Plant Cell Walls

Plant cell walls have been relevant for human survival throughout evolution, from cell walls recognised as an essential ingredient in human and livestock nutrition, to their use in energy generation, construction, tool making, paper and clothing. This plant-generated material is at the centre of a myriad of human activities, and it represents the world's most abundant natural resource for fuel, fibre, food and fodder. *Plant Cell Walls: Research Milestones and Conceptual Insights* provides an overview of the key discoveries of hundreds of years of plant cell wall research. With chapter contributions from prominent scientists in the cell wall field, this book provides a comprehensive treatment of plant cell wall research, accompanied by a historical overview to illustrate how concepts have evolved, and how progress has been enabled by emerging technological advances.

Plant Cell Walls: Research Milestones and Conceptual Insights elaborates on the translation of research to application in biotechnology and agriculture, and highlights its relevance for climate change mitigation and adaptation. It will be a key resource for plant cell biologists, biochemists and geneticists.

Plant Cell Walls

Research Milestones and Conceptual Insights

Edited by

Anja Geitmann

CRC Press
Taylor & Francis Group
Boca Raton London New York

CRC Press is an imprint of the
Taylor & Francis Group, an **informa** business

Designed cover image: Anja Geitmann

First edition published 2024
by CRC Press
2385 NW Executive Center Drive, Suite 320, Boca Raton FL 33431

and by CRC Press
4 Park Square, Milton Park, Abingdon, Oxon, OX14 4RN

CRC Press is an imprint of Taylor & Francis Group, LLC

ISBN: 9781032013213 (hbk)
ISBN: 9781032013626 (pbk)
ISBN: 9781003178309 (ebk)

DOI: 10.1201/9781003178309

Typeset in Times
by codeMantra

Contents

Preface

Anja Geitmann

Plants provide us with the energy and oxygen to sustain our lives, clothe us, produce building materials and renewable fuel, and inspire our built forms through their exquisite shapes, texture, water properties and engineering design. Plants also play a crucial role in addressing the most important challenge facing Society in the twenty-first century—climate change—due to their ability to capture gaseous carbon dioxide and transform it into polymers that immobilise carbon into a solid state. Understanding the ways in which plants convert carbon into polymers, the roles of these polymers for plant form and functioning, and the ways in which these polymers can subsequently be utilised by humans, are the underlying goals of a research field that focuses on the very material that composes most of the carbon-rich biomass of plants—the cell walls. This book illustrates the state-of-the-art of this sector of plant biology and biotechnology and highlights major technical and conceptual advances that propelled the field to its current state.

The cell wall envelopes plant cells, regulates how the cells grow and interconnects cells providing the extracellular continuum (the apoplast) that enables communication throughout the plant. Plant development is thus strongly influenced by the way in which the cell wall controls tissue architecture and organ morphogenesis. Understanding plant functioning at the macroscopic scale, therefore, requires interrogating structural processes happening at the microscopic molecular and cellular scales during wall morphogenesis and development.

A significant milestone in the visual investigation of plants at a cellular level was their description by early microscopists in the seventeenth century. Tracheids—cells displaying beautifully helical cell wall thickenings shown on the cover of this book—were observed by Marcello Malpighi (1628–1694) in a snapped branch of a chestnut tree. They reminded him of insect tracheae, structures he had studied intensely (West 2013). Malpighi's detailed observations of plant anatomy (Malpighi 1675, 1679) inspired contemporary and future studies. The actual term *'cell'* had, of course, been coined by Robert Hooke (1635–1703) as documented in his book *Micrographia* (Hooke 1665). Upon viewing cross sections of cork tissue in which thick walls surround small, rectangular openings (remnants of protoplasts in the now dead phellem tissue), Hooke was reminded of the monks' cells in a monastery (*cella* in Latin). Another prolific plant anatomist during that era was Nehemiah Grew (1641–1712). His detailed descriptions of plant structures in *Anatomy of Plants* (Grew 1682) contain the first known microscopic description of pollen, a cell type with stunningly beautiful cell wall features. The nineteenth and a greater part of the twentieth centuries were characterised by the amazing work of the organic chemists who set about establishing the structures of the major wall polymers (polysaccharides, lignins, suberins and glycoproteins) driven primarily due to their biotechnological significance to humans; an incredible achievement given that linkage-specific enzymes were largely unknown and analytical technologies (such as nuclear magnetic resonance (NMR)/mass spectrometry) only became available in the later part of the twentieth century. In parallel, improved imaging methods allowed researchers to look deeper into the structural features of the cell wall seizing technological advances such as 3D imaging, super resolution visualisation and live observation of molecular dynamics using fluorescent protein technologies. The later part of the twentieth century resulted in a step change in the field with the applications of modern molecular biological and genetic techniques, which together with the lure of plants as biofuels and an increasing realisation that the wall was central to plant growth and development, saw a massive injection of funds and a re-kindling of interest in the cell wall field by a new generation of scientists.

Today, research on plant cell walls is performed by numerous laboratories around the world; yet, compared with many other life science domains, the community is surprisingly small and lacks sufficient disciplinary depth given the centrality of walls to plant growth, development and biotechnological applications. With a total of 70 authors, a substantial contingent of this community contributed to the 25 chapters of this book. Many more gave their time as peer reviewers. Through everyone's combined efforts, this book has become both a cross-section of modern research disciplines related to plant cell walls and a retrospective into the history of the research field as most chapters highlight milestones that led to the knowledge we have today. Many chapters also outline how the evolution in scientific method and progress in experimental technology enabled and propelled scientific insight in the cell wall field.

Setting the scene with a large overview over the cell wall research field, Chapter 1 provides the entry point to this book. Doblin and colleagues introduce the different polysaccharides composing plant cell walls, methods to detect them, and experimental tools to manipulate their synthesis *in planta*. Their chapter provides the overall context, alludes to the structural diversity and elaborates on human uses of cell wall material.

The conversion of carbon dioxide into plant cell wall polymers proceeds through the production of sugar molecules, a process for which solar energy is harnessed through photosynthesis. From these sugars, various types of polysaccharides are synthesised either within the cellular endomembrane system or by enzymes located in the plasma membrane. Among the synthetic enzymes are glycosyltransferases (Chapter 2). The transport of sugars to sites of polymer synthesis is executed by sugar transporters

(Chapter 3). Once polysaccharides are deposited at the outside of the cell, they continue to be modified by an array of different enzymes (Chapter 4). This applies, among others, to pectin, one of the principal components of the primary plant cell wall (Chapter 5). Such modifications and remodelling characterise the differentiation of tissues with highly specific functionality (Chapters 6 and 7).

The molecular complexity of plant cell wall polysaccharides is staggering and poses a significant challenge to structural investigations. A multitude of methods is used to analyse polysaccharides, including solution and solid state NMR (Chapter 8). How such methods have advanced our understanding of the wall composition, synthesis, remodelling and degradation in the grasses is elaborated in Chapter 9. To link particular polysaccharides to cell and tissue functioning, information on their precise location is required and among the principal methods used for their detection are antibody-based labelling strategies (Chapter 10). Insight into the molecular structure of the cell wall has informed our conceptual understanding of the interactions between cell wall polymers (Chapter 11). Chapter 12 proposes the interpretation of the cell wall structure–function relationship as a 'glycomic code'.

Comparisons across the phylogenetic tree have enabled tracing the evolutionary pressures that may have led to the diversification in cell wall molecular composition and structure (Chapters 9 and 13). An important development during the evolution of land plants was the impermeabilisation of cell walls to improve the control of evaporation from plant tissues (Chapter 14). Another adaptation to terrestrial life was the development of stomates—openings in the plant surface that facilitate gas exchange between the outside and the inside of the organ (McAdam et al. 2021). The guard cells composing the stomates are able to change shape reversibly, a feature they share with selected other plant cells and tissues and that is governed by cell wall properties (Chapter 15).

Next to polysaccharide-modifying enzymes, the cell wall has other types of proteins, including hydroxyproline-rich glycoproteins, such as arabinogalactan proteins (AGPs). Among other functions, AGPs are involved in plants' response to stress (Chapter 16) and in their interaction with microbes (Chapter 17). Stress response and defence are complex processes involving intracellular signalling cascades and cell wall remodelling (Chapters 18–20). The response to stress is mediated in a large variety of ways including by affecting the deposition of secondary cell walls and the development of wood tissue designed to provide increased support to plant organs (Chapter 21). A highly performant adaptation to drought stress, unique to resurrection plants, is their ability to dry out almost entirely while preventing irreversible damage to their cell structures (Chapter 22).

In plant tissues, cells are connected to each other at the interface between adjacent cell walls—the middle lamella (Chapter 23). During secondary wall deposition, this thin layer (Zamil and Geitmann 2017) becomes part of the compound middle lamella, a structure of particular focus in the context of plant biomass utilisation since it tends to be highly lignified thus making the material recalcitrant to microbial or enzymatic deconstruction (Chapter 24). The middle lamella is formed after mitosis when the growing cell plate starts dividing the daughter cells. The assembly of this newly developing wall in the centre of the dividing cell during cytokinesis is one of the most distinguishing features in the cell biology of plants (Chapter 25).

In science, we all stand on the shoulders of giants as we try to add our part to the collective knowledge. Most chapters in this book reference historical milestones in the field, thus highlighting the work of numerous key members of the plant cell wall research community who may not be with us anymore. Among these is one of the reviewers of this book. Simon McQueen-Mason did not only contribute as a peer-reviewer, his own body of work is cited in no less than ten chapters. He tragically died in a bicycle accident shortly before the 2022 Cell Wall Conference in East Lansing, Michigan. His passing was a shock to the community as the group gathered for the first time after a pandemic-related two-year hiatus. May this book be a salute to this eminent cell wall researcher and the many others who preceded us and who contributed to the advancement in the field in so many ways.

To those colleagues who wrote chapters and/or accepted review assignments, I express my huge appreciation for enabling me to make this a fully peer-reviewed book. Thank you for the hard work in putting together and critiquing the manuscripts and thank you for everyone's patience with multiple rounds of editorial requests and queries. The scope of the book was to go beyond a summary of the current state-of-the-art by documenting the scientific history of the cell wall research field. Even with 25 chapters, such a work could never be exhaustive, and it isn't, but it provides a wide overview of many research endeavours that have allowed us to advance our understanding of plant functioning.

REFERENCES

Grew N (1682) *Anatomy of Plants with and Idea of a Philosophical History of Plants*. W. Rawlins, London.

Hooke R (1665) *Micrographia: or Some Physiological Descriptions of Minute Bodies Made by Magnifying Glasses with Observations and Inquiries Thereupon*. The Royal Society, London.

Malpighi M (1675) *Anatomes Plantarum Pars Prima*. Philosophical Transactions of the Royal Society, London.

Malpighi M (1679) *Anatomes Plantarum Pars Altera*. Philosophical Transactions of the Royal Society, London.

McAdam SAM, Duckett JG, Sussmilch FC, Pressel S, Renzaglia KS, Hedrich R, Brodribb TJ, and Merced A (2021) Stomata: the holey grail of plant evolution. *American Journal of Botany* 108:366–371.

West JB (2013) Marcello Malpighi and the discovery of the pulmonary capillaries and alveoli. *American Journal of Physiology-Lung Cellular and Molecular Physiology* 304:L383–L390.

Zamil MS, and Geitmann A (2017) The middle lamella-more than a glue. *Physical Biology* 14:015004.

Biography

Anja Geitmann investigates the role of the cell wall in plant reproduction and development. Her research focuses on the cellular processes that lead to the formation of plant organs and functional tissues. She holds the Canada Research Chair in Biomechanics of Plant Development and leads an interdisciplinary team of biologists and engineers. Dr. Geitmann's research combines cell biology with engineering methodology and mathematical concepts to reveal how the cell wall deforms during cell growth and to detect the forces involved. Her team has developed sophisticated experimental and computational tools enabling researchers to study questions such as how the pollen tube delivers the sperm cells during reproduction in plants and how plant cells change shape to generate tissues with various functionalities.

Dr. Geitmann is the editor of the books *Pollen and Pollen Tube Biology* (2020), *Plant Biomechanics - From Structure to Function at Multiple Scales* (2018) and *Cell Biology of Plant and Fungal Tip Growth* (2001). She published 130+ journal articles and book chapters, and she has supervised more than 100 trainees. She has organised numerous international conferences and workshops including a NATO Workshop on 'Tip Growth', joint meetings between the American and Canadian Societies of Plant Biology, the *9th International Plant Biomechanics Conference* and international workshops on 'Cell Wall Mechanics' and 'Multiscale Modeling of Cell Wall Mechanics and Growth in Walled Cells' hosted by the Banff International Research Station. She is frequently solicited to serve on the scientific advisory committees of international conferences on the plant cell wall.

In 2013–2015, Dr. Geitmann served as President of the Microscopical Society of Canada and she was the President of the Canadian Society of Plant Biologists in 2015–2017. She served as the President of the International Association of Plant Reproduction Research in 2018–2022, and she is a member of the editorial and advisory boards of multiple scientific journals including *Cell*, *Plant Physiology*, and *Journal of Experimental Botany*.

Anja Geitmann obtained her PhD in Environmental Biology in 1997 from the University of Siena (Italy), following undergraduate and graduate studies in biology at the University of Constance (Germany), Oregon State University (USA) and Stockholm University (Sweden). Between 1997 and 2001, she performed postdoctoral research at the *Université Laval*, Québec, and at the University of Wageningen, The Netherlands. From 2001 to 2015, she was Professor at the Department of Biological Sciences of the *Université de Montréal* and Scientist at the *Institut de recherche en biologie végétale*. She taught Plant Anatomy, Microscopy and Scientific Communication at undergraduate and graduate levels. In 2015, Dr. Geitmann was appointed as the Dean of the Faculty of Agricultural and Environmental Sciences and Associate Vice-Principal (Macdonald Campus) of McGill University, Montreal, Canada.

Contributors

Dhika Amanda
Norwegian University of Science and Technology
Trondheim, Norway

Charles T. Anderson
The Pennsylvania State University
University Park, Pennsylvania

Melani A. Atmodjo
University of Georgia
Athens, Georgia

Antony Bacic
La Trobe University
Bundoora, Australia
and
Zhejiang Agriculture and Forestry University
Hangzhou, China

Marcos S. Buckeridge
University of São Paulo
São Paulo, Brasil

Vincent Bulone
The University of Adelaide
South Australia, Australia

Nicholas C. Carpita
Biosciences Center
National Renewable Energy Laboratory
Golden, Colorado

Romain Castilleux
Swedish University of Agricultural Sciences
Umeå, Sweden

Felice Cervone
Sapienza Università di Roma
Rome, Italy

Mingqin Chang
University of California Davis
Davis, California

Sílvia Coimbra
Universidade do Porto
Porto, Portugal
and
LAQV Requimte
Porto, Portugal

Thomas Michael Curry
Complex Carbohydrate Research Centre (CCRC)
University of Georgia
Athens, Georgia

Giulia De Lorenzo
Sapienza Università di Roma
Rome, Italy

Monika S. Doblin
La Trobe University
Bundoora, Australia
and
Zhejiang Agriculture and Forestry University
Hangzhou, China

Georgia Drakakaki
University of California Davis
Davis, California

Azeddine Driouich
Université de Rouen Normandie
Mont-Saint-Aignan, France

Berit Ebert
The University of Melbourne
Melbourne, Australia
and
Ruhr Universität Bochum
Bochum, Germany

Jill M. Farrant
University of Cape Town
Rondebosch, South Africa

Simone Ferrari
Sapienza Università di Roma
Rome, Italy

Maria João Ferreira
Universidade do Porto
Porto, Portugal
and
LAQV Requimte
Porto, Portugal

Raquel Figueiredo
Universidade do Porto
Porto, Portugal
and
LAQV Requimte
Porto, Portugal

Geoffrey B. Fincher
The University of Adelaide
South Australia, Australia

Stephen C. Fry
The University of Edinburgh
Edinburgh, United Kingdom

Anja Geitmann
McGill University
Montreal, Quebec, Canada

Eleni Giannoutsou
National and Kapodistrian University of Athens
Athens, Greece

Sarah Giulietti
Sapienza Università di Roma
Rome, Italy

Giovanna Gramegna
Sapienza Università di Roma
Rome, Italy

Michael G. Hahn
University of Georgia
Athens, Georgia

Thorsten Hamann
Norwegian University of Science and Technology
Trondheim, Norway

Herman Höfte
University Paris Saclay, INRAE
Versailles, France

Michael C. Jarvis
Glasgow University
Glasgow, Scotland, United Kingdom

Pradeepa Jayawardhane
University of Georgia
Athens, Georgia

Kim L. Johnson
La Trobe University
Bundoora, Australia
and
Zhejiang Agriculture and Forestry University
Hangzhou, China

Panagiotis Kalaitzis
Mediterranean Agronomic Institute of Chania
Chania, Greece

Anna Konkina
Mediterranean Agronomic Institute of Chania
Chania, Greece

Vincenzo Lionetti
Sapienza Università di Roma
Rome, Italy

Ana Lopes
Universidade do Porto
Porto, Portugal
and
BioISI
Porto, Portugal

Riccardo Lorrai
Sapienza Università di Roma
Rome, Italy

Yingxuan Ma
La Trobe University
Bundoora, Australia
and
Nanjing Forestry University
Nanjing, China

Rawen Ben Malek
Université Paris-Saclay
Versailles, France
and
Martin Luther University
Halle-Wittenberg, Germany

Alban Mariette
Max-Planck Institute of Molecular Plant Physiology
Potsdam, Germany
and
Ruhr Universität Bochum
Bochum, Germany

Lucia Marti
Sapienza Università di Roma
Rome, Italy

Maureen C. McCann
Biosciences Center
National Renewable Energy Laboratory
Golden, Colorado

Ewa J. Mellerowicz
Swedish University of Agricultural Sciences
Umeå, Sweden

Debra Mohnen
University of Georgia
Athens, Georgia

John P. Moore
Stellenbosch University
Matieland, South Africa

Diana Moreira
Universidade do Porto
Porto, Portugal
and
LAQV Requimte
Porto, Portugal

Grégory Mouille
Université Paris-Saclay, INRAE
Versailles, France

Eric Nguema-Ona
Centre Mondial de l'innovation - Roullier
Saint Malo, France

Lazar Novaković
La Trobe University
Bundoora, Australia
University of Melbourne
Parkville, Australia
and
University of Leeds
Leeds, United Kingdom

Ariel Orellana
Universidad Andrés Bello
Santiago, Chile

Ana Marta Pereira
Universidade do Porto
Porto, Portugal
and
LAQV Requimte
Porto, Portugal

Luís Gustavo Pereira
Universidade do Porto
Porto, Portugal
and
Sustainable Agrifood Production Research Centre
Porto, Portugal

Daniela Pontiggia
Sapienza Università di Roma
Rome, Italy

Zoë A. Popper
University of Galway
Galway, Ireland

Peter M. Ray
Stanford University
Palo Alto, California

Kay Schneitz
Technical University of Munich
Freising, Germany

Lukas Schreiber
University of Bonn
Bonn, Germany

Julian G. Schwerdt
The University of Adelaide
South Australia, Australia

Jessy Silva
LAQV Requimte
Porto, Portugal
and
Universidade do Minho
Braga, Portugal

Rosalie Sinclair
University of California Davis
Davis, California

Pramod Sivan
KTH Royal Institute of Technology
Stockholm, Sweden

Kiran Suresh
University of Bonn
Bonn, Germany

Kristen Thorne
University of Georgia
Athens, Georgia

Peter Ulvskov
Copenhagen University
Frederiksberg, Denmark

Breeanna R. Urbanowicz
Complex Carbohydrate Research Centre (CCRC)
University of Georgia
Athens, Georgia

Maïté Vicré
Université de Rouen Normandie
Mont-Saint-Aignan, France

Francisco Vilaplana
KTH Royal Institute of Technology
Stockholm, Sweden

Kris Vissenberg
University of Antwerp
Antwerp, Belgium
and
Hellenic Mediterranean University
Chania, Greece

Sebastian Wolf
Eberhard Karls Universität Tübingen
Tübingen, Germany

Viktoria V. Zeisler-Diehl
University of Bonn
Bonn, Germany

Abbreviations

ΔUA	Unsaturated uronic acid (4-deoxy-β-L-*threo*-hex-4-enopyranuronic acid)
$[Ca^{2+}]_{cyt}$	Cytoplasmic free calcium ion
%	Percentage
OH	Hydroxyl radical
ΔG	Change in free energy
1D	One-dimensional
2D	Two-dimensional
3D	Three-dimensional
3D-dSTORM	Three-dimensional direct stochastic optical reconstruction microscopy
4-OMe GlcA	4-O-Methyl-glucuronic acid
4CL	4-Coumarate:CoA ligase
4D	Four-dimensional
5-OH-G	5-Hydroxyguaiacyl
A	dehydro-L-ascorbic acid
ABA	Abscisic acid
ABC	ATP-binding cassette
ABP	Actin binding protein
Ac	Acetyl
ADP	Adenosine diphosphate
ADPG	Arabidopsis dehiscence zone polygalacturonase
AFEX	Ammonia-fibre expansion
AFM	Atomic force microscope/microscopy
AFM-IR	Atomic force microscopy-infrared spectroscopy
AG	Arabinogalactan
AG-I	Arabinogalactan type I
AG-II	Arabinogalactan type II
AGA	Apiogalacturonan
AGM	AG glucuronic acid methyltransferase
AGP	Arabinogalactan protein
AGX	Arabinoglucuronoxylan
AH_2	L-Ascorbic acid
AHA	Autoinhibited plasma membrane proton (H^+) ATPases
AI	Artificial intelligence
AIR	Alcohol insoluble residue
AIS	Alcohol insoluble solids
ala	Alanine
AM	Arbuscular mycorrhiza
AMSA	Array-based microfibril surface assessment
ANJ	Anjea
ANX	Anxur
AOS	Allene oxide synthase
AP	Apiogalacturonan
APAP	Supramolecular structure comprising arabinoxylan, pectin and AGP
APAP1	Arabinoxylan pectin arabinogalactan protein 1
API	Apiose
ApiT	Apiosyltransferase
Ara	Arabinose
Arabidopsis	Arabidopsis thaliana
ARAD	Arabinan deficient
Ara*f*	Arabinofuranose
Ara*p*	Arabinopyranose
AraT	Arabinosyltransferase
ARF	Auxin response factor
ARF GEF	ADP-ribosylation factor guanine-nucleotide exchange factor
arg	Arginine
asn	Asparagine
Asp	Asparagine
AT	Acetyltransferase
At	Arabidopsis thaliana
AtFUT1	*Arabidopsis thaliana* Xyloglucan fucosyltransferase 1
AtKIN13A	Arabidopsis thaliana Kinesin-13A
ATP	Adenosine triphosphate
AtRGIL6	Arabidopsis thaliana RG-LYASE6
AtUTr	*Arabidopsis thaliana* UDP-galactose transporter
AX	Arabinoxylan
BAHD	Benzyl alcohol, anthocyanin, *N*-hydroxycinnamoyl, deacetylvindoline acyl-CoA transferase
BAK	BRI1-associated kinase
BAR	Bio-analytic resource
BASL	Breaking of asymmetry in the stomatal lineage
BBE-like	Berberine bridge enzyme-like
BC	Border cell
BcMF	Brassica campestris Male fertility
BFA	Brefeldin A
BG	Beta-glucanase
BGAL	β-galactosidase
bHLH	Basic helix loop helix
BiFC	Bimolecular fluorescence complementation
BL	Brassinolide
BLC	Border-like cell
BR	Brassinosteroid
BRI1	Brassinosteroid-insensitive 1
BRX	Brevis radix
BUP	Buddha's paper seal
BY-2	Bright yellow 2
C3′H	Cinnamate 3′-hydroxylase
C4H	Cinnamate 4-hydroxylase
Ca^{2+}	Calcium ion
CAD	Cinnamyl alcohol dehydrogenase
CalS	Callose synthase
CAP	$[Ca^{2+}]_{cyt}$-associated protein
CaZy	Carbohydrate active enzymes
CAZymes	Carbohydrate active enzymes
CBI	Cellulose biosynthesis inhibition
CBL	Calcineurin B-like protein

CBM	Carbohydrate-binding module
CCH	Calcium channel
CCR	Cinnamoyl CoA reductase
CCRC	Complex Carbohydrate Research Center
CCV	Clathrin-coated vesicles
CD	Cellodextrin
cDNA	Copy DNA
CDPK	Calcium-dependent protein kinase
CDTA	Cyclohexanediaminetetraacetic acid
CDZ	Cortical division zone
CELA	Cellulose synthase A (bacterial)
CES	Cellulase synthase
CESA/CesA	Cellulose synthase A
CEV	Constitutive expression of VSP
CHOR	Chorus
CID	High energy collision induced dissociation
CKX	Cytokinin oxidase/dehydrogenase
Cl^-	Chloride ion
CLV	Clavata
CML	Calmoduline-like protein (Chapter 18), Compound middle lamella (Chapters 21 and 24)
CMP	Cytidine monophosphate
Co-IP	Co-immunoprecipitation
CO_2	Carbon dioxide
CoA	CoenzymeA
CoMPP	Comprehensive microarray polymer profiling
COMT	Caffeic acid *O*-methyltransferase
COOH	Carboxy
COP	Coat protein
COR	Cold-related
COS	Chitosan oligosaccharides
COSY	Correlation spectroscopy
CP	Cross Polarisation
CPAM	Cell plate assembly matrix
CPC	Caprice
CPK	Calcium-dependent protein kinase
CRK	Cys-rich receptor-like kinase
CRLK	Calcium/calmodulin-regulated receptor-like kinase
CrRLK	*Catharanthus roseus* receptor-like kinase
CrRLK1L	*Catharanthus roseus* receptor-like kinase 1-like
Cryo-ET	Cryo-electron tomography
CRZ	Calcineurin-responsive zinc finger
CSC	Cellulose synthase complex
CSI	Cellulose synthases interacting
CSL/Csl	Cellulose synthase like
CSLD	Cellulose synthase-like D
CSP	Chelated soluble fraction
CtecII	Commercial enzyme cocktail from Novozymes
CVY	Curvy
CWD	Cell wall damage
CWDE	Cell wall degrading enzyme
CWI	Cell wall integrity
CXE	Cellulose:xyloglucan endotransglucosylase (activity)
CYC	Cyclin
cys	Cysteine
d.w.	Dry weight
Da	Dalton
DAMP	Damage-associated molecular pattern
DARR	Dipolar assisted rotational resonance
DCB	2,6-Dichlorobenzonitrile
ddPCR	Droplet digital polymerase chain reaction
DEK	Defective kernel
DFRC	Derivatisation followed by reductive cleavage
DFS	Donnan-free space
Dha	3-Deoxy-D-*lyxo*-heptulosaric acid
DHB	2,5-Dihydroxybenzoic acid
DM	Degree of methylesterification
DMSO	Dimethyl sulphoxide
***d*N**	Rate of nonsynonymous substitution
DNA	Desoxyribonucleic acid
DNP	Dynamic nuclear polarisation
DP	Degree of polymerisation
DP-S	Degree of polymerisation with substitution tolerance
DPA	Days post anthesis
***d*S**	Rate of synonymous substitution
DSC	Differential scanning calorimetry
DUF	Domain of unknown function
$\dot{E}$	Rate of extension of cell wall or of entire cell
E2F	E2 promoter binding factor
EC	Enzyme Commission
ECD	Extracellular domain
ECH	Echidna
ECM	Extracellular matrix
EC	β-xylanase
Ecto-M	Ectomycorrhiza
EDTA	Ethylenediaminetetraacetic acid
EE	Early endosome
EELS	Image-electron energy loss spectroscopy
EFR	Elongation factor-TU receptor
EG	Endo glucanase
ELI	Ectopic lignification
ELISA	Enzyme-linked immunosorbent assay
EM	Electron microscopy
EMBL	European Molecular Biology Laboratory
EMBL-EBI	EMBL's European Bioinformatics Institute
EMS	Ethyl methanesulphonate
endo-EC	Endo-1, 4-β-xylanase
endo-PG	Endo-polygalacturonase
ENODL	Early nodulin-like protein
EPG	Endopolygalacturonase
EPR	Electron paramagnetic resonance
ER	Endoplasmic reticulum
ERU	Erulus
ES7	Endosidin 7
ESI-MS	Electrospray ionisation-mass spectrometry
EST	Expressed sequence tag
ETI	Effector-triggered immunity
EXAD	Extensin arabinose deficient
EXGT	Endoxyloglucan transferase

exo-PG	Exo-polygalacturonase
EXP	Expansin
EXPA	Alpha-expansin
EXPO	Exocyst-positive organelle
EXT	Extensin
F5H	Ferulate 5-hydroxylase
FAS	Fasciclin
FC	Fusicoccin
FCC	Fusiform cambial cell
FEM	Finite element modelling
FER	Feronia
FESEM	Field emission scanning electron microscope/ microscopy
FIB	Focused ion beam
FIND-IT	Fast identification of nucleotide variants by DigITal PCR
FITC	Fluorescein isothiocyanate
FLA	Fasciclin-like arabinogalactan-protein
FLA4/ SOS5	Fasciclin-like AGP 4/Salt overly sensitive 5
FLS	Flagellin-sensing
FRA	Fragile fibre
FRAP	Fluorescence recovery after photobleaching
FT-IR	Fourier-transform infra-red spectroscopy
Fuc	Fucose
Fuc*p*	Fucopyranose unit
FUT	Fucosyltransferase
FVS	Fusion of Golgi-derived vesicles stage
G	Free energy
G-	Gelatinous (Chapter 21), Guaiacyl units (Chapter 24)
G-fibre	Gelatinous fibre
G-layer	Gelatinous layer
G/S	Guaiacyl/syringyl
Gal	Galactose
GalA	Galacturonic acid
Gal*p*	Galactopyranose unit
Gal*p*A	Galactopyranuronic acid unit
GALS	Galactan synthase
GALT	Galactosyltransferase
GAS	GPI anchor addition sequence
GAUT	Galacturonosyltransferase
GAX	Glucuronoarabinoxylan
GC	Gas-liquid chromatography
GC-MS	Gas chromatography–mass spectrometry
GDP	Guanosine diphosphate
GEF	Guanine exchange factor
GFP	Green fluorescent protein
GFT	GDP-fucose transporter
gg	gauche-gauche
GGLT	GDP-L-galactose transporter
GGM	Galactoglucomannan
GH	Glycoside hydrolase
GIPC	Glycosyl inositol phosphorylceramide
GL2	Glabra 2
Glc	Glucose
GlcA	Glucuronic acid
GLCAT/ GlcAT	Glucuronosyltransferase
GlcNAc	N-Acetylglucosamine
Glc*p*	Glucopyranose unit
Glc*p*A	Glucopyranuronic acid unit
gln	Glutamine
glu	Glutamic acid
gly	Glycine
GM	Galactomannan (Chapter 2), Glucomannan (Chapter 21), Genetically modified (Chapters 7 and 9)
GMGT	Galactomannan-specific galactosyltransferase
GONST	Golgi nucleotide sugar transporter
GPI	Glycosylphosphatidylinositol
GPI-AP	Glycosylphosphatidylinositol-anchored protein
GPI-T	GPI transamidase complex
GRP	Glycine-rich protein
GSL	Glucan synthase like
GT	Glycosyltransferase
gt	gauche-trans
GT	Glycosyltransferase
GTP	Guanosine triphosphate
GUX	Glucuronyl substitution of xylan
GX	Glucuronoxylan
GXM/ GXMT	Glucuronoxylan methyl transferase
GXXG	Xyloglucan hexasaccharide ($Xyl_2.Glc_4$)
H-	*p*-Hydroxyphenyl units
H_2O	Water
HAMK	Heat-activated map-kinase
HCT	Hydroxycinnamoyl-Coenzyme A shikimate/ quinate hydroxycinnamoyl transferase
HEK	Human embryonic kidney
HERK	Hercules receptor kinase
HETCOR	Heteronuclear correlation
HG	Homogalacturonan
HG:GalAT	Homogalacturonan:galacturonosyltransferase
HGT	Horizontal gene transfer
his	Histidine
HM	Heteromannan
HMBC	Heteronuclear multiple bond correlation
HMQC	Heteronuclear multiple-quantum coherence
HOG	High osmolarity glycerol
HPAT	Hydroxyproline *O*-arabinosyltransferase
HPF	High pressure freezing
HPGT	Hyp *O*-galactosyltransferase
HPLC	High-performance liquid chromatography
HPTG	Hydroxyproline-*O*-galactosyltransferase
HRGP	Hydroxyproline-rich glycoprotein
HSP	Heat-shock protein
HTG	Hetero-trans-β-glucanase
HTP	High throughput
Hyp	Hydroxyproline
IAA	Indole-3-acetic acid
ile	Isoleucine
IMK2	Inflorescence meristem kinase 2

INADEQUATE	Incredible natural abundance double quantum transfer experiment
INEPT	Insensitive nuclei enhanced by polarisation transfer
IRX	Irregular xylem
IUBMB	International Union of Biochemistry and Molecular Biology
JA	Jasmonic acid
JAGGER	Arabinogalactan protein 4
JAR	Jasmonic acid resistant
JBEI	Joint BioEnergy Institute
JIM5	John Innes monoclonal 5
K^+	Potassium ion
K_{cat}	Catalytic rate constant
KD	Knock-down
Kdo	3-Deoxy-D-*manno*-octulosonic acid
KN1	Knotted1
KNOLLE	Syntaxin-related protein KNOLLE
L	Load
LC-MS	Liquid chromatography–mass spectrometry
LCC	Lignin-carbohydrate complex
LeRLK	L-type lectin-like receptor-like kinase
leu	Leucine
LLG1	Lorelei-like GPI-AP1
LM7	Leeds monoclonal 7
L_p	Hydraulic conductivity
LPMO	Lytic polysaccharide monooxygenase
LRE	Lorelei
LRR	Leucine-rich repeat
LRR-RLK	Leucine-rich repeat receptor-like kinase
LRX	Leucine-rich repeat extensin
LTN	Leaf tip necrosis
LYCH	Lucifer yellow carbohydrazide
lys	Lysine
mAbs	Monoclonal antibodies
MALDI-MSI	Matrix-assisted laser desorption/ionisation mass spectrometry imaging
MALDI-TOF/MS	Matrix-assisted laser desorption/ionisation-time-of-flight mass spectrometry
MAMP	Microbe-associated molecular pattern
Man	Mannose
Man*p*	Mannopyranose unit
MAP	Mitogen-activated protein
MAPK	Mitogen-activated protein kinase
MAP	Microtubule associated protein
MAS	Magic-angle spinning
MC4	Metacaspase 4
MCA	Mating pheromone induced death 1 complementing activity
MDP	Microtubule-destabilising protein
MDS	Medos
Me-GlcA	(4-*O*-methyl)-glucuronosyl unit
Me-HG	Methyl esterified homogalacturonan
MeFuc	2-*O*-methylfucose
MeGlc*p*A	4-*O*-methyl-glucopyranuronic acid unit
MEKK	Stress-induced MAPK kinase kinase
MeT	Methyltransferase
met	Methionine
MeXyl	2-*O*-methylxylose
MF	Microfibril
MFA	Microfibril angle
Mg^{2+}	Magnesium ion
MGAT	N-acetylglucosaminyltransferase
mGLCA	Methylated glucuronic acid
MGP4	Male gametophyte defective 4
MID	Mating-induced death
MIK	Male discoverer 1-interacting receptor-like kinase
miRNA	Micro-ribonucleic acid
MKK	Mitogen-activated protein kinase (MAPK) kinase
ML	Maximum likelihood
MLG	(1,3-1,4)-β-D-glucan
MLG	Mixed-linkage glucan
MLR	Malectin-like receptor kinase
MMNO-DMSO	N-methylmorpholine N-oxide/dimethyl sulphoxide
mnn2-2	*Kluyveromyces lactis* mutant deficient in GlcNAc in glycoproteins
MOCA	Monocation-induced $[Ca^{2+}]_{cyt}$-induced
MOMT	Monolignol 4-*O*-methyltransferase
MoO_4^{2-}	Molybdate ion
MPa	Megapascal
MPK	Mitogen-activated protein kinase
MS	Mass spectrometry
MSCS	Mechano-responsive channel of small conductance
MSL	MSCS-like
MSR	Mannan-synthesis-related
MT	Microtubule
MTH	Mannan transglycosylase hydrolase
MTL	MID two-like protein
MTR1	Microspore and tapetum regulator 1
MUCI	Mucilage-related
mum2	mucilage-modified 2, a galactosidase mutant
MUR	Murus
mur1	*Murus1 Arabidopsis thaliana* mutant deficient in GDP-D-mannose-4,6-dehydratase
MXE	Mixed-linkage β-glucan:xyloglucan endotransglucosylase
MYA	Million years ago
MYB	Myb-like DNA binding-domain transcription factor
NO_3^-	Nitrate ion
N_2	Nitrogen
NA	Numerical aperture
NAC	NAC domain transcription factor; derived from the names of the three genes first described as containing the domain, namely *NAM (no apical meristem), ATAF1,2* and *CUC2 (cup-shaped cotyledon)*

NACK1	Kinesin-like protein
NaCl	Sodium chloride
NDP	Nucleoside diphosphate
NDR1	Non-race-specific disease resistance 1
NET	Neutrophil extracellular trap
NH_2	Amino
NH_4^+	Ammonium ion
NIA	Nitrate reductase
nm	Nanometer
NMP	Nucleoside monophosphate
NMR	Nuclear magnetic resonance
NO	Nitric oxide
NOESY	Nuclear Overhauser enhancement spectroscopy
NRE	Non-reducing end
NST	NAC secondary wall thickening promoting factor (Chapter 1), Nucleotide sugar transporter (Chapters 2 and 3)
NW	Normal wood
ø	Extensibility of cell wall
O_2	Oxygen
OCH	Outer chain elongation
OE	Overexpression
OG	Oligogalacturonide
OH^-	Hydroxide anion
OLIMP	Oligo mass profiling
OSCA	Ca^{2+}-permeable channel induced Ca^{2+} increase
OsELA3	*Oryza sativa* Early nodulin-like AGP 3
OST	Oligosaccharyltransferase
P	Phosphate
P	Turgor pressure
P4H	Prolyl-4-hydroxylase
PAC	PRP–AGP-containing Cys
PACE	Polysaccharide analysis by gel electrophoresis
PAE	Pectin acetylesterase
PAGE	Polyacrylamide gel electrophoresis
PAGR	Pectic arabinogalactan synthesis-related
PAL	Phenylalanine ammonia lyase
PAM1	Phage-display monoclonal 1
PAMP	Pathogen-associated molecular pattern
PC	Post-chlorite
PCNA	Proliferating cell nuclear antigen
PCR	Polymerase chain reaction
PD	Plasmodesmata
PDCA	Pyridine 2,4–dicarboxylate
PDSD	Proton-driven spin diffusion
PEP	Plant elicitor peptide
PEPR	Plant elicitor peptide receptor
PERK	Proline-rich extensin-like receptor kinase
PFS	Planar fenestrated sheet
PG	Polygalacturonase
PG2a	Polygalacturonase 2a
PGA	Polygalacturonic acid
PGase	Polygalacturonanase
PGIP	Polygalacturonase-inhibiting protein
phe	Phenylalanine
PI	Propidium iodide
PIP5K	Phosphatidylinositol-4-phosphate 5-kinase
PKC	Protein kinase C
PL	Pectate lyase
PL1	Polysaccharide lyase 1 family
PM	Plasma membrane
pm	Picometer
PME	Pectin methylesterase
PMEI	Pectin methylesterase inhibitor
PNL	Pectin lyase
POK	Phragmoplast orienting kinesin
PPB	Preprophase band
PR	Pathogenesis-related
PRC	Procuste
pro	Proline
PROPEP	Precursor of peptide
Proton ATPase	Adenosine-triphosphate-dependent proton pump
PRP	Proline-rich protein
PRR	Pattern recognition receptor
PRX	Peroxidase (Chapter 6), Protein kinase associated with BRX (Chapter 19)
PT1	Pto-interacting protein 1
PTI	Pattern-triggered immunity
PTO	Serine/threonine protein kinase
PW	Primary wall
QKY	Quirky
QTL	Quantitative trait locus
QUA	Quasimodo
R	Resistance
Ras	Rat sarcoma
RAB/Rab	Ras-associated binding
RALF	Rapid alkalinization factor
RAY	Reduced arabinose yariv
RB	Retinoblastoma
RBOHD	Respiratory burst oxidase homolog d
RCC	Ray cambial cell
REDOR	Rotational echo double resonance
REN	Rop1 Enhancer
RET	Root extracellular trap
RFDR	Radiofrequency-driven recoupling
RG	Rhamnogalacturonan
RG-I	Rhamnogalacturonan I
RG-II	Rhamnogalacturonan II
RGA	Repressor of gibberellic acid
RGase	Rhamnogalacturonase
RGD	Arginine-glycine-aspartic acid
RGGAT	Rhamnogalacturonan I:Galacturonosyl-transferase
RGL	Rhamnogalacturonan lyase
RGP	Reversibly glycosylated protein
RGXT	Rhamnogalacturonan II:Xylosyltransferase
Rha	Rhamnose
RHO	Family of small signalling G proteins
RHM1	Rhamnose synthase1
RK	Receptor kinase

RLK	Receptor-like kinase
RLP44	Receptor-like protein 44
RNA	Ribonucleic acid
RNAi	RNA interference
RNS	Root nodulation symbiosis
ROCK	Repressor of cytokinin deficiency
ROP	Rho-related protein from plants
ROS	Reactive oxygen species
RRA	Reduced residual arabinose
RRT	Rhamnogalacturonan I:Rhamnosyl-transferase
RT	Rhamnosyltransferase
RWA	Reduced wall acetylation
S-	Syringyl units
S1P	SITE-1 protease
S2i	S2 inner layer
SA	Salicylic acid
SAMK	Stress-activated MAP-kinase
SAR	Systemic acquired resistance
SAUR	Small auxin upregulated RNA gene family
SBT	Subtilisin
SCaBP	SOS3 and SOS3-like calcium binding protein
SCAMP2	Secretory carrier membrane protein 2
SCD	Stomatal cytokinesis-defective
SCM	Scrambled
SCOOP	Serin-rich endogenous peptide
SCW	Secondary cell wall
SELEX	Systematic evolution of ligands by exponential enrichment
SEM	Scanning electron microscope/microscopy
ser	Serine
SGT	Serine galactosyltransferase
SHR	Short root
SHYG	Speedy hyponastic growth
SNARE	Soluble N-ethylmaleimide-sensitive factor protein attachment protein receptor
SND	Secondary wall-associated NAC domain protein
SO_4^{2-}	Sulphate ion
SOK	Soseki
SOS	Salt overly sensitive
SPCH	Speechless
SPR	Spiral
SR	Sulphorhodamine
SUB	Strubbelig
SuSy	Sucrose synthase
SW	Secondary wall
SYP61	Syntaxin of plants 61
T-DNA	Transfer-DNA
TAN1	Tangled 1
TAPG1	Tomato abscission-specific PG
TAXI	*Triticum aestivum* xylanase inhibitor
TBG4	β-galactanase
TBL	Trichome birefringence-like
TC	Terminal complex
TCA	Trichloroacetic acid
TCH	Touch
TEM	Transmission electron microscope/microscopy
TEMPO	2,2,6,6-Tetramethylpyperidine-1-oxyl
TFA	Trifluoroacetic acid
tg	trans-gauche
TGN	Trans-Golgi network
THE1	Theseus1
thr	Threonine
TLC	Thin-layer chromatography
TM	Transmembrane
TMD	Transmembrane domain
TMK1	Transmembrane kinase 1
TMV	Tobacco mosaic virus
TN	Tubular network
TOCSY	Total correlation spectroscopy
TOF	Time-of-flight
TOR	Target of rapamycin
TRAPP	Transport protein particle
trp	Tryptophan
TTG1	Transparent testa glabra 1
TTS	Transmitting tissue-specific protein
TVN	Tubular-vesicular network
TW	Tension wood
tyr	Tyrosine
UAFT	UDP-arabinofuranose transporter
UDP	Uridine diphosphate
UDPG	Uridine diphosphate glucose
UGE	UDP-glucose epimerase
UGNT	UDP-N-acetylglucosamine transporter
URGT	UDP-rhamnose/UDP-galactose transporter
UTP	Uridine triphosphate
UUAT	UDP-glucuronic acid/UDP-galacturonic acid transporter
UV	Ultraviolet
UXT	UDP-xylose transporter
val	Valine
VAMP	Vesicle-associated membrane protein
VIGS	Virus induced gene silencing
V_m	Characteristic molar volume
VND	Vascular-related NAC domain
VRG4	Vanadate-resistant *Saccharomyces cerevisiae* mutant 4 deficient in transport of GDP-mannose
VXXG	Xyloglucan octasaccharide ($Xyl_4.Glc_4$)
WAK	Wall-associated kinase
WER	Werewolf
WFS	Water free space
WSC	Wall stress-responsive component
WSP	Water soluble fraction
XAT	Xylan arabinosyltransferase
XAX	Xylosyl arabinosyl substitution of xylan
XEG	Xyloglucan-specific ß-endoglucanase
XEGIP	Xyloglucan endoglucanase inhibitor protein
XEH	Xyloglucan endo-hydrolase
XET	Xyloglucan endo-transglycosylase
XG	Xyloglucan
XGA	Xylogalacturonan
XGD	Xylogalacturonan deficient
XIP	Xylanase inhibitor protein

XLT	XyG "L"-sidechain transferase
XOAT	Xyla *O*-acetyltransferase
XST	Xyloglucan "S"-sidechain transferase
XTH	Xyloglucan endotransglycosylase/hydrolase
XUT	Xyloglucan-specific galacturonosyltransferase
XXT	Xyloglucan xylosyltransferase
XXXG	Xyloglucan heptasaccharide ($Xyl_3.Glc_4$)
XyG	Xyloglucan
Xyl	Xylose
Xyl*p*	Xylopyranose unit
XylT	Xylosyltransferase
XYN10	Xylanase of family GH10
XYS	Xylan synthase
XYXT	Xylan xylosyltransferase
Y	Yield threshold
YH	Yeast-hybrid
β-Gal	β-Galactosidase
β-Yariv	β-Glycosyl Yariv
µm	Micrometre
π	Osmotic potential (potential pressure)

1 Plant Cell Walls – Past, Present, and Future

Monika S. Doblin
La Trobe University

Yingxuan Ma
La Trobe University
Nanjing Forestry University

Lazar Novaković
La Trobe University
University of Leeds

Antony Bacic
La Trobe University

Kim L. Johnson
La Trobe University

INTRODUCTION

The 21st century has thrown up enormous challenges, and we are living during a period of profound change—the first pandemic since the Spanish Flu of 1918, global climate change as a consequence of fossil fuel use and a human population explosion—in which these events are transforming our societies, agricultural production systems and manufacturing practises. In these uncertain times, there is nothing more certain than 'business as usual is not an option'. The necessary transition away from a fossil fuel–based economy to one based upon renewables has drawn attention back to the importance of plants and their walls. Plants are our most abundant renewable bioresource and are the major source of energy (heat/biofuels), functional foods/feed, bioinspired materials and biopolymer building blocks to replace those based on fossil fuels. Perhaps not since the early hunter–gatherer period have plants been more important in human history for our very existence and that of the diversity of life on Earth. In addition to the traditional drivers in plant biotechnology, we must now place sustainability 'front & centre' of our solutions. How we will feed the growing population, estimated to reach approximately 10 billion by 2050, with sustainable and nutritious food (and fuels) is vital to the survival of our planet and life on Earth as we know it. The projected population increase of 25% in the next 30 years will require the production of 70% more calories as diets change. Promotion of healthy plant-based diets as a preventative medicine for human health and reducing the impact of human activities through more sustainable yet intensified plant cultivation will be part of the solution.

The wall is a polysaccharide-rich structure that surrounds the protoplast of almost all plant cells. It performs a myriad of functions, providing a protective barrier against the environment including pests and pathogens; is a reservoir of signalling molecules to perceive, transduce and respond to both external and internal cues; defines cellular morphology and therefore tissue and plant architecture; and is essential for growth.

Ontogenetically, the wall originates at the cell plate or phragmoplast at cell division (Chapter 25). During growth, the primary wall is laid down between the middle lamella, the junction between cells formed during the later stages of cytokinesis, and the plasma membrane (PM), and in some cells, subsequent differentiation and addition of material can occur at the cessation of growth to form a secondary cell wall (SCW) that may or may not be lignified. During plant growth, the thin, flexible primary wall (approximately 100 nm) must preserve its structural integrity to be able to withstand the hydrostatic forces generated by turgor pressure that drives growth, its thickness maintained with the continuous deposition of new wall material. The primary wall is highly hydrated (~60% of wet weight) comprising an organic phase (~40% of wet weight) that is polysaccharide-rich (~90%) with some (glyco)proteins (~10%). In contrast, the SCW that must endure enormous compressive forces is a dehydrated ('hydrophobic') structure comprised of approximately two-thirds largely unbranched polysaccharides (~50% cellulose and 50% non-cellulosic polysaccharides ('hemicelluloses')) and one-third lignin.

DOI: 10.1201/9781003178309-1

While primary walls are deposited in all cells, SCWs are laid down only in specialised cell types, for instance in xylem elements, tracheids and fibres of the vasculature. The SCW layer, deposited between the PM and primary wall, thickens the wall considerably and offers plants greater mechanical support, and is thus considered an adaptive feature that allowed their colonisation of land (Johnson et al. 2017b).

The proposed basic structure of primary walls is of cellulose microfibrils (MFs) embedded in a 'matrix' composed of hemicelluloses (mainly xyloglucans (XGs) in dicots, gymnosperms and non-commelinid monocots; glucuronoarabinoxylans (GAXs) and (1,3;1,4)-β-D-glucans (MLGs) in commelinid monocots, such as grasses), pectins and (glyco) proteins (Doblin et al. 2010). The thickened walls of cells that contain SCWs are comprised largely of cellulose and hemicelluloses (glucuronoxylans (GXs) and heteromannans (HMs) in dicots/gymnosperms and non-commelinid monocots; GAXs and HMs in commelinid monocots) along with other minor components, including glycoproteins. Lignin, a complex hydrophobic polymer comprised of phenylpropanoids building blocks, is also incorporated into many SCWs.

Structural diversity, organisation and biological function are also displayed within and between other major taxa (algae/fungi/bacteria) with prominent cell walls that are essential to their survival. For example, algal species have distinct walls from terrestrial plants with a predominance of charged (sulphated) polysaccharides in their cell walls (red algae (agars and carrageenans), green algae (ulvans) and brown algae (fucoidans)). The biosynthesis and structure of sulphated polysaccharides is highly complex and beyond the scope of this chapter, but they have been extensively reviewed elsewhere (for example, see Usov 2011; Deniaud-Bouet et al. 2017; Wahlström et al. 2020). Similarly, fungal cell walls are rich in chitin, mannoproteins and β-(3,6)-glucans (Xie and Lipke 2010), and bacterial walls (of both Gram-positive and Gram-negative) consist of complex peptidoglycans (Cabeen and Jacobs-Wagner 2005; Radkov et al. 2018).

Thus, the composition of walls varies depending on the plant species, cell type, tissue, developmental stage and history of responses to (a)biotic stresses, each factor with the potential to modify the type and distribution of constituent polysaccharides/(glyco)proteins within individual cells. That is, both macro- and micro-heterogeneities are observed.

In this opening chapter, we will first provide examples of the extensive utilisation of primary and secondary walls by society, a potted history of the major 'step changes' (Figure 1.1) that have fundamentally transformed our knowledge of walls, followed by an attempt to predict their future roles!

CELL WALLS AND SOCIETIES

Feeding Communities

One of the essential uses of plant tissues is as a food (energy) and dietary fibre source. A range of plant species are consumed in human diets in the form of fruits, seeds, leafy shoots, underground organs such as roots, and flowers. Of the over 400,000 known flowering plant species, less than 1% are consumed as part of today's diets, considerably less than estimates of edible species (Ulian et al. 2020). Today, ~30 domesticated species comprise a significant portion of our dietary diversity, and only three cereal grains—rice, wheat and maize—contribute to more than half of the world's calorie intake (FAO 2010). Plant food and its production systems have changed significantly over the period since domestication began ~8,000–10,000 years ago, being strongly geared towards increased size of harvested organs and greater ease of collection. Commonly, domestication has led to a higher harvest index (ratio of harvested to total biomass) through modified partitioning of photosynthates.

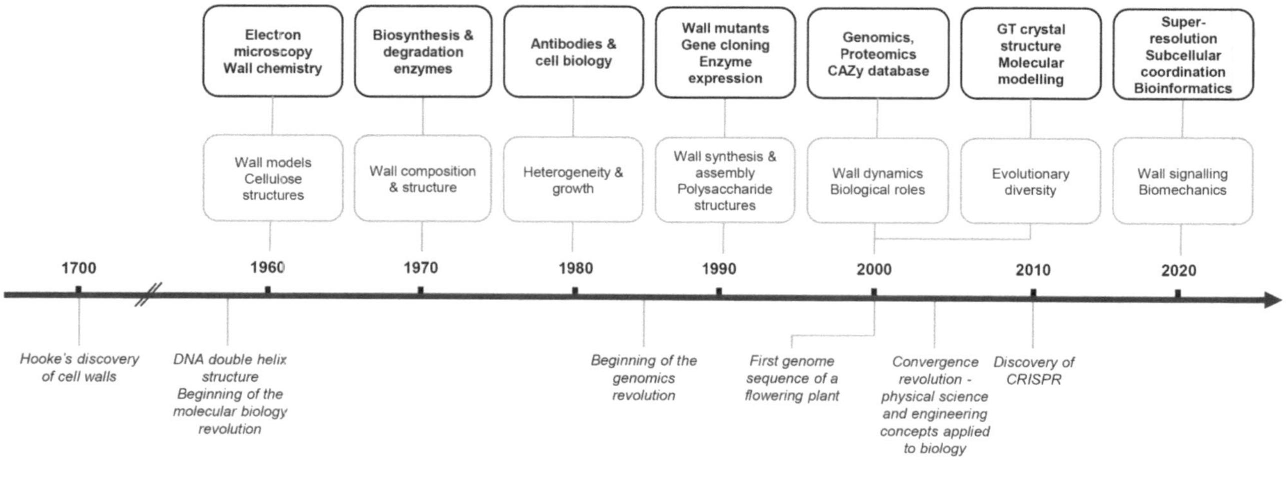

FIGURE 1.1 Timeline highlighting tools and techniques that led to 'step changes' in our understanding of plant walls since their discovery by Hooke in 1665.

impacting both internal stores of carbohydrates such as starch, with attendant, unwitting effects on proteins and fats, as well as walls.

In addition to meeting energy needs, walls play an important function as the major source of dietary fibre, the non-digestible component of plant tissues, in human (and production animal) diets. Dietary fibre components are classified as soluble (hemicelluloses such as XGs, galactomannans and MLGs, pectins, resistant starches and oligosaccharides such as fructans and galacto-oligosaccharides) or insoluble (cellulose and lignin), but depending on the species, tissue source, age and degree of post-harvest processing, many of these polymers can be in either form when consumed and can vary in structure and molecular size. Humans do not possess the digestive enzymes to hydrolyse wall polysaccharides; instead, they are fermented by microbiota in the large intestine. A significant body of evidence has accumulated demonstrating the many health benefits of consuming fruits and vegetables and whole grain cereal foods as a source of dietary fibre (prebiotics) for optimal human health. Dietary fibre exerts physiological influence throughout the gastrointestinal tract via a number of mechanisms, briefly described below. Additionally, wall-based polysaccharides from plant, fungal and algal sources are used extensively as texture modifiers (emulsifiers and thickening agents) in the food processing industry and contribute to the organoleptic properties of most plant-based foods.

Connecting Communities

SCWs often make up the majority of plant biomass, particularly in woody dicots and gymnosperms, and some of the most important applications include their use as a fuel source for heat and mechanical work, the almost infinite applications of wood and paper products, and spinnable fibres for cloth and rope (Doblin et al. 2010; Johnson et al. 2018). Wood has long been used for household items and construction due to its many favourable features, including being renewable, low in energy to produce and an excellent thermal and electrical insulator. Our current understanding suggests that the major factors influencing wood quality relate to the location, number and distribution of SCW cell types within tissues. In softwood species, long, thin tracheids for water and nutrient transport and strength are the most common cell types with small, radially distributed parenchyma cells also present. In hardwoods, large vessel elements form the pores that create the characteristic wood 'grain'. Wall thickness, cell length and wall composition also have important effects on wood properties.

In textile fibres, SCWs are usually cellulose-rich and are desired for their softness. The development of textiles and their many applications is thought to have started with hemp as early as 10,000 years ago. Plant fibres are broadly defined as a cell type that is long, thick walled and possessing high tensile strength (Albersheim et al. 2010 and references therein). Inevitably, you have regularly used products with such fibres, dressing in cotton clothing that is derived from epidermal fibres of seeds, wiping your feet on your coir doormat from coconut fruit fibres, using a reusable jute bag made from stem fibres for your shopping and tied a strong knot with a manila rope made of leaf fibres. Beyond the very 'functional' applications of walls, they have made significant contributions to societies through communication media, art and culture. Beginning with papyrus sheets made from *Cyperus papyrus*, papermaking facilitated documentation of stories, military communication, money and trade. The written word enabled a reading and education culture to be established; book making, libraries and postal services gave insight into the lives of others, broadening views and influencing politics and societies. Similarly, art has used canvas, made of cotton, linen or hemp, for paintings that have depicted war, poverty, love and hope. Representing the diverse properties of wood, musical instruments, wooden sculptures and carvings have added further richness to societies.

These selective examples emphasise the importance of walls to human survival and well-being, providing the context for why they are so necessary to understand. In the following sections, we will provide a historical context of the research and methodological breakthroughs that have enabled us to begin to unravel wall structure and function.

GETTING TO KNOW CELL WALLS

Wall Models

Models are generally useful for conceptualising the experimental evidence of the day and providing guidance on the gaps in our knowledge and how these might be tackled. The sophistication of wall models, elaborated upon in further detail in Chapter 11, reflects the paradigms of their times, the synthesis of accumulated knowledge and often the 'bias' of their creators! In any event, since plants contain around 40 different cell types and walls determine their functional specialisation, it is unlikely that any single model will suffice.

The ontogeny of the wall was captured schematically by Wardrop and Bland (1959) with a particular focus on the multi-lamellate patterning of the SCW visualised by the alternating patterns of deposited cellulose MFs, contrasting with the random orientation of cellulose MFs in the primary wall (Figure 1.2a). This depiction was largely the product of observations from electron (transmission and scanning) microscopy (see section on imaging below).

Since that time, wall models have mostly focused on dicot primary walls, largely of *Arabidopsis thaliana*, the most comprehensively studied plant to date. Some fundamental principles have endured since the first model of Keegstra et al. (1973) almost 50 years ago of a suspension-cultured sycamore cell wall. For example, the biphasic structure of primary walls comprises two to three 'layers' of cellulose MFs embedded in a gel-like matrix of pectins, hemicelluloses and proteins (Figure 1.2b). Additionally, early models were influenced by the prevailing view that the wall was

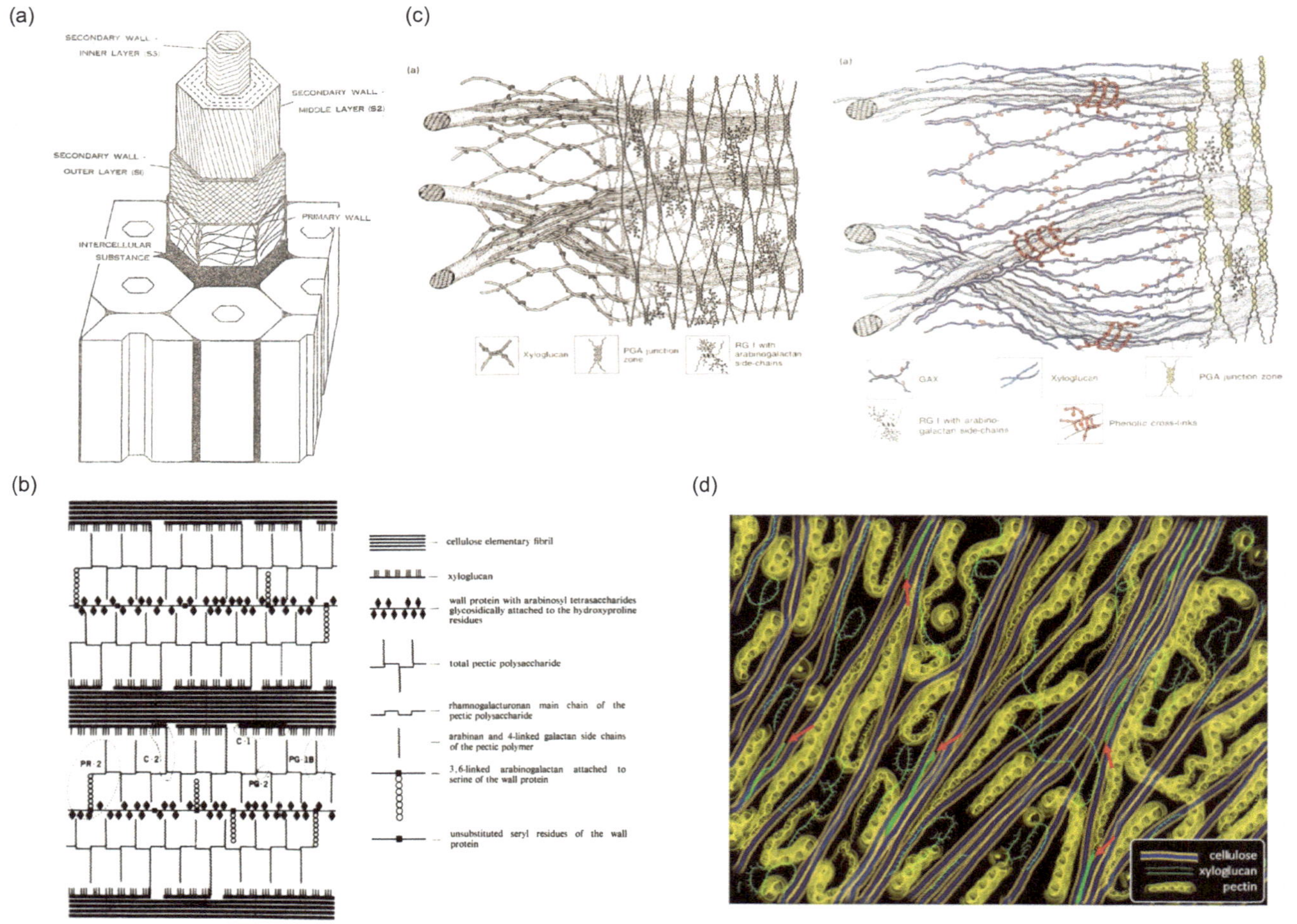

FIGURE 1.2 Wall models over the ages. (a) Schematic representation of the cellulose MF orientation in the primary (P) and secondary (S1, S2, S3) wall layers of a xylem fibre cell or tracheid. Reprinted with permission from Wardrop and Bland (1959). (b) Earliest molecular model of a primary wall based on data from sycamore suspension culture primary walls. Reprinted with permission from Keegstra et al. (1973). (c) Diagrammatic representation of type 1 (left; dicots/gymnosperms/non-commelinid monocots) and type 2 (right; commelinid monocots) primary walls. Reprinted with permission from Carpita and Gibeaut (1993). (d) Conceptual depiction of structural features of a dicot primary cell wall. Cellulose MFs are rods with hydrophobic (blue) and hydrophilic (orange) faces; XG (green) found as both coiled and extended conformations; and pectins (yellow) depicted as coiled structures. Red arrows point to cellulose-XG-cellulose junctions that are sites of wall loosening by bifunctional endoglucanases. Reprinted with permission from Cosgrove (2018).

essentially fulfilling a structural role that could accommodate growth processes. Thus, Keegstra et al. (1973) envisaged the wall as a 'single macromolecule' with covalent cross-links between the hemicelluloses and pectins. This aspect of the wall model has proved irreconcilable with the ability to solubilise various types and proportions of pectic and non-cellulosic polysaccharides from the wall with reagents that only disrupt non-covalent interactions (e.g., chelating and chaotropic reagents), although numerous covalent cross-links have been demonstrated between pectic polysaccharides (e.g., polysaccharide–protein, protein–protein borate-diester bonds of rhamnogalacturonan II (RGII)). The discovery of a cross-linking network involving arabinogalactan protein (AGP) at the core covalently linked to xylan and pectic polysaccharides, named ARABINOXYLAN PECTIN ARABINOGALACTAN PROTEIN 1 (APAP1) (Tan et al. 2013), is the most compelling experimental evidence for a 'single' macromolecular wall complex for at least some proportion of a wall.

Displaying enormous prescience, Keegstra et al. (1973) also predicted a non-covalent interaction (H-bonding) between XGs and cellulose MFs to enable movement by a 'non-enzymically catalysed creep, that is, by the xyloglucan moving like an inchworm along the cellulose fibre' to enable extension growth induced by auxin acidification of the apoplast predicted by Cleland (1971). How the timescales for wall 'creep' matched known biological growth rates was difficult to conceptualise at the time. It was not until some 20 years later that this could be explained with the discovery of 'expansins' by McQueen-Mason et al. (1992) in Daniel Cosgrove's laboratory. Although the molecular mechanism of expansin action remains unresolved, it is known that they are non-catalytic proteins capable of disrupting H-bonds at acidic pH (see Cosgrove (2018) and references therein).

Work emerging in the late 1970s/early 1980s was indicating that primary walls of land plants could be distinguished on their polysaccharide composition and that a simple hydroxycinnamic acid marker, an esterified ferulic acid (and less abundantly *p*-coumaric acid), on AXs was a rapid and diagnostic (autofluorescence) marker. This was summarised in a review by Bacic et al. (1988), where it was proposed that the Embryophyta could be distinguished into two groups based on this wall feature: dicots/gymnosperms/non-commelinid monocots and the commelinid monocots; the latter group having wall-bound ferulic acid. Further differences exist in the abundance of the matrix phase polysaccharides with the first group being rich in pectic polysaccharides and XGs, whereas the second being depleted in both these polysaccharides but containing an abundance of GAXs and some MLGs.

In the 1990s, Carpita and Gibeaut (1993) diagrammatically represented these two groups coining them as type 1 and type 2 walls (Figure 1.2c), groupings that have persisted to this day. This concept was extremely helpful, but it is somewhat redundant as subsequent surveys, particularly using antibody profiling techniques (Moller et al. 2008; Pattathil et al. 2010), indicate a continuum of wall structures rather than clearly delineated wall types (Harris et al. 1997; Harris 2006; Fangel et al. 2012).

This century has seen a remarkable transformation of our understanding of wall assembly and organisation but most importantly its increasing significance in regulating plant growth and development leading to a realisation that any useful model will require a systems-based approach that integrates biophysical, developmental and genetic knowledge (Somerville et al. 2004). Technological advances in imaging and biophysical techniques (see below) have enabled the probing of how primary (largely dicot) wall structures and molecular interaction of (and between) polysaccharides (and proteins) define their biomechanical/functional properties. The outputs have led to new paradigms in wall models and an understanding of how the wall, as a biomechanical construct and cell surface sensor, modulates both plant growth and material properties (Johnson et al. (2018); see the section on biophysical properties below and references therein). This has necessarily led to refinement in wall models and highlighted another complex and interacting network, the wall–PM continuum, as a cell wall integrity (CWI) sensor that likely also exerts control over wall composition (and hence function) (Wolf et al. 2012; Doblin et al. 2014; Vaahtera et al. 2019; Gigli-Bisceglia et al. 2020); see also below.

The initial thinking that the cellulose–hemicellulose (XG) network resists tension while the co-extensive, intermeshed pectin network resists compression and sheering forces has needed re-defining. Previous models depicted that only XGs interacted and cross-linked the cellulose MFs, and this acted as the major load-bearing network. Evidence from an *Arabidopsis* mutant (*xxt1 xxt2*) that lacks detectable XG and displays only a relatively mild growth phenotype (Cavalier et al. 2008), and that commelinid monocot walls are XG-poor and grow normally, indicate that there must be other wall polymers that can interact with cellulose MFs to maintain functional integrity. Furthermore, the *xxt1 xxt2* mutant has altered MF patterning and is more highly aligned and bundled than in the wildtype (Xiao et al. 2016). Nuclear magnetic resonance (NMR) studies of a 'never dried' wall preparation show pectins make the majority of contacts with MFs and may, therefore, be the primary mechanical tethers (Wang et al. 2015). *In vitro* studies using biophysical techniques (X-ray scattering, NMR and small angle neutron scattering (SANS)) to probe bacterial cellulose, XG and pectin hydrogel composites demonstrate that: (i) XG plays a role in separating MFs and modifying crystalline cellulose structure during cellulose crystallisation/assembly, (ii) homogalacturonans (HGs) and the galactan/arabinan side chains of rhamnogalacturonan I (RGI) interact extensively and largely irreversibly with the surfaces of MFs without impacting crystallinity (Lin et al. 2016; Lopez-Sanchez et al. 2017) and (iii) a minor pectic fraction of either HG or neutral side chains of RGI interacts irreversibly with MFs during assembly (Lopez-Sanchez et al. 2016, 2017) without impacting crystallinity (Lopez-Sanchez et al. 2015) and influences mechanical properties through poroelastic effects.

These and other studies have been integrated into a biomechanical 'hotspots' model (Figure 1.2d; Cosgrove (2018) and references therein) in which pectin–cellulose MF interactions are prevalent and XG makes contact with limited sites that are specifically targeted by expansins to allow wall loosening, that is, wall stress relaxation (wall relaxation acid-induced growth). The 'rigidity' of the wall is also influenced by wall pH that can impact non-covalent cross-linking of pectins through the regulation of Ca^{2+} cross bridges, as well as an array of wall-located enzymic proteins that can be active at either low pHs (e.g., polygalacturonases) or higher pHs (e.g., pectin methylesterases (PMEs)) (Höfte and Voxeur 2017). Wall pH is regulated by PM-bound ATPases whose activity is modulated by hormones (auxin and brassinosteroids). Cosgrove (2018) has distinguished 'loosening'/softening caused by expansins from that of lytic enzymes—the former does not impact the tensile strength of the wall and facilitates extension growth whereas the latter likely affect rheological (visco-elastic) properties. That is, many processes can affect 'softening/wall stiffness' but only expansins can allow the wall to grow! Given that certain pectic polysaccharides and heteroxylans, albeit via their hydrophilic face (compared to XGs that interact with MFs through cellulose's hydrophobic face) can interact with MFs, it seems highly likely that mechanisms other than XG hotspots may also act as biomechanical determinants of tensile strength in walls. What role the CWI sensors, some of which, such as the PM-bound Wall-Associated Kinases (WAKs) that are covalently cross-linked to pectins (Kohorn 2016), play in modulating wall properties and how these nano-scale mechanisms operate at the meso-scale in these 'smart biomaterials' is the next challenge (see below).

When devising wall models, the molecular properties of water (for a review, see Dumais and Forterre (2012)) and its distribution throughout the wall is a further consideration that undoubtedly impacts wall properties during growth. There is considerable accumulated evidence, both *in vivo* and *in vitro*, that the hydration/dehydration process of walls/pectic polysaccharides are not entirely reversible and can contribute to the (altered) mechanical properties of walls. Thus, for example, pectin-induced changes in wall mechanics underlie organ initiation in *Arabidopsis* (Peaucelle et al. 2011) and are implicated in affecting the flexibility of the wall during water deficit stress (Moore et al. 2008) and during extreme desiccation in resurrection plants (Kuroki et al. 2019). Chapters 5, 6, 22 and 23 provide a more detailed discussion of the structure–function relationships of pectins and their role in plant growth.

Structural Techniques

It is worth being reminded that oligo/polysaccharides are secondary gene products, and as such, they are not template-synthesised like DNA/RNA/proteins. There is no sequence template that allows one to read the sequence of an oligomeric sugar molecule. To further complicate structure determination, there is no defined length (degree of polymerisation (DP)), although we know that, for example, cellulose chains are shorter and more heterogeneous in primary walls (DP 2,000–6,000) than from SCWs (DP ~14,000). Cells are likely to possess some as yet unknown regulatory processes that control DP as well as a plethora of non-glycosyl substituents (methyl/acetyl/phosphoryl, etc.). All these factors contribute to making the sequencing of a polysaccharide the most challenging of any class of biological macromolecules. As a consequence, the complete sequence of any plant polysaccharide has not been described with rare exceptions, for example, homopolymers with no branching nor non-glycosyl substituents (cellulose) or laminarin, a small branched algal (1,3)-β-glucan (Read et al. 1996). At best, one has to be content with an average structure derived from a multi-pronged approach. It is neither the brief nor is it the scope of this chapter to comprehensively review the myriad of approaches and techniques used to structurally characterise glycans, but a good starting point for polysaccharide and glycoproteins can be found in Pettolino et al. (2012) and Oxley et al. (2004). However, remarkable progress has occurred since the pioneering work of chemists in the late 1800s/early 1900s in polysaccharide structure determination and is briefly overviewed below.

The available tools/methods for structural determination of oligo-/polysaccharides and glycoproteins can be divided into three broad categories: physical, chemical and enzymatic. The first two (physical and chemical) have historically been the workhorse tools and have benefited from enormous advances in chromatographic separation technologies, particularly high pH anion exchange chromatography combined with pulsed amperometric detection (HPAEC-PAD) and instrumentation (particularly mass spectrometry (MS) and NMR spectroscopy). Traditionally, enzymatic approaches lagged, but the advent of molecular biology in the late 20th/early 21st centuries culminated in the availability of easily purified polysaccharide-degrading linkage specific enzymes (Bauer et al. 2006). Coupling of enzymatic approaches with the physical methods (particularly MS) has led to extraordinary insights into the structure of wall polysaccharides and their diversity between cells, tissues, organs and different plant species. This was most elegantly demonstrated by Markus Pauly's group using OLIgo-saccharide Mass Profiling (OLIMP) (Obel et al. 2009; Günl et al. 2010).

The most used and informative chemical technique has undoubtedly been methylation analysis (for a comprehensive review, see Pettolino et al. (2012)). Neutral (to preserve non-glycosyl base-labile substituents) and base-catalysed (using NaOH, most popular and simplest) methylations combined with prior carboxyl reductions for acidic polysaccharides and subsequent acid hydrolysis, reduction and acetylation can provide monosaccharide and linkage composition data in a single analysis by gas chromatography-MS (GC-MS) detection. For oligosaccharides and some small polysaccharides (e.g., laminarin, see Read et al. (1996)), methylation followed by soft ionisation MS approaches (electrospray ionisation (ESI) and matrix-assisted laser desorption (MALDI)) provides both molecular weight and sequence information.

MS and NMR are undoubtedly the two most powerful and used physical techniques as they provide considerable structural information with either very low (MS) or zero (NMR) consumption of material. Advances in NMR instrumentation have greatly contributed to glycan structure elucidation, particularly solution state NMR and more recently the advances in solid state NMR technologies in their various guises have emerged as indispensable in resolving the molecular architecture and intermolecular packing of polysaccharides in both primary and secondary walls (see Zhu et al. (2021a) and Chapter 8 for a review).

Antibodies (polyclonal and monoclonal) first generated for the wall field by the Roberts/Knox/Hahn/Stone laboratories in the 1990s/2000s revolutionised our understanding of wall organisation at the cellular level (see below and Chapter 10 for more details). Antibodies were used to profile polysaccharides in walls to observe spatiotemporal patterning and their solvent-solubilised fractions by techniques such as CoMPP (Comprehensive Microarray Polymer Profiling; (Moller et al. 2007, 2008)) and glycan profiling (Pattathil et al. 2010). Antibodies became even more powerful tools in structure determination following the landmark work of the Pfrengle laboratory building on decades of chemo-enzymic synthesis of oligosaccharides. Using automated chemical synthesis and synthetic glycan microarrays allowed epitopes recognised by long-used antibodies to be characterised (Ruprecht et al. 2017). This also led to the development of a glycan microarray-based assay for plant glycosyltransferases (Ruprecht et al. 2020).

Imaging the Possibilities

Walls, the most distinguishable structure of plant cells, were discovered in 1665 when Robert Hooke used a compound optical microscope to look through thin sections of cork, making plant cells the first cells imaged and called such (Figure 1.3a and b). The thick SCWs, composed mostly of suberin, made the cork cells easily observable. They reminded Hooke of monk cells in monasteries (Latin *cella*) from which the name for the basic unit of life is derived.

Imaging technologies progressed slowly for centuries so the wall remained little more than a thick line defining cell shape with limited knowledge of cellular structures. Improvements in organic and analytical chemistry during the nineteenth century enabled deciphering the polysaccharide building blocks of the wall, cellulose, pectins and hemicelluloses (Braconnot 1825; Payen 1838; Schulze 1892), thereby enabling microscopists to interpret images of the ultrastructure of the wall.

Clues of wall ultrastructure were initially obtained using different modes of electron microscopy (EM). Transmission EM (TEM) helped resolve the structure of cellulose, showing that it forms long unbranched MFs (Frey-Wyssling et al. 1948). Major advances in understanding cellulose MF structure were made by imaging walls of the green alga *Valonia* (Figure 1.3c) (Preston et al. 1948; Preston and Ripley 1954). Numerous crystalline properties of cellulose MFs, including the distinction between Iα and Iβ cellulose allomorphs (Hebert and Müller 1974), were discovered by applying various combinations of diffracted electron beams (Ogawa et al. 2019). Initial TEM imaging had predicted 36 glucan chains, six per subunit of the 'rosette' (cellulose synthase complex (CSC)), in one elementary MF (Herth 1983). However, these results did not concur with data generated using wide angle X-ray scattering and NMR, which suggested an elementary MF comprises either of 18 or 24 glucan chains (Newman et al. 2013). This was later validated with computational modelling (Oehme et al. 2015). Finally, improved freeze–fracture TEM and advances in image analysis software confirmed that the basic MF contains 18, not 36 glucan chains (Nixon et al. 2016), thus fundamentally changing our understanding of the CSC. The structure of the cellulose synthase (CESA) catalytic subunit was also investigated using cryo-EM, which provided an insight into CESA oligomerisation and catalytic mechanism (Purushotham et al. 2020; Qiao et al. 2021).

Advances in light and fluorescence imaging coupled with the development of fluorescent molecules and fluorophore conjugates that target different types of polysaccharides (DeVree et al. 2021) enabled tissue- and cellular-level examination of the distribution of wall molecules. Dyes such as Calcofluor White (detecting a broad range of β-glucans), used extensively by the textile industry, was used to detect cellulose and are still widely used today (Herth and Schnepf 1980). These fluorophores enabled the basic properties of the wall (thickness, shape, cellulose distribution and orientation, etc.) to be revealed. Later, more specific molecules such as Pontamine Fast Scarlet 4B and the Carbohydrate Binding Module 3a (CBM3a) were developed and applied (Figure 1.3d) (Lehtiö et al. 2003; Anderson et al. 2009; Liesche et al. 2013).

Further advances were made using various modalities of EM. Scanning electron microscopy (SEM) and field emission SEM (FESEM) enabled high-resolution imaging of cell surfaces, thus enabling characterisation of wall features of the outer organ layer or the inner face of walls of cells cut open. FESEM and related methodologies enabled quantification of cellulose MF length, width, orientation and other structural properties (Figure 1.3e) (Du et al. 2020). Further advances, such as cryo-EM and high-pressure freezing, which reduces sample damage during preparation due to ice crystal formation, enabled samples to be viewed in near native state and with increased resolution (Studer et al. 2008; Cressey and Callaway 2017). For example, electron tomography of high-pressure frozen *Arabidopsis* meristematic cells in different stages of mitosis has provided an insight into the assembly of the phragmoplast (Seguí-Simarro et al. 2004). Focused ion beam (FIB) microscopy, similar to SEM except that an ion beam is used instead of electrons, enabled the imaging of internal tissue layers (Bassim et al. 2014). FIB combined with electron cryotomography was used to image the layers of onion bulb walls showing that the ~90° alternations of MF angle were preserved in internal lamellae of the wall, throughout the onion bulb scales (Nicolas et al. 2021).

An array of different polysaccharide-specific antibodies (monoclonal and polyclonal) were established during the 1980s–2000s and remain some of the best tools to determine high-resolution localisation of wall components (Roberts et al. 1985; Puhlmann et al. 1994; Knox 1997; DeVree et al. 2021). For example, a study of tobacco endosperm cells using confocal laser scanning microscopy in which three different RGI components, arabinan, galactan and the RGI backbone were labelled (Figure 1.3f) showed the heterogeneity in the distribution of different RGI domains at the tissue and cell levels. Fluorescence-based imaging of walls contributed to a major shift in viewing the wall as a dynamic structure. For example, pectin dynamics, as well as their structural and developmental roles were further uncovered using other fluorescent dyes (e.g., propidium iodide) and fluorescently tagged oligosaccharides, such as the chitosan COS-488 probe (Mravec et al. 2014). These probes have also been used to understand the role of mechanical forces and pectins in *Arabidopsis* epidermal pavement cell development (Altartouri et al. 2019), to understand wall dynamics in growing pollen tubes (Rounds et al. 2011), and in root hairs during growth (Schoenaers et al. 2018). Wall imaging has also benefited from recent innovations, such as metabolic labelling using click-chemistry (Anderson and Wallace 2012), fluorescently tagged precursors for *in muro* labelling (Dumont et al. 2016) and lipophilic dyes (Stocker and Timmer 2013). Raman spectroscopy

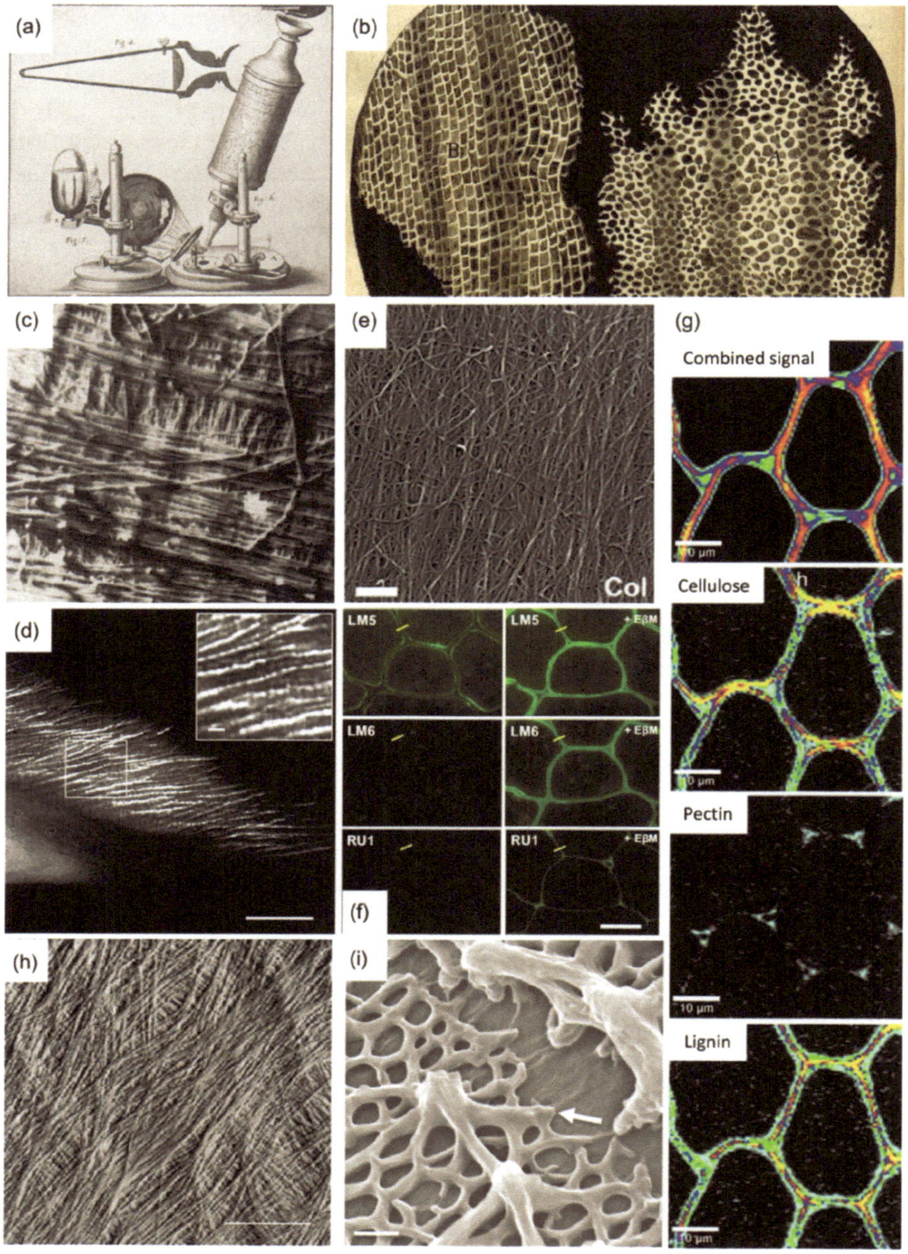

FIGURE 1.3 Revealing wall ultrastructure on different length scales using an array of imaging techniques. Microscope used by Robert Hooke (a) to image cork cells (b), thus visualising cells and walls for the first time. Adapted from Hooke (1665). (c) TEM of green alga *Valonia ventricose* wall, revealing the multi-lamellate arrangement of the cellulose MFs. Scale bar = 1 μm. Modified from Preston et al. (1948). (d) Pontamine Fast Scarlet 4B staining showing cellulose MFs in onion epidermal cells imaged using super-resolution confocal microscopy. Scale bar = 10 μm, 1 μm in inserts. Adapted from Liesche et al. (2013). (e) FESEM image of *Arabidopsis* hypocotyl walls providing high resolution images with clear details of cellulose MFs, enabling quantification of different cellulose structural properties, such as MF orientation, thickness, fibre density and interaction with other polymers. Scale bar = 200 nm. Reprinted with permission from Xiao et al. (2016). (f) Confocal micrographs using polysaccharide-specific antibodies to reveal (sub) cellular heterogeneity in the distribution of RGI: LM5 and LM6 visualising galactan and arabinan-rich areas, respectively, and RU1 the RG-I backbone, before (left panels) and after (right panels) endo-β-mannanase (+EβM) demasking treatment in tobacco endosperm walls. Scale bar = 10 μm. Adapted from Lee et al. (2013). (g) Non-invasive investigation of cell wall composition using Raman confocal microscopy in collenchyma cells of *Phormium tenax* leaves. Colour denotes relative abundance and location of each polymer (cellulose/pectin/lignin) in both a tissue/cell context. Scale bar = 10 μm. Adapted from Richter et al. (2011). (h) AFM image of 'never-dried' epidermal wall of *Allium cepa* (onion) bulbs. AFM gives high resolution images, with minimal physical disruption of the sample, enabling quantification of various parameters such as cellulose MF orientation, thickness, fibre density and interaction with other polymers. Scale bar = 500 nm. Reprinted with permission from Zhang et al. (2014). (i) FESEM image of HG polygonal lattice deposits on the surface of the external wall layer of the unicellular alga *P. margaritaceum* unveils the structure of pectin polymers. Scale bar = 150 nm. Adapted from Domozych et al. (2014).

has also provided a valuable insight into the distribution of chemical components in native cell walls (see Figure 1.3g) (Richter et al. 2011).

An indirect method yielding molecular insight into how the wall is constructed and regulated has come from advances in live cell confocal imaging of various wall-related proteins (e.g., biosynthesis enzymes, regulatory proteins including membrane trafficking components, and elements of the cytoskeleton) expressed as fluorescent fusion proteins (see below). Dynamics of cellulose synthesis was elucidated using *Arabidopsis* fluorescent reporter lines *pCESA3:GFP:CESA3* and *pCESA6:YFP:CESA6* crossed with various microtubule/actin filaments markers, in various mutant backgrounds, and imaged with spinning disc confocal microscopy (Paradez et al. 2006; Persson et al. 2007; Crowell et al. 2009; Gutierrez et al. 2009; Fujita et al. 2011; Sampathkumar et al. 2013). Confocal microscopy has determined relationships between the velocity of CSC movement in the PM and rate of cellulose synthesis; the faster the CSCs move across the PM, the more the cellulose produced (Gutierrez et al. 2009; Fujita et al. 2011; Vain et al. 2014; Watanabe et al. 2018). Use of different *Arabidopsis* mutants crossed with fluorescently tagged cytoskeleton markers has shown that CSCs are physically connected to cortical microtubules (Gu and Somerville 2010; Bringmann et al. 2012) and move along them during cellulose synthesis. Microtubules (MTs) were also shown to position CSCs prior to delivery to and recycling from PM via a microtubule-associated cellulose synthase compartment termed SmaCCs/MASCs, as well as influence the degree of cellulose crystallinity (Paredez et al. 2006; Crowell et al. 2009; Gutierrez et al. 2009). Live cell confocal imaging has also shown that the actin cytoskeleton plays an important role both in delivery of CSCs to the PM and in regulation of their biosynthetic properties, such as their velocity and resident time at the PM (i.e., turnover) (Sampathkumar et al. 2013).

The combined application of high-resolution microscopy using live and fixed tissue with molecular biology approaches has begun to expose the underlying processes by which the spatiotemporal heterogeneity of wall components is elaborated during biosynthesis, secretion and/or post-deposition, although these processes are yet to be fully understood (see Chapters 4, 7, 14 and 24 for further details). Many insights into where wall constituents are made within the secretory system and how these molecules are trafficked to and from the PM have been revealed, although integration of the biosynthesis and trafficking fields has been somewhat lacking to date. The consensus appears to be that the majority of the numerous motile Golgi stacks in plants contribute to cell wall matrix polysaccharide and glycoprotein biosynthesis, likely using a combination of sequential sugar addition and enzyme complexes. In most cases, cargo reaches the PM through the conventional secretory pathway via the trans-Golgi network (TGN) regulated by YPT/RAB GTPase interacting proteins YIP4a/YIP4b in a complex with ECHIDNA, a protein of unknown function (Zhang and Staehelin 1992; Somerville et al. 2004; Young et al. 2008; Gendre et al. 2011, 2013; Hoffmann et al. 2021). Some notable exceptions have been detected, for instance glycosylphosphatidyl-inositol (GPI)-anchored AGPs (Poulsen et al. 2014) and some resident wall proteins, for example, PMEI1, PGIP2 (De Caroli et al. 2011, 2015) that have been shown to be secreted through unconventional pathways (Ding et al. 2012; Robinson et al. 2016; van de Meene et al. 2017), including exocyst-positive organelles (EXPOs). The exocyst is a conserved octameric protein complex found in yeast, mammals and plants and is critical for specification of the site of vesicle docking and fusion, likely acting as a PM tethering complex (Synek et al. 2006; Žárský et al. 2009, 2013). Whereas each exocyst protein is encoded by a single gene in yeast and mammals, paralog number is expanded in plants, particularly the Exo70 subunit that contains a C-terminal lipid binding region. It has been proposed that the Exo70 paralogs in plants serve as landmarks for exocyst vesicle targeting which may initiate and then maintain specific membrane domains for delivery and/or recycling both in polarised cells such as pollen tubes and root hairs, and also in cells undergoing diffuse growth. Thus, cargo delivery by EXPOs provides a mechanism by which the spatiotemporal patterning of wall deposition may be precisely controlled (Žárský et al. 2009, 2013).

High-end microscopy techniques including fluorescence recovery after photobleaching (FRAP) have also been used to dissect post-deposition processes. For example, in the case of lignin, it appears that the localised secretion and/or anchoring of oxidative enzymes such as laccases and peroxidases control when and where lignin is polymerised in specific wall domains such as thickened SCWs, cell corners and the middle lamella (Schuetz et al. 2014; Yi Chou et al. 2018; Hoffmann et al. 2020). Thus, protein anchoring via covalent attachment to wall constituents within distinct regions is another means by which plants can regulate localised wall structure and architecture to high resolution.

Atomic Force Microscopy (AFM), a non-destructive method yielding nanometer resolution images with minimum sample preparation, has also brought important knowledge advances. AFM confirmed earlier TEM findings relating to cellulose structure, for example, MF diameter and their cross-laminate organisation in onion and maize walls (Figure 1.3h). Due to its large and robust fibrils, cellulose is the wall polymer best observable using current methods. Hemicelluloses and pectins have proven more difficult to visualise since they are branched, generally non-fibrillar, interact with many other polymers, are of smaller size and mass, and are highly hydrated forming the gel matrix phase of the wall (Zhang et al. 2016b, 2017). In specific systems, some of the gel-forming polysaccharides, such as HG, have been visualised *in vivo/in muro*. For example, the single-cell alga *Penium margaritaceum* has a multi-layered outer wall that has an intricate polygonal HG lattice sitting on the cell surface above a cellulose layer (Figure 1.3i) (Domozych et al. 2014). In parallel, super-resolution microscopy techniques have also proven insightful with regard to wall structure and architecture. Cryo-SEM and three-dimensional direct stochastic optical reconstruction microscopy (3D-dSTORM) have revealed existence of HG nano-filaments in anticlinal cell walls of *Arabidopsis* epidermal pavement cells (Haas et al. 2020). MLG and AX have been imaged *in vivo* within the walls of *Lolium multiflorum* suspension-cultured cells, but simultaneously imaging hemicellulose and pectin polymers *in vivo* and *in muro* using either EM- or AFM-based methods remains challenging.

Genetics and Molecular Biology

Our understanding of wall structure, biosynthesis and function advanced rapidly when genetics and molecular biology were introduced into the wall field in the 1990s/2000s. Mutant collections in *Arabidopsis* and later rice (*Oryza sativa*) were instrumental to the identification of many genes underlying wall-deficient phenotypes such as root/hypocotyl swelling, dwarfism and irregular xylem (Figure 1.4), a process that was made significantly easier by the availability of sequenced genomes. The advent of polymerase chain reaction (PCR), a key technique in gene cloning, followed by *in planta* testing, largely using *Agrobacterium*-mediated plant transformation, has underpinned gene functional analysis. In more recent years, further genomic and RNA sequencing (microarrays and other high-throughput technologies), bioinformatics including comparative analyses between species, proteomics and protein structural analyses have driven key discoveries of the proteins responsible for wall biosynthesis, the biological functions of wall components, conservation/divergence of structure and function

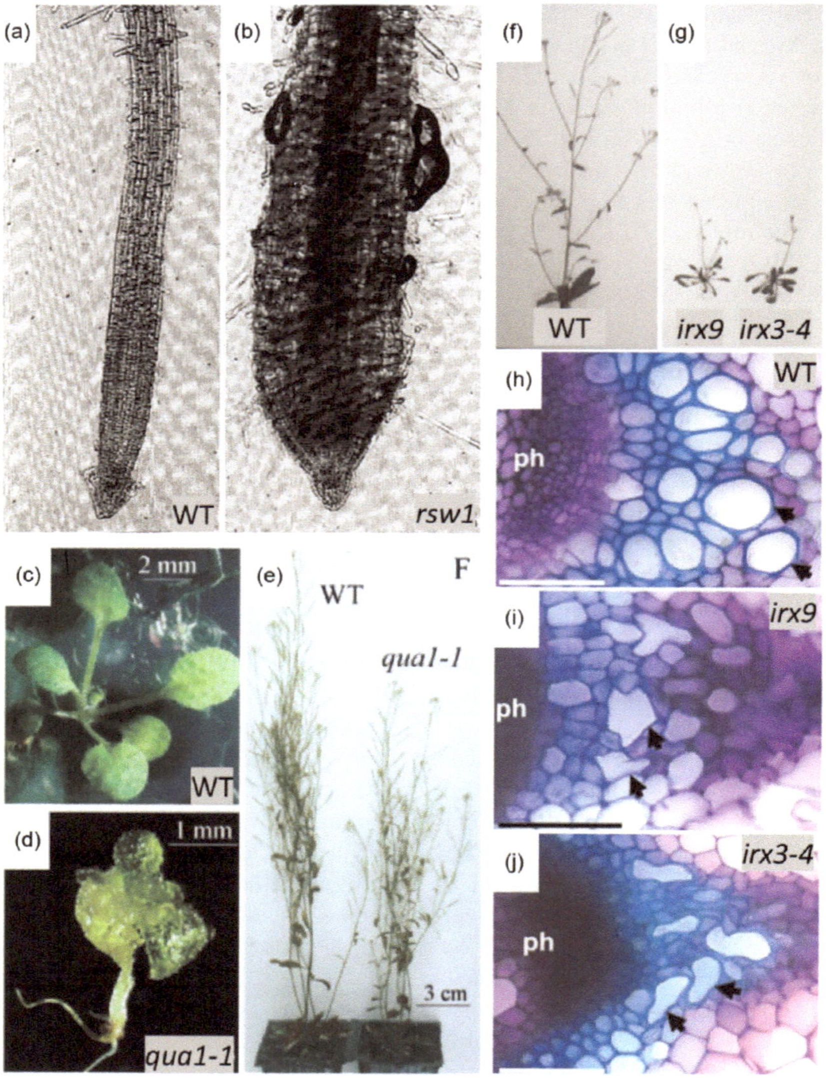

FIGURE 1.4 Phenotypes of *Arabidopsis* wall mutant plants compared to wildtype (WT). (a–b) Root swelling phenotype observed in *rswl* (*cesal*) mutant plant (a) compared to WT (b). Adapted from Baskin et al. (1992). (c–e) Pectin synthesis mutant *qual-1* causes dwarfism and defective leaf phenotypes compared to WT. Adapted from Bouton et al. (2002). (f–j) Xylan backbone synthase (*irx9*) and secondary wall *CESA* (*irx3–4*) mutant plants have a dwarfism phenotype with irregular xylem (*irx*) morphology in stems compared to WT. Adapted from Brown et al. (2005).

through plant evolution and wall dynamics (see Chapters 2, 3, 9 and 13 for additional information).

Cloning of the first wall synthesis genes stemmed from decades of biochemical investigation of enzyme activities using plant species/tissues rich in specific wall constituents. For example, plant *CESA* genes were first cloned from cotton (*Gossypium hirsutum*) fibres during active SCW development (Pear et al. 1996). Close homologs of cotton *CESAs* were then cloned from *Arabidopsis* including *RSW1* (*AtCESA1*) and *IRX3* (*AtCESA7*), respectively (Turner and Somerville 1997; Arioli et al. 1998). AGP protein backbone cDNAs were also cloned from pear (*Pyrus communis*) suspension culture cells and *Nicotiana alata* style tissue in the early 1990s (Chen et al. 1994; Du et al. 1994). Cloning of glycosyl transferase (GT) genes greatly advanced when genome and transcript sequencing data became available for a wide variety of plants in the 2000s. Comprehensive GT gene lists were compiled and categorised into families with the CAZy database (www.cazy.org) launched online in 1998 (Drula et al. 2021). This database now includes ~600 predicted wall biosynthesis enzymes from *Arabidopsis* and has provided an informative map for cloning GT genes and biochemical activity analyses to verify function.

The simultaneous development of next-generation sequencing and bioinformatics tools brought wall research into a new era. In 2000, the first plant genome became available with the sequencing of *Arabidopsis* (The Arabidopsis Genome Initiative). The role of *Arabidopsis* as a plant research model for the next two decades was established, including for the investigation of walls. It has been predicted that of the approximately 30,000 genes in *Arabidopsis*, ~10% relate to wall processes (Liepman et al. 2010). Genome sequencing coupled with gene annotation has introduced a voluminous amount of information about wall-related genes, revealing a higher complexity than expected. For example, the plant-specific Hydroxyproline-Rich Glycoprotein (HRGPs) family thought initially to only contain extensins (Lamport 1977) is now recognised as a protein superfamily that includes very diverse extensins, AGPs and proline-rich proteins (PRPs) including hybrid and chimeric variants (Showalter et al. 2010; Johnson et al. 2017a). Microarrays, followed by RNA sequencing, have been widely used to identify genes involved in regulating wall synthesis, modification and signalling pathways. For example, by analysing microarray data, a variety of genes co-expressed with secondary wall *CESAs* were identified as potential candidates involved in SCW synthesis, including laccase, COBRA-like 4, fasciclin-like AGP (FLA) 11 and 12 and KNAT7 (Persson et al. 2005), although the precise role of their encoded proteins has proven challenging to resolve. With genome and transcriptome sequencing data available from a range of plant species, comparative analyses were able to be undertaken. The *Cellulose synthase-like F (CSLF), H* and *J* subfamilies, for example, were found to be grass-specific and various isoforms have been shown to be (1,3;1,4)-β-D-glucan synthases (Burton et al. 2006; Doblin et al. 2009; Schwerdt et al. 2015; Little et al. 2018). Comparison of wall genes between algae, mosses and higher plants made possible by the 1,000 Plant Transcriptome Initiative (onekp.com) also revealed the conservation and expansion of wall gene numbers such as HRGPs and transcription factors regulating wall synthesis (Xu et al. 2014; Hervé et al. 2016; Johnson et al. 2017b).

Largely in parallel to gene cloning, approaches to induce single nucleotide polymorphisms or small deletions, such as physical/chemical induced gene mutations, as well as methods that disrupt gene function, T-DNA insertions and later RNA interference (RNAi), have proven highly valuable tools for investigation of wall synthesis processes. Between the 1970s and 1990s, a number of mutants were identified that disrupted (largely secondary) wall composition and growth including the brown midrib mutants with defects in lignin in maize (*Zea mays* L.) and sorghum (*Sorghum bicolor* L.) (Barriere et al. 2004), defects in cotton (*Gossypium hirsutum* L.) fibre elongation (Kohel et al. 1974) and barley (*Hordeum vulgare* L.) brittle culm mutants (Kokubo et al. 1991). It was in the 1990s with the establishment of resources in *Arabidopsis* such as T-DNA collections, mapping populations and genome sequencing initiatives that the underlying genes causing wall defects could be uncovered on a larger scale (O'Malley and Ecker 2010). Evidence of the biological importance of cellulose for plant development was shown by investigation of the *Arabidopsis rsw1* mutant, a chemically induced mutant with defects in *AtCESA1* (Figure 1.4a and b) (Baskin et al. 1992; Arioli et al. 1998). Other *CESAs* were functionally verified using mutants including *ixr1* (*AtCESA3*), *ixr2/prc1* (*AtCESA6*), *irx1* (*AtCESA8*), *irx3* (*AtCESA7*) (Figure 1.4f–j) and *irx5* (*AtCESA4*) (Turner and Somerville 1997; Taylor et al. 1999, 2000; Scheible et al. 2001; Brown et al. 2005), together with *cesa2*, *cesa5* and *cesa9* mutants, revealing that three different CESA isoforms form primary and SCW CSCs (Desprez et al. 2007; Persson et al. 2007). The important function of pectin for plant growth and development was supported by analysis of a T-DNA insertion mutant *quasimodo1* (*qua1-1*), which disrupted *QUASIMODO1/GAUT8* that encodes a putative galacturonosyltransferase, and led to stunted plants with reduced cell adhesion and defects in pectin synthesis (Figure 1.4c–e) (Bouton et al. 2002). Xylan backbone synthesis genes were also identified and verified by analysing T-DNA insertion mutants including *irx9/irx9L*, *irx10/irx10L* and *irx14/irx14L* (Figure 1.4f–j) (Brown et al. 2007, 2009; Pena et al. 2007; Wu et al. 2009, 2010). Some of the key transcription factors regulating SCW development including VASCULAR-RELATED NAC-DOMAIN (VND) 1–7 and NAC SECONDARY WALL THICKENING PROMOTING FACTOR 1 and 3 (NST1 and SND1/NST3) were identified and verified using mutant and overexpression plants (Kubo et al. 2005; Zhong et al. 2006; Mitsuda et al. 2007). The discovery of transcription factors regulating major transitions in wall development enabled a suite of other advances for understanding wall biosynthesis. Combined with genome editing tools, a useful VND7 system was developed which can induce protoxylem development at the plant tissue surface (Yamaguchi et al. 2010). The development of CRISPR-Cas9 approaches to edit wall biosynthesis genes is expected to more efficiently and accurately reveal function in a range of plant species.

Application of protein–protein interaction techniques such as yeast-hybrid (YH), co-immunoprecipitation (co-IP) and bimolecular fluorescence complementation (BiFC) have enabled investigations of both biosynthetic GT complexes as well as their cognate interactor proteins (e.g., Gu and Somerville 2010; Atmodjo et al. 2011; Zeng et al. 2016), allowing more molecular players to be revealed and in part validating gene co-expression analyses, particularly in the case of cellulose biosynthesis (Ruprecht et al. 2016). Proteomic studies that enable systematic, qualitative and quantitative analysis of proteins, for example, enriched sub-cellular organellar fractions, in addition to their post-translational modifications (e.g., Nühse et al. 2004; Dunkley et al. 2006; Drakakaki et al. 2012; Parsons et al. 2012) have also contributed to our understanding of walls, especially CSCs and their regulatory networks. Y1H was used to investigate protein–protein interactions involved in the regulation of SCW development (Taylor-Teeples et al. 2015). Golgi-localised proteins that may be involved in wall synthesis and GPI-anchored proteins that may regulate this process or potentially wall sensing at the PM were able to be identified by coupling sub-cellular fractionation and proteomics methods (Borner et al. 2003; Parsons et al. 2012). Co-IP and BiFC have been widely used to identify proteins/protein complexes and pathways involved in wall synthesis and signalling, together with the sensitive MS-based proteomics methods that have provided vast datasets to identify novel wall synthesis proteins, analysis of protein complexes and to predict the post-translational modifications involved in wall synthesis and dynamics (McBride et al. 2017; Mergner et al. 2020).

Biophysical Properties

The wall is now viewed as a 'smart' material with its own biophysical (also known as mechanical) properties, such as elasticity, stiffness and adhesive properties, that together with sensing capacity are determinants of plant tissue and organ morphogenesis and growth (Altartouri et al. 2019; Zhang et al. 2019). Historically, stress patterns in plants were observed by biologists in the 1800s such as Julius von Sachs and Wilhelm Hofmeister who incorporated tissue tension into concepts of organ growth with elegant studies cutting sunflower stems and deducing which layers were under tension/compression (Kutschera and Niklas 2018). In the early 1900s, D'Arcy Thompson used mathematics to study the mechanics of morphogenesis, pioneering the study of biophysics and modelling (Thompson 1917). The role of mechanics and stress in plant growth continued to be investigated during the subsequent decades yet was largely overshadowed by the focus on hormones and the wave of molecular technologies. Of note are the studies by Paul Green in the 1960s who identified microtubules as directing the orientation of cellulose MFs in *Nitella* (Green 1962, 1968). MTs have since been shown to orientate in response to stress patterns using cell biology tools, such as labelling of MTs (Hamant et al. 2008; Sampathkumar et al. 2014), MT and wall mutants (Ambrose et al. 2011; Uyttewaal et al. 2012; Belteton et al. 2021) combined with mathematical modelling of stress dynamics and AFM (Sampathkumar et al. 2014; Bidhendi and Geitmann 2019).

In parallel with the developments in cell biology techniques, the integration of methods and technologies developed in materials science and engineering for mechanical and biophysical characterisation of molecules and materials (see Figure 1.1) to the wall have informed how mechanical behaviour on the nano-scale influences morphogenesis on a macro-scale. This convergence of disciplines in the late 20th/early 21st centuries resulted in a paradigm shift in the conceptual thinking of how we view and interpret the structure and function of biological structures. Here, we briefly overview how some of these developments that have influenced our understanding of wall biology. This section should be read in conjunction with the earlier sections on structural techniques and imaging where coverage of other physical technologies (e.g., NMR) is discussed.

Early studies of wall mechanical properties were performed using tensile testing. In tensile tests, a sample is mounted on an instrument called an extensometer, and stretched in either one direction (uniaxial) or in two directions (biaxial) (Bidhendi and Geitmann 2019). Uniaxial tensile tests, such as wall creep tests, are most often used for testing of biological materials mostly because of a simple experimental setup, geometry of the sample and the ease of calculating mechanical properties (see Figure 1.5a) (Bidhendi and Geitmann 2018, 2019). Wall creep tests mimic the growth conditions in plant cells, by applying a constant force to the sample resembling the one caused by turgor pressure. In particular, creep tests have helped us understand acid pH-induced growth of plant cells (Cosgrove 1989; Durachko and Cosgrove 2009), the roles (Park and Cosgrove 2011; Zhang et al. 2019), as well as the molecular behaviour of wall polymers (Marga et al. 2005) and of specific wall regions during growth (Wang et al. 2021).

In parallel, extensometers have been coupled with confocal microscopes to follow the behaviour of different fluorescently labelled structures, such as proteins, organelles or the cytoskeleton during live tissue/organ deformation and tracking of mechanical properties at cellular resolution (Robinson et al. 2017). For example, an automated confocal micro-extensometer (Robinson et al. 2017) was used to observe behaviour of fluorescently labelled MTs in hypocotyls during tensile stress tests. The study confirmed results obtained by other, more indirect methods, such as cell ablation, that MTs orient perpendicularly to the direction of the highest mechanical stress in the tissue (Robinson and Kuhlemeier 2018). These approaches enable tracking the direct molecular output (i.e., microtubule dynamics) that mechanical stress has on cell dynamics.

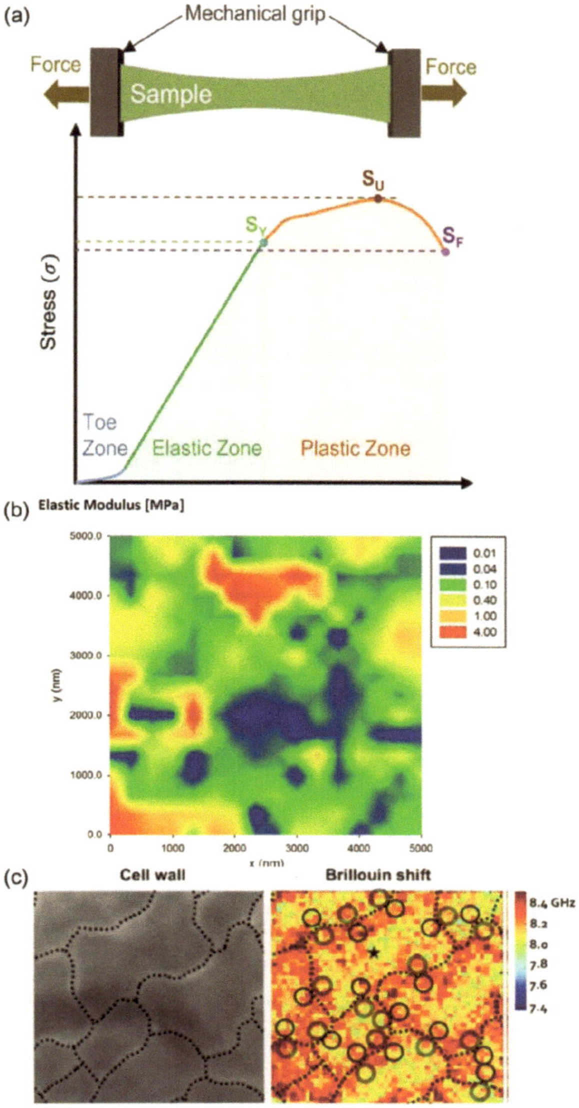

FIGURE 1.5 Assessing mechanical properties of the plant cell wall using different techniques. (a) Stylised stress–strain graph for the loading portion of tensile testing. Wall creep tests performed using uniaxial deformation of plant tissue give valuable information about wall growth and its mechanical behaviour. Walls experience non-linear deformation, meaning that different nano-regions of the wall will deform at different intensities and rates. Modified from Bidhendi and Geitmann (2019). (b) Nano-indentation performed using AFM-enabled understanding of mechanical properties of the wall on micro- and nano-scales. Image represents distribution of Young's elastic modulus on a 5×5 μm patch of wall of *Lolium multiflorum* protoplast regenerating a wall. Analysis demonstrated high heterogeneity of cell wall mechanical properties, which was correlated with the wall biochemical composition. Areas of the cell wall with the lower elastic modulus are softer, and vice versa. Modified from Yakubov et al. (2016). (c) Analysis of the mechanical properties of the neck and lobe wall regions of *Arabidopsis* epidermal pavement cells. Brillouin shift is given in GHz with higher values indicating a higher longitudinal modulus, which gives information on visco-elastic properties of the material. Pavement cell necks have higher stiffness than lobes. Adapted from Altartouri et al. (2019).

AFM has emerged as a popular method to ascertain the different mechanical properties of plant organs, tissues, cells or even isolated walls. AFM was previously mentioned as a powerful method for non-invasive imaging of native walls (see Figure 1.3h and the above section on imaging), but AFM is a highly multifunctional tool. AFM nano-probes, that is, cantilevers fitted with tips of variable size and shape, can also be used to mechanically characterise materials, using nano-indentation methods (Gavara 2017). Materials are commonly characterised using apparent elastic modulus (Young's modulus), which determines the degree of elastic deformation of a material when it is subjected to compression on its outer boundary (Gavara 2017) (Figure 1.5b).

One of the major contributions to our knowledge of wall growth was made using AFM combined with tensile test results and wall degrading enzymes on onion wall monolayers as a model system, revealing the contribution of various wall components in the wall response to different aspects of mechanical stress (Zhang et al. 2019). Responses of the wall to indentation depends on the elastic compliance of HG and bendability of the cellulose MFs, while wall responses to creep and stretching are influenced by selective wall loosening at certain points (biomechanical hotspots; see also Figure 1.2d) and compliances of the cellulose MF networks (Cosgrove 2014; Zhang et al. 2019).

Microheterogeneity of wall mechanical properties were observed in AFM studies of *L. multiflorum* suspension-cultured cells and leaves of *A. thaliana* (Yakubov et al. 2016). AFM was used to determine the distribution of apparent elastic modulus values on the surface of the cells and leaves, by constructing the so-called 'force maps', which enabled the distribution of mechanical properties of the wall over the analysed cell or tissue region on micro- and nano-scales to be determined (Figure 1.5b). These ranges of wall elastic properties also correspond to those observed in a study by Zhang and co-workers (2019) in onion epidermal wall monolayers. This suggests that heterogeneous distribution of cellulose MFs, pectins and other components throughout the wall contributes to the observed variability of the apparent elastic modulus values.

Another interesting method for assessment of wall mechanical properties is Brillouin acoustic confocal microscopy (Figure 1.5c). This method enables calculation of a longitudinal modulus giving information on the viscoelastic properties of the material. Analysis of *Arabidopsis* cotyledon pavement cells has shown that pavement cell necks have higher stiffness than lobes (Altartouri et al. 2019). This method, although not widely used in plant biology, does not require sample preparation, and has the potential to combine with Raman microscopy and other imaging techniques to assist in correlating mechanical properties of the wall with biochemical composition and/or put it into appropriate developmental/genetic context.

Synthetic wall models have also been used to understand how different components of the wall contribute to its overall mechanical status and behaviour (Johnson et al. 2018). For example, when the bacterium *Komagataeibacter xylinus* is grown in static cultures, it synthesises and extrudes cellulose into the culture medium to form a pellicle to assist with flotation. Adding various combinations and concentrations of hemicelluloses and/or pectic polysaccharides into *K. xylinus* cultures under different environmental conditions results in the production of polysaccharide composites (Mikkelsen et al. 2020). These composites could then be tested using AFM, tensile tests or various biophysical methods, providing information about the role of various polysaccharides in determining mechanical status and behaviour of composite materials.

Walls in Their Environments

From the earliest stages of growth where the radicle extends through the seed coat, modification of the wall is essential throughout plant development. Constant feedback loops between the wall and other cellular compartments are starting to be uncovered and support the wall being an active, integrated organelle (see the above section on wall models, Figure 1.2). Adaptation of the wall is also a key response to environmental conditions, such as cold, drought and pathogens. Monitoring and activation of cellular processes in response to wall properties involves cell surface proteins including CWI sensors covered in detail in Chapters 18 and 19. These may interact with wall polysaccharides, proteins, small molecules, ions and more to 'sense' the status of the wall (Figure 1.6).

Although much has been learned about the architecture and biophysical properties of the wall (see the above sections on imaging and biophysical properties as well as Chapter 15), gaps remain about the molecular mechanisms of CWI sensing during growth (Voiniciuc et al. 2018). For example, turgor generates stresses within the wall and stress patterns will differ depending on the cell size, shape and context (Mosca et al. 2017; Cosgrove 2018). It has become

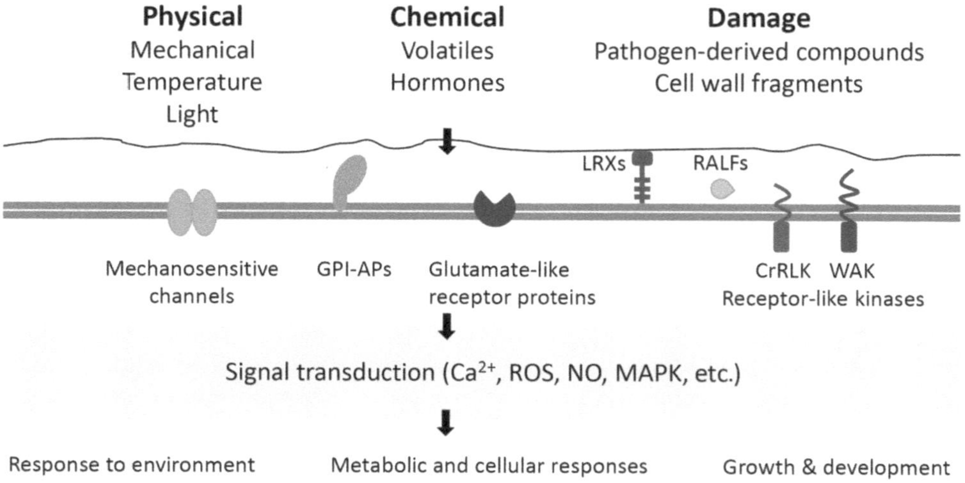

FIGURE 1.6 Schematic overview of the cell wall-plasma membrane-cytoplasmic continuum highlighting the different types of environmental signals detected and transduced.

clear that heterogeneity exists within walls with stiffer and softer regions, and these differences also generate local stresses. Questions remain as to whether the wall is laid down in such a way that controlled expansion follows or if directional growth and shape changes occur in response to stress patterns in the wall (see Chapter 6). While both are likely to contribute, constant feedback must occur to maintain integrity.

Experiments of the past have been revisited and there is a surge of activity to deliver greater understanding and appreciation of the importance of wall mechanics in regulating growth (Trinh et al. 2021). Elegantly shown by pectin mutants (Verger et al. 2018), what is emerging is that walls, as the connection between cells, are crucially important for generating tissue tension and this stress is in turn involved in modifying walls in response, maintaining integrity. The connections between the wall and cellular processes involve CWI pathways (Wolf et al. 2012; Vaahtera et al. 2019). CWI monitors are proposed to act by binding wall polysaccharides that have likely altered or become present during stress, leading to conformational changes in the cell surface proteins that then trigger intercellular downstream signalling cascades. A number of reviews have covered this topic in detail (see Novakovic et al. 2018; Vaahtera et al. 2019; Gigli-Bisceglia et al. 2020) and here we will provide a selected overview with Chapters 18 and 19 in this volume providing further information.

Clues to the existence of CWI sensors came in the 1990s from studies of the compensatory changes in wall composition in response to wall damage from chemicals (Shedletzky et al. 1992), biosynthesis mutants (Anterola and Lewis 2002) and pathogens (Mohnen and Hahn 1993). A breakthrough came with identification of the receptor-like kinase (RLK) THESEUS1 (THE1) that when mutated suppressed the growth defects of a cellulose-deficient mutant (Hematy et al. 2007). CWI mechanisms in plants have since been shown to include cell surface proteins that interact with wall polysaccharides and stretch-activated ion channels that respond to changes in the PM, similar to CWI pathways in yeast (Wolf et al. 2012; Vaahtera et al. 2019; Gigli-Bisceglia et al. 2020). Pectin has been identified as a key integrity 'signal'. This most likely results from its complex and heterogeneous chemical structure that can be readily modified, for example, with esterification, acetylation and sidechain variation, resulting in either making them more/less accessible, allowing cross-linking or degradation (De Lorenzo et al. 2019; Shin et al. 2021). It should be noted that the primary wall of dicots is assumed to be pectin-rich, whereas grasses have relatively lower levels. Even small amounts of pectin may be sufficient to trigger responses (Wu et al. 2020), and it is likely that sensing mechanisms involve matrix components other than pectin as well as mechanosensitive channels. A large number of RLKs exist in plants and could be correlated with the diversity of wall compositions and architectures; however, the ligands of relatively few RLKs have been identified. In addition to the *Catharanthus roseus* RLK (CrRLK) family to which THE1 belongs, WAKs that can bind pectins are involved in both expansion and other stress-related pathways (Kohorn 2016). Cell surface glycoproteins, such as GPI-anchored AGPs, also have potential to interact with wall carbohydrates such as pectins and arabinoxylans (Tan et al. 2013) and transmit signals of wall properties (see the above section on wall models).

Members of the CrRLKs are some of the best characterised in stress responses (Yang et al. 2021; Zhu et al. 2021b). Containing malectin-like domains proposed to interact with oligosaccharides, FERONIA (FER) can bind pectins *in vitro* and also mediate roles in pattern-triggered immunity (PTI) and development through binding RAPID ALKALINISATION FACTORS (RALFs) (Shih et al. 2014; Stegmann et al. 2017; Feng et al. 2018) (see Chapter 20 for a more detailed discussion of PTI). FER plays roles in pollen tube growth and integrity (Huck 2003), mechanoperception in roots (Shih et al. 2014) and salinity tolerance (Feng et al. 2018). THE1 is also proposed to detect both wall fragments and secondary messengers formed during stress conditions as ligands as well as RALFs (Merz et al. 2017; Gonneau et al. 2018). Another CrRLK, ANXUR, has also been implicated in CWI in pollen tubes by regulating ROS production and Ca^{2+} influx to maintain wall integrity at the tip during elongation (Boisson-Dernier et al. 2013). The leucine-rich repeat extensin (LRX) family of wall glycoproteins are also emerging as candidates in CWI (Herger et al. 2019). These multidomain glycoproteins include an extensin domain that can be *O*-glycosylated with short chains of arabinose as well as tyrosine motifs with potential to cross-link by intra- and intermolecular linkages via dityrosine (Hohmann et al. 2017). The leucine-rich repeat (LRR) domain is found in organisms of all kingdoms and implicated in binding a range of proteins—from small peptides to large proteins and potentially non-proteinaceous small molecular compounds. LRXs have been found to bind RALF peptides (Mecchia et al. 2017) via the LRR domain, with crystal structures of LRX2/8-RALF4 complexes showing a dimeric arrangement (Moussu et al. 2020). The similarity of *lrx3 lrx4 lrx5* triple mutants with *fer-4* led to further investigation and identified a physical interaction between LRXs and FER (Dünser et al. 2019). The extensin domain was found to be essential for LRX function in root hairs leading to suggestions that its tight interaction with the wall forms the link with PM-located CWI receptors (Herger et al. 2019).

Mechanically dependent Ca^{2+} influxes have long been associated with changes in membrane tension and the role of mechanosensitive channels in wall remodelling and cell growth in many different tissues throughout plant growth is becoming increasingly apparent (Malivert et al. 2018; Hartmann et al. 2021). MATING PHEROMONE INDUCED DEATH 1 COMPLEMENTING ACTIVITY channels have been proposed to be Ca^{2+} channels with MCA2 shown to be activated by membrane tension (Yoshimura et al. 2021). MCAs are implicated in isoxaben-induced wall damage and production of jasmonic acid (Engelsdorf et al. 2018) and their roles during stress-induced wall remodelling will be an interesting area for further investigation.

A connection between membrane tension and CWI is evident in the reliance of a rapidly mechanically activated Ca^{2+} channel on DEFECTIVE KERNEL 1 (DEK1) and has been likened to the Piezo pathways in animals (Guerringue et al. 2018; Malivert et al. 2018). DEK1 is a plant-specific member of the calpain superfamily of Ca^{2+} activated Cys-proteases that regulates wall composition and thickness (Amanda et al. 2016, 2017). Interestingly, plants with reduced levels of DEK1 show defects in cell adhesion and epidermal integrity and suggest a key role during plant development in the feedback loop between tissue tension and integrity (Verger et al. 2018).

There remain many unanswered questions related to the pathways involved in stress sensing and growth. Many factors are involved in processing the CWI signals, including Ca^{2+}, nitric oxide and ROS (Figure 1.6; Hamann 2015). The relationship between MT response to stress and CWI is unclear. A pectin-FER pathway involving ROP6 has been proposed to regulate MT reorganisation (Tang et al. 2021) yet a suite of mechanical tests on *fer* and other mechanosensor candidate mutants showed FER and MT responses to stress act independently (Malivert et al. 2021). This fascinating area of research will no doubt have decades more of discoveries.

Abiotic stresses, such as drought, heat and cold (freezing), impact the water potential of the cell and therefore turgor pressure. For example, cold stress can lead to the formation of ice crystals in the apoplastic space, and these crystals then cause rearrangement of wall architecture, which exerts increased mechanical strain on the cell membrane, and in extreme cases, cell rupture (Wei et al. 2021). Cold stress has been shown to induce a decrease in pectin levels, as well as in enzyme activity associated with pectins in leaves of maize (Bilska-Kos et al. 2017) and *Arabidopsis* (Alonso-Serra et al. 2020), whereas pectin levels increased in *Brassica napus* (Solecka et al. 2008) and pea (Baldwin et al. 2014). Structural and architectural changes in other matrix components that could modulate the mechanical and rheological properties of walls and increase freezing tolerance include MLGs in maize (Domon et al. 2013), XGs in *Arabidopsis* (Takahashi et al. 2020, 2021) and GAXs in winter wheat (Willick et al. 2018). The response of plants to cold stress is therefore likely to differ depending on the degree and extent of cold/freezing, wall composition and the tissues affected. Elucidating the mechanism(s) awaits identification of wall-related gene expression changes, their transcriptional regulation and the CWI sensors involved.

In the case of metal/ion toxicities, such as that of high soil salinity, monovalent ions such as Na^+ and K^+ can cause displacement of Ca^{2+} ions and disruption of the pectin 'egg box' matrix structure (Byrt et al. 2018) and Ca^{2+} bound by glycoproteins such as AGPs (Lopez-Hernandez et al. 2020). Salt stress-induced changes in wall mechanical properties are proposed to be interpreted by CWI sensors (Liu et al. 2021a). FER, together with its co-receptor LORELEI-like GPI-AP1 (LLG1), is implicated in sensing salt-induced disruption of pectin cross-linking and initiation of signalling (Feng et al. 2018). The FER-dependent signalling pathway leads to a cell-specific increase in Ca^{2+} that aids growth recovery stage and prevents cell bursting under high salinity (Feng et al. 2018). Maintenance of cellulose synthesis during salt stress has been shown to be important for tolerance and involves CESAs and interacting proteins such as CSI1 (Zhang et al. 2016a) as well as MT-associated proteins CC1 and CC2 (Endler et al. 2015). AGPs are implicated in salt stress tolerance as shown by the *Arabidopsis* salt overly sensitive mutant *sos5* (*fla4*) (Shi et al. 2003) and increase of glucuronic acid side chains in seagrasses (Pfeifer et al. 2020) proposed to bind Ca^{2+} (Lopez-Hernandez et al. 2020) that may mediate increased salt tolerance (see Chapters 16, 17 and 22 for further information). A rice OsWAK112 was shown to negatively regulate salt stress responses by inhibiting ethylene production (Lin et al. 2021) and WAKs have also been shown to function during heavy metal stress (Novakovic et al. 2018).

Biotic stress caused by pathogens results in wall damage by wall degrading enzymes produced and secreted by pathogenic fungi and bacteria and include cellulases, hemicellulases and pectinases (Allwood et al. 2021). Plants use several strategies to counter-attack, including formation of localised appositions (termed papillae) in an attempt to 'wall off' the invading pathogens. Plants sense pathogens through pathogen-associated molecular patterns (PAMPs) and damage-associated molecular patterns (DAMPs) (Vaahtera et al. 2019; Allwood et al. 2021). PTI responses are likely to be coordinated with CWI sensing through both wall damage and altered biomechanics and potentially overlapping signalling components. THE1 and FER are implicated in biotic stress due to influencing resistance responses to *Botrytis cinerea* and *Pseudomonas syringae*, respectively (Qu et al. 2017; Guo et al. 2018). The versatility of FER interactions with RALFs and RLKs has been shown to be important for immunity responses in *Arabidopsis*, rice and soybean (reviewed in Zhu et al. 2021b). A THE1 dependent pathway in response to wall damage was shown to be reduced by PTI receptors (plant elicitor peptide receptors, PEPR1,2) and this suppression could be alleviated when PTI signalling downstream of PEPR1 and PEPR2 was impaired (Engelsdorf et al. 2018). CWI pathways are therefore likely to complement PTI signalling pathways.

The pectin fragments oligogalacturonides (OGs) have been shown to act as ligands to trigger plant resistance to a variety of pathogens (Benedetti et al. 2015). A study of OGs present during infection of *Arabidopsis* by the fungus *B. cinerea* revealed the large diversity of OGs present, with many being acetyl- and methyl-esterified and produced by fungal pectin lyases (Voxeur et al. 2019). A systematic analysis of wall mutants to different pathogens supports the key roles of walls in mediating abiotic stress responses (Molina et al. 2021). WAKs have been shown to bind oligogalacturonide fragments and a wealth of data is solidifying their role in plant defence (Stephens et al. 2021). WAK4 in rose is involved in resistance to *B. cinerea* (Liu et al. 2021b; Stephens et al. 2021), TaWAK2A-800 (Guo et al. 2021) and

TaWAK-6D (Qi et al. 2021) in resistance to *Fusarium* Head Blight and Sharp Eyespot in wheat and many other examples (reviewed in Stephens et al. 2021). Integration of CWI sensors and cytoplasmic pathways with other cell surface monitoring, such as PTI, are strategies developed by plants to fine-tune adaptation of walls based on the specific context (Hamann 2015; Vaahtera et al. 2019; Gigli-Bisceglia et al. 2020).

WALLS OF THE FUTURE

Sustainable Biomaterials

A new trend is currently taking the building sector by storm with structures such as Mjøstårnet in Norway, Forté in Melbourne and W350 in Japan. This amazing new material is 'wood'! Wooden buildings are having a resurgence and being reinvented into skyscrapers as the construction industry strives for a more sustainable footprint. Interestingly, the secret to these new building materials is a cross-laminate structure creating additional strength, much like the structure of the SCWs that make up woody or fibrous tissues. Plant biomaterials, especially the vast amount of SCWs (wood) from plants, are major renewable biomaterials. Plant products continue to inspire innovation in traditional applications (see the 'Cell walls and societies' section in addition to Chapters 12, 14, 17, 21 and 24) and increasingly are viewed as a source of building blocks (monomers), produced from renewable feedstocks such as corn, sugar beet and wheat for biodegradable biopolymer production to replace those based on fossil fuels (de Vries et al. 2021).

Innovations in building materials are still being developed with materials such as 'hempcrete' that mix woody hemp (*Cannabis sativa*) stems with lime to make a lightweight insulation material and 'ecoceramics' from cork with uses including in the aeronautical industry as part of the protective heat shield in missiles and spacecraft (Pullar and Novais 2017). Processing walls into pulp and paper has historically involved harsh chemicals and a large energy input that are not aligned with biomaterials as environmentally friendly. Improvements in processing, bleaching and biotechnology applications to change lignin structures have made paper-based and other biomaterials excellent alternatives to plastics and non-renewables (de Vries et al. 2021; Jedrzejczak et al. 2021).

Biological and bioinspired materials in material science are at the forefront of innovation by material scientists and engineers as they strive for novel sustainable materials with enhanced functionalities. Nature has evolved materials with properties and mechanisms that are beyond the current know-how of the engineering-materials industry (for a review, see Wang et al. 2020). A number of these have gained inspiration from the properties of plant walls and their surfaces. For example, the function of superwettability, including superhydrophobicity and self-cleaning, is being informed by the shape (topography) and chemistry of cuticular structures on surfaces of plant organs such as lotus leaves and petals, respectively. Lightweight and high strength material design are inspired by the structure and organisation of SCWs in the tissues in stems (wood) – thus, bioinspired polymeric woods have better axial compressive properties and density and are able to be fine-tuned to balance weight with strength through control of the microstructure and fabrication parameters. The inherent biocompatibility and bioactivity of biological macromolecules, particularly in relation to wall-derived polysaccharides, have seen their widespread application in biomedicine as hydrogels and in the encapsulation of drugs for targeted delivery in therapeutics. For example, the use of cellulose nanofibres with switchable permeability in the design and fabrication of microcapsules (Paulraj et al. 2017). Industrial polymers, such as bioplastics (bio-PP, bio-PE or bio-PET), are increasingly being synthesised from renewable plant-based sources. Some bioplastics are obtained by processing directly from natural biopolymers including polysaccharides (e.g., starch and cellulose from plants, chitosan from fungi and alginate from brown algae), while others are chemically synthesised from sugar derivatives (e.g., lactic acid) from plants or biologically generated by fermentation of sugars.

A pre-requirement for future utilisation of plant wall–based biomaterials are sustainable plant production systems. Extreme weather caused by climate change and pests can divert plants into defence status that reduces their growth, wall synthesis and quality (Vaahtera et al. 2019). A better understanding of the wall's role in plant stress resistance would help us secure plant growth and wall material production (see the section on walls in their environments). The diversity of plant responses, incorporating genetics and multiple environmental stresses, can lead to walls with different mechanical properties and chemical compositions, which can be a limiting step for the utilisation of materials such as wood for industry and construction. For example, control of reaction wood (Gardiner et al. 2014) and standardisation of woody wall structure and composition would greatly assist the vast utilisation of plant stems.

Breeding plants, either by gene editing or marker-assisted selection, to produce germplasm with woody stems faster while maintaining high quality will be a key objective in the future. Although some fast-growing species have been utilised such as in *Eucalyptus*, poplar and bamboo, their wood quality for wider utilisation is usually limited. Our understanding at the molecular level of the driving force of wall thickening and regulation pathways is currently insufficient to assist the design of fast-growing wood types producing walls with the desired industrial qualities.

Additionally, the number of steps needed to bring the raw materials to a product, harvesting, transport, processing, waste and recycling, are key considerations. Future walls should be designed to enable efficient processing of wall raw materials to the end-product and reduce waste during processing and after consumption. These can include *de novo* customisation of wall material (Gardiner et al.

2014). A good example for raw material customisation is genome-edited trees with reduced lignin content for easier paper making and bioenergy production (de Vries et al. 2021). Genome-edited poplar (*Populus tremula*×*Populus alba*) trees with down-regulated cinnamoyl-CoA reductase (CCR) expression showed increased saccharification efficiency (Van Acker et al. 2014). With increased understanding of walls and by use of genome editing tools, we would expect in the future more *de novo* edited plants that provide specialised wall materials and reduce processing steps.

To make full use of the wonderful opportunities of walls, it is essential we also explore the immense diversity present throughout algal and plant evolution. Maintaining the available diversity through resources such as seedbanks will ensure that we can explore future potential. For example, an extremely small set of dicot plant walls have been examined in the level of detail to understand the composition and architecture. Grasses that represent our largest food source are poorly described in comparison to dicots. Understanding of the evolutionary transitions and adaptations walls have made to different environments will allow exploration of key features that are consistent as well as different between organisms, tissues and cells. Plant environment interactions also include a wealth of symbionts and beneficial microbes crucial to growth and wall components such as in root exudates are known to be involved in these interactions.

Determinants of Human Health

Walls play an important function as the major source of dietary fibre, the non-digestible component of plant tissues, in human (and production animal) diets (see the 'Feeding Communities' section above). Dietary fibre exerts physiological influence throughout the gastrointestinal tract via mechanisms of physical structuring of digesta, modulation of digestive processes regulating transit time and by acting as an energy source for microbial fermentation (Williams et al. 2017, 2019). These mechanisms relate to the physicochemical characteristics of dietary fibre such as its water-holding capacity ('dispersibility'), viscosity, binding ability, absorptive capacity, faecal bulking capacity and fermentability (Gidley and Yakubov 2019; Williams et al. 2019). The health benefits of dietary fibre are attributed to a combination of rheological and biophysical behaviours, its function as a matrix and the biochemical traits of its polymer constituents. The main benefits associated with dietary fibre fermentability relate to its effects on the maintenance of the diversity and function of the gut microbiota, and the associated beneficial end-products of the fermentation process such as short chain fatty acids (Williams et al. 2011, 2017, 2019; Gorham et al. 2017; Grant et al. 2019). What is emerging from various *in vitro* fermentation and animal/human studies is that while polymer solubility has an influence, it is their fermentability that is critical with both chemical composition and wall architecture playing a role (Low et al. 2015; Williams et al. 2017, 2019; Lu et al. 2020). How and to what extent these wall features can be harnessed to drive better health outcomes within our multicultural communities is a considerable future challenge.

Walls also have other functions, for example, in encapsulation, as they form an impassable (physical and diffusion) barrier to the release of nutrients such as lipids and secondary metabolites (Dhital et al. 2016; Bhattarai et al. 2018, 2021; Grant et al. 2019). Walls are also matrices for nutrient binding, in case of polyphenolics (e.g., Padayachee et al. 2013) and digestive enzymes. For example, a combination of encapsulation and binding has been shown to result in almost complete resistance to *in vitro* digestion of starch and protein from intact cells isolated from legume cotyledons (Dhital et al. 2016; Bhattarai et al. 2021), possibly a key factor in the nutritional benefit of legumes in the human diet (Johnson et al. 2018). Transport has also been shown to be impacted whereby walls within multicellular tissue pieces can act as a diffusion barrier to molecular movement into or out of cells (Dhital et al. 2016; Grundy et al. 2016). Solubilisation of polysaccharides during either food processing and/or digestion can also slow enzyme and nutrient movement due to local viscosity changes as well as potentially acting as an accessible substrate for binding of micronutrients or enzymes (Zhang et al. 2015; Padayachee et al. 2017). Understanding the elements that control the extent of polysaccharide solubilisation from walls (wall architecture) and tissues (plant architecture) will provide targets for either plant selection or tailored food processing to modify digestion properties for optimal nutrient bioavailability (Williams et al. 2017; Johnson et al. 2018).

CONCLUSIONS

Moving through the 21st century, humankind must solve some enormous challenges in providing food and other resources to support the projected population increases in a more volatile climate on less agronomically productive land. There will be increasing pressure to use renewable resources—plants—to meet these needs. Much is still to be learnt as to how we can maintain if not improve plant productivity under increasing environmental stress, in addition to enhancing their utility for the myriad of functions plants are used for within human societies and natural ecosystems. Plant breeding and biotechnological engineering of plants for improved growth, yield and end-use properties will include significant modifications to plant walls. Further research is required to sufficiently understand how plant molecular structure from nano- to meso-scale is controlled such that this information can be harnessed to manipulate plant tissue structures for specific utilisation and re-utilisation. For example, in plant-based foods, the objective is to deliver macro- and micronutrients at appropriate rates and sites within the gastrointestinal tract but also to control host responses such as satiety and microbiome ecology to support health and wellness. Though much has been learnt about plant wall composition, the micro-/macro-heterogeneity of wall composition within and between species, material properties of walls and the biosynthesis of its components in the last 50 years, further knowledge must be revealed to deliver the tailored raw ingredients for our food, fibre, biomaterials and fuel requirements from plants while retaining their capacity for productive growth. Because of their importance, many domesticated species have been utilised as experimental models within the wall field. Yet, there still remains a significant gap in our understanding of the magnitude and types of wall changes that have accompanied this selection process and the consequence of these on plant function. Also highlighted is the enormous diversity of plant, algal and fungal taxa that are yet to be investigated with the potential to underpin the future demand for plant-based foods. To fulfil these societal needs, we will also need to understand the role of the wall as an environmental sensor, the feedback loops that control wall composition and how composition influences material properties so that we can use the myriad of molecular genetic approaches to generate 'designer walls/fit-for-purpose walls' on demand (Doblin et al. 2014; Johnson et al. 2018). We forecast the plant wall biology field to continue to flourish, ushering in a new era of intensified research with development and the application of new tools and technologies to meet this challenge.

ACKNOWLEDGEMENTS

YM acknowledges the support of The University of Melbourne Research Scholarship and the Albert Shimmins Fund for a writing-up award. LN acknowledges the support of The University of Melbourne Research Scholarship, Max Planck Society and Norma Hilda Schuster neé Swift Scholarship. KLJ was supported by a La Trobe Research Focus Area grant 2000004372. AB, MSD and KLJ would like to acknowledge a start-up grant from La Trobe University and from Zhejiang A&F University for the Sino-Australia Plant Cell Wall Research Centre. We acknowledge the meticulous editorial skills of Ms Andrea Gallo and Dr Frazer Thorpe for assistance with figure design.

REFERENCES

Albersheim P, Darvill AG, Roberts K, Sederoff R, Staehelin A (2010) *Plant Cell Walls*. New York: Garland Science.

Allwood JW, Williams A, Uthe H, van Dam NM, Mur LAJ, Grant MR, Petriacq P (2021) Unravelling plant responses to stress-the importance of targeted and untargeted metabolomics. *Metabolites* 11:558.

Alonso-Serra J, Shi X, Peaucelle A, Rastas P, Bourdon M, Immanen J, Takahashi J, Koivula H, Eswaran G, Muranen S, Help H, Smolander OP, Su C, Safronov O, Gerber L, Salojarvi J, Hagqvist R, Mahonen AP, Helariutta Y, Nieminen K (2020) ELIMAKI locus is required for vertical proprioceptive response in birch trees. *Current Biology* 30:589–599 e585.

Altartouri B, Bidhendi AJ, Tani T, Suzuki J, Conrad C, Chebli Y, Liu N, Karunakaran C, Scarcelli G, Geitmann A (2019) Pectin chemistry and cellulose crystallinity govern pavement cell morphogenesis in a multi-step mechanism. *Plant Physiology* 181:127–141.

Amanda D, Doblin MS, Galletti R, Bacic A, Ingram GC, Johnson KL (2016) DEFECTIVE KERNEL1 (DEK1) regulates cell walls in the leaf epidermis. *Plant Physiology* 172:2204–2218.

Amanda D, Doblin MS, Galletti R, Bacic A, Ingram GC, Johnson KL (2017) Regulation of cell wall genes in response to DEFECTIVE KERNEL1 (DEK1)-induced cell wall changes. *Plant Signaling and Behavior* 12:e1345405.

Ambrose C, Allard JF, Cytrynbaum EN, Wasteneys GO (2011) A CLASP-modulated cell edge barrier mechanism drives cell-wide cortical microtubule organization in Arabidopsis. *Nature Communications* 2:430.

Anderson CT, Carroll A, Akhmetova L, Somerville C (2009) Real-time imaging of cellulose reorientation during cell wall expansion in Arabidopsis roots. *Plant Physiology* 152:787–796.

Anderson CT, Wallace IS (2012) Illuminating the wall: using click chemistry to image pectins in Arabidopsis cell walls. *Plant Signaling and Behavior* 7:661–663.

Anterola AM, Lewis NG (2002) Trends in lignin modification: a comprehensive analysis of the effects of genetic manipulations/mutations on lignification and vascular integrity. *Phytochemistry* 61:221–294.

Arioli T, Peng LC, Betzner AS, Burn J, Wittke W, Herth W, Camilleri C, Hofte H, Plazinski J, Birch R, Cork A, Glover J, Redmond J, Williamson RE (1998) Molecular analysis of cellulose biosynthesis in Arabidopsis. *Science* 279:717–720.

Atmodjo MA, Sakuragi Y, Zhu X, Burrell AJ, Mohanty SS, Atwood JA, 3rd, Orlando R, Scheller HV, Mohnen D (2011) Galacturonosyltransferase (GAUT)1 and GAUT7 are the core of a plant cell wall pectin biosynthetic homogalacturonan:galacturonosyltransferase complex. *Proceedings of the National Academy of Sciences of the United States of America* 108:20225–20230.

Bacic A, Harris PJ, Stone BA (1988) Structure and function of plant cell walls. In *The Biochemistry of Plants*, ed. Preiss J, 297–371. New York: Academic Press.

Baldwin L, Domon JM, Klimek JF, Fournet F, Sellier H, Gillet F, Pelloux J, Lejeune-Henaut I, Carpita NC, Rayon C (2014) Structural alteration of cell wall pectins accompanies pea development in response to cold. *Phytochemistry* 104:37–47.

Barriere Y, Ralph J, Mechin V, Guillaumie S, Grabber JH, Argillier O, Chabbert B, Lapierre C (2004) Genetic and molecular basis of grass cell wall biosynthesis and degradability. II. Lessons from brown-midrib mutants. *Comptes Rendus: Biologies* 327:847–860.

Baskin TI, Betzner AS, Hoggart R, Cork A, Williamson RE (1992) Root morphology mutants in *Arabidopsis thaliana*. *Functional Plant Biology* 19:427–437.

Bassim N, Scott K, Giannuzzi LA (2014) Recent advances in focused ion beam technology and applications. *MRS Bulletin* 39:317–325.

Bauer S, Vasu P, Persson S, Mort AJ, Somerville CR (2006) Development and application of a suite of polysaccharide-degrading enzymes for analyzing plant cell walls. *Proceedings of the National Academy of Sciences of the United States of America* 103:11417–11422.

Belteton SA, Li W, Yanagisawa M, Hatam FA, Quinn MI, Szymanski MK, Marley MW, Turner JA, Szymanski DB (2021) Real-time conversion of tissue-scale mechanical forces into an interdigitated growth pattern. *Nature Plants* 7:826–841.

Benedetti M, Pontiggia D, Raggi S, Cheng ZY, Scaloni F, Ferrari S, Ausubel FM, Cervone F, De Lorenzo G (2015) Plant immunity triggered by engineered in vivo release of oligogalacturonides, damage-associated molecular patterns. *Proceedings of the National Academy of Sciences of the United States of America* 112:5533–5538.

Bhattarai RR, Dhital S, Mense AL, Gidley MJ, Shi Y-C (2018) Intact cellular structure in cereal endosperm limits starch digestion *in vitro*. *Food Hydrocolloids* 81:139–148.

Bhattarai RR, Dhital S, Williams BA, Yang HJ, Mikkelsen D, Flanagan BM, Gidley MJ (2021) In vitro fermentation of legume cells and components: effects of cell encapsulation and starch/protein interactions. *Food Hydrocolloids* 113:106538.

Bidhendi AJ, Geitmann A (2018) Finite element modeling of shape changes in plant cells. *Plant Physiology* 176:41–56.

Bidhendi AJ, Geitmann A (2019) Methods to quantify primary plant cell wall mechanics. *Journal of Experimental Botany* 70:3615–3648.

Bilska-Kos A, Solecka D, Dziewulska A, Ochodzki P, Jonczyk M, Bilski H, Sowinski P (2017) Low temperature caused modifications in the arrangement of cell wall pectins due to changes of osmotic potential of cells of maize leaves (Zea mays L.). *Protoplasma* 254:713–724.

Boisson-Dernier A, Lituiev DS, Nestorova A, Franck CM, Thirugnanarajah S, Grossniklaus U (2013) ANXUR receptor-like kinases coordinate cell wall integrity with growth at the pollen tube tip via NADPH oxidases. *PLOS Biology* 11:e1001719.

Borner GHH, Lilley KS, Stevens TJ, Dupree P (2003) Identification of glycosylphosphatidylinositol-anchored proteins in Arabidopsis. A proteomic and genomic analysis. *Plant Physiology* 132:568–577.

Bouton S, Leboeuf E, Mouille G, Leydecker MT, Talbotec J, Granier F, Lahaye M, Hofte H, Truong HN (2002) Quasimodo1 encodes a putative membrane-bound glycosyltransferase required for normal pectin synthesis and cell adhesion in Arabidopsis. *The Plant Cell* 14:2577–2590.

Braconnot H (1825) Recherches sur un nouvel acide universellement répandu dans tous les vegetaux. *Annales de chimie et de physique* 2:173–178.

Bringmann M, Landrein B, Schudoma C, Hamant O, Hauser MT, Persson S (2012) Cracking the elusive alignment hypothesis: the microtubule-cellulose synthase nexus unraveled. *Trends in Plant Science* 17:666–674.

Brown DM, Goubet F, Vicky WWA, Goodacre R, Stephens E, Dupree P, Turner SR (2007) Comparison of five xylan synthesis mutants reveals new insight into the mechanisms of xylan synthesis. *The Plant Journal* 52:1154–1168.

Brown DM, Zeef LAH, Ellis J, Goodacre R, Turner SR (2005) Identification of novel genes in Arabidopsis involved in secondary cell wall formation using expression profiling and reverse genetics. *The Plant Cell* 17:2281–2295.

Brown DM, Zhang ZN, Stephens E, Dupree P, Turner SR (2009) Characterization of IRX10 and IRX10-like reveals an essential role in glucuronoxylan biosynthesis in Arabidopsis. *The Plant Journal* 57:732–746.

Burton RA, Wilson SM, Hrmova M, Harvey AJ, Shirley NJ, Stone BA, Newbigin EJ, Bacic A, Fincher GB (2006) Cellulose synthase-like CslF genes mediate the synthesis of cell wall (1,3;1,4)-beta-D-glucans. *Science* 311:1940–1942.

Byrt CS, Munns R, Burton RA, Gilliham M, Wege S (2018) Root cell wall solutions for crop plants in saline soils. *Plant Science* 269:47–55.

Cabeen MT, Jacobs-Wagner C (2005) Bacterial cell shape. *Nature Reviews Microbiology* 3:601–610.

Carpita NC, Gibeaut DM (1993) Structural models of primary cell walls in flowering plants: consistency of molecular structure with the physical properties of the walls during growth. *The Plant Journal* 3:1–30.

Cavalier DM, Lerouxel O, Neumetzler L, Yamauchi K, Reinecke A, Freshour G, Zabotina OA, Hahn MG, Burgert I, Pauly M, Raikhel NV, Keegstra K (2008) Disrupting Two *Arabidopsis thaliana* xylosyltransferase genes results in plants deficient in xyloglucan, a major primary cell wall component. *The Plant Cell* 20:1519–1537.

Chen CG, Pu ZY, Moritz RL, Simpson RJ, Bacic A, Clarke AE, Mau SL (1994) Molecular cloning of a gene encoding an arabinogalactan-protein from pear (*Pyrus communis*) cell suspension culture. *Proceedings of the National Academy of Sciences of the United States of America* 91:10305–10309.

Cleland R (1971) Cell wall extension. *Annual Review of Plant Physiology* 22:197–222.

Cosgrove DJ (1989) Characterization of long-term extension of isolated cell walls from growing cucumber hypocotyls. *Planta* 177:121–130.

Cosgrove DJ (2014) Re-constructing our models of cellulose and primary cell wall assembly. *Current Opinion in Plant Biology* 22:122–131.

Cosgrove DJ (2018) Diffuse growth of plant cell walls. *Plant Physiology* 176:16–27.

Cressey D, Callaway E (2017) Cryo-electron microscopy wins chemistry Nobel. *Nature* 550:167–167.

Crowell EF, Bischoff V, Desprez T, Rolland A, Stierhof YD, Schumacher K, Gonneau M, Hofte H, Vernhettes S (2009) Pausing of Golgi bodies on microtubules regulates secretion of cellulose synthase complexes in Arabidopsis. *The Plant Cell* 21:1141–1154.

De Caroli M, Lenucci MS, Di Sansebastiano GP, Dalessandro G, De Lorenzo G, Piro G (2011) Protein trafficking to the cell wall occurs through mechanisms distinguishable from default sorting in tobacco. *The Plant Journal* 65:295–308.

De Caroli M, Lenucci MS, Manualdi F, Dalessandro G, De Lorenzo G, Piro G (2015) Molecular dissection of Phaseolus vulgaris polygalacturonase-inhibiting protein 2 reveals the presence of hold/release domains affecting protein trafficking toward the cell wall. *Frontiers in Plant Science* 6: 660.

De Lorenzo G, Ferrari S, Giovannoni M, Mattei B, Cervone F (2019) Cell wall traits that influence plant development, immunity, and bioconversion. *The Plant Journal* 97:134–147.

de Vries L, Guevara-Rozo S, Cho M, Liu LY, Renneckar S, Mansfield SD (2021) Tailoring renewable materials via plant biotechnology. *Biotechnology for Biofuels* 14:167.

Deniaud-Bouet E, Hardouin K, Potin P, Kloareg B, Herve C (2017) A review about brown algal cell walls and fucose-containing sulfated polysaccharides: cell wall context, biomedical properties and key research challenges. *Carbohydrate Polymers* 175:395–408.

Desprez T, Juraniec M, Crowell EF, Jouy H, Pochylova Z, Parcy F, Höfte H, Gonneau M, Vernhettes S (2007) Organization of cellulose synthase complexes involved in primary cell wall synthesis in *Arabidopsis thaliana*. *Proceedings of the National Academy of Sciences of the United States of America* 104:15572–15577.

DeVree BT, Steiner LM, Głazowska S, Ruhnow F, Herburger K, Persson S, Mravec J (2021) Current and future advances in fluorescence-based visualization of plant cell wall components and cell wall biosynthetic machineries. *Biotechnology for Biofuels* 14:78.

Dhital S, Bhattarai RR, Gorham J, Gidley MJ (2016) Intactness of cell wall structure controls the in vitro digestion of starch in legumes. *Food & Function* 7:1367–1379.

Ding Y, Wang J, Wang J, Stierhof Y-D, Robinson DG, Jiang L (2012) Unconventional protein secretion. *Trends in Plant Science* 17:606–615.

Doblin MS, Johnson KL, Humphries J, Newbigin EJ, Bacic A (2014) Are designer walls a realistic aspiration or will the plasticity of the plant's metabolism win out? *Current Opinion in Biotechnology* 26:108–114.

Doblin MS, Pettolino FA, Bacic A (2010) Plant cell walls: the skeleton of the plant world. *Functional Plant Biology* 37:357–381.

Doblin MS, Pettolino FA, Wilson SM, Campbell R, Burton RA, Fincher GB, Newbigin E, Bacic A (2009) A barley cellulose synthase-like CSLH gene mediates (1,3;1,4)-beta-D-glucan synthesis in transgenic Arabidopsis. *Proceedings of the National Academy of Sciences of the United States of America* 106:5996–6001.

Domon JM, Baldwin L, Acket S, Caudeville E, Arnoult S, Zub H, Gillet F, Lejeune-Henaut I, Brancourt-Hulmel M, Pelloux J, Rayon C (2013) Cell wall compositional modifications of Miscanthus ecotypes in response to cold acclimation. *Phytochemistry* 85:51–61.

Domozych DS, Sørensen I, Popper ZA, Ochs J, Andreas A, Fangel JU, Pielach A, Sacks C, Brechka H, Ruisi-Besares P, Willats WGT, Rose JKC (2014) Pectin metabolism and assembly in the cell wall of the charophyte green alga *Penium margaritaceum*. *Plant Physiology* 165:105–118.

Drakakaki G, van de Ven W, Pan S, Miao Y, Wang J, Keinath NF, Weatherly B, Jiang L, Schumacher K, Hicks G, Raikhel N (2012) Isolation and proteomic analysis of the SYP61 compartment reveal its role in exocytic trafficking in *Arabidopsis*. *Cell Research* 22:413–424.

Drula E, Garron ML, Dogan S, Lombard V, Henrissat B, Terrapon N (2021) The carbohydrate-active enzyme database: functions and literature. *Nucleic Acids Research*. doi: 10.1093/nar/gkab1045.

Du H, Simpson RJ, Moritz RL, Clarke AE, Bacic A (1994) Isolation of the protein backbone of an arabinogalactan-protein from the styles of *Nicotiana alata* and characterization of a corresponding cDNA. *The Plant Cell* 6:1643–1653.

Du J, Kirui A, Huang S, Wang L, Barnes WJ, Kiemle SN, Zheng Y, Rui Y, Ruan M, Qi S, Kim SH, Wang T, Cosgrove DJ, Anderson CT, Xiao C (2020) Mutations in the pectin methyltransferase QUASIMODO2 influence cellulose biosynthesis and wall integrity in Arabidopsis. *The Plant Cell* 32:3576–3597.

Dumais J, Forterre Y (2012) "Vegetable Dynamicks": the role of water in plant movements. *Annual Review of Fluid Mechanics* 44:453–478.

Dumont M, Lehner A, Vauzeilles B, Malassis J, Marchant A, Smyth K, Linclau B, Baron A, Mas Pons J, Anderson CT, Schapman D, Galas L, Mollet JC, Lerouge P (2016) Plant cell wall imaging by metabolic click-mediated labelling of rhamnogalacturonan II using azido 3-deoxy-D-manno-oct-2-ulosonic acid. *The Plant Journal* 85:437–447.

Dunkley TP, Hester S, Shadforth IP, Runions J, Weimar T, Hanton SL, Griffin JL, Bessant C, Brandizzi F, Hawes C, Watson RB, Dupree P, Lilley KS (2006) Mapping the Arabidopsis organelle proteome. *Proceedings of the National Academy of Sciences of the United States of America* 103:6518–6523.

Dünser K, Gupta S, Herger A, Feraru MI, Ringli C, Kleine-Vehn J (2019) Extracellular matrix sensing by FERONIA and leucine-rich repeat Extensins controls vacuolar expansion during cellular elongation in *Arabidopsis thaliana*. *EMBO Journal* 38(7):e100353.

Durachko DM, Cosgrove DJ (2009) Measuring plant cell wall extension (creep) induced by acidic pH and by alpha-expansin. *Journal of Visualized Experiments*. doi: 10.3791/1263:e1263.

Endler A, Kesten C, Schneider R, Zhang Y, Ivakov A, Froehlich A, Funke N, Persson S (2015) A mechanism for sustained cellulose synthesis during salt stress. *Cell* 162:1353–1364.

Engelsdorf T, Gigli-Bisceglia N, Veerabagu M, McKenna JF, Vaahtera L, Augstein F, Van der Does D, Zipfel C, Hamann T (2018) The plant cell wall integrity maintenance and immune signaling systems cooperate to control stress responses in *Arabidopsis thaliana*. *Science Signaling* 11 (536):eaao3070.

Fangel J, Ulvskov P, Knox JP, Mikkelsen M, Harholt J, Popper Z, Willats W (2012) Cell wall evolution and diversity. *Frontiers in Plant Science* 3:152.

FAO (2010) *The Second Report on the State of the World's Plant Genetic Resources for Food and Agriculture* Rome, Italy: FAO.

Feng W, Kita D, Peaucelle A, Cartwright HN, Doan V, Duan Q, Liu MC, Maman J, Steinhorst L, Schmitz-Thom I, Yvon R, Kudla J, Wu HM, Cheung AY, Dinneny JR (2018) The FERONIA receptor kinase maintains cell-wall integrity during salt stress through Ca^{2+} signaling. *Current Biology* 28:666-675 e665.

Frey-Wyssling A, Mühlethaler K, Wyckoff RWG (1948) Mikrofibrillenbau der pflanzlichen Zellwände. *Experientia* 4:475–476.

Fujita M, Himmelspach R, Hocart CH, Williamson RE, Mansfield SD, Wasteneys GO (2011) Cortical microtubules optimize cell-wall crystallinity to drive unidirectional growth in Arabidopsis. *The Plant Journal* 66:915–928.

Gardiner BB, Saranpää P, Gril, J (2014) *The Biology of Reaction Wood*. Berlin, Heidelberg: Springer.

Gavara N (2017) A beginner's guide to atomic force microscopy probing for cell mechanics. *Microscopy Research and Technique* 80:75–84.

Gendre D, McFarlane HE, Johnson E, Mouille G, Sjödin A, Oh J, Levesque-Tremblay G, Watanabe Y, Samuels L, Bhalerao RP (2013) Trans-golgi network localized ECHIDNA/Ypt interacting protein complex is required for the secretion of cell wall polysaccharides in *Arabidopsis*. *The Plant Cell* 25:2633–2646.

Gendre D, Oh J, Boutté Y, Best JG, Samuels L, Nilsson R, Uemura T, Marchant A, Bennett MJ, Grebe M, Bhalerao RP (2011) Conserved Arabidopsis ECHIDNA protein mediates trans-Golgi-network trafficking and cell elongation. *Proceedings of the National Academy of Sciences of the United States of America* 108:8048–8053.

Gidley MJ, Yakubov GE (2019) Functional categorisation of dietary fibre in foods: beyond 'soluble' vs 'insoluble'. *Trends in Food Science & Technology* 86:563–568.

Gigli-Bisceglia N, Engelsdorf T, Hamann T (2020) Plant cell wall integrity maintenance in model plants and crop species-relevant cell wall components and underlying guiding principles. *Cellular and Molecular Life Sciences* 77:2049–2077.

Gonneau M, Desprez T, Martin M, Doblas VG, Bacete L, Miart F, Sormani R, Hematy K, Renou J, Landrein B, Murphy E, Van De Cotte B, Vernhettes S, De Smet I, Hofte H (2018) Receptor kinase THESEUS1 is a rapid alkalinization factor 34 receptor in Arabidopsis. *Current Biology* 28:2452-2458 e2454.

Gorham JB, Kang S, Williams BA, Grant LJ, McSweeney CS, Gidley MJ, Mikkelsen D (2017) Addition of arabinoxylan and mixed linkage glucans in porcine diets affects the large intestinal bacterial populations. *European Journal of Nutrition* 56:2193–2206.

Grant LJ, Mikkelsen D, Ouwerkerk D, Klieve AV, Gidley MJ, Williams BA (2019) Whole fruit pulp (mango) and a soluble fibre (pectin) impact bacterial diversity and abundance differently within the porcine large intestine. *Bioactive Carbohydrates and Dietary Fibre* 19:100192.

Green PB (1962) Mechanism for plant cellular morphogenesis. *Science* 138:1404–1405.

Green PB (1968) Growth physics in *Nitella*: a method for continuous in vivo analysis of extensibility based on a micromanometer technique for turgor pressure. *Plant Physiology* 43:1169–1184.

Grundy MML, Edwards CH, Mackie AR, Gidley MJ, Butterworth PJ, Ellis PR (2016) Re-evaluation of the mechanisms of dietary fibre and implications for macronutrient bioaccessibility, digestion and postprandial metabolism. *British Journal of Nutrition* 116:816–833.

Gu Y, Somerville C (2010) Cellulose synthase interacting protein: a new factor in cellulose synthesis. *Plant and Signaling Behavior* 5:1571–1574.

Guerringue Y, Thomine S, Frachisse JM (2018) Sensing and transducing forces in plants with MSL10 and DEK1 mechanosensors. *FEBS Letters* 592:1968–1979.

Günl M, Gille S, Pauly M (2010) OLIgo mass profiling (OLIMP) of extracellular polysaccharides. *Journal of Visualized Experiments* 40:e2046.

Guo F, Wu T, Xu G, Qi H, Zhu X, Zhang Z (2021) TaWAK2A-800, a wall-associated kinase, participates positively in resistance to fusarium head blight and sharp eyespot in wheat. *International Journal of Molecular Sciences* 22(21):11493.

Guo H, Nolan TM, Song G, Liu S, Xie Z, Chen J, Schnable PS, Walley JW, Yin Y (2018) FERONIA receptor kinase contributes to plant immunity by suppressing jasmonic acid signaling in *Arabidopsis thaliana*. *Current Biology* 28:3316–3324 e3316.

Gutierrez R, Lindeboom JJ, Paredez AR, Emons AM, Ehrhardt DW (2009) *Arabidopsis* cortical microtubules position cellulose synthase delivery to the plasma membrane and interact with cellulose synthase trafficking compartments. *Nature Cell Biology* 11:797–806.

Haas KT, Wightman R, Meyerowitz EM, Peaucelle A (2020) Pectin homogalacturonan nanofilament expansion drives morphogenesis in plant epidermal cells. *Science* 367:1003–1007.

Hamann T (2015) The plant cell wall integrity maintenance mechanism–a case study of a cell wall plasma membrane signaling network. *Phytochemistry* 112:100–109.

Hamant O, Heisler MG, Jönsson H, Krupinski P, Uyttewaal M, Bokov P, Corson F, Sahlin P, Boudaoud A, Meyerowitz EM, Couder Y, Traas J (2008) Developmental patterning by mechanical signals in *Arabidopsis*. *Science* 322:1650–1655.

Harris PJ (2006) Non-cellulosic polysaccharides in plant cell walls: their diversity, organization and approaches to understanding their functions. In *The Hemicelluloses Workshop 2005*, eds. Entwistle KM, Walker JCF, 36–53, Christchurch: University of Canterbury.

Harris PJ, Kelderman MR, Kendon MF, McKenzie RJ (1997) Monosaccharide compositions of unlignified cell walls of monocotyledons in relation to the occurrence of wall-bound ferulic acid. *Biochemical Systematics and Ecology* 25(2):167–179.

Hartmann FP, Tinturier E, Julien JL, Leblanc-Fournier N (2021) Between stress and response: function and localization of mechanosensitive Ca^{2+} channels in herbaceous and perennial plants. *International Journal of Molecular Sciences* 22(20):11043.

Hebert JJ, Müller LL (1974) An electron diffraction study of the crystal structure of native cellulose. *Journal of Applied Polymer Science* 18:3373–3377.

Hematy K, Sado PE, Van Tuinen A, Rochange S, Desnos T, Balzergue S, Pelletier S, Renou JP, Hofte H (2007) A receptor-like kinase mediates the response of *Arabidopsis* cells to the inhibition of cellulose synthesis. *Current Biology* 17:922–931.

Herger A, Dunser K, Kleine-Vehn J, Ringli C (2019) Leucine-rich repeat extensin proteins and their role in cell wall sensing. *Current Biology* 29:R851–R858.

Herth W (1983) Arrays of plasma-membrane "rosettes" involved in cellulose microfibril formation of *Spirogyra*. *Planta* 159:347–356.

Herth W, Schnepf E (1980) The fluorochrome, calcofluor white, binds oriented to structural polysaccharide fibrils. *Protoplasma* 105:129–133.

Hervé C, Siméon A, Jam M, Cassin A, Johnson KL, Salmeán AA, Willats WGT, Doblin MS, Bacic A, Kloareg B (2016) Arabinogalactan proteins have deep roots in eukaryotes: identification of genes and epitopes in brown algae and their role in *Fucus serratus* embryo development. *New Phytologist* 209:1428–1441.

Hoffmann N, Benske A, Betz H, Schuetz M, Samuels AL (2020) Laccases and peroxidases co-localize in lignified secondary cell walls throughout stem development. *Plant Physiology* 184:806–822.

Hoffmann N, King S, Samuels AL, McFarlane HE (2021) Subcellular coordination of plant cell wall synthesis. *Developmental Cell* 56:933–948.

Höfte H, Voxeur A (2017) Plant cell walls. *Current Biology* 27:R865–R870.

Hohmann U, Lau K, Hothorn M (2017) The structural basis of ligand perception and signal activation by receptor kinases. *Annual Review of Plant Biology* 68:109–137.

Hooke R (1665) *Micrographia: Or Some Physiological Descriptions of Minute Bodies made by Magnifying Glasses with Observations and Inquiries thereupon.* London: Printed by Jo. Martyn, and Ja. Allestry: Royal Society.

Huck N (2003) The *Arabidopsis* mutant *feronia* disrupts the female gametophytic control of pollen tube reception. *Development* 130:2149–2159.

Jedrzejczak P, Collins MN, Jesionowski T, Klapiszewski L (2021) The role of lignin and lignin-based materials in sustainable construction - a comprehensive review. *International Journal of Biological Macromolecules* 187:624–650.

Johnson KL, Cassin AM, Lonsdale A, Bacic A, Doblin MS, Schultz CJ (2017a) Pipeline to Identify Hydroxyproline-Rich Glycoproteins. *Plant Physiology* 174:886–903.

Johnson KL, Cassin AM, Lonsdale A, Wong GKS, Soltis DE, Miles NW, Melkonian M, Melkonian B, Deyholos MK, Leebens-Mack J, Rothfels CJ, Stevenson DW, Graham SW, Wang XM, Wu SX, Pires JC, Edger PP, Carpenter EJ, Bacic A, Doblin MS, Schultz CJ (2017b) Insights into the evolution of hydroxyproline-rich glycoproteins from 1000 plant transcriptomes. *Plant Physiology* 174:904–921.

Johnson KL, Gidley M, Bacic A, Doblin MS (2018) Cell wall biomechanics: a tractable challenge in manipulating plant cell walls "fit for purpose"! *Current Opinion in Biotechnology* 49:163–171.

Keegstra K, Talmadge KW, Bauer WD, Albersheim P (1973) The structure of plant cell walls. *Plant Physiology* 51:188–196.

Knox JP (1997) The use of antibodies to study the architecture and developmental regulation of plant cell walls. *International Review of Cytology* 171:79–120.

Kohel RJ, Quisenberry JE, Benedict CR (1974) Fiber elongation and dry weight changes in mutant lines of cotton. *Crop Science* 14: 471–474.

Kohorn BD (2016) Cell wall-associated kinases and pectin perception. *Journal of Experimental Botany* 67:489–494.

Kokubo A, Sakurai N, Kuraishi S, Takeda K (1991) Culm brittleness of barley (*Hordeum vulgare* L.) mutants is caused by smaller number of cellulose molecules in cell wall. *Plant Physiology* 97:509–514.

Kubo M, Udagawa M, Nishikubo N, Horiguchi G, Yamaguchi M, Ito J, Mimura T, Fukuda H, Demura T (2005) Transcription switches for protoxylem and metaxylem vessel formation. *Genes & Development* 19:1855–1860.

Kuroki S, Tsenkova R, Moyankova D, Muncan J, Morita H, Atanassova S, Djilianov D (2019) Water molecular structure underpins extreme desiccation tolerance of the resurrection plant *Haberlea rhodopensis. Scientific Reports* 9:3049.

Kutschera U, Niklas KJ (2018) Julius Sachs (1868): the father of plant physiology. *American Journal of Botany* 105:656–666.

Lamport DT (1977) Structure, biosynthesis and significance of cell wall glycoproteins. In *The Structure, Biosynthesis, and Degradation of Wood*, eds Loewus FA, Runeckles VC, 79–115. New York: Springer New York.

Lee KJD, Cornuault V, Manfield IW, Ralet M-C, Paul Knox J (2013) Multi-scale spatial heterogeneity of pectic rhamnogalacturonan I (RG-I) structural features in tobacco seed endosperm cell walls. *The Plant Journal* 75:1018–1027.

Lehtiö J, Sugiyama J, Gustavsson M, Fransson L, Linder M, Teeri TT (2003) The binding specificity and affinity determinants of family 1 and family 3 cellulose binding modules. *Proceedings of the National Academy of Sciences of the United States of America* 100:484–489.

Liepman AH, Wightman R, Geshi N, Turner SR, Scheller HV (2010) Arabidopsis - a powerful model system for plant cell wall research. *The Plant Journal* 61:1107–1121.

Liesche J, Ziomkiewicz I, Schulz A (2013) Super-resolution imaging with pontamine fast scarlet 4BS enables direct visualization of cellulose orientation and cell connection architecture in onion epidermis cells. *BMC Plant Biology* 13:226.

Lin D, Lopez-Sanchez P, Gidley M (2016) Interactions of pectins with cellulose during its synthesis in the absence of calcium. *Food Hydrocolloids* 52:57–68.

Lin W, Wang Y, Liu X, Shang JX, Zhao L (2021) OsWAK112, A wall-associated kinase, negatively regulates salt stress responses by inhibiting ethylene production. *Frontiers in Plant Science* 12:751965.

Little A, Schwerdt JG, Shirley NJ, Khor SF, Neumann K, O'Donovan LA, Lahnstein J, Collins HM, Henderson M, Fincher GB, Burton RA (2018) Revised phylogeny of the cellulose synthase gene superfamily: insights into cell wall evolution. *Plant Physiology* 177:1124–1141.

Liu J, Zhang W, Long S, Zhao C (2021a) Maintenance of cell wall integrity under high salinity. *International Journal of Molecular Sciences* 22(6):3260.

Liu X, Wang Z, Tian Y, Zhang S, Li D, Dong W, Zhang C, Zhang Z (2021b) Characterization of wall-associated kinase/wall-associated kinase-like (WAK/WAKL) family in rose (*Rosa chinensis*) reveals the role of *RcWAK4* in Botrytis resistance. *BMC Plant Biology* 21:526.

Lopez-Hernandez F, Tryfona T, Rizza A, Yu XL, Harris MOB, Webb AAR, Kotake T, Dupree P (2020) Calcium binding by arabinogalactan polysaccharides is important for normal plant development. *The Plant Cell* 32:3346–3369.

Lopez-Sanchez P, Martinez-Sanz M, Bonilla MR, Wang D, Gilbert EP, Stokes JR, Gidley MJ (2017) Cellulose-pectin composite hydrogels: intermolecular interactions and material properties depend on order of assembly. *Carbohydrate Polymers* 162:71–81.

Lopez-Sanchez P, Martinez-Sanz M, Bonilla MR, Wang D, Walsh CT, Gilbert EP, Stokes JR, Gidley MJ (2016) Pectin impacts cellulose fibre architecture and hydrogel mechanics in the absence of calcium. *Carbohydrate Polymers* 153:236–245.

Lopez-Sanchez P, Schuster E, Wang D, Gidley MJ, Strom A (2015) Diffusion of macromolecules in self-assembled cellulose/hemicellulose hydrogels. *Soft Matter* 11:4002–4010.

Low DY, Williams BA, D'Arcy BR, Flanagan BM, Gidley MJ (2015) In vitro fermentation of chewed mango and banana: particle size, starch and vascular fibre effects. *Food & Function* 6:2464–2474.

Lu S, Flanagan BM, Williams BA, Mikkelsen D, Gidley MJ (2020) Cell wall architecture as well as chemical composition determines fermentation of wheat cell walls by a faecal inoculum. *Food Hydrocolloids* 107:105858.

Malivert A, Erguvan O, Chevallier A, Dehem A, Friaud R, Liu M, Martin M, Peyraud T, Hamant O, Verger S (2021) FERONIA and microtubules independently contribute to mechanical integrity in the *Arabidopsis* shoot. *PLOS Biology* 19:e3001454.

Malivert A, Hamant O, Ingram G (2018) The contribution of mechanosensing to epidermal cell fate specification. *Current Opinion in Genetics & Development* 51:52–58.

Marga F, Grandbois M, Cosgrove DJ, Baskin TI (2005) Cell wall extension results in the coordinate separation of parallel microfibrils: evidence from scanning electron microscopy and atomic force microscopy. *The Plant Journal* 43:181–190.

McBride Z, Chen DL, Reick C, Xie J, Szymanski DB (2017) Global analysis of membrane-associated protein oligomerization using protein correlation profiling. *Mol Cell Proteomics* 16:1972–1989.

McQueen-Mason S, Durachko DM, Cosgrove DJ (1992) Two endogenous proteins that lnduce cell wall extension in plants. *The Plant Cell* 4:1425–1433.

Mecchia MA, Santos-Fernandez G, Duss NN, Somoza SC, Boisson-Dernier A, Gagliardini V, Martinez-Bernardini A, Fabrice TN, Ringli C, Muschietti JP, Grossniklaus U (2017) RALF4/19 peptides interact with LRX proteins to control pollen tube growth in *Arabidopsis. Science* 358:1600–1603.

Mergner J, Frejno M, List M, Papacek M, Chen X, Chaudhary A, Samaras P, Richter S, Shikata H, Messerer M, Lang D, Altmann S, Cyprys P, Zolg DP, Mathieson T, Bantscheff M, Hazarika RR, Schmidt T, Dawid C, Dunkel A, Hofmann T, Sprunck S, Falter-Braun P, Johannes F, Mayer KFX, Jurgens G, Wilhelm M, Baumbach J, Grill E, Schneitz K,

Schwechheimer C, Kuster B (2020) Mass-spectrometry-based draft of the *Arabidopsis* proteome. *Nature* 579:409–414.

Merz D, Richter J, Gonneau M, Sanchez-Rodriguez C, Eder T, Sormani R, Martin M, Hematy K, Hofte H, Hauser MT (2017) T-DNA alleles of the receptor kinase THESEUS1 with opposing effects on cell wall integrity signaling. *Journal of Experimental Botany* 68:4583–4593.

Mikkelsen D, Lopez-Sanchez P, Wang D, Gidley MJ (2020) Formation of cellulose-based composites with hemicelluloses and pectins using *Komagataeibacter* fermentation. *Methods in Molecular Biology (Clifton, N.J.)* 2149:73–87.

Mitsuda N, Iwase A, Yamamoto H, Yoshida M, Seki M, Shinozaki K, Ohme-Takagi M (2007) NAC transcription factors, NST1 and NST3, are key regulators of the formation of secondary walls in woody tissues of *Arabidopsis. The Plant Cell* 19:270–280.

Mohnen D, Hahn MG (1993) Cell wall carbohydrates as signals in plants. *Seminars in Cell Biology* 4:93–102.

Molina A, Miedes E, Bacete L, Rodriguez T, Melida H, Denance N, Sanchez-Vallet A, Riviere MP, Lopez G, Freydier A, Barlet X, Pattathil S, Hahn M, Goffner D (2021) Arabidopsis cell wall composition determines disease resistance specificity and fitness. *Proceedings of the National Academy of Sciences of the United States of America* 118(5):e2010243118.

Moller I, Marcus SE, Haeger A, Verhertbruggen Y, Verhoef R, Schols H, Ulvskov P, Mikkelsen JD, Knox JP, Willats W (2008) High-throughput screening of monoclonal antibodies against plant cell wall glycans by hierarchical clustering of their carbohydrate microarray binding profiles. *Glycoconjugate Journal* 25:37–48.

Moller I, Sørensen I, Bernal AJ, Blaukopf C, Lee K, Øbro J, Pettolino F, Roberts A, Mikkelsen JD, Knox P, Bacic A, Willats WGT (2007) High-throughput mapping of cell wall polymers within and between plants using novel microarrays. *The Plant Journal.* doi: 10.1111/j.1365-313X.2007.03114.x:1118-1128.

Moore JP, Farrant JM, Driouich A (2008) A role for pectin-associated arabinans in maintaining the flexibility of the plant cell wall during water deficit stress. *Plant Signaling & Behavior* 3:102–104.

Mosca G, Sapala A, Strauss S, Routier-Kierzkowska AL, Smith RS (2017) On the micro-indentation of plant cells in a tissue context. *Physical Biology* 14:015003.

Moussu S, Broyart C, Santos-Fernandez G, Augustin S, Wehrle S, Grossniklaus U, Santiago J (2020) Structural basis for recognition of RALF peptides by LRX proteins during pollen tube growth. *Proceedings of the National Academy of Sciences of the United States of America* 117:7494–7503.

Mravec J, Kračun SK, Rydahl MG, Westereng B, Miart F, Clausen MH, Fangel JU, Daugaard M, Van Cutsem P, De Fine Licht HH, Höfte H, Malinovsky FG, Domozych DS, Willats WGT (2014) Tracking developmentally regulated post-synthetic processing of homogalacturonan and chitin using reciprocal oligosaccharide probes. *Development* 141:4841–4850.

Newman RH, Hill SJ, Harris PJ (2013) Wide-angle x-ray scattering and solid-state nuclear magnetic resonance data combined to test models for cellulose microfibrils in mung bean cell walls. *Plant Physiology* 163:1558–1567.

Nicolas W, Fäßler F, Meyerowitz E, Jensen G (2021) Peaking into the plant cell wall using cryo-FIB milling and electron cryo-tomography. *Microscopy and Microanalysis* 27:3024–3026.

Nixon BT, Mansouri K, Singh A, Du J, Davis JK, Lee J-G, Slabaugh E, Vandavasi VG, O'Neill H, Roberts EM, Roberts AW, Yingling YG, Haigler CH (2016) Comparative structural and computational analysis supports eighteen cellulose synthases in the plant cellulose synthesis complex. *Scientific Reports* 6:28696.

Novakovic L, Guo T, Bacic A, Sampathkumar A, Johnson KL (2018) Hitting the wall-sensing and signaling pathways involved in plant cell wall remodeling in response to abiotic stress. *Plants (Basel)* 7(4):89.

Nühse TS, Stensballe A, Jensen ON, Peck SC (2004) Phosphoproteomics of the Arabidopsis plasma membrane and a new phosphorylation site database. *The Plant Cell* 16:2394–2405.

Obel N, Erbenc V, Schwarza T, Kühnel S, Fodora A, Pauly M (2009) Microanalysis of plant cell wall polysaccharides. *Molecular Plant* 2:922–932.

Oehme DP, Downton MT, Doblin MS, Wagner J, Gidley MJ, Bacic A (2015) Unique aspects of the structure and dynamics of elementary Iβ cellulose microfibrils revealed by computational simulations. *Plant Physiology* 168:3–17.

Ogawa Y, Chanzy H, Putaux J-L (2019) Transmission electron microscopy of cellulose. Part 1: historical perspective. *Cellulose* 26:5–15.

O'Malley RC, Ecker JR (2010) Linking genotype to phenotype using the Arabidopsis unimutant collection. *The Plant Journal* 61:928–940.

Oxley D, Currie GC, Bacic A (2004) Analysis of carbohydrate from glycoproteins. In *Purifying Proteins for Proteomics: A Laboratory Manual*, ed Simpson RJ, 579–636. New York: Cold Spring Harbor Cold Spring Harbor Laboratory Press.

Padayachee A, Day L, Howell K, Gidley MJ (2017) Complexity and health functionality of plant cell wall fibers from fruits and vegetables. *Critical Reviews in Food Science and Nutrition* 57:59–81.

Padayachee A, Netzel G, Netzel M, Day L, Mikkelsen D, Gidley MJ (2013) Lack of release of bound anthocyanins and phenolic acids from carrot plant cell walls and model composites during simulated gastric and small intestinal digestion. *Food & Function* 4:906–916.

Paradez A, Wright A, Ehrhardt DW (2006) Microtubule cortical array organization and plant cell morphogenesis. *Current Opinion in Plant Biology* 9:571–578.

Paredez AR, Somerville CR, Ehrhardt DW (2006) Visualization of cellulose synthase demonstrates functional association with microtubules. *Science* 312:1491–1495.

Park YB, Cosgrove DJ (2011) Changes in cell wall biomechanical properties in the xyloglucan-deficient *xxt1/xxt2* mutant of Arabidopsis *Plant Physiology* 158:465–475.

Parsons HT, Christiansen K, Knierim B, Carroll A, Ito J, Batth TS, Smith-Moritz AM, Morrison S, McInerney P, Hadi MZ, Auer M, Mukhopadhyay A, Petzold CJ, Scheller HV, Loque D, Heazlewood JL (2012) Isolation and proteomic characterization of the Arabidopsis golgi defines functional and novel components involved in plant cell wall biosynthesis. Plant Physiology *159:12–26.*

Pattathil S, Avci U, Baldwin D, Swennes AG, McGill JA, Popper Z, Bootten T, Albert A, Davis RH, Chennareddy C, Dong R, O'Shea B, Rossi R, Leoff C, Freshour G, Narra R, O'Neil M, York WS, Hahn MG (2010) A comprehensive toolkit of plant cell wall glycan-directed monoclonal antibodies. *Plant Physiology* 153:514–525.

Paulraj T, Riazanova AV, Yao K, Andersson RL, Mullertz A, Svagan AJ (2017) Bioinspired layer-by-layer microcapsules based on cellulose nanofibers with switchable permeability. *Biomacromolecules* 18:1401–1410.

Payen A (1838) Mémoire sur la composition du tissu propre des plantes et du ligneux. *Comptes Rendus* 7:1052–1056.

Pear JR, Kawagoe Y, Schreckengost WE, Delmer DP, Stalker DM (1996) Higher plants contain homologs of the bacterial *celA* genes encoding the catalytic subunit of cellulose synthase. *Proceedings of the National Academy of Sciences of the United States of America* 93:12637–12642.

Peaucelle A, Braybrook SA, Le Guillou L, Bron E, Kuhlemeier C, Hofte H (2011) Pectin-induced changes in cell wall mechanics underlie organ initiation in *Arabidopsis*. *Current Biology* 21:1720–1726.

Pena MJ, Zhong RQ, Zhou GK, Richardson EA, O'Neill MA, Darvill AG, York WS, Ye ZH (2007) *Arabidopsis irregular xylem8* and *irregular xylem9*: implications for the complexity of glucuronoxylan biosynthesis. *The Plant Cell* 19:549–563.

Persson S, Paredez A, Carroll A, Palsdottir H, Doblin M, Poindexter P, Khitrov N, Auer M, Somerville CR (2007) Genetic evidence for three unique components in primary cell-wall cellulose synthase complexes in *Arabidopsis*. *Proceedings of the National Academy of Sciences of the United States of America* 104:15566–15571.

Persson S, Wei HR, Milne J, Page GP, Somerville CR (2005) Identification of genes required for cellulose synthesis by regression analysis of public microarray data sets. *Proceedings of the National Academy of Sciences of the United States of America* 102:8633–8638.

Pettolino F, Walsh C, Fincher GB, Bacic A (2012) Determining the polysaccharide composition of plant cell walls. *Nature Protocols* 7:1590–1607.

Pfeifer L, Shafee T, Johnson KL, Bacic A, Classen B (2020) Arabinogalactan-proteins of *Zostera marina* L. contain unique glycan structures and provide insight into adaption processes to saline environments. *Scientific Reports* 10:8232.

Poulsen CP, Dilokpimol A, Mouille G, Burow M, Geshi N (2014) Arabinogalactan glycosyltransferases target to a unique subcellular compartment that may function in unconventional secretion in plants. *Traffic* 15:1219–1234.

Preston RD, Nicolai E, Reed R., Millard A (1948) An electron microscope study of cellulose in the wall of *Valonia ventricosa*. *Nature* 162:665–667.

Preston RD, Ripley GW (1954) Electron diffraction diagrams of cellulose micro-fibrils in *Valonia*. *Nature* 174:76–77.

Puhlmann J, Bucheli E, Swain MJ, Dunning N, Albersheim P, Darvill AG, Hahn MG (1994) Generation of monoclonal antibodies against plant cell-wall polysaccharides. I. characterization of a monoclonal antibody to a terminal [alpha]-(1,2)-linked fucosyl-containing epitope. *Plant Physiology* 104:699–710.

Pullar RC, Novais RM (2017) Cork-based biomimetic ceramic 3-DOM foams. *Materials Today* 20:45–46.

Purushotham P, Ho RY, Zimmer J (2020) Architecture of a catalytically active homotrimeric plant cellulose synthase complex. *Science* 369:1089–1094.

Qi H, Guo F, Lv L, Zhu X, Zhang L, Yu J, Wei X, Zhang Z (2021) The wheat wall-associated receptor-like kinase TaWAK-6D mediates broad resistance to two fungal pathogens *Fusarium pseudograminearum* and *Rhizoctonia cerealis*. *Frontiers in Plant Science* 12:758196.

Qiao Z, Lampugnani ER, Yan XF, Khan GA, Saw WG, Hannah P, Qian F, Calabria J, Miao YS, Gruber G, Persson S, Gao YG (2021) Structure of *Arabidopsis* CESA3 catalytic domain with its substrate UDP-glucose provides insight into the mechanism of cellulose synthesis. *Proceedings of the National Academy of Sciences of the United States of America* 118(11):e2024015118

Qu S, Zhang X, Song Y, Lin J, Shan X (2017) THESEUS1 positively modulates plant defense responses against *Botrytis cinerea* through GUANINE EXCHANGE FACTOR4 signaling. *Journal of Integrative Plant Biology* 59:797–804.

Radkov AD, Hsu YP, Booher G, VanNieuwenhze MS (2018) Imaging bacterial cell wall biosynthesis. *Annual Review of Biochemistry* 87:991–1014.

Read SM, Currie G, Bacic A (1996) Analysis of the structural heterogeneity of laminarin by electrospray ionisation-mass spectrometry. *Carbohydrate Research* 281:187–201.

Richter S, Müssig J, Gierlinger N (2011) Functional plant cell wall design revealed by the Raman imaging approach. *Planta* 233:763–772.

Roberts K, Phillips J, Shaw P, Grief C, Smith E (1985) An immunological approach to the plant cell wall. In *Biochemistry of Plant Cell Walls*, ed Hillman CTBa JR, 125–154. Cambridge, UK: Cambridge University Press.

Robinson DG, Ding Y, Jiang L (2016) Unconventional protein secretion in plants: a critical assessment. *Protoplasma* 253:31–43.

Robinson S, Huflejt M, Barbier de Reuille P, Braybrook SA, Schorderet M, Reinhardt D, Kuhlemeier C (2017) An automated confocal micro-extensometer enables in vivo quantification of mechanical properties with cellular resolution. *The Plant Cell* 29:2959–2973.

Robinson S, Kuhlemeier C (2018) Global compression reorients cortical microtubules in *Arabidopsis* hypocotyl epidermis and promotes growth. *Current Biology* 28:1794-1802.e1792.

Rounds CM, Lubeck E, Hepler PK, Winship LJ (2011) Propidium iodide competes with Ca^{2+} to label pectin in pollen tubes and Arabidopsis root hairs. *Plant Physiology* 157:175–187.

Ruprecht C, Bartetzko MP, Senf D, Dallabernadina P, Boos I, Andersen MCF, Kotake T, Knox JP, Hahn MG, Clausen MH, Pfrengle F (2017) A synthetic glycan microarray enables epitope mapping of plant cell wall glycan-directed antibodies. *Plant Physiology* 175:1094–1104.

Ruprecht C, Bartetzko MP, Senf D, Lakhina A, Smith PJ, Soto MJ, Oh H, Yang JY, Chapla D, Varon Silva D, Clausen MH, Hahn MG, Moremen KW, Urbanowicz BR, Pfrengle F (2020) A glycan array-based assay for the identification and characterization of plant glycosyltransferases. *Angewandte Chemie International Edition* 59:12493–12498.

Ruprecht C, Mendrinna A, Tohge T, Sampathkumar A, Klie S, Fernie AR, Nikoloski Z, Persson S, Mutwil M (2016) FamNet: a framework to identify multiplied modules driving pathway expansion in plants. *Plant Physiology* 170:1878–1894.

Sampathkumar A, Gutierrez R, McFarlane HE, Bringmann M, Lindeboom J, Emons A-M, Samuels L, Ketelaar T, Ehrhardt DW, Persson S (2013) Patterning and lifetime of plasma membrane-localized cellulose synthase is dependent on actin organization in Arabidopsis interphase cells. *Plant Physiology* 162:675–688.

Sampathkumar A, Krupinski P, Wightman R, Milani P, Berquand A, Boudaoud A, Hamant O, Jonsson H, Meyerowitz EM (2014) Subcellular and supracellular mechanical stress prescribes cytoskeleton behavior in *Arabidopsis* cotyledon pavement cells. *eLife* 3:e01967.

Scheible WR, Eshed R, Richmond T, Delmer D, Somerville C (2001) Modifications of cellulose synthase confer resistance to isoxaben and thiazolidinone herbicides in *Arabidopsis Ixr1* mutants. *Proceedings of the National Academy of Sciences of the United States of America* 98:10079–10084.

Schoenaers S, Balcerowicz D, Breen G, Hill K, Zdanio M, Mouille G, Holman TJ, Oh J, Wilson MH, Nikonorova N, Vu LD, De Smet I, Swarup R, De Vos WH, Pintelon I, Adriaensen D, Grierson C, Bennett MJ, Vissenberg K (2018) The auxin-regulated CrRLK1L kinase *ERULUS* controls cell wall composition during root hair tip growth. *Current Biology* 28:722–732 e726.

Schuetz M, Benske A, Smith RA, Watanabe Y, Tobimatsu Y, Ralph J, Demura T, Ellis B, Samuels AL (2014) Laccases direct lignification in the discrete secondary cell wall domains of protoxylem *Plant Physiology* 166:798–807.

Schulze E (1892) Zur chemie der pflanzlichen zellmembranen. *Hoppe-Seyler's Zeitschrift für Physiologische Chemie* 16:387–438.

Schwerdt JG, MacKenzie K, Wright F, Oehme D, Wagner JM, Harvey AJ, Shirley NJ, Burton RA, Schreiber M, Halpin C, Zimmer J, Marshall DF, Waugh R, Fincher GB (2015) Evolutionary dynamics of the cellulose synthase gene superfamily in grasses. *Plant Physiology* 168:968–983.

Seguí-Simarro JM, Austin JR, 2nd, White EA, Staehelin LA (2004) Electron tomographic analysis of somatic cell plate formation in meristematic cells of Arabidopsis preserved by high-pressure freezing. *The Plant Cell* 16:836–856.

Shedletzky E, Shmuel M, Trainin T, Kalman S, Delmer D (1992) Cell wall structure in cells adapted to growth on the cellulose-synthesis inhibitor 2,6-dichlorobenzonitrile : a comparison between two dicotyledonous plants and a graminaceous monocot. *Plant Physiology* 100:120–130.

Shi HZ, Kim Y, Guo Y, Stevenson B, Zhu JK (2003) The Arabidopsis *SOS5* locus encodes a putative cell surface adhesion protein and is required for normal cell expansion. *The Plant Cell* 15:19–32.

Shih HW, Miller ND, Dai C, Spalding EP, Monshausen GB (2014) The receptor-like kinase FERONIA is required for mechanical signal transduction in *Arabidopsis* seedlings. *Current Biology* 24:1887–1892.

Shin Y, Chane A, Jung M, Lee Y (2021) Recent advances in understanding the roles of pectin as an active participant in plant signaling networks. *Plants* 10(8):1712.

Showalter AM, Keppler B, Lichtenberg J, Gu D, Welch LR (2010) A bioinformatics approach to the identification, classification, and analysis of hydroxyproline-rich glycoproteins. *Plant Physiology* 153:485–513.

Solecka D, Zebrowski J, Kacperska A (2008) Are pectins involved in cold acclimation and de-acclimation of winter oil-seed rape plants? *Annals of Botany* 101:521–530.

Somerville C, Bauer S, Brininstool G, Facette M, Hamann T, Milne J, Osborne E, Paredez A, Persson S, Raab T, Vorwerk S, Youngs H (2004) Toward a systems approach to understanding plant cell walls. *Science* 306:2206–2211.

Stegmann M, Monaghan J, Smakowska-Luzan E, Rovenich H, Lehner A, Holton N, Belkhadir Y, Zipfel C (2017) The receptor kinase FER is a RALF-regulated scaffold controlling plant immune signaling. *Science* 355:287–289.

Stephens C, Hammond-Kosack KE, Kanyuka K (2021) WAKsing plant immunity, waning diseases. *Journal of Experimental Botany*. doi: 10.1093/jxb/erab422.

Stocker BL, Timmer MS (2013) Chemical tools for studying the biological function of glycolipids. *ChemBioChem* 14:1164–1184.

Studer D, Humbel BM, Chiquet M (2008) Electron microscopy of high pressure frozen samples: bridging the gap between cellular ultrastructure and atomic resolution. *Histochemistry and Cell Biology* 130:877–889.

Synek L, Schlager N, Eliás M, Quentin M, Hauser M-T, Zárský V (2006) AtEXO70A1, a member of a family of putative exocyst subunits specifically expanded in land plants, is important for polar growth and plant development. *The Plant Journal* 48:54–72.

Takahashi D, Johnson KL, Hao P, Tuong T, Erban A, Sampathkumar A, Bacic A, Livingston DP, 3rd, Kopka J, Kuroha T, Yokoyama R, Nishitani K, Zuther E, Hincha DK (2021) Cell wall modification by the xyloglucan endotransglucosylase/hydrolase XTH19 influences freezing tolerance after cold and sub-zero acclimation. *Plant, Cell & Environment* 44:915–930.

Takahashi D, Zuther E, Hincha DK (2020) Analysis of changes in plant cell wall composition and structure during cold acclimation. *Methods in Molecular Biology (Clifton, N.J.)* 2156:255–268.

Tan L, Eberhard S, Pattathil S, Warder C, Glushka J, Yuan C, Hao Z, Zhu X, Avci U, Miller JS, Baldwin D, Pham C, Orlando R, Darvill A, Hahn MG, Kieliszewski MJ, Mohnen D (2013) An *Arabidopsis* cell wall proteoglycan consists of pectin and arabinoxylan covalently linked to an arabinogalactan protein. *The Plant Cell* 25:270–287.

Tang W, Lin W, Zhou X, Guo J, Dang X, Li B, Lin D, Yang Z (2021) Mechano-transduction via the pectin-FERONIA complex activates ROP6 GTPase signaling in *Arabidopsis* pavement cell morphogenesis. *Current Biology*. doi: 10.1016/j.cub.2021.11.031.

Taylor NG, Laurie S, Turner SR (2000) Multiple cellulose synthase catalytic subunits are required for cellulose synthesis in Arabidopsis. *The Plant Cell* 12:2529–2539.

Taylor NG, Scheible WR, Cutler S, Somerville CR, Turner SR (1999) The *irregular xylem3* locus of Arabidopsis encodes a cellulose synthase required for secondary cell wall synthesis. *The Plant Cell* 11:769–779.

Taylor-Teeples M, Lin L, de Lucas M, Turco G, Toal TW, Gaudinier A, Young NF, Trabucco GM, Veling MT, Lamothe R, Handakumbura PP, Xiong G, Wang C, Corwin J, Tsoukalas A, Zhang L, Ware D, Pauly M, Kliebenstein DJ, Dehesh K, Tagkopoulos I, Breton G, Pruneda-Paz JL, Ahnert SE, Kay SA, Hazen SP, Brady SM (2015) An *Arabidopsis* gene regulatory network for secondary cell wall synthesis. *Nature* 517:571–U307.

Thompson DA (1917) *On Growth and Form*. Cambridge: Cambridge University Press.

Trinh DC, Alonso-Serra J, Asaoka M, Colin L, Cortes M, Malivert A, Takatani S, Zhao F, Traas J, Trehin C, Hamant O (2021) How mechanical forces shape plant organs. *Current Biology* 31:R143–R159.

Turner SR, Somerville CR (1997) Collapsed xylem phenotype of Arabidopsis identifies mutants deficient in cellulose deposition in the secondary cell wall. *The Plant Cell* 9:689–701.

Ulian T, Diazgranados M, Pironon S, Padulosi S, Liu U, Davies L, Howes M-JR, Borrell JS, Ondo I, Pérez-Escobar OA, Sharrock S, Ryan P, Hunter D, Lee MA, Barstow C, Łuczaj Ł, Pieroni A, Cámara-Leret R, Noorani A, Mba C, Nono Womdim R, Muminjanov H, Antonelli A, Pritchard HW, Mattana E (2020) Unlocking plant resources to support food security and promote sustainable agriculture. *Plants People Planet* 2:421–445.

Usov AI (2011) Polysaccharides of the red algae. *Advances in Carbohydrate Chemistry and Biochemistry* 65:115–217.

Uyttewaal M, Burian A, Alim K, Landrein B, Borowska-Wykret D, Dedieu A, Peaucelle A, Ludynia M, Traas J, Boudaoud A, Kwiatkowska D, Hamant O (2012) Mechanical stress acts via katanin to amplify differences in growth rate between adjacent cells in Arabidopsis. *Cell* 149:439–451.

Vaahtera L, Schulz J, Hamann T (2019) Cell wall integrity maintenance during plant development and interaction with the environment. *Nature Plants* 5:924–932.

Vain T, Crowell EF, Timpano H, Biot E, Desprez T, Mansoori N, Trindade LM, Pagant S, Robert S, Höfte H, Gonneau M, Vernhettes S (2014) The cellulase KORRIGAN is part of the cellulose synthase complex. *Plant Physiology* 165:1521–1532.

Van Acker R, Leple JC, Aerts D, Storme V, Goeminne G, Ivens B, Legee F, Lapierre C, Piens K, Van Montagu MCE, Santoro N, Foster CE, Ralph J, Soetaert W, Pilate G, Boerjan W (2014) Improved saccharification and ethanol yield from field-grown transgenic poplar deficient in cinnamoyl-CoA reductase. *Proceedings of the National Academy of Sciences of the United States of America* 111:845–850.

van de Meene AML, Doblin MS, Bacic A (2017) The plant secretory pathway seen through the lens of the cell wall. *Protoplasma* 254:75–94.

Verger S, Long Y, Boudaoud A, Hamant O (2018) A tension-adhesion feedback loop in plant epidermis. *eLife* 7:e34460.

Voiniciuc C, Pauly M, Usadel B (2018) Monitoring polysaccharide dynamics in the plant cell wall. *Plant Physiology* 176:2590–2600.

Voxeur A, Habrylo O, Guenin S, Miart F, Soulie MC, Rihouey C, Pau-Roblot C, Domon JM, Gutierrez L, Pelloux J, Mouille G, Fagard M, Hofte H, Vernhettes S (2019) Oligogalacturonide production upon *Arabidopsis thaliana-Botrytis cinerea* interaction. *Proceedings of the National Academy of Sciences of the United States of America* 116:19743–19752.

Wahlström N, Nylander F, Malmhäll-Bah E, Sjövold K, Edlund U, Westman G, Albers E (2020) Composition and structure of cell wall ulvans recovered from *Ulva spp.* along the Swedish west coast. *Carbohydrate Polymers* 233:115852.

Wang D, Lin L, Fu F (2021) The difference of creep compliance for wood cell wall CML and secondary S2 layer by nanoindentation. *Mechanics of Time-Dependent Materials* 25:219–230.

Wang T, Park YB, Cosgrove DJ, Hong M (2015) Cellulose-pectin spatial contacts are inherent to never-dried arabidopsis primary cell walls: evidence from Solid-State Nuclear Magnetic Resonance. *Plant Physiology* 168:871–884.

Wang Y, Naleway SE, Wang B (2020) Biological and bioinspired materials: structure leading to functional and mechanical performance. *Bioactive Materials* 5:745–757.

Wardrop AB, Bland DE (1959) Process of lignification in woody plants. *Proceedings of the 4th International Congress of Biochemistry* 2(Synop. II.):92–116.

Watanabe Y, Schneider R, Barkwill S, Gonzales-Vigil E, Hill JL, Jr., Samuels AL, Persson S, Mansfield SD (2018) Cellulose synthase complexes display distinct dynamic behaviors during xylem transdifferentiation. *Proceedings of the National Academy of Sciences of the United States of America* 115:E6366–E6374.

Wei X, Liu S, Sun C, Xie G, Wang L (2021) Convergence and divergence: signal perception and transduction mechanisms of cold stress in *Arabidopsis* and rice. *Plants* 10(9):1864..

Williams BA, Grant LJ, Gidley MJ, Mikkelsen D (2017) Gut fermentation of dietary fibres: physico-chemistry of plant cell walls and implications for health. *International Journal of Molecular Sciences* 18(10):E2203.

Williams BA, Mikkelsen D, Flanagan BM, Gidley MJ (2019) "Dietary fibre": moving beyond the "soluble/insoluble" classification for monogastric nutrition, with an emphasis on humans and pigs. *Journal of Animal Science and Biotechnology* 10:45.

Williams BA, Mikkelsen D, le Paih L, Gidley MJ (2011) In vitro fermentation kinetics and end-products of cereal arabinoxylans and (1,3;1,4)-β-glucans by porcine faeces. *Journal of Cereal Science* 53:53–58.

Willick IR, Takahashi D, Fowler DB, Uemura M, Tanino KK (2018) Tissue-specific changes in apoplastic proteins and cell wall structure during cold acclimation of winter wheat crowns. *Journal of Experimental Botany* 69:1221–1234.

Wolf S, Hematy K, Hofte H (2012) Growth control and cell wall signaling in plants. *Annual Review of Plant Biology* 63:381–407.

Wu AM, Hornblad E, Voxeur A, Gerber L, Rihouey C, Lerouge P, Marchant A (2010) Analysis of the Arabidopsis IRX9/IRX9-L and IRX14/IRX14-L pairs of glycosyltransferase genes reveals critical contributions to biosynthesis of the hemicellulose glucuronoxylan. *Plant Physiology* 153:542–554.

Wu AM, Rihouey C, Seveno M, Hornblad E, Singh SK, Matsunaga T, Ishii T, Lerouge P, Marchant A (2009) The Arabidopsis IRX10 and IRX10-LIKE glycosyltransferases are critical for glucuronoxylan biosynthesis during secondary cell wall formation. *The Plant Journal* 57:718–731.

Wu X, Bacic A, Johnson KL, Humphries J (2020) The role of *Brachypodium distachyon* wall-associated kinases (WAKs) in cell expansion and stress responses. *Cells* 9(11):2478.

Xiao C, Zhang T, Zheng Y, Cosgrove DJ, Anderson CT (2016) Xyloglucan deficiency disrupts microtubule stability and cellulose biosynthesis in Arabidopsis, altering cell growth and morphogenesis. *Plant Physiology* 170:234–249.

Xie X, Lipke PN (2010) On the evolution of fungal and yeast cell walls. *Yeast* 27:479–488.

Xu B, Ohtani M, Yamaguchi M, Toyooka K, Wakazaki M, Sato M, Kubo M, Nakano Y, Sano R, Hiwatashi Y, Murata T, Kurata T, Yoneda A, Kato K, Hasebe M, Demura T (2014) Contribution of NAC transcription factors to plant adaptation to land. *Science* 343:1505–1508.

Yakubov GE, Bonilla MR, Chen H, Doblin MS, Bacic A, Gidley MJ, Stokes JR (2016) Mapping nano-scale mechanical heterogeneity of primary plant cell walls. *Journal of Experimental Botany* 67:2799–2816.

Yamaguchi M, Goue N, Igarashi H, Ohtani M, Nakano Y, Mortimer JC, Nishikubo N, Kubo M, Katayama Y, Kakegawa K, Dupree P, Demura T (2010) Vascular-related NAC-Domain6 and vascular-related NAC-Domain7 effectively induce transdifferentiation into xylem vessel elements under control of an induction system. *Plant Physiology* 153:906–914.

Yang H, Wang D, Guo L, Pan H, Yvon R, Garman S, Wu HM, Cheung AY (2021) Malectin/Malectin-like domain-containing proteins: a repertoire of cell surface molecules with broad functional potential. *The Cell Surface* 7:100056.

Yi Chou E, Schuetz M, Hoffmann N, Watanabe Y, Sibout R, Samuels AL (2018) Distribution, mobility, and anchoring of lignin-related oxidative enzymes in Arabidopsis secondary cell walls. *Journal of Experimental Botany* 69:1849–1859.

Yoshimura K, Iida K, Iida H (2021) MCAs in Arabidopsis are Ca^{2+}-permeable mechanosensitive channels inherently sensitive to membrane tension. *Nature Communications* 12:6074.

Young RE, McFarlane HE, Hahn MG, Western TL, Haughn GW, Samuels AL (2008) Analysis of the golgi apparatus in Arabidopsis seed coat cells during polarized secretion of pectin-rich mucilage. *The Plant Cell* 20:1623–1638.

Žárský V, Cvrčková F, Potocký M, Hála M (2009) Exocytosis and cell polarity in plants - exocyst and recycling domains. *New Phytologist* 183:255–272.

Žárský V, Kulich I, Fendrych M, Pečenková T (2013) Exocyst complexes multiple functions in plant cells secretory pathways. *Current Opinion in Plant Biology* 16:726–733.

Zeng W, Lampugnani ER, Picard KL, Song L, Wu AM, Farion IM, Zhao J, Ford K, Doblin MS, Bacic A (2016) Asparagus IRX9, IRX10, and IRX14A are components of an active xylan backbone synthase complex that forms in the golgi apparatus. *Plant Physiology* 171:93–109.

Zhang D, Williams BA, Mikkelsen D, Li X, Keates HL, Lisle AT, Collins HM, Fincher GB, Bird AR, Topping DL, Gidley MJ, Bryden WL (2015) Soluble arabinoxylan alters digesta flow and protein digestion of red meat-containing diets in pigs. *Nutrition* 31:1141–1147.

Zhang GF, Staehelin LA (1992) Functional compartmentation of the golgi apparatus of plant cells 1: immunocytochemical analysis of high-pressure frozen- and freeze-substituted sycamore maple suspension culture cells. *Plant Physiology* 99:1070–1083.

Zhang SS, Sun L, Dong X, Lu SJ, Tian W, Liu JX (2016a) Cellulose synthesis genes CESA6 and CSI1 are important for salt stress tolerance in Arabidopsis. *Journal of Integrative Plant Biology* 58:623–626.

Zhang T, Mahgsoudy-Louyeh S, Tittmann B, Cosgrove DJ (2014) Visualization of the nanoscale pattern of recently-deposited cellulose microfibrils and matrix materials in never-dried primary walls of the onion epidermis. *Cellulose* 21:853–862.

Zhang T, Tang H, Vavylonis D, Cosgrove DJ (2019) Disentangling loosening from softening: insights into primary cell wall structure. *The Plant Journal* 100:1101–1117.

Zhang T, Vavylonis D, Durachko DM, Cosgrove DJ (2017) Nanoscale movements of cellulose microfibrils in primary cell walls. *Nat Plants* 3:17056.

Zhang T, Zheng Y, Cosgrove DJ (2016b) Spatial organization of cellulose microfibrils and matrix polysaccharides in primary plant cell walls as imaged by multichannel atomic force microscopy. *The Plant Journal* 85:179–192.

Zhong RQ, Demura T, Ye ZH (2006) SND1, a NAC domain transcription factor, is a key regulator of secondary wall synthesis in fibers of *Arabidopsis*. *The Plant Cell* 18:3158–3170.

Zhu JY, Agarwal UP, Ciesielski PN, Himmel ME, Gao R, Deng Y, Morits M, Österberg M (2021a) Towards sustainable production and utilization of plant-biomass-based nanomaterials: a review and analysis of recent developments. *Biotechnology for Biofuels* 14:114.

Zhu S, Fu Q, Xu F, Zheng H, Yu F (2021b) New paradigms in cell adaptation: decades of discoveries on the CrRLK1L receptor kinase signalling network. *New Phytologist* 232:1168–1183.

2 Plant Cell Wall Glycosyltransferases

From Sequence to Structure and Function

Alban Mariette
Max-Planck Institute of Molecular Plant Physiology
Ruhr Universität Bochum

Thomas Michael Curry and Breeanna R. Urbanowicz
University of Georgia

Berit Ebert
The University of Melbourne
Ruhr Universität Bochum

INTRODUCTION

In both plants and animals, polysaccharides have long been recognised as important storage materials and energy sources (Nikaido, 1971; Beck and Ziegler, 1989; Robyt, 2001). Beyond these crucial functions, more recent studies have demonstrated that they also play a myriad of biological roles, including cell signalling and communication. In plants, polysaccharides are particularly important as they make up most of the cell wall that surrounds each cell, and due to the structural complexity of the polysaccharides, estimations suggest 10% or more of the plant genome is dedicated to building the wall (Carpita and McCann, 2015). Despite the importance of polysaccharides for plant growth and development and our reliance on them for food, textiles, paper, timber and more, it was only relatively recently that scientists discovered the crucial roles that glycosyltransferases play. The coordinated action of these enzymes results in the template-independent synthesis of cell wall carbohydrates, with final structures dependent on enzyme activity, availability of sugar donor and acceptor molecules and their organellar microenvironments. Biosynthetic glycosyltransferases are found in different cellular compartments and while those active at the plasma membrane directly incorporate polysaccharides into the wall during assembly, matrix polysaccharides are typically made by Golgi-localised enzymes and secreted to the plant cell wall via exocytosis. In Arabidopsis, around 1.7% of the genome encodes for glycosyltransferases with over 573 distributed among 42 GT families, highlighting their importance. This chapter therefore focuses on glycosyltransferases that catalyse formation of glycosidic linkages to produce the complex carbohydrates present in plant cell walls.

LELOIR GLYCOSYLTRANSFERASES

The term glycosyltransferase (GT; EC.2.4.*x*.*y*) specifies enzymes that use an activated sugar containing a phosphate leaving group as a donor, catalysing the transfer of a glycosyl residue from this donor onto an acceptor substrate. Donor sugar substrates are most often nucleoside diphosphate sugars (NDP-sugars); however, nucleoside monophosphate sugars such as cytidine 5′-monophospho-3-deoxy-D-*manno*-2-octulosonic acid (CMP-Kdo), lipid phosphates, e.g., dolichol phosphate oligosaccharides and unsubstituted phosphate are also used. NDP-sugars were discovered in 1950 by Luis L. Leloir and his colleagues Ranwel Caputto, Carlos E. Cardini, Raúl Trucco and Alejandro C. Paladini (Caputto et al., 1950). These early works from Leloir focused on the utilisation of galactose (Gal) in yeast and subsequently led to the isolation of uridine diphosphate glucose (UDP-Glc, UDPG) (Caputto et al., 1950). Todd and colleagues confirmed the structure 5 years later (Todd, 1954). Following the yeast study, many other nucleotide sugars were detected in animals, plants and bacteria (Hassid et al., 1959). Ginsburg et al. isolated a mix of UDP-Glc, UDP-galactose (UDP-Gal), UDP-xylose (UDP-Xyl) and UDP-arabinose (UDP-Ara) from mung bean (*Phaseolus aureus*) (Ginsburg et al., 1956), while Solms and Hassid identified UDP-*N*-acetyl-D-glucosamine (UDP-GlcNAc) and UDP-glucuronic acid (UDP-GlcA) (Solms and Hassid, 1957). For more detail on NDP-sugars and how they are translocated by nucleotide sugar transporters (NSTs) from the cytosol into the Golgi apparatus for use in lumenal polysaccharide biosynthesis pathways, see Chapter 3 in this volume.

The studies by Leloir and others changed long-standing perceptions of carbohydrate biosynthesis, which in the

DOI: 10.1201/9781003178309-2

1940s was still thought to involve the transglycosylation of sugar phosphates (Cori et al., 1939; Hanes, 1940), and established the central role of NDP-sugars as substrates for (the now so-called) Leloir glycosyltransferases (Bean and Hassid, 1955; Cardini et al., 1955; Leloir and Cardini, 1955; Cabib and Leloir, 1958). In 1953, evidence promoting the idea that nucleotide sugars serve as glycosyl donors came from studies undertaken by Dutton and Storey (1953) and Leloir and Cabib (1953). Dutton and Storey demonstrated the transfer reaction where UDP-GlcA acts as donor for the biosynthesis of phenyl β-D-glucopyranosiduronic acid with enzymes from liver extracts, while Leloir and Cabib described the case in which UDP-Glc was found to act as a glucosyl donor in the formation of trehalose phosphate (Dutton and Storey, 1953; Leloir and Cabib, 1953). A role for UDP-Glc as a sugar donor had also been independently suggested by Buchanan et al. (1952) and Kalckar (1954).

Once Leloir and colleagues had discovered one transferase reaction, they soon found more. Using wheat germ extracts they identified two enzymes. One of which was involved in the formation of sucrose, a disaccharide consisting of α-Glc and β-fructose connected through an α-1; β-2 linkage (Cardini et al., 1955), and the other produced sucrose phosphate (Leloir and Cardini, 1955), confirming Buchanan's hypothesis that sucrose is formed in plants from UDP-Glc and fructose-1-phosphate. However, it was not until 1957 that Glaser and Brown (1957) showed that NDP-sugars are involved in the biosynthesis of more complex polysaccharides by confirming the transfer of *N*-acetylglucosamine (GlcNAc) from UDP-GlcNAc to hyaluronic acid and chitin using enzymes present in mould extracts (Glaser and Brown, 1957). In the following years, UDP-Glc was shown to be the sugar donor for callose synthesis, while xylan was obtained with UDP-Xyl in reactions incubated with plant extracts (Feingold et al., 1958, 1959). Around the same time, Glaser (1957) worked with the cellulose-producing bacterium *Acetobacter xylinum* (recently reclassified as *Gluconacetobacter xylinus*) and reported transfer of Glc from UDP-Glc to cellulose (Glaser, 1957). These early studies marked the beginning of GT discovery and highlight the pivotal role of NDP-sugars in complex carbohydrate biosynthesis. Since then, many more GT-catalysed reactions utilising nucleotide sugar donors were detected during the era of biochemistry starting in the 1960s.

We now know that non-cellulosic polysaccharides are synthesised in the Golgi apparatus, while cellulose and callose are made at the plasma membrane. It was established relatively early that enzymes involved in synthesis of matrix polysaccharides are associated with the Golgi membrane (Northcote, 1969). Multiple studies presented evidence that subcellular fractions obtained by centrifugation possess GT activity in both the Golgi apparatus and endoplasmic reticulum (ER) (Schachter et al., 1970; Carey and Hirschberg, 1980) and that microsomes contained biochemically active, integral membrane proteins following centrifugation (Diesperger et al., 1974; Fujiki et al., 1982). Therefore, since the late 1960s, biosynthetic activities of Golgi-localised GTs were measured using microsomal membrane preparations, marking a key era of carbohydrate biochemistry. Most often, microsomal membranes were prepared from seedlings, coleoptiles, hypocotyls or suspension-cultured cells since they could be easily obtained in large quantities and because wall biosynthesis occurs more rapidly in these tissues at early growth stages. In order to measure GT activity, *in vitro* assays were developed where microsomal membrane preparations were incubated with radiolabeled nucleotide sugars, such as UDP-[^{14}C]Glc, followed by measurement of incorporation of the radiolabel [^{14}C] into ethanol-insoluble polysaccharides. With this method, Golgi membrane-linked synthesis of the α-1,4 galacturonic acid (GalA) backbone of homogalacturonan (HG) (Villemez et al., 1965) and the β-1,4-glucan backbone of xyloglucan (XG) (Ray et al., 1969) were measured.

Despite these advances, until 1995, no GTs involved in plant cell wall biosynthesis had been purified, nor had a gene encoding one been isolated (Delmer and Amor, 1995). That all changed rapidly with the era of functional genomics. A major milestone was the cloning and characterisation of the cellulose synthase genes from cotton, identified based on their homology to the bacterial celA genes (Pear et al., 1996). A few years later, a ground-breaking study of cellulose synthase genes in Arabidopsis using forward genetics paved the way towards the analysis of *cellulose synthase* (*cesa*) mutants (Arioli et al., 1998). In 1999, molecular genetics was coupled with biochemistry and heterologous protein expression to demonstrate the synthesis of specific glycosidic linkages by a cell wall GT, identifying the XG side chain FUCOSYLTRANSFERASE 1 (FUT1) (Perrin et al., 1999) and a galactomannan (GM)-specific galactosyltransferase (GMGT) (Edwards et al., 1999).

Early studies promoted a now antiquated dogma of carbohydrate biosynthesis that transfer of each specific sugar in a specific glycosidic linkage would require the action of a single, high specificity GT. This idea prompted Hagopian and Eylar (1968) to propose the 'one enzyme – one linkage' hypothesis (Hagopian and Eylar, 1968), which basically argues that the number of glycosidic linkages equals the number of GTs present in an organism. At the time, this hypothesis provided a plausible explanation as to why there are so many GTs in plants. However, while evidence supporting this hypothesis has been demonstrated for a multitude of GTs over the past several decades, and it might hold true as a broader general concept, there are several studies suggesting a more complex relationship between glycosidic linkage and GT, and many of these exceptions are highlighted throughout this chapter. The availability of nucleotide and amino acid sequences of GTs have shown that recombinant proteins harbouring a single catalytic domain are indeed able to catalyse the transfer of more than one sugar from distinct uridine diphospho-sugar (UDP) precursors, as has been shown in some notable plant GT examples in XG synthesis by XYLOGLUCAN XYLOSYLTRANSFERASE 1 (XXT1) (Ehrlich et al., 2021), pectic β-1,4 galactan synthesis by GALACTAN SYNTHASE 1 (GALS1) (Laursen et al., 2018) and glucomannan and glucoxylan synthesis by CELLULOSE SYNTHASE LIKE (CSL) A and CSL-F proteins, respectively (Liepman et al., 2005; Little et al., 2019).

The First Genetic Approaches to Connect Genes with Cell Wall Biosynthesis

Genetic approaches have been extensively used to identify mutants with altered cell wall composition, commonly referred to as chemical phenotypes or 'chemotypes'. The first mutants known to impact cell wall polysaccharide components were the *brittle culm* mutants found in maize, rice and barley. The barley *brittle culm* mutants contain less cellulose and demonstrate a decreased breaking strength (Kokubo et al., 1989, 1991). Some of the first mutants with altered cell wall composition were identified using a classical 'forward' genetics approach driven by pioneering plant cell wall research in the lab of Chris Somerville. In the 1990s, Reiter, Chapple and Somerville laboriously analysed the neutral monosaccharide composition of leaves, which were enriched in primary walls of an ethyl methanesulfonate (EMS)-mutagenised population of Arabidopsis. These studies led to the identification of 38 heritable murus (*mur*) loci with altered cell wall polysaccharide composition (Reiter et al., 1993, 1997). Eleven *mur* mutants (*mur1-mur11*) were selected for further study (Reiter et al., 1997), and eventually resulted in identification of GTs involved in XG biosynthesis, including MURUS2 (MUR2) and MUR3, encoding a XG α-1,2-fucosyltransferase and a β-1,2-galactosyltransferase, respectively (Vanzin et al., 2002; Madson et al., 2003).

Inspired by Kokubo's work on *brittle culm* mutants, Turner and Somerville (1997) screened ~1,600 EMS mutant plants for altered xylem cell morphology through meticulous microscopic examination of hand-cut sections of toluidine blue–stained Arabidopsis stems. One Hundred plants were selected for microscopic analysis from a prescreen of ~4,600 plants for altered stem strength. This approach yielded a collection of mutant plants with irregular, collapsed-looking xylem elements, which subsequently were named *irregular xylem* (*irx*) mutants (Turner and Somerville, 1997). Similar genetic screens have also revealed the *fragile fibre* (*fra*) mutants in Arabidopsis (Zhong et al., 2001). While *irx1*, *irx3* and *irx5* demonstrated defects in CELLULOSE SYNTHASE A (CESA) subunits involved in secondary cell wall synthesis (Taylor et al., 1999, 2000), other *irx* and *fra* mutants were linked to cellulose, xylan and lignin deposition or signal-transduction (Jones et al., 2001; Szyjanowicz et al., 2004; Zhong et al., 2005; Brown et al., 2007; Pena et al., 2007). Due to the complexity of pectins, the identification of specific mutants lagged significantly behind enzymes involved in the synthesis of their neutral counterparts in the wall, such as XG. For example, the original *mur* screen relied on the analysis of alditol acetate derivatives from cell wall hydrolysates and only considered neutral sugars, and not uronic acids, thereby likely missing many pectin-specific chemotypes.

As transfer-DNA (T-DNA) insertional mutant collections grew in popularity (Samson et al., 2002; Alonso et al., 2003), many studies began using a combination of both forward and reverse genetic approaches. Identification of one of the initial pectin-specific GTs was based on the knowledge that pectin organisation is important for plant growth and especially cell adhesion. For example, the *quasimodo1* (*qua1-1*) mutant, eventually found to encode a HG α-1,4 galacturonosyltransferase from glycosyltransferase family 8 (GT8) (Sterling et al., 2006), was identified through a systematic screening of the Versailles T-DNA insertional mutant population to identify misshapen plants with potential cell adhesion phenotypes (Bouton et al., 2002). The analysis of cell wall sugars from *qua1-1* mutants showed that neutral sugars were present at similar levels in both mutant and control plants, while GalA content was significantly lower in *qua1-1*.

The full-genome sequence of Arabidopsis was published in 2000 (Kaul et al., 2000) and has undergone 11 annotation releases to date. By early 2000, microarray technology made large-scale transcript analyses possible. These approaches took advantage of publicly available microarray datasets, and several studies demonstrated that genes with similar function are often transcriptionally correlated (Stuart et al., 2003; Ihmels et al., 2004), but did not investigate plant species. In 2005, the first gene expression map of Arabidopsis development was generated using Affymetrix ATH1 arrays that contained probes representing a little more than 80% of known genes (Schmid et al., 2005). These data and others provided the plant science community with access to comprehensive sets of gene expression data for the first time and jump-started the use of computational pipelines relying on systems biology to enable the identification of cell wall–related genes. In 2005, two pioneering studies were published that used co-expression networks to successfully identify additional genes involved in secondary cell wall formation and cellulose biosynthesis in Arabidopsis (Brown et al., 2005; Persson et al., 2005). Later, large-scale biology efforts to elucidate plant cell wall biosynthetic pathways leveraged gene co-expression analysis to identify transcriptional relationships of cell wall–related genes in monocots and dicots (Mitchell et al., 2007; Oikawa et al., 2010), which led to the identification of some grass-specific genes involved in cell wall synthesis. These approaches and numerous databases, including GENEVESTIGATOR®, the Bio-Analytic Resource (BAR; Toufighi et al., 2005) and PlaNet (Mutwil et al., 2011) have proven invaluable for identifying new cell wall–related genes for a number of cell wall polymers (e.g., cellulose) and specific tissues or cell types (e.g., secondary cell walls and seed coat mucilage).

THE RISE OF CAZy CLASSIFICATION OF GTs

With rapid advances in the ability to identify genes and gene products encoding GTs, it soon became clear that classification schemes beyond those available in the literature were needed. Firstly, GTs can be categorised by both their structure and mechanism of action. Mechanistically, they are either inverting or retaining based on the anomeric carbon linkage of the newly synthesised glycosidic bond with respect to the linkage of the donor substrate (Figure 2.1) (Lairson et al., 2008; Moremen and Haltiwanger, 2019). In other words, a GT that uses an α-linked nucleotide sugar donor substrate is considered retaining if it produces an α-linked polysaccharide product, while it would be inverting if it produces a β-linked product. For example, β-1,4-D-xylan is synthesised from UDP-α-D-xylose by xylan synthase (XYS) enzymes classified as members of the GT47 family, which are inverting GTs.

Second, GTs (EC 2.4.x.y) are classified by Enzyme Commission numbers (EC number) according to standards set by the International Union of Biochemistry and Molecular Biology (IUBMB; https://iubmb.org/) and are based solely on substrate specificity: EC 2.4.1.y, hexosyltransferases; EC 2.4.2.y, pentosyltransferases; and EC 2.4.99.y, transferring other glycosyl groups. As the era of genomics unfolded, it became clear that correlations between EC number and amino acid sequence were often absent. Throughout the 1990s, pioneering work on sequence-based classification of carbohydrate-active enzymes (CAZymes), initially focused on glycoside hydrolases (GH), was initiated by Bernard Henrissat (Henrissat, 1991; Henrissat and Bairoch, 1993, 1996), recognising that structural features of enzymes play a considerable role in their function. In 1997, Campbell, Davies, Bulone and Henrissat classified the first set of known GTs into 27 sequence-derived families based on analysis of amino acid sequence similarities (Campbell et al., 1997). In 1999, Henrissat and Coutinho established the CAZy database (http://www.cazy.org) (Cantarel et al., 2009; Lombard et al., 2014), and by 2003 the classification of GTs by Coutinho and colleagues encompassed 65 distinct sequence-derived families (Coutinho et al., 2003). CAZypedia (http://www.cazypedia.org) was initiated in 2007 to work in tandem with the CAZy database, functioning as a curated, online encyclopaedia for CAZymes (CAZypedia Consortium 2018) where biocurators create pages that provide a wealth of information on different families. To this day, the CAZy database is continuously expanded with new families that are added when the sequence and biochemical characterisation of a minimum of one member is published in a peer-reviewed scientific journal. In 2021, the CAZy database underwent a large overhaul intended to facilitate access to information and aid taxonomic navigation (Drula et al., 2021). At the time of this publication, 849,449 GT sequences, classified into 114 families (GT1-GT114), have been curated and 298 solved GT structures (all species) are present in the database (Drula et al., 2021).

As structural information became available, GTs were also classified based on the structure of their catalytic domains. Most GTs adopt one of three common folds: GT-A, GT-B or GT-C. Both GT-A and GT-B fold enzymes are considered Leloir GTs, employing nucleotide sugars as donors, while GT-C fold enzymes are multipass transmembrane proteins that use polyisoprenoid sugar donors for glycan extension. GT-A enzymes typically have one Rossmann-like fold, characterised by alternating α-helices and β-sheets (Figure 2.2). The majority of GT-A enzymes, but not all, are divalent cation dependent and coordinate the NDP-sugar donor via a metal ion interaction with carboxylate residues in the characteristic active site Asp (asparagine)-X-Asp (DxD) motif. Recently, Taujale et al. used computational tools to map out the evolution of GT-A enzymes, defining a minimal structural unit that is expanded by substrate-binding loops, to understand how these enzymes evolved diverse activities (Taujale et al., 2020). GT-B fold enzymes are generally more variable in structure, especially for the acceptor substrate-binding domain, and do not have well-established landmark catalytic site features (i.e., DxD) or employ enzyme-bound metals in catalysis that can be used for predictive biology. GT-B enzymes contain two linked Rossmann-like domains where one is predominately involved in donor recognition while the second, more variable domain contains acceptor-binding subsites (Lairson et al., 2008; Moremen and Haltiwanger, 2019). In GT-B enzymes, the Rossmann-like domains come

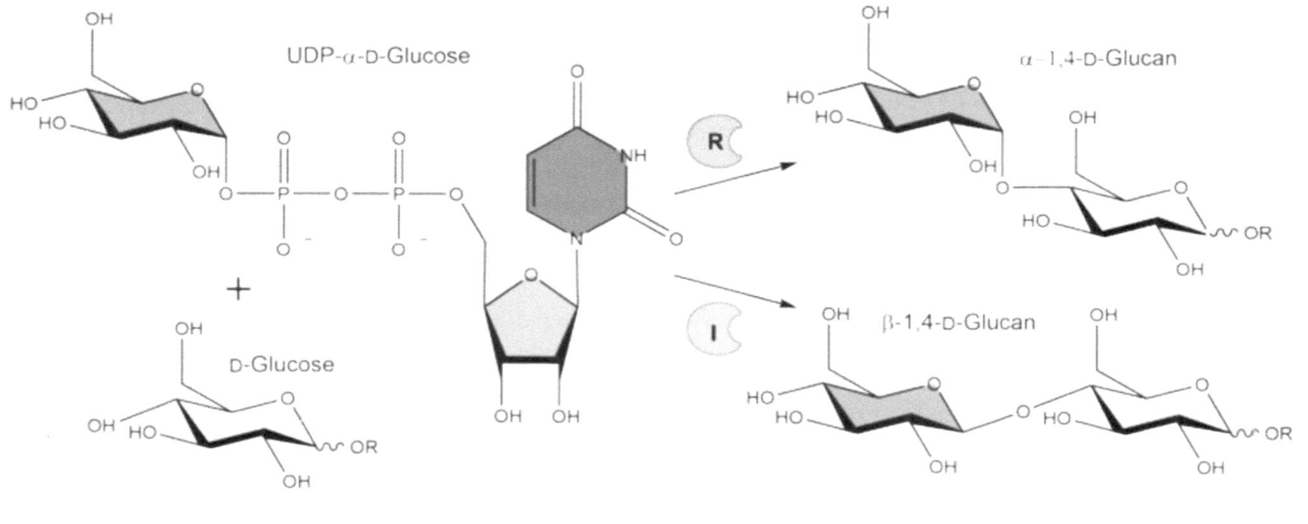

FIGURE 2.1 Schematic of the substrates and saccharide products of inverting (I) or retaining (R) glycosyltransferases using UDP-α-D-Glc as a donor.

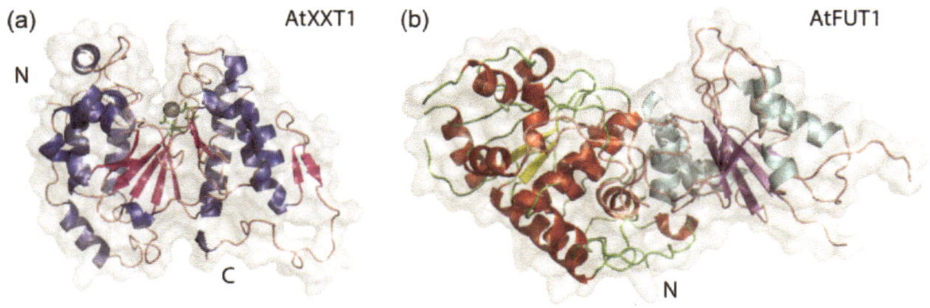

FIGURE 2.2 Overall GT folds depicted by representative plant enzymes involved in xyloglucan (XG) biosynthesis. Solved secondary structures are displayed as cartoon models with transparent surfaces with the amino- (N) and carboxy- (C)-terminal ends of each indicated where visible. (a) The GT-A fold is represented by the retaining XG α-1,6-xylosyltransferase (AtXXT1; PDB 6BSU) from GT34. Helixes are blue and beta strands are magenta, residues that form the DxD motif are shown in green, and the coordinated Mn^{2+} ion is a grey sphere. (b) The GT-B fold is represented by the inverting XG α-1,2-fucosyltransferase 1 from GT37 (AtFUT1; PDB 5KWK) from *Arabidopsis thaliana*. The N- and C-terminal Rossmann-like folds are depicted by red and cyan helixes with yellow and magenta beta strands, respectively.

together, forming the active site and are (almost exclusively) cation independent (Albuquerque-Wendt et al., 2019). All known plant cell wall GTs, which have the structure of their catalytic domains determined, adopt either a GT-A or GT-B fold. The fold and mechanism of a GT are not predictive of each other and both GT-A and GT-B enzymes can be either retaining or inverting.

The catalytic properties of GTs are often difficult to understand without structural information. Experimentally determined structural information can provide a wealth of biological and mechanistic information on an enzyme; unfortunately, this is lacking for the majority of enzymes. For decades, researchers in the cell wall field have attempted the structural characterisation of GTs involved in wall synthesis but long remained unsuccessful. This changed only recently when two groups independently achieved the scientific breakthrough of solving plant GT crystal structures for enzymes involved in cell wall synthesis. The reports provided the crystal structure of FUT1, also one of the first cell wall biosynthetic GTs discovered and shed light on important information about the mechanisms of substrate-binding and catalytic activity of the enzyme (Rocha et al., 2016; Urbanowicz et al., 2017). A year later, XXT1 was also structurally characterised (Culbertson et al., 2018). In ground-breaking work published in 2020, Purushotham, Ho and Zimmer determined the structure of a trimeric *Populus tremula x tremuloides* cellulose synthase (PttCESA8) complex by cryo-electron microscopy (cryo-EM) associated with a nascent polysaccharide, shedding light on how plant cellulose synthase complexes (CSC) actually assemble (Purushotham et al., 2020).

The 2020s marked the era of protein structure modelling. The development of new machine learning approaches using deep neural networks, which involve the use of multiple, or even thousands, of neural network layers focused on pattern recognition to classify and order information (Sharma and Singh, 2017; Samek et al., 2021; Baek and Baker, 2022), have rapidly and dramatically changed the game for structural biology workflows. The year 2021 saw the release of both the Baker lab's RoseTTAFold and Deepmind's Alphafold(2) for prediction of native protein structures from their amino acid sequences (Baek et al., 2021; Jumper et al., 2021). In partnership with the European Bioinformatics Institute (EMBL-EBI), Deepmind utilised Alphafold to computationally predict and release protein structures from the human proteome (Tunyasuvunakool et al., 2021) as well as 20 other organisms, including key plant species such as soybean, rice, maize, and Arabidopsis (https://alphafold.ebi.ac.uk/). Thus, deep learning–based protein structural models of virtually any GT encoded by the genomes of these plant species can be readily downloaded and visualised (Figure 2.3); one key caveat is that they do not contain any predicted bound substrates. Next, the Baker lab went a step further and developed a protein interaction pipeline by incorporating both RoseTTAFold and Alphafold, which use coevolution to evaluate and predict protein–protein interactions of gene products encoded by the eukaryotic genome of *Saccharomyces cerevisiae* (Humphreys et al., 2021). It will be exciting as the field moves forward to see how these rapidly advancing tools impact the way we study plant biosynthetic pathways.

Functional Characterisation of GTs—Now That We Have Identified Them, How Do We Confirm Their Activity?

To date, many cell wall GTs have been cloned and various strategies have been developed aimed at their identification and characterisation, including enzyme enrichment and biochemical analysis, molecular genetics approaches and targeted 'omics' for candidate selection.

Membrane Preparations and Enzyme Enrichment

Microsomal expression is one of the most common methods utilised for GT characterisation, facilitating the *in vitro* characterisation of membrane-bound proteins and protein complexes. This method entails the lysis of cells and subsequent enrichment of microsomes derived from the desired

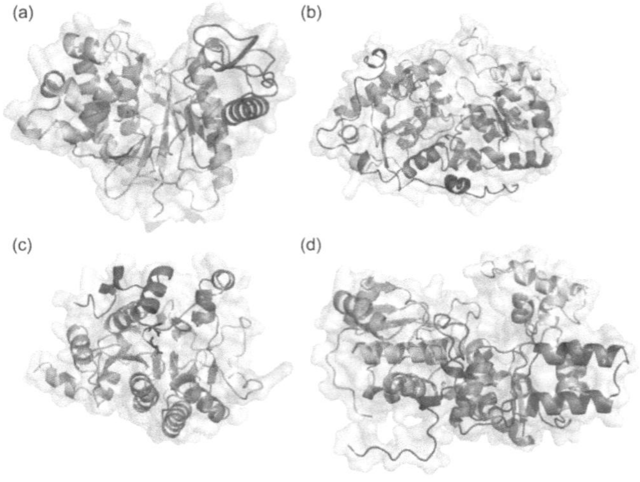

FIGURE 2.3 Alphafold predicted structures for four GTs. (a) AtHPAT1- GT95. (b) AtIRX10L- GT47. (c) AtRGXT1- GT77. (d) AtRRT1- GT106.

membrane. For GTs expressed in plant expression systems, it is usually Golgi-derived microsomes that are enriched due to the Golgi being the site of non-cellulosic cell wall polysaccharide biosynthesis (Amos and Mohnen, 2019). While microsomal preparations have been widely used for the study of Golgi-localised GTs, an issue with these approaches is the possibility of native protein interference. Golgi-enriched microsomes contain native Golgi-localised GTs that can interact or form complexes with other native proteins or the heterologously expressed target enzyme and thereby impact its functional analysis. Microsomes are therefore an important, but often inadequate, stand-alone approach to conclude a direct effect of an individual GT. Similarly, microsomes are often prepared from plants overexpressing GTs, resulting in the desired protein existing at a much higher level than normal, which can also impact protein function. However, GTs can also be heterologously expressed in species that do not produce the target polysaccharide. For example, a milestone achievement was expression of the mixed-linkage glucan (MLG) synthesising CSLF6 in Arabidopsis, resulting in the production of MLG (Burton et al., 2006), which does not naturally occur in this species. Such results provide increased confidence that product formation is a direct effect of the overexpressed protein.

Heterologous Expression of GTs

The gold standard for enzyme characterisation is *in vitro* assays with pure, soluble, recombinant protein. Unfortunately, for plant GTs, this is typically easier said than done. One issue that scientists face when attempting heterologous expression and purification of recombinant GTs is that eukaryotic GTs overwhelmingly contain at least a single transmembrane domain (TMD). Many GTs, including all members of the CESA/CSL superfamily, are multipass transmembrane proteins where the transmembrane region is directly involved in catalytic activity (Daras et al., 2021). Others, such as the GT43 enzymes involved in xylan backbone synthesis, have a single transmembrane region that anchors the soluble catalytic region in the Golgi or even lack one entirely (Wu et al., 2009). For this second group of GTs, expression of truncated proteins with intact catalytic regions, but lacking a TMD, is an approach that can alleviate issues with membrane association (Prabhakar et al., 2020). Heterologous expression in the bacteria *Escherichia coli* is a simple and popular method of producing protein, but plant GTs often express poorly or lack activity when purified from bacteria (Welner et al., 2017). One reason for this is that *E. coli* lacks an *N*-glycosylation pathway. Many plant proteins, including GTs, are post-translationally glycosylated in the plant secretory pathway (Strasser et al., 2021). The absence of glycans can be detrimental to both folding and stability, and therefore, the *in vitro* activity of a protein (Imperiali and O'Connor, 1999). Eukaryotic expression systems are therefore preferred for GT purification despite the increased cost and complexity. Good eukaryotic systems for plant GT expression are HEK293 cells (Prabhakar et al., 2020), yeast (Petersen et al., 2009; Voiniciuc et al., 2019) and insect cells (Rocha et al., 2016; Purushotham et al., 2020). However, some GTs purified from eukaryotic expression systems will still exhibit poor activity *in vitro*. While there are myriad reasons for poor activity, a common concern is complex-dependent function. Complex-forming GTs have been demonstrated to function poorly without their binding partners (Atmodjo et al., 2011). Additionally, some putative GTs provide a structural or scaffold forming function rather than a catalytic one. Overexpression of such scaffold forming proteins *in vivo* can even enhance the activity of binding partners by proxy, giving the impression of direct catalytic activity. These issues can cause uncertainty about GT activity *in vitro* and raise a number of questions: Is this protein inactive due to suboptimal expression? Is it only active in a complex? Is it a catalytically inactive scaffold forming member of a complex? For these reasons, *in vitro* approaches are best done in tandem with *in vivo* methods including but not limited to mutant analysis and RNAi.

Substrate Issues

Another problem that arises with *in vitro* approaches is the increasing demand to obtain well-characterised substrates for GTs. While some donor nucleotide sugars can be purchased commercially, they are often prohibitively expensive, particularly those with more niche applications. Furthermore, the incredible diversity of cell wall polysaccharides makes it extremely challenging to identify the exact molecular structures that serve as acceptors for each highly specific GT. Simple acceptor substrates are becoming more readily available from commercial suppliers such as Megazyme (https://www.megazyme.com), but access to more complex substrates such as RG-II derivatives can be a major hindrance to *in vitro* work. Some plant species synthesise unusual cell wall polysaccharides that can be used as acceptor substrates. For instance, tamarind seed storage XG lacks fucose and has been isolated and used as an acceptor substrate, which allowed for the demonstration of FUT1 activity (Perrin et al., 1999). Unsubstituted saccharide backbones often exhibit solubility issues, including oligosaccharides of β-1,4 glucose or xylose that are used as acceptor substrates for XG and xylan modifying enzymes, respectively. While the degree of polymerisation (DP) of XG and xylan can reach into the hundreds *in vivo*, the rapid addition of substituents to nascent backbones maintains solubility and inhibits aggregation (Berglund et al., 2020). Without these substituents present, acceptor substrates are limited to small sizes, typically a DP of six or less for cello- and xylooligosaccharides. However, GTs often preferentially modify longer substrates and will therefore display little or no activity on shorter oligosaccharides. An alternative to the use of natural polysaccharides as described above are chemically synthesised acceptors. These have the advantage that they are well-defined and highly pure in comparison to oligosaccharides obtained from digests of natural compounds (Kinnaert et al., 2017; Pfrengle, 2017) although synthesis can be elaborate and time-consuming.

In Vitro Biochemical Characterisation

In the past, transferase activity assays aiming at the biochemical characterisation of plant cell wall GTs often relied on the preparation of microsomal membranes, e.g., from leaves of *Nicotiana benthamiana* (tobacco) plants transiently overexpressing the enzyme or Arabidopsis mutant plants (discussed above), combined with radiolabeled donor substrates. In these experiments, microsomal membranes are purified from infiltrated or mutant plants and used directly in transferase assays with NDP-sugars as substrates where the sugar moiety is labelled, either uniformly or on selected carbons with a ^{14}C or ^{3}H radioisotope. For example, UDP-[^{14}C]GlcA and xylohexaose were used as substrates to demonstrate the activity of the GLUCURONIC ACID SUBSTITUTION OF XYLAN 1 (GUX1) enzyme (Rennie et al., 2012). Following an incubation period, excess radiolabel that has not been incorporated is removed by filtration. This can be done by passing the reaction mixture through anion exchange resin when neutral monosaccharides are tested; commonly self-made mini columns in syringes filled with the chromatographic matrix are used. In the case of negatively charged monosaccharides, i.e., uronic acids, this technique cannot be used as those would be retained on the column. Alternatively, paper chromatography (Fry, 2000), thin-layer chromatography (Fry, 2000) and high-performance anion exchange chromatography (HPAEC) (Rautengarten et al., 2019) are applied to separate the labelled products as has been used to study galacturonosyltransferases and glucuronosyltransferases (Doong and Mohnen, 1998; Rennie et al., 2012). Radiolabelled products are measured by scintillation counting, which directly corresponds to the level of enzyme activity. While this approach has been successfully applied to characterise several cell wall GTs, e.g., including galactan synthases of the GT92 family (Liwanag et al., 2012; Ebert et al., 2018), it is not ideal for a couple of reasons. Firstly, not all NDP-sugars are available in their radio-labelled form, which limits the number of substrates that can be tested. In addition, given the safety risks that are associated with radioactivity, attempts have been made to avoid the use of radioactive substrates. As an alternative, assays have been developed where the use of fluorescent oligosaccharide acceptors and substrate separation by HPAEC or reverse phase high-performance liquid chromatography (HPLC) is coupled to fluorescence detection. This approach has been successfully used to probe the activities of several plant GTs (Akita et al., 2002; Ishii et al., 2002; Konishi et al., 2006; Lee et al., 2012; Urbanowicz et al., 2014; Xuan et al., 2021). However, one drawback is that each run can take between 20 and 60 minutes, limiting the number of samples that can be processed in a given time period. The method is also limited to enzymes that form water-soluble products since water-insoluble polysaccharides cannot be separated.

Characterisation of Saccharide Products

GTs carry out bisubstrate reactions, and thus it is important to carefully analyse the saccharide products to confirm their structure. A multitude of approaches have been developed, some of which are discussed below; however, a detailed description of carbohydrate analytical techniques is beyond the scope of this chapter. Thus, we instead recommend some excellent review and method articles for additional information. A biochemical method for the structural analysis of GT products pioneered by the Dupree lab is polysaccharide analysis by gel electrophoresis (PACE) (Goubet et al., 2002). In this method, polysaccharides are hydrolysed by specific GH enzymes, the released oligosaccharides fluorescently labelled and then separated on large format polyacrylamide gels and detected by fluorescence imaging (Goubet et al., 2002). There is also a range of chemical techniques available including glycosidic linkage analysis (Pettolino et al., 2012), nuclear magnetic resonance (NMR) spectroscopy (Yao et al., 2021) and mass spectrometry (MS) (Harvey, 2021).

Glycosidic linkage analysis is another useful technique to characterise the products of GT assays. Briefly, all free hydroxyl groups are methylated followed by hydrolysis of the methylated. Finally the released partially methylated sugars are reduced to yield partially methylated alditol acetate derivatives that can be separated, identified and quantified by gas chromatography–mass spectrometry (GC-MS) (Pettolino et al., 2012) or liquid chromatography–mass spectrometry (LC-MS) (Galermo et al., 2018). While the method enables the identification of the linkage position of each individual sugar in a polysaccharide, it is limited in that the technique cannot reveal either the position or anomeric configuration of these residues in a given polymer. Another drawback is that it is a qualitative rather than a quantitative method because standards for each individual monosaccharide derivative, while very important, are difficult to obtain or unavailable.

MS is a powerful tool for the analysis of oligosaccharide products of GTs, or polysaccharide digestion products after treatment with linkage specific GHs. Oligosaccharide mass profiling (OLIMP) of cell wall polysaccharides by matrix-assisted laser desorption/ionisation time-of-flight mass spectrometry (MALDI-TOF/MS) is a mass spec technique that utilises a laser to ionise samples embedded in a crystalline chemical matrix (Gunl et al., 2011). For MALDI-TOF/MS, the most commonly used matrix is 2,5-dihydroxybenzoic acid (DHB), but a variety of specialised matrices are available (Harvey, 1999, 2021). Benefits of MALDI-TOF/MS are that it is a rapid and simple approach to identify and structurally compare carbohydrates based on mass. Limitations to the technique are that it is only semiquantitative and cannot provide direct structural data (linkage, composition) for a saccharide product. Therefore, when characterising GTs, MALDI-TOF/MS is most useful when there is a putative or documented function associated with the GT, and often requires a secondary technique for confirmation of product structure. For an in-depth review of MALDI-TOF/MS, see Harvey (2021). This technique has been used to study several GTs involved in the addition of neutral and acidic monosaccharides. Some notable examples include transfer of fucose by AtFUT1 (Urbanowicz et al., 2017), xylan backbone synthesis by XYS (Urbanowicz et al., 2014), transfer of xylose by XG xylosyltransferases (Ade et al., 2014) and synthesis of HG by galacturonosyltransferases (Amos et al., 2018).

An essential analytical method that is frequently used to define the anomeric configurations (α- or β-) and patterns of glycosidic linkages as well as the sequence of linkages present in polysaccharides is NMR. Most often a combination of 1- and 2D NMR spectra, chemical shifts and coupling constants are used to investigate the structures of oligo- and polysaccharides. 1D NMR (^{1}H and ^{13}C NMR) and increasingly 2D NMR spectroscopic analysis is utilised for characterisation of polysaccharides. One issue with NMR data is that they are often highly complex and difficult to interpret, and it is therefore recommended to compare and validate new data to published spectra. Applications of this technique for structural characterisation of saccharide products to investigate the regio- or stereo-specificity of enzymes include transfer of fucose by AtFUT1 (Urbanowicz et al., 2017), xylan backbone synthesis by XYS enzymes (Smith et al., 2022), real-time NMR analysis of regiospecific xylan acetylation by xylan *O*-acetyltransferases (Urbanowicz et al., 2014; Lunin et al., 2020) and recent studies on xylan arabinosylation (Zhong et al., 2021). Biophysical procedures are also available for the characterisation of polysaccharides, e.g., X-ray diffraction is used for crystalline polysaccharides (Rongpipi et al., 2018). In general, it should be noted that the method of choice for substrate characterisation should be selected depending on the type and amount of polysaccharide that is synthesised in the *in vitro* assay.

High-Throughput Methods for the Biochemical Characterisation of GTs

To accelerate GT discovery and overcome common issues reported for traditional GT assays, new high-throughput (HTP) methods allowing rapid screens of NDP-sugar donor and acceptor specificity are currently being developed. One HTP method is based on the synthesis of UDP byproducts, which can be detected by luminescence and allow the rapid determination of the donor NDP-sugar for a specific GT. When the donor substrate is determined, the GT can be probed with the specific donor and a range of available acceptors, thus significantly reducing the screening efforts. HTP methods have many advantages over traditional methods including being harmless, robust, cost-effective and reproducible. The commercially available UDP-Glo™ Glycosyltransferase Assay (Promega, https://www.promega.de/en/) is one such HTP method that is increasingly used for GT discovery. Many GTs will even hydrolyse NDP-sugar donors in the absence of acceptor substrates, facilitating characterisation even without the knowledge of the relevant acceptor substrate (Sheikh et al., 2017).

Recently, a collaborative effort between Breeanna Urbanowicz's and Fabian Pfrengle's groups achieved a major technical breakthrough with the development of an HTP platform for the analysis of cell wall GT enzymes (Ruprecht et al., 2020). This analytical platform uses glycan arrays with chemically synthesised and natural glycans printed on them, and its application has been shown to successfully identify GT acceptor substrates (Ruprecht et al., 2020). These important technical developments are supported by genomic tools like the Joint BioEnergy Institute (JBEI) GT collection, a clone collection of more than 500 Arabidopsis and rice GTs (Lao et al., 2014; Xuan et al., 2021). In the future, glycan array technology in combination with genetics tools and advancements in enzyme expression and chemical synthesis should significantly accelerate GT discovery.

Overview of Characterised GTs of Known Function Discussed by Cell Wall Polymer

As discussed previously, unique linkages in a polysaccharide typically require a specific GT. Plants have an incredible diversity of polysaccharides in their cell walls, which is reflected in the sheer number of GTs encoded in their genomes. This chapter will cover current knowledge and provide a general overview about plant GTs (Table 2.1); however, it should be noted that there are still many unknowns about plant GTs and limitations to GT studies in particular when it comes to lineage-specific substrate discrimination, sampling of species and the use of Arabidopsis as a model system. In addition to GTs, there are also numerous other proteins including methyltransferases and acetyltransferases that are involved in the addition of crucial non-glycosyl substituents to cell wall polysaccharides; however, they will not be covered in this chapter.

TABLE 2.1
Plant Cell Wall Biosynthetic GTs with Known Function Organised by Polymer

Enzyme or Family	Enzyme Function	Polymer	CAZy GT Family	Fold	Mechanism	Activity Biochemically Verified by
CESA	β-1,4 glucan synthase	Cellulose	2	GT-A	Inverting	Cho et al. (2015)
GSL	β-1,3 glucan synthase	Callose	48	ND	Inverting	N/A
CSLC	β-1,4 glucan synthase	Xyloglucan	2	GT-A	Inverting	Cocuron et al. (2007)
XXT	α-1,6 xylosyl transferase	Xyloglucan	34	GT-A	Retaining	Faik et al. (2002), Cavalier and Keegstra (2006), Vuttipongchaikij et al. (2012), Culbertson et al. (2016), Ruprecht et al. (2018)
MUR3	β-1,2 galactosyl transferase	Xyloglucan	47	GT-B	Inverting	Madson et al. (2003)
FUT1 (MUR2)	α-1,2 fucosyl transferase	Xyloglucan	37	GT-A	Inverting	Perrin et al. (1999), Faik et al. (2000), Vanzin et al. (2002), Ciceron et al. (2016), Rocha et al. (2016), Urbanowicz et al. (2017)
XLT2	β-1,2 galactosyl transferase	Xyloglucan	47	GT-B	Inverting	N/A
XST	α-1,2 arabinosyl transferases	Xyloglucan	47	GT-B	Inverting	N/A
XUT1	β-1,2 galacturonosyl transferase	Xyloglucan	47	GT-B	Inverting	N/A
XYS1/IRX10-L	β-1,4 xylosyl transferase	Xylan	47	GT-B	Inverting	Zeng et al. (2010, 2016), Lee et al. (2012), Jensen et al. (2014, 2018), Urbanowicz et al. (2014), Jiang et al. (2016), Wang et al. (2022)
GUX	α-1,2 glucuronosyl transferase	Xylan	8	GT-A	Retaining	Lee et al. (2012), Rennie et al. (2012)
XAT1	α-1,3 arabinosyl transferase	Xylan	61	GT-B	Inverting	Anders et al. (2012)
XAX1	*β-1,2 xylosyl transferase; #hydroxycinnamoyl α -1,3-arabinosyl transferase	Xylan	61	GT-B	Inverting	*Chiniquy et al. (2012), #Feijao et al. (2021)
CSLA	β-1,4 mannan synthase; β-1,4 glucan synthase	Gluco mannan	2	GT-A	Inverting	Dhugga et al. (2004), Liepman et al. (2005), Goubet et al. (2009), Voiniciuc et al. (2019)
CSLD	*β-1,4 mannan synthase; #β-1,4 glucan synthase#	Mannan	2	GT-A	Inverting	*Verhertbruggen et al. (2011), *Yin et al. (2011), #Yang et al. (2020)
GMGT/GalT	α-1,6 galactosyl transferase	Mannan	34	GT-A	Retaining	Edwards et al. (1999), Voiniciuc et al. (2015)
MSR	ND	(Gluco) mannan	106	ND	Inverting	N/A

(Continued)

TABLE 2.1 (*Continued*)
Plant Cell Wall Biosynthetic GTs with Known Function Organised by Polymer

Enzyme or Family	Enzyme Function	Polymer	CAZy GT Family	Fold	Mechanism	Activity Biochemically Verified by
CSLF (Note: glucoxylan activity only detected for *HvCslF3* and *HvCslF10*)	*β-1,3; 1,4 glucan synthase; #1,4-β glucoxylan synthase	Mixed-Linkage Glucan; Glucoxylan	2	GT-A	Inverting	*Burton et al. (2006), *Vega-Sanchez et al. (2012), *Jobling (2015), *Kim et al. (2015), *Dimitroff et al. (2016), *Little et al. (2018), #Little et al. (2019)
CSLH	β-1,3; 1,4 glucan synthase	Mixed-Linkage Glucan	2	GT-A	Inverting	Doblin et al. (2009), Little et al. (2018)
CSLJ	β-1,3; 1,4 glucan synthase	Mixed-Linkage Glucan	2	GT-A	Inverting	Little et al. (2018)
GAUT	α-1,4 galacturonosyl transferase	Homo galacturonan	8	GT-A	Retaining	Sterling et al. (2006), Atmodjo et al (2011), Amos et al. (2018), Biswal et al. (2018), Voiniciuc et al. (2018)
XGD1	β-1,3 xylosyl transferase	Xylo galacturonan	47	GT-B	Inverting	Jensen et al. (2008)
ARAD	α-1,5 arabinosyl transferase	Rhamno galacturonan I	47B	GT-B	Inverting	N/A
RRT1	α-1,4 rhamnosyl transferase	Rhamno galacturonan I	106	ND	Inverting	Uehara et al. (2017), Takenaka et al (2018)
GALS	*β-1,4 galactan synthase; #α-? arabinosyl Transferase	Galactan	92	GT-A	Inverting	*Liwanag et al. (2012), *Ebert et al (2018), #Laursen et al. (2018)
RGXT	α-1,3 Xylosyl transferase	Rhamno galacturonan II	77	ND	Retaining	Egelund et al. (2006, 2008), Petersen et al. (2009), Liu et al. (2011)
SGT1	Serine *O*-α-galactosyl transferase	Extensins	96	ND	Retaining	Saito et al. (2014)
HPAT	Hydroxyproline *O*- β-arabinofuranosyl transferase	Extensins	95	ND	Retaining	Ogawa-Ohnishi et al. (2013)
RRA	β-1,2-Arabinosyl transferase	Extensins	77A	ND	Retaining	N/A
XEG113	β-1,2-Arabinosyl transferase	Extensins	77	ND	Retaining	N/A
EXAD1	α-1,3 Arabinosyl transferase	Extensins	47B	GT-B	Inverting	N/A
RAY1	β-Arabinofuranosyl transferase	Arabinogalactan proteins	77	ND	Retaining	N/A
HPGT	Hydroxyproline-*O*-galactosyltransferase	Arabinogalactan proteins	31	GT-A	Inverting	Ogawa-Ohnishi and Matsubayashi (2015)
GALT2–6 (Note: GALT1 is a non-cell wall GT)	Hydroxyproline-*O*-galactosyltransferase	Arabinogalactan proteins	31	GT-A	Inverting	Basu et al. (2013), Basu, Tian et al. 2015), Basu, Wang et al. (2015)
FUT4, FUT6	α-1,2 Fucosyl transferase	Arabinogalactan proteins	37	GT-B	Inverting	Wu et al. (2010)
GALT29A	β-1,6 Galactosyl transferase	Arabinogalactan proteins	29	GT-A	Inverting	Dilokpimol et al. (2014)
GALT31A	*β-1,6 Galactosyl transferase; #β-1,3 galactosyl transferase#	Arabinogalactan proteins	31	GT-A	Inverting	*Geshi et al. (2013), #Ruprecht et al. (2020)
GlcAT14	β-1,6 Glucuronosyl transferase	Arabinogalactan proteins	14	GT-A	Inverting	Knoch et al. (2013), Dilokpimol and Geshi (2014)
At1g77810	β-1,3 Galactosyl transferase	Arabinogalactan proteins	31	GT-A	Inverting	Qu et al. (2008)

(*Continued*)

TABLE 2.1 (*Continued*)
Plant Cell Wall Biosynthetic GTs with Known Function Organised by Polymer

Enzyme or Family	Enzyme Function	Polymer	CAZy GT Family	Fold	Mechanism	Activity Biochemically Verified by
KNS4/UPEX1	β-1,3 Galactosyl transferase	Arabinogalactan proteins and/or pectin	31	GT-A	Inverting	Suzuki et al. (2017)
GALT8	β-1,3 Galactosyl transferase	Arabinogalactan proteins and/or pectin	31	GT-A	Inverting	Narciso et al. (2021)
GhGALT1	β-1,3 Galactosyl transferase	Arabinogalactan proteins	31	GT-A	Inverting	N/A
AtCDI	L-Galactosyl transferase	RG-II	Unk/ND	ND	ND	Peng et al. (2021)

Cellulose Synthase

Cellulose is the most abundant saccharide component of the cell wall and on Earth. It is composed of unsubstituted β-1,4 linked glucose chains that assemble into microfibrils that are now thought to comprise 18 chains in plants (Hill et al., 2014). Plant cellulose is synthesised by GT2 CESA enzymes (Pear et al., 1996). The Arabidopsis genome, as a base reference, encodes ten CESA genes annotated as CESA1–10. Differential expression of CESA genes is related to the tissue and type of cell wall being synthesised (Hill et al., 2014). Cellulose synthase enzymes are multi-pass transmembrane proteins that form a pore in the plasma membrane through which β-1,4-glucan chains are synthesised and translocated into the apoplastic space in a processive manner (Sethaphong et al., 2013; Purushotham et al., 2020). Cellulose microfibrils are synthesised by CSC that consist of multiple isoforms of CESAs, which vary for primary and secondary wall synthesis, and adopt a rosette shape in the plasma membrane (Kimura et al., 1999). In Arabidopsis, the primary cell wall CSC consists of CESA1, CESA3 and CESA6-like proteins and it is speculated that any of the CESA6-like proteins can fill the CESA6-like position. CESA4, CESA7 AND CESA8 participate in the secondary cell wall CSC, with only CESA4 being able to homodimerise. The existence of mixed CSCs holding both primary and secondary CESAs has been suggested; however, it remains to be shown. Notably CESA3 seems to be functionally related to CESA7 while CESA1 is functionally related to CESA8.

Callose Synthase

Callose is a polymer that is not always present in the wall, being deposited in specialised walls or wall-associated structures. Therefore, callose is often forgotten, which is surprising given its crucial role in many individual stages of development (Stone 1992). These include cell plate formation (Samuels et al., 1995; Verma and Hong, 2001), sieve pore development and regulation, plasmodesmata regulation (Iglesias and Meins, 2000), functional megaspore selection, pollen development (McCormick, 1993; Enns et al., 2005) and pollen tube growth, as well as in response to abiotic and biotic stresses (Jacobs et al., 2003; Nishimura et al., 2003; Dong et al., 2008). In plants, callose is a linear β-glucan that is composed of hundreds of β-1,3 linked glucose residues with some 1,6 linked branches. When compared to cellulose, callose is rather amorphous and forms helical structures. Callose is reportedly synthesised by so-called GLUCAN-SYNTHASE-LIKE (GSL) proteins belonging to the GT48 family although direct biochemical evidence remains to be provided. The GSL proteins have been identified by their similarity to proteins encoded by the yeast *FK506 hypersensitivity (FKS)* genes, which are believed to be involved in β-1,3 glucan synthesis (Dijkgraaf et al., 2002). In Arabidopsis, the GSL family of proteins consists of 12 members distributed across two main clades (Hong et al., 2001). It has been speculated that the individual members of the GSL family mediate callose synthesis in different tissues and/or under different environmental conditions (Jacobs et al., 2003). GSL1, GSL2, GSL 6, GSL 8 and GSL10 are involved in fertility and cell division, while GSL5, GSL7 and GSL12 have been associated with structural reinforcement (Ellinger and Voigt, 2014). The functions of GLS3, GLS4, GLS9 and GLS11 remain to be elucidated. If, and to what extent, the different GSLs form oligomers remains also to be shown; however, a callose synthase complex holding a monomeric GTPase, UDP-glucose transferase (UGT), annexin and sucrose synthase that may interact with the hydrophilic, intracellular loop of the GSLs has been suggested (Stone, 2006). For the sake of brevity, callose synthesis will not be covered in detail in this review. In terms of the biochemical characterisation of the GT48 enzymes, there have not been many significant advances made over the course of the past 10 years, and the current state of the proposed pathway(s) for callose synthesis have been critically discussed by Stone et al. (2010) and Nedukha (2015).

Hemicellulose Biosynthetic GTs

Compared to cellulose, hemicelluloses are more complex structures that contain a range of sugar substituents that require a diverse repertoire of GTs to synthesise. Hemicelluloses have a β-1,4 linked backbone, with the exception of mixed-linkage glucan (MLG), which has β-1,3 linkages interspersed in the backbone (Scheller and Ulvskov, 2010). For XG, glucomannan, and MLG, these backbones are synthesised by multipass transmembrane GT2 enzymes of the CSL family (Scheller and Ulvskov, 2010).

Xyloglucan: XG GTs are some of the most well studied matrix polysaccharide synthetic enzymes. The β-1,4 glucan backbone of XG is synthesised by members of the CSLC family (Cocuron et al., 2007; Schultink et al., 2014), and is further extended by side chains initiated with α-1,6-d-Xyl. At least 19 different side chains have been identified across different plant phyla and are referred to using a single-letter nomenclature originally described by Fry et al. (1993) and most recently expanded by Tuomivaara et al. (2015). For simplicity, we will cover the most common side chains: unsubstituted Glc is denoted by the letter G, Glc appended Xyl is termed X, X further substituted by β-D-Gal is called L and Glc substituted with α-l-Fuc-(1,2)-β-D-Gal-(1,2)-α-d-Xyl is termed F. The X, L and F side chains are nearly universally distributed across all plant taxa (Schultink et al., 2014). In Arabidopsis, α-1,6 linked xylosyl residues are added by XYLOGLUCAN XYLOSYLTRANSFERASE (XXT) enzymes in the GT34 family. Of the five XXT enzymes, XXT1, XXT2, and to a lesser extent, XXT5 appear to be the most active and ubiquitously expressed xylosyltransferases (Cavalier and Keegstra, 2006; Culbertson et al., 2016). The addition of a β-1,2 galactose residue on the second and third xylose of the XXXG motif is conducted by GT47 members XYLOGLUCAN L-SIDE CHAIN GALACTOSYLTRANSFERASE POSITION-2 (XLT) 2 and MURUS (MUR) 3, respectively (Reiter et al., 1997; Madson et al., 2003; Jensen et al., 2012). The α-1,2 fucose residue on the F side chain is added by GT37 member, FUT1 (Perrin et al., 1999; Faik et al., 2000). Interspecies variance of XG biosynthetic GTs has been demonstrated experimentally. Tomato (*Solanum lycopersicum*) expresses XYLOGLUCAN 'S'-SIDECHAIN TRANSFERASE (XST) 1 and XST2, homologs of XLT2, which can produce α-1,2 arabinosylated S side chains upon overexpression in Arabidopsis, a species that normally does not produce arabinosylated XG (Schultink et al., 2013). There is also evidence of tissue-specific variance of XG GTs. The Arabidopsis XG-specific galacturonosyltransferase XUT1, expressed only in root hairs, is responsible for the addition of the β-1,2 galacturonic acid to xylose, forming the Y side chain (Pena et al., 2012). Due to the diversity of XG side chains, many more XG biosynthetic GTs remain uncharacterised and future studies will be required across several species to uncover them.

Mannan: The backbones of mannan and glucomannan were proposed to be synthesised by the CSLA and CSLD enzyme families (Dhugga et al., 2004; Goubet et al., 2009; Verhertbruggen et al., 2011). The extent to which the activities of these families overlap is uncertain. There is evidence that they may be differentially expressed in a developmental stage-specific or tissue-specific manner (Verhertbruggen et al., 2011). Notably, some CSLA enzymes have been shown to have glucosyltransferase and mannosyltransferase activity, indicating that single enzymes can synthesise the mixed sugar glucomannan backbone (Liepman et al., 2005; Goubet et al., 2009). The biosynthetic activity of CSLD enzymes has long remained controversial with different activities being proposed for these proteins. Recently, Yang et al. (2020) demonstrated unequivocally that Arabidopsis CSLD3 has β-1,4-glucosyltransferase activity. Both galactomannan and galactoglucomannan are decorated with α-1,6 galactose residues. The GT34 Arabidopsis enzyme MUCILAGE-RELATED (MUCI) 10 has been reported as a glucomannan galactosyltransferase and homologous enzymes in other species have mannan galactosyltransferase activity (Edwards et al., 1999; Voiniciuc et al., 2015). The MANNAN-SYNTHESIS-RELATED (MSR) proteins are also thought to be involved in (gluco)mannan synthesis (Wang et al., 2013). The proteins are now part of the GT106 family, whose only characterised members are rhamnosyltransferases; however, their functions have so far proven difficult to elucidate and GT activity of MSR has not yet been detected by itself. However, recently Voiniciuc and colleagues demonstrated that MSR proteins assist CSLAs and function as cofactors for both the elongation and the composition of plant (gluco)mannan (Voiniciuc et al., 2019).

Mixed-linkage β-1,3/1,4-glucan: MLG lacks sugar side chains; however, backbone synthesis is still not entirely understood. Three CSL families have been implicated in MLG synthesis: CSLF, CSLH and CSLJ (Little et al., 2018). Enzymes from each of these families have the demonstrated capability to synthesise MLG in Arabidopsis without co-expression of enzymes from the other families (Burton et al., 2006; Doblin et al., 2009; Little et al., 2018). In addition, not all plants that synthesise MLG encode genes from all three families (Little et al., 2018). Therefore, the individual contributions of the MLG synthesising members of these groups remain unknown and compelling questions regarding their action still need to be explored.

Xylan: The xylan backbone is a unique hemicellulose in that it is not synthesised by GT2 enzymes, but instead by GT47 enzymes that are active in the lumen of the Golgi (Wu et al., 2009). Interestingly, several enzymes associated with xylan backbone synthesis have redundant homologs. IRX 10 and its homolog IRX10L/XYLAN SYNTHASE-1 (XYS1) have been identified as the catalytic GTs involved in xylan backbone biosynthesis (Jensen et al., 2014; Urbanowicz et al., 2014). The GT43 proteins IRX9, IRX9L, IRX14 and IRX14L have also been implicated in xylan backbone synthesis (Brown et al., 2007; Lee et al., 2007; Wu et al., 2010). However, IRX9 and IRX14 lack known catalytic activities. The canonical GT-A fold catalytic DxD motif is absent in IRX9, suggesting it is incapable of being enzymatically active (Ren et al., 2014). Despite this, IRX9 or IRX14 loss-of-function plants are deficient in xylan and have severe growth defects, which prompt the idea that they might form a complex (Zeng et al., 2010; Ren et al., 2014; Rennie and Scheller, 2014). Many dicot glucuronoxylans have a unique reducing end oligosaccharide, also known as sequence 1 (York and O'Neill, 2008). Two GT8 enzymes and a GT47 enzyme are believed to be involved in its synthesis (Brown et al., 2007; Pena et al., 2007). Mutants lacking the function of these genes contain reduced total xylan content but the contribution of sequence 1 to xylan biosynthesis is not known.

The xylan backbone is heavily decorated with sugar substituents. In Arabidopsis, one in eight xylan backbone residues are substituted with glucuronic acid (GlcA) or methylated GlcA (Brown et al., 2007). Arabidopsis encodes five GT8 GUX enzymes, some of which are responsible for the addition of α-1,2 GlcA residues onto the xylan backbone (Mortimer et al., 2010). Both GUX1 and GUX2 are responsible for secondary cell wall glucuronidation of xylan (Rennie et al., 2012). GlcA residues added to the xylan backbone by GUX1 are spaced evenly and exhibit wide spacing whereas GUX2 adds GlcA irregularly with tight spacing between residues (Bromley et al., 2013). GUX3 is believed to be involved in the synthesis of a low abundance primary cell wall xylan that features pentose-GlcA disaccharide side chains (Mortimer et al., 2015). GUX4 has also been reported to have glucuronosyltransferase activity but its role *in vivo* and any functional differences compared to the other GUX enzymes are not known (Rennie et al., 2012). Monocot xylan is notable for the prevalence of pentose substituents. Enzymes from the GT61 family have been identified as transferases for three prominent pentose substituents. XYLAN ARABINOSYLTRANSFERASE (XAT) 1 and XAT2 add α-1,3 arabino*furanose* to the xylan backbone (Anders et al., 2012). More recently, the Ye lab heterologously expressed 20 GT61 proteins in the Arabidopsis *gux1/2/3* triple mutant and in HEK293 cells and confirmed that they mediate 3-*O*-arabinosylation of xylan (Zhong et al., 2021). XYLAN XYLOSYLTRANSFERASE (XYXT) 1 transfers xylose directly to the xylan backbone with a β-1,2 linkage (Zhong et al., 2018). Further, XYLOSYL ARABINOSYL SUBSTITUTION OF XYLAN (XAX) 1 was reported to add β-1,2 xylose to the α-1,3 arabino*furanose* product of XAT (Chiniquy et al., 2012). However, recent data based on detailed analysis of rice *xax1* mutants by the Dupree lab suggests XAX1 may be involved in the direct transfer of hydroxycinnamic acid modified arabinosyl substitutions to the xylan backbone, and not in xylosyl transfer to arabinosyl substituents (Feijao et al., 2021). Thus, the function of this enzyme remains enigmatic, and limitations in the availability of substrates will hinder biochemical confirmation.

Pectin Biosynthetic GTs

Current models suggest that there are several types of native pectic polymers in the plant cell wall: HG, xylogalacturonan, apiogalacturonan (some species), rhamnogalacturonan I (RGI), rhamnogalacturonan II (RG-II) and the xylan- and pectin-containing arabinogalactan protein APAP1. However, that has yet to be confirmed to be present in the wall (for details and schematic models on pectin structure, see Chapter 5).

Homogalacturonan: The HG backbone is synthesised by the GT8 GALACTURONOSYLTRANSFERASE (GAUT) family of enzymes. HG synthesis has been documented for GAUT1, GAUT4 and GAUT11 (Sterling et al., 2006; Biswal et al., 2018; Voiniciuc et al., 2018). How the *in vivo* roles of these enzymes differ and what roles most other members of the large GAUT family play is still unknown. GAUT7, however, is known to form a complex with GAUT1 despite not having HG biosynthesis activity itself. In a very recent study, HG synthesis has also been shown for GAUT 10, GAUT13 and GAUT14. The study further showed that *de novo* synthesis of HG may be the major function of GAUT13 and GAUT14 (Engle et al., 2022) and that different GAUTs synthesise HG at different rates. This would be consistent with the hypothesis that individual GAUTs involved in HG synthesis have distinct roles, both scaffold forming and catalytically functional, for different pectic polymers that contain HG (Atmodjo et al., 2011, 2013; Biswal et al., 2018). Therefore, it is also possible that GAUT enzymes are responsible for RG-II backbone synthesis (Biswal et al., 2018). The Arabidopsis GT47 enzyme XYLOGALACTURONAN DEFICIENT 1 (XGD1) has β-1,3 xylosyltransferase activity on HG backbone substrates, producing xylogalacturonan (Jensen et al., 2008).

RG-I: Only relatively recently, the enzymes that add the rhamnose component of the repeating disaccharide backbone were identified as the founding members of the GT106 RG-I RHAMNOSYLTRANSFERASE (RRT) family (Uehara et al., 2017). Five members of the RRT family have rhamnosyltransferase activity, and they are predominantly expressed in different tissues in Arabidopsis (Takenaka et al., 2018; Wachananawat et al., 2020). Another member of the Arabidopsis GT106 family is PECTIC ARABINOGALACTAN SYNTHESIS-RELATED (PAGR). Although activity could not be demonstrated for PAGR, mutant data support its involvement in the biosynthesis of RG-I arabinogalactans (AGP) (Stonebloom et al., 2016). The enzyme responsible for adding the galacturonic acid (GalA) unit of the repeating disaccharide remains unknown at the time of this publication. The GT92 GALS family of enzymes is involved in the synthesis of RG-I side chain galactan (Liwanag et al., 2012). Triple mutant *gals1/gals2/gals3* plants had no defect in the abundance of RG-I GalA branching, indicating that another GT is likely responsible for initiating galactan side chain synthesis (Ebert et al., 2018). Interestingly, GALS1 appears to be a key exception to the 'One-Linkage-One-Enzyme' dogma and is capable of capping β-1,4 galactan chains by adding arabino*pyranose* (Laursen et al., 2018). Arabinan also occurs as RG-I side chains and is thought to be synthesised by the GT47 proteins ARABINAN DEFICIENT 1 (ARAD1) and ARAD2 (Harholt et al., 2006). Direct evidence of arabino*furanosyl* transferase activity has not been demonstrated yet; however, ARAD1 and ARAD2 have been shown to form a complex *in vivo* and ARAD1 loss-of-function plants that display a significant reduction in α-1,5 arabinan (Harholt et al., 2006, 2012).

Rhamnogalacturonan II (*RG-II*): Due to its complexity, RG-II biosynthesis is the black box of plant polysaccharide biosynthesis. If we assume that one GT activity is required for the formation of each different glycosidic linkage, at least 22 different enzymes would be needed to synthesise its core saccharide structure (Bar-Peled et al., 2012). In addition, due to the crucial nature of RG-II in plant development, specific genetic knockouts are hypothesised to be largely lethal making reverse genetics approaches that have been utilised for other pathways nearly impossible for this polymer. Thus, only two verified activities have been identified in addition to a slightly dramatic conflict regarding another candidate. For the conventional side of RG-II biosynthesis, three GT77 enzymes, RHAMNOGALACTURONAN XYLOSYLTRANSFERASE (RGXT) 1, RGXT2, and RGXT3 are responsible for the addition of an α-1,3 xylose to the fucose on RG-II chain A (Egelund et al., 2006, 2008). Recently, it was shown that the enzyme encoded by a Fe deficiency–induced gene (*At1g64980*) referred to as *Cdi* (as it was initially identified as a Cd^{2+}-induced gene) likely catalyses the transfer of GDP-L-Gal (Peng et al., 2021). However, CDI is not listed in the CAZY annotation of the Arabidopsis genome as a characterised or non-characterised GT at the time of publication, so no GT family information is available. The dramatic side of RG-II research began in 2002 when a report appeared that described the tobacco nolac-H18 (nonorganogenic callus with loosely attached cells) mutant, which had four distinct T-DNA insertions and a cell separation phenotype. Analysis of nolac-H18 cell wall material indicated that it lacked the terminal GlcA-Gal disaccharide on side chain A of RG-II, which in turn affected the ability of RG-II to form dimers (Iwai et al., 2002). Iwai et al. identified a candidate gene in *Nicotiana plumbaginifolia* using T-DNA tagging and showed it was NpGUT1, a member of the GT47 family, which is associated with the observed loss of GlcA in the mutant (Iwai et al., 2002). However, in 2008, a detailed report from Alan Marchant's group later showed that the Arabidopsis orthologs of NpGUT1 are IRX10 and IRX10-L and provided very strong data for a role for these enzymes in glucuronoxylan and not RG-II formation (Wu et al., 2009). Five years later, IRX10/10-L were biochemically shown by two independent labs to specifically catalyse formation of the xylan backbone (Jensen et al., 2014; Urbanowicz et al., 2014), confirming Wu's work and proving beyond a doubt that these enzymes are involved in xylan synthesis.

Overall, pectins are among the most complex polysaccharides found in nature and only a fraction of the GTs expected to be involved in their synthesis are identified and characterised to date, given that it likely requires 25–35 enzymes to make RG-II alone, especially considering the complexity of the non-glycosyl substituents present on this polymer. Thus, it will be crucial to further pursue the many unanswered questions related to this enigmatic cell wall polymer. For more information on pectin, see Chapter 5 by Debra Mohnen in this volume.

Arabinogalactan Protein and Extensin Biosynthetic GTs

GTs also catalyse the many glycosidic linkages found in cell wall proteins. Several GTs involved in AGP polysaccharide biosynthesis are known. There are two families of GT31 enzymes that have been implicated in AGP galactan backbone initiation. HYDROXYPROLINE *O*-GALACTOSYLTRANSFERASES (HPGT) 1–3 and AGP galactosyltransferases (GALT) 2–6 all have demonstrated β-1,4 galactosyltransferase activity in microsomal preparations (Basu et al., 2013; Basu, Tian et al., 2015; Basu, Wang et al., 2015; Ogawa-Ohnishi and Matsubayashi, 2015). These eight GALTs add the first galactose to the hydroxyproline residue on the core protein. Despite sharing a CAZy family, these enzymes are dissimilar and cluster in distinct clades within the GT31 family but appear to produce the same product (Basu et al., 2013; Basu, Tian et al., 2015; Basu, Wang et al., 2015; Ogawa-Ohnishi and Matsubayashi, 2015).

Four additional members of the GT31 family have been identified as putative AGP backbone GALTs, namely, KAONASHI4/UNEVEN PATTERN OF EXINE1 (KNS4/UPEX1, At1g33430, Suzuki et al., 2017), GALT8 (At1g22015, Narciso et al., 2021), cotton (*Gossypium hirsutum*), GhGALT1 (Qin et al., 2017) and plausibly also At1g77810 (Qu et al., 2008). The β-1,6 galactan side chains have been reported to be elongated by the GT31 family member GALT31A (Geshi et al., 2013). The GT29 enzyme GALT29A was found to form a complex with GALT31A that resulted in increased galactosyltransferase activity compared to either enzyme alone. The complex was also able to add β-1,6 galactose to the galactan backbone, suggesting that this is also an activity of GALT29A (Dilokpimol et al., 2014). However, more recently Ruprecht and colleagues obtained surprising results using a glycan array-based GT assay. When they incubated the array with UDP-6-N_3-Gal and AtGALT31A, they found AtGALT31A was able to galactosylate substituted and unsubstituted β-1,3-galactan oligosaccharides rather than purely β-1,6-linked galactan oligosaccharides, indicating β-1,3- rather than β-1,6-galactosylation activity for this enzyme (Ruprecht et al., 2020).

Two members of the GT37 family FUT4 and FUT6 are responsible for the non-redundant addition of α-1,2 fucose residues to AGP branches (Wu et al., 2010). These enzymes are in the same family as the XG-acting FUT1, representing a rare example of different members of an enzyme family acting on different cell wall components. The GlcAT14 family of GT14 enzymes have β-1,6 glucuronosyltransferase activity on both AGP side chains and backbone galactose (Knoch et al., 2013; Dilokpimol and Geshi, 2014; Xuan et al., 2021). Finally, the REDUCED ARABINOSE YARIV (RAY)1 enzyme belonging to GT family 77 has been reported to be a putative AGP β-arabinosyltransferase (Gille et al., 2013). Evidence for the proposed activity is well supported by mutant studies; however, β-linked arabinose (Ara) is rarely (e.g., (Tryfona et al., 2010)) reported in AGPs, and thus questions remain whether RAY1 indeed functions in AGP biosynthesis. However, recent work on degradation of RG-I by the common human gut bacterium *Bacteroides thetaiotaomicron* showed that the enzyme BT0349 was able to cleave β-L-Ara from an arabinan-derived oligosaccharide (Luis et al., 2018), suggesting that β-linked Ara is present in RG-I. Interestingly, BT0349 became the founding member of GH146. For more information on structure, synthesis and function of AGPs, see Chapter 16 of this volume by Silvia Coimbra.

Most of the nine predicted extensin GTs have been identified. These include the SERINE GALACTOSYLTRANSFERASE1 (SGT1), which adds single galactose residues to serine (Ser) residues in serin-hydroxyprolin (Hyp) Ser-Hyp-Hyp-Hyp-Hyp sequences (Saito et al., 2014). When this enzyme was finally added to the CAZy database, it became part of the GT96 family, which has been recently predicted to adopt a completely novel GT fold using new structural modelling tools driven by machine learning (Taujale et al., 2021). The HYDROXYPROLINE *O*-ARABINOSYLTRANSFERASES (HPATs) mediate the addition of single Ara residues to Hyp residues in Ser-Hyp-Hyp-Hyp-Hyp sequences (Ogawa-Ohnishi et al., 2013) and although related to GT8 enzymes, HPATs have been grouped in GT95. The GT77 family enzymes RESIDUAL ARABINOSE (RRA) 1–3 and XYLOGLUCAN-SPECIFIC ß-ENDOGLUCANASE (XEG) 113 catalyse the addition of the second and third Ara residue in a β-1,2 linkage to the extensin arabino-oligosaccharide (Egelund et al., 2007; Gille et al., 2009; Velasquez et al., 2011). Most recently EXTENSIN ARABINOSE DEFICIENT (ExAD) has been identified, which adds the fourth Ara residue with an α-1,3 linkage. In Arabidopsis, ExAD is the only gene present in the GT47 family (Moller et al., 2017). While extensins are typically substituted with three to four Ara residues, some also have five, thus at least one more GT needs to be identified in future studies.

GT ORGANISATION AND COMPLEX FORMATION

Apart from cellulose synthases of the GT2 family and callose synthases of the GT48 family, most plant cell wall biosynthetic GTs are active in the Golgi apparatus. In the Golgi, cell wall matrix polysaccharides are produced, cell wall proteins are glycosylated and the CSCs are assembled. These crucial Golgi functions are supported by the unique morphological features of this organelle. Light microscopy and later transmission electron microscopy (EM) revealed a reticulum-like structure composed of successive flattened disc-shaped cisternae typical for a maturing organelle, with membranes assembling and maturing in a *cis* to *trans* direction (Golgi, 1989; Fullekrug and Nilsson, 1998). On top of the surrounding vesicles, tubular structures linking the different cisternae were observed at the edge of Golgi (Rambourg and Clermont, 1990). Early works on mammalian Golgi in the 1970s and 1980s involving subcellular fractionation of cellular membranes and ultrastructural localisation with immunoelectron microscopy supported a view that glycosylation events were strictly static and sequentially organised throughout the organelle (Roth, 1997). It was hypothesised that a distinct set of GTs acted on a specific substrate during its migration from the *cis* to *trans* Golgi (Kornfeld and Kornfeld, 1985; Rothman and Orci, 1990; Nilsson et al., 2009), which thereby functionally defined the different Golgi cisternae (Figure 2.4). At the same time, similar observations were made in plants using antibodies targeting glycosylated products (Moore et al., 1991; Zhang and Staehelin, 1992). These further supported the so-called compartmentation model and the hypothesis that distinct Golgi cisternae are involved in the production of specific polysaccharides and glycoproteins.

The compartmentation model was first challenged when numerous GTs including *N*-acetylglucosaminyltransferase 1 (GlcNAc-T1), galactosyltransferase I (GT1), sialyltransferase I (ST1), mannosidase II (Man II) and N-acetylgalactosaminyltransferase (GalNAc-T) 1–3 involved in mammalian *N*-glycosylation were localised along a gradient across the Golgi stacks (Nilsson et al., 1993; Rabouille et al., 1995; Rottger et al., 1998). It was proposed that these gradients could allow for only so much overlap between enzymes and reactions, and that they could easily be taking place in the same Golgi compartment (Nilsson et al., 2009). Two relatively recent studies report similar findings. One approach applied free-flow electrophoresis to successfully separate endomembrane compartments from Arabidopsis cell cultures, revealing the distribution of proteins, including GTs and glycans across the Golgi (Parsons et al., 2019). The results of this study hint at functional compartmentalisation among resident Golgi proteins and distinct features for *cis*, middle or late compartments (Parsons et al., 2019). The authors propose that a continuum of transmembrane features, rather than discrete rules, guide proteins to earlier or later locations within the Golgi stack. Complementary work performed by Meents et al. (2019) used live-cell imaging and transmission electron microscopy (TEM) to study xylan biosynthesis and showed that IRX9 is most abundant in the ring of the inner margins of medial cisternae while the xylan product is more concentrated in the trans cisternal margins. This is evidence that GTs occupy specific domains in the Golgi suggesting that in addition to being distributed from *cis*- to *trans*, GTs also have a preferential localisation in the cisternae of plants. However, whether this is a general feature of GT/cargo organisation in the Golgi remains unknown.

Conflicting models of GT organisation need to be considered in light of two models related to Golgi function: the cisternal maturation model and the vesicular transport model (Figure 2.4) (de Graffenried and Bertozzi, 2004). The cisternal maturation model proposes that the cisternae

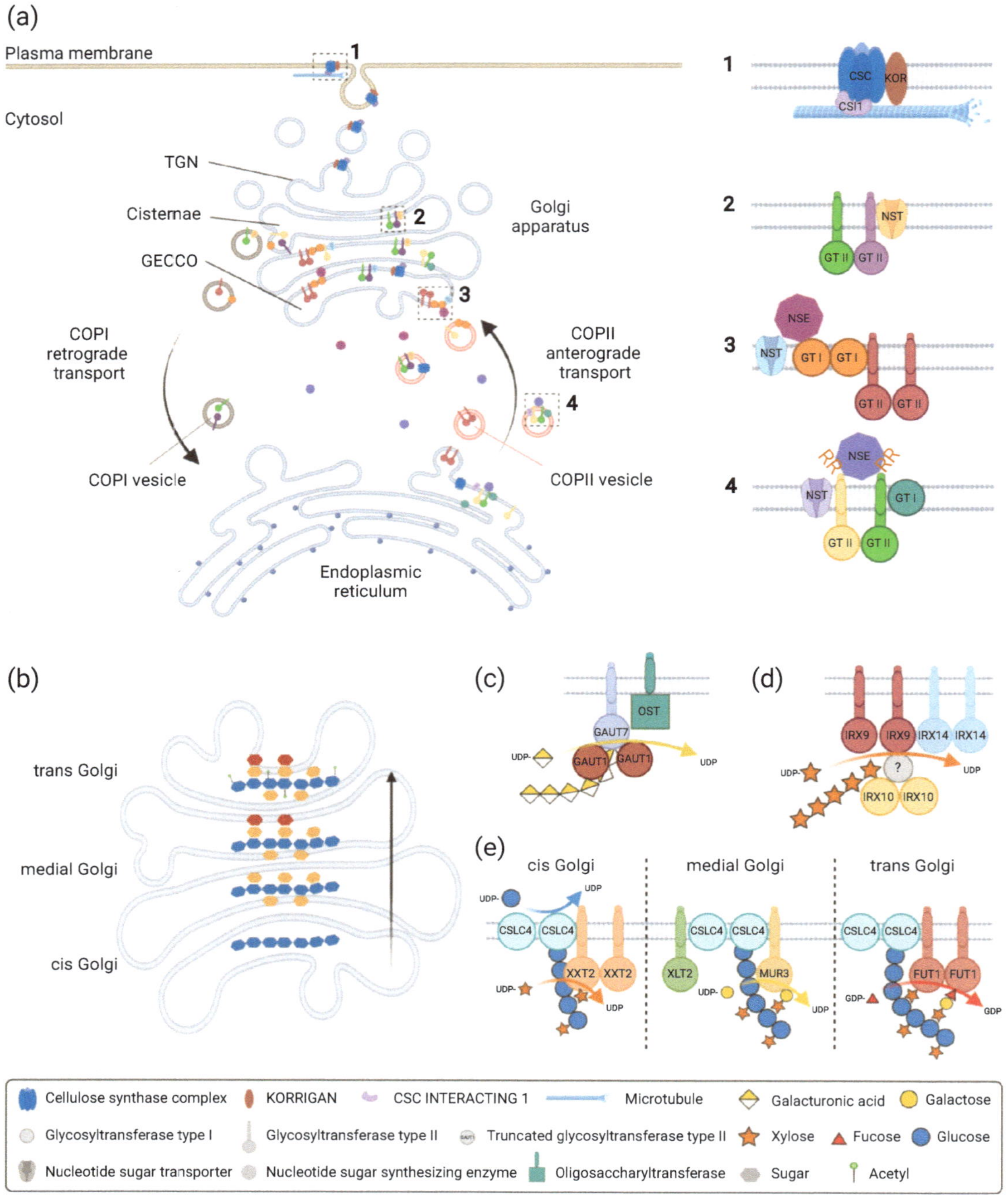

FIGURE 2.4 Polysaccharide biosynthesis in the Golgi apparatus. (a) Models for Golgi protein trafficking. Protein trafficking involves the following compartments: Endoplasmic Reticulum (ER), Golgi Entry Core Compartment (GECCO), Golgi and post-Golgi trafficking with the Trans-Golgi Network (TGN) and the plasma membrane (PM). Pre-Golgi trafficking involves COPII-anterograde transport as well as COPI-retrograde transport. The GECCO is independent of COPII machinery and may represent an intermediate compartment between the ER and Golgi. Schematic representation of known or proposed protein complexes involved in cell wall polysaccharide biosynthesis: (1) Cellulose synthase complex (CSC) with possibly interacting proteins CSC INTERACTING 1 (CSI1) and KORRIGAN (KOR) (McFarlane et al., 2014), (2) hetero and homodimerisation of type II GTs, (3) hetero and homodimersation of type II and type I GTs, (4) hetero and homodimerisation of type II and type I GTs and ER retrieval motif masking (Jiang et al., 2016). Complexes holding GTs and nucleotide-sugar synthesising enzymes (NSE) have been reported, while interactions with nucleotide sugar transporters (NST) that import the GT substrates are very likely (and shown in the schematic model), they remain to be experimentally demonstrated. (b) Sequential polysaccharide biosynthesis across sub-Golgi compartments. Schematics of the (c) pectin biosynthetic complex showing the GAUT1:GAUT7 interaction and possible interactions with subunits of the oligosaccharyltransferase (OST) (Atmodjo et al., 2011), (d) xylan biosynthetic complex (Zeng et al., 2010, 2016) and (e) xyloglucan biosynthetic complex (Chou et al., 2015).

themselves move from the *cis* to the *trans* face of the Golgi. Meaning that the cargo undergoes processing in cisternal compartments that mature in composition through the accumulation of *medial* and *trans* enzymes that move in vesicles from later to earlier cisternae (retrograde traffic). By contrast, the vesicular model established by Farquhar and Palade in 1998 (Farquhar and Palade, 1998) is based on the idea that each cisternae remains in one place with the same enzyme composition, and that the cargo moves in vesicles from *cis* to *trans* cisternae (anterograde traffic). Today, no clear consensus has been reached; however, most scientists agree that likely aspects of both models are true. Evidence supporting the cisternal maturation model argues for a requirement of the efficient sorting of GTs to anchor them in specific cisternae, excluding them from vesicles and thus separating them from the cargo (de Graffenried and Bertozzi, 2004). Different models have been proposed for the retention of Golgi enzymes. In the 'kin recognition model', self-assembled oligomers or 'kin' complexes of enzyme homodimers were suggested to ensure the proper localisation of GTs working in the same pathway (Nilsson et al., 1993). This is distinct from the 'membrane thickness model' that assumes a lipid gradient, caused by gradually increasing levels of cholesterol and sphingolipids across the Golgi cisternae, is the driving force for protein sorting and retention (Bretscher and Munro, 1993).

Finally, the 'retrograde transport model' is based on the observation that COAT PROTEIN (COP) I-coated vesicles are involved in intra-Golgi transport and retrograde transport of proteins between Golgi and ER, which might allow the retention of GTs despite Golgi maturation because they compete for the entry into retrograde COPI vesicles (Glick et al., 1997). Different proteins, e.g., Vps74p, have been found to regulate retrograde transport of GTs by COPI vesicles in mammalian and yeast systems (Hsu et al., 2013). In the 'retrograde transport model', the rate of anterograde transport of GTs within COPII-coated vesicles and the rate of retrograde transport by COPI-coated vesicles determines the localisation of GTs (Melancon et al., 1987; Malhotra et al., 1989; Banfield, 2011). The carboxy-terminal 'KKXX' and 'K/HDEL' signals required for ER-protein retrograde transport from the Golgi are well-described; however, no signal sequences maintaining the sub-Golgi localisation of GTs have been identified to date (Banfield, 2011). This has also been hypothesised for plant GTs (Gimeno-Ferrer et al., 2017); however, direct proof remains to be shown.

While it is most probable that the 'kin recognition model' and the 'membrane thickness model' at least partially coexist in the Golgi, a lipid-driven exclusion of proteins is rather unlikely (Nilsson et al., 2009). This is supported by observations that the diffusion rate of lipids is higher than that of membrane proteins and experimental evidence that protease treatment, but not cholesterol depletion, triggers a thickening of the lipid membrane across the Golgi stacks. It is more likely that proteins would drive the partitioning of lipids in the Golgi membrane than the other way around (de Graffenried and Bertozzi, 2004; Ubarretxena-Belandia and Tate, 2004; Nilsson et al., 2009). Differences in the length of TMDs are implicated in enzyme sorting and have been described between plasma membrane- and Golgi-localised proteins (Rayner and Pelham, 1997). This might be a general feature of protein-based lipid partitioning within the endomembrane system, which creates exclusion domains specific to proteins with different target localisations (Sharpe et al., 2010; Banfield, 2011). Such protein driven lipid partitioning would be consistent with the observation that GT mutants can have an abnormal endomembrane organisation and/or Golgi/TGN structure, which has been observed for *irx9* and *mur3* mutants, indicating that Golgi GTs might play an important role in the maintenance of cisternae structure (Tamura et al., 2005; Meents et al., 2019). However, cause of these endomembrane organisation phenotypes remains unknown.

The fact that GTs participating in the biosynthesis of the same polysaccharide localise to the same sub-Golgi compartment suggests that multiple enzymes work together and that different enzymes with the same targets might compete. Thus, raising questions regarding how enzyme specificity is guaranteed and whether certain substrates and linkages are favoured over others? This argues for the fact that mechanisms need to be in place to separate enzyme reactions and allow for efficient polysaccharide biosynthesis. Protein–protein interactions have been reported to play a crucial role for the organisation of Golgi GTs involved in glycan synthesis in both mammalian and yeast systems and similarly such interactions could also be a common principle of polysaccharide biosynthesis in the plant Golgi (Oikawa et al., 2013). Indeed, recent reports support the idea that multiprotein complexes participate in the efficient synthesis of plant cell wall polysaccharides and proteoglycans (Zeng et al., 2010, 2016; Chou et al., 2015).

In plants, not only homodimers and heterodimers but also multiprotein complexes have been described to be involved in the biosynthesis of most cell wall polymers (Figure 2.4). This stimulated discussions that protein complex formation is a general requirement for proper GT function (Oikawa et al., 2013). For example, Arabidopsis GAUT (Atmodjo et al., 2011), ARAD (Harholt et al., 2012) and REVERSIBLY GLYCOSYLATED PEPTIDES (RGPs, GT75) (Rautengarten et al., 2011) complexes play a role in pectin biosynthesis, while XG biosynthesis relies on complexes involving Arabidopsis XXTs and CSLC4 (Chou et al., 2012). Some of these complexes and their implications for GT activity and organisation in the endomembrane system will be discussed as part of this chapter (Figure 2.4).

The CSC Complexes

In plants, CESA proteins from the GT2 family are likely the best studied complex-forming cell wall GTs. The CSC complex was first observed in 1972 and described as 'granules' in two filamentous green algae: Cladophora and Chaetomorpha (Robinson et al., 1972). In 1976, using freeze-fracture TEM, Brown and colleagues observed similar 'granules' in the prokaryote *Acetobacter xylinum* (now

Gluconacetobacter xylinus), the unicellular green algae Oocystis and in roots of corn seedlings. They subsequently proposed that they were the enzymatic component responsible for cellulose synthesis (Brown and Montezinos, 1976; Mueller et al., 1976). Later, the interaction between CESA subunits was shown to be required for proper cellulose synthesis through Arabidopsis mutant studies (Taylor et al., 2003). Interestingly, however, *in vitro* a single CESA protein is capable of synthesising linear chains of β-1,4-glucose when heterologously expressed in *Pichia* (Purushotham et al., 2016). The structure of the CSC has been part of many debates in the past, but recent analyses support the hypothesis that it is composed of six CESA trimers that together produce a bundle of 18 β-1,4 glucan chains that form a cellulose microfibril (Nixon et al., 2016). It is also uncertain if the rosettes are composed of a homotrimer or heterotrimer of CESA proteins (Hill et al., 2014). Recently, Jochen Zimmer's lab made a step towards the structural characterisation of the CSC and suggested six heterogeneous CESA trimers form a single plant CSC (Purushotham et al., 2020). Here, the interfaces formed by residues interacting between PttCESA8 homotrimer subunits are nearly identical in heterooligomers, which might explain the favoured formation of these species *in vivo* (Timmers et al., 2009; Purushotham et al., 2020). Furthermore, it could be speculated that the apparent competition between CESA homodimers and heterodimers observed in rosette structures is caused by a spatial separation of the heterodimers forming in the Golgi and competing homodimers that have formed in the ER. The CSC has now been extensively studied in Arabidopsis and other species, and over the past decades additional proteins participating in the formation of cellulose as part of the CSC have been identified. For example, interactions between CESAs and TRACHERARY ELEMENT DIFFERENTIATION-RELATED6 (TED6), CELLULOSE SYNTHASE INTERACTING1 (CSI1) and KORRIGAN1 (KOR1) have been demonstrated using protein interaction screens (Endo et al., 2009; Gu et al., 2010; Mansoori et al., 2014). Several additional proteins that may interact with the CSC have been identified using an immunoprecipitation approach in poplar (Song et al., 2010). However, not all those proteins have been verified and fully functionally characterised yet. Finally, CESA complexes have been observed in the Golgi and trans-Golgi network where they form from proteins produced in the ER, but the exact extent that the endomembrane domain contributes to complex formation remains unclear (Gutierrez et al., 2009; Zhang et al., 2016).

Pectin Biosynthesising Protein Complexes

The best studied complex involving Golgi GTs is likely the pectin HG biosynthesis complex, formed by GAUT enzymes from the GT8 family. GAUT1 was first isolated from a galacturonosyltransferase active protein fraction in Arabidopsis (Sterling et al., 2006) and later demonstrated to be in a complex with GAUT7 and at least 12 additional proteins that are noncovalently associated to form a larger complex (Atmodjo et al., 2011). In this complex, the association of GAUT7 and GAUT1 allows the retention of GAUT1 in the Golgi despite proteolysis of its N-terminal TMD in Arabidopsis (Atmodjo et al., 2011). Instead, GAUT7 anchors soluble GAUT1 to the Golgi membrane (Atmodjo et al., 2011; Amos et al., 2018). This interaction is mediated by both disulphide covalent interactions and non-covalent interactions. The interaction between GAUT1 and GAUT7 is responsible for the two-phase synthesis of the HG backbone similar to the mechanisms described for the galactofuranosyltransferase 2 (GlfT2) from *Mycobacterium tuberculosis* (Levengood et al., 2011; Amos et al., 2018). A model for the synthesis of the HG backbone involving the dimeric structure has been proposed by Amos et al. (2018). In this model, GAUT7 is required for the activity of GAUT1 to synthesise high-molecular weight HG according to a distributive mechanism (Amos et al., 2018). Interestingly, putative methyltransferases have also been identified as part of the GAUT complex, promoting the idea that GTs form complexes with methyltransferases so that concurrent methylation of the pectin backbone can occur (Atmodjo et al., 2011). In addition to methyltransferases, subunits of the oligosaccharyltransferase (OST) were also found in the GAUT complex, which supports the domain synthesis model for the synthesis of pectins, involving *en bloc* transfer of pectic polysaccharides to the backbone (Atmodjo et al., 2011, 2013; Kim and Brandizzi, 2016). For more information about pectin biosynthesis, see Chapter 5 by Debra Mohnen in this volume.

The Xyloglucan Biosynthetic Complex

The synthesis of XG engages complexes involving both type I and type II GTs in the Golgi. In Arabidopsis, CSLC4 is responsible for the biosynthesis of the β-1,4 glucan backbone (Cocuron et al., 2007), which is then decorated with xylose residues by the XXT enzymes forming the repeating XXXG, which represents three xylosylated Glc (X) residues followed by an unsubstituted Glc (G). Other CSLCs are also involved in synthesising the XG backbone (Kim et al., 2020); however, apart from CLSC4, no other CSLC has been demonstrated to participate in a complex. Numerous other modifications are added to the main XXXG core structure. For example, Xyl residues can be galactosylated (L) to form the XLXG or XXLG structures by XLT2 and MUR3, respectively (Madson et al., 2003; Jensen et al., 2012), and FUT1 (Perrin et al., 1999) fucosylates the second galactosyl (L) residue in XLLG/XXLG subunits to form the F side chain. Bimolecular fluorescence complementation (BiFC) and co-immunoprecipitation (Co-IP) were used to demonstrate the organisation of these GTs in a multi-enzyme biosynthetic complex, which could allow the sequential synthesis of a fully substituted XG molecule (Chou et al., 2012, 2015). Such organisation of GTs in a multi-enzyme complex allows the efficient substitution of the backbone while it is synthesised and simultaneously the solubility of the developing polysaccharide is maintained in the Golgi lumen. This seems to be an important requirement of XG biosynthesis as the precipitation of non-substituted XGs in the Golgi lumen has been proposed as the cause of a number of mutant phenotypes impacting the XG biosynthesis pathway (Kong et al., 2015; Pauly and Keegstra, 2016).

The Xylan Biosynthetic Complex

In Arabidopsis, two GT43 family members IRX9 and IRX14 and the GT47 family enzyme IRX10 are involved in xylan backbone synthesis and likely part of a xylan biosynthetic complex (Zeng et al., 2010; Rennie and Scheller, 2014). IRX9-like (IRX9-L), IRX10-L and IRX14-L are related genes that appear to encode functionally redundant paralogs (Brown et al., 2007; Lee et al., 2007). Evidence that xylan synthesising enzymes work synergistically together came from a study where IRX9 and IRX14 when overexpressed together demonstrated higher xylosyltransferase activity (Lee et al., 2012). Later, immunoprecipitation of a wheat (*Triticum aestivum*) glucuronoarabinoxylan active complex containing TaGT43–4 (wheat homolog of IRX14), TaGT47-13 (wheat homolog of IRX10 and IRX10-L) and TaGT75-4 (wheat homolog of Arabidopsis RGP) provided definite experimental evidence that xylan synthesising complexes exist *in vivo* (Zeng et al., 2010; Jiang et al., 2016). Similarly, in asparagus (*Asparagus officinalis*), AoIRX9, AoIRX10 and AoIRX14A were shown to participate in a xylan synthesis complex using a combination of Co-IP and BiFC (Zeng et al., 2016). The function of IRX14 in this complex is uncertain, but it maintains an intact catalytic motif and appears to hold some catalytic or substrate-binding function (Ren et al., 2014). Although no reports indicating a xylan synthesis complex exists in Arabidopsis, recent studies of IRX9 revealed its specific localisation to the inner margins of the medial Golgi cisternae (Meents et al., 2019), and it is tempting to speculate that it is held in this specific location because it is part of a complex formed in this particular Golgi subdomain. Supporting this hypothesis, a recent investigation of the wheat xylan synthesis complex suggests that complex formation is required for proper localisation of xylan-biosynthesising GTs in the Golgi allowing the masking of COPI-recognition sequences required for ER retrieval (Jiang et al., 2016). The authors found a conserved di-arginine (RR) and a long sequence motif that they suggested to control forward trafficking and Golgi accumulation of the complexes (Jiang et al., 2016). Future studies should reveal whether such complex-dependent localisation in the Golgi or ER, through the availability of recognition sequences for the COPI retrieval system, is a more general feature of GT localisation *in planta*.

GT Homo and Heterodimerisation and Domains Involved in Complex Formation

In mammals, GT homodimers and heterodimers have been found to be ER and Golgi specific, respectively. Experimental evidence suggests that heterodimers are preferentially formed in the Golgi and that their disassembly occurs during the Golgi to ER retrograde transport (Hassinen and Kellokumpu, 2014). In addition, it has been demonstrated that some GT heterodimers form pH-dependent complexes and several homodimers that involve covalent disulphide bond formation are dependent on the redox balance of the ER/Golgi endomembrane system (Kellokumpu, 2019). Unfortunately, almost nothing is known about the impact of the pH in the ER and Golgi, redox balance, and cation concentration on GT function in plants. Future studies will require novel tools to test those parameters. For example, a Golgi targeted pH sensor would allow monitoring the Golgi pH (Deschamps et al., 2020).

Apart from CesAs and CSLCs, most of the Golgi GTs are type II membrane proteins and harbour a short N-terminal cytosolic domain, a TMD, a linker or stem domain and a globular catalytic domain facing the Golgi lumen. Different GT domains have been implicated in complex formation, indicating that multiple proteins could be required to form and maintain complex specificity and stability (Grabenhorst and Conradt, 1999). In plants, so far only a few GT domains have been found to be involved in protein–protein interactions, i.e., the catalytic domains of GAUT1 and GAUT7 (Atmodjo et al., 2011) or ARAD1 and ARAD2 (Harholt et al., 2012), while in the mammalian system, the TMD and the stem of GTs have also been shown to participate in complex formation (Yamaguchi and Fukuda, 1995). Such domain diversity might be a requirement to allow the concurrent existence of multiple complexes forming distinct catalytic units (Kellokumpu et al., 2016). The evidence supporting the formation of homodimers and heterooligomers, however, raises the question if competition exists between multiple binding sites in a single complex. Even though the stoichiometry of most GT complexes has not yet been characterised, preliminary data for the XG synthesising XXTs indicate that the binding domains involved in homodimerisation are different from domains involved in heteromerisation (Chou et al., 2012). Thus, the existence of higher order complexes would theoretically be possible. However, at this point, the dynamics and stability of GT complex formation remains unresolved, in particular when considering the impact of environments with different pHs and reducing conditions, which will likely affect the strength of such interactions (Kellokumpu, 2019).

In light of functional relevance, the role of GT homodimers is difficult to explain; in some cases, their pH insensitivity might indicate a requirement for trafficking within the endomembrane system or their formation in the absence of a suitable heteromeric partner. In mammalian cells, the pH dependency of GT heterodimer assembly in the Golgi has been speculated to be responsible for altered glycosylation patterns observed in cancer cells (Hassinen et al., 2011). Nonetheless, other heterodimers are insensitive to Golgi pH, but respond to the oxidative status of the cisternae, suggesting that similar to the elusive 'localisation signal for a distinct cisternae', different protein interactions can be mediated or depend on the different physio-chemical properties of Golgi cisternae (Kellokumpu, 2019). GT heterodimers in the Golgi have been proposed to regulate the activity of interacting partners, maintain their localisation and allow the efficient synthesis of long-chain polysaccharides (Jiang et al., 2016; Amos et al., 2018). However, considering the numerous and diverse evidence that pertains to the role of these complexes, it is difficult to derive any general rules about the effects of complex formation on the different properties of the GTs involved. In this area of GT research, there are currently more questions than answers and identifying the mechanisms underlying GT complex formation are of particular interest for future studies.

The collaborative function of GTs with other GTs, but also with other cell wall synthesising machinery or Golgi-localised membrane proteins, has been demonstrated or proposed. In plants, some of the polymer synthesising protein complexes contain NDP-sugar interconverting enzymes in addition to GTs from other families (Zeng et al., 2010; Jiang et al., 2016). Similarly, NSTs have also been proposed to be part of such complexes in plants (Seifert, 2004, 2018; Hoffmann et al., 2021; Mariette et al., 2021). In the mammalian system, complexes that contain both NSTs and GTs have already been reported. Such complex formation was suggested when a UDP-Gal transporter was partially retained in the ER while co-expressed with an ER-localised ceramide galactosyltransferase in CHOLec8 cells, suggesting that this complex facilitates the specific delivery of UDP-Gal in the ER directly when it is produced (Sprong et al., 2003). This is the earliest example of a complex that allows the specific formation of a biosynthetic pathway in a subdomain of the endomembrane system when it is required by a cell (Sprong et al., 2003; de Graffenried and Bertozzi, 2004). The mammalian *N*-acetyl-glucosaminyltransferases (MGATs) and the UDP-GlcNAc transporter SLC35A3 have also been shown to form multi-enzyme/multi-transporter complexes in the Golgi (Khoder-Agha et al., 2019). These complexes allow efficient synthesis of *N*-glycans by maintaining close proximity for substrate supply to the GT forming the glycosidic linkage. Similar complexes have been suggested in plants. For example, the tight association of the cytosolic RGPs (GT75) with the Golgi apparatus has been hypothesised to be mediated by an interaction with UDP-ARABINOFURANOSE TRANSPORTERS (UA*f*Ts) that facilitate the transport of arabino*furanose* precursors across the Golgi membrane (Rautengarten et al., 2011, 2017; Mariette et al., 2021). Similarly, the association of the UDP-GLUCOSE EPIMERASE (UGE) with the Golgi membrane could be explained by an interaction of UGE4 and the transporters delivering UDP-Gal (Seifert, 2004). Such organisation of nucleotide sugar interconversion enzymes, NSTs and GTs in one complex might be a common design feature of specific glycosylation complexes in eukaryotes, which would support targeted delivery of substrates to specific enzymes (Seifert, 2004, 2018). However, further studies must be conducted to elucidate if complexes composed of GTs, NSTs and/or interconverting enzymes indeed exist in plants. Further, studies of the dynamics of such complexes should help uncover mechanisms that control the delivery and turn-over of cell wall biosynthetic complexes.

Given that the collaborative function of GTs and other players involved in glycosylation is likely a general feature of cell wall synthesising machinery, complex formation could be speculated to work as a concentrative mechanism to group polysaccharide synthesis enzymes into larger 'multi' complexes, similar to what has been observed for complexes holding transferases involved in anterograde trafficking in mitochondria (Brickley et al., 2011). However, such complexes are not necessary in the cisternal maturation model of the Golgi and have not been observed yet. Indeed, they might not be possible because of structural limitations, and future studies must illuminate whether they exist or not in the Golgi.

FUTURE PERSPECTIVES

Over the past 60 years, the field of GT research has seen pioneering discoveries and many innovative technical developments (Table 2.2). Nonetheless, we are still far away from fully understanding the details of GT function in cell wall polysaccharide assembly and regulation *in vivo*. Many aspects of these enzymes, their activity and organisation as well as their roles in the biosynthesis and regulation of carbohydrate structures must still be elucidated. A major bottleneck for our understanding of GTs is that experimentally determined structural data are missing for most plant cell wall GTs. However, the recent structural characterisation of three cell wall synthesising GT families and innovative tools like Alphafold will bring new insights into the mechanisms underlying the function of GTs through precise protein structure predictions in combination with experimental validation. Other aspects that should be addressed are how protein–protein interactions and the enzyme domains participating in those interactions impact polysaccharide biosynthesis. For example, by combining MS, activity assays, crosslinking, EM, and computational modelling, new insights into both domain structure and plasticity of protein complexes could be gained in future.

In addition, many questions in relation to specific Golgi environments and how these might affect GT localisation and function remain unresolved, but new knowledge in this area will significantly contribute to our understanding of mechanisms underpinning the function of these enzymes in future. Another approach, one-pot multi-enzyme synthesis, is gaining interest in the broader glycobiology field as a solution for substrate accessibility issues and as a resource for glycan synthesis (Muthana et al., 2015; Li et al., 2017). Our increasing understanding of plant cell wall GTs makes this an interesting opportunity for expansion of these techniques into plant polysaccharides. Finally, the identification of the myriad of the enzymes that are required for the biosynthesis of all known glycosidic linkages present in plants will be a challenging task for cell wall researchers. Despite significant advances, the identities and biochemical

TABLE 2.2
Landmarks in the History of Cell Wall Glycosyltransferases

1950	Discovery of NDP-sugars by Luis L. Leloir and his colleagues Ranwel Caputto, Carlos E. Cardini, Raúl Trucco and Alejandro C. Paladini.	Caputto et al. (1950)
1953	Evidence promoting the idea that nucleotide sugars serve as glycosyl donors emerges.	Dutton and Storey (1953), Leloir and Cabib (1953)
1956	Isolation of UDP-glucose, UDP-galactose, UDP-xylose and UDP-arabinose from mung bean.	Ginsburg et al. (1956)
1957	NDP-sugars were demonstrated to be involved in the biosynthesis of complex polysaccharides by confirming the transfer of *N*-acetylglucosamine from UDP-GlcNAc to hyaluronic acid and chitin.	Glaser and Brown (1957)
1957	Cellulose-producing bacterium *Acetobacter xylinum* (now *Gluconacetobacter xylinus*) was reported to transfer glucose from UDP-Glc to cellulose.	Glaser (1957)
1959	UDP-Glc and UDP-Xyl were shown to be the sugar donors for callose and xylan biosynthesis, respectively.	Feingold et al. (1958), Feingold et al. (1959)
1965	Synthesis of the α-1,4 galacturonic acid backbone of homogalacturonan (HG) is shown to be associated with Golgi membranes.	Villemez et al. (1965)
1968	The 'one enzyme–one linkage' hypothesis was proposed.	Hagopian and Eylar (1968)
1969	Synthesis of the β-1,4-glucan backbone of xyloglucan (XG) was measured.	Ray et al. (1969)
1989	The *brittle culm* mutants in maize, rice and barley impacting cell wall polysaccharide components were identified.	Kokubo et al. (1989), Kokubo et al. (1991)
1993	Rise of the MURUS. An analysis of the neutral monosaccharide composition of leaves of an EMS-mutagenised population of Arabidopsis identifies a fucose-deficient *mur* mutant.	Reiter et al. (1993)
1996	Cloning and characterisation of the cellulose synthase genes from cotton.	Pear et al. (1996)
1997	*Mur2* and *mur3* mutants were shown to be involved in xyloglucan (XG) biosynthesis.	Reiter et al. (1997), Vanzin et al. (2002), Madson et al. (2003)
1997	EMS mutant screen for altered xylem cell morphology yielded *irregular xylem* (*irx*) mutants.	Turner and Somerville (1997)
1997	Campbell, Davies, Bulone, and Henrissat classified the first set of known GTs into 27 sequence-derived families based on amino acid sequence similarities.	Campbell et al. (1997)
1998	Forward genetics allowed the analysis of the *cellulose synthase* (*cesa*) mutants.	Arioli et al. (1998)
1999	Molecular genetics was coupled with biochemistry and heterologous protein expression to demonstrate the synthesis of specific glycosidic linkages by a cell wall GT.	Perrin et al. (1999), Edwards et al. (1999)
2002	The pectin-specific *quasimodo1* mutant was identified.	Bouton et al. (2002)
2005	Gene co-expression analysis was established as a tool to identify additional cellulose and secondary cell wall synthesis genes.	Brown et al. (2005), Persson et al. (2005)
2007	Initiation of CAZypedia (http://www.cazypedia.org)	CAZypedia Consortium (2018)
2014	Identification and biochemical characterisation of xylan synthases (XYS) and confirmation that they are members of GT47, and not GT2.	Urbanowicz et al. (2014), Jensen et al. (2014)
2016	The crystal structure of FUT1 was solved.	Rocha et al. (2016), Urbanowicz et al. (2017)
2017	Structural characterisation of XXT1.	Culbertson et al. (2018)
2020	Structural characterisation of a trimeric *Populus tremula x tremuloides* cellulose synthase (PttCESA8) complex.	Purushotham et al. (2020)
2021	CAZy database underwent a large overhaul to facilitate access to information and aid taxonomic navigation.	Drula et al. (2021)
2021	Release of the Baker lab's RoseTTAFold and Deepmind's Alphafold(2) for prediction of native protein structures from their amino acid sequences.	Baek et al. (2021), Jumper et al. (2021)

activities of most GTs required for wall synthesis have yet to be identified and/or functionally characterised at the time of publication. This is particularly true when we consider the known diversity of cell wall polysaccharide structures across the plant kingdom and how little we really know about the true scope of what is out there. Nonetheless, with the recent development of sophisticated approaches for enzyme expression, HTP biochemical assays using glycan arrays, new plant models and advanced molecular genetics and phenotypic analysis techniques, many new GTs are expected to be discovered and functionally characterised over the next decades.

ACKNOWLEDGEMENTS

A.M. is supported by a Scholarship of the International Max-Planck Research School 'Primary Metabolism and Plant Growth' and acknowledges support from the Megan Klemm Postgraduate Award at the University of Melbourne. T.C. and B.U. are supported by the Center for Bioenergy

Innovation (CBI), a U.S. Department of Energy Bioenergy Research Center supported by the Office of Biological, Environmental Research in the DOE Office of Science. B.U. also acknowledges the National Science Foundation under grant no. 1817697 and the U.S. Department of Energy, Office of Science, Basic Energy Sciences [DE-SC0015662 and DE-SC0008472] to the Complex Carbohydrate Research Center. B.E. acknowledges the Inaugural Botany Foundation Fellowship 2020 from the University of Melbourne Botany Foundation.

REFERENCES

Ade CP, Bemm F, Dickson JM, Walter C, Harris PJ (2014) Family 34 glycosyltransferase (GT34) genes and proteins in Pinus radiata (radiata pine) and Pinus taeda (loblolly pine). *Plant Journal* 78:305–318.

Akita K, Ishimizu T, Tsukamoto T, Ando T, Hase S (2002) Successive glycosyltransfer activity and enzymatic characterization of pectic polygalacturonate 4-alpha-galacturonosyltransferase solubilized from pollen tubes of Petunia axillaris using pyridylaminated oligogalacturonates as substrates. *Plant Physiology* 130:374–379.

Albuquerque-Wendt A, Hutte HJ, Buettner FFR, Routier FH, Bakker H (2019) Membrane Topological Model of Glycosyltransferases of the GT-C Superfamily. *International Journal of Molecular Sciences* 20(19):4842. https://doi.org/10.3390/ijms20194842

Alonso JM, Stepanova AN, Leisse TJ, Kim CJ, Chen H, Shinn P, Stevenson DK, Zimmerman J, Barajas P, Cheuk R, Gadrinab C, Heller C, Jeske A, Koesema E, Meyers CC, Parker H, Prednis L, Ansari Y, Choy N, Deen H, Geralt M, Hazari N, Hom E, Karnes M, Mulholland C, Ndubaku R, Schmidt I, Guzman P, Aguilar-Henonin L, Schmid M, Weigel D, Carter DE, Marchand T, Risseeuw E, Brogden D, Zeko A, Crosby WL, Berry CC, Ecker JR (2003) Genome-wide insertional mutagenesis of Arabidopsis thaliana. *Science* 301:653–657.

Amos RA, Mohnen D (2019) Critical review of plant cell wall matrix polysaccharide glycosyltransferase activities verified by heterologous protein expression. *Frontiers in Plant Science* 10:915.

Amos RA, Pattathil S, Yang JY, Atmodjo MA, Urbanowicz BR, Moremen KW, Mohnen D (2018) A two-phase model for the non-processive biosynthesis of homogalacturonan polysaccharides by the GAUT1:GAUT7 complex. *Journal of Biological Chemistry* 293:19047–19063.

Anders N, Wilkinson MD, Lovegrove A, Freeman J, Tryfona T, Pellny TK, Weimar T, Mortimer JC, Stott K, Baker JM, Defoin-Platel M, Shewry PR, Dupree P, Mitchell RA (2012) Glycosyl transferases in family 61 mediate arabinofuranosyl transfer onto xylan in grasses. *The Proceedings of the National Academy of Sciences USA* 109:989–993.

Arioli T, Peng L, Betzner AS, Burn J, Wittke W, Herth W, Camilleri C, Hofte H, Plazinski J, Birch R, Cork A, Glover J, Redmond J, Williamson RE (1998) Molecular analysis of cellulose biosynthesis in Arabidopsis. *Science* 279:717–720.

Atmodjo MA, Hao Z, Mohnen D (2013) Evolving views of pectin biosynthesis. *Annual Review of Plant Biology* 64:747–779.

Atmodjo MA, Sakuragi Y, Zhu X, Burrell AJ, Mohanty SS, Atwood JA, 3rd, Orlando R, Scheller HV, Mohnen D (2011) Galacturonosyltransferase (GAUT)1 and GAUT7 are the core of a plant cell wall pectin biosynthetic homogalacturonan:galacturonosyltransferase complex. *The Proceedings of the National Academy of Sciences USA* 108:20225–20230.

Baek M, Baker D (2022) Deep learning and protein structure modeling. *Nature Methods* 19:13–14.

Baek M, DiMaio F, Anishchenko I, Dauparas J, Ovchinnikov S, Lee GR, Wang J, Cong Q, Kinch LN, Schaeffer RD, Millan C, Park H, Adams C, Glassman CR, DeGiovanni A, Pereira JH, Rodrigues AV, van Dijk AA, Ebrecht AC, Opperman DJ, Sagmeister T, Buhlheller C, Pavkov-Keller T, Rathinaswamy MK, Dalwadi U, Yip CK, Burke JE, Garcia KC, Grishin NV, Adams PD, Read RJ, Baker D (2021) Accurate prediction of protein structures and interactions using a three-track neural network. *Science* 373:871–876.

Banfield DK (2011) Mechanisms of protein retention in the Golgi. *Cold Spring Harbor Perspectives in Biology* 3:a005264.

Bar-Peled M, Urbanowicz BR, O'Neill MA (2012) The synthesis and origin of the pectic polysaccharide rhamnogalacturonan II - insights from nucleotide sugar formation and diversity. *Frontiers in Plant Science* 3:92.

Basu D, Liang Y, Liu X, Himmeldirk K, Faik A, Kieliszewski M, Held M, Showalter AM (2013) Functional identification of a hydroxyproline-o-galactosyltransferase specific for arabinogalactan protein biosynthesis in Arabidopsis. *Journal of Biological Chemistry* 288:10132–10143.

Basu D, Tian L, Wang W, Bobbs S, Herock H, Travers A, Showalter AM (2015) A small multigene hydroxyproline-O-galactosyltransferase family functions in arabinogalactan-protein glycosylation, growth and development in Arabidopsis. *BMC Plant Biology* 15:295.

Basu D, Wang W, Ma S, DeBrosse T, Poirier E, Emch K, Soukup E, Tian L, Showalter AM (2015) Two hydroxyproline galactosyltransferases, GALT5 and GALT2, function in arabinogalactan-protein glycosylation, growth and development in Arabidopsis. *PLoS One* 10:e0125624.

Bean RC, Hassid WZ (1955) Assimilation of C14O2 by a photosynthesizing red alga, iridophycus flaccidum. *Journal of Biological Chemistry* 212:411–425.

Beck Ea, Ziegler P (1989) Biosynthesis and degradation of starch in higher plants. *Annual Review of Plant Physiology and Plant Molecular Biology* 40:95–117.

Berglund J, Mikkelsen D, Flanagan BM, Dhital S, Gaunitz S, Henriksson G, Lindstrom ME, Yakubov GE, Gidley MJ, Vilaplana F (2020) Wood hemicelluloses exert distinct biomechanical contributions to cellulose fibrillar networks. *Nature Communications* 11:4692.

Biswal AK, Atmodjo MA, Pattathil S, Amos RA, Yang X, Winkeler K, Collins C, Mohanty SS, Ryno D, Tan L, Gelineo-Albersheim I, Hunt K, Sykes RW, Turner GB, Ziebell A, Davis MF, Decker SR, Hahn MG, Mohnen D (2018) Working towards recalcitrance mechanisms: increased xylan and homogalacturonan production by overexpression of GAlactUronosylTransferase12 (GAUT12) causes increased recalcitrance and decreased growth in Populus. *Biotechnology for Biofuels* 11:9.

Bouton S, Leboeuf E, Mouille G, Leydecker MT, Talbotec J, Granier F, Lahaye M, Hofte H, Truong HN (2002) QUASIMODO1 encodes a putative membrane-bound glycosyltransferase required for normal pectin synthesis and cell adhesion in Arabidopsis. *Plant Cell* 14:2577–2590.

Bretscher MS, Munro S (1993) Cholesterol and the Golgi apparatus. *Science* 261:1280–1281.

Brickley K, Pozo K, Stephenson FA (2011) N-acetylglucosamine transferase is an integral component of a kinesin-directed mitochondrial trafficking complex. *Biochimica et Biophysica Acta* 1813:269–281.

Bromley JR, Busse-Wicher M, Tryfona T, Mortimer JC, Zhang Z, Brown DM, Dupree P (2013) GUX1 and GUX2 glucuronyltransferases decorate distinct domains of glucuronoxylan with different substitution patterns. *Plant Journal* 74:423–434.

Brown DM, Goubet F, Wong VW, Goodacre R, Stephens E, Dupree P, Turner SR (2007) Comparison of five xylan synthesis mutants reveals new insight into the mechanisms of xylan synthesis. *Plant Journal* 52:1154–1168.

Brown DM, Zeef LA, Ellis J, Goodacre R, Turner SR (2005) Identification of novel genes in Arabidopsis involved in secondary cell wall formation using expression profiling and reverse genetics. *Plant Cell* 17:2281–2295.

Brown RM, Jr., Montezinos D (1976) Cellulose microfibrils: visualization of biosynthetic and orienting complexes in association with the plasma membrane. *The Proceedings of the National Academy of Sciences USA* 73:143–147.

Buchanan JGB, Benson, AA, Bradley DF, Calvin M, Daus LL, Goodman M, Hayes P, Lynch VH, Norris LT, Wilson AT (1952) Phosphorus metabolism. *John Hopkins Baltimore* 11:440.

Burton RA, Wilson SM, Hrmova M, Harvey AJ, Shirley NJ, Medhurst A, Stone BA, Newbigin EJ, Bacic A, Fincher GB (2006) Cellulose synthase-like CslF genes mediate the synthesis of cell wall (1,3;1,4)-beta-D-glucans. *Science* 311:1940–1942.

Cabib E, Leloir LF (1958) The biosynthesis of trehalose phosphate. *Journal of Biological Chemistry* 231:259–275.

Campbell JA, Davies GJ, Bulone V, Henrissat B (1997) A classification of nucleotide-diphospho-sugar glycosyltransferases based on amino acid sequence similarities. *Biochemical Journal* 326 (Pt 3):929–939.

Cantarel BL, Coutinho PM, Rancurel C, Bernard T, Lombard V, Henrissat B (2009) The Carbohydrate-Active EnZymes database (CAZy): an expert resource for Glycogenomics. *Nucleic Acids Research* 37:D233–238.

Caputto R, Leloir LF, Cardini CE, Paladini AC (1950) Isolation of the coenzyme of the galactose phosphate-glucose phosphate transformation. *Journal of Biological Chemistry* 184:333–350.

Cardini CE, Leloir LF, Chiriboga J (1955) The biosynthesis of sucrose. *Journal of Biological Chemistry* 214:149–155.

Carey DJ, Hirschberg CB (1980) Kinetics of glycosylation and intracellular transport of sialoglycoproteins in mouse liver. *Journal of Biological Chemistry* 255:4348–4354.

Carpita NC, McCann MC (2015) Characterizing visible and invisible cell wall mutant phenotypes. *Journal of Experimental Botany* 66:4145–4163.

Cavalier DM, Keegstra K (2006) Two xyloglucan xylosyltransferases catalyze the addition of multiple xylosyl residues to cellohexaose. *Journal of Biological Chemistry* 281:34197–34207.

Chiniquy D, Sharma V, Schultink A, Baidoo EE, Rautengarten C, Cheng K, Carroll A, Ulvskov P, Harholt J, Keasling JD, Pauly M, Scheller HV, Ronald PC (2012) XAX1 from glycosyltransferase family 61 mediates xylosyltransfer to rice xylan. *The Proceedings of the National Academy of Sciences USA* 109:17117–17122.

Cho SH, Du J, Sines I, Poosarla VG, Vepachedu V, Kafle K, Park YB, Kim SH, Kumar M, Nixon BT (2015) In vitro synthesis f cellulose microfibrils by a membrane protein from protoplasts of the non-vascular plant Physcomitrella patens. *Biochemical Journal* 470:195–205.

Chou YH, Pogorelko G, Young ZT, Zabotina OA (2015) Protein-protein interactions among xyloglucan-synthesizing enzymes and formation of Golgi-localized multiprotein complexes. *Plant & Cell Physiology* 56:255–267.

Chou YH, Pogorelko G, Zabotina OA (2012) Xyloglucan xylosyltransferases XXT1, XXT2, and XXT5 and the glucan synthase CSLC4 form Golgi-localized multiprotein complexes. *Plant Physiology* 159:1355–1366.

Ciceron F, Rocha J, Kousar S, Hansen SF, Chazalet V, Gillon E, Breton C, Lerouxel O (2016) Expression, purification and biochemical characterization of AtFUT1, a xyloglucan-specific fucosyltransferase from Arabidopsis thaliana. *Biochimie* 128–129:183–192.

Cocuron JC, Lerouxel O, Drakakaki G, Alonso AP, Liepman AH, Keegstra K, Raikhel N, Wilkerson CG (2007) A gene from the cellulose synthase-like C family encodes a beta-1,4 glucan synthase. *The Proceedings of the National Academy of Sciences USA* 104:8550–8555.

Cori CF, Schmidt G, Cori GT (1939) The synthesis of a polysaccharide from glucose-1-phosphate in muscle extract. *Science* 89:464–465.

Coutinho PM, Deleury E, Davies GJ, Henrissat B (2003) An evolving hierarchical family classification for glycosyltransferases. *Journal of Molecular Biology* 328:307–317.

Culbertson AT, Ehrlich JJ, Choe JY, Honzatko RB, Zabotina OA (2018) Structure of xyloglucan xylosyltransferase 1 reveals simple steric rules that define biological patterns of xyloglucan polymers. *The Proceedings of the National Academy of Sciences USA* 115:6064–6069.

Culbertson AT, Smith AL, Cook MD, Zabotina OA (2016) Truncations of xyloglucan xylosyltransferase 2 provide insights into the roles of the N- and C-terminus. *Phytochemistry* 128:12–19.

Daras G, Templalexis D, Avgeri F, Tsitsekian D, Karamanou K, Rigas S (2021) Updating insights into the catalytic domain properties of plant cellulose synthase (CesA) and cellulose synthase-like (Csl) proteins. *Molecules* 26(14):4335. doi: 10.3390/molecules26144335

de Graffenried CL, Bertozzi CR (2004) The roles of enzyme localisation and complex formation in glycan assembly within the Golgi apparatus. *Current Opinion in Cell Biology* 16:356–363.

Delmer DP, Amor Y (1995) Cellulose biosynthesis. *Plant Cell* 7:987–1000.

Deschamps A, Colinet AS, Zimmermannova O, Sychrova H, Morsomme P (2020) A new pH sensor localized in the Golgi apparatus of Saccharomyces cerevisiae reveals unexpected roles of Vph1p and Stv1p isoforms. *Scientific Reports* 10:1881.

Dhugga KS, Barreiro R, Whitten B, Stecca K, Hazebroek J, Randhawa GS, Dolan M, Kinney AJ, Tomes D, Nichols S, Anderson P (2004) Guar seed beta-mannan synthase is a member of the cellulose synthase super gene family. *Science* 303:363–366.

Diesperger H, Muller CR, Sandermann H, Jr. (1974) Rapid isolation of a plant microsomal fraction by Mg2+--precipitation. *FEBS Letters* 43:155–158.

Dijkgraaf GJ, Abe M, Ohya Y, Bussey H (2002) Mutations in Fks1p affect the cell wall content of beta-1,3- and beta-1,6-glucan in Saccharomyces cerevisiae. *Yeast* 19:671–690.

Dilokpimol A, Geshi N (2014) Arabidopsis thaliana glucuronosyltransferase in family GT14. *Plant Signaling and Behavior* 9:e28891.

Dilokpimol A, Poulsen CP, Vereb G, Kaneko S, Schulz A, Geshi N (2014) Galactosyltransferases from Arabidopsis thaliana in the biosynthesis of type II arabinogalactan: molecular interaction enhances enzyme activity. *BMC Plant Biology* 14:90.

Dimitroff G, Little A, Lahnstein J, Schwerdt JG, Srivastava V, Bulone V, Burton RA, Fincher GB (2016) (1,3;1,4)-beta-glucan biosynthesis by the CSLF6 enzyme: position and flexibility of catalytic residues influence product fine structure. *Biochemistry* 55:2054–2061.

Doblin MS, Pettolino FA, Wilson SM, Campbell R, Burton RA, Fincher GB, Newbigin E, Bacic A (2009) A barley cellulose synthase-like CSLH gene mediates (1,3;1,4)-beta-D-glucan synthesis in transgenic Arabidopsis. *The Proceedings of the National Academy of Sciences USA* 106:5996–6001.

Dong X, Hong Z, Chatterjee J, Kim S, Verma DP (2008) Expression of callose synthase genes and its connection with Npr1 signaling pathway during pathogen infection. *Planta* 229:87–98.

Doong RL, Mohnen D (1998) Solubilization and characterization of a galacturonosyltransferase that synthesizes the pectic polysaccharide homogalacturonan. *Plant Journal* 13:363–374.

Drula E, Garron ML, Dogan S, Lombard V, Henrissat B, Terrapon N (2021) The carbohydrate-active enzyme database: functions and literature. *Nucleic Acids Research* 50(D1):D571–D577. doi:10.1093/nar/gkab1045

Dutton GJ, Storey ID (1953) The isolation of a compound of uridine diphosphate and glucuronic acid from liver. *Biochemical Journal* 53:xxxvii–xxxviii.

Ebert B, Birdseye D, Liwanag AJM, Laursen T, Rennie EA, Guo X, Catena M, Rautengarten C, Stonebloom SH, Gluza P, Pidatala VR, Andersen MCF, Cheetamun R, Mortimer JC, Heazlewood JL, Bacic A, Clausen MH, Willats WGT, Scheller HV (2018) The three members of the Arabidopsis glycosyltransferase family 92 are functional beta-1,4-galactan synthases. *Plant & Cell Physiology* 59:2624–2636.

Edwards ME, Dickson CA, Chengappa S, Sidebottom C, Gidley MJ, Reid JS (1999) Molecular characterisation of a membrane-bound galactosyltransferase of plant cell wall matrix polysaccharide biosynthesis. *Plant Journal* 19:691–697.

Egelund J, Damager I, Faber K, Olsen CE, Ulvskov P, Petersen BL (2008) Functional characterisation of a putative rhamnogalacturonan II specific xylosyltransferase. *FEBS Letters* 582:3217–3222.

Egelund J, Obel N, Ulvskov P, Geshi N, Pauly M, Bacic A, Petersen BL (2007) Molecular characterization of two Arabidopsis thaliana glycosyltransferase mutants, rra1 and rra2, which have a reduced residual arabinose content in a polymer tightly associated with the cellulosic wall residue. *Plant Molecular Biology* 64:439–451.

Egelund J, Petersen BL, Motawia MS, Damager I, Faik A, Olsen CE, Ishii T, Clausen H, Ulvskov P, Geshi N (2006) Arabidopsis thaliana RGXT1 and RGXT2 encode Golgi-localized (1,3)-alpha-D-xylosyltransferases involved in the synthesis of pectic rhamnogalacturonan-II. *Plant Cell* 18:2593–2607.

Ehrlich JJ, Weerts RM, Shome S, Culbertson AT, Honzatko RB, Jernigan RL, Zabotina OA (2021) Xyloglucan xylosyltransferase 1 displays promiscuity toward donor substrates during in vitro reactions. *Plant & Cell Physiology* 62(12):1890–1901. doi:10.1093/pcp/pcab114

Ellinger D, Voigt CA (2014) Callose biosynthesis in Arabidopsis with a focus on pathogen response: what we have learned within the last decade. *Annals of Botany* 114:1349–1358.

Endo S, Pesquet E, Yamaguchi M, Tashiro G, Sato M, Toyooka K, Nishikubo N, Udagawa-Motose M, Kubo M, Fukuda H, Demura T (2009) Identifying new components participating in the secondary cell wall formation of vessel elements in zinnia and Arabidopsis. *Plant Cell* 21:1155–1165.

Engle KA, Amos RA, Yang JY, Glushka J, Atmodjo MA, Tan L, Huang C, Moremen KW, Mohnen D (2022) Multiple Arabidopsis galacturonosyltransferases synthesize polymeric homogalacturonan by oligosaccharide acceptor-dependent or de novo synthesis. *Plant Journal* 109:1441–1456.

Enns LC, Kanaoka MM, Torii KU, Comai L, Okada K, Cleland RE (2005) Two callose synthases, GSL1 and GSL5, play an essential and redundant role in plant and pollen development and in fertility. *Plant Molecular Biology* 58:333–349.

Faik A, Bar-Peled M, DeRocher AE, Zeng W, Perrin RM, Wilkerson C, Raikhel NV, Keegstra K (2000) Biochemical characterization and molecular cloning of an alpha-1,2-fucosyltransferase that catalyzes the last step of cell wall xyloglucan biosynthesis in pea. *Journal of Biological Chemistry* 275:15082–15089.

Faik A, Price NJ, Raikhel NV, Keegstra K (2002) An Arabidopsis gene encoding an alpha-xylosyltransferase involved in xyloglucan biosynthesis. *The Proceedings of the National Academy of Sciences USA* 99:7797–7802.

Farquhar MG, Palade GE (1998) The Golgi apparatus: 100 years of progress and controversy. *Trends in Cell Biology* 8:2–10.

Feijao C, Morreel K, Anders N, Tryfona T, Busse-Wicher M, Kotake T, Boerjan W, Dupree P (2022) Hydroxycinnamic acid-modified xylan side chains and their cross-linking products in rice cell walls are reduced in the Xylosyl arabinosyl substitution of xylan 1 mutant. *Plant Journal* 109(5):1152–1167. doi: 10.1111/tpj.15620

Feingold DS, Neufeld EF, Hassid WZ (1958) Synthesis of a beta-1, 3-linked glucan by extracts of Phaseolus aureus seedlings. *Journal of Biological Chemistry* 233:783–788.

Feingold DS, Neufeld EF, Hassid WZ (1959) Xylosyl transfer catalyzed by an asparagus extract. *Journal of Biological Chemistry* 234:488–489.

Fry S (2000) *The Growing Plant Cell Wall: Chemical and Metabolic Analysis*. Reprint Edition. Caldwell, NJ: The Blackburn Press.

Fry SC, York WS, Albersheim P, Darvill A, Hayashi T, Joseleau J-P, Kato Y, Lorences EP, Maclachlan GA, McNeil M, Mort AJ, Grant Reid JS, Seitz HU, Selvendran RR, Voragen AGJ, White AR (1993) An unambiguous nomenclature for xyloglucan-derived oligosaccharides. *Physiologia Plantarum* 89:1–3.

Fujiki Y, Hubbard AL, Fowler S, Lazarow PB (1982) Isolation of intracellular membranes by means of sodium carbonate treatment: application to endoplasmic reticulum. *Journal of Cell Biology* 93:97–102.

Fullekrug J, Nilsson T (1998) Protein sorting in the Golgi complex. *Biochimica et Biophysica Acta* 1404:77–84.

Galermo AG, Nandita E, Barboza M, Amicucci MJ, Vo TT, Lebrilla CB (2018) Liquid chromatography-tandem mass spectrometry approach for determining glycosidic linkages. *Analytical Chemistry* 90:13073–13080.

Geshi N, Johansen JN, Dilokpimol A, Rolland A, Belcram K, Verger S, Kotake T, Tsumuraya Y, Kaneko S, Tryfona T, Dupree P, Scheller HV, Hofte H, Mouille G (2013) A galactosyltransferase acting on arabinogalactan protein glycans is essential for embryo development in Arabidopsis. *Plant Journal* 76:128–137.

Gille S, Hansel U, Ziemann M, Pauly M (2009) Identification of plant cell wall mutants by means of a forward chemical genetic approach using hydrolases. *The Proceedings of the National Academy of Sciences USA* 106:14699–14704.

Gille S, Sharma V, Baidoo EE, Keasling JD, Scheller HV, Pauly M (2013) Arabinosylation of a Yariv-precipitable cell wall polymer impacts plant growth as exemplified by the Arabidopsis glycosyltransferase mutant ray1. *Molecular Plant* 6:1369–1372.

Gimeno-Ferrer F, Pastor-Cantizano N, Bernat-Silvestre C, Selvi-Martinez P, Vera-Sirera F, Gao C, Perez-Amador MA, Jiang L, Aniento F, Marcote MJ (2017) alpha2-COP is involved in early secretory traffic in Arabidopsis and is required for plant growth. *Journal of Experimental Botany* 68:391–401.

Ginsburg V, Hassid WZ, Stumpf PK (1956) The isolation of uridine diphosphate derivatives of D-glucose, D-galactose, D-xylose, and L-arabinose from mung bean seedlings. *Journal of Biological Chemistry* 223:977–983.

Glaser L (1957) The enzymic synthesis of cellulose by Acetobacter xylinum. *Biochimica et Biophysica Acta* 25:436.

Glaser L, Brown DH (1957) The enzymatic synthesis in vitro of hyaluronic acid chains. *The Proceedings of the National Academy of Sciences USA* 41:253–260.

Glick BS, Elston T, Oster G (1997) A cisternal maturation mechanism can explain the asymmetry of the Golgi stack. *FEBS Letters* 414:177–181.

Golgi C (1989) On the structure of the nerve cells of the spinal ganglia. 1898. *Journal of Microscopy (Oxford)* 155:9–14.

Goubet F, Barton CJ, Mortimer JC, Yu X, Zhang Z, Miles GP, Richens J, Liepman AH, Seffen K, Dupree P (2009) Cell wall glucomannan in Arabidopsis is synthesised by CSLA glycosyltransferases, and influences the progression of embryogenesis. *Plant Journal* 60:527–538.

Goubet F, Jackson P, Deery MJ, Dupree P (2002) Polysaccharide analysis using carbohydrate gel electrophoresis: a method to study plant cell wall polysaccharides and polysaccharide hydrolases. *Analytical Biochemistry* 300:53–68.

Grabenhorst E, Conradt HS (1999) The cytoplasmic, transmembrane, and stem regions of glycosyltransferases specify their in vivo functional sublocalization and stability in the Golgi. *Journal of Biological Chemistry* 274:36107–36116.

Gu Y, Kaplinsky N, Bringmann M, Cobb A, Carroll A, Sampathkumar A, Baskin TI, Persson S, Somerville CR (2010) Identification of a cellulose synthase-associated protein required for cellulose biosynthesis. *The Proceedings of the National Academy of Sciences USA* 107:12866–12871.

Gunl M, Kraemer F, Pauly M (2011) Oligosaccharide mass profiling (OLIMP) of cell wall polysaccharides by MALDI-TOF/MS. *Methods in Molecular Biology* 715:43–54.

Gutierrez R, Lindeboom JJ, Paredez AR, Emons AM, Ehrhardt DW (2009) Arabidopsis cortical microtubules position cellulose synthase delivery to the plasma membrane and interact with cellulose synthase trafficking compartments. *Nature Cell Biology* 11:797–806.

Hagopian A, Eylar EH (1968) Glycoprotein biosynthesis: the basic protein encephalitogen from bovine spinal cord, a receptor for the polypeptidyl: N-acetylgalactosaminyl transferase from bovine submaxillary glands. *Archives of Biochemistry and Biophysics* 126:785–794.

Hanes CS (1940) Enzymic Synthesis of Starch from Glucose-I-Phosphate. *Nature* 145:348–349.

Harholt J, Jensen JK, Sorensen SO, Orfila C, Pauly M, Scheller HV (2006) ARABINAN DEFICIENT 1 is a putative arabinosyltransferase involved in biosynthesis of pectic arabinan in Arabidopsis. *Plant Physiology* 140:49–58.

Harholt J, Jensen JK, Verhertbruggen Y, Sogaard C, Bernard S, Nafisi M, Poulsen CP, Geshi N, Sakuragi Y, Driouich A, Knox JP, Scheller HV (2012) ARAD proteins associated with pectic Arabinan biosynthesis form complexes when transiently overexpressed in planta. *Planta* 236:115–128.

Harvey DJ (1999) Matrix-assisted laser desorption/ionization mass spectrometry of carbohydrates. *Mass Spectrometry Reviews* 18:349–450.

Harvey DJ (2021) Analysis of carbohydrates and glycoconjugates by matrix-assisted laser desorption/ionization mass spectrometry: an update for 2017-2018. *Mass Spectrometry Reviews* 42:227–431. https://doi.org/10.1002/mas.21721

Hassid WZ, Neufeld EF, Feingold DS (1959) Sugar nucleotides in the interconversion of carbohydrates in higher plants. *The Proceedings of the National Academy of Sciences USA* 45:905–915.

Hassinen A, Kellokumpu S (2014) Organizational interplay of Golgi N-glycosyltransferases involves organelle microenvironment-dependent transitions between enzyme homo- and heteromers. *Journal of Biological Chemistry* 289:26937–26948.

Hassinen A, Pujol FM, Kokkonen N, Pieters C, Kihlstrom M, Korhonen K, Kellokumpu S (2011) Functional organization of Golgi N- and O-glycosylation pathways involves pH-dependent complex formation that is impaired in cancer cells. *Journal of Biological Chemistry* 286:38329–38340.

Henrissat B (1991) A classification of glycosyl hydrolases based on amino acid sequence similarities. *Biochemical Journal* 280(Pt 2):309–316.

Henrissat B, Bairoch A (1993) New families in the classification of glycosyl hydrolases based on amino acid sequence similarities. *Biochemical Journal* 293 (Pt 3):781–788.

Henrissat B, Bairoch A (1996) Updating the sequence-based classification of glycosyl hydrolases. *Biochemical Journal* 316 (Pt 2):695–696.

Hill JL, Jr., Hammudi MB, Tien M (2014) The Arabidopsis cellulose synthase complex: a proposed hexamer of CESA trimers in an equimolar stoichiometry. *Plant Cell* 26:4834–4842.

Hoffmann N, King S, Samuels AL, McFarlane HE (2021) Subcellular coordination of plant cell wall synthesis. *Developmental Cell* 56:933–948.

Hong Z, Delauney AJ, Verma DP (2001) A cell plate-specific callose synthase and its interaction with phragmoplastin. *Plant Cell* 13:755–768.

Hsu JW, Chang LC, Jang LT, Huang CF, Lee FJ (2013) The N-terminus of Vps74p is essential for the retention of glycosyltransferases in the Golgi but not for the modulation of apical polarized growth in Saccharomyces cerevisiae. *PLoS One* 8:e74715.

Humphreys IR, Pei J, Baek M, Krishnakumar A, Anishchenko I, Ovchinnikov S, Zhang J, Ness TJ, Banjade S, Bagde SR, Stancheva VG, Li XH, Liu K, Zheng Z, Barrero DJ, Roy U, Kuper J, Fernandez IS, Szakal B, Branzei D, Rizo J, Kisker

C, Greene EC, Biggins S, Keeney S, Miller EA, Fromme JC, Hendrickson TL, Cong Q, Baker D (2021) Computed structures of core eukaryotic protein complexes. *Science* 374:eabm4805.

Iglesias VA, Meins F, Jr. (2000) Movement of plant viruses is delayed in a beta-1,3-glucanase-deficient mutant showing a reduced plasmodesmatal size exclusion limit and enhanced callose deposition. *Plant Journal* 21:157–166.

Ihmels J, Bergmann S, Barkai N (2004) Defining transcription modules using large-scale gene expression data. *Bioinformatics* 20:1993–2003.

Imperiali B, O'Connor SE (1999) Effect of N-linked glycosylation on glycopeptide and glycoprotein structure. *Current Opinion in Chemical Biology* 3:643–649.

Ishii T, Ichita J, Matsue H, Ono H, Maeda I (2002) Fluorescent labeling of pectic oligosaccharides with 2-aminobenzamide and enzyme assay for pectin. *Carbohydrate Research* 337:1023–1032.

Iwai H, Masaoka N, Ishii T, Satoh S (2002) A pectin glucuronyltransferase gene is essential for intercellular attachment in the plant meristem. *The Proceedings of the National Academy of Sciences USA* 99:16319–16324.

Jacobs AK, Lipka V, Burton RA, Panstruga R, Strizhov N, Schulze-Lefert P, Fincher GB (2003) An arabidopsis callose synthase, GSL5, is required for wound and papillary callose formation. *Plant Cell* 15:2503–2513.

Jensen JK, Busse-Wicher M, Poulsen CP, Fangel JU, Smith PJ, Yang JY, Pena MJ, Dinesen MH, Martens HJ, Melkonian M, Wong GK, Moremen KW, Wilkerson CG, Scheller HV, Dupree P, Ulvskov P, Urbanowicz BR, Harholt J (2018) Identification of an algal xylan synthase indicates that there is functional orthology between algal and plant cell wall biosynthesis. *New Phytologist* 218:1049–1060.

Jensen JK, Johnson NR, Wilkerson CG (2014) Arabidopsis thaliana IRX10 and two related proteins from psyllium and Physcomitrella patens are xylan xylosyltransferases. *Plant Journal* 80:207–215.

Jensen JK, Schultink A, Keegstra K, Wilkerson CG, Pauly M (2012) RNA-Seq analysis of developing nasturtium seeds (Tropaeolum majus): identification and characterization of an additional galactosyltransferase involved in xyloglucan biosynthesis. *Molecular Plant* 5:984–992.

Jensen JK, Sorensen SO, Harholt J, Geshi N, Sakuragi Y, Moller I, Zandleven J, Bernal AJ, Jensen NB, Sorensen C, Pauly M, Beldman G, Willats WG, Scheller HV (2008) Identification of a xylogalacturonan xylosyltransferase involved in pectin biosynthesis in Arabidopsis. *Plant Cell* 20:1289–1302.

Jiang N, Wiemels RE, Soya A, Whitley R, Held M, Faik A (2016) Composition, assembly, and trafficking of a wheat xylan synthase complex. *Plant Physiology* 170:1999–2023.

Jobling SA (2015) Membrane pore architecture of the CslF6 protein controls (1-3,1-4)-b-glucan structure. *Science Advances* 1:e1500069

Jones L, Ennos AR, Turner SR (2001) Cloning and characterization of irregular xylem4 (irx4): a severely lignin-deficient mutant of Arabidopsis. *Plant Journal* 26:205–216.

Jumper J, Evans R, Pritzel A, Green T, Figurnov M, Ronneberger O, Tunyasuvunakool K, Bates R, Zidek A, Potapenko A, Bridgland A, Meyer C, Kohl SAA, Ballard AJ, Cowie A, Romera-Paredes B, Nikolov S, Jain R, Adler J, Back T, Petersen S, Reiman D, Clancy E, Zielinski M, Steinegger M, Pacholska M, Berghammer T, Silver D, Vinyals O, Senior AW, Kavukcuoglu K, Kohli P, Hassabis D (2021) Applying and improving alphafold at CASP14. *Proteins* 89:1711–1721.

Kalckar HM (1954) The enzymes of uridine polyphosphates. *Science* 119:479–480.

Kaul S, Koo HL, Jenkins J, Rizzo M, Rooney T, Tallon LJ, Feldblyum T, Nierman W, Benito MI, Lin XY, Town CD, Venter JC, Fraser CM, Tabata S, Nakamura Y, Kaneko T, Sato S, Asamizu E, Kato T, Kotani H, Sasamoto S, Ecker JR, Theologis A, Federspiel NA, Palm CJ, Osborne BI, Shinn P, Conway AB, Vysotskaia VS, Dewar K, Conn L, Lenz CA, Kim CJ, Hansen NF, Liu SX, Buehler E, Altafi H, Sakano H, Dunn P, Lam B, Pham PK, Chao Q, Nguyen M, Yu GX, Chen HM, Southwick A, Lee JM, Miranda M, Toriumi MJ, Davis RW, Wambutt R, Murphy G, Dusterhoft A, Stiekema W, Pohl T, Entian KD, Terryn N, Volckaert G, Salanoubat M, Choisne N, Rieger M, Ansorge W, Unseld M, Fartmann B, Valle G, Artiguenave F, Weissenbach J, Quetier F, Wilson RK, de la Bastide M, Sekhon M, Huang E, Spiegel L, Gnoj L, Pepin K, Murray J, Johnson D, Habermann K, Dedhia N, Parnell L, Preston R, Hillier L, Chen E, Marra M, Martienssen R, McCombie WR, Mayer K, White O, Bevan M, Lemcke K, Creasy TH, Bielke C, Haas B, Haase D, Maiti R, Rudd S, Peterson J, Schoof H, Frishman D, Morgenstern B, Zaccaria P, Ermolaeva M, Pertea M, Quackenbush J, Volfovsky N, Wu DY, Lowe TM, Salzberg SL, Mewes HW, Rounsley S, Bush D, Subramaniam S, Levin I, Norris S, Schmidt R, Acarkan A, Bancroft I, Quetier F, Brennicke A, Eisen JA, Bureau T, Legault BA, Le QH, Agrawal N, Yu Z, Martienssen R, Copenhaver GP, Luo S, Pikaard CS, Preuss D, Paulsen IT, Sussman M, Britt AB, Selinger DA, Pandey R, Mount DW, Chandler VL, Jorgensen RA, Pikaard C, Juergens G, Meyerowitz EM, Theologis A, Dangl J, Jones JDG, Chen M, Chory J, Somerville MC, In AG (2000) Analysis of the genome sequence of the flowering plant Arabidopsis thaliana. *Nature* 408:796–815.

Kellokumpu S (2019) Golgi pH, ion and redox homeostasis: how much do they really matter? *Frontiers in Cell and Developmental Biology* 7:93.

Kellokumpu S, Hassinen A, Glumoff T (2016) Glycosyltransferase complexes in eukaryotes: long-known, prevalent but still unrecognized. *Cellular and Molecular Life Sciences* 73:305–325.

Khoder-Agha F, Sosicka P, Escriva Conde M, Hassinen A, Glumoff T, Olczak M, Kellokumpu S (2019) N-acetylglucosaminyltransferases and nucleotide sugar transporters form multi-enzyme-multi-transporter assemblies in Golgi membranes in vivo. *Cellular and Molecular Life Sciences* 76:1821–1832.

Kim SJ, Brandizzi F (2016) The plant secretory pathway for the trafficking of cell wall polysaccharides and glycoproteins. *Glycobiology* 26:940–949.

Kim SJ, Chandrasekar B, Rea AC, Danhof L, Zemelis-Durfee S, Thrower N, Shepard ZS, Pauly M, Brandizzi F, Keegstra K (2020) The synthesis of xyloglucan, an abundant plant cell wall polysaccharide, requires CSLC function. *The Proceedings of the National Academy of Sciences USA* 117:20316–20324.

Kim SJ, Zemelis S, Keegstra K, Brandizzi F (2015) The cytoplasmic localization of the catalytic site of CSLF6 supports a channeling model for the biosynthesis of mixed-linkage glucan. *Plant Journal* 81:537–547.

Kimura S, Laosinchai W, Itoh T, Cui X, Linder CR, Brown RM, Jr. (1999) Immunogold labeling of rosette terminal cellulose-synthesizing complexes in the vascular plant vigna angularis. *Plant Cell* 11:2075–2086.

Kinnaert C, Daugaard M, Nami F, Clausen MH (2017) Chemical synthesis of oligosaccharides related to the cell walls of plants and algae. *Chemical Reviews* 117:11337–11405.

Knoch E, Dilokpimol A, Tryfona T, Poulsen CP, Xiong G, Harholt J, Petersen BL, Ulvskov P, Hadi MZ, Kotake T, Tsumuraya Y, Pauly M, Dupree P, Geshi N (2013) A beta-glucuronosyltransferase from Arabidopsis thaliana involved in biosynthesis of type II arabinogalactan has a role in cell elongation during seedling growth. *Plant Journal* 76:1016–1029.

Kokubo A, Kuraishi S, Sakurai N (1989) Culm strength of barley: correlation among maximum bending stress, cell wall dimensions, and cellulose content. *Plant Physiology* 91:876–882.

Kokubo A, Sakurai N, Kuraishi S, Takeda K (1991) Culm brittleness of barley (Hordeum vulgare L.) mutants is caused by smaller number of cellulose molecules in cell wall. *Plant Physiology* 97:509–514.

Kong Y, Pena MJ, Renna L, Avci U, Pattathil S, Tuomivaara ST, Li X, Reiter WD, Brandizzi F, Hahn MG, Darvill AG, York WS, O'Neill MA (2015) Galactose-depleted xyloglucan is dysfunctional and leads to dwarfism in Arabidopsis. *Plant Physiology* 167:1296–1306.

Konishi T, Ono H, Ohnishi-Kameyama M, Kaneko S, Ishii T (2006) Identification of a mung bean arabinofuranosyltransferase that transfers arabinofuranosyl residues onto (1, 5)-linked alpha-L-arabino-oligosaccharides. *Plant Physiology* 141:1098–1105.

Kornfeld R, Kornfeld S (1985) Assembly of asparagine-linked oligosaccharides. *Annual Review of Biochemistry* 54:631–664.

Lairson LL, Henrissat B, Davies GJ, Withers SG (2008) Glycosyltransferases: structures, functions, and mechanisms. *Annual Review of Biochemistry* 77:521–555.

Lao J, Oikawa A, Bromley JR, McInerney P, Suttangkakul A, Smith-Moritz AM, Plahar H, Chiu TY, Gonzalez Fernandez-Nino SM, Ebert B, Yang F, Christiansen KM, Hansen SF, Stonebloom S, Adams PD, Ronald PC, Hillson NJ, Hadi MZ, Vega-Sanchez ME, Loque D, Scheller HV, Heazlewood JL (2014) The plant glycosyltransferase clone collection for functional genomics. *Plant Journal* 79:517–529.

Laursen T, Stonebloom SH, Pidatala VR, Birdseye DS, Clausen MH, Mortimer JC, Scheller HV (2018) Bifunctional glycosyltransferases catalyze both extension and termination of pectic galactan oligosaccharides. *Plant Journal* 94:340–351.

Lee C, O'Neill MA, Tsumuraya Y, Darvill AG, Ye ZH (2007) The irregular xylem9 mutant is deficient in xylan xylosyltransferase activity. *Plant & Cell Physiology* 48:1624–1634.

Lee C, Zhong R, Ye ZH (2012) Arabidopsis family GT43 members are xylan xylosyltransferases required for the elongation of the xylan backbone. *Plant & Cell Physiology* 53:135–143.

Leloir LF, Cabib E (1953) The enzymic synthesis of trehalose phosphate1. *Journal of the American Chemical Society* 75:5445–5446.

Leloir LF, Cardini CE (1955) The biosynthesis of sucrose phosphate. *Journal of Biological Chemistry* 214:157-165.

Levengood MR, Splain RA, Kiessling LL (2011) Monitoring processivity and length control of a carbohydrate polymerase. *Journal of the American Chemical Society* 133:12758–12766.

Li S, Wang S, Fu X, Liu XW, Wang PG, Fang J (2017) Sequential one-pot multienzyme synthesis of hyaluronan and its derivative. *Carbohydrate Polymers* 178:221–227.

Liepman AH, Wilkerson CG, Keegstra K (2005) Expression of cellulose synthase-like (Csl) genes in insect cells reveals that CslA family members encode mannan synthases. *The Proceedings of the National Academy of Sciences USA* 102:2221–2226.

Little A, Lahnstein J, Jeffery DW, Khor SF, Schwerdt JG, Shirley NJ, Hooi M, Xing X, Burton RA, Bulone V (2019) A novel (1,4)-beta-linked glucoxylan is synthesized by members of the cellulose synthase-like F gene family in land plants. *ACS Central Science* 5:73–84.

Little A, Schwerdt JG, Shirley NJ, Khor SF, Neumann K, O'Donovan LA, Lahnstein J, Collins HM, Henderson M, Fincher GB, Burton RA (2018) Revised phylogeny of the cellulose synthase gene superfamily: insights into cell wall evolution. *Plant Physiology* 177:1124-1141.

Liu XL, Liu L, Niu QK, Xia C, Yang KZ, Li R, Chen LQ, Zhang XQ, Zhou Y, Ye D (2011) Male gametophyte defective 4 encodes a rhamnogalacturonan II xylosyltransferase and is important for growth of pollen tubes and roots in Arabidopsis. *Plant Journal* 65:647–660.

Liwanag AJ, Ebert B, Verhertbruggen Y, Rennie EA, Rautengarten C, Oikawa A, Andersen MC, Clausen MH, Scheller HV (2012) Pectin biosynthesis: GALS1 in Arabidopsis thaliana is a beta-1,4-galactan beta-1,4-galactosyltransferase. *Plant Cell* 24:5024–5036.

Lombard V, Golaconda Ramulu H, Drula E, Coutinho PM, Henrissat B (2014) The carbohydrate-active enzymes database (CAZy) in 2013. *Nucleic Acids Research* 42:D490–495.

Luis AS, Briggs J, Zhang X, Farnell B, Ndeh D, Labourel A, Basle A, Cartmell A, Terrapon N, Stott K, Lowe EC, McLean R, Shearer K, Schuckel J, Venditto I, Ralet MC, Henrissat B, Martens EC, Mosimann SC, Abbott DW, Gilbert HJ (2018) Dietary pectic glycans are degraded by coordinated enzyme pathways in human colonic Bacteroides. *Nature Microbiology* 3:210–219.

Lunin VV, Wang HT, Bharadwaj VS, Alahuhta M, Pena MJ, Yang JY, Archer-Hartmann SA, Azadi P, Himmel ME, Moremen KW, York WS, Bomble YJ, Urbanowicz BR (2020) Molecular mechanism of polysaccharide acetylation by the arabidopsis xylan O-acetyltransferase XOAT1. *Plant Cell* 32:2367–2382.

Madson M, Dunand C, Li X, Verma R, Vanzin GF, Caplan J, Shoue DA, Carpita NC, Reiter WD (2003) The MUR3 gene of Arabidopsis encodes a xyloglucan galactosyltransferase that is evolutionarily related to animal exostosins. *Plant Cell* 15:1662–1670.

Malhotra V, Serafini T, Orci L, Shepherd JC, Rothman JE (1989) Purification of a novel class of coated vesicles mediating biosynthetic protein transport through the Golgi stack. *Cell* 58:329–336

Mansoori N, Timmers J, Desprez T, Alvim-Kamei CL, Dees DC, Vincken JP, Visser RG, Hofte H, Vernhettes S, Trindade LM (2014) KORRIGAN1 interacts specifically with integral components of the cellulose synthase machinery. *PLoS One* 9:e112387.

Mariette A, Kang HS, Heazlewood JL, Persson S, Ebert B, Lampugnani ER (2021) Not just a simple sugar: arabinose metabolism and function in plants. *Plant & Cell Physiology* 62(12):1791–1812. https://doi.org/10.1093/pcp/pcab087

McCormick S (1993) Male gametophyte development. *Plant Cell* 5:1265–1275.

McFarlane HE, Doring A, Persson S (2014) The cell biology of cellulose synthesis. *Annual Review of Plant Biology* 65:69–94.

Meents MJ, Motani S, Mansfield SD, Samuels AL (2019) Organization of xylan production in the golgi during secondary cell wall biosynthesis. *Plant Physiology* 181:527–546.

Melancon P, Glick BS, Malhotra V, Weidman PJ, Serafini T, Gleason ML, Orci L, Rothman JE (1987) Involvement of GTP-binding "G" proteins in transport through the Golgi stack. *Cell* 51:1053–1062.

Mitchell RA, Dupree P, Shewry PR (2007) A novel bioinformatics approach identifies candidate genes for the synthesis and feruloylation of arabinoxylan. *Plant Physiology* 144:43–53.

Moller SR, Yi X, Velasquez SM, Gille S, Hansen PLM, Poulsen CP, Olsen CE, Rejzek M, Parsons H, Yang Z, Wandall HH, Clausen H, Field RA, Pauly M, Estevez JM, Harholt J, Ulvskov P, Petersen BL (2017) Identification and evolution of a plant cell wall specific glycoprotein glycosyl transferase, ExAD. *Scientific Reports* 7:45341.

Moore PJ, Swords KM, Lynch MA, Staehelin LA (1991) Spatial organization of the assembly pathways of glycoproteins and complex polysaccharides in the Golgi apparatus of plants. *Journal of Cell Biology* 112:589–602.

Moremen KW, Haltiwanger RS (2019) Emerging structural insights into glycosyltransferase-mediated synthesis of glycans. *Nature Chemical Biology* 15:853–864.

Mortimer JC, Faria-Blanc N, Yu X, Tryfona T, Sorieul M, Ng YZ, Zhang Z, Stott K, Anders N, Dupree P (2015) An unusual xylan in Arabidopsis primary cell walls is synthesised by GUX3, IRX9L, IRX10L and IRX14. *Plant Journal* 83:413–426.

Mortimer JC, Miles GP, Brown DM, Zhang Z, Segura MP, Weimar T, Yu X, Seffen KA, Stephens E, Turner SR, Dupree P (2010) Absence of branches from xylan in Arabidopsis gux mutants reveals potential for simplification of lignocellulosic biomass. *The Proceedings of the National Academy of Sciences USA* 107:17409–17414.

Mueller SC, Brown RM, Jr., Scott TK (1976) Cellulosic microfibrils: nascent stages of synthesis in a higher plant cell. *Science* 194:949–951.

Muthana MM, Qu J, Xue M, Klyuchnik T, Siu A, Li Y, Zhang L, Yu H, Li L, Wang PG, Chen X (2015) Improved one-pot multienzyme (OPME) systems for synthesizing UDP-uronic acids and glucuronides. *Chemical Communications* 51:4595-4598.

Mutwil M, Klie S, Tohge T, Giorgi FM, Wilkins O, Campbell MM, Fernie AR, Usadel B, Nikoloski Z, Persson S (2011) PlaNet: combined sequence and expression comparisons across plant networks derived from seven species. *Plant Cell* 23:895–910.

Narciso JO, Zeng W, Ford K, Lampugnani ER, Humphries J, Austarheim I, van de Meene A, Bacic A, Doblin MS (2021) Biochemical and functional characterization of GALT8, an Arabidopsis GT31 beta-(1,3)-Galactosyltransferase that influences seedling development. *Frontiers in Plant Science* 12:678564.

Nedukha OM (2015) Callose: localization, functions, and synthesis in plant cells. *Cytology and Genetics* 49:49–57.

Nikaido Ha H, W.Z. (1971) Biosynthesis of saccharides from glycopyranosyl esters of nucleoside pyrophosphates sugar nucleotides. *Advances in Carbohydrate Chemistry and Biochemistry*, 26, 351–483.

Nilsson T, Au CE, Bergeron JJ (2009) Sorting out glycosylation enzymes in the Golgi apparatus. *FEBS Letters* 583:3764–3769.

Nilsson T, Pypaert M, Hoe MH. Slusarewicz P, Berger EG, Warren G (1993) Overlapping distribution of two glycosyltransferases in the Golgi apparatus of HeLa cells. *Journal of Cell Biology* 120:5–13.

Nishimura MT, Stein M, Hou BH, Vogel JP, Edwards H, Somerville SC (2003) Loss of a callose synthase results in salicylic acid-dependent disease resistance. *Science* 301:969–972.

Nixon BT, Mansouri K, Singh A, Du J, Davis JK, Lee JG, Slabaugh E, Vandavasi VG, O'Neill H, Roberts EM, Roberts AW, Yingling YG, Haigler CH (2016) Comparative structural and computational analysis supports eighteen cellulose synthases in the plant cellulose synthesis complex. *Scientific Reports* 6:28696.

Northcote DH (1969) The synthesis and metabolic control of polysaccharides and lignin during the differentiation of plant cells. *Essays in Biochemistry* 5:89–137.

Ogawa-Ohnishi M, Matsubayashi Y (2015) Identification of three potent hydroxyproline O-galactosyltransferases in Arabidopsis. *Plant Journal* 81:736–746.

Ogawa-Ohnishi M, Matsushita W, Matsubayashi Y (2013) Identification of three hydroxyproline O-arabinosyltransferases in Arabidopsis thaliana. *Nature Chemical Biology* 9:726–730.

Oikawa A, Joshi HJ, Rennie EA, Ebert B, Manisseri C, Heazlewood JL, Scheller HV (2010) An integrative approach to the identification of Arabidopsis and rice genes involved in xylan and secondary wall development. *PLoS One* 5:e15481.

Oikawa A, Lund CH, Sakuragi Y, Scheller HV (2013) Golgi-localized enzyme complexes for plant cell wall biosynthesis. *Trends in Plant Science* 18:49–58.

Parsons HT, Stevens TJ, McFarlane HE, Vidal-Melgosa S, Griss J, Lawrence N, Butler R, Sousa MML, Salemi M, Willats WGT, Petzold CJ, Heazlewood JL, Lilley KS (2019) Separating Golgi proteins from CIS to trans reveals underlying properties of cisternal localization. *Plant Cell* 31:2010–2034.

Pauly M, Keegstra K (2016) Biosynthesis of the plant cell wall matrix polysaccharide xyloglucan. *Annual Review of Plant Biology* 67:235–259.

Pear JR, Kawagoe Y, Schreckengost WE, Delmer DP, Stalker DM (1996) Higher plants contain homologs of the bacterial celA genes encoding the catalytic subunit of cellulose synthase. *The Proceedings of the National Academy of Sciences USA* 93:12637–12642.

Pena MJ, Kong Y, York WS, O'Neill MA (2012) A galacturonic acid-containing xyloglucan is involved in Arabidopsis root hair tip growth. *Plant Cell* 24:4511–4524.

Pena MJ, Zhong R, Zhou GK, Richardson EA, O'Neill MA, Darvill AG, York WS, Ye ZH (2007) Arabidopsis irregular xylem8 and irregular xylem9: implications for the complexity of glucuronoxylan biosynthesis. *Plant Cell* 19:549–563.

Peng JS, Zhang BC, Chen H, Wang MQ, Wang YT, Li HM, Cao SX, Yi HY, Wang H, Zhou YH, Gong JM (2021) Galactosylation of rhamnogalacturonan-II for cell wall pectin biosynthesis is critical for root apoplastic iron reallocation in Arabidopsis. *Molecular Plant* 14:1640–1651.

Perrin RM, DeRocher AE, Bar-Peled M, Zeng W, Norambuena L, Orellana A, Raikhel NV, Keegstra K (1999) Xyloglucan fucosyltransferase, an enzyme involved in plant cell wall biosynthesis. *Science* 284:1976–1979.

Persson S, Wei H, Milne J, Page GP, Somerville CR (2005) Identification of genes required for cellulose synthesis by regression analysis of public microarray data sets. *The Proceedings of the National Academy of Sciences USA* 102:8633–8638.

Petersen BL, Egelund J, Damager I, Faber K, Jensen JK, Yang Z, Bennett EP, Scheller HV, Ulvskov P (2009) Assay and heterologous expression in Pichia pastoris of plant cell wall type-II membrane anchored glycosyltransferases. *Glycoconjugate Journal* 26:1235–1246.

Pettolino FA, Walsh C, Fincher GB, Bacic A (2012) Determining the polysaccharide composition of plant cell walls. *Nature Protocols* 7:1590–1607.

Pfrengle F (2017) Synthetic plant glycans. *Current Opinion in Chemical Biology* 40:145–151.

Prabhakar PK, Wang HT, Smith PJ, Yang JY, Barnes WJ, Pena MJ, Moremen KW, Urbanowicz BR (2020) Heterologous expression of plant glycosyltransferases for biochemistry and structural biology. *Methods in Cell Biology* 160:145–165.

Purushotham P, Cho SH, Diaz-Moreno SM, Kumar M, Nixon BT, Bulone V, Zimmer J (2016) A single heterologously expressed plant cellulose synthase isoform is sufficient for cellulose microfibril formation in vitro. *The Proceedings of the National Academy of Sciences USA* 113:11360–11365.

Purushotham P, Ho R, Zimmer J (2020) Architecture of a catalytically active homotrimeric plant cellulose synthase complex. *Science* 369:1089–1094.

Qin LX, Chen Y, Zeng W, Li Y, Gao L, Li DD, Bacic A, Xu WL, Li XB (2017) The cotton beta-galactosyltransferase 1 (GalT1) that galactosylates arabinogalactan proteins participates in controlling fiber development. *Plant Journal* 89:957–971.

Qu Y, Egelund J, Gilson PR, Houghton F, Gleeson PA, Schultz CJ, Bacic A (2008) Identification of a novel group of putative Arabidopsis thaliana beta-(1,3)-galactosyltransferases. *Plant Molecular Biology* 68:43–59.

Rabouille C, Hui N, Hunte F, Kieckbusch R, Berger EG, Warren G, Nilsson T (1995) Mapping the distribution of Golgi enzymes involved in the construction of complex oligosaccharides. *Journal of Cell Science* 108 (Pt 4):1617–1627.

Rambourg A, Clermont Y (1990) Three-dimensional electron microscopy: structure of the Golgi apparatus. *European Journal of Cell Biology* 51:189–200.

Rautengarten C, Birdseye D, Pattathil S, McFarlane HE, Saez-Aguayo S, Orellana A, Persson S, Hahn MG, Scheller HV, Heazlewood JL, Ebert B (2017) The elaborate route for UDP-arabinose delivery into the Golgi of plants. *The Proceedings of the National Academy of Sciences USA* 114:4261–4266.

Rautengarten C, Ebert B, Herter T, Petzold CJ, Ishii T, Mukhopadhyay A, Usadel B, Scheller HV (2011) The interconversion of UDP-arabinopyranose and UDP-arabinofuranose is indispensable for plant development in Arabidopsis. *Plant Cell* 23:1373–1390.

Rautengarten C, Heazlewood JL, Ebert B (2019) Profiling cell wall monosaccharides and nucleotide-sugars from Plants. *Current Protocols in Plant Biology* 4:e20092.

Ray PM, Shininger TL, Ray MM (1969) Isolation of beta-glucan synthetase particles from plant cells and identification with Golgi membranes. *The Proceedings of the National Academy of Sciences USA* 64:605–612.

Rayner JC, Pelham HR (1997) Transmembrane domain-dependent sorting of proteins to the ER and plasma membrane in yeast. *EMBO Journal* 16:1832–1841.

Reiter WD, Chapple CC, Somerville CR (1993) Altered growth and cell walls in a fucose-deficient mutant of Arabidopsis. *Science* 261:1032–1035.

Reiter WD, Chapple C, Somerville CR (1997) Mutants of Arabidopsis thaliana with altered cell wall polysaccharide composition. *Plant Journal* 12:335–345.

Ren Y, Hansen SF, Ebert B, Lau J, Scheller HV (2014) Site-directed mutagenesis of IRX9, IRX9L and IRX14 proteins involved in xylan biosynthesis: glycosyltransferase activity is not required for IRX9 function in Arabidopsis. *PLoS One* 9:e105014.

Rennie EA, Hansen SF, Baidoo EE, Hadi MZ, Keasling JD, Scheller HV (2012) Three members of the Arabidopsis glycosyltransferase family 8 are xylan glucuronosyltransferases. *Plant Physiology* 159:1408–1417.

Rennie EA, Scheller HV (2014) Xylan biosynthesis. *Current Opinion in Biotechnology* 26:100–107.

Robinson DG, White RK, Preston RD (1972) Fine structure of swarmers of Cladophora and Chaetomorpha: III. Wall synthesis and development. *Planta* 107:131–144.

Robyt JF (2001) Polysaccharides: energy storage. *eLS*. https://onlinelibrary.wiley.com/action/showCitFormats?doi=10.1038%2Fnpg.els.0000700

Rocha J, Ciceron F, Lerouxel O, Breton C, de Sanctis D (2016) The galactoside 2-alpha-L-fucosyltransferase FUT1 from Arabidopsis thaliana: crystallization and experimental MAD phasing. *Acta Crystallographica Section F: Structural Biology Communications* 72:564–568.

Rongpipi S, Ye D, Gomez ED, Gomez EW (2018) Progress and opportunities in the characterization of cellulose - an important regulator of cell wall growth and mechanics. *Frontiers in Plant Science* 9:1894.

Roth J (1997) Topology of glycosylation in the Golgi apparatus. In EG Berger, J Roth, eds., *The Golgi Apparatus*, pp 131–161. Basel: Birkhäuser Basel.

Rothman JE, Orci L (1990) Movement of proteins through the Golgi stack: a molecular dissection of vesicular transport. *FASEB Journal* 4:1460–1468.

Rottger S, White J, Wandall HH, Olivo JC, Stark A, Bennett EP, Whitehouse C, Berger EG, Clausen H, Nilsson T (1998) Localization of three human polypeptide GalNAc-transferases in HeLa cells suggests initiation of O-linked glycosylation throughout the Golgi apparatus. *Journal of Cell Science* 111 (Pt 1):45–60.

Ruprecht C, Bartetzko MP, Senf D, Lakhina A, Smith PJ, Soto MJ, Oh H, Yang JY, Chapla D, Varon Silva D, Clausen MH, Hahn MG, Moremen KW, Urbanowicz BR, Pfrengle F (2020) A glycan array-based assay for the identification and characterization of plant glycosyltransferases. *Angewandte Chemie, International Edition* 59:12493–12498.

Ruprecht C, Dallabernardina P, Smith PJ, Urbanowicz BR, Pfrengle F (2018) Analyzing Xyloglucan Endotransglycosylases by Incorporating Synthetic Oligosaccharides into Plant Cell Walls. *ChemBioChem* 19:793–798.

Saito F, Suyama A, Oka T, Yoko OT, Matsuoka K, Jigami Y, Shimma YI (2014) Identification of novel peptidyl serine alpha-galactosyltransferase gene family in plants. *Journal of Biological Chemistry* 289:20405–20420.

Samek W, Montavon G, Lapuschkin S, Anders CJ, Muller KR (2021) Explaining deep neural networks and beyond: a review of methods and applications. *Proceedings of the IEEE* 109:247–278.

Samson F, Brunaud V, Balzergue S, Dubreucq B, Lepiniec L, Pelletier G, Caboche M, Lecharny A (2002) FLAGdb/FST: a database of mapped flanking insertion sites (FSTs) of Arabidopsis thaliana T-DNA transformants. *Nucleic Acids Research* 30:94–97.

Samuels AL, Giddings TH, Jr., Staehelin LA (1995) Cytokinesis in tobacco BY-2 and root tip cells: a new model of cell plate formation in higher plants. *Journal of Cell Biology* 130:1345–1357.

Schachter HV, Jabbal I, Hudgin RL, Pinteric L, McGuire EJ, Roseman S (1970) Intracellular localization of liver sugar nucleotide glycoprotein glycosyltransferases in a Golgi-rich fraction. *Journal of Biological Chemistry* 245:1090–1100.

Scheller HV, Ulvskov P (2010) Hemicelluloses. *Annual Review of Plant Biology* 61:263–289.

Schmid M, Davison TS, Henz SR, Pape UJ, Demar M, Vingron M, Scholkopf B, Weigel D, Lohmann JU (2005) A gene expression map of Arabidopsis thaliana development. *Nature Genetics* 37:501–506.

Schultink A, Cheng K, Park YB, Cosgrove DJ, Pauly M (2013) The identification of two arabinosyltransferases from tomato reveals functional equivalency of xyloglucan side chain substituents. *Plant Physiology* 163:86–94.

Schultink A, Liu L, Zhu L, Pauly M (2014) Structural diversity and function of xyloglucan sidechain substituents. *Plants (Basel)* 3:526–542.

Seifert GJ (2004) Nucleotide sugar interconversions and cell wall biosynthesis: how to bring the inside to the outside. *Current Opinion in Plant Biology* 7:277–284.

Seifert GJ (2018) Mad moves of the building blocks - nucleotide sugars find unexpected paths into cell walls. *Journal of Experimental Botany* 69:905–907.

Sethaphong L, Haigler CH, Kubicki JD, Zimmer J, Bonetta D, DeBolt S, Yingling YG (2013) Tertiary model of a plant cellulose synthase. *The Proceedings of the National Academy of Sciences USA* 110:7512–7517.

Sharma P, Singh A (2017) Era of deep neural networks: a review. *2017 8th International Conference on Computing, Communication and Networking Technologies (Icccnt)*, Dehli.

Sharpe HJ, Stevens TJ, Munro S (2010) A comprehensive comparison of transmembrane domains reveals organelle-specific properties. *Cell* 142:158–169.

Sheikh MO, Halmo SM, Patel S, Middleton D, Takeuchi H, Schafer CM, West CM, Haltiwanger RS, Avci FY, Moremen KW, Wells L (2017) Rapid screening of sugar-nucleotide donor specificities of putative glycosyltransferases. *Glycobiology* 27:206–212.

Smith PJ, Curry TM, Yang J-Y, Barnes WJ, Ziegler SJ, Mittal A, Moremen KW, York WS, Bomble YJ, Peña MJ, Urbanowicz BR (2022) Enzymatic synthesis of xylan microparticles with tunable morphologies. *ACS Materials Au* 2(4):440–452.

Solms J, Hassid WZ (1957) Isolation of uridine diphosphate N-acetylglucosamine and uridine diphosphate glucuronic acid from mung bean seedlings. *Journal of Biological Chemistry* 228:357–364.

Song D, Shen J, Li L (2010) Characterization of cellulose synthase complexes in Populus xylem differentiation. *New Phytologist* 187:777–790.

Sprong H, Degroote S, Nilsson T, Kawakita M, Ishida N, van der Sluijs P, van Meer G (2003) Association of the Golgi UDP-galactose transporter with UDP-galactose:ceramide galactosyltransferase allows UDP-galactose import in the endoplasmic reticulum. *Molecular Biology of the Cell* 14:3482–3493.

Sterling JD, Atmodjo MA, Inwood SE, Kumar Kolli VS, Quigley HF, Hahn MG, Mohnen D (2006) Functional identification of an Arabidopsis pectin biosynthetic homogalacturonan galacturonosyltransferase. *The Proceedings of the National Academy of Sciences USA* 103:5236–5241.

Stone BA (1992) Chemistry and physiology of higher plant 1,3-β-glucanase (callose). In BA Stone AE Clark, eds., *Chemistry and Biology of (1-3)-β-Glucans*. Bundoora: La Trobe University Press.

Stone BA (2006) *Callose and Related Glucans*. eLS, (Ed.). Chichester: John Wiley & Sons, Ltd Online library. doi:10.1038/npg.els.0004111

Stone BA, Hrmova M, Burton RA, Fincher GB (2010) *Biosynthesis of Plant Cell Wall and Related Polysaccharides by Enzymes of the GT2 and GT48 Families*. Oxford: John Wiley & Sons Ltd.

Stonebloom S, Ebert B, Xiong G, Pattathil S, Birdseye D, Lao J, Pauly M, Hahn MG, Heazlewood JL, Scheller HV (2016) A DUF-246 family glycosyltransferase-like gene affects male fertility and the biosynthesis of pectic arabinogalactans. *BMC Plant Biology* 16:90.

Strasser R, Seifert G, Doblin MS, Johnson KL, Ruprecht C, Pfrengle F, Bacic A, Estevez JM (2021) Cracking the "Sugar Code": a snapshot of N- and O-glycosylation pathways and functions in plants cells. *Frontiers in Plant Science* 12:640919.

Stuart JM, Segal E, Koller D, Kim SK (2003) A gene-coexpression network for global discovery of conserved genetic modules. *Science* 302:249–255.

Suzuki T, Narciso JO, Zeng W, van de Meene A, Yasutomi M, Takemura S, Lampugnani ER, Doblin MS, Bacic A, Ishiguro S (2017) KNS4/UPEX1: a type II Arabinogalactan beta-(1,3)-galactosyltransferase required for pollen exine development. *Plant Physiology* 173:183–205.

Szyjanowicz PM, McKinnon I, Taylor NG, Gardiner J, Jarvis MC, Turner SR (2004) The irregular xylem 2 mutant is an allele of korrigan that affects the secondary cell wall of Arabidopsis thaliana. *Plant Journal* 37:730–740.

Takenaka Y, Kato K, Ogawa-Ohnishi M, Tsuruhama K, Kajiura H, Yagyu K, Takeda A, Takeda Y, Kunieda T, Hara-Nishimura I, Kuroha T, Nishitani K, Matsubayashi Y, Ishimizu T (2018) Pectin RG-I rhamnosyltransferases represent a novel plant-specific glycosyltransferase family. *Nature Plants* 4:669–676.

Tamura K, Shimada T, Kondo M, Nishimura M, Hara-Nishimura I (2005) KATAMARI1/MURUS3 Is a novel golgi membrane protein that is required for endomembrane organization in Arabidopsis. *Plant Cell* 17:1764–1776.

Taujale R, Venkat A, Huang LC, Zhou Z, Yeung W, Rasheed KM, Li S, Edison AS, Moremen KW, Kannan N (2020) Deep evolutionary analysis reveals the design principles of fold A glycosyltransferases. *Elife* 9:e54532. https://doi.org/10.7554/eLife.54532

Taujale R, Zhou Z, Yeung W, Moremen KW, Li S, Kannan N (2021) Mapping the glycosyltransferase fold landscape using interpretable deep learning. *Nature Communications* 12:5656.

Taylor NG, Howells RM, Huttly AK, Vickers K, Turner SR (2003) Interactions among three distinct CesA proteins essential for cellulose synthesis. *The Proceedings of the National Academy of Sciences USA* 100:1450–1455.

Taylor NG, Laurie S, Turner SR (2000) Multiple cellulose synthase catalytic subunits are required for cellulose synthesis in Arabidopsis. *Plant Cell* 12:2529–2540.

Taylor NG, Scheible WR, Cutler S, Somerville CR, Turner SR (1999) The irregular xylem3 locus of Arabidopsis encodes a cellulose synthase required for secondary cell wall synthesis. *Plant Cell* 11:769–780.

The CAZypedia Consortium (2018) Ten years of CAZypedia: a living encyclopedia of carbohydrate-active enzymes. *Glycobiology* 28(1): 3–8. https://doi.org/10.1093/glycob/cwx089

Timmers J, Vernhettes S, Desprez T, Vincken JP, Visser RG, Trindade LM (2009) Interactions between membrane-bound cellulose synthases involved in the synthesis of the secondary cell wall. *FEBS Letters* 583:978–982.

Todd AR (1954) Bakerian lecture. The chemistry of the nucleotides. *Proceedings of the Royal Society of London. Series A, Mathematical and Physical Sciences* 226:70–82.

Toufighi K, Brady SM, Austin R, Ly E, Provart NJ (2005) The botany array resource: e-northerns, expression angling, and promoter analyses. *Plant Journal* 43:153–163.

Tryfona T, Liang HC, Kotake T, Kaneko S, Marsh J, Ichinose H, Lovegrove A, Tsumuraya Y, Shewry PR, Stephens E, Dupree P (2010) Carbohydrate structural analysis of wheat flour arabinogalactan protein. *Carbohydrate Research* 345:2648–2656.

Tunyasuvunakool K, Adler J, Wu Z, Green T, Zielinski M, Zidek A, Bridgland A, Cowie A, Meyer C, Laydon A, Velankar S, Kleywegt GJ, Bateman A, Evans R, Pritzel A, Figurnov M, Ronneberger O, Bates R, Kohl SAA, Potapenko A, Ballard AJ, Romera-Paredes B, Nikolov S, Jain R, Clancy E, Reiman D, Petersen S, Senior AW, Kavukcuoglu K, Birney E, Kohli P, Jumper J, Hassabis D (2021) Highly accurate protein structure prediction for the human proteome. *Nature* 596:590–596.

Tuomivaara ST, Yaoi K, O'Neill MA, York WS (2015) Generation and structural validation of a library of diverse xyloglucan-derived oligosaccharides, including an update on xyloglucan nomenclature. *Carbohydrate Research* 402:56–66.

Turner SR, Somerville CR (1997) Collapsed xylem phenotype of Arabidopsis identifies mutants deficient in cellulose deposition in the secondary cell wall. *Plant Cell* 9:689–701.

Ubarretxena-Belandia I, Tate CG (2004) New insights into the structure and oligomeric state of the bacterial multidrug transporter EmrE: an unusual asymmetric homo-dimer. *FEBS Letters* 564:234–238.

Uehara Y, Tamura S, Maki Y, Yagyu K, Mizoguchi T, Tamiaki H, Imai T, Ishii T, Ohashi T, Fujiyama K, Ishimizu T (2017) Biochemical characterization of rhamnosyltransferase involved in biosynthesis of pectic rhamnogalacturonan I in plant cell wall. *Biochemical and Biophysical Research Communications* 486:130–136.

Urbanowicz BR, Bharadwaj VS, Alahuhta M, Pena MJ, Lunin VV, Bomble YJ, Wang S, Yang JY, Tuomivaara ST, Himmel ME, Moremen KW, York WS, Crowley MF (2017) Structural, mutagenic and in silico studies of xyloglucan fucosylation in Arabidopsis thaliana suggest a water-mediated mechanism. *Plant Journal* 91:931–949.

Urbanowicz BR, Pena MJ, Moniz HA, Moremen KW, York WS (2014) Two Arabidopsis proteins synthesize acetylated xylan in vitro. *Plant Journal* 80:197–206.

Vanzin GF, Madson M, Carpita NC, Raikhel NV, Keegstra K, Reiter WD (2002) The mur2 mutant of Arabidopsis thaliana lacks fucosylated xyloglucan because of a lesion in fucosyltransferase AtFUT1. *The Proceedings of the National Academy of Sciences USA* 99:3340–3345.

Vega-Sanchez ME, Verhertbruggen Y, Christensen U, Chen X, Sharma V, Varanasi P, Jobling SA, Talbot M, White RG, Joo M, Singh S, Auer M, Scheller HV, Ronald PC (2012) Loss of Cellulose synthase-like F6 function affects mixed-linkage glucan deposition, cell wall mechanical properties, and defense responses in vegetative tissues of rice. *Plant Physiology* 159:56–69.

Velasquez SM, Ricardi MM, Dorosz JG, Fernandez PV, Nadra AD, Pol-Fachin L, Egelund J, Gille S, Harholt J, Ciancia M, Verli H, Pauly M, Bacic A, Olsen CE, Ulvskov P, Petersen BL, Somerville C, Iusem ND, Estevez JM (2011) O-glycosylated cell wall proteins are essential in root hair growth. *Science* 332:1401–1403.

Verhertbruggen Y, Yin L, Oikawa A, Scheller HV (2011) Mannan synthase activity in the CSLD family. *Plant Signaling and Behavior* 6:1620–1623.

Verma DP, Hong Z (2001) Plant callose synthase complexes. *Plant Molecular Biology* 47:693–701.

Villemez CL, Lin TY, Hassid WZ (1965) Biosynthesis of the polygalacturonic acid chain of pectin by a particulate enzyme preparation from Phaseolus aureus seedlings. *The Proceedings of the National Academy of Sciences USA* 54:1626–1632.

Voiniciuc C, Dama M, Gawenda N, Stritt F, Pauly M (2019) Mechanistic insights from plant heteromannan synthesis in yeast. *The Proceedings of the National Academy of Sciences USA* 116:522–527.

Voiniciuc C, Engle KA, Gunl M, Dieluweit S, Schmidt MH, Yang JY, Moremen KW, Mohnen D, Usadel B (2018) Identification of key enzymes for pectin synthesis in seed mucilage. *Plant Physiology* 178:1045–1064.

Voiniciuc C, Schmidt MH, Berger A, Yang B, Ebert B, Scheller HV, North HM, Usadel B, Gunl M (2015) MUCILAGE-RELATED10 produces galactoglucomannan that maintains pectin and cellulose architecture in Arabidopsis seed mucilage. *Plant Physiology* 169:403–420.

Vuttipongchaikij S, Brocklehurst D, Steele-King C, Ashford DA, Gomez LD, McQueen-Mason SJ (2012) Arabidopsis GT34 family contains five xyloglucan alpha-1,6-xylosyltransferases. *New Phytologist* 195:585–595.

Wachananawat B, Kuroha T, Takenaka Y, Kajiura H, Naramoto S, Yokoyama R, Ishizaki K, Nishitani K, Ishimizu T (2020) Diversity of pectin rhamnogalacturonan I rhamnosyltransferases in glycosyltransferase family 106. *Frontiers in Plant Science* 11:997.

Wang H, Yang H, Wen Z, Gao C, Gao Y, Tian Y, Xu Z, Liu X, Persson S, Zhang B, Zhou Y (2022) Xylan-based nanocompartments orchestrate plant vessel wall patterning. *Nature Plants* 8:295–306.

Wang Y, Mortimer JC, Davis J, Dupree P, Keegstra K (2013) Identification of an additional protein involved in mannan biosynthesis. *Plant Journal* 73:105–117.

Welner DH, Shin D, Tomaleri GP, DeGiovanni AM, Tsai AY, Tran HM, Hansen SF, Green DT, Scheller HV, Adams PD (2017) Plant cell wall glycosyltransferases: high-throughput recombinant expression screening and general requirements for these challenging enzymes. *PLoS One* 12:e0177591.

Wu AM, Rihouey C, Seveno M, Hornblad E, Singh SK, Matsunaga T, Ishii T, Lerouge P, Marchant A (2009) The Arabidopsis IRX10 and IRX10-LIKE glycosyltransferases are critical for glucuronoxylan biosynthesis during secondary cell wall formation. *Plant Journal* 57:718–731.

Wu Y, Williams M, Bernard S, Driouich A, Showalter AM, Faik A (2010) Functional identification of two nonredundant Arabidopsis alpha(1,2)fucosyltransferases specific to arabinogalactan proteins. *Journal of Biological Chemistry* 285:13638–13645.

Xuan L, Zhang J, Lu W, Gluza P, Ebert B, Kotake T, Lu M, Zhang Y, Clausen MH, Johnson KL, Doblin MS, Heazlewood JL, Bacic A, Song L, Zeng W (2021) A pipeline towards the biochemical characterization of the Arabidopsis GT14 family. *International Journal of Molecular Sciences* 22(3):1360. https://doi.org/10.3390/ijms22031360

Yamaguchi N, Fukuda MN (1995) Golgi retention mechanism of beta-1,4-galactosyltransferase. Membrane-spanning domain-dependent homodimerization and association with alpha- and beta-tubulins. *Journal of Biological Chemistry* 270:12170–12176.

Yang J, Bak G, Burgin T, Barnes WJ, Mayes HB, Pena MJ, Urbanowicz BR, Nielsen E (2020) Biochemical and genetic analysis identify CSLD3 as a beta-1,4-Glucan synthase that functions during plant cell wall synthesis. *Plant Cell* 32:1749–1767.

Yao HY, Wang JQ, Yin JY, Nie SP, Xie MY (2021) A review of NMR analysis in polysaccharide structure and conformation: progress, challenge and perspective. *Food Research International* 143:110290.

Yin L, Verhertbruggen Y, Oikawa A, Manisseri C, Knierim B, Prak L, Jensen JK, Knox JP, Auer M, Willats WG, Scheller HV (2011) The cooperative activities of CSLD2, CSLD3, and CSLD5 are required for normal Arabidopsis development. *Molecular Plant* 4:1024–1037.

York WS, O'Neill MA (2008) Biochemical control of xylan biosynthesis - which end is up? *Current Opinion in Plant Biology* 11:258–265.

Zeng W, Jiang N, Nadella R, Killen TL, Nadella V, Faik A (2010) A glucurono(arabino)xylan synthase complex from wheat contains members of the GT43, GT47, and GT75 families and functions cooperatively. *Plant Physiology* 154:78–97.

Zeng W, Lampugnani ER, Picard KL, Song L, Wu AM, Farion IM, Zhao J, Ford K, Doblin MS, Bacic A (2016) Asparagus IRX9, IRX10, and IRX14A are components of an active xylan backbone synthase complex that forms in the golgi apparatus. *Plant Physiology* 171:93–109.

Zhang GF, Staehelin LA (1992) Functional compartmentation of the Golgi apparatus of plant cells : immunocytochemical analysis of high-pressure frozen- and freeze-substituted sycamore maple suspension culture cells. *Plant Physiology* 99:1070–1083.

Zhang Y, Nikolovski N, Sorieul M, Vellosillo T, McFarlane HE, Dupree R, Kesten C, Schneider R, Driemeier C, Lathe R, Lampugnani E, Yu X, Ivakov A, Doblin MS, Mortimer JC, Brown SP, Persson S, Dupree P (2016) Golgi-localized STELLO proteins regulate the assembly and trafficking of cellulose synthase complexes in Arabidopsis. *Nature Communications* 7:11656.

Zhong R, Cui D, Phillips DR, Sims NT, Ye ZH (2021) Functional analysis of GT61 glycosyltransferases from grass species in xylan substitutions. *Planta* 254:131.

Zhong R, Cui D, Phillips DR, Ye ZH (2018) A novel rice xylosyltransferase catalyzes the addition of 2-O-xylosyl side chains onto the xylan backbone. *Plant & Cell Physiology* 59:554–565.

Zhong R, Pena MJ, Zhou GK, Nairn CJ, Wood-Jones A, Richardson EA, Morrison WH, Darvill AG, York WS, Ye ZH (2005) Arabidopsis fragile fiber8, which encodes a putative glucuronyltransferase, is essential for normal secondary wall synthesis. *Plant Cell* 17:3390–3408.

Zhong R, Burk DH, Ye ZH (2001) Fibers. A model for studying cell differentiation, cell elongation, and cell wall biosynthesis. *Plant Physiology* 126(2):477–479. https://doi org/10.1104/pp.126.2.477

3 Nucleotide Sugars and the Transporters That Put Them at the Right Place

Ariel Orellana
Universidad Andrés Bello

INTRODUCTION

The plant cell wall is one of the most abundant biomass components. Most of it is composed of monosaccharides assembled into different polymers by enzymes present within the plant cell. The building blocks for these processes are nucleotide sugars. These are the substrates for glycosyltransferases, which create the glycosidic bonds that join the monosaccharides to assemble the polysaccharides, determining their conformation. This chapter focuses on the nucleotide sugars and how these molecules are produced by the plant cell and then brought to the Golgi apparatus to synthesise hemicelluloses, pectin, and proteoglycans, important components of the plant cell wall.

DISCOVERY OF NUCLEOTIDE SUGARS

In the 1950s, seminal work towards our understanding of carbohydrate metabolism was generated by several research groups. Among these, Leloir and co-workers' contribution at Fundación Campomar in Buenos Aires, Argentina, is one of the most significant. This brilliant effort was recognised in 1970 when Luis F. Leloir was awarded The Nobel Prize in Chemistry "for his discovery of sugar nucleotides and their role in the biosynthesis of carbohydrates." Cardini et al. (1950) reported the structure for UDP-glucose (Figure 3.1), and a whole new area emerged in the biochemistry of carbohydrates. Through the years, new nucleotide sugars were discovered in living organisms, and by 1963 more than 40 had been described (Cabib 1963). In plants, 30 nucleotide sugars have been described (Bar-Peled and O'Neill, 2011), and at least half of them are involved in the biosynthesis of plant cell wall polysaccharides.

NUCLEOTIDE SUGAR METABOLISM

Plants capture CO_2 from the atmosphere and transform it into sugars. In addition, solar energy is captured by photosynthesis and converted into high-energy bonds through the synthesis of adenosine triphosphate (ATP), the energy currency of living organisms. In general terms, ATP is involved in the phosphorylation of sugars. Then, uridine triphosphate (UTP) or guanosine triphosphate (GTP) provides the nucleoside monophosphate moiety to form the nucleoside diphosphate sugar known as nucleotide sugar (Figure 3.2).

Sugars on nucleotide sugars can be converted into other sugars, leading to different nucleotide sugars, which are used in different polysaccharide biosynthetic processes. For example, UDP-glucose is the initial substrate for a series of reactions, as can be seen in Figure 3.3. Hence, a very intricate series of metabolic pathways are involved in synthesising the nucleotide sugars required for the biosynthesis of cell wall polysaccharides. A detailed and comprehensive analysis of these pathways can be seen in the work of Bar-Peled and O'Neill (2011). This chapter focuses on nucleotide sugars and the transporter proteins that bring them into the Golgi apparatus, the organelle where the biosynthesis of plant cell wall polysaccharides takes place. However,

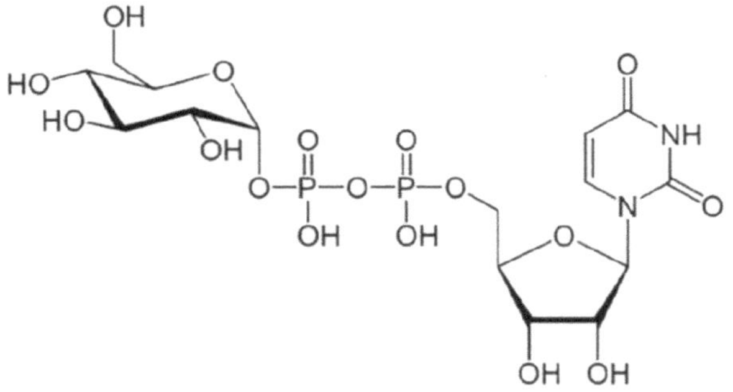

FIGURE 3.1 Uridine diphosphate glucose (UDP-glucose). Nucleotide sugars are formed of four basic structures, the nitrogenous base, the pyrimidine uracil in UDP-glucose, a pentose (ribose), a pyrophosphate group that at physiological pH is negatively charged, and the sugar (glucose). Uracil and ribose form the nucleoside uridine, and together with the pyrophosphate form the nucleoside diphosphate (NDP) or uridine diphosphate (UDP) in UDP-glucose.

DOI: 10.1201/9781003178309-3

$$(1)\ \text{Sugar} + \text{ATP} \rightarrow \text{Sugar-P} + \text{ADP}$$
$$(2)\ \text{Sugar-P} + \text{UTP} \rightarrow \text{UDP-Sugar} + \text{P-P}_i$$

FIGURE 3.2 Sugars are phosphorylated using adenosine triphosphate (ATP) to form the sugar phosphate. This phosphorylated sugar is enzymatically converted into the UDP-sugar which generates a high-energy bond between the sugar and the UDP, energy that is necessary for the transfer reaction of the sugar into the acceptor.

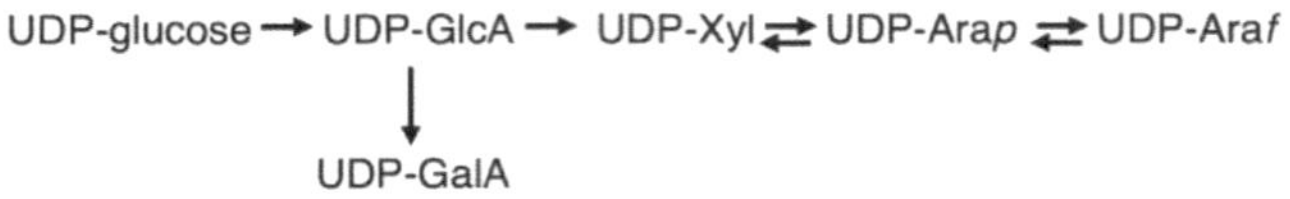

FIGURE 3.3 Nucleotide sugars can be converted into other nucleotide sugars. Here is depicted the conversion of UDP-glucose to UDP-glucuronic acid, a reaction that is not reversible. In turn, UDP-glucuronic acid is converted into UDP-galacturonic acid, UDP-xylose, UDP-arabinopyranose, and UDP-arabinofuranose.

it is important to mention that nucleotide sugars in plants are also important for protein glycosylation, glycolipid biosynthesis, and small molecule glycosylation. Finally, ADP-glucose is fundamental for starch biosynthesis.

TOWARDS IDENTIFYING THE NUCLEOTIDE SUGARS AS THE SUBSTRATES FOR THE BIOSYNTHESIS OF PLANT CELL WALL POLYSACCHARIDES

Feingold et al. (1958) reported that ^{14}C-labelled 1,3-linked ß-glucan (callose) was formed when extracts of *Phaseolus aureus* were incubated with UDP-^{14}C-Glc, indicating that the sugar moiety from this nucleotide sugar was utilised to make a sugar polymer in plants. Furthermore, despite some initial suggestions that GDP-glucose was involved in cellulose biosynthesis, it was finally shown that UDP-glucose is involved in the biosynthesis of cellulose (Ordin and Hall, 1968). These were the initial steps onto a road that concluded that nucleotide sugars are essential components to build the plant cell wall. Today, many glycosyltransferases involved in plant cell wall polysaccharides have been identified. To get a detailed description on glycosyltransferases involved in plant cell wall polysaccharide biosynthesis, see the Chapter 2 in this book.

WHERE DO PLANT CELL WALL POLYSACCHARIDE BIOSYNTHETIC REACTIONS TAKE PLACE?

Northcote and Pickett-Heaps (1966) provided evidence that the Golgi apparatus plays a role in the biosynthesis of cell wall polysaccharides. Furthermore, Ray et al. (1969) showed that subcellular fractions containing Golgi membranes produced β-glucan from UDP-glucose and GDP-glucose, thus indicating that this organelle is involved in the biosynthesis of plant cell wall polysaccharides. On the other hand, evidence obtained by electron microscopy suggested that cellulose is synthesised at the plasma membrane; thus, it became clear that biosynthesis of the plant cell wall polysaccharides occurs at the Golgi apparatus and the plasma membrane.

THE PURIFICATION AND CHARACTERISATION OF THE ENZYMES INVOLVED IN THE BIOSYNTHESIS OF THE CELL WALL: THE GLYCOSYLTRANSFERASES

Early studies by Northcote, Bolwell, and many others had identified several glycosyltransferases activities but had not purified/identified the enzymes. One of the crucial issues in identifying these enzymes was setting reliable assays to test their actual activity. Furthermore, glycosyltransferases located in the Golgi apparatus are low abundant proteins, and thus, their purification became quite tricky. On the other hand, transcriptome and genome information was becoming available in the mid-nineties, and comparison with other glycosyltransferases identified in bacteria, yeast, and animal cells became a way to identify putative glycosyltransferases involved in polysaccharide biosynthesis. Furthermore, proteomic approaches isolating Arabidopsis Golgi membranes enriched fractions were also utilised to advance the identification of proteins playing a role in this process (Dunkley et al., 2006).

The first studies towards identifying the enzymes involved in biosynthesis of plant cell wall polysaccharides came from the identification of cDNA sequences from cotton, homologous to bacterial *CELA* genes that encode the catalytic subunit of CELLULOSE SYNTHASE (Pear et al., 1996). Then, genetics and map-based cloning proved to be a powerful tool to identify the first gene involved in cellulose biosynthesis in plants (Arioli et al., 1998). A cellulose-deficient temperature-sensitive mutant in Arabidopsis led to the identification of a gene that could complement the mutant and restore the normal synthesis of crystalline cellulose. Compelling evidence indicated that this gene codes for the catalytic subunit of CELLULOSE SYNTHASE.

These pioneering studies were important in the identification of the proteins and genes involved in cellulose biosynthesis, and today cellulose synthases engaged in primary and secondary cell walls are known, as well as some of the partners that are involved in regulating the biosynthesis of cellulose in plants (Polko and Kieber, 2019).

The first plant cell wall polysaccharide-specific glycosyltransferase gene to be characterised was the Arabidopsis

Xyloglucan Fucosyltransferase (AtFUT1) (Perrin et al., 1999). The reasons to succeed in this effort were the availability of the sugar donor GDP-Fuc and an acceptor substrate (tamarind xyloglucan and xyloglucan oligosaccharides), which allowed to set a specific and reliable enzymatic assay, along with a tremendous effort in protein purification, that led to the identification of peptides, whose sequence was used to design PCR primers to amplify a cDNA sequence encoding the gene. This approach, combined with the heterologous expression of the AtFUT1 cDNA in animal cells to confirm that the identified gene contained information for encoding a protein with the enzymatic activity, proved to be successful. Subsequently, many genes encoding glycosyltransferases involved in hemicellulose, pectin, and arabinogalactan proteins biosynthesis have been identified and some of their activities demonstrated. A comprehensive analysis of this topic can be found in a review published by Amos and Mohnen (2019).

THE STRUCTURE OF THE GLYCOSYLTRANSFERASES LEADS TO ISSUES IN THE TOPOLOGY OF REACTIONS INVOLVED IN THE BIOSYNTHESIS OF PLANT CELL WALL POLYSACCHARIDES

Analyses of the amino acid sequence of the Golgi-localised glycosyltransferases predict that most of them have a transmembrane domain close to the N-terminus of the sequence (Type II membrane bound proteins). Furthermore, in the Golgi apparatus can also be found a class of glycosyltransferases known as cellulose synthase-like (CSL) proteins with multiple transmembrane domains. However, there are a few examples of proteins that are not membrane bound either because they have no transmembrane domain like in IRX-10 (Wu et al., 2009) or because the transmembrane domain is cleaved *in vivo* as part of the processing of some proteins synthesised in the endoplasmic reticulum like in GAUT1 (Atmodjo et al., 2011). Most likely they are retained in the Golgi by interaction with other molecules present in that organelle.

Studies to identify the active site of the glycosyltransferases using protease-inactivation assays in Golgi vesicles showed that both the Xyloglucan Fucosyltransferase (FUT1) and the α-1,4-galacturonosyltransferase (GAUT1) involved in pectin biosynthesis have their catalytic domain facing the lumen of the Golgi apparatus (Wulff et al. 2000; Sterling et al., 2001). These results clearly showed that the polymers' biosynthesis occurs in the Golgi apparatus' lumen. However, at that time, most nucleotide sugars were believed to be synthesised in the cytosol of plant cells; hence, this finding raised the question how the substrates cross the Golgi membrane. Muñoz et al. (1996) provided evidence for the transport of nucleotide sugars; thus, these transporters became another player in the biosynthetic process leading to the plant cell wall biosynthesis.

A similar topological issue was raised in the case of the mechanism of cellulose biosynthesis given that UDP-glucose is located in the cytosol, but cellulose microfibrils are deposited outside the plasma membrane. Today, the crystalline structure of cellulose synthase from bacteria and extensive modelling of the process led to the widely accepted view that the elongating polymer crosses the membrane through a pore present in the protein (McNamara et al., 2015). Along the same line of thinking, CSL proteins (glycosyltransferases located in the Golgi membrane) may utilise the nucleotide sugars directly from the cytosol (Daras et al., 2021); therefore, no transport of nucleotide sugars would be required in this case. However, experimental evidence suggests that the catalytic domain of CSLA9, a protein involved in the biosynthesis of mannan (Liepman et al., 2005), and CSLC4, a protein involved in the biosynthesis of ß-1,4 glucan from xyloglucan (Cocuron et al., 2007), may have opposite orientations in the Golgi membrane (Davis et al., 2010). Although this result supports the idea that nucleotide sugar may need to cross the Golgi membrane to provide the substrate to some CSL proteins, further experiments are needed to confirm whether CSLs can take the substrate from the cytosol or in some cases may need the nucleotide sugar to be available in the Golgi lumen.

NUCLEOTIDE SUGARS UTILISED TO BUILD THE PLANT CELL WALL ARE SYNTHESISED IN THE CYTOPLASM AND THE GOLGI APPARATUS

UDP-glucose, which is transformed into other UDP-sugars, is made in the cytoplasm; nucleotide sugars derived directly from this UDP-sugar, such as UDP-galactose, UDP-glucuronic acid, and UDP-rhamnose, are also produced in the cytoplasm. However, some nucleotide sugars, such as UDP-galacturonic, UDP-arabinopyranose, and UDP-xylose, are synthesised in the Golgi apparatus. The precursor for the synthesis of these three nucleotide sugars is UDP-glucuronic acid, which is produced in the cytoplasm. Therefore, as part of the metabolic pathway leading to these nucleotide sugars, UDP-glucuronic acid must enter the Golgi lumen. On the other hand, UDP-glucuronic acid is a precursor for the synthesis of UDP-apiose in the cytoplasm and is also utilised to synthesise UDP-xylose in this compartment. Thus, UDP-glucuronic acid is a critical intermediate for the synthesis of several nucleotide sugars used in the biosynthesis of plant cell wall polysaccharides. Furthermore, it is interesting to notice that UDP-glucuronic acid can be used as an intermediate in nucleotide sugars biosynthesis both in the cytoplasm and the lumen of the Golgi apparatus. The synthesis of UDP-xylose from UDP-glucuronic acid in the cytoplasm and the Golgi lumen is intriguing because UDP-xylose is utilised to synthesise polysaccharides within the Golgi, but it is synthesised in both compartments (Harper and Bar-Peled, 2002).

Another intriguing process regarding the compartmentalisation in the synthesis of nucleotide sugars occurs in the interconversion of UDP-arabinopyranose and UDP-arabinofuranose (Konishi et al., 2007). Several polysaccharides contain arabinose in their structures, but the most abundant structural isomer is the furanose form; then, it is relevant to identify the pathway that leads to the synthesis of this nucleotide sugar. UDP-arabinofuranose is produced by the epimerisation of UDP-arabinopyranose, which is synthesised in the lumen of the Golgi apparatus; however, the UDP-Ara*p*/UDP-Ara*f* mutase is cytosolic, although it appears to be closely associated with the Golgi. This topological issue implies that UDP-arabinopyranose must be transported from the Golgi lumen to the cytoplasm (Rautengarten et al., 2011). This arrangement is a rather unusual phenomenon since there are no other known examples of metabolites leaving the Golgi apparatus as part of a metabolic pathway. Finally, among UDP-sugars, UDP-glcNac is an important nucleotide sugar for N-glycosylation of proteins, and it is present in fair amounts in plants (Ito et al., 2014). Even though N-acetylglucosamine is not present in polysaccharide, there are numerous reports describing the presence of N-glycoproteins in the plant cell wall (Ruiz-May et al., 2012) and some of them have been identified as glycoside hydrolases (Liu et al., 2021); thus, N-glycosylation of proteins may play a role in the modification of plant cell wall polysaccharides.

GDP-mannose, in a similar fashion to UDP-glucose, is the source for other nucleotide sugars, but also is needed in polysaccharide biosynthesis as well as in protein and lipid glycosylation. GDP-fucose is produced from GDP-mannose in a three-step mechanism, and a mutant involved in this process (*mur1*) exhibits alterations in cell wall polysaccharides and developmental abnormalities (Reiter et al., 1993; Bonin et al., 1997). GDP-mannose is also the substrate for the GDP-D-mannose 3,5-epimerase that leads to GDP-L-galactose, and the mutation of this gene affects the cell wall (Gilbert et al., 2009). GDP-glucose instead is synthesised from GTP and glucose-1-P (Barber, 1966), and even though it was proposed to play a role in cellulose biosynthesis, and it is a substrate for the biosynthesis of some hemicelluloses, less is known about the genes involved in its biosynthesis.

3-Deoxy-d-manno-2-octulosonic acid (KDO) is a sugar present in the cell wall (York et al., 1985), and Kobayashi et al. (2011) reported that KDO is activated as CMP-KDO by the gene product of At1g53000.

NUCLEOTIDE SUGAR TRANSPORTERS

As we mentioned above, the topology of the Golgi glycosyltransferases indicated that nucleotide sugars must be available for the biosynthetic reactions in the lumen of the Golgi apparatus. Furthermore, biochemical approaches provided evidence that different nucleotide sugars (UDP-Glc, GDP-Fuc, UDP-GalA, and UDP-Gal) were incorporated into Golgi vesicles (Muñoz et al., 1996; Wulff et al., 2000; Sterling et al., 2001; Norambuena et al., 2002), suggesting that transporters for nucleotide sugars should be present in the Golgi membrane.

Nucleotide sugar transporters were first biochemically characterised in animal cells. CMP-sialic acid is the donor used to add sialic acid into glycoproteins in animal cells. Carey et al. (1980) isolated sealed, right-side-in Golgi vesicles to demonstrate that CMP-sialic acid is incorporated into the Golgi. The transporter for this nucleotide sugar was the first to be biochemically characterised. Carlos Hirschberg provided an exciting reflection of the process that led him and his co-workers to identify this class of proteins (Hirschberg, 2018).

Furthermore, a series of elegant experiments demonstrated that nucleotide sugar transporters work as antiporters, using the monophosphate form of the nucleotide as the antiporter molecule. It is worth noting that no energy source is required to transport nucleotide sugars (Capasso and Hirschberg, 1984). In the case of UDP- or GDP-sugars, the transfer of the sugar catalysed by the glycosyltransferases release nucleoside diphosphate (NDP), which needs to be cleaved of to produce the nucleoside monophosphate (NMP) that is used in the antiporter mechanism. The hydrolysis of the NDP occurs by a nucleoside diphosphatase (NDPase), an enzyme located in the Golgi lumen (Abeijon et al., 1993; Neckelmann and Orellana, 1998). The mutation of this NDPase in yeast leads to a decrease in the N- and O-glycosylation of proteins and glycolipids (Abeijon et al., 1993), confirming the need to produce NMP, the antiporter molecule for the transport of nucleotide sugars.

THE IDENTIFICATION OF THE GENES ENCODING NUCLEOTIDE SUGAR TRANSPORTERS

The identification of the *Kluyveromyces lactis* mutant (*mnn2-2*) with impaired activity in the transport of UDP-glcNac (Abeijon et al., 1996a) was instrumental for the identification of the first gene coding for a nucleotide sugar transporter. Complementation of the *mnn2-2* mutant with a genomic library facilitated the identification of a DNA sequence containing an open reading frame that was predicted to encode a hydrophobic protein (328 amino acids), with 5–8 transmembrane domains (Abeijon et al., 1996b). Homologous sequences were found in *S. cerevisiae* and *C. elegans*, suggesting that orthologous sequences encoding nucleotide sugar transporters were present in these species.

A similar approach (complementing cells deficient in nucleotide sugar transport activity) was utilised to identify a UDP-galactose transporter in animal cells (Miura et al., 1996). Hence, an open reading frame predicting a protein of 393 amino acids was discovered. This protein also was highly hydrophobic, and eight transmembrane domains were predicted.

Glycoproteins in *Saccharomyces cerevisiae* contain high-mannose oligosaccharides rather than complex or hybrid N-glycans as they occur in animal cells. These highly mannosylated proteins, are modified in the Golgi apparatus, and transport of GDP-mannose is needed in this process. The identification of mutants with lower mannose levels led to the identification of different mutants (*VRG*) that are more sensitive to vanadate. One of these mutants (*vrg4*) exhibited a deficiency in the transport of GDP-mannose (Dean et al., 1997), and this deficiency was partially restored by a plasmid bearing the *VRG4* gene; thus, a gene encoding a GDP-mannose transporter was identified.

The identification of these genes, the progress in the sequencing efforts of the *Arabidopsis thaliana* genome, and the availability of an extensive collection of expressed sequence tags (ESTs) enabled the identification of homologous sequences to nucleotide sugar transporter genes in plants. The search for homologous sequences to *VRG4* in Arabidopsis led to the identification of Golgi Nucleotide Sugar Transporter 1 (GONST1). *GONST1* complemented the *Saccharomyces cerevisiae vrg4* mutant (Baldwin et al., 2001). Further analyses showed that GONST1 could transport GDP-mannose, GDP-fucose, and GDP-glucose in proteoliposomes *in vitro* (Mortimer et al., 2013). Later on, the concentration of substrate at which half of the maximum rate of transport is achieved (Km) for GDP-mannose and GDP-fucose were estimated to be 17 and 76 µM, respectively (Rautengarten et al., 2016). However, cell wall polysaccharides are not altered in *gonst1* plants based on monosaccharide analysis of cell wall material; instead, this mutant exhibits a deficiency of mannose in glycosphingolipids (Mortimer et al., 2013), indicating that GONST1 is involved in glycolipid biosynthesis but not in polysaccharide biosynthesis. Recent studies revealed that GONST2 is a functional homolog of GONST1 (Jing et al., 2021), and GONST1 and 2 are both involved in transporting GDP-mannose in *Arabidopsis thaliana*.

The search in the Arabidopsis genome for homologous sequences to the UDP-galactose transporter cloned from animal cells led to the identification of several sequences exhibiting some degree of homology to the UDP-galactose transporter. AtUTR1 was identified as a UDP-galactose/UDP-glucose transporter (Norambuena et al., 2002). This gene was able to complement UDP-galactose transporter-deficient mutants in MDCK cells. Also, UDP-galactose and UDP-glucose transport was observed in microsomal fractions obtained from yeast expressing AtUTR1. However, the subcellular localisation analysis showed that the protein is located in the endoplasmic reticulum (Reyes et al., 2006). The study of mutants suggested that AtUTR1 is involved in the cellular response to endoplasmic reticulum stress. Then, it is likely that somehow, AtUTR1 is involved in protein folding in the endoplasmic reticulum but not in polysaccharide biosynthesis. Furthermore, Reyes et al. (2010) appear to suggest that AtUTR1 and 3 are more important for transport of UDP-Glc into the endoplasmic reticulum.

The identification of these few genes encoding nucleotide sugar transporters in plants was essential to discover homologous sequences in the Arabidopsis genome; thus, through bioinformatic analyses, around 50 sequences coding for putative nucleotide sugar transporters were discovered, and a gene family for these proteins emerged in plants (Figure 3.4). However, the challenge was still to identify the activity of these putative nucleotide sugar transporter genes and to try to identify some unique NSTs expected to be present in plants such as transporters for UDP-rhamnose, UDP-arabinose, UDP-xylose, and UDP-uronic acid transporter among others.

For many years, the assessment of the activity of the nucleotide sugar transporters was performed using only radio-labelled substrates. However, several nucleotide sugars utilised in the biosynthesis of plant cell wall polysaccharides were not available in a radio-labelled form to perform the assays. Then, the establishment of a new methodology to assess the activity of the putative nucleotide sugar transporters was necessary; thus, a new procedure was set, and it became a breakthrough towards the identification of transmembrane proteins playing a role in polysaccharides biosynthesis in plants. This new assay used non-radioactive nucleotide sugars, which were separated by HPLC and detected by mass spectrometry (Rautengarten et al., 2014). The genes encoding the putative transporters were heterologously expressed, the recombinant proteins incorporated into proteoliposomes, and transport assays were performed. However, many nucleotide sugars utilised for polysaccharide biosynthesis in plants were not commercially available; therefore, nucleotide sugars had to be synthesised. Fortunately, genes for many enzymes involved in the biosynthesis of nucleotide sugars had been identified, and procedures to express functional recombinant proteins that can be used to produce the desired NDP-sugar were developed to synthesise those nucleotide sugars that were not commercially available. Therefore, the use of this new method to assess the activity of the nucleotide sugar transporters, together with a systematic biochemical analysis of several putative genes, allowed the characterisation of new nucleotide sugar transporters (Figure 3.5), which are described below.

UDP-RHAMNOSE/UDP-GALACTOSE TRANSPORTERS

UDP-rhamnose is a critical substrate for rhamnogalacturonan I (RG-I) and rhamnogalacturonan II (RG-II) biosynthesis, and this nucleotide sugar must be mobilised from the cytosol to the Golgi lumen. Six Arabidopsis genes encoding proteins that can transport UDP-rhamnose *in vitro* were identified (Figure 3.4) (Rautengarten et al., 2014). All six proteins were located in the Golgi apparatus upon heterologous expression of GFP-fused proteins. The assessment of the activity of the gene products indicated that they transport UDP-rhamnose as well as UDP-galactose; however, their apparent Km for UDP-rhamnose was in the range of 17–87 µM, whereas the apparent Km for UDP-galactose was in the range of 68–196 µM, indicating that they have a higher affinity for UDP-rhamnose than UDP-galactose. In addition, UDP-galactose competed with the UDP-rhamnose transport activity, suggesting that they share the same transport mechanism. No significant transport activity for other nucleotide sugar was reported. Both activities

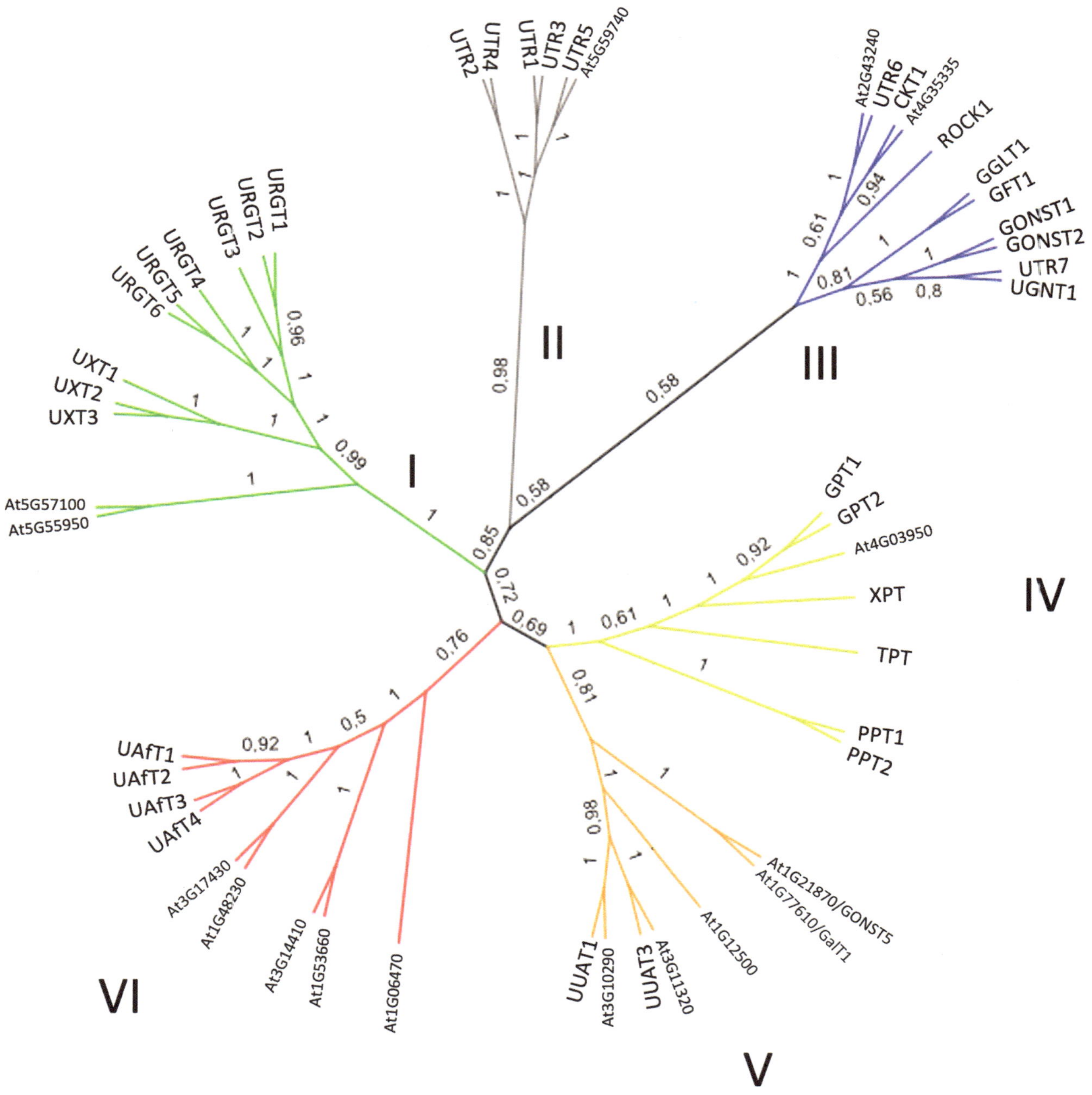

FIGURE 3.4 Nucleotide sugar transporter/triose phosphate translocator gene family. This family is represented by six clades where clade IV contains the triose phosphate translocator genes. Genes in clades II and III were initially identified by homologous sequences to yeast and animal cells. Clades I, III, V, and VI contain genes encoding NSTs involved in the supply of substrates needed for the biosynthesis of different plant cell wall polysaccharides. Biochemical evidence of their activity *in vitro* has been obtained for most of them, except for *UUAT3*, *CKT1*, and *GGLT1*. Figure adapted from Rautengarten et al., 2014.

required UMP as an antiporter molecule. How URGTs transport nucleotide sugars that exhibit significative differences in their sugar moieties remains to be analysed at the structural level of these proteins.

URGT genes show different levels of expression in the plant organs. *URGT1* is highly expressed throughout plant development, and it shows the highest level of expression of the URGTs. Interestingly, T-DNA insertional mutant lines in *URGT1* show a slight decrease in galactose in leaves but no changes in rhamnose or other monosaccharides. Furthermore, overexpression of *URGT1* led to an increase in galactose but not in rhamnose. This increase likely

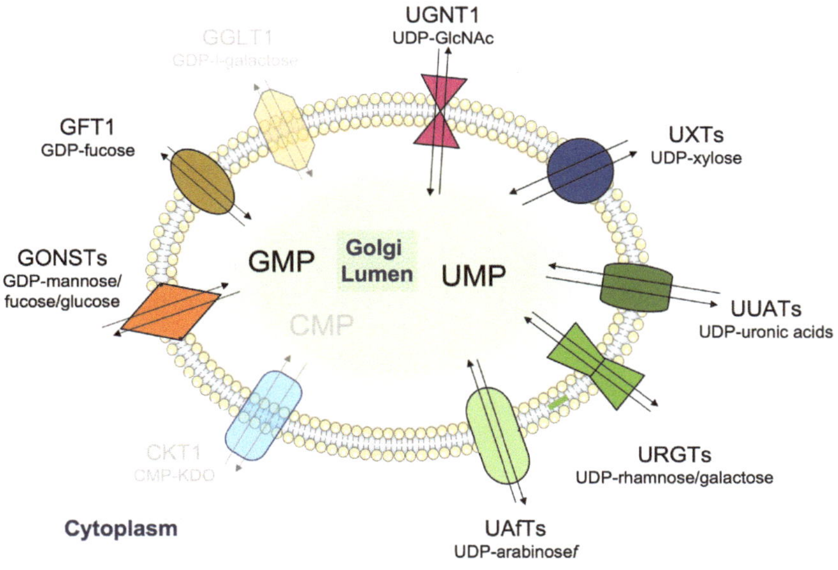

FIGURE 3.5 Nucleotide sugar transporters identified in the plant Golgi apparatus. They work as antiporters using UMP, GMP, or CMP as the exchanger, depending on the base of the nucleotide sugar. According to structural information, they are dimers in the membrane. CKT1 and GGLT1 are depicted in a transparent fashion because there is compelling evidence for their role but no functional evidence of their activities has been yet provided.

occurred in galactan but not in xyloglucan (Rautengarten et al., 2014); thus, it seems that URGT1 has a primary role in galactan biosynthesis in rosette leaves. On the other hand, *URGT2* is highly expressed in seeds, and it is known that seed mucilage contains a high amount of unbranched RG-I, composed of a linear polymer of alternating rhamnose and galacturonic acid. This polysaccharide accounts for most of the mucilage surrounding the hydrated seed (Arsovski et al., 2010). Mutations in *URGT2* exhibited lower contents of rhamnose in mucilage but also lower contents of galacturonic acid; indicating that there is a lower content of RG-I in the mucilage of this mutant. URGT1 and URGT2 exhibit 89.4% identity and 93.4% similarity at the protein level; then, it is intriguing that mutants in these genes have different cell wall chemotypes. Celiz-Balboa et al. (2020) addressed that question by using a promoter swapping approach; thus, URGT1 was expressed in *urgt2* plants under the control of the *URGT2* promoter. Interestingly, these transgenic plants have a wildtype chemotype. On the other hand, expression of URGT2 in *urgt1* plants under the control of the *URGT1* promoter led to transgenic plants that also exhibit a wildtype chemotype. Hence, the *in vivo* role of URGT1 and URGT2 seems to be modulated by their transcriptional expression, suggesting that these nucleotide sugar transporters may be involved in different pathways depending on which cellular context they are expressed in.

Mutants in *URGT2* exhibited around 20% less rhamnose and galacturonic acid in seed mucilage; therefore, it is quite likely that other URGT transporters must be involved in the biosynthesis of RG-I in seeds. In order to identify other URGTs playing a role in the biosynthesis of this polysaccharide, a systematic analysis of URGT mutants was carried out, and the content of rhamnose and galacturonic acid was assessed in each of them. The results of analysing individual mutants indicated that URGT2, URGT4, and URGT6 are involved in RG-I biosynthesis (Saez-Aguayo et al., 2021). On the other hand, mutants in *URGT1*, *URGT3*, and *URGT5* showed no differences in rhamnose and galacturonic acid content in seed mucilage suggesting that they are not involved in RG-I biosynthesis.

To analyse further the role of URGT2, URGT4, and URGT6 in RG-I biosynthesis, double and triple mutants were obtained. The results showed that shorter RG-I polymers were present in the mucilage from the triple mutants, suggesting that a lower availability of UDP-Rha, provided by URGT2, URGT4, and URGT6, has an impact in the length of this polysaccharide. Intriguingly and unexpectedly, the content of xylose was raised in the mucilage; then, since it was previously reported that xylan chains are associated with RG-I (Ralet et al., 2016), the results suggest that mucilage from the triple mutant contains higher number of RG-I chains but are shorter in length. This increase in xylose showed that some compensation phenomena were probably triggered in the mutants; thus, a transcriptomic analysis of the *URGT2* mutant was performed to analyse this further (Parra-Rojas et al., 2019). The results showed that 3,149 genes exhibited some degree of misregulation. Thirty-seven genes of that list corresponded to genes already known to play a role in mucilage biosynthesis. In addition, *URGT4* and other genes coding for other nucleotide sugar transporters were up-regulated, likely to compensate for the lack of *URGT2*. Hence, we need to be aware that phenotypes in mutants may be due to the impact of the knock-out or knock-down gene and the effect of other genes that are misregulated to compensate for the mutation present in the plant.

UDP-XYLOSE TRANSPORTERS

As outlined above, UDP-xylose can be synthesised in the cytoplasm and the Golgi lumen. Furthermore, it has been established that the UDP-xylose synthesised in the cytoplasm is utilised for xylan biosynthesis. Then, transporters for UDP-xylose must be located in the Golgi membrane. Three genes encoding UDP-xylose transporters were described (Figure 3.4) (Ebert et al., 2015). UXT2 and UXT3 are located in the Golgi apparatus, whereas UXT1 is found both in the Golgi and the endoplasmic reticulum. The Km for these transporters is between 39 and 58 μM. They are fairly specific for UDP-xylose, although they can transport minor amounts of UDP-Ara*p* in proteoliposomes preloaded with UMP.

To analyse the role of UXTs *in vivo*, two mutant lines were obtained for *UXT1* and analysed. Monosaccharide composition analysis of the plants' upper and lower stem sections showed lower levels of xylose compared to the wildtype. Additional studies focused on xylose-containing polysaccharides showed that xylan decreased in both mutant lines compared to wildtype plants (Zhao et al., 2018). These results were consistent with previous experiments that showed that UDP-xylose synthases located in the cytosol, but not in the Golgi, are involved in providing the substrate for xylan biosynthesis (Kuang et al., 2016). Interestingly, the abundance of xyloglucan, another cell wall polysaccharide that contains high levels of xylose, did not change in the *UXT1* mutants, suggesting that this transporter is not involved in the biosynthesis of xyloglucan (Ebert et al., 2015). On the other hand, no differences were observed in glycoproteins, suggesting that the addition of xylose into glycoproteins does not depend on the UXTs.

The results obtained for the role of UXTs, which seems to be more focused on xylan biosynthesis and not in another xylose-containing polysaccharide such as xyloglucan, suggest that some sort of compartmentalisation is taking place in the feeding of UDP-xylose for the biosynthesis of xylan and another polysaccharide such as xyloglucan. However, a higher Km for UDP-xylose of the xylosyltransferases involved in xylan biosynthesis, in comparison to the other xylosyltransferases, may also explain the phenotype. Overall, the evidence suggests that *UXT1* is necessary for xylan biosynthesis; however, it is less clear that these transporters are involved in the biosynthesis of other xylose-containing polysaccharides.

UDP-URONIC ACID TRANSPORTERS

UDP-glucuronic and UDP-galacturonic acid are essential substrates in the plant cell wall biosynthesis. UDP-glucuronic acid is the precursor of other important nucleotide sugars involved in polysaccharide biosynthesis such as UDP-xylose, UDP-arabinose, and UDP-apiose. In addition, it is the precursor for the biosynthesis of UDP-galacturonic, one of the most important substrates required in pectin biosynthesis. As outlined above, UDP-arabinopyranose is synthesised in the Golgi apparatus from UDP-xylose, and it is well documented that some of the synthesis of UDP-xylose, occurring from the decarboxylation of UDP-glucuronic acid occurs in the Golgi apparatus. Therefore, UDP-glucuronic must enter into the Golgi lumen. Furthermore, UDP-glucuronic acid is also utilised in the biosynthesis of glucuronoxylan, and it is also needed for the biosynthesis of RG-II and arabinogalactan proteins in the Golgi lumen. Also, UDP-glucuronic is required to synthesise UDP-galacturonic acid, a process catalysed by UDP-GlcA 4-epimerase, an enzyme located in the Golgi lumen; thus, the transport of UDP-glucuronic acid becomes a critical step in the biosynthesis of hemicelluloses and pectins. However, likely, UDP-galacturonic is also synthesised in the cytoplasm by a salvage pathway. Then, UDP-galacturonic acid may also need to be transported into the Golgi lumen since galacturonosyltransferases are located in that compartment.

The external layer of Arabidopsis seed mucilage is highly enriched in RG-I; therefore, it is quite likely that transporters that provide the nucleotide sugars that form the backbone of galacturonic acid and rhamnose should be expressed in the epidermal cells of the seed coat. The assessment of the number of genes encoding for nucleotide sugar transporters expressed in the epidermal cells allowed the detection of 21 genes. Ten of them were already known, and one belonged to the triose phosphate translocator family. Mutants were obtained for the remaining ten genes, and their contribution to mucilage formation was analysed. In addition to *URGT2*, already mentioned above, the gene At5g04160 emerged as one exhibiting a clear phenotype with reduced mucilage content (Saez-Aguayo et al., 2017). This gene was named *UUAT1* (Figure 3.4) and encoded for a protein that is in the Golgi apparatus and transports both UDP-glucuronic acid and UDP-galacturonic acid in proteoliposomes. The apparent Km is 1.5 mM for UDP-glucuronic acid using UMP as an antiporter molecule, which is very high compared to the affinity of other nucleotide sugar transporters for their substrates. This transporter also exhibits some transport activity towards UDP-arabinopyranose, but no other substrate seems to be transported by UUAT1.

Three mutant lines for *UUAT1* were analysed. The results indicated that in addition to the lower content of galacturonic acid observed in seed mucilage, a lower level of rhamnose was detected, confirming that the synthesis of RG-I was affected in the mutants. In addition, less arabinose was observed in roots and pollen tubes, supporting the hypothesis that transport of UDP-glucuronic acid is required for the synthesis of UDP-arabinose within the Golgi apparatus. Unexpectedly, homogalacturonan's methylation level was increased in the mutant, indicating that compensation mechanisms were activated in the mutant. Studies on the misregulation of genes observed in the *URGT2* mutant support the hypothesis that altering mucilage biosynthesis leads to changes in the expression of genes involved in this process. *UUAT1* is significantly overexpressed in the urgt2 mutant, suggesting that the impairment in the supply of UDP-rhamnose produces an increase in the expression of the transporter for UDP-glucuronic acid (Parra-Rojas et al., 2019). Furthermore, overexpression of *UUAT3* is also observed in the *urgt2* mutant, suggesting that UUAT3 may also be involved in the transport of UDP-glucuronic acid. Additional evidence was provided indicating that mutants in *uuat3* contain lower levels of galacturonic acid and rhamnose in soluble mucilage. These results suggest that *UUAT3* and *UUAT1* are involved in the transport of UDP-glucuronic acid for the biosynthesis of RG-I.

UDP-ARABINOFURANOSE TRANSPORTERS

Arabinose is present in two forms, the pyranose and the furanose forms. However, the most abundant form of arabinose present in cell wall polysaccharides is the arabinofuranose form. The biosynthesis of UDP-arabinofuranose occurs by the epimerisation of UDP-arabinopyranose, a reaction catalysed by a mutase, an enzyme located in the cytoplasm. However, UDP-arabinopyranose is synthesised in the lumen of the Golgi apparatus; thus, transport of both forms of UDP-arabinose must occur but in different directions. The UDP-arabinopyranose must leave the Golgi apparatus to the cytosol, whereas UDP-arabinofuranose must enter the Golgi apparatus. Therefore, there should be two types of transporters for UDP-arabinose (Temple et al. 2016).

The systematic search for NSTs present in the Golgi apparatus and involved in the biosynthesis of cell wall polysaccharides led to the discovery of four Golgi-localised UDP-arabinofuranose transporters from *Arabidopsis thaliana* (Figure 3.4). The apparent Km for these transporters was between 7 and 10 μM.

Arabinose is widely distributed and an abundant sugar in the plant cell wall. The expression of the genes for the four UAfTs showed that they are expressed throughout the plant except for UAfT1, which is mainly expressed in flowers. This result suggests a likely redundancy in the functioning of these transporters. Indeed, experiments to determine the functioning of these transporters in plants using mutants revealed that only mutants in UAfT4 exhibited a decrease in arabinose in leaves (Rautengarten et al., 2017). Therefore, a different strategy was utilised to examine the role of UAfTs in plants; thus, the genes were overexpressed in Arabidopsis wildtype plants. The overexpression of these genes resulted in morphological and biochemical changes such that plants overexpressing UAfT3 and UAfT4 became dwarfs compared to wildtype plants. Furthermore, an increase in arabinose content was also observed in rosette leaves. In addition, the analysis of the glycome profiles of these plants using monoclonal antibodies provided clear evidence that overexpression of UAfT3 and UAfT4 led to higher content of arabinan in the pectic fraction. These results confirm that UAfT3 and UAfT4 are involved in the biosynthesis of Ara-containing polymers in plants.

Interestingly, analyses of the misregulation of genes in the *URGT2* mutant showed that *UAfT2* is highly expressed in the mutant compared to wildtype plants (Parra-Rojas et al., 2019). Changes in the arabinose content accompanied this behaviour, but intriguingly they were higher in the adherent mucilage, the most inner layer, but lower in the soluble mucilage, the most external layer. Further analysis of two allelic mutant lines of *UAfT2* showed reduced levels of arabinose in the adherent mucilage; however, no change in soluble mucilage was observed, although increases in rhamnose and galacturonic acid were detected as well as in other monosaccharides, suggesting that arabinose-containing polymers may play a role in the adherence of mucilage to the seed coat.

GDP-FUCOSE TRANSPORTER

Fucose (6-deoxy L-galactose) is a sugar present in polysaccharides such as xyloglucan, RG-II, and glycoproteins. The substrate for fucosylation reactions is GDP-fucose, a nucleotide sugar derived from GDP-mannose. This interconversion requires the combined activities of the GDP-D-mannose-4,6-dehydratase and the GDP-4-keto-6-deoxy-D-mannose-3,5-epimerase-4-reductase. Mutations in the first enzyme produce a severe reduction of fucose in Arabidopsis, and plants are dwarf, and fucose is replaced by L-galactose (Zablackis et al., 1996).

The interconversion of GDP-mannose to GDP-fucose occurs in the cytoplasm, and Wulff et al. (2000) showed that GDP-fucose is transported into the Golgi apparatus. The search for genes encoding GDP-fucose transporters led to the identification of At5g19980 as the gene encoding a protein capable to transport this nucleotide sugar in the Golgi apparatus (Rautengarten et al., 2016). This gene, named *GFT1* (Figure 3.4), was formerly known as *GONST4*, since it was first identified as part of the *GONST1* homologous sequences (Baldwin et al., 2001; Handford et al., 2004). Despite the fact that GONST1 exhibited activity for GDP-Man, GDP-Glc, GDP-Fuc, and GDP-L-Gal in proteoliposome assays, mutants in *GONST1* have no changes in fucose; instead, they exhibit changes in GIPC glycosylation, suggesting that GONST1 provides GDP-mannose for glycolipid biosynthesis (Mortimer et al. 2013). The *in vitro* analysis of GFT1 activity showed that this transporter preferentially transports GDP-fucose with an apparent Km = 7 μM. The transport activity using GDP-mannose or GDP-glucose was not different from the control; however, using GDP-galactose as substrate, a rather low but significative activity over the control was observed (Rautengarten et al., 2016). Whether the transport of GDP-galactose is a meaningful activity *in vivo* remains to be tested.

The *GFT1* gene is expressed throughout the plant, and the protein is located in the Golgi. In addition, it is known that a reduction in GDP-fucose produces severe phenotypes in the plant; therefore, since no other transporter exhibited a specific role in the transport of GDP-fucose, it could be expected that an impairment in GFT1 could have significant effects on the plant. To assess the role of *GFT1 in vivo*, since no T-DNA insertional plants were identified, an RNA interference approach was developed. Plants exhibited different growth phenotypes, with increased developmental abnormalities correlating with lower contents of *GFT1* transcript. These phenotypes are reminiscent of those found in *mur1* plants, which have a deficiency in the GDP-D-mannose-4,6-dehydratase, the enzyme involved in the first step in the biosynthesis of GDP-fucose (Bonin et al., 1997). The content of fucose in the cell wall from RNAi lines plants was reduced, correlating with lower levels of *GFT1* expression; thus, plants with low expression of *GFT1* had no detectable fucosylated oligosaccharides derived from xyloglucan, lower contents of fucose in glycoproteins as well as in the pectic fraction. In conclusion, GFT1 plays an essential role in plant growth, and the developmental abnormalities observed in the lines with lower contents of *GFT1* transcript are likely due to lower levels of fucose in RG-II, as it occurs in *mur1* plants (Rautengarten et al., 2016).

GDP-L-GALACTOSE TRANSPORTER

GDP-L-galactose is required for the biosynthesis of RG-II, a polysaccharide we already mentioned which plays an important role in development. It is synthesised from GDP-mannose in the cytoplasm (Gilbert et al., 2009) and L-galactose is transferred to the A side chain of RG-II in the lumen of the Golgi. Sechet et al. (2018) provided genetic evidence that *GONST3* is involved in GDP-L-galactose transport and renamed the gene as Golgi GDP-L-galactose transporter1 (GGLT1) (Figure 3.4). *GGLT1*-silenced plants exhibit a reduction of L-galactose in RG-II, affecting the ability of RG-II to form dimers. The reduction in *GGLT1* also correlates with a dwarf phenotype, like *mur1* plants, suggesting that the RG-II structure is impaired. Despite the lack of a functional characterisation, all the evidence suggests that GGLT1is a transporter for GDP-L-galactose.

A PUTATIVE CMP-KDO TRANSPORTER

The eight-carbon sugar acid 3-deoxy-D-manno-2-octulosonic acid (KDO) is present in RG-II and it seems to be essential for the role of this polysaccharide in plant development (Kobayashi et al., 2011). KDO is similar to N-acetylneuraminic acid (sialic acid, SA), a monosaccharide found in mammalian cells, and it is activated to CMP-KDO (Kobayashi et al., 2011) as it occurs with CMP-sialic acid. As mentioned before, the transport of CMP-sialic acid into the Golgi was the first transport activity ever reported, the gene has been cloned and the crystal structure of the protein solved (Ahuja and Whorton, 2019). Intriguingly, orthologous sequences to the gene encoding for the mammalian CMP-sialic acid transporter gene are present in plant genomes. In addition, different lines of evidence have confirmed that these plant orthologue sequences can efficiently transport CMP-sialic acid both *in vitro* and *in vivo* (Bakker et al., 2008; Takashima et al., 2009; Nji et al., 2019). Since plants do not contain sialic acid, the question is "what is the actual function of these transporters in plants?" Given the similarity between sialic acid and KDO, it has been proposed that they transport CMP-KDO (Figure 3.5), an interesting hypothesis that remains to be tested.

UDP-GLcNAc TRANSPORTER

N-Acetylglucosamine is an important monosaccharide in glycoproteins, but it is not present in plant cell wall polysaccharides. The substrate donor for the incorporation of this sugar into glycoproteins is UDP-N-acetylglucosamine, which is utilised by the enzyme N-acetylglucosaminyltransferase located in the Golgi apparatus, organelle where the modification of the N-linked oligosaccharide takes place (Strasser, 2022). A Golgi-localised UDP-glcNac transporter (UGNT1) was identified and characterised (Figure 3.4) (Ebert et al., 2018). UGNT1 shows high specificity for UDP-glcNac and an apparent Km of 9 μM. Mutants in this gene exhibited altered content of complex and hybrid N-glycans, which is expected if the supply of UDP-glcNAc into the Golgi apparatus is impaired. Furthermore, glycolipids containing N-acetylglucosamine were affected in the *UGNT1* mutant, thus, providing strong evidence that UDP-glcNac supplied by this transporter is needed both for proteins and lipids glycosylation. On the other hand, Niemann et al. (2015) reported a protein (ROCK1) that was able to transport UDP-glcNac and UDP-galNac *in vitro*. However, this protein was located in the endoplasmic reticulum, and the analyses of mutants in this gene did not alter the N-glycosylation of proteins compared to wild type (Ebert et al., 2018); therefore, its function in the endoplasmic reticulum remains unclear.

THE STRUCTURE OF NUCLEOTIDE SUGAR TRANSPORTERS

Parker and Newstead (2017) provided the first insights into the structure of the nucleotide sugar transporters by determining the crystal structure of the GDP-mannose transporter from yeast (VRG4) to a 3.2 Å resolution. This protein consisted of 10 transmembrane domains' α-helices connected by short loops. The even number of transmembrane domains determine that the N-terminus and the C-terminus are on the same side of the membrane, likely in the cytosolic face of the Golgi apparatus.

The analysis of the crystal structure in a state where GDP-mannose was bound revealed separate binding pockets for the sugar and the nucleotide moieties and consensus motifs for the mannose (GALNK) and the guanine (FYNN) in the transmembrane domains TM9 and TM7, respectively. The crystal structure also revealed that VRG4 is a dimer where both monomers face the membrane in the same way. This result confirms previous findings that nucleotide sugar transporters, such as GDP-fucose and UDP-N-acetylgalactosamine, are dimers (Parker and Newstead, 2017). Furthermore, lipids seem to play an important role in the dimerisation of VRG4. Even more, short chain lipids seem to regulate VRG4 activity (Parker et al., 2019).

CMP-sialic acid is not present in plants but is a very important nucleotide sugar in animal cells. CMP-sialic acid is transported into the Golgi lumen by a transporter. The crystal structure for this protein was also determined in complex with CMP and CMP-Sia at 2.6 and 2.8 Å resolution, respectively. The analysis of the structure of the CMP-sialic acid transporter revealed that the CMP and sialic acid moieties have different binding pockets. The presence of separate binding pockets for the nucleotide and the sugar is a feature also found in the GDP-mannose transporter VRG4. Interestingly, as mentioned above, plants (*Zea mays*) have a nucleotide sugar transporter that shares 80%–92% to the mammalian CMP-sialic acid transporter. The crystal structure of this protein was determined in complex with CMP and utilised to study the exchange of CMP-sialic acid and CMP (Nji et al., 2019). Likely, this transporter may be used to transport CMP-KDO in plants; then, the results could be extrapolated to understand the transport of this nucleotide sugar in plants.

FUTURE CHALLENGES

Significant progress in the identification of nucleotide sugar transporters involved in cell wall biosynthesis has been made in the last 10 years. Yet, a few of them are still to be discovered (UDP-apiose, UDP-arabinopyranose, to mention some of them). Furthermore, there are monosaccharides present in RG-II, which play a role in plant development, such as Dha (2-keto-3-deoxy-D-lyxo-heptulosaric acid) and aceric acid (3-C-carboxy-5-deoxy-L-xylose), whose activated forms remain unidentified and thus, transporters for them, that remain to be discovered, might also be required (Bar-Peled et al., 2012). Since there are still members of the NST gene family that have not been characterised, it is likely that they may be involved in the transport of these nucleotide sugars.

Regarding the mechanism of action, there are questions regarding how nucleotide sugar transporters can transport different substrates that have differences in their sugar moieties like the URGTs for instance. Likely, the analysis of their protein structure, along with the introduction of point mutations coupled to functional analysis, should lead to answers in this regard. On the other hand, the importance of making dimers and the role of lipids in this process emerged as an interesting regulatory mechanism that needs to be addressed in NSTs other than VRG4.

Plant cell wall polysaccharides have large molecular masses and are often highly branched, and they likely need to be synthesised in a very coordinated manner. The example of changing the expression of *URGT1* and *URGT2* by swapping their promoters (Celiz-Balboa et al., 2020) suggests that channelling of substrate into different polysaccharides depends upon the cellular context where the polymers are synthesised. Then, one of the questions is how the different players, glycosyltransferases, and acceptors are coordinated with the NSTs. One possibility is that the formation of protein complexes may expedite the biosynthetic process. Of course, there are questions regarding specificity that remained unanswered. Still, the crystal structure of some of them will help to perform molecular dynamics and understand the mechanisms for many of them. In this regard, tools like AlphaFold will be very valuable to provide insights in the protein structure of nucleotide sugar transporters.

ACKNOWLEDGEMENT

I would like to acknowledge current and former students/postdocs who contributed at different stages of the work in my laboratory.

REFERENCES

Abeijon C, Mandon EC, Robbins PW, Hirschberg CB (1996a) A mutant yeast deficient in Golgi transport of uridine diphosphate N-acetylglucosamine. *Journal of Biological Chemistry*, 271: 8851–8854.

Abeijon C, Robbins PW, Hirschberg, CB (1996b). Molecular cloning of the Golgi apparatus uridine diphosphate-N-acetylglucosamine transporter from *Kluyveromyces lactis. Proceedings of the National Academy of Sciences*, 93: 5963–5968.

Abeijon C, Yanagisawa K, Mandon EC, Häusler A, Moremen K, Hirschberg CB, Robbins PW (1993). Guanosine diphosphatase is required for protein and sphingolipid glycosylation in the Golgi lumen of Saccharomyces cerevisiae. *The Journal of Cell Biology*, 122: 307–323.

Ahuja S, Whorton MR (2019). Structural basis for mammalian nucleotide sugar transport. *eLife*, 8: e45221.

Amos RA, Mohnen D (2019). Critical review of plant cell wall matrix polysaccharide glycosyltransferase activities verified by heterologous protein expression. *Frontiers in Plant Science*, 10: 915.

Arioli T, Peng L, Betzner AS, Burn J, Wittke W, Herth W, ... Williamson RE (1998). Molecular analysis of cellulose biosynthesis in Arabidopsis. *Science*, 279: 717–720.

Arsovski AA, Haughn GW, Western TL (2010). Seed coat mucilage cells of *Arabidopsis thaliana* as a model for plant cell wall research. *Plant Signaling & Behavior*, 5: 796–801.

Atmodjo MA, Sakuragi Y, Zhu X, Burrell AJ, Mohanty SS, Atwood III JA, ... Mohnen D (2011). Galacturonosyltransferase (GAUT) 1 and GAUT7 are the core of a plant cell wall pectin biosynthetic homogalacturonan: galacturonosyltransferase complex. *Proceedings of the National Academy of Sciences*, 108: 20225–20230.

Bakker H, Routier F, Ashikov A, Neumann D, Bosch D, Gerardy-Schahn R. (2008). A CMP-sialic acid transporter cloned from Arabidopsis thaliana. *Carbohydrate Research*, 343: 2148–2152.

Baldwin TC, Handford MG, Yuseff MI, Orellana A, Dupree P (2001). Identification and characterization of GONST1, a Golgi-localized GDP-mannose transporter in Arabidopsis. *The Plant Cell*, 13: 2283–2295.

Barber GA (1966). GDP-glucose pyrophosphorylase from peas. *Methods in Enzymology*, 8: 266–268.

Bar-Peled M, O'Neill MA (2011). Plant nucleotide sugar formation, interconversion, and salvage by sugar recycling. *Annual Review of Plant Biology*, 62: 127–155.

Bar-Peled M, Urbanowicz BR, O'Neill MA (2012). The synthesis and origin of the pectic polysaccharide rhamnogalacturonan II-insights from nucleotide sugar formation and diversity. *Frontiers in Plant Science*, 3: 92.

Bonin CP, Potter I, Vanzin GF, Reiter WD (1997). The *MUR1* gene of Arabidopsis thaliana encodes an isoform of GDP-D-mannose-4, 6-dehydratase, catalyzing the first step in the de novo synthesis of GDP-L-fucose. *Proceedings of the National Academy of Sciences*, 94: 2085–2090.

Cabib E (1963). Carbohydrate metabolism. *Annual Review of Biochemistry*, 32: 321–354.

Capasso JM, Hirschberg CB (1984). Mechanisms of glycosylation and sulfation in the Golgi apparatus: evidence for nucleotide sugar/nucleoside monophosphate and nucleotide sulfate/nucleoside monophosphate antiports in the Golgi apparatus membrane. *Proceedings of the National Academy of Sciences*, 81: 7051–7055.

Cardini CE, Paladini AC, Caputto R, Leloir LF (1950). Uridine diphosphate glucose: the coenzyme of the galactose-glucose phosphate isomerization. *Nature*, 165: 191–192.

Carey DJ, Sommers LW, Hirschberg CB (1980). CMP-N-acetylneuraminic acid: isolation from and penetration into mouse liver microsomes. *Cell*, 19: 597–605.

Celiz-Balboa J, Largo-Gosens A, Parra-Rojas JP, Arenas-Morales V, Sepulveda-Orellana P, Salinas-Grenet H, Saez-Aguayo S, Orellana A (2020). Functional interchangeability of nucleotide sugar transporters URGT1 and URGT2 reveals that *urgt1* and *urgt2* cell wall chemotypes depend on their spatio-temporal expression. *Frontiers in Plant Science*, 11: 594544.

Cocuron JC, Lerouxel O, Drakakaki G, Alonso AP, Liepman AH, Keegstra K, ... Wilkerson CG (2007). A gene from the cellulose synthase-like C family encodes a β-1,4 glucan synthase. *Proceedings of the National Academy of Sciences*, 104: 8550–8555.

Daras G, Templalexis D, Avgeri F, Tsitsekian D, Karamanou K, Rigas S (2021). Updating insights into the catalytic domain properties of plant cellulose synthase (CESA) and cellulose synthase-like (CSL) Proteins. *Molecules*, 26: 4335.

Davis J, Brandizzi F, Liepman AH, Keegstra K (2010). Arabidopsis mannan synthase CSLA9 and glucan synthase CSLC4 have opposite orientations in the Golgi membrane. *The Plant Journal*, 64: 1028–1037.

Dean N, Zhang YB, Poster JB (1997). The *VRG4* gene is required for GDP-mannose transport into the lumen of the Golgi in the yeast, *Saccharomyces cerevisiae*. *Journal of Biological Chemistry*, 272: 31908–31914.

Dunkley TP, Hester S, Shadforth IP, Runions J, Weimar T, Hanton SL,... Lilley KS (2006). Mapping the Arabidopsis organelle proteome. *Proceedings of the National Academy of Sciences*, 103: 6518–6523.

Ebert B, Rautengarten C, Guo X, Xiong G, Stonebloom S, Smith-Moritz AM, ... Scheller HV (2015). Identification and characterization of a Golgi-localized UDP-xylose transporter family from Arabidopsis. *The Plant Cell*, 27: 1218–1227.

Ebert B, Rautengarten C, McFarlane HE, Rupasinghe T, Zeng W, Ford K,... Heazlewood JL (2018). A Golgi UDP-GlcNAc transporter delivers substrates for N-linked glycans and sphingolipids. *Nature Plants*, 4: 792–801.

Feingold DS, Neufeld EF, Hassid WZ (1958). Synthesis of a β-1,3-linked glucan by extracts of *Phaseolus aureus* seedlings. *Journal of Biological Chemistry*, 233: 783–788.

Gilbert L, Alhagdow M, Nunes-Nesi A, Quemener B, Guillon F, Bouchet B, ... Baldet P (2009). GDP-D-mannose 3, 5-epimerase (GME) plays a key role at the intersection of ascorbate and non-cellulosic cell-wall biosynthesis in tomato. *The Plant Journal*, 60: 499–508.

Handford MG, Sicilia F, Brandizzi F, Chung JH, Dupree P (2004). *Arabidopsis thaliana* expresses multiple Golgi-localised nucleotide-sugar transporters related to GONST1. *Molecular Genetics and Genomics*, 272: 397–410.

Harper AD, Bar-Peled M (2002). Biosynthesis of UDP-xylose. Cloning and characterization of a novel Arabidopsis gene family, UXS, encoding soluble and putative membrane-bound UDP-glucuronic acid decarboxylase isoforms. *Plant Physiology*, 130: 2188–2198.

Hirschberg CB (2018). My journey in the discovery of nucleotide sugar transporters of the Golgi apparatus. *Journal of Biological Chemistry*, 293: 12653–12662.

Ito J, Herter T, Baidoo EE, Lao J, Vega-Sánchez ME, Smith-Moritz AM,... Heazlewood JL (2014). Analysis of plant nucleotide sugars by hydrophilic interaction liquid chromatography and tandem mass spectrometry. *Analytical Biochemistry*, 448: 14–22.

Jing B, Ishikawa T, Soltis N, Inada N, Liang Y, Murawska G, ... Mortimer JC (2021). The *Arabidopsis thaliana* nucleotide sugar transporter GONST2 is a functional homolog of GONST1. *Plant Direct*, 5: e00309.

Kobayashi M, Kouzu N, Inami A, Toyooka K, Konishi Y, Matsuoka K, Matoh T (2011). Characterization of Arabidopsis CTP: 3-deoxy-D-manno-2-octulosonate cytidylyltransferase (CMP-KDO synthetase), the enzyme that activates KDO during rhamnogalacturonan II biosynthesis. *Plant and Cell Physiology*, 52: 1832–1843.

Konishi T, Takeda T, Miyazaki Y, Ohnishi-Kameyama M, Hayashi T, O'Neill MA, Ishii T (2007). A plant mutase that interconverts UDP-arabinofuranose and UDP-arabinopyranose. *Glycobiology*, 17: 345–354.

Kuang B, Zhao X, Zhou C, Zeng W, Ren J, Ebert B, ... Wu AM (2016). Role of UDP-glucuronic acid decarboxylase in xylan biosynthesis in Arabidopsis. *Molecular Plant*, 9: 1119–1131.

Liepman AH, Wilkerson CG, Keegstra K (2005). Expression of cellulose synthase-like (Csl) genes in insect cells reveals that CslA family members encode mannan synthases. *Proceedings of the National Academy of Sciences*, 102: 2221–2226.

Liu Y, Ma L, Cao D, Gong Z, Fan J, Hu H, Jin X (2021). Investigation of cell wall proteins of *C. sinensis* leaves by combining cell wall proteomics and N-glycoproteomics. *BMC Plant Biology*, 21: 1–15.

McNamara JT, Morgan JL, Zimmer J (2015). A molecular description of cellulose biosynthesis. *Annual Review of Biochemistry*, 84: 895–921.

Miura N, Ishida N, Hoshino M, Yamauchi M, Hara T, Ayusawa D, Kawakita M (1996). Human UDP-galactose translocator: molecular cloning of a complementary DNA that complements the genetic defect of a mutant cell line deficient in UDP-galactose translocator. *The Journal of Biochemistry*, 120: 236–241.

Mortimer JC, Yu X, Albrecht S, Sicilia F, Huichalaf M, Ampuero D,... Dupree P (2013). Abnormal glycosphingolipid mannosylation triggers salicylic acid-mediated responses in Arabidopsis. *The Plant Cell*, 25: 1881–1894.

Muñoz P, Norambuena L, Orellana A (1996). Evidence for a UDP-glucose transporter in Golgi apparatus-derived vesicles from pea and its possible role in polysaccharide biosynthesis. *Plant Physiology*, 112: 1585–1594.

Neckelmann G, Orellana A (1998). Metabolism of uridine 5′-diphosphate-glucose in Golgi vesicles from pea stems. *Plant Physiology*, 117: 1007–1014.

Niemann MC, Bartrina I, Ashikov A, Weber H, Novák O, Spíchal L, ... Werner T (2015). Arabidopsis ROCK1 transports UDP-GlcNAc/UDP-GalNAc and regulates ER protein quality control and cytokinin activity. *Proceedings of the National Academy of Sciences*, 112: 291–296.

Nji E, Gulati A, Qureshi AA, Coincon M, Drew D (2019). Structural basis for the delivery of activated sialic acid into Golgi for sialyation. *Nature Structural & Molecular Biology*, 26: 415–423.

Norambuena L, Marchant L, Berninsone P, Hirschberg CB, Silva H, Orellana A (2002). Transport of UDP-galactose in plants: identification and functional characterization of AtUTR1, an *Arabidopsis thaliana* UDP-galactose/UDP-glucose transporter. *Journal of Biological Chemistry*, 277: 32923–32929.

Northcote DH, Pickett-Heaps JD (1966). A function of the Golgi apparatus in polysaccharide synthesis and transport in the root-cap cells of wheat. *Biochemical Journal*, 98: 159–167.

Ordin L, Hall MA (1968). Cellulose synthesis in higher plants from UDP glucose. *Plant Physiology*, 43: 473.

Parker JL, Corey RA, Stansfeld PJ, Newstead S (2019). Structural basis for substrate specificity and regulation of nucleotide sugar transporters in the lipid bilayer. *Nature Communications*, 10: 1–10.

Parker JL, Newstead S (2017). Structural basis of nucleotide sugar transport across the Golgi membrane. *Nature*, 551: 521–524.

Parra-Rojas JP, Largo-Gosens A, Carrasco T, Celiz-Balboa J, Arenas-Morales V, Sepúlveda-Orellana P, ... Orellana A (2019). New steps in mucilage biosynthesis revealed by analysis of the transcriptome of the UDP-rhamnose/UDP-galactose transporter 2 mutant. *Journal of Experimental Botany*, 70: 5071–5088.

Pear JR, Kawagoe Y, Schreckengost WE, Delmer DP, Stalker DM (1996). Higher plants contain homologs of the bacterial celA genes encoding the catalytic subunit of cellulose synthase. *Proceedings of the National Academy of Sciences*, 93: 12637–12642.

Perrin RM, DeRocher AE, Bar-Peled M, Zeng W, Norambuena L, Orellana A, Raikhel NV, Keegstra K (1999) Xyloglucan fucosyltransferase, an enzyme involved in plant cell wall biosynthesis. *Science* 284:1976–1979.

Polko JK, Kieber JJ (2019). The regulation of cellulose biosynthesis in plants. *The Plant Cell*, 31: 282–296.

Ralet MC, Crépeau MJ, Vigouroux J, Tran J, Berger A, Sallé C, Granier F, Botran L, North HM (2016). Xylans provide the structural driving force for mucilage adhesion to the Arabidopsis seed coat. *Plant Physiology*, 171: 165–178.

Rautengarten C, Birdseye D, Pattathil S, McFarlane HE, Saez-Aguayo S, Orellana A, ... Ebert B (2017). The elaborate route for UDP-arabinose delivery into the Golgi of plants. *Proceedings of the National Academy of Sciences*, 114: 4261–4266.

Rautengarten C, Ebert B, Herter T, Petzold CJ, Ishii T, Mukhopadhyay A, ... Scheller HV (2011). The interconversion of UDP-arabinopyranose and UDP-arabinofuranose is indispensable for plant development in Arabidopsis. *The Plant Cell*, 23: 1373–1390.

Rautengarten C, Ebert B, Liu L, Stonebloom S, Smith-Moritz AM, Pauly M, ... Heazlewood JL (2016). The Arabidopsis Golgi-localized GDP-L-fucose transporter is required for plant development. *Nature Communications*, 7: 1–10.

Rautengarten C, Ebert B, Moreno I, Temple H, Herter T, Link B, ... Mortimer JC (2014). The Golgi localized bifunctional UDP-rhamnose/UDP-galactose transporter family of Arabidopsis. *Proceedings of the National Academy of Sciences*, 111: 11563–11568.

Ray PM, Shininger TL, Ray MM (1969). Isolation of β-glucan synthetase particles from plant cells and identification with Golgi membranes. *Proceedings of the National Academy of Sciences*, 64: 605–612.

Reiter WD, Chapple CC, Somerville CR (1993). Altered growth and cell walls in a fucose-deficient mutant of Arabidopsis. *Science*, 261: 1032–1035.

Reyes F, León G, Donoso M, Brandizzi F, Weber AP, Orellana A (2010). The nucleotide sugar transporters AtUTr1 and AtUTr3 are required for the incorporation of UDP-glucose into the endoplasmic reticulum, are essential for pollen development and are needed for embryo sac progress in *Arabidopsis thaliana*. *The Plant Journal*, 61: 423–435.

Reyes F, Marchant L, Norambuena L, Nilo R, Silva H, Orellana A (2006). AtUTr1, a UDP-glucose/UDP-galactose transporter from *Arabidopsis thaliana*, is located in the endoplasmic reticulum and up-regulated by the unfolded protein response. *Journal of Biological Chemistry*, 281: 9145–9151.

Ruiz-May E, Kim SJ, Brandizzi F, Rose JK (2012). The secreted plant N-glycoproteome and associated secretory pathways. *Frontiers in Plant Science*, 3: 117.

Saez-Aguayo S, Parra-Rojas JP, Sepúlveda-Orellana P, Celiz-Balboa J, Arenas-Morales V, Sallé C, ... Orellana A (2021). Transport of UDP-rhamnose by URGT2, URGT4, and URGT6 modulates rhamnogalacturonan-I length. *Plant Physiology*, 185: 914–933.

Saez-Aguayo S, Rautengarten C, Temple H, Sanhueza D, Ejsmentewicz T, Sandoval-Ibañez O, ... Orellana A. (2017). UUAT1 is a Golgi-localized UDP-uronic acid transporter that modulates the polysaccharide composition of Arabidopsis seed mucilage. *The Plant Cell*, 29: 129–143.

Sechet J, Htwe S, Urbanowicz B, Agyeman A, Feng W, Ishikawa T, ... Mortimer JC (2018). Suppression of Arabidopsis GGLT 1 affects growth by reducing the L-galactose content and borate cross-linking of rhamnogalacturonan-II. *The Plant Journal*, 96: 1036–1050.

Sterling JD, Quigley HF, Orellana A, Mohnen D (2001). The catalytic site of the pectin biosynthetic enzyme α-1, 4-galacturonosyltransferase is located in the lumen of the Golgi. *Plant Physiology*, 127: 360–371.

Strasser R (2022). Recent developments in deciphering the biological role of plant complex N-glycans. *Frontiers in Plant Science*, 13: 897549.

Takashima S, Seino J, Nakano T, Fujiyama K, Tsujimoto M, Ishida N, Hashimoto Y (2009). Analysis of CMP-sialic acid transporter-like proteins in plants. *Phytochemistry*, 70: 1973–1981.

Temple H, Saez-Aguayo S, Reyes FC, Orellana A (2016). The inside and outside: topological issues in plant cell wall biosynthesis and the roles of nucleotide sugar transporters. *Glycobiology*, 26: 913–925.

Wu AM, Rihouey C, Seveno M, Hörnblad E, Singh S K, Matsunaga T, ... Marchant A (2009). The Arabidopsis IRX10 and IRX10-LIKE glycosyltransferases are critical for glucuronoxylan biosynthesis during secondary cell wall formation. *The Plant Journal*, 57: 718–731.

Wulff C, Norambuena L, Orellana A (2000). GDP-fucose uptake into the Golgi apparatus during xyloglucan biosynthesis requires the activity of a transporter-like protein other than the UDP-glucose transporter. *Plant Physiology*, 122: 867–878.

York WS, Darvill AG, McNeil M, Albersheim P (1985). 3-Deoxy-D-manno-2-octulosonic acid (KDO) is a component of rhamnogalacturonan II, a pectic polysaccharide in the primary cell walls of plants. *Carbohydrate Research*, 138: 109–126.

Zablackis E, York WS, Pauly M, Hantus S, Reiter WD, Chapple CC, ... Darvill A (1996). Substitution of L-fucose by L-galactose in cell walls of Arabidopsis *mur1*. *Science*, 272: 1808–1810.

Zhao X, Liu N, Shang N, Zeng W, Ebert B, Rautengarten C,... Wu AM (2018). Three UDP-xylose transporters participate in xylan biosynthesis by conveying cytosolic UDP-xylose into the Golgi lumen in Arabidopsis. *Journal of Experimental Botany*, 69: 1125–1134.

4 Enzymic and Non-Enzymic Remodelling of Cell-Wall Polysaccharides

Stephen C. Fry
The University of Edinburgh

INTRODUCTION: ACTIVITY VERSUS ACTION

The plant cell wall is principally composed of diverse polysaccharides, classified as matrix polymers (pectin and hemicelluloses) plus microfibrillar cellulose. Contrary to former assumptions, the wall is not an inert 'cardboard box'—its structural polymers undergo various interesting and important chemical changes *in vivo*, such as during organ growth, fruit ripening and abscission, and these changes may be promoted or inhibited by endogenous signals (hormones, etc.) and by environmental stimuli (Albersheim et al. 2010). Intimately involved in such metabolism within the wall, numerous enzymes exist in the apoplast (the extra-protoplasmic fluid permeating the cell wall): some freely soluble in the apoplastic fluid (thus extractable by low-salt buffers, such as during a vacuum-infiltration–centrifugation protocol), some ionically bound to acidic wall polysaccharides (thus extractable with salt solutions), and some probably covalently bound to structural polymers of the cell wall and thus very difficult to extract. In plant cell-suspension cultures, the culture medium is effectively a greatly extended apoplast; therefore, such cultures are very convenient for obtaining freely soluble apoplastic enzymes (collected simply by filtering or centrifuging the culture) and ionically bound wall enzymes (collected in the same way but after addition of salt to the medium).

Early work on the role of cell walls in controlling cell expansion had focused on their biophysical properties (plasticity or extensibility) with less emphasis on chemical properties (Frey-Wyssling 1959; Pilet 1971; Preston 1974) and less still on wall enzymes. Nevertheless, Pilet (1971) was already drawing attention to the presence and possible physiological roles of wall-localised enzyme activities, including phenol oxidase, auxin oxidase, ascorbate oxidase, peroxidase, ATPase and phosphatase, leading him to the concept of a 'living cell wall'. Importantly, certain wall-polysaccharide-attacking hydrolase activities were being documented, such as 'hemicellulases' (such as β-glucanases), pectin methylesterase and polygalacturonase (PG). Preston (1974) concluded that

> the ultimate effect as far as [auxin-induced] growth is concerned must presumably be the passage into the cell wall, or activation within the wall, of a substance active in bond breaking and making, and it could be envisaged that this should be an enzyme. There is, however, as yet no idea what this enzyme (if any) might be.

Preston further speculated that 'the effects of all such degrading enzymes are irreversible whereas auxin effects are reversible so that none of them can be a candidate unless some parallel "repairing" system is also present'—perhaps prescient of the discovery of the 'cutting and pasting' abilities of xyloglucan endotransglucosylase (XET) activity (see later). His parting shot was,

> In any event, the primary wall is clearly a complex system delicately constructed and under delicately balanced regulatory control. There can be no longer any doubt that its mechanical properties are of concern in growth only because it is part of the living system, a cell organelle rather than a dead excretory product.
>
> **(Preston 1974).**

This point of view established the excitement of documenting the wall's enzyme activities—subsequent discoveries being documented in the present chapter. Since 1974, numerous enzyme activities have been found in the cell wall, many of them potentially capable of acting on wall polysaccharides and loosening the cell wall to enable cell growth.

Wall enzymes possess a wide range of activities, as assayed enzymologically *in vitro* after extraction. The wall contains some enzymes that serve in the *de novo* synthesis of certain non-polysaccharide polymers, especially cutin, suberin and lignin, and cellulose is manufactured by enzymes located in the plasma membrane. However, the wall's matrix polysaccharides (except callose) are synthesised intraprotoplasmically and then released, essentially fully formed, into the wall: this chapter focuses on the subsequent metabolic fate of matrix polysaccharides in the walls of living cells. It will be shown later that such post-secretion metabolism is only partly enzymic, and that there are additional non-enzymic reactions modifying certain wall polysaccharides.

Wall enzymes are produced as a result of transcription (mRNA synthesis) and translation (protein synthesis).

DOI: 10.1201/9781003178309-4

In order to contribute to wall metabolism, the enzymes then need to be secreted into the apoplast. Transcription, translation and secretion are necessary, but not sufficient, for claiming any biological 'contribution'. A tissue's transcriptome and proteome can suggest such contributions, but not prove them: e.g., an mRNA may not be translated; an enzymic protein may not be secreted into the wall; and a secreted enzyme may lack activity (e.g., it may require post-translational modifications). It certainly cannot be assumed without additional evidence that an mRNA observed in the transcriptome leads to any kind of structural cell-wall remodelling *in vivo*.

Furthermore, an important distinction is drawn between enzyme activity and enzyme action. Even an 'active' enzyme located in the cell wall at the right time and the right place may not necessarily be catalysing any reaction (Prakash et al. 2012). The wall's pH may be inappropriate for the enzyme, an inhibitor may be present or a necessary activator may be absent. An 'active' enzyme is one which, after extraction, can be demonstrated enzymologically to be capable of catalysing a given reaction on a given wall-derived polysaccharide *in vitro*. Activity is measured in katals under optimised conditions *in vitro*; but such conditions may not prevail *in vivo*. In contrast, action is what can be observed *in vivo*, in the 'rough and tumble' of the wall of a living cell (Fry 2004). Enzyme action is the gold standard in this field, demonstrating reactions that do occur *in vivo*, not simply reactions that *might* occur. Direct evidence for enzyme action can be provided by analysis of changes in wall polysaccharide chemistry during development. The present chapter carefully distinguishes activity and action, and explores both.

HYDROLASE ACTIVITIES ASSAYED *IN VITRO*

What they are: A hydrolase is an enzyme catalyses hydrolysis—cutting a chemical bond (X–Y) by adding H_2O across it:

$$X - Y + H_2O \rightarrow X - OH + H - Y$$

Such a reaction is almost always irreversible. Whether it is the X or the Y group that ends up with the O atom from the water molecule depends on the chemical nature of X and Y, and on the enzyme's specificity.

Known wall-localised hydrolases are capable of cleaving numerous types of glycosidic bonds (such as in polysaccharides and oligosaccharides), peptide bonds (in proteins and glycoproteins), carboxy-ester bonds (such as in cutin, pectin, *O*-acetylesters and *O*-feruloylesters) and phosphate esters. This chapter focuses on hydrolases that cut glycosidic bonds (i.e. glycosylhydrolases or glycoside hydrolases) in matrix polysaccharides. In this case, it is the former glycosyl group (=glycoside; i.e. not the aglycone or leaving group) that inherits the O atom from H_2O. [A rule of thumb is that the 'glycosyl' group in this context is the sugar ring whose carbon in the C–O–C linkage to be cleaved is attached to two O atoms. Note that, although etymologically 'aglycone' means 'not sugar', chemically speaking an aglycone can itself be a sugar: e.g., both lactose and methyl β-galactoside can be represented as galactosyl–Y, with the aglycone 'Y' being glucose and methanol, respectively.]

A glycoside hydrolase cuts its polysaccharide substrate either in mid-chain, such enzymes being described here as glycanases (= endo-glycoside hydrolases), or at a non-reducing terminus (NRT), described here as glycosidases (= exo-glycoside hydrolases). Very few enzymes (none from plants?) cleave preferentially at the reducing terminus (RT). A very convenient class of substrates for assaying both these broad types of glycoside hydrolases is medium-sized oligosaccharides, typically hexasaccharides, many of which are available commercially. A glycanase cleaves a hexasaccharide in mid-chain, producing, e.g., two trisaccharide molecules. Conversely, a glycosidase cleaves a hexasaccharide at the NRT, producing a monosaccharide plus a pentasaccharide. If both types of glycoside hydrolase are present in a crude plant extract, it is often possible to monitor both of them simultaneously: rapid release of the monosaccharide indicating glycosidase action and rapid trisaccharide production (before the glycosidase has pared the hexasaccharide down the trisaccharide by sequential removal of monosaccharides), indicating glycanase action.

Glycanase Activities Assayed *In Vitro*

Plant cell walls contain glycanase activities that catalyse the mid-chain cleavage of numerous different polysaccharides (Table 4.1). A survey of such activities in extracts of diverse plants is given in GHATAbase (https://fry.bio.ed.ac.uk/GHATAbase.html) (GlycosylHydrolase And Transglycosylase Activity database), derived from Franková and Fry (2011).

Oligosaccharides serving as useful substrates for known glycanases include β-(1→4)-xylohexaose, β-(1→4)-glucohexaose [= cellohexaose], β-(1→3)-glucohexaose [= laminarihexaose], β-(1→4)-mannohexaose, α-(1→5)-arabinohexaose and α-(1→4)-galacturonohexaose. The exception to this rule is substrates for glycanases that cleave xyloglucan since such enzymes require a minimum backbone of ~8 glucose (Glc) residues bearing ~6 side-chain xylose (Xyl) residues for optimal activity.

Glycanases, by virtue of their ability to cleave polysaccharides in mid-chain, are widely assumed to play roles in loosening the cell wall, such as facilitating cell expansion, fruit softening and leaf abscission (Ayour et al. 2021; Wang et al. 2020; Forlani et al. 2019; Defilippi et al. 2018; Lohani et al. 2004; Witasari et al. 2019; Duan et al. 2008). Glycanase activity assays always need to be carefully evaluated to check that they provide data specifically on the reaction of interest. Importantly, published enzyme assays have often not satisfactorily distinguished between certain activities. For example, PG activity in fruit extracts was often assayed as *in vitro* production of 'galacturonic acid' (GalA) equivalents (total reducing groups) from the substrate homogalacturonan (Figueroa et al. 2010; Villarreal

TABLE 4.1
A Selection of Plant Glycanase Activities Potentially Targeting Cell-Wall Polysaccharides

Enzyme Activity	Potential Substrates in the Primary Cell Wall
Endo-α-1,4-polygalacturonase (EPG)	Homogalacturonan (de-esterified domains)
• β-1,4-Mannanase	Mannans and hetero-mannans
• β-1,4-Glucanase (cellulase)	Cellulose, mixed-linkage (1→3),(1→4)-β-glucan (MLG)
• Xyloglucanase (xyloglucan endohydrolase; XEH)	Xyloglucan
• β-1,4-Xylanase	Xylans and hetero-xylans
• α-1,5-Arabinanase	Neutral side-chains of RG-I
• β-1,4-Galactanase	Neutral side-chains of RG-I
• Rhamnogalacturonase	RG-I. Cleaves GalA-(1→2)-Rha bonds, not Rha-(1→4)-GalA
• β-1,3-Glucanase (laminarinase)	Callose
• Lichenase	MLG

et al. 2009; Zhou et al. 2015, based on an influential paper by Gross 1982); however, reducing groups can be generated from homogalacturonan by endo-polygalacturonase (EPG), α-galacturonidase (= exo-PG) and pectate lyase (PL), so these three enzyme activities would not be distinguished in this assay.

The use of RG-I as substrate in place of homogalacturonan allows the assay of rhamnogalacturonanase, which catalyses hydrolytic endo- or exo-cleavage of a GalA–Rha bond in RG-I, where Rha=rhamnose. Its activity has been reported in fruits of tomato, apple and grape (Gross et al. 1995) and in carrot roots (Stratilová et al. 1998).

Glycosidase Activities Assayed *In Vitro*

The terminal sugar residue cleaved off by a glycosidase may be located either at the non-reducing end of the polysaccharide's backbone or at the end of a side-chain. In principle, it is not usually the enzyme that decides which of these two options operates; rather, it depends on the structure of the substrate polysaccharide. For example, β-D-galactosidase can shave off terminal β-D-Gal (where Gal=galactose) residues present as a side-chain of RG-I, and it can also slightly shorten the backbone of β-galactan by removing a single β-D-Gal residue from the NRT. However, plant α-xylosidase appears to be able only to act on the single α-Xyl residue at the NRT of xyloglucan; it cannot cleave any of the numerous terminal α-Xyl residues that occur as side-chains of xyloglucan (O'Neill et al. 1989).

Plants possess glycosidases that can cleave most of the non-reducing terminal sugar residues that exist in plant cell-wall polysaccharides (Table 4.2), including α-arabinosidase (Kamiyoshihara et al. 2018; Figueroa et al. 2010). A survey of such activities in extracts of diverse plants is given in GHATAbase (https://fry.bio.ed.ac.uk/GHATAbase.html), derived from Franková and Fry (2011). The principal exceptions appear to be the terminal sugar residues of RG-II, a polysaccharide domain, which seems to be stable *in planta* and only cleaved by certain microbial enzymes (Ndeh et al. 2017); these termini include β-L-Ara (where Ara=arabinose), α-L-Rha, α-L-Gal, α-D-MeXyl (where MeXyl=2-*O*-methylxylose) and α-L-MeFuc (where MeFuc=2-*O*-methylfucose).

EVIDENCE FOR HYDROLASE ACTION *IN VIVO*

Three major strands of evidence can be adduced to show that glycanases or glycosidases are at work *in vivo*: (i) a change may occur in the chemistry of the target polysaccharide between two developmental stages (for instance, the polysaccharide may decrease in molecular weight or may lose certain sugar residues during fruit ripening); (ii) a labelled exogenous substrate, infiltrated into the living plant's apoplast, may decrease in molecular weight; and (iii) manipulation of the expression of the gene encoding the enzyme of interest may result in changes in wall or tissue properties. Approach (ii) demonstrates that enzymes exist in the cell wall that can, *in vivo*, hydrolyse the oligosaccharide used as a model of the polysaccharide of interest, though not necessarily that they exhibit such action on the endogenous polysaccharide substrates themselves.

Evidence for Glycanase Action *In Vivo*

Strand (i) evidence. Discernible decreases in molecular weight of wall polysaccharides *in vivo* probably indicate backbone cleavage because side-chain trimming would not cause any major size change. Examples of target polysaccharides that have been tracked in this way during plant development include xyloglucan, especially during auxin-induced wall loosening (Labavitch and Ray 1974; Nishitani and Masuda 1981), mixed-linkage β-glucan (MLG) (Hatfield and Nevins 1984) and homogalacturonan (Ortiz et al. 2010). Surprisingly, there is little evidence for decreases in molecular weight of cellulose or xylan. Related to this approach of monitoring changes in polysaccharides during development, it is of interest that some hemicelluloses are 'mobilised' in the leaves of deciduous trees prior to abscission, potentially salvaging carbon (Hoch 2007)—an observation deserving more detailed characterisation.

Strand (ii) evidence. Labelled exogenous substrates employed as tracers are not usually polysaccharides with

TABLE 4.2
A Selection of Plant Glycosidase Activities Potentially Targeting Cell-Wall Polysaccharides

Enzyme Activity	Potential Substrates in the Primary Cell Wall
• β-D-Glucosidase	Xyloglucan (after attack by α-D-xylosidase), MLG, callose, cellulose
• Cellobiohydrolase	Cellulose [releases the disaccharide cellobiose, not the monosaccharide]
• α-D-Xylosidase	Xyloglucan (single residue at non-reducing terminus of backbone)
• β-D-Xylosidase	(Hetero)xylans
• β-D-Glucuronidase	RG-II?
• β-D-Mannosidase	(Hetero)mannans
• α-D-Galactosidase	Galactomannans
• β-D-Galactosidase	RG-I side-chains, galactans
• α-L-Fucosidase	Xyloglucan oligosaccharides
• α-L-Arabinosidase	Heteroxylans, RG-I side-chains
• α-D-Galacturonidase (exo-PG)	Homogalacturonan (de-esterified)

molecular weights similar to those naturally occurring in the cell-wall architecture. More routinely, we infiltrate oligosaccharide models (typically, hexa- to nonasaccharides) of the polysaccharides of interest. Such oligosaccharides can be radiolabelled (with ^3H or ^{14}C) or fluorescently labelled (such as with sulphorhodamine, SR) [both are available commercially; Edipos (https://fry.bio.ed.ac.uk/edipos.html)], and the decrease in their molecular weight is then monitored by thin-layer chromatography (TLC) or high-pressure liquid chromatography (HPLC) of samples extracted from the test tissue at various intervals after feeding (Franková and Fry 2011). Observed decreases in molecular weight can be due to glycanase or glycosidase action: the former would be indicated by an immediate large decrease in molecular weight, such as:

$$Xyl_6 + H_2O \rightarrow 2\ Xyl_3$$

whereas the latter would be indicated by a stepwise decrease ($Xyl_6 \rightarrow Xyl_5 \rightarrow Xyl_4 \rightarrow Xyl_3 \rightarrow Xyl_2$), accompanied by a continual release of the monosaccharide (Franková and Fry 2011).

Strand (iii) evidence. Extractable EPG activity from strawberry is low, but downregulating its EPG (or PL; see below) genes delayed softening (Posé et al. 2015). Downregulating EPG mRNA expression in tomato and strawberry produced discrepant data—in tomato, PG activity could be 99% blocked without affecting softening, whereas in strawberry downregulation of PG did hinder softening (Atkinson et al. 2012; Quesada et al. 2009). Downregulating a cellulase gene in strawberry was also able to delay softening (Lee and Kim 2011), suggesting that fruit softening is not dictated solely by a single enzymic reaction.

Evidence for Glycosidase Action *In Vivo*

Strand (i) evidence. Evidence for loss of side-chains, likely to indicate glycosidase action *in vivo*, comes from observed decreases in the β-D-Gal and α-L-Ara content of pectin during fruit ripening. This process itself is unlikely to rapidly change the polysaccharide's molecular weight, but it may alter the biophysical properties (solubility, etc.). It would be difficult to discern the α-xylosidase-catalysed loss of the single non-reducing terminal α-Xyl from xyloglucan because there is in principle only one such 'target' residue in a whole xyloglucan molecule of, typically, >1,000 sugar residues. Nevertheless, such loss could be highly significant biologically because the single terminal α-Xyl is the potential acceptor substrate site for the action of XET action (see below). Another approach to looking for glycosidase action would be to detect the released monosaccharides; however, it would be difficult to prove that these arose directly from polysaccharide hydrolysis rather than being produced *de novo* by protoplasmic metabolism and sugar secretion; in addition, any monosaccharides released may well be short-lived *in vivo*, being scavenged by the cells for respiration or recycling to new NDP-sugars (Sharples and Fry 2007).

Strand (ii) evidence. Labelled exogenous oligosaccharides that are being attacked primarily by glycosidases will undergo a progressive stepwise decrease in molecular weight. Usually, the 'label' is located solely at the reducing end of the oligosaccharide: for instance, it is ^3H in a reductively tritiated hexasaccharide-alditol such as Xyl-Xyl-Xyl-Xyl-Xyl-[^3H]xylitol, or it is SR introduced by reductive amination to make Xyl-Xyl-Xyl-Xyl-Xyl-xylitol-SR. Thus, progressive attack from the non-reducing end gives a clear-cut stepwise breakdown, uncomplicated by the (undetectable) monosaccharides that are liberated:

Xyl_5-[^3H]xylitol + $H_2O \rightarrow$ Xyl (unlabelled) + Xyl_4-[^3H]xylitol (1st labelled product)
Xyl_4-[^3H]xylitol + $H_2O \rightarrow$ Xyl (unlabelled) + Xyl_3-[^3H]xylitol (2nd labelled product)
Xyl_3-[^3H]xylitol + $H_2O \rightarrow$ Xyl (unlabelled) + Xyl_2-[^3H]xylitol (3rd labelled product)
Xyl_2-[^3H]xylitol + $H_2O \rightarrow$ Xyl (unlabelled) + Xyl-[^3H]xylitol (last labelled product)

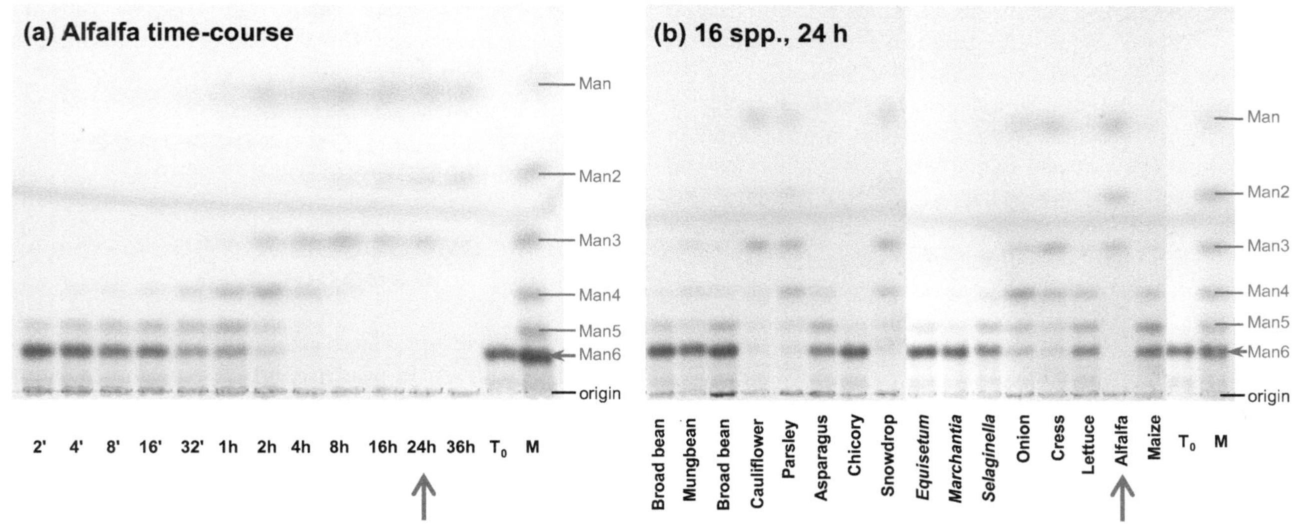

FIGURE 4.1 Evidence for β-mannosidase and β-mannanase activities in crude plant extracts. The images show TLCs of the reaction products formed when various plant extracts were incubated with 1.5 mM Man_6 [(1→4)-β-D-mannohexaose] at 20°C in a buffer at pH 4.7. The TLCs were run in butan-1-ol/acetic acid/water (2:1:1), and then stained with thymol/H_2SO_4, which detects carbohydrates. In (a), an extract from alfalfa was incubated with Man_6 for various time periods (2' = 2 minutes,..... 36 h = 36 hours). In (b), extracts from 16 different plant species were incubated with Man_6 under the same conditions for 24 hours. On each TLC, sample 'T_0' is an enzyme-free control (commercial Man_6 with no added enzyme). Sample 'M' is a manno-oligosaccharide ladder. As expected, the 24-hour track in (a) is almost identical to track 15 in (b). TLC (a) shows clear evidence for β-mannosidase: as time progresses, the pathway $Man_6 \rightarrow Man_5 \rightarrow Man_4 \rightarrow Man_3 \rightarrow Man_2$ unfolds, with monosaccharide mannose (Man) starting to appear from the outset. If all the extracts in (b) showed only β-mannosidase, they would all give profiles closely resembling one of the tracks in (a), and 15 of the 16 do so. For example, the broad bean extract (1) in (b) resembles the 4-min alfalfa sample in (a), and the cress extract (13) resembles the 4-hour alfalfa sample, showing that the β-mannosidase activity in the alfalfa extract was ~6× that in the cress and ~360× that in the broad bean extract. Thus, 15 of the 16 extracts are qualitatively similar (exhibiting principally β-mannosidase activity) though very different quantitatively. The exception is the mung bean extract (2), which shows appreciable Man_3 even though a large amount of the starting material Man_6 remains intact—a pattern not matched at any time-point in (a), and indicating that endo-hydrolysis reactions were occurring such as $Man_6 \rightarrow 2\ Man_3$, catalysed by β-mannanase. The Man_6 substrate in this experiment may be regarded as a model of the hemicellulose mannan. (Adapted from Franková and Fry (2011).)

Often the dimer ('last labelled product' above) is not further attacked by the glycosidase under investigation.

Even if a glycanase and a glycosidase are both acting simultaneously on the same substrate, it is often possible to discern them both in the same analysis, since the glycanase action

Xyl_5-[^{3}H]xylitol + H2O → Xyl_3 + Xyl_2-[^{3}H]xylitol (1st labelled glycanase product)

results in an early production of a labelled trimer *at the same time* as the glycosidase is generating a labelled pentamer:

Xyl_5-[^{3}H]xylitol + H2O → Xyl + Xyl_4-[^{3}H]xylitol (1st labelled glycosidase product).

Similar information can be obtained with unlabelled oligosaccharides such as mannohexaose (Figure 4.1) (Franková and Fry 2011).

Strand (iii) evidence. Antisense downregulation of β-galactosidase gene expression in strawberry can delay softening (Paniagua et al. 2016), suggesting a possible role for this enzyme in fruit ripening. Furthermore, downregulation of α-xylosidase gene expression in arabidopsis seedlings resulted in an abnormal accumulation of the heptasaccharide XXXG (a substrate of α-xylosidase) in the seedlings' medium, and concomitant growth defects, suggesting that the enzyme normally exhibits action *in vivo* (Sampedro et al. 2010).

TRANSGLYCOSYLASE ACTIVITIES ASSAYED *IN VITRO*

A transglycosylase is an enzyme that cuts a glycosidic bond (X–Y) and then transfers the 'X' moiety onto a different molecule, Z

$$X - Y + Z \rightarrow X - Z + Y$$

through a double-displacement mechanism. In a transglycosylation reaction, the compound whose bond is broken (X–Y) is called the donor substrate, Z is the acceptor substrate and Y is the leaving group. It could be considered that glycosidases and glycanases (see above), which catalyse hydrolysis

$$X - Y + H_2O \rightarrow X - OH + H - Y,$$

are special forms of 'transglycosylase' in which 'Z' is H_2O.

In principle, the group on the acceptor substrate, to which a new glycosidic bond is made, could be any nucleophile such as –OH, –SH or $-NH_2$, depending on the enzyme's substrate specificity, but in the examples to be discussed here it will always be an –OH group on a sugar residue, usually at the non-reducing end of a polysaccharide.

Transglycosylation reactions can be freely reversible or essentially irreversible. Irreversible ones are involved in the *de novo* biosynthesis of cell-wall polysaccharides and the donor substrate in that case is an NDP-sugar (X is a monosaccharide residue, and Y is a nucleoside diphosphate (NDP); usually UDP but in some cases GDP), and the acceptor substrate Z is a growing polysaccharide chain. For instance, in the case of xylan synthase, the reaction is:

$$Xyl - UDP \ + \ Xyl_n \ \rightarrow \ Xyl_{n+1} \ + \ UDP$$

(Petrik et al. 2020). For polysaccharide biosynthesis to continue, the released UDP is then converted back to Xyl–UDP (usually termed UDP-Xyl), at the expense of cellular energy, ultimately ATP (Fry 2000); thus, the reactions are driven irreversibly towards xylan biosynthesis. Such transglycosylases are membrane-localised: usually in Golgi membranes, but in the plasma membrane in the case of cellulose synthase (Purushotham et al. 2016) and callose synthase (Cai et al. 2011). Such membrane-localised enzymes are not further discussed in this chapter.

Other transglycosylases, discussed in more detail below, are located in the cell wall and can modify wall polysaccharides post-synthetically. Such enzymes catalyse reversible reactions, where X–Y and Z are both polysaccharides, and no nucleotides are involved (Franková and Fry 2013).

Transglycanase Activities Assayed *In Vitro*

The transferred fragment (X) of the donor substrate polysaccharide can be either a string of sugar residues (i.e. an oligo- or polysaccharide) or single monosaccharide residue. This distinction is thus comparable with that met in the case of hydrolases: there are transglycanases and transglycosidases, comparable with glycanases and glycosidases, respectively. However, another question, which must be taken into account in the case of transglycosylase activities, is whether the acceptor substrate (Z) resembles or differs from the leaving group (Y): if similar, the enzyme is exhibiting a homo-transglycosylase activity; if qualitatively different, it is showing a hetero-transglycosylase activity. Both homo- and hetero-transglycanases exist in the plant cell wall, though there seems to be no clear evidence for dedicated hetero-transglycosidases.

The best-known transglycanase in plant cell walls is a homo-transglycanase activity called XET (Rose et al. 2002). Logically, it could also be termed xyloglucan:xyloglucan xyloglucanotransferase, though in practice this term is rarely used. XET activity was found in the primary cell walls of all land-plants tested (Fry et al. 1992), and it can effectively 'cut and paste' xyloglucan chains. XET activity cleaves one xyloglucan chain (the donor substrate, shown below as ... ⊖⊖⊖●●●...) and then re-attaches the '⊖⊖⊖' part of the donor to a neighbouring xyloglucan chain (the acceptor substrate, ⊙⊙⊙...) and releases the '●●●...' part of the donor substrate (the leaving group). The net effect can be summarised as:

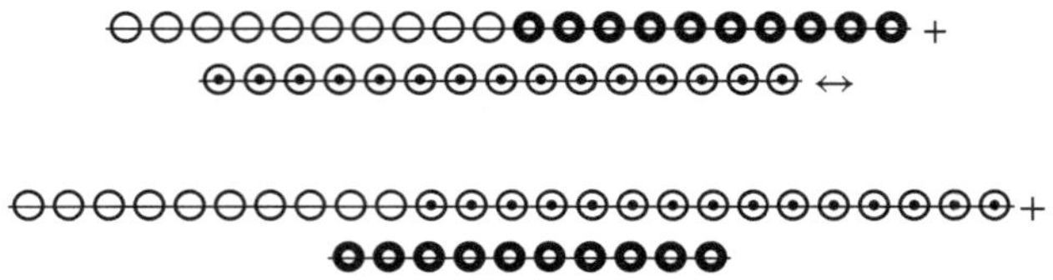

where each circle represents a single repeat unit (oligosaccharide) of xyloglucan. In principle, all the repeat-units can be chemically identical. This 'cutting and pasting' process may reversibly loosen the wall, favouring cell expansion (Rose et al. 2002; Fry et al. 1992). Alternatively or additionally, this type of reaction may help to assemble the cell-wall architecture by bonding formerly separate polysaccharide molecules together (Thompson and Fry 2001)—albeit at the cost of cleaving one of them. The reaction depicted may be difficult to detect when both the donor and the acceptor substrates are polysaccharides (since the reaction is essentially 'polysaccharide+polysaccharide ↔ polysaccharide+polysaccharide'), so it is usual to employ a labelled oligosaccharide as acceptor substrate when trying to monitor it (Fry et al. 1992; Fry 1997). This strategy exploits the fact that transglycanases can often use oligosaccharides as acceptor substrate, even though the donor substrate usually has to be a larger polysaccharide.

For 'labelling' the oligosaccharide acceptor substrate, either radioactive or fluorescent tags can be used. The radioactive tag is usually ^3H, introduced into an oligosaccharide at its RT by reaction with radioactive borohydride (NaB^3H_4) (Hetherington and Fry 1993). A very sensitively detectable fluorescent tag is SR (Miller et al. 2007). A range of relevant ^3H- and SR-labelled oligosaccharides is available from Edipos. The labelling does not affect the NRT of the xyloglucan oligosaccharide, i.e. the site recognised by XET-active enzymes and related hetero-transglycanases.

Besides XET, other homo-transglycanase activities exist in the plant cell wall. These act on β-(1→4)-xylan and β-(1→4)-mannan, i.e. the backbones of heteroxylans (such as glucuronoarabinoxylan; Franková and Fry 2011; Johnston et al. 2013) and heteromannans (such as galactomannan; Schröder et al. 2009). Such activities can be assayed by the use of [^{3}H]xylohexaitol and [^{3}H]mannohexaitol, respectively, with non-labelled (hetero)xylans and (hetero)mannans as donors; the diagnostic reaction product is then a ^3H-labelled polysaccharide, distinguishable from the radioactive hexasaccharide by any of a variety of techniques—for instance, the polysaccharide but not the hexasaccharide is insoluble in 75% ethanol (Franková and Fry 2015). An alternative strategy for assaying transglycanases is to use a medium-sized, non-labelled oligosaccharide such as a hexasaccharide, as both donor and acceptor substrate. The reaction catalysed is then, for example,

$$2\ \text{XXXXXX} \leftrightarrow \text{XXXXXXXXX} + \text{XXX}$$

where *X* could be a xylose residue (Franková and Fry 2011). This approach is comparable to the method described above for assaying hydrolysis. The key difference is that a transglycanase creates a product bigger than the starting material(s), a size change that is easily observed by chromatography (such as TLC or HPLC).

Enzymes that exhibit transglycanase activity are in some cases 'dedicated' transglycanases, i.e. cannot also hydrolyse the donor substrate at appreciable rates. Most proteins with XET activity fall into this category. Some, however, can also catalyse xyloglucan hydrolysis—in some cases preferentially (Shi et al. 2015). For this reason, the family of XET-active proteins and close relatives are termed XTHs (xyloglucan endotransglucosylase/hydrolases) (Rose et al. 2002). Arabidopsis has 33 *XTH* genes (Nishitani 2005), three of which appear to preferentially hydrolyse xyloglucan rather than transglycosylating it (Baumann et al. 2007). All currently known proteins that exhibit transxylanase and transmannanase activity can catalyse both transglycosylation and hydrolysis, and the transglycosylation:hydrolysis ratio rises with increasing acceptor substrate concentration (Schröder et al. 2006; Franková and Fry 2011, 2012b, 2021).

XTHs catalyse a 2-step reaction in which part of the donor xyloglucan chain becomes covalently attached to a glutamate residue in the enzyme via a glycosyl-ester bond, and the xyloglucan–enzyme conjugate thus formed later transfers the xyloglucan moiety onto the acceptor substrate (another xyloglucan chain or oligosaccharide thereof). The xyloglucan–enzyme conjugate is remarkably long-lived (in the order of 1 h; Sulová et al. 1998), potentially giving it time to 'find' another xyloglucan chain in the cell wall to use as its acceptor substrate, completing the transglycosylation process and releasing the enzyme for another round.

Besides the homo-transglycanase activities discussed above, there is also considerable interest in hetero-transglycanase activities. An early example was an activity (MLG:xyloglucan endotransglucosylase; MXE) where the donor is MLG and the acceptor is a xyloglucan oligosaccharide, creating a 'hybrid' MLG–xyloglucan product. This activity was first noted in a barley XTH protein (Hrmova et al. 2007), which exhibited high XET activity and low MXE activity (MXE:XET ≈ 0.002). Thereafter, enzyme extracts from fern-related plants of the genus *Equisetum* (horsetails) were found to preferentially exhibit MXE activity (MXE:XET ratio ≈ 7) (Simmons et al. 2015). The responsible enzyme was subsequently discovered via its gene, and found to possess the additional ability catalysing a cellulose:xyloglucan endotransglucosylase (CXE) hetero-transglycosylation reaction at a high rate (Simmons et al. 2015). The protein was termed hetero-trans-β-glucanase (HTG) and is of interest since it can covalently graft a cellulose chain to a xyloglucan hemicellulose, potentially contributing to cell-wall assembly in *Equisetum* and applicable biotechnologically for functionalising cellulose, creating cellulose–xyloglucan bonds. Certain arabidopsis XTHs (specifically AtXTH3) have also been shown to possess CXE in addition to XET activity (Shinohara et al. 2017). Furthermore, a *Tropaeolum* XTH [TmXET(6.3)] is able to utilise hydroxyethylcellulose as donor and cello-oligosaccharides as acceptor (Stratilová et al. 2010), and some AtXTHs can also use the artificial cellulose derivatives hydroxyethylcellulose and water-soluble cellulose acetate as donors (Maris et al. 2011). The range of donor:acceptor permutations known to be tolerated by diverse transglycanases is steadily being expanded (Seven et al. 2021).

Other hetero-transglycanase activities have been found in land-plants (Kosík et al. 2010; Kovacová and Farkaš 2016) and in charophytic algae (Franková and Fry 2021), with xyloglucan, MLG, xylans and mannans as donors and acceptors in various permutations.

Transglycosidase Activities Assayed *In Vitro*

Corresponding to glycosidase activities, there are also transglycosidase activities in the plant cell wall. These activities transfer sugar residues, one at a time, from one polysaccharide (or oligosaccharide) to another. So far, all transglycosidase activities have turned not to be 'dedicated' transglycosylases as they can also catalyse hydrolysis. The transglycosylation:hydrolysis ratio rises with increasing acceptor substrate concentration because the oligosaccharide used as acceptor then competes with H_2O (Franková and Fry 2013). Examples are trans-α-xylosidase (Franková and Fry 2012a), trans-β-xylosidase (Franková and Fry 2011), trans-β-galactosidase (Franková and Fry 2012b) and trans-β-glucosidase (Crombie et al. 1998).

A convenient assay for demonstrating such enzyme activities employs a xyloglucan oligosaccharide possessing [^{3}H]xylose residues (such as the heptasaccharide [*Xyl*-^{3}H]XXXG) as donor and an amino-labelled (therefore cationic) but non-radioactive acceptor, such as XXXG-NH_2 [in which the reducing terminal glucose moiety has been artificially converted to 1-amino-1-deoxyglucitol ('glucamine')] (Franková and Fry 2012a). The product is both radiolabelled and cationic, thus clearly demonstrating inter-molecular transfer of a [^{3}H]xylose residue (Figure 4.2). The cationic products can be separated from any others by electrophoresis or ion-exchange chromatography. Similar strategies can be employed for other transglycosidase activities.

EVIDENCE FOR TRANSGLYCOSYLASE ACTION ASSAYED *IN VIVO*

Application of exogenous substrates is a highly effective means of testing for transglycosylase action *in vivo*. Indeed, the first evidence for the existence of XET activity came from an experiment in which a radiolabelled xyloglucan oligosaccharide was fed to living *Rosa* cell cultures and was found to become covalently attached to endogenous xyloglucan by the action of endogenous XET activity (Baydoun and Fry 1989).

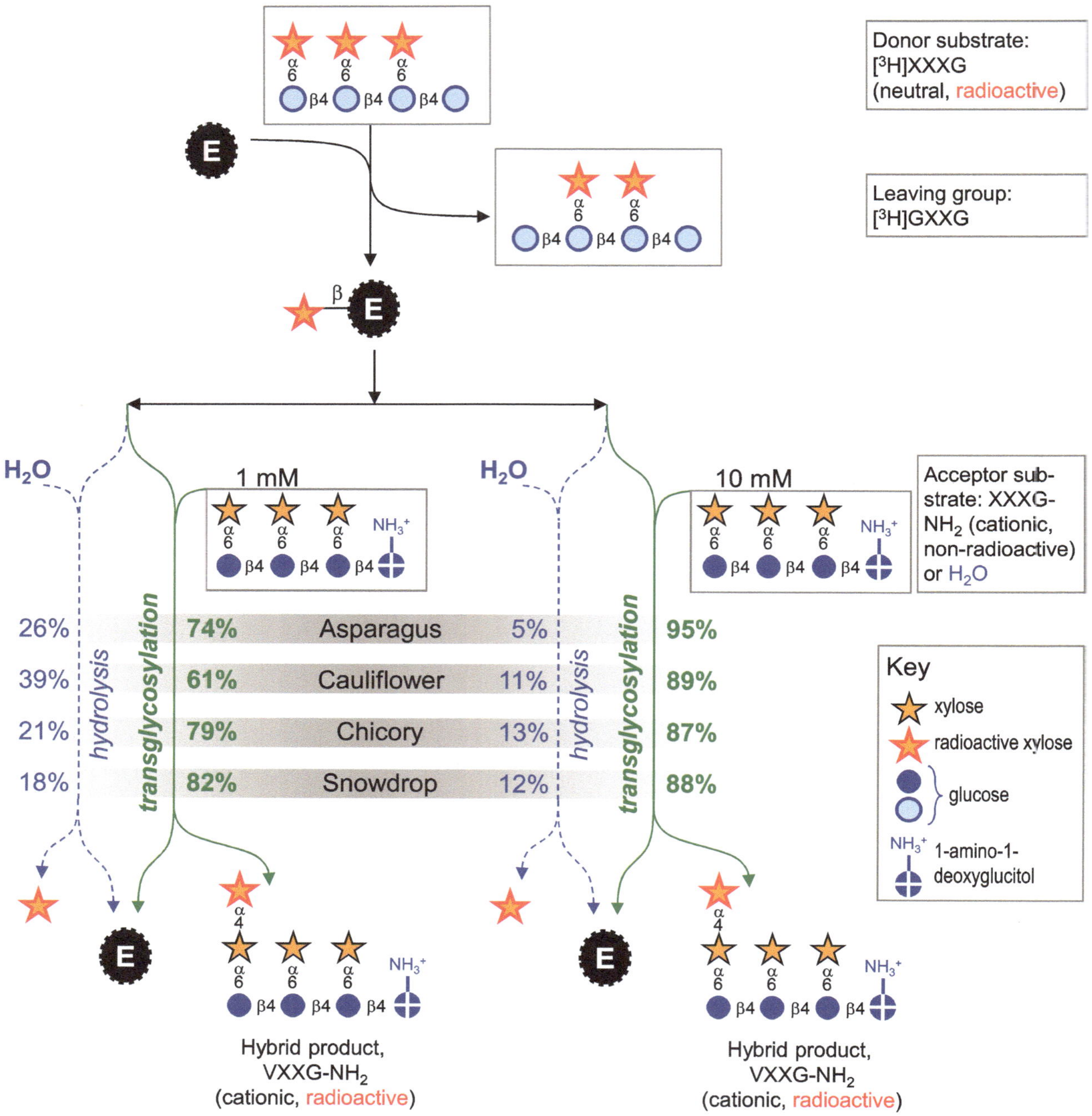

FIGURE 4.2 Hydrolysis/transglycosylation competition during trans-α-xylosidase action. The enzyme (E) cuts a [^{3}H]xylose residue off the donor substrate (top), forming a [^{3}H]xylosyl–enzyme bond. This then reacts with H_2O (hydrolysis, dashed arrows) or with XXXG-NH_2 (transglycosylation, solid arrows). Only transglycosylation produces a cationic, radioactive product (which can be isolated by electrophoresis and assayed by scintillation-counting); only hydrolysis generates free [^{3}H]xylose. Measured hydrolysis:transglycosylation ratios are indicated for four plant enzyme sources supplied with [*Xyl*-^{3}H]XXXG plus either 1 mM (left) or 10 mM (right) XXXG-NH_2. The higher concentration of acceptor substrate leads to a higher proportion of transglycosylation. (Adapted from Franková and Fry (2012b).)

Evidence for Transglycanase Action Assayed *In Vivo*

XET action *in vivo* can be probed by the feeding of plant tissues with xyloglucan oligosaccharides labelled with either ^{3}H (Baydoun and Fry 1989) or SR (Vissenberg et al. 2001; Nishikubo et al. 2007). After a brief incubation time (permitting enzyme action), the remaining unreacted oligosaccharides are washed out of the tissue, for instance with acidified 75% ethanol, and the unextractable labelled material (enzymic reaction product) is monitored. The SR labelling approach has the advantage over ^{3}H labelling in that it permits easy visualisation of the location of XET action since the reaction product is a fluorescent polysaccharide within the cell wall. However, ^{3}H-labelled products are more easily quantified.

In the case of HTG action, there is the additional challenge of determining which of the three potential donor substrates present in the cell wall (xyloglucan, MLG or cellulose) was being selected by the enzyme *in vivo*. This can be addressed by experiments in which the cell walls are examined after each of the three steps:

a. After washing with acidified alcohol only, which removes unreacted labelled oligosaccharides (XET, MXE plus CXE products are thereby visualised);
b. Subsequently, digested with lichenase (a commercially available MLG endo-hydrolase), which digests and removes MLG (only XET and CXE products are then still present);
c. Subsequently, washed with 6M NaOH at 37°C, which solubilises all hemicelluloses but not cellulose (so only CXE products now remain).

By difference between a, b and c, the location (and, with ^{3}H-labelling, quantities) of each of the three activities of HTG in *Equisetum* can be separately determined (Herburger et al. 2020).

This general approach involves an endogenous XTH enzyme acting on an endogenous polysaccharide donor substrate together with an exogenous, labelled oligosaccharide acceptor substrate. Thus, two of the three players are natural constituents of the plant cell wall, and the result confirms that XTH exhibits action on xyloglucan within a given cell wall (or even a specific part thereof). However, this approach makes the assumption that the intermediary xyloglucan–enzyme conjugate that is formed can normally 'find' the NRT of an endogenous xyloglucan molecule to use as an acceptor substrate with which to consummate the transglycosylation process. It should be emphasised that the mere presence of an enzyme protein (such as an XTH or HTG) in the plant cell wall, together with the known presence of the appropriate substrate(s) (such as xyloglucan), does not guarantee that the enzyme performs any action *in vivo*. A possible reason why an active enzyme may not exhibit action *in vivo* could be the absence of a necessary promoter. Plants do indeed contain intriguing heat-stable polymers that can activate XTH proteins (Nguyen-Phan and Fry 2019).

An alternative experiment that does not rely on the above assumption that the transglycanase can 'find' a suitable acceptor substrate in the cell wall of a living cell has been developed by use of dual isotopic labelling (Thompson and Fry 2001). Heterotrophic plant cell-suspension cultures are pre-grown for approximately a week in a medium whose principal carbon source is 'heavy', non-radioactive glucose (containing ^{2}H and/or ^{13}C). The cells then contain 'heavy' polysaccharides. Subsequently, at a particular time-point, the cultured cells are transferred into a medium based on ordinary 'light' glucose (with ^{1}H and ^{12}C) and fed a tiny trace of radioactive [^{3}H]arabinose (too small a quantity to affect the density of polysaccharides). The radioactive ('hot') arabinose is quickly taken up by the cells and efficiently incorporated into UDP-[^{3}H]arabinose and UDP-[^{3}H]xylose (Sharples and Fry 2007), from which some of the tritium is incorporated into newly synthesised xyloglucan chains. The cell walls thus now contain two populations of xyloglucan: 'old, heavy, cold' chains plus 'young, light, hot' chains. XET action in the walls of the living cells then permits cutting and pasting between these two populations to create some 'half-heavy, hot' chains, which can be distinguished from the *de novo* synthesised 'light, hot' chains by isopycnic centrifugation in a density gradient of caesium trifluoroacetate (Thompson and Fry 2001). The key advantage of this approach is that all three players (enzyme, and donor and acceptor substrates) are endogenous, unambiguously demonstrating inter-polymeric transglycosylation (transglycanase action) *in vivo*.

Evidence for Transglycosidase Action Assayed *In Vivo*

There is as yet little experimental proof of transglycosidase action on wall polysaccharides *in vivo*. Evidence could in principle be obtained from experiments in which oligosaccharides are fed to living plant cells and observed to increase in molecular weight by one sugar at a time, as has been observed with crude plant extracts *in vitro* (Franková and Fry 2012b). Such exogenous oligosaccharides could be radioactively or fluorescently labelled, facilitating the detection of the enlarged products without interference from co-occurring natural oligosaccharides.

Given that crude plant extracts can exhibit trans-α-xylosidase activity *in vitro*, it seems likely that trans-α-xylosidase action will be detectable *in vivo* if appropriate experiments are conducted. The reaction product formed from exogenous labelled XXXG (heptasaccharide) would be GXXG (hexasaccharide) plus VXXG (octasaccharide), where V is a xyloglucan building block comprising α-D-Xyl-(1→4)-α-D-Xyl-(1→6)-D-Glc at the non-reducing extremity of the acceptor substrate. That is a highly strategic location, as it is the single acceptor substrate site for XET, MXE or CXE action, and a 'V' unit at that site in place of an 'X' would be expected to influence acceptor substrate function. VXXG has yet to be tested as an XET acceptor substrate.

LYASE ACTIVITIES ASSAYED *IN VITRO*

A lyase is an enzyme that cleaves a bond, leaving two products, one of which now possesses a double-bond. If the substrate is a polysaccharide, the net result is in effect very similar to that achieved by a glycanase, *viz.*, irreversible cutting of the polysaccharide chain at one mid-chain bond.

Two lyase activities have been well documented in the plant cell wall: PL and rhamnogalacturonan lyase (RGL). These two activities have in common that they cleave a '.....sugar-(1→4)-α-D-GalA.....' bond within a polysaccharide chain, leaving the '.....sugar' moiety as a new (ordinary) RT, and generating a new non-reducing terminal unsaturated uronic acid residue ('ΔUA.....') in place of the 'GalA.....'. The full name of the ΔUA residue is 4-deoxy-β-L-*threo*-hex-4-enopyranuronosyl. [Note that although this residue starts out as an 'α-D-' residue in pectin, the lyase product is a 'β-L-' residue despite the fact that the anomeric carbon (C-1 in this case) has not been altered; this is because of the rules of carbohydrate nomenclature. The ΔUA product would be the same whether the lyase was acting on a GalA or a glucuronic acid (GlcA), hence the term ΔUA (meaning 'unsaturated uronic acid') is used rather than ΔGalA or ΔGlcA.] The ΔUA moiety absorbs ultra-violet (UV) at about 235 nm (Fuchs 1965) owing to its double-bond, and the lyase-catalysed reaction can therefore readily be followed as an increase in UV absorbance (Zhou et al. 2016). Also, the product can be seen on a chromatogram or paper electrophoretogram as a dark (UV-absorbing) spot under a 254-nm UV lamp.

Pectate Lyase Activity Assayed *In Vitro*

The bond cleaved by PL is .....α-D-GalA-(1→4)-α-D-GalA....., the repeat unit within the pectic domain homogalacturonan. The products are therefore one homogalacturonan fragment with a simple GalA RT (.....GalA-GalA-GalA) and another with a ΔUA NRT (ΔUA-GalA-GalA-GalA.....), which thus absorbs UV. [Note that reducing and non-reducing termini are conventionally placed at the right- and left-hand ends of a carbohydrate sequence.] PL activity is extractable from various plant tissues including strawberry fruit (Zhou et al. 2016). Plant PLs appear to attack homogalacturonan chains only at non-methylesterified residues, so existing methylester groups in the substrate may need to be removed by pectin methylesterase action (or, artificially, by alkali, such as 1 M Na_2CO_3) prior to PL activity. Some microorganisms possess pectin lyase activities, which do cleave methylesterified GalA residues (Albersheim et al. 1960).

Rhamnogalacturonan Lyase Activity Assayed *In Vitro*

The backbone of the pectic domain rhamnogalacturonan-I (RG-I) comprises

α-L-Rha-(1→4)-α-D-GalA-(1→2)-α-L-Rha-(1→4)-α-D-GalA-

The bond in this backbone that is cleaved by RGL is '.....α-L-Rha-(1→4)-α-D-GalA.....', not '.....GalA-(1→2)-Rha.....'. It is therefore expected that the two products of RGL activity will be one RG-I fragment with a simple Rha RT (i.e. .....GalA-Rha-GalA-Rha) and another with a ΔUA NRT (ΔUA-Rha-GalA-Rha.....) (Molina-Hidalgo et al. 2013; Mendez-Yanez et al. 2020; Yang et al. 2020; Kozlova et al. 2017). RGL can be assayed by the increase in A_{235} due to the formation of ΔUA residues when RG-I is used as a substrate (Ochoa-Jimenez et al. 2018).

In RG-I, roughly half the Rha residues may bear neutral side-chains rich in D-Gal and/or L-Ara (Mohnen 2008). Little is known about the influence of these on RGL activity, or how close the RGL enzyme can get to an Ara/Gal-rich side-chain and still be able to cleave the RG-I backbone. RG-I lacks methylester groups, although it can have acetylester groups on some of the backbone GalA residues: the influence of these on RGL activity is also unclear.

EVIDENCE FOR LYASE ACTION ASSAYED *IN VIVO*

The genetic manipulation of PL and RGL activities has been reported to influence fruit ripening (Jimenez-Bermudez et al. 2002; Marin-Rodriguez et al. 2002; Wang et al. 2018), suggesting that the encoded proteins exert action on fruit cell walls *in vivo*.

The author's laboratory recently obtained direct evidence for lyase action *in vivo* by detecting cell-wall 'fingerprints' in the shape of the diagnostic ΔUA non-reducing termini on oligosaccharides released from fruit cell walls by the commercial enzyme mixture, Driselase (Figure 4.3) (Al Hinai et al. 2021). Oligosaccharides possessing ΔUA differ from all other Driselase-generated cell-wall fragments that possess only standard uronic acid and/or neutral sugar residues in that the ΔUA residue has an unusually low pK_a value, causing migration towards the anode during electrophoresis even in a very low-pH buffer (pH 2.0). At this pH, GalA residues are only slightly ionised, but ΔUA residues are still appreciably anionic. In principle, the ΔUA-containing fingerprint fragments could also be isolated by anion-exchange chromatography instead of electrophoresis, though this has not yet been reported in practice.

Evidence for Pectate Lyase Action Assayed *In Vivo*

In a search for evidence of recent PL action *in vivo*, fruit cell walls were isolated and then digested *in vitro* with either pure EPG (after treatment with Na_2CO_3 which removes methylester groups) or with Driselase. EPG digests homogalacturonan to a mixture of mono-, di and trisaccharides ($GalA_{1,2,3}$), whereas Driselase converts it almost quantitatively to the monosaccharide, GalA (and removes the methylester groups if any). Where they encounter a ΔUA residue, these enzymes stall: EPG yields a mixture of a dimer (ΔUA-GalA) and a trimer (ΔUA-$GalA_2$), whereas Driselase yields only the dimer. The EPG or Driselase digestion products were then resolved by paper electrophoresis, and the ΔUA-$GalA_n$ products are the fastest-migrating substances present. For further purification and quantification, the ΔUA-$GalA_n$ zones were eluted from the electrophoretogram and then, even if present at very low concentrations, concentrated and re-run by TLC followed by thymol staining (Figure 4.4; Al Hinai et al. 2021).

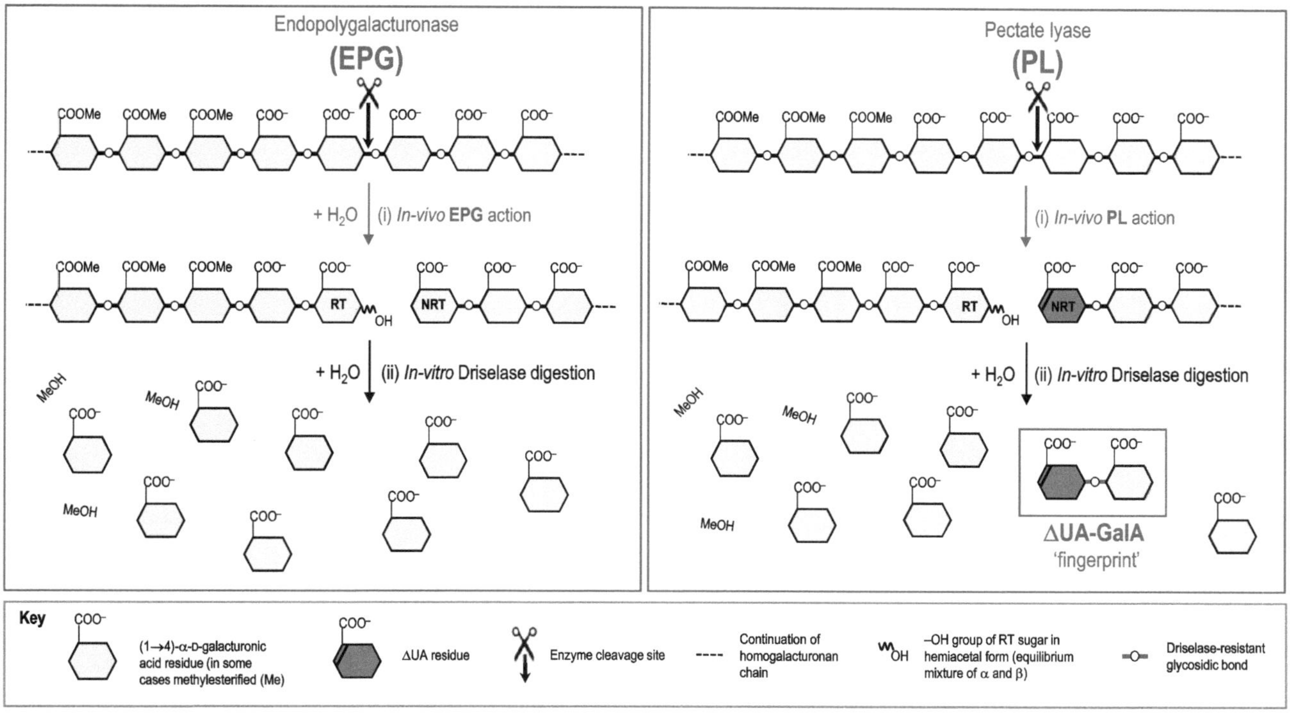

FIGURE 4.3 Endo-polygalacturonase (EPG) and pectate lyase (PL) action on homogalacturonan *in vivo* followed by *in-vitro* Driselase digestion. *Left:* EPG attacking a (1→4)-glycosidic bond between two neighbouring GalA residues in homogalacturonan by hydrolysis, producing a new reducing terminus (RT) and a new saturated non-reducing terminus (NRT). Subsequent *in-vitro* digestion with Driselase cleaves the remaining methyl ester groups (releasing methanol; MeOH) and the whole chain of homogalacturonan to yield monomeric GalA. *Right:* PL attacking the same substrate, producing an NRT and a new unsaturated NRT (ΔUA residue) by β-elimination. Subsequent digestion with Driselase cleaves the remaining methyl ester groups and the whole chain of HG to GalA monomers plus the dimer, ΔUA-GalA, the unique PL action fingerprint. (Adapted from Al Hinai et al. (2021).)

EVIDENCE FOR RHAMNOGALACTURONAN LYASE ACTION ASSAYED *IN VIVO*

In principle, evidence for *in vivo* RGL action might be sought by a comparable approach to that used for PL action if the cell walls are treated *in vitro* with an enzyme that hydrolyses RG-I. The predicted products such as ΔUA-Rha-GalA-Rha would also have a high electrophoretic mobility due to the low pK_a of the ΔUA residue.

Additional indirect evidence for RGL action *in vivo* is the finding of the disaccharide ΔUA-(1→2)-Rha ('lepidimoic acid') in seed-coat exudate, such as in cress (Yamada et al. 2010; Iqbal et al. 2016). Since RG-I is the only plant constituent known to contain the GalA-(1→2)-Rha linkage, it is likely that lepidimoic acid arises *in vivo* by lyase action on RG-I. However, the alternative possibility has not been excluded that lepidimoic acid is synthesised *de novo* as a disaccharide.

HYDROXYL RADICALS CLEAVING WALL POLYSACCHARIDES

The hydroxyl radical (OH) is the most reactive known molecule (Halliwell 2007). It is one of the reactive oxygen species (ROS) and can attack essentially all organic molecules, including polysaccharides, resulting in a range of rather random reactions, including oxidative scission that results in the cleaving of a polysaccharide chain (Schuchmann and von Sonntag 1978; Fry 1998; Vreeburg and Fry 2005). If $^{\bullet}OH$ can be generated *in vivo* in the apoplast, it is expected to be capable of cleaving wall polysaccharides and thereby loosening the cell wall. The $^{\bullet}OH$ radical is widely regarded as biologically 'hazardous' as it can oxidise DNA (causing mutation) and phospholipids (causing membrane leakiness) (Halliwell 2007). However, these adverse effects would only come into play if the $^{\bullet}OH$ radicals were present on or in the protoplast. The half-life of an $^{\bullet}OH$ radical in a biological milieu is so short (in the order of 1 ns) that it would be unable to diffuse more than about 1 nm from its site of production before reacting with an organic molecule (Griffiths and Lunec 1996). Thus, if generated within the primary cell wall (thickness typically ~100 nm), $^{\bullet}OH$ would be very likely to react with a wall polysaccharide before reaching the plasma membrane, let alone the DNA in the nucleus.

OH Radicals could be generated in the cell wall by non-enzymic Fenton reactions, such as:

$$Cu^{+} + H_2O_2 {}^{\bullet}\leftarrow OH + OH^{-} + Cu^{2+}$$

$$Fe^{2+} + H_2O_2 \rightarrow {}^{\bullet}OH + OH^{-} + Fe^{3+}$$

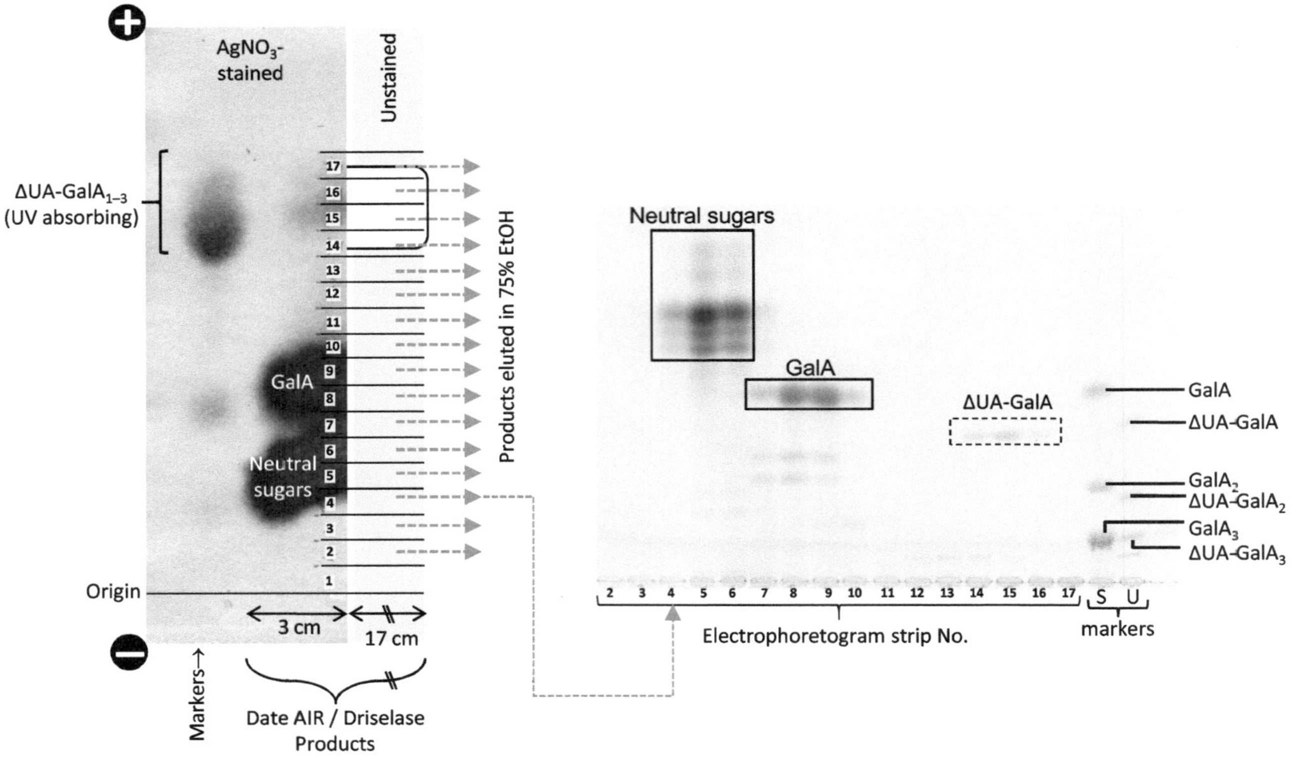

FIGURE 4.4 Detecting pectate lyase fingerprints in digests of date fruit cell walls by paper electrophoresis and TLC. Date cell walls (25 mg AIR) were digested in Driselase, which hydrolyses most cell-wall polysaccharides to monosaccharides but cannot cleave the ΔUA-(1→4)-GalA bond. *Left:* The products were loaded as a 20-cm streak on Whatman No. 3 paper and electrophoresed in a pH 2 buffer. The left-hand 3-cm fringe of the streak-loading (plus markers, which were run alongside it) were then stained with $AgNO_3$, visualising the sugars (brown spots). The remaining 17 cm of the streak-loading, only part of which is depicted (in grey), was not stained, but green/white shading indicates the position of a UV-absorbing band attributable to ΔUA residues in the Driselase digestion products. The unstained portion was cut into 17 1×17-cm strips as shown, and the digestion products were eluted from the paper with 75% ethanol. *Right:* Eluates from strips 2–17 were run by TLC in butan-1-ol/acetic acid/water (2:1:1) alongside marker mixtures and stained with thymol. Marker mixtures were **S**, saturated oligogalacturonides; **U**, unsaturated oligogalacturonides. Neutral sugars remain near the origin of the electrophoretogram and are seen on the TLC plate in fractions 4–6; the monosaccharide GalA is slightly anionic under these electrophoresis conditions (pH 2) and is seen in fractions 7–10; and ΔUA-GalA is more strongly anionic (because of the low pK_a of the ΔUA residue) and is seen in fractions 14–16. The thymol-stained spots on the TLC plate are then quantified on the digital image, such as by use of ImageJ. (Adapted from Al Hinai et al. (2021).)

which are feasible because the apoplast contains H_2O_2 as well as Cu^{2+} and Fe^{3+} and the necessary electron donors (especially secreted ascorbate) to reduce a proportion of these to Cu^+ and Fe^{2+}, respectively, required for Fenton reactions (Fry 1998; Liskay et al. 2003). Thus, OH is most likely to be produced on a transition metal ion, itself bound to a wall polymer (Walker and Webb 1981), exactly positioned to attack that polymer molecule or a near neighbour (Fry et al. 2002).

Non-Enzymic Scission of Polysaccharides *In Vitro*

Modelled on the above mechanism of OH production, the process can readily be mimicked *in vitro* in a solution containing H_2O_2, Cu^{2+} and ascorbate. Indeed, in an aerobic environment, it is unnecessary to add the H_2O_2 since ascorbate (represented below as AH_2) can non-enzymically reduce O_2 to make H_2O_2:

$$AH_2 + O_2 \rightarrow A + H_2O_2$$

where 'A' is dehydroascorbic acid. If the solution also contains a polysaccharide, such as xyloglucan or pectin, the polymer will quickly decrease in molecular weight owing to oxidative scission. This scission can easily be monitored as the polysaccharide solution loses viscosity (Fry 1998). Polysaccharide scission can also be monitored by gel-permeation chromatography, for instance on a Sepharose column.

Evidence for OH Action *In Vivo*

Evidence of this type of reaction occurring in the cell wall *in vivo* has been obtained by detection of chemical fingerprints in cell-wall polysaccharides. The relatively 'random' reactions inflicted by OH on polysaccharides include not only oxidative scission but also the oxidation of some of the sugar residues' C–OH groups to form oxo groups (C=O; i.e. aldehydes or ketones), creating products known as glycosulose residues (Schuchmann and von Sonntag 1978). These can be detected in alcohol-insoluble residues (crude cell-wall preparations) of plant tissues by virtue of the reactivity of oxo groups with either of the following:

a. Radiolabelled borohydride (NaB^3H_4), which reduces the >C=O group to a >C^3H–OH group (Miller and Fry 2001; Fry et al. 2002). The radiolabelled product thus formed can then be released from the mid-chain of the OH-attacked polysaccharide by acid or enzymic hydrolysis in the form of a tritiated monosaccharide aldose (not alditol, which is the product formed from an ordinary RT). In some cases, the ^{3}H-labelled monosaccharide obtained is an unusual one (such as [^{3}H]ribose or [^{3}H]lyxose, which are not known wall polysaccharide components), confirming their origin from an •OH-generated ketone group. This approach has been applied to wall polysaccharides from ripening fruits (Fry et al. 2001).
b. Fluorescent reagents, such as aminoacridone, whose –NH_2 groups add to the C=O group in the presence of $NaCNBH_3$, stably introducing a fluorescent tag. The tagged sugar residue can then be released (still attached to one or more neighbouring sugar residues) by digestion of the wall polysaccharides, for instance with Driselase, and detected by chromatography or electrophoresis (Vreeburg et al. 2014). This approach has been successfully applied to fruit AIR samples prepared at different stages of ripening (Figure 4.5): an increase in labelled products accompanied the ripening of many fruit species—especially true fruits, derived from the ovary-wall, rather than false fruits such as apples and strawberries (Airianah et al. 2016). This provides evidence for attack of wall polysaccharides by OH in ripening fruits.

Probing for Apoplastic OH Radicals *In Vivo*

Hydroxyl radicals are too short-lived to be collected for analysis (Griffiths and Lunec 1996), but evidence for their presence *in vivo* can be sought by the use of reporter molecules that react with OH to produce recognisable products (Vreeburg and Fry 2005). In the context of this chapter, it

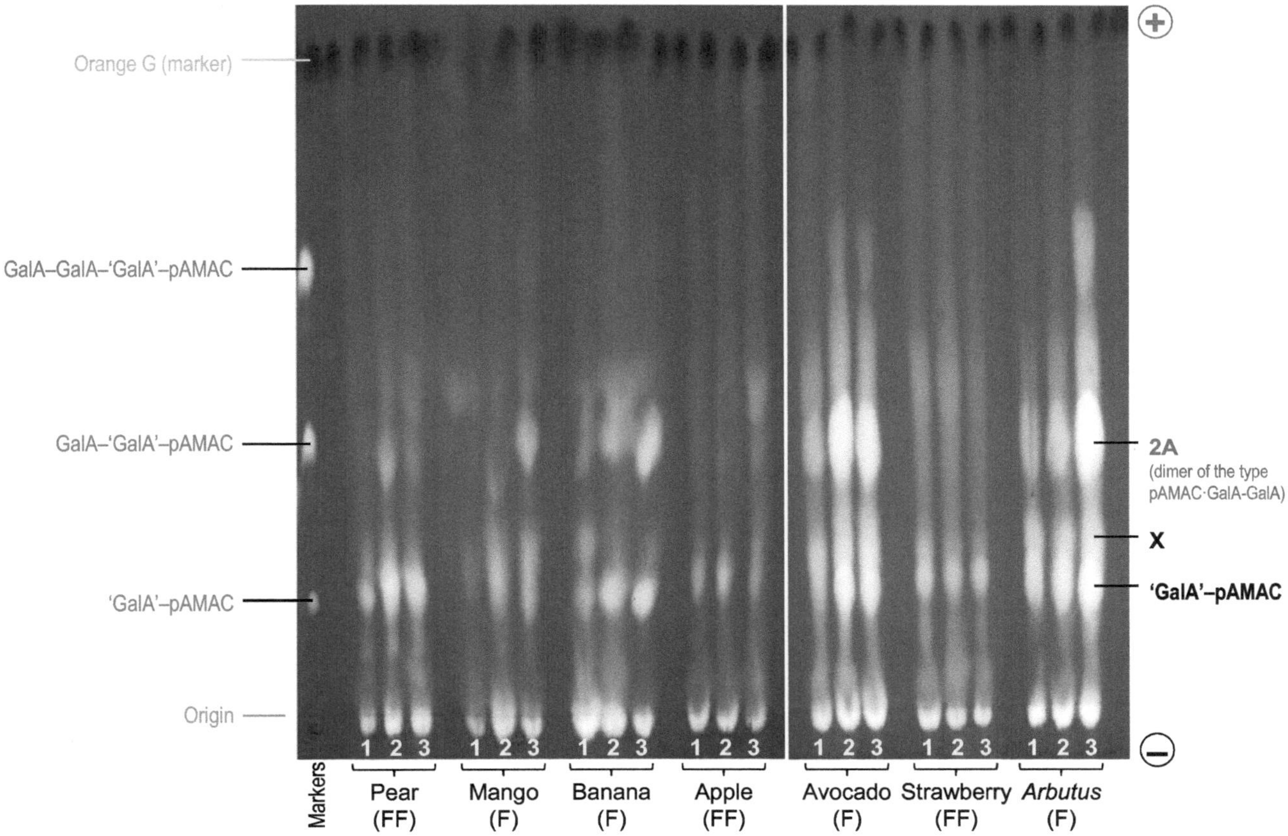

FIGURE 4.5 Evidence for hydroxyl radical attack on fruit cell walls *in vivo*. Fruit cell walls, harvested at three stages of ripening (**1–3**), were labelled with isopropylaminoacridone (pAMAC) *in vitro*, then Driselase-digested, which hydrolyses most of the cell-wall polysaccharide sugar residues, but not those located immediately adjacent to a pAMAC group, to monosaccharides. The pAMAC-labelled fragments thus generated were subjected to paper electrophoresis in buffer at pH 6.5. Each electrophoretogram loading contained the products from 20 mg fresh weight of fruit tissue. The Driselase-resistant disaccharide spot **2A** contains fluorescent fingerprint compound(s) that indicate mid-chain oxo groups introduced by *in vivo* OH radical attack. This fingerprint product increases during fruit softening, especially in the true fruits (F) rather than the false fruits (FF). (Adapted from Airianah et al. (2016).)

is important to demonstrate that any •OH radicals detected were located in the cell wall rather than protoplasm. Thus, the reporter molecule used must be confined to the apoplast and unable to pass through the plasma membrane. This proviso can be ensured if the reporter is sufficiently hydrophilic.

One set of potentially useful reporters in this respect is provided by compounds with a ^3H-labelled benzoyl moiety attached to a hydrophilic compound such as polyallylamine or an oligopeptide (Gly_3 or Lys_5-methylester) (Fry et al. 2002; Miller and Fry 2004). Some of these probes, such as *N*-[^3H]benzoyl-polyallylamine, are cationic (Figure 4.6) and therefore have the additional advantage of ionically bonding to acidic polysaccharides and thus specifically reporting OH radicals in the immediate vicinity (<1 nm) of, say, a pectin molecule. Although •OH radicals react with essentially any organic compound, they have a particular affinity for aromatics such as the benzoyl group. The products formed by the reaction of •OH with [^3H]benzoyl groups include ^3H_2O, which is easily distinguished from all other labelled compounds present, for instance by its volatility or by its failure to bind to a small ion-exchange or affinity chromatography column to which the reported molecule does bind. This method of probing for apoplastic •OH radicals *in vivo* is awaiting more widespread exploitation.

BORON-BRIDGING OF RHAMNOGALACTURONAN-II

Boron is an essential element in plant nutrition (Wimmer et al. 2019). Plants obtain their boron in the form of un-ionised boric acid [$B(OH)_3$], which is soluble in the soil water. Its best-established role in plants is as a covalent cross-link between rhamnogalacturonan-II (RG-II) domains, with the B atom forming a diester with a specific apiose residue in each of two participating RG-II molecules (Ishii et al. 1999; O'Neill et al. 2004).

Borate esters and diesters can be reversibly formed with most carbohydrates, as demonstrated by the paper electrophoretic mobility of 'neutral' carbohydrates in the presence of borate buffer (Weigel 1963; Fry 2000). However, the efficient formation of these ester bonds is in most cases dependent on an unphysiologically high pH, such as 9.4, although with certain sugars, boron binding is also detectable at lower (physiological) pH values—such sugars are those with a *cis*-diol on a furanose ring, such as ribose (in RNA) and apiose. However, the formation of all such borate esters is freely reversible, even at optimal pH values, which is not the case with boron-bridging of RG-II. There is clearly something very special about the RG-II–B–RG-II bridge, which remains stable at pH values as low as 3.0; not readily mimicked in other apiose-containing substances.

The precise role of RG-II's boron bridges is difficult to define. By cross-linking neighbouring pectin chains, cross-linking might be involved in the assembly of the cell-wall's architecture. However, the cross-linking of wall polymers might be expected to tighten the wall and thus to correlate negatively with cell expansion. In fact, however, prevention of boron-bridging (either by boron starvation or by genetic modification of the RG-II's structure) results in diminished cell expansion (O'Neill et al. 2004). Nevertheless, B is essential for plant life, and it clearly cross-links RG-II domains—presumed to be its principal botanical role.

In Vitro Non-Enzymic Cross-Linking of RG-II with Boric Acid

Simply mixing boric acid with pure RG-II *in vitro* leads to only a small proportion of the RG-II molecules becoming B-bridged (regardless of pH), whereas newly synthesised RG-II very quickly becomes firmly B-bridged *in vivo* (see next section). It is evident that some additional factor present *in vivo* is necessary. No evidence could be found for a hypothetical enzyme ('boryltransferase') creating boron bridges (Chormova et al. 2014). However, it was found that certain inorganic cations can greatly promote B-bridging of RG-II: the most effective were Pb^{2+} and Sr^{2+}. Also effective was Ca^{2+}, but only at considerably higher concentrations (O'Neill et al. 2004). These cations are proposed to act as 'chaperones', helping to bring two negatively charged (thus mutually repulsive) RG-II molecules close enough together to be bridged (Chormova and Fry 2016). Since Pb^{2+} and Sr^{2+} are not essential elements for plants, a survey was conducted for more natural, organic, chaperones.

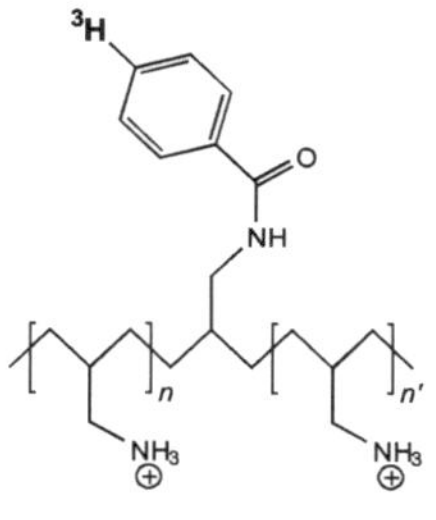

FIGURE 4.6 *N*-[^3H]Benzoyl-polyallylamine, a probe suitable for detection of OH radicals in the cell wall *in vivo*. The tritium-labelled benzoyl group reacts avidly with •OH, releasing tritiated water, which can easily be quantified (Fry et al. 2002). The cationic backbone (polyallylamine) is too hydrophilic to pass through the plasma membrane, confining the exogenous probe to the apoplast. The positive charges may help the probe to specifically attach to anionic cell-wall components, especially pectins, thus revealing •OH radicals generated in the immediate vicinity of pectin molecules.

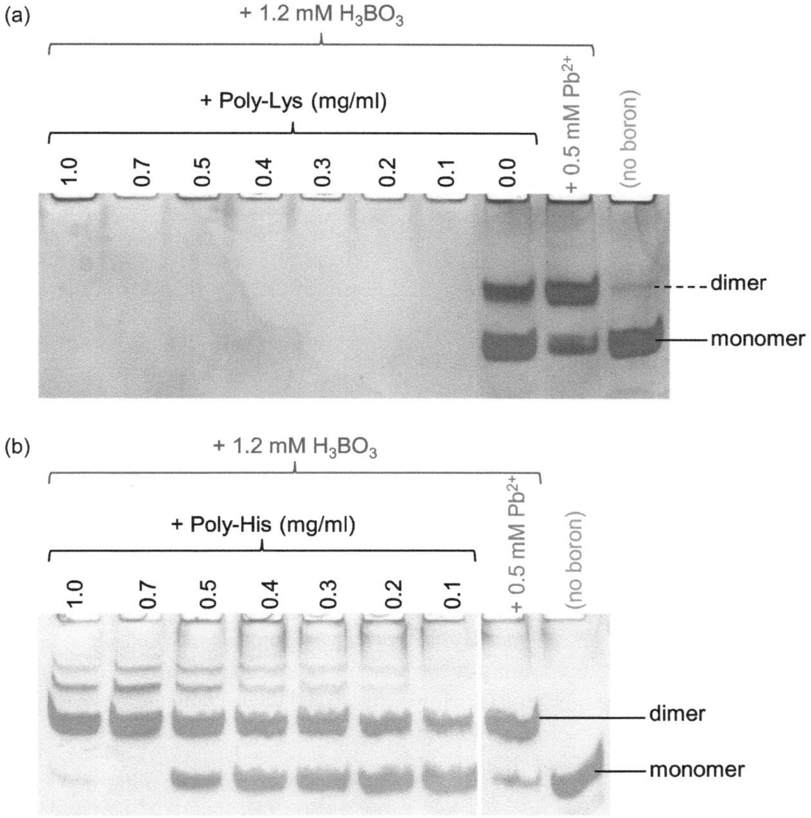

FIGURE 4.7 Polylysine and polyhistidine have different 'chaperoning' effects on rhamnogalacturonan-II. Monomeric RG-II (0.1 mg/mL; 20 μM) was incubated at 20°C for 24 hours with a dilution series of (a) polylysine and (b) polyhistidine, with and without the other additives indicated. Later, the samples were subjected to polyacrylamide gel electrophoresis, and the gel was silver-stained. The right-hand lane shows the monomeric RG-II with no additives. In the presence of poly-Lys, the RG-II was lost from the electrophoretogram (a) because the chaperone remained attached to the RG-II under the conditions of electrophoresis (thus, the poly-Lys–RG-II complex had a net positive charge and migrated towards the cathode—upwards in the picture). This did not happen with poly-His, which chaperones RG-II reversibly. In the presence of poly-His and boric acid, the RG-II was dimerised. Lead ions, supplied as $Pb(NO_3)_2$, also serve as an excellent non-biological chaperone. (Adapted from Chormova and Fry (2016).)

Boron-bridging (dimerisation) of RG-II was monitored by polyacrylamide gel electrophoresis (PAGE), the RG-II being detected either by silver-staining or by tracing its previously incorporated radioactivity (Chormova et al. 2014). Monomeric RG-II migrates faster than the dimer, which is retarded by the sieving effect of the polyacrylamide gel. The most effective organic chaperones turned out to be certain artificial polypeptides, especially polyhistidine, which can reversibly bond to RG-II and strongly promote dimerisation in the presence of boric acid (Figure 4.7). The reversibility of the ionic interaction was shown by the fact that electrophoresis of a polyhistidine–RG-II mixture produced an anionic RG-II band identical with that given by pure RG-II (whereas polylysine formed a more permanent ionic complex with RG-II that reversed out of the polyacrylamide gel towards the cathode) (Chormova and Fry 2016).

Polyhistidine does not occur naturally in plants, but there are histidine-rich (glyco)proteins such as certain extensins and arabinogalactan–proteins (AGPs) that *in vitro* can serve the same role as artificial polyhistidine (Chormova and Fry 2016). Certain glycosylinositol-phosphorylceramides (glycolipids) present in plant membranes have also been reported to possess chaperone-like properties, enhancing RG-II dimerisation in the presence of boric acid *in vitro* (Voxeur and Fry 2014).

Timing of Boron-Bridging of RG-II *In Vivo*

The kinetics of RG-II dimerisation *in vivo* were monitored in experiments with *Rosa* cell-suspension cultures that had been acclimated to grow in the absence of boron and which thus produced only monomeric RG-II domains. Re-addition of boric acid to these cells caused only slow accumulation of RG-II dimers over 24 hours, and without any loss of the pre-existing monomers, suggesting that only those RG-II newly synthesised after the re-addition of boron could be dimerised. Supporting this conclusion, *Rosa* cultures that had been rendered incapable of making new RG-II domains (for instance by carbon starvation, respiratory inhibitors or freezing) were unable to dimerise their pre-existing RG-II monomers even in the presence of H_3BO_3 (Chormova et al. 2014). Therefore, it appears that RG-II is normally boron-bridged in living cells during polysaccharide synthesis or secretion, but not once the RG-II domains have been deposited in the cell wall. In conclusion, in cultured *Rosa* cells, RG-II domains have a brief 'window of opportunity' for boron-bridging intraprotoplasmically or during secretion, but secretion into the apoplast is a point of no return beyond which additional boron-bridging does not readily occur.

CONCLUSIONS

Strategies have been presented for assaying diverse reactions that can modify wall-related polysaccharides *in vitro*, under artificial conditions that may mimic the apoplastic conditions of living cells. This approach provides valuable information about the types of reaction that might theoretically occur *in vivo* and thereby build and/or loosen the cell wall, with consequences for cell expansion, fruit ripening, abscission, etc. However, to go from 'might theoretically occur *in vivo*' to 'occur *in vivo*' is a key additional hurdle, and this chapter has emphasised strategies for achieving this. A valuable approach to this challenge is to monitor the chemical changes that wall polysaccharides undergo during the various stages of plant growth and development, and specifically to document chemical 'fingerprints' that diagnose polysaccharide reactions that have recently occurred *in vivo*.

REFERENCES

Airianah OB, Vreeburg RAM, Fry SC (2016) Pectic polysaccharides are attacked by hydroxyl radicals in ripening fruit: evidence from a fluorescent fingerprinting method. *Annals of Botany* 117:441–455.

Al Hinai TZS, Vreeburg RAM, Mackay CL, Murray L, Sadler IH, Fry SC (2021) Fruit softening: evidence for pectate lyase action *in vivo* in date (*Phoenix dactylifera*) and rosaceous fruit cell walls. *Annals of Botany* 128:511–525.

Albersheim P, Darvill A, Roberts K, Sederoff R, Staehelin A (2010). *Plant Cell Walls. From Chemistry to Biology*. New York: Garland Science.

Albersheim P, Neukom H, Deuel H (1960) Über die Bildung von ungesättigten Abbauprodukten durch ein pektinabbauendes Enzym. *Helvetica Chimica Acta* 43:1422–1426.

Atkinson RG, Sutherland PW, Johnston SL, Gunaseelan K, Hallett IC, Mitra D, Brummell DA, Schröder R, Johnston JW, Schaffer RJ (2012) Down-regulation of polygalacturonase1 alters firmness, tensile strength and water loss in apple (*Malus × domestica*) fruit. *BMC Plant Biology* 12:129. https://doi.org/10.1186/1471-2229-12-129.

Ayour J, Le Bourvellec C, Gouble B, Audergon JM, Benichou M, Renard CMGC (2021) Changes in cell wall neutral sugar composition related to pectinolytic enzyme activities and intra-flesh textural property during ripening of ten apricot clones. *Food Chemistry* 339:Article No 128096. https://doi.org/10.1016/j.foodchem.2020.128096.

Baumann MJ, Eklof JM, Michel G, Kallas AM, Teeri TT, Czjzek M, Brumer H (2007) Structural evidence for the evolution of xyloglucanase activity from xyloglucan endo-transglycosylases: biological implications for cell wall metabolism. *Plant Cell* 19:1947–1963.

Baydoun EA- H, Fry SC (1989). *In vivo* degradation and extracellular polymer-binding of xyloglucan nonasaccharide, a natural anti-auxin. *Journal of Plant Physiology* 134:453–459.

Cai G, Faleri C, Del Casino C, Emons AMC, Cresti M (2011) Distribution of callose synthase, cellulose synthase, and sucrose synthase in tobacco pollen tube is controlled in dissimilar ways by actin filaments and microtubules. *Plant Physiology* 155:1169–1190.

Chormova D, Fry SC (2016) Boron bridging of rhamnogalacturonan-II is promoted *in vitro* by cationic chaperones, including polyhistidine and wall glycoproteins. *New Phytologist* 209:241–251.

Chormova D, Messenger DJ, Fry SC (2014) Boron bridging of rhamnogalacturonan-II, monitored by gel electrophoresis, occurs during polysaccharide synthesis and secretion, but not post-secretion. *The Plant Journal* 77:534–546.

Crombie HJ, Chengappa S, Hellyer A, Reid JSG (1998) A xyloglucan oligosaccharide-active, transglycosylating β-D-glucosidase from the cotyledons of nasturtium (*Tropaeolum majus* L.) seedlings - purification, properties and characterization of a cDNA clone. *The Plant Journal* 15:27–38.

Defilippi BG, Ejsmentewicz T, Covarrubias MP, Gudenschwager O, Campos-Vargas R (2018) Changes in cell wall pectins and their relation to postharvest mesocarp softening of hass avocados (*Persea americana* Mill.). *Plant Physiology and Biochemistry* 128:142–151.

Duan XW, Cheng GP, Yang E, Yi C, Ruenroengklin N, Lu WJ, Luo YB, Jiang YM (2008) Modification of pectin polysaccharides during ripening of postharvest banana fruit. *Food Chemistry* 111:144–149.

Figueroa CR, Rosli HG, Civello PM, Martinez GA, Herrera R, Moya-Leon MA (2010) Changes in cell wall polysaccharides and cell wall degrading enzymes during ripening of *Fragaria chiloensis* and *Fragaria × ananassa* fruits. *Scientia Horticulturae* 124:454–462.

Forlani S, Masiero S, Mizzotti C (2019) Fruit ripening: the role of hormones, cell wall modifications, and their relationship with pathogens. *Journal of Experimental Botany* 70:2993–3006.

Franková L, Fry SC (2011) Phylogenetic variation in glycosidases and glycanases acting on plant cell wall polysaccharides, and the detection of transglycosidase and trans-β-xylanase activities. *The Plant Journal* 67:662–681.

Franková L, Fry SC (2012a). Trans-α-xylosidase, a widespread enzyme activity in plants, introduces (1→4)-α-D-xylobiose side-chains into xyloglucan structures. *Phytochemistry*:78, 29–43.

Franková L, Fry SC (2012b). Trans-α-xylosidase and trans-β-galactosidase activities, widespread in plants, modify and stabilise xyloglucan structures. *The Plant Journal* 71:45–60.

Franková L, Fry SC (2013) Darwin review: biochemistry and physiological roles of enzymes that 'cut and paste' plant cell-wall polysaccharides. *Journal of Experimental Botany* 64:3519–3550.

Franková L, Fry SC (2015) A general method for assaying homo- and hetero-transglycanase activities that act on plant cell-wall polysaccharides. *Journal of Integrative Plant Biol* 57:411–428.

Franková L, Fry SC (2021) Hemicellulose-remodelling transglycanase activities from charophytes: towards the evolution of the land-plant cell wall. *The Plant Journal* 108:7–28.

Frey-Wyssling A (1959). *Die Pflanzliche Zellwand*. Berlin: Springer.

Fry SC (1997) Novel 'dot-blot' assays for glycosyltransferases and glycosylhydrolases: optimization for xyloglucan endotransglycosylase (XET) activity. *The Plant Journal* 11:1141–1150.

Fry SC (1998). Oxidative scission of plant cell wall polysaccharides by ascorbate-induced hydroxyl radicals. *Biochemical Journal* 332:507–15.

Fry SC (2000). *The Growing Plant Cell Wall: Chemical and Metabolic Analysis*. Reprint Edition, Caldwell, NJ: The Blackburn Press.

Fry SC (2004) Tansley review: primary cell wall metabolism: tracking the careers of wall polymers in living plant cells. *New Phytologist* 161:641–675.

Fry SC, Dumville JC, Miller JG (2001). Fingerprinting of polysaccharides attacked by hydroxyl radicals *in vitro* and in the cell walls of ripening pear fruit. *Biochemical Journal* 357:729–735.

Fry SC, Miller JG, Dumville JC (2002) A proposed role for copper ions in cell wall loosening. *Plant and Soil* 247:57–67.

Fry SC, Smith RC, Renwick KF, Martin DJ, Hodge SK, Matthews KJ (1992). Xyloglucan endotransglycosylase, a new wall-loosening enzyme activity from plants. *Biochemical Journal* 282:821–828.

Fuchs A (1965) The trans-eliminative breakdown of Na-polygalacturonate by *Pseudomonas fluorescens*. *Antonie Van Leeuwenhoek* 31:323–340.

Griffiths HR, Lunec J (1996) Investigating the effects of oxygen free radicals on carbohydrates in biological systems. In *Free Radicals: A Practical Approach*, eds. Punchard NA, Kelly FJ (pp. 185–200). Oxford: IRL Press.

Gross KC (1982) A rapid and sensitive spectrophotometric method for assaying polygalacturonase using 2-cyanoacetamide. *HortSci* 17:933–934.

Gross KC, Starrett, DA, Chen H (1995) Rhamnogalacturonase, α-galactosidase, and β-galactosidase: potential roles in fruit softening. *Acta Horticulturae* 398:121–130.

Halliwell B (2007) Oxidative stress and cancer: have we moved forward? *Biochemical Journal* 401:1–11.

Hatfield R, Nevins DJ (1984) Characterization of an endo-β-D-glucanase from corn cell walls. *Plant Physiology* 75:61–61.

Herburger L, Franková L, Pičmanová M, Loh JW, Valenzuela-Ortega M, Meulewaeter F, Hudson A, French CE, Fry SC (2020) Hetero-trans-β-glucanase produces cellulose-xyloglucan covalent bonds in the cell walls of structural plant tissues and is stimulated by expansin. *Molecular Plant* 13:1047–1062.

Hetherington PR, Fry SC (1993). Xyloglucan endotransglycosylase activity in carrot cell suspensions during cell elongation and somatic embryogenesis. *Plant Physiology* 103:987–992.

Hoch G (2007) Cell wall hemicelluloses as mobile carbon stores in non-reproductive plant tissues. *Functional Ecology* 21:823–834.

Hrmova, M, Farkaš V, Lahnstein J, Fincher GB (2007) A barley xyloglucan xyloglucosyl transferase covalently links xyloglucan, cellulosic substrates, and (1,3;1,4)-β-D-glucans. *Journal of Biological Chemistry* 282:12951–12962.

Iqbal A, Miller JG, Murray L, Sadler IH, Fry SC (2016) The pectic disaccharides lepidimoic acid and β-D-xylopyranosyl-(1→3)-D-galacturonic acid occur in cress-seed exudate but lack allelochemical activity. *Annals of Botany* 117:607–623.

Ishii T, Matsunaga T, Pellerin P, O'Neill MA, Darvill AG, Albersheim P (1999) The plant cell wall polysaccharide rhamnogalacturonan ii self-assembles into a covalently crosslinked dimer. *Journal of Biological Chemistry* 274:13098–13104.

Jimenez-Bermudez S, Redondo-Nevado J, Munoz-Blanco J, Caballero JL, Lopez-Aranda JM, Valpuesta V, Pliego-Alfaro F, Quesada MA, Mercado JA (2002) Manipulation of strawberry fruit softening by antisense expression of a pectate lyase gene. *Plant Physiology* 128:751–759.

Johnston SL, Prakash R, Chen NJ, Kumagai MH, Turano HM, Cooney JM, Atkinson RG, Paull RE, Cheetamun R, Bacic A, Brummell DA, Schröder R (2013) An enzyme activity capable of endotransglycosylation of heteroxylan polysaccharides is present in plant primary cell walls. *Planta* 237:173–187.

Kamiyoshihara Y, Mizuno S, Azuma M, Miyohashi F, Yoshida M, Matsuno J, Takahashi S, Abe S, Shiba H, Watanabe K, Inoue H, Tateishi A (2018) Opposite accumulation patterns of two glycoside hydrolase family 3 α-L-arabinofuranosidase proteins in avocado fruit during ripening. *Horticulture Journal* 87:430–435.

Kosík O, Auburn RP, Russell S, Stratilova E, Garajova S, Hrmova M, Farkaš V (2010) Polysaccharide microarrays for high-throughput screening of transglycosylase activities in plant extracts. *Glycoconjugate Journal* 27:79–87.

Kovacová K, Farkaš V (2016) Two variants of a high-throughput fluorescent microplate assay of polysaccharide endotransglycosylases. *Applied Biochemistry and Biotechnology* 178:1652–1665.

Kozlova LV, Mokshina NE, Nazipova AR, Gorshkova TA (2017) Systemic use of limping enzymes in plant cell walls. *Russian Journal of Plant Physiology* 64:808–821.

Labavitch JM, Ray PM (1974) Relationship between promotion of xyloglucan metabolism and induction of elongation by indoleacetic-acid. *Plant Physiology* 54:499–502.

Lee YK, Kim IJ (2011) Modulation of fruit softening by antisense suppression of endo-β-1,4-glucanase in strawberry. *Molecular Breeding* 27:375–383.

Liskay A, Kenk B, Schopfer P (2003) Evidence for the involvement of cell wall peroxidase in the generation of hydroxyl radicals mediating extension growth. *Planta* 217:658–667.

Lohani S, Trivedi PK, Nath P (2004) Changes in activities of cell wall hydrolases during ethylene-induced ripening in banana: effect of 1-MCP, ABA and IAA. *Postharvest Biology and Technology* 31:119–126.

Marin-Rodriguez MC, Orchard J, Seymour GB (2002) Pectate lyases, cell wall degradation and fruit softening. *Journal of Experimental Botany* 53:2115–2119.

Maris A, Kaewthai N, Eklöf JM, Miller JG, Brumer H, Fry SC, Verbelen J-P, Vissenberg K (2011) Differences in enzymic properties of five recombinant xyloglucan endotransglucosylase/ hydrolase (XTH) proteins of *Arabidopsis thaliana*. *Journal of Experimental Botany* 62:261–271.

Mendez-Yanez A, Gonzalez M, Carrasco-Orellana C, Herrera R, Moya-Leon MA (2020) Isolation of a rhamnogalacturonan lyase expressed during ripening of the Chilean strawberry fruit and its biochemical characterization. *Plant Physiology and Biochemistry* 146:411–419.

Miller JG, Farkaš V, Sharples SC, Fry SC (2007) *O*-Oligosaccharidyl-1-amino-1-deoxyalditols as intermediates for fluorescent labelling of oligosaccharides. *Carbohydrate Research* 342:44–54.

Miller JG, Fry SC (2001). Characteristics of xyloglucan after attack by hydroxyl radicals. *Carbohydrate Research* 332:389–403.

Miller JG, Fry SC (2004) *N*-[^{3}H]Benzoylglycylglycylglycine as a probe for hydroxyl radicals. *Analytical Biochemistry* 335:126–134.

Mohnen D (2008) Pectin structure and biosynthesis. *Current Opinion in Plant Biology* 11:266–277.

Molina-Hidalgo FJ, Franco AR, Villatoro C, Medina-Puche L, Mercado JA, Hidalgo MA, Monfort A, Caballero JL, Munoz-Blanco J, Blanco-Portales R (2013) The strawberry (*Fragaria ananassa*) fruit-specific rhamnogalacturonate lyase 1 (*FaRGLYASE1*) gene encodes an enzyme involved in the degradation of cell-wall middle lamellae. *Journal of Experimental Botany* 64:1471–1483.

Ndeh D, Rogowski A, Cartmell A, Luis AS, Basle A, Gray J, Venditto I, Briggs J, Zhang XY, Labourel A, Terrapon N, Buffetto F, Nepogodiev S, Xiao Y, Field RA, Zhu YP, O'Neill MA, Urbanowicz BR, York WS, Davies GJ, Abbott DW, Ralet MC, Martens EC, Henrissat B, Gilbert HJ (2017) Complex pectin metabolism by gut bacteria reveals novel catalytic functions. *Nature* 544:65–70.

Nguyen-Phan CT, Fry SC (2019) Functional and chemical characterisation of XAF: a heat-stable plant polymer that activates xyloglucan endotransglucosylase/hydrolase (XTH). *Annals of Botany* 124:131–147.

Nishikubo N, Awano T, Banasiak A, Bourquin V, Ibatullin F, Funada R, Brumer H, Teeri TT, Hayashi T, Sundberg B, Mellerowicz EJ (2007) Xyloglucan endo-transglycosylase (XET) functions in gelatinous layers of tension wood fibers in poplar—a glimpse into the mechanism of the balancing act of trees, *Plant and Cell Physiology* 48:843–855.

Nishitani K (2005) Division of roles among members of the *xth* gene family in plants. *Plant Biosystems* 139:98–101.

Nishitani K, Masuda Y (1981) Auxin-induced changes in the cell-wall structure - changes in the sugar compositions, intrinsic-viscosity and molecular-weight distributions of matrix polysaccharides of the epicotyl cell-wall of *Vigna angularis*. *Physiologia Plantarum* 52:482–494.

O'Neill MA, Ishii T, Albersheim P, Darvill AG (2004) Rhamnogalacturonan II : structure and function of a borate cross-linked cell wall pectic polysaccharide. *Annual Review of Plant Biology* 55:109–139.

O'Neill R, Albersheim P, Darvill AG (1989) Purification and characterization of a xyloglucan oligosaccharide-specific xylosidase from pea seedlings. *Journal of Biological Chemistry* 260:20430–20437.

Ochoa-Jimenez VA, Berumen-Varela G, Burgara-Estrella A, Orozco-Avitia JA, Ojeda-Contreras AJ, Trillo-Hernandez EA, Rivera-Dominguez M, Troncoso-Rojas R, Baez-Sanudo R, Datsenka T, Handa AK, Tiznado-Hernandez ME (2018) Functional analysis of tomato rhamnogalacturonan lyase gene *solyc11g011300* during fruit development and ripening. *Journal of Plant Physiology* 231:31–40.

Ortiz A, Seymour GB, Tucker GA, Lara I (2010) Cell wall disassembly during the melting phase of softening in 'snow queen' nectarines. *Postharvest Biology and Technology* 58:88–92.

Paniagua C, Blanco-Portales R, Barcelo-Munoz M, Garcia-Gago JA, Waldron KW, Quesada MA, Munoz-Blanco J, Mercado JA (2016) Antisense down-regulation of the strawberry β-galactosidase gene *Fa beta gal4* increases cell wall galactose levels and reduces fruit softening. *Journal of Experimental Botany* 67:619–631.

Petrik DL, Tryfona T, Dupree P, Anderson CT (2020) *Bdgt43b2* functions in xylan biosynthesis and is essential for seedling survival in *Brachypodium distachyon*. *Plant Direct* 4: e00216 https://doi.org/10.1002/pld3.216.

Pilet P-E (1971) *Les Parois Cellulaires*. Paris: Doin.

Posé S, Kirby AR, Paniagua C, Waldron KW, Morris VJ, Quesada MA, Mercado JA (2015) The nanostructural characterization of strawberry pectins in pectate lyase or polygalacturonase silenced fruits elucidates their role in softening. *Carbohydrate Polymers* 132:134–145.

Prakash R, Johnston SL, Boldingh HL, Redgwell RJ, Atkinson RG, Melton LD, Brummell DA, Schröder R (2012) Mannans in tomato fruit are not depolymerized during ripening despite the presence of endo-β-mannanase. *Journal of Plant Physiology* 169:1125–1133.

Preston RD (1974) *The Physical Biology of Plant Cell Walls*. London: Chapman and Hall.

Purushotham P, Cho SH, Diaz-Moreno SM, Kumar M, Nixon BT, Bulone V, Zimmer J (2016) A single heterologously expressed plant cellulose synthase isoform is sufficient for cellulose microfibril formation *in vitro*. *Proceedings of the National Academy of Sciences* 113:11360–11365.

Quesada MA, Blanco-Portales R, Pose S, Garcia-Gago JA, Jimenez-Bermudez S, Munoz-Serrano A, Caballero JL, Pliego-Alfaro F, Mercado JA, Munoz-Blanco J (2009) Antisense down-regulation of the *FaPG1* gene reveals an unexpected central role for polygalacturonase in strawberry fruit softening. *Plant Physiology* 150:1022–1032.

Rose JKC, Braam J, Fry SC, Nishitani K (2002) The XTH family of enzymes involved in xyloglucan endotransglucosylation and endohydrolysis: current perspectives and a new unifying nomenclature. *Plant and Cell Physiology* 43:1421–1435.

Sampedro J, Pardo B, Gianzo C, Guitián E, Revilla G, Zarra I (2010). Lack of α-xylosidase activity in arabidopsis alters xyloglucan composition and results in growth defects. *Plant Physiology* 154:1105–1115.

Schröder R, Atkinson RG, Redgwell RJ (2009) Re-interpreting the role of endo-β-mannanases as mannan endotransglycosylase/hydrolases in the plant cell wall. *Annals of Botany* 104:197–204.

Schröder R, Wegrzyn TF, Sharma NN, Atkinson RG (2006) LeMAN4 endo-β-mannanase from ripe tomato fruit can act as a mannan transglycosylase or hydrolase. *Planta* 224:1091–1102.

Schuchmann MN, von Sonntag C (1978) The effect of oxygen on the OH-radical-induced scission of the glycosidic linkage of cellobiose. *International Journal of Radiation Biology* 34:397–400.

Seven M, Cem Derman Ü, Harvey AJ (2021) Enzymatic characterization of ancestral/group-IV clade xyloglucan endotransglycosylase/hydrolase enzymes reveals broad substrate specificities. *The Plant Journal* 106:1660–1673.

Sharples SC, Fry SC (2007) Radio-isotope ratios discriminate between competing pathways of cell wall polysaccharide and RNA biosynthesis in living plant cells. *The Plant Journal* 52:252–262.

Shi YZ, Zhu XF, Miller JG, Gregson T, Zheng SJ, Fry SC (2015) Distinct catalytic capacities of two aluminium-repressed *Arabidopsis thaliana* xyloglucan endotransglucosylase/hydrolases, XTH15 and XTH31, heterologously produced in *Pichia*. *Phytochemistry* 112:160–169.

Shinohara N, Sunagawa N, Tamura S, Yokoyama R, Ueda M, Igarashi K, Nishitani K (2017) The plant cell-wall enzyme AtXTH3 catalyses covalent cross-linking between cellulose and cello-oligosaccharide. *Scientific Reports* 7:46099. https://doi.org/10.1038/srep46099.

Simmons TJ, Mohler KE, Holland C, Goubet F, Franková L, Houston DR, Hudson AD, Meulewaeter F, Fry SC (2015) Hetero-trans-β-glucanase, an enzyme unique to *Equisetum* plants, functionalises cellulose. *The Plant Journal* 83:753–769.

Stratilová E, Ait-Mohand F, Rehulka P, Garajova S, Flodrová D, Rehulková H, Farkaš V (2010) Xyloglucan endotransglycosylases (XETs) from germinating nasturtium (*Tropaeolum majus*) seeds: isolation and characterization of the major form. *Plant Physiology and Biochemistry* 48:207–215.

Stratilová E, Markovic O, Dzurová M, Maloviková A, Capek P, Omelková J (1998) The pectolytic enzymes of carrots. *Biologia* 53:731–738.

Sulová Z, Takácová M, Steele NM, Fry SC, Farkaš V (1998) Xyloglucan endotransglycosylase: evidence for the existence of a relatively stable glycosyl-enzyme intermediate. *Biochemical Journal* 330:1475–1480.

Thompson JE, Fry SC (2001). Restructuring of wall-bound xyloglucan by transglycosylation in living plant cells. *The Plant Journal* 26:23–34.

Villarreal NM, Martinez GA, Civello PM (2009) Influence of plant growth regulators on polygalacturonase expression in strawberry fruit. *Plant Science* 176:749–757.

Vissenberg K, Verbelen J-P, Fry SC (2001) Root hair initiation is coupled to a highly localized increase of xyloglucan endotransglycosylase action in *Arabidopsis* roots. *Plant Physiology* 127:1125–1135.

Voxeur A, Fry SC (2014) Glycosylinositol phosphorylceramides (GIPCs) from *Rosa* cell cultures are boron-bridged in the plasma membrane and form complexes with rhamnogalacturonan II. *The Plant Journal* 79:139–149.

Vreeburg RAM, Airianah OB, Fry SC (2014) Fingerprinting of hydroxyl radical-attacked polysaccharides by *N*-isopropyl 2-aminoacridone labelling. *Biochemical Journal* 463:225–237.

Vreeburg RAM, Fry SC (2005) Reactive oxygen species in cell walls. In *Antioxidants and Reactive Oxygen Species in Plants*, ed. Smirnoff N (pp. 215–249). Oxford: Blackwell.

Walker CD, Webb J (1981) Copper in plants: forms and behaviour. In *Copper in Soils and Plants*, eds. Loneragan JF, Robson AD, Graham RD (pp. 189–212. Sydney: Academic Press.

Wang D, Yeats TH, Uluisik S, Rose JKC, Seymour GB (2018) Fruit softening: revisiting the role of pectin. *Trends in Plant Science* 23:302–310.

Wang K, Li TT, Chen SQ, Li YL, Rashid A (2020) The biochemical and molecular mechanisms of softening inhibition by chitosan coating in strawberry fruit (*Fragaria × ananassa*) during cold storage. *Scientia Horticulturae* 271:109483 https://doi.org/10.1016/j.scienta.2020.109483.

Weigel H (1963) Paper electrophoresis of carbohydrates. *Advances in Carbohydrate Chemistry* 18:61–96.

Wimmer MA, Abreu I, Bell RW, Bienert MD, Brown PH, Dell B, Fujiwara T, Goldbach HE, Lehto T, Mock HP, von Wiren N, Bassil E, Bienert GP (2019) Boron: an essential element for vascular plants. A comment on Lewis (2019) 'boron: the essential element for vascular plants that never was'. *New Phytologist* 226:1232–1237.

Witasari LD, Huang F-C, Hoffmann T, Rozhon W, Fry SC, Schwab W (2019) Higher expression of the strawberry xyloglucan endotransglucosylase/hydrolase genes *FvXTH9* and *FvXTH6* accelerates fruit ripening. *The Plant Journal* 100:1237–1253. https://doi.org/10.1111/tpj.14512.

Yamada K, Hondo M, Shigemori H, Hirose K, Hasegawa K (2010) Physiological effects of lepidimoic acid (a stimulatory allelochemical) on cucumber cotyledons. *Allelopathy Journal* 25:497–502.

Yang HB, Benatti MR, Karve RA, Fox A, Meilan R, Carpita NC, McCann MC (2020) Rhamnogalacturonan-I is a determinant of cell-cell adhesion in poplar wood. *Plant Biotechnology Journal* 18:1027–1040.

Zhou HC, Li G, Zhao X (2016) Comparative analysis of pectate lyase in relation to softening in strawberry fruits. *Canadian Journal of Plant Science* 96:604–612.

Zhou HC, Li G, Zhao X, Li LJ (2015) Comparative analysis of polygalacturonase in the fruit of strawberry cultivars. *Genetics and Molecular Research* 14:12776–12787.

5 Reconsidering Pectin Structure
A Historical Perspective Informed by Studies of the Biosynthetic Enzymes

Debra Mohnen, Melani A. Atmodjo, and Pradeepa Jayawardhane
University of Georgia

INTRODUCTION

In 1990, we initiated a long-term research project to study pectin synthesis. Our initial aim was to identify the enzymes that synthesise the major pectic polysaccharide homogalacturonan (HG). The prevailing view at the time, and one that we held when we began our studies of the pectin biosynthetic enzymes, was of pectin as a family of three polysaccharides: HG (Figure 5.1), rhamnogalacturonan I (RG-I) (Figures 5.2 and 5.3) and the substituted galacturonans including the complex rhamnogalacturonan II (RG-II) (Figure 5.4) and, in aquatic plants and specific tissues, apiogalacturonan (AGA) (Figure 5.5) and xylogalacturonan (XGA) (Figure 5.6), respectively (Albersheim et al. 1996; Mohnen et al. 1996; Mohnen 1999, 2008; Ridley et al. 2001). It was not known at the time whether the HG, RG-I and RG-II were covalently connected as a single network via their backbones, or alternatively if they existed as two or more distinct heteropolymers. However, since cleavage of cell walls with endopolygalacturonase (EPG, which fragments HG) resulted in the release of RG-I and RG-II polymers from the wall, pectin was frequently depicted in cell wall models as a network of RG-I and RG-II covalently connected by HG glycan regions (domains) (Mohnen 2008). Our original goal had been to study pectin synthesis. We had not intended or expected to redefine pectin structure. However, now with more than 30 years of results on the enzymatic products of different HG GALACTURONOSYLTRANSFERASES (GAUTs), their enzymatic properties and the phenotypes of GAUT mutants and transgenic plants down-regulated or knocked-out in *GAUT* gene expression, *we conclude that the prevailing view of pectin structure is inadequate to explain the results that have been obtained.* Thus, a reevaluation of pectin structure is in order.

This chapter begins with a summary of the history of the discovery of pectin and studies of its structure. The goal is to recognise the early researchers in the field and review if and how the past data address the following questions.

- How did we arrive at the view of pectin as a family of three pectic glycans: HG, RG-I and RG-II?
- How much of that view is justified by definitive chemical/structural data?
- How are these three pectic glycans covalently connected to each other in the wall?
- How many different covalently connected pectic heteroglycans exist in the wall?
- Does homogenous, independent HG exist in the cell wall, or is it always synthesised as part of a heteroglycan?

(a) 4)-α-D-GalA-(1,4)-α-D-GalA-(1,4)-α-D-GalA-(1,4)-α-D-GalA-(1,4)-α-D-GalA-(1,4)-α-D-GalA-(1, (*n*)

(b) 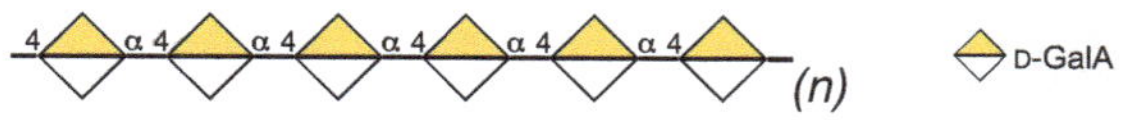

FIGURE 5.1 Representative structure of homogalacturonan (HG). (a) Anomeric linkage and glycosyl residue content. (b) Glycobiology symbol nomenclature (https://www.ncbi.nlm.nih.gov/glycans/snfg.html) representation. HG is a homoglycan of α-D-1,4-GalA that can be methylesterified at the C-6 carboxyl and, to a lesser extent, acetylated at O-2 or O-3 (not depicted in figure). HG is often found covalently connected to rhamnogalacturonan I (RG-I) and may also be substituted with xylose, apiose or more complex side branches to comprise xylogalacturonan, apiogalacturonan and rhamnogalacturonan II (RG-II) that as a group are referred to as substituted galacturonans. It is not known if all these pectic glycans are covalently connected into a single polymer. It is also not known how much HG, if any, is synthesised as a separate homoglycan polysaccharide, or rather, if it is always synthesised as part of a larger glycan or glycoconjugate complex. HG can also be part of a glycoconjugate, such as the Arabinoxylan Pectin Arabinogalactan Protein APAP1 (Tan et al. 2013). Although it is commonly stated that HG is synthesised in a highly methylesterified state, experimental confirmation of the degree of esterification during synthesis has not been determined and the variation in the degree of esterification of newly synthesised HG is not known. (*n*) Degrees of polymerisation (DPs) of 100 or more for HG in the cell wall have been reported (Thibault et al. 1993) and DPs of over 500 for HG enzymatically *in vitro* synthesised by GAUTs (Amos et al. 2018).

DOI: 10.1201/9781003178309-5

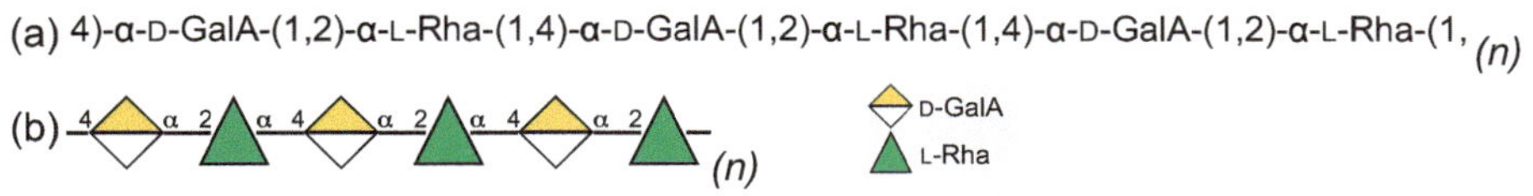

FIGURE 5.2 Representative structure of the backbone of rhamnogalacturonan I (RG-I). (a) Anomeric linkage and glycosyl residue content. (b) Glycobiology symbol nomenclature (https://www.ncbi.nlm.nih.gov/glycans/snfg.html) representation. RG-I has a disaccharide repeat heteroglycan [α-1,4-D-GalA-α-1,2-L-Rha] backbone. (*n*) The degree of polymerisation of the RG-I backbone appears to vary in different types of tissues with values of 30–300 being reported (Ropartz and Ralet 2020). *In vitro* synthesised RG-I backbone has lengths of up to DP 130 (Amos et al. 2022).

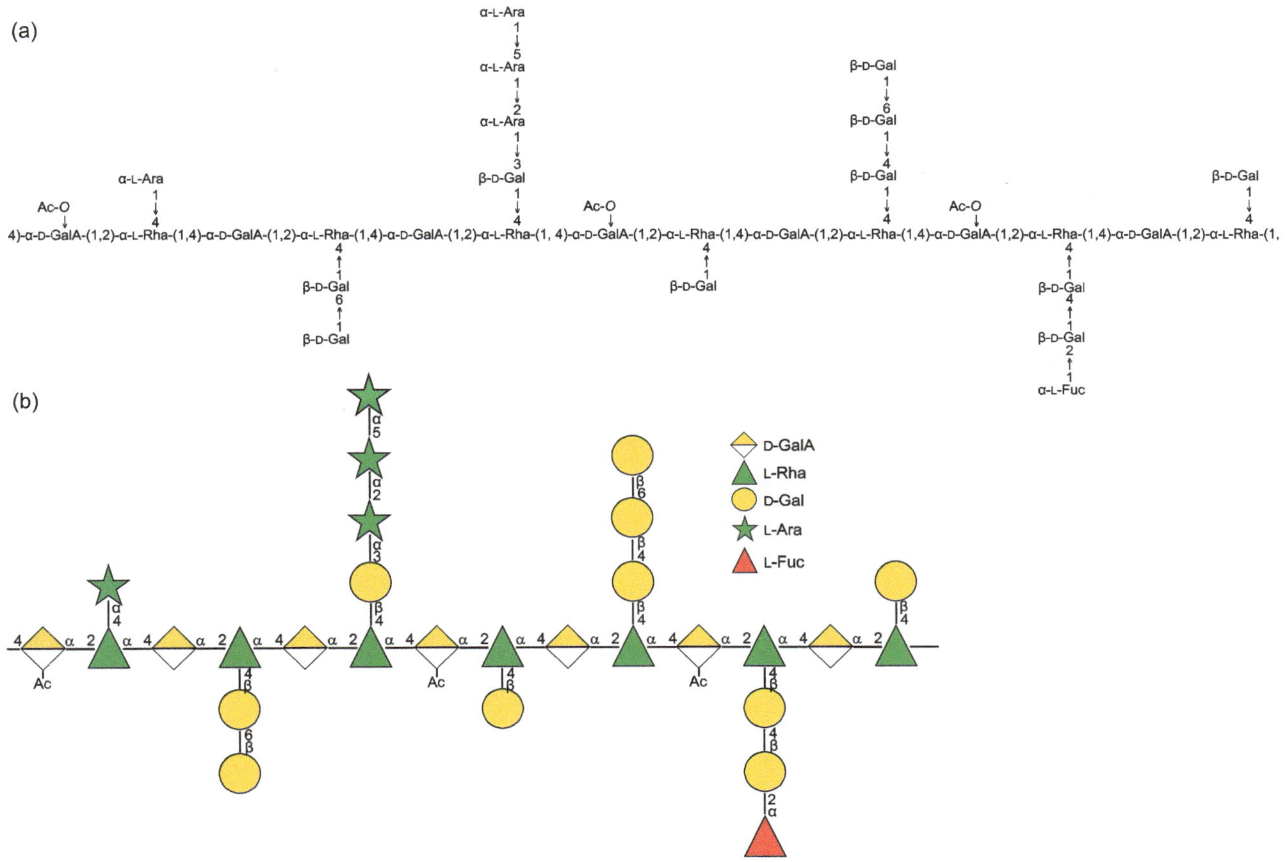

FIGURE 5.3 Representative structure of rhamnogalacturonan I (RG-I) with representative side chains. (a) Anomeric linkage and glycosyl residue content. (b) Glycobiology symbol nomenclature (https://www.ncbi.nlm.nih.gov/glycans/snfg.html) representation. RG-I is branched at up to 90% of the rhamnosyl residues with individual arabinose or galactose residues or more complex arabinan, galactan or arabinogalactan side branches. Arabinogalactan side branches on RG-I are sometimes referred to as Type I or Type II arabinogalactans. RG-I can be acetylated at O-3 of the GalA in the backbone. RG-I is often found covalently connected to HG and/or xylogalacturonan, and has been shown to be a part of a glycoconjugate, such as the Arabinoxylan Pectin Arabinogalactan Protein (APAP) 1 (Tan et al. 2013). Note that the exact locations of the side chains and acetate groups along the RG-I backbone are not known, and the specific oligosaccharide structures depicted are representative.

- Does homogenous, independent RG-I exist in the cell wall, or is it always synthesised as part of a heteroglycan or proteoglycan?
- What is the evidence that the so-called Type II arabinogalactans (i.e. β-1,3 and β-1,6-linked galactans with α-1–5-linked arabinan) are present in RG-I?
- What is the evidence that pectic glycans are covalently connected to proteins or to hemicellulose?
- Could at least part of the pectic matrix consist of a covalently linked "net" akin in function to bacterial peptidoglycan? [Such a proposition was originally proposed by Lamport (1970) and supported by the cell wall model by Keegstra et al. (1973), Albersheim et al. (1973), Chambat et al. (1984), Qi et al. (1995), and others.]
- Are there components of the pectins or pectic arabinogalactan proteins (AGPs) that are synthesised

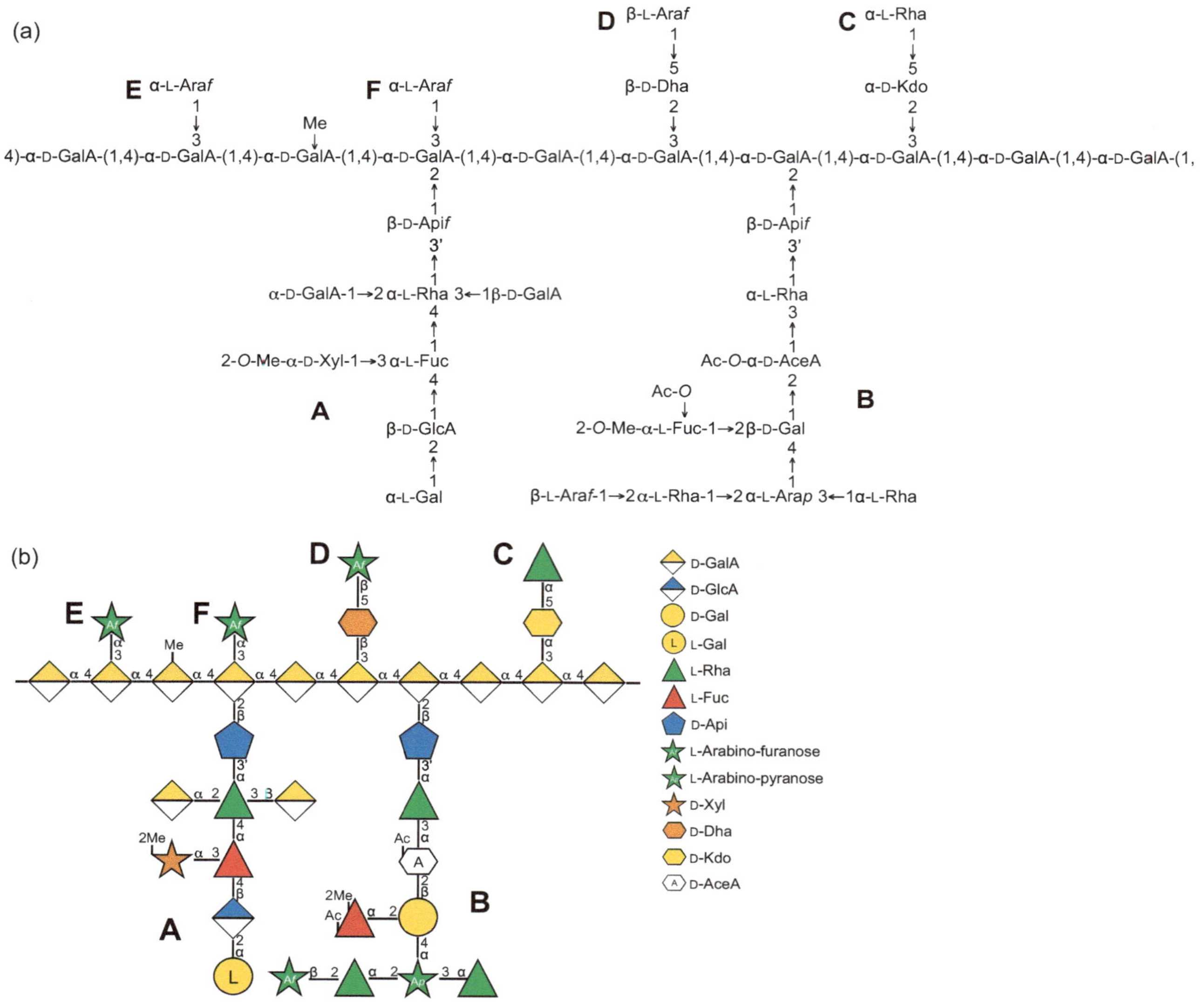

FIGURE 5.4 Representative structure of rhamnogalacturonan II (RG-II). (a) Anomeric linkage and glycosyl residue content. (b) Glycobiology symbol nomenclature (https://www.ncbi.nlm.nih.gov/glycans/snfg.html) representation. Structural data are based on Ndeh et al. (2017). Note that the exact locations of the side chains along the RG-II backbone are not known.

separately within the endomembrane system and covalently connected in the wall by transglycosidases, or are the complex heteroglycans and proteoglycans all synthesised within the endomembrane system?

The chapter goes on to compare the prevailing view of pectin structure in the plant with the properties of characterised pectin biosynthetic enzymes and their products. We propose that addressing many of the questions posed above will require *in vitro* reconstitution of pectin using pure heterologously expressed recombinant enzymes or by *in vivo* synthesis of pectic polymers and glycoconjugates in non-plant systems. Knowledge of the enzyme kinetics, acceptor and substrate specificities, products and protein complexes will provide a critical foundation for generation of hypotheses for pectin structure in the plant and for mechanisms for how that structure is generated.

When we began our study of pectin biosynthetic enzymes, we did not have enough information to critically test pectin structure models. However, the data now allow us to revisit the models and test and refine them. The current data lead us to propose that a change in the way we discuss and view pectin structure is required. Specifically, we put forward that *pectin is an assortment of unique pectic glycans, heteroglycans and glycoconjugates with unique HG and rhamnogalacturonan domains and that to fully understand pectin function one must know what the full polymer(s) is that is affected in any given mutant or is the product of any specific gene(s).* We propose that different pectic glycans (e.g. HG, RG-I, RG-II) function as structural domains in multiple complex heteropolymers that have unique roles in cell wall architecture.

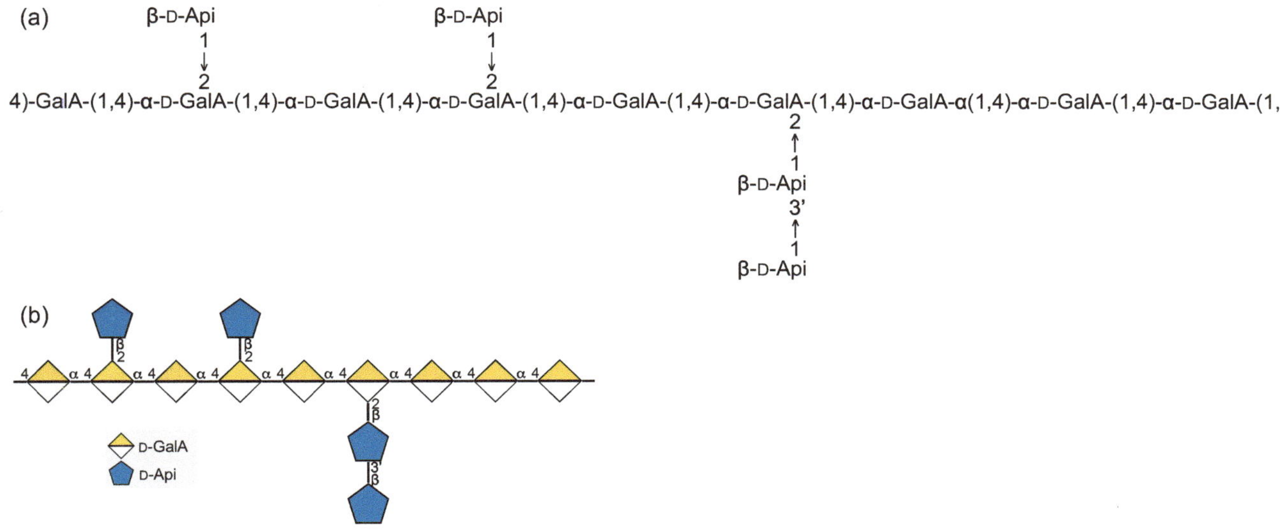

FIGURE 5.5 Representative structure of apiogalacturonan (AGA). (a) Anomeric linkage and glycosyl residue content. (b) Glycobiology symbol nomenclature (https://www.ncbi.nlm.nih.gov/glycans/snfg.html) representation.

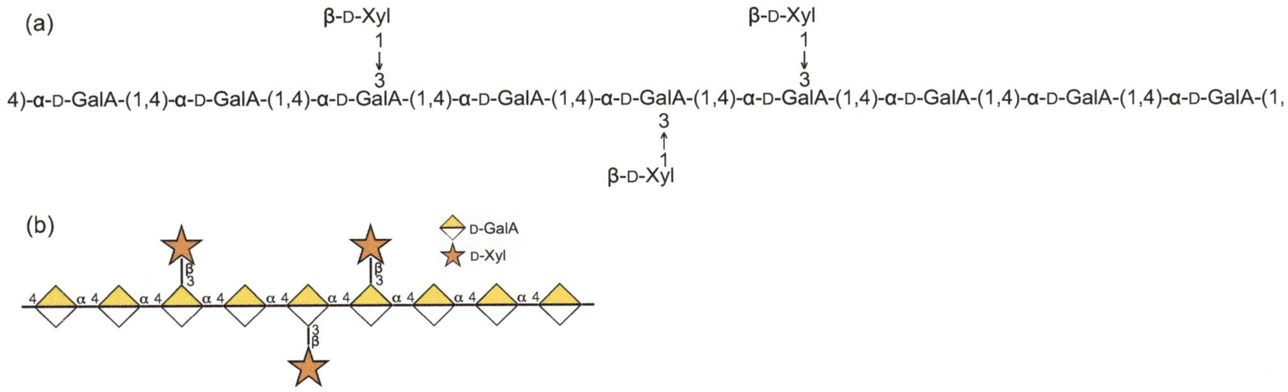

FIGURE 5.6 Representative structure of xylogalacturonan (XGA). (a) Anomeric linkage and glycosyl residue content. (b) Glycobiology symbol nomenclature (https://www.ncbi.nlm.nih.gov/glycans/snfg.html) representation.

EARLY HISTORY OF RESEARCH ON PECTIN STRUCTURE – 1825 TO 1973

The name pectin (pectique) from the Greek "pēktikós" meaning congealed was first used by Braconnot in 1825 to describe the viscous, acidic extract with gelling properties that could be isolated from many fruits and vegetables, including plant tubers (Braconnot 1825; Pilnik 1990; Willats et al. 2006; Ropartz and Ralet 2020). A long history of structural characterisation studies ensued, some aimed at understanding the structure of pectin from commercial crops, particularly citrus peel and apple pomace (BeMiller 1986; Willats et al. 2006), and others focused on pectin from diverse plant species. These two types of studies aimed, respectively, to understand the structure of commercial food pectin, and the structure of pectin in plant cell walls and during biogenesis. Most analyses used pectin extracted, initially by chemical extraction and later by enzymatic fragmentation, from insoluble cell walls. The picture of pectin obtained from these early studies was thus heavily influenced by information from commercial citrus and apple and from pectic oligosaccharides generated from pectic cell wall polymers by chemical and/or enzymatic fractionation. Since these data represented both mixtures of products and only a portion of the pectic material that was present in the walls, i.e., that which was extracted and recovered by the methods used, understanding of the full intact polymeric structure of pectin in the plant and of the number and types of diverse pectin-containing structures in the cell walls was incomplete. This lack of understanding continues to beset the field today.

In the early 1900s, pectin was identified as a substance rich in galacturonic acid (GalA) (Ehrlich 1917; Smolenski 1923) and shortly thereafter as a polymer chain of GalA that also contained neutral sugars (reviewed in Kertesz (1951); Naseri et al. (2008); Ropartz and Ralet (2020)).

In his informative 1951 book entitled "The Pectic Substances", Kertesz provided a detailed history of the numerous attempts by multiple chemists to decipher what the industrial colloid pectin was. Kertesz summarised the state of knowledge when he concluded that pectins were polyuronides composed mostly of GalA but that two schools of thought existed with regard to their structure (Kertesz 1951). Some researchers thought pectins were composed entirely of GalA, while others held that neutral sugars including arabinose (Ara), galactose (Gal) and rhamnose (Rha) were attached to chains of GalA. Kertesz noted that "The characteristics and composition of pectic preparations depend on the source material and on the method of preparation" (Kertesz 1951). This is a theme that reoccurs throughout this review and remains a limitation in our understanding of the "structure of pectin". We do not understand the full structural diversity of pectin in any specific cell or tissue type. As noted above, two overarching goals drove, and still drive, many researchers who study pectin, those interested in the commercial uses of pectin (particularly in the food industry) for whom most attention focuses on pectin as a polymer of α-1,4-linked GalA for which functionality is often associated with the amount of methylesterification of the GalA residues (reviewed in Willats et al. (2006)) and those interested in the *in planta* biogenesis and function of pectin. The latter researchers somewhat earlier appreciated that neutral sugars were also part of pectin and had roles in its structure and function.

Kertesz reported that "Two groups of polysaccharides almost invariably accompany the pectic substances in plants and in crude preparations". These were "arabans" (later called arabinans) and galactans. While at that time many held that these were not pectin but rather incidentally co-isolated, Kertesz did conclude that "Although there is considerable likelihood that they are in some manner related to the pectic substances, at the present time there is no creditable theory which would clearly state their genetical relationship to the polyuronides" (Kertesz 1951). We now know (see below) that there are two types of arabinans and galactans, so-called Type I (i.e. containing β-1,4-linked galactans) and Type II (i.e. containing β-1,3; β-1,6-linked galactans) (Fincher et al. 1983; Gibeaut and Carpita 1994; Hijazi et al. 2014) and that they are indeed associated with pectin. However, we remain uncertain as to how they are associated with pectin and whether Type II arabinogalactans are attached to the RG-I backbone, associated with RG-I containing proteoglycans (Tan et al. 2013), or both, although the bulk of available data tend to support the third option.

The view of what pectin is has changed over time. For example, Aspinall in his 1973 review of plant cell walls highlighted several transitions in thinking regarding pectin, including a period when pectin was considered as a "pectic triad" of galacturonan, galactan and arabinan (Aspinall 1973). This shifted to a view that pure galacturonans occurred only infrequently but rather were associated with some amount of neutral sugars, although the full polymeric structure of pectin(s) in the wall remained unclear. To this day, we still do not know how much pure HG, if any, exists in plant cell walls, although we know that homo-polymeric HG can be synthesised *in vitro* by pure enzymes in a non-methylesterified form (Sterling et al. 2006; Atmodjo et al. 2011; Amos et al. 2018; Engle et al. 2022) and that over the years a handful of researchers claimed to have purified "pure" HG (e.g. Chambat and Joseleau (1980)). Curiously, although Aspinall published some of the earliest data identifying pectic oligosaccharide fragments with GalA linked to Rha (Aspinall et al. 1967b,c), he did not describe such rhamnogalacturonan in his 1973 review within the context of cell wall pectin. Rather, he stated that "the role of L-rhamnose may be assessed separately, since as far as known in pectins, this sugar occurs only in the interior chains, whereas residues of all other neutral sugars are encountered in the exterior chains only" (Aspinall 1973). This is an intriguing statement that touches on aspects of architectural organisation of pectin in the wall, an area for which we still have little understanding. It is important to note the major contributions Aspinall made to the early studies of pectin (and what would eventually be known as RG-I structure). At the time, he carried out his seminal studies and identified what he called the "acidic polysaccharide complex from soybean cotyledon meal" (Aspinall et al. 1967a–c); it was not clear how the various oligosaccharide fragments identified would eventually be found to be connected into a cell wall polymer. However, among the fragments he discovered, already in 1967, were fragments of the GalA-Rha disaccharide repeat backbone of RG-I and fragments of the RG-I side chains including β-1,4 galactan chains branched at O-3 with 1,5-linked arabinan (Aspinall et al. 1967a–c).

Finally, in his 1951 book, Kertesz tackled the confusing issue of terminology regarding different pectin fractions and structures by reintroducing the term "protopectin", which he defined as "the water-insoluble parent pectic substance which occurs in plants and which, upon restricted hydrolysis, yields pectic acid" (Kertesz 1951). At that time, protopectin was believed to be insoluble due to association with cellulose, presence of polyvalent ions, large size or a combination thereof. Regardless of the mechanism by which some pectin is held tightly in the wall, the recognition that an appreciable amount of pectin exists in cell walls in an insoluble form underscored the importance of distinguishing between the properties of soluble pectin isolated from the walls by water or mild solvent extracts versus the insoluble form of pectin in the plant wall. Kertesz's evaluation of the state of understanding of protopectin at that time rings true today: "The nature of protopectin is one of the most intricate problems in pectin chemistry and there is no theory at this time which would explain its behaviour and properties to our complete satisfaction" (Kertesz 1951). The structure and function of that surprisingly large portion of pectin that is held very tightly within the wall remains unclear today and is a critical question for future studies (Biswal et al. 2018a,b) (see below).

HISTORY OF PECTIN PRIMARY STRUCTURE RESEARCH – 1973 TO 1990 AND BEYOND

A major advance in our understanding of pectin structure occurred in the 1970s and 1980s with the development of methods to determine the glycosyl residue composition, linkage and anomeric configuration of pectic oligosaccharide fragments from pectic cell wall preparations from diverse plants and tissues. The use of such methods enabled the determination of the primary structure (sequence) of pectic oligosaccharides released from cell walls by treatment with pectin-specific hydrolytic or lytic enzymes or via specific chemical fragmentation. These advances provided specific pectic oligosaccharide structures that could be classified into categories such as: HG backbone fragments, RG-I backbone fragments, substituted HG (i.e. HG backbone oligosaccharides with one or more covalently attached side chain sugars), RG-I side chains (i.e. oligosaccharides released from the RG-I backbone), RG-HG covalently connected backbone regions, RG-II and non-sugar substitutions on pectin (i.e. methylesterification, methyletherification, acetylation, feruloylation, etc.). While it is outside the scope of this review to summarise the immense amounts of glycosyl residue composition, anomeric configuration and linkage data published over the years based on analyses of pectic oligosaccharides recovered from diverse plants and plant tissues, Table 5.1 provides examples of some of the structural data that contributed to leaps in our understanding. More details about the oligosaccharides identified can be found in several reviews (O'Neill et al. 1990; Mohnen 2002; Caffall and Mohnen 2009; Atmodjo et al. 2013; Kaczmarska et al. 2022).

TABLE 5.1
Representative Pectic Oligosaccharides Identified in RG-I-Enriched Preparations.

Oligosaccharide Identified	Reference
4-β-D-Gal*p*-1,4-β-D-Gal*p*-(L-Ara-1,5-L-Ara-1,3-)-1,4-β-D-Gal*p*-1,4-β-D-Gal*p*-1	Aspinall et al. (1967a)
β-D-Xyl*p*-1,3-D-GalA	Aspinall et al. (1967b)
Gal*p*A1,4-Gal*p*A,1,2-Rha	Aspinall et al. (1967c)
GalA-1,2-Rha*p*-1,2-Rha*p*	Aspinall et al. (1967c)
GalA-1,2-Rha-1,4-GalA-1,2-Rha-1,4-GalA-1,4-GalA [HG on reducing end of RG-I]	Talmadge et al. (1973)
3-O-(3-O-feruloyl-α-L-Ara*p*)-L-Ara	Fry (1982)
4-O-(6-O-feruloyl-β-D-Gal*p*)-D-Gal	Fry (1982)
GalA-1,2-Rha-1,4-GalA-1,2-Rha-1,4-GalA-	Lau et al. (1985)
-4-D-GalA*p*-1,6-D-Gal-1-	Lau et al. (1985)
β-D-Galp-1,4-L-Rhaol	Lau et al. (1987)
Ara-1,5-Araol	Lau et al. (1987)
Gal-Galol	Lau et al. (1987)
Ara-1,4-Rhaol	Lau et al. (1987)
α-L-Ara*f*-1,3-β-D-Gal*p*-1,4-L-Rhaol	Lau et al. (1987)
β-D-Gal*p*-1,6-β-D-Gal*p*-1,4-L-Rhaol	Lau et al. (1987)
β-D-Gal*p*-1,4-β-D-Gal*p*-1,4-L-Rhaol	Lau et al. (1987)
Ara-Ara-Araol	Lau et al. (1987)
Gal-Gal-Galol	Lau et al. (1987)
Ara-Ara-Rhaol	Lau et al. (1987)
Fuc-Ara-Rhaol	Lau et al. (1987)
α-L-Ara*f*-1,2-α-L-Ara*f*-1,3-β-D-Gal*p*-1,4-L-Rhaol	Lau et al. (1987)
β-D-Gal*p*-1,6-β-D-Gal*p*-1,4-β-D-Gal*p*-1,4-L-Rhaol	Lau et al. (1987)
α-L-Fuc*p*-1,2-β-D-Gal*p*-1,4-β-D-Gal*p*-1,4-L-Rhaol	Lau et al. (1987)
Ara-Ara-Ara-Araol	Lau et al. (1987)
Gal-Gal-Gal-Galol	Lau et al. (1987)
l-Ara*f*-1,5-L-Ara*f*-1,2-L-Ara*f*-1,3-D-Gal*p*-1,4-L-Rhaol	Lau et al. (1987)
Gal-Gal-Gal-Gal-Galol	Lau et al. (1987)
Ara-Ara-Ara-Ara-Araol	Lau et al. (1987)
GalA-1,4-GalA Also small amounts of 3-linked and 3,6-linked galactose	Sun et al. (1987)
Rha-1,4-GalA	Sun et al. (1987)
GalA-1,2-Rha	Sun et al. (1987)
Gal-1,4-Rha	Sun et al. (1987)
Gal-1,4-Gal	Sun et al. (1987)
Ara-1,5-Ara	Sun et al. (1987)

(Continued)

TABLE 5.1 (*Continued*)
Representative Pectic Oligosaccharides Identified in RG-I-Enriched Preparations.

Oligosaccharide Identified	Reference
Ara-1,5-Ara-1,5-Ara	Sun et al. (1987)
Terminal Ara	Sun et al. (1987)
Terminal Gal	Sun et al. (1987)
Ara-Galol	Thomas et al. (1989)
Gal-Ara-Rhaol	Thomas et al. (1989)
Gal-Gal-Ara-Rhaol	Thomas et al. (1989)
Ara-Gal-Rhaol	Thomas et al. (1989)
α-L-Ara-β-D-Gal-1,4-Rhaol	Thomas et al. (1989)
β-D-Gal-1,4-α-L-Rha-Galol	Thomas et al. (1989)
Ara-Araol	Thomas et al. (1989)
Gal-Galol	Thomas et al. (1989)
β-D-Gal-1,4-Rhaol	Thomas et al. (1989)
Ara-1,3-Gal-Rhaol	Thomas et al. (1989)
Ara-Ara-1,3-Gal-Rhaol	Thomas et al. (1989)
Ara-Ara-Ara-1,3-Gal-Rhaol	Thomas et al. (1989)
Gal-Gal-Rhaol	Thomas et al. (1989)
Ara-Rhaol	Thomas et al. (1989)
Rha-l,2-Rhaol	Gao et al. (1990)
Rha-l,4-Rhaol	Gao et al. (1990)
Ara-l,4-Rhaol	Gao et al. (1990)
Gal-1,4-Rhaol	Gao et al. (1990)
Gal-l,6-Galol	Gao et al. (1990)
GlcA-l,4-Rhaol	Gao et al. (1990)
Rha-Rhaol	Gao et al. (1990)
HexA-Rha-Rhaol	Gao et al. (1990)
α-Rha*p*-1,4-α-Gal*p*A-1,2-α-Rha*p*-1,4-α-Gal*p*A	Schols and Voragen (1994)
α-Rha*p*-1,4-α-Gal*p*A-1,2-α-Rha*p*-(β-Gal*p*-1,4)-1,4-α-Gal*p*A	Schols and Voragen (1994)
β-Gal*p*-1,4-α-Rha*p*-1,4-α-Gal*p*A-1,2-α-Rha*p*-1,4-α-Gal*p*A	Schols and Voragen (1994)
β-Gal*p*-1,4-α-Rha*p*-1,4-α-Gal*p*A-1,2-α-Rha*p*-(β-Gal*p*-1,4)-1,4-α-Gal*p*A	Schols and Voragen (1994)
α-Rha*p*-1,4-α-Gal*p*A-1,2-α-Rha*p*-1,4-α-Gal*p*A-1,2-α-Rha*p*-1,4-α-Gal*p*A	Schols and Voragen (1994)
β-Gal*p*-1,4-α-Rha*p*-1,4-α-Gal*p*A-1,2-α-Rha*p*-(β-Gal*p*-1,4)-1,4-α-Gal*p*A-1,2-α-Rha*p*-1,4-α-Gal*p*A	Schols and Voragen (1994)
β-Gal*p*-1,4-α-Rha*p*-1,4-α-Gal*p*A-1,2-α-Rha*p*-1,4-α-Gal*p*A-1,2-α-Rha*p*-(β-Gal*p*-1,4)-1,4-α-Gal*p*A	Schols and Voragen (1994)
β-Gal*p*-1,4-α-Rha*p*-1,4-α-Gal*p*A-1,2-α-Rha*p*-(β-Gal*p*-1,4)-1,4-α-Gal*p*A-1, 2-α-Rha*p*-(β-Gal*p*-1,4)-1,4-α-Gal*p*A	Schols and Voragen (1994)
-4-α-D-Gal*p*A-1,2-α-L-Rhap-1- (DP-6–20)	Renard et al. (1997)
β-D-Glc*p*A-1,3-α-D-Gal*p*A-1,2-α-L-Rha*p*-1,4-α-D-Gal*p*A-1,2-L-Rha*p*	Renard et al. (1999)
α-D-Gal*p*A-1,2-α-L-Rha*p*-1,4-α-D-Gal*p*A-(β-D-Glc*p*A-1,3)-1,2-L-Rha*p*	Renard et al. (1999)
α-Rha*p*-(β-Gal*p*-1,4)-1,4-α-Gal*p*A-1,2-α-Rha*p*-(α-Ara*f*-1,3-β-Gal*p*-(4,1-α-Ara*f*)-1,4)-1,4-α-Gal*p*A	Zheng and Mort (2008)
3-O-acetylated-1,2,4-α-L-Rha	Sengkhamparn et al. (2009)
terminal-α-D-Gal	Sengkhamparn et al. (2009)

For each glycosyl sequence: reducing end is on the right, and the sequence in parentheses indicates covalent linkage to the sugar residue to the left.

Early Definition of Rhamnogalacturonan

Knowledge of the primary structure of pectic oligosaccharides from different plants, tissues and pectic preparations during the 1970s to the 1990s began to provide at least partial answers to some key questions regarding pectin primary structure. These included questions regarding the nature of the backbone of pectin. In the early 1970s, the term rhamnogalacturonan was used to refer to a proposed heteropolymer containing HG that was interspersed with one or more 2-linked rhamnosyl residues (Talmadge et al. 1973). There was no distinction in the earliest models between rhamnogalacturonan and HG, rather rhamnogalacturonan was considered as the entire polymer containing relatively modestly sized HG with degrees of polymerisation (DPs) 4–8 interspersed by occasional individual or dimeric dimers of Rha-GalA (Albersheim et al. 1973; Talmadge et al. 1973; Albersheim 1975). In those early cell wall models, it was recognised based on space-filling calotte models and work from bacterial polysaccharides with α-1,2-Rha linkages that the presence of 2-linked Rha in the RG-I backbone would result in a zigzagged conformation of the backbone (Albersheim et al. 1973).

The early models also provided fractionation evidence (not structurally confirmed) for the possible covalent attachment between xyloglucan and acidic pectin. Although this concept was not widely accepted, it is noteworthy that multiple researchers over the years have come to the same conclusion using diverse co-fractionation and *in vivo* labelling techniques (Barrett and Northcote 1965; Thompson and Fry 2000; Brett et al. 2005; Cumming et al. 2005; Abdel-Massih et al. 2007; Popper and Fry 2008). This topic is discussed more below, but it is important to note here since the concept/model of a cell wall that included covalent connections between pectins, hemicelluloses and proteins was proposed early on, and remains an unresolved but important question. Definitive studies to prove or disprove such covalent connections are critical for formulation of cell wall models. For history on conceptual models of plant cell walls, see Chapter 11 in this book.

Rhamnogalacturonan Becomes RG-I and Confirmation of the Rha-GalA Disaccharide Repeat Backbone

Peter Albersheim's group carried out a series of studies in the 1970s–1990s using sycamore (*Acer pseudoplatanus*) suspension culture cell walls treated with a purified EPG to solubilise pectin from the walls. It was known at that time that a large amount of pectin could be released from cell walls by treatment with EPG, strongly indicating that the released pectins had been covalently linked to HG (Figure 5.1). While this method did not address the nature of the starting pectic polymer from which the soluble material was released or how the released pectin had been connected to HG, the process did yield large amounts of pectic material that could be used for primary sequence determination.

From the analyses of this EPG-released pectic material over the years, Albersheim and colleagues discovered that the cell wall contained two different types of Rha-containing pectic polymers. This was first demonstrated in the landmark study by Darvill et al. (1978) when they identified a new rhamnose-containing pectic polysaccharide that was different from the previously identified rhamnogalacturonan (i.e. the original previous polymer was thought to consist of moderately sized HG interspersed by one or two Rha-GalA disaccharides). The new pectic polymer was therefore named rhamnogalacturonan II (RG-II) and the previously identified rhamnogalacturonan was named rhamnogalacturonan I (RG-I). Shortly after the discovery of RG-II (which is described later in this review), Lau et al. (1985) used liquid chromatography-mass spectrometry (LC-MS), gas chromatography-mass spectrometry (GC-MS) and nuclear magnetic resonance (NMR) spectroscopy to conclusively show that RG-I had a strictly GalA-Rha repeating backbone of alternating -4-α-D-Gal*p*A-1,2-α-L-Rha*p*-1- (Figure 5.2). The current definition of rhamnogalacturonan as RG-I thus began with the Lau et al. (1985) study which built off the work of McNeil et al. (1980), both showing that ~50% of the Rha in the backbone was substituted at the O-4 position with side chains (thought to be arabinans and galactans). Importantly, Lau et al. (1985) were able to obtain glycosyl sequence data for 21 unique oligosaccharides released from purified RG-I by acid hydrolysis and concluded that no discernable pattern could be discerned for the distribution of the branched versus unbranched Rha in the RG-I backbone. Such results support a possible stochastic mechanism of side branch synthesis, at least in sycamore suspension culture cells.

Taken together, the results supported a model for how the early pectic fragments described by Aspinall (Aspinall et al. 1967a,b) and others fit into larger pectic structures. A quote from Albersheim's 1973 review summarised the general significance of the work: "The structures proposed for the sycamore pectic polymers are generally consistent with the previous structural studies of pectin polysaccharides. Rhamnogalacturonans appear to be a common feature of all pectic polysaccharides" (Albersheim et al. 1973). Thus, starting in the 1970s until the present it has been accepted that at least some HG (Figure 5.1) in pectin is covalently associated with RG-I in a single polymeric structure that consists of HG and RG-I connected via their backbones and with RG-I having Gal and Ara side chains on ~50% of the rhamnose residues (Figure 5.3).

Side Chains of RG-I: Type, Linkage to the Backbone and Positioning

A question that researchers sought to answer in the early 1970s was whether all the neutral sugars that co-fractionated with pectins were linked to RG-I, and if so, how they were linked. A challenge associated with this question was the observation that arabinans, galactans and arabinogalactans sometimes, but not always, fractionated as if these were covalently linked to pectin. In a study of EPG-released RG-I from sycamore suspension cell walls, McNeil et al. (1980) concluded that RG-I accounted for 7% of the starting dry mass of the cell wall, had a molecular weight of ~200,000 Da and was composed of primarily D-GalA, L-Ara, D-Gal, L-Rha with smaller amounts of fucose (L-Fuc), glucose (D-Glc) and xylose (D-Xyl). Since it would be expected that non-covalently bound proteoglycans such as AGPs would have been removed because the starting walls had been washed with 500 mM potassium phosphate prior to EPG treatment, the presence of large amounts of Gal and Ara provided evidence that arabinans and galactans were part of the RG-I structure. Furthermore, the linkage data confirmed that the bulk of the Gal had linkages associated with the β-galactans and the Ara with 5-linked, 3,5-linked and 2,3,5-linked arabinans. Subsequent work by Lau et al. (1985) on RG-I that had been deesterified prior to EPG treatment to remove residual HG confirmed the alternating disaccharide backbone of RG-I with a composition of equimolar amounts of GalA (17%) and Rha (16%) and large amounts of L-Ara (32%) and D-Gal (31%) as well as 2% L-Fuc. It is noteworthy that although the bulk of galactosyl linkages identified were those expected for Type I β-1,4-linked galactans, the presence of some 3,6-linked galactan, and 6-linked and 3-linked galactan reported by both McNeil et al. (1980) and Lau et al. (1985) supports the presence of some Type II arabinogalactans in the RG-I, or in RG-I AGP glycoconjugates, as would be later discovered.

In their defining 1987 paper, Lau et al. (1987) selectively cleaved off the GalA residues in the RG-I backbone of RG-I purified from sycamore suspension culture cell walls by treatment of walls with EPG and separation by ion exchange and size exclusion chromatography. They then dissolved the RG-I in ethylenediamine and treated it for 1 hour with lithium to generate RG-I side branch oligosaccharides with a reduced Rha (rhamnitol) on the reducing end. Analysis of these side chains by LC-MS, GC-MS and proton NMR provided glycosyl residue information on the oligosaccharides that had been directly attached to the RG-I backbone Rha. This procedure enabled the researchers to identify RG-I side chains ranging from 1 to 14 residues (see Table 5.1). The bulk of the identified backbone rhamnitol residues had 4-linked β-Gal that was further branched either at the 3-position by α-L-Ara*f*, at the 6-position by β-D-Gal or at the 4-position by β-D-Gal. Some side chains contained only β-1,4-linked D-galactans while others had additional residues including α-1,2-linked L-Ara*f*, β-1,6-linked D-Gal, α-1,2-linked L-Fuc*p* or 1,5-linked L-Ara*f* (see Table 5.1). Evidence for Ara linked directly to the Rha in the backbone was also obtained (Lau et al. 1987). These results provided some of the earliest data for the structure of side chains connected to the RG-I backbone (Figure 5.3). Lau and colleagues could not rule out that some GalA or GlcA residues may have been present in the RG-I side chains or that some Ara or Gal may have been attached to the GalA in the backbone, an area of remaining uncertainty. Based on their total data, Lau and colleagues proposed several RG-I side chain families, which included (i) β-1,4-galactan side chains, (ii) single initial β-Gal residues elongated by α-1,3-linked L-Ara*f* alone or (iii) these further elongated by α-1,2-linked L-Ara*f*. The latter was also found (iv) terminated with α-1,5-linked L-Ara*f*. Finally, some β-1,4 galactan side branches were terminated with α-L-Fuc*p*.

In a follow-up paper, cell walls from maize and rice suspension culture cells were EPG-solubilised and the resulting RG-I enriched material selectively cleaved with lithium in ethylenediamine (Thomas et al. 1989). The structure of the resulting oligosaccharide products was determined and compared to those previously identified from sycamore RG (Lau et al. 1987). Like sycamore, the EPG-solubilised material from the grass species was rich in D-GalA, L-Ara, D-Gal and L-Rha with lesser amounts of Xyl. Unlike sycamore, no Fuc was recovered. While many of the same oligosaccharides were found in both the dicot and monocot grass species (Thomas et al. 1989), seven new RG-I side chain residues were identified (see Table 5.1), including several more oligosaccharides showing Ara attached to the rhamnitol representing the RG-I backbone.

At about the same time, work by multiple researchers identified characteristics of pectic fractions that contained rhamnogalacturonan. However, it is important to note that some researchers at that time were still using the broader earlier definition of rhamnogalacturonan that meant pectic preparations containing both HG and what we now call RG-I. For example, the rhamnogalacturonan described by Sun et al. (1987), which was said to be ~22% methylesterified, was almost certainly a heteroglycan of HG and RG-I. Thus, care must be taken when reading the pectin literature in the 1980s regarding characteristics attributed to "rhamnogalacturonan" in order to decipher which results likely arose from HG glycan regions in the material studied versus characteristics resulting from more purified RG-I-rich material. Sun et al. (1987) reported oligosaccharide fragment data from the neutral sugar-rich pectic material from tobacco, results confirming the RG-I GalA-Rha disaccharide repeat backbone as well as confirming neutral sugar side chains at C-4 of Rha that were rich in β-D-Gal and contained terminal and 5-linked α-L-Ara residues. However, these authors were unable to identify whether the arabinosyl side chains were directly linked to the backbone Rha or rather linked to Gal in the side chains. Interestingly, Sun et al. (1987) also found small amounts of 3-linked and 3,6-linked galactosyl residues. The presence of such linkages, which are now called Type II arabinogalactans, in pectin and rhamnogalacturonan-rich preparations suggests that such glycans may indeed represent a small percentage of the side chains of RG-I, although the possibility that pectic AGPs are generally often part of RG-I rich fractions (Tan et al. 2023) leaves the question unresolved as to whether such glycans exist as side chains of the RG-I backbone, as a part of pectic AGPs such as Arabinoxylan Pectin Arabinogalactan Protein 1 (APAP1) (Tan et al. 2013) or both.

Pilnik and colleagues carried out numerous studies over the years on pectins isolated from fruits such as apple (De Vries et al. 1981), providing important information on characteristics of pectin important to the pectin food industry. These included studies showing that pectin was often highly (60%–70%) methylesterified and that at least a portion of pectin was rich in neutral sugars including Gal, Rha, Ara, Xyl and Glc. The separation of pectins present in alcohol insoluble residues (AIR, i.e., cell walls) by sequential sodium acetate, ethylenediaminetetraacetic acid (EDTA), ammonium oxalate and dilute acid identified pectin fractions with varying amounts of neutral sugars and led to the observation that there was a discontinuous distribution of neutral sugars along the pectin chain(s) which the authors referred to as smooth and hairy regions (De Vries et al. 1981). The smooth and hairy regions corresponded to HG

(Figure 5.1) and the more neutral sugar containing regions (which would eventually be recognised as RG-I) (Figures 5.2 and 5.3), respectively. In 1982, based on studies of pectic fractions released by pectate and pectin lyase treatment of walls, De Vries et al. (1982) concluded that 90% of the GalA in pectic fractions from apple was present as HG and that the neutral sugar side chains in the pectin were arranged in blocks of the so-called "hairy regions". While the authors reported that the GalA in such hairy regions was highly methylesterified, it is not clear from that work whether the methylesterification was associated with HG covalenty attached to the rhamnogalacturonan, or rather, was a property of the rhamnogalacturonan backbone itself. Taken together, these results solidified the existence of HG (Figure 5.1) and RG-I (Figure 5.3) regions in pectin, and suggested that at least in extracted apple fruit pectin, HG was more abundant than RG-I.

In 1990, our understanding of the structure of the pectic polysaccharides was reviewed by O'Neill et al. (1990). This review made the compelling argument that pectins were composed of only three polysaccharides: HG (Figure 5.1), RG-I (Figures 5.2 and 5.3) and RG-II (Figure 5.4) (although the possibility that AGA and XGA might represent additional substituted galacturonans was also presented). Noteworthy points included: (i) It was not known how the pectic polysaccharides were distributed within the plant cell wall; (ii) The location of acetyl groups on the pectic polysaccharides was not known, although GalA in RG-I had been shown to be O-acetylated on O-3 (Komalavilas and Mort 1989); (iii) Evidence indicated that all plants contained HG (α-1,4-linked GalA), that HG was highly methylesterified, but the overall length(s) of HG were unknown (although DPs of 20 or more had been recovered); (iv) Since RG-I isolated from cells eluted as a single peak on BioGel Agarose size exclusion chromatography whose leading edge was enriched in Ara residues while the latter edge was enriched in Rha and GalA, the data suggested that different populations of RG-I existed with different side chains, i.e., "RG-I is not uniformly substituted with side chains" (O'Neill et al. 1990); (v) The major sugars in RG-I were Ara, Gal, GalA and Rha with small amounts of Fuc (and, we note, with Xyl frequently present); (vi) In sycamore suspension culture walls, most/all Rha in RG-I was linked to O-4 of GalA, all Rha was substituted at O-2 with GalA, 50% of Rha was substituted at O-4 with side branches and no regular pattern of distribution of Rha side chains was detected such that neighbouring Rha residues along the GalA-Rha backbone could be branched, unbranched or one branched and the other not; (vii) Sugars attached O-4 to Rha in RG-I from sycamore included terminal Ara, terminal Gal, 3-linked Gal, 4-linked Gal, 6-linked Gal, 2,6-linked Gal, 3,6-linked Gal and further glycosylation of the substituted Gal (see Table 5.1). [We note: the identification of the 3,6-linked Gal could support the existence of some Type II arabinogalactans as side branches of RG-I]; (viii) It had not been established whether some side chains of RG-I contained GalA; (ix) Intriguingly, ~50% of pectin in sycamore suspension culture cell walls was solubilised by polygalacturonase treatment and about half of the remaining pectins were solubilised from the remaining walls by sodium carbonate at 1°C and 20°C. However, while much of the sugar composition was the same for the two solubilised RG-I populations, work by Ishii et al. (1989) revealed that the density of side-branching was different in the sodium carbonate–solubilised RG-I populations, with one of those RG-I populations having less dense branching and one having greater branching than the polygalacturonase solubilised RG-I. [We note: these results support the existence of a diversity of RG-I structures (or regions) within the wall]; (x) The precise structure and patterning of the RG-I side chains was not known and it was not known if all or some of the arabinans, galactans and arabinogalactans were associated with RG-I; and (xi) The structure of RG-II (which was determined) appeared relatively conserved across plants although the precise order of the side branches was not known. It is important to stress that a statement from the O'Neill et al. (1990) review remains true today:

> It remains to be determined whether homogalacturonan, RG-I, RG-II and the arabans, galactans, and arabinogalactans are part of a single pectin polysaccharide primary cell-wall complex or whether some of them exist as individual polysaccharides or at different locations in the cell wall.
>
> (O'Neill et al. 1990)

Voragen and colleagues in an extensive set of papers from the 1980s onward characterised the hairy region of pectic polysaccharides from diverse fruits and vegetables. In 1990, they began studies of a so-called "modified hairy region", which represented a high molecular weight pectic fraction isolated upon liquefaction of apple pomace using an assortment of pectinases, hemicellulases and cellulases (Schols et al. 1990b). This modified hairy region was shown in this work to be a highly branched RG-I (Figure 5.3) with Ara-rich side chains. Importantly, the authors used the term "modified hairy region" to make clear that the material studied was modified from the larger starting polymer present in the original apple tissue. The characterised "modified hairy region" was rich in Ara, Rha, Xyl, Gal and GalA, with a 3.4 GalA/Rha ratio (suggesting approximately equal amounts of HG and RG-I). The RG-I enriched fraction was 42% methylesterified and 60% acetylated on the GalAs with the authors' resulting RG-I structure model showing methyl esters on GalA in the HG or the XGA regions of HG and the GalA acetylation indicated on GalAs in both the HG and RG-I backbones. Arabinans in the RG-I model were represented as attached both to the Gal attached to the RG-I backbone and also attached directly to the Rha in the backbone. Thus, by 1990, Voragen's work noted the presence of XGA regions of HG (i.e. HG substituted with single xylose residues on some of the GalA residues), confirmed the premise that HG and RG-I were connected via their backbones, acknowledged the role of HG methylesterification and RG-I acetylation and supported the view of arabinans as both components of arabinogalactan side chains and as direct side chains of the RG-I backbone (Voragen et al. 2009) (although confirmatory linkage data for the latter were not provided).

In a noteworthy follow-up study, Schols and Voragen (1994) characterised modified hairy region fractions from apple, pear, potato, carrot, leek and onion isolated by liquefaction enzyme treatment of the starting tissues using both their newly discovered rhamnogalacturonase (RGase) from *Aspergillus aculeatus* (Schols et al. 1990a) and multiple chemical characterisation techniques. Significantly, they showed that RG-I from these different species had similar characteristics. For example, the major sugars identified in the glycosyl residue compositions were Rha, Ara, Gal and GalA, although the amount of Ara appeared to be species specific with pear having 40 mol% Ara while onion had only 7 mol% Ara. There were also differences in the less abundant sugars with pear and apple having much greater Xyl levels than the other species. All the modified hairy regions had GalA/Rha ratios ranging from 1.6 to 2.3, indicating that the RG-I preparations contained approximately 30%–65% HG, assuming equimolar amounts of GalA and Rha in the RG-I backbone regions. Whether the HG was attached to the reducing or non-reducing end of the RG-I backbone was not investigated. The degree of acetylation of the RG-I-enriched material ranged from 33% for pear to 90% for potato. The relative amount of Gal versus Ara varied depending upon the species. Size exclusion chromatography indicated that all species contained three (and possibly four) RG-I size populations, although the relative abundances of the different populations varied depending upon the tissue/species. Importantly, the results led these authors to conclude that the modified pectic hairy regions (i.e. RG-I fractions) were comparable across different fruit and vegetable tissues, and fell into three or four size ranges.

In a subsequent study to probe the existence of different RG-I populations in apple, Voragen and colleagues compared the RG-I recovered when apple alcohol insoluble solids (AIS) (cell walls) were extracted sequentially with cold or hot buffers, chelators or with alkaline solutions (to deesterify the polymers) (Schols et al. 1995). Using these methods, they recovered 75% of the GalA from the AIS (suggesting that 25% of the GalA remained in the walls). They further reported that the cold buffer, hot buffer, chelator and alkaline extracts represented 10%, 7%, 13% and 10% of the mass of the starting AIS, with the first three extracts containing 64% of the GalA in the AIS. These results suggest that the bulk of HG is relatively easily solubilised from cell walls. The authors also found that the pectins contained more Rha and other neutral sugars when more severe extraction conditions were used. This result likely signifies that RG-I containing wall structures are, in general, more tightly held in the wall than at least some/most of the HG and/or HG-containing polymers, a conclusion that was also recently reached by Perzon et al. (2020) in their analysis of polysaccharides tightly bound to sugar beet nanocellulose nanofibres using the new technique of Array-based Microfibril Surface Assessment (AMSA).

In a study comparing the RG-I backbone of deesterified apple, beet and citrus pectins, Renard et al. (1995) identified 4-α-D-Gal*p*A-1,2-α-L-Rha*p*-1 oligomers with DPs from 6 to 20 (i.e. 10 Rha residues and 10 GalA residues) by NMR analysis of material recovered following ion-exchange chromatography, again confirming the strict disaccharide repeat backbone of RG-I. Significantly, in a subsequent study to determine whether exceptions to the alternating GalA-Rha structure of the RG-I backbone could be found (Renard et al. 1997), only strict alternating GalA-Rha repeating structures were identified by the methods used. Two years later in studies of sugar beet the researchers (Renard et al. 1999) found that ~1 out of 72 GalA residues in the RG-I backbone was substituted by GlcA at the O-3 position of GalA (see Table 5.1). How general this GlcA substitution is in other plants needs to be established.

Sengkhamparn et al. (2009) in studies of hot buffer soluble and chelator soluble RG-I from okra found that the hot buffer soluble RG-I was highly branched and acetylated, with both the GalA and the Rha being acetylated. The RG-I in the chelator soluble fraction (which was more HG-rich than the hot buffer soluble fraction) had longer Gal containing side chains. These results support the existence of different populations of RG-I and suggest that RG-I side chain characteristics, at least in okra, may depend on the pectic environment in which the RG-I resides. Another unique characteristic of the okra pod RG-I was the presence of α-Gal (see Table 5.1) as terminal residues in some of the neutral side chains (Sengkhamparn et al. 2009). Zheng and Mort (2008) in studies of RG-I purified from cotton (*Gossypium hirsutum*) cell walls identified and characterised a novel RG-I oligosaccharide via TOtal Correlation SpectroscopY (TOCSY), COrrelation SpectroscopY (COSY), Heteronuclear Multiple-Quantum Coherence (HMQC), Heteronuclear Multiple Bond Correlation (HMBC) and Nuclear Overhauser Enhancement SpectroscopY (NOESY) NMR spectroscopy as an RG-I backbone tetramer with the two Rha being β4-galactosylated and with one of the Gal residues having α-Ara at both the O-3 and O-4 positions. Importantly, these results show that Rha residues in consecutive GalA-Rha repeats in the RG-I backbone can be branched. Studies from potato indicated the presence of β-1,3-linked Gal within the more common β-1,4-galactan side chains of RG-I (Hinz et al. 2005) and provide evidence that β-1,4-galactan may be covalently linked to the non-reducing end of α-1,5-arabinan in the RG-I side chains and that α-1,5-arabinan may be covalently connected to the non-reducing end of β-1,4-galactan side chains of potato (Øbro et al. 2004). These results indicate that some RG-I branches can be both large and complex.

Although pectins appear to have rather similar characteristics across species, some species such as sugar beet and spinach have high levels of ferulic acid, which is at least in part attached to terminal Gal and Ara residues in RG-I (Fry 1983). Renard and Jarvis (1999) proposed that the high ferulic acid content and acetylation level of sugar beet pectin might enable sugar beet pectin to associate with cellulose, thereby fulfilling a role normally carried out by hemicellulose that was thought to be of low abundance in sugar beet primary walls. To test if sugar beet pectins (or other wall components) provided such a structural role, ^{13}C-solid state NMR was used to investigate polymer mobility and domain size and to identify polymers in the sugar beet root walls that were closely associated with cellulose in a manner normally expected for hemicellulose (Renard and Jarvis 1999). As this topic is of interest today, it is noteworthy that the authors did not obtain evidence for an association of arabinans or galacturonan with cellulose. To the contrary, they found an absence of signals for these polymers in delayed-contact experiments, indicating that arabinans or galacturonans were not sufficiently closely associated with the cellulose microfibrils to make them more rigid. The authors concluded that these pectins, rather than having a cross-linking function with cellulose in the wall, filled the pores in the wall network and acted as plasticisers and water-binding agents (Renard and Jarvis 1999).

Non-Sugar Modifications of HG and RG-I

Methylesterification and acetylation of HG: In all naturally isolated pectins, the HG is partially methylesterified at the C6 carboxyl, with the degree of methylesterification almost certainly a result of the initial degree of methylesterification of the HG when it is synthesised, and the subsequent deesterification of the polymer in the wall by pectin methylesterases. Although it is often stated that HG is synthesised in a highly methylesterified form, this has never been rigorously determined and remains an important question, since the HG biosynthetic glycosyltransferases (GTs), the GAUTs, can synthesise non-methylesterified HG of high molecular mass *in vitro* (Amos et al. 2018; Engle et al. 2022). Thus, HG polymer synthesis does not require methylesterification of the product, at least *in vitro*.

Many industrial researchers are interested in the HG portion of pectin due to its gelling properties, which are influenced by the degree of HG esterification (Willats et al. 2006). In addition to partial methylesterification, HG may have acetate esters at O-2 or O-3. HG with stretches of 10 or more non-esterified GalA residues makes it sensitive to calcium cross-linking (Voragen et al. 2009). The methylester and acetylester content of HG influences the gel forming and solubility properties of pectin, with commercial sources such as sugar beet, sunflower and potato having prohibitively high acetyl ester levels for industrial purposes in the food industry (BeMiller 1986). For industrial utilisation, pectins with a degree of methylesterification of >50% may be named high-methoxyl pectins and those with methylesterification content <50% may be named low-methoxyl pectins (BeMiller 1986; Willats et al. 2006). Multiple HG chains can interact with each other and lead to gelling in two manners. Regions of HG with low methylester content have abundant, ionised, negatively charged carbonate groups that can ionically interact with calcium (or other cations) to form HG-calcium-HG gels. Regions of unesterified GalA on adjacent HG chains can cooperatively bind calcium to form the so-called "egg-box" junction zones with the chains in a 2_1 conformation (Grant 1973; Rees 1982). Alternatively, HG with high methylester content will gel due to hydrogen bonding and hydrophobic forces between the methoxy groups in the presence of high concentrations of co-solute such as sugar (e.g. sucrose) under acidic conditions (protonating the carboxyl groups). Due to interest in the gelling properties of pectins, attributes of pectin that affect gelling have over time been identified and have been given names denoting gelling state or likelihood. For example, pectin preparations that contain HG with no, or very little, methylester content are sometimes referred to as pectic acid, while those with higher methylester content are referred to as pectinic acid. Salts of pectinic acids may be referred to as pectinates. These terms should be kept in mind when reading the older and commercially directed pectin literature. Proton NMR, X-ray diffraction and mathematical models suggest that HG GalA takes the $^4C_1$ conformation and that pectinic acid, pectic acid and sodium and calcium pectates have a right-handed 3_1 helical three-fold screw axis conformation (Walkinshaw and Arnott 1981a,b; BeMiller 1986).

A common statement found in most pectin reviews is that HG in the plant is synthesised in a highly methylesterified form and becomes deesterified in the wall by the action of pectin methylesterase(s). However, although commonly believed, there is no definitive evidence to support this and the degree of methylesterification of HG during synthesis *in vivo* is an area that requires experimental verification. It is quite possible that the methylesterification state of HG prior to insertion in the wall may differ in different cell types and/or states of development. The fact that polymeric HG can be synthesised *in vitro* in a non-methylesterified state by both *de novo* and acceptor-dependent synthesis, and that HG can be synthesised in a non-methylesterified form by multiple GAUTs including the GAUT1:GAUT7 complex, GAUT10, GAUT11, GAUT13 and GAUT14 (Amos et al. 2018; Engle et al. 2022) clearly shows that HG-synthesising catalysts have the ability to synthesise non-methylesterified HG. Thus, critical questions that need to be resolved are the amount and variation of the methylesterification state(s) of HG synthesised in the endomembrane system and at the time of its insertion into the wall. It is reasonable to hypothesise that different HGs are synthesised *in vivo* in different methylesterification states and that these impact the biological activity of the pectin.

HG may carry acetate esters at O-2 or O-3. Rombouts and Thibault (1986b) showed that methyl and acetyl groups were randomly distributed in HG. Ishii (1997) identified in Driselase-treated tuber cell walls, acetates at the O-3 position of HG oligosaccharides. More recently, Voxeur et al. (2019) using a high-performance size exclusion chromatography-MS approach to analyse HG oligosaccharides (oligogalacturonides) released from Arabidopsis leaves during *Botrytis cinerea* infection revealed both methylesterification and acetylesterification of HG. These results make clear that studies into the regulation and patterning of these HG non-sugar modifications during plant growth and development are warranted.

DRISELASE

A commercial mixture of enzymes including cellulases, hemicellulases, pectinases and others, often used to fragment/remove plant cell walls.

Uronyl esters: – There have been multiple intriguing suggestions over the years that pectins, and particularly GalA of the HG regions of pectin, may be linked via O-D-galacturonoyl ester cross-links to GalA in other HG chains or to other hydroxyl containing compounds/polymers (Brown and Fry 1993; Gelineo-Albersheim et al. 2001; Voragen et al. 2009). It has also been suggested that pectin methylesterases (which reside in very large gene families) might catalyse such proposed trans-esterification reactions between methylesterified HG and substrate – OH groups (Gelineo-Albersheim et al. 2001). However, it has been challenging to test this hypothesis. Brown and Fry (1993), using ^{14}C-uronate-labelled cell walls from spinach suspension cultures showed that ~5% of the GalA in spinach cell wall HG was ester-linked at O-6 to undefined hydrophobic alcohols, providing evidence for the existence of uronyl esters, at least in spinach. Research is needed to determine what role uronyl esters have in wall structure.

Methylesterification of RG-I: Jarvis and colleagues evaluated the pectic polysaccharides in 3-day-old flax seedlings as potential early markers of fibre differentiation (Rihouey et al. 1995). They reported that cyclohexanediaminetetraacetic acid (CDTA)-extracted RG-I-containing-fractions had degrees of acetylation ranging from 37% to 83% and concluded that some of the methylesterification was likely associated with RG-I in addition to HG. The question of whether or not RG-I is methylesterified requires experimental verification.

Acetylation of RG-I: Mort and colleagues showed using NMR analysis that GalA-Rha disaccharides fragmented from carrot, cotton, tobacco and tomato RG-I-enriched cell wall material by anhydrous hydrogen fluoride treatment were acetylated at the O-3 position of GalA (Komalavilas and Mort 1989). Ishii (1997) working with Driselase-treated potato tuber walls determined that GalA in the RG-I backbone was acetylated at O-2 and O-3, a result also found for sycamore cell walls by Lerouge et al. (1993). These results provided strong evidence that much of the acetylation described for RG-I occurs on GalA in the backbone. Interestingly, Sengkhamparn et al. (2009) in studies of hot buffer and chelator soluble RG-I from okra pods identified both GalA and Rha as being acetylated with the acetylation of Rha being at O-3.

Feruloyl esters in RG-I: Using spinach suspension culture cells that contain large amounts of ferulic acid, Fry (1982) identified feruloylated glycosyl residues in pectin. Digestion of spinach cell walls with the commercial enzyme cocktail "Driselase", which includes among others an RG-I hydrolase, released >60% of the cell wall ferulic acid in two oligosaccharides. These were identified as 3-O-(3-O-feruloyl-α-L-Ara*p*)-L-Ara and 4-O-(6-O-feruloyl-β-D-Gal*p*)-D-Gal (Fry 1982). The presence of the feruloyl on Ara*p* is noteworthy as the bulk of Ara in the cell wall is in the furanose form, suggesting specificity of this process to specific structural moieties of RG-I.

Selvendran (1985) showed that pectic fractions from sugar beet had ester-linked hydroxycinnamic acids associated with arabinan-rich pectic fractions. During the same period, this observation was independently confirmed by Rombouts and Thibault (1986a) who showed that sugar beet pectins contained ferulic acid, leading Selvendran to conclude that at least some sugar beet pectins had ester-linked ferulic and/or coumaric acid on terminal (e.g. Ara) RG-I side chain sugar residues (Selvendran 1985). Selvendran further proposed that in apples and sugar beets, both phenolic-ester linkages and phenolic-ether linkages may cross-link pectic polymers in some tissues. This was based in part on the observation that chlorite/acetic acid treatment of sugar beet released more pectin than oxalate depectination or saponication with base.

Rombouts and Thibault reported that sugar beet pectin has high methylester (>60%), acetylester (~30%) and feruloylester content (between 2% and 3% in acid and alkali soluble fractions) (Rombouts and Thibault 1986b). Subsequently they showed that methyl and acetyl groups were randomly distributed in the HG while the feruloyl groups were located in the RG-I (Rombouts and Thibault 1986a). Although the authors did not identify the specific location of the feruloyl groups, they concluded that they were located "in the hairy fragments" (in the RG-I region) and were "probably linked to the neutral sugar side-chains" (Rombouts and Thibault 1986a). Thus, the results of Rombouts and Thibault agreed with those of Steven Fry (1983) who provided evidence that feruloyl esters in spinach pectin occur on the non-reducing terminal residues of Gal and Ara.

Probing the Existence of Pure HG in the Cell Wall

Due to interest in the gelling properties of citrus pectin, and because commercial citrus pectin (isolated by acid extraction of peels) has a high (~80%) GalA content (Daas et al. 2000), it is frequently thought that pure HG is an abundant component of pectin. However, this notion has not been substantiated. Aspinall and colleagues, after studying pectin from numerous types of plants concluded that "The presently available evidence suggests that pure galacturonan devoid of neutral sugar residues are of infrequent natural occurrence" (Aspinall et al. 1970). The only source of possibly pure HG that has been reported is from sunflower heads (Zitko and Bishop 1965, 1966); however, as noted below, it is unlikely that the sources studied were pure HG.

Zitko and Bishop (1965) are often cited as having identified pure HG in sunflower. However, inspection of their paper shows that upon extracting pectin from sunflower heads using ammonium oxalate:oxalic acid, the initial sunflower pectin was ~72% GalA, i.e., not pure HG. Subsequent base and acid treatments led to a yield of ~78% pectinic acid, which at that point appeared to be highly enriched for HG. This analysis suggests that the starting pectin was not pure HG. Chambat and Joseleau (1980) also reported the isolation of pure HG from *Rosa glauca* suspension culture cells. Interestingly, they reported that some HG remained in the insoluble material following sequential ammonium oxalate, EDTA and sodium hydroxide extractions but was hydrolysed from the insoluble cellulose-rich material by acid hydrolysis (Chambat and Joseleau 1980). They concluded that insoluble HG with a DP >100 was associated with cellulose in the wall via acid-labile linkages. In a follow-up study, Chambat and colleagues using two cell wall extraction procedures (EDTA/ammonium oxalate and alkaline) showed that while significant amounts of pectin and hemicellulose were extracted from *Rosa glauca* cells, more than half of the cell wall components remained insoluble and were associated with cellulose (Chambat et al. 1984). Using the crystalline cellulose chaotropic solvent system N-methylmorpholine N-oxide/dimethyl sulphoxide (MMNO-DMSO), the authors were able to solubilise the entire cell wall. A comparison of three different extraction procedures led them to conclude that at least two "blocks" of polysaccharides exist in the cell wall that are not held together by covalent bonds and, of relevance to this review, the authors concluded that their data supported a cell wall model in which covalent glycosidic bonds exist between xyloglucan and pectin (e.g. a pectin with HG and RG-I) causing the pectin to be tightly bound to cellulose via hydrogen bonding by xyloglucan (Chambat et al. 1984). This research provides another example in which studies to understand the structure of the entire cell wall led researchers to conclude that covalent connections exist between pectin and other wall polymers. *A critical unresolved question in pectin structure is the extent to which such covalent connections between pectin and hemicellulose (or other constituents) exist in the wall, what their exact structure is, how they are synthesised and the role that such covalent connections have in wall architecture.*

More often when researchers attempted to isolate pure HG, they obtained something akin to the pectinic acid purified from apple by Barrett and Northcote (1965), where the sample contained 13% neutral sugars and 87% GalA. Specifically, the pectinic acid was composed of neutral sugars 9.3% Ara, 1.4% Gal, 1.2% Rha, 0.88% Xyl and trace amounts of 2-O-methyl-Xyl and 2-O-methyl-Fuc. Based on our current understanding of pectin structure, it is tempting to suggest that the pectinic acid consisted of RG-II within an HG backbone. If true, this would indicate that a population of HG:RG-II may exist separately from RG-I.

Indeed, as noted by McNeil et al. (1980),

> No pure homogalacturonan has been isolated from primary cell walls without treatments that are likely to cleave bonds. However, there is strong evidence that such molecules are present in primary cell walls, although they may be covalently attached to other cell wall polymers.

This important point highlights the need to determine whether or not pure HG exists in the cell wall. The fact that pure HG can be synthesised *in vitro* by multiple GAUTs makes it clear that cells have catalysts that can produce this homopolymer in high molecular weight forms (Amos et al. 2018; Engle et al. 2022). What is less clear is whether *in vivo* they synthesise HG that is not covalently connected to another pectin or other wall glycan. Atomic force microscopic analyses of CDTA extracts from ripe strawberry fruit cell walls provided some evidence for the presence of linear HG chains (Posé et al. 2012), suggesting that inclusion of high resolution scanning probe microscopy may shed light onto this area.

Length and Positioning of RG-I and HG Glycan Domains along the Backbone

Although structural research from 1973 onwards identified primary structures of some of the pectic glycans, the question of how the pectic glycans are linked together into one or more pectic heteroglycan and/or glycoconjugate structures remains unresolved. For example, it has been shown that pectins contained regions of α-1,4-D-galacturonan (referred to over time as, among others, polyuronides, polygalacturonic acid, pectate and HG) and regions abundant in Rha and GalA with β-1,4-D-galactan and α-1,5-L-arabinan side chains (a polymer eventually named RG-I). However, how these fit together into one or more polymeric structures and how many different heteroglycan structures are present in cell walls have not been resolved.

The available evidence indicates that at least some RG-I and HG are covalently linked via their backbone. Multiple studies have shown that RG-I can be isolated from cell walls by cleavage with polygalacturonase. Furthermore, oligosaccharide sequence data confirm a linkage between RG-I and HG sequences (see Table 5.1). Importantly, we also recently identified at least two enzymes that elongate the RG-I backbone on the non-reducing end with HG. What is not clear is how long HG (or RG-I) domains are in the plant, how much variation there is in the length of the HG (or RG-I domains), and in how many different polymers/glycoconjugates these domains reside.

It is interesting to note that studies of transgenics expressing RG-I degrading enzymes lead to the same conclusion that HG and RG-I are connected via their backbones. The examination of transgenics in which *in planta* cleavage of the RG-I backbone (Oomen et al. 2002), of β-1,4-galactan (Oxenbøll Sørensen et al. 2000), and of α-1,5-arabinan side chains (Skjot et al. 2002) upon *in planta* expression of the enzymes originally inspired the proposition that HG could be attached as side chains of RG-I (Vincken et al. 2003). However, attempts to support this hypothesis through the isolation of a diagnostic oligosaccharide instead led to the isolation of an RG-I backbone fragment with HG attached on the backbone at the non-reducing end (Coenen et al. 2007), confirming that HG and RG-I are connected via their backbones and providing the first structural evidence for a covalent linkage of HG to RG-I.

Thibault et al. (1993) reported that HG has a length of 72–100 GalA residues based on controlled and partial acid hydrolysis of pectin from apple, beet and citrus. The length of the HG that is associated with RG-I appears to vary depending on the source of the cell wall material analysed. For example, Talmadge et al. (1973) concluded that in the materials solubilised by EPG treatment of sycamore suspension cells, the remaining GalA that was associated with RG-I consisted of regions of HG of ~8 residues. However, the length of the HG that was digested by EPG was not established. A length of 25 for HG in acid hydrolysed citrus pectin has been calculated (see below). The meaning of the different HG lengths reported by different investigators is unclear and may be due to the methods used or to differences in the biological materials analysed. One is tempted to try to draw biological and/or structural meaning from the data. For example, the observation that acid-catalysed hydrolysis of pectin (which hydrolyses the rhamnosyl glycosidic bond) yields HG oligosaccharides of DP ~25 in citrus may indicate that HG regions of 25 residues are associated with RG-I glycan regions (BeMiller 1986). However, subsequent work from Ralet and colleagues (Yapo et al. 2007) using multi-angle laser light scattering and viscometry found the DP of HG extracted from citrus peels by mild acid hydrolysis after deesterification ranged from 81 to 117. This was true in each of HG preparations obtained upon sequential extraction of citrus peels with water, oxalate, hot dilute acid and cold dilute base. Such results from pectin isolated directly from a plant tissue and which compared the length of HG in the different polymers extracted suggest that at least in lemon peel, HG of DP ~100 may be a primary component of the pectin extracted under these conditions. It should be noted, however, that the amount and length of the HG in the pectin that was not extracted by the methods used, i.e., pectin that is held tightly in the wall, was not evaluated and such pectin could have different characteristics (see below).

Ropartz and Ralet (2020) observed that the DP of RG-I appears to vary depending on the plant. For example, 300 backbone repeats have been reported for RG-I from sycamore suspension culture cells (McNeil et al. 1984), whereas reports of 60 repeats for RG-I from citrus albedo, 120 repeats for RG-I from sugar beet and three different RG-I disaccharide repeat lengths of 15, 28 and 100 have been reported for soybean (Nakamura et al. 2001). Yapo et al. (2007) reported DPs of 34–40 for RG-I from citrus peel. Selvendran in his 1985 review concluded that pectin polymers in the primary cell wall appear to have more branched RG-I with longer side chains than pectin in the middle lamellae, supporting a control of RG-I length and a function for RG-I side branches (Selvendran 1985).

Rhamnogalacturonan II (RG-II)

In their landmark 1978 paper, Darvill et al. (1978) in Albersheim's lab identified a novel complex pectic polysaccharide that was named RG-II, since it was the second pectic polysaccharide identified that contained Rha (after the rhamnogalacturonan described above with the Rha-GalA backbone, i.e., RG-I). Digestion of sycamore suspension culture cells with EPG followed by sequential ion exchange and size exclusion chromatographies yielded a single peak of material that constituted 3%–4% of the primary cell wall (Darvill et al. 1978). Glycosyl residue composition identified 10 sugars in the polymer including apiose (Api), 2-O-methyl-Xyl and 2-O-methyl-Fuc. Of the 20 different linkages, some included the rare 2-linked GlcA, 3,4-linked Fuc and 3-linked Rha residues.

Determining the structure of RG-II (Figure 5.4) took dedicated work by many researchers over many years and led to the identification of additional unusual sugars in RG-II including 3-deoxy-D-*manno*-octulosonic acid (Kdo), 3-deoxy-D-*lyxo*-heptulosaric acid (Dha) and aceric acid. Further results indicated that the sugar residues in RG-II existed in both furanose and pyranose ring forms, β and α configurations as well as with some methyl, acetyl or methylester substituents (Hervé du Penhoat et al. 1999). Glycosyl linkage and NMR analyses of purified RG-II preparations showed that the RG-II was composed of a 1,4-linked α-D-Gal*p*A residue (i.e. HG) backbone (Stevenson et al. 1988b) with four unique oligosaccharide side chains (A-D) (Spellman et al. 1983; York et al. 1985; Melton et al. 1986; Stevenson et al. 1988a,b; Puvanesarajah et al. 1991; Whitcombe et al. 1995; Vidal et al. 2000) and two single arabinosyl side chains (E-F) (Ndeh et al. 2017). However, the precise order of attachment of the side chains to the HG backbone remains to be determined.

Despite the complexity of RG-II, most studies suggest that the RG-II structure is highly conserved throughout the plant kingdom (Bar-Peled et al. 2012). There are also, however, some reports claiming the existence of potential variation in RG-II side chain monosaccharide composition and alterations in methylation status within and across RG-II in different plant species (Pabst et al. 2013).

The initial studies on the significance of borate for plant growth and development and the observations of a correlation between the amount of RG-II and the boron requirement suggested an interaction between borate and pectin (Matoh et al. 1993, 1996). Subsequently, several studies revealed that two RG-II molecules are cross-linked by a borate diester bond between the apiosyl residue of side chain A in each of the two RG-II molecules and that the dimer contained 3' and 2,3,3'-linked apiosyl residues, whereas the monomer contained only 3'-linked apiosyl residues (Pellerin et al. 1996; Ishii et al. 1999; Kobayashi et al. 1999). There is strong evidence that RG-II dimerisation is required for plant growth (reviewed in Bar-Peled et al. (2012)).

Since RG-II dimers are generally studied by purification of RG-II from cell walls treated with EPG, the full structure of the parent RG-II containing polymer(s) is not known. For example, whether the RG-II is cross-linked to an HG:RG-II heteroglycan, one or more HG:RG-II:HG:RG-I heteroglycans or a pectic AGP-type AGP-RG-I:HG:RG-II glycoconjugate (e.g. proteoglycan) is not known. The question of how RG-II (or any of the pectins) fits into cell wall architecture is a major unresolved question in plant cell wall research. Furthermore, it is important to realise that while numerous mutant studies show that RG-II boron diester cross-linking is critical for plant cell walls and for plant growth, the notion that RG-II dimers exist as unique independent glycans is likely not supported since RG-II cannot be isolated from cell walls without prior EPG treatment.

COVALENT CONNECTIONS BETWEEN PECTINS AND OTHER GLYCANS AND/OR PROTEINS

The fact that HG, RG-I and RG-II are solubilised by EPG treatment has led to the hypothesis that they are all covalently linked together as a single macromolecule. Knowledge, however, of how many unique heteroglycan polymers exist and how large and structurally diverse they are, has not been resolved (Kaczmarska et al. 2022). Likewise, although data from numerous studies have supported the hypothesis that at least some pectins are connected to protein and/or to hemicellulose (i.e. xyloglucan and/or xylan), only recently has a structural connection between pectin and a hemicellulose been confirmed with the identification of the covalent connections between a short xylan and RG-I in the pectic AGP proteoglycan APAP1 (Tan et al. 2013). Our ability to understand pectin structure and elucidate its function requires a more complete understanding of the structure of the pectic glycans and heteroglycans and their associations with proteins and non-cellulosic polysaccharides.

COVALENT LINKAGES BETWEEN PECTIN AND PROTEINS

In his review of pectic–protein complexes, Selvendran (1985) compiled data from multiple species and provided evidence that pectin–protein complexes are genuine and not artefacts. He concluded that they had the following characteristics. The pectin:protein complexes occur in cell walls of parenchymatous tissues [this is significant since such cells have primary walls that have the ability to expand]. Selvendran also found the proteins in the complexes to be rich in hydroxyproline and serine. Relatedly, Andrew Mort and colleagues provided evidence for pectin–protein cross-links in cotton that appeared to be between RG-I and extensin (Qi et al. 1995). More recently, Tan et al. (2013) confirmed a covalent connection between the arabinogalactan region of an AGP protein (AGP57C) with RG-I in Arabidopsis suspension culture cells. The arabinogalactan domain of the AGP was also covalently linked to a short xylan chain and the RG-I was further linked with HG and with a side chain of xylan. This pectic AGP known as APAP1 is the most fully characterised of the pectin–protein covalent cross-links. It is likely not the only one present in plant cell walls. Importantly, the presence in a single proteoglycan of xylan which can hydrogen bond to cellulose and of HG which can ionically crosslink with other HG provides a mechanism for cross-linking across the plant cell wall and has implications for cell growth and wall integrity (Tan et al. 2013; Biswal et al. 2018a,b). Taken together, the evidence for the existence of pectin–protein covalent complexes in the cell wall is strong and growing. The recent demonstration that RG-I-AGP protein glycoconjugate was the major component in RG-I isolated from cell walls by EPG digestion and size exclusion chromatography [the classical method for isolating RG-I] provides strong support for the hypothesis that most RG-I in plants is covalently linked to protein (Tan et al. 2023). Future work should be targeted at combining molecular genetic, imaging, spectroscopic and biochemical approaches to elucidate the function of such complexes in cell wall architecture and in plant cell function.

COVALENT LINKAGES BETWEEN PECTIN AND XYLOGLUCAN

As noted above, in their early cell wall model, Keegstra et al. (1973) proposed a covalent connection between pectin and xyloglucan via the arabinogalactan side chains of rhamnogalacturonan. Although such a covalent connection has not yet been structurally confirmed, multiple studies since then have provided supporting evidence for a connection between pectin and xyloglucan (reviewed in Brett et al. (2005)). For example, Femenia et al. (1999) in studies of 0.5 M KOH extracts of cauliflower stems provided evidence for possible covalent interactions between pectic–xylan–xyloglucan during secondary cell wall thickening in stem tissues. Thompson and Fry (2000) in studies of rose suspension culture cells provided multiple lines of evidence that ~30% of the xyloglucan was present in covalent complexes with pectin. Although the work stopped short of providing the specific chemical linkage between the pectin and xyloglucan, the results led the authors to propose that xyloglucan may have been attached to arabinogalactan side chains of pectin. Abdel-Massih et al. (2003) and Cumming et al. (2005) provided evidence that a pectin–xyloglucan complex may form in the Golgi. In studies with pea epicotyl membrane fractions synthesising radiolabelled β-1,4-galactan, evidence for an association of xyloglucan with *in vivo* synthesised β-1,4-galactan attached to a pectic RG-I:HG complex was obtained. The evidence included sensitivity of the product to cleavage by both xyloglucan and pectin-specific enzymes. However, structural proof for the specific pectin–xyloglucan covalent linkage was not provided. In a follow-up study using radiolabelled uridine 5'-diphosphate (UDP)-GlcA to label HG, Abdel-Massih et al. (2007) provided evidence that the synthesis of HG and β-1,4-galactosylated pectin may be temporally separate events in the Golgi, and that for at least the initially produced HG labelled in this experiment, it was not immediately associated with the xyloglucan. These results support a possible cross-linking later in the process during movement through the endomembrane system. Such *in vivo* Golgi-localised synthesis studies are an important tool for probing heteroglycan and glycan-protein association events. Additional tools that could provide insight into these interactions include labelling techniques with fluorescent probes and antibodies (e.g. anti-cell wall antibodies (Larsen et al. 2006; Knox 2008; Pattathil et al. 2012)). Finally, working with Arabidopsis cell suspensions, Popper and Fry (2008) tested two hypotheses for the mechanism for formation of a pectin–xyloglucan crosslink, i.e., transglycosylation versus within the endomembrane system. Their intriguing pulse-labelling results using [^{3}H]Ara indicated that xyloglucan–pectin bonds form within the cell and that ~50% of the xyloglucan produced by Arabidopsis suspension culture cells is synthesised on a pectin primer.

Chambat et al. (1984) in studies of suspension-cultured rose cells used graded acid hydrolysis of the cell walls to provide evidence for a role for HG interconnections with xyloglucans and arabinogalactans. They further suggested that these covalently associated polysaccharides remained associated with cellulose even after alkaline extraction of the wall. It is possible that such a structure could have an architectural function in the cell wall.

Finally, Selvendran (1985) in studies of proteoglycan complexes found in apple, cabbage, carrot and runner bean cell walls identified hot-water-soluble pectic polysaccharides with characteristics of pectic-hemicellulose complexes. Some of these pectic complexes appeared associated with xyloglucan, others with xylan. Selvendran concluded that while the bulk of the xyloglucan was not covalently linked to pectin, a small but significant amount existed in a covalent association with pectin.

Covalent Linkages between Pectin and Xylan

Waldron and Selvendran (1992) in studies of the cell wall material remaining after sequential extraction of asparagus cell walls with CDTA, Na_2CO_3 and KOH provided evidence for the existence of xylan–pectic polysaccharide complexes. As noted above, the pectic AGP APAP1 has xylan covalently attached at two points in the proteoglycan: the reducing end of a short xylan chain is attached to the O-5 position of Ara in the arabinogalactan glycan of the AGP and a different xylan chain is attached via its reducing end to a Rha in the RG-I backbone that is covalently elongated on the AGP (Tan et al. 2013). These linkages have been confirmed by sequence and NMR analyses. Ralet et al. (2016) provided biochemical evidence for attachment of xylan to RG-I in Arabidopsis seed mucilage, although NMR confirmation of the sequence and covalent linkage has not yet been reported.

PROBING CELL WALL ARCHITECTURE

Talmadge et al. (1973), in their first paper analysing sycamore suspension culture cells reported that treatment of the cell walls with EPG from *Colletotrichum lindemuthianum* increased the ability to degrade the wall by purified endoglucanase and pronase. These results suggested that the cleaved HG had structurally impeded access of these enzymes to other wall polymers and, importantly, indicated a region-specific architecture and multi-polymer association within the wall. In addition, the observation that EPG treatment released 50% of the total pectic polysaccharides and 75% of the GalA in the wall, suggested that at least two populations of pectin exist in the wall, one that can be solubilised by EPG fragmentation of HG and one that cannot. The existence of pectin that is tightly bound in the wall and perhaps associated with cellulose has repeatedly been reported. For example, when Arabidopsis leaf cell walls were extensively extracted by treatments with EPG, sodium carbonate, 1M KOH and 4M KOH, the final remaining cell wall residue contained the following mol% of non-cellulosic sugars: 3% Rha, 11% Man, 5% Ara, 6% Xyl, 3% Gal and 8% GalA (Zablackis et al. 1995). These sugars are consistent with a pectic RG-I-enriched architectural structure. It is noteworthy and deserving of further study that the portion of pectin that can be solubilised by EPG contains both HG and RG-I, while the pectin that is held tightly into the wall structure is enriched in RG-I (Biswal et al. 2015, 2018a,b).

A further significant observation made by Schols et al. (1995) is that pectins that were more difficult to extract from cell walls had RG-I with more substituted side chains than the more easily extracted pectin. Upon extracting the AIS from apple fruit with EDTA/oxalate and mild alkali, the authors found that more RG-I oligosaccharides with more Gal substituted side chains was released with the more severe extraction treatments. Evidence that different types of RG-I side branches are not randomly distributed has been obtained from studies of apple and ginseng pectin treated with rhamnogalacturonan hydrolase (reviewed in Ropartz and Ralet (2020)). These results suggest a regulation of RG-I structure and an association with function in wall architecture.

Finally, Perzon et al. (2020) in a recent study of sugar beet cellulose nanofibres found evidence for tightly bound HG and RG-I on the cellulose nanofibres. Using a method the researchers developed called AMSA, which included a collection of anti-cell wall glycan-epitope-specific monoclonal antibodies, the researchers detected polysaccharides present on cellulose nanofibres or exposed by enzyme treatment of the nanofibres. The authors concluded that most cellulose nanofibres are masked by pectin or hemicellulose and that RG-I, specifically, tightly interacts with cellulose.

Taken together, the results to date lead to the following conclusions.

1. Our basic understanding of pectin structure is mostly associated with readily soluble pectins.
2. By extrapolation, we are largely uninformed about the structure of insoluble pectin.
3. For many years, the data have indicated that there is little free HG (galacturonan) in cell walls, rather,

HG is usually associated with some amount of Gal, Ara, Rha and Xyl residues, suggesting that the wall is composed of either a single interconnected HG:RG-I polymer, or that multiple HG:RG-I polymers/glycoconjugates exist. We favour the second hypothesis.

Abundance of Pectin in Different Cell Wall Fractionations and What it Tells us about Pectin Structure

The work of many investigators over the years has shown that the primary cell walls of dicots and non-grass monocots are rich in pectin, cellulose and hemicellulose (particularly xyloglucans), whereas grass primary wall is rich in cellulose and hemicellulose (particularly acidic arabinoxylan) with lesser amounts of pectin (Selvendran 1985). It has also long been known that "only about 50%–90% of the total pectic substances present in the walls is extractable" and that the "insoluble pectic material is always found in close association with other cell wall polymers, particularly the α-cellulose fraction" (from Selvendran (1985)). What is this insoluble pectin? Has additional evidence been obtained since 1985 to better understand why up to 50% of pectin (depending on the species and cell type) remains tightly bound to the wall? Recent studies to determine the mass yield of different types of wall polymers by sequential extraction of poplar and switchgrass cell walls (i.e. walls from dicots and grasses) provide information about the possible nature of the pectic polymers that are held tightly in the wall, and about the quantitative distribution of these in different tissues and plants.

We previously studied the cell walls of wild-type and transgenic poplar wood and switchgrass tillers down-regulated in or overexpressing different HG biosynthetic *GAUT* genes. We thus had available multiple sets of cell wall glycosyl residue composition and wall mass yield data from sequential extracts of wild-type walls from four sets of experiments (Biswal et al. 2015, 2018a, b) for an analysis of easily extractable and tightly bound pectin. The sequential extracts commonly used in wall studies include ammonium oxalate, sodium carbonate, 1M KOH, 4M KOH, sodium chlorite, 4M KOH post-chlorite (PC) and the final insoluble wall pellet. The first two fractions are commonly considered pectin-rich. The next two are considered enriched for hemicellulose. The sodium chlorite extract is believed to contain polysaccharides that had been bound to lignin, and the 4M KOH PC fraction is thought to contain those glycans that can be extracted once lignin is removed.

To estimate the amount of HG and RG-I in these extracts, we evaluated the amount of Rha and GalA recovered in each extract and in the final insoluble pellet. It has been repeatedly stated in publications over the years that HG is the most abundant pectic glycan. We have made this statement ourselves (Mohnen 2008; Atmodjo et al. 2013). However, this statement has been based largely on the content of pectins in wall fractions obtained by salt and sodium carbonate extraction of cell walls, or from food-quality pectin extracted from apples and citrus. The remaining pectin in the wall has not been adequately considered.

The first four data rows in Tables 5.2 and 5.3 show composition and yield data for Rha and GalA in the sequential ammonium oxalate, sodium carbonate, 1M KOH, 4M KOH, sodium chlorite, 4M KOH PC extracts and in the remaining insoluble wall residue from poplar and switchgrass, respectively. Since both Rha and GalA are present in the reducing end of xylan in dicots and non-graminceous monocots, the amount of Rha and GalA that may have been present in the xylan reducing end sequence was removed from the calculations for poplar by subtracting out a 1% xylan equivalent of Rha and GalA from those fractions likely to contain xylan (see Tables 5.2 and 5.3). What can immediately be seen is that for both GalA and Rha, the bulk of these sugars reside not in the ammonium oxalate and sodium carbonate fractions (those traditionally studied as enriched in pectin), rather in poplar the bulk of GalA and Rha is in the final insoluble pellet and in the 1M KOH fraction (Tables 5.2 and 5.3). In switchgrass, there were also significant amounts of these sugars in the sodium chlorite and 4M KOH PC fractions. Importantly, if the amount of GalA required for the RG-I backbone is calculated (Table 5.4), it becomes abundantly clear that in switchgrass and poplar only the ammonium oxalate and sodium carbonate fractions from poplar samples have sufficient GalA to justify the conclusion that HG is the most abundant pectin. In fact, if one considers the amounts of GalA and Rha in all the wall extracts from poplar and switchgrass, it is immediately evident that in both dicot wood and grass walls, the most abundant pectin is RG-I. Equimolar amounts of GalA and Rha are needed for the RG-I backbone. *Equal amounts of HG and RG-I are present when the GalA/Rha ratio in a given fraction is three.* While some caution must be taken in comparing mass yield data for GalA since, as a uronic acid it is generally underrepresented in composition analyses, nonetheless the data indicate that the commonly expressed statement that HG is the most abundant pectic glycan needs to be reconsidered, at least when considering cell walls from wood dicot tissue or from grasses. A further noteworthy result is the large amount of pectin (~30%–60%) that remains in the insoluble pellet. What is this pectin? What is its structure? What is its function?

We further considered whether more actively growing dicot tissues, known to contain a high percentage of primary walls, might contain different amounts of HG versus RG-I and/or whether they would be distributed differently in the cell wall fractions compared to poplar and switchgrass. To analyse these, we adapted the cell wall fractionation data from a study in our lab using exponentially growing wild-type Arabidopsis suspension culture cells to compare these actively growing cells with the poplar and switchgrass samples (Tables 5.2 and 5.3). Several things become immediately clear. Arabidopsis suspension culture cells produce much more ammonium oxalate and sodium carbonate-extractable pectin. The high GalA

TABLE 5.2
Location of Pectin-Associated GalA in Different Fractions of Extracted AIR and Final Insoluble Pellet from Wild-Type Switchgrass Tillers, Poplar Stems and Arabidopsis Suspension Culture Cells

		GalA (µg/mg AIR) in the Designated Extracts from Wild-Type Samples									
Plant Source	Reference	Ammonium Oxalate	Sodium Carbonate	1M KOH	4M KOH	Sodium Chlorite	4M KOH PC	Insoluble	Total	GalA in Ammonium Oxalate + Sodium Carbonate	GalA in Insoluble Residue
Switchgrass tillers	Biswal et al. (2018a)	0.29	0.19	1.4[b]	0.3[b]	0.54[b]	0.54[b]	1.69[b]	4.95	9.7%	34.1%
Poplar stem	Biswal et al. (2018a)	0.63	0.28	1.01 (0.44)[a]	0.19 (0.06)[a]	1.69 (1.66)[a]	0.73 (0.57)[a]	1.87 (1.77)[a]	5.4	16.9%	32.8%
	Biswal et al. (2015)	0.52	0.26	1.21 (0.66)[a]	0.21 (0.06)[a]	1.00 (0.97)[a]	0.60 (0.45)[a]	1.52 (1.44)[a]	5.32	14.7%	27.1%
	Biswal et al. (2018b)	0.54	0.27	1.52 (0.99)[a]	0.24 (0.51)[a]	1.04 (1.01)[a]	0.70 (0.54)[a]	1.62 (1.54)[a]	5.93	13.7%	26.0%
Arabidopsis suspension culture cells	Atmodjo and Mohnen (unpublished)	28.2	1.01	0.26[b]	2.07[b]	2.61[b]	2.92[b]	1.25[b]	38.32	76.2%	3.3%

[a] 1% of xylan mass from poplar was subtracted from GalA values in 1M KOH, 4M KOH, sodium chlorite and 4M KOH post-chlorite (PC) fractions and in the insoluble pellet due to the presence of GalA in the xylan reducing end sequence (Rha-Gal-Xyl) of dicots and non-grass monocots. Since xylan has a DP of ~100, 1% of xylan mass in respective fractions was subtracted from the GalA value (shown in red fonts in parentheses).

[b] 1% of xylose content was not subtracted from GalA in Arabidopsis because suspension culture cells have primary walls that lack xylan and in switchgrass since grass xylan does not have the reducing end sequence.

TABLE 5.3
Location of Pectin-Associated Rha in Different Fractions of Extracted AIR and Final Insoluble Pellet from Wild-Type Switchgrass Tillers, Poplar Stems, and Arabidopsis Suspension Culture Cells

		Rha (µg/mg AIR) in the Designated Extracts from Wild-Type Samples									
Plant Source	Reference	Ammonium Oxalate	Sodium Carbonate	1 M KOH	4 M KOH	Sodium Chlorite	4 M KOH PC	Insoluble	Total	Rha in Ammonium Oxalate + Sodium Carbonate	Rha in Insoluble Residue
Switchgrass tillers	Biswal et al. (2018a)	0.19	0.10	0.74[b]	0.32[b]	1.12[b]	0.71[b]	0.33[b]	3.51	8.3%	9.4%
Poplar stem	Biswal et al. (2018a)	0.13	0.04	1.30 (0.73)[a]	0.22 (0.09)[a]	0.83 (0.8)[a]	0.82 (0.66)[a]	2.20 (2.1)[a]	4.55	3.7%	46.2%
	Biswal et al. (2015)	0.10	0.04	0.87 (0.32)[a]	0.21 (0.06)[a]	0.59 (0.56)[a]	0.04 (0)[a]	1.80 (1.72)[a]	2.80	5.0%	61.4%
	Biswal et al. (2018b)	0.10	0.04	1.00 (0.47)[a]	0.25 (0.11)[a]	0.67 (0.64)[a]	0.51 (0.35)[a]	2.06 (1.98)[a]	3.69	3.8%	53.7%
Arabidopsis suspension culture cells	Atmodjo and Mohnen (unpublished)	1.00	0.22	0.08[b]	0.59[b]	0.53[b]	1.02[b]	0.26[b]	3.70	32.0%	21.3%

[a] 1% of xylan mass from poplar was subtracted from Rha values in the 1M KOH, 4M KOH, sodium chlorite and 4M KOH post-chlorite (PC) fractions and in the insoluble pellet due to the presence of Rha in the xylan reducing end sequence (Rha-Gal-Xyl) of dicots and non-grass monocots. Since xylan has a DP of ~100, 1% of xylan mass in respective fractions was subtracted for Rha values (shown in red fonts in parentheses).

[b] 1% of xylose content was not subtracted from Rha in Arabidopsis because suspension culture cells have primary walls that lack xylan and in switchgrass since grass xylan does not have the reducing end sequence.

TABLE 5.4
GalA/Rha Ratio and HG/RG-I Content in Different Fractions of Extracted AIR and Final Insoluble Pellet from Wild-Type Switchgrass Tillers, Poplar Stem and Arabidopsis Suspension Culture Cells

		GalA/Rha Ratio (μg/mg AIR) in the Designated Extracts from Wild-Type Samples								
Plant Source	Reference	Ammonium Oxalate	Sodium Carbonate	1M KOH	4M KOH	Sodium Chlorite	4M KOH PC	Insoluble	Total	HG/RG-I Ratio for Total AIR[a]
Switchgrass tiller	Biswal et al. (2018a)	0.29/0.19=1.53	0.19/0.10=1.9	1.40/0.74=1.89	0.3/0.32=0.94	0.54/1.12=0.48	0.54/0.71=0.76	1.69/0.33=5.12	4.95/3.51=1.4	0.2
Poplar stem	Biswal et al. (2018a)	0.63/0.13=4.85	0.28/0.04=7	0.44/0.73=0.63	0.06/0.09=0.67	1.66/0.8=2.08	0.57/0.66=0.86	1.77/2.1=0.84	5.4/4.55=1.2	0.1
	Biswal et al. (2015)	0.52/0.10=5.2	0.26/0.04=6.5	0.66/0.32=2.1	0.06/0.06=1	0.97/0.56=1.7	0.45/0 =?	1.44/1.72=0.84	5.32/2.8=1.9	0.45
	Biswal et al. (2018b)	0.54/0.1=5.4	0.27/0.04=6.75	0.99/0.47=2.1	0.51/0.11=4.6	1.01/0.64=1.5	0.54/0.35=1.5	1.54/1.98=0.78	5.93/3.69=1.6	0.3
Arabidopsis suspension culture cells	Atmodjo and Mohnen (unpublished)	28.2/1.0=28.2	1.01/0.22=4.6	0.26/0.08=3.25	2.07/0.59=3.5	2.61/0.53=4.9	2.92/1.02=2.9	1.25/0.26=4.8	38.3/3.7=10.4	4.7

Note: In poplar, the amount of GalA and Rha was corrected in the last four fractions and in solid residue by subtracting 1% mass xylan due to reducing end sequence.

[a] Equimolar amounts of GalA and Rha are needed for RG-I.

For equimolar amounts of HG and RG-I, the GalA/Rha ratio would need to be: (1 Rha equivalent + 1 GalA equivalent) + 2 GalA equivalents = **3/1 = 3**

Since in poplar only the ammonium oxalate and sodium carbonate fractions have a GalA/Rha ratio of 3 or greater, only these fractions have more HG than RG-I. For switchgrass, only the insoluble pellet has more HG than RG-I. Conversely, in the primary wall–rich Arabidopsis fractions, all fractions except the 4M KOH PC have more HG than RG-I, with particularly high HG levels in the ammonium oxalate fraction and in the insoluble residual pellet.

Based on total amounts of GalA and Rha in the AIR from the shown lines (after subtracting reducing end Rha and GalA values), it is clear that HG is NOT the most abundant pectic glycan in poplar stems or in switchgrass tillers. RG-I is the most abundant pectic glycan in both switchgrass and poplar. Conversely, in cell walls from Arabidopsis suspension culture cells, HG is the most abundant pectic glycan, particularly in the ammonium oxalate fraction

FORMULA: HG/RG-I content = [amount of GalA – (1x amount of Rha)] / (2x amount of Rha)

Example: HG/RG-I content for switchgrass ammonium oxalate fraction: [0.63 – 0.13] / (2×0.13) = 1.9

The GalA/Rha ratios for the total amount of GalA and Rha in the AIR data shown ranged from 1.2 to 1.9 for poplar and switchgrass, while for Arabidopsis it was 10.4.

* A GalA/Rha ratio of 1.2 means GalA:Rha of 1.2 : 1 ==> HG/RG-I = (1.2 – 1) / 2 = 0.1

Thus, RG-I in this calculation is ten times more abundant than HG.

* A GalA/Rha ratio of 1.9 means GalA:Rha of 1.9 : 1 ==>. HG/RG-I = (1.9 – 1) / 2 = 0.45

Thus, RG-I in this calculation is 2.2 times more abundant than HG.

* A GalA/Rha ratio of 10.4 means GalA:Rha of 10.4 : 1 ==> HG/RG-I = (10.4 – 1)/2 = 4.7

Thus, from this calculation, HG in Arabidopsis cells is 4.7 times more abundant than RG-I.

These numbers need to be taken with some caution, since GalA as an uronic acid is likely underrepresented. Nonetheless, the commonly expressed statement that HG is the most abundant pectic glycan needs to be qualified for the type of tissue. For Arabidopsis suspension cell walls, unlike walls from poplar wood and switchgrass tillers, HG is indeed the most abundant pectic glycan.

content indicates that this pectin is highly enriched in HG. Notably, in these dicot suspension culture cells, HG is the most abundant pectin in almost all cell wall fractions (Table 5.4). While there remains much to glean from these data, we propose that a critical major question is what is/are the structures of the pectin in these different wall fractions and in the final pellet. Are they "pure" pectins? Are they cross-linked heteroglycans or glycoconjugates? Is the pectin in the final pellet part of a wall architectural net? Is the greater relative amount of HG to RG-I a function of cell type, developmental state or both?

WHAT WE LEARN FROM A REVIEW OF PECTIN STRUCTURE RESEARCH

- *How did we arrive at the view of pectin as a family of three pectic glycans: HG, RG-I and RG-II?* Chemical/structure data from multiple laboratories over the past 100+ years have confirmed the central structural role of glycans with either an HG or repeating GalA-Rha (RG-I) backbone as central to what we call pectin. The degree of variation in the pattern and extent of single sugar versus oligosaccharide side chain substitutions of the backbone remains unclear, as does the frequency and positioning of RG-II and other substituted galacturonans such as AGA and XGA. The way these different types of pectic glycans are organised in the wall is not clear. Perhaps, Arabidopsis seed mucilage with its relatively unsubstituted RG-I may be a model system to study RG-I and HG positioning along a heteropolymer, although the importance of that organisation for pectin in non-mucilage cells in not clear.
- *How are these three pectic glycans covalently connected to each other in the wall?* We still do not know how HG, RG-I and RG-II are connected in the wall, other than that a large volume of data show that EPG digestion of HG releases RG-I and RG-II. The identification of an enzyme that can extend RG-I with HG (Amos and Mohnen, unpublished results) opens the use of *in vitro* reconstitution to aid in elucidating this. However, the initial backbone branching GTs still need to be identified to enable broader *in vitro* reconstitution studies of heteroglycan pectin synthesis.
- *How many different covalently connected pectic heteroglycans exist in the wall?* Data from multiple labs showing different types/frequencies of RG-I branching and the existence of "populations" of RG-I during chromatography suggests that there may be different patterns and frequencies of RG-I branching. However, since to see these populations the pectin is first fragmented either chemically or enzymatically, it remains unclear how many populations exist in any given cell wall type. This is a critical area of needed future research.
- *Does homogenous, independent HG exist in the cell wall, or is it always synthesised as part of a heteroglycan?* The sum of the accumulated data would suggest that if HG exists as a homogenous separate polysaccharide, it is in very low abundance.
- *Does homogenous, independent RG-I exist in the cell wall, or is it always synthesised as part of a heteroglycan?* The data would suggest that, with the possible exception of Arabidopsis seed mucilage, the answer to this is no.
- *What is the evidence that the so-called Type II arabinogalactans (i.e. β-1,3 and β-1,6-linked galactans with α-1–5 linked arabinan) are present in RG-I?* Multiple researchers over the years have identified oligosaccharide side branches and linkages that suggest that small amounts of the so-called Type II arabinogalactans may be covalently linked to the RG-I backbone. However, the possibility that the RG-I preparations contained pectic AGP and that the identified linkages arose from the AGP part of the preparation remains strong, particularly in light of the recent report that most classically isolated RG-I is attached to AGP (Tan et al. 2023). A critical question that needs to be addressed is whether all RG-I is initially synthesised as pectic AGPs. If so, the next question is how much of the original pectic AGPs function as a proteoglycan (possibly GPI-anchored to the plasma membrane (Tan et al. 2013)) versus as a soluble pectic AGP or pectic glycan that is enzymatically cleaved from the membrane-bound AGP and thereby functions freely in the wall.
- *What is the evidence that pectic glycans are covalently connected to proteins or to hemicellulose?* There have been multiple reports indicating that at least a fraction of pectin is attached to proteins and to hemicellulosic xyloglucan and xylan. Tan et al. (2013) with the discovery of APAP1 showed conclusively that some/much of RG-I is attached to at least one AGP protein and to xylan. Structural linkage confirmation of an attachment of xyloglucan to pectin is lacking, but the large number of reports from multiple laboratories over the past >40 years makes it likely that such a linkage exists.
- *Could at least part of the pectic matrix consist of a covalently linked "net" akin in function to bacterial peptidoglycan?* The authors of this review find the prior data supporting this hypothesis to be substantial. We hypothesise that there is a yet-to-be-defined "net-like" structure in plant cell walls that has fundamental roles in plant cell growth and function. The finding that herbicide-resistant cells that are almost devoid of cellulose but enriched in pectin can grow and divide and that such cells

have fibrous-looking cross-linked walls would appear to provide support for such a net-like structure (Shedletzky et al. 1990; Wells et al. 1994). From our review of pectic structure above based on data from multiple groups, we hypothesise that such a pectin-rich net-like structure has an RG-I-rich core and includes HG (and possibly RG-II) domains and pectic AGP.

- *Are there components of the pectins or pectic AGPs that are synthesised separately within the endomembrane system and covalently connected in the wall by transglycosidases, or are the complex heteroglycans and proteoglycans all synthesised within the endomembrane system?* Although this issue needs to be resolved, the available data support the hypothesis that pectic AGPs are synthesised in the Golgi membrane and then transported to the wall.

PECTIN SYNTHESIS AND HOW IT INFORMS PECTIN STRUCTURE

Glycosidic linkage data indicate that the different structural domains of the pectic polysaccharides (HG, RG-I, RG-II, XGA and AGA) are covalently linked to each other. However, the biosynthetic mechanisms behind their interconnections and where and how cross-linking occurs remains unknown. The pectin biosynthesis process is hypothesised to proceed, like all biological polymer synthesis, through three stages: initiation, elongation and termination. A comprehensive understanding of those stages for the synthesis of the different pectic domains is lacking, as is the identity of many of the pectin biosynthetic genes/enzymes. That said, considerable progress has been made on the identity and biochemical activity of the pectin biosynthetic enzymes and this has informed our understanding of pectin structure.

Immunocytochemical research using anti-pectin antibodies, subcellular fractionation and topology studies indicate that pectin is synthesised in the Golgi, is modified and assembled as it moves through different Golgi cisternae and the trans-Golgi network and is transported via vesicles to the cell wall, where further *in muro* modification and re-modelling may occur during and after deposition in the walls (Atmodjo et al. 2013; Anderson 2016; Sinclair et al. 2018; Gu and Rasmussen 2022; Sinclair et al. 2022). However, there are very few details known about this process.

The current view of pectin biosynthesis (Caffall and Mohnen 2009; Harholt et al. 2010; Atmodjo et al. 2013; Amos and Mohnen 2019; Shin et al. 2021; Engle et al. 2022) is that the coordinated action of multiple GTs leads to the transfer of specific glycosyl residues from activated nucleotide sugar donors to oligosaccharide/polysaccharide/proteoglycan acceptors, although some of the GAUTs have also been shown to catalyse *de novo* synthesis of HG using UDP-GalA only (Amos et al. 2018; Engle et al. 2022). Methyltransferases (MeTs) and acetyltransferases (ATs) are also critical for catalysis of methylesterification, methyletherification and acetylation of specific sugar acceptors (Caffall and Mohnen 2009). Pectin GTs are generally specific for their nucleotide sugar donor and oligosaccharide acceptor substrates, although Arabidopsis GALACTAN SYNTHASE 1 (AtGALS1) has been shown to be bifunctional in catalysing both the transfer of Gal from UDP-α-D-Gal and arabinopyranose from UDP-β-L-Ara*p* to β-1,4-galactan chains (Laursen et al. 2018). Below we summarise the known pectin biosynthetic transferases and when applicable, how results of analyses of these enzymes and the products produced inform our understanding of pectin structure.

BIOSYNTHESIS OF HG

Early biochemical studies of pectin biosynthetic enzymes revealed that HG:galacturonosyltransferase (HG:GalAT) activity transferred GalA from UDP-GalA onto the growing HG polymer and the enzymatic activity was identified in microsomal fractions, suggesting that HG is made in the endomembrane system (reviewed in Mohnen (2002); Caffall and Mohnen (2009)). Subsequent studies using isolated Golgi localised this enzyme activity to the Golgi with the catalytic site in the Golgi lumen (Sterling et al. 2001; Ohashi et al. 2017). Relevant to this review, size exclusion chromatography of the product synthesised in tobacco microsomes and in detergent solubilised microsomes incubated with UDP-[^{14}C]GalA showed that HG:GalAT synthesised an ~105 kDa product that was only ~40% esterified, indicating that HG is not necessarily synthesised in a highly methylated form (Doong et al. 1995; Doong and Mohnen 1998). The results agree both with the early observation by Kauss et al. (1969) that the rate of HG synthesis is independent of the methylesterification process and with the more recent results of Amos et al. (2018) and Engle et al. (2022) showing that HG of greater than >100 kDa is synthesised *in vitro* by multiple GAUTs in a non-methylesterified form. These results call into question the frequently made statement that HG is synthesised in a highly methylesterified form. This assumption likely depends on the specific cell type, state of development and biosynthetic enzymes expressed in the cell, since there is no evidence based on the HG biosynthetic catalysts that methylesterification is required for HG synthesis. The question of what the degree of HG methylesterification (or acetylation) is during synthesis and how that is regulated requires future research.

In 2002, the reduced amounts of GalA in Arabidopsis QUASIMODO1 (QUA1, At3g25140) mutants compared to wild type suggested that the gene might encode an HG:GalAT (Bouton et al. 2002; Orfila et al. 2005); however, no confirmatory biochemical evidence has been obtained to substantiate this possibility.

The first enzymatically demonstrated HG biosynthetic enzyme, GALACTURONOSYLTRANSFERASE 1 (GAUT1), is an α-1,4-GalAT that catalyses the addition of (1,4)-linked α-D-GalA residues onto the non-reducing end of HG. It was identified via purification of the protein from Arabidopsis suspension-cultured cells followed by MS sequencing and subsequently immunoprecipitation (Sterling et al. 2006). Based on BLAST analyses, a family of 25 genes comprising 15 GAUT and 10 GAUT-like (GATL) with high sequence similarity was identified, providing genetic and biochemical tools for further studies of pectin biosynthetic enzymes.

Subsequently, GAUT7, a member of the GAUT family that co-purified with GAUT1 was shown to anchor the catalytic GAUT1 in the Golgi. Immunoprecipitation and bimolecular fluorescence complementation studies revealed that these Arabidopsis proteins function as a GAUT1:GAUT7 heteromeric enzyme complex of two GAUT1 subunits and one GAUT7 subunit held together by disulphide and non-covalent bonds (Atmodjo et al. 2011). The GAUT1:GAUT7 complex synthesises high molecular weight HG in both an acceptor-dependent and independent manner (i.e. via *de novo* synthesis) (Amos et al. 2018). These results make the GAUT1:GAUT7 complex a strong candidate for functioning in both complex heteroglycan and homopolymer syntheses. Recently, our characterisation of the *gaut1* mutant indicates that it may function in the synthesis of a complex pectic polymer held tightly into the cell wall. Once GAUT1 and the GAUT family were identified as encoding proven and putative HG biosynthetic genes, proteomic studies confirmed the presence of the GAUT proteins in the Golgi (Nikolovski et al. 2012) as did fluorescent protein fusion studies of the GAUTs (Atmodjo et al. 2011; Lund et al. 2020).

The transient expression of Arabidopsis, poplar and switchgrass GAUT4 homologs in *Nicotiana benthamiana* and the study of knockdown (KD) expression of *GAUT4* in poplar and switchgrass revealed that *GAUT4* encodes an HG:GalAT that is required for production of the bulk of HG and RG-II in these tree and grass species (Biswal et al. 2018a), confirming the central and conserved role of pectin in diverse plants. Downregulation of GAUT4 resulted in increased cell size and biomass yield in both species, strongly implicating a role of pectin in cell expansion. Importantly, for this review, the finding that GAUT4 appears to encode the HG biosynthetic enzyme that synthesises the bulk of HG and RG-II pushed into centre stage the question of what the other HG biosynthetic GAUTs were synthesising. Multiple GAUTs, such as GAUT1 and GAUT8, are required for plant growth as mutants of these genes cause severe plant dwarfism (Bouton et al. 2002) (Atmodjo and Mohnen, unpublished). If they are not synthesising the bulk of HG, what HG do they synthesise? These results lead us to hypothesise that different GAUTs synthesise HG glycan domains in unique cell wall polymers.

Prior to the analysis of the phenotypes of the *GAUT4*-KD plants, poplar with both KD (Biswal et al. 2015) and overexpression (OE) (Biswal et al. 2018b) of GAUT12 had been generated and analysed. The *GAUT12*-KD poplar, like the *GAUT4*-KD poplar, had increased cell size and biomass yield, but for a different reason. *GAUT12*-KD plants, like the previously studied *GAUT12/IRX8* mutants (Peña et al. 2007; Persson et al. 2007; Hao et al. 2014; Hao and Mohnen 2014), had reduced xylan and a reduction in a subfraction of HG (Biswal et al. 2015). A detailed analysis of the changes in cell wall composition and linkage in the poplar *GAUT12*-KD (Biswal et al. 2015) and *GAUT12*-OE plants (Biswal et al. 2018b) led us to conclude that the evidence best supported the hypothesis that GAUT12 synthesises a subfraction of HG that is covalently linked to xylan and required for xylan synthesis. These results further support the hypothesis that different GAUTs synthesise unique HG domains in unique wall polymers.

The hypothesis that different GAUTs synthesise unique HG domains in different polymers was further supported by the study of GAUT11 and its role in Arabidopsis mucilage production (Voiniciuc et al. 2018a). Arabidopsis mutated in *gaut11* produced reduced amounts of seed mucilage with reduced HG content (Caffall et al. 2009; Voiniciuc et al. 2018a). Expression of recombinant Arabidopsis GAUT11 in human embryonic kidney (HEK293) cells and assays of enzyme activity identified GAUT11 as an HG:GalAT (Voiniciuc et al. 2018a). In a follow-up study designed to express the entire GAUT family in HEK293 cells in an effort to determine if all family members were HG:GalATs, three additional *GAUT* family genes, *GAUTs 10*, *13* and *14* were also shown to encode HG:GalATs that synthesised polymeric HG *in vitro* (Engle et al. 2022). The same study revealed that GAUT13 and GAUT14 could also *de novo* synthesise (initiate) HG in the presence of UDP-GalA and in the absence of exogenous HG acceptors, an activity which had previously been demonstrated for the GAUT1:GAUT7 complex. Comparison of the relative activities of those GAUTs with confirmed HG:GalAT activity showed that the GAUT1:GAUT7 complex and GAUTs 13 and 14 were the most enzymatically active, GAUT1 alone had intermediate activity and GAUTs 10 and 11 had the lowest relative activity in *in vitro* catalysis studies.

One hypothesis that can be formed from these results is that GAUTs 1:7, 13 and 14 function when rapidly synthesised and long HG is required, while GAUTs 10 and 11 may synthesise short HG domains during the synthesis of more complex polymers. *In vitro* reconstitution of complex pectin synthesis will be one means to test this hypothesis. Also, although GAUTs 1, 1:7, 10, 11, 13 and 14 have confirmed HG:GalAT activity, it is not yet known whether GAUTs other than GAUT1:GAUT7 function in protein complexes. Recently, Lund et al. (2020) provided evidence that GAUTs 5 and 6 can serve as alternative anchors to hold GAUT1 in the Golgi, a function that was further supported by GAUT co-expression in HEK293 cells (Engle et al. 2022). While none of the other GAUTs showed evidence for GAUT:GAUT co-expression, the possibility that they function with other GTs, MeTs or ATs has not been tested. Future experiments aimed at identifying additional pectin biosynthetic complexes and at probing the location in the endomembrane system at which the GAUTs and other pectin biosynthetic enzymes function are important areas of future research.

It has been proposed that the cooperative action of GTs and MeTs in the Golgi leads to the synthesis and transport to the wall of highly esterified HG (Driouich et al. 2012). The methylesterification of the C-6 carboxyl of GalA residues is catalysed by HG-methyltransferase (HG-MeT). HG-MeT activity has been detected in Golgi-enriched fractions and isolated microsomes from tobacco (*Nicotiana tabacum*), and has been localised to the Golgi with the catalytic site identified in the Golgi lumen (Goubet and Mohnen 1999; Ridley et al. 2001). Based on bioinformatics studies and sugar composition analyses of cell walls from loss or gain of function Arabidopsis mutants, Cotton Golgi-Related 3 (CGR3, At5g65810) and QUASIMODO 3 (QUA3, At4g00740) have been identified as putative HG-MeTs (Held et al. 2011; Miao et al. 2011) but proof of enzymatic activity is needed to substantiate the biochemical activity of the proteins. In this regard, it is interesting that QUA3 was previously identified as a GAUT1:GAUT7 complex transiently interacting protein (Atmodjo et al. 2011), leading to the hypothesis that QUA3 may interact with GAUT1:GAUT7 during HG synthesis, possibly as an HG:MeT. This should be tested. In 2020, QUASIMODO2 (QUA2) (Mouille et al. 2007) from *Arabidopsis thaliana* was heterologously expressed and shown to have HG:MeT activity, making it the first confirmed HG:MeT (Du et al. 2020).

HG may be partially O-acetylated at O-2 or O-3 of GalA, but no HG (or RG-I) AT has yet been biochemically confirmed. Pauly and Scheller (2000) provided enzymatic evidence that microsomal membranes from potato suspension culture cells catalyse the transfer of ^{14}C-acetate from ^{14}C-AcetylCoA to multiple cell wall polysaccharides including RG-I, and in subsequent work identified Reduced Wall Acetylation (RWA) proteins as being required for cell wall acetylation in Arabidopsis. However, these proteins are unlikely to catalyse the acetate transfer (Manabe et al. 2013). More recently, candidate pectin ATs have been identified (Stranne et al. 2018; Chiniquy et al. 2019) and with active interest from multiple research groups in pectin acetylation, it is likely that HG (and RG-I) ATs will be identified in the near future.

Biosynthesis of Substituted Galacturonans

Xylogalacturonan (XGA) biosynthesis: Biosynthesis of XGA (Figure 5.6) is catalysed by XGA:xylosyltransferases (XGA:XylT). Composition analysis of mutants, transformation-based molecular complementation studies and heterologous protein expression studies followed by *in vitro* demonstration of enzyme activity identified Arabidopsis XYLOGALACTURONAN DEFICIENT1 (XGD1, At5g33290) as a Golgi-localised XGA:XylT (Jensen et al. 2008).

Apiogalacturonan (AGA) biosynthesis: There are currently no reports identifying an AGA:apiosyltransferase (AGA:ApiT) that synthesises AGA (Figure 5.5). As Api or apiobiose is a glycosyl residue that also exists as β-Api*f* in two of four side branches of RG-II, it remains to be determined whether the same ApiT synthesises RG-II and AGA.

Biosynthesis of RG-II Side Chains

RG-II appears to have an evolutionary conserved structure within the plant kingdom. It is composed of complex side chains (A-D) and two arabinosyl short chains (E-F) with 12 different types of glycosyl residues including methylesterified, methyletherified and O-acetylated forms and nearly 20 different linkages (Bar-Peled et al. 2012; Ndeh et al. 2017). Evidence suggests that two RG-II molecules are cross-linked to one another by borate diester bonds (Kobayashi et al. 1996; Bar-Peled et al. 2012). It is not known whether the complex side chains of RG-II are synthesised by GTs that transfer individual glycosyl residues onto the growing non-reducing end of RG-II molecules or whether side chains of RG-II are synthesised separately as oligosaccharides and subsequently transferred onto HG.

Owing to the structural complexity of RG-II, the cooperative activities of approximately 20 GTs including RG-II:apiosyltransferase (RG-II:ApiT), RG-II:xylosyltransferase (RG-II:XylT), RG-II:glucuronosyltransferase (RG-II:GlcAT), as well as transferases for GalA, Gal, Rha, Ara, Dha, Fuc, Kdo and Aceric Acid are required for the biosynthesis of side chains in RG-II (Voxeur et al. 2012). Furthermore, RG-II:methyltransferase (RG-II:MeT) and RG-II:acetyltransferase (RG-II: AT) are also required for the modification of glycosyl residues. Reverse genetic strategies to identify candidate GTs in RG-II biosynthesis has been challenging, as mutations appear to lead to either strong developmental defects or lethality. Furthermore, identification of putative GTs through sequence homologies with well characterised GTs from other organisms has been limited, since RG-II is composed of many types of glycosyl residues with unusual ring forms or configurations that are rare in nature (Voxeur et al. 2012). To date, only a handful of enzymes have been characterised and enzymatically demonstrated for functionality in RG-II biosynthesis, as described below.

RG-II:MeT and RG-II:AT: Although side chain residues of RG-II contain several methyl and acetyl groups such as 2-O-methyl-Fuc, 2-O-methyl-Xyl, 3-O-acetylaceric acid and acetylated methyl Fuc, the genes encoding RG-II side chain-specific MeTs and ATs have not yet been identified. Also, the location of catalytic activities of methylation and acetylation, and whether those methyl and acetyl groups are added to the nucleotide sugar before or after being added to RG-II remains to be investigated.

RG-II: xylosyltransferases (RG-II:XylT): Recombinant protein expression followed by biochemical enzymatic analyses, NMR spectroscopy and fluorescence-based subcellular localisation studies revealed RGXT1 (At4g01770) and RGXT2 (At4g01750) from *Arabidopsis thaliana* as Golgi-localised (1,3)-α-D-XylT involved in the xylosylation of the internal L-fucose of the side chain A of RG-II, yielding an α-(1,3)-linkage between the xylose and fucose moiety (Egelund et al. 2006). In a subsequent study, the third member of the RGXT family, RGXT3 (At1g56550) from *A. thaliana*, was characterised as an RG-II:XylT identifying it as also catalysing the transfer of D-xylose from UDP-α-D-Xyl onto methyl α-L-Fuc (Egelund et al. 2008). Later, Liu et al. (2011) reported MALE GAMETOPHYTE DEFECTIVE 4 (MGP4, At4g01220) as a putative XylT. Through heterologous protein expression and activity assays, its enzymatic function as a XylT that transfers D-Xyl onto L-Fuc during the biosynthesis of side chain A in RG-II was confirmed. Thus, MGP4 has been designated as RGXT4, the fourth member of RGXT family (Liu et al. 2011).

Biosynthesis of RG-I

Recently progress has been made in the synthesis of the disaccharide repeat Rha-GalA backbone of RG-I (Takenaka et al. 2018; Amos et al. 2022) that opens the door to understanding RG-I synthesis and structure. The GalA residues in RG-I backbone may be acetylated at O-3 or O-2 and the Rha residues can be substituted at O-4, and to lesser extents at O-3 with single residues or with β-1,4-galactans, α-1,5-arabinans or arabinogalactan side chains (Yapo et al. 2007). Both Type I and Type II arabinogalactans have been identified as components of RG-I (see above). Some Type II arabinogalactans may be associated with mannoglycans or AGPs (Clarke et al. 1979). Immunocytochemical evidence based on RG-I-specific antibodies suggests that the branching pattern, length and linkages of RG-I side branches vary in a cell-type, plant species and developmental specific manner (Thomas et al. 1989; Willats et al. 2001). Subcellular localisation studies with RG-I backbone and side chain-specific antibodies suggest synthesis occurs in the early and late Golgi compartments, respectively (Wachananawat et al. 2020). However, whether RG-I is synthesised independently of HG, requires an existing HG backbone as primer or is synthesised onto the arabinogalactan region of AGPs (Tan et al. 2013) remains to be determined. The biosynthesis of the RG-I side chains requires the cooperative action of numerous GTs.

RG-I:Rhamnosyltransferase (RG-I:RT): Over the last few years, seed coat mucilage in *Arabidopsis thaliana*, which is mainly composed of unbranched RG-I, has been identified as a promising tissue for studies of plant cell wall biosynthetic enzymes. One gene/enzyme identified as highly expressed in Arabidopsis seed mucilage is RRT1 (At5g15740), which has been shown to transfer Rha from UDP-β-L-Rha to RG-I oligosaccharides (Takenaka et al. 2018). This important achievement was the first enzymatically proven RG-I backbone biosynthetic enzyme. Subcellular localisation studies using GFP fusion protein confirmed the presence of RRT1 in the Golgi. RRT1 belongs to a 10-member gene family in Arabidopsis, and in the same study (Takenaka et al. 2018) three other genes RRT2, RRT3, and RRT4 (At3g02250, At2g03280 and At1g14020, respectively) were shown to have the same catalytic activity. Transcript expression studies revealed the highest level of expression in seed mucilage, leaf and flowers, dry seeds and pollen tubes for RRT1, RRT2, RRT3 and RRT4, respectively. The results suggest that RG-I synthesis in specific organs may be governed by specific RRTs, with possible cooperation from other RRTs.

Subsequently, based on mutants and recombinant protein expression studies, Wachananawat et al. (2020) reported AtFRB1/RRT8 (At5g01100) to also be an RG-I:RT. In the same study, recombinant protein expression and *in vitro* activity assays of AtRRT1–AtRRT4 orthologs from *Marchantia polymorpha* (MpRRT1–Mapoly0033s0138 and MpRRT3–Mapoly0014s149, respectively) showed that they had RG-I:RT activity. The question remains as to whether RRTs function independently or in complexes composed of rhamnosyltransferase, galacturonosyltransferase or other enzymes.

RG-I:Galacturonosyltransferase (RG-I:GalAT): Recently, based on heterologous protein expression and *in vitro* activity assays using RG-I acceptor oligosaccharides, MUCILAGE RELATED 70 (MUCI70, At1g28240) was enzymatically shown to be a RG-I:GalAT and renamed RGGAT1 to reflect its catalytic activity (Amos et al. 2022). Furthermore, the combined activities of RGGAT1 and RRT4 led to the production of RG-I via elongation of RG-I acceptors *in vitro*. The results show that RGGAT1/MUCI70 is a RG-I:GalAT that functions with RRT to synthesise the alternating [-4-α-D-Gal*p*A-1,2-α-L-Rha*p*-1-] repeat backbone of RG-I. Since like the GAUTs, RGGAT1 is a member of a multigene family, it remains to be shown which, if any, of the remaining family members have RGGAT activity and if so, how they differ in their activities and their production of the RG-I backbone. Importantly, the ability to synthesise both the RG-I and HG backbone enables *in vitro* reconstitution experiments that will be critical to explore the breadth of pectin structures that can be synthesised. Preliminary results show that enzymes exist that are able to elongate the RG-I backbone with HG (Amos and Mohnen, unpublished results), opening up the possibility of complex pectic heteroglycan *in vitro* synthesis. A combination of *in vitro* reconstitution of complex pectin synthesis with molecular genetic, transcriptomics, proteomics and *in vivo* labelling will be required to tease apart the different types of pectin heteroglycans that exist in plants and to determine pectin structure–function relationships.

RG-I:Arabinosyltransferase (RG-I:AraT): L-Ara in RG-I side chains exists in multiple linkages, most often in the furanose ring form, although some terminal Ara exists in the pyranose form. Early biosynthetic studies conducted using beans revealed that AraT activity is primarily found in the Golgi and to a lesser extent in the endoplasmic reticulum (Bolwell and Northcote 1983). Initial efforts to investigate AraT activity involved in RG-I biosynthesis was challenging as multiple polysaccharides including pectin, hemicellulose and proteoglycans contain Ara. Furthermore, due to the presence of both Ara*p* and Ara*f* in RG-I side chains, the uncertainty of the required nucleotide sugar substrates also confounded biosynthetic studies of RG-I:AraT (Nunan and Scheller 2003). Differential gene expression, mutant data, immunohistochemical analysis and transformation-based molecular complementation studies along with compositional and linkage analysis of RG-I isolates suggest that ARABINAN DEFICIENT 1 (ARAD1, At2g35100) from *Arabidopsis thaliana* may be an arabinan α-1,5-AraT (Harholt et al. 2006). In a separate study, ARAD2 (At5g44930) was also identified as a putative RG-I:α-1,5-AraT with a non-redundant function to ARAD1, and an interaction between ARAD1 and ARAD2 as both homodimers and heterodimers were demonstrated, implying their function as complexes during biosynthesis (Harholt et al. 2012). However, biochemical proof of enzymatic activity is required to confirm, or not, the role of ARAD1 and ARAD2 in RG-I synthesis.

RG-I:Galactosyltransferase (RG-I:GalT): Early biosynthetic studies conducted using microsomal preparations from mung beans identified β-1,4-GalT and β-1,3-GalT activities (McNab et al. 1968; Panayotatos and Villemez 1973). Microsomal preparations from potato suspension-cultured cells were shown to contain β-1,4-GalT activities that elongated RG-I side chains containing β-1,4-galactans (Geshi et al. 2000). Later, through subcellular organelle fractionation studies, the location of catalytic activity of RG-I:GalT was shown to be in the lumen of the Golgi (Geshi et al. 2004). Based on differential expression, composition and linkage analysis of *Arabidopsis thaliana* mutants, β-1,4-GalT activity was demonstrated for three GALACTAN SYNTHASE (GALS) genes: GALS1-At2g33570, GALS2-At5g44670 and GALS3-At4g20170 (Liwanag et al. 2012). Subcellular localisation studies revealed that all three GALS proteins are Golgi-localised and biochemical analyses have confirmed that they successively add β-1,4-Gal residues onto existing β-1,4-galactans of the RG-I backbone (Ebert et al. 2018). However, the mechanism by which side chains with varying DPs are synthesised remains poorly understood and the enzyme(s) involved in the initial Gal substitution that initiates β-1,4-galactan side chain formation on the RG-I backbone remains to be discovered.

A subsequent study demonstrated the potential of AtGALS1 to both catalyse the β-1,4-Gal extension of β-(1,4) galactan as well as to terminate the galactan side chain by the addition of Ara*p* (Laursen et al. 2018). The bifunctionality of AtGALS1 suggests that a dedicated GT for each glycosyl linkage may not necessarily be required for the biosynthesis of structurally diverse polysaccharides. The authors of this study proposed that, based on the availability of substrates, complex cell wall polysaccharides could potentially be biosynthesised by fewer enzymes than initially anticipated (Mohnen 2008; Caffall and Mohnen 2009). More studies on diverse GTs are required to test this proposal.

RG-I:Acetyltransferase (RG-I:AT): RG-I acetylation studies conducted using microsomal membranes from suspension-cultured potato cells identified RG-I:AT activity (Pauly and Scheller 2000). Based on differential gene expression, mutant studies and composition analysis data, TRICHOME BIREFRINGENCE LIKE 10 (TBL10, At3g06080) from *Arabidopsis thaliana* was proposed as a putative RG-I O-AT (Stranne et al. 2018). However, biochemical data are required to confirm this hypothesis.

RG-I:Methyltransferase (RG-I:MeT): Some evidence has been presented that RG-I may undergo methylation (An et al. 1994). The RG-I:MeTs and the exact positions of methylesterification or methyletherification remain to be identified.

CURRENT PROPOSED MINIMUM SET OF PECTIC POLYMERS BASED ON STRUCTURAL AND BIOSYNTHETIC DATA

Based on our current knowledge of pectin structure and the products synthesised by the GAUTs, we propose a minimum set of five HG-containing glycans, heteroglycans and glycoconjugates in the plant cell wall (Figure 5.7). We believe this number may be low and think it likely that additional complex heteroglycans exist.

1. *HG:* We know HG can be synthesised *in vitro* as a pure homopolymer by all the enzymatically active GAUTs identified (Amos et al. 2018; Engle et al. 2022). We do not know if this homopolymer exists to any considerable extent as a homopolymer in the wall, or rather, if it exists as a heteropolymer.
2. *HG:RG-II:HG*: We know that RG-II has "HG tails" and that KD expression of GAUT4 reduces both HG and RG-II by >65% in poplar and switchgrass (Biswal et al. 2018a). Thus, we know that an HG:RG-II:HG heteropolymer exists. We do not know if it exists as shown, or rather, as a part of a larger heteroglycan (i.e. with RG-I).
3. *HG:RG-I:HG:* We know that RG-I can be cleaved from the wall by EPG, and there are some data showing RG-I linked at the non-reducing and the reducing end to HG. We do not know if it exists as shown, or rather, as part of a larger heteroglycan (i.e. with RG-II) or part of a pectic AGP.
4. *Pectic AGP:* We know that APAP1 exists with the arabinogalactan portion elongated by RG-I, with the RG-I attached to HG, and with xylan attached to both the RG-I and the arabinogalactan portion.
5. *HG required for xylan synthesis:* Studies of *GAUT12* KD and overexpression poplar, as well as *gaut12/irx8* Arabidopsis mutants, show reduced (in the KDs and mutants) or increased (in the overexpression lines) HG. Without GAUT12-synthesised HG, xylan levels are reduced. We have results from one unpublished study that HG:GalAT activity is reduced in microsomes from *gaut12* mutants. However, we have not been able to obtain *in vitro* enzyme activity for GAUT12. Thus, this structure remains hypothetical, but our combined data lead us to list it as minimum structure 5.

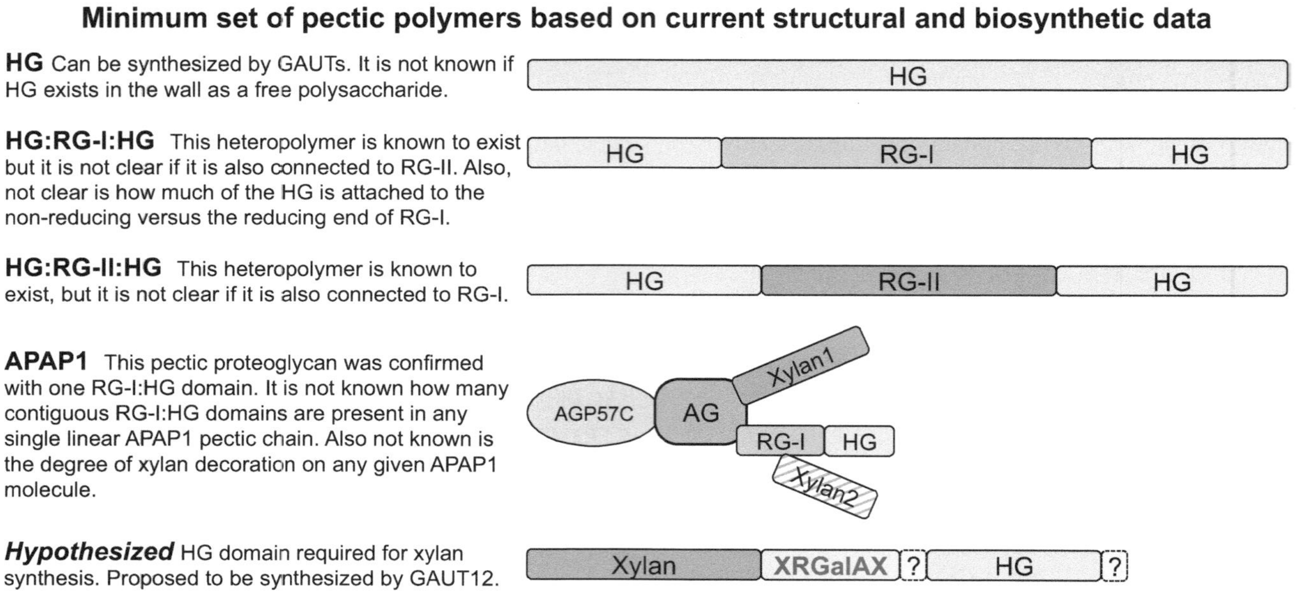

FIGURE 5.7 Representations of different possible pectin domain organisations in different heteroglycans and glycoconjugates. The top four structures are known to exist. The bottom structure is hypothesised (Biswal et al. 2018b).

CONCLUSIONS: CHALLENGES AND FUTURE DIRECTIONS IN STUDYING PECTIN STRUCTURE, SYNTHESIS AND FUNCTION

We conclude that future pectin research must include the development of methods to interrogate pectin structure both within the endomembrane system and in the wall. This will require development and use of multiple advanced nanoscale imaging techniques (Voiniciuc et al. 2018b; Haas et al. 2020) (see advanced imaging in Chapter 6) including the utilisation of anti-cell wall antibodies and carbohydrate-binding domains (Pattathil et al. 2015; Kračun et al. 2017; Perzon et al. 2020) to probe the architecture of the wall (see Chapter 10 for monoclonal antibody techniques). Also needed are microanalytics to determine cell-specific wall structures and their modifications in response to environmental and developmental changes. *In vitro* reconstitution approaches using pure and characterised biosynthetic enzymes to probe the boundaries of what wall biosynthetic catalysts can produce will be necessary to confirm structural studies. Structural and biosynthetic data are essential to enable robust interpretations of molecular genetic experiments and the formulation of rigorous testable hypotheses for pectin structure/function relationships based on the phenotypes of pectin mutants. Without understanding the diverse and different populations of pectic glycans and heteroglycans in any given cell and/or tissue, and knowing which one(s) is/are affected in a given mutant, biologically meaningful interpretations of results from molecular genetic experiments will be restricted and potentially wrong. In summary, an understanding of pectin structure is essential to determine the roles of pectin in cell wall architecture, plant growth, development and response to biological and environmental stresses.

ACKNOWLEDGEMENTS

This review was supported in part by funding from the U.S Department of Energy, Office of Science, Basic Energy Sciences, Chemical Sciences, Geosciences and Biosciences Division, under award #DE-SC0015662 and from The Center for Bioenergy Innovation, a U.S. Department of Energy Research Center supported by the Office of Biological and Environmental Research in the DOE Office of Science. We thank Peter Ulvskov for careful and critical reading of the manuscript and helpful comments. We also thank our colleagues at the CCRC and all the cell wall enthusiasts for innumerable discussions that continue to fertilise and challenge our thinking.

REFERENCES

Abdel-Massih RM, Baydoun EAH, Brett CT (2003) *In vitro* biosynthesis of 1,4-β-galactan attached to a pectin-xyloglucan complex in pea. *Planta* 216:502–511.

Abdel-Massih RM, Rizkallah HD, Al-Din RS, Baydoun EA, Brett CT (2007) Nascent pectin formed in Golgi apparatus of pea epicotyls by addition of uronic acids has different properties from nascent pectin at the stage of galactan elongation. *Journal of Plant Physiology* 164:1–10.

Albersheim P (1975) The walls of growing plant cells. *Scientific American* 232:80–95.

Albersheim P, Bauer WD, Keegstra K, Talmadge KW (1973) The structure of the wall of suspension-cultured sycamore cells. In *Biogenesis of Plant Cell Wall Polysaccharides*, ed. Loewus F, 117–147. New York: Academic Press.

Albersheim P, Darvill AG, O'Neill MA, Schols HA, Voragen AGJ (1996) An hypothesis: the same six polysaccharides are components of the primary cell walls of all higher plants. In *Pectins and Pectinases*, eds. Visser J and Voragen AGJ, 47–55. Amsterdam: Elsevier.

Amos RA, Atmodjo MA, Huang C, Gao Z, Venkat A, Taujale R, Kannan N, Moremen KW, Mohnen D (2022) Polymerization of the backbone of the pectic polysaccharide rhamnogalacturonan I. *Nature Plants* 8:1289–1303.

Amos RA, Mohnen D (2019) Critical review of plant cell wall matrix polysaccharide glycosyltransferase activities verified by heterologous protein expression. *Frontiers of Plant Science* 10:915.

Amos RA, Pattathil S, Yang JY, Atmodjo MA, Urbanowicz BR, Moremen KW, Mohnen D (2018) A two-phase model for the non-processive biosynthesis of homogalacturonan polysaccharides by the GAUT1:GAUT7 complex. *Journal of Biological Chemistry* 293:19047–19063.

An J, O'Neill MA, Albersheim P, Darvill AG (1994) Isolation and structural characterization of β-D-glucosyluronic acid and 4-O-methyl β-D-glucosyluronic acid-containing oligosaccharides from the cell-wall pectic polysaccharide, rhamnogalacturonan I. *Carbohydrate Research* 252:235–243.

Anderson CT (2016) We be jammin': an update on pectin biosynthesis, trafficking and dynamics. *Journal of Experimental Botany* 67:495–502.

Aspinall GO (1973) Carbohydrate polymers of plant cell walls. In *Biogenesis of Plant Cell Wall Polysaccharides*, ed. Loewus F, 95–115. New York: Academic Press.

Aspinall GO, Begbie R, Hamilton A, Whyte JN (1967a) Polysaccharides of soy-beans. III. Extraction and fractionation of polysaccharides from cotyledon meal. *Journal of the Chemical Society C* 11:1065–1070.

Aspinall GO, Cottrell IW, Egan SV, Morrison IM, Whyte JN (1967b) Polysaccharides of soy-beans. IV. Partial hydrolysis of the acidic polysaccharide complex from cotyledon meal. *Journal of the Chemical Society C* 11:1071–1080.

Aspinall GO, Cottrell IW, Molloy JA, Uddin M (1970) Lemon-peel pectin. III. Fractionation of pectic acids from lemon-peel and lucerne. *Canadian Journal of Chemistry* 48:1290–1295.

Aspinall GO, Hunt K, Morrison IM (1967c) Polysaccharides of soy-beans. V. Acidic polysaccharides from the hulls. *Journal of the Chemical Society C* 11:1080–1086.

Atmodjo MA, Hao Z, Mohnen D (2013) Evolving views of pectin biosynthesis. *Annual Review of Plant Biology* 64:747–779.

Atmodjo MA, Sakuragi Y, Zhu X, Burrell AJ, Mohanty SS, Atwood JA, 3rd, Orlando R, Scheller HV, Mohnen D (2011) Galacturonosyltransferase (GAUT)1 and GAUT7 are the core of a plant cell wall pectin biosynthetic homogalacturo nan:galacturonosyltransferase complex. *Proceedings of the National Academy of Sciences USA* 108:20225–20230.

Bar-Peled M, Urbanowicz BR, O'Neill MA (2012) The synthesis and origin of the pectic polysaccharide rhamnogalacturonan II - insights from nucleotide sugar formation and diversity. *Frontiers in Plant Science* 3:92.

Barrett AJ, Northcote DH (1965) Apple fruit pectic substances. *Biochemical Journal* 94:617–627.

BeMiller JN (1986) An introduction to pectins: structure and properties. In *Chemistry and Function of Pectins*, eds. Fishman ML and Jen JJ, 2–12. Washington, DC: American Chemical Society.

Biswal AK, Atmodjo MA, Li M, Baxter HL, Yoo CG, Pu Y, Lee YC, Mazarei M, Black IM, Zhang JY, Ramanna H, Bray AL, King ZR, LaFayette PR, Pattathil S, Donohoe BS, Mohanty SS, Ryno D, Yee K, Thompson OA, Rodriguez M, Jr., Dumitrache A, Natzke J, Winkeler K, Collins C, Yang X, Tan L, Sykes RW, Gjersing EL, Ziebell A, Turner GB, Decker SR, Hahn MG, Davison BH, Udvardi MK, Mielenz JR, Davis MF, Nelson RS, Parrott WA, Ragauskas AJ, Neal Stewart C, Jr., Mohnen D (2018a) Sugar release and growth of biofuel crops are improved by downregulation of pectin biosynthesis. *Nature Biotechnology* 36:249–257.

Biswal AK, Atmodjo MA, Pattathil S, Amos RA, Yang X, Winkeler K, Collins C, Mohanty SS, Ryno D, Tan L, Gelineo-Albersheim I, Hunt K, Sykes RW, Turner GB, Ziebell A, Davis MF, Decker SR, Hahn MG, Mohnen D (2018b) Working towards recalcitrance mechanisms: increased xylan and homogalacturonan production by overexpression of GAlactUronosylTransferase12 (GAUT12) causes increased recalcitrance and decreased growth in *Populus*. *Biotechnology for Biofuels* 11:9.

Biswal AK, Hao Z, Pattathil S, Yang X, Winkeler K, Collins C, Mohanty SS, Richardson EA, Gelineo-Albersheim I, Hunt K, Ryno D, Sykes RW, Turner GB, Ziebell A, Gjersing E, Lukowitz W, Davis MF, Decker SR, Hahn MG, Mohnen D (2015) Downregulation of *GAUT12* in *Populus deltoides* by RNA silencing results in reduced recalcitrance, increased growth and reduced xylan and pectin in a woody biofuel feedstock. *Biotechnology for Biofuels* 8:41.

Bolwell GP, Northcote DH (1983) Arabinan synthase and xylan synthase activities of *Phaseolus vulgaris*. Subcellular localization and possible mechanism of action. *Biochemical Journal* 210:497–507.

Bouton S, Leboeuf E, Mouille G, Leydecker MT, Talbotec J, Granier F, Lahaye M, Höfte H, Truong HN (2002) *QUASIMODO1* encodes a putative membrane-bound glycosyltransferase required for normal pectin synthesis and cell adhesion in Arabidopsis. *Plant Cell* 14:2577–2590.

Braconnot H. (1825) Recherches sur un nouvel acide universellement répandu dans tous les végétaux. *Annales de Chimie et de Physique* 28:173–178.

Brett CT, Baydoun E-H, Abdel-Massih RM (2005) Pectin-xyloglucan linkages in type I primary cell walls of plants. *Plant Biosystems* 139:54–59.

Brown JA, Fry SC (1993) Novel O-D-galacturonoyl esters in the pectic polysaccharides of suspension-cultured plant cells. *Plant Physiology* 103:993–999.

Caffall KH, Mohnen D (2009) The structure, function, and biosynthesis of plant cell wall pectic polysaccharides. *Carbohydrate Research* 344:1879–1900.

Caffall KH, Pattathil S, Phillips SE, Hahn MG, Mohnen D (2009) *Arabidopsis thaliana* T-DNA mutants implicate *GAUT* genes in the biosynthesis of pectin and xylan in cell walls and seed testa. *Molecular Plant* 2:1000–1014.

Chambat G, Barnoud F, Joseleau JP (1984) Structure of the primary cell walls of suspension-cultured *Rosa glauca* cells: I. Polysaccharides associated with cellulose. *Plant Physiology* 74:687–693.

Chambat G, Joseleau JP (1980) Isolation and characterization of a homogalacturonan in the primary walls of *rosa* cells cultures *in vitro*. *Carbohydrate Research* 85:C10–C12.

Chiniquy D, Underwood W, Corwin J, Ryan A, Szemenyei H, Lim CC, Stonebloom SH, Birdseye DS, Vogel J, Kliebenstein D, Scheller HV, Somerville S (2019) PMR5, an acetylation protein at the intersection of pectin biosynthesis and defense against fungal pathogens. *Plant Journal* 100:1022–1035.

Clarke A, Anderson R, Stone B (1979) Form and function of arabinogalactans and arabinogalactan-proteins. *Phytochemistry* 18:521–540.

Coenen G, Bakx E, Verhoef R, Schols H, Voragen A (2007) Identification of the connecting linkage between homo- or xylogalacturonan and rhamnogalacturonan type I. *Carbohydrate Polymers* 70:224–235.

Cumming CM, Rizkallah HD, McKendrick KA, Abdel-Massih RM, Baydoun EA, Brett CT (2005) Biosynthesis and cell-wall deposition of a pectin-xyloglucan complex in pea. *Planta* 222:546–555.

Daas PJ, Voragen AG, Schols HA (2000) Characterization of non-esterified galacturonic acid sequences in pectin with endo-polygalacturonase. *Carbohydrate Research* 326:120–129.

Darvill AG, McNeil M, Albersheim P (1978) Structure of plant cell walls: VIII. A new pectic polysaccharide. *Plant Physiology* 62:418–422.

De Vries JA, Rombouts FM, Voragen AGJ, Pilnik W (1982) Enzymic degradation of apple pectins. *Carbohydrate Polymers* 2:25–33.

De Vries JA, Voragen AGJ, Rombouts FM, Pilnik W (1981) Extraction and purification of pectins from Alcohol Insoluble Solids from ripe and unripe apples. *Carbohydrate Polymers* 1:117–127.

Doong RL, Liljebjelke K, Fralish G, Kumar A, Mohnen D (1995) Cell-free synthesis of pectin (identification and partial characterization of polygalacturonate 4-α-galacturonosyltransferase and its products from membrane preparations of tobacco cell-suspension cultures). *Plant Physiology* 109:141–152.

Doong RL, Mohnen D (1998) Solubilization and characterization of a galacturonosyltransferase that synthesizes the pectic polysaccharide homogalacturonan. *The Plant Journal* 13:363–374.

Driouich A, Follet-Gueye ML, Bernard S, Kousar S, Chevalier L, Vicré-Gibouin M, Lerouxel O (2012) Golgi-mediated synthesis and secretion of matrix polysaccharides of the primary cell wall of higher plants. *Frontiers in Plant Science* 3:79.

Du J, Kirui A, Huang S, Wang L, Barnes WJ, Kiemle SN, Zheng Y, Rui Y, Ruan M, Qi S, Kim SH, Wang T, Cosgrove DJ, Anderson CT, Xiao C (2020) Mutations in the pectin methyltransferase QUASIMODO2 influence cellulose biosynthesis and wall integrity in Arabidopsis. *Plant Cell* 32:3576–3597.

Ebert B, Birdseye D, Liwanag AJM, Laursen T, Rennie EA, Guo X, Catena M, Rautengarten C, Stonebloom SH, Gluza P, Pidatala VR, Andersen MCF, Cheetamun R, MortiVmer JC, Heazlewood JL, Bacic A, Clausen MH, Willats WGT, Scheller HV (2018) The three members of the Arabidopsis Glycosyltransferase Family 92 are functional β-1,4-galactan synthases. *Plant and Cell Physiology* 59:2624–2636.

Egelund J, Damager I, Faber K, Olsen C-E, Ulvskov P, Petersen BL (2008) Functional characterisation of a putative rhamnogalacturonan II specific xylosyltransferase. *FEBS Letters* 582:3217–3222.

Egelund J, Petersen BL, Motawia MS, Damager I, Faik A, Olsen CE, Ishii T, Clausen H, Ulvskov P, Geshi N (2006) *Arabidopsis thaliana RGXT1* and *RGXT2* encode Golgi-localized (1,3)-α-D-xylosyltransferases involved in the synthesis of pectic rhamnogalacturonan-II. *Plant Cell* 18:2593–2607.

Ehrlich F (1917) Die pektinstoffe, ihre konstitution und bedeutung. *Chemiker Zeitung* 41:197–200.

Engle KA, Amos RA, Yang JY, Glushka J, Atmodjo MA, Tan L, Huang C, Moremen KW, Mohnen D (2022) Multiple Arabidopsis galacturonosyltransferases synthesize polymeric homogalacturonan by oligosaccharide acceptor-dependent or *de novo* synthesis. *Plant Journal* 109:1441–1456.

Femenia A, Rigby NM, Selvendran RR, Waldron KW (1999) Investigation of the occurrence of pectic-xylan-xyloglucan complexes in the cell walls of cauliflower stem tissues. *Carbohydrate Polymers* 39:151–164.

Fincher GB, Stone BA, Clarke AE (1983) Arabinogalactan-proteins: structure, biosynthesis, and function. *Annual Review of Plant Physiology* 34:47–70.

Fry SC (1982) Phenolic components of the primary cell wall. Feruloylated disaccharides of D-galactose and L-arabinose from spinach polysaccharide. *Biochemical Journal* 203:493–504.

Fry SC (1983) Feruloylated pectins from the primary cell wall: their structure and possible functions. *Planta* 157:111–123.

Gao QP, Kiyohara H, Yamada H (1990) Further structural studies of anti-complementary acidic heteroglycans from the leaves of *Panax ginseng* C.A. Meyer. *Carbohydrate Research* 196:111–125.

Gelineo-Albersheim I, Darvill A, Albersheim P. (2001) Structural studies of pectin. In *Cell Walls' 01: 9th International Cell Wall Meeting, Toulouse.* Auzeville, France: Université Paul Sabatier, 183. *(Abstract).*

Geshi N, Jørgensen B, Scheller HV, Ulvskov P (2000) In vitro biosynthesis of 1,4-β-galactan attached to rhamnogalacturonan I. *Planta* 210:622–629.

Geshi N, Jorgensen B, Ulvskov P (2004) Subcellular localization and topology of β(1→4)galactosyltransferase that elongates β(1→4)galactan side chains in rhamnogalacturonan I in potato. *Planta* 218:862–868.

Gibeaut DM, Carpita NC (1994) Biosynthesis of plant cell wall polysaccharides. *FASEB Journal* 8:904–915.

Goubet F, Mohnen D (1999) Solubilization and partial characterization of homogalacturonan-methyltransferase from microsomal membranes of suspension-cultured tobacco cells. *Plant Physiology* 121:281–290.

Grant GT (1973) Biological interactions between polysaccharides and divalent cations: the egg-box model. *FEBS Letters* 32:195–198.

Gu Y, Rasmussen CG (2022) Cell biology of primary cell wall synthesis in plants. *The Plant Cell* 34:103–128.

Haas KT, Wightman R, Meyerowitz EM, Peaucelle A (2020) Pectin homogalacturonan nanofilament expansion drives morphogenesis in plant epidermal cells. *Science* 367:1003–1007.

Hao Z, Avci U, Tan L, Zhu X, Glushka J, Pattathil S, Eberhard S, Sholes T, Rothstein GE, Lukowitz W, Orlando R, Hahn MG, Mohnen D (2014) Loss of Arabidopsis *GAUT12/IRX8* causes anther indehiscence and leads to reduced G lignin associated with altered matrix polysaccharide deposition. *Frontiers in Plant Science* 5:357.

Hao Z, Mohnen D (2014) A review of xylan and lignin biosynthesis: foundation for studying Arabidopsis *irregular xylem* mutants with pleiotropic phenotypes. *Critical Reviews in Biochemistry and Molecular Biology* 49:212–241.

Harholt J, Jensen JK, Sorensen SO, Orfila C, Pauly M, Scheller HV (2006) ARABINAN DEFICIENT 1 is a putative arabinosyltransferase involved in biosynthesis of pectic arabinan in Arabidopsis. *Plant Physiology* 140:49–58.

Harholt J, Jensen JK, Verhertbruggen Y, Søgaard C, Bernard S, Nafisi M, Poulsen CP, Geshi N, Sakuragi Y, Driouich A, Knox JP, Scheller HV (2012) ARAD proteins associated with pectic Arabinan biosynthesis form complexes when transiently overexpressed *in planta*. *Planta* 236:115–128.

Harholt J, Suttangkakul A, Vibe Scheller H (2010) Biosynthesis of pectin. *Plant Physiology* 153:384–395.

Held MA, Be E, Zemelis S, Withers S, Wilkerson C, Brandizzi F (2011) CGR3: a Golgi-localized protein influencing homogalacturonan methylesterification. *Molecular Plant* 4:832–844.

Hervé du Penhoat C, Gey C, Pellerin P, Perez S (1999) An NMR solution study of the mega-oligosaccharide, rhamnogalacturonan II. *Journal of Biomolecular NMR* 14:253–271.

Hijazi M, Velasquez SM, Jamet E, Estevez JM, Albenne C (2014) An update on post-translational modifications of hydroxyproline-rich glycoproteins: toward a model highlighting their contribution to plant cell wall architecture. *Frontiers in Plant Science* 5:395.

Hinz SW, Verhoef R, Schols HA, Vincken J-P, Voragen AG (2005) Type I arabinogalactan contains β-D-Galp-(1→3)-β-D-Galp structural elements. *Carbohydrate Research* 340:2135–2143.

Ishii T (1997) O-Acetylated oligosaccharides from pectins of potato tuber cell walls. *Plant Physiology* 113:1265–1272.

Ishii T, Matsunaga T, Pellerin P, O'Neill MA, Darvill A, Albersheim P (1999) The plant cell wall polysaccharide rhamnogalacturonan II self-assembles into a covalently cross-linked dimer. *Journal of Biological Chemistry* 274:13098–13104.

Ishii T, Thomas J, Darvill A, Albersheim P (1989) Structure of plant cell walls: XXVI. The walls of suspension-cultured sycamore cells contain a family of rhamnogalacturonan-I-like pectic polysaccharides. *Plant Physiology* 89:421–428.

Jensen JK, Sørensen SO, Harholt J, Geshi N, Sakuragi Y, Møller I, Zandleven J, Bernal AJ, Jensen NB, Sørensen C, Pauly M, Beldman G, Willats WGT, Scheller HV (2008) Identification of a xylogalacturonan xylosyltransferase involved in pectin biosynthesis in Arabidopsis. *Plant Cell* 20:1289–1302.

Kaczmarska A, Pieczywek PM, Cybulska J, Zdunek A (2022) Structure and functionality of rhamnogalacturonan I in the cell wall and in solution: a review. *Carbohydrate Polymers* 278:118909.

Kauss H, Swanson AL, Arnold R, Odzuck W (1969) Biosynthesis of pectic substances Localisation of enzymes and products in a lipid-membrane complex. *Biochimica et Biophysica Acta (BBA) - General Subjects* 192:55–61.

Keegstra K, Talmadge KW, Bauer WD, Albersheim P (1973) The structure of plant cell walls: III. A model of the walls of suspension-cultured sycamore cells based on the interconnections of the macromolecular components. *Plant Physiology* 51:188–197.

Kertesz ZI (1951) *The Pectic Substances*. New York: Interscience Publishers.

Knox JP (2008) Revealing the structural and functional diversity of plant cell walls. *Current Opinion in Plant Biology* 11:308–313.

Kobayashi M, Matoh T, Azuma J (1996) Two chains of rhamnogalacturonan II are cross-linked by borate-diol ester bonds in higher plant cell walls. *Plant Physiology* 110:1017–1020.

Kobayashi M, Nakagawa H, Asaka T, Matoh T (1999) Borate-rhamnogalacturonan II bonding reinforced by Ca^{2+} retains pectic polysaccharides in higher-plant cell walls. *Plant Physiology* 119:199–204.

Komalavilas P, Mort AJ (1989) The acetylation of O-3 of galacturonic acid in the rhamnose-rich portion of pectins. *Carbohydrate Research* 189:261–272.

Kračun SK, Fangel JU, Rydahl MG, Pedersen HL, Vidal-Melgosa S, Willats WGT (2017) Carbohydrate microarray technology applied to high-throughput mapping of plant cell wall glycans using comprehensive microarray polymer profiling (CoMPP). In *High-Throughput Glycomics and Glycoproteomics*, eds. Lauc G and Wuhrer M, 147–165. New York: Humana Press.

Lamport DTA (1970) Cell wall metabolism. *Annual Review of Plant Physiology* 21:235–270.

Larsen K, Thygesen MB, Guillaumie F, Willats WG, Jensen JK (2006) Solid-phase chemical tools for glycobiology. *Carbohydrate Research* 341:1209–1234.

Lau JM, McNeil M, Darvill AG, Albersheim P (1985) Structure of the backbone of rhamnogalacturonan I, a pectic polysaccharide in the primary cell walls of plants. *Carbohydrate Research* 137:111–125.

Lau JM, McNeil M, Darvill AG, Albersheim P (1987) Treatment of rhamnogalacturonan I with lithium in ethylenediamine. *Carbohydrate Research* 168:245–274.

Laursen T, Stonebloom SH, Pidatala VR, Birdseye DS, Clausen MH, Mortimer JC, Scheller HV (2018) Bifunctional glycosyltransferases catalyze both extension and termination of pectic galactan oligosaccharides. *Plant Journal* 94:340–351.

Lerouge P, O'Neill MA, Darvill AG, Albersheim P (1993) Structural characterization of endo-glycanase-generated oligoglycosyl side chains of rhamnogalacturonan I. *Carbohydrate Research* 243:359–371.

Liu XL, Liu L, Niu QK, Xia C, Yang KZ, Li R, Chen LQ, Zhang XQ, Zhou Y, Ye D (2011) *MALE GAMETOPHYTE DEFECTIVE 4* encodes a rhamnogalacturonan II xylosyltransferase and is important for growth of pollen tubes and roots in Arabidopsis. *Plant Journal* 65:647–660.

Liwanag AJM, Ebert B, Verhertbruggen Y, Rennie EA, Rautengarten C, Oikawa A, Andersen MCF, Clausen MH, Scheller HV (2012) Pectin biosynthesis: GALS1 in *Arabidopsis thaliana* is a β-1,4-galactan β-1,4-galactosyltransferase. *Plant Cell* 24:5024–5036.

Lund CH, Stenbæk A, Atmodjo MA, Rasmussen RE, Moller IE, Erstad SM, Biswal AK, Mohnen D, Mravec J, Sakuragi Y (2020) Pectin synthesis and pollen tube growth in Arabidopsis involves three GAUT1 Golgi-anchoring proteins: GAUT5, GAUT6, and GAUT7. *Frontiers in Plant Science* 11:585774.

Manabe Y, Verhertbruggen Y, Gille S, Harholt J, Chong SL, Pawar PM, Mellerowicz EJ, Tenkanen M, Cheng K, Pauly M, Scheller HV (2013) Reduced wall acetylation proteins play vital and distinct roles in cell wall O-acetylation in Arabidopsis. *Plant Physiology* 163:1107–1117.

Matoh T, Ishigaki K-i, Ohno K, Azuma J-i (1993) Isolation and characterization of a boron-polysaccharide complex from radish roots. *Plant and Cell Physiology* 34:639–642.

Matoh T, Kawaguchi S, Kobayashi M (1996) Ubiquity of a borate-rhamnogalacturonan II complex in the cell walls of higher plants. *Plant and Cell Physiology* 37:636–640.

McNab JM, Villemez CL, Albersheim P (1968) Biosynthesis of galactan by a particulate enzyme preparation from *Phaseolus aureus* seedlings. *Biochemical Journal* 106:355–360.

McNeil M, Darvill AG, Albersheim P (1980) Structure of plant cell walls: X. Rhamnogalacturonan I, a structurally complex pectic polysaccharide in the walls of suspension-cultured sycamore cells. *Plant Physiology* 66:1128–1134.

McNeil M, Darvill AG, Fry SC, Albersheim P (1984) Structure and function of the primary cell walls of plants. *Annual Review in Biochemistry* 53:625–663.

Melton LD, McNeil M, Darvill AG, Albersheim P, Dell A (1986) Structural characterization of oligosaccharides isolated from the pectic polysaccharide rhamnogalacturonan II. *Carbohydrate Research* 146:279–305.

Miao Y, Li H-Y, Shen J, Wang J, Jiang L (2011) QUASIMODO 3 (QUA3) is a putative homogalacturonan methyltransferase regulating cell wall biosynthesis in Arabidopsis suspension-cultured cells. *Journal of Experimental Botany* 62:5063–5078.

Mohnen D (1999) Biosynthesis of pectins and galactomannans. In *Comprehensive Natural Products Chemistry Vol. 3. Carbohydrates and Their Derivatives Including Tannins, Cellulose, and Related Lignins*, ed. Pinto BM, 497–527. Oxford: Elsevier.

Mohnen D (2002) Biosynthesis of pectins. In *Pectins and Their Manipulation*, eds. Seymour GB and Knox JP, 52–98. Oxford: Blackwell Publishing.

Mohnen D (2008) Pectin structure and biosynthesis. *Current Opinion in Plant Biology* 11:266–277.

Mohnen D, Doong RL, Liljebjelke K, Fralish G, Chan J (1996) Cell free synthesis of the pectic polysaccharide homogalacturonan. In *Progress in Biotechnology*, eds. Visser J and Voragen AGJ, 109–126. Wageningen: Elsevier.

Mouille G, Ralet MC, Cavelier C, Eland C, Effroy D, Hématy K, McCartney L, Truong HN, Gaudon V, Thibault JF, Marchant A, Höfte H (2007) Homogalacturonan synthesis in *Arabidopsis thaliana* requires a Golgi-localized protein with a putative methyltransferase domain. *Plant Journal* 50:605–614.

Nakamura A, Furuta H, Maeda H, Nagamatsu Y, Yoshimoto A (2001) Analysis of structural components and molecular construction of soybean soluble polysaccharides by stepwise enzymatic degradation. *Bioscience, Biotechnology, and Biochemistry* 65:2249–2258.

Naseri AT, Thibault J-F, Ralet-Renard M-C (2008) Citrus pectin: structure and application in acid dairy drinks. In *Tree and Forestry Science and Biotechnology 2, Special Issue 1*. eds. Benkeblia N, Tennant, P, 60–70. Boca Raton, FL: Global Science Books.

Ndeh D, Rogowski A, Cartmell A, Luis AS, Baslé A, Gray J, Venditto I, Briggs J, Zhang X, Labourel A, Terrapon N, Buffetto F, Nepogodiev S, Xiao Y, Field RA, Zhu Y, O'Neill MA, Urbanowicz BR, York WS, Davies GJ, Abbott DW, Ralet M-C, Martens EC, Henrissat B, Gilbert HJ (2017) Complex pectin metabolism by gut bacteria reveals novel catalytic functions. *Nature* 544:65–70.

Nikolovski N, Rubtsov D, Segura MP, Miles GP, Stevens TJ, Dunkley TP, Munro S, Lilley KS, Dupree P (2012) Putative glycosyltransferases and other plant Golgi apparatus proteins are revealed by LOPIT proteomics. *Plant Physiology* 160:1037–1051.

Nunan KJ, Scheller HV (2003) Solubilization of an arabinan arabinosyltransferase activity from mung bean hypocotyls. *Plant Physiology* 132:331–342.

Øbro J, Harholt J, Scheller HV, Orfila C (2004) Rhamnogalacturonan I in *Solanum tuberosum* tubers contains complex arabinogalactan structures. *Phytochemistry* 65:1429–1438.

O'Neill M, Albersheim P, Darvill A (1990) The pectic polysaccharides of primary cell walls. In *Methods in Plant Biochemistry*, Vol. 2, ed. Dey PM, 415–441. London: Academic Press Limited.

Ohashi T, Jinno J, Inoue Y, Ito S, Fujiyama K, Ishimizu T (2017) A polygalacturonase localized in the Golgi apparatus in *Pisum sativum*. *Journal of Biochemistry* 162:193–201.

Oomen RJ, Doeswijk-Voragen CH, Bush MS, Vincken JP, Borkhardt B, van den Broek LA, Corsar J, Ulvskov P, Voragen AG, McCann MC, Visser RG (2002) *In muro* fragmentation of the rhamnogalacturonan I backbone in potato (*Solanum tuberosum* L.) results in a reduction and altered location of the galactan and arabinan side-chains and abnormal periderm development. *Plant Journal* 30:403–413.

Orfila C, Sørensen SO, Harholt J, Geshi N, Crombie H, Truong HN, Reid JS, Knox JP, Scheller HV (2005) *QUASIMODO1* is expressed in vascular tissue of *Arabidopsis thaliana* inflorescence stems, and affects homogalacturonan and xylan biosynthesis. *Planta* 222:613–622.

Oxenbøll Sørensen S, Pauly M, Bush M, Skjøt M, McCann MC, Borkhardt B, Ulvskov P (2000) Pectin engineering: modification of potato pectin by *in vivo* expression of an endo-1,4-β-D-galactanase. *Proceedings of the National Academy of Sciences USA* 97:7639–7644.

Pabst M, Fischl RM, Brecker L, Morelle W, Fauland A, Köfeler H, Altmann F, Léonard R (2013) Rhamnogalacturonan II structure shows variation in the side chains monosaccharide composition and methylation status within and across different plant species. *Plant Journal* 76:61–72.

Panayotatos N, Villemez CL (1973) The formation of a β-(1→4)-D-galactan chain catalysed by a *Phaseolus aureus* enzyme. *Biochemical Journal* 133:263–271.

Pattathil S, Avci U, Miller JS, Hahn MG (2012) Immunological approaches to plant cell wall and biomass characterization: glycome profiling. In *Biomass Conversion: Methods and Protocols*, ed. Himmel ME, 61–72. Totowa, NJ: Humana Press.

Pattathil S, Avci U, Zhang T, Cardenas CL, Hahn MG (2015) Immunological approaches to biomass characterization and utilization. *Frontiers in Bioengineering and Biotechnology* 3:173.

Pauly M, Scheller HV (2000) O-Acetylation of plant cell wall polysaccharides: identification and partial characterization of a rhamnogalacturonan O-acetyl-transferase from potato suspension-cultured cells. *Planta* 210:659–667.

Pellerin P, Doco T, Vidal S, Williams P, Brillouet JM, O'Neill MA (1996) Structural characterization of red wine rhamnogalacturonan II. *Carbohydrate Research* 290:183–197.

Peña MJ, Zhong R, Zhou GK, Richardson EA, O'Neill MA, Darvill AG, York WS, Ye ZH (2007) Arabidopsis *irregular xylem8* and *irregular xylem9*: implications for the complexity of glucuronoxylan biosynthesis. *Plant Cell* 19:549–563.

Persson S, Caffall KH, Freshour G, Hilley MT, Bauer S, Poindexter P, Hahn MG, Mohnen D, Somerville C (2007) The Arabidopsis *irregular xylem8* mutant is deficient in glucuronoxylan and homogalacturonan, which are essential for secondary cell wall integrity. *Plant Cell* 19:237–255.

Perzon A, Kračun SK, Jørgensen B, Ulvskov P (2020) Array-based microfibril surface assessment (AMSA): a method for probing surface-exposed polysaccharides on cellulose nanofibres. *Cellulose* 27:8635–8651.

Pilnik W (1990) Pectin-a many splendoured thing. In *Gums and Stabilisers for the Industry*, eds. Phillips GO, Williams PA, and Wedlock DJ, 313–326. Oxford: Oxford University Press.

Popper ZA, Fry SC (2008) Xyloglucan-pectin linkages are formed intra-protoplasmically, contribute to wall-assembly, and remain stable in the cell wall. *Planta* 227:781–794.

Posé S, Kirby AR, Mercado JA, Morris VJ, Quesada MA (2012) Structural characterization of cell wall pectin fractions in ripe strawberry fruits using AFM. *Carbohydrate Polymers* 88:882–890.

Puvanesarajah V, Darvill AG, Albersheim P (1991) Structural characterization of two oligosaccharide fragments formed by the selective cleavage of rhamnogalacturonan II: evidence for the anomeric configuration and attachment sites of apiose and 3-deoxy-2-heptulosaric acid. *Carbohydrate Research* 218:211–222.

Qi X, Behrens BX, West PR, Mort AJ (1995) Solubilization and partial characterization of extensin fragments from cell walls of cotton suspension cultures. Evidence for a covalent cross-link between extensin and pectin. *Plant Physiology* 108:1691–1701.

Ralet MC, Crepeau MJ, Vigouroux J, Tran J, Berger A, Salle C, Granier F, Botran L, North HM (2016) Xylans provide the structural driving force for mucilage adhesion to the Arabidopsis seed coat. *Plant Physiology* 171:165–178.

Rees DA (1982) Polysaccharide conformation in solutions and gels - recent results on pectins. *Carbohydrate Polymers* 2:254–263.

Renard CMGC, Crépeau M-J, Thibault J-F (1995) Structure of the repeating units in the rhamnogalacturonic backbone of apple, beet and citrus pectins. *Carbohydrate Research* 275:155–165.

Renard CMGC, Crépeau MJ, Thibault JF (1999) Glucuronic acid directly linked to galacturonic acid in the rhamnogalacturonan backbone of beet pectins. *European Journal of Biochemistry* 266:566–574.

Renard CM, Jarvis MC (1999) A cross-polarization, magic-angle-spinning, ^{13}C-nuclear-magnetic-resonance study of polysaccharides in sugar beet cell walls. *Plant Physiology* 119:1315–1322.

Renard CM, Lahaye M, Mutter M, Voragen FG, Thibault JF (1997) Isolation and structural characterisation of rhamnogalacturonan oligomers generated by controlled acid hydrolysis of sugar-beet pulp. *Carbohydrate Research* 305:271–280.

Ridley BL, O'Neill MA, Mohnen D (2001) Pectins: structure, biosynthesis, and oligogalacturonide-related signaling. *Phytochemistry* 57:929–967.

Rihouey C, Morvan C, Borissova I, Jauneau A, Demarty M, Jarvis M (1995) Structural features of CDTA-soluble pectins from flax hypocotyls. *Carbohydrate Polymers* 28:159–166.

Rombouts FM, Thibault J-F (1986a) Enzymic and chemical degradation and the fine structure of pectins from sugar-beet pulp. *Carbohydrate Research* 154:189–203.

Rombouts FM, Thibault J-F (1986b) Feruloylated pectic substances from sugar-beet pulp. *Carbohydrate Research* 154:177–187.

Ropartz D, Ralet MC (2020) Pectin structure. In *Pectin: Technological and Physiological Properties*, ed. Kontogiorgos V, 17–36. Cham: Springer International Publishing.

Schols HA, Geraeds CCJM, Searle-van Leeuwen MF, Kormelink FJM, Voragen AGJ (1990a) Rhamnogalacturonase: a novel enzyme that degrades the hairy regions of pectins. *Carbohydrate Research* 206:105–115.

Schols HA, Posthumus MA, Voragen AGJ (1990b) Structural features of hairy regions of pectins isolated from apple juice produced by the liquefaction process. *Carbohydrate Research* 206:117–129.

Schols HA, Vierhuis E, Bakx EJ, Voragen AG (1995) Different populations of pectic hairy regions occur in apple cell walls. *Carbohydrate Research* 275:343–360.

Schols HA, Voragen AGJ (1994) Occurrence of pectic hairy regions in various plant cell wall materials and their degradability by rhamnogalacturonase. *Carbohydrate Research* 256:83–95.

Selvendran RR (1985) Developments in the chemistry and biochemistry of pectic and hemicellulosic polymers. *Journal of Cell Science Supplement* 2:51–88.

Sengkhamparn N, Bakx EJ, Verhoef R, Schols HA, Sajjaanantakul T, Voragen AG (2009) Okra pectin contains an unusual substitution of its rhamnosyl residues with acetyl and α-linked galactosyl groups. *Carbohydrate Research* 344:1842–1851.

Shedletzky E, Shmuel M, Delmer DP, Lamport DT (1990) Adaptation and growth of tomato cells on the herbicide 2,6-dichlorobenzonitrile leads to production of unique cell walls virtually lacking a cellulose-xyloglucan network. *Plant Physiol* 94:980–987.

Shin Y, Chane A, Jung M, Lee Y (2021) Recent advances in understanding the roles of pectin as an active participant in plant signaling networks. *Plants* 10:1712.

Sinclair R, Hsu G, Davis D, Chang M, Rosquete M, Iwasa JH, Drakakaki G (2022) Plant cytokinesis and the construction of new cell wall. *FEBS Letters* 596:2243–2255.

Sinclair R, Rosquete MR, Drakakaki G (2018) Post-Golgi trafficking and transport of cell wall components. *Frontiers in Plant Science* 9:1784.

Skjot M, Pauly M, Bush MS, Borkhardt B, McCann MC, Ulvskov P (2002) Direct interference with rhamnogalacturonan I biosynthesis in Golgi vesicles. *Plant Physiology* 129:95–102.

Smolenski K (1923) Pectins. *Rocznigi Chemki* 3:86–152.

Spellman MW, McNeil M, Darvill AG, Albersheim P, Dell A (1983) Characterization of a structurally complex heptasaccharide isolated from the pectic polysaccharide rhamnogalacturonan II. *Carbohydrate Research* 122:131–153.

Sterling JD, Atmodjo MA, Inwood SE, Kumar Kolli VS, Quigley HF, Hahn MG, Mohnen D (2006) Functional identification of an Arabidopsis pectin biosynthetic homogalacturonan galacturonosyltransferase. *Proceedings of the National Academy of Sciences USA* 103:5236–5241.

Sterling JD, Quigley HF, Orellana A, Mohnen D (2001) The catalytic site of the pectin biosynthetic enzyme α-1,4-galacturonosyltransferase is located in the lumen of the Golgi. *Plant Physiology* 127:360–371.

Stevenson TT, Darvill AG, Albersheim P (1988a) 3-deoxy-D-lyxo-2-heptulosaric acid, a component of the plant cell-wall polysaccharide rhamnogalacturonan-II. *Carbohydrate Research* 179:269–288.

Stevenson TT, Darvill AG, Albersheim P (1988b) Structural features of the plant cell-wall polysaccharide rhamnogalacturonan-II. *Carbohydrate Research* 182:207–226.

Stranne M, Ren Y, Fimognari L, Birdseye D, Yan J, Bardor M, Mollet JC, Komatsu T, Kikuchi J, Scheller HV, Sakuragi Y (2018) TBL10 is required for O-acetylation of pectic rhamnogalacturonan-I in *Arabidopsis thaliana*. *Plant Journal* 96:772–785.

Sun HH, Wooten JB, Ryan WS, Bokelman GH, Åman P (1987) Structural characterization of a tobacco rhamnogalacturonan. *Carbohydrate Polymers* 7:143–158.

Takenaka Y, Kato K, Ogawa-Ohnishi M, Tsuruhama K, Kajiura H, Yagyu K, Takeda A, Takeda Y, Kunieda T, Hara-Nishimura I, Kuroha T, Nishitani K, Matsubayashi Y, Ishimizu T (2018) Pectin RG-I rhamnosyltransferases represent a novel plant-specific glycosyltransferase family. *Nature Plants* 4:669–676.

Talmadge KW, Keegstra K, Bauer WD, Albersheim P (1973) The structure of plant cell walls: I. The macromolecular components of the walls of suspension-cultured sycamore cells with a detailed analysis of the pectic polysaccharides. *Plant Physiology* 51:158–173.

Tan L, Eberhard S, Pattathil S, Warder C, Glushka J, Yuan C, Hao Z, Zhu X, Avci U, Miller JS, Baldwin D, Pham C, Orlando R, Darvill A, Hahn MG, Kieliszewski MJ, Mohnen D (2013) An Arabidopsis cell wall proteoglycan consists of pectin and arabinoxylan covalently linked to an arabinogalactan protein. *The Plant Cell* 25:270–287.

Tan L, Zhang L, Black I, Glushka J, Urbanowicz B, Heiss C, Azadi P (2023) Most of the rhamnogalacturonan-I from cultured Arabidopsis cell walls is covalently linked to arabinogalactan-protein. *Carbohydrate Polymers* 301:120340.

Thibault JF, Renard CMGC, Axelos MAV, Roger P, Crépeau M-J (1993) Studies of the length of homogalacturonic regions in pectins by acid hydrolysis. *Carbohydrate Research* 238:271–286.

Thomas JR, Darvill AG, Albersheim P (1989) Rhamnogalacturonan I, a pectic polysaccharide that is a component of monocot cell-walls. *Carbohydrate Research* 185:279–305.

Thompson JE, Fry SC (2000) Evidence for covalent linkage between xyloglucan and acidic pectins in suspension-cultured rose cells. *Planta* 211:275–286.

Vidal S, Doco T, Williams P, Pellerin P, York WS, O'Neill MA, Glushka J, Darvill AG, Albersheim P (2000) Structural characterization of the pectic polysaccharide rhamnogalacturonan II: evidence for the backbone location of the aceric acid-containing oligoglycosyl side chain. *Carbohydrate Research* 326:277–294.

Vincken JP, Schols HA, Oomen RJFJ, McCann MC, Ulvskov P, Voragen AGJ, Visser RGF (2003) If homogalacturonan were a side chain of rhamnogalacturonan I. Implications for cell wall architecture. *Plant Physiology* 132:1781–1789.

Voiniciuc C, Engle KA, Günl M, Dieluweit S, Schmidt MH, Yang JY, Moremen KW, Mohnen D, Usadel B (2018a) Identification of key enzymes for pectin synthesis in seed mucilage. *Plant Physiology* 178:1045–1064.

Voiniciuc C, Pauly M, Usadel B (2018b) Monitoring polysaccharide dynamics in the plant cell wall. *Plant Physiology* 176:2590–2600.

Voragen AGJ, Coenen G-J, Verhoef RP, Schols HA (2009) Pectin, a versatile polysaccharide present in plant cell walls. *Structural Chemistry* 20:263–275.

Voxeur A, André A, Breton C, Lerouge P (2012) Identification of putative rhamnogalacturonan-II specific glycosyltransferases in Arabidopsis using a combination of bioinformatics approaches. *PLoS One* 7:e51129.

Voxeur A, Habrylo O, Guénin S, Miart F, Soulié MC, Rihouey C, Pau-Roblot C, Domon JM, Gutierrez L, Pelloux J, Mouille G, Fagard M, Höfte H, Vernhettes S (2019) Oligogalacturonide production upon *Arabidopsis thaliana-Botrytis cinerea* interaction. *Proceedings of the National Academy of Sciences USA* 116:19743–19752.

Wachananawat B, Kuroha T, Takenaka Y, Kajiura H, Naramoto S, Yokoyama R, Ishizaki K, Nishitani K, Ishimizu T (2020) Diversity of pectin rhamnogalacturonan I rhamnosyltransferases in Glycosyltransferase Family 106. *Frontiers in Plant Sciences* 11:997.

Waldron KW, Selvendran RR (1992) Cell wall changes in immature Asparagus stem tissue after excision. *Phytochemistry* 31:1931–1940.

Walkinshaw MD, Arnott S (1981a) Conformation and interactions of pectins. I. X-ray diffraction analyses of sodium pectate in neutral and acidified forms. *Journal of Molecular Biology* 153:1055–1073.

Walkinshaw MD, Arnott S (1981b) Conformations and interactions of pectins. II. Models for junction zones in pectinic acid and calcium pectate gels. *Journal of Molecular Biology* 153:1075–1085.

Wells B, McCann M, Shedletzky E, Delmer D, Roberts K (1994) Structural features of cell walls from tomato cells adapted to grow on the herbicide 2, 6-dichlorobenzonitrile. *Journal of Microscopy* 173:155–164.

Whitcombe AJ, O'Neill MA, Steffan W, Albersheim P, Darvill AG (1995) Structural characterization of the pectic polysaccharide, rhamnogalacturonan-II. *Carbohydrate Research* 271:15–29.

Willats WGT, Knox JP, Mikkelsen JD (2006) Pectin: new insights into an old polymer are starting to gel. *Trends in Food Science & Technology* 17:97–104.

Willats WGT, McCartney L, Mackie W, Knox JP (2001) Pectin: cell biology and prospects for functional analysis. *Plant Molecular Biology* 47:9–27.

Yapo BM, Lerouge P, Thibault J-F, Ralet M-C (2007) Pectins from citrus peel cell walls contain homogalacturonans homogenous with respect to molar mass, rhamnogalacturonan I and rhamnogalacturonan II. *Carbohydrate Polymers* 69:426–435.

York WS, Darvill AG, McNeil M, Albersheim P (1985) 3-Deoxy-D-manno-2-octulosonic acid (KDO) is a component of rhamnogalacturonan II, a pectic polysaccharide in the primary cell walls of plants. *Carbohydrate Research* 138:109–126.

Zablackis E, Huang J, Muller B, Darvill AG, Albersheim P (1995) Characterization of the cell-wall polysaccharides of *Arabidopsis thaliana* leaves. *Plant Physiology* 107:1129–1138.

Zheng Y, Mort A (2008) Isolation and structural characterization of a novel oligosaccharide from the rhamnogalacturonan of *Gossypium hirsutum* L. *Carbohydrate Research* 343:1041–1049.

Zitko V, Bishop CT (1965) Fractionation of pectins from sunflowers, sugar beets, apples, and citrus fruits. *Canadian Journal of Chemistry* 43:3206–3214.

Zitko V, Bishop CT (1966) Structure of a galacturonan from sunflower pectic acid. *Canadian Journal of Chemistry* 44:1275–1282.

6 Cell Wall-Related Mechanisms Underlying Plant Cell Expansion

Kris Vissenberg
University of Antwerp
Hellenic Mediterranean University

Herman Höfte
University Paris Saclay, INRAE, AgroParisTech, Institut Jean-Pierre Bourgin

INTRODUCTION

The cells of prokaryotes, fungi, oomycetes, many protists, algae and plants are all surrounded by strong cell walls. A central question is how the walls of such cells manage to expand while resisting the force of the cell's turgor pressure and preventing rupture. As nicely reviewed by Dumais (2013), different solutions for this conundrum can be found in nature, two of which allow irreversible cell wall expansion (see Box 6.1): (i) growth by processive intussusception, which refers to cell wall area increase driven by the insertion of cell wall elements in a regularly arranged monolayer, as for instance observed in prokaryotes and (ii) chemorheological growth, which is defined as the chemically mediated flow of polymeric materials under stress. Here, the deformation of the wall depends on the rate of chemical reactions leading to the breakage of load-bearing cross-links within the network, as observed in plant cell walls.

The current consensus is that plant cell expansion is driven by the turgor pressure and limited by the extensibility of the cell wall, which relaxes through a chemorheological mechanism (see Box 6.1). This view has evolved over time. Early landmark publications (Nägeli 1864; Sachs 1874) proposed that primary cell walls grow by intussusception. An alternative view (Strasburger 1882) proposed that wall material is deposited through apposition onto the inner cell wall surface. Since such a mechanism on its own does not explain cell wall expansion, an additional process was needed to account for cell enlargement. During the same period there was the discovery of an 'influence', which moves from the site of light perception to the site of bending during phototropic bending of canary grass coleoptiles (Darwin 1880). This "influence" was later identified as auxin by Went (1928), which, together with observations by Cholodny (1926), led to the Cholodny–Went theory that states that during tropic responses, asymmetric distribution of auxin throughout the organ stimulates differential growth responses, which results in organ bending (Went and Thimann 1937). This in turn led to the formulation of the acid growth theory (Rayle and Cleland 1970), in which auxin triggers an acidification of the apoplast, promoting relaxation of the cell wall, supposedly through pH-sensitive cell wall hydrolytic activities (see Box 6.2). This was clearly in line with a chemorheological growth model, also adopted by Lockhart (1965), who formalised turgor-driven cell expansion, with the now famous formula $E = \varphi(P - Y)$, where E is the rate of extension, φ the plastic compliance, P the turgor pressure and Y the yield threshold and where Y drops and φ increases with decreasing pH. Finally, the

BOX 6.1: PROCESSIVE INTUSSUSCEPTION VS CHEMORHEOLOGY

Processive intussusception is illustrated by the growth of the sacculus of gram-negative bacteria (Höltje 1998). This structure is a peptidoglycan monolayer with an extremely regular structure consisting of long glycan chains, which support circumferential stress, interconnected by short peptide chains, which support axial stress. Surface deformation is controlled by a protein complex that concomitantly removes peptide bonds, depolymerises one template glycan chain and synthesises three new glycan chains (Figure 6.1a). The regularity of the cell wall structure is such that the "shape" of the new material determines the deformation anisotropy of the cell wall (Figure 6.1b). The much less regular plant cell walls are thought to undergo chemorheological growth (Figure 6.1c), which differs from intussusception in that the deformation is dictated by the anisotropic relaxation (the anisotropy results from the anisotropy of both wall structure (Figure 6.1d) and the stresses that are exerted upon the wall (Figure 6.1e)) upon the removal of load-bearing bonds. The newly inserted wall material is thought to be amorphous or sufficiently flexible to passively fill in the voids created by the removal of load-bearing cross-links, and thus does not contribute to orienting the deformation (Dumais 2013).

DOI: 10.1201/9781003178309-6

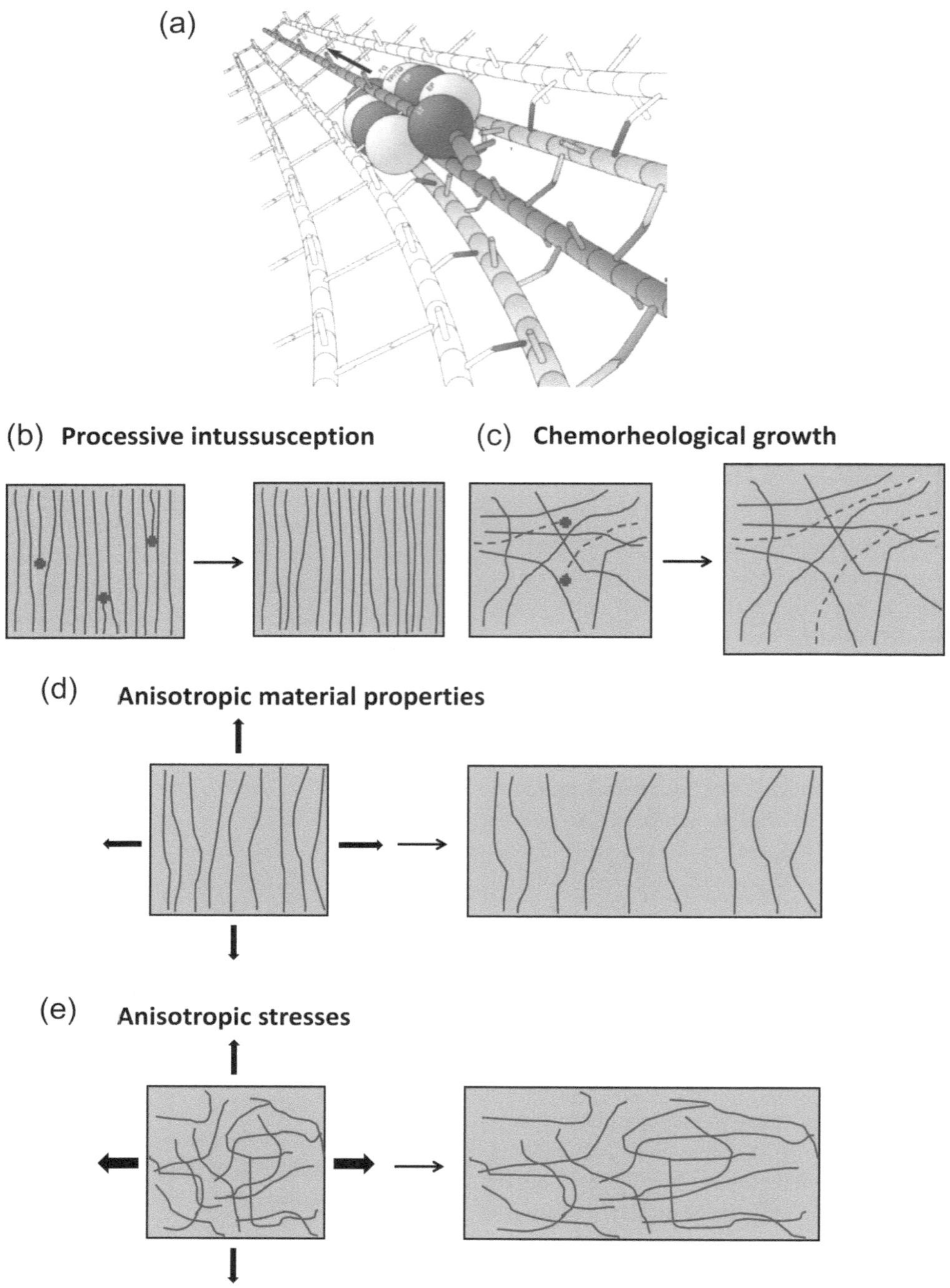

FIGURE 6.1 Models explaining (an)isotropic growth. (Adapted from Dumais (2013) with permission.) (a) The three-for-one model of peptidoglycan wall assembly. The wall assembly machinery travels along one docking glycan strand, and depolymerises the docking strand as it progresses forward. Meanwhile, it is synthesising three new strands to replace it, resulting in a net growth of two glycan chains. (b) Intussusception in a cell wall with ordered cell wall material. New material is deposited between existing material, separating former neighbouring material and inducing cell wall growth in one direction during the process of intussusception. (Brown crosses and dotted lines depict the deposition of new cell wall material by synthase proteins/complexes, also in C). (c) Chemorheological growth of a cell wall with randomly organised cell wall material. New material becomes deposited and concomitantly chemical bonds between cell wall components are broken and reformed, leading to turgor-driven isotropic growth. (d) Anisotropic growth. Cell wall deformation is anisotropic when a cell wall with anisotropic material properties undergoes isotropic stress. (The thickness of the arrows reflects the level of stress in that direction, also in E). (e) Anisotropic growth. Cell wall deformation is anisotropic when an isotropic cell wall is subjected to anisotropic stress.

BOX 6.2: AUXIN-CONTROLLED CELL EXPANSION: AN INTERPLAY BETWEEN PH AND CELL WALL MODIFICATIONS

Since the Cholodny–Went theory (Went and Thimann 1937), we now know that during tropic and other growth responses (Estelle 1996; Ma and Li 2019; Fendrych et al. 2016), auxin levels are positively correlated with cell elongation and auxin is redistributed through the action of different auxin transport proteins to the cells in the zone that will elongate (Friml et al. 2002; Noh et al. 2001; Rakusová et al. 2011).

Curiously, however, auxin can also elicit opposite growth responses, depending on the organ or even the process within the same organ. For instance, in line with Cholodny–Went (Du et al. 2020), lower auxin content or signalling in mutant Arabidopsis roots correlates with shorter epidermal cells and higher cell wall pH (Barbez et al. 2017). In contrast, in the same roots during the gravitropic bending response, higher auxin levels correlate with the *inhibition* of cell elongation at the lower side of the root (Band et al. 2012). Similarly, external auxin supply inhibits cell elongation (Růžička et al. 2007; Swarup et al. 2007) and promotes surface alkalinisation in the root's elongation zone (Staal et al. 2011). Such opposite responses also occur within etiolated Arabidopsis hypocotyls, where auxin application within 20 minutes activates plasma membrane-localised H^+-ATPases leading to cell wall acidification and increased cell elongation (Fendrych et al. 2016), but where, during apical hook formation, auxin accumulation correlates with the inhibition of cell elongation at the future inner side of the hook (Liao et al. 2015). It should be noted, however, that no matter the growth response to auxin, cell elongation rate inversely correlates with the cell wall pH in accordance with the acid growth theory.

It thus seems that auxin, depending on the context, can trigger either acidification or alkalinisation of the cell wall. Lin et al. (2021) recently explained this dual action by showing that two antagonistically acting auxin signalling pathways converge on the apoplastic pH: the cell surface-located TRANSMEMBRANE KINASE1 (TMK1), which interacts with and mediates the phosphorylation of plasma membrane H^+-ATPases for apoplast acidification and the stronger intracellular canonical auxin signalling that promotes net cellular H^+ influx and apoplast alkalinisation. The picture is more complex, however, since TMK1 can also mediate slower transcriptional responses. Indeed, Cao et al. (2019) showed that the differential growth response during apical hook formation also depends on TMK1, where local auxin accumulation in cells at the inner side of the hook leads to cleavage of the C-terminus of TMK1, which then stabilises IAA32 and IAA34 proteins that are specific regulators of auxin signalling (by repressing auxin response transcription factor (ARF) activity).

The study of apical hook formation provided additional insights into the link between auxin levels, cell wall pH and cell expansion. Indeed, Jonsson et al. (2021) observed that higher auxin levels at the inner side of the hook correlate with higher methylesterification levels of the pectin homogalacturonan (HG). Using mutants and transgenics, the authors show that auxin prevents HG de-methylesterification, but interestingly, those changes in HG methylesterification alter the auxin response domain through a so far unidentified feedback loop. We will discuss below the potential relevance of pectin de-methylesterification for the control of cell elongation. The same group also provided evidence that xyloglucan (XG) is required to limit cell growth rates at the inner side of the hook, and that XG is required for the auxin maxima in the inner side cells by transcriptional regulation of auxin transporters through ARF2 (Aryal et al. 2020). Auxin therefore controls apical hook formation through changes in cell wall chemistry, and in turn, the cell wall provides mechanochemical feedback towards auxin.

identification of expansins, proteinaceous agents responsible for acid-promoted relaxation of stretched cell wall preparations (McQueen-Mason et al. 1992), provided a biochemical correlate for φ and acid growth.

CHEMORHEOLOGICAL CELL WALL EXPANSION

As also discussed in Chapters 1 and 11 in this book, much of the literature is focused on the nature of the cross-links and on how they are broken to allow relaxation of the network. In the most advanced model, in terms of experimental validation and modelling, the load-bearing network of the primary cell wall consists of cellulose microfibrils, which make lateral contacts at a small number of locations (Zhang et al. 2021). Recent studies, using AFM (atomic force microscopy), cryo-focused ion beam milling and cryo-electron tomography (Nicolas et al. 2022) provided new insights into the cell wall architecture. These studies used the outer epidermal cell walls of onion bulbs as a model, which are up to 7 μm thick and consist of multiple lamellae, each ~40 nm thick. Each lamella consists of fields of aligned microfibrils, the overall orientation of which alternates in the successive lamellae between −45° and +45° relative to the cell's long axis. The fibrils are surrounded by a reticulated matrix, referred to by the authors as «meshing», which consists primarily of HG pectins. This meshing forms patches with no preferential orientation as well as segments in between the aligned fibres. Meshing was most abundant at the inner surface of the cell wall, and levels abruptly dropped off around 1 μm into the wall. AFM combined with coarse grain modelling on walls before and after stretching, showed that tensile mechanics of the cell wall are dominated by complex movements of cellulose microfibrils (bundling, sliding), with no significant contribution from the matrix (Zhang et al. 2021). At least a subset of the lateral contact sides between microfibrils are thought to be sandwiched together by tightly bound XG chains (at least in non-Commelinoid species) (see Chapter 1 for a description of the cell wall polysaccharides). The latter cross-links were referred to as "biomechanical hotspots" (Cosgrove 2018). The existence of a XG fraction closely associated with cellulose was shown by Pauly et al. (1999) and later by Vissenberg et al. (2005), who used an *in vivo* XG endotransglycosylase (XET) assay, in which fluorescently labelled XG-oligosaccharides are covalently incorporated into existing XG by XET action. Upon removal of these incorporated XG-oligosaccharides by a XG-specific β-endoglucanase (XEG), a substantial portion of the XG remained in the cell wall, suggesting that they were in close contact with cellulose and hence protected against XEG enzyme action. The existence of such hybrid XG-cellulose structures, recalcitrant to XEG-treatment, was later confirmed by the discovery of a class of specialised fungal β-endoglucanases (members of the glycosyl hydrolase family 12, which is absent in plants) that specifically removes these structures while leaving the bulk of the "unprotected" XG intact (Park and Cosgrove 2012). If pathogenic fungi have evolved such highly specialised enzymes, one can expect that the structures that they target have some critical role in the cell wall's architecture. Interestingly, Cosgrove and colleagues indeed showed that this class of β-endoglucanases caused a sustained deformation (creep) of isolated cell walls in an extensometer, which was not the case for XEGs (Park and Cosgrove 2012), such as those used by Vissenberg et al. (2005). In an elegant series of experiments using AFM, they could also show at the nanoscale that these fungal enzymes promoted the relaxation of stressed walls by removing connections between individual microfibrils (Zhang et al. 2017). The same group also identified expansins as endogenous creep-promoting proteins, suggesting that they also disrupt these biomechanical hotspots, but through a poorly understood, non-enzymatic mechanism (McQueen-Mason et al. 1992; Wang, Yuan et al. 2013). Interestingly, expansin activity requires low pH (≤ 5), consistent with the acid growth theory. Expansins also promote cell expansion of intact cells (see references in Marowa et al. 2016), as shown by ectopic administration on tomato apical meristems (Pien et al. 2001) or by overexpression in Arabidopsis stems and rosettes (Cho and Cosgrove 2000; Goh et al. 2014).

An essential *in vivo* role for expansins in cell elongation was pinpointed by the demonstration that root hair growth was reduced in loss-of-function mutants for the root hair–specific expansin *OsEXPA17* in rice (ZhiMing et al. 2011) or in RNAi lines for *AtEXPA7* in Arabidopsis (Lin et al. 2011), that leaves and petioles were smaller and often twisted in antisense lines for *AtEXPA10* (Cho and Cosgrove 2000), that pollen tube growth was reduced in double *atexpa4/atexpb5* mutants and that the root meristem size was reduced in a mutant for *AtEXPA4* (Liu et al. 2021).

It should be noted that creep measurements in general are carried out on isolated and often heat-inactivated fragments from cucumber or Arabidopsis hypocotyls, mung bean epicotyls, oat coleoptiles or the above-mentioned onion peels (Suslov and Vissenberg 2018). The cells in these tissues are unique in that cell expansion occurs uncoupled from cell wall deposition, perhaps as an adaptation allowing rapid growth. Indeed, the outer epidermal cell walls of young cells are multilayered and very thick (>1 μm, up to 7 μm for onion cell walls) (Zhang et al. 2021; Suslov et al. 2009), in comparison to much thinner walls (35–50 nm) described for instance for anticlinal cell walls of pavement cells (Haas et al. 2020a). A study using Arabidopsis hypocotyls showed that such thick walls are deposited during a slow growth phase, after which they undergo significant thinning during a rapid growth phase (Refrégier et al. 2004). The use of such tissues has allowed the assessment of the remodelling capacity of multilayered walls without having to worry about the replenishment of new cell wall material. The observation of the uncoupling of wall synthesis and cell elongation led Bonner to conclude in 1934 (Bonner 1934) that, "Increase of wall area due primarily to active intussusception of new material is therefore excluded, at least in the case of the Avena coleoptile". Subsequent studies using similar materials were often less careful in avoiding extrapolation to plant cell expansion in general.

Indeed, the problem is more complex for the majority of expanding cell walls, in which wall deposition and cell expansion are tightly coupled. This is illustrated by the anticlinal (perpendicular to the organ surface) walls of Arabidopsis epidermal pavement cells, which showed, on cryo-scanning electron microscopy (SEM) sections, a thickness of 35–50 nm on each side of the middle lamella (Haas et al. 2020a). Such walls can thus accommodate not more than a single cellulose layer (estimated to have a thickness of around 40 nm as determined by field emission scanning electron microscopy (FESEM) and AFM in dried onion cell walls (Zhang et al. 2013)) or a few cellulose layers, if we assume that these walls have a denser architecture than onion outer cell walls. As discussed below, expansion of such walls cannot occur without a continuous insertion of new wall material.

Very few studies have addressed in plants how new cell wall material is inserted into the structure and how critical cross-links are formed. Ray (1967) fed oat coleoptile segments with Glucose-^{3}H in the presence or absence of auxin and used electron microscope radioautography to observe in which parts of the cell wall cellulose and matrix polysaccharides were incorporated. Consistent with the more recent observation of the activity of cellulose synthase complexes in the plasma membrane (McFarlane et al. 2014), they observed that cellulose was deposited close to the plasma membrane, whereas hemicellulose also penetrated into deeper cell layers. Anderson et al. (2012) used click chemistry upon feeding Arabidopsis seedlings with alkynylated fucose to show the incorporation of newly synthesised polymers, presumably fucose-containing rhamnogalacturonan I (RG-I), into the outer cell wall of root epidermis cells. They observed deposits of the polymer aligned with cellulose microfibrils and reorientation of these polymers associated with cell elongation consistent with the multinet growth theory by Roelofsen and Houwink (1953), the above-mentioned AFM observations by Zhang et al. (2017) and the results from coarse grain modelling (Zhang et al. 2021). More detailed observations were reported for the Charophyte *Chara corallina* (Boyer 2016). Charophytes are freshwater green algae, which descended from a common ancestor with land plants (Domozych et al. 2016). This proximity is reflected in a very similar composition of the cell walls (Popper et al. 2011). In Chara, calcium-pectate cross-links rather than XG-cellulose bonds appear to be the load-bearing structures of the cell wall (Boyer 2016). A large body of evidence shows that these cell walls expand as a result of the insertion of pectate, which chelates calcium from load-bearing bonds, thus causing the relaxation of the cell wall and the formation of new load-bearing bonds. It should be noted that in this system, the insertion of new wall material is sufficient to promote wall expansion, which is reminiscent of intussusception (Figure 6.1b). The difference, however, is that in Dumais' view, the flexible pectin only adopts its structure by occupying the void created upon the turgor-driven relaxation of the cell wall network after removal of load-bearing cross-links (Dumais 2013) and hence does not orient the deformation like the insertion of new material in bacterial cell walls for instance.

In what follows, we will first discuss the role of pectin metabolism in cell expansion and then compare models for cell wall expansion in tip-growing and diffusely growing cells.

PECTIN STRUCTURE–FUNCTION RELATIONSHIPS, THE TIP OF THE ICEBERG?

As discussed in detail in Chapter 5 in this book, pectins are characterised by the presence of galacturonic acid (GalA) and constitute the most diverse family of cell wall polysaccharides in plants. Among those, HG, a linear polymer of α-1,4-linked GalA is, in many cell types of non-Commelinoid species, the most abundant cell wall polymer. For instance, it represents 20% of the cell wall dry mass in Arabidopsis leaf tissue (Zablackis et al. 1995) and even over 50% in the outer walls of the onion epidermis (Wilson et al. 2021). HGs are found as part of different block-copolymers (HG-RG-I, HG-xylogalacturonan, HG-RG-II the proteoglycan ARABINOXYLAN PECTIN ARABINOGALACTAN PROTEIN1 (APAP1; Tan et al. 2013)) and perhaps also as homogeneous independent HG polymers (Chapter 5). The function of this structural modularity is not understood in most cases.

HGs are unique in the plant cell wall in that their charge patterns are precisely regulated. Indeed, they are thought to be secreted in a methylesterified form to become selectively de-esterified by cell wall–associated pectin methylesterases (PMEs) (see Chapter 5 for a detailed discussion on HG methylesterification). This generates negatively charged residues in the polymer at pH values above the pKa of the polymer (which can range from 3.5 to 5 depending on its charge density; Ralet et al. 2001).

Why does polymer charge matter that much? Firstly, anionic polymers, such as pectins but also glucuronated arabinogalactan proteins (AGPs) (Lopez-Hernandez et al. 2020), trap counterions and thus may affect the apoplastic pH, the proton motive force and ion currents across the plasma membrane (Grignon and Sentenac 1991; Sentenac and Grignon 1981), which control turgor and morphogenesis (Weisenseel and Kicherer 1981; Hoffmann et al. 2020). In addition, pectins, like many other polymers, form cross-linked networks that contain water in their interstices. The water retention capacity increases dramatically with increasing polymer charge, going from water-polymer ratios of 5:1 for uncharged polymers, up to 3,000:1 for charged polymers (Osada and Gong 1993). To understand this, one should keep in mind that the volume of an ionised polymer gel is determined by its osmotic pressure, which is the sum of three forces: (i) the rubber elasticity, (ii) the polymer–polymer affinity and (iii) the counterion pressure (Tanaka 1981). The rubber elasticity derives from the elasticity of the individual polymer strands. The magnitude of this pressure depends on the temperature, whereas the sign depends only on the volume of the gel (positive for a condensed gel, with polymer ends held close together or negative for a swollen gel, with polymer ends stretched far apart). The polymer–polymer affinity depends on the solubility of the polymers. Poor solubility will tend to draw polymers together creating a negative pressure. This pressure is temperature insensitive and decreases with increasing gel volume. The counterion pressure is the osmotic pressure resulting from counterions sequestered by the polymer's charges (Gibbs-Donnan effect). This pressure increases with temperature and is positive since it tends to draw water into the polymer's interior and push its exterior parts away from one another. Changes in the "environment" (temperature, pH, ionic strength, calcium concentration, mechanical stresses, etc.) of the gel will affect the three parameters in independent and often discontinuous ways. For instance, increasing the temperature will increase the osmotic pressure of the gel due to the increased rubber elasticity and counterion pressure. The resulting volume increase will lead to a decrease in polymer–polymer affinity, which, beyond a certain threshold, will again be counteracted by the negative pressure due to the stretching of the polymers. This leads to complex emergent behaviours, including discontinuous changes in volume (phase transitions) in response to even minute changes in the "environment" or the composition of the gel (Tanaka 1981) (Figure 6.2a and b). So, it is easy to understand how even subtle chemical changes (e.g. de-methylesterification,

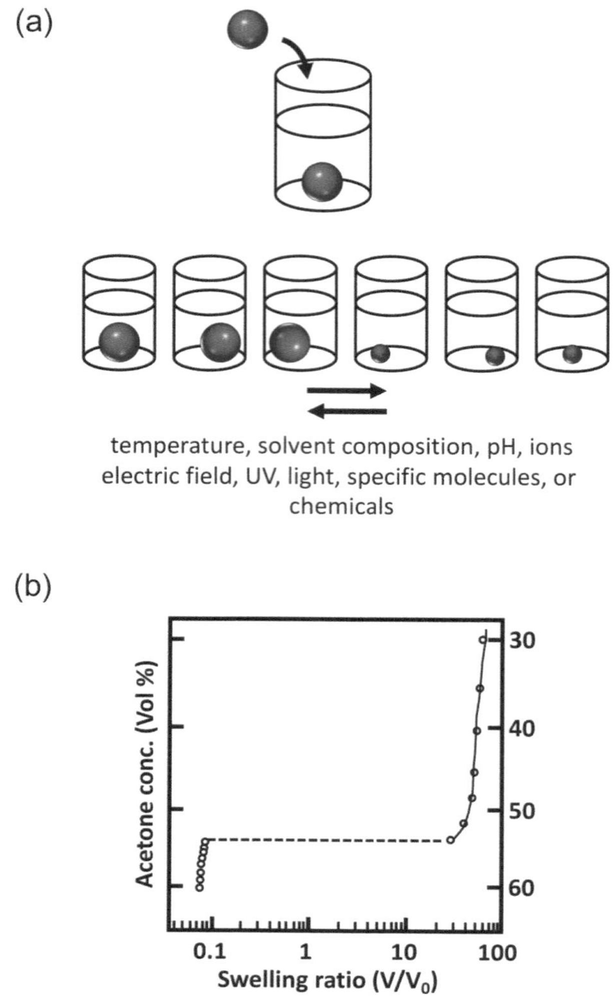

FIGURE 6.2 Phase separation in ionised polymer gels. (Adapted from Tanaka et al. 1992.) (a) Large volume transitions (here in acrylamide gel beads) are triggered by subtle changes in the environment. From left to right: gradually increasing changes in environmental conditions lead to abrupt changes in the volume of the bead. (b) Large volume transition in a gel induced by minute changes in the hydrophobicity of the medium.

de-acetylation, polymer cleavage) that affect one or more of these three parameters can strongly affect the hydration capacity of the gel and hence that of the cell wall.

This idea is nicely illustrated by studies of the mucilage that is released from the Arabidopsis seed coat upon seed imbibition. The released portion of the mucilage is a hydrogel consisting mainly of RG-I (Šola et al. 2019). Genetic studies showed that the removal of one or more galactose residues from a RG-I side chain by the β-galactosidase MUM2 is essential for mucilage release (Macquet et al. 2007). The authors provided evidence that galactoses in RG-I side chains, when oxidised by a putative galactose oxidase, encoded by the *RUBY* gene, are able to form hemi-acetal cross-links with neighbouring sugars during dehydration (Šola et al. 2019). Mucilage swelling therefore appears to depend on preventing cross-linking through the enzymatic removal of galactose residues from RG-I side chains.

To understand the *in vivo* function of pectins, it should be kept in mind that their concentration in the cell wall is an order of magnitude higher than in the gels used in most *in vitro* physicochemical studies. Such concentrated gels can have tensile moduli close to those of plant cell walls (Zsivanovits et al. 2004), suggesting that, besides cellulose, pectins can significantly contribute to the mechanical properties of the cell wall. In addition, uni-axially oriented HG chains can form crystalline fibrils. This was shown in *vitro* using X-ray diffraction on oriented HG preparations of high Degree of Methylesterification (DM), gelified at low pH in the presence of sucrose, or calcium cross-linked low DM HG, where the two types of fibrils showed different crystal units (Palmer et al. 1947; Sterling 1957; Walkinshaw and Arnott 1981a, 1981b). Pectin fibrils were also shown to exist *in vivo* (Figure 6.3). For instance, electron microscopy and X-ray diffraction revealed predominantly axially oriented pectin fibrils in fresh collenchyma cell walls of *Petasites vulgaris* petioles (Roelofsen and Kreger 1951) (Figure 6.3a). "Non-cellulosic fibrillar galacturonic acid" was also observed *in vivo*, pointing out from the cell wall into the extracellular medium of liquid plant cell cultures (Leppard et al. 1971) (Figure 6.3b), in the inner (cytoplasmic-facing) cell wall surface of these cells, or in root epidermal cell

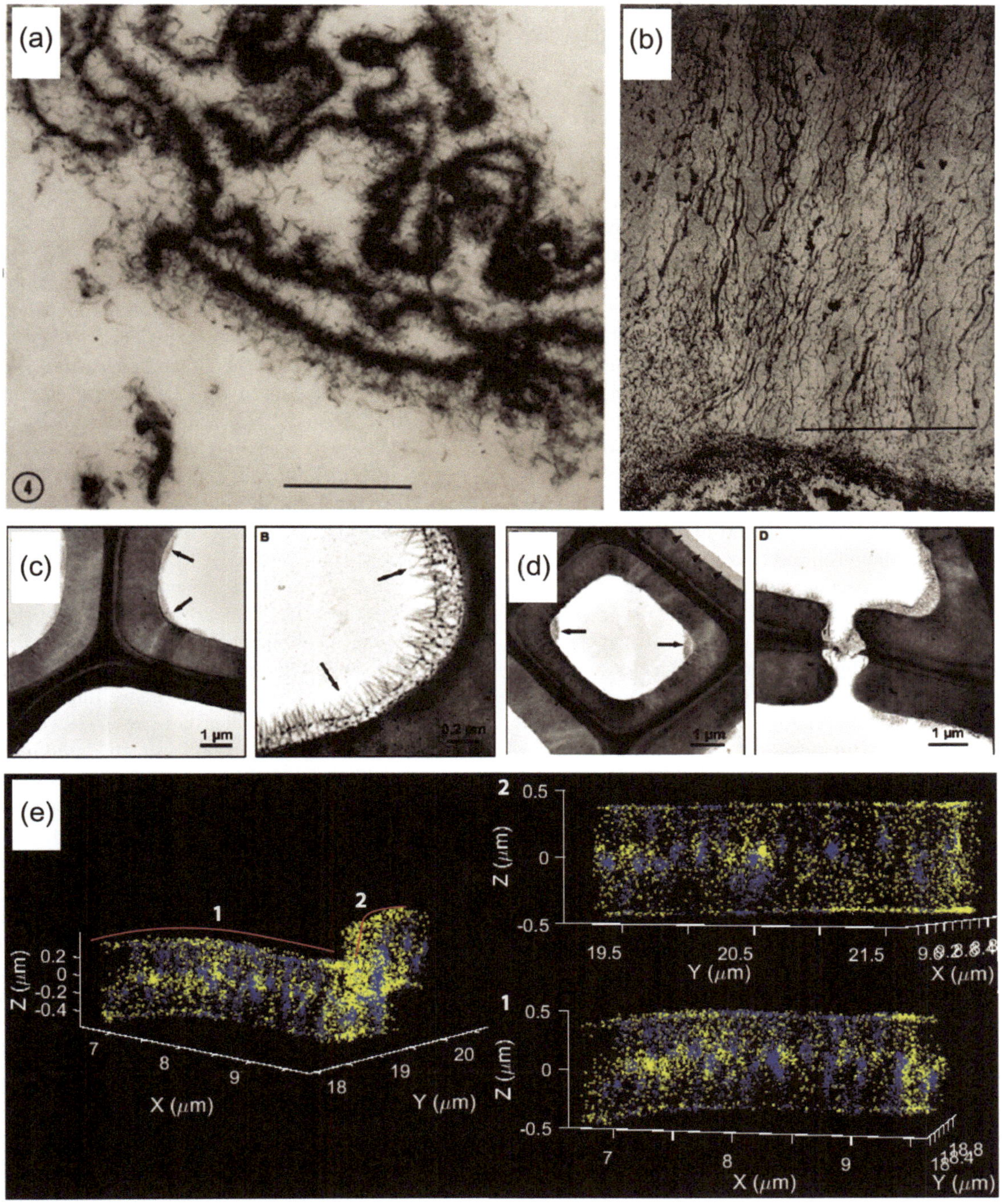

FIGURE 6.3 Visualisation of fibrillar pectin. (a) Rope-like aggregate at the cell wall surface of *Phaseolus* culture cells, as visualised by TEM. (From Leppard et al. (1971), with permission.) (b) Non-cellulosic fibrillar galacturonic acid pointing out from the cell wall into the extracellular medium of liquid plant cell cultures, as visualised by TEM. (From Leppard and Colvin (1972), with permission.) (c and d) TEM images of pectin fibrils in the lumen of poplar xylem fibre cells. (From Arend et al. (2008), with permission.) (d) TEM images of pectin fibrils in the lumen of poplar xylem fibre cells. (From Arend et al. (2008), with permission.) (e) Super-resolution microscopy of the anticlinal cell walls of pavement cells in the Arabidopsis leaf epidermis labelled with antibodies against methylesterified HG (LM20, blue) and against de-methylesterified HG (2F4, yellow), respectively. Vertically oriented labelling patterns presumably corresponding to filamentous structures can be observed. (From Haas et al. (2020a)).

walls (Figure 6.3c). These structures with a diameter in the order of 20 nm were initially christened "lignofibrils", due to their weak UV absorption properties and phloroglucinol-HCl staining. The UV absorption, however, could, besides lignin, also come from tyrosine-rich proteins such as polycationic extensin-related proteins, which can form polyelectrolyte complexes with pectate (Cannon et al. 2008; Valentin et al. 2010). This may go some way to explain why these fibres were reported to be largely resistant to pectinases (Leppard et al. 1971). Pectin fibrils were also observed in the lumen of poplar xylem fibre cells (Arend et al. 2008) (Figure 6.3d). More recently, super-resolution microscopy revealed uni-axially oriented HG in the anticlinal cell walls of pavement cells in the Arabidopsis epidermis (Haas et al. 2020a; Figure 6.3e). The significance of this for cell expansion will be discussed below.

A last, so far poorly explored aspect relevant for the regulation of cell expansion is how substitution and charge patterns of HGs may govern interactions with other polysaccharides or proteins. Well-studied examples for other polysaccharides are the periodic (glucuronic acid (GlcA), me-GlcA or acetyl) substitution patterns on xylan backbones that condition cellulose binding (Grantham et al. 2017) or, in the animal extracellular matrix, the sulphate substitution patterns of the glycosaminoglycan heparin that govern, often highly specific, interactions with proteins (Capila and Linhardt 2002). HGs can also be acetylated or xylosylated (at position 2 or 3) and can have random or blockwise charge patterns. In addition, specific repeated substitution patterns (such as $GalA_4MeAc$) could be inferred from the profiling of oligogalacturonides released from the cell wall by a fungal pectate lyase (Voxeur et al. 2019). Such patterns are the hallmarks of specific PME activities, which remain to be characterised.

PMEs are encoded by large gene families (66 members in Arabidopsis thaliana) and, in vascular plants, PME activity can be modulated by endogenous proteinaceous PME inhibitors (PMEIs), which are encoded by equally large families (71 in *A. thaliana*) (Wolf et al. 2009; Wang et al. 2013). In addition, demethylation exposes the HG to the activity of endogenous polygalacturonases (PGs) and pectate lyases (PLs), which require demethylated substrates (Hocq et al. 2017). The few plant PMEs that have been studied so far have neutral to basic pH optima and processive activities, generating blockwise charge patterns. At least some of these PMEs can lose processivity in certain conditions (low pH; Hocq et al. 2021) or in the presence of Ca^{2+} (Vincent et al. 2013) thus generating non-clustered patterns. It will be interesting to see to what extent HG charge and substitution patterns can govern interactions with cell wall proteins. A recent study (Francoz et al. 2019) revealed that a specific PMEI (PMEI6) is required for the generation of a binding site for a cell wall peroxidase (PRX36), presumably by altering the activity of an unknown PME, at a specific location within the outer cell wall of the Arabidopsis seed coat. At this specific location, the peroxidase changes the cell wall properties generating brittle spots that selectively rupture upon hydration, facilitating the release of the seed mucilage. Many other cell wall proteins have basic surfaces and bind to pectins as shown *in vitro* (Carpin et al. 2001; Jamet et al. 2008) and *in vivo* (Wang et al. 2015). This is expected to contribute not only to the nanoscale distribution of cell wall–modifying activities but also to the formation of polyelectrolyte complexes that may also locally impact the hydration, porosity and rheology of cell walls. For instance, the interaction with basic extensins caused dehydration of pectin layers *in vitro* (Valentin et al. 2010). Similar effects can be expected from the interaction with other polycations such as endogenous polyamines (Messiaen and Van Cutsem 1999) or chitosan present in the walls of endophytic (Noorifar et al. 2021) or pathogenic fungi (Baker et al. 2007).

TIP GROWTH

Tip growth is mainly studied in pollen tubes and root hairs. Therefore, there is much literature dealing with discoveries in both cell systems. It has been known since 1897 that growing pollen tubes contain moving granular protoplasm (Potter 1897). Years later, Castle (1958) concluded that the apical regions of tip-growing cells may have the highest relative elemental rates of elongation known. Following this, the oldest models of tip growth zone activities were developed by Green (1973), while in the same period electrophysiological experiments showed that calcium enters the growing pollen tube tip, with other positive ions, mainly protons, leaving at the more basal region (Weisenseel et al. 1975). The importance of calcium was backed up with the visualisation of an internal Ca^{2+}-gradient in pollen tubes (Jaffe et al. 1975) and the observations that raising calcium in the tubes leads to growth arrest, but not of cell wall build-up, resulting in a thicker cell wall (Picton and Steer 1983) while its deprivation or the use of a calcium-channel blocker results in localised swelling at the tube tip (Reiss and Herth 1985). Later on Li et al. (1994) observed a periodically arranged pattern of esterified and unesterified pectin in pollen tubes, which was linked to growth velocity fluctuations by (Pierson et al. 1996). Further correlations between the oscillations of pollen tube growth, and those of calcium-ions and protons were described by (Pierson et al. 1996; Feijó et al. 1999), respectively. In the early years of the 21st century, localised cell wall remodelling by expansin and XTH in tip-growing root hairs was shown by Baluška et al. (2000) and Vissenberg et al. (2001), while the discovery that Rapid Alkalinisation Factors (RALFs) regulate extracellular pH and growth of root hairs (Wu et al. 2007) opened up the findings that members of the *Catharanthus roseus* superfamily of receptor-like kinases (RLKs) control cell wall integrity at the pollen tube tip (Boisson-Dernier et al. 2009) and in root hairs (Duan et al. 2010). Only in 2018, Schoenaers et al. (2018) made a link between cell wall integrity sensing by one of these RLKs, ERULUS (ERU), and the oscillations of cell wall pectin methylesterification degree and the resulting growth of root hairs.

Like in Chara, pectin insertion thus also appears to drive the expansion of tip-growing pollen tubes and root hairs. In such cells, growth is restricted to the extreme apex and depends on the fusion of vesicles containing new cell wall material with the apical plasma membrane (Bove et al. 2008; Ketelaar et al. 2008; Balcerowicz et al. 2015). This results in polar growth that is highly responsive to local changes in the cell's environment (Schoenaers et al. 2017).

Valuable insights into growth mechanisms in these cells came from the observation of growth rate oscillations, which, as detailed below, are accompanied by oscillations with the same temporal or spatial frequency as a number of other cellular parameters (Kroeger and Geitmann 2012; Damineli et al. 2017). These include exocytosis rate, cell wall modification and thickness, Rho-like GTPase (ROP) activity, apical F-actin presence, cytoplasmic pH and calcium gradients, apoplastic pH and Reactive Oxygen Species (ROS) levels (Hepler et al. 2001; Hwang et al. 2010; Yan et al. 2009; Schoenaers et al. 2017). Such oscillations are also reflected in the periodic cell wall deposition patterns that were reported in tobacco pollen tubes (Geitmann et al. 1995; Derksen et al. 2011). McKenna et al. (2009) monitored PME-GFP levels in the cell wall of pollen tubes, as a marker for exocytosis. They concluded that exocytosis oscillates with the growth rate, where peaks in exocytosis precede periodic growth bursts. (A critical note here is that GFP is less fluorescent at lower pH (Kneen et al. 1998), hence PME-GFP fluorescence oscillations might also reflect apoplastic pH oscillations). Cell wall thickness was also shown to oscillate, also with its maxima preceding the growth rate maxima (Figure 6.4a). Interestingly, variations in the amount of deposited cell wall material correlated with the extent of the subsequent growth rate increases. This suggested that cell wall deposition determines the elongation rate of these cells (McKenna et al. 2009) (Figure 6.4b), in contrast to the above-mentioned "biomechanical hotspot" model for diffusely growing cells, in which cell wall deposition somehow passively follows cell wall relaxation. A role for cytosolic calcium oscillations in the control of oscillations in exocytosis, cell wall thickness and growth rate was explored in a model by Kroeger et al. (2008).

The fusion of vesicles with the plasma membrane is controlled by members of the syntaxin family of soluble N-ethyl-maleimide sensitive factor attachment protein receptors (SNAREs; Rodriguez-Furlán et al. 2016), the exocyst complex (Hála et al. 2008) and in part by members of the plant ROP family. ROP activity oscillates in pollen tubes with the same frequency as tip growth oscillations (Hwang et al. 2010), and its apical location depends on the action of ROP1 ENHANCERs (RENs), proteins that are located subapically and help to remove ROPs from the plasma membrane using clathrin-mediated endocytosis (Li et al. 2018). ROPs interact with diverse effectors and mediate actin dynamics, microtubule organisation (and thus vesicle trafficking, reviewed in Nibau et al. 2006; Yalovsky et al. 2008), ROS production (Carol et al. 2005) and the formation of a tip-focused Ca^{2+} gradient (Li et al. 1999), and are therefore considered as 'hubs' for signal integration during tip growth.

The requirement for pectin deposition for proper tip growth was suggested by the observation that root hair growth was inhibited in mutants (for two distinct UDP-GalA biosynthetic enzymes) with reduced pectin levels (Pang et al. 2010) or in wild type upon pectate lyase treatment (Park et al. 2011). Next to deposition, PME-driven pectin modification (see Box 6.2) is also critical for pollen tube and root hair growth. In pollen tubes, PME activity is essential for growth, as shown by the curved, irregularly shaped and stunted pollen tubes with reduced elongation rates upon inactivation of *PPME1 or VANGUARD1* (*VGD1*), which both encode pollen-specific PMEs (Tian et al. 2006; Jiang et al. 2005). The involvement of PME activity in proper tip growth of pollen tubes was modelled using finite element modelling (FEM) and also experimentally validated (Fayant et al. 2010). In root hairs, genetic validation for a role of PME activity in elongation awaits the identification of *pme*-mutants with a root hair phenotype, but Schoenaers et al. (2018) were already able to show that inhibiting PME activity with a catechin extract also affected root hair growth. Interestingly, PME activity also appears to oscillate in the tip of growing root hairs (Schoenaers et al. 2018) or pollen tubes (Rounds et al. 2011). This was suggested by the oscillations in the staining intensity of propidium iodide, a dye that binds to anionic pectin. These oscillations may mirror oscillations in PME activity, which, given the high pH optimum of the bulk of PME activity (Hocq et al. 2017), is expected to be controlled by the oscillations in surface pH, reported for pollen tubes (Hepler et al. 2006) and root hairs (Monshausen et al. 2007) (Box 6.2).

In line with this, the activity of most other cell wall–modifying enzymes is also highly pH-dependent. The activity of expansins is promoted in an acidic environment (McQueen-Mason et al. 1992), whereas recombinantly produced XG endo-transglycosylase-hydrolases (XTHs) show widely diverging pH optima (Maris et al. 2009, 2011). Recombinant PLs have alkaline pH optima (Ali et al. 2015), while PGs are more active at acidic pH (Hocq et al. 2017). In this respect, it needs to be mentioned that in alkaline conditions, PME activity on methylesterified pectin generates protons, and hence contributes to cell wall acidification (Moustacas et al. 1991), in concert with plasma membrane-localised H^+-ATPases (AHAs). In pollen tubes, AHAs are specifically localised in the shank and they have been shown to be the drivers for cytosolic pH oscillations (Hoffmann et al. 2020). Indeed, mutants in AHAs show reduced H^+ and anion fluxes around the pollen tube tip, slower cytosolic pH oscillations, shallower cytosolic pH gradients and reduced growth rates. Root hair and pollen tube elongation therefore seem to be in line with the acid growth theory (Box 6.2) (Rayle and Cleland 1992), where cell wall pH oscillations cause periodic activation of PMEs at high pH in antiphase with the activation of expansins at low pH (Cosgrove 2000).

Based on the calcium-pectate exchange mechanism previously proposed for Chara, Rojas et al. (2011) developed a computational model for pollen tube growth (Figure 6.4c). This model posits that cell wall deposition promotes cell elongation and that cell elongation inhibits cell wall deposition through a negative feedback loop. It also implies that cell expansion ceases when all available pectin is incorporated into the load-bearing cell wall layer and no new material is available. This model was able to predict pollen tube shape, trajectories of cell wall material at the growing surface and even growth oscillations, which occur above a certain elongation rate, where cell expansion outruns cell wall deposition. The model clearly indicates that in the case of tip-growing cells, wall expansion is limited by deposition and/or insertion of material. Evidently, turgor pressure provides the driving force for growth in this model (Kroeger et al. 2011; Dumais 2021).

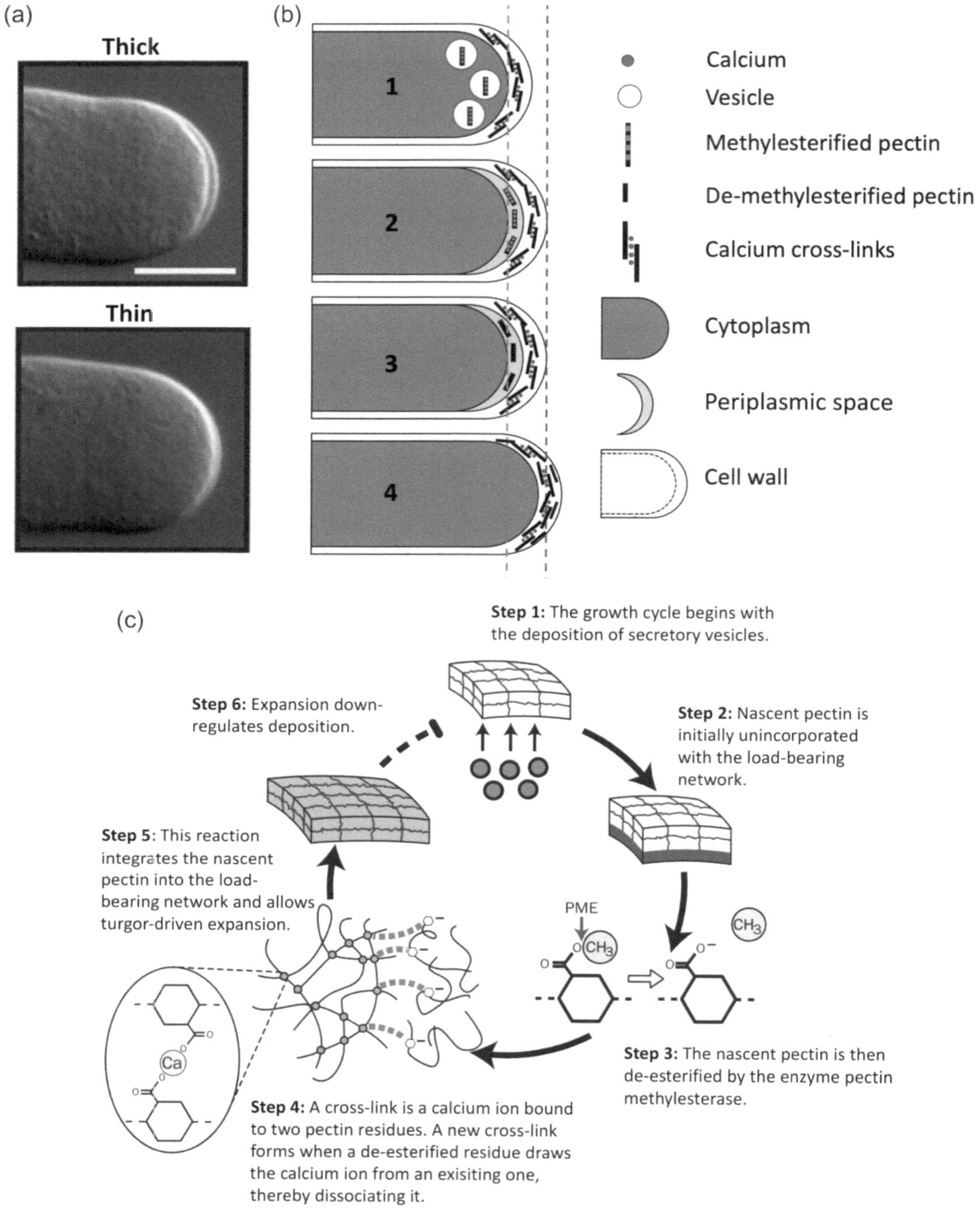

FIGURE 6.4 Calcium-pectate exchange model to explain tip growth. (a) Two Differential interference contrast microscopy-images taken at different moments during oscillating pollen tube tip growth. The upper image shows a thick and the lower image shows a thin apical cell wall. (Images from McKenna et al. (2009, with permission.) (b) Schematic representation of events happening at the tip during pollen tube tip growth. (i) Vesicles containing highly methylesterified pectin move towards the tip. The cell wall contains load-bearing calcium-pectate cross-links. (ii) Methylesterified pectin is secreted into the inner cell wall layers, creating a periplasmic space, pushing the cell wall a bit forward and creating the thick cell wall appearance (see panel (a), upper image). (iii) PME activity de-methylesterifies the newly deposited pectin. (iv) This de-methylesterified pectin competes for calcium-ions with the older de-methylesterified and load-bearing pectins and becomes integrated in the existing pectin matrix, emptying the periplasmic space and thinning the cell wall (see panel (a), lower image). By dissociating the calcium-bonds and recreating new bonds with the young de-methylesterified pectin, the older pectins can yield under turgor pressure and are pushed aside, leading to cell wall growth. The cycle can now restart from (i). (Modified after McKenna et al. (2009), with permission.) (c) More detailed view on the events explained in (b) happening at the pollen tube tip. (From Rojas et al. (2011), with permission.) Load-bearing cross-links are altering positions in the wall, from the old pectin located more in the outer cell wall to the newly incorporated pectins more to the inner side of the wall. The model is based on three key concepts: (i) cross-link dissociation allows turgor pressure to stretch (and thin) the cell wall, (ii) newly deposited methylesterified and uncross-linked pectin serves as a reservoir of material that waits to be incorporated into the cross-linked network by dissociating existing cross-links, after being de-methylesterified by PMEs and (iii) cross-links reform at the same rate as they are broken, but in a relaxed state.

The calcium-pectate exchange model does not address the role in tip growth of cellulose-XG remodelling described in the biomechanical hotspot model. As discussed below, however, this cannot be ignored given the strong root hair growth phenotypes of mutants for cellulose or XG synthesis or expansins.

Cellulose deposition is essential for maintaining cell wall integrity as shown by the growth inhibition and/or rupture of root hairs treated with a cellulase or the cellulose synthesis inhibitor 2,6-dichlorobenzonitrile (DCB) (Park et al. 2011). Mutant analysis and localisation experiments suggest that proteins other than those involved in cellulose synthesis in primary walls (CESA1, CESA3 and CESA6) synthesise cellulose at the root hair tip (Endler and Persson 2011; Park et al. 2011). *CELLULOSE SYNTHASE LIKE (CSL) D2* and *CSLD3* (*KOJAK*) are likely candidates. Indeed their mutants produce abnormal root hairs that rupture early (Favery et al. 2001) or late (Bernal et al. 2008) during development or which show a swollen base, wavy growth or occasional branches. Evidence for a role of CSLD3 in cellulose synthesis came from the observation that a chimeric gene encoding CSLD3 with the catalytic domain of CESA6 complemented the *csld3/kjk* root hair phenotype (Galway et al. 2011, Park et al. 2011). In addition, recent data confirmed that CSLD3 is indeed a UDP-glucose-dependent (1–4)-β-glucan synthase that can be purified as higher-order complexes with similar ultrastructural features as those formed by CESA6 proteins (Yang et al. 2020). The fact that CSLDs are active where plasma membrane-associated cortical microtubules are generally absent (in the root hair tip or at the new cell plate) suggests that they contribute to the production of cellulose structures different from the long microfibrils generated by microtubule-associated cellulose synthase complexes. It will be interesting to see whether the shorter disordered cellulose-like fibrils that can be observed in the tip of growing root hairs are produced by CLSD complexes (Akkerman et al. 2012; Newcomb and Bonnett 1965).

Xyloglucan is also essential for root hair growth. Indeed, mutants for XG 6-XYLOSYLTRANSFERASES (*XXTs*) (*xxt1/xxt2*, *xxt1/xxt2/xxt5)* lack detectable XG and have short root hairs with swollen bases (Cavalier et al. 2008; Zabotina et al. 2008). Disruption of the synthesis of a GalA-containing XG (in the *ROOT HAIR SPECIFIC8/XYLOGLUCAN-SPECIFIC GALACTURONOSYLTRANSFERASE1* (*RHS8/XUT1*) mutant) results in shorter root hairs (Peña et al. 2012), which, however, do not rupture. The same is true for root hairs treated with a xyloglucanase (Park et al. 2011).

Expansins are also required for normal root hair growth as shown by the shorter root hair phenotypes of RNAi lines for *EXPANSIN A7* (*EXPA7*) in Arabidopsis (Lin et al. 2011), or an *OsEXPA17* mutant in rice (ZhiMing et al. 2011). In addition, application of expansin to growing root hairs triggered, at low concentrations, radial swelling of the tip and transient growth inhibition or, at higher concentrations, bursting (Cosgrove et al. 2002).

XTHs act on the same cellulose/XG network. Although XG endotransglycosylase activity (XET) was observed at the tip and in the shank of root hairs (Vissenberg et al. 2000, 2001, 2003), no *xth* mutants with a detectable root hair phenotype have been described so far, possibly due to redundancy.

Another critical oscillating parameter during root hair growth is the production of ROS in the apoplast. Indeed, mutations in *ROOT HAIR DEFECTIVE* (*RHD2*) encoding a plasma membrane NADPH oxidase strongly decreased ROS accumulation and caused stunted roots with no or very short root hairs (Foreman et al. 2003). Different interpretations of this phenotype have been formulated. Foreman et al. (2003) proposed that extracellular ROS is required for the activation of hyperpolarisation-activated Ca^{2+}-channels in the plasma membrane, which form the cytoplasmic calcium gradient that is required for root hair growth and acts in the negative feedback regulation of ROP activity (Yan et al. 2009). Monshausen et al. (2007) proposed an alternative interpretation, in which the lack of root hairs in *rhd2* is the result of premature bursting due to reduced apoplastic ROS levels. In this view, cell wall expansion in root hairs reflects a balancing act between growth-promoting (e.g. acid growth) and growth-restricting activities, where apoplastic ROS promotes wall strengthening through oxidative cross-linking of cell wall extensins (Schnabelrauch et al. 1996; Velasquez et al. 2011). Support for this view comes from the observation that growth at higher pH, which slows down cell expansion, restored root hair growth in the *rhd2* mutant (Monshausen et al. 2007). The essential role for extensins (Velasquez et al. 2011; 2015a,b; Draeger et al. 2015; Møller et al. 2017) but also of other structural proteins such as PRP1, PRP3 (Bernhardt and Tierney 2000), PRPL1 (Boron et al. 2014) and AGPs (Mollet et al. 2002; Marzec et al. 2015) in cell wall assembly of root hairs was shown by genetic and pharmacological studies.

Together, these observations show that understanding tip growth requires integration of the growth-promoting "calcium-pectate exchange" and "biomechanical hotspot" models but also growth-restricting activities involving oxidative cross-linking of cell wall proteins.

Given the importance and highly dynamic character of the cell wall at the growing root hair and pollen tube tip, the status of the cell wall must be constantly monitored to prevent (premature in case of the pollen tubes) bursting of the cell. This information on cell wall status must therefore feed back to the growth-controlling processes. RLKs of the *Catharanthus roseus* subfamily (CrRLK1Ls), such as FERONIA (FER) (Duan et al. 2010) and ERULUS (ERU) (Schoenaers et al. 2018), were suggested to function in the perception of cell wall status in root hairs by binding to specific cell wall ligands, and they are indeed related to growth control through activity changes of the apical plasma membrane-localised ROPs (Molendijk et al. 2001; Jones et al. 2002). Our current understanding states that ligand-bound RLKs activate ROPs, which then initiate NADPH-oxidase-dependent ROS production, leading to changes in calcium-channel activity and changes in the cell wall (Duan et al. 2010). It would be interesting to see which ligands serve as cell wall status indicators, and whether vesicle fusion and new cell wall deposition are controlled by RLK activation, especially in the light of the above-mentioned calcium-pectate exchange model to explain tip growth, where RLKs could act in the control of wall deposition and expansion (Rojas et al. 2011; Dumais 2021). Schoenaers et al. (2018) have indeed shown that ERU is involved in the control of cell wall PME activity dynamics, since *eru* root hairs have elevated PME activity levels, reflected in increased de-methylesterified pectin abundance and in propidium iodide fluorescence intensity oscillations with a lower frequency and higher amplitude.

DIFFUSE CELL EXPANSION

In contrast to the extremely polarised expansion in tip-growing cells, the growth of cells embedded in tissues is considered "diffuse", i.e., not restricted to specific subcellular locations. The distinction between tip growth and diffuse growth, however, is not always that clear (Geitmann and Ortega 2009). Indeed, the "diffuse" growth pattern can be either homogeneous or non-homogeneous, referred to as "homotropic" and "heterotropic", respectively. In the latter case, local foci with increased growth rate, reminiscent of tip growth, create irregular deformation patterns, underlying the large variety of cell shapes that can be found in plants (Figure 6.5).

One well-studied example of heterotropic growth is that of the formation of the lobes of puzzle-shaped pavement cells in the epidermis of Arabidopsis leaves or cotyledons (Figure 6.5a). In these cells, the appearance of inhomogeneities in cell wall deformation patterns depends, like in root hairs and pollen tubes, on ROP signalling modules, which not only promote the formation of specific membrane domains that locally organise the cytoskeleton and secretory events (Liu et al. 2021) but also other processes crucial for growth, such as protein translation (through TARGET OF RAPAMYCIN (TOR) signalling upon activation by the plant hormone auxin) (Schepetilnikov et al. 2017).

This raises the question of how the position of these ROP domains is determined. Two recent studies (Bidhendi et al. 2019; Belteton et al. 2021) showed that stress patterns may provide initial positional cues. Bidhendi et al. (2019) developed 3D Finite Element (FE) models of pressurised epidermal pavement cells (Figure 6.5a). This revealed that highest stress occurs along the anticlinal (perpendicular to the surface) cell walls. In addition, the periclinal (parallel to the surface) walls that bulge out pull the anticlinal walls (AWs) inward, which causes the latter to buckle. Although the buckling only creates a small deformation, it would be sufficient as a symmetry-breaking event for the lobe formation. The outgrowth of the lobe would then require a feedback mechanism involving the alignment of cortical microtubules along the stress patterns as has been observed in other systems (Hamant et al. 2008, 2019). Transfacial microtubules following AWs and radiating out from the neck region underneath the periclinal walls along the predicted stress patterns were indeed observed. These microtubules guide the deposition of cellulose microfibrils, which are expected to stiffen the cell wall (Figure 6.5b). The authors also observed an accumulation of de-methylesterified pectin at the neck region, which they expected to also contribute to the stiffening of the cell wall through the formation of Ca^{2+}-cross-links. The modelling showed that lobe outgrowth required not only growth restriction at the neck but also growth promotion, through a so far unknown mechanism, between the necks. Belteton et al. (2021) came to similar conclusions using a similar FE (finite element) model. They also used time lapse imaging to observe the transient formation of transfacial microtubule arrays and the cellulose synthase trajectories at predicted stress hotspots at the neck side of the future lobe. The authors proposed that the newly deposited cellulose microfibrils create a "patch of anisotropic wall". The subsequent local anisotropic cell wall deformation, perpendicular to the microfibril orientation, would then drive lobe formation (Belteton et al. 2021).

These studies provide a nice example of how mechanical stress patterns, predicted from a geometrical model, may actively initiate (Bidhendi et al. 2019) and/or provide initial positional cues (Bidhendi et al. 2019; Belteton et al. 2021) for local cell wall deformation patterns, which may be sharpened and reinforced by ROP-mediated feedback signalling. These models are implicitly based on the Lockhart model in which turgor drives and wall relaxation controls cell expansion (Lockhart 1965). An alternative driving force was proposed for the formation of the natural cell wall ingrowths, or furrows, observed in a number of cell types (Borowska-Wykręt and Kwiatkowska 2018) (Figure 6.5c and d). The authors propose that ingrowth is driven by the insertion and swelling of the matrix between the oriented microfibrils in the anticlinal cell walls, selectively on the convex side of the lobe. In this respect, the insertion and expansion of cell wall material may also be a driving force for the formation of Rhizobium infection threads (Rae et al. 1992) or fungal infection structures that progress against the turgor pressure of the host cell (Hückelhoven 2007).

A potential molecular mechanism for turgor-independent cell wall deformation was suggested by a recent study on Arabidopsis pavement cells (Haas et al. 2020a). This study used super-resolution microscopy with anti-HG antibodies to reveal that HG, instead of forming an amorphous gel, can show a striking degree of organisation, at least in anticlinal cell walls. Indeed, the epitopes for a subset of these antibodies (2F4, LM20 and LM19) formed filaments paralleling the cellulose microfibrils. Stacked parallel chains of HG are expected to form semi-crystalline fibrils, such as those that have been reported previously (Figure 6.3). Interestingly, other anti-HG antibodies (JIM7) did not reveal such filaments (Haas et al. 2020b). This indicates that firstly, the observed filaments were not artefactual due, for instance, to the ordering of an amorphous matrix constrained by the microfibrils and secondly, different HG populations with a distinct mesoscale organisation co-exist in the same cell wall. The study also showed a higher proportion of low/high DM HG in the wall on the convex side of the lobe. Given

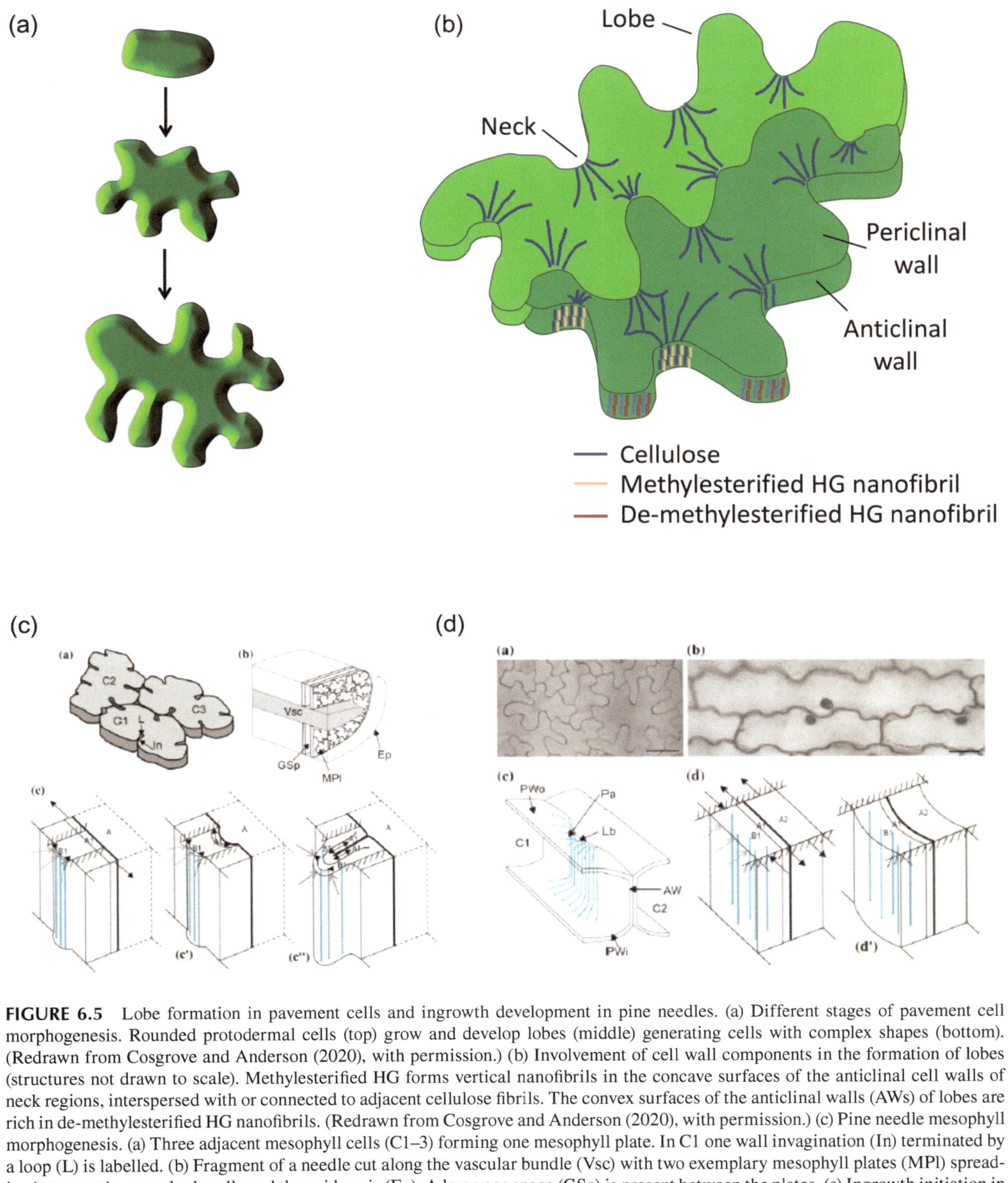

FIGURE 6.5 Lobe formation in pavement cells and ingrowth development in pine needles. (a) Different stages of pavement cell morphogenesis. Rounded protodermal cells (top) grow and develop lobes (middle) generating cells with complex shapes (bottom). (Redrawn from Cosgrove and Anderson (2020), with permission.) (b) Involvement of cell wall components in the formation of lobes (structures not drawn to scale). Methylesterified HG forms vertical nanofibrils in the concave surfaces of the anticlinal cell walls of neck regions, interspersed with or connected to adjacent cellulose fibrils. The convex surfaces of the anticlinal walls (AWs) of lobes are rich in de-methylesterified HG nanofibrils. (Redrawn from Cosgrove and Anderson (2020), with permission.) (c) Pine needle mesophyll morphogenesis. (a) Three adjacent mesophyll cells (C1–3) forming one mesophyll plate. In C1 one wall invagination (In) terminated by a loop (L) is labelled. (b) Fragment of a needle cut along the vascular bundle (Vsc) with two exemplary mesophyll plates (MPl) spreading between the vascular bundle and the epidermis (Ep). A large gas space (GSp) is present between the plates. (c) Ingrowth initiation in one of two contacting cell walls due to the swelling of the local wall thickening (B1 in c) is followed by the expansion of the ingrowth assisted by the stresses generated by the B1 layer (c', c"). (From Borowska-Wykręt and Kwiatkowska (2018), with permission.) (d) Jigsaw puzzle–shaped epidermal cells. (a and b) Images showing optical paradermal sections of the leaf epidermis in Arabidopsis (a) and maize (b). Scale bars 20 μm. (c) At the site of lobe (Lb) formation, cellulose microfibrils (blue/grey lines) form bundles in the AW of cell 1 (C1), into which lumen the wall is bending. The cellulose microfibril bundles extend radiating into the outer (PWo) and inner periclinal walls (PWi) of C1. Pads (Pa) are formed at the junctions between the anticlinal and periclinal walls. (d) Asymmetric swelling in one of the single walls comprising a compound wall fragment with fixed ends leads to the buckling of the wall (B1, A1, A2 in d). As a result, a lobe is initiated (d'). (From Borowska-Wykręt and Kwiatkowska (2018), with permission.)

the above-mentioned increased distance between polymers in crystals of low DM HG relative to those of high DM HG, the authors proposed a mechanism in which HG de-methylesterification induces a phase transition causing the cell wall to swell. Since this occurred preferentially on one side of the cell wall, a lobe was expected to form. Indeed, lobe formation could be simulated with a FE model based on this mechanism. The causality between PME activity, the HG phase transition and lobe formation was suggested by the inhibition of all these events by the inducible overexpression of a PMEI in transgenic Arabidopsis. In addition, inducible overexpression of a PME promoted the accumulation of low DM HG on both sides of the wall, increased the distance between fibrils observed by SEM and promoted cell expansion while preventing lobe formation. The latter was expected given the lack of asymmetry in the connected opposing cell walls.

It should be noted that lobe formation occurs during the expansion of the entire cell. This requires differential expansion rates across an already expanding wall. If we assume that the expansion of the straight walls is also, at least in part, driven by pectin insertion and swelling, lobe formation could arise from enhanced activation of PME activity on the convex side but also from the inhibition of de-methylesterification on the concave side. It will therefore be interesting to see whether the local cell wall pH differs between the two opposing cell walls at the concave and convex sides, since it would affect the activities of PMEs and other growth regulators.

Together, the study of lobe and furrow formation suggests that besides turgor-driven cell wall deformation, a turgor-independent driving force for cell wall deformation based on the conversion of chemical energy into mechanical energy associated with the de-methylesterification of HG may contribute to cell expansion.

NEW FOOD FOR THOUGHT

In summary, three models for cell expansion are currently being proposed in different contexts: (i) the biomechanical hotspot model for acid growth in diffusely growing cells, (ii) the Ca^{2+}-pectate cross-link exchange model for Chara and tip-growing cells and (iii) the expanding pectin filament model for lobe formation in pavement cells. The question is whether these models explain cell expansion only in certain cell wall types or whether they are part of an integrated processing system in which overlapping, synergistic or antagonistic chemo-mechanical phenomena and control points modulate cell wall expansion, with some aspects becoming dominant in some cell types/developmental stages/environmental conditions and others being completely absent in other instances. The above discussions showed that the latter possibility is the more likely one. Indeed, we discussed the requirement for XG-cellulose cross-link removal in tip-growing root hairs, in addition to Ca^{2+}-pectate cross-link exchange. Inversely, in diffusely growing cells, genetic evidence showed a key growth-promoting role for pectin de-methylesterification in a variety of contexts (Peaucelle et al. 2008, 2011, 2015; Andres-Robin et al. 2018; Haas et al. 2020a; Wachsman et al. 2020; Jonsson et al. 2021; Stefanowicz et al. 2021), suggesting that Ca^{2+}-pectate exchange and/or pectin swelling also occurs in these cells, although more complex scenarios involving signalling or PME-dependent cell wall acidification impinging on cellulose-XG remodelling cannot be excluded. In pavement cells, FE modelling showed that cell wall deformation driven by swelling pectins would create additional stresses in particular in external periclinal cell walls (Haas et al., 2020a), and perhaps also between younger and older cell wall layers, which would require relaxation for instance through XG-cellulose cross-link removal. Finally, pectin swelling upon de-methylesterification might also contribute to turgor-driven expansion. For instance, cell wall swelling was observed in Arabidopsis root epidermis cells upon auxin or brassinosteroid treatment (Elgass et al. 2009).

The study of tip growth showed the importance of the surface pH and ROS oscillations that contribute to expansion and cross-linking cycles leading to the formation of periodic cell wall deposition patterns. Similar processes could explain the deposition of layered cell walls of diffusely growing cells, although such oscillations would be more difficult to observe.

The observation of oriented HG chains and pectin fibrils raises the question how the HG chains can reach their uni-axial orientation. Is this due to steric constraints imposed by closely packed, similarly oriented, microfibrils or does it reflect an interaction with similarly oriented cellulose? Pectin–cellulose interactions have indeed been observed *in muro* using solid-state NMR on never-dried cell wall preparations (Wang et al. 2015). Also, fractionation studies revealed a pectin fraction strongly bound to the cellulosic residue. Interestingly, this fraction and not the chelator-extractable pectin fraction had disappeared in a seedling lethal mutant for the GalA transferase GAUT1 (Chapter 5 in this book), suggesting the existence of a specialised biosynthetic pathway. How pectin is bound to cellulose is not known, but it may be through xylan (Ralet et al. 2016), arabinan or galactan side chains (Zykwinska et al. 2005) of RG-I or even through covalent links that have been observed between HG and XG (Stratilová et al. 2020) or cellulose (Broxterman and Schols 2018), respectively. Alternative orientating mechanisms are not excluded, such as self-assembly through a nematic crystalline phase (Sanchez et al. 2012; Saw et al. 2018) or perhaps through exocytosis by vesicular-tubular structures reminiscent of those observed in neurons or in suberising plant cells (De Bellis et al. 2022).

Finally, the presence of uni-axially oriented HG filaments, the swelling of which contributes to the growth direction of single-layered anticlinal cell walls, corresponds more to a growth mechanism by intussusception than chemorheological growth. The above-mentioned statement of Bonner in 1934 (Bonner 1934) definitely cannot be extrapolated to all cell/tissue types or species!

ACKNOWLEDGEMENTS

The authors acknowledge the financial support by the University of Antwerp (BOF-DOCPRO4) and the Research Foundation Flanders (FWO Post-doc fellowship and Grant G013023N) to KV and ANR project ANR-20-CE11–0008, HOMEOWALL to HH. The IJPB benefits from the support of Saclay Plant Sciences-SPS (ANR-17-EUR-0007).

REFERENCES

Akkerman M, Franssen-Verheijen MAW, Immerzeel P, Den Hollander L, Schel JHN, Emons AMC (2012) Texture of cellulose microfibrils of root hair cell walls of *Arabidopsis thaliana*, *Medicago truncatula*, and *Vicia sativa*. *Journal of Microscopy* 247:60–67.

Ali S, Søndergaard CR, Teixeira S, Pickersgill RW (2015) Structural insights into the loss of catalytic competence in pectate lyase activity at low pH. *FEBS Letters* 589:3242–3246.

Anderson CT, Wallace IS, Somerville CR (2012) Metabolic click-labeling with a fucose analog reveals pectin delivery, architecture, and dynamics in Arabidopsis cell walls. *Proceedings of the National Academy of Sciences of the USA* 109:1329–1334.

Andres-Robin A, Reymond MC, Dupire A, Battu V, Dubrulle N, Mouille G, Lefebvre V, Pelloux J, Boudaoud A, Traas J, Scutt CP, Monéger F (2018) Evidence for the regulation of gynoecium morphogenesis by *ETTIN* via cell wall dynamics. *Plant Physiology* 178:1222–1232.

Arend M, Muninger M, Fromm J (2008) Unique occurrence of pectin-like fibrillar cell wall deposits in xylem fibres of poplar. *Plant Biology* 10:763–770.

Aryal B, Jonsson K, Baral A, Sancho-Andres G, Routier-Kierzkowski A-L, Kierzkowski D, Bhalerao R (2020) Interplay between cell wall and auxin mediates the control of differential cell elongation during apical hook development. *Current Biology* 30:1733–1739.

Baker LG, Specht CA, Donlin MJ, Lodge JK (2007) Chitosan, the deacetylated form of chitin, is necessary for cell wall integrity in Cryptococcus neoformans. *Eukaryotic Cell* 6:855–867.

Balcerowicz D, Schoenaers S, Vissenberg K (2015) Cell fate determination and the switch from diffuse growth to planar polarity in Arabidopsis root epidermal cells. *Frontiers in Plant Science* 6:1–13.

Baluška F, Salaj J, Mathur J, Braun M, Jasper F, **Šamaj** J, Chua NH, Barlow PW, Volkmann D (2000) Root hair formation: F-actin-dependent tip growth is initiated by local assembly of profilin-supported F-actin meshworks accumulated within expansin-enriched bulges. *Developmental Biology* 227: 618–632.

Band LR, Wells DM, Larrieu A, Sun J, Middleton AM, French AP, Brunoud G, Mendocilla Sato E, Wilson MH, Péret B, Oliva M, Swarup R, Sairanen I, Parry G, Ljung K, Beeckman T, Garibaldi JM, Estelle M, Owen MR, Vissenberg K, Hodgman TC, Pridmore TP, King JR, Vernoux T, Bennett MJ (2012) Root gravitropism is regulated by a transient lateral auxin gradient controlled by a tipping-point mechanism. *Proceedings of the National Academy of Sciences of the USA* 109:4668–4673.

Barbez E, Dünser K, Gaidora A, Lendl T, Busch W (2017) Auxin steers root cell expansion via apoplastic pH regulation in *Arabidopsis thaliana*. *Proceedings of the National Academy of Sciences of the USA* 114:4884–4893.

Belteton SA, Li W, Yanagisawa M, Hatam FA, Quinn MI, Szymanski MK, Marley MW, Turner JA, Szymanski DB (2021) Real-time conversion of tissue-scale mechanical forces into an interdigitated growth pattern. *Nature Plants* 7:989.

Bernal AJ, Yoo C-M, Mutwil M, Jensen JK, Hou G, Blaukopf C, Sørensen I, Blancaflor EB, Scheller HV, Willats WGT (2008) Functional analysis of the cellulose synthase-like genes *CSLD1*, *CSLD2*, and *CSLD4* in tip-growing Arabidopsis cells. *Plant Physiology* 148:1238–1253.

Bernhardt C, Tierney ML (2000) Expression of AtPRP3, a proline-rich structural cell wall protein from Arabidopsis, is regulated by cell-type-specific developmental pathways involved in root hair formation. *Plant Physiology* 122:705–714.

Bidhendi AJ, Altartouri B, Gosselin FP, Geitmann A (2019) Mechanical stress initiates and sustains the morphogenesis of wavy leaf epidermal cells. *Cell Reports* 28:1237–1250.

Boisson-Dernier A, Roy S, Kritsas K, Grobei MA, Jaciubek M, Schroeder JI, Grossniklaus U (2009) Disruption of the pollen-expressed FERONIA homologs ANXUR1 and ANXUR2 triggers pollen tube discharge. *Development* 136:3279–3288.

Bonner J (1934) Studies on the growth hormone of plants. V. The relation of cell elongation to cell wall formation. *Proceedings of the National Academy of Sciences of the USA* 20:393–397.

Boron AK, Orden JV, Markakis MN, Mouille G, Adriaensen D, Verbelen J-P, Höfte H, Vissenberg K (2014) Proline-rich protein-like PRPL1 controls elongation of root hairs in Arabidopsis thaliana. *Journal of Experimental Botany* 65:5485–5495.

Borowska-Wykręt D, Kwiatkowska D (2018). Folding, wrinkling, and buckling in plant cell walls. In *Plant Biomechanics*, eds Geitmann A, Gril J, 209–233. Cham: Springer.

Bove J, Vaillancourt B, Kroeger J, Hepler PK, Wiseman PW, Geitmann A (2008) Magnitude and direction of vesicle dynamics in growing pollen tubes using spatiotemporal image correlation spectroscopy and fluorescence recovery after photobleaching. *Plant Physiology* 147:1646–1658.

Boyer JS (2016) Enzyme-less growth in Chara and terrestrial plants. *Frontiers in Plant Science* 7:866.

Broxterman SE, Schols HA (2018) Interactions between pectin and cellulose in primary plant cell walls. *Carbohydrate Polymers* 192:263–272.

Cannon MC, Terneus K, Hall Q, Tan L, Wang Y, Wegenhart BL, Chen L, Lamport DT, Chen Y, Kieliszewski MJ (2008) Self-assembly of the plant cell wall requires an extensin scaffold. *Proceedings of the Nationall Academy of Sciences of the USA* 105:2226–2231.

Cao M, Chen R, Li P, Yu Y, Zheng R, Ge D, Zheng W, Wang X, Gu Y, Gelova Z, Friml J, Zhang H, Liu R, He J, Xu T (2019) TMK1-mediated auxin signalling regulates differential growth of the apical hook. *Nature* 568:240–243.

Capila I, Linhardt RJ (2002) Heparin-protein interactions. *Angewandte Chemie International Edition* 41:391–412.

Carol RJ, Takeda S, Linstead P, Durrant MC, Kakesova H, Derbyshire P, Drea S, Zarsky V, Dolan L (2005) A RhoGDP dissociation inhibitor spatially regulates growth in root hair cells. *Nature* 438:1013–1016.

Carpin S, Crévecoeur M, de Meyer M, Simon P, Greppin H, Penel C (2001) Identification of a Ca^{2+}-pectate binding site on an apoplastic peroxidase. *The Plant Cell* 13:511–520.

Castle ES (1958) The topography of tip growth in a plant cell. *Journal of General Physiology* 41:913–925.

Cavalier DM, Lerouxel O, Neumetzler L, Yamauchi K, Reinecke A, Freshour G, Zabotina OA, Hahn MG, Burgert I, Pauly M, Raikhel NV, Keegstra K (2008) Disrupting two Arabidopsis thaliana xylosyltransferase genes results in plants deficient in xyloglucan, a major primary cell wall component. *The Plant Cell* 20:1519–1537.

Cho H-T, Cosgrove DJ (2000) Altered expression of expansin modulates leaf growth and pedicel abscission in *Arabidopsis thaliana*. *Proceedings of the National Academy of Sciences of the USA* 97:9783–9788.

Cholodny N (1926) Beiträge zur analyse der geotropischen reaktion. *Jahrbücher für Wissenschaftliche Botanik* 65:447–459.

Cosgrove DJ (2000) Loosening of plant cell walls by expansins. *Nature* 407:321–326.

Cosgrove DJ (2018) Diffuse growth of plant cell wall. *Plant Physiology* 176:16–27.

Cosgrove DJ, Anderson CT (2020) Plant cell growth: do pectins drive lobe formation in Arabidopsis pavement cells? *Current Biology* 30:R660–662.

Cosgrove DJ, Li LC, Cho H-T, Hoffmann-Benning S, Moore RC, Blecker D (2002) The growing world of expansins. *Plant and Cell Physiology* 43:1436–1444.

Damineli DSC, Portes MT, Feijó JA (2017) Oscillatory signatures underlie growth regimes in Arabidopsis pollen tubes: computational methods to estimate tip location, periodicity, and synchronization in growing Cells. *Journal of Experimental Botany* 68:3267–3281.

Darwin CR (1880) *The Power of Movement in Plants*. London: Murray J.

De Bellis D, Kalmbach L, Marhavy P, Daraspe J, Geldner N, Barberon M (2022) Extracellular vesiculo-tubular structures associated with suberin deposition in plant cell walls. *Nature Communications* 13:1489.

Derksen J, Janssen GJ, Wolters-Arts M, Lichtscheidl I, Adlassnig W, Ovecka M, Doris F, Steer M (2011) Wall architecture with high porosity is established at the tip and maintained in growing pollen tubes of Nicotiana tabacum. *The Plant Journal* 68:495–506.

Domozych DS, Popper ZA, Sørensen I (2016) Charophytes: evolutionary giants and emerging model organisms. *Frontiers in Plant Science* 7:1–8.

Draeger C, Fabrice TN, Gineau E, Mouille G, Kuhn BM, Moller I, Abdou MT, Frey B, Pauly M, Bacic A, Ringli C (2015) Arabidopsis leucine-rich repeat extensin (LRX) proteins modify cell wall composition and influence plant growth. *BMC Plant Biology* 15:1–11.

Du M, Spalding ES, Gray WM (2020) Rapid auxin-mediated cell expansion. *Annual Review of Plant Biology* 71:379–402.

Duan Q, Kita D, Li C, Cheung AY, Wu HM (2010) FERONIA receptor-like kinase regulates RHO GTPase signaling of root hair development. *Proceedings of the National Academy of Sciences of the USA* 107:17821–17826.

Dumais J (2013) Modes of deformation of walled cells. *The Journal of Experimental Botany* 64:4681–4695.

Dumais J (2021) Mechanics and hydraulics of pollen tube growth. *New Phytologist* 232:549–1565.

Elgass K, Caesar K, Schleifenbaum F, Stierhof YD, Meixner AJ, Harter K (2009) Novel application of fluorescence lifetime and fluorescence microscopy enables quantitative access to subcellular dynamics in plant cells. *PLoS One* 4:e5716.

Endler A, Persson S (2011) Cellulose synthases and synthesis in Arabidopsis. *Molecular Plant* 4:199–211.

Estelle M (1996) Plant tropisms: the ins and outs of auxin. *Current Biology* 6:1589–1591.

Favery B, Ryan E, Foreman J, Linstead P, Boudonk K, Steer M, Shaw P, Dolan L (2001) KOJAK encodes a cellulose synthase-like protein required for root hair cell morphogenesis in *Arabidopsis*. *Genes and Development* 15:79–89.

Fayant P, Girlanda O, Chebli Y, Aubin CEE, Villemure I, Geitmann A (2010) Finite element model of polar growth in pollen tubes. *The Plant Cell* 22:2579–2593.

Feijó JA. Sainhas J, Hackett GR, Kunkel JG, Hepler PK (1999) Growing pollen tubes possess a constitutive alkaline band in the clear zone and a growth-dependent acidic tip. *Journal of Cell Biology* 144:483–496.

Fendrych M, Leung J, Friml J (2016) TIR1/AFB-Aux/IAA auxin perception mediates rapid cell wall acidification and growth of Arabidopsis hypocotyls. *Elife* 14:e19048.

Foreman J, Demidchik V, Bothwell JH, Mylona P, Miedema H, Torres MA, Linstead P, Costa S, Brownlee C, Jones JD, Davies JM, Dolan L (2003) Reactive oxygen species produced by NADPH oxidase regulate plant cell growth. *Nature* 422:442–446.

Francoz E, Ranocha P, Le Ru A, Martinez Y, Fourquaux I, Jauneau A, Dunand C, Burlat V (2019) Pectin Demethylesterification generates platforms that anchor peroxidases to remodel plant cell wall domains. *Developmental Cell* 48:261–276.

Friml J, Benková E, Blilou I, Wisniewska J, Hamann T, Ljung K, Woody S, Sandberg G, Scheres B, Jürgens G, Palme K (2002) AtPIN4 mediates sink-driven auxin gradients and root patterning in Arabidopsis. *Cell* 108:661–673.

Galway ME, Eng RC, Schiefelbein JW, Wasteneys GO (2011) Root hair-specific disruption of cellulose and xyloglucan in *AtCSLD3* mutants, and factors affecting the post-rupture resumption of mutant root hair growth. *Planta* 233:985–999

Geitmann A, Li YQ, Cresti M (1995) Ultrastructural immunolocalization of periodic pectin depositions in the cell wall of Nicotiana tabacum pollen tubes. *Protoplasma* 187:168–171.

Geitmann A, Ortega JKE (2009) Mechanics and modeling of plant cell growth. *Trends in Plant Science* 14:467–478.

Goh H-H, Sloan J, Malinowski R, Fleming A (2014) Variable expansin expression in Arabidopsis leads to different growth responses. *Journal of Plant Physiology* 171:329–339.

Grantham NJ, Wurman-Rodrich J, Terrett OM, Lyczakowski JJ, Stott K, Luga D, Simmons TJ, Durand-Tardiff M, Brown SP, Dupree R, Busse-Wicher M, Dupree P (2017) An even pattern of xylan substitution is critical for interaction with cellulose in plant cell walls. *Nature Plants* 3:859–865.

Green PB (1973) Morphogenesis of the cell and organ axis-biophysical models. *Brookhaven Symposium in Biology* 25:166–190.

Grignon C, Sentenac H (1991) pH and ionic conditions in the apoplast. *Annual Review of Plant Physiology and Plant Molecular Biology* 42:103–128.

Haas KT, Rivière M, Wightman R, Peaucelle A. (2020b) Multitarget immunohistochemistry for confocal and super-resolution imaging of plant cell wall polysaccharides. *Bio-Protocol* 10: e3783.

Haas KT, Wightmann R, Meyerowitz EM, Peaucelle A (2020a) Pectin homogalacturonan nanofilament expansion drives morphogenesis in plant epidermal cells. *Science* 367:1003–1007.

Hála M, Cole R, Synek L, Drdová E, Pečenková T, Nordheim A, Lamkemeyer T, Madlung J, Hochholdinger F, Fowler JE, Žárský V (2008) An exocyst complex functions in plant cell growth in Arabidopsis and tobacco. *The Plant Cell* 20:1330–1345.

Hepler PK, Lovy-Wheeler A, McKenna ST, Kunkel JG (2006) Ions and pollen tube growth. *Plant Cell Monographs* 3:47–69.

Hepler PK, Vidali L, Cheung AY (2001) Polarized growth in higher plants. *Annual Review of Cell and Developmental Biology* 17:159–187.

Hocq L, Pelloux J, Lefebvre V (2017) Connecting homogalacturonan-type pectin remodeling to acid growth. *Trends in Plant Science* 22: 20–29.

Hocq L, Habrylo O, Voxeur A, Pau-Roblot C, Safran J, Sénéchal F, Fournet F, Bassard S, Battu V, Demailly H, Tovar JC, Pilard S, Marcelo P, Savary BJ, Mercadante D, Njo MF, Beeckman T, Boudaoud A, Pelloux J, Lefèbvre V (2021) Arabidopsis AtPME2 has a pH-dependent processivity and control cell wall mechanical properties. *BioRXiv*. https://doi.org/10.1101/2021.03.03.433777.

Hoffmann RD, Portes MT, Olsen LI, Damineli DSC, Hayashi M, Nunes CO, Pedersen JT, Lima PT, Campos C, Feijó JA, Palmgren M (2020) Plasmamembrane H+-ATPases sustain pollen tube growth and fertilization. *Nature Communications* 11:2395.

Höltje JV (1998) Growth of the stress-bearing and shape-maintaining murein sacculus of Escherichia coli. *Microbiology and Molecular Biology Reviews* 62:181–203.

Hückelhoven R (2007) Cell wall-associated mechanisms of disease resistance and susceptibility. *Annual Review of Phytopathology* 45:101–127.

Hwang JU, Wu G, Yan A, Lee YL, Grierson CS, Yang Z (2010) Pollen-tube tip growth requires a balance of lateral propagation and global inhibition of Rho-Family GTPase activity. *Journal of Cell Science* 123:340–350.

Jaffe LA, Weisenseel MH, Jaffe LF (1975) Calcium accumulations within the growing tips of pollen tubes. *The Journal of Cell Biology* 67:488–492.

Jamet E, Boudart G, Borderies G, Charmont S, Lafitte C, Rossignol M, Canut H, Pont-Lezica R (2008) Isolation of plant cell wall proteins. *Series: Methods in Molecular Biology* 425:187–201.

Jiang L, Yang SL, Xie LF, Puah CS, Zhang XQ, Yang WC, Sundaresan V, Ye D (2005) VANGUARD1 encodes a pectin methylesterase that enhances pollen tube growth in the Arabidopsis style and transmitting tract. *The Plant Cell* 2005 17:584–596.

Jones MA, Shen JJ, Fu Y, Li H, Yang Z, Grierson CS (2002) The Arabidopsis Rop2 GTPase is a positive regulator of both root hair initiation and tip growth. *The Plant Cell* 14:763–776.

Jonsson K, Lathe RS, Kierzkowksi D, Routier-Kierzskowski A-L, Hamant O, Bhalerao R (2021) Mechanochemical feedback mediates tissue bending required for seedling emergence. *Current Biology* 31:1154–1164.

Ketelaar T, Galway ME, Mulder BM, Emons AMC (2008) Rates of exocytosis and endocytosis in Arabidopsis root hairs and pollen tubes. *Journal of Microscopy* 231:265–273.

Kneen M, Farinas J, Li Y, Verkman AS (1998) Green fluorescent protein as a noninvasive intracellular pH indicator. *Biophysical Journal* 74:1591–1599.

Kroeger JH, Geitmann A (2012) The pollen tube paradigm revisited. *Current Opinion in Plant Biology* 15:618–624.

Kroeger JH, Geitmann A, Grant M (2008) Model for calcium dependent oscillatory growth in pollen tubes. *Journal of Theoretical Biology* 253:363–374.

Kroeger JH, Zerzour R, Geitmann A (2011) Regulator or driving force? The role of turgor pressure in oscillatory plant cell growth. *PLoS ONE* 6:e18549.

Leppard GG, Colvin JR (1972) Electron-opaque fibrils and granules in and between the cell walls of higher plants. *Journal of Cell Biology* 53:635–708.

Leppard GG, Colvin JR, Rose D, Martin SM (1971) Lignofibrils on the external cell wall surface of cultured plant cells. *Journal of Cell Biology* 50:63–80.

Li H, Lin Y, Heath RM, Zhu MX, Yang Z (1999). Contr'ol of pollen tube tip growth by a Rop GTPase-dependent pathway that leads to tip-localized calcium influx. *The Plant Cell* 11:1731–1742.

Li H, Luo N, Wang W, Liu Z, Chen J, Zhao L, Tan L, Wang C, Qin Y, Li C, Xu T, Yang Z (2018) The REN4 rheostat dynamically coordinates the apical and lateral domains of Arabidopsis pollen tubes. *Nature Communications* 9:2573.

Li YQ, Chen F, Linskens HF, Cresti M (1994) Distribution of unesterified and esterified pectins in cell walls of pollen tubes of flowering plants. *Sexual Plant Reproduction* 7:145–152.

Liao C-Y, Smet W, Brunoud G, Yoshida S, Vernoux T, Weijers D (2015) Reporters for sensitive and quantitative measurement of auxin response. *Nature Methods* 12:207–210.

Lin C, Choi H-S, Cho H-T (2011) Root hair-specific EXPANSIN A7 is required for root hair elongation in *Arabidopsis*. *Molecular Cells* 31:393–397.

Lin W, Zhou W, Tang W, Takahashi K, Pan X, Dai J, Ren H, Zhu X, Pan S, Zheng H, Gray WM, Xu T, Kinoshita T, Yang Z (2021) TMK-based cell-surface auxin signaling activates cell-wall acidification. *Nature* 599:278–282.

Liu W, Xu L, Lin H, Cao J (2021) Two expansin genes, *AtEXPA4* and *AtEXPB5*, are redundantly required for pollen tube growth and *AtEXPA4* is involved in primary root elongation in *Arabidopsis thaliana*. *Genes* 12:249.

Lockhart JA (1965) An analysis of irreversible plant cell elongation. *Journal of Theoretical Biology* 8:264–275.

Lopez-Hernandez F, Tryfona T, Rizza A, Yu XL, Harris MOB, Webb AAR, Kotake T, Dupree P (2020) Calcium binding by arabinogalactan polysaccharides is important for normal plant development. *The Plant Cell* 32:3346–3369.

Ma L, Li G (2019) Auxin-dependent cell elongation during the shade avoidance response. *Frontiers in Plant Science 10*:914.

Macquet A, Ralet M-C, Loudet O, Kronenberger J, Mouille G, Marion-Poll A, North HM (2007) A naturally occurring mutation in an Arabidopsis accession affects a beta-D-galactosidase that increases the hydrophilic potential of rhamnogalacturonan I in seed mucilage. *The Plant Cell* 19:3990–4006.

Maris A, Kaewthai N, Eklöf JM, Miller JG, Brumer H, Fry SC, Verbelen J-P, Vissenberg K (2011) Characterization of five recombinant xyloglucan endotransglucosylase/hydrolase (XTH) proteins of Arabidopsis reveals specific enzymatic properties. *Journal of Experimental Botany* 62:261–271.

Maris A, Suslov D, Fry SC, Verbelen J-P, Vissenberg K (2009) Enzymic characterization of two recombinant xyloglucan endotransglucosylase/hydrolase (XTH) proteins of Arabidopsis and their effect on root growth and cell wall extension. *Journal of Experimental Botany* 60:3959–3972.

Marowa P, Ding A, Kong Y (2016) Expansins: roles in plant growth and potential applications in crop development. *Plant Cell Reports* 35:949–965.

Marzec M, Szarejko I, Melzer M (2015) Arabinogalactan proteins are involved in root hair development in Barley. *Journal of Experimental Botany* 66:1245–1257.

McFarlane HE, Doring A, Persson S (2014) The cell biology of cellulose synthesis. *Annual Review of Plant Biology* 65:69–94.

McKenna ST, Kunkel JG, Bosch M, Rounds CM, Vidali L, Winship LJ, Hepler PK (2009) Exocytosis precedes and predicts the increase in growth in oscillating pollen tubes. *The Plant Cell* 21:3026–3040.

McQueen-Mason S, Durachko DM, Cosgrove DJ (1992) Two endogenous proteins that induce cell wall extension in plants. *The Plant Cell* 4:1425–1433.

Messiaen J, Van Cutsem P (1999) Polyamines and pectins. II. Modulation of pectic-signal transduction. *Planta* 208:247–256.

Molendijk AJ, Bischoff F, Rajendrakumar CS, Friml J, Braun M, Gilroy S, Palme K (2001) Arabidopsis thaliana Rop GTPases are localized to tips of root hairs and control polar growth. *The EMBO Journal* 20:2779–2788.

Møller SR, Yi X, Velásquez SM, Gille S, Hansen PML, Poulsen CP, Olsen CE, Rejzek M, Parsons H, Yang Z, Wandall HH, Clausen H, Field RA, Pauly M, Estevez JM, Harholt J, Ulvskov P, Petersen BL (2017) Identification and evolution of a plant cell wall specific glycoprotein glycosyl transferase, ExAD. *Scientific Reports* 7:1–16.

Mollet JC, Kim S, Jauh GY, Lord EM (2002) Arabinogalactan proteins, pollen tube growth, and the reversible effects of Yariv phenylglycoside. *Protoplasma* 219:89–98.

Monshausen GB, Bibikova TN, Messerli MA, Shi C, Gilroy S (2007) Oscillations in extracellular pH and reactive oxygen species modulate tip growth of Arabidopsis root hairs. *Proceedings of the National Academy of Sciences of the USA* 104:20996–21001.

Moustacas AM, Nari J, Borell M, Noat G, Ricard J (1991) Pectin methylesterase: metal ions and plant cell wall extension. The role of metal ions in plant cell wall extension. *Journal of Biochemistry* 279:351–354.

Nägeli C (1864) **Über den inneren Bau der vegetabilischen Zellmembranen**. *Sitzungsberichte Der Königl, Bayer Akademie Der Wissenschaften Zu München* 1:282–323.

Newcomb EH, Bonnett HT (1965) Cytoplasmic microtubule and wall microfibril orientation in root hairs of radish. *Journal of Cell Biology* 27:575–589.

Nibau C, Wu HM, Cheung AY (2006) RAC/ROP GTPases: 'hubs' for signal integration and diversification in plants. *Trends in Plant Science* 11:309–315.

Nicolas WJ, Fassler F, Dutka P, Schur FKM, Jensen G, Meyerowitz E, Jensen G (2022) Cryo-electron tomography of the onion cell wall shows bimodally oriented cellulose fibers and reticulated homogalacturonan networks. *Current Biology* 32:1–15.

Noh B, Murphy AS, Spalding EP (2001) Multidrug resistance-like genes of Arabidopsis required for auxin transport and auxin-mediated development. *The Plant Cell* 13:2441–2454.

Noorifar N, Savoian MS, Ram A, Lukito Y, Hassing B, Weikert TW, Moerschbacher BM, Scott B (2021) Chitin deacetylases are required for *Epichloë festucae* endophytic cell wall remodeling during establishment of a mutualistic symbiotic interaction with *Lolium perenne*. *Molecular Plant Microbe Interactions* 34:1181–1192.

Osada Y, Gong JP (1993) Stimuli-responsive polymer gels and their application to chemomechanical systems. *Progress in Polymer Science* 18:187–226.

Palmer KJ, Merrill RC, Owens HS, Ballantyne M (1947) An X-ray diffraction investigation of pectinic and pectic acids. *Journal of Physical Chemistry* 51:710–720.

Pang CY, Wang H, Pang Y, Xu C, Jiao Y, Qin YM, Western TL, Yu SX, Zhu YX (2010) Comparative proteomics indicates that biosynthesis of pectic precursors is important for cotton fiber and Arabidopsis root hair elongation. *Molecular and Cellular Proteomics* 9:2019–1033.

Park S, Szumlanski AL, Gu FW, Guo F, Nielsen E (2011) A role for CSLD3 during cell-wall synthesis in apical plasma membranes of tip-growing root-hair cells. *Nature Cell Biology* 13:973–980.

Park YB, Cosgrove DJ (2012) A revised architecture of primary cell walls based on biomechanical changes induced by substrate-specific endoglucanases. *Plant Physiology* 158:1933–1943.

Pauly M, Albersheim P, Darvill A, York WS (1999) Molecular domains of the cellulose/xyloglucan network in the cell walls of higher plants. *The Plant Journal* 20:629–639.

Peaucelle A, Louvet R, Johansen JN, Höfte H, Laufs P, Pelloux J, Mouille G (2008) Arabidopsis phyllotaxis is controlled by the methyl-esterification status of cell-wall pectins. *Current Biology* 18:1943–1948.

Peaucelle A, Braybrook S, Guillou L, Bron E, Kuhlemeier C, Höfte H (2011). Pectin-induced changes in cell wall mechanics underlie organ initiation in Arabidopsis. *Current Biology* 21:1720–1726.

Peaucelle A, Wightman R, Höfte H (2015) The control of growth symmetry breaking in the arabidopsis hypocotyl. *Current Biology* 25:1746–1752.

Peña MJ, Kong Y, York WS, O'Neill MA. (2012) A galacturonic acid-containing xyloglucan is involved in Arabidopsis root hair tip growth. *The Plant Cell* 24:4511–4524.

Picton JM, Steer MW (1983) Evidence for the role of Ca^{2+} ions in tip extension in pollen tubes. *Protoplasma* 115:11–17.

Pien S, Wyrzykowska J, McQueen-Mason SJ (2001) local expression of expansin induces the entire process of leaf development and modifies cell shape. *Proceedings of the National Academy of Sciences of the USA* 98:11812–11817.

Pierson ES, Miller DD, Callaham DA, Van Aken J, Hackett G, Hepler PK (1996) Tip-localized calcium entry fluctuates during pollen tube growth. *Developmental Biology* 174:160–173.

Popper ZA, Michel G, Hervé C, Domozych DS, Willats WGT, Tuohy MG, Kloareg B, Stengel DB (2011) Evolution and diversity of plant cell walls: from algae to flowering plants. *Annual Review of Plant Biology* 62:567–590.

Potter HB (1897) Streaming movements of the protoplasm in pollen of flowers. *Nature* 1446:248.

Rakusová H, Gallego-Bartolomé J, Vanstraelen M, Robert HS, Alabadí D, Blázquez MA, Benková E, Friml J (2011) Polarization of PIN3-dependent auxin transport for hypocotyl gravitropic response in *Arabidopsis thaliana*. *The Plant Journal* 67:817–826.

Rae AL, Bonfante-Pasolo P, Brewin NJ (1992) Structure and growth of infection threads in the legume symbiosis with *Rhizobium leguminosarum*. *The Plant Journal* 2:385–395.

Rakusová H, Gallego-Bartolomé J, Vanstraelen M, Robert HS, Alabadí D, Blázquez MA, Benková E, Friml J (2011) Polarization of PIN3-dependent auxin transport for hypocotyl gravitropic response in *Arabidopsis thaliana*. *The Plant Journal* 67:817–826.

Ralet M C, Dronnet V, Buchholt H C, Thibault J F (2001) Enzymatically and chemically de-esterified lime pectins: characterisation, polyelectrolyte behaviour and calcium binding properties. *Carbohydrate Research* 336:117–125.

Ralet M-C, Crépeau M-J, Vigouroux J, Tran J, Berger A, Sallé C, Granier F, Botran L, North H (2016) Xylans provide the structural driving force for mucilage adhesion to the Arabidopsis seed coat. *Plant Physiology* 171:165–178.

Ray PM (1967) Radioautographic study of cell wall deposition in growing plant cells. *The Journal of Cell Biology* 35:659–674.

Rayle DL, Cleland RE (1970) Enhancement of wall loosening and elongation by acid solutions. *Plant Physiology* 46:250–253.

Rayle DL, Cleland RE (1992) The acid growth theory of auxin-induced cell elongation is alive and well. *Plant Physiology* 99:1271–1274.

Refrégier G, Pelletier S, Jaillard D, Höfte H (2004) Interaction between wall deposition and cell elongation in dark-grown hypocotyl cells in Arabidopsis. *Plant Physiology* 135:959–968.

Reiss HD, Herth W (1985) Nifedipine-sensitive calcium channels are involved in polar growth of Lily pollen tubes. *Journal of Cell Science* 76:247–254.

Rodriguez-Furlán C, Salinas-Grenet H, Sandoval O, Recabarren C, Arraño-Salinas P, Soto-Alvear S, Orellana A, Blanco-Herrera F (2016) The root hair specific SYP123 regulates the localization of cell wall components and contributes to Rizhobacterial priming of induced systemic resistance. *Frontiers in Plant Science* 7:1–14.

Roelofsen PA, Houwink AL (1953) Architecture and growth of the primary cell wall in some plant hairs and in the phycomyces sporangiophore. *Acta Botanica Néerlandica* 2:218–125.

Roelofsen PA, Kreger DR (1951) The sub-microscopic structure of pectin in collenchyma cell-walls. *Journal of Experimental Botany* 2:332–343.

Rojas ER, Hotton S, Dumais J (2011) Chemically mediated mechanical expansion of the pollen tube cell wall. *Biophysical Journal* 101:1844–1853.

Rounds CM, Lubeck E, Hepler PK, Winship LJ (2011) Propidium iodide competes with Ca^{2+} to label pectin in pollen tubes and Arabidopsis root hairs. *Plant Physiology* 157:175–187.

Růžička K, Ljung K, Vanneste S, Podhorská R, Beeckman T, Friml J, Benková E (2007) Ethylene regulates root growth through effects on auxin biosynthesis and transport-dependent auxin distribution. *The Plant Cell* 19:2197–2212.

Sachs J (1874) *Lehrbuch der Botanik nach dem gegenwärtigen Stand der Wissenschaft*. Leipzig: Verlag von Wilhelm Engelmann.

Sanchez T, Chen DT, DeCamp SJ, Heymann M, Dogic Z (2012) Spontaneous motion in hierarchically assembled active matter. *Nature* 491:431–434.

Saw TB, Xi W, Ladoux B, Lim CT (2018) Biological tissues act as active nematic liquid crystals. *Advanced Materials* 30:1802579.

Schepetilnikov M, Makarian J, Srour O, Geldreich A, Yang Z, Chicher J, Hammann P, Ryabova LA (2017) GTPase ROP2 binds and promotes activation of target of rapamycin, TOR, in response to auxin. *The EMBO Journal* 36:886–903.

Schnabelrauch LS, Kieliszewski MJ, Upham BL, Alizedeh H, Lamport DTA (1996) Isolation of pI 4.6 extensin peroxidase from tomato cell suspension cultures and identification of Val-Tyr-Lys as putative intermolecular cross-link site. *The Plant Journal* 9:477–489.

Schoenaers S, Balcerowicz D, Breen G, Hill K, Zdanio M, Mouille G, Holman TJ, Oh J, Wilson MH, Nikonorova N, Vu LD, De Smet I, Swarup R, De Vos WH, Pintelon I, Adriaensen D, Grierson C, Bennett MJ, Vissenberg K (2018) The auxin-regulated CrRLK1L kinase ERULUS controls cell wall composition during root hair tip growth. *Current Biology* 28:722–732.

Schoenaers S, Balcerowicz D, Vissenberg K (2017) Molecular mechanisms regulating root hair tip growth: a comparison with pollen tubes. In *Pollen Tip Growth: From Biophysical Aspects to Systems Biology*, eds Obermeyer G, Feijó JA, 167–243. Berlin, Heidelberg, New York: Springer.

Sentenac H, Grignon C (1981) A model for predicting ionic equilibrium concentrations in cell walls. *Plant Physiology* 68:415–419.

Šola K, Gilchrist EJ, Ropartz D, Wang L, Feussner I, Mansfield SD, Ralet MC, Haughn GW (2019) RUBY, a putative galactose oxidase, influences pectin properties and promotes cell-to-cell adhesion in the seed coat epidermis of Arabidopsis. *The Plant Cell* 31: 809–831.

Staal M, De Cnodder T, Simon D, Vandenbussche F, van Der Straeten D, Verbelen J-P, Vissenberg K (2011) Apoplastic alkalinization is instrumental for the inhibition of cell elongation in the Arabidopsis root by the ethylene precursor 1-aminocyclopropane-1-carboxylic acid. *Plant Physiology* 155:2049–2055.

Stefanowicz K, Szymanska-Chargot M, Truman W, Walerowski P, Olszak M, Augustyniak A, Kosmala A, Zdunek A, Malinowski R (2021) *Plasmodiophora brassicae*-triggered cell enlargement and loss of cellular integrity in root systems are mediated by pectin demethylation. *Frontiers in Plant Science* 12:711838.

Sterling C. (1957) Structure of oriented gels of calcium polyuronates. *Biochimica et Biophysica Acta* 26:186–197.

Strasburger E (1882) *Ueber den Bau und das Wachstum der Zellhäute*. Jena: Verlag von Gustav Fischer.

Stratilová B, Sestak S, Mravec J, Garajová S, Pakanová Z, Vadinová K, Kucerová D, Kozmon S, Schwerdt JG, Shirley N, Stratilová E, Hrmova M (2020) Another building block in the plant cell wall: Barley xyloglucan xyloglycosyl transferases link covalently xyloglucan and anionic oligosaccharides derived from pectin. *The Plant Journal* 104:752–767.

Suslov D, Verbelen J-P, Vissenberg K (2009) Onion epidermis as a new model to study the control of growth anisotropy in higher plants. *Journal of Experimental Botany* 60:4175–4187.

Suslov D, Vissenberg K (2018) Cell wall expansion as viewed by the creep method. In *Plant Biomechanics - From Structure to Function at Multiple Scales*, eds Geitmann A, Gril J, 305–320. Berlin Heidelberg New York: Springer Verlag.

Swarup R, Perry P, Hagenbeek D, van Der Straeten D, Beemster GTS, Sandberg G, Bhalerao R, Ljung K, Bennett MJ (2007) Ethylene upregulates auxin biosynthesis in Arabidopsis seedlings to enhance inhibition of root cell elongation. *The Plant Cell* 19:2186–2196.

Tan L, Eberhard S, Pattathil S, Warder C, Glushka J, Yuan C, Hao Z, Zhu X, Avci U, Miller JS, Baldwin D, Pham C, Orlando R, Darvill A, Hahn MG, Kieliszewski MJ, Mohnen D (2013) An Arabidopsis cell wall proteoglycan consists of pectin and arabinoxylan covalently linked to an arabinogalactan pProtein. *The Plant Cell* 25:270–287.

Tanaka T (1981) Gels. *Scientific American* 244:124–138.

Tanaka T, Annaka M, Ilmain F, Ishii K, Kokufuta E, Suzuki A, Tokita M (1992) Phase transitions of gels. In *Mechanics of Swelling*, ed Karalis TK, 683–703. Berlin, Heidelberg: NATO ASI Series Springer.

Tian GW, Chen MH, Zaltsman A, Citovsky V (2006) Pollen-specific pectin methylesterase involved in pollen tube growth. *Developmental Biology* 294:83–91.

Valentin R, Cerclier C, Geneix N, Aguie-Beghin V, Gaillard C, Ralet MC, Cathala B (2010) Elaboration of extensin-pectin thin film model of primary plant cell wall. *Langmuir* 26:9891–9898.

Velasquez SM, Ricardi MM, Dorosz JG, Fernandez PV, Nadra AD, Pol-Fachin L, Egelund J, Gille S, Harholt J, Ciancia M, Verli H, Pauly M, Bacic A, Olsen CE, Ulvskov P,

Petersen BL, Somerville C, Iusem ND, Estevez JM (2011) O-glycosylated cell wall proteins are essential in root hair growth. *Science* 332:1401–1403.

Velasquez SM, Marzol E, Borassi C, Pol-Fachin L, Ricardi MM, Mangano S, Juarez SPD, Salgado Salter JD, Dorosz JG, Marcus SE, Knox JP, Dinneny JR, Iusem ND, Verli H, Estevez JM (2015a) Low sugar is not always good: impact of specific O-glycan defects on tip growth in Arabidopsis. *Plant Physiology* 168:808–813.

Velasquez SM, Ricardi MM, Poulsen CP, Oikawa A, Dilokpimol A, Halim A, Mangano S, Juarez SPD, Marzol E, Salgado Salter JD, Dorosz JG, Borassi C, Möller SR, Buono R, Ohsawa Y, Matsuoka K, Otegui MS, Scheller HV, Geshi N, Petersen BL, Iusem ND, Estevez JM (2015b) Complex regulation of prolyl-4-hydroxylases impacts root hair expansion. *Molecular Plant* 8:734–746.

Vincent RRR, Mansel BW, Kramer A, Kroy K, Williams MAK (2013) Micro-rheological behaviour and nonlinear rheology of networks assembled from polysaccharides from the plant cell wall. *New Journal of Physics* 15:35002.

Vissenberg K, Martinez-Vilchez IM, Verbelen J-P, Miller JG, Fry SC (2000) In vivo colocalisation of xyloglucan endotransglycosylase activity and its donor substrate in the elongation zone of Arabidopsis roots. *The Plant Cell* 12:1229–1237.

Vissenberg K, Fry SC, Verbelen J-P (2001) Root hair initiation is coupled to a highly localized increase of xyloglucan endotransglycosylase action in Arabidopsis roots. *Plant Physiology* 127:1125–1135.

Vissenberg K, Van Sandt V, Fry SC, Verbelen J-P (2003) Xyloglucan endotransglucosylase action is high in the root elongation zone and in the trichoblasts of all vascular plants from *Selaginella* to *Zea mays*. *Journal of Experimental Botany* 54:335–344.

Vissenberg K, Fry SC, Pauly M, Höfte H, Verbelen J-P (2005) XTH acts at the microfibril-matrix interface during cell elongation. *Journal of Experimental Botany* 56:673–683.

Voxeur A, Habrylo O, Guénin S, Miart F, Soulié M-C, Rihouey C, Pau-Roblot C, Domon J-M, Gutierrez L, Pelloux J, Mouille G, Fagard M, Höfte H, Vernhettes S (2019) Oligogalacturonide production upon Arabidopsis thaliana-Botrytis cinerea interaction. *Proceedings of the National Academy of Sciences of the USA* 116:19743–19752.

Wachsman G, Zhang J, Moreno-Risueno MA, Anderson CT, Benfey PN (2020) Cell wall remodeling and vesicle trafficking mediate the root clock in *Arabidopsis*. *Science* 370:819–823.

Walkinshaw MD, Arnott S (1981a) Conformations and interactions of pectins. *Journal of Molecular Biology* 153:1075–1085.

Walkinshaw MD, Arnott S (1981b) Conformations and interactions of pectins. *Journal of Molecular Biology* 153:1055–1073.

Wang M, Yuan D, Gao W, Li Y, Tan J, Zhang X (2013) A comparative genome analysis of PME and PMEI families reveals the evolution of pectin metabolism in plant cell walls. *PLoS ONE* 8:e72082.

Wang T, Park YB, Caporini MA, Rosay M, Zhong L, Cosgrove DJ, Hong M (2013) Sensitivity-enhanced solid-state NMR detection of expansin's target in plant cell walls. *Proceedings of the National Academy of Sciences of the USA* 110:16444–16449.

Wang T, Park YB, Cosgrove DJ, Hong M (2015) Cellulose-pectin spatial contacts are inherent to never-dried Arabidopsis primary cell walls: evidence from solid-state nuclear magnetic resonance. *Plant Physiology* 168:871–884.

Weisenseel M, Kicherer RM (1981) Ionic currents as control mechanism in cytomorphogenesis. *Cell Biology Monographs* 8: 379–399.

Weisenseel M, Nuccitelli R, Jaffe LF (1975) Large electrical currents traverse growing pollen tubes. *Journal of Cell Biology* 66:556–567.

Went FW (1928) Wuchsstoff und Wachstum. *Recueil des travaux botaniques néerlandais* 25:1–116.

Went FW, Thimann KV (1937) *Phytohormones*. New York: Macmillan Press Ltd.

Wilson LA, Deligey F, Wang T, Cosgrove DJ (2021) Saccharide analysis of onion outer epidermal walls. *Biotechnology for Biofuels* 14:66.

Wolf S, Mouille G, Pelloux J (2009) Homogalacturonan methyl-esterification and plant development. *Molecular Plant* 2:851–860.

Wu J, Kurten EL, Monshausen G, Hummel GM, Gilroy S, Baldwin IT (2007) NaRALF, a peptide signal essential for the regulation of root hair tip apoplastic pH in Nicotiana attenuata, is required for root hair development and plant growth in native soils. *The Plant Journal* 52:877–890.

Yalovsky S, Bloch D, Sorek N, Kost B (2008) Regulation of membrane trafficking, cytoskeleton dynamics, and cell polarity by ROP/RAC GTPases. *Plant Physiology* 147:1527–1543.

Yan A, Xu G, Yang ZB (2009) Calcium participates in feedback regulation of the oscillating ROP1 Rho GTPase in pollen tubes. *Proceedings of the National Academy of Sciences of the USA* 106:22002–22007.

Yang J, Bak G, Burgin T, Barnes WJ, Mayes HB, Peña MJ, Urbanowicz BR, Nielsen E (2020) Biochemical and genetic analysis identify CSLD3 as a beta-1,4-glucan synthase that functions during plant cell wall synthesis. *The Plant Cell* 32:1749–1767.

Zablackis E, Huang J, Muller B, Darvill AG, Albersheim P (1995) Characterization of the cell-wall polysaccharides of Arabidopsis thaliana leaves. *Plant Physiology* 107:1129–1138.

Zabotina OA, van de Ven WTG, Freshour G, Drakakaki G, Cavalier D, Mouille G, Hahn MG, Keegstra K, Raikhel H (2008) Arabidopsis XXT5 gene encodes a putative α-1,6-xylosyltransferase that is involved in xyloglucan biosynthesis. *The Plant Journal* 56:101–115.

Zhang T, Mahgsoudy-Louyeh S, Tittmann B, Cosgrove DJ (2013) Visualization of the nanoscale pattern of recently-deposited cellulose microfibrils and matrix materials in never-dried primary walls of the onion epidermis. *Cellulose* 21:853–862

Zhang T, Vavylonis D, Durachko DM, Cosgrove DJ (2017) Nanoscale movements of cellulose microfibrils in primary cell walls. *Nature Plants* 3:17056.

Zhang Y, Yu J, Wang X, Durachko DM, Zhang S, Cosgrove DJ (2021) Molecular insights into the complex mechanics of plant epidermal cell walls. *Science* 711:706–711.

ZhiMing Y, Bo K, XiaoWei H, ShaoLei M, YouHuang B, WoNa D, Ming C, Hyung-Taeg C, Ping W (2011) Root hair-specific expansins modulate root hair elongation in rice. *The Plant Journal* 66:725–734.

Zabotina OA, van de Ven WTG, Freshour G, Drakakaki G, Cavalier D, Mouille G, Hahn MG, Keegstra K, Raikhel H (2008) Arabidopsis XXT5 gene encodes a putative α-1,6-xylosyltransferase that is involved in xyloglucan biosynthesis. *The Plant Journal* 56:101–115.

Zsivanovits G, MacDougall AJ, Smith AC, Ring SG (2004) Material properties of concentrated pectin networks. *Carbohydrate Research* 339:1317–1322.

Zykwinska AW, Ralet MCJ, Garnier CD, Thibault J-FJ (2005) Evidence for in vitro binding of pectin side chains to cellulose. *Plant Physiology* 139:397–407.

7 Role of the Cell Wall in the Regulation of Fruit Ripening

Panagiotis Kalaitzis
Mediterranean Agronomic Institute of Chania

Eleni Giannoutsou
National and Kapodistrian University of Athens

Anna Konkina
Mediterranean Agronomic Institute of Chania

INTRODUCTION

Fruits and vegetables constitute the basis of a healthy human diet and, to satisfy supply chain needs and consumer demands, it is important to be able to preserve high-quality fresh fruits for extended periods of time. Quality of fresh fruits is characterised by various properties. A key quality component is texture, which is associated with cell wall modifications in the context of fruit softening. Based on the degree of softening during ripening, fleshy fruits are categorised into two groups (Bourne 1979; Paniagua et al. 2014): those that soften significantly as they ripen, acquiring a melting texture and therefore cannot be stored for a long period of time (e.g. strawberry, tomato, peach and apricot), and those whose ripening is accompanied by moderate softening while retaining their crispness and can be stored for extended periods of time (e.g. apple, pear and watermelon).

Fleshy fruits are characterised as climacteric or non-climacteric, depending on whether they produce autocatalytic ethylene or not, respectively (Cherian et al. 2014; Klee and Giovannoni 2011). Tomato, apple, pear, peach, kiwi, mango, avocado and banana are climacteric and are characterised by an increase in respiration and ethylene production at the onset of ripening. Strawberry, olive, blackcurrant, cucumber, grape, grapefruit, pepper and citrus are considered non-climacteric, and ethylene production is not induced (Cherian et al. 2014). Exogenous ethylene promotes ripening in climacteric fruits, while it might induce a rise in the respiration rate of non-climacteric fruits. Thus, ethylene is considered an initial and essential factor of fruit ripening.

There are three widely studied tomato fruit mutants, the *Cnr*, *rin* and *nor* whose investigation has contributed significantly to the characterisation of fruit ripening regulation (Li et al. 2021). The tomato *Cnr* is a mutant exhibiting ripe fruits of yellow colour with a loss of cell-to-cell adhesion in pericarp tissue, which affects fruit texture and softening (Manning et al. 2006). This mutation was attributed to hypermethylation of the promoter region of the transcription factor SBP-box protein (Manning et al. 2006). The tomato *rin* mutant shows a significant inhibition of fruit ripening (Vrebalov et al. 2002). The fruits do not exhibit the climacteric increase in ethylene production and respiration rate, while a delay in softening, flavour development and carotenoid accumulation is observed (Vrebalov et al. 2002). The *rin* mutation was localised on a MADS-box transcription factor and the phenotype was attributed to loss of function of this gene (Vrebalov et al. 2002). The Rin MADS-box is considered a master regulator of tomato fruit ripening. The tomato *nor* mutant affects ethylene biosynthesis, carotenoids production and fruit softening and is caused by a two base pair deletion in the coding sequence of a NAC transcription factor (Seymour et al. 2013).

Fruit ripening is a highly coordinated developmental programme, which comprises various, dramatic changes in colour, flavour, texture and aroma, pigments and sugars accumulation as well as chloroplasts differentiation into chromoplasts (Klee and Giovannoni 2011). One of the first changes that occur during fruit ripening is the dissolution of the middle lamella according to electron microscopy observations (Brummell 2006; Fry 2003). This layer between cells is rich in pectins, whose dissolution leads to the formation of intercellular spaces that appear triangular on tissue cross sections (see Chapter 23). The second step in the fruit softening process is a progressive degradation of the cell wall polysaccharides and an alteration of the linkages connecting the various components (Brummell 2006; Cosgrove 1997). The progressive dismantling causes a softening of the cell wall, and in some species, the cells walls are swollen and intercellular adhesion is further decreased (Hallett et al. 1992; Redgwell et al. 1992). To ascertain the changes that occur during softening at the cell wall of the fleshy fruits, a reminder of the primary cell wall architecture is essential.

DOI: 10.1201/9781003178309-7

FRUIT RIPENING AND CELL WALL STRUCTURAL COMPONENTS

Atomic force microscopy (AFM) is a scanning probe microscopy technique based on surface topography sensing and is used to study the nanostructure of cell wall polysaccharides isolated from fruits (Morris and Sharma 2010). AFM provides magnification similar to electron microscopy without using lenses or beam irradiation, requires minimal sample preparation and can be conducted in air or aqueous matrix (Kirby et al. 1995; Morris and Sharma 2010). Three different methods are used for cell wall polysaccharide imaging: the contact mode, the non-contact mode and the tapping mode (Morris and Sharma 2010). The contact mode refers to the direct contact of the AFM tip to the sample leading to high-contrast images (Morris and Sharma 2010). In the non-contact method, the sample is in the air, while there is a distance between the sample and the tip leading to loss of resolution (Morris and Sharma 2010). In the tapping mode, the probe is in contact with the sample intermittently leading to higher contrast images and lower damage of the sample (McIntire and Brant 1997).

AFM has been applied on cell walls of various fruit crops, such as tomato (*Solanum lycopersicum* L.) (Kirby et al. 2008, Round et al. 2010, Xin et al. 2010), strawberry (*Fragaria ananassa*) (Posé et al. 2012, Chen et al. 2011), peach (*Prunus Persian* L. Batsch.) (Yang et al. 2006a, andb, 2009, Zhang et al. 2012), cherry (*Prunus pseudocerasus* L.) (Lai et al. 2013, Zhang et al. 2008) and apricot (*Prunus armeniaca* L.) (Chen et al. 2013).

The use of AFM has led to substantial new insights into the cell wall structure (Round et al. 2001, 2010; Paniagua, Kirby et al. 2017) such as better understanding of polysaccharide structure and composition during fruit ripening. This in turn has contributed to the elucidation of mechanistic details in softening-induced pectin dissolution and depolymerisation. AFM imaging allows the visualisation of plant polysaccharides at the molecular level, including linear chains, branches, stiff chains, aggregates, helix and other structures. Various polysaccharides including pectin, xyloglucan, arabinoxylan and starch were analysed by AFM (Wang and Nie 2019).

The plant cell wall analysis involves the use of extractans of gradual astringency for fractionation of cell wall components (Posé et al. 2019). The use of a weak alkali such as sodium carbonate produces the corresponding sodium carbonate soluble (SSP) fraction enriched in pectins covalently bound to the cell wall including ramnogalacturonan-I (RG-I) (Posé et al. 2019). The use of chelating agents such as imidazole and trans-1,2-diaminocyclohexane-N, N, N', N'-tetraacetic acid (CDTA) produces the chelated soluble fraction (CSP) enriched in ionically bound pectins such as homogalacturonan (HG) (Round et al. 2001). There is also the water soluble fraction (WSP), which comprises freely soluble pectins in the apoplast (Round et al. 2001).

The length of pectin polymers usually ranges between 20 and 1000 nm, while longer chains were observed in fruits such as peaches (Yang et al. 2009) and ripe strawberries (Posé et al. 2012). HG is the most abundant pectin polymer. It is linear and comprises 100–200 galacturonic acid units (GalA). Analytical approaches indicated that fruit HG polymers comprise 72–100 galacturonic acid residues and are approximately 100 nm long (Thibault et al. 1993), while AFM studies indicated a molecular length at the range of 75–100 nm (Carpita and Gibeaut 1993). In unripe tomato fruits, HG chains of approximately 320 residues were also determined (Round et al. 2010).

RG-I comprises rhamnose and galacturonic acid with side chains of arabinogalactans (AGPs), arabinans or galactans of various lengths starting from single and extending up to 50 residues (Yapo 2011). RG-I length was proposed to be around 500 glycosyl residues with a molecular weight of 200 kDa (McNeil et al. 1980) while in unripe tomatoes it was estimated at a molecular weight of 400 kDa (Round et al. 2010). Rhamnogalacturonan-II (RG-II) is a polymer with branches comprising 9 galacturonic acids with four side chains made from more than 20 different glycosyl residues (Matsunaga et al. 2004). In the cell wall, RG-II is found as dimers cross-linked by borate diesters (Chormova et al. 2014) and is considered to be present in the cell wall but not in the middle lamella (Vincken et al. 2003). It is important to note that pectins might not only consist of HGs, RG-Is and RG-IIs but also might comprise a wider assemblage of heteroglycans and glycoconjugates (Chapter 5).

Analysis of pectin polymers and complexes isolated from the pericarp of tomato has led to the identification of branched structures not corresponding to galactan or arabinan side chains of RG-I (Round et al. 2010). Sodium carbonate soluble pectins extracted from strawberry fruits showed susceptibility to a fungal polygalacturonase, suggesting that they likely correspond to HGs (Paniagua, Santiago-Doménech et al. 2017). The fact that these branches might be attached to HGs and not to a RG-I backbone suggest that, opposite to what was thought, HGs might not be linear and uniform, but represent a branching structure which might interlink with the cellulose microfibrils network (Vincken et al. 2003). In fruits, de-esterified HGs were immuno-detected mainly at cellular junctions and cell corners. Fluorescence signal was also observed at the centre of the cell wall, but only in ripe apple fruits (Leszczuk et al. 2018b). The presence of non-methylesterified HGs in cellular junctions enables calcium cross-linking leading to gel formation that stiffens the wall material. The degree of methyl-esterification could therefore be a tool to define cell adhesion properties as was proposed for tomato pericarp (Guillon et al. 2017).

In apple fruit, label for methylesterified HGs was found throughout the cell wall, at the plasma membrane and in the cytoplasm in immature fruit, whereas only the central cell wall was labelled in the ripe fruit (Leszczuk et al. 2018b). Interestingly, the temporal patterns of changes in the abundance of HG and AGP during fruit ripening differed significantly. HG signal intensity was relatively constant in green, red ripe fruit and during postharvest storage although changes were observed in the degree of methyl-esterification. The AGP signal on the other hand showed strong changes over time (Leszczuk et al. 2018b). The overall AGP immunofluorescence signal increased during fruit ripening and decreased long after postharvest storage (Leszczuk et al. 2018b). These data suggest specific and distinct temporal and spatial variation in AGPs' content and methylesterified HGs during fruit growth and ripening in apple.

One example of glycophospatidylinositol(GPI)-anchored protein is the Arabidopsis AGP31, which was identified in the cell walls of etiolated hypocotyls (Hijazi et al. 2012). AGP31 comprises a PAC (PRP–AGP-containing Cys) domain, which does not contain AGs and was shown to interact with RG-I branches comprising galactans (Hijazi et al. 2012), suggesting similar interactions in fruit ripening affecting cell wall structure and function.

The height and the width of pectin chains ranges between 0.5 and 2 nm (Paniagua et al. 2014) and 15–230 nm in different fruits, respectively (Posé et al. 2019). In addition, various widths of pectin chains were observed in peach ranging from 54 to 91 nm (Zhang et al. 2012). The length of pectin molecules and the height and width of their branches are important components that determine the structure and rigidity of cell walls during ripening. The structure of pectins, which was extensively characterised during fruit ripening by AFM, indicated specific differences in structural properties at different stages (Zdunek et al. 2020, Posé et al. 2019). In most of the fruits, such as tomato, cherry, peach and strawberry, pectin length, width and degree of branching decrease during ripening (Zdunek et al. 2020; Posé et al. 2019; Paniagua, et al. 2017). Fruit ripening in tomato fruit initiates at the breaker stage and gradually progresses to turning pink, light red, red ripe and overripe stage. Higher frequency of wider pectin chains was observed at the turning stage of tomato fruit ripening compared to the light red ripe stage (Xin et al. 2010). In addition, at the light red ripe stage, more linear and single chains were detected (Xin et al. 2010). The ratio of galacturonic acid (GalA) to the sum of arabinans and galactans is an indicator of HG and RG-I content in fruit pectins and a marker of pectin side branching. The higher the ratio, the higher the HG content compared to the side chains of RG-I (Liu et al. 2023). Tomato cultivars with higher ratios and as a result lower branching were those with softer red ripe fruits such as Microtome and Ailsa Craig (Liu et al. 2023). This decrease in branching of RG-Is at the red ripe ripening stage decreases cell adhesion leading to an increase in tomato fruit softening (Liu et al. 2023).

Ripe strawberry fruits exhibited shorter pectin chains compared to longer chains in unripe fruits, while branched polymers were detected in both stages with higher frequency at ripe fruit tissue (Paniagua et al. 2014). Peach cultivars with crisp fruit texture exhibited longer pectins with a higher number of branches compared to cultivars with soft texture (Yang et al. 2009; Paniagua et al. 2014). These data indicate that a decrease in the structural complexity of pectins in terms of branching and length during ripening progression affects cell adhesion and enhances fruit softening.

THE ROLE OF PECTIN IN CELL WALL MODELS

The primary cell wall comprises cellulose microfibrils embedded in the matrix, which is composed of pectin and hemicellulose polysaccharides. Several models have been proposed to explain how these polymers interact to form the primary cell wall. In the "covalent network" model, first described by Keegstra et al. (1973), xyloglycans (XyG) are linked with a galactan side chain of RG-I pectin, while the other main crosslink is between the glycan part of AGPs and RG-I (Chapter 11). A specific role for HGs was not considered in this model. In the "sticky network" or "tethering network" model (Chapter 11; Carpita and Gibeaut 1993; Hayashi 1989), pectins form an extensive but separate network, by forming the "egg-box" structures, characterised by chains of de-esterified HGs that are cross-linked by calcium ions, leading to cell wall matrix stiffening not related to any cellulose–pectin interactions. This cross-linking of HG chains via calcium is particularly significant in regulating cellular adhesion, since the middle lamella has a high content of de-esterified HGs (Willats et al. 2001), which are also present at the sites delimiting the intercellular spaces formed in three-way cell junction sites (Giannoutsou et al. 2013). In "the multi-coat" model (Talbott and Ray 1992; Brummell 2006), hemicelluloses are thought to form layers that coat each cellulose microfibrils, with each layer successively less firmly bound. Pectin molecules are attached, by the formation of layers, to the glycan-coated microfibrils. In this way, cellulose microfibrils are indirectly connected by interactions between hemicelluloses and pectin layers. In the "hybrid model", microfibrils are directly connected to xyloglycans, while pectins interact between the layers (Ha et al. 1997).

Recent studies have challenged these models and suggest that pectins are directly connected to cellulose microfibrils, and would thus be implicated much more directly in defining the mechanical properties of cell wall compared to previous assumptions (Wang et al. 2018). Recently, it was shown that cellulose microfibrils can bind to RG-I arabinan and galactan side chains *in vitro* with affinities that resemble xyloglucan (Zykwinska et al. 2018). Lin et al. (2015, 2016) have proven that arabinan and galactan derived from pectins can bind to bacterial cellulose during its synthesis. In addition, solid-state nuclear magnetic resonance (ssNMR) spectroscopy provided new data on the field of cell wall component interconnection. In the *Arabidopsis esk1* mutant, the lower levels of acetylation deregulate the position-specific activity of the xylan glucuronosyltransferase (GUX1) (Grantham et al. 2017; Simmons et al. 2016). This causes the loss of the even pattern of glucuronic acid on the xylan, and it prevents docking of the xylan on the cellulose fibrils (Grantham et al. 2017; Simmons et al. 2016). It was also shown by NMR microscopy that pectins interact extensively with cellulose microfibrils, while cellulose–xyloglycan contacts are not as common as originally thought (Dick-Perez et al. 2011; Wang et al. 2012, 2015). The most striking result of the above studies is that

25%–50% of the cellulose chains are tightly connected with pectins, while some pectins are highly dynamic, supposedly filling the spaces between microfibrils. Also, HG and pectin galactan chains seem to crosslink nascent cellulose microfibrils. These data suggest that the role of pectins in cellulose microfibril cross-linking might be more important considering that previously the main molecules responsible for the interconnection of microfibrils were thought to be xyloglucans.

Further evidence for this concept is provided by the observation that an *Arabidopsis* mutant deficient in xyloglucan exhibited a phenotype similar to that of wild type without any effect on the mechanical properties of cell wall despite the fact that cellulose and xyloglucan were considered very important for bearing the load of cell expansion process (Cavalier et al. 2008). Moreover, the failure of a specific xyloglycan endoglucanase to cause hypocotyl cell wall creep (Park and Cosgrove 2012) has also supported the reconsideration of the role of xyloglucans as molecules mediating the formation of "hot spots" in close contact with cellulose microfibrils ("hot-spot model", Chapter 11). In a recent study, two allelic *Arabidopsis* mutants, *qua2* and *tsd2*, with defects in pectin biosynthesis were used to investigate the molecular mechanisms that functionally link pectin and cellulose in the cell wall and the effects of pectin deficiency on cellulose, the cytoskeleton and cell wall and tissue integrity (Du et al. 2020). In *qua2* and *tsd2* mutants, the absence of a pectin methyltransferase, QUA2, hampers the biosynthesis of pectic HG and decreases its accumulation in the cell wall. The consequent alteration of cell wall structure negatively impacts the activity of plasma membrane–localised cellulose synthase complexes (SCSs). Loss of cell wall integrity destabilises cortical microtubules, resulting in the loss of the aligned deposition of cellulose in the cell wall. As pectin is reduced in the mutants, the orientation of cellulose microfibrils that interact with pectins is changed, concomitant with increased molecular mobility of both cellulose and pectin. These new data strongly point to a reconsideration of the role of pectins, since the new findings suggest a closer relation between pectins and cellulose microfibrils than previously thought (Park and Cosgrove 2012; Du et al. 2020).

FRUIT SOFTENING PROGRAMME

The ripening of fleshy fruits is a developmental programme highly coordinated at the physiological, biochemical and transcriptional level and strictly regulated by hormones, mainly ethylene as well as abscisic acid (ABA), jasmonic acid (JA) and brassinosteroids (BR) (Li et al. 2021; Klee and Giovannoni 2011). Fruit ripening might be considered an organised disorganisation leading to senescence and organ death, while the ultimate aim is to ensure seed dispersal by attracting frugivores through the fruit attributes of colour, flavour, aroma and texture.

Fruit softening has been under investigation for a long time focusing mainly on mechanisms that allow regulating softening progression without altering flavour, aroma, colour, and nutritional and quality attributes. The molecular basis of fruit softening is rather complicated and mainly depends on the structural features of cell wall which maintain turgor pressure and regulate cell–cell adhesion (Seymour et al. 2013). Presumed predominate causes for fruit softening are cell wall modifications and osmosis in fruit tissues (Seymour et al. 2013) but precise mechanisms are still not fully understood. Most studies on fruit softening were performed on the model plant tomato (*Solanum lycopersicum*) because of the available genetic resources, its importance as a crop and the strong association of softening with the ripening process (Seymour et al. 2013).

The flesh of tomato fruit consists primarily of parenchyma cells endowed with thin, primary, unlignified walls and large vacuoles. Fruit firmness is regulated by the structural features of the parenchyma cells, such as size, turgor and wall thickness (Toivonen and Brummell 2008), as well as by cell adhesion, shape and packing characteristics (Harker et al. 1997). Parenchyma cells of the mesocarp are extensively modified during ripening. Their shape and size change, their cell turgor pressure is reduced due to an increase in osmotic solutes outside the plasma membrane and water loss takes place (Brummell 2006; Shackel et al. 1991; Harker and Sutherland 1993). This reduction of the turgor pressure leads to a decline in the pressure exerted on the cell wall leading to changes in textural properties of the fruit. The tomato cultivar DFD (delayed fruit deterioration) exhibits a significant delay in fruit softening of several months despite the fact that all the fruit softening related changes that occur during ripening are similar to wild type fruits (Saladié et al. 2007). The expression of cell wall hydrolases considerably increases, while the cell-to-cell adhesion properties of pericarp tissue decrease (Saladié et al. 2007), indicating that the observed fruit softening delay cannot be attributed to a delay in pectin solubilisation and depolymerisation caused by the cell degrading enzymes. Instead, the slower softening may be attributed to a reduced water loss due to transpiration and an increase in turgor of pericarp cells which might result from an altered structure and composition of the cuticle (Romero and Rose 2019; Saladié et al. 2007; Lara et al. 2014). These data indicate that fruit cuticles and particularly those which are thick and well developed might directly regulate transpiration rate and fruit firmness (Romero and Rose 2019).

One of the initial changes in ripening fruits that can only be observed in the electron microscope is the dissolution of the middle lamella (Brummell 2006; Fry 2003). This layer between cells is rich in pectins, whose dissolution leads to the formation of intercellular spaces that appear triangular in cross sections. The second step that follows during the fruit softening process is a progressive degradation of the cell wall polysaccharides and an alteration of the linkages connecting the various components (Brummell 2006; Cosgrove 1997). The progressive dismantling causes softening of the cell wall, and in some species, the cell walls are swollen and intercellular adhesion is further decreased (Hallett et al. 1992; Redgwell et al. 1992).

Fruit ripening is characterised by the secretion of various cell wall matrix degrading enzymes in the cell wall, the synergistic action of which contributes to pectin polymer degradation and to the subsequent fruit cell wall softening. Pectin solubilisation takes place in most plant species and pectin depolymerisation takes place in a smaller number of species, while both modifications occur during ripening. During solubilisation, an increase in the content of the pectins that are loosely bound to the cell wall occurs accompanied by a decrease in the amount of covalently bound pectins (Paniagua et al. 2014). This facilitates the entrance of water into intermicrofibrillar spaces (Redgwell et al. 1997), resulting in the swelling of the cell walls, increasing substrate accessibility to various hydrolytic enzymes (Brummell 2006; Posé et al. 2019). Those fruits that show high or moderate pectin solubilisation are those that soften significantly, while those with low or undetectable solubilisation display a crisp texture (apple, watermelon) (Mercado et al. 2011).

Less degraded, thicker and branched pectin molecules were found to be related to increased firmness and favourable texture in pear fruit (Zdunek et al. 2014). In fruit softening, pectin molecules become more weakly attached to the extracellular matrix (Brummell 2006; Wang et al. 2018) while pectin degradation results in an increase in free Ca^{2+} in the apoplast and the loss of cell–cell adhesion (Ortiz et al. 2011; reviewed in Gao et al. 2019). In apple, calcium infiltration delayed harvesting and storage, while the loss of middle lamella calcium was shown to facilitate softening (Stow 1993). Calcium action in fruit development and ripening was extensively reviewed in Ferguson (1984) and Gao et al. (2019). Calcium mainly contributes to fruit softening by affecting cell wall–degrading enzymatic activities including pectinesterase, cellulase, polygalacturonase, β-galactosidase and pectin lyase.

Depolymerisation of chelated and/or covalently bound pectins also leads to solubilisation, but that occurs later at the ripening procedure (Mercado et al. 2011; Posé et al. 2015). The pectin fraction extracted with sodium carbonate is rich in RG-I and considered covalently bound to the primary cell wall, while the side chains comprise neutral sugars such as arabinose and galactose (Posé et al. 2019). Arabinans and galactans of RG-I have the capacity for wide range interactions forming a temporary matrix, which might control both the structural integrity and polymer organisation in cell walls. Apart from the solubilisation of the pectins, the loss of side chain neutral sugars is a ripening-related process in all fruit species (Gross and Sams 1984; Goulao and Oliveira 2008). Enzymatic removal of galactose from galactan side chains of pectins was often related to primary solubilisation effects that are later followed by major significant fruit softening processes (Ortiz et al. 2011). Two enzymes are considered responsible for neutral sugar removal during fruit ripening, β-galactosidase and α-L-arabinofuranosidase (Goulao et al. 2007). High β-galactosidase activity was observed during apple fruit ripening, while an increase in α-L-arabinofuranosidase activity was detected in overripe fruits of apples and pears (Goulao et al. 2007). In addition, rhamnogalacturonan lyase activity was associated with the softening of tomato fruit ripening while silencing of this mRNA in strawberry resulted in a decrease in dissolution of middle lamellae (Kaczmarska et al. 2021).

PECTIN-DEGRADING AND OTHER ENZYMES

Fruit softening is controlled by the concerted action of various cell wall–modifying enzymes, such as PGs, PMEs, endo-β-(1,4)-glucanases, β-galactosidases and expansins. The enzymatic activity of the enzymes involved is accurately and differentially controlled and differs greatly between species (Klee and Giovannoni 2011). Cell wall–modifying enzymes can be clustered in two main categories according to the substrate: pectolytic and non-pectolytic enzymes. Pectolytic enzymes are: a) endo- or exo-PGs, PLs, PMEs, pectin acetylesterases (PAEs), β-galactosidases or α-L-arabinofuranosidases. The pectin-related enzymes can cleave or modify the polysaccharide backbone or remove neutral sugars from side chains. The non-pectolytic enzymes comprise endo-1,4-β-glucanases (EGs), endo-1,4-β-xylanases (endo-ECs), β-xylanases (ECs), xyloglycan endotransglycosylase/hydrolases (XTH) and expansins (EXPs) (Garcia-Romera and Fry 1994; Goulao and Oliveira 2008).

PME removes the methyl groups from methylesterified HGs while endo-PG hydrolyses these HGs by cleaving α-(1,4)-galacturonic acid linkages and exo-PG cleaves the terminal galacturonic acid residues of HGs (Wang et al. 2018). Galactanase cleaves the β-(1–4) galactose linkages in pectin galactans found in side chains of RG-I and β-galactosidase cleaves terminal, non-reducing galactosyl residues from cell wall pectins (Smith and Gross 2000). The α-L-arabinofuranosidase removes terminal arabinofuranosyl residues from pectins and hemicelluloses (Wang et al. 2018).

It was reported that PME activity increased during fruit growth and particularly at the mature green stage, and it decreased thereafter (Gaffe et al. 1994). Others reported an increase in enzymatic activity as well as gene expression in tomato fruit ripening (Tucker et al. 1992) or up-regulation until the turning stage followed by a decrease (Harriman et al. 1991; Pressey and Avants 1982). A minor retardation of tomato fruit softening was demonstrated after silencing of PME (Phan et al. 2007). Similar results were obtained after silencing of a β-galactosidase mRNA (Smith et al. 2002). Further, it was shown that pectin solubilisation is a complicated process controlled by several classes of pectin-degrading enzymes. HG that is secreted to the apoplast is highly methylated, and once *in muro* it is subjected to PME activity, which removes the methyl ester groups. The lower degree of methylation then allows further cleavage by PL and PG.

PLs are known to randomly cleave α-1,4-galacturonic acids of HG by a β-elimination mechanism while it was also shown to target un-esterified pectins at three-way cell junctions as well as middle lamellae regions in tomato (Uluisik et al. 2016). The available data on silencing show that pectin solubilisation in tomato fruits by PL made possible subsequent depolymerisation by PG (Wang et al. 2018). The significant and impressive effect on tomato fruit softening was later demonstrated by silencing PL (Uluisik et al. 2016; Yang et al. 2017). The control of tomato softening was altered without any effect on the ripening process, such as yield or weight, ethylene biosynthesis, colour or total soluble solids. Interestingly, Uluisik et al. (2016) stated that the tricellular junctions are considered major sites for PL activity. A comparative study on PL, polygalacturonase 2a (PG2a) and β-galactanase (TBG4) using CRISPR mutants in tomato confirmed the leading role of PL in fruit softening (Wang et al. 2019). In strawberry, silencing of PL also resulted in the reduction of softening progression during ripening (Santiago-Doménech et al. 2008). In strawberry fruits with silenced PL, a decrease in pectin solubility and polymerisation of polyuronides was observed, as well as ripening-induced loss of cell–cell adhesion. These data suggest that PL is involved in pectin degradation processes in primary wall and middle lamella.

Expansins are known to catalyse loosening of cell walls through acid-mediated growth without showing lytic activity of wall polymers (Cosgrove 2020, 2022). It was suggested that at sites where cellulose microfibrils are attached through xyloglycan monolayers, the EXPs are active and disrupt non-covalent binding (Cosgrove 2022). This might provide available space for cellulases to approach cellulose microfibrils and use them as substrate (Cosgrove 2022). The subfamily of α-expansins comprises genes highly expressed during fruit ripening in apple, strawberry and tomato (Valenzuela-Riffo et al. 2020; Chen et al. 2022; Brummell et al. 1999). Among the members of the tomato expansin gene family, Expansin 1 (EXP1) is highly up-regulated during fruit ripening and, therefore, its expression was altered in stable transgenic lines by silencing and overexpression constructs (Brummell et al. 1999). Fruit softening was accelerated in the overexpression lines which was associated with extensive depolymerisation only of hemicelluloses but not of polyuronides (Brummell et al. 1999). The down-regulation of EXP1 in the silencing lines resulted in a decrease in softening rate, which gradually diminished to levels of control fruits during ripening progression accompanied by a significant inhibition of polyuronide depolymerisation but not of structural hemicelluloses (Brummell et al. 1999). However, it was suggested recently that α-expansins are involved in fruit softening based on data obtained from the overexpression of an apple α-expansin in tomato, resulting in acceleration of fruit softening (Chen et al. 2022). Moreover, an α-expansin, *Fragaria ananassa* EXPA5, is up-regulated during fruit ripening in strawberry and is related to fruit softening according to its expression pattern and molecular dynamic simulation, which showed that this α-expansin was forming stable complex with cellulose (Valenzuela-Riffo et al. 2020).

Taking into consideration the new data on closer relation between pectins and cellulose microfibrils, the possible mode of action of PL might be the attack on the branched HGs as their structure analysis by AFM revealed. Therefore, PL might be required for pectin solubilisation and depolymerisation, which results in the decrease of de-esterified HGs from tricellular junctions allowing the subsequent degradation of HG and RG-I by PG2a and TBG4 during ripening (Wang et al. 2019). According to this new perspective of the cell wall remodelling during fruit softening, expansin activity might improve the access of pectins to enzymatic degradation. Furthermore, the action of galactosidases and α-L-arabinofuranosidases might facilitate the access of the other cell wall–degrading enzymes (PMEs, PLs and PGs) to the pectic polysaccharide backbones (Wang et al. 2018). The action of the above enzymes can directly alter the cell wall biomechanics by decreasing cellulose microfibril tethering, which is mediated by pectins, as suggested by new data on pectin-cellulose microfibrils contacts.

THE TOMATO FRUIT ENDO-POLYGALACTURONASE STORY

The objective of this section is a historical perspective on fruit softening–associated cell wall hydrolases mainly affecting the structure of pectins in relation to cell adhesion properties and middle lamella dissolution (Figure 7.1).

The endo-PG will be the main focus considering that it is the first cell wall hydrolase, which was extensively characterised mainly in tomato fruit ripening using various biotechnological approaches leading to the introduction of the first genetically modified (GM) vegetable on the market in 1994 (Figure 7.1), while recent reports indicated the central role of PL in fruit softening without interfering with other developmental programmes regulating flavour, aroma and nutritional value (Wang et al. 2019).

In 1967, a close correlation between the firmness of four tomato genotypes and PG activity suggested that this enzyme might control the loss of firmness (Hobson 1967) while in 1978, it was proposed that PG activity inhibition might be the primary genetic event in *rin* and *nor* mutants (Tigchelaar et al. 1978) (Figure 7.1).

The interest in endo-PG could also be attributed to the significant increase in gene expression during ripening, which coincides with the climacteric increase in ethylene production (Grierson, Slater, Maunders et al. 1985; DellaPenna et al. 1986). In 1981, biochemical and molecular evidence supported the involvement of gene expression in fruit ripening (Grierson et al. 1981), followed by the cloning, identification and expression analysis of endo-PG due to enzyme *de novo* synthesis and participation in cell wall dissolution (Crookes and Grierson 1983) (Figure 7.1). The involvement of endo-PG in pectin degradation was also associated with softening in pear fruits (Bartley et al. 1982).

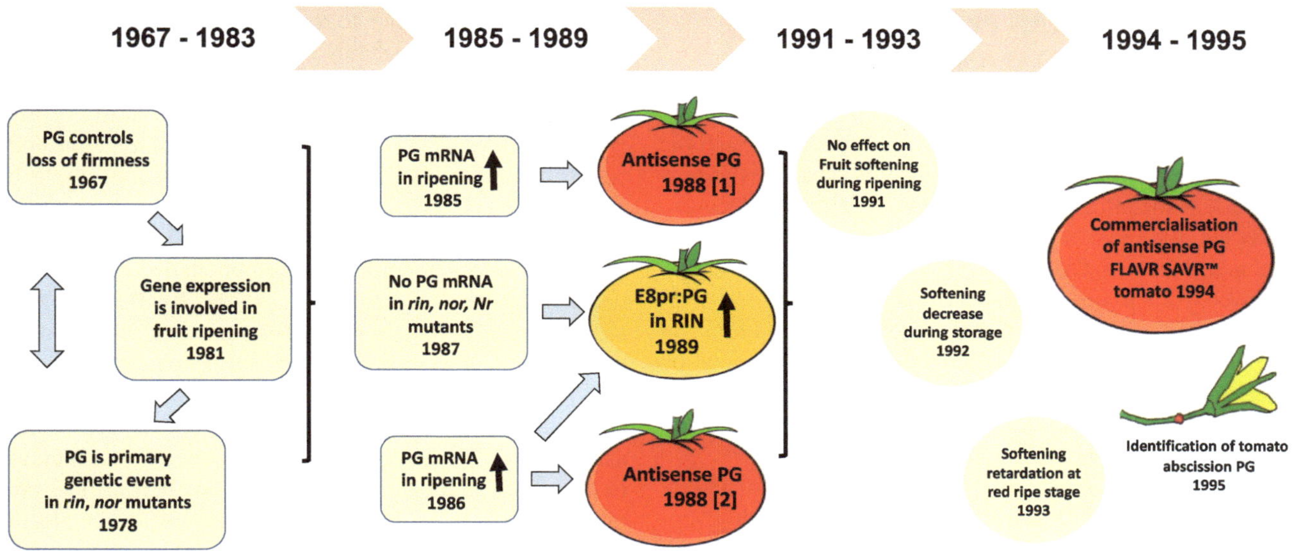

FIGURE 7.1 Milestones of tomato endo-polygalacturonase (endo-PG) identification, characterisation and commercialisation in chronological order. The publications of significant findings for endo-PG are presented starting in 1967 up to 1995, from left to right in the illustration. Publications are the following: Hobson (1967), Grierson et al. (1981), Tigchelaar et al. (1978), Grierson, Slater, Speirs et al. (1985), DellaPenna et al. (1987), DellaPenna et al. (1986), Sheehy et al. (1988), Giovannoni et al. (1989), Smith et al. (1988), Fischer et al. (1991), Kramer et al. (1992), Carrington et al. (1993), Kramer and Redenbaugh (1994) and Kalaitzis et al. (1995).

Further biochemical studies indicated an increase of PG mRNA during ripening (Grierson, Slater, Speirs et al. 1985), which was preceded by ethylene production (Grierson and Tucker 1983) (Figure 7.1). Moreover, lower levels of PG mRNA were detected in fruit tissue of *rin* mutant, which lacks ethylene production during ripening despite the fact that PG expression and protein levels were always associated with climacteric ethylene (DellaPenna et al. 1987). No detectable PG expression levels were observed in *nor* and *Nr* mutants (DellaPenna et al. 1987) (Figure 7.1).

This interest in tomato fruit PG involvement in fruit softening was enhanced by the identification of a 1,9 kb tomato fruit PG mRNA after screening a ripe tomato fruit cDNA expression library (DellaPenna et al. 1986) while another group in the same period reported the isolation of a tomato fruit PG transcript producing a polypeptide of different molecular weight (Slater et al. 1985) (Figure 7.1). However, both groups identified the same tomato fruit PG cDNA since there is only one tomato fruit-specific endo-PG. The expression of this 1,9 kb fruit PG was significantly increased during tomato ripening by at least 2,000-fold, while no expression was detected in immature green fruits, suggesting developmental regulation (DellaPenna et al. 1986) (Figure 7.1). In 1995, a tomato abscission-specific PG, TAPG1, was cloned and characterised (Kalaitzis et al. 1995) while few years later another two tomato abscission PGs, TAPG2 and TAPG4, were identified (Kalaitzis et al. 1997).

In August 1988, tomato was transformed with a tomato fruit PG antisense construct, resulting in suppression of PG transcript levels in ripe fruits accompanied by a decrease in PG protein levels (Smith et al. 1988) (Figure 7.1). However, no delay in fruit softening was observed in the antisense PG transgenic lines, indicating the complexity of pectin degradation process (Smith et al. 1988). Few months later, Sheehy et al. (1988) reported the suppression of PG mRNA by using also an antisense approach, which resulted in substantial reduction in PG transcript levels and enzymatic activity, while no effect on lycopene accumulation was observed assuming that a delay either in ripening or/and softening might have had an effect on lycopene accumulation. The E8 is a fruit-specific, ethylene responsive, negative regulator of ethylene production and developmentally regulated gene belonging to the family of 2-oxoglutarate dependent dioxygenases (Deikman et al. 1992; Kneissl and Deikman 1996). The E8 promoter has been extensively used to direct expression of genes during fruit ripening since its expression pattern follows the climacteric ethylene production pattern. Moreover, 1 year later, the PG cDNA was expressed under the control of the E8 gene promoter in the *rin* mutant in order to investigate the role of PG in fruit softening and ripening (Giovannoni et al. 1989) (Figure 7.1). Although the expression of PG increased the degradation of polyuronide, it did not induce either fruit softening progression nor lycopene accumulation and ethylene production (Giovannoni et al. 1989), indicating the limitations of PG's regulatory role in the overall developmental programme of ripening.

The quality characteristics and postharvest behaviour of endo-PG antisense transgenic tomato fruits were extensively evaluated (Smith et al. 1990a,b; Schuch et al. 1991) including not only processing and firmness but also disease resistance (Kramer et al. 1992) (Figure 7.1). These endo-PG antisense tomato fruits were introduced in the market with the commercial name FLAVR SAVR in 1994, while the first tomato abscission PG was identified in 1995 (Kalaitzis et al. 1995). The FLAVR SAVR is considered the first transgenic fruit vegetable introduced in the market as a commercial product.

ARABINOGALACTAN PROTEINS AND PROLYL-4-HYDROXYLASES IN FRUIT SOFTENING AND RIPENING

Most of the cell wall proteins are translocated through the secretory pathway and therefore undergo post-translational alterations including glycosylation. The cell wall glycoproteins mainly undergo *N*- and/or *O*-glycosylation as well as the addition of a GPI-anchor (Tegl and Nidetzky 2020). *O*-glycosylation occurs on hydroxyl groups of hydroxyproline (Hyp) (Gorres and Raines 2010). The addition of the glycan part by *N*- or *O*-glycosylation is determined by specific glycosylation motifs (Kieliszewski 2001; Tan et al. 2003; Seifert 2020), indicating that molecules that undergo hydroxylation might also be *O*-glycosylated.

AGPs are extremely highly glycosylated proteins of the cell wall and belong to the hydroxyproline-rich glycoproteins superfamily (HRGPs), a diversified group of cell wall glycoproteins (Chapter 16; Seifert and Roberts 2007). This glycosylation comprises the addition of arabinose and galactose to non-contiguous hydroxyprolines during nascent glycopeptide synthesis and is further extended with the attachment of α-arabinose, α-fucose, β-(methyl) glucuronic acid and α-rhamnose which is catalysed by several galactosyltransferases such as *O*-β-galactosyltransferases, β-1,3-galactosyltransferases, α-rhamnosyl transferases, β-1,6-galactosyltransferases, α-fucosyltransferases, α-arabinosyltransferases and β-glucuronosyltransferases in the Golgi network (Chapter 16; Showalter and Basu 2016). Proline hydroxylation and the subsequent glycosylation of AGPs are integral parts of their structure which determine their functional significance and as a result their physiological role in plethora of developmental programmes (Strasser et al. 2021; Perrakis et al. 2019). Alterations in the levels of proline hydroxylation occurrence perturb the subsequent glycosylation, and thus the secondary structures of AGPs, leading either to their degradation or to incomplete structure and probably function (Perrakis et al. 2019; Fragkostefanakis et al. 2012). Proline hydroxylation is catalysed by proteins encoded by the gene family of prolyl 4 hydroxylases (P4Hs). A proline hydroxylation code was proposed predicting which proline residues might be hydroxylated (Duruflé et al. 2017).

Several AGPs are up-regulated during tomato fruit ripening, indicating potential involvement in this developmental programme (Leszczuk et al. 2020a, 2023). During the red ripe stage of tomato fruit ripening, two classical AGPs, AGP4 and AGP5, showed significant up-regulation by 20- and 5-fold, respectively (Leszczuk et al. 2020a). Moreover, two additional classical AGPs, AGP10-like and AGP9-like, as well as AGP13 and FLA17 also exhibited minor increase in transcript abundance (Leszczuk et al. 2020a). The promoter sequences of AGP4, AGP5 and AGP13 contain the E4-cis acting element initially identified in the tomato fruit E4 gene, which is known to be regulated by ripening suggesting possible regulation of tomato fruit AGPs by ripening (Leszczuk et al. 2020a).

Subsequently, the ripening-induced P4Hs participating in the hydroxylation of these AGPs might be involved in the regulation of ripening. In addition, suppression or silencing of P4Hs by using VIGS (virus-induced gene silencing) resulted either in a decrease in content or in a complete suppression of AGP biosynthesis in tomato fruits according to western blot analysis with broad spectrum AGPs-bound epitopes (Fragostefanakis et al. 2014). Moreover, VIGS-induced silencing of two P4Hs, SlP4H5 and SlP4H7, in tomato fruits resulted in ripening delay (Figure 7.2). Several areas of pink pericarp tissue were observed in red ripe fruits, which were injected at 15 days post anthesis with the pTRV:P4H5 and pTRV:P4H7 silencing constructs (Figure 7.2). These data might indicate involvement of P4Hs in tomato fruit ripening progression.

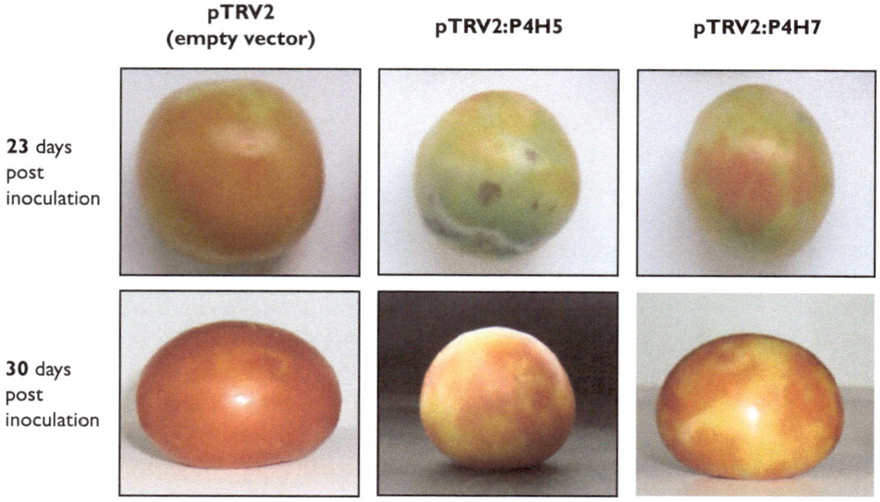

FIGURE 7.2 Effect of virus-induced gene silencing of SlP4H5 and SlP4H7 in tomato fruit ripening progression. Visual representation of tomato fruits after 23 days post inoculation with the pTRV2:P4H5 and pTRV2:P4H7 constructs. Ripening retardation was observed in various pericarp areas with both silencing constructs.

The presence and involvement of AGPs in fruit tissue and during fruit ripening were reported since 2000 and 2010, respectively (Qin et al. 2000; Kalaitzis et al. 2010). Two Type-II AGPs were purified from fruit tissue of *Lycium chinense* (Qin et al. 2000), while AGP epitopes were detected in the light red stage of tomato fruit, indicating expression during ripening (Kalaitzis et al. 2010). In addition, pyridine 2,4-dicarboxylate (PDCA), a putative inhibitor of prolyl 4 hydroxylases (P4Hs), decreased AGP epitopes in a dose-dependent manner, indicating involvement in AGP content (Kalaitzis et al. 2010). Protein glycosylation was also reported to participate in fruit ripening (Méndez-Yañez et al. 2017, 2021; Zhang et al. 2020).

There are tomato mutants, such as *fab*, *fin* and *fab2*, which show enlarged shoot meristem size, increased inflorescence branching and fasciated flowers with higher number of flower organs as well as larger fruits due to the presence of additional carpels (Xu et al. 2015). The *fab*, *fin* and *fab2* encode the Clavata 1 receptor kinase (CLV1), the hydroxyproline *O*-arabinosyltransferase 3 (HPAT3) and an arabinosyltransferase acting after the HPATs, respectively (Xu et al. 2015). These genes are all involved in the Clavata (CLV) signalling pathway in shoot meristems (Xu et al. 2015). This pathway maintains meristem size by controlling the population of stem cells through a regulatory loop between CLAVATA and WUSCHEL genes (Stahl and Simon 2010; Schoof et al. 2000). The Clavata 3 (CLV3) glycopeptide binds to leucine-rich repeat receptor complexes, which comprise also the receptor kinase CLV1 initiating a cascade that negatively regulates the expression of WUSCHEL, while WUSCHEL enhances the expression of CLV3 (Fletcher et al. 1999).

The arabinosylation of SlCLV3 by HPAT3 is required for regular function suggesting that the activity of specific tomato prolyl 4 hydroxylase(s) might also be considered essential for SlCLV3 activity since proline *O*-hydroxylation precedes arabinosylation of glycopeptides. These results indicate that mutations in genes encoding proteins involved in the biosynthesis of secreted glycopeptides such as CLV3 might result in important agronomical traits such as fruit size and flower number affecting productivity in tomato.

Biochemical studies using β-glycosyl Yariv (β-Yariν), a synthetic phenyl glycoside that binds to AGPs, suggested a role for AGPs in plant cell expansion (Willats and Knox 1996). The involvement of tomato AGPs in plant growth was also demonstrated indirectly by using PDCA (pyridine 2,4-dicarboxylic acid), a 2-oxoglutarate analogue, which might putatively block the activity of P4Hs and consequently *O*-glycosylation of the HRGPs (Fragkostefanakis et al. 2018). This PDCA-induced decrease in tomato seedling root and hypocotyl length was observed to be dose dependent and was accompanied by a down-regulation in AGP epitopes. This phenotype was attributed to changes in the expansion of root and hypocotyl cells (Fragkostefanakis et al. 2018). The use of PDCA in tomato just after pollination and fruit setting might affect either cell division or cell expansion of fruit pericarp cell layers inducing changes in fruit growth in tomato or other fruits.

The characterisation of fruit ripening–related AGPs was recently reported mainly at the structural and spatio-temporal level. AGPs and pectins are dynamic, structural and regulatory constituents of cell wall and the effect on fruit structure has been demonstrated (Moore et al. 2014; Sala et al. 2017; Leszczuk et al. 2018a, 2019b). The localisation and distribution of AGPs is dynamic and might be determined by a mechanism yet to be determined, and therefore should not be considered random (Wu et al. 2017; Leszczuk et al. 2018a, 2019a; Parra et al. 2020). The spatio-temporal distribution of AGPs was investigated in apple, tomato (Leszczuk et al. 2019b,c, 2020a), banana (Wu et al. 2017) and *Lycium barbarum* L. (Bao et al. 2016).

The apple fruit AGPs were extracted using the β-D-Glucosyl Yariv reagent and then after lyophilisation were analysed for monosaccharide composition with HPLC (Leszczuk et al. 2020b). The apple AGPs consisted mainly of arabinose, galactose, glucose, galacturonic acid and xylose, while the polypeptide comprises only the 3.15%–4.58% of total molecular mass (Leszczuk et al. 2020b). Green apple fruit AGPs contained significantly higher amounts of galactose and much lower arabinose compared to red fruit and postharvest stored fruits, while the arabinose increased progressively during ripening and after long postharvest storage (Leszczuk et al. 2020b). In the same report, the green fruits contained lower amount of galacturonic acid, which is in accordance with the increase in uronic acids' content in ripe and postharvest stored fruits compared to green fruits. A previous study in *Lycium chinense* fruit tissue AGPs indicated higher amounts of sugar chains of galactose and arabinose at a ratio of 3:1 compared to the protein constituent (Qin et al. 2000) which is similar to the carbohydrate content of other plant organ AGPs (Showalter 2001). Smaller amounts of fucose, mannose and rhamnose were also detected in the apple fruit tissues (Leszczuk et al. 2020b).

Preliminary data indicate higher amount of Yariv reagent precipitated AGPs in ripe apple fruit pericarp tissue compared to green fruit and after a period of postharvest storage (Leszczuk et al. 2020a). AGPs identified by LM2 showed increased abundance in red ripe fruit compared to green and stored fruits (Leszczuk et al. 2020a). The AGP epitopes recognised by LM14, which comprise arabinose- and galactose-rich units in the AGPs glycan structure, exhibited lower signal intensity (Leszczuk et al. 2020a). Label with JIM13 indicated a temporal variation not only in the total content but also in the molecular mass of AGPs depending on the stage of ripening and postharvest storage (Leszczuk et al. 2020a). Higher content of JIM13-labelled AGPs were detected in the ripe fruit according to western blot analysis (Leszczuk et al. 2020a). Similar data were observed in ripe berries of *Vitis vinifera* (Moore et al. 2014) and in tomato pericarp tissue (Fragostefanakis et al. 2014).

Immunolocalisation showed evenly distributed AGPs as a line in the cell wall near to plasma membrane region, while this pattern was disrupted in the green apple fruit showing distribution in the entire cell wall area (Leszczuk et al. 2020a, 2018a). After postharvest storage for 1 month, the signal intensity of JIM13 decreased considerably (Leszczuk et al. 2020a). Immunolocalisation of various AGP epitopes indicated the highest AGP accumulation at the plasma membrane border with the cell wall, while arabinan epitopes showed localisation also at the inner cell wall close to plasma membrane (Leszczuk et al. 2018a). Similar results were observed during tomato fruit ripening; most of the AGP epitopes were detected in areas lining the cell wall and the plasma membrane, while treatment of fruits with a mixture of pectinolytic enzymes induced remodelling of AGP epitopes due to the degradation of carbohydrate moieties (Leszczuk et al. 2018b). It was suggested that AGPs might not participate in intercellular adhesion in fruit tissue, particularly in apple fruit, due to their apparent absence in tricellular junctions (Leszczuk et al. 2019a, 2020a).

The interaction of AGPs with additional cell wall constituents, particularly with pectins and arabinoxylans, was investigated by Tan et al. (2013). It was shown that AGP polypeptides were covalently associated with matrix polysaccharides, such as pectin and hemicellulose.. This cell wall structure was called ARABINOXYLAN PECTIN ARABINOGALACTAN PROTEIN1 (APAP1) and is not in agreement with previous models which suggest that AGPs do not interact with pectin and hemicellulose polysaccharide networks (Tan et al., 2013). Such connection of AGPs with pectins and arabinoxylans might be taken into consideration in terms of physiological significance in fruit softening (Leszczuk 2020a).

The calcium content in fruit AGPs was lower in green apple fruits and higher by 8-fold in ripe and after long postharvest storage fruits, while there was equal distribution in parenchymatic and epidermal as well as sub-epidermal cells (Leszczuk et al. 2020b). In addition, the galacturonic acid increased in AGPs by 3-fold during ripening which might explain the higher accumulation of calcium by AGPs considering the ability of carboxyl and hydroxyl groups to interact with calcium ions in accordance to the egg-box model (Braccini and Pérez 2001). This Ca^{2+}-driven cross-linking of AGPs galacturonic acid with pectic acids (Huang et al. 2016) might influence the textural properties of fruit tissue by increasing the stiffness of the cell wall.

A previously proposed model suggested that the calcium ions are released from AG glycans into the cytoplasm and initiate downstream calcium-mediated signalling (Lamport and Várnai 2013; Lamport et al. 2014, 2020; Lopez-Hernandez et al. 2020). Thus, Lamport et al. (2014) proposed that at pH ~5,5, AGPs bind calcium with a $K_{dissociation}$ of approximately 6 μM in the plasma membrane. This calcium is released at $pk_{association}$ of approximately 3 μM leading to dissociation of AGP–Ca^{2+} carboxylates when auxin-dependent plasma membrane H^+-ATPases decrease the periplasmic pH. The AGPs in this model act as external dynamic Ca^{2+} repository, coordinating plant growth via an AGP–Ca^{2+} oscillator (Lopez-Hernandez et al. 2020). It was also shown that double or triple mutants of *Arabidopsis* AG β-glycuronyltransferases (GlcATs) exhibited a strong reduction in gluronidation, resulting in lower calcium binding capacity and consequently in phenotypes characterised by plant sizes reduced by 30%, deformed leaves and less branched leaf trichomes (Lopez-Hernandez et al. 2020).

AFM analysis of green and red apple fruit nanostructure of AGPs indicated irregular and extended shapes of aggregates in the cell wall ultrastructure as well as smaller individual molecules compared to postharvest storage fruits (Leszczuk et al. 2020a). The AGP molecules in green fruit were of larger size compared to red and postharvest storage fruits, while the aggregates were more compact in the postharvest fruits. The size of the aggregates varied between 75 and 150 nm, while the shape was also altered towards a decrease in aggregation as ripening was progressing (Leszczuk et al. 2020a). Scanning electron microscopy (SEM) of AGPs extracted from apple fruit also revealed globular aggregates comprised of small-size granules of various shapes (Leszczuk et al. 2020b). A single granule diameter was estimated to be within 100–200 nm, while each aggregate showed a diameter of 2–3 nm (Leszczuk et al. 2020b). After HPLC purification and western blot analysis, spherical nanoparticles comprising AGPs were also identified, which possess significant adhesive properties forming interactions with pectic polysaccharides through calcium-driven electrostatic interactions (Huang et al. 2016). This is another line of evidence indicating interactions between pectins and AGP molecules which might be involved in fruit ripening regulation. FT-IR (Infrared spectroscopy – Fourier transform) analysis indicated a specific fingerprint with unique AGPs wavenumbers in apple fruit tissue which might be used as an indicator of AGPs' structure shifts in different fruit ripening stages (Leszczuk et al. 2020a). Overall, the structural analysis of apple fruit AGPs showed changes in diameter, shape of aggregates, amount and molecular mass, indicating strictly regulated biosynthesis leading to specific structure and as a result function during fruit ripening.

CONCLUSIONS

The roles of pectic HGs, RG-I and II are more important in the softening process of fleshy fruit cell walls than what was thought few years ago. Tools such as AFM provide important structural information on the alterations in length, width and side chains of pectins during different stages of fruit ripening. These structural changes were associated with cell wall–acting enzymatic activities of PL, endo-PG and β-galactosidases as well as of non-pectolytic enzymes such as EXPs. The physiological significance of these enzymes in fruit softening requires functional studies including either silencing or overexpression of the mRNAs encoding these enzymes and extensive cell wall analysis and particularly AFM. PL might be considered important for softening in tomato fruits, while different pectolytic enzymes might have different significance for ripening and softening progression in fruits other than tomato. The data suggest that targeting pectin metabolism might result in successful regulation of fruit softening. The extension of shelf-life of fruits and the delay of softening is first priority, although it will be important to consider the preservation if not the enhancement of nutritional value of horticultural commodities, which might be considered as a challenging task. There are indications that the cell wall glycoproteins comprising mainly AGPs and extensins and their biosynthetic components such as prolyl-4-hydroxylases might be considered significant for the regulation of fruit ripening processes including fruit softening. Most of these glycoproteins and their biosynthetic components are large gene families. Therefore, the experimental approaches should take into consideration the possibility of redundant function among genes within the gene family. In this context, the creation of an interactome between glycoproteins and their biosynthetic components such as P4Hs might provide information on the identification of target genes for either overexpression and/or silencing in order to investigate fruit ripening and softening.

ACKNOWLEDGEMENTS

This work has been financed by the European Regional Development Fund of the European Union and Greek national funds through the Operational Competitiveness, Entrepreneurship and Innovation, under the call RESEARCH-CREATE-INNOVATE (project code: T2EDK-01332: n-Tomatomics-DEVELOPMENT OF NEW TOMATO CULTIVARS BY USING OMICS TECHNOLOGIES).

REFERENCES

Bao H, Zheng G, Qi GL, Wang J (2016) Cellular localization and levels of arabinogalactan proteins in *Lycium barbarum* fruit. *Pakistan Journal of Botany* 48:1951–1963.

Bartley IM, Knee M, Casimir MA (1982) Fruit softening I. Changes in cell wall composition and endo-polygalacturonase in ripening pears. *Journal of Experimental Botany* 33:1248–1255.

Bourne MC (1979) Texture of temperate fruits. *Texture Studies* 10:25–44.

Braccini I, Pérez, S (2001) Molecular basis of Ca^{2+}-induced gelation in alginates and pectins: the egg-box model revisited. *Biomacromolecules* 2:1089–1096.

Brummell DA (2006) Cell wall disassembly in ripening fruit. *Functional Plant Biology* 33:103–119.

Brummell DA, Harpster MH, Civello PM, Palys JM, Bennett AB, Dunsmuir P (1999) Modification of expansin protein abundance in tomato fruit alters softening and cell wall polymer metabolism during ripening. *The Plant Cell* 11:2203–2216.

Carpita NC, Gibeaut DM (1993) Structural models of primary cell walls in flowering plants: Consistency of molecular structure with the physical properties of the walls during growth. *The Plant Journal* 3:1–30.

Carrington CS, Greve LC, Labavitch JM (1993) Cell wall metabolism in ripening fruit (VI. Effect of the antisense polygalacturonase gene on cell wall changes accompanying ripening in transgenic tomatoes). *Plant Physiology* 103(2):429–434.

Cavalier DM, Lerouxel O, Neumetzler L, Yamauchi K, Reinecke A, Freshour G, Zabotina OA, Hahn MG, Burgert I, Pauly M, Raikhel NV, Keegstra K (2008) Disrupting two *Arabidopsis thaliana* xylosyltransferase genes results in plants deficient in xyloglucan, a major primary cell wall component. *Plant Cell* 20:1519–1537.

Chen FS, Liu H, Yang HS, Lai SJ, Cheng XL, Xin Y, Yang B, Hou HJ, Yao YZ, Zhang SB, Bu GH, Deng Y (2011) Quality attributes and cell wall properties of strawberries (*Fragaria ananassa* Duch.) under calcium chloride treatment. *Food Chemistry* 126:450–459.

Chen Y, Chen F, Lai S, Yang H, Liu H, Liu K, Bu G, Deng Y (2013) *In vitro* study of the interaction between pectinase and chelate-soluble pectin in postharvest apricot fruits. *European Food Research and Technology* 237:987–993.

Chen Y, Xie B, An X, Ma R, Zhao D, Cheng C, Li, Zhou J, Kang G, Zhang Y (2022) Overexpression of the apple expansin-like gene MdEXLB1 accelerates the softening of fruit texture in tomato. *Journal of Integrative Agriculture* 21:3578–3588.

Cherian S, Figueroa CR, Nair H (2014) 'Movers and shakers' in the regulation of fruit ripening: a cross-dissection of climacteric versus non-climacteric fruit. *Journal of Experimental Botany* 65: 4705–4722.

Chormova D, Messenger DJ, Fry SC (2014) Boron bridging of rhamnogalacturonan-II, monitored by gel electrophoresis, occurs during polysaccharide synthesis and secretion but not post-secretion. *Plant Journal* 77:534–546.

Cosgrove DJ (1997) Relaxation in a high-stress environment: the molecular bases of extensible cell walls and cell enlargement. *Plant Cell* 9:1031–1041.

Cosgrove DJ (2020) Non-enzymatic action of expansins. *Journal of Biological Chemistry* 295:6782.

Cosgrove DJ (2022) Building an extensible cell wall. *Plant Physiology* 189:1246–1277.

Crookes PR, Grierson D (1983) Ultrastructure of tomato fruit ripening and the role of polygalacturonase isoenzymes in cell wall degradation. *Plant Physiology* 72:1088–1093.

Deikman J, Kline R, Fischer RL (1992) Organization of ripening and ethylene regulatory regions in a fruit-specific promoter from tomato (*Lycopersicon esculentum*). *Plant Physiology* 100:2013–2017.

Dellapenna D, Alexander DC, Bennett AB (1986) Molecular cloning of tomato fruit polygalacturonase: Analysis of polygalacturonase mRNA levels during ripening. *Proceedings of the National Academy of Science, USA* 83:6420–6424.

Dellapenna D, Kates DS, Bennett AB (1987) Polygalacturonase gene expression in Rutgers, *rin*, *nor*, and *Nr* tomato fruits. *Plant Physiology* 85:502–507.

Dick-Perez M, Zhang Y, Hayes J, Salazar A, Zabotina OA, Hong M (2011) Structure and interactions of plant cell-wall polysaccharides by two- and three-dimensional magic-angle-spinning solid-state NMR. *Biochemistry* 50:989–1000.

Du J, Kirui A, Huang S, Wang L, Barnes WJ, Kiemle SN, Zheng Y, Rui Y, Ruan M, Qi S, Kim SH, Wang T, Cosgrove DJ, Anderson CT, Xiao C (2020) Mutations in the pectin methyltransferase QUASIMODO2 influence cellulose biosynthesis and wall integrity in Arabidopsis. *The Plant Cell* 32:3576–3597.

Duruflé H, Hervé V, Balliau T, Zivy M, Dunand C, Jamet E (2017) Proline hydroxylation in cell wall proteins: Is it yet possible to define rules? *Frontiers in Plant Science* 8:1802.

Ferguson IB (1984) Calcium in plant senescence and fruit ripening. *Plant Cell and Environonment* 7:477–489.

Fischer RL, Bennett AB (1991) Role of cell wall hydrolases in fruit ripening. *Annual Review of Plant Physiology and Plant Molecular Biology* 42:675–703.

Fletcher JC, Brand U, Running MP, Simon R, Meyerowitz EM (1999) Signaling of cell fate decisions by CLAVATA3 in Arabidopsis shoot meristems. *Science* 283:1911–1914.

Fragkostefanakis S, Dandachi F, Kalaitzis P (2012) Expression of arabinogalactan proteins during tomato fruit ripening and in response to mechanical wounding, hypoxia and anoxia. *Plant Physiology and Biochemistry* 52:112–118.

Fragostefanakis S, Sedeek EMK, Raad M, Zaki SM, Kalaitzis P (2014) Virus induced gene silencing of three putative prolyl 4-hydroxylases enhances plant growth in tomato (*Solanum lycopersicum*) *Plant Molecular Biology.* doi: 10.1007/s11103-014-0197-6

Fragkostefanakis S, Kaloudas D, Kalaitzis P (2018) Pyridine 2,4-dicarboxylic acid suppresses tomato seedling growth. *Frontiers in Chemistry* 6:3.

Fry LW (2003) Toward a theory of spiritual leadership. *The Leadership Quarterly* 14:693–727.

Gaffe J, Tieman DM, Handa AK (1994) Pectin methylesterase isoforms in tomato (*Lycopersicon esculentum*) tissue: effects of expression of a pectin methylesterase antisense gene. *Plant Physiology* 105:199–203.

Gao Q, Xiong T, Li X, Chen W, Zhu X (2019) Calcium and calcium sensors in fruit development and ripening. *Science Horticulturae* 253:412–421.

Garcia-Romera I, Fry SC (1994) Absence of transglycosylation with oligogalacturonides in plant cells. *Phytochemistry* 35:67–72.

Giannoutsou E, Sotiriou P, Apostolakos P, Galatis B (2013) Early local differentiation of the cell wall matrix defines the contact sites in lobed mesophyll cells of *Zea mays*. *Annals of Botany* 112:1067–1081.

Giovannoni JJ, DellaPenna D, Bennett AB, Fischer RL (1989) Expression of a chimeric polygalacturonase gene in transgenic rin (ripening inhibitor) tomato fruit results in polyuronide degradation but not fruit softening. *Plant Cell* 1:53–63.

Gorres KL, Raines RT (2010) Prolyl 4-hydroxylase. *Critical Reviews in Biochemistry and Molecular Biology* 45:106–124.

Goulao LF, Oliveira CM (2008) Cell wall modification during fruit ripening: when a fruit is not the fruit. *Trends in Food Science & Technology* 19:4 -25.

Goulao LF, Santos J, de Sousa I, Oliveira CM (2007) Patterns of enzymatic activity of cell wall-modifying enzymes during growth and ripening of apples. *Postharvest Biology and Technology* 43:307–318.

Grantham NJ, Wurman-Rodrich J, Terrett OM, Lyczakowski JJ, Stott K, Iuga D, Simmons TJ, Durand-Tardif M, Brown SP, Dupree R, Busse-Wicher M, Dupree P (2017) An even pattern of xylan substitution is critical for interaction with cellulose in plant cell walls. *Nature Plants* 3:859–865.

Grierson D, Slater A, Maunders M, Crookes PR, Tucker GA, Schuch W, Edwards K (1985) Regulation of the expression of tomato fruit ripening genes: the involvement of ethylene. In *Ethylene and Plant Development*, eds. Roberts JA, Tucker GA, 147–161. Butterworths, London.

Grierson D, Slater A, Speirs J, Tucker GA (1985) The appearance of polygalacturonase mRNA in tomatoes: one of a series of changes in gene expression during development and ripening. *Planta* 163:263–271.

Grierson D, Tucker GA (1983) Timing of ethylene and polygalacturonase synthesis in relation to the control of tomato fruit ripening. *Planta* 157:174–179.

Grierson D, Tucker GA, Robertson NG (1981) The molecular biology of ripening. In *Biochemistry of Fruit and Vegetables*, eds. Friend J, 179–191. London: Academic Press.

Gross KC, Sams CE (1984) Changes in cell wall neutral sugar composition during fruit ripening: a species survey. *Phytochemistry* 23:2457–2461.

Guillon F, Moïse A, Quemener B, Bouchet B, Devaux MF, Alvarado C, Lahaye M (2017) Remodeling of pectin and hemicelluloses in tomato pericarp during fruit growth. *Plant Science* 257:48–62.

Ha HC, Woster PM, Yager JD, Casero RA (1997) The role of polyamine catabolism in polyamine analogue-induced programmed cell death. *Proceedings of the National Academy of Sciences, USA* 94: 11557–11562.

Hallett IC, MacRae EA, Wegrzyn TF (1992) Changes in kiwifruit cell-wall ultrastructure and cell packing during postharvest ripening. *International Journal of Plant Sciences* 153:49–60.

Harker FR, Redgwell RJ, Hallet IC, Murray SH (1997) Texture of fresh fruit. *Horticultural Reviews* 20:121–224.

Harker FR, Sutherland PW (1993) Physiological changes associated with fruit ripening and the development of mealy texture during storage of nectarines. *Postharvest Biology and Technology* 2:269–277.

Harriman RW, Tieman DM, Handa AK (1991) Molecular cloning of tomato pectin methylesterase gene and its expression in rutgers, ripening inhibitor, nonripening, and Never ripe tomato fruits. *Plant Physiology* 97:80–87.

Hayashi T (1989) Xyloglucans in the primary cell wall. *Annual Review of Plant Physiology and Plant Molecular Biology* 40:139–168.

Hijazi M, Canada E, Jamet E (2012) Characterization of the Arabinogalactan Protein 31 (GP31) of *Arabidopsis thaliana*. *Journal of Biological Chemistry* 287:9623–9632.

Hobson GE (1967) The effects of alleles at the "Never ripe" locus on the ripening of tomato fruit. *Phytochemistry* 6:1337–1341.

Huang Y, Wang Y, Tan L, Sun L, Petrosino J, Cui MZ, Hao F, Zhang M (2016) Nanospherical arabinogalactan proteins are a key component of the high-strength adhesive secreted by English ivy. *Proceedings of the National Academy of Sciences* 113:3193–3202.

Kaczmarska A, Pieczywek PM, Cybulska J, Zdunek A (2021) Structure and functionality of Rhamnogalacturonan I in the cell wall and in solution: A review. *Carbohydrate Polymers* 2781:118909.

Kalaitzis P, Fragkostefanakis S, Dandachi F, Siomos AS (2010) The role of prolyl 4-hydroxylases in tomato fruit ripening. *Acta Horticulturae* 877:999–1004.

Kalaitzis P, Koehler SM, Tucker ML (1995) Cloning of a tomato polygalacturonase expressed in abscission. *Plant Molecular Biology* 28: 647–656.

Kalaitzis P, Solomos T, Tucker ML (1997) Three different polygalacturonases are expressed in tomato leaf and flower abscission, each with a different temporal expression pattern. *Plant Physiology* 113:1303–1308.

Keegstra K, Talmadge KW, Bauer WD, Albersheim P (1973) The structure of plant cell Walls: III. A model of the walls of suspension-cultured sycamore cells based on the interconnections of the macromolecular components. *Plant Physiology* 51:188–197.

Kieliszewski MJ (2001) The latest hype on Hyp-O-glycosylation codes. *Phytochemistry* 57:319–323.

Kirby AR, Gunning AP, Morris VJ, Ridout MJ (1995) Observation of the helical structure of the bacterial polysaccharide acetan by atomic force microscopy. *Biophysical Journal* 68:360–363.

Kirby AR, MacDougall AJ, Morris VJ (2008) Atomic force microscopy of tomato and sugar beet pectin molecules. *Carbohydrate Polymers* 71:640–647.

Klee HJ, Giovannoni JJ (2011) Genetics and control of tomato fruit ripening and quality attributes. *Annual Review of Genetics* 45:41–59.

Kneissl ML, Deikman J (1996) The tomato E8 gene influences ethylene biosynthesis in fruit but not in flowers. *Plant Physiology* 112:537–547.

Kramer MG, Redenbaugh K (1994) Commercialization of a tomato with an antisense polygalacturonase gene: The FLAVR SAVR™ tomato story. *Euphytica* 79, 293–297.

Kramer M, Sanders R, Bolkan H, Waters C, Sheeny RE, Hiatt WR (1992) Postharvest evaluation of transgenic tomatoes with reduced levels of polygalacturonase: processing, firmness and disease resistance. *Postharvest Biology and Technology* 1:241–255.

Lai S, Chen F, Zhang L, Yang H, Deng Y, Yang B (2013) Nanostructural difference of water-soluble pectin and chelate-soluble pectin among ripening stages and cultivars of Chinese cherry. *Natural Product Research* 27:379–385.

Lamport DTA, Tan L, Held M, Kieliszewski MJ (2020) Phyllotaxis turns over a new leaf-A new hypothesis. *International Journal of Molecular Sciences* 21:1145.

Lamport DTA, Várnai P (2013) Periplasmic arabinogalactan glycoproteins act as a calcium capacitor that regulates plant growth and development. *New Phytology* 197:58–64.

Lamport DTA, Varnai P, Seal CE (2014) Back to the future with the AGP-Ca^{2+} flux capacitor. *Annals of Botany* 114:1069–1085.

Lara I, Belge B, Goulao LF (2014) The fruit cuticle as a modulator of postharvest quality. *Postharvest Biology and Technology* 87: 103 -112.

Leszczuk A, Chylinska M, Zdunek A (2019a) Distribution of arabinogalactan proteins and pectins in the cells of apple (*Malus×domestica*) fruit during post-harvest storage. *Annals of Botany* 123:47–55.

Leszczuk A, Chylińska M, Zdunek A (2019b) Enzymes and vitamin C as factors influencing the presence of arabinogalactan proteins (AGPs) in *Solanum lycopersicum* fruit. *Plant Physiology and Biochemistry* 139:681–690.

Leszczuk A, Chylińska M, Zięba E, Skrzypek T, Szczuka E, Zdunek A (2018b) Structural network of arabinogalactan proteins (AGPs) and pectins in apple fruit during ripening and senescence processes. *Plant Science* 275:36–48.

Leszczuk A, Cybulska J, Skrzypek T, Zdunek A (2020b) Properties of arabinogalactan proteins (AGPs) in apple (*Malus×domestica*) fruit at different stages of ripening. *Biology (Basel)* 9:1–12.

Leszczuk A, Kalaitzis P, Blazakis KN, Zdunek A (2020a) The role of arabinogalactan proteins (AGPs) in fruit ripening - a review. *Horticulture Research* 7:176.

Leszczuk A, Kalaitzis P, Kulik J, Zdunek A (2023) Review: structure and modifications of arabinogalactan proteins (AGPs). *BMC Plant Biology* 23:45.

Leszczuk A, Pieczywek PM, Gryta A, Frąc M, Zdunek A (2019c) Immunocytochemical studies on the distribution of arabinogalactan proteins (AGPs) as a response to fungal infection in *Malus* x *domestica* fruit. *Scientific Reports* 9:17428.

Leszczuk A, Szczuka E, Wydrych J, Zdunek A (2018a) Changes in arabinogalactan proteins (AGPs) distribution in apple (*Malus* x *domestica*) fruit during senescence. *Postharvest Biology and Technology* 138:99–106.

Li S, Chen K, Grierson D (2021) Molecular and hormonal mechanisms regulating fleshy fruit ripening. *Cells* 10:1136.

Lin D, Lopez-Sanchez P, Gidley MJ (2015) Binding of arabinan or galactan during cellulose synthesis is extensive and reversible. *Carbohydrate Polymers* 126:108–121.

Lin D, Lopez-Sanchez P, Gidley MJ (2016) Interactions of pectins with cellulose during its synthesis in the absence of calcium. *Food Hydrocolloids* 52:57–68.

Liu D, Zhou W, Zhong Y, Xie X, Liu H, Huang H, Wang Q, Xiao G (2023) Involvement of branched RG-I pectin with hemicellulose in cell-cell adhesion of tomato during fruit softening. *Food Chemistry* 413:135574.

Lopez-Hernandez F, Tryfona T, Rizza A, Yu XL, Harris MOB, Webb AAR, Kotake T, Dupree P (2020) Calcium binding by arabinogalactan polysaccharides is important for normal plant development. *Plant Cell* 32:3346–3369.

Manning K, Tör M, Poole M, Hong Y, Thompson AJ, King GJ, Giovannoni JJ, Seymour GB (2006) A naturally occurring epigenetic mutation in a gene encoding an SBP-box transcription factor inhibits tomato fruit ripening. *Nature Genetics* 38:948–952.

Matsunaga E, Tauszig-Delamasure S, Monnier PP, Mueller BK, Strittmatter SM, Mehlen P, Chédotal A (2004) RGM and its receptor neogenin regulate neuronal survival. *Nature Cell Biology* 6:749–755.

McIntire TM, Brant DA (1997) Imaging of individual biopolymers and supramolecular assemblies using noncontact atomic force microscopy. *Biopolymers* 42:133–146.

McNeil M, Darvill AG, Albersheim P (1980) Structure of plant cell walls: X. Rhamnogalacturonan I, a structurally complex pectic polysaccharide in the walls of suspension-cultured sycamore cells. *Plant Physiology* 66:1128–1134.

Méndez-Yañez Á, Beltrán D, Campano-Romero C, Molinett S, Herrera R, Moya-León MA, Morales-Quintana L (2017) Glycosylation is important for FcXTH1 activity as judged by its structural and biochemical characterization. *Plant Physiology and Biochemistry* 119:200–210.

Méndez-Yañez Á, Ramos P, Morales-Quintana L (2021) Role of glycoproteins during fruit ripening and seed development. *Cells* 10:2095.

Mercado JA, Pliego-Alfaro F, Quesada MA (2011) Fruit shelf life and potential for its genetic improvement. In *Breeding for Fruit Quality*, eds. Jenks MA, Bebeli PJ, 81–104. John Wiley & Sons, Oxford.

Moore JP, Fangel JU, Willats WGT, Vivier MA (2014) Pectic-β(1,4)-galactan, extensin and arabinogalactan-protein epitopes differentiate ripening stages in wine and table grape cell walls. *Annals of Botany* 114:1279–1294.

Morris VB, Sharma CP (2010) Enhanced in-vitro transfection and biocompatibility of l-arginine modified oligo (-alkylaminosiloxanes)-graft-polyethylenimine. *Biomaterials* 31:8759–8769.

Ortiz A, Graell J, Lara I (2011) Cell wall-modifying enzymes and firmness loss in ripening 'Golden Reinders' apples: a comparison between calcium dips and ULO storage. *Food Chemistry* 128:1072–107.

Paniagua C, Kirby AR, Gunning AP, Morris VJ, Matas AJ, Quesada MA (2017) Unravelling the nanostructure of strawberry fruit pectins by endo-polygalacturonase digestion and atomic force microscopy. *Food Chemistry* 224:270–279.

Paniagua C, Posé S, Morris VJ, Kirby AR, Quesada MA, Mercado JA (2014) Fruit softening and pectin disassembly: an overview of nanostructural pectin modifications assessed by atomic force microscopy. *Annals of Botany* 114:1375–1383.

Paniagua C, Santiago-Doménech N, Kirby AR, Gunning AP, Morris VJ, Quesada MA (2017) Structural changes in cell wall pectins during strawberry fruit development. *Plant Physiology and Biochemistry* 118:55–63.

Park YB, Cosgrove DJ (2012) A revised architecture of primary cell walls based on biomechanical changes induced by substrate-specific endoglucanases. *Plant Physiology* 158:1933–1943.

Parra R, Paredes MA, Labrador J, Nunes C, Coimbra MA, Fernandez-Garcia N, Olmos E, Gallardo M, Gomez-Jimenez MC (2020) Cell wall composition and ultrastructural immunolocalization of pectin and arabinogalactan protein during *Olea europaea* L. Fruit abscission. *Plant Cell Physiology* 61:814–825.

Perrakis A, Bita CE, Arhondakis S, Krokida A, Mekkaoui K, Denic D, Blazakis KN, Kaloudas D, Kalaitzis P (2019) Suppression of a prolyl 4 hydroxylase results in delayed abscission of overripe tomato fruits. *Frontiers in Plant Science* 10:348.

Phan TD, Bo W, West G, Lycett GW, Tucker GA (2007) Silencing of the major salt-dependent isoform of pectinesterase in tomato alters fruit softening. *Plant Physiology* 144:1960–1967.

Posé S, Kirby AR, Mercado JA, Morris VJ, Quesada MA (2012) Structural characterization of cell wall pectin fractions in ripe strawberry fruits using AFM. *Carbohydrate Polymers* 88:882–889.

Posé S, Kirby AR, Paniagua C, Waldron KW, Morris VJ, Quesada M (2015) The nanostructural characterization of strawberry pectins in pectate lyase or polygalacturonase silenced fruits elucidates their role in softening. *Carbohydrate Polymers* 132:134–145.

Posé S, Paniagua C, Matas AJ, Gunning AP, Morris VJ, Quesada MA, Mercado JA (2019) A nanostructural view of the cell wall disassembly process during fruit ripening and postharvest storage by atomic force microscopy. *Trends in Food Science and Technology* 87:47–58.

Pressey R, Avants JK (1982) Solubilization of cell walls by tomato polygalacturonases: effects of pectinesterases. *Journal of Food Biochemistry* 6:57–74.

Qin X, Yamauchi R, Aizawa K, Inakuma T, Kat K (2000) Isolation and characterization of arabinogalactan-protein from fruit of *Lycium Chinense* Mill. *Journal of Applied Glycoscience* 47:155–161.

Redgwell RJ, Macrae E, Hallett I, Fischer M, Perry J, Harker R (1997) *In vivo* and *in vitro* swelling of cell walls during fruit ripening. *Planta* 203:162–173.

Redgwell RJ, Melton LD, Brasch DJ (1992) Cell wall dissolution in ripening kiwifruit (*Actinidia deliciosa*): solubilisation of the pectic polymers. *Plant Physiology* 98:71–81.

Romero P, Rose JKC (2019) A relationship between tomato fruit softening, cuticle properties and water availability. *Food Chemistry* 295:300–310.

Round AN, Rigby NM, MacDougall AJ, Morris VJ (2010) A new view of pectin structure revealed by acid hydrolysis and atomic force microscopy. *Carbohydrate Research* 345:487 -497.

Round AN, Rigby NM, MacDougall AJ, Ring SG, Morris VJ (2001) Investigating the nature of branching in pectin by atomic force microscopy and carbohydrate analysis. *Carbohydrate Research* 331:337–342.

Sala K, Malarz K, Barlow PW, Kurczyńska EU (2017) Distribution of some pectic and arabinogalactan protein epitopes during *Solanum lycopersicum* (L.) adventitious root development. *BioMed Center Plant Biology* 17:25.

Saladié M, Matas AJ, Isaacson T, Jenks MA, Goodwin SM, Niklas KJ, Xiaolin R, Labavitch JM, Shackel KA, Fernie AR, Lytovchenko A, O'Neill MA, Watkins CB, Rose JK (2007) A reevaluation of the key factors that influence tomato fruit softening and integrity. *Plant Physiology* 144:1012–1028.

Santiago-Doménech N, Jiménez-Bemúdez S, Matas AJ, Rose JK, Muñoz-Blanco J, Mercado JA, Quesada MA (2008) Antisense inhibition of a pectate lyase gene supports a role for pectin depolymerization in strawberry fruit softening. *Journal of Experimental Botany* 59:2769–2779.

Schoof H, Lenhard M, Haecker A, Mayer KFX, Jürgens G, Laux T (2000) The stem cell population of Arabidopsis shoot meristems is maintained by a regulatory loop between the CLAVATA and WUSCHEL genes. *Cell* 100:635–644.

Schuch W, Kanczler J, Robertson D, Hobson G, Tucker G, Grierson D, Bright S, Bird C (1991) Fruit quality characteristics of transgenic tomato fruit with altered polygalacturonase activity. *Horticultural Science* 26:1517–1520.

Seifert GJ (2020) On the potential function of type II arabinogalactan *O*-glycosylation in regulating the fate of plant secretory proteins. *Frontiers in Plant Science* 11:1–11.

Seifert GJ, Roberts K (2007) The biology of arabinogalactan proteins. *Annual Review of Plant Biology* 58:137–161.

Seymour GB, Chapman NH, Chew BL, Rose JKC (2013) Regulation of ripening and opportunities for control in tomato and other fruits. *Plant Biotechnology Journal* 11:269–278.

Shackel KA, Greve C, Labavitch JM, Ahmadi H (1991) Cell turgor changes associated with ripening in tomato pericarp tissue. *Plant Physiology* 97:814–816.

Sheehy RE, Kramer M, Hiatt WR (1988) Reduction of polygalacturonase activity in tomato fruit by antisense RNA. *Proceedings of the National Academy of Sciences* 85:8805–8809.

Showalter AM (2001) Arabinogalactan-proteins: structure, expression and function. *Cellular and Molecular Life Sciences CMLS* 58:1399–1417.

Showalter AM, Basu D (2016) Extensin and arabinogalactan-protein biosynthesis: glycosyltransferases, research challenges, and biosensors. *Frontiers in Plant Science* 7:814.

Simmons TJ, Mortimer JC, Bernardinelli OD, Pöppler AC, Brown SP, deAzevedo ER, Dupree R, Dupree P (2016) Folding of xylan onto cellulose fibrils in plant cell walls revealed by solid-state NMR. *Nature Communications* 7:13902.

Slater A, Mauders MJ, Edwards K, Schuch W, Grierson D (1985) Isolation and characterization of cDNA clones for tomato polygalacturonase and other ripening-related proteins. *Plant Molecular Biology* 5:137–147.

Smith CJS, Watson CF, Bird CR, Ray J, Schuch W, Grierson D (1990a) Expression of a truncated tomato polygalacturonase gene inhibits expression of the endogenous gene in transgenic plants. *Molecular and General Genetics 224*:477–481

Smith CJS, Watson CF, Morris PC, Bird CR, Seymour GB, Gray JE, Arnold C, Tucker GA, Schuch W, Harding S (1990b) Inheritance and effect on ripening of antisense polygalacturonase genes in transgenic tomatoes. *Plant Molecular Biology* 14:369–379.

Smith CJS, Watson CF, Ray J, Bird CR, Morris PC, Schuch W, Grierson D (1988) Antisense RNA inhibition of polygalacturonase gene expression in transgenic tomatoes. *Nature* 334:724–726.

Smith DL, Abbott JA, Gross KC (2002) Down-regulation of tomato beta-galactosidase 4 results in decreased fruit softening. *Plant Physiology* 129:1755–1762.

Smith DL, Gross KC (2000) A family of at least seven β-galactosidase genes is expressed during tomato fruit development. *Plant Physiology* 123:1173–1184.

Stahl Y, Simon R (2010) Plant primary meristems: shared functions and regulatory mechanisms. *Current Opinion in Plant Biology* 13:53–58.

Stow J (1993) Effect of calcium ions on apple fruit softening during storage and ripening. *Postharvest Biology and Technology* 3:1–9.

Strasser R, Seifert G, Doblin MS, Johnson KL, Ruprecht C, Pfrengle F, Bacic A, Estevez JM (2021) Cracking the "Sugar Code": a snapshot of *N*- and *O*-glycosylation pathways and functions in plants cells. *Frontiers in Plant Science* 12:640919.

Talbott LD, Ray PM (1992) Molecular size and separability features of pea cell wall polysaccharides: implications for models of primary wall structure. *Plant Physiology* 98:357–368.

Tan L, Eberhard S, Pattathil S, Warder C, Glushka J, Yuan C, Hao Z, Zhu X, Avci U, Miller JS, Baldwin D, Pham C, Orlando R, Darvill A, Hahn MG, Kieliszewski MJ, Mohnen D (2013) An Arabidopsis cell wall proteoglycan consists of pectin and arabinoxylan covalently linked to an arabinogalactan protein. *Plant Cell* 25:270–287.

Tan L, Leykam JF, Kieliszewski MJ (2003) Glycosylation motifs that direct arabinogalactan addition to arabinogalactan-proteins. *Plant Physiology* 132:1362–1369.

Tegl G, Nidetzky B (2020) Leloir glycosyltransferases of natural product C-glycosylation: structure, mechanism and specificity. *Biochemical Society Transactions* 48:1583–1598.

Thibault JF, Renard CMGC, Axelos MAV, Roger P, Crepeau MJ (1993) Studies of the length of homogalacturonic regions in pectins by acid hydrolysis. *Carbohydrate Research* 238: 271–286.

Tigchelaar EC, Mcglasson WB, Franklin MJ (1978) Natural and ethephon-stimulated ripening of F1 hybrids of the *ripening inhibitor (rin)* and *non-ripening (nor)* mutants of tomato (*Lycopersicon esculentum* Mill.). *Functional Plant Biology* 5:449–456.

Toivonen PMA, Brummell DA (2008) Biochemical bases of appearance and texture changes in fresh-cut fruit and vegetables. *Postharvest Biology and Technology* 48:1–14.

Tucker GA, Seymour GB, Bundick Y, Robertson D, Hall L, Smith CJS, Grierson D, Bird CR and Schuch W (1992) Use of antisense RNA technology to study pectin degradation in tomato fruit. *New Zealand Journal of Crop and Horticultural Science* 20:119–124.

Uluisik S, Chapman NH, Smith R, Poole M, Adams G, Gillis RB, Besong TM, Sheldon J, Stiegelmeyer S, Perez L, Samsulrizal N, Wang D, Fisk ID, Yang N, Baxter C, Rickett D, Fray R, Blanco-Ulate B, Powell AL, Harding SE, Craigon J, Rose JK, Fich EA, Sun L, Domozych DS, Fraser PD, Tucker GA, Grierson D, Seymour GB (2016) Genetic improvement of tomato by targeted control of fruit softening. *Nature Biotechnology* 34:950–952.

Valenzuela-Riffo F, Parra-Palma C, Ramos P, Morales-Quintana L (2020) Molecular and structural insights into FaEXPA5, an alpha-expansin protein related with cell wall disassembly during ripening of strawberry fruit. *Plant Physiology and Biochemistry* 154:581–589.

Vincken JP, Schols HA, Oomen RJFJ, McCann MC, Ulvskov P, Voragen AGJ, Visser RGF (2003) If homogalacturonan were a side chain of rhamnogalacturonan I. Implications for cell wall architecture. *Plant Physiology* 132:1781–1789.

Vrebalov J, Ruezinsky D, Padmanabhan V, White R, Medrano D, Drake R, Schuch W, Giovannoni J (2002) A MADS-Box gene necessary for fruit ripening at the tomato Ripening-Inhibitor (Rin) locus. *Science* 296: 343–346.

Wang D, Yeats TH, Uluisik S, Rose JKC, Seymour GB (2018) Fruit softening: revisiting the role of pectin. *Trends in Plant Science* 23:302–310.

Wang J, Nie S (2019) Application of atomic force microscopy in microscopic analysis of polysaccharide. *Trends in Food Science & Technology* 87:35–46.

Wang RF, Tavano ECD, Lammers M, Martinelli AP, Angenent GC, de Maagd RA (2019) Re-evaluation of transcription factor function in tomato fruit development and ripening with CRISPR/Cas9-mutagenesis. *Scientific Reports* 9:1696.

Wang T, Williams JK, Schmidt-Rohr K, Hong M (2015) Relaxation-compensated difference spin diffusion NMR for detecting 13C-13C long-range correlations in proteins and polysaccharides. *Journal of Biomolecular NMR* 61:97–107.

Wang T, Zabotina O, Hong M (2012) Pectin-cellulose interactions in the Arabidopsis primary cell wall from two-dimensional magic-angle-spinning solid-state nuclear magnetic resonance. *Biochemistry* 51:9846–9856.

Willats WGT, Knox JP (1996) A role for arabinogalactan-proteins in plant cell expansion: evidence from studies on the interaction of β-glucosyl Yariv reagent with seedlings of *Arabidopsis thaliana. The Plant Journal* 9: 919–925.

Willats WGT, McCartney L, Mackie W, Knox JP (2001) Pectin: cell biology and prospects for functional analysis. *Plant Molecular Biology* 47:9–27.

Wu Y, Fan W, Li X, Chen H, Takáč T, Šamajová O, Fabrice MR, Xie L, Ma J, Šamaj J, Xu C (2017) Expression and distribution of extensins and AGPs in susceptible and resistant banana cultivars in response to wounding and *Fusarium oxysporum. Scientific Reports* 7:42400.

Xin Y, Chen F, Yang H, Zhang P, Deng Y, Yang B (2010) Morphology, profile and role of chelate-soluble pectin on tomato properties during ripening. *Food Chemistry* 121:372–380.

Xu C, Liberatore KL, MacAlister CA, Huang Z, Chu YH, Jiang K, Brooks C, Ogawa-Ohnishi M, Xiong G, Pauly M, Van Eck J, Matsubayashi Y, van der Knaap E, Lippman ZB (2015) A cascade of arabinosyltransferases controls shoot meristem size in tomato. *Nature Genetics* 47:784–792.

Yang H, Chen F, An H, Lai S (2009) Comparative studies on nanostructures of three kinds of pectins in two peach cultivars using atomic force microscopy. *Postharvest Biology and Technology* 51:391–398.

Yang H, Lai S, An H, Li Y (2006a) Atomic force microscopy study of the ultrastructural changes of chelate-soluble pectin in peaches under controlled atmosphere storage. *Postharvest Biology and Technology* 39:75–83.

Yang HS, Feng GP, An HJ, Li YF (2006b) Microstructure changes of sodium carbonate-soluble pectin of peach by AFM during controlled atmosphere storage. *Food Chemistry* 94:179–192.

Yang L, Huang W, Xiong F, Xian Z, Su D, Ren M, Li Z (2017) Silencing of SlPL, which encodes a pectate lyase in tomato, confers enhanced fruit firmness, prolonged shelf-life and reduced susceptibility to grey mould. *Plant Biotechnology Journal* 15:1544–1555.

Yapo BM (2011) Rhamnogalacturonan-I: A structurally puzzling and functionally versatile polysaccharide from plant cell walls and mucilages. *Polymer Reviews* 51:391–413.

Zdunek A, Kozioł A, Pieczywek PM, Cybulska J (2014) Evaluation of the nanostructure of pectin, hemicellulose and cellulose in the cell walls of pears of different texture and firmness. *Food Bioprocess Technology* 7:3525–3535.

Zdunek A, Pieczywek PM, Cybulska J (2020) The primary, secondary, and structures of higher levels of pectin polysaccharides. *Comprehensive Reviews in Food Science and Food Safety* 20:1101–1117.

Zhang LF, Chen F, An H, Yang H, Sun X, Guo X, Li L (2008) Physicochemical properties, firmness, and nanostructures of sodium carbonate-soluble pectin of 2 Chinese cherry cultivars at 2 ripening stages. *Journal of Food Science* 73:17–22.

Zhang LF, Yang H, Chen FS, Ye XQ, Xin-Yuan S, Liu DH, Yang B, Hongjie A, Yun D (2012) Effects of temperature and cultivar on nanostructural changes of water-soluble and chelate-soluble pectin in peaches. *Carbohydrate Polymers* 87:816–821.

Zhang X, Tang H, Du H, Liu Z, Bao Z, Shi Q (2020) Comparative N-glycoproteome analysis provides novel insights into the regulation mechanism in tomato (*Solanum lycopersicum* L.) during fruit ripening process. *Plant Science* 293:110413.

Zykwinska A, Tripon-Le Berre L, Sinquin C, Ropartz D, Rogniaux H, Colliec-Jouault S, Delbarre-Ladrat C (2018) Enzymatic depolymerization of the GY785 exopolysaccharide produced by the deep-sea hydrothermal bacterium *Alteromonas infernus*: Structural study and enzyme activity assessment. *Carbohydrate Polymers* 188:101–107.

8 NMR Studies on Cellulose and Plant Cell Walls

Michael C. Jarvis
Glasgow University

INTRODUCTION

NMR (nuclear magnetic resonance) spectroscopy is more informative about molecular and supramolecular cell-wall structure than any other single experimental method. NMR relaxation experiments are also a key source of information about the rigidity or otherwise of cell-wall components, including water. NMR methods have made a large contribution towards our understanding of cell-wall structure since the 1980s, especially during the past decade, and the continuing development of new NMR methods is likely to have a substantial impact in the next few years.

The author's early research on cell walls was before the structure of such insoluble materials could be determined by NMR. Polysaccharides were studied then using techniques like methylation analysis (Talmadge et al. 1973), traditionally applied after each polymer had been extracted and rigorously purified in solution (Jarvis and Duncan 1976; Jarvis et al. 1981). Cellulose, of course, is not extractable and was studied by crystallographic methods usually intended for large perfect crystals, leading to frustration because cellulose microfibrils are neither large nor perfectly crystalline (e.g. Chanzy et al. 1979). Lignin is not very extractable either and studying lignin was restricted to a few determined specialists (e.g. Nimz et al. 1981). With the methods then available, it was not possible to determine the structures of cell-wall polymers *in situ*. For that and other reasons, it was very difficult to connect the structure of the intact cell wall to its functions in the living plant (Green et al. 1964; Nishitani and Masuda, 1980). Solid-state NMR methods have provided a route towards that kind of understanding.

NMR methods originated in the 1930s and 1940s as a branch of nuclear physics (Andrew, 1984; Becker, 1993). Atomic nuclei have magnetic properties that make them interact with an external magnetic field. Because the interaction is a quantum effect, it undergoes transitions with discrete changes in energy and therefore with characteristic frequencies (Ernst et al. 1997). These transitions can be induced by a resonating electromagnetic field. The resonant frequencies are in the radiofrequency region of the spectrum. They are very slightly modulated by the magnetic shielding of any nucleus by the electrons around it, including bonding electrons; that is, by the local chemistry (Ernst et al. 1997). Decades of development by numerate physical chemists, working with or for the instrument manufacturers, transformed this last observation into a powerful technology for determining the structures of organic molecules in solution (Ernst, 1992).

From its inception until the 1970s, NMR was a mainly solution-state form of spectroscopy. Cell walls, even when composed principally of water, are not solutions in the relevant sense: their polymers are too rigid to give usable spectra for structural analysis by conventional solution-state NMR procedures. The development of effective solid-state NMR methods around 1980 made it possible, however, to record worthwhile ^{13}C NMR spectra from cellulose and from intact cell walls (reviewed by Atalla and VanderHart 1999). Among the first significant advances in biological chemistry using solid-state ^{13}C NMR were the discoveries that cellulose from higher plants contains crystalline and disordered chains differing in conformation (Horii et al. 1983, 1984) and that fully crystalline native cellulose has two distinguishable structures (Atalla and VanderHart 1984), which we now know as cellulose Iα and Iβ.

These early solid-state ^{13}C NMR experiments included two key technical innovations (Schmidt-Rohr and Spiess 1994). One was magic-angle spinning (MAS), to alleviate the extreme spectral broadening that arises from dispersed orientation of molecules with respect to the NMR magnetic field (Yannoni 1982). The orientations of small molecules in solution are randomised by thermal tumbling, but in a solid they are fixed on the NMR timescale. By mathematics rather than magic, spinning the solid sample at an angle of 54.74° to the magnetic field, the angle between the diagonal of a cube and its edge, simulates the random effect of molecular tumbling and greatly improves the resolution of the spectra. The magic angle of 54.74° is not unique to NMR (Sturcova et al. 2003), but its practical significance is greater than in other forms of spectroscopy. The other key innovation was cross-polarisation (CP) from abundant magnetised protons to the less abundant ^{13}C, together with high-power proton decoupling, to improve the signal intensity of the ^{13}C spectra (Yannoni 1982).

These CP-MAS (Cross-polarisation with magic-angle spinning) experiments comprised the basic NMR approach to cell-wall characterisation for more than two decades and led to many advances in our understanding of

DOI: 10.1201/9781003178309-8

cellulose (Atalla and VanderHart 1999) and cell-wall structure (Leary et al. 1986, Jarvis and McCann 2000), even though the spectral resolution available was still far below what was then achieved routinely in solution-state NMR. Magnetic relaxation and spin-diffusion experiments, timing the decay or redistribution of ^{1}H or ^{13}C magnetisation, proved to be useful for measuring the rigidity of individual polymers within the overall cell-wall structure and for determining the spatial distance between cell-wall domains (Jarvis and McCann 2000).

A step change in the capability of NMR came when, following earlier experiments on isolated cellulose (Sakellariou et al. 2003, Witter et al., 2006, Hesse-Ertelt et al., 2008), the research groups of Hong in the USA (Perez Garcia et al. 2011; Wang et al. 2012, 2014) and later the Duprees in England (Dupree et al. 2015) adopted multi-dimensional solid-state NMR methods that provide ways round the problem of limited resolution for intact cell walls. More recently, other NMR spectroscopists have taken up multidimensional methods and added further experimental innovations (e.g. Gao and Mortimer 2020; Yuan et al. 2021), a trend that clearly has some way still to proceed. There have also been developments in the gentle solubilisation of cell-wall polymer complexes for high-resolution analysis by solution-state NMR (Belesov et al. 2020; Liu et al. 2020; Koso et al. 2020; Wang et al. 2020), and in characterising the state of water in cell walls by proton broadline NMR (Grunin et al. 2020; Li and Zhao 2020).

Historically, solid-state NMR technology has not been easy for non-spectroscopists to assimilate, and disconnects arose between the parts of the cell-wall research community specialising in NMR, crystallography and other approaches (Jarvis and McCann 2000). In this chapter, there will be an attempt to bridge these disconnects and survey the results from NMR methods against the backdrop of data from other analytical methods. The NMR methodology is not the focus: it has been well reviewed elsewhere (e.g. Reif et al. 2021), and the emphasis here is on what has been learned about cell walls from NMR.

LOOKING BACK: EARLY 1D ^{13}C NMR EXPERIMENTS ON PLANT CELL WALLS

CP-MAS NMR experiments on intact cell walls were first carried out in the 1990s (Jarvis and Apperley, 1990; Newman 1992; Newman et al. 1994; Larsson et al. 1999), with the aim—in this author's mind, at least—of avoiding the need to destroy the interconnections that keep the cell-wall polymers insoluble in the native state: interconnections that permit growth in primary walls but allow secondary walls to support tall trees. Typical one-dimensional (1D) ^{13}C spectra of primary and woody cell walls are shown in Figure 8.1. They show peaks from C4 and C6 of 'crystalline' cellulose at 89 and 65 parts per million (ppm) and from less ordered cellulose at 84 and 62 ppm (Newman et al. 1994). After some controversy, the conformational difference responsible for separating both C4 and C6 peaks from these two cellulose forms was ascribed to the rotation of the exocyclic (C6) group (Newman 1999), with the trans-gauche (*tg*) conformation (approx. 89 ppm/65 ppm) predominating in the crystalline form as in the Iα and Iβ cellulose allomorphs and the gauche-trans (*gt*) conformation, perhaps with some gauche-gauche (*gg*), in the less ordered form (approx. 84 ppm/62 ppm) (Vietor et al. 2002). These C4 and C6 chemical shifts are in reasonable agreement with Horii et al. (1983), who found that in model saccharides the *gt* conformation gave a ^{13}C chemical shift intermediate between those of the *tg* and *gg* conformations.

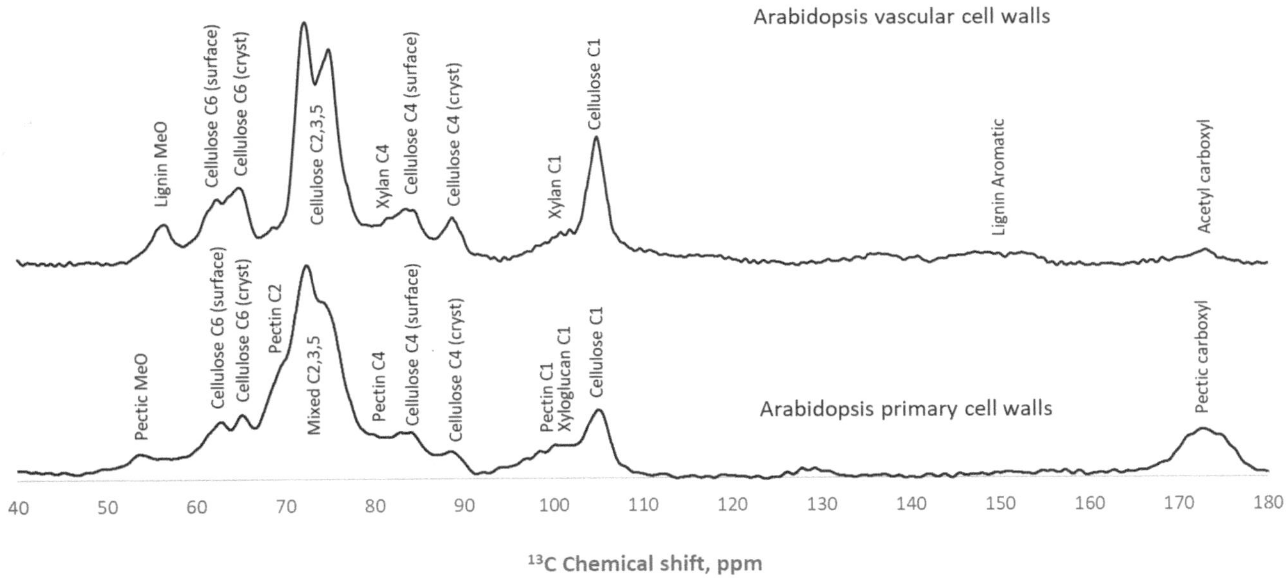

FIGURE 8.1 Solid-state CP-MAS ^{13}C spectra of primary and xylem cell walls isolated from wild-type Arabidopsis hypocotyls and flowering stems. For experimental details, see Ha et al. (2002) and MacKinnon et al. (2006).

It was generally assumed that the 'crystalline' cellulose was a mixture of the known Iα and Iβ forms (Newman et al. 1996, Sturcova et al. 2004), perhaps with additional, partly ordered chains (Wickholm et al. 1998). With hindsight, this assumption was unjustified (Wang, Yang et al. 2016): hydrothermal conversion of native cellulose to the Iβ form causes quite large changes to the 1D ^{13}C spectra as shown in Figure 8.2, which is based on a data set used by Sturcova et al. (2004).

The spectral contributions of C2, C3 and C5 of cellulose and other polymers are jumbled together in the 72–76 ppm region, but peaks or shoulders can be assigned to primary-wall pectic galacturonans and arabinans, methoxyl and acetyl groups and the methyl groups of rhamnose and fucose (Figure 8.1). Hemicellulose peaks are quite hard to distinguish from those of cellulose and pectins (Newman et al. 1994; Ha et al. 1998; Altaner et al. 2006). In spectra from lignified cell walls, the aromatic carbon peaks from lignin are dispersed between 120 and 160 ppm (Altaner et al. 2006), but the lignin aliphatic carbon peaks (Nimz et al. 1981) are obscured under the polysaccharide signals.

The spectral resolution is very low by comparison with solution-state spectra, or with spectra from fully crystalline solids: compare with the sucrose spectrum in Figure 8.3. This observation suggests that the limited resolution of cell-wall spectra is linked to the intrinsic disorder of the cell wall, consistent with the lack of improved resolution in higher-field spectrometers and with some limited improvement on hydrating the cell walls and by tuning the CP and ^{1}H decoupling conditions. A group of pectic chains including β(1,4')-D-galactans and branched arabinans, undetectable under CP conditions optimised for the cell wall as a whole, gave notably higher-resolution spectra (Figure 8.4) in hybrid MAS experiments without CP but incorporating features of solution-state experiments (Ha et al. 1996; Thimm et al. 2002; Ha et al. 2005). Monomer residues

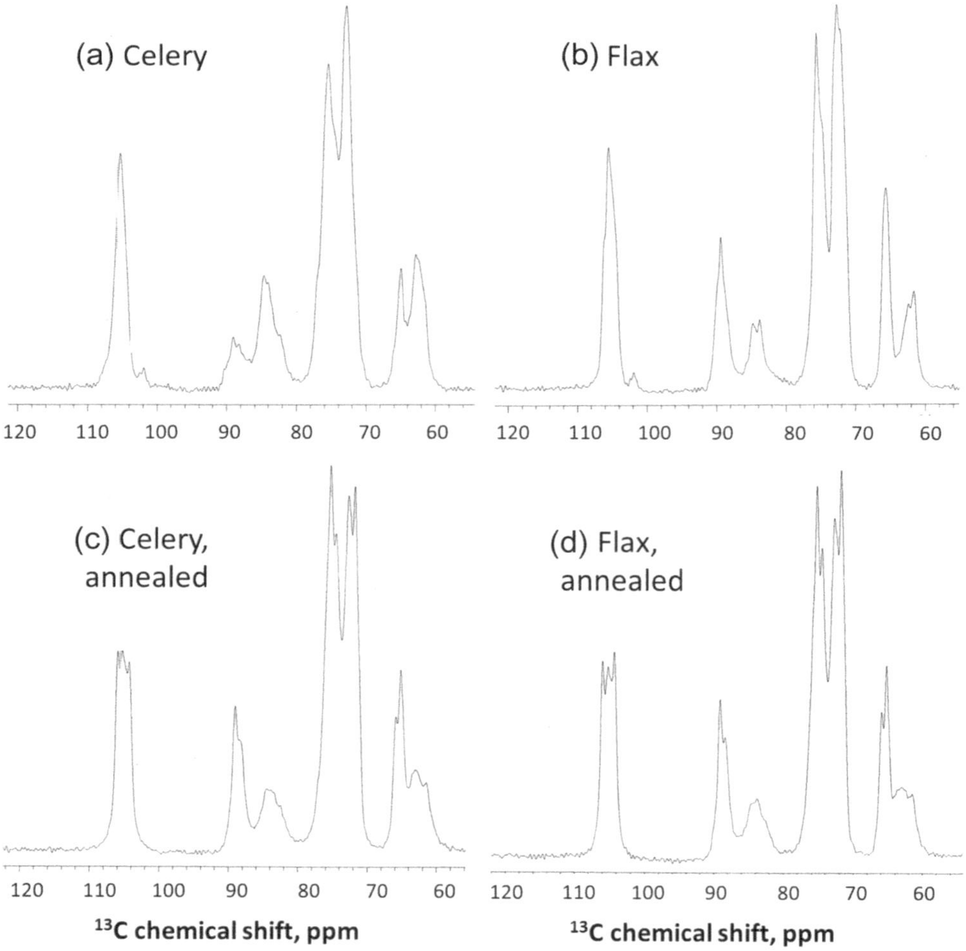

FIGURE 8.2 Solid-state CP-MAS ^{13}C spectra of primary cell walls isolated from celery collenchyma and fibre cell walls isolated from flax, before and after hydrothermal ('annealing') treatment to convert other forms of cellulose to cellulose Iβ. Even for flax cellulose, which was generally considered to be dominated by the Iβ allomorph in its native state, the changes in the C1, C4 and C6 signals are readily evident. This observation demonstrates that the crystalline fraction of native cellulose microfibrils in higher plants is not a simple mixture of the Iα and Iβ forms, although that was not appreciated at the time. The celery cellulose also became dominated by the C6 *tg* conformation, probably indicating an increase in microfibril diameter and tensile performance. For experimental details, see Sturcova et al. (2004).

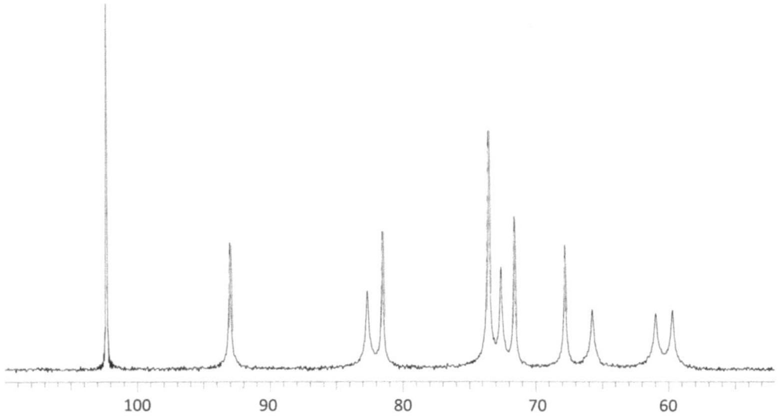

FIGURE 8.3 Solid-state CP-MAS ^{13}C spectrum of crystalline sucrose (icing sugar) collected under similar conditions to the cell-wall spectra shown in Figure 8.1. The increased resolution is attributed to crystalline order.

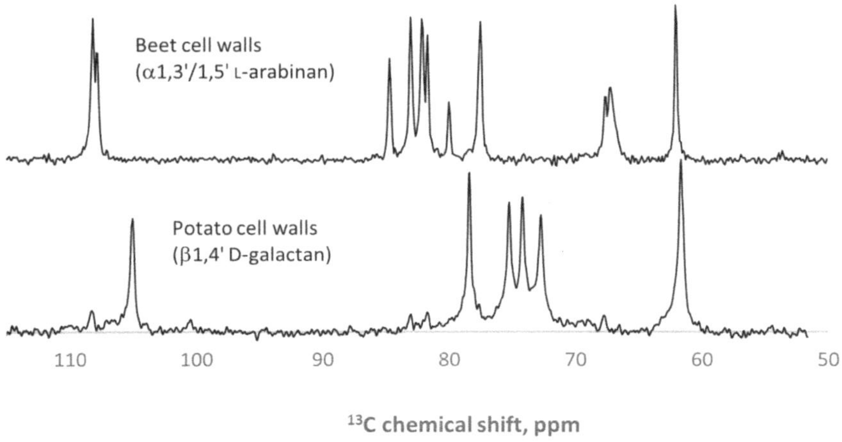

FIGURE 8.4 Direct-polarisation MAS ^{13}C spectra of hydrated beet and potato cell walls, collected with WALZ-16 ^{1}H decoupling as for solution NMR. Only the 'tethered solute' α(1,3'/1,5')-L-arabinan (beet) and β(1,4')-D-galactan (potato) chains of rhamnogalacturonan I are represented in these hybrid solution/solid-state spectra. The increased resolution compared to conventional CP-MAS spectra (c.f. Figure 8.1) is attributed to time-averaged orientational motion. For experimental details, see Ha et al. (2005).

within these very flexible chains were assumed to approximate the motion of small molecules in solution. The term 'tethered solutes' was coined to describe the solution-like behaviour of these flexible primary-wall polymer chains (Ha et al. 2005). Nothing similar was observed for secondary cell walls. With suitable adjustments, all solid components of cell walls can be observed by solid-state NMR: an unforeseen advantage for intractable polymers like cutin and suberin (e.g. Ha et al. 1997).

CP-MAS spectra are not rigorously regarded as quantitative, because the efficiency of the CP step varies according to the number and mobility of the protons attached to each carbon atom. CP efficiency is particularly low for carboxyl and substituted aromatic carbons, which lack covalently bonded protons. However semi-quantitative data on cell-wall composition can be extracted if peaks from carboxyl, aromatic and methyl carbons are excluded (Newman 1999). Microfibril dimensions were estimated by several research groups on the assumption that the 89 and 65 ppm peaks, similar to crystalline cellulose signals, represented C4 and C6 in core cellulose chains and the 84 and 62 ppm peaks represented C4 and C6 in surface cellulose chains with greater mobility and sensitivity to hydration, so that a nominal surface: interior ratio could be calculated (Wickholm et al. 1998; Newman 1999; Vietor et al. 2002). The resulting estimates of microfibril diameter correlated roughly with those obtained by electron microscopy (EM) or the application of the Scherrer equation to X-ray diffraction patterns (Fernandes et al. 2011; Newman et al. 2013): microfibrils from most cell walls of higher plants appeared to be approximately 3 nm in diameter, too small to be consistent with the 36-chain model then accepted for the elementary fibrils synthesised by a single terminal complex (Jarvis 2018). In contrast when few non-cellulosic polysaccharides are present as in cotton (Kirui et al. 2019) flax (Thomas et al. 2013) or tension wood (Leppanen et al. 2011), somewhat larger (5–7 nm) microfibrils are formed, probably by fusion of several newly polymerised ~3 nm units laid down in parallel, and the surface: interior ratio is smaller. There is an implication that non-cellulosic polymers help to keep individual microfibrils from fusing into larger units immediately after they are formed (Jarvis 2018). It now seems likely that higher plants normally synthesise cellulose in 18-chain units (Nixon et al. 2016), which would have dimensions of about 2.4×3.0 nm if arranged as in cellulose Iβ. These dimensions would be slightly increased if there is a small degree of local post-synthesis fusion or when bound hemicelluloses contribute (Kumar and Turner 2015; Jarvis 2018; Thomas et al. 2020)

POLYMER FLEXIBILITY FROM EARLY NMR RELAXATION EXPERIMENTS

In a suitably designed relaxation experiment, each NMR signal decays (relaxes) at a rate that depends on local thermal motion (Newman 1992). In a ^{13}C spin-lattice relaxation experiment, for example, relaxation depends on MHz-frequency thermal motion of the carbon giving rise to the signal. As another example, in a CP-MAS ^{1}H rotating-frame spin-lattice relaxation experiment, kHz-frequency motions within the pool of nearby polymer or water protons that transfer magnetisation to a carbon determine the relaxation rate of the ^{13}C signal. Thus, ^{13}C or ^{1}H relaxation can be exploited to probe the rigidity of different domains within the cell wall, on different spatial and time scales (Ha et al. 1998; Vietor et al. 2002).

From proton relaxation experiments on primary cell walls, mobile domains dominated by hydrated pectins could be distinguished from a rigid domain containing cellulose (Ha et al. 1996, 1997; Hediger et al. 1999; Thimm et al. 2002). Lignin in secondary cell walls was nearly as rigid as cellulose, probably due to cross-linking, while hemicelluloses, so far as their signals could be identified with the resolution available in 1D experiments, had lower overall rigidity (Altaner et al. 2006).

More localised information on mobility can be derived from ^{13}C relaxation experiments, because spin diffusion is minimal between the widely spaced ^{13}C nuclei at natural ^{13}C abundance. The original assignment of the spectral signatures of 'crystalline' and less ordered cellulose to interior and surface chains was based on the greater rigidity of the interior chains as estimated from ^{13}C relaxation experiments without CP (Horii et al. 1984). Further experiments of this type showed that C6 of cellulose was much more mobile than the other carbons and that the *gt* and *gg* conformers were more mobile than *tg* (Vietor et al. 2002).

With hindsight, it is not clear how far these early experiments on polymer mobility answered the questions that were being posed at the time. There was a strong interest in the nature of the cell-wall loosening that permits growth under turgor stress, and in the mechanical deformation of wood, but there was not much clear thinking about the relationship between such mechanical properties and the thermal mobility that could be measured by NMR. In the 1990s, there was also much current interest in the extent to which parts of the primary cell wall formed 3D bonded networks that restricted otherwise soluble polymer chains from escaping into solution (Jarvis and McCann 2000). A single covalent bond to an insoluble domain, or a few covalent cross-links to let a polymer form an infinite network, can confer insolubilisation with little restriction of the thermal mobility measured by NMR relaxation times, as the example of the 'tethered solute' pectic chains demonstrates (Ha et al. 2005). Such sparse covalent cross-links were not likely to be directly visible in the 1D solid-state NMR spectra, and it has never been easy to distinguish their consequences from the insolubilisation and rigidification that come from more numerous but weaker linkages such as hydrogen bonds.

POLYMER LOCATIONS FROM 1D SPIN-DIFFUSION EXPERIMENTS

Differential magnetisation of rigid and mobile domains within the cell wall, created in a ^{13}C or ^{1}H relaxation experiment, can be allowed to re-equilibrate by spin diffusion over millisecond or longer timescales (Newman 1992). Spin diffusion follows Fick's Law with a diffusion coefficient that can be predicted from first principles (Schmidt-Rohr and Spiess 1994). On that basis, it is possible to estimate distances, on the sub-nanometre to nanometre scale, between domains differing in rigidity (Newman 1992). The information obtained on cell walls fell short of imaging, but with appropriate diffusion models it was possible to reconstruct the radial distribution of chain types around the microfibril axis and to draw some conclusions about the distribution of wood polymers (Altaner et al. 2006), even within the limitations of the 1D proton spin-diffusion experiments that were in general use before about 2010.

At the same time it was becoming possible, although technically challenging, to image the internal structure of cell walls at 1–10 nm resolution by EM (Xu et al. 2007). It was emerging that there was significant variation in polymer composition across the thickness of both primary and woody cell walls, that microfibrils in wood were loosely bundled into aggregates sometimes called macrofibrils (Fahlen and Salmen 2005; Xu et al. 2007) and that along their length, lignin was inhomogeneously distributed (Terashima et al. 2009). The results of the NMR relaxation and ^{1}H spin-diffusion experiments on conifer cell walls were broadly consistent with these models and showed that lignin and acetylated hemicelluloses occupied the space between the macrofibrils, and their constituent microfibrils were partially in direct contact with one another, so that there was incomplete penetration of water into the macrofibril interior (Newman 1992; Altaner et al. 2006; Hult et al. 2000). The spin-diffusion data also supported a model in which the less ordered cellulose chains were mostly so close to the crystalline cellulose component that the microfibril structure could be approximately described as concentric, with a crystalline core and a less ordered surface

(Newman 1991; Altaner et al. 2006). The concentric disposition of ordered and disordered chains was disputed by crystallographers, who interpreted the axial broadening of the cellulose diffraction patterns in terms of ordered and disordered domains alternating along the microfibril (Andersson et al. 2003). This interpretation in its simplest form was inconsistent with the NMR data. Later it became apparent that short regions of disorder do exist along the microfibril axis and are potential hydrolysis sites in nanocellulose manufacture (Thygesen and Gierlinger 2013), but that these domains probably do not account for a large proportion of the mass of disordered cellulose (Jarvis 2018). It was proposed that cellulose chains at microfibril interfaces within wood macrofibrils give a broad C4 signal distinct from the pair of narrow peaks assigned to exposed surfaces (Hult et al. 2000).

Similar but more hydration-dependent aggregation was observed in celery collenchyma cell walls, which resemble typical dicot primary cell walls in polymer composition (Kennedy et al. 2007); however, these appear to be atypical of primary walls with respect to microfibril aggregation (Jarvis 2018).

These early ^{1}H relaxation and spin-diffusion experiments showed that there were domains of greater and less rigidity outside the microfibrils, but with limited spectral resolution it was difficult to locate each of the known cell-wall polymers or to link the information to EM images with any precision (Altaner et al. 2006). More precise identification had to await the additional resolution provided by 2D NMR. It should be noted in passing that 2D spectra can be extracted from any 1D experiment that is repeated under slightly varying conditions, such as classic Goldman–Shen spin-diffusion experiments, either by a statistical correlation approach (Noda 1993, Thomas et al. 2021) or through other forms of data processing (Fernandes et al. 2011). The key innovation, however, was true 2D solid-state ^{13}C NMR experiments with magnetisation transferred between defined nuclei, as described below.

MULTIDIMENSIONAL NMR AND THE STRUCTURES OF CELLULOSE MICROFIBRILS

Before about 2000, multidimensional experiments had become routine in solution NMR but were not commonly implemented on solid samples. In the early 2000s, there were limited pioneering trials of 2D solid-state NMR experiments on cellulose, particularly in the laboratory of Emsley in France (Hediger et al. 2002; Sakellariou et al. 2003). After about 2010, the application of multidimensional NMR methods (Box 1) to primary cell walls by Hong's group (Perez Garcia et al. 2011; Wang et al. 2012), and later to secondary cell walls by Dupree and co-workers (2015), caused a re-thinking of cell-wall structures and especially, a reassessment of how microfibrils were assembled.

Box 1: Abbreviations for NMR Experiments

^{13}C-^{13}C through-space correlation
- PDSD Proton-Driven Spin Diffusion
- RFDR Radiofrequency-Driven Recoupling
- DARR Dipolar-Assisted Rotational Resonance

^{13}C-^{13}C through-bond correlation
- INEPT Insensitive Nuclei Enhanced by Polarization Transfer
- INADEQUATE Incredible Natural Abundance DoublE QUAntum Transfer Experiment

^{13}C-^{1}H correlation
- HETCOR HETeronuclear CORrelation
- REDOR Rotational Echo DOuble Resonance

Signal enhancement
- DNP Dynamic Nuclear Polarization

There are a number of ways in which ^{13}C NMR experiments can be designed to give 2D or multidimensional outputs. Most commonly, the basis is proximity, so that cross peaks in the 2D spectrum appear between signals from pairs of ^{13}C atoms that are sufficiently close together (Figure 8.5). This has two advantages. First, with the data spread out in two dimensions, the number of signals that can be resolved is increased by one to two orders of magnitude even when there is no improvement in the resolution in each dimension. Secondly, the proximity information makes it possible to distinguish many or all of the ^{13}C signals from carbons in any one monosaccharide residue: in some experiments, they can be traced in sequence around the monosaccharide ring (Figure 8.6) (Wang et al. 2012).

The main disadvantage, at least initially, was that the lower signal: noise ratio inherent in 2D NMR meant that the experiments needed to be carried out on ^{13}C-enriched cell walls. Growing ^{13}C-enriched Arabidopsis plants is relatively straightforward but time-consuming. Obtaining ^{13}C-enriched cell walls from other species, such as conifer trees, is more challenging.

Given ^{13}C-enriched cell walls, there were a number of solid-sate NMR experiments already used in protein structure determination (Hong et al. 2012) that gave a proximity-based 2D output in different ways, responding to either through-space (including intermolecular) or through-bond proximity. The PDSD, RFDR (Wang et al. 2012) and DARR experiments (Dupree et al. 2015) (Box 1) rely on through-space proximity between ^{13}C atoms. In these experiments, the ^{13}C spectrum is on both vertical and horizontal axes, and the 'mixing' time allowed for spin diffusion between ^{13}C nuclei can be adjusted to optimise the threshold spatial separation that will generate an off-diagonal cross peak. When a short mixing time is chosen, cross peaks are restricted to pairs of carbons that are separated in space by about 0.15 nm, the length of a single covalent bond. When the mixing time is increased, cross peaks start to be detectable from carbons further apart or

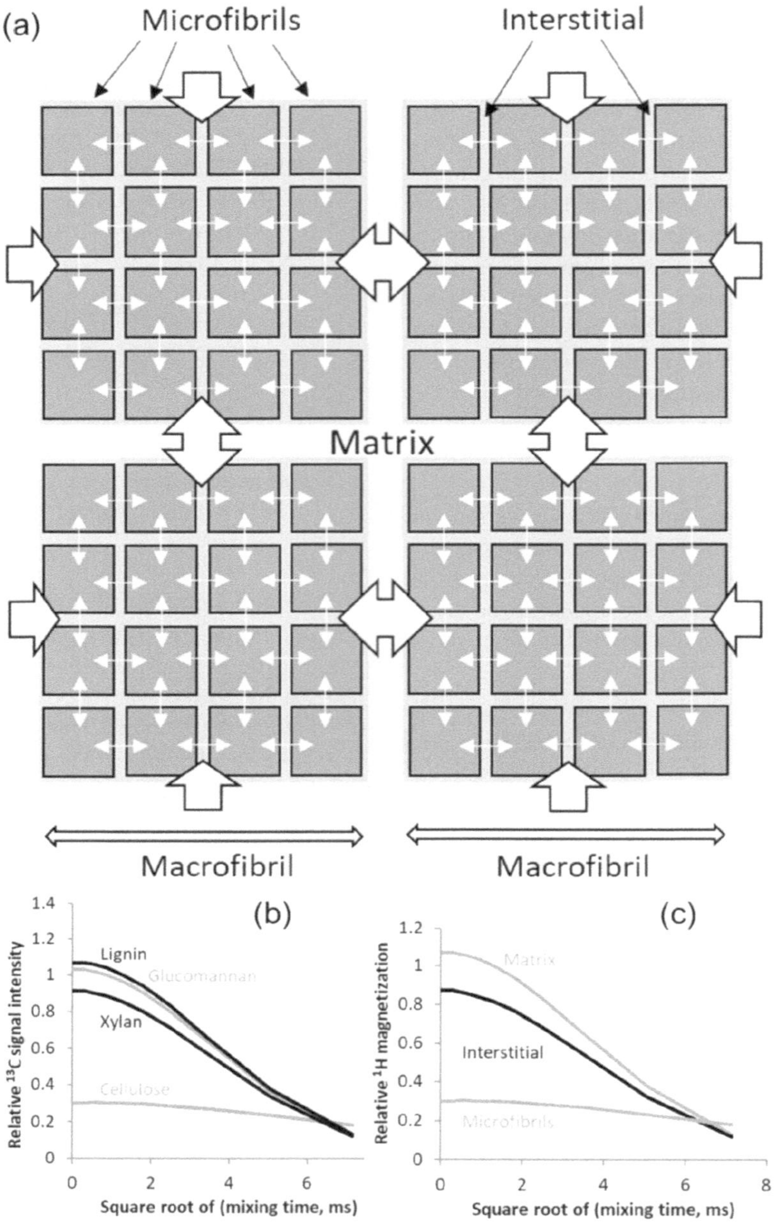

FIGURE 8.5 Modelling of proton spin diffusion, as measured in 1D experiments of the Goldman–Shen type on Sitka spruce cell walls. (a) Modelled structure showing four macrofibrils, each of 16 microfibrils. (b). Spin-diffusion curves for individual polymers, fitted approximately to the experimental data of Altaner et al. (2006). (c) Spin diffusion curves for spatial domains, derived by finite element analysis of spin diffusion within the structure shown in A and the behaviour of each polymer in B. For experimental details, see Altaner et al. (2006). A 2D presentation of the output from this experiment is shown in Fernandes et al. (2011).

in adjacent molecules, but the maximum separation that will permit cross peaks is only about 1 nm, considerably less than the maximum separation distances that can be probed in 1D proton spin-diffusion experiments. The refocused CP-INADEQUATE 2D experiment (Sakellariou et al. 2003; Hesse-Ertelt et al. 2008; Wang et al. 2014) and related through-bond (*J*-coupling) experiments (Wang et al. 2012, Dupree et al. 2015, Kirui et al. 2019) specifically give cross peaks from pairs of ^{13}C atoms connected by a covalent C-C bond. In these double-quantum experiments, the normal CP-MAS spectrum is on the horizontal axis and the sum of the chemical shifts for a bonded pair of ^{13}C nuclei is on the vertical axis.

To illustrate how these 2D experiments are used, in the diagrammatic spectrum shown in Figure 8.6, both dimensions show the ^{13}C spectrum. With a short mixing time only adjacent carbons give off-axis cross peaks Thus, for any one form of cellulose, there are cross peaks linking C1 with C2, C2 with C3 and so on. Then, even if the resolution of the spectrum in each dimension is no greater than in a 1D experiment, it becomes feasible to identify the spectral contributions of individual polymers in much more detail

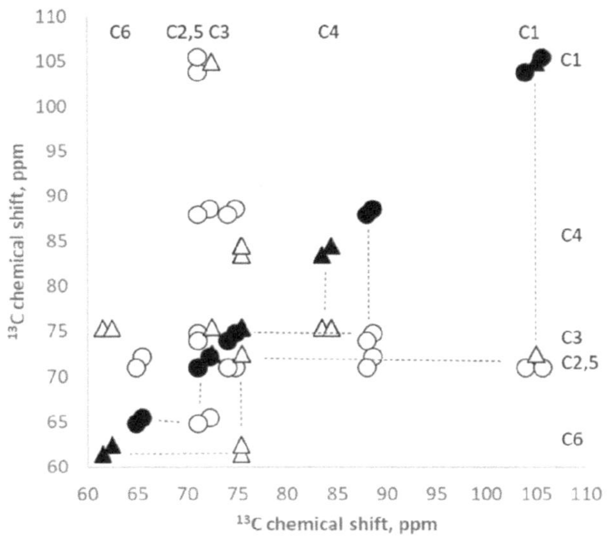

FIGURE 8.6 Assigning signals in a diagrammatic 2D ^{13}C spectrum of cellulose, such as would be collected in a through-space correlation experiment with a short mixing time that gives cross peaks only at one-bond separation distances. Triangles refer to ordered cellulose with C6 *tg*. Circles refer to less ordered surface cellulose with C6 *gt*. The cross peaks also allow signals that overlap in the 1D spectrum (on the diagonal) to be clearly resolved. Dupree et al. (2015) describe ^{13}C-^{13}C RFDR experiments with outputs of this kind.

than was ever possible in 1D. Lengthening the mixing time leads to additional cross peaks between carbons in adjacent polymer chains (Figure 8.7).

The ~3 nm microfibrils of higher-plant cellulose that have been studied by multidimensional NMR contain about seven identifiable types of glucosyl residue (Wang et al. 2014; Wang, Yang et al. 2016; Terrett et al. 2019). It would be tempting to say 'seven types of cellulose chain', but caution is needed because each cellulose chain has glucosyl residues facing in alternate directions (Nishiyama et al. 2012), so that if the environments of the two sides of the chain differ, the hydrogen-bonding patterns of consecutive glucosyl residues will also differ (Jarvis, 2018) and their contributions to the NMR spectrum may be distinguishable.

Two of the seven types of glucosyl unit can be identified with the 'surface' cellulose known from 1D ^{13}C NMR (Wang, Yang et al. 2016; Dupree et al. 2015): they have the approximate *gt* conformation at C6 (Phyo et al. 2018) giving a C4 doublet signal centred on 84 ppm, they show moderate C6 mobility (Wang et al. 2015) and they are spatially associated with water, consistent with a location on exposed surfaces (Wang, Yang et al. 2016).

Modelling studies (Kubicki et al. 2018) suggest that the [110] and [1–10] planes occupy much of the microfibril surface in higher plants. On these hydrophilic surfaces, the alternating glucosyl units with C6 facing inwards and outwards are in different environments (Wickholm et al. 1998). Their NMR signal intensities are easiest to reconcile with plausible microfibril structures if both inward- and outward-facing C6 groups are predominantly in the *gt* conformation (Phyo et al. 2018; Wang, Yang et al. 2016). It is not clear why: hydrogen-bonding schemes on the two sides of a cellulose chain do not interact, although there are some possibilities for stereoelectronic interactions between C6 conformations in successive glucosyl residues.

The *gt* conformation does not permit inter-residue H-bonding from O6'H to O2. Instead, the orientation of the H-bonding from both of these positions, as shown by Fourier-transform infrared (FTIR) spectroscopy (Fernandes et al. 2011), is redirected outwards towards water or other microfibrils. The inward-facing *gt* conformation, with O6 being a hydrogen-bond donor, would imply that each surface chain is attached to the subsurface chain in the same sheet by a H-bond pattern different from that found in cellulose Iβ. There is some freedom of rotation around the *gt* conformation, and some *gg* may also be present (Fernandes et al. 2011). Hydrogen bonding to surface chains in an adjacent sheet might also be permitted (Simmons et al. 2016).

In 1D ^{13}C spectra of softwood cell walls with their cellulose microfibrils aggregated into larger units, it had been suggested that cellulose surface chains accessible to water could be distinguished from inaccessible surfaces in contact with other microfibrils, the latter giving broadly dispersed C4 signals (Wickholm et al. 1998). While the presence of both accessible and inaccessible surfaces is consistent with other evidence (Fahlen and Salmen 2005; Xu et al. 2007; Thomas et al. 2021), a separate NMR assignment for inaccessible surfaces does not seem to be supported (Ling et al. 2019) by more recent 2D spectra, which have the two narrow C4 signals near 84 ppm assigned to the C6 *gt* surface residues (Dupree et al. 2015; Wang, Yang et al. 2016).

In the native state of the cell wall, cellulose–cellulose contacts due to microfibril aggregation have water interposed to varying extents that can be inferred from the decrease in centre-to-centre spacing on dehydration of secondary cell walls: small-angle neutron scattering (SANS) experiments show that the spacing is particularly sensitive to dehydration in bamboo, a model for grasses (Thomas et al. 2015), less sensitive in conifer wood (Fernandes et al.

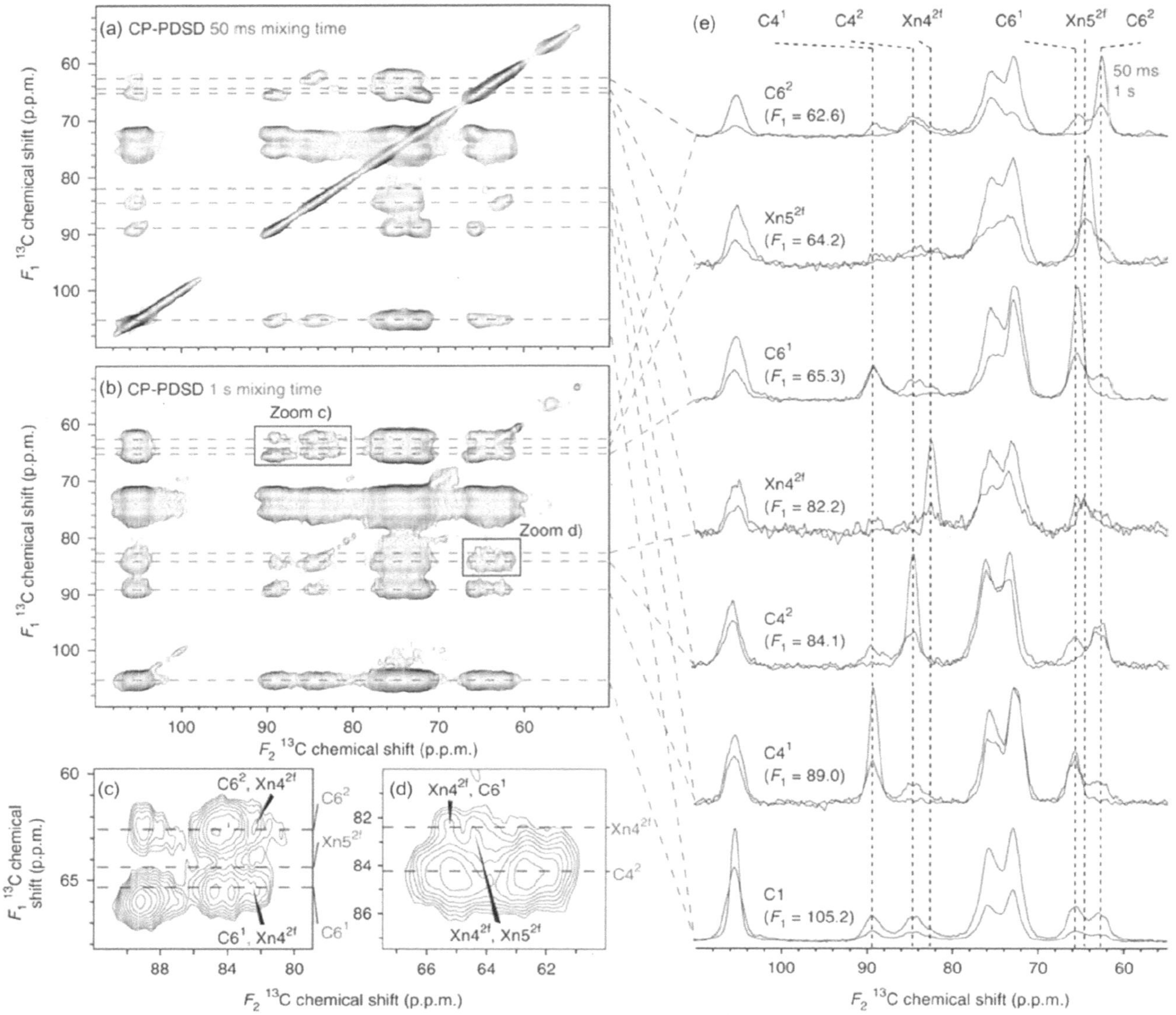

FIGURE 8.7 2D DARR spectra and corresponding 1D slices from Arabidopsis stem cell walls (Simmons et al. 2016). With kind assistance from the authors.

2011; Thomas et al. 2021) and hardly affected at all in dicot wood (Langan et al. 2014; Thomas et al. 2014). A recent study on radiata pine wood (Cresswell et al. 2021) showed that dehydration from the never-dried state led to conformational changes at C6 of cellulose and in glucomannans, together with well-defined spatial redistribution of the glucomannans and lignin. Xylans were less affected. There was a small increase in the C6 *tg* conformation of cellulose (Cresswell et al. 2021), but not enough to be consistent with the idea that the transition to the *gt* and other C6 conformations at cellulose surfaces is simply due to hydration: non-hydrated contact with other microfibrils may also destabilise the *tg* conformation.

The other five types of glucosyl residue identifiable in the 2D spectra have C6 in approximately the *tg* conformation, as in crystalline cellulose (Dupree et al. 2015, Wang et al. 2014; Wang, Yang et al. 2016). The C6 *tg* glucosyl residues identifiable in the corresponding 2D spectra differ from both of the two distinct cellulose Iβ glucosyl units, and from the two distinct cellulose Iα glucosyl units (Wang and Hong 2016). Cotton and flax cellulose are crystallographically closer to cellulose Iβ and their 2D spectra are more similar, but still not identical, to the spectra of crystalline cellulose Iβ (Kirui et al.2019).

In Arabidopsis primary cell walls, four of the five types of glucosyl residues with C6 *tg* are tentatively assigned to subsurface locations, although it is not yet clear which surface each underlies. The first 2D spin-diffusion experiments on Arabidopsis cell walls showed close proximity of the C6 *tg* residues to the C6 *gt* residues, consistent with their presence in the same segments of the microfibrils (Wang et al. 2014; Wang, Yang et al. 2016) as in the 'concentric' model proposed from 1D ^{13}C NMR, although one study of sorghum cell walls did not show this (Gao et al. 2020).

An unexpected feature of the early 2D spin-diffusion experiments was that many of the C6 *tg* types of glucosyl residue, which would have been assumed to be enclosed in the crystalline core of the microfibril, showed close spatial associations with hemicelluloses, pectins, water and lignin (Wang et al. 2014, 2015; Dupree et al. 2015). This caused consternation, with the implication that the prevailing concentric models for the microfibril structure might be inadequate. To examine this anomaly in more detail, it is necessary to look at the way in which non-cellulosic polymers associate with cellulose.

CELLULOSE: HEMICELLULOSE BINDING

The classic example is the glucuronoarabinoxylan in angiosperm secondary cell walls, including hardwoods, where a regularly alternating pattern of substituents allows hydrogen bonding to the hydrophilic [110] and [1–10] faces of cellulose along the inner edge of the flat-ribbon (2_1 helical) xylan chain (Busse-Wicher et al. 2014; Falcoz-Vigne et al. 2017; Grantham et al. 2017). The outer edge carries bulky acetyl and glucuronosyl substituents. Binding to hydrophilic microfibril surfaces has been clearly demonstrated by NMR spin-diffusion experiments (Simmons et al. 2016) and is reduced on hydration (Dupree et al. 2015) and in mutants where the alternating substitution pattern is disrupted (Lyczakowski et al. 2017). Unbound xylan remains in the coiled, roughly 3_1 helical conformation found in solution (Simmons et al. 2016). Because the xylan chain has been shown to bind to cellulosic glucosyl units with C6 *tg*, its presence appears to force the cellulosic surface into a conformation like that of the subsurface chains in an uncoated microfibril (Terrett et al. 2019). The absence of C6 on xylose precludes one option for an inward-facing H-bond to cellulose and leaves a gap between the xylan and cellulose chains, possibly occupied by water (Jarvis 2018; Cresswell et al. 2021).

In softwoods, the xylan component binds to cellulose surfaces with C6 *tg*, in very much the same way as in angiosperms, even though its alternating substituents do not include acetyl groups (Terrett et al. 2019; Thomas et al. 2020; Cresswell et al. 2021). Modelling suggests that softwood xylans can also bind in a flat-ribbon conformation to the hydrophobic microfibril surfaces, although not in exactly the same position as a cellulose chain would occupy in the Iβ lattice (Martinez-Abad et al. 2017). The arabinoxylans of grasses and cereals appear to adopt similar flat-ribbon chain conformations and to contain domains capable of binding to cellulose surfaces (Martinez-Sanz et al. 2017; Zhang et al. 2019), but being more highly branched than dicot xylans they may bind in a less ordered way (Wang et al. 2014; Martinez-Sanz et al. 2017; Duan et al. 2021). In never-dried sorghum cell walls, the arabinoxylans adopted primarily the 3_1 helical conformation and were in proximity to C6 *gt* rather than C6 *tg* cellulose surfaces (Gao et al. 2020).

Galactoglucomannans, which are more abundant than xylans in softwood cell walls, can apparently adopt flat-ribbon 2_1 helical conformation as in crystalline linear mannans (Martinez-Abad et al. 2020, Terrett et al. 2019). In that conformation, non-acetylated galactoglucomannan segments bind to cellulose surfaces (Terrett et al. 2019; Cresswell et al. 2021). The underlying cellulose has both C6 *tg* and C6 *gt* conformations (Terrett et al. 2019). Other, irregularly acetylated galactoglucomannan chain segments remain detached (Martinez-Abad et al. 2020) in irregularly coiled conformations (Terrett et al. 2019).

In the primary cell walls of dicot species, xyloglucans appear to be in spin-diffusion proximity to cellulose surfaces to a lesser degree than pectins (Wang and Hong 2016), but the enzymatic removal of the xyloglucans exposes binding sites for cellulose-binding proteins that are considered to attach along the narrow, hydrophobic faces of microfibrils (Wang et al. 2013). Hemicellulose binding to the hydrophobic [200] microfibril face presumably involves a mechanism with some similarity to the inter-sheet binding in cellulose. Molecular dynamics simulations suggest that the free energy difference between binding to hydrophobic and to hydrophilic faces is slight (Martinez-Abad et al. 2020). That being the case, it seems advisable to draw conclusions from direct experimental evidence of the binding location rather than from simulations. For example, there is evidence that in softwoods, acetylated and non-acetylated galactoglucomannans bind in proximity to cellulose monomers with different NMR signatures (Terrett et al. 2019), although these are not yet assigned to specific faces of the microfibril.

Are there common features in microfibril structure across all higher plants? Similar, although not identical, sets of about seven cellulose signals have been observed in Arabidopsis secondary cell walls (Simmons et al. 2016), in cell walls of the grass *Brachypodium* (Duan et al. 2021) and in softwood cell walls (Terrett et al. 2019). Such similarities would be expected for microfibril segments to which no non-cellulosic polymers were attached. However different non-cellulosic polymers have been shown to associate with hydrophilic surfaces of microfibrils in each case and as discussed above, binding of each kind of non-cellulosic polymer is associated with the C6 *tg* conformation in the underlying cellulose chain. Differences between taxa in the spectral signatures of at least some of the C6 *tg* glucosyl units might then be expected. These differences seem to be quite small, a puzzle remaining to be resolved.

LIGNIN–CELLULOSE ASSOCIATION

In lignified *Arabidopsis* cell walls, at least a small amount of lignin is in close contact with cellulose domains having C6 either *tg* or *gt*, *tg* being more abundant (Dupree et al. 2015). Close contacts between syringyl lignin and both *tg* and *gt* cellulose domains were also detected in poplar and eucalyptus, but lignin–cellulose contacts were less abundant in grasses (Kirui et al. 2022). It has been suggested that lignin binds to the hydrophobic [200] faces of cellulose in a manner resembling the tyrosine residues of cellulose-binding proteins (Jarvis 2018). If the 200 faces are two or more chains wide, they may have C6 in the *tg* conformation anyway, as in the microfibril interior.

More abundant contacts have been found between lignins and xylans, both in the flat-ribbon conformation that binds to cellulose (Kirui et al 2022) and in the coiled conformation that does not (Fernando et al. 2020; Kirui et al. 2022). The guaiacyl lignins of softwoods were associated with acetylated galactoglucomannans as well as xylans (Terrett et al. 2019; Kirui et al. 2022). Lignin–xylan and lignin–galactoglucomannan associations may involve covalent linkages (Terrett and Dupree 2019; Zhang et al. 2021). There remains some uncertainty as to how much lignin lies within bundles of microfibrils (Xu et al. 2007) and how much is associated with hemicelluloses between bundles, but close associations of lignin with cellulose seem to increase in the order grasses and Arabidopsis < hardwoods < softwoods (Kirui et al. 2022).

PECTIN–CELLULOSE ASSOCIATION

The ^{13}C spin-diffusion experiments of Perez Garcia et al. (2011) and Wang et al. (2012, 2015) have demonstrated clear proximity of pectic galacturonans and possibly rhamnogalacturonan I to cellulose domains, including cellulose domains with C6 *tg*. The pectin–cellulose interactions diminished on reducing the pH to 4.0 and thus suppressing pectic ionisation (Phyo et al. 2019).

The nature of the pectin–cellulose binding is unknown, and may depend on different principles from hemicellulose–cellulose binding. None of the diverse chains making up the pectic group of polysaccharides has any known conformational similarity to cellulose. The C4 chemical shift of the galacturonans located close to cellulose, about 80 ppm (Wang et al. 2012), suggests that these approximate to a 3_1 helical conformation (Jarvis and Apperley 1995) with a trisaccharide repeat distance of 1.3 nm compared to the cellobiose repeat distance of 1.04 nm, so that any favourable binding conformation at one point on the microfibril cannot be repeated longitudinally. Localised or misaligned associations therefore seem more likely, and what stabilises them is unclear, nor is it clear what might induce the adoption of the C6 *tg* conformation (Wang et al. 2014) by the underlying cellulose.

Summarising, where non-cellulosic polymers associate with cellulose surfaces the underlying glucosyl units are commonly found to have the *tg* conformation at C6, in contrast to the *gt* conformation that predominates at exposed, hydrated surfaces and perhaps at interfaces between microfibrils. It may be the hydrophobic [200] cellulose surfaces that underlie lignin chains, but glucuronoarabinoxylans, at least some galactoglucomannans and pectins, all appear to convert the underlying cellulose chains at the [110]/[1–10] surfaces into the *tg* conformation, by a means presently unknown. Looking at this puzzle from the other side, it is not clear why cellulose surfaces interacting with water or with other microfibrils should favour the *gt* conformation with its predominant hydrogen-bond donor character, on both the inner and the outer sides of the cellulose chains.

^{1}H-^{13}C AND ^{2}H-^{13}C HETERONUCLEAR CORRELATION NMR IN THE SOLID STATE

Most solid-state ^{1}H NMR experiments lead to broadline spectra in which protons in different molecular environments cannot be resolved. Information on proton mobility can nevertheless be derived in broadline experiments such as WISE (Schmidt-Rohr et al. 1992). For example, wood contains highly mobile water in the cell lumina and at least two pools of less mobile, non-freezing water within the cell walls. These can be distinguished in broadline ^{1}H relaxation experiments (Cox et al. 2010). Hediger et al. (1999, 2002) described ways to correlate ^{1}H broadline spectra with the ^{13}C spectra either through space (dipolar coupling) or through C-H covalent bonds (*J*-coupling). Rapid polarisation transfer between water and onion cellulose suggested that the water was not fully excluded from the microfibril structure (Hediger et al. 2002). Similar dipolar-coupled experiments have since been used for Arabidopsis primary cell walls (White et al. 2014) showing close interaction of water and pectins; and for cotton cellulose (Taylor et al. 2008; Garvey et al. 2019) showing how hydration influences the mobility of surface chains.

A different type of ^{1}H-^{13}C correlation experiment was used by Phyo et al. (2018) to estimate the distance between the protons on C1 and C6, and hence the C6 conformation. The ^{1}H spectrum does not appear in this 2D experiment: proton spin diffusion is sandwiched within a series of forward and reverse CP steps. The results for primary cell walls from Arabidopsis and the grass *Brachypodium*, together with analysis of spin-diffusion rates from water to cellulose, rigorously confirmed the conclusion of Horii et al. (1984) that interior chains have the *tg* conformation at C6. Surface chains were shown to adopt predominantly the *gt* conformation (Phyo et al. 2018), but with a small proportion of *gg* as reported in earlier 1D ^{13}C relaxation experiments (Vietor et al. 2002).

A similar type of heteronuclear experiment might provide a solution to the problem of correlating ^{13}C signals between successive monosaccharide residues, as in solution-state glycosyl sequencing (Gray et al. 2019). In cellulose-like chain conformations, the C1–C4' distance spanning the glycosidic linkage and the C1–C4 distance spanning the pyranose ring are both in the range of 0.24–0.28 nm and are not readily distinguishable, but the H1–H4' distance is potentially short enough to bridge the ^{13}C correlations from one monosaccharide to the next.

Heteronuclear 2D ^{1}H-^{13}C correlation experiments with high resolution in the proton dimension have extensive potential. The chemical shift of CH protons is a probe of stereoelectronic effects such as the anomeric effect, while that of hydroxyl protons is strongly dependent on hydrogen-bond strength. Hydrogen-bond strength (and orientation) can also be derived by FTIR spectroscopy, but the assignment of FTIR signals to specific hydroxyls has been controversial (Altaner et al. 2014; Thomas et al. 2021). We attempted such heteronuclear NMR experiments in 2001, but we could not make them work for cellulosic materials on the spectrometers then available. With the use of higher

MAS rates (Phyo and Hong 2019), it becomes possible to obtain well-resolved ^{1}H spectra and to carry out a variety of 2D and 3D heteronuclear correlation experiments, giving ^{13}C spectra in one or two dimensions and ^{1}H spectra in the remaining dimension, with the option of ^{1}H detection. For fully rigid solids, ultrafast MAS rates of at least 100 kHz are required to obtain usable resolution in the ^{1}H dimension, but the thermal mobility of matrix polysaccharides in hydrated,^{13}C-enriched primary cell walls makes such experiments practicable at more modest, although still technically demanding, MAS rates of 50–60 kHz. In a form of heteronuclear RFDR experiment, Phyo and Hong (2019) observed cross peaks with OH as well as CH protons.

Recently, Yuan et al. (2021) applied ^{1}H-^{13}C heteronuclear correlation experiments, with a MAS rate of 150 kHz and without ^{13}C enrichment, to cotton cellulose and spruce and maple wood. The minor *gg* conformer of surface cellulose in spruce wood was distinguishable by its C4H ^{1}H chemical shift (Yuan et al. 2021).

Deuterium NMR is a very different technique from proton NMR, as ^{2}H is a quadrupolar nucleus. Deuterium spectra are not site-resolved but encode information on molecular motions (Lemke et al. 2012; Lindh et al. 2017), particularly motions of large amplitude in disordered systems. Gelenter et al. (2017) described 3D experiments, with the ^{2}H spectrum in the third dimension, on bacterial cellulose and Arabidopsis primary cell walls. Fast isomerisation at C6 in surface chains of bacterial cellulose was observed, suggesting that the *gt* and *gg* conformations are time-averaged. Fast large-amplitude motions were also observable in the pectin-rich matrix of primary cell walls (Gelenter et al. 2017). An off-resonance ^{2}H/^{13}C REDOR experiment made it possible to obtain sensitive and specific information on motions of CH groups in bacterial cellulose (Gelenter et al. 2021).

ENHANCING THE SENSITIVITY OF NMR EXPERIMENTS ON CELL WALLS

The insensitivity of ^{13}C solid-state NMR lies behind many practical problems, including the need for ^{13}C enrichment prior to most 2D experiments and the difficulty of detecting sparse structural motifs like covalent cross-links between different polymers. The duration of an NMR experiment, i.e., the number of repetitions (scans) needed to accumulate an adequate averaged spectrum, depends on the square of the signal:noise ratio and rapidly becomes more than the instrument time available if the signal:noise ratio is insufficient. Packing more sample into a wide MAS rotor can help, but is limited by centrifugal forces and inhomogeneity of the magnetic field. The inherent insensitivity of NMR methodology is made worse, for cell walls, by the fact that peaks in the ^{13}C NMR spectrum are rather broad and therefore not very high.

It is possible, although quite challenging, to enhance the sensitivity of MAS experiments by dynamic nuclear polarisation (DNP-MAS) from a paramagnetic doping radical (Gauto et al. 2021). This approach was adopted in an attempt to locate the site at which the protein expansin binds and permits controlled, turgor-driven loosening of the cell-wall structure during growth: ^{15}N-labelled expansin was located at a site with xyloglucan and a specific C6 *tg* cellulose domain, in Arabidopsis (Wang et al. 2013), but the binding site was different in maize (Wang, Chen et al. 2016).

DNP sensitivity enhancement has also been used to avoid the need for ^{13}C enrichment, for example, in 2D NMR analyses of cotton (Kirui et al. 2019), maize (Kang et al. 2019) and rice cell walls (Zhao et al. 2021), for cellulose synthesised *in vitro* (Deligey et al. 2022) and in studies of chemical valorisation of biomass (Perras et al. 2017, Viger-Gravel et al. 2019). The ability of 2D DNP-MAS experiments to resolve large numbers of signals from complex systems is illustrated by the lignin spectra (Sapouna and Lawoko 2021), where cross peaks from the diverse aliphatic interconnections between monomers are clearly distinguishable from the carbohydrate signals that obscure them in the 1D spectra (Fernando et al. 2020). At natural ^{13}C abundance, there is little through-space spin diffusion between ^{13}C nuclei, so that through-bond ^{13}C-^{13}C correlation experiments such as INADEQUATE are preferred.

An alternative way to gain sensitivity is to revert to one spectral dimension: not in the simple 1D CP-MAS experiments used formerly but by recording a 1D slice of a 2D correlation experiment containing cross peaks to a single signal of interest (Wang et al. 2012; Dupree et al. 2015). The utility of this approach has been explored by Addison et al. (2020). They did not attempt to obtain 1D spectra at natural abundance but used the extra sensitivity of 1D DARR difference experiments to construct detailed spin-diffusion curves, confirming the spatial association of lignin and xylan in ^{13}C-enriched poplar cell walls.

SOLUTION NMR AFTER SOLUBILISATION OF CELL-WALL POLYMERS

Solution NMR methods have of course been used for many years to study isolated, purified cell-wall polysaccharides. A number of more recent developments, however, have aimed at the total solubilisation of the cell wall followed by simultaneous NMR characterisation of its polymer components. A number of solvent systems have been used, e.g., DMSO/pyridine after ball-milling (Kim and Ralph 2010, Abbas et al. 2020); ionic liquids after partial acetylation (Liu et al. 2020); and acetonitrile following permethylation (Ndukwe et al. 2020). Ideally, only non-covalent bonding would be disrupted during the solubilisation step and any covalently bonded networks should remain intact, but ball-milling and permethylation certainly break some covalent bonds and the partial acetylation conditions used by Liu et al. (2020) might cleave any intermolecular ester linkages by transesterification. These experiments are nevertheless useful in allowing powerful and time-efficient solution-state NMR methods to be brought to bear on the structures of individual polymers. An alternative approach is partial solubilisation in hot water followed by hybrid solution/solid-state NMR characterisation of both dissolved and suspended polymers (Ning et al. 2020).

LOOKING FORWARD—SOME OPEN QUESTIONS

Some predictions about the utility of NMR methods can be made with a degree of confidence. Recent developments like DNP and faster MAS, reducing the dependence on ^{13}C enrichment, are making cell-wall NMR more accessible and easier to integrate with other directions of investigation such as molecular genetics and biomass processing. It seems reasonable to expect that a full description of microfibril structure will soon emerge, including which cellulose chain types are located where and how primary- and secondary-wall microfibrils differ.

Looking further ahead would need a crystal ball. In the author's (probably biased) opinion, understanding how cell walls function—where their mechanical properties come from, and how the time-dependent mechanical properties of primary cell walls permit and control growth—will be possible only with substantial inputs from NMR experiments.

Unlike vibrational spectroscopy and scattering methods (Thomas et al 2021), solid-state MAS NMR experiments cannot feasibly be carried out under controlled mechanical loading. There are in fact very large mechanical loads imposed by centrifugal force inside the MAS rotor, but it has never been feasible to pack cell-wall samples into the rotor in such a way that the direction of loading is uniform, controllable and useful: for example, with the microfibrils radially oriented (we tried, in the 1990s). It may also be surmised that large oscillating forces (phonons) are generated within cell-wall samples as a piezoelectric response to radiofrequency (RF) proton decoupling fields. Nevertheless, the structural information emerging from NMR studies will be key to understanding how cell walls respond to force, when combined with observations under strain by other methods. NMR mobility measurements will also be relevant, but it has not been fully appreciated that in composite materials like cell walls, stiffness and strength have a complex and poorly understood relationship to the thermal mobility that can be measured by NMR relaxation: indeed, the relationship between chemical or enzyme kinetics as normally described, based on thermal energy, and alternatively based on a sustained mechanical force, has only rarely been investigated by any means in any system (Tinoco and Bustamante 2002).

To understand the relationship between structure and mechanical function, we need to know more about how cell-wall polymers are attached together at the scale of nanometers to tens of nanometers. For example, we know very little about what shear-resisting linkages join one microfibril (with or without a hemicellulosic coating) to the next, adhering, microfibril within an aggregate. Nor do we know much about how shear is resisted across wider gaps, such as the lignin–hemicellulose region between macrofibrils in wood or the cellulose-anchored pectin phase in primary walls, although it seems likely that in every case hydrogen-bonded polymer interfaces are involved (Ye et al. 2020; Thomas et al. 2021). Within cellulose (and probably other polysaccharide) chains (Jarvis and Apperley 1995; Loerbroks et al. 2013; Berglund et al. 2016; Yang et al. 2018), stereoelectronic factors have an incompletely understood influence on conformation and molecular deformation under stress (Altaner et al. 2014; Wohlert et al. 2022).

In the author's opinion, ^{1}H-^{13}C (and ^{2}H-^{13}C) heteronuclear correlation experiments with ultrafast MAS (Phyo et al. 2019; Yuan et al. 2021) hold particular promise for these questions to be addressed. We have come to assume that hydrogen-bond topology cannot be accessed directly, only sketched indirectly by inference from the structure of the carbon/oxygen framework. Identifying hydroxyl protons (Phyo et al. 2019) has potential for answering direct questions about hydrogen-bonding patterns that are not currently being asked. The chemical shifts of CH protons provide insights into stereoelectronic effects, the other poorly understood source of the stability and mechanical strength of cellulose (Wohlert et al. 2022).

Open questions abound. Why is the *gt* conformation, with its outward-facing hydrogen bonds, preferred on hydrated and interface cellulose surfaces? What is the free energy contribution of displacing water molecules from microfibril surfaces during these processes, if water is structured (or de-structured) by the topology of hydrogen-bond donors and acceptors on the surface? How strong is the co-operativity when hydrogen bonds line up head-to-tail, for example in the centre chains of cellulose Iβ with C6 *tg* (Nishiyama et al. 2002; Jarvis 2003)? Sufficient chemical shift resolution in the proton dimension plus spatial information would permit a more direct view of hydrogen-bonding schemes and stereoelectronic patterns, and a potential route to answering questions such as these.

ACKNOWLEDGEMENTS

Almost all the author's NMR research was in collaboration with D.C. Apperley (University of Durham) whose many inputs are gratefully acknowledged. The author also thanks a long series of research students and postdocs at Glasgow University, and the UK BBSRC, EPSRC and SFC for funding. J. Sugiyama (Kyoto University) is thanked for the hydrothermal treatment of the cellulose samples shown in Figure 8.3 and R. Dupree (University of Warwick) for Figure 8.7 and for enjoyable discussions on modern NMR.

REFERENCES

Abbas M, Peszlen I, Shi R, Kim H, Katahira R, Kafle K, Xiang Z, Huang X, Min D, Mohamadamin M, Yang C, Dai X, Yan X, Park S, Li Y, Kim SH, Davis M, Ralph J, Sederoff RR, Chiang VL, Li Q (2020) Involvement of CesA4, CesA7-A/B and CesA8-A/B in secondary wall formation in *Populus trichocarpa* wood. *Tree Physiology* 40:73–89. doi:10.1093/treephys/tpz020.

Addison B, Stengel D, Bharadwaj VS, Happs RM, Doeppke C, Wang T, Bomble YJ, Holland GP, Harman-Ware AE (2020) Selective one-dimensional C-13-C-13 spin-diffusion solid-state nuclear magnetic resonance methods to probe spatial arrangements in biopolymers including plant cell walls, peptides, and spider silk. *Journal of Physical Chemistry B* 124:9870–9883. doi:10.1021/acs.jpcb.0c07759.

Altaner CM, Apperley DC, Jarvis MC (2006) Spatial relationships between polymers in Sitka spruce: Proton spin-diffusion studies. *Holzforschung* 60:665–673. doi:10.1515/hf2006.112.

Altaner CM, Thomas LH, Fernandes AN, Jarvis MC (2014) How cellulose stretches: synergism between covalent and hydrogen bonding. *Biomacromolecules* 15:791–798. doi:10.1021/bm401616n.

Andersson S, Serimaa R, Paakkari T, Saranpaa P, Pesonen E (2003) Crystallinity of wood and the size of cellulose crystallites in Norway spruce (*Picea abies*). *Journal of Wood Science* 49:531–537. doi:10.1007/s10086-003-0518-x.

Andrew ER (1984) A historical review of NMR and its clinical applications. *British Medical Bulletin* 40:115–119.

Atalla RH, Vanderhart DL (1984) Native cellulose - a composite of 2 distinct crystalline forms. *Science* 223:283–285. doi:10.1126/science.223.4633.283.

Atalla RH, VanderHart DL (1999) The role of solid state C-13 NMR spectroscopy in studies of the nature of native celluloses. *Solid State Nuclear Magnetic Resonance* 15:1–19. doi:10.1016/s0926-2040(99)00042-9.

Becker ED (1993) A brief history of nuclear magnetic resonance. *Analytical Chemistry* 65:295–302.

Belesov AV, Ladesov AV, Pikovskoi II, Faleva AV, Kosyakov DS (2020) Characterization of ionic liquid lignins isolated from Sprucewood with 1-butyl-3-methylimidazolium acetate and methyl sulfate and their binary mixtures with DMSO. *Molecules* 25. doi:10.3390/molecules25112479.

Berglund J, d'Ortoli TA, Vilaplana F, Widmalm G, Bergenstrahle-Wohlert M, Lawoko M, Henriksson G, Lindstrom M, Wohlert J (2016) A molecular dynamics study of the effect of glycosidic linkage type in the hemicellulose backbone on the molecular chain flexibility. *The Plant Journal* 88:56–70. doi:10.1111/tpj.13259.

Busse-Wicher M, Gomes TCF, Tryfona T, Nikolovski N, Stott K, Grantham NJ, Bolam DN, Skaf MS, Dupree P (2014) The pattern of xylan acetylation suggests xylan may interact with cellulose microfibrils as a twofold helical screw in the secondary plant cell wall of *Arabidopsis thaliana*. *The Plant Journal* 79:492–506. doi:10.1111/tpj.12575.

Chanzy H, Imada K, Mollard A, Vuong R, Barnoud F (1979) Crystallographic aspects of sub-elementary cellulose fibrils occurring in the wall of rose cells cultured in vitro. *Protoplasma* 100:303–316. doi:10.1007/bf01279318.

Cox J, McDonald PJ, Gardiner BA (2010) A study of water exchange in wood by means of 2D NMR relaxation correlation and exchange. *Holzforschung* 64:259–266. doi:10.1515/hf.2010.036.

Cresswell R, Dupree R, Brown SP, Pereira CS, Skaf MS, Sorieul M, Dupree P, Hill S (2021) Importance of water in maintaining softwood secondary cell wall nanostructure. *Biomacromolecules* 22:4669–4680. doi:10.1021/acs.biomac.1c00937.

Deligey F, Frank MA, Cho SH, Kirui A, Mentinck-Vigier F, Swulius MT, Nixon BT, Wang T (2022) Structure of in vitro-synthesized cellulose fibrils viewed by cryo-electron tomography and ^{13}C natural-abundance dynamic nuclear polarization solid-state NMR. *Biomacromolecules*. Published online ahead of print. doi:10.1021/acs.biomac.1c01674.

Duan P, Kaser SJ, Lyczakowski JJ, Phyo P, Tryfona T, Dupree P, Hong M (2021) Xylan structure and dynamics in native *Brachypodium* grass cell walls investigated by solid-state NMR spectroscopy. *ACS Omega* 6:15460–15471. doi:10.1021/acsomega.1c01978.

Dupree R, Simmons TJ, Mortimer JC, Patel D, Iuga D, Brown SP, Dupree P (2015) Probing the molecular architecture of *Arabidopsis thaliana* secondary cell walls using two- and three-dimensional C-13 solid state nuclear magnetic resonance spectroscopy. *Biochemistry* 54:2335–2345. doi:10.1021/bi501552k.

Ernst RR (1992) Nuclear magnetic resonance Fourier transform spectroscopy (Nobel Lecture). *Angewandte Chemie* 31:805–823. doi: 10.1002/anie.199208053.

Ernst RR, Bodenhausen G, Wokaun A (1997) *Principles of Nuclear Magnetic Resonance in One and Two Dimensions*. Oxford: Oxford Science Publications. ISBN 978-0198556473.

Fahlen J, Salmen L (2005) Pore and matrix distribution in the fiber wall revealed by atomic force microscopy and image analysis. *Biomacromolecules* 6:433–438. doi:10.1021/bm040068x.

Falcoz-Vigne L, Ogawa Y, Molina-Boisseau S, Nishiyama Y, Meyer V, Petit-Conil M, Mazeau K, Heux L (2017) Quantification of a tightly adsorbed monolayer of xylan on cellulose surface. *Cellulose* 24:3725–3739. doi:10.1007/s10570-017-1401-z.

Fernandes AN, Thomas LH, Altaner CM, Callow P, Forsyth VT, Apperley DC, Kennedy CJ, Jarvis MC (2011) Nanostructure of cellulose microfibrils in spruce wood. *Proceedings of the National Academy of Sciences of the USA* 108:E1195–E1203. doi:10.1073/pnas.1108942108.

Fernando LD, Zhao W, Widanage MCD, Mentink-Vigier F, Wang T (2020) Solid-state NMR and DNP investigations of carbohydrates and cell-wall biomaterials. *Emagres* 9:251–258. doi:10.1002/9780470034590.emrstm1624.

Gao Y, Lipton AS, Wittmer Y, Murray DT, Mortimer JC (2020) A grass-specific cellulose-xylan interaction dominates in sorghum secondary cell walls. *Nature Communications* 11. doi:10.1038/s41467-020-19837-z.

Gao Y, Mortimer JC (2020) Unlocking the architecture of native plant cell walls via solid-state nuclear magnetic resonance. In: Anderson CT, Haswell ES, Dixit R (eds.). *Methods in Cell Biology*, vol. 160, pp. 121–143. doi:10.1016/bs.mcb.2020.05.001.

Garvey CJ, Simon GP, Whittaker AK, Parker IH (2019) Moisture-activated dynamics on crystallite surfaces in cellulose. *Colloid and Polymer Science* 297:521–527. doi:10.1007/s00396-018-04464-4.

Gauto D, Dakhlaoui O, Marin-Montesinos I, Hediger S, De Paepe G (2021) Targeted DNP for biomolecular solid-state NMR. *Chemical Science* 12:6223–6237. doi:10.1039/d0sc06959k.

Gelenter MD, Chen KJ, Hong M (2021) Off-resonance C-13-H-2 REDOR NMR for site-resolved studies of molecular motion. *Journal of Biomolecular NMR* 75:335–345. doi:10.1007/s10858-021-00377-7.

Gelenter MD, Wang T, Liao S-Y, O'Neill H, Hong M (2017) H-2-C-13 correlation solid-state NMR for investigating dynamics and water accessibilities of proteins and carbohydrates. *Journal of Biomolecular NMR* 68:257–270. doi:10.1007/s10858-017-0124-7.

Grantham NJ, Wurman-Rodrich J, Terrett OM, Lyczakowski JJ, Stott K, Iuga D, Simmons TJ, Durand-Tardif M, Brown SP, Dupree R, Busse-Wicher M, Dupree P (2017) An even pattern of xylan substitution is critical for interaction with cellulose in plant cell walls. *Nature Plants* 3:859–865. doi:10.1038/s41477-017-0030-8.

Gray CJ, Migas LG, Barran PE, Pagel K, Seeberger PH, Eyers CE, Boons G-J, Pohl NLB, Compagnon I, Widmalm G, Flitsch SL (2019) Advancing solutions to the carbohydrate sequencing challenge. *Journal of the American Chemical Society* 141:14463–14479. doi:10.1021/jacs.9b06406.

Green PB (1964) Cell walls and the geometry of plant growth. *Brookhaven Symposia in Biology* 16:203–217.

Grunin YB, Grunin LY, Schiraya VY, Ivanova MS, Masas DS (2020) Cellulose-water system's state analysis by proton nuclear magnetic resonance and sorption measurements. *Bioresources and Bioprocessing* 7. doi:10.1186/s40643-020-00332-8.

Ha MA, Apperley DC, Evans BW, Huxham M, Jardine WG, Vietor RJ, Reis D, Vian B, Jarvis MC (1998) Fine structure in cellulose microfibrils: NMR evidence from onion and quince. *The Plant Journal* 16:183–190. doi:10.1046/j.1365-313x.1998.00291.x.

Ha MA, Evans BW, Jarvis MC, Apperley DC, Kenwright AM (1996) CP-MAS NMR of highly mobile hydrated biopolymers: Polysaccharides of *Allium* cell walls. *Carbohydrate Research* 288:15–23. doi:10.1016/0008-6215(96)00095-x.

Ha MA, Jardine WG, Jarvis MC (1997) Solid-state C-13 NMR of cell walls in wheat bran. *Journal of Agricultural and Food Chemistry* 45:117–119. doi:10.1021/jf960311m.

Ha MA, MacKinnon IM, Sturcova A, Apperley DC, McCann MC, Turner SR, Jarvis MC (2002) Structure of cellulose-deficient secondary cell walls from the irx3 mutant of *Arabidopsis thaliana*. *Phytochemistry* 61:7–14. doi:10.1016/s0031-9422(02)00199-1.

Ha MA, Vietor RJ, Jardine GD, Apperley DC, Jarvis MC (2005) Conformation and mobility of the arabinan and galactan side-chains of pectin. *Phytochemistry* 66:1817–1824. doi:10.1016/j.phytochem.2005.06.001.

Hediger S, Emsley L, Fischer M (1999) Solid-state NMR characterization of hydration effects on polymer mobility in onion cell-wall material. *Carbohydrate Research* 322:102–112. doi:10.1016/s0008-6215(99)00195-0.

Hediger S, Lesage A, Emsley L (2002) A new NMR method for the study of local mobility in solids and application to hydration of biopolymers in plant cell walls. *Macromolecules* 35:5078–5084. doi:10.1021/ma020065h.

Hesse-Ertelt S, Witter R, Ulrich AS, Kondo T, Heinze T (2008) Spectral assignments and anisotropy data of cellulose I-alpha: C-13-NMR chemical shift data of cellulose I-alpha determined by INADEQUATE and RAI techniques applied to uniformly C-13-labeled bacterial celluloses of different *Gluconacetobacter xylinus* strains. *Magnetic Resonance in Chemistry* 46:1030–1036. doi:10.1002/mrc.2311.

Hong M, Zhang Y, Hu F (2012) Membrane protein structure and dynamics from NMR spectroscopy. In: Johnson MA, Martinez TJ (eds.). *Annual Review of Physical Chemistry*, vol 63, pp. 1–24. doi:10.1146/annurev-physchem-032511-143731.

Horii F, Hirai A, Kitamaru R (1983) Solid-state C-13 NMR study of conformations of oligosaccharides and cellulose - conformation of CH_2 group about the exo-cyclic C-C bond. *Polymer Bulletin* 10:357–361. doi:10.1007/bf00281948.

Horii F, Hirai A, Kitamaru R (1984) CP MAS C-13 NMR Study of spin relaxation phenomena of cellulose containing crystalline and noncrystalline components. *Journal of Carbohydrate Chemistry* 3:641–662. doi:10.1080/07328308408057922.

Hult EL, Larsson PT, Iversen T (2000) A comparative CP/MAS C-13-NMR study of cellulose structure in spruce wood and kraft pulp. *Cellulose* 7:35–55. doi:10.1023/a:1009236932134.

Jarvis MC (2003) Chemistry - Cellulose stacks up. *Nature* 426:611–612. doi:10.1038/426611a.

Jarvis MC (2018) Structure of native cellulose microfibrils, the starting point for nanocellulose manufacture. *Philosophical Transactions of the Royal Society A-Mathematical Physical and Engineering Sciences* 376. doi:10.1098/rsta.2017.0045.

Jarvis MC, Apperley DC (1990) Direct observation of cell-wall structure in living plant tissues by solid-state C-13 NMR spectroscopy. *Plant Physiology* 92:61–65. doi:10.1104/pp.92.1.61.

Jarvis MC, Apperley DC (1995) Chain conformation in concentrated pectic gels - evidence from C-13 NMR. *Carbohydrate Research* 275:131–145. doi:10.1016/0008-6215(95)00033-p.

Jarvis MC, Duncan HJ (1976) Aquilinan, an acidic polysaccharide from *Pteridium aquilinum*. *Phytochemistry* 15:171–174. doi:10.1016/s0031-9422(00)89078-0.

Jarvis MC, Hall MA, Threlfall DR, Friend J (1981) The polysaccharide structure of potato cell walls - chemical fractionation. *Planta* 152:93–100. doi:10.1007/bf00391179.

Jarvis MC, McCann MC (2000) Macromolecular biophysics of the plant cell wall: concepts and methodology. *Plant Physiology and Biochemistry* 38:1–13. doi:10.1016/s0981-9428(00)00172-8.

Kang X, Kirui A, Widanage MCD, Mentink-Vigier F, Cosgrove DJ, Wang T (2019) Lignin-polysaccharide interactions in plant secondary cell walls revealed by solid-state NMR. *Nature Communications* 10. doi:10.1038/s41467-018-08252-0.

Kennedy CJ, Sturcova A, Jarvis MC, Wess TJ (2007) Hydration effects on spacing of primary-wall cellulose microfibrils: a small angle X-ray scattering study. *Cellulose* 14:401–408. doi:10.1007/s10570-007-9129-9.

Kim H, Ralph J (2010) Solution-state 2D NMR of ball-milled plant cell wall gels in DMSO-d(6)/pyridine-d(5). *Organic & Biomolecular Chemistry* 8:576–591. doi:10.1039/b916070a.

Kirui A, Ling Z, Kang X, Widanage MCD, Mentink-Vigier F, French AD, Wang T (2019) Atomic resolution of cotton cellulose structure enabled by dynamic nuclear polarization solid-state NMR. *Cellulose* 26:329–339. doi:10.1007/s10570-018-2095-6.

Kirui A, Zhao W, Deligey F, Yang H, Kang X, Mentink-Vigier F, Wang T (2022) Carbohydrate-aromatic interface and molecular architecture of lignocellulose. *Nature Communications* 13. doi:10.1038/s41467-022-28165-3.

Koso T, del Cerro DR, Heikkinen S, Nypelo T, Buffiere J, Perea-Buceta JE, Potthast A, Rosenau T, Heikkinen H, Maaheimo H, Isogai A, Kilpelainen I, King AWT (2020) 2D Assignment and quantitative analysis of cellulose and oxidized celluloses using solution-state NMR spectroscopy. *Cellulose* 27:7929–7953. doi:10.1007/s10570-020-03317-0.

Kubicki JD, Yang H, Sawada D, O'Neill H, Oehme D, Cosgrove D (2018) The shape of native plant cellulose microfibrils. *Scientific Reports* 8. doi:10.1038/s41598-018-32211-w.

Kumar M, Turner S (2015) Plant cellulose synthesis: CESA proteins crossing kingdoms. *Phytochemistry* 112:91–99. doi:10.1016/j.phytochem.2014.07.009.

Langan P, Petridis L, O'Neill HM, Pingali SV, Foston M, Nishiyama Y, Schulz R, Lindner B, Hanson BL, Harton S, Heller WT, Urban V, Evans BR, Gnanakaran S, Ragauskas AJ, Smith JC, Davison BH (2014) Common processes drive the thermochemical pretreatment of lignocellulosic biomass. *Green Chemistry* 16:63–68. doi:10.1039/c3gc41962b

Larsson PT, Hult EL, Wickholm K, Pettersson E, Iversen T (1999) CP/MAS C-13-NMR spectroscopy applied to structure and interaction studies on cellulose I. *Solid State Nuclear Magnetic Resonance* 15:31–40. doi:10.1016/s0926-2040(99)00044-2.

Leary GJ, Morgan KR, Newman RH, Samuelsson B, Westermark U (1986) A CP-MAS NMR comparison of wood fractions from spruce. *Holzforschung* 40:221–224. doi:10.1515/hfsg.1986.40.4.221.

Lemke CH, Dong RY, Michal CA, Hamad WY (2012) New insights into nano-crystalline cellulose structure and morphology based on solid-state NMR. *Cellulose* 19:1619–1629. doi:10.1007/s10570-012-9759-4.

Leppanen K, Bjurhager I, Peura M, Kallonen A, Suuronen JP, Penttila P, Love J, Fagerstedt K, Serimaa R (2011) X-ray scattering and microtomography study on the structural changes of never-dried silver birch, European aspen and hybrid aspen during drying. *Holzforschung* 65:865–873. doi:10.1515/hf.2011.108.

Li X, Zhao Z (2020) Time domain-NMR studies of average pore size of wood cell walls during drying and moisture adsorption. *Wood Science and Technology* 54:1241–1251. doi:10.1007/s00226-020-01209-x.

Lindh EL, Terenzi C, Salmen L, Furo I (2017) Water in cellulose: evidence and identification of immobile and mobile adsorbed phases by H-2 MAS NMR. *Physical Chemistry Chemical Physics* 19:4360–4369. doi:10.1039/c6cp08219j.

Ling Z, Wang T, Makarem M, Cintron MS, Cheng HN, Kang X, Bacher M, Potthast A, Rosenau T, King H, Delhom CD, Nam S, Edwards JV, Kim SH, Xu F, French AD (2019) Effects of ball milling on the structure of cotton cellulose. *Cellulose* 26:305–328. doi:10.1007/s10570-018-02230-x.

Liu X, Zhu R, Chen T, Song P, Lu F, Xu F, Ralph J, Zhang X (2020) Mild acetylation and solubilization of ground whole plant cell walls in EmimAc: a method for solution-state NMR in DMSO-d(6). *Analytical Chemistry* 92:13101–13109. doi:10.1021/acs.analchem.0c02124.

Loerbroks C, Rinaldi R, Thiel W (2013) The electronic nature of the 1,4-beta-glycosidic bond and its chemical environment: DFT insights into cellulose chemistry. *Chemistry-a European Journal* 19:16282–16294. doi:10.1002/chem.201301366.

Lyczakowski JJ, Wicher KB, Terrett OM, Faria-Blanc N, Yu X, Brown D, Krogh KBRM, Dupree P, Busse-Wicher M (2017) Removal of glucuronic acid from xylan is a strategy to improve the conversion of plant biomass to sugars for bioenergy. *Biotechnology for Biofuels* 10. doi:10.1186/s13068-017-0902-1.

MacKinnon IM, Sturcova A, Sugimoto-Shirasu K, His I, McCann MC, Jarvis MC (2006) Cell-wall structure and anisotropy in *procuste*, a cellulose synthase mutant of *Arabidopsis thaliana*. *Planta* 224:438–448. doi:10.1007/s00425-005-0208-6.

Martinez-Abad A, Berglund J, Toriz G, Gatenholm P, Henriksson G, Lindstrom M, Wohlert J, Vilaplana F (2017) Regular motifs in xylan modulate molecular flexibility and interactions with cellulose surfaces. *Plant Physiology* 175:1579–1592. doi:10.1104/pp.17.01184.

Martinez-Abad A, Jimenez-Quero A, Wohlert J, Vilaplana F (2020) Influence of the molecular motifs of mannan and xylan populations on their recalcitrance and organization in spruce softwoods. *Green Chemistry* 22:3956–3970. doi:10.1039/d0gc01207f.

Martinez-Sanz M, Pettolino F, Flanagan B, Gidley MJ, Gilbert EP (2017) Structure of cellulose microfibrils in mature cotton fibres. *Carbohydrate Polymers* 175:450–463. doi:10.1016/j.carbpol.2017.07.090.

Ndukwe IE, Black I, Heiss C, Azadi P (2020) Evaluating the utility of permethylated polysaccharide solution NMR data for characterization of insoluble plant cell wall polysaccharides. *Analytical Chemistry* 92:13221–13228. doi:10.1021/acs.analchem.0c02379.

Newman RH (1992) Nuclear magnetic resonance study of spatial relationships between chemical components in wood cell walls. *Holzforschung* 46:205–210. doi:10.1515/hfsg.1992.46.3.205.

Newman RH (1999) Estimation of the lateral dimensions of cellulose crystallites using C-13 NMR signal strengths. *Solid State Nuclear Magnetic Resonance* 15:21–29. doi:10.1016/s0926-2040(99)00043-0.

Newman RH, Davies LM, Harris PJ (1996) Solid-state C-13 nuclear magnetic resonance characterization of cellulose in the cell walls of *Arabidopsis thaliana* leaves. *Plant Physiology* 111:475–485. doi:10.1104/pp.111.2.475.

Newman RH, Ha MA, Melton LD (1994) Solid-state C-13 NMR investigation of molecular ordering in the cellulose of apple cell walls. *Journal of Agricultural and Food Chemistry* 42:1402–1406. doi:10.1021/jf00043a002.

Newman RH, Hill SJ, Harris PJ (2013) Wide-angle X-Ray scattering and solid-state nuclear magnetic resonance data combined to test models for cellulose microfibrils in mung bean cell walls. *Plant Physiology* 163:1558–1567. doi:10.1104/pp.113.228262.

Nimz HH, Robert D, Faix O, Nemr M (1981) C-13 NMR spectra of lignins. 8. Structural differences between lignins of hardwoods, softwoods, grasses and compression wood. *Holzforschung* 35:16–26. doi:10.1515/hfsg.1981.35.1.16.

Ning P, Lane D, Soong R, Schmidig D, Frei T, De Castro P, Kovacevic I, Graf S, Wegner S, Busse F, Struppe J, Fey M, Stronks H, Monette M, Simpson MJ, Simpson AJ (2020) Comprehensive multiphase NMR-a powerful tool to understand and monitor molecular processes during biofuel production. *ACS Sustainable Chemistry & Engineering* 8:17551–17564. doi:10.1021/acssuschemeng.0c07205.

Nishitani K, Masuda Y (1980) Modifications of cell wall polysaccharides during auxin-induced growth in Adzuki bean epicotyl segments. *Plant and Cell Physiology* 21:169–181.

Nishiyama Y, Johnson GP, French AD (2012) Diffraction from nonperiodic models of cellulose crystals. *Cellulose* 19:319–336. doi:10.1007/s10570-012-9652-1.

Nishiyama Y, Langan P, Chanzy H (2002) Crystal structure and hydrogen-bonding system in cellulose 1 beta from synchrotron X-ray and neutron fiber diffraction. *Journal of the American Chemical Society* 124:9074–9082. doi:10.1021/ja0257319.

Nixon BT, Mansouri K, Singh A, Du J, Davis JK, Lee J-G, Slabaugh E, Vandavasi VG, O'Neill H, Roberts EM, Roberts AW, Yingling YG, Haigler CH (2016) Comparative structural and computational analysis supports eighteen cellulose synthases in the plant cellulose synthesis complex. *Scientific Reports* 6. doi:10.1038/srep28696.

Noda I (1993) Generalized 2-dimensional correlation method applicable to infrared, Raman and other types of spectroscopy. *Applied Spectroscopy* 47:1329–1336. doi:10.1366/0003702934067694.

Perez Garcia M, Zhang Y, Hayes J, Salazar A, Zabotina OA, Hong M (2011) Structure and interactions of plant cell-wall polysaccharides by two- and three-dimensional magic-angle-spinning solid-state NMR. *Biochemistry* 50:989–1000. doi:10.1021/bi101795q.

Perras FA, Luo H, Zhang X, Mosier NS, Pruski M, Abu-Omar MM (2017) Atomic-level structure characterization of biomass pre- and post-lignin treatment by dynamic nuclear polarization-enhanced solid-state NMR. *Journal of Physical Chemistry A* 121:623–630. doi:10.1021/acs.jpca.6b11121.

Phyo P, Gu Y, Hong M (2019) Impact of acidic pH on plant cell wall polysaccharide structure and dynamics: insights into the mechanism of acid growth in plants from solid-state NMR. *Cellulose* 26:291–304. doi:10.1007/s10570-018-2094-7.

Phyo P, Hong M (2019) Fast MAS H-1-C-13 correlation NMR for structural investigations of plant cell walls. *Journal of Biomolecular NMR* 73:661–674. doi:10.1007/s10858-019-00277-x.

Phyo P, Wang T, Yang Y, O'Neil H, Hong M (2018) Direct determination of hydroxymethyl conformations of plant cell wall cellulose using H-1 polarization transfer solid-state NMR. *Biomacromolecules* 19:1485–1497. doi:10.1021/acs.biomac.8b00039.

Reif B, Ashbrook SE, Emsley L, Hong M (2021) Solid-state NMR spectroscopy. *Nature Reviews Methods Primers* 1. doi:10.1038/s43586-020-00002-1.

Sakellariou D, Brown SP, Lesage A, Hediger S, Bardet M, Meriles CA, Pines A, Emsley L (2003) High-resolution NMR correlation spectra of disordered solids. *Journal of the American Chemical Society* 125:4376–4380. doi:10.1021/ja0292389.

Sapouna I, Lawoko M (2021) Deciphering lignin heterogeneity in ball milled softwood: unravelling the synergy between the supramolecular cell wall structure and molecular events. *Green Chemistry* 23:3348–3364. doi:10.1039/d0gc04319b.

Schmidt-Rohr K, Clauss J, Spiess HW (1992) Correlation of structure, mobility and morphological information in heterogeneous polymer materials by 2-dimensional wide-line-separation NMR spectroscopy. *Macromolecules* 25:3273–3277. doi:10.1021/ma00038a037.

Schmidt-Rohr K, Spiess H, (1994) *Multidimensional Solid-State NMR and Polymers*. Elsevier ISBN: 9780080925622.

Simmons TJ, Mortimer JC, Bernardinelli OD, Poppler A-C, Brown SP, DeAzevedo ER, Dupree R, Dupree P (2016) Folding of xylan onto cellulose fibrils in plant cell walls revealed by solid-state NMR. *Nature Communications* 7. doi:10.1038/ncomms13902.

Sturcova A, His I, Apperley DC, Sugiyama J, Jarvis MC (2004) Structural details of crystalline cellulose from higher plants. *Biomacromolecules* 5:1333–1339. doi:10.1021/bm034517p.

Sturcova A, His I, Wess TJ, Cameron G, Jarvis MC (2003) Polarized vibrational spectroscopy of fiber polymers: Hydrogen bonding in cellulose II. *Biomacromolecules* 4:1589–1595. doi:10.1021/bm034295v.

Talmadge KW, Keegstra K, Bauer WD, Albersheim P (1973) Structure of plant cell walls. I Macromolecular components of walls of suspension-cultured sycamore cells with a detailed analysis of pectic polysaccharides. *Plant Physiology* 51:158–173. doi:10.1104/pp.51.1.158.

Taylor RE, French AD, Gamble GR, Himmelsbach DS, Stipanovic RD, Thibodeaux DP, Wakelyn PJ, Dybowski C (2008) H-1 and C-13 solid-state NMR of *Gossypium barbadense* (Pima) cotton. *Journal of Molecular Structure* 878:177–184. doi:10.1016/j.molstruc.2007.08.006.

Terashima N, Kitano K, Kojima M, Yoshida M, Yamamoto H, Westermark U (2009) Nanostructural assembly of cellulose, hemicellulose, and lignin in the middle layer of secondary wall of ginkgo tracheid. *Journal of Wood Science* 55:409–416. doi:10.1007/s10086-009-1049-x.

Terrett OM, Dupree P (2019) Covalent interactions between lignin and hemicelluloses in plant secondary cell walls. *Current Opinion in Biotechnology* 56:97–104. doi:10.1016/j.copbio.2018.10.010.

Terrett OM, Lyczakowski JJ, Yu L, Iuga D, Franks WT, Brown SP, Dupree R, Dupree P (2019) Molecular architecture of softwood revealed by solid-state NMR. *Nature Communications* 10:4978–4978. doi:10.1038/s41467-019-12979-9.

Thimm JC, Burritt DJ, Sims IM, Newman RH, Ducker WA, Melton LD (2002) Celery (*Apium graveolens*) parenchyma cell walls: cell walls with minimal xyloglucan. *Physiologia Plantarum* 116:164–171. doi:10.1034/j.1399-3054.2002.1160205.x.

Thomas LH, Altaner CM, Forsyth VT, Mossou E, Kennedy CJ, Martel A, Jarvis MC (2021) Nanostructural deformation of high-stiffness spruce wood under tension. *Scientific Reports* 11. doi:10.1038/s41598-020-79676-2.

Thomas LH, Altaner CM, Jarvis MC (2013) Identifying multiple forms of lateral disorder in cellulose fibres. *Journal of Applied Crystallography* 46:972–979. doi:10.1107/s002188981301056x.

Thomas LH, Forsyth VT, Martel A, Grillo I, Altaner CM, Jarvis MC (2014) Structure and spacing of cellulose microfibrils in woody cell walls of dicots. *Cellulose* 21:3887–3895. doi:10.1007/s10570-014-0431-z.

Thomas LH, Forsyth VT, Martel A, Grillo I, Altaner CM, Jarvis MC (2015) Diffraction evidence for the structure of cellulose microfibrils in bamboo, a model for grass and cereal celluloses. *BMC Plant Biology* 15. doi:10.1186/s12870-015-0538-x.

Thomas LH, Martel A, Grillo I, Jarvis MC (2020) Hemicellulose binding and the spacing of cellulose microfibrils in spruce wood. *Cellulose* 27:4249–4254. doi:10.1007/s10570-020-03091-z.

Thygesen LG, Gierlinger N (2013) The molecular structure within dislocations in *Cannabis sativa* fibres studied by polarised Raman microspectroscopy. *Journal of Structural Biology* 182:219–225. doi:10.1016/j.jsb.2013.03.010.

Tinoco I, Bustamante C (2002) The effect of force on thermodynamics and kinetics of single molecule reactions. *Biophysical Chemistry* 101:513–533. doi:10.1016/s0301-4622(02)00177-1.

Vietor RJ, Newman RH, Ha MA, Apperley DC, Jarvis MC (2002) Conformational features of crystal-surface cellulose from higher plants. *The Plant Journal* 30:721–731. doi:10.1046/j.1365-313X.2002.01327.x.

Viger-Gravel J, Lan W, Pinon AC, Berruyer P, Emsley L, Bardet M, Luterbacher J (2019) Topology of pretreated wood fibers using dynamic nuclear polarization. *Journal of Physical Chemistry C* 123:30407–30415. doi:10.1021/acs.jpcc.9b09272.

Wang T, Chen Y, Tabuchi A, Cosgrove DJ, Hong M (2016) The target of beta-expansin EXPB1 in maize cell walls from binding and solid-state NMR studies. *Plant Physiology* 172:2107–2119. doi:10.1104/pp.16.01311.

Wang T, Hong M (2016) Solid-state NMR investigations of cellulose structure and interactions with matrix polysaccharides in plant primary cell walls. *Journal of Experimental Botany* 67:503–514. doi:10.1093/jxb/erv416.

Wang T, Park YB, Caporini MA, Rosay M, Zhong L, Cosgrove DJ, Hong M (2013) Sensitivity-enhanced solid-state NMR detection of expansin's target in plant cell walls. *Proceedings of the National Academy of Sciences of the USA* 110:16444–16449. doi:10.1073/pnas.1316290110.

Wang T, Park YB, Cosgrove DJ, Hong M (2015) Cellulose-pectin spatial contacts are inherent to never-dried Arabidopsis primary cell walls: evidence from solid-state nuclear magnetic resonance. *Plant Physiology* 168:871–884. doi:10.1104/pp.15.00665.

Wang T, Salazar A, Zabotina OA, Hong M (2014) Structure and dynamics of *Brachypodium* primary cell wall polysaccharides from two-dimensional C-13 solid-state nuclear magnetic resonance spectroscopy. *Biochemistry* 53:2840–2854. doi:10.1021/bi500231b.

Wang T, Yang H, Kubicki JD, Hong M (2016) Cellulose structural polymorphism in plant primary cell walls investigated by high-field 2D solid-state NMR spectroscopy and density functional theory calculations. *Biomacromolecules* 17:2210–2222. doi:10.1021/acs.biomac.6b00441.

Wang T, Zabotina O, Hong M (2012) Pectin-cellulose interactions in the Arabidopsis primary cell wall from two-dimensional magic-angle-spinning solid-state nuclear magnetic resonance. *Biochemistry* 51:9846–9856. doi:10.1021/bi3015532.

Wang W, Cai J, Xu Z, Zhang Y, Niu F, Gao M, Yang J (2020) Structural characteristics of plant cell wall elucidated by solution-state 2D NMR spectroscopy with an optimized procedure. *Green Processing and Synthesis* 9:650–663. doi:10.1515/gps-2020-0049.

White PB, Wang T, Park YB, Cosgrove DJ, Hong M (2014) Water-polysaccharide interactions in the primary cell wall of *Arabidopsis thaliana* from polarization transfer solid-state NMR. *Journal of the American Chemical Society* 136:10399–10409. doi:10.1021/ja504108h.

Wickholm K, Larsson PT, Iversen T (1998) Assignment of non-crystalline forms in cellulose I by CP/MAS C-13 NMR spectroscopy. *Carbohydrate Research* 312:123–129. doi:10.1016/s0008-6215(98)00236-5.

Witter R, Sternberg U, Hesse S, Kondo T, Koch F-T, Ulrich AS (2006) C-13 chemical shift constrained crystal structure refinement of cellulose I-alpha and its verification by NMR anisotropy experiments. *Macromolecules* 39:6125–6132. doi:10.1021/ma052439n.

Wohlert M, Benselfelt T, Wagberg L, Furo I, Berglund LA, Wohlert J (2022) Cellulose and the role of hydrogen bonds: not in charge of everything. *Cellulose* 29:1–23. doi:10.1007/s10570-021-04325-4.

Xu P, Donaldson LA, Gergely ZR, Staehelin LA (2007) Dual-axis electron tomography: a new approach for investigating the spatial organization of wood cellulose microfibrils. *Wood Science and Technology* 41:101–116. doi:10.1007/s00226-006-0088-3.

Yang H, Wang T, Oehme D, Petridis L, Hong M, Kubicki JD (2018) Structural factors affecting C-13 NMR chemical shifts of cellulose: a computational study. *Cellulose* 25:23–36. doi:10.1007/s10570-017-1549-6.

Yannoni CS (1982) High resolution NMR in solids - the CPMAS experiment. *Accounts of Chemical Research* 15:201–208. doi:10.1021/ar00079a003.

Ye D, Rongpipi S, Kiemle SN, Barnes WJ, Chaves AM, Zhu C, Norman VA, Liebman-Pelaez A, Hexemer A, Toney MF, Roberts AW, Anderson CT, Cosgrove DJ, Gomez EW, Gomez ED (2020) Preferred crystallographic orientation of cellulose in plant primary cell walls. *Nature Communications* 11. doi:10.1038/s41467-020-18449-x.

Yuan EC- Y, Huang S-J, Huang H-C, Sinkkonen J, Oss A, Org M-L, Samoson A, Tai H-C, Chan JCC (2021) Faster magic angle spinning reveals cellulose conformations in woods. *Chemical Communications* 57:4110–4113. doi:10.1039/d1cc01149a.

Zhang K, Liu Y, Cui S, Xie Y (2021) Elucidation of the structure of lignin-carbohydrate complexes in *Ginkgo* CW-DHP by ^{13}C-^{2}H dual isotope tracer. *Molecules (Basel, Switzerland)* 26. doi:10.3390/molecules26195740.

Zhang L, Gao C, Mentink-Vigier F, Tang L, Zhang D, Wang S, Cao S, Xu Z, Liu X, Wang T, Zhou Y, Zhang B (2019) Arabinosyl deacetylase modulates the arabinoxylan acetylation profile and secondary wall formation. *Plant Cell* 31:1113–1126. doi:10.1105/tpc.18.00894.

Zhao W, Kirui A, Deligey F, Mentink-Vigier F, Zhou Y, Zhang B, Wang T (2021) Solid-state NMR of unlabeled plant cell walls: high-resolution structural analysis without isotopic enrichment. *Biotechnology for Biofuels* 14:14–14. doi:10.1186/s13068-020-01858-x.

9 Adoption of Evolving Technologies to Define Cell Wall Biology in the Grasses

A Brief but Unfinished History

Vincent Bulone, Julian G. Schwerdt, and Geoffrey B. Fincher
The University of Adelaide

INTRODUCTION

Advances in our understanding of plant cell wall biology share a parallel history with the development and refinement of new experimental technologies. Here, we focus on emerging technologies that have been progressively adapted over the last 50 years or so to incrementally enhance our knowledge of wall composition, synthesis, remodelling and degradation in the grasses. The grasses (family Poaceae) are of particular importance because they include the common cereal species that provide the majority of dietary carbohydrate for human societies, together with grass-based fodder and pasture species, which are of key nutritional importance for major livestock species and for native herbivores. More recently, crop and other plant residues have been recognised as a valuable and abundant source of renewable biofuels; cell walls and their partially degraded products account for a large proportion of these residues. Furthermore, many of the dietary polysaccharides that originate in walls of grasses and other plants have been implicated in lowering the risk of contracting serious human diseases, including type II diabetes, colorectal cancer and coronary heart disease (Collins et al. 2010; Zhang et al. 2015).

Cell walls of the grasses exhibit compositional characteristics that differ significantly from walls of dicotyledonous plants. In both cases, overall wall structure involves cellulosic microfibrils embedded in matrices of non-cellulosic polysaccharides and lignin. In some cases, there may be covalent cross-links between these components (Tan et al. 2013), but this is not always so. It is generally held that the cellulosic microfibrils and lignin are essential for providing tensile and compressive strength, respectively. Walls have traditionally been classified as primary walls, which are deposited during cell expansion, and secondary walls, which are thickened and usually lignified walls that are deposited once cell growth has ceased. However, there is not usually a sharp cut-off point between primary and secondary cell wall biosynthesis.

The major differences between the grasses and the eudicots lie in the composition of the non-cellulosic polysaccharide components of the wall (Carpita and Gibeaut 1993). In the eudicots, a complex array of pectic polysaccharides and xyloglucans account for a large proportion of these components. Heteroxylans usually make up a much smaller proportion of the non-cellulosic polysaccharide constituents in walls of the eudicots. In contrast, walls of the grasses generally have much lower levels of pectic polysaccharides and xyloglucans, but much higher levels of heteroxylans. Indeed, the heteroxylans appear to be the 'core' non-cellulosic polysaccharides of cell walls in the grasses. However, walls of the grasses often, but not always, contain (1,3;1,4)-β-glucans; these polysaccharides are not found in eudicots. Mannans and glucomannans also occur in relatively low concentrations in walls of the grasses.

Pectic polysaccharides can be found transiently in walls during the early stages of grain development (Chateigner-Boutin et al. 2014) and are important wall constituents in vegetative tissues of rice and switchgrass (Biswal et al. 2018; Zhang et al. 2020; Ohara et al. 2021). Although pectic polysaccharides are generally present at relatively low levels in walls of the grasses, they are likely to play a general role in intercellular adhesion, as well as other roles in specific tissues at specific developmental stages (Amos and Mohnen 2019; Ohara et al. 2021). Lignin is found in the 'secondarily thickened' walls of certain cell types that are present in all plant species, including the grasses (Iiyama et al. 1990), but will not be discussed further here.

So, in this review, the emphasis will be focused on advances in our knowledge of the biology of heteroxylans and (1,3;1,4)-β-glucans in the grasses and how new technologies have enabled these advances over the last 50–60 years. We start with the overlapping technological contributions of the chemical and biochemical eras, where walls were isolated for compositional analysis and for early investigations into the enzymes and pathways involved in the biosynthesis and degradation of individual components of the wall. Monoclonal antibody technologies enabled the generation of specific antibodies for wall polysaccharides or enzymes of interest, and could be applied to the identification of specific polysaccharides or proteins in gels, tissue sections

DOI: 10.1201/9781003178309-9

and isolated wall preparations (Knox 2008; Pattathil et al. 2010). Antibody microarrays for the analysis of wall polysaccharides (comprehensive microarray polymer profiling; CoMPP) quickly followed and allowed high-throughput, semi-quantitative mapping of cell wall polysaccharides (Moller et al. 2006; Pattathil et al. 2015). Concomitantly, the development and enhanced resolution of imaging methods, including Fourier-transform infra-red (FT-IR), Raman and nuclear magnetic resonance (NMR) spectroscopy, high-performance anion-exchange chromatography (HPAEC) and matrix-assisted laser desorption/ionisation mass spectrometry imaging (MALDI–MSI) have been central to the gains in our knowledge of cell wall biology (McCann et al. 2007; Chapters 1 and 8). Tissue culture, initially focused on the eudicots *Acer pseudoplaatanus* (sycamore maple) and *Nicotiana benthamiana* (tobacco), and the monocot *Lolium multiflorum* (Italian ryegrass) enabled the rapid generation of relatively high levels of walls and associated enzymes for analysis.

The emergence of disruptive molecular biology technologies in the 1980s provided a generational advance in our ability to address wall biology, initially in the characterisation of hydrolytic enzymes and later in the description of the more challenging glycosyl transferases. *In vitro* synthesis, isolation and nucleotide sequencing of cDNAs and genes enabled the complete amino acid sequences of enzymes to be defined and greatly accelerated the application of structural biology to enzymes involved in the metabolism of key wall polysaccharides. A manual of common molecular biology protocols, such as cloning methods and Northern and Southern hybridisation, was assembled by Sambrook et al. (1989) and was crucial in arming biochemists with these new technologies. According to Google Scholar, this manual has been cited close to 200,000 times, as of November 2021. Heterologous expression systems for the identification and characterisation of candidate genes were developed in barley and other grass species amenable to tissue culture, while transformation techniques and mutagenesis provided new experimental approaches for proof-of-function testing. Atomic level resolution of protein 3D structures subsequently enabled the molecular mechanisms of key enzymes to be described. The genomics era then evolved over several years, from the sequencing and mapping of expressed sequence tags (ESTs), through sequencing of larger genomic fragments cloned in yeast and bacterial artificial chromosomes and finally to the generation of complete sequences of plant monocot genomes (e.g. IRGSP 2005; Paterson et al. 2009; Schnable et al. 2009; Vogel et al. 2010; Mascher et al. 2017; IWGSC 2018). The co-evolution of dramatically increased computing power and bioinformatics technologies led to our ability to identify large numbers of genes, gene transcripts and proteins in single experiments (Chen et al. 2014; Zhang et al. 2016), while evolving laser dissection methods (Obel et al. 2009) are continually improving the resolution with which these experiments can be undertaken. As a result, large, publicly accessible databases have been generated.

These technologies simultaneously provided new approaches to defining the evolution of wall polysaccharides at the molecular level, which has been of interest for many years (Scheller and Ulvskov 2010) and, as noted above, several important resources have been made publicly available, through both the private and public sectors, to support our research activities. In many cases, the technological advances have been incremental and sustained in nature, but certain radical new technologies have led to the so-called disruptive changes that have dramatically altered our experimental approaches to the elucidation of unanswered questions in wall biology. Concomitantly, these disruptive technologies have greatly enhanced knowledge in the field and our ability to transform research advances into practical applications. Finally, major knowledge gaps in our understanding of wall biology in the grasses remain but continually evolving technologies are providing new approaches through which these might be addressed. Opportunities presented by genetic engineering and new gene-editing technologies will also be assessed.

CHEMICAL STRUCTURES OF MAJOR WALL POLYSACCHARIDES OF THE GRASSES

Analytical Methods

The first step required for quantitative compositional analyses of walls is to extract from plant tissues a wall preparation that is largely free of contaminating starch, proteins and low molecular weight material. Early methods for the isolation of cell walls often involved the use of aqueous media, but this resulted in errors due to the loss of water-soluble wall polysaccharides and to the depolymerisation of wall components during the purification procedure. The subsequent use of 70% ethanol as a purification medium helped minimise the loss of water-soluble material and simultaneously inhibited hydrolytic enzymes in tissue lysates. Starch and intracellular proteins were removed enzymatically and/or physically (Mares and Stone 1973; Fincher 1975). This fundamental principle is still used today to rapidly prepare the starch-free, alcohol-insoluble residue (AIR) fraction from plant tissue homogenates; the AIR fraction consists mainly of wall components.

Polysaccharides of the wall preparations were subsequently fractionated according to their solubility in water, dilute alkali and more concentrated alkaline solutions. The final alkali-insoluble residue normally consisted mainly of cellulose, although small amounts of xylan and mannan were often present (Mares and Stone 1973; Fincher 1975). Special pre-treatment involving sodium borodeuteride was used for the analysis of uronic acid residues in polysaccharides. The next step was to hydrolyse the polysaccharides with dilute acid and quantitate the alditol acetate or trimethylsilyl derivatives of released monosaccharides by gas liquid chromatography (GC) or high-performance liquid chromatography (HPLC) and hence to deduce the types of polysaccharides present in the wall fraction.

The linkage types in the polysaccharides analysed in this manner were investigated using enzymes of known linkage specificity. However, a breakthrough chemical technology for monosaccharide linkage analysis (Hakomori 1964) used the methyl-sulphinyl (dimsyl) carbanion in in dimethyl sulphoxide (DMSO) to ionise the free hydroxyl groups of the polysaccharide. These ionised hydroxyls are methylated with methyl iodide, the permethylated polysaccharide is hydrolysed and the partially volatile alditol acetate derivatives of released permethylated monosaccharides are generated with acetic anhydride prior to GC analysis (Björndal et al. 1970). However, bulk generation of the methyl-sulphinyl carbanion, which is unstable and potentially explosive in the presence of oxygen, meant that this method was not always amenable to standard biochemistry laboratories. To overcome this problem, Ciucanu and Kerek (1984) subsequently modified the procedure by using solid NaOH in DMSO to ionise the free hydroxyl groups. In due course, mass spectrometry (MS) linked with GC greatly enhanced structural information generated from the wall polysaccharide derivatives.

The ongoing evolution of these analytical technologies over several decades has culminated in sophisticated, rapid and reliable analytical methods for wall polysaccharides (Biswal et al. 2017). These methods have revealed a high degree of heterogeneity in the chemistry and structure of cell wall polysaccharides (Burton et al. 2010). Detailed protocols for core methodologies are available (e.g. Pettolino et al. 2012). New methods and the refinement of existing technologies, such as MALDI-Time of Flight (ToF)-Mass Spectrometry (MS), MALDI-ToF/ToF high energy collision-induced dissociation (CID) and Polysaccharide Analysis by Carbohydrate gel Electrophoresis (PACE) have been used to revise and update earlier structures of plant polysaccharides, including arabinoxylans (Brown et al. 2009; Hrmova et al. 2007). Solid-state NMR has been used to examine polysaccharide-polysaccharide interactions in cell walls *in situ* (Höfte and Voxeur 2017; Turner and Kumar 2018; Terrett et al. 2019). Specialised laboratories have also been established to service the needs of the cell wall community. These include, *inter alia*, the Complex Carbohydrate Research Center (CCRC) at the University of Georgia (UGA) (Box 9.1).

Structures of (1,3;1,4)-β-Glucans

The (1,3;1,4)-β-glucans are widespread but not ubiquitous in cell walls of the grasses. The polysaccharide commonly consists of about 30% (1,3)-β-linked and 70% (1,4)-β-linked glucopyranosyl residues, arranged in an unbranched, unsubstituted chain of glucosyl residues. The predominant structural feature of the polysaccharide involves cellotriosyl and cellotetraosyl residues linked by single (1,3)-β-linkages. However, the relative proportion of the two linkage types varies between and within species. The insertion of (1,3)-linkages into what is essentially a 'cellulosic' backbone introduces flexible 'kinks' into the chain and renders (1,3;1,4)-β-glucans much more soluble than cellulosic polymers. The variations in the proportions of the two linkage types can be broadly described through the analysis of DP3:DP4 ratios, where DP denotes the degree of polymerisation. The predominant arrangement of the two linkage types involves single (1,3)-β-glucosyl residues separated by two or three adjacent (1,4)-β-glucosyl residues. However, up to 10% of the polysaccharide consists of longer blocks of adjacent (1,4)-linkages, ranging from four to more than 10 (1,4)-linkages (Woodward et al. 1983). The structural features can be quantitated following hydrolysis of the polysaccharide with specific (1,3;1,4)-β-glucan endohydrolases, which only hydrolyse (1,4)-β-linkages that are immediately adjacent to (and towards the reducing terminus of the polysaccharide) a (1,3)-β-linkage. The released oligosaccharides are predominantly 3-O-β-cellobiosyl-d-glucose (DP3) and 3-O-β-cellotriosyl-d-glucose (DP4), which can be readily separated and quantitated by HPLC or related methods for the calculation of DP3:DP4 ratios.

Examination of the distribution of (1,3)-linkages in the polysaccharide has revealed both non-random and random structural features. The observation that adjacent (1,3)-β-glucosyl residues are rarely if ever found in (1,3;1,4)-β-glucans of the grasses (Buliga et al. 1986) indicates that the distribution of these linkage types is non-random, given that (1,3)-β-glucosyl residues constitute about 30% of the total linkages in the chain. If their distribution were random, blocks of two or three adjacent (1,3)-linkages would be commonly found in a polymer of this size. In contrast to the non-random distribution of (1,3)-β-glucosyl

BOX 9.1: THE COMPLEX CARBOHYDRATE RESEARCH CENTER

The CCRC was founded at the UGA in 1985 to address the national need for a centre devoted to increasing knowledge of the structures and functions of complex carbohydrates. Peter Albersheim and Alan Darvill (and their team from the University of Colorado) were recruited by UGA to lead the CCRC. Peter Albersheim had spent 21 years as a professor of biochemistry at the University of Colorado in Boulder, prior to taking up his position at UGA. The CCRC became and remains a fertile spawning ground for excellent carbohydrate chemists and biochemists working on plant cell walls. It now offers a wide range of analytical services and equipment for carbohydrate analysis and molecular modelling, together with nucleotide-sugar substrates and a comprehensive range of monoclonal antibodies against cell wall constituents. These invaluable activities have significantly accelerated our understanding of cell wall biology, in both the grasses and eudicots, over many years.

residues, Markov chain analyses of the water-soluble (1,3;1,4)-β-glucan from barley showed that the distribution of the blocks of two and three adjacent (1,4)-β-glucosyl residues was indeed random (Staudte et al. 1983). The distribution pattern of the longer chains of adjacent (1,4)-linkages has not been defined.

Lazaridou and Biliaderis (2007) recognised the link between the DP3:DP4 ratios of cereal (1,3;1,4)-β-glucans and the solubility of the polysaccharide. Given that the (1,3)-β-glucosyl residues will form a molecular kink in the (1,3;1,4)-β-glucan chain, if these residues are regularly spaced in the polysaccharide one would expect the polymers to more easily align and that their solubility in aqueous solutions would thereby be reduced. As noted above, the blocks of two and three adjacent (1,4)-β-glucosyl residues are randomly arranged, so the (1,3)-β-glucosyl residues will not be regularly spaced in the polysaccharide. Thus, the polysaccharide will be relatively soluble. However, at very high (e.g. *Rhyncosporium*) or very low (e.g. *Equisetum*) DP3:DP4 ratios, the DP3 or the DP4 component will predominate, respectively. In both cases, the (1,3)-β-glucosyl residues will become more regularly spaced and the polysaccharide chains will more easily align to form insoluble complexes.

This link between (1,3;1,4)-β-glucan fine structure and solubility is crucially important in the commercial use of cereals in such applications as renewable biofuel production, human health and nutrition, and malting and brewing (Lazaridou and Biliaderis 2007; Collins et al. 2010).

Structures of Heteroxylans

As introduced above, there has been a significant evolutionary shift in the polysaccharide composition of walls in the grasses, which are manifested by higher heteroxylan contents and lower pectin contents compared with walls of eudicots. Heteroxylans constitute 25% and 70% of walls from the barley starchy endosperm and aleurone (Fincher 1975; Bacic and Stone 1981), respectively. This can be compared with just a few per cent in the primary walls of most eudicots (Zablackis et al. 1995; Burton and Fincher 2014). For example, cell wall polysaccharides of *Arabidopsis thaliana* leaves consist of just 4% heteroxylan (Zablackis et al. 1995).

Heteroxylans are characterised by a main chain of (1,4)-linked β-D-xylopyranosyl residues, which may or may not be substituted with single α-L-arabinofuranosyl residues at C(O)3 and/or C(O)2. In some cases, both C(O)3 and C(O)2 are substituted. Uronic acid residues and a variety of short oligosaccharide chains can also be linked to the xylan backbone, while phenolic acid residues, commonly ferulic acid but sometimes *p*-coumaric acid, can be covalently linked to the arabinofuranosyl residues (Harris and Hartley 1976). The ferulic acid residues appear to be involved in oxidative cross-linking of arabinoxylan molecules or in the covalent linkage of arabinoxylan molecules to lignin (Fincher and Stone 2004). Uronic acids are usually low in abundance in heteroxylans from cereal grains, which are therefore commonly described simply as arabinoxylans. In developing barley grain, newly synthesised arabinoxylans have a high degree of arabinosyl substitution, which subsequently decreases (Wilson et al. 2006). This is consistent with the detection in the early developing grain of transcripts encoding arabinoxylan arabinofuranohydrolase (AXAH) (Zhang et al. 2016), which is capable of removing arabinosyl residues from the polysaccharide. The degree of substitution and the nature of main chain substituents are key determinants of the physical and molecular properties of heteroxylans, including their solubility and ability to form non-covalent associations with other cell wall components. High degrees of substitution sterically inhibit xylan main chain associations. The lower the degree of substitution of xylan backbones, the greater their tendency to self-align to form insoluble complexes and the greater their ability to associate with other wall constituents, such as cellulose (Terrett et al. 2019). Thus, xylose:arabinose ratios provide an indication of the solubility of arabinoxylans in the grasses.

During this early period in which the structural analysis of wall polysaccharides was a central activity in many biochemistry laboratories, important new resources became available. Enzymes used to define polysaccharide structures, enzyme substrates, purified wall polysaccharides and assay kits were developed and represented a huge contribution to the wall community (Box 9.2).

EARLY FRUSTRATIONS IN THE BIOCHEMISTRY OF WALL POLYSACCHARIDE SYNTHESIS

During the biochemistry and early molecular biology eras of the 1970s, the 1980s and the 1990s, leaders of several of the best biochemistry laboratories around the world attempted to purify cellulose synthase, (1,3;1,4)-β-glucan synthase and (1,3)-β-glucan synthase enzymes, and thereby to obtain partial amino acid sequence information. These enzymes were known to be type II integral membrane proteins that generally lost activity on dissociation from membranes; this presented a central problem for biochemical approaches. The common first step was to prepare membrane fractions from cells with relatively low levels of secondary thickening, such as tissue-cultured *Acer pseudoplatanus* (sycamore maple) and *Lolium multiflorum* (Italian ryegrass) cells (e.g. Smith and Stone 1973; Okuda et al. 1993; Becker et al. 1964). The membrane preparations, in many cases enriched in a particular fraction (e.g. Golgi or plasma membrane), were incubated with a glucosyl donor, typically UDP-[^{14}C]-glucose, and the radioactive products of the assay mixture were analysed for their monosaccharide and linkage compositions, using enzymic and/or chemical analyses. Incorporation of the [^{14}C]-label into polymeric material was observed, but sugar-nucleotide interconverting enzymes in the membrane preparations in many cases caused the epimerisation of the UDP-[^{14}C]-glucose into UDP-[^{14}C]-galactose and

BOX 9.2: MEGAZYME INTERNATIONAL

Barry McCleary obtained his PhD degree in 1975 under the supervision of Don Matheson, a carbohydrate chemist, at the University of Sydney. After postdoctoral positions in both the public and private sectors, Barry decided to follow his commercial interests through the preparation and sale of carbohydrates and enzymes of interest to the cell wall community. He founded Megazyme Australia in 1988, and it is rumoured that one of his first products was barley (1,3;1,4)-β-glucan, which he extracted in the garage of his home in Sydney (McCleary, 1988). Megazyme was re-located to Cork in Ireland in 1996 and, with its recent North American partner Neogen, has a long history of providing a wide range of assay kits, carbohydrates, enzymes and enzyme substrates that have proven invaluable for the development of plant cell wall biology, especially for the cereals and food industries. A key secret of Megazyme's success lies with Barry's continuing involvement with the development and refinement of his products and his open statements on potential difficulties with methods or potential minor contaminating components in products. Megazyme has also sponsored the international Bruce Stone Award, 'to recognise outstanding contributions to our understanding of cell wall and more general carbohydrate biochemistry' for many years through a medal and an honorarium.

thence into other sugar-nucleotides. The re-distribution of the [^{14}C]-label presented difficult interpretative problems, as did difficulties in obtaining enriched membrane preparations (e.g. Golgi membranes) that were completely free of membranes from other subcellular organelles (Sadowski et al. 2008). Furthermore, the membrane preparations from monocotyledons invariably contained cellulose synthases, (1,3;1,4)-β-glucan synthases and (1,3)-β-glucan synthases, which competed for the UDP-[^{14}C]-glucose substrate. The synthases usually had different kinetic properties, so adjustment of UDP-[^{14}C]-glucose concentrations in the assay mixture was attempted to favour the synthesis of the different polysaccharides (Meikle et al. 1991). But there was seldom a clear-cut answer and after many years this approach had provided no amino acid sequence information for even partially purified cellulose, (1,3;1,4)-β-glucan or (1,3)-β-glucan synthase enzymes and hence very few insights into their biochemical properties or the genes that encoded them. Similar frustrations were experienced by groups attempting to purify xylan synthases using similar biochemical technologies (Dalessandro and Northcote 1981; Bolwell and Northcote 1983; Porchia and Scheller 2008). Thus, the identification of enzymes that mediated wall polysaccharide synthesis in the grasses remained elusive for many years and awaited the emergence of molecular biology technologies before further progress was possible, as described below.

CHARACTERISING THE CELLULOSE SYNTHASE GENE SUPERFAMILY

Despite the considerable effort invested in the identification and purification of plant cellulose synthases, these enzymes could not be purified or characterised through classical biochemical approaches. However, Pear et al. (1996) used emerging molecular biology techniques to show that amino acid sequences derived from cotton and rice cDNA clones shared three homologous regions with bacterial cellulose synthase (*celA*) enzymes (Wong et al. 1990; Saxena et al. 1990), which had been purified to homogeneity. Proof-of-function was quickly provided by Arioli et al. (1998) and Taylor et al. (1999), who showed that the *RSW1* and *IRX3* loci of Arabidopsis mediated cellulose synthesis.

Phylogeny and Evolution

Richmond and Somerville (2000) applied emerging bioinformatics techniques to demonstrate that the Arabidopsis genome sequence contained a group of closely related cellulose synthase genes, which they designated *CesA*, and a larger group of 'cellulose synthase-like' genes (*CslA*, *CslB*, *CslC*, *CslD*, *CslE* and *CslG*) that together constituted the cellulose synthase gene 'superfamily'. In a similar analysis of the newly published rice genome sequence, Hazen et al. (2002) revealed additional groups of *Csl* genes (*CslF*, *CslH*) that appeared to be specific for monocots.

The genes in the cellulose synthase gene superfamily are classified as GT2 glycosyl transferases (Lombard et al. 2014; Box 9.3) and, with few exceptions, contain a conserved QxxRW amino acid sequence motif that is involved in UDP-sugar-binding at the catalytic site. At the molecular level, the phylogeny and evolutionary dynamics of genes within the cellulose synthase gene superfamily have been continually upgraded, as more plant genome sequences become available (Popper et al. 2011; Yin et al. 2014; Schwerdt et al. 2015; Little et al. 2018). The most recent version of the phylogenetic tree, derived from the data of Little et al. (2018), is shown in Figure 9.1. Because of the likely dual endosymbiotic origin of the superfamily, no direct connection is shown between the two major evolutionary groups (CslA/CslC and CslF/CslA/CesA/CslB/H/CslJ/M/CslG/CslE) in Figure 9.1.

The cell wall polysaccharides synthesised by the various synthase clades or sub-groups have been partly but not completely defined. Several but not all *CesA* genes have been shown to mediate cellulose synthesis; *CslA* genes were implicated in mannan biosynthesis in a landmark paper from Dhugga et al. (2004), which was quickly confirmed by Liepman et al. (2005). Similarly, *CslC* was shown to be involved in xyloglucan biosynthesis (Cocuron et al. 2007; Kim et al. 2020), *CslD* genes in single-stranded cellulose synthesis, particularly in cells that undergo tip growth (Doblin et al. 2001; Favery et al. 2001; Wang et al. 2001; Yang et al. 2020), and *CslF* and *CslH* genes in (1,3;1,4)-β-glucan synthesis (Burton et al. 2006; Doblin et al. 2009). The *CslJ* subfamily also mediates (1,3;1,4)-β-glucan synthesis (Little et al. 2018).

At this stage, it is important to stress that not all the genes in a particular subfamily can be assumed to mediate the synthesis of the same or similar wall polysaccharide. Thus, it has not yet been demonstrated that every *CesA* gene in a particular species encodes a cellulose synthase, if all *CslA* genes encode mannan synthases or if all *CslF* genes encode (1,3;1,4)-β-glucan synthases. For example, the barley *CslF3* gene was recently shown to encode a (1,4)-β-xyloglucan synthase. This polysaccharide contains both (1,4)-β-glucosyl and (1,4)-β-xylosyl residues at ratios ranging from 1.5:1 to 5:1 in an unbranched polysaccharide chain (Little et al. 2019).

The phylogenetic data summarised in Figure 9.1 (Schwerdt et al. 2015; Little et al. 2018) have revealed a number of evolutionary features of the cellulose synthase gene superfamily in angiosperms, which can be summarised as follows:

- The *CslA*/*CslC* clade is highly diverged from the *CesA* and other *Csl* subfamilies and may have evolved from a separate endosymbiotic transfer event.
- At least some members of the three Poaceae-specific subfamilies (*CslF*, *CslH* and *CslJ*) mediate (1,3;1,4)-β-glucan synthesis; the ability to synthesise (1,3;1,4)-β-glucans appears to have evolved several times.
- The *CslF* subfamily is monophyletic with the *CslD* clade and is only found in the graminids and restiids; this indicates its very recent evolutionary origin.
- The *CslJ* clade forms a reciprocally monophyletic monocot-eudicot grouping with a previously unreported eudicot lineage, *CslM*, which is not capable of directing (1,3;1,4)-β-glucan synthesis; this relationship is repeated in the monocot (1,3;1,4)-β-glucan synthase *CslH* family and its eudicot sister group, *CslB*; the function of the *CslB* clade is currently unknown.

Selection Pressure

Selection pressure on individual genes or on specific nucleotides within genes can be estimated through the ratio of the rate of nonsynonymous substitutions (*d*N), which result in changes of amino acid residues in the encoded proteins, and the rate of synonymous substitutions (*d*S), where no amino acid changes are observed. A *d*N:*d*S ratio greater than 1 broadly indicates that positive selection has occurred, a ratio of about 1 suggests stationary/stabilising selection and ratios that approach zero indicate purifying (negative) selection pressure (Yang and Bielawski 2000).

Schwerdt et al. (2015) investigated selection pressure patterns across the *CesA*, *CslD*, *CslF*, *CslE*, *CslJ* and *CslH* subfamilies and found relatively low *d*N:*d*S values for most clades. Genes with low *d*N:*dS* ratios are likely to encode enzymes that direct the synthesis of polysaccharides that are advantageous and, conversely, that alteration of these genes might be disadvantageous to the plant (Tan et al. 2015). Nevertheless, some *CesA*, *CslF* and other cellulose synthase superfamily genes showed positive selection, particularly following recent gene duplication events, and these genes might generate new enzyme specificities or adaptive changes to gene expression profiles (Schwerdt et al. 2015).

BIOSYNTHESIS OF HETEROXYLANS IN THE GRASSES

The identification of xylan synthase genes has faced many challenges over several decades. Attempts to purify xylan synthases from plant homogenates and membrane fractions were largely unsuccessful (Dallasandro and Northcote 1981; Bolwell and Northcote 1983; Scheller and Ulvskov 2010). Subsequent overexpression and knock-down experiments, using transcript profiling, bioinformatics and wall analyses of wheat endosperm and *irregular xylem* (*irx*) mutants of Arabidopsis, implicated GT43 and GT47 genes in the synthesis of the xylan backbone (Pena et al. 2007; Persson et al. 2007; Brown et al. 2009; Jiang et al. 2016). Genes from the GT8 and GT61 families have been implicated in the addition of α-glucuronic acid and α-4-O-methylglucuronic acid residues to the heteroxylan of Arabidopsis (Mortimer et al. 2010; Zeng et al. 2010) and in the addition of α-l-arabinosyl residues to the wheat arabinoxylan (Mitchell et al. 2007). Lunin et al. (2020) solved the 3D structure of a xylan *O*-acetyltransferase from Arabidopsis and were able to describe the precise molecular mechanism of catalysis of the enzyme.

Interpretative difficulties of data generated in these experiments are attributable to the presence of candidate genes in large gene families, complicating pleiotropic effects during proof-of-function analyses, and our limited success in developing reliable biochemical assays for membrane-bound enzymes expressed in heterologous systems. Furthermore, it is difficult to identify suitable plant species for gain-of-function approaches, which are often more easily interpreted than partial loss-of-function experiments through the analysis of walls from mutant lines. Thus, while it was possible to use eudicot plants that have no (1,3;1,4)-β-glucans in their walls to test candidate genes for (1,3;1,4)-β-glucan synthesis (Burton et al. 2006), most eudicot and monocot plants have at least some heteroxylan in their walls and therefore do not provide a 'clean background' for gain-of-function approaches for the identification of xylan synthase genes.

Although direct biochemical proof for the functions of putative xylan synthase enzymes was usually lacking,

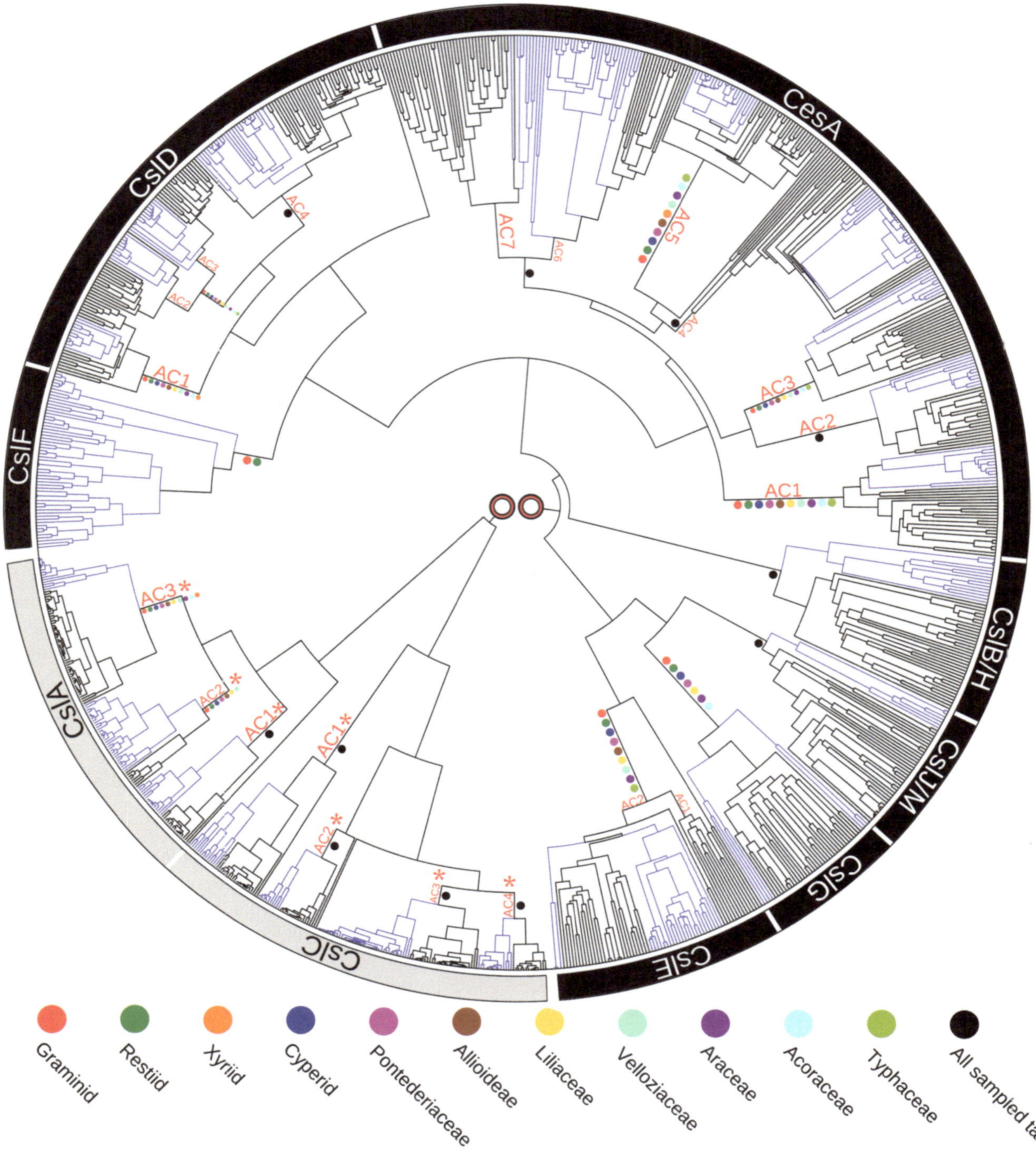

FIGURE 9.1 Phylogeny of the cellulose synthase gene superfamily of angiosperms. Midpoint rooted, ultrametric best-known maximum likelihood (ML) tree generated from species with fully sequenced genomes. Blue branches indicate monocots and black branches indicate eudicots. Clades existing before the monocot-eudicot divergence are shown as Ancestral Clusters (AC). The central red rings indicate putative independent endosymbiotic transfer events. Asterisks indicate ancestral clusters that are not well supported by the ML analyses. Red rings indicate putative independent endosymbiotic transfer events. Coloured dots indicate transcriptome data placed on the ML tree. Figure and legend reproduced from Little et al. (2018).

the ability to express candidate genes in heterologous systems nevertheless provided a new approach through which the specificity of the encoded enzymes could be defined. Jensen et al. (2014) and Urbanowicz et al. (2014) showed that the *IRX10* GT47 gene, when expressed in heterologous systems, mediated the addition of several xylosyl residues to oligoxyloside acceptors. However, it was necessary to use incubation times of 10–96h, so catalytic rate constants (k_{cat}) appeared to be quite low and the DP values of the products were relatively short. It is possible that the synthases behave in a two-phase manner, as is observed with the homogalacturonan synthases (GAUT), which catalyse elongation at a slow rate initially but, at a critical DP, elongation rates increase rapidly (Amos and Mohnen 2019). Jensen et al. (2018) subsequently showed that an *IRX10* orthologue from a charophytic green alga, when expressed transiently in a HEK293-F foetal kidney cell line, generated an enzyme capable of directing the synthesis of (1,4)-β-xylan chains with an average DP of 25 within 4 hours. Pellny et al. (2012) reported that GT47 transcripts, which are orthologues of the Arabidopsis *IRX10* gene, were the most abundant in developing wheat endosperm, followed by GT61 transcripts, while Jiang et al. (2016) used proteomics technologies to show that the xylan synthase complex of wheat mainly contained the products of GT47 and GT61 gene expression.

IDENTIFICATION OF (1,3;1,4)-β-GLUCAN SYNTHASE GENES

Following a gap of close to 20 years, emerging genomics technologies enabled the biology of (1,3;1,4)-β-glucan and (1,3)-β-glucan synthase enzymes and genes to be re-visited in the early 2000s. With respect to the (1,3;1,4)-β-glucan synthases of the grasses, there were three major breakthroughs that enabled the unanswered questions to be addressed. These included the publication of a draft rice genome sequence through a collaborative project coordinated by the International Rice Genome Sequencing Project (IRGSP 2005; de Vries 2000), the use of expressed sequence tags (EST) to map phenotypes of interest in barley (Han et al. 1995) and the discovery of genetic synteny, or gene collinearity, in cereals (Gale and Devos 1998). Although there was no genome sequence available for barley at this stage, syntenic similarities were discovered between cereal species, through which gene order along a particular region of a cereal genome often corresponded to gene order along homologous regions of other cereal genomes. The collinearity was not perfect, because during the course of cereal genome evolution, chromosomes were fragmented, re-arranged and recombined (Gale and Devos 1998). Nevertheless, local synteny could still be seen in these 'shuffled' genomes.

The starting information was provided by a molecular marker linkage map that identified the genetic positions (quantitative trait loci; QTLs) of genes controlling (1,3;1,4)-β-glucan content in barley grain (Han et al. 1995). The QTLs for (1,3;1,4)-β-glucan content were located on chromosomes 2 and 5. The stronger of the two was on chromosome 2 and this was the target used for the identification of (1,3;1,4)-β-glucan synthase genes (Burton et al. 2006). The chromosome 2 QTL peak was very broad and there were probably hundreds of genes under that peak. Two flanking ESTs, each about half way up each side of the chromosome 2 QTL peak, were chosen and their sequences were compared with the newly released rice genome sequence. The two barley sequence orthologues were found on chromosome 8 of rice and it was reasoned that, if the property of synteny held up in those regions, the (1,3;1,4)-β-glucan synthase gene should reside somewhere between those two rice sequences. After manually scanning through this large region of the rice genome sequence, a group of *cellulose synthase-like* genes (*CslF*) **was** found close to one end of the interval defined in rice by the two barley EST sequences (Burton et al. 2006).

Thus, candidate genes for the (1,3;1,4)-β-glucan synthase enzymes were identified. Now for their proof-of-function. This proved to be relatively straightforward because (1,3;1,4)-β-glucans are found almost exclusively in monocotyledons. This enabled a clean gain-of-function approach, through which *CslF* genes cloned from rice were transformed into the eudicot *Arabidopsis thaliana*, which contains no (1,3;1,4)-β-glucan in its walls and no *CslF* genes in its genome. Using highly sensitive and specific antibodies against (1,3;1,4)-β-glucan, it was soon shown that the Arabidopsis plants transformed with the rice *CslF* genes did indeed synthesise (1,3;1,4)-β-glucan in their cell walls. To double-check the specificity of the antibodies, sections for microscopy were pre-treated with purified (1,3;1,4)-β-glucan endohydrolase; this treatment removed subsequent antibody binding (Burton et al. 2006).

The barley genome sequence (Mascher et al. 2017) subsequently revealed that the *HvCslF* gene family consists of 12 genes and one pseudogene (Burton et al. 2008; Schreiber et al. 2014). These genes are distributed across the genome, although there is a conserved *CslF* gene cluster on chromosome 2H of barley; that cluster is also evident in syntenic positions on other cereal genomes (Burton et al. 2008; Schreiber et al. 2014). The *CslF6* gene is the most highly transcribed member of the *CslF* family in developing barley grain, although *CslF9* transcripts are abundant early in endosperm development (Burton et al. 2008). The role of the *CslF6* gene in (1,3;1,4)-β-glucan synthesis in barley has been confirmed, but the unequivocal demonstration that the CslF9 protein directs (1,3;1,4)-β-glucan synthesis has proved more elusive (Garcia-Gimenez et al. 2020).

A similar approach was used to demonstrate that a barley *CslH* gene also mediated the synthesis of (1,3;1,4)-β-glucan in transgenic Arabidopsis and in this case sufficient polysaccharide was deposited in the walls to enable its extraction and its unequivocal identification by enzymic analysis (Doblin et al. 2009). Later, it was shown that the *HvCslJ* genes also encode (1,3;1,4)-β-glucan synthases (Little et al. 2018).

EVOLUTIONARY DYNAMICS OF (1,3;1,4)-β-GLUCAN SYNTHASES IN THE GRASSES

The distribution of (1,3;1,4)-β-glucan in walls has been compared with the distribution of *CslF, CslH* and *CslJ* genes in commelinid and non-commelinid species of the monocots through the use of specific antibodies against (1,3;1,4)-β-glucan, linked with transmission electron microscopy, and confirmed where possible through biochemical assays (Little et al. 2018). This revealed that *CslF*, *CslH* and *CslJ* genes are found outside the Poaceae family. The *CslF* genes are detected in both graminids and restiids. In contrast, the *CslH* and *CslJ* genes are distributed more widely across commelinids and non-commelinids. The immunocytochemical assay detected (1,3;1,4)-β-glucans in the graminids, restiids, xyrids, cyperids and bromeliads, and in more distantly related commelinids and non-commelinids (Little et al. 2018).

Thus, (1,3;1,4)-β-glucans have a much wider distribution across the monocots than previously believed. Little et al. (2018) suggested that (1,3;1,4)-β-glucan synthases have evolved convergently on several occasions or possibly that synthesis of this polysaccharide has been gained and lost several times during evolution of the cellulose synthase gene superfamily. Further evidence for the multiple evolution of (1,3;1,4)-β-glucan synthases (Harris and Fincher 2009) lies in the detection of these polysaccharides outside the angiosperms, namely, in lower plants such as the *Equisetum* spp. horsetail ferns (Trethewey et al. 2005; Fry et al. 2008; Sørensen et al. 2008), bryophytes (Popper and Fry 2003), the pathogenic fungus of barley, *Rhynchosporium secalis* (Pettolino et al. 2009), brown, green and red algae (Lechat et al. 2000; Eder et al. 2008; Popper and Tuohy 2010) and the lichen *Cetraria islandica* (Stone and Clarke 1992; Wood et al. 1994).

In a broader metabolic context, it is worth noting that the (1,3;1,4)-β-glucan synthase genes probably co-evolved with the (1,3;1,4)-β-glucan endohydrolases (Bulone et al. 2019). Initially, cellulases would have enabled plants to recover or remove the newly evolved (1,3;1,4)-β-glucans in cell walls, and enzymes that would more specifically hydrolyse these polysaccharides probably evolved later. Comparative structural analyses provided strong evidence that the (1,3;1,4)-β-glucanases evolved from the widely distributed and abundant (1,3)-β-glucanases (Varghese et al. 1994; Bulone et al. 2019).

The development of bioinformatics technologies was crucial for these advances in our understanding of wall biology, enzymology, genetics and evolution. The CAZy database of Carbohydrate-Active enZYmes evolved from classifications of enzymes involved in the synthesis and metabolism of complex carbohydrates, including those found in plant cell walls (http://www.cazy.org/; Lombard et al. 2014; Box 9.3).

BOX 9.3: THE CARBOHYDRATE-ACTIVE ENZYMES DATABASE (CAZY)

This database was launched online by Bernard Henrissat at the University of Marseille in 1998 (http://www.cazy.org/; Lombard et al. 2014). It provides an encyclopaedic family classification of glycoside hydrolases and transferases, polysaccharide lyases, carbohydrate esterases, carbohydrate-binding modules and auxiliary activities. As an indicator of the importance of this database to the international scientific community, over the last 6 years (up until July 2021) over 1 million hits were recorded for CAZY, from more than 380,000 users. That is an average of about 63,500 users and 74,000 sessions per annum. In addition, each visitor accessed 10 pages (on average), which means that more than 10 million pages have been visited or downloaded during the last 6 years. In 2006, Bruce Stone suggested that an encyclopaedia of carbohydrate-active enzymes might be a useful complement for CAZY for the cell wall community and in 2007 Harry Brumer began setting up CAZYpedia online, with help from Steve Withers and Bernard Henrissat, and with input on specific enzymes provided by experts in the field. CAZYpedia follows a similar overall structure as CAZY. These online databases provide an invaluable resource for the plant cell wall community.

MECHANISM OF ACTION OF (1,3;1,4)-β-GLUCAN SYNTHASES

There are two central questions to be addressed in explaining the molecular mechanism of action of (1,3;1,4)-β-glucan synthases. Firstly, are both (1,3)-β- and (1,4)-β-linkages inserted into the elongating polysaccharide by a single enzyme or are additional enzymes required? Secondly, how are the single (1,3)-β- and (1,4)-β-linkages arranged in a non-random fashion while the cellotriosyl and cellotetraosyl oligomers are arranged at random (Burton and Fincher 2009), as described above?

At this stage, the weight of evidence lies with the suggestion that a single enzyme, for example, CslF6, is capable of inserting both (1,3)-β- and (1,4)-β-linkages (Buckeridge et al. 1999; Kim et al. 2015). There are biological precedents for this possibility. Many mammalian and other glycosyl transferases can catalyse the synthesis of polysaccharides that contain different linkage types and/or different glycosyl residues. However, in many of these cases, the glycosyl residues and linkages are arranged in a strictly repeating fashion. For example, the bacterial SIII pneumococcal polysaccharide is composed of a repeating sequence of (1,3)-β-linked cellobiuronic acid residues. Reduction of the uronic acid residues yields a (1,3;1,4)-β-glucan with strictly alternating (1,3)- and (1,4)-linkages that has been used to examine the specificity of various (1,3;1,4)-β-glucanases (Anderson and Stone 1975). Similarly, hyaluronan is composed of a repeating disaccharide of N-acetylglucosamine (GlcNAc) and glucuronic acid (GlcA) that has alternating (1,3)- and (1,4)-linkages and is synthesised by a single oligomeric enzyme (Weigel and DeAngelis 2007). To our knowledge, none of these types of polysaccharide exhibit the non-random and random characteristics typical of the (1,3;1,4)-β-glucans from the grasses.

In a breakthrough paper, Morgan et al. (2013) published the 3D crystal structure of a bacterial CesA enzyme. This crystal structure provided a template for homology modelling of related enzymes, including the barley CslF6 enzyme, for which the complete amino acid sequence had been obtained from corresponding cDNAs (Burton et al. 2006). The model of the CSLF6 enzyme suggests that a large cytoplasmic region is anchored into the membrane by five membrane spanning helices. These helices form a pore across the membrane through which the nascent (1,3;1,4)-β-glucan chain is extruded; the dispositions of the active site QxxRW and TED motifs were also modelled (Dimitroff et al. 2016).

The homology models allowed the identification of sections of the CslF6 enzyme that control the DP3:DP4 ratio of the nascent polysaccharide chain and the total amount of (1,3;1,4)-β-glucan synthesised (Jobling 2015; Dimitroff et al. 2016). The *CslF6* genes from *Sorghum bicolor* and *Hordeum vulgare*, which mediate the formation of significantly different levels of (1,3;1,4)-β-glucan that also differ in their DP3:DP4 ratios, were expressed heterologously in *Nicotiana benthamiana*. Chimeric cDNA constructs with interchanged sections of the two *CslF6* genes identified regions of the enzymes important for total activity and DP3:DP4 ratios (Figure 9.2) (Dimitroff et al. 2016). At this stage, it is not known if the CslF enzymes form oligomeric complexes of the type observed with cellulose synthases (Höfte and Voxeur, 2017; Turner and Kumar, 2018; Jarvis, 2018).

Although a complete understanding of the catalytic mechanism of (1,3;1,4)-β-glucan synthases is not yet available, Purushotham et al. (2022) successfully applied structural biology technologies to define the 3D structure of the enzyme and were able to explain how a single, monomeric enzyme can insert both (1,3)- and (1,4)-linkages into the elongating polysaccharide chain. The potentially transformative use of artificial intelligence (AI) to accurately predict 3D structures of proteins at the atomic level, based solely on their amino acid sequences, is without doubt accelerating the achievement of this goal (Callaway 2021; Jumper et al. 2021). In this technology, the machine learning, neural network model, AlphaFold, provides a computational tool that can rapidly predict protein structures with atomic accuracy, even where no homologous structure is known. The technology was developed by a consortium involving DeepMind, Google and Alphabet. It has been validated through the provision of accurate models of all proteins in the human proteome, which will be made publicly available (Jumper et al. 2021).

SUBCELLULAR SITES OF (1,3;1,4)-β-GLUCAN AND ARABINOXYLAN SYNTHESIS

Traditionally, synthesis of the non-cellulosic polysaccharides of cell walls has been considered to occur in the Golgi apparatus (Carpita and McCann 2010; Dhugga 2012; Pauly et al. 2013). However, when Wilson et al. (2006) probed sections of barley coleoptiles, developing endosperm and barley cell cultures with (1,3;1,4)-β-glucan-specific antibodies, they observed heavy labelling over the walls but little if any labelling of adjacent Golgi cisternae. Subsequent immunocytochemical studies with antibodies against the CslF6 and CslH1 (1,3;1,4)-β-glucan synthase enzymes showed a concentration of the CslF6 enzyme in plasma membrane–enriched fractions; the CslH1 enzyme did not appear to be enriched in the plasma membrane fraction (Wilson et al. 2015).

In contrast, Kim et al. (2015) used live imaging and immuno-electron microscopy to trace a yellow fluorescent protein fusion of the *Brachypodium distachyon* CslF6 protein in tobacco and yeast heterologous systems and showed unequivocally that the enzyme was localised to the Golgi apparatus, where (1,3;1,4)-β-glucan synthesis was also observed. Topology predictions suggested that the catalytic domain of the enzyme is located in the cytoplasm, which would be consistent with either a Golgi or plasma membrane location for synthesis of the polysaccharide. In maize coleoptiles, (1,3;1,4)-β-glucan was detected in both cell walls and the Golgi (Carpita and McCann 2010). Consistent with this result, Okekeogbu et al. (2019) detected CslF proteins in highly enriched Golgi membranes from maize.

The cell biologists who have published these apparently conflicting results are world class experts in their fields. This begs the question as to whether both conclusions are correct, namely, that (1,3;1,4)-β-glucans can be synthesised either in the Golgi or at the plasma membrane. Enzymes derived from a range of different genes have been implicated in (1,3;1,4)-β-glucan synthesis and the polysaccharide is synthesised in several different tissues in different species of the grasses. Tissue-specific, isoenzyme-specific or species-specific differences in gene expression might occur. Wilson et al. (2015) reported that the CslF6 and CslH1 enzymes, both of which mediate (1,3;1,4)-β-glucan synthesis, are found in plasma membrane–enriched and plasma membrane–depleted fractions, respectively. The UDP-glucose substrate would be available at both sites. At least half of the total sucrose synthase (SuSy) enzyme, which generates UDP-glucose from sucrose, is tightly associated with the plasma membrane in developing cotton fibres (*Gossypium hirsutum*) and could therefore deliver the UDP-glucose substrate to either callose or cellulose synthases that are located in the plasma membrane (Amor et al. 1995). The SuSy could also deliver the UDP-glucose substrate if (1,3;1,4)-β-glucan synthases were indeed located in the plasma membrane. As noted above, Okekeogbu et al. (2019) detected CslF proteins in enriched Golgi fractions from maize, but CesA proteins were also seen in high abundance in these Golgi membrane preparations, and it was reasonably concluded that the CesA proteins were in transit to the plasma membrane, where cellulose synthesis occurs. With respect to the Golgi-associated CslF proteins, Okekeogbu et al. (2019) noted that "an association of CslF proteins with CesA transport to the PM is possible". Is it

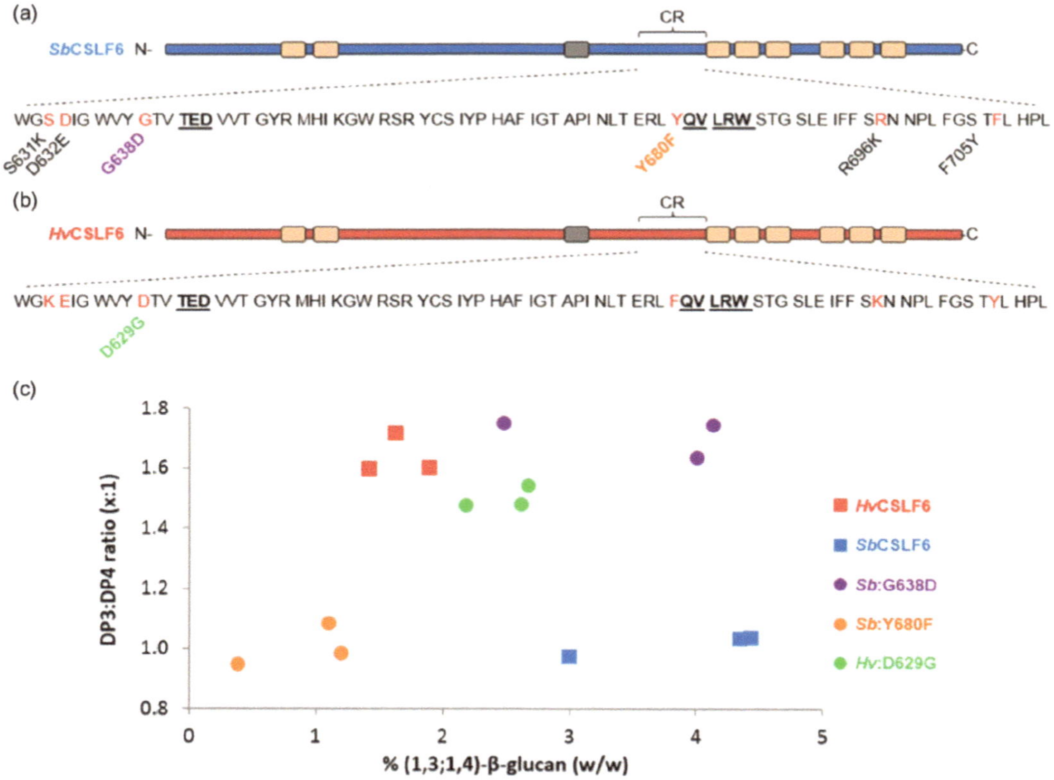

FIGURE 9.2 Structures and levels of (1,3;1,4)-β-glucan synthesised by *CSLF6* mutation constructs expressed in *N. benthamiana* leaves. The primary structures of the sorghum SbCSLF6 (a) and the barley HvCSLF6 (b) showing predicted amphipathic helices (brown boxes), the 'extra region of 55 amino acid residues' (grey box) and the catalytic region (CR), for which the amino acid sequence is expanded. Conserved motifs are in bold font and underlined. Red residues indicate differences between SbCSLF6 and HvCSLF6. Details of specific point mutation constructs are underneath. The amount (% w/w) and structure (x:1) of (1,3;1,4)-β-glucan produced by the constructs is plotted (c). Three spots of the same colour represent the values generated by biological triplicates for a particular construct. Figure and legend from Dimitroff et al. (2016).

therefore possible that different CslF enzymes are directed to and activated in both the Golgi and the plasma membrane in response to specific spatio-temporal mechanisms?

It is also tempting to question whether other synthases of non-cellulosic wall polysaccharides might also be located only in the plasma membrane or in the Golgi, or in both the plasma membrane and the Golgi. Is there any connection here with observations that papillae formed in the periplasmic space of barley leaves infected with a fungal pathogen contain high levels of arabinoxylan, together with high levels of callose and cellulose (Chowdhury et al. 2014), as described in detail later in this article?

These observations present some challenges to our conventional wisdom that non-cellulosic polysaccharides are synthesised exclusively in the Golgi, but the ever-increasing resolution of immunocytochemical and related technologies should allow us to investigate other possibilities in the immediate future. New confocal, super-resolution and electron microscopies, together with cytochemical labelling, have recently been used to define the sequence and timing of wall polysaccharide synthesis in walls (van de Meene et al. 2017; 2021). As pointed out by these authors, rapid advances in imaging techniques at the light, fluorescence and electron microscopy levels are allowing resolution to the nanometer level and new molecular probes are enabling the characterisation of supramolecular complexes *in situ*. Emerging atomic force microscopy will allow structure–function relationships to be defined (van de Meene et al. 2021). These technologies will further advance our knowledge of the locations of cell wall polysaccharide synthases in the immediate future.

HYDROLYSIS OF WALL POLYSACCHARIDES IN THE GRASSES

Despite the experimental barriers that slowed the identification and characterisation of the membrane-bound polysaccharide synthases responsible for (1,3;1,4)-β-glucan and arabinoxylan biosynthesis, great advances were achieved during the biochemical era in our understanding of the corresponding hydrolytic enzymes, which are generally soluble and therefore amenable to traditional purification technologies. Purification allowed the determination of NH_2-terminal amino acid sequences, which enabled the identification of corresponding cDNAs and genes early in the molecular biology era. DNA sequencing technologies allowed the complete amino acid sequences of these hydrolases to be defined, which were prerequisite for solving their 3D structures at atomic resolution, and hence their molecular mechanisms of hydrolysis. In this section, we present current information on the hydrolytic enzymes and encoding genes that mediate the degradation of arabinoxylans and (1,3;1,4)-β-glucans from the grasses.

Arabinoxylans

Here is an example of how the emergence of genomics and associated transcriptomics technologies forced us to radically change our thinking on previously accepted 'conventional wisdom'. Early work on arabinoxylan degradation was focused on the endo-(1,4)-β-xylanases from germinated wheat or barley. The enzymes could be purified from grain extracts or, in many cases, from isolated aleurone layers treated with the phytohormone gibberellic acid. Benjavongkulchai and Spencer (1986) purified a family GH10 endo-(1,4)-β-xylanase from isolated barley aleurone layers and reported an apparent molecular weight of 34,000 Da, while Slade et al. (1989) separated and purified three endo-(1,4)-β-xylanases from extracts of germinated barley grain. However, the latter authors reported apparent molecular weights of 41,000 Da for each of the three isoforms. Using the NH_2-terminal amino acid sequences of these three 41,000 Da enzymes, Banik et al. (1996) isolated two cDNAs from germinated barley but reported difficulties in identifying the position of the NH_2-terminal amino acid and the expected signal peptide, and could therefore not explain the observed discrepancy in the molecular weights of the purified enzymes.

Caspers et al. (2001) subsequently solved this inconsistency. They showed that the barley endo-(1,4)-β-xylanase primary translation product is a 61,500 Da cytoplasmic proenzyme with NH_2- and COOH-terminal pro-peptides that facilitate correct folding and activation of the precursor. Progressive removal of fragments from both the NH_2- and COOH-termini of the primary translation product resulted firstly in the active 41,000 Da intermediate form, which was further processed to generate the final, active 34,000 Da enzyme. Release of the enzyme from barley aleurone layers coincided with the loss of membrane integrity associated with programmed cell death (Bethke et al. 1999; Caspers et al. 2001), consistent with earlier observations by Slade et al. (1989).

The central GH10 domain and the 52 amino acid COOH-terminal of the barley endo-(1,4)-β-xylanase correspond to the active 41,000 Da enzyme purified by Slade et al. (1989), following the removal of the approximately 200 amino acid NH_2-terminal CBM22 domain by cysteine endopeptidases (Caspers et al. 2001). Subsequent removal of the COOH-terminal region results in the formation of the final 34,000 Da enzyme. Van Campenhout et al. (2007) reported that the unprocessed 61,500 Da barley aleurone endo-(1,4)-β-xylanase isoenzyme X-I was also active, while Caspers et al. (2001) could barely detect activity in the 61,500 Da protein. Nevertheless, the apparent discrepancy between the results of Benjavongkulchai and Spencer (1986) and Slade et al. (1989) is most probably explained on the basis of differences in the timing of enzyme extraction with respect to the stage of aleurone function and hence the timing of xylanase processing. Slade et al. (1989) purified the intermediate, active 41,000 Da form from extracts of whole grain 5 days after the initiation of germination, while Benjavongkulchai and Spencer (1986) purified the 34,000 Da enzyme from isolated aleurone layers treated with GA_3 for 3 days. Aleurone function would appear to be more advanced in the latter experimental system.

For 30 years or more, it was accepted that the barley (1,4)-β-xylanase gene family consisted of two or possibly three members. But when Betts et al. (2021) re-visited the barley (1,4)-β-xylanase gene family, armed with the barley genome sequence (Mascher et al. 2017) and new transcriptomics technologies, it was revealed that the family actually consisted of at least 15 genes. Furthermore, while earlier homology models of the 3D structure of the X-1 (1,4)-β-xylanase isoform revealed the 3D structure of one form of the enzyme (Banik et al. 1996), Betts et al. (2021) refined these models and showed that the enzymes consisted of a number of conserved domains (Figure 9.3). All enzymes contain one or more NH_2-terminal CBM22 or CBM22-like domains, with a linker peptide to a central catalytic domain, followed by COOH-terminal β-sandwich sequences (Betts et al. 2021). The CBM22 region has the characteristics of a xylan-binding domain (http://www.cazy.org/; Lombard et al. 2014) and is probably involved in binding the proenzyme to arabinoxylans in the cell wall. The GH10 domain of 288 amino acid residues corresponds to the fully processed, active, 34 kDa enzyme identified by Benjavongkulchai and Spencer (1986) and Caspers et al. (2001). It forms a $(\beta/\alpha)_8$ barrel fold that is typical of many cell wall polysaccharide endohydrolases, insofar as it has, right across the enzyme, a deep polysaccharide-binding groove that enables the hydrolysis of internal linkages in an essentially random fashion. Along this polysaccharide-binding groove are evenly spaced hydrophobic amino acid residues that constitute the binding subsites of individual glycosyl residues of the polysaccharide. The catalytic acid and catalytic nucleophile residues (usually aspartic acid or glutamic acid residues) are positioned adjacent to the glycosidic linkage that is hydrolysed (Hrmova et al. 1995; Banik et al. 1996; Betts et al. 2021).

Hydrolytic removal of the arabinosyl residues of the arabinoxylans of barley grain, together with the degradation of oligo-β-D-xylosides released by the endo-xylanase, is catalysed by both α-L-arabinofuranosidases and bifunctional arabinofuranosidase/xylosidase enzymes (Lee et al. 2001, 2003). An additional class of xylanolytic enzymes contains the AXAHs, which remove arabinosyl residues from the backbone of arabinoxylan molecules and therefore alter the physical properties of the polysaccharides and potentially the wall as a whole (Laidlaw et al. 2012). It appears that nascent arabinoxylans are heavily substituted with arabinosyl residues (Gibeaut et al. 2005; Wilson et al. 2015) and that AXAHs might be important in stripping arabinosyl

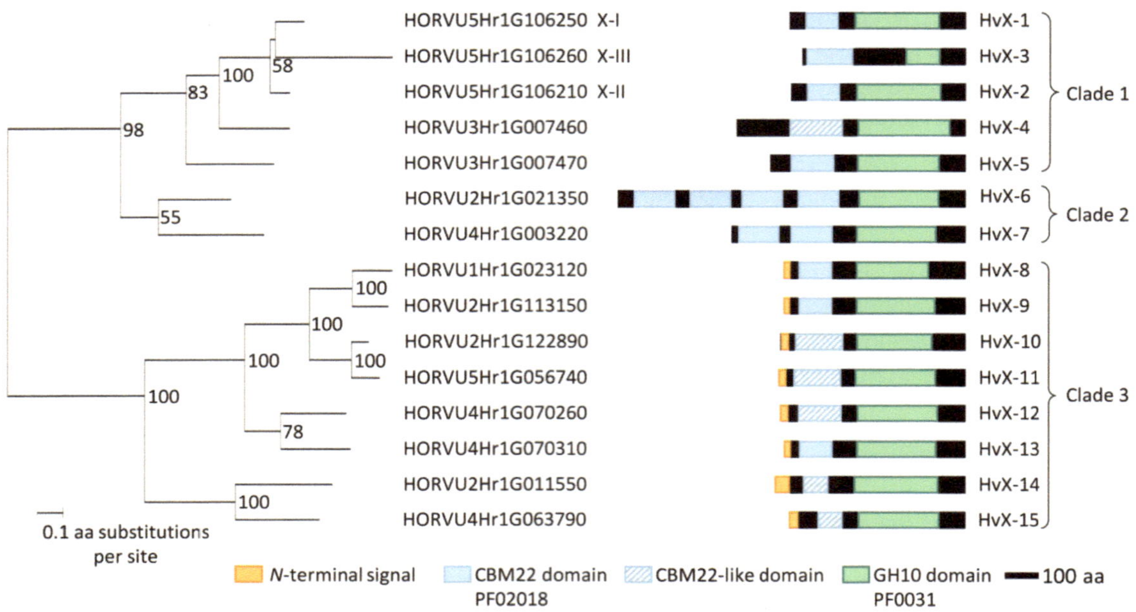

FIGURE 9.3 Structure of the conserved xylanases enzyme modularity in barley. Note that three phylogenetically different clades of genes can be distinguished. Barley gene annotations are indicated, bootstrap support values (%) are shown on branch points (%) and predicted protein domains are shown as coloured blocks (from Betts et al. 2021).

residues from the main chain xylan backbone (Zhang et al. 2016). This would serve to enhance non-covalent hydrogen bonding between the heteroxylans themselves and with other wall polysaccharides, such as cellulose. This, in turn, would provide the extra toughness and strength required by the walls, either as the increasing weight of the developing plant placed additional loads on the walls, or in the event that the plant was attacked by potentially pathogenic microorganisms. However, co-localisation analyses showed that the interactions between cellulose and arabinoxylans actually decreased slightly over time in walls regenerating on tissue-cultured *Lolium multiflorum* cells (van de Meene et al. 2021). Similarly, Laidlaw et al. (2012) were unable to demonstrate the involvement of AXAH enzymes in the remodelling of barley endosperm cell walls during development. The role of these enzymes in remodelling arabinoxylans therefore remains unclear.

Betts et al. (2021) also showed that, in addition to the 15 (1,4)-β-xylanase genes, six β-xylosidase and five or six AXAH genes are present in the barley genome; the chromosomal locations of these genes have been defined. The spatio-temporal expression patterns of the genes were investigated in germinated grain, young seedlings, vegetative tissues and in developing grain. These expression patterns are summarised in Figure 9.4. Transcripts of several genes are detected in most tissues, while a smaller number of both the (1,4)-β-xylanase and xylosidase genes are expressed at much higher levels (Betts et al. 2021). Additional enzymes would be required to remove uronic acid residues, the constituents of short oligosaccharide chains that are found in certain heteroxylans and the ferulic acid residues that are attached to the arabinosyl residues (Mitchell et al. 2007). The relatively large number of putative (1,4)-β-xylanase genes, compared with endo-(1,3;1,4)-β-glucanase genes, presumably accommodates the broader distribution of heteroxylans in different tissues, coupled with the need for independent regulation of these genes during vegetative and reproductive growth.

An obvious question here is 'why do all the barley (1,4)-β-xylan endohydrolases have this relatively complex modular structure?'. Clues may be provided by data from germinated barley grain. As noted above, the (1,4)-β-xylanases are synthesised *de novo* in the aleurone layer but are not released into the starchy endosperm until late in the process of secretion of degradative enzymes from the aleurone (Fincher 1989). The aleurone cell wall consists of a thick outer layer, through which enzymes are released via channels formed in the outer wall layer. A much thinner inner layer of the aleurone wall remains intact until programmed cell death and is presumably crucial for cell integrity until the function of the aleurone layer is completed. The CBM22 domain might bind the enzyme to the cell wall in an inactive form until it is released during programmed cell death (Fincher 1989). Given the various (1,4)-β-xylanases of barley all have similar modular structures, it might be argued that each of these enzymes is likely to be released in an active form only after programmed cell death and that they therefore function mainly in the recovery of constituent sugars of heteroxylans in senescing tissues. The immediately apparent flaw in this argument is that the walls of the starchy endosperm, which contain up to 20% arabinoxylan, are degraded following germination, before the aleurone (1,4)-β-xylanases are released. Does this mean that (1,4)-β-xylanases are also synthesised in the developing starchy endosperm in an inactive form and activated in the starchy endosperm following the initiation of germination? The transcript data of Betts et al. (2021), which show that some (1,4)-β-xylanase transcripts are detectable in developing grain, are consistent with this possibility (Figure 9.4).

Thus, emerging genomics and bioinformatics technologies have revealed that enzymic degradation of heteroxylans is a much more complicated process, involving many more genes and enzymes, coupled with various post-translational modifications of the nascent enzymes, than previously thought.

(1,3;1,4)-β-Glucans

In comparison with the multiple enzymes required for the complete hydrolysis of heteroxylans, relatively few enzymes are needed to completely depolymerise (1,3;1,4)-β-glucan to glucose. In the case of barley, these include two family GH17 (1,3;1,4)-β-glucan endohydrolases (Woodward and Fincher 1982), several exo-hydrolases from the broad specificity family GH3 β-d-glucan glucohydrolase group of enzymes (Hrmova and Fincher 1998; Hrmova et al. 2002) and a range of family GH1 β-d-glucosidases that may or may not be specifically involved in the hydrolysis of oligosaccharides released by the endohydrolases. Crystal structures of representatives of each of these classes of enzymes have allowed their molecular mechanisms to be defined in detail at the atomic level. Cloned cDNAs and genes allowed the complete amino acid sequences of the enzymes to be defined (Fincher et al. 1986; Hrmova et al. 1996) and hence molecular biology technologies contributed to the resolution of their 3D structures via X-ray crystallography. The barley (1,3;1,4)-β-glucan endohydrolase isoenzyme EII adopts a typical $(\beta/\alpha)_8$ barrel conformation that is characterised by a broad substrate-binding groove that extends across the enzyme surface and binds the polysaccharide, and therefore enables hydrolysis of internal (1,4)-β-linkages (Varghese et al. 1994). This barley enzyme has the same specificity as those described above that are used to define the DP3:DP4 ratios of (1,3;1,4)-β-glucans. The structural biology analyses indicate that the (1,3;1,4)-β-glucan endohydrolases evolved from the more widely distributed (1,3)-β-glucan endohydrolases (Varghese et al. 1994).

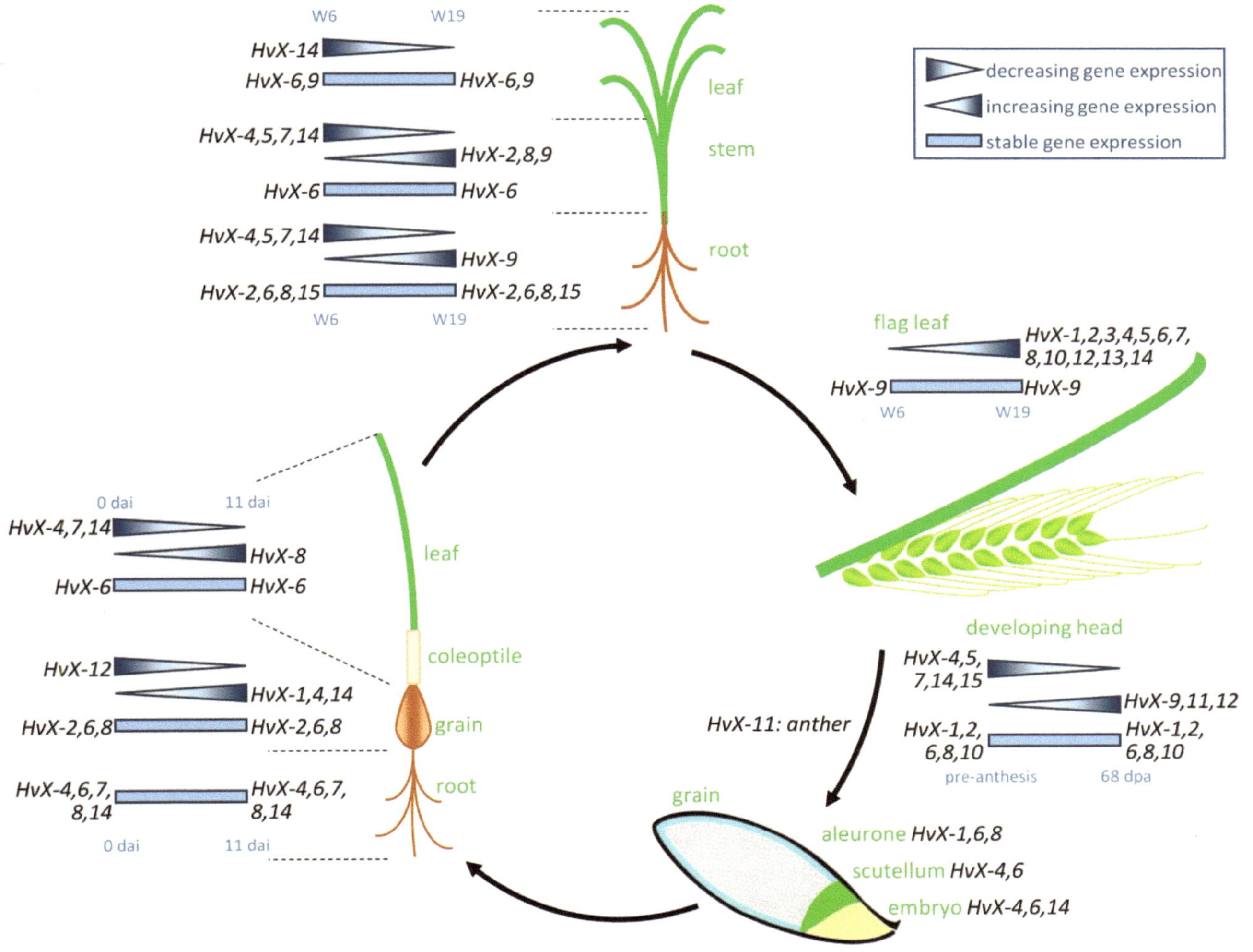

FIGURE 9.4 Transcript patterns of the 15 (1,4)-β-xylanase genes during barley development. This figure is from Betts et al. (2021). X-1 to X-15 are the designations of the xylanase genes; dai denotes 'days after imbibition'; W denotes week; dpa denotes days post-anthesis.

In contrast, the barley family GH3 β-d-glucan glucohydrolase isoenzyme ExoI is an exohydrolase in which the non-reducing terminus of the polysaccharide is bound in a short, dead-end pocket on the surface of the enzyme (Varghese et al. 1999). This enzyme has a broad substrate specificity insofar as it can hydrolyse (1,3)-, (1,4)- or (1,6)-β-glucosyl linkages (Hrmova et al. 1996). The broad specificity resides in the disposition of two tryptophan residues at the mouth of the substrate-binding pocket. These residues bind the penultimate glucosyl residue of the substrate and their relatively large surface area, and hydrophobic nature means that the positioning of the penultimate glucosyl residue, which depends on the linkage type, is somewhat flexible and does not need to be absolutely fixed; this binding flexibility explains the broad specificity of the enzyme (Varghese et al. 1999). In a recent breakthrough, Stratilová et al. (2020) demonstrated that this β-d-glucan glucohydrolase is capable of processive catalysis, whereby multiple hydrolytic events release the terminal glucosyl residue of the substrate without the enzyme losing contact with its oligo- or polymeric substrates. β-Glucosidases might also participate in the final degradation of (1,3;1,4)-β-oligoglucosides to glucose, but there are very large numbers of these enzymes in most plant tissues and the role of any particular enzyme in (1,3;1,4)-β-glucan depolymerisation has not been confirmed.

UNRAVELLING THE FUNCTIONS OF XTHS

The xyloglucan endo-transglycosylase/hydrolase (XTH) group of enzymes, which are members of the relatively diverse GH16 family (Lombard et al. 2014; Viborg et al. 2019), can be subdivided into xyloglucan endo-transglycosylases (XETs; EC 2.4.1.207) and xyloglucan endohydrolases (XEHs; EC 3.2.1.151) (Behar et al. 2018). The XTH enzymes derived their names from multiple early observations that the enzymes could hydrolyse xyloglucan and subsequently transfer the non-reducing terminal fragment of the hydrolytic products to another xyloglucan molecule or to xyloglucan oligosaccharides; these were XETs (Farkaš et al. 1992; Fry et al. 1992; Nishitani and Tominaga 1992). It was later found that the XTH enzymes only catalysed the hydrolysis of xyloglucans (XEHs). The different specificities reside in an approximately 70 amino acid COOH-terminal extension of the enzymes, where sequence differences determine whether the enzyme will be a hydrolase or a transglycosylase (Hrmova et al. 2009; Behar et al. 2018; Stratilová et al. 2020).

Given that the cell walls of barley contain little if any xyloglucan, it was somewhat of a surprise not only to find abundant levels of the XET enzyme in extracts of young barley seedlings (Hrmova et al. 2009) but also to detect highly abundant XET mRNA transcripts in many tissues and 36 XET or XET-like genes in the barley genome (Stratilová et al. 2020). Mohler et al. (2012) detected (1,3;1,4)-β-glucan:xyloglucan endotransglycosylase (MXE) activity in the horsetail fern (*Equisetum*) but not in extracts of barley seedlings. However, a series of investigations has suggested that the enzymes might not be acting exclusively on xyloglucans. It has been demonstrated, *in vitro*, that purified XETs from both barley and nasturtium can covalently link xyloglucan and cellulosic donor substrates to a variety of acceptor oligosaccharides derived from (1,3;1,4)-β-glucans, cellulose, xylans and pectic polysaccharides (Hrmova et al. 2007; Stratilová et al. 2019, 2020).

These observations strongly suggest that XETs in the grasses play an important role in the architecture and remodelling of cell walls, and that xyloglucans may not necessarily be involved in these processes. Perhaps, the enzymes are involved in the covalent cross-linking of at least some (1,3;1,4)-β-glucan, cellulose and even arabinoxylan. These hetero-transglycosylation cross-linking processes might play a role in toughening and strengthening the wall during normal growth and development, or in response to pathogen attack.

LONG- AND SHORT-TERM STORAGE ROLES FOR WALL POLYSACCHARIDES

In eudicots, cell wall (1,4)-β-mannans, which may contain (1,4)-β-glucosyl residues in their main chain backbones (glucomannans) and which also can be substituted with (1,6)-α-galactosyl residues (galactoglucomannans), are abundant in thick cell walls in the endosperm of seeds of the Leguminosae and several other plant families (Meier and Reid 1982; Buckeridge 2010). Enzymes and genes involved in both the synthesis and the degradation of these mannan storage polysaccharides following germination have been characterised extensively (Reid and Bewley, 1979). In addition, xyloglucans act as storage polysaccharides in the seeds of nasturtium, tamarind and related plants (Edwards et al. 1985). Walls in the cotyledons of legumes and related eudicots contain storage xylo-(1,4)-β-glucans and arabino-(1,4)-β-galactans (Crawshaw and Reid 1984; Tine et al. 2006; Buckeridge 2010). It is possible that these cell wall storage polysaccharides perform dual functions in the cell, namely, as structural components of the walls and as sources of monosaccharides for the developing seedling. Or perhaps, these polysaccharides do not have an obligatory structural role in the wall? In this connection, Sowinski et al. (2022) have shown that a mutant of Arabidopsis containing no xyloglucan in its walls has an essentially normal phenotype, but in an apparent compensation for the absence of xyloglucan, the mutant has enhanced levels of homogalacturonans and glucomannans in its walls.

The same phenomenon is observed in the grasses, where evidence is mounting that the (1,3;1,4)-β-glucans in the starchy endosperm of grain can provide a secondary source of metabolisable glucose during germination. Firstly, not all walls in different species of the Poaceae or in individual tissues of a single species contain (1,3;1,4)-β-glucan. This suggests that the polysaccharide is not essential for wall structure. Secondly, it was calculated long ago that the (1,3;1,4)-β-glucan in the endosperm of barley grain accounted for 18.5% of the total glucose available for seedling development after germination and that this was likely to be a significant component of the glucose used for seedling growth (Morrall and Briggs 1978).

The grain of *Brachypodium distachyon*, a grass species of the Pooideae subfamily of the Poaceae, consists of about 45% w/w (1,3;1,4)-β-glucan and only about 6% by weight of starch (Guillon et al. 2012). The former is located in thick cell walls of the starchy endosperm and appears to accumulate as a result of a reduced capacity for starch synthesis during grain development (Trafford et al. 2013). These values may be compared with grain (1,3;1,4)-β-glucan contents in barley and oats (4%–10%, w/w) and wheat, maize and sorghum (0.06%–1% w/w); starch contents in these grains are typically about 65% (Fincher and Stone 2004). Thus, the (1,3;1,4)-β-glucans are quantitatively more important than starch as a source of glucose for the developing *Brachypodium distachyon* seedling following germination of the grain. Members of the Bromeae tribe have starch:(1,3;1,4)-β-glucan ratios midway between these extremes (Trafford et al. 2013). The inverse relationship between starch and (1,3;1,4)-β-glucan contents was confirmed through targeted CRISP/Cas9 gene editing of barley *CslF* genes by Garcia-Gimenez et al. (2020).

In addition to their role in long-term carbohydrate storage, a possible role for (1,3;1,4)-β-glucans as a secondary, short-term source of metabolisable glucose came from light/dark experiments on young barley seedlings, where the (1,3;1,4)-β-glucan in walls of leaves disappeared when plants were placed in the dark but reappeared once the lights were turned on again (Roulin et al. 2002). Gibeaut and Carpita (1991) also reported the transient nature and turnover of (1,3;1,4)-β-glucans in maize coleoptiles. Might it be possible that other wall polysaccharides (such as xyloglucans and heteromannans) in leaves of the species mentioned above also act as a short-term storage of metabolisable sugars? The investigation of this possibility would only require exposure of vegetative tissues to simple light/dark intervals, accompanied by wall analyses of these tissues.

In barley, the use of wall-located (1,3;1,4)-β-glucans as a source of metabolisable glucose provides a number of potential advantages to the plant, especially when the deposition or mobilisation of carbohydrate reserves is required in a short time frame. For example, when unfavourable climatic conditions such as water stress occur during grain-filling, it would be advantageous for the plant to quickly store as much of the incoming carbohydrate as possible. A single (1,3;1,4)-β-glucan synthase enzyme would deposit the incoming glucose, derived from sucrose, directly into the cell wall and thereby bypass the more time-consuming starch synthesis process, which requires multiple enzymes and membrane translocations. This possibility is consistent with the data of Coles (1979), who observed elevated levels of (1,3;1,4)-β-glucan in developing barley grain placed under heat and moisture stress. Similarly, if the plant required extra glucose in a short period during normal growth and development or in periods of darkness, the plant could shortcut the relatively complex starch mobilisation process through the rapid depolymerisation of wall bound (1,3;1,4)-β-glucans to glucose; this process requires a single (1,3;1,4)-β-glucan endohydrolase and a single β-glucan exohydrolase.

WALLS AS A BARRIER TO PATHOGEN INVASION

Terrestrial plants are constantly under attack from fungal pathogens and the cell wall is one of the first lines of defence against such attacks. Although plants have evolved tough surface layers, these have naturally occurring pores (stomata) and in many cases they can be disrupted through a variety of mechanical or enzymic forces. At some stage, the penetrating pathogen will attempt to break through the walls into living cells. In one early defence response, the cell wall is reinforced through the deposition of dome-shaped papillae in the periplasmic space, near the site of attempted penetration by the pathogen. Papillae represent physical and chemical barriers that either prevent penetration or delay the infection process so that other defence strategies can be initiated (Zeyen and Bushnell 1979; Albersheim et al. 2011; Hückelhoven 2014). In some cases, papillae develop quickly enough to block fungal penetration, but in other cases the response appears to be too slow and the fungus successfully penetrates the cell wall. The major component of papillae has traditionally been identified as the (1,3)-β-glucan, callose, which is easily detected in tissue sections using the aniline blue fluorochrome (Stone et al. 1984; Jacobs et al. 2003). Autofluorescence studies suggested that phenolic acids might also be present (Aist 1976). It should be noted that callose is not normally considered to be an integral component of plant cell walls, but is rather deposited in the periplasmic space following mechanical perturbation of the plasma membrane–cell wall interface by osmotic, mechanical or other stresses, or is synthesised transiently at the cell plate during cell division (Clarke and Stone 1992).

Isolation of intact papillae for detailed compositional analysis remains technically difficult, but the emergence of new immunocytochemical technologies using monoclonal antibodies to detect a wide range of wall polysaccharides *in situ* (Knox 2008; Pattathil et al. 2010) allowed the composition of papillae to be described in much more detail (Chowdhury et al. 2014). Somewhat surprisingly the immunocytochemical experiments showed that when leaves of barley (*Hordeum vulgare*) were inoculated with conidia of the powdery mildew fungus (*Blumeria graminis* f. sp. *hordei*), the papillae that formed in the periplasmic space contained, in addition to callose, relatively high levels of cellulose and arabinoxylan (Figure 9.5); the latter appeared to be feruloylated (Chowdhury et al. 2014). Thus, cell wall polysaccharides were being deposited in the papillae. Effective papillae, which successfully blocked penetration by the fungus, contained higher concentrations of callose, cellulose and arabinoxylan than ineffective papillae, which were successfully penetrated by the fungal penetration peg

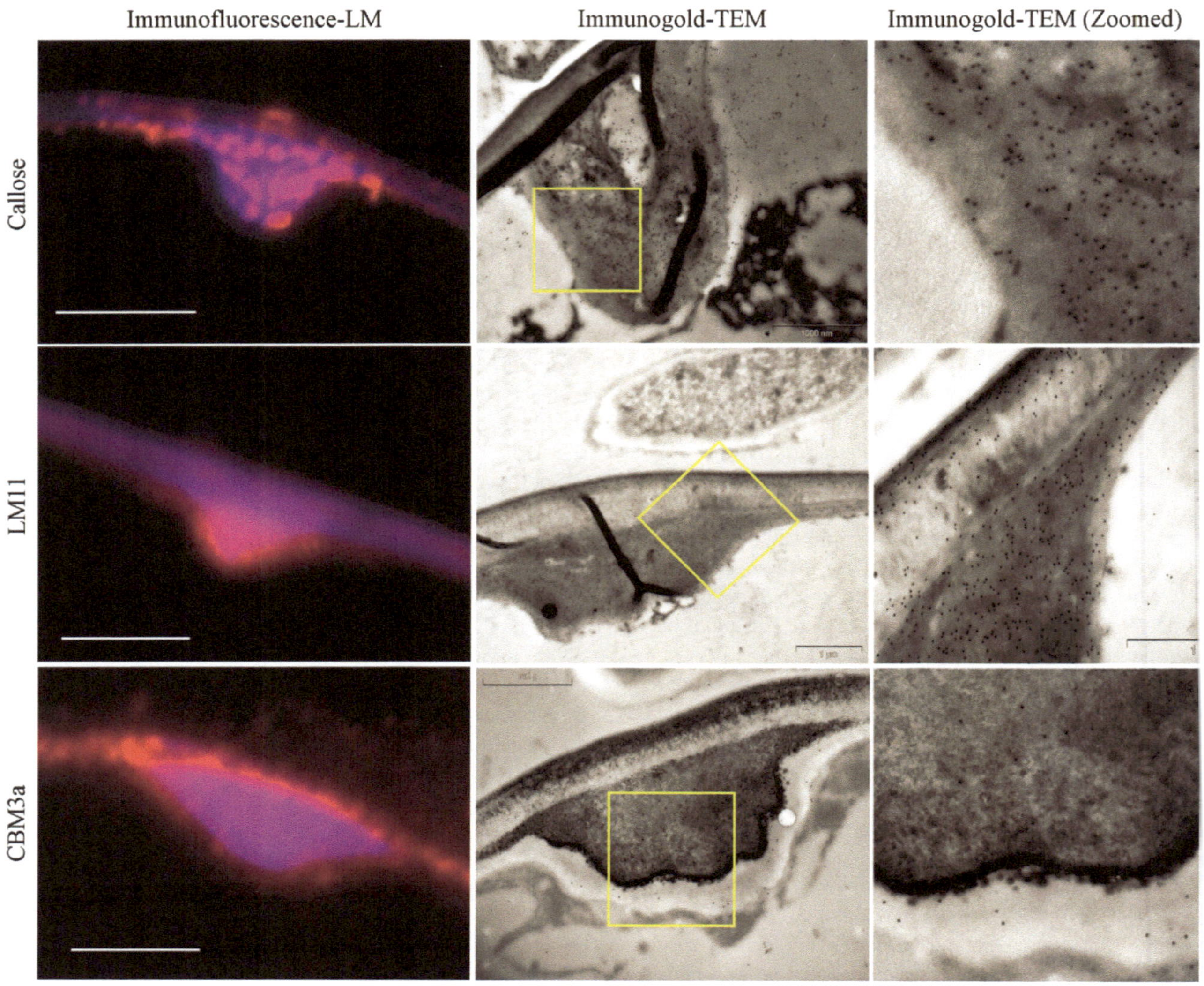

FIGURE 9.5 Deposition patterns of polysaccharides in papillae in barley. These data are related to the formation of effective and ineffective papillae during the barley/*Blumeria graminis* f. sp. *hordei* (Bgh) interaction. Transverse sections of barley papillae formed against Bgh at 24 hours after inoculation labelled with callose-, arabinoxylan- (LM11) and crystalline cellulose-specific probes. First column; red fluorescence, secondary antibodies attached to respective primary antibodies; blue fluorescence, phenolic acid-associated autofluorescence; scale bar = 5 µm. Second column, immunogold labelled transmission electron micrograph, scale bar = 1 µm. Third column, magnified view of boxed area from second column. The figure and legend are reproduced from Chowdhury et al. (2014).

(Figure 9.6; Chowdhury et al. 2014). Transcript profiling later showed that members of key GT43 and GT47 gene families reported to be involved in heteroxylan biosynthesis were up-regulated in the epidermal layer of barley leaves during powdery mildew infection (Chowdhury et al. 2017). Douchkov et al. (2016) reported that, in transgenic barley plants, silencing the *HvCslD2* gene demonstrated a role for this gene as an important defence response to both quantitative and non-host resistance. The discovery of cellulose and arabinoxylans as constituents of effective papillae provides new target genes for the enhancement of disease resistance in important plant species.

High-throughput transcriptomics analyses were also central to recent advances in our understanding of the role of cell walls in plant–pathogen interactions. For example, Aditya et al. (2015) showed that infection of barley roots with the cereal cyst nematode (CCN; *Heterodera avenae*) resulted in the deposition of higher levels of (1,3;1,4)-β-glucan in cell walls around the feeding sites in resistant varieties, compared with susceptible varieties, while callose levels were similar in both susceptible and resistant varieties. These higher levels of (1,3;1,4)-β-glucan were associated with increases in *HvCslF10* transcripts and decreases in *HvCslF3* and *HvCslF9* transcripts in the resistant varieties (Aditya et al. 2015). However, the precise roles of these transcripts and (1,3;1,4)-β-glucan, and the molecular mechanisms underlying the different responses to nematode attack, have not yet been defined.

The sites of synthesis of the various polysaccharides found in papillae deposited in response to pathogen attack remain a subject for future research. The callosic and cellulosic components of the papillae can be synthesised at

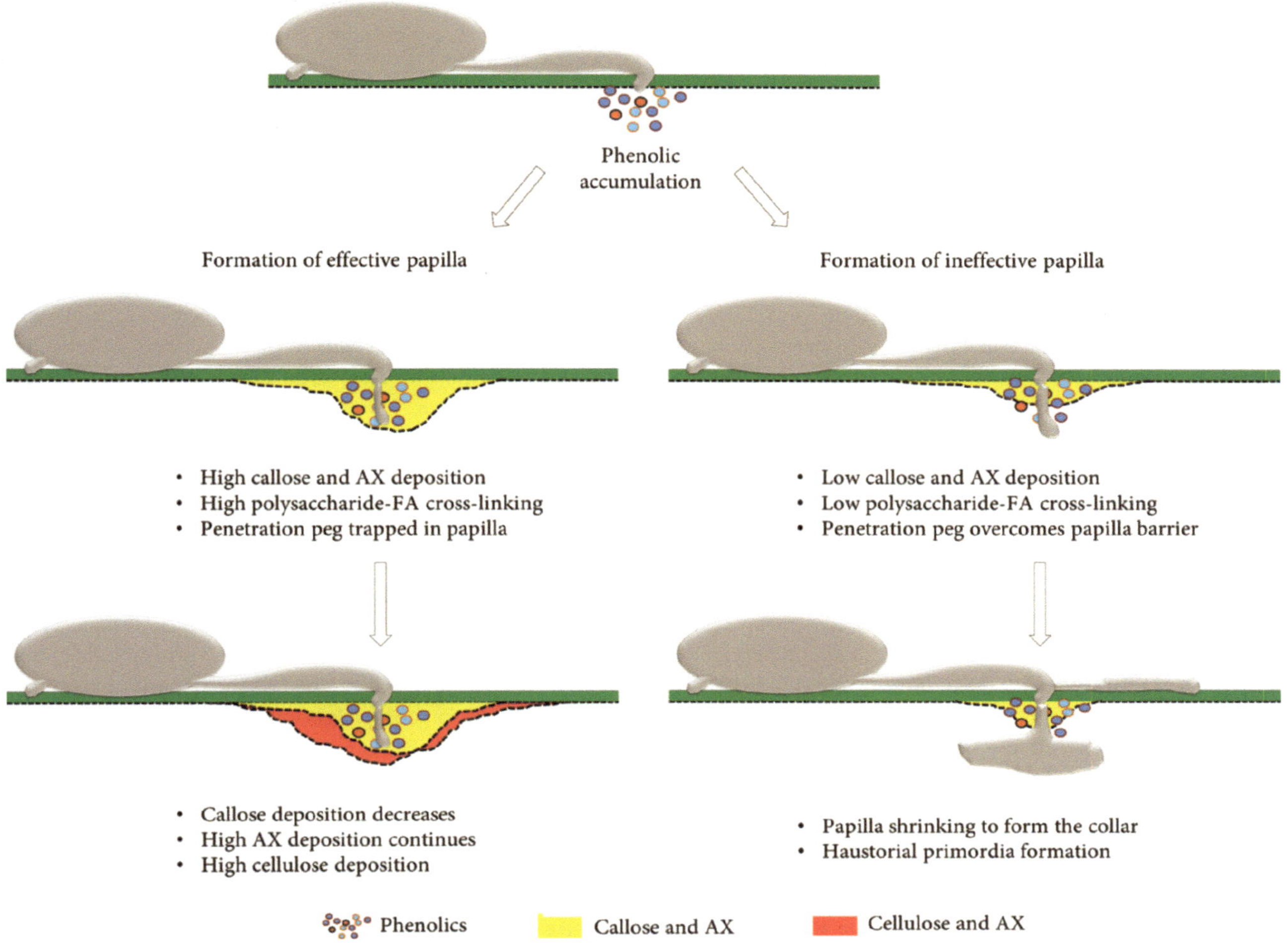

FIGURE 9.6 Model of polysaccharide and phenolic acid deposition in effective (EP) and ineffective (IP) papillae following Inoculation of young barley leaves with *Blumeria graminis* f. sp. *hordei*. Reproduced from Chowdhury et al. (2014). AX denotes arabinoxylan, FA denotes ferulic acid.

the plasma membrane, presumably in response to local signals. But are the arabinoxylans delivered by relatively slow, conventional Golgi vesicle secretory pathways, or via pre-existing xylan synthases at the plasma membrane? In a more general sense, Houston et al. (2016) proposed that cell wall remodelling during both abiotic and biotic stresses involve a common set of genes and transcription factors, and have presented a comprehensive list of genes that might be involved in wall remodelling during pathogen attack and/or abiotic stress. For example, GT61 genes, which have been implicated in arabinoxylan synthesis, and GT75 genes, which encode UDP-Ara mutases that generate UDP-arabinofuranose for arabinoxylan synthesis, are believed to be broadly involved in defence responses (Houston et al. 2016).

EMERGING TRANSLATIONAL TECHNOLOGIES FOR WALL POLYSACCHARIDE APPLICATIONS

Cell walls from the grasses are attracting a good deal of interest for the production of renewable biofuels and for their role in human health and nutrition. Thus, cereal crop residues and specialist grasses are being used for bioethanol and other biofuel production, as political pressure on phasing out the use of fossil fuels gains pace around the world. In these plant materials, cell wall components represent the most important energy source, whether they are used for bioethanol production or for production of oils through technologies such as hydrothermal liquefaction. These applications of cell walls are described in detail in other chapters of this book.

In human health and nutrition applications, the arabinoxylans and the (1,3;1,4)-β-glucans from cell walls of cereal grains are not digested by human small intestinal enzymes and are therefore important components of dietary fibre. Dietary fibre, especially soluble dietary fibre, is now recognised as an effective dietary component for lowering the risk of contracting many of the world's most serious human diseases, including colorectal cancer, cardiovascular disease, type II diabetes and diverticulitis (Collins et al. 2010; Burton and Fincher 2014). Properties of non-cellulosic cell wall polysaccharides that are implicated in improving human health include their fine structure and associated physicochemical properties, in particular their solubility and viscosity in aqueous media. Thus, the two most important wall polysaccharides for human health in cereal grains are the (1,3;1,4)-β-glucans and the arabinoxylans. As discussed above, the solubilities of these polysaccharide are determined in part by the degree of substitution of arabinoxylans or by the DP3:DP4 ratios of (1,3;1,4)-β-glucans.

Experiments with pigs fed diets enriched in soluble arabinoxylan from wheat endosperm showed that this material provides health benefits associated with the efficiency of digestion and satiety; these are important for prevention of obesity and related health problems (Zhang et al. 2015). In the case of (1,3;1,4)-β-glucans, overexpression of the barley *CslF6* gene resulted in an increase of more than 50% in dietary fibre (Burton et al. 2011).

In another application, high levels of (1–3,1–4)-β-glucans in extracts of malted barley are traditionally viewed as undesirable in the brewing industry, where they are associated with slow filtration rates and with the formation of hazes in the final beer product (Bamforth 2009). The levels of (1,3;1,4)-β-glucans can be either increased or decreased in barley grain and their DP3:DP4 ratios can be altered, depending on the required outcome and application (Burton et al. 2011; Jobling 2015; Dimitroff et al. 2016). It is also feasible to manipulate the composition and chemistry of walls in the vegetative tissues of grasses, with a view to maximising yields and speed of biofuel production from crop residues.

Knowledge of the fine structures of these wall polysaccharides, coupled with the identification of genes encoding enzymes involved in their metabolism and with transformation technologies, allow us to manipulate their levels and properties in transgenic plants. New CRISPR/Cas9-based gene-editing technologies are being increasingly used for gene replacement and site-specific mutagenesis of selected genes in plants, including the grasses (Zhang et al. 2020; Tang et al. 2020). As noted above, Garcia-Gimenez et al. (2020) used this technology to manipulate the levels of (1,3;1,4)-β-glucans in barley grain.

However, translation of this increasing cell wall polysaccharide knowledge into commercial processes or valuable crop species for the realisation of their potential benefits is not always straightforward. For example, the release of genetically modified (GM) crop plants faces time-consuming and expensive regulatory barriers in many countries and the transgenic material can seldom be generated in elite varieties, because many valuable varieties are notoriously difficult to grow in culture and hence to transform. Despite the promise of the CRISPR/Cas9 technologies, they have been plagued by off-target edits (Zhang et al. 2020), and they usually require a transformation step and are classified as GM in the European Union (Callaway 2018) and other jurisdictions.

A new technology has the potential to greatly facilitate the translation of our knowledge of cell wall polysaccharides of the grasses into valuable new crops. The method is termed FIND-IT ('Fast Identification of Nucleotide variants by DigITal PCR') and allows the ultrafast mining of genome diversity in any plant or microorganism (Knudsen et al. 2022). The genomic diversity can result from natural variation found in germplasm collections or elite breeding lines, or can be induced by chemical mutagens. The technology is based upon droplet digital polymerase chain reaction (ddPCR), which is 1,000-fold more sensitive than conventional PCR, and allows libraries of 500,000 variants to be screened for a specific, predetermined (targeted) variant within 10 working days, to single nucleotide resolution. The technology has been validated through the rapid isolation of targeted gene knockouts, targeted variants with specified amino acid changes in their encoded proteins and targeted variants that alter gene expression levels through changes in microRNA (miRNA) and transcription factor binding sites (Knudsen et al. 2022). The technology requires no cloning or transformation and isolated variants are therefore not classified as GM material. Because no tissue culture or transformation steps are required, the technology can be applied to any elite crop variety. The capability of the procedure to rapidly screen very large populations means that, in the case of chemically induced variant libraries, relatively gentle mutagenesis protocols can be used to reduce the mutation load of the libraries and this in turn minimises or completely eliminates the requirement for subsequent backcrossing in crop breeding programmes. The authors have already generated large variant libraries in barley, wheat, rapeseed, oat, yeast and bacteria (Knudsen et al. 2022). Although the examples chosen by Knudsen et al. (2022) were focused on crop yield improvements necessary to feed a world population of 9 billion by 2050, they also included examples of variants of the barley *CslF6* (1,3;1,4)-β-glucan synthase gene that have the potential to improve the production efficiency of the malting and brewing industries.

Thus, if we wish to reduce cross-linking of wall components to facilitate rapid wall degradation for biofuel production, if we want to increase the proportion of soluble polysaccharides to enhance human health benefits, if we want to increase the proportion of insoluble polysaccharides to reduce problems caused by soluble components in the brewing process, or if we want to alter wall composition and structure in other ways, our existing knowledge of the biochemistry and genetics of wall constituents can be combined with this new ultrafast variant generation and screening technology to produce translational benefits in a very short time frame. Furthermore, the technology can be used to generate a large number of base substitutions in a single targeted gene and to quickly isolate variants for analysis. This 'saturation' approach targeted to variants of a predetermined, single gene, for example, could be used to unravel the molecular mechanisms of action of the CslF6 (1,3;1,4)-β-glucan synthase enzyme.

In a more general sense, Langridge and Waugh (2019) estimate that there are more than 7 million crop plant accessions in global gene banks, harbouring both naturally occurring and induced genetic variation that is considered a key resource for crop improvement in the immediate future. However, devising rapid strategies for efficiently screening out valuable genes or alleles and hence harnessing the potential of the genetic diversity embodied in these collections have frustrated crop improvement scientists. This new FIND-IT technology will remove this limitation and greatly accelerate the meeting of crop improvement targets required to feed the burgeoning world population in the face of changing climatic conditions (Knudsen et al. 2022).

UNANSWERED QUESTIONS AND FUTURE DIRECTIONS

The steady and at times spectacular progress in our understanding of the biochemistry, cell biology and genetics of non-cellulosic cell wall polysaccharides in the grasses over the last 50 years has been enabled by a parallel stream of emerging technologies and their progressive refinement. Nevertheless, a number of unanswered questions remain and are likely to form the focus of our research activities in the immediate future. The background to these unanswered questions has been described in detail in the sections above, but the questions are listed in summary below:

- Have we identified all the genes that mediate the processive biosynthesis of the (1,4)-β-xylan backbone of cereal arabinoxylans? Might some be members of the GT2 family of glycosyl transferases?
- What roles do the abundant XET enzymes play in grass species that contain very low levels of xyloglucans in their walls? Are they involved in cross-linking and remodelling of walls? Is covalent cross-linking of different wall polysaccharides more prevalent than previously thought?
- What molecular mechanisms control how (1,3;1,4)-β-glucan synthase enzymes incorporate (1,3)- and (1,4)-β-glucosyl residues into the elongating (1,3;1,4)-β-glucan chain?
- Are the current inconsistencies in our understanding of the subcellular site of synthesis of (1,3;1,4)-β-glucans explained by the synthesis of these and other non-cellulosic wall polysaccharides both in the Golgi apparatus and at the plasma membrane?
- Why do the large number of (1,4)-β-xylan endohydrolases all have a relatively complex modular structure that suggests cell wall binding and progressive release and activation of the enzymes?
- Can we use currently emerging technologies to exploit natural or induced genetic variation for the improvement and translation of our knowledge of cell wall biology?

Although these questions have challenged us for many years (Burton and Fincher 2012), they can now be considered as a list of testable hypotheses that, together with the latest technologies, can guide our research activities in the future. Increasing demand for renewable biofuel production and our increased interest in dietary contributions to human health, together with the imperative of feeding the burgeoning global population, will help drive us towards an increasing understanding of cell wall biology and its applications.

ACKNOWLEDGEMENTS

Our work has been supported continuously by the Australian Research Council and the Grains Research and Development Corporation over the period covered in this review. The experimental work has been performed by many bright and enthusiastic postgraduate students, postdoctoral scientists and senior academic colleagues. In particular, we acknowledge the exceptional technical skills and contributions of Rachel Burton, Maria Hrmova and Neil Shirley, and the key role played by Bruce Stone in training many of our current cell wall biologists in Australia. Our interactions with these groups and individual scientists culminated in the formation of the Australian Research Council's Centre of Excellence in Plant Cell Walls in 2011, through which we developed exciting research programmes with Tony Bacic, Monika Doblin and their colleagues in Melbourne, and Mike Gidley's team at the University of Queensland.

REFERENCES

Aditya J, Lewis J, Shirley NJ, Tan H-T, Henderson M, Fincher GB, Burton RA, Mather DE, Tucker MR (2015) The dynamics of cereal cyst nematode infection differ between susceptible and resistant barley cultivars and lead to changes in (1,3;1,4)-β-glucan levels and *HvCslF* gene transcript abundance. *New Phytologist* 207: 135–147.

Aist JR (1976) Papillae and related wound plugs of plant-cells. *Annual Review of Phytopathology* 14: 145–163.

Albersheim P, Darvill A, Roberts K, Sederoff R, Staehelin A (2011) Cell walls and plant-microbe interactions (2011) In: *Plant Cell Walls: From Chemistry to Biology*, pp. 319–363. New York, NY: Garland Science, Taylor & Francis.

Amor Y, Haigler CH, Johnson S, Wainscott M, Delmer DP (1995) A membrane-associated form of sucrose synthase and its potential role in synthesis of cellulose and callose in plants. *Proceedings of the National Academy of Sciences (USA)* 92: 9353–9357.

Amos RA, Mohnen D (2019) Critical review of plant cell wall matrix polysaccharide glycosyltransferase activities verified by heterologous protein expression. *Frontiers of Plant Science* 10: 915.

Anderson MA, Stone BA (1975) A new substrate for investigating the specificity of β-glucan hydrolases. *FEBS Letters* 52: 202–207.

Arioli T, Peng L, Betzner AS, Burn J, Wittke W, Herth W, Camilleri C, Hofte H, Plazinski J, Birch R, Cork A, Glover J, Redmond J, Williamson RE (1998) Molecular analysis of cellulose biosynthesis in Arabidopsis. *Science* 279: 717–720.

Bacic A, Stone BA (1981) Chemistry and organization of aleurone cell wall components from wheat and barley. *Functional Plant Biology* 8: 475–495.

Bamforth C (2009) *Beer: Tap into the Art and Science of Brewing*. Oxford: Oxford University Press.

Banik M, Garrett TPJ, Fincher GB (1996) Molecular cloning of cDNAs encoding (1→4)-β-xylan endohydrolases from the aleurone layer of germinated barley (*Hordeum vulgare*). *Plant Molecular Biology* 31: 1163–1172.

Becker GE, Hui PA, Albersheim P (1964) Synthesis of extracellular polysaccharide by suspensions of *Acer pseudoplatanus* cells. *Plant Physiology* 39: 913–920.

Behar H, Graham SW, Brumer H (2018) Comprehensive cross-genome survey and phylogeny of glycoside hydrolase family 16 members reveals the evolutionary origin of EG 16 and XTH proteins in plant lineages. *Plant Journal* 95: 1114–1128.

Benjavongkulchai E, Spencer MS (1986) Purification and characterization of barley aleurone xylanase. *Planta* 169: 415–419.

Bethke PC, Lonsdale JE, Fath A, Jones RL (1999) Hormonally regulated programmed cell death in barley aleurone cells. *Plant Cell*, 11, 1033–1045.

Betts NS, Collins HM, Shirley NJ, Cuesta-Seijo JA, Schwerdt JG, Phillips RJ, Finnie C, Fincher GB, Dockter C, Skadhauge B, Bulone V (2021) Identification and spatio-temporal expression analysis of barley genes that encode putative modular xylanolytic enzymes. *Plant Science* 308: 110792.

Biswal AK, Atmodjo MA, Li M, Baxter HL, Yoo CG, Pu Y, Lee YC, Mazarei M, Black IM, Zhang JY, Ramanna H (2018) Sugar release and growth of biofuel crops are improved by down-regulation of pectin biosynthesis. *Nature Biotechnology* 36: 249–257.

Biswal AK, Tan L, Atmodjo MA, DeMartini J, Gelineo-Albersheim I, Hunt K, Black IM, Mohanty SS, Ryno D, Wyman CE, Mohnen D (2017) Comparison of four glycosyl residue composition methods for effectiveness in detecting sugars from cell walls of dicot and grass tissues. *Biotechnology for Biofuels and Bioproducts* 10: 182.

Björndal H, Hellerqvist CG, Lindberg B, Svensson S (1970). Gas-liquid chromatography and mass spectrometry in methylation analysis of polysaccharides. *Angewandte Chemie International Edition in English*, 9: 610–619.

Bolwell GP, Northcote DH (1983) Arabinan synthase and xylan synthase activities of Phaseolus vulgaris. Subcellular localization and possible mechanism of action. *Biochemical Journal* 210: 497–507.

Brown DM, Zhang Z, Stephens E, Dupree P, Turner SR (2009). Characterization of *IRX10* and *IRX10-like* genes reveals an essential role in glucuronoxylan biosynthesis in Arabidopsis. *Plant Journal* 57: 732–746.

Buckeridge MS (2010) Seed cell wall storage polysaccharides: models to understand cell wall biosynthesis and degradation. *Plant Physiology* 154: 1017–1023.

Buckeridge MS, Vergara CE, Carpita NC (1999) The mechanism of synthesis of a mixed-linkage (1,3);(1,4)-β-D-glucan in maize. Evidence for multiple sites of glucosyl transfer in the synthase complex. *Plant Physiology* 120: 1105–1116.

Buliga GS, Brandt DA, Fincher GB (1986) The sequence statistics and solution conformation of a barley (1→3,1→4)-β-D-glucan. *Carbohydrate Research* 157: 139–156.

Bulone V, Schwerdt JG, Fincher GB (2019) Co-evolution of enzymes involved in plant cell wall metabolism in the grasses. *Frontiers in Plant Science* 10: 1009.

Burton RA, Collins HM, Kibble NAJ, Smith JA, Shirley NJ, Jobling SA, Henderson M, Singh RR, Pettolino F, Wilson SM, Bird AR, Topping DL, Bacic A, Fincher GB (2011) Over-expression of specific *HvCslF* cellulose synthase-like genes in transgenic barley increases the levels of cell wall (1,3;1,4)-β-D-glucans and alters their fine structure. *Plant Biotechnology Journal* 9: 117–135.

Burton RA, Fincher GB (2009) (1,3;1,4)-β-D-Glucans in cell walls of the Poaceae, lower plants and fungi: a tale of two linkages. *Molecular Plant* 2: 873–882.

Burton RA, Fincher GB (2012) Current challenges in cell wall biology in the cereals and grasses. *Frontiers in Plant Physiology* 3: 130.

Burton RA, Fincher GB (2014) Plant cell wall engineering: applications in biofuel production and improved human health. *Current Opinions in Biotechnology* 26: 79–84.

Burton RA, Gidley MJ, Fincher GB (2010) Heterogeneity in the chemistry, structure and function of plant cell walls. *Nature Chemical Biology* 6: 724–732.

Burton RA, Jobling SA, Harvey AJ, Shirley NJ, Mather DE, Bacic A, Fincher GB (2008) The genetics and transcriptional profiles of the cellulose synthase-like *HvCslF* gene family in barley (*Hordeum vulgare* L.). *Plant Physiology* 146: 1821–1833.

Burton RA, Wilson SM, Hrmova M, Harvey AJ, Shirley NJ, Medhurst A, Stone BA, Newbigin EJ, Bacic A, Fincher GB (2006) Cellulose synthase-like *CslF* genes mediate the synthesis of cell wall (1,3;1,4)-β-D-glucans. *Science* 311: 1940–1942.

Callaway E (2018) CRISPR plants now subject to tough GM laws in European Union. *Nature* 560: 16–17.

Callaway E (2021) DeepMind's AI predicts structures for a vast trove of proteins. *Nature* 595: 635.

Carpita NC, Gibeaut DM (1993). Structural models of primary cell walls in flowering plants: consistency of molecular structure with the physical properties of the walls during growth. *The Plant Journal* 3: 1–30.

Carpita NC, McCann MC (2010) The maize mixed-linkage (1→3),(1→4)-β-D-glucan polysaccharide is synthesized at the Golgi membrane. *Plant Physiology* 153: 1362–1371.

Caspers MP, Lok F, Sinjorgo K, Van Zeijl MJ, Nielsen KA, Cameron-Mills V (2001) Synthesis, processing and export of cytoplasmic endo-β-1,4-xylanase from barley aleurone during germination. *Plant Journal* 26: 191–204.

Chateigner-Boutin A-L, Bouchet B, Alvarado C, Bakan B, Guillon F (2014) The wheat grain contains pectic domains exhibiting specific spatial and development-associated distribution. *PLoS One* 9: e89620.

Chen J, Zeng B, Zhang M, Xie S, Wang G, Hauck A, Lai J (2014) Dynamic transcriptome landscape of maize embryo and endosperm development. *Plant Physiology* 166: 252–264.

Chowdhury J, Henderson M, Schweizer P, Burton RA, Fincher GB, Little A (2014) Differential accumulation of callose, arabinoxylan and cellulose in nonpenetrated versus penetrated papillae on leaves of barley infected with *Blumeria graminis* f. sp. *Hordei*. *New Phytologist* 204: 650–660.

Chowdhury J, Lueck S, Rajaraman J, Douchkov D, Shirley NJ, Schwerdt JG, Schweizer P, Fincher GB, Burton RA, Little A (2017) Altered expression of genes implicated in xylan biosynthesis affects penetration resistance against powdery mildew. *Frontiers in Plant Science*. doi: 10.3389/fpls.2017.00445.

Ciucanu I, Kerek F (1984) A simple and rapid method for the permethylation of carbohydrates. *Carbohydrate Research* 131: 209–217.

Clarke AE, Stone BA (1992) *Chemistry and Biology of 1,3-β-Glucans*. Victoria, Australia: La Trobe University Press.

Cocuron JC, Lerouxel O, Drakakaki G, Alonso AP, Liepman AH, Keegstra K, Raikhel N, Wilkerson CG (2007) A gene from the cellulose synthase-like C family encodes a beta-1,4 glucan synthase. *Proceedings of the National Academy of Sciences* 104: 8550–8555.

Coles G (1979) Relationship of mixed-link beta-glucan accumulation to accumulation of free sugars and other glucans in the developing barley endosperm. *Carlsberg Research Communications* 44: 439–453.

Collins HM, Burton RA, Topping DL, Liao M-L, Bacic A, Fincher GB (2010) Variability in the fine structures of non-cellulosic cell wall polysaccharides from cereal grains: potential importance in human health and nutrition. *Cereal Chemistry* 87: 272–282.

Crawshaw LA, Reid JSG (1984) Changes in cell-wall polysaccharides in relation to seedling development and the mobilisation of reserves in the cotyledons of *Lupinus angustifolius* cv. Unicrop. *Planta* 160: 449–454.

Dalessandro G, Northcote DH (1981) Xylan synthetase activity in differentiated xylem cells of sycamore trees (*Acer pseudoplatanus*). *Planta* 151: 53–60.

De Vries GE (2000) Monsanto shares rice genome data. *Trends in Plant Science* 5: 277.

Dhugga KS (2012) Biosynthesis of non-cellulosic polysaccharides of plant cell walls. *Phytochemistry* 74: 8–19.

Dhugga KS, Barreiro R, Whitten B, Stecca K, Hazebroek J, Randhawa GS, Dolan M, Kinney AJ, Tomes D, Nichols S, Anderson P (2004) Guar seed beta-mannan synthase is a member of the cellulose synthase super gene family. *Science* 303: 363–366.

Dimitroff G, Little A, Lahnstein J, Schwerdt JG, Shirley NJ, Burton RA, Fincher GB (2016) (1,3;1,4)-β-Glucan biosynthesis by the CslF6 enzyme: Position and flexibility of catalytic residues influence product fine structure. *Biochemistry* 55: 2054–2061.

Doblin MS, De Melis L, Newbigin E, Bacic A, Read SM (2001). Pollen tubes of *Nicotiana alata* express two genes from different beta-glucan synthase families. *Plant Physiology* 125: 2040–2052.

Doblin MS, Pettolino F, Wilson SM, Campbell R, Burton RA, Fincher GB, Newbigin E, Bacic A (2009) A barley cellulose synthase-like *CSLH* gene mediates (1,3;1,4)-β-D-glucan synthesis in transgenic Arabidopsis. *Proceedings of the National Academy of Sciences USA* 106: 5996–6001.

Douchkov D, Lueck S, Hensel G, Kumlehn J, Rajaraman J, Johrde A, Doblin MS, Beahan CT, Kopischke M, Fuchs R, Lipka V, Niks RE, Bulone V, Chowdhury J, Little A, Burton RA, Bacic A, Fincher GB, Schweizer P (2016) The barley (*Hordeum vulgare*) *cellulose synthase-like D2* gene (*HvCslD2*) mediates penetration resistance to host-adapted and nonhost isolates of the powdery mildew fungus. *New Phytologist* 212: 421–433.

Eder M, Tenhaken R, Driouich A, Lütz-Meindl U (2008). Occurrence and characterization of arabinogalactan-like proteins and hemicelluloses in *Micrasterias* (streptophyta). *Journal of Phycology* 44: 1221–1234.

Edwards M, Dea ICM, Bulpin PV, Reid JSG (1985) Xyloglucan (amyloid) mobilisation in the cotyledons of *Tropaeolum majus* L. seeds following germination. *Planta* 163: 133–140.

Farkaš V, Sulová Z, Stratilová E, Hanna R, Maclachlan GA (1992) Cleavage of xyloglucan by nasturtium seed xyloglucanase and transglycosylation to xyloglucan subunit oligosaccharides. *Archives of Biochemistry and Biophysics* 298: 365–370.

Favery B, Ryan E, Foreman J, Linstead P, Boudonck K, Steer M, Shaw P, Dolan L (2001) KOJAK encodes a cellulose synthase-like protein required for root hair cell morphogenesis in Arabidopsis. *Genes and Development* 15: 79–89.

Fincher GB (1975) Morphology and chemical composition of barley endosperm cell walls. *Journal of the Institute of Brewing* 81: 116–122.

Fincher GB (1989) Molecular and cellular biology associated with endosperm mobilization in germinated cereal grains. *Annual Review of Plant Physiology and Plant Molecular Biology* 40: 305–346.

Fincher GB, Lock PE, Morgan MM, Lingelbach K, Wettenhall REH, Mercer JFB, Brandt A, Thomsen KK (1986) Primary structure of the (1→3,1→4)-β-D-glucan 4-glucanohydrolase from barley aleurone. *Proceedings of the National Academy of Sciences USA* 83: 2081–2085.

Fincher GB, Stone BA (2004) Chemistry of non-starch polysaccharides from cereal grains. In: *Encyclopedia of Grain Science* (Wrigley C, Corke H, Walker CE, eds), vol. 1, pp. 206–223. Elsevier Academic Press.

Fry SC, Nesselrode BHWA, Miller JG, Mewburn BR (2008) Mixed-linkage (1,3;1,4)-β-D-glucan is a major hemicellulose of *Equisetum* (horsetail) cell walls. *New Phytologist* 179: 104–115.

Fry SC, Smith R, Renwick K, Martin D, Hodge S, Matthews K (1992) Xyloglucan endotransglycosylase, a new wall-loosening enzyme activity from plants. *Biochemical Journal* 282: 821–828.

Gale MD, Devos KM (1998) Plant comparative genetics after 10 years. *Science* 282: 656–659.

Garcia-Gimenez G, Barakate A, Smith P, Stephens J, Khor S, Doblin M, Hao P, Bacic A, Fincher GB, Burton RA, Waugh R, Tucker M, Houston K (2020) Targeted mutation of barley (1,3;1,4)-β-glucan synthases reveals complex relationships between the storage and cell wall polysaccharide content. *Plant Journal* 104: 1009–1022.

Gibeaut DM, Carpita NC (1991) Tracing cell wall biogenesis in intact cells and plants: Selective turnover and alteration of soluble and cell wall polysaccharides in grasses. *Plant Physiology*, 97: 551–561.

Gibeaut DM, Pauly M, Bacic A, Fincher GB (2005). Changes in cell wall polysaccharides in developing barley (*Hordeum vulgare*) coleoptiles. *Planta* 221: 729–738.

Guillon F, Larré C, Petipas F, Berger A, Moussawi J, Rogniaux H, Santoni A, Saulnier L, Jamme F, Miquel M, Lepiniec L, Dubreucq B (2012) A comprehensive overview of grain development in *Brachypodium distachyon* variety Bd21. *Journal of Experimental Botany* 63: 739–755.

Hakomori S-I (1964) A rapid permethylation of glycolipid and polysaccharide catalyzed by methylsulfinyl carbanion in dimethyl sulfoxide. *Journal of Biochemistry* 55: 205–208.

Han F, Ullrich SE, Chirat S, Menteur S, Jestin L, Sarrafi A, Hayes PM, Jones BL, Blake TK, Wesenberg DM, Kleinhofs A, Kilian A (1995) Mapping of β-glucan content and β-glucanase activity loci in barley grain and malt. *Theoretical and Applied Genetics* 91: 921–927.

Harris PJ, Fincher GB (2009) Distribution, fine structure and function of (1,3;1,4)-β-glucans in the grasses and other taxa. In: *Chemistry, Biochemistry, and Biology of (1,3)-β-Glucans and Related Polysaccharides* (Stone BA, Bacic T, Fincher GB, eds), pp. 621–654. Philadelphia, PA: Elsevier Inc.

Harris PJ, Hartley RD (1976) Detection of bound ferulic acid in cell walls of the Gramineae by ultraviolet fluorescence microscopy. *Nature* 259: 508–510.

Hazen SP, Scott-Craig JS, Walton JD (2002) Cellulose synthase-like genes of rice. *Plant Physiology* 128: 336–340.

Höfte H, Voxeur A (2017) Plant cell walls. *Current Biology* 27(17): R865–870.

Houston K, Tucker MR, Chowdhury J, Shirley NJ, Little A (2016) The plant cell wall: a complex and dynamic structure as revealed by the responses of genes under stress conditions. *Frontiers of Plant Science* 7: 984.

Hrmova M, De Gori R, Smith BJ, Fairweather JK, Driguez H, Varghese JN, Fincher GB (2002) Structural basis for broad substrate specificity in higher plant β-D-glucan glucohydrolases. *Plant Cell* 14: 1033–1052.

Hrmova M, Farkas V, Harvey AJ, Lahnstein J, Wischmann B, Kaewthai N, Ezcurra I, Teeri TT, Fincher GB (2009) Substrate specificity and catalytic mechanism of a xyloglucan xyloglucosyl transferase HvXET6 from Barley (*Hordeum vulgare* L.). *FEBS Journal* 276: 437–456.

Hrmova M, Farkas V, Lahnstein J, Fincher GB (2007) A barley xyloglucan xyloglucosyl transferase covalently links xyloglucan, cellulosic substrates and (1,3;1,4)-β-D-glucans. *Journal of Biological Chemistry* 282: 12951–12962.

Hrmova M, Fincher GB (1998) Barley β-D-glucan exohydrolases. Substrate specificity and kinetic properties. *Carbohydrate Research* 305: 209–221.

Hrmova M, Garrett TPJ, Fincher GB (1995) Subsite affinities and disposition of catalytic amino acids in the substrate-binding region of barley 1,3-ß-glucanases: implications in plant-pathogen interactions. *Journal of. Biological Chemistry* 270: 14556–14563.

Hrmova M, Harvey AJ, Wang J, Shirley NJ, Jones GP, Stone BA, Høj PB, Fincher GB (1996) Barley β-D-glucan exohydrolases with β-glucosidase activity. Purification, characterization and determination of primary structure from a cDNA clone. *Journal of Biological Chemistry* 271: 5277–5286.

Hückelhoven R (2014) The effective papilla hypothesis. *New Phytologist* 204: 438–440.

Iiyama K, Lam TBT, Stone BA (1990). Phenolic acid bridges between polysaccharides and lignin in wheat internodes. *Phytochemistry* 29: 733–737.

International Rice Genome Sequencing Project, Sasaki T (2005) The map-based sequence of the rice genome. *Nature* 436: 793–800. https://doi.org/10.1038/nature03895

International Wheat Genome Sequencing Consortium (IWGSC), Appels R, Eversole K, Stein N, Feuillet C, Keller B, ... Singh NK (2018). Shifting the limits in wheat research and breeding using a fully annotated reference genome. *Science* 361: 7191.

Jacobs AK, Lipka V, Burton RA, Panstruga R, Strizhov N, Schulze-Lefert P, Fincher GB (2003) An *Arabidopsis thaliana* callose synthase, GSL5, is required for wound and papillary callose formation. *Plant Cell* 15: 2503–2513.

Jarvis MC (2018). Structure of native cellulose microfibrils, the starting point for nanocellulose manufacture. *Philosophical Transactions of the Royal Society A: Mathematical, Physical and Engineering Sciences*, 376(2112): 20170045.

Jensen JK, Busse-Wicher M, Poulsen CP, Fangel JU, Smith PJ, Yang J-Y, Peña MJ, et al. (2018) Identification of an algal xylan synthase indicates that there is functional orthology between algal and plant cell wall biosynthesis. *New Phytologist* 218: 1049–1060.

Jensen JK, Johnson NR, Wilkerson CG (2014). *Arabidopsis thaliana* IRX10 and two related proteins from psyllium and *Physcomitrella patens* are xylan xylosyltransferases. *Plant Journal* 80: 207–215.

Jiang N, Wiemels RE, Soya A, Whitley R, Held M, Faik A. (2016). Composition, assembly, and trafficking of a wheat xylan synthase complex. *Plant Physiology* 170: 1999–2023.

Jobling SA (2015) Membrane pore architecture of the CslF6 protein controls (1-3,1-4)-β-glucan structure. *Science Advances* 1(5): e1500069.

Jumper J, Evans R, Pritzel A, Green T, Figurnov M, Ronneberger O, Tunyasuvunakool K, Bates R, Žídek A, Potapenko A, Bridgland A, Meyer C, Kohl SAA, Ballard AJ, Cowie A, Romera-Paredes B, Nikolov S, Jain R, Adler J, Back T, Petersen S, Reiman D, Clancy E, Zielinski M, Steinegger M, Pacholska M, Berghammer T, Bodenstein S, Silver D, Vinyals O, Senior AW, Kavukcuoglu K, Kohli P, Hassabis D (2021) Highly accurate protein structure prediction with AlphaFold. *Nature* 596: 583–589.

Kim SJ, Chandrasekar B, Rea AC, Danhof L, Zemelis-Durfee S, Thrower N, Shepard ZS, Pauly M, Brandizzi F, Keegstra K (2020) The synthesis of xyloglucan, an abundant plant cell wall polysaccharide, requires CSLC function. *Proceedings of the National Academy of Sciences USA* 117: 20316–20324.

Kim SJ, Zemelis S, Keegstra K, Brandizzi F (2015) The cytoplasmic localization of the catalytic site of CSLF6 supports a channeling model for the biosynthesis of mixed-linkage glucan. *Plant Journal* 81: 537–547.

Knox JP (2008) Revealing the structural and functional diversity of plant cell walls. *Current Opinion in Plant Biology* 11: 306–313.

Knudsen S, Wendt T, Dockter C, Thomsen HC, Rasmussen M, Jørgensen ME, Lu Q, Voss C, Murozuka E, Østerberg J, Harholt J, Braumann I, Cuesta-Seijo JA, Bodevin S, Petersen LT, Carciofi M, Pedas PR, Husum JO, Simmelsgaard MT, Nielsen K, Jensen MK, Møller LA, Gojkovic Z, Striebeck A, Lengeler K, Fennessy RT, Katz M, Sanchez RG, Solodovnikova N, Forster J, Olsen O, Møller BL, Fincher GB, Skadhauge B (2022) FIND-IT: accelerated trait development for a green evolution. *Science Advances* 8: eabq2266.

Laidlaw HKC, Lahnstein J, Burton RA, Fincher GB, Jobling SA (2012) Analysis of the arabinoxylan arabinofuranohydrolase gene family in barley does not support their involvement in the remodelling of endosperm cell walls during development. *Journal of Experimental Botany* 63: 3031–3045.

Langridge P, Waugh R (2019) Harnessing the potential of germplasm collections. *Nature Genetics* 51: 200–201.

Lazaridou A, Biliaderis CG (2007) Molecular aspects of cereal β-glucan functionality: Physical properties, technological applications and physiological effects. *Journal of Cereal Science* 46: 101–118.

Lechat H, Amat M, Mazoyer J, Buleon A, Lahaye M (2000) Structure and distribution of glucomannan and sulfated glucan in the cell walls of the red alga *Kappaphycus alvarezii* (gigartinales, rhodophyta). *Journal of Phycology* 36: 891–902.

Lee RC, Burton RA, Hrmova M, Fincher GB (2001) Barley arabinoxylan arabinofuranohydrolases: purification, characterization and determination of primary structures from cDNA clones. *Biochemical Journal* 356: 181–189.

Lee RC, Hrmova M, Burton RA, Lahnstein J, Fincher GB (2003) Bifunctional family 3 glycoside hydrolases from barley with α-L-arabinofuranosidase and β-D-xylosidase activity: characterization, primary structures and COOH-terminal processing. *Journal of Biological Chemistry* 278: 5377–5387.

Liepman AH, Wilkerson CG, Keegstra K (2005). Expression of cellulose synthase-like (*Csl*) genes in insect cells reveals that *CslA* family members encode mannan synthases. *Proceedings of the National Academy of Sciences USA* 102: 2221–2226.

Little A, Lahnstein J, Jeffery DW, Khor SF, Schwerdt JG, Shirley NJ, Hooi M, Xing X, Burton RA, Bulone V (2019) A novel (1,4)-β-linked glucoxylan is synthesized by members of the cellulose synthase-like F gene family in land plants. *ACS Central Science* 5: 73–84.

Little A, Schwerdt JG, Shirley NJ, Khor SF, Neumann K, O'Donovan LA, Lahnstein J, Collins HM, Henderson M, Fincher GB, Burton RA (2018) Revised phylogeny of the cellulose synthase gene superfamily: insights into cell wall evolution. *Plant Physiology* 177: 1124–1141.

Lombard V, Golaconda Ramulu H, Drula E, Coutinho PM, Henrissat B (2014) The carbohydrate-active enzymes database (CAZy) in 2013. *Nucleic Acids Research* 42: D490–D495.

Lunin VV, Wang H-T, Bharadwaj VS, Alahuhta M, Peña MJ, Yang J-Y, Archer-Hartmann SA, Azadi P, Himmel ME, Moremen KW, York WS, Bomble YJ, Urbanowicz BR (2020) Molecular mechanism of polysaccharide acetylation by the arabidopsis xylan O-acetyltransferase XOAT1. *Plant Cell* 32: 2367–2382.

Mares DJ, Stone BA (1973) Studies on wheat endosperm I. Chemical composition and ultrastructure of the cell walls. *Australian Journal of Biological Sciences* 26: 793–812.

Mascher M, Gundlach H, Himmelbach A, Beier S, Twardziok S, Wicker T, Radchuk V, Dockter C. Hedley P, Russell J, Bayer M, Ramsay L, Liu H, Haberer G, Zhang X-Q, Zhang Q, Barrero R, Li L, Taudien S, Groth M, Felder M, Hastie A, Šimková H, Staňková H, Vrána J, Chan S, Muñoz-Amatriaín M, Ounit R, Wanamaker S, Bolser D, Colmsee C, Schmutzer T, Aliyeva-Schnorr L, Grasso S, Tanskanen J, Chailyan A, Sampath D, Heavens D, Clissold L, Cao S, Chapman B, Dai F, Han Y, Li H, Li X, Lin C, McCooke J, Tan C, Wang P, Wang S, Yin S, Zhou G, Poland J, Bellgard M, Borisjuk L, Houben A, Doležel J, Ayling S, Lonardi S, Kersey P, Langridge P, Muehlbauer G, Clark M, Caccamo M, Schulman A, Mayer K, Platzer M, Close T, Scholz U, Hansson M, Zhang G, Braumann I, Spannagl M, Li C, Waugh R, Stein N (2017) A chromosome conformation capture ordered sequence of the barley genome. *Nature* 544: 427–433.

McCann MC, Defernez M, Urbanowicz BR, Tewari JC, Langewisch T, Olek A, Wells B, Wilson RH, Carpita NC (2007) Neural network analyses of infrared spectra for classifying cell wall architectures. *Plant Physiology* 143: 1314–1326.

McCleary BV (1988) Purification of (1→3),(1→4)-β-D-glucan from barley flour. *Methods in Enzymology* 160: 511–514.

Meier H, Reid JSG (1982) Reserve polysaccharides other than starch in higher plants. In *Encyclopedia of Plant Physiology, New Series 13A: Plant Carbohydrates I* (Loewus FA, Tanner W, eds), pp. 418–471. Berlin: Springer.

Meikle PJ, Ng KF, Johnson E, Hoogenraad NJ, Stone BA (1991) The beta-glucan synthase from Lolium multiflorum. Detergent solubilization, purification using monoclonal antibodies, and photoaffinity labeling with a novel photoreactive pyrimidine analogue of uridine 5'-diphosphoglucose. *Journal of Biological Chemistry* 266: 22569–22581.

Mitchell RAC, Dupree P, Shewry PR (2007) A novel bioinformatics approach identifies candidate genes for the synthesis and feruloylation of arabinoxylan. *Plant Physiology* 144: 43–53.

Mohler KE, Simmons TJ, Fry SC (2012) Mixed-linkage glucan:xyloglucan endotransglucosylase (MXE) re-models hemicelluloses in Equisetum shoots but not in barley shoots or Equisetum callus. *New Phytologist* 197: 111–122.

Moller I, Sørensen I, Bernal AJ, Blaukopf C, Lee K, Øbro J, Pettolino F, Roberts A, Mikkelsen JD, Knox JP, Bacic A, Willats WGT (2006) High-throughput mapping of cell-wall polymers within and between plants using novel microarrays. *Plant Journal* 50: 1118–1128.

Morgan JL, Strumillo J, Zimmer J (2013). Crystallographic snapshot of cellulose synthesis and membrane translocation. *Nature* 493: 181–186.

Morrall P, Briggs DE (1978) Changes in cell wall polysaccharides of germinating barley grains. *Phytochemistry* 17: 1495–1502.

Mortimer JC, Miles GP, Brown DM, Zhang Z, Segura MP, Weimar T, et al. (2010). Absence of branches from xylan in Arabidopsis *gux* mutants reveals potential for simplification of lignocellulosic biomass. *Proceedings of the National Academy of Sciences USA* 107: 17409–17414.

Nishitani K, Tominaga R (1992) Endo-xyloglucan transferase, a novel class of glycosyltransferase that catalyzes transfer of a segment of xyloglucan molecule to another xyloglucan molecule. *Journal of Biological Chemistry* 267: 21058–21064.

Obel N, Erben V, Schwarz T, Kühnel S, Fodor A, Pauly M (2009). Microanalysis of plant cell wall polysaccharides. *Molecular Plant* 2: 922–932.

Ohara T, Takeuchi H, Sato J, Nakamura A, Ichikawa H, Yokoyama R, Nishitani K, Minami E, Satoh S, Iwai H (2021). Structural alteration of rice pectin affects cell wall mechanical strength and pathogenicity of the rice blast fungus under weak light conditions. *Plant and Cell Physiology* 62: 641–649.

Okekeogbu IO, Pattathil S, González Fernández-Niño SM, Aryal UK, Penning BW, Lao J, Heazlewood JL, Hahn MG, McCann MC, Carpita NC (2019) Glycome and proteome components of Golgi membranes are common between two angiosperms with distinct cell-wall structures. *The Plant Cell* 31: 1094–1112.

Okuda K, Li L, Kudlicka K, Kuga S, Brown RM Jr (1993). β-Glucan synthesis in the cotton fiber (I. Identification of β-1, 4-and β-1, 3-glucans synthesized *in vitro*). *Plant Physiology* 101: 1131–1142.

Paterson AH, Bowers JE, Bruggmann R, Dubchak I, Grimwood J, Gundlach H, ... Rokhsar DS (2009). The *Sorghum bicolor* genome and the diversification of grasses. *Nature*, 457: 551–556.

Pattathil S, Avci U, Baldwin D, Swennes AG, McGill JA, Popper Z, Bootten T, Albert A, Davis RH, Chennareddy C, Dong R, O'Shea B, Rossi R, Leoff C, Freshour G, Narra R, O'Neil M, York WS, Hahn MG (2010) A comprehensive toolkit of plant cell wall glycan-directed monoclonal antibodies. *Plant Physiology* 153: 514–525.

Pattathil S, Hahn MG, Dale BE, Chundawat SP (2015). Insights into plant cell wall structure, architecture, and integrity using glycome profiling of native and AFEXTM-pretreated biomass. *Journal of Experimental Botany* 66: 4279–4294.

Pauly M, Gille S, Liu L, Mansoori N, de Souza A, Schultink A, Xiong G (2013) Hemicellulose biosynthesis. *Planta* 238: 627–642.

Pear JR, Kawagoe Y, Schreckengost WE, Delmer DP, Stalker DM (1996). Higher plants contain homologs of the bacterial *Cela* genes encoding the catalytic subunit of cellulose synthase. *Proceedings of the National Academy of Sciences USA* 93: 12637–12642.

Pellny TK, Lovegrove A, Freeman J, Tosi P, Love CG, Knox JP, Shewry PR, Mitchell RAC (2012) Cell Walls of developing wheat starchy endosperm: comparison of composition and RNA-Seq transcriptome. *Plant Physiology* 158: 612–627.

Pena MJ, Zhong R, Zhou G-K, Richardson EA, O'Neill MA, Darvill AG, et al. (2007). Arabidopsis irregular xylem8 and irregular xylem9: implications for the complexity of glucuronoxylan biosynthesis. *Plant Cell* 19: 549–563.

Persson S, Caffall KH, Freshour G, Hilley MT, Bauer S, Poindexter P, et al. (2007). The Arabidopsis *irregular xylem8* mutant is deficient in glucuronoxylan and homogalacturonan, which are essential for secondary cell wall integrity. *Plant Cell* 19: 237–255.

Pettolino FA, Sasaki I, Turbic A, Wilson SM, Bacic A, Hrmova M, Fincher GB (2009) Hyphal cell walls from the plant pathogen *Rhynchosporium secalis* contain (1,3/1,6)-β-D-glucans, galacto- and rhamnomannans, (1,3;1,4)-β-D-glucans and chitin. *FEBS Journal* 276: 4122–4133.

Pettolino FA, Walsh C, Fincher GB, Bacic A (2012) Determining the polysaccharide composition of plant cell walls. *Nature Protocols* 7: 1590–1607.

Popper ZA, Fry SC (2003). Primary cell wall composition of bryophytes and charophytes. *Annals of Botany* 91: 1–12.

Popper ZA, Herve MG, Domozych DS, Willats WGT, Tuohy MG, Kloareg B, Stengel DB (2011) Evolution and diversity of plant cell walls: from algae to flowering plants. *Annual Review of Plant Biology* 62: 567–590.

Popper ZA, Tuohy MG (2010). Beyond the green: understanding the evolutionary puzzle of plant and algal cell walls. *Plant Physiology* 153: 373–383.

Porchia AC, Scheller HV (2008) Arabinoxylan biosynthesis: identification and partial characterization of β-1,4-xylosyltransferase from wheat. *Physiologia Plantarum* 110: 350–356.

Purushotham P, Ho R, Yu L, Fincher GB, Bulone V, Zimmer J (2022) Mechanism of mixed linkage glucan biosynthesis by barley cellulose synthase-like CslF6 (1,3;1,4)-β-glucan synthase. *Science Advances* 8(45): eadd1596.

Reid JSG, Bewley JD (1979) A dual role for the endosperm and its galactomannan reserve in the germinative physiology of fenugreek (*Trigonellafoenun-graecum* L.), an endospermic leguminous seed. *Planta* 147: 145–150.

Richmond TA, Somerville CR (2000) The cellulose synthase superfamily. *Plant Physiology* 124: 495–498.

Roulin S, Buchala AJ, Fincher GB (2002) Induction of (1→3,1→4)-β-glucan hydrolases in leaves of dark-incubated barley seedlings. *Planta* 215: 51–59.

Sadowski PG, Groen AJ, Dupree P, Lilley KS (2008). Sub-cellular localization of membrane proteins. *Proteomics*, 8: 3991–4011.

Sambrook J, Fritsch EF, Maniatis T (1989) *Molecular Cloning: A Laboratory Manual*. Cold Spring Harbor: CAB Direct.

Saxena IM, Lin FC, Brown RM Jr (1990) Cloning and sequencing of the cellulose synthase catalytic subunit gene of *Acetobacter xylinum*. *Plant Molecular Biology* 15: 673–683.

Scheller HV, Ulvskov P (2010) Hemicelluloses. In: *Annual Review of Plant Biology* (Merchant S, Briggs WR, Ort D eds), vol. 61, pp. 263–289.

Schnable PS, Ware D, Fulton RS, Stein JC, Wei F, Pasternak S, ... Presting GG (2009). The B73 maize genome: complexity, diversity, and dynamics. *Science* 326: 1112–1115.

Schreiber M, Wright F, MacKenzie K, Hedley PE, Schwerdt JG, Little A, Burton RA, Fincher GB, Marshall D, Waugh R, Halpin C (2014) The barley genome sequence assembly reveals three additional members of the *CslF* (1,3;1,4)-β-glucan synthase gene family. *Plos One 9*: e90888.

Schwerdt JG, MacKenzie K, Wright F, Oemhe D, Wagner JM, Harvey AJ, Shirley NJ, Burton RA, Wright F, MacKenzie K, Schreiber M, Halpin C, Zimmer J, Marshall DF, Waugh R, Fincher GB (2015) Evolutionary dynamics of the cellulose synthase gene superfamily ingrasses. *Plant Physiology* 168: 968–983.

Slade AM, Høj PB, Morrice NA, Fincher GB (1989) Purification and characterization of three (1,4)-β-D-xylan endohydrolases from germinated barley. *European Journal of Biochemistry* 185: 533–539.

Smith MM, Stone BA (1973). β-Glucan synthesis by cell-free extracts from *Lolium multiflorum* endosperm. *Biochimica et Biophysica Acta General Subjects* 313: 72–94.

Sørensen I, Pettolino FA, Wilson SM, Doblin MS, Johansen B, Bacic A, Willats WGT (2008). Mixed-linkage (1,3;1,4)-beta-D-glucan is not unique to the Poales and is an abundant component of *Equisetum arvense* cell walls. *Plant Journal* 54: 510–521.

Sowinski EE, Westman BM, Redmond CR, Kong Y, Olek AT, Olek J, McCann MC, Carpita NC (2022) Lack of xyloglucan in the cell walls of the Arabidopsis xxt1/xxt2 mutant results in specific increases in homogalacturonan and glucomannan. *The Plant Journal* 110: 212–227.

Staudte RG, Woodward JR, Fincher GB, Stone BA (1983) Water-soluble (1→3),(1→4)-β-D-glucans from barley (*Hordeum vulgare*) endosperm. III. Distribution of cellotriosyl and cellotetraosyl residues. *Carbohydrate Polymers* 3: 299–312.

Stone BA, Clarke AE (1992). *Chemistry and Biology of (1,3)-β-D-Glucans*. Victoria, Australia: La Trobe University Press.

Stone BA, Evans NA, Bonig I, Clarke AE (1984) The application of Sirofluor, a chemically defined fluorochrome from aniline blue for the histochemical detection of callose. *Protoplasma* 122: 191–195.

Stratilová B, Firáková Z, Klaudiny J, Šesták S, Kozmon S, Strouhalová D, Garajová S, Ait-Mohand F, Horváthová A, Farkaš V, Stratilová E, Hrmova M (2019) Engineering the acceptor substrate specificity in the xyloglucan endotransglycosylase TmXET6.3 from nasturtium seeds (*Tropaeolum majus* L.). *Plant Molecular Biology* 100: 181–197.

Stratilová B, Šesták S, Mravec J, Garajová S, Pakanová Z, Vadinová K, Kučerová D, Kozmon S, Schwerdt JG, Shirley NJ, Stratilová E, Hrmova M (2020) Another building block in the plant cell wall: barley xyloglucan xyloglucosyl transferases link covalently xyloglucan and anionic oligosaccharides derived from pectin. *Plant Journal* 104: 752–767.

Tan H-T, Shirley NJ, Singh RR, Henderson M, Dhugga KS, Mayo GM, Fincher GB, Burton RA (2015) Powerful regulatory systems and post-transcriptional gene silencing resist increases in cellulose content in cell walls of barley. *BMC Plant Biology* 15: 62.

Tan L, Eberhard S, Pattathil S, Warder C, Glushka J, Yuan C, Hao Z, Zhu X, Avci U, Miller JS, Baldwin D, Pham C, Orlando R, Darvill A, Hahn MG, Kieliszewski MJ, Mohnen D (2013) An Arabidopsis cell wall proteoglycan consists of pectin and arabinoxylan covalently linked to an arabinogalactan protein. *Plant Cell* 25: 270–287.

Tang, X, Sretenovic S, Ren Q, Jia X, Li M, Fan T, Yin D, Xiang S, Guo Y, Liu L, Zheng X, Qi Y, Zhang Y (2020) Plant prime editors enable precise gene editing in rice cells. *Molecular Plant* 13: 667–670.

Taylor NG, Scheible WR, Cutler S, Somerville CR, Turner SR (1999) The irregular xylem 3 locus of *Arabidopsis* encodes a cellulose synthase gene required for secondary cell wall synthesis. *Plant Cell* 11: 769–780.

Terrett OM, Lyczakowski JJ, Yu L, Iuga, D, Franks WT, Brown SP, Dupree R, Dupree P (2019) Molecular architecture of softwood revealed by solid-state NMR. *Nature Communications* 10: 4978.

Tine MAS, Silva CO, Lima DU, Carpita NC, Buckeridge MS (2006) Fine structure of a mixed-oligomer storage xyloglucan from seeds of *Hymenaea courbaril*. *Carbohydrate Polymers* 66: 444–454.

Trafford K, Haleux P, Henderson M, Parker M, Shirley NJ, Tucker MR, Fincher GB, Burton RA (2013) Grain development in *Brachypodium* and other grasses: interactions between cell expansion, starch deposition and cell wall synthesis. *Journal of Experimental Botany* 64: 5033–5047.

Trethewey JAK, Campbell LM, Harris PJ (2005) (1,3;1,4)-beta-D-glucans in the cell walls of the poales (*sensu lato*): an immunogold labeling study using a monoclonal antibody. *American Journal of Botany* 92: 1660–1674.

Turner S, Kumar M. (2018) Cellulose synthase complex organization and cellulose microfibril structure. *Philosophical Transactions of the Royal Society A: Mathematical, Physical and Engineering Sciences* 376(2112): 20170048.

Urbanowicz BR, Peña MJ, Moniz HA, Moremen KW, York WS (2014). Two Arabidopsis proteins synthesize acetylated xylan *in vitro*. *Plant Journal* 80: 197–206.

Van Campenhout S, Pollet A, Bourgois TM, Rombouts S, Beaugrand J, Gebruers K, De Backer E, Courtin CM, Delcour JA, Volckaert G (2007) Unprocessed barley aleurone endo-β-1,4-xylanase X-I is an active enzyme. *Biochemical and Biophysical Research Communications* 356: 799–804.

van de Meene AML, Doblin MS, Bacic A (2017) The plant secretory pathway seen through the lens of the cell wall. *Protoplasma* 254: 75–94.

van de Meene AML, McAloney L, Wilson SM, Zhou J, Zeng W, McMillan P, Bacic A, Doblin MS (2021) Interactions between cellulose and (1,3;1,4)-β-glucans and arabinoxylans in the regenerating wall of suspension culture cells of the ryegrass *Lolium multiflorum*. *Cells* 10: 127.

Varghese JN, Garrett TPJ, Colman PM, Chen L, Høj PB, Fincher GB (1994) The three-dimensional structures of two plant β-glucan endohydrolases with distinct substrate specificities. *Proceedings of the National Academy of Sciences USA* 91: 2785–2789.

Varghese JN, Hrmova M, Fincher GB (1999) Three-dimensional structure of a barley β-D-glucan exohydrolase; a family 3 hydrolase. *Structure* 7: 179–190.

Viborg AH, Terrapon N, Lombard V, Michel G, Gurvan M, Czjzek M, Henrissat B, Brumer H (2019) A subfamily roadmap for functional glycogenomics of the evolutionarily diverse Glycoside Hydrolase Family 16 (GH16). *Journal of Biological Chemistry* 294: 15973–15986.

Vogel J, Garvin D, Mockler TC, Schmutz J, Rokhsar D, et al. (2010). Genome sequencing and analysis of the model grass Brachypodium distachyon. *Nature* 463: 763–768.

Wang X, Cnops G, Vanderhaeghen R, De Block S, Van Montagu M, Van Lijsebettens M (2001). AtCSLD3, a cellulose synthase-like gene important for root hair growth in Arabidopsis. *Plant Physiology* 126: 575–586.

Weigel PH, DeAngelis PL (2007). Hyaluronan synthases: a decade-plus of novel glycosyltransferases. *Journal of Biological Chemistry* 282: 36777–36781.

Wilson SM, Burton RA, Doblin MS, Stone BA, Newbigin EJ, Fincher GB, Bacic A (2006). Temporal and spatial appearance of wall polysaccharides during cellularization of barley (*Hordeum vulgare*) endosperm. *Planta* 224: 655–667.

Wilson SM, Ho YY, Lampugnani ER, van de Meene AML, Bain MP, Bacic A, Doblin MS (2015) Determining the subcellular location of synthesis and assembly of the cell wall polysaccharide (1,3;1,4)-β-D-glucan in grasses. *Plant Cell* 27: 754–771.

Wong HC, Fear AL, Calhoon RD, Eichinger GH, Mayer R, Amikam D, Benziman M, Gelfand DH, Meade JH, Emerick AW (1990) Genetic organization of the cellulose synthase operon in *Acetobacter xylinum*. *Proceedings of the National Academy of Sciences USA* 87: 8130–8134.

Wood PJ, Weisz J, Blackwell BA (1994) Structural studies of (1→3),(1→4)-β-D-glucans by13C-nuclear magnetic resonance spectroscopy and by rapid analysis of cellulose-like regions using high-performance anion-exchange chromatography of oligosaccharides released by lichenase. *Cereal Chemistry* 71: 301–307.

Woodward JR, Fincher GB (1982) Substrate specificities and kinetic properties of two (1→3),(1→4)-β-D-glucan endohydrolases from germinating barley (*Hordeum vulgare*). *Carbohydrate Research* 106: 111–122.

Woodward JR, Fincher GB, Stone BA (1983) Water-soluble (1→3),(1→4)-β-D-glucans from barley (*Hordeum vulgare*) endosperm. II. Fine structure. *Carbohydrate Polymers* 3: 207–225.

Yang J, Bak G, Burgin T, Barnes WJ, Mayes HB, Peña MJ, Urbanowicz BR, Nielsen E (2020) Biochemical and genetic analysis identify CSLD3 as a beta-1,4-glucan synthase that functions during plant cell wall synthesis. *Plant Cell* 32: 1749–1767.

Yang Z, Bielawski JP (2000) Statistical methods for detecting molecular adaptation. *Trends in Ecology and Evolution* 15: 496–503.

Yin Y, Johns MA, Cao H, Rupani M (2014) A survey of plant and algal genomes and transcriptomes reveals new insights into the evolution and function of the cellulose synthase superfamily. *BMC Genomics* 15: 260.

Zablackis E, Huang J, Muller B, Darvill AG, Albersheim P (1995). Characterization of the cell-wall polysaccharides of *Arabidopsis thaliana* leaves. *Plant Physiology* 107: 1129–1138.

Zeng W, Jiang N, Nadella R, Killen TL, Nadella V, Faik A (2010) A glucurono(arabino)xylan synthase complex from wheat contains members of the GT43, GT47, and GT75 families and functions cooperatively. *Plant Physiology* 154: 78–97.

Zeyen RJ, Bushnell WR (1979) Papilla response of barley epidermal cells caused by *Erysiphe graminis*: rate and method of deposition determined by microcinematography and transmission microscopy. *Canadian Journal of Botany* 57: 898–913.

Zhang D, Williams BA, Mikkelsen D, Li X, Keates HL, Lisle AT, Collins HM, Fincher GB, Bird AR, Topping DL, Gidley MJ, Bryden WL (2015) Soluble arabinoxylan alters digesta flow and protein digestion of red meat-containing diets in pigs. *Nutrition* 31: 1141–1147.

Zhang R, Tucker MR, Burton RA, Shirley NJ, Little A, Morris J, Milne L, Houston K, Hedley PE, Waugh R, Fincher GB (2016). The dynamics of transcript abundance during cellularisation of developing barley endosperm. *Plant Physiology* 170: 1549–1565.

Zhang Y, Pribil M, Palmgren M, Gao CA (2020) CRISPR way for accelerating improvement of food crops. *Nature Food* 1: 200–205.

10 Plant Cell Wall Glycan-Directed Monoclonal Antibodies

Kristen Thorne, Breeanna R. Urbanowicz, and Michael G. Hahn
University of Georgia

INTRODUCTION

Plant cell walls play a ubiquitous role in physiological processes such as cellular growth and morphogenesis, reproductive development, and environmental and immune defence (Chapters 1, 11). As such, their structures are complex and depend on the interactions between a diverse set of macromolecules, predominantly glycans, proteins, and in secondary walls, phenolics. Keith Roberts and colleagues stated some three decades ago that "models of cell wall architecture have been Neoclassical in their simplicity but it is now clear that the cell wall has a complexity that is more typically Byzantine" (McCann et al. 1992). This account persists to this day, as it remains a challenge to uncover the molecular architecture and regulatory mechanisms of plant cell walls. Thus, it is critical that we develop and apply enabling tools that can be used to refine and expand upon the following key questions. Which polymers are where and what polymers are in proximity to each other? How are polymer synthesis, location, and proximity modulated during plant development? How do the latter change in response to biotic and abiotic stimuli and/or stress?

As described elsewhere in this volume, plant cell wall polymers are structurally and functionally diverse. The availability of highly specific and selective probes for individual cell wall components, and specifically for cell wall glycans, is critical to answering the questions raised above. Scientists have long been able to generate such probes for proteins of interest, which has dramatically increased our functional understanding of the targeted proteins in diverse biological processes. Proteins have the advantage that they are usually non-repetitive polymers with a unique structure, facilitating, for example, the production of highly specific and selective antibodies, even in polyclonal sera. Further, proteins are direct gene products, allowing for the generation of tagged versions of target proteins *in vivo* to study their dynamics and interactions. Glycans, in contrast, are not direct products of genes, but rather are synthesised by assemblages of gene products, precluding genetic approaches to tagging individual polysaccharides. Furthermore, glycans are frequently repetitive in their structures and some glycan structural features can be found in multiple macromolecular contexts. This largely precludes the use of polyclonal antisera and antibodies for studies of glycans, as the selectivity and specificity of such probes is almost impossible to determine in most cases. Keith Roberts and colleagues, almost four decades ago, recognised that "monoclonal antibodies form ideal probes for both structural and functional studies of plant cell walls", and further that "monoclonal antibodies against short sugar sequences hold out the promise of providing us with the appropriate specificity … [to define] … the biological function of the bewildering variety of biochemical structures found in the plant cell wall" (Roberts et al. 1985). This is because monoclonal antibodies (mAbs) are, by definition, monospecific, given that they are the product of a clonal cell line derived from a single antibody producing cell taken from an immunised animal (Zola 1987; Goding 1996). Thus, each antibody will bind to a specific glycan structure (epitope) in the plant cell wall.

The first cell wall glycan–directed antibodies came onto the scene over 50 years ago when Valerie Vreeland generated a polyclonal antibody against an alginate fraction from the brown alga, *Fucus distichus* (Vreeland 1970). The first cell wall glycan–directed monoclonal antibodies arrived about 14 years later, and were directed against glycans in Fucus and Chlamydomonas (Table 10.1). Over the next 40 years, about a dozen laboratories world-wide have generated more than 200 distinct mAbs that recognise diverse glycan sub-structures, or epitopes, within plant cell wall polysaccharides (Knox 2008; Pattathil et al. 2010) (Figure 10.2). Initial screening of these antibodies against plant cell wall glycan preparations from diverse plants demonstrated that the mAb library includes groups of antibodies against most of the major non-cellulosic plant cell wall glycan families (Figure 10.1). It is clear from these initial screens that each of the groups of mAbs contain multiple sub-sets of antibodies that appear to have different specificities based on their differing binding intensities to the various glycan preparations within each glycan class. Furthermore, while some of the groups of mAbs are highly selective for a given glycan class, the mAbs are not always polymer selective, depending on the plant of origin for a particular glycan (see examples highlighted by green boxes in Figure 10.1). Indeed, the galactan- and arabinogalactan-directed mAbs show binding to a broad diversity pectic and arabinogalactan glycans (Figure 10.1). Recent advances in synthetic and enzymatic methods to generate pure oligosaccharides based on, or related to, cell wall polysaccharides have permitted

DOI: 10.1201/9781003178309-10

TABLE 10.1
Milestones for the Generation and Use of Antibodies Against Cell Wall Glycan Epitopes

Year	Event	Reference(s)
1970	First generation of a cell wall glycan–directed antibody probe; polyclonal antibody against alginate in brown algae (Fucus).	Vreeland (1970)
1984	First generation of monoclonal antibodies against algal cell wall glycan epitopes (Fucus and Chlamydomonas).	Smith et al. (1984); Vreeland et al. (1984)
1986	First generation and use of cell wall glycan–directed antibody probes in higher plants; polyclonal antibodies against xyloglucan and RG-I epitopes in suspension-cultured plant cells.	Moore et al. (1986)
1989	First generation and use of a cell wall glycan–directed monoclonal antibody probe against de-esterified homogalacturonan in higher plants.	Vandenbosch et al. (1989)
1990	Demonstration of domain-specific distribution of a de-esterified homogalacturonan epitope in higher plant walls.	Knox et al. (1990)
1994	Generation of a monoclonal antibody against a fucosylated epitope in xyloglucans.	Puhlmann et al. (1994)
1996	Use of monoclonal antibodies to demonstrate domain-specific and asymmetric distribution of several glycan epitopes in the walls of single cells.	Freshour et al. (1996)
1996	Use of a monoclonal antibody against a rhamnogalacturonan I epitope as a cell-specific marker in a plant developmental mutant.	Di Laurenzio et al. (1996)
2001	First generation and use of a plant mannan–directed monoclonal antibody.	Pettolino et al. (2001)
2004	First generation and characterisation of a plant xylan–directed monoclonal antibody.	Ordaz-Ortiz et al. (2004)
2007	First use of monoclonal antibodies and CBMs for comparative glycomics of plant cell walls.	Moller et al. (2007)
2008	First use of monoclonal antibody against backbone of rhamnogalacturonan I for immunolocalisation.	Young et al. (2008)
2010	Generation of a large and diverse collection of monoclonal antibodies against higher plant cell wall glycan epitopes.	Pattathil et al. (2010)
2012	Use of oligosaccharide arrays to define epitopes recognised by cell wall glycan–directed monoclonal antibodies.	Pedersen et al. (2012)
2017	First use of microarrays and cell wall glycan–directed monoclonal antibodies for genome-wide association studies.	Wood et al. (2017)

the detailed characterisation of the binding specificities of well over half of the mAbs in the world-wide collection to date. We will summarize these epitopes in the following sections of this review. All of the plant glycan-directed mAbs are available to the research community from several suppliers world-wide (Table 10.2), as well as from individual laboratories that generated the antibodies.

A variety of immunogens have been used to stimulate the production of antibody producing cells in immunised animals, such as mice or rats, that could be captured and immortalised by hybridoma technology (Zola 1987; Goding 1996). These immunogens include cell wall glycoproteins (Smith et al. 1984), membrane preparations (e.g. Hahn et al. 1987; Vandenbosch et al. 1989), cell wall preparations (Pattathil et al. 2010), plant polysaccharide preparations (e.g. Liners et al. 1989; Meikle et al. 1991; Puhlmann et al. 1994), and plant cell wall–derived oligosaccharides (e.g. Jones et al. 1997; Willats et al. 1998). Of these, the cell wall preparations proved to be the least effective in stimulating an immune response in the injected animals (Pattathil et al. 2010). Free glycans, because of their inherent repetitive structural features, are not immunogenic and frequently give rise to an IgM-dominated immune response in animals; indeed, a sizable percentage of the current plant glycan–directed mAb collection is of the IgM isotype. However, when coupled to protein, glycans tend to be immunodominant, as exemplified by the frequency with which arabinogalactan-directed antibodies are obtained when plant membrane preparations are used as immunogen (Anderson et al. 1984; Norman et al. 1986; Hahn et al. 1987; Knox et al. 1989, 1991; Pennell et al. 1989, 1991). Thus, most attempts to generate plant cell wall glycan–directed antibodies have used polysaccharides or oligosaccharides that have been covalently linked to or complexed with protein carriers to maximally stimulate the immune response of the injected animals. Available evidence suggests that the use of protein-coupled oligosaccharides leads to the generation of mAbs directed towards the non-reducing end of the oligosaccharides, while the use of protein-coupled polysaccharide immunogens tends to yield mAbs directed at internal epitopes on those polysaccharides (Ruprecht et al. 2017). Monoclonal antibodies typically bind to their epitopes with high affinity ($K_{D\sim}10^{-6}$ M), which makes them very sensitive reagents for detecting and quantitating the molecules to which they bind (Pattathil et al. 2015). Their affinity, combined with the epitope-specific nature of mAbs, as opposed, for example, to that of non-specific polysaccharide stains, is advantageous in studies focused on cell wall regulation, as the presence or absence of an epitope can be indicative of fine-scale alterations to the biochemical composition of distinct polymers, and hence, cell walls (Figure 10.2).

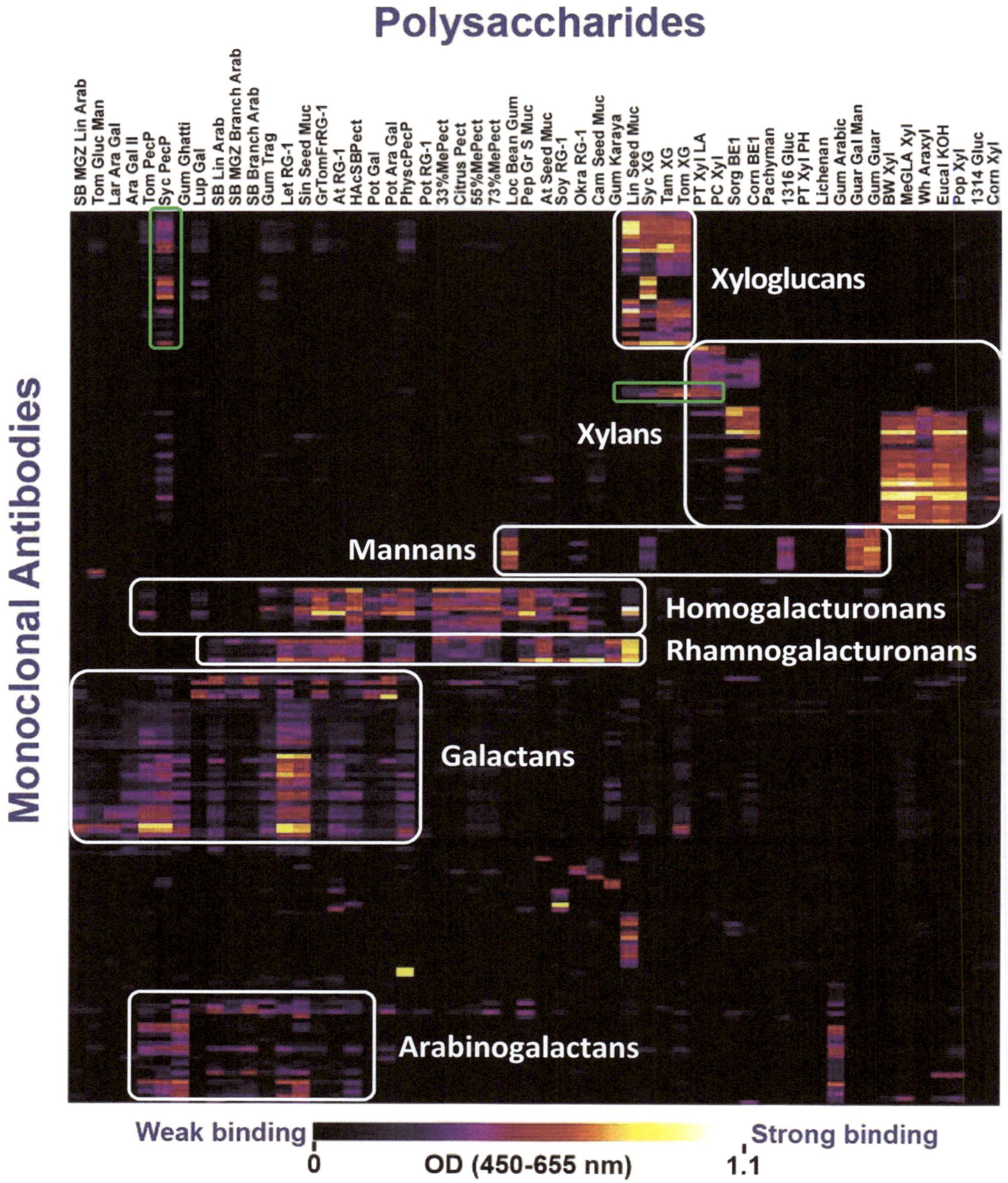

FIGURE 10.1 The world-wide collection of plant cell wall glycan-directed mAbs contains sets of antibodies against most of the major classes of plant cell wall glycans. The mAb collection was screened by ELISA against a panel of 54 structurally known plant cell wall glycan preparations from diverse plants (Pattathil et al., 2010). The binding strengths are depicted in a black-red-bright yellow color scheme where maximum and no binding are denoted by bright yellow and black colors, respectively (color gradient bar below the heat-map). The antibodies were then grouped according to the patterns of their binding activities against this polysaccharide panel (white boxes). Green boxes denote examples of sets of antibodies that exhibit binding to more than one class of plant cell wall glycan. Also note the extensive overlap in binding to different glycan classes by antibodies against galacturonans, galactans, and arabinogalactans.

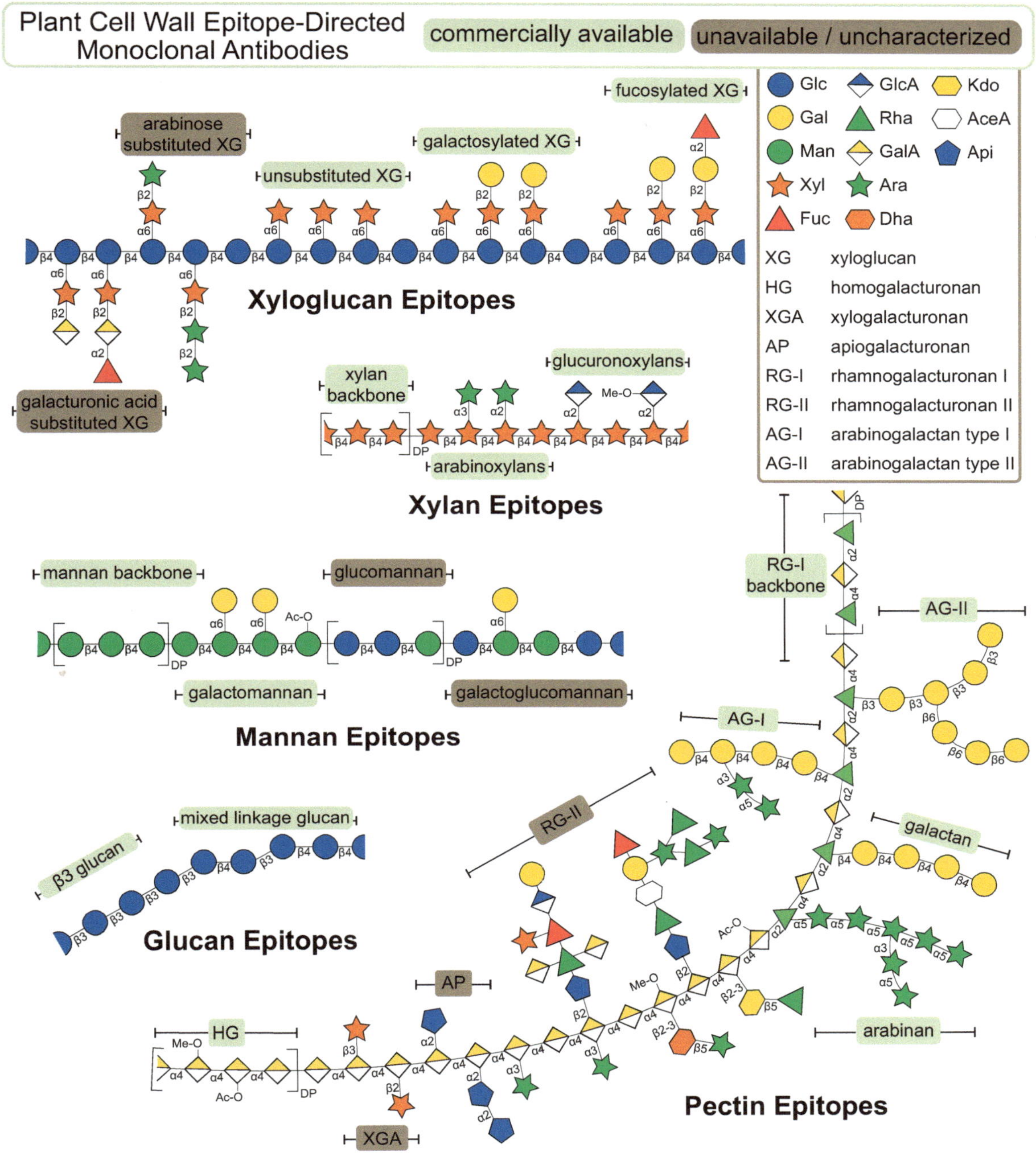

FIGURE 10.2 Schematic of common epitopes found in plant cell walls. Glycan-directed monoclonal antibodies target a wide range of known structural epitopes within the major classes of cell wall glycans, including xyloglucans, xylans, mannans, glucans, pectins, and arabinogalactan polymers. The epitope descriptions shaded in green and grey indicate the respective presence or absence of commercially available mAbs for immunological studies. The epitopes depicted in this figure do not encompass all potential epitopes, but rather emphasise the structural diversity within and between plant cell wall polymers.

TABLE 10.2
Commercial Suppliers for Plant Cell Wall Glycan-Directed Monoclonal Antibodies

Source	Antibodies Available	Web Link
CarboSource	All CCRC, JIM and MAC series	https://carbosource.uga.edu/
Kerafast	All LM and selected CCRC series	https://www.kerafast.com/
AgriSera	Selected CCRC series	https://www.agrisera.com/
BioSupplies Australia	LAMP, BG1, BGM C6, and selected LM series	http://www.biosupplies.com.au/

Timeless in their utility as biological imaging tools, mAbs are notably versatile in their application; the evolution of imaging and analysis methods has yielded a multitude of novel and improved light microscopy techniques. Immunohistochemical analysis with mAbs prevails as one of the most popular approaches for investigating the abundance, distribution, and interactions of plant cell wall polysaccharides, providing fundamental insights into the many functions of cell wall micro- and macrostructures. In this review, we will focus on the diversity of plant cell wall glycan–directed mAbs that have been generated and discuss what is known, and remains enigmatic, about their epitope specificities in a polymer-by-polymer manner. Furthermore, we will describe key advances that have had important implications for our ability to use mAbs as tools to distinguish distinct glycan epitopes and facilitate the characterisation of structurally heterogeneous plant cell walls.

HEMICELLULOSE

Xyloglucan-Directed Antibodies

Xyloglucans (XGs) are hemicellulosic polysaccharides present in the primary cell walls of all vascular plants. The biochemical composition of XGs has been shown to vary between different plant families, cell and tissue types, and developmental morphologies (Burton et al. 2010; Muszyński et al. 2015; Voiniciuc et al. 2018). XGs consist of a β-1,4-linked glucan backbone that is highly decorated with α-linked xylosyl residues in a regular pattern that can vary to some extent in different plant species. Advances in chemical analysis have revealed the structure of 24 naturally occurring XG side-chains (Pauly and Keegstra 2016), which are defined by further substitution of these xylosyl residues. Taken together, the known diversity of this polymer suggests that there is an important correlation between XG structure and its functional properties in plant cell walls. Here, we will discuss the epitope specificities of the currently available XG-directed antibodies (Figure 10.3). For simplicity's sake, we refer to more complex XG structures according to a nomenclature based on one-letter designations (Fry et al. 1993; Tuomivaara et al. 2015). The letter G represents an unsubstituted glucosyl backbone residue and the letter X represents a backbone glucosyl residue with an α-linked xylosyl residue. The letter D denotes an α-linked arabinosyl residue attached to the xylosyl residue, while the letters L and F denote the side-chains where the xylosyl residue is extended by addition of a β-linked galactosyl (L) residue and where that Gal is further extended by an α-linked fucosyl substituent (F), respectively.

Epitope: Unsubstituted Xyloglucan (X-XG)

The mAbs CCRC-M100 and CCRC-M103 target internal XG epitopes consisting of the β-1,4-D-glucan backbone with at least two sequential α-1,6-D-xylopyranose (Xyl*p*) residues (XX-XG). The binding specificities of each mAb have been characterised through enzyme-linked immunosorbent assay (ELISA)-based screens and glycan microarray mapping against plant-derived and synthetic plant cell wall oligosaccharides. Hierarchical clustering analysis of the ELISA-based polysaccharide screen (Pattathil et al. 2010) concluded that CCRC-M100 and CCRC-M103 recognise non-fucosylated XG. In a subsequent screen against synthetic XG oligosaccharides (Ruprecht et al. 2017), CCRC-M100 and CCRC-M103 bound preferentially to oligosaccharides with an unsubstituted glucosyl backbone residue next to the xylosyl substituent towards the non-reducing end; substitution of any xylosyl residue in the oligosaccharide with another sugar prevents recognition by either of these two antibodies. The mAb LM15 was originally generated to target XXXG motifs of XG (Marcus et al. 2008), but screens using chemically synthesised oligosaccharides or oligosaccharides prepared by enzymatic digestion of XG polymers provided a refinement of the recognised epitope structure. LM15 is instead directed towards the non-reducing end of XGs and requires either the second to last or the last glucosyl residue to be substituted with xylose (Ruprecht et al. 2017). LM15 will bind to any XG oligosaccharide carrying either GX or XX at the non-reducing terminus, regardless of substituents present or absent on the other xylosyl residues. The mAb LM25 has a similar binding profile to LM15, but is not bothered by the presence of galactosyl substituents on some of the xylosyl residues, although not all possible substituted oligosaccharides have been tested (Pedersen et al. 2012). The observed tolerances for substitutions on adjacent xylosyl residues suggest that these substituted xylosyl residues are not part of the epitope structure recognised by these antibodies.

Epitope: Galactosylated Xyloglucan (L-XG)

From the 25 mAbs initially characterised to bind to non-fucosylated natural XGs (Pattathil et al. 2010), two subgroups have been partially characterised to recognise XG side-chains that contain a galactosyl residue β-1,2-linked to xylose, referred to as the L side-chain. The refinement of these binding profiles was accomplished through successive studies, many of which used innovative approaches such as chemically synthesised galactose–substituted XG oligosaccharides. For example, a microarray using fragments generated from automated glycan assembly revealed that a single galactosyl substitution (L-XG) was sufficient for the binding of CCRC-M87, CCRC-M88, CCRC-M89, CCRC-M93, CCRC-M95, CCRC-M101, and CCRC-M104 (Dallabernardina et al. 2017). The other subgroup, consisting of CCRC-M50, CCRC-M52, CCRC-M53, CCRC-M55, and CCRC-M96, appears to target a larger epitope that encompasses the L side-chain and additional residues (XLX-XG), although the exact structural requirements for binding are unknown.

One other antibody that appears to be directed to the galactosylated side-chains of XGs has a binding pattern that is distinct from the two groups of antibodies discussed in the previous paragraph. CCRC-M48 has an epitope that includes the non-reducing terminal X side-chain and the L side-chain closest to the reducing terminus (XnLG);

Polymer Class	Monoclonal Antibody		Validated Epitope Structure M = minimum E = encompassed	Animal
Xyloglucan - XG	CCRC-M100	XX-XG	M 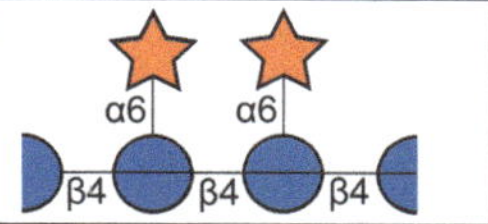	Mouse
	CCRC-M103			Mouse
	CCRC-M87	L-XG	M 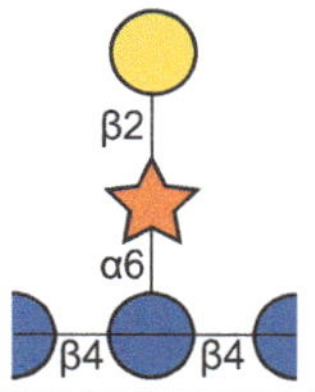	Mouse
	CCRC-M88			Mouse
	CCRC-M89			Mouse
	CCRC-M93			Mouse
	CCRC-M95			Mouse
	CCRC-M101			Mouse
	CCRC-M104			Mouse
	CCRC-M50	XLX-XG	E 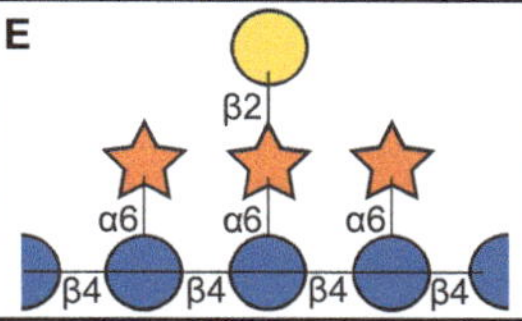	Mouse
	CCRC-M52			Mouse
	CCRC-M53			Mouse
	CCRC-M55			Mouse
	CCRC-M96			Mouse
	CCRC-M1	α-1,2-Fuc	M 	Mouse
	CCRC-M102	F-XG	E 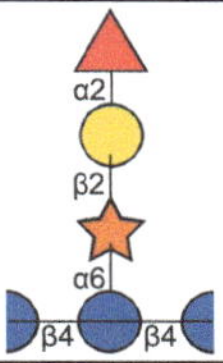	Mouse
	CCRC-M39	FF-XG	E 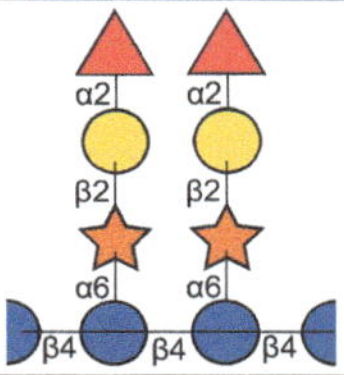	Mouse
	CCRC-M84			Mouse
	CCRC-M106			Mouse
	CCRC-M48	XnLG-XG	M 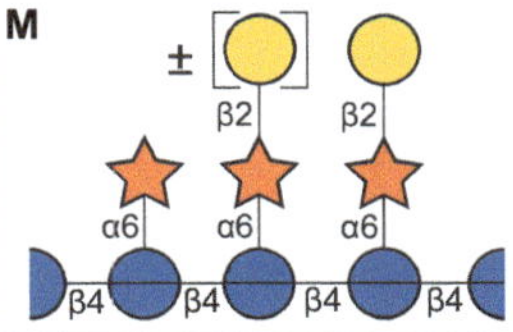	Mouse
	LM15	GX/XX-XG (NRE)	M 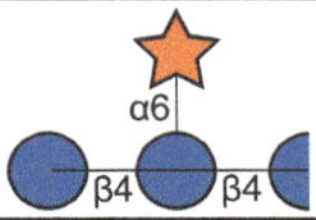	Rat
	LM25			
	CCRC-M51	$XXXG_n$ n≥2	E 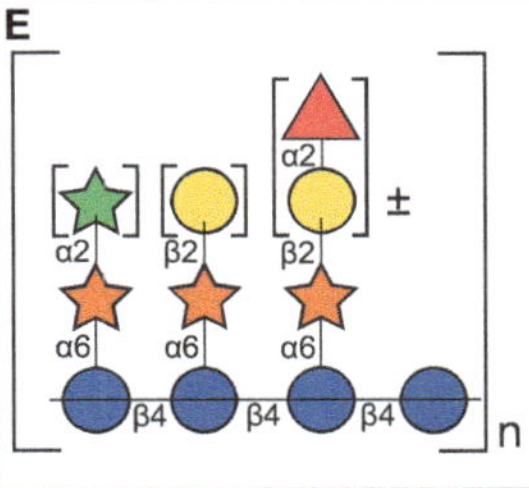	Mouse
	CCRC-M58			Mouse
	CCRC-M99			Mouse
	CCRC-M49	$XXXG_n$ n≥3		Mouse
	CCRC-M54			Mouse
	CCRC-M57			Mouse
	CCRC-M86			Mouse

FIGURE 10.3 Xyloglucan-directed monoclonal antibodies. Experimentally validated epitope structures for xyloglucan-directed mAbs are shown. The epitope descriptions follow a nomenclature based on one-letter designations (Fry et al. 1993; Tuomivaara et al. 2015). The representative schematics of each epitope are labelled with M or E. M indicates that the schematic shown depicts the minimum structure required for recognition. E indicates that the minimum epitope structure is encompassed in the depicted schematic, but additional information is needed for further refinement. NRE=non-reducing end.

substitutions of the middle xylosyl residue appear not to be necessary for or interfere with antibody binding (Tuomivaara et al. in preparation).

Epitope: Fucosylated Xyloglucan (F-XG)

The mAbs CCRC-M1, CCRC-M39, CCRC-M84, CCRC-M102, and CCRC-M106 target XG epitopes that contain α-1,2-linked fucosyl residues extended from galactosyl substituents, termed the F side-chain (Pattathil et al. 2010; Pattathil et al. 2015). While a single fucosyl residue is sufficient for recognition by most of these mAbs, CCRC-M39, CCRC-M84, and CCRC-M106 appear to preferentially bind to di-fucosylated XG (XFFG) (Tuomivaara et al. in preparation). CCRC-M1 recognises terminal α-fucosyl residues linked 1,2 to galactosyl residues and was the first plant cell wall–directed mAb to be thoroughly characterised with respect to epitope specificity (Puhlmann et al. 1994). The binding of CCRC-M1 can be completely inhibited in competitive immunoassays with mM concentrations of free fucose (Freshour et al. 1996; Brennan and Harris 2011), suggesting that the fucosyl residue is the predominant feature of the epitope recognised by this antibody. Terminal fucosyl residues are present in XGs, rhamnogalacturonan I (RG-I) from maize and sycamore maple, and arabinogalactans in some plants. However, CCRC-M1 is selective for fucosylated XGs in most plants that have been studied to date. *In vitro* experiments supported this conclusion, as the binding of CCRC-M1 was abolished after pre-incubation with fucosylated XG oligosaccharides (Freshour et al. 2003; Brennan and Harris 2011). Further confirmation was provided in immunolabelling studies of *fut1* and *mur1* mutant plants, which respectively lack fucosylated and galactosylated XG but retain these substituents on other wall polymers, as CCRC-M1 failed to bind to RG-I in these Arabidopsis mutants (Freshour et al. 2003; Perrin et al. 2003; Brennan and Harris 2011). With a comprehensive understanding of the specificity of CCRC-M1, the scientific community has been able to observe the dynamic localisation of its epitope *in planta* and explore potential physiological functions of fucosylated XGs (Shevell et al. 2000; Freshour et al. 2003; Sandquist et al. 2010; Brennan and Harris 2011; Zabotina et al. 2012; Lampugnani et al. 2013; Cornuault et al. 2014; Kong et al. 2015; Leroux et al. 2015; Brennan et al. 2019; Livingston et al. 2021; Molina et al. 2021; Velasquez et al. 2021; Petrova et al. 2022).

Epitope: Xyloglucan Multimers (XXXGn)

Oligosaccharide screens against multimers of $XXXG_n$ and $XLLG_n$ uncovered a group of seven XG-directed mAbs with an apparent chain length requirement for epitope recognition (Tuomivaara et al. in preparation). CCRC-M49, CCRC-M57, and CCRC-M86 comprise a subgroup that binds multimers of at least $n=3$, also referred to as S3 subunits. CCRC-M51, CCRC-M54, CCRC-M58, and CCRC-M99 can recognise slightly shorter XG chains, and preferentially bind multimers of at least $n=2$, also referred to as S2 subunits. These mAbs have yet to be further characterised, which illustrates an important consideration for the interpretation of data from polysaccharide- and oligosaccharide-based microarrays. While these tools can uncover critical information regarding the binding specificities of mAbs, the library of available epitope structures for screening does not cover the full range of length and side-chain diversity present *in vivo*. Rigorous experimental designs that employ biochemical analysis, glycan-based arrays, and immunolabelling techniques are therefore essential to enhance the utility of mAbs.

The breadth and diversity of the available collection of XG-directed antibodies has allowed exploration of the diversity of XG glycoforms, demonstrating that different XGs are deposited in different tissue types (Brennan and Harris 2011; Leroux et al. 2015; Brennan et al. 2019), different cell types (Freshour et al. 2003; Cavalier et al. 2008; Zabotina et al. 2008; Leroux et al. 2015; Petrova et al. 2022), and different wall domains both in relation to surrounding cells (Freshour et al. 1996; Cavalier et al. 2008; Peña et al. 2012; Zabotina et al. 2012) and with respect to sub-domains across a single wall (Freshour et al. 1996; Sandquist et al. 2010; Donaldson and Knox 2012). These observations emphasise the need to use multiple mAbs against diverse XG epitopes in order to obtain a complete picture of XG localisation patterns and to gain deeper insights into XG function in plant cell wall biology.

Xylan-Directed Antibodies

Xylans are a class of hemicellulosic polysaccharides that are present in great abundance in secondary plant cell walls. Based on a backbone of β-1,4-linked xylosyl residues, xylans can be substituted with *O*-acetyl, α-arabinofuranosyl, β-xylosyl, α-glucuronosyl, and/or 4-*O*-methyl-α-glucuronosyl residues. The type and degree of substitution patterning varies across different plant tissues and species, which contributes to the diverse utility of this polymer in industrial and pharmaceutical applications (Doering et al. 2012; Rennie and Scheller 2014; Anderson anjd Kieber 2020). The tunable design of xylan derivatives requires a better understanding of how the functional properties of this polymer are controlled in the assembly of the plant cell wall. As a result, numerous mAbs have been developed to study the organisation and regulation of various xylan structures (Figure 10.4).

Epitope: Xylan Backbone

Several sets of xylan-directed antibodies recognise the β-1,4-linked xylosyl backbone of this polymer. These and other xylan-directed antibodies were initially broadly categorised based on hierarchical clustering of their binding profiles to plant xylan preparations (Moller et al. 2008; Pattathil et al. 2010). Glycan microarray experiments using linear and substituted xylan oligosaccharides of up to a degree of polymerisation (DP) of eight have distinguished individual mAbs and their corresponding epitopes according to chain length requirement and tolerance

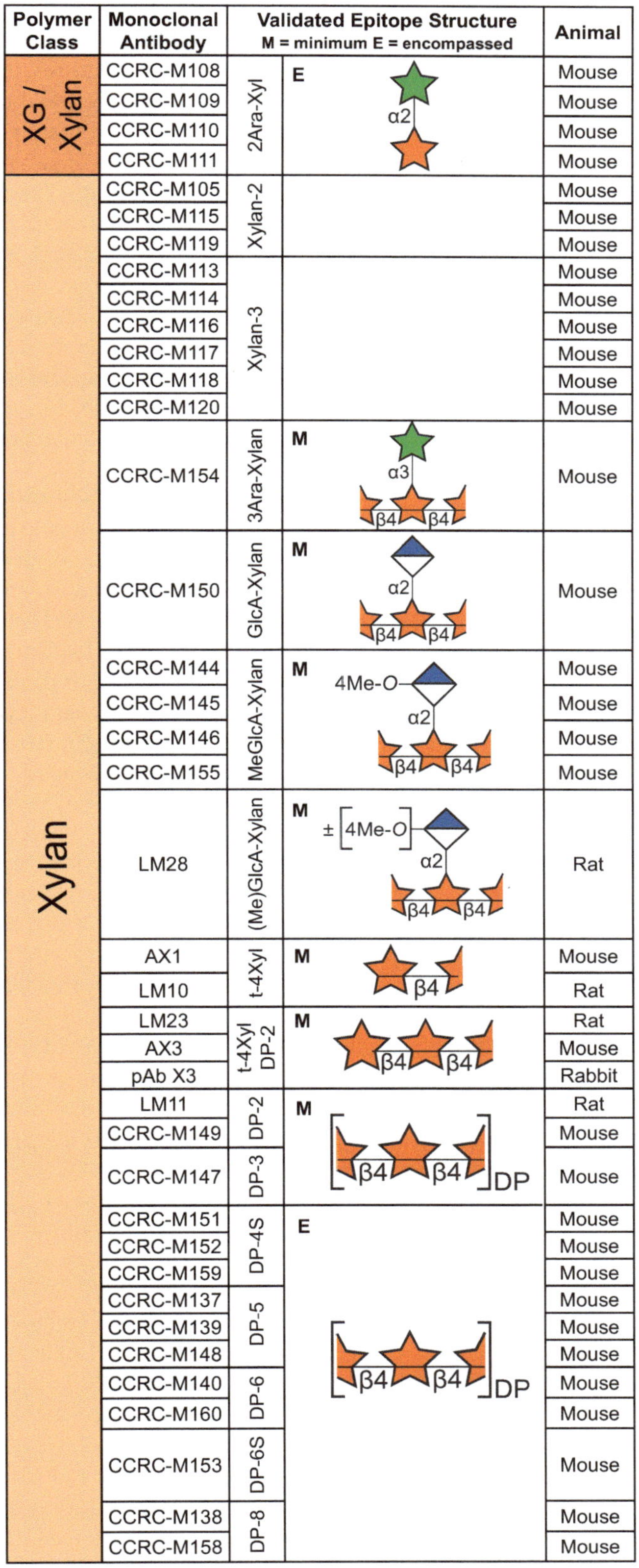

FIGURE 10.4 Xylan-directed monoclonal antibodies. Experimentally validated epitope structures for xylan-directed mAbs are shown. t-4Xyl indicates a binding specificity for the non-reducing terminus of the β-1,4-linked xylosyl backbone. DP, or degrees of polymerisation, designates the minimum number of unsubstituted backbone residues that are required for antibody recognition. S, such as in DP-4S, indicates that the presence of substituents on the backbone residues does not interfere with antibody recognition. The representative schematics of each epitope are labelled with M or E. M indicates that the schematic shown depicts the minimum structure required for recognition. E indicates that the minimum epitope structure is encompassed in the depicted schematic, but additional information is needed for further refinement.

for substituents (Pedersen et al. 2012; Schmidt et al. 2015; Ruprecht et al. 2017). To responsibly interpret microarray analyses, it must be acknowledged that the structures available for oligosaccharide-based screens do not completely represent the diverse xylan compositions in the plant cell wall. This is particularly important when binding profiles indicate a preference for longer chains of the xylan backbone, as there is a positive correlation between the DP and potential substitution pattern of an oligosaccharide. Nonetheless, the xylan backbone–directed antibodies can be classified into three groups. The first includes mAbs that will bind only to unsubstituted stretches of the backbone, while the second group tolerates side-chain substituents. The third group of antibodies binds to the non-reducing terminus of the backbone. The first two groups are further subdivided based on the minimum length of the backbone that is required for antibody recognition.

The mAbs LM11 (McCartney et al. 2005), CCRC-M149 (Pattathil et al. 2010), and CCRC-M147 (Pattathil et al. 2010) recognise short chains of β-1,4-linked xylosyl residues with low tolerance for substitution. Glycan microarray analyses indicate that the minimum epitope of LM11 and CCRC-M149 consists of a β-1,4-linked xylosyl disaccharide (Pedersen et al. 2012; Schmidt et al. 2015; Ruprecht et al. 2017). The binding profile of CCRC-M147 differed slightly in these oligosaccharide screens, and suggests a requirement for a β-1,4-linked xylosyl trisaccharide for optimal recognition.

CCRC-M137, CCRC-M139, and CCRC-M148 target unsubstituted xylan backbones of DP=5 or greater. CCRC-M140 and CCRC-M160 target somewhat longer unsubstituted backbone regions with a DP of six or larger, while CCRC-M138 and CCRC-M158 require at least eight unsubstituted xylosyl residues. For these longer epitopes, it should be noted that the number of xylan oligosaccharides of this length that are available is quite limited, especially regarding substitution. Hence, further refinement of these epitopes may occur as additional compounds become available. Integration of xylan-degrading enzymes with binding assays has proven to be a valuable approach for overcoming the limitations of chemical synthesis. For example, CCRC-M148 was able to bind certain arabinoxylans after treatment with arabinofuranosidases, ultimately leading to a refinement in chain length requirement and substitution tolerance of this antibody (Schmidt et al. 2015; Ruprecht et al. 2017).

Several xylan backbone–directed antibodies tolerate backbone substituents. These include CCRC-M151, CCRC-M152, and CCRC-M159, which require a minimum backbone length of four xylosyl residues. CCRC-M153 also appears to accept backbone substituents but requires six xylosyl backbone residues. Each of these antibodies binds to linear xylooligosaccharides of the required DP but will also bind to xylosyl oligosaccharides carrying a single substituent.

Lastly, there is a group of five antibodies that appear to bind to the non-reducing termini of xylan backbones. The mAbs AX1 (Guillon et al. 2004; Ordaz-Ortiz et al. 2004) and LM10 (McCartney et al. 2005) display strong binding to all synthetic oligosaccharides except those with an arabinose substituent on the non-reducing terminal backbone xylosyl residue. Enzymatic removal of this arabinosyl residue restored the binding capability of LM10 and thus established that the epitope for AX1 and LM10 is an unsubstituted terminal xylose at the non-reducing end of xylan oligosaccharides. The mAbs LM23 (Ruprecht et al. 2017), AX3 (Guillon et al. 2004; Ordaz-Ortiz et al. 2004), and pAb X3 (Guillon et al. 2004) are slightly more selective probes that also target the non-reducing end of xylan. In contrast to AX1 and LM10, LM23, AX3, and X3 require two unsubstituted terminal xylose residues for effective binding. AX3 also appears to preferentially bind arabinoxylans, while AX1, LM10, and LM23 are capable of binding both arabinoxylans and glucuronoxylans.

Epitope: Xylan Arabinosyl Substituents

Arabinoxylans represent a major dietary fibre component in cereal grains and are formed by the addition of α-1,2- and/or α-1,3-linked arabinofuranosyl residues to the xylan backbone (Ebringerová and Heinze 2000; Rennie and Scheller 2014). The degree of substitution, in which xylose residues may be un-, mono-, or disubstituted, influences the charge and solubility of arabinoxylan polysaccharides, resulting in varied arabinoxylan structure and distribution throughout primary and secondary cell walls (Burton et al. 2010). Hierarchical clustering analysis from an ELISA-based polysaccharide screen partially characterised two clades of mAbs that recognise low- or high-substituted arabinoxylans (Pattathil et al. 2010). Subsequent glycan-based microarrays against synthetic oligosaccharides enabled the refinement of the epitope specificities of a number of these mAbs (Schmidt et al. 2015; Ruprecht et al. 2017).

One set of antibodies comprised of CCRC-M108, CCRC-M109, CCRC-M110, and CCRC-M111, each of which were generated using xylan preparations from *Phormium* spp. (New Zealand flax), recognise linear chains of β-1,4-linked xylan that are substituted with an α-1,2-linked arabinosyl substituent. In contrast to 3-linked arabinoxylans, 2-linked arabinoxylans are less densely substituted to promote hydrogen-bonding interactions with unsubstituted xylan chains or other cell wall polysaccharides (Lee et al. 2001). These antibodies also cross-react with arabinosylated epitopes in some XGs, and exemplify an epitope-selective, rather than a polymer-selective, binding profile (Figure 10.1). CCRC-M108 can be distinguished from the other three antibodies in this group because it recognises an XDDG XG oligosaccharide, while the others only bind to larger XG oligosaccharides. Further experiments will be necessary to characterise the XG structure(s) that these antibodies bind to.

One antibody, CCRC-M154, binds exclusively to xylan oligosaccharides carrying one or more substituents consisting of single α-1,3-linked arabinofuranosyl residues (Schmidt et al. 2015; Ruprecht et al. 2017). This antibody does not bind to linear unsubstituted xylan backbones, nor does it bind to oligosaccharides where the side-chains consist of two arabinosyl residues. Further support for the selectivity and specificity of CCRC-M154 towards α-1,3-linked arabinofuranosyl residues linked to xylans was obtained from studies using arabinoxylan-specific arabinofuranosidases (McCleary et al. 2015) together with *in vitro* and *in planta* binding experiments, in which enzymatic removal of arabinose decorations was sufficient to abolish CCRC-M154 binding.

Epitope: Xylan Glucuronosyl Substituents

Glucuronoxylans, the predominant hemicellulose in the secondary walls of dicots, are substituted with α-1,2-linked glucuronic acid (GlcA) or 4-*O*-methyl glucuronic acid (MeGlcA). The backbone residues of glucuronoxylans contain finely regulated patterns of GlcA and MeGlcA that can be distinguished according to the spacing between each side-chain (Bromley et al. 2013; Grantham et al. 2017). The spacing and proportion of each substituent contributes to the molecular structure and physicochemical properties of individual xylans and thus varies across different plants and tissues (Ebringerová 2005; Doering et al. 2012; Rennie and Scheller 2014). Several mAbs that exclusively recognise unmethylated or methylated glucuronoxylans have been characterised by ELISA-based polysaccharide screens and synthetic glycan microarrays (Pattathil et al. 2010; Schmidt et al. 2015; Ruprecht et al. 2017). The mAb CCRC-M150 specifically recognises xylan oligosaccharides containing unmethylated α-1,2-linked GlcA, while CCRC-M144, CCRC-M145, CCRC-M146, and CCRC-M155 require methyl-etherification at *O*-4 to bind. The mAb LM28 binds to both methylated and unmethylated GlcA side-chains, as enzymatic removal of the α-1,2-GlcA substituents leads to significant binding inhibition. Furthermore, screening cell wall extracts from mutant plants deficient in MeGlcA (e.g. Urbanowicz et al. 2012; Li et al. 2013) generated a negligible change in total signal, indicating that methylation status does not influence target recognition by LM28 (Cornuault et al. 2015).

Epitope: Incomplete Characterisation

Two additional sets of xylan-directed mAbs have been generated; these are referred to in the literature as Xylan-2 and Xylan-3 antibodies (Pattathil et al. 2010). The Xylan-2 group, consisting of CCRC-M105, CCRC-M115, and CCRC-M119, is highly selective for the Phormium xylans against which they were generated (Pattathil et al. 2010). To date, these antibodies have not been observed to bind to xylan preparations from any other plant. The Xylan-3 antibodies, CCRC-M113, CCRC-M114, CCRC-M116, CCRC-M117, and CCRC-M118, bind to xylans from monocot grasses, but do not bind to xylans from dicotyledonous plants or to xylans from non-Commelinoid monocots (Figure 10.1). Thus, these two groups of antibodies are unique among the plant glycan–directed antibodies in that they are selective for epitopes found only in specific plant taxa. Unfortunately, no detailed information about the epitopes recognised by these two sets of antibodies has been obtained. The unique epitope recognised by these antibodies apparently is not represented within the available library of xylan oligosaccharides used in ELISA-based screens and glycan microarrays.

Mannan-Directed Antibodies

Heteromannans are hemicellulosic polysaccharides present in the primary and secondary walls of all angiosperms (Selvendran and Ryden 1990; Silva et al. 2011). These structurally diverse polymers can be distinguished by backbone composition and pattern of substitution. Mannans and glucomannans are based on a β-1,4-linked backbone of mannose residues, the latter of which is randomly interspersed with glucose residues. Galactomannans and galactoglucomannans arise from the presence of α-1,6-galactose substituents, with large variations in mannose to galactose ratios in different plant tissues and species (Ebringerová et al. 2005; Gilbert 2010; Prajapati et al. 2013). Recent advancements in the structural characterisation of these polymers have suggested that heteromannans play critical roles in cell wall structure, carbohydrate storage, and metabolic networks (Liepman et al. 2007; Yin et al. 2011; Voiniciuc et al. 2015, 2018; Yu et al. 2018). Currently, there are a limited number of mAbs available for the recognition of mannan polysaccharides (Figure 10.5). To gain a comprehensive understanding of this polymer's functional versatility, it is essential that additional tools for studying the structural dynamics of mannan continue to be developed.

Epitope: Heteromannans

The mAb BGM C6 (also referred to as BS-400–4 in some publications) targets β-1,4-linked mannopyranosyl residues within mannan and galactomannan polysaccharides (Pettolino et al. 2001). The minimum epitope required for binding was identified as a mannose tetrasaccharide in indirect and competitive ELISA screens (Pettolino et al. 2001), as well as in glycan microarrays (Pedersen et al. 2012). BGM C6 will bind to galactomannans from a variety of plant species, probably because unsubstituted stretches of the backbone are available in the polysaccharides for antibody binding (Pettolino et al. 2001). Two other mannan-directed mAbs, LM21 and LM22, have also been used frequently for the identification of β-1,4-linked mannan polysaccharides and bind a wide range of heteromannans. ELISA-based screens and glycan microarrays against mannan-, glucomannan-, and galactomannan-derived oligosaccharides have determined that each mAb is capable of recognising mannan and glucomannan polymers (Marcus et al. 2010; Pedersen et al. 2012). The binding profiles differ in the recognition of galactomannan, as only LM21 can effectively bind polymers appended with α-1,6-galactose substituents. LM22 differs from BGM C6 in that it will bind to a β-1,4-linked manno-disaccharide (Pedersen et al. 2012), while BGM C6 requires a longer oligosaccharide for binding (Pettolino et al. 2001). While LM21 and LM22 lack the specificity to distinguish between heteromannans that consist of a mannan or glucomannan backbone structure, they continue to serve as important tools for studying the spatial

Polymer Class	Monoclonal Antibody	Validated Epitope Structure M = minimum E = encompassed		Animal
Mannan	CCRC-M70	galactomannan-1	E 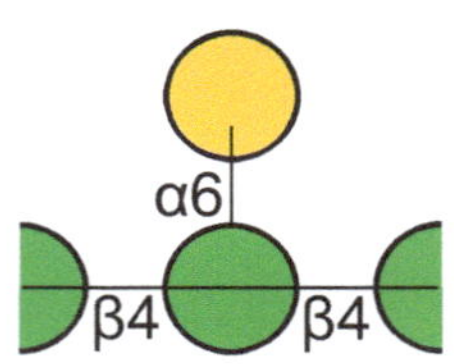	Mouse
	CCRC-M74			Mouse
	CCRC-M75			Mouse
	CCRC-M166	galactomannan-2		Mouse
	CCRC-M168			Mouse
	CCRC-M174			Mouse
	CCRC-M175			Mouse
	LM21	β-4 mannan DP-3S	M 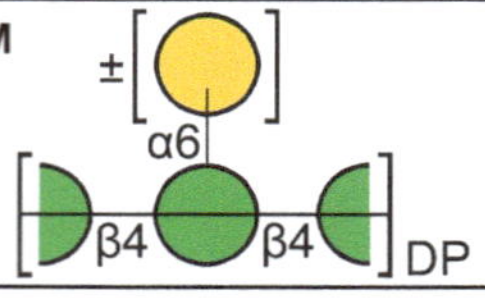	Rat
	LM22	β-4 mannan DP-2	M 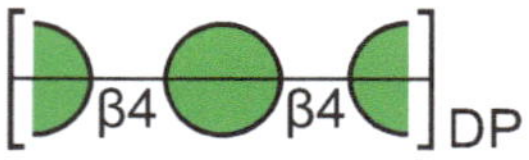	Rat
	BGM C6	β-4 mannan DP-4		Mouse
	CCRC-M169	acetylated mannan	M 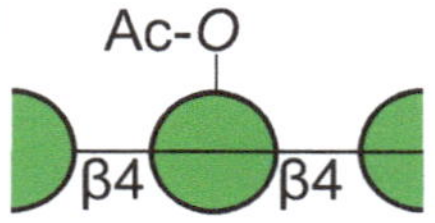	Mouse
	CCRC-M170			Mouse

FIGURE 10.5 Mannan-directed monoclonal antibodies. Experimentally validated epitope structures for mannan-directed mAbs are shown. DP, or degrees of polymerisation, designates the minimum number of unsubstituted backbone residues that are required for antibody recognition. S, such as in DP-4S, indicates that the presence of substituents on the backbone residues do not interfere with antibody recognition. The representative schematics of each epitope are labelled with M or E. M indicates that the schematic shown depicts the minimum structure required for recognition. E indicates that the minimum epitope structure is encompassed in the depicted schematic, but additional information is needed for further refinement.

distribution of heteromannans within the macromolecular architecture of the cell wall, particularly in distinct plants, tissues, and developmental stages (Marcus et al. 2010; Hernandez-Gomez et al. 2015; Muszyński et al. 2015; Yu et al. 2018; Petrova et al. 2022).

Heteromannans are often substituted at *O*-2 and *O*-3 with acetyl groups, with the degree of acetylation corresponding to the gel-like properties of the polymer (Capek et al. 2002; Ebringerová et al. 2005). CCRC-M169 and CCRC-M170 have been characterised and found to have a feature not yet seen within the library of plant cell wall–directed probes, as both recognise acetylated mannans (Zhang et al. 2014; Pattathil et al. 2015). Removal of the acetyl esters on the mannans leads to loss of antibody recognition (Zhang et al. 2014). These mAbs have been used to investigate the structural and functional role of acetylated mannan in the plant cell wall (Yu et al. 2014; Molina et al. 2021), and have also demonstrated important utility in the research of mannan-directed esterases (Zhang et al. 2014).

Epitope: Galactomannans

Galactomannans are heterogeneous in both composition and function; the degree of α-galactose substitution directly influences the ability to interact with surrounding polymers, and thus can control the rheological properties of the cell wall (Buckeridge et al. 2000; Prajapati et al. 2013). Hierarchical clustering analysis from ELISA-based screens has identified two distinct clades of mAbs with the ability to recognise galactomannans isolated from natural plant sources, although the binding profiles of each remain only partially characterised at this point (Pattathil et al. 2010). The first group, comprised of CCRC-M70, CCRC-M74, and CCRC-M75, displayed a preference for galactomannan with mannose to galactose ratios of around 1.5:1. The second group of mAbs displayed a similar, but less selective binding profile. CCRC-M166, CCRC-M168, CCRC-M174, and CCRC-M175 were able to recognise galactomannan with mannose to galactose ratios of around 1.5-4:1, and additionally showed weak binding to sycamore maple XG and mixed-linkage glucans. Immunolabelling studies using these mAbs has uncovered unique galactomannan patterning across distinct cells and tissues in developing plants (Hall et al. 2013; Muszyński et al. 2015; Molina et al. 2021), although it is essential that novel mAbs for galactomannan epitopes continue to be developed and characterised in detail. The mannose to galactose ratio of galactomannan can range from 1:1 to 20:1 and as many as 33% of the backbone residues can be substituted by acetyl groups (Capek et al. 2002; Ebringerová et al. 2005). A library of target-specific mAbs that accommodate this diversity would facilitate important research focused on the functional regulation of mannans in plants.

β-Glucan-Directed Antibodies

Antibodies against two types of glucan polymers, mixed-linkage glucans and callose, found in plant cell walls have been generated and characterised (Figure 10.6). Mixed-linkage (1,3)(1,4)-β-D-glucans belong to a family of hemicellulosic matrix polysaccharides that are believed to undergo extensive structural modification after synthesis to contribute to the dynamic mechanical properties of the cell wall (Ebringerová et al. 2005; Burton et al. 2010). While hemicellulose polymers are typically modified with various sugar and non-sugar substituents, such as in XG and xylan, unbranched mixed-linkage glucans are assembled by introducing irregularly spaced β-1,3-linkages within a backbone composed predominantly of β-1,4-linked glucosyl residues. The distribution of β-1,3-linkages directly affects the structural conformation and solubility of a single mixed-linkage polysaccharide chain, and if multiple polysaccharide chains differ in this distribution, they are less likely to aggregate together within the cell wall (Lazaridou and Biliaderis 2007; Fincher 2009). The physicochemical properties of this polymer are likely to be regulated by plants to support distinct morphologies in the primary wall. Mixed-linked glucans are well-known dietary components due to their high abundance in grass and cereal species, and thus are quite relevant in numerous nutritional and physiological applications, which have generated industry-wide interest in the ability to modulate the functional properties of this polymer (Buckeridge et al. 2004; Lazaridou and Biliaderis 2007).

Callose is rapidly synthesised and degraded in specialised tissues with developmental and environmental implications in plants (Anderson and Kieber 2020). Specifically, callose is one of the first polysaccharides to be deposited to the cell wall in developing plants and is the dominant component of growing pollen tubes (Li et al. 1999; Doblin et al. 2001). Furthermore, callose is frequently deposited at sites of infection by plant pathogens as part of the plant's disease defence responses (Benhamou 1995).

Epitope: Mixed-Linkage Glucans

BG1 is currently the only available mAb for the recognition of β-(1,3)(1,4)-linked glucan oligosaccharides. Competitive inhibition assays have characterised BG1 to have a high affinity to mixed-linkage glucooligosaccharides with little to no recognition of β-(1,3)-linked glucans (Meikle et al. 1994). Immunolabelling studies have demonstrated that BG1 can detect mixed-linkage glucan that comprises less than 0.1% of the cell wall (Burton et al. 2006). While this level of specificity would be unattainable through biochemical or enzymatic analyses, the utilisation of BG1 has uncovered many intricate details regarding the subcellular location of mixed-linkage glucans during cell wall synthesis (Burton et al. 2006; Fincher 2009; Carpita and McCann 2010; Wilson et al. 2015) and plant development (Carpita et al. 2001; Trethewey and Harris 2002; Kozlova et al. 2014; Petrova et al. 2022).

Epitope: β-1,3-Linked Glucans

An additional mAb that targets the β-1,3-linked glucan, callose, has been grouped here, although this polymer does not have an equatorial β-(1→4)-linked backbone structure and thus is not considered a hemicellulose (Scheller and Ulvskov 2010). The mAb LAMP, sometimes referred to as BS-400–2, recognises a β-1,3-glucan epitope and has been extensively characterised through competitive inhibition

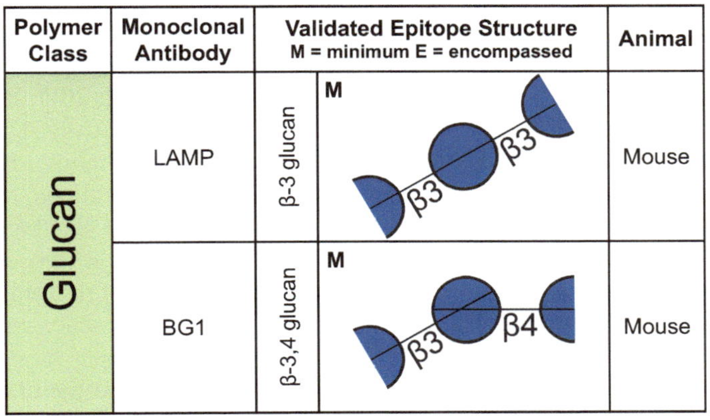

FIGURE 10.6 Glucan-directed monoclonal antibodies. Experimentally validated epitope structures for glucan-directed mAbs are shown. The representative schematics of each epitope are labelled with M or E. M indicates that the schematic shown depicts the minimum structure required for recognition. E indicates that the minimum epitope structure is encompassed in the depicted schematic, but additional information is needed for further refinement.

assays, ELISA-based screens, and glycan arrays (Meikle et al. 1991; Moller et al. 2008; Pedersen et al. 2012). Early uses of LAMP aimed to identify the functional role of callose in developmental tissues (Brown et al. 1997; Nickle and Meinke 1998; Nishikawa et al. 2005; Wilson et al. 2006), and instead, revealed an accumulation of callose at plasmodesmal channels and thus a potential role in cell-to-cell communication (Salnikov et al. 2003; Ruan et al. 2004). Together with the identification of the enzymes responsible for callose synthesis, deposition, and degradation (Schlüpmann et al. 1994; Cui et al. 2001; Doblin et al. 2001), the accomplishments of callose-directed immunolocalisation have likely contributed to almost three decades of continued interest in this polysaccharide. Modern uses of LAMP are notably more versatile due to the established functions of callose in organ development and plant environmental responses. LAMP has been used in studies of callose deposition in various plant species (Avci et al. 2013; Raimundo et al. 2017) and is currently a widely used tool for elucidating the effects of callose on the permeability of plasmodesmal channels under pathogen infection and mechanical wounding stress (Cui and Lee 2016; Amsbury et al. 2018).

Hemicellulose-Directed Antibodies: Recent Advancements and Prospects

This library of hemicellulose-directed antibodies can collectively recognise most of the common hemicellulosic epitopes found in plant cell walls. XGs are one of the most abundant hemicellulosic polysaccharides in the primary cell wall of dicots. mAbs raised against XG polymers and oligosaccharides as antigens are some of the most well-studied and used mAbs to date, and antibodies binding to all common side-chains have been identified. XGs can also be substituted with arabinose and galacturonic acid residues and acetyl groups, although mAbs that target these modifications have yet to be developed (Julian and Zabotina 2022). Several of the described mAbs have been used to observe the complex formation and localisation of XG polysaccharides within the Golgi (Chevalier et al. 2010; Parsons et al. 2019), and will continue to serve as important tools for studying the regulation and quality assurance of XG biosynthesis.

In contrast to XG, xylans are often associated with the secondary cell walls of both monocots and dicots. Glycome profiling with xylan-directed mAbs has been extensively used to study cell wall structural changes in industrial applications, particularly in the characterisation, pretreatment, and deconstruction of the world's biomass (Alonso-Simón et al. 2010; DeMartini et al. 2011; de Souza et al. 2013; Liu et al. 2014; Socha et al. 2014; Pattathil et al. 2015). Immunolabelling studies have been used to monitor the spatial and temporal regulation of this polymer and uncover the diverse patterns associated with xylan biosynthesis and deposition into the cell wall (Hao et al. 2014; Marcotuli et al. 2016; Peralta et al. 2017; Meents et al. 2019; Melida et al. 2020; Velasquez et al. 2021). Xylans, and most other plant cell wall polysaccharides, can be acetylated to varying degrees, but there are insufficient immunocytochemical methods for locating *O*-acetyl substituents. Specifically, there is a critical need for novel mAbs that target acetylated xylans for observing how the location(s) of this modification contributes to the structural and functional diversity of cell walls. Studies using plants carrying mutations that affect xylan structure have implicated xylan acetylation in numerous developmental and immune responses (Xiong et al. 2013; Urbanowicz et al. 2014; Escudero et al. 2017), and additionally revealed complex interactions between xylans and other critical cell wall components, such as cellulose and pectins (Tan et al. 2013; Cornuault et al. 2015; Grantham et al. 2017; Broxterman and Schols 2018).

To date, the only characterised mAbs that target an acetylated plant cell wall glycan are the mannan-directed mAbs CCRC-M169 and CCRC-M170. Heteromannans are widely distributed across the plant kingdom and are abundant in algae, bryophytes, and some land plants (Voiniciuc 2022). The occurrence and structure of heteromannans are incredibly variable and immunolabelling studies have been used to identify cell-specific patterns of heteromannans in several plant species (Yang et al. 2016; Guedes et al. 2017; Chavan et al. 2021; Petrova et al. 2022). Heteromannan acetylation influences the physicochemical properties of this polymer, and thus could have important implications in the application of mannans in the pharmaceutical, food, and textile industries (Singh et al. 2018). Therefore, the ability to specifically locate acetylated mannans can be used to enhance targeted genomics efforts focused on the identification of the genes responsible for mannan acetylation and deacetylation (Yu et al. 2014; Zhang et al. 2014; Voiniciuc et al. 2015).

As has been demonstrated with mannan- and β-glucan-directed antibodies, even a small number of well-characterised mAbs can be used to observe how the location or abundance of an epitope may correlate to dynamic cell wall reconfigurations. However, the utility of a single probe is not likely to expand beyond localisation studies. For example, mixed-linkage glucans, which are present in bacterial and fungal species and monocot plants (Fincher 2009), have been suggested to trigger pattern recognition receptors in immune response and disease resistance signalling pathways (Salmeán et al. 2017; Barghahn et al. 2021; Rebaque et al. 2021). The mAb BG1 is insufficient for the identification and differentiation of potential immunogenic fragments; therefore, additional mixed-linkage glucan-directed antibodies that can distinguish between heterogeneous chain compositions are necessary to investigate the functional significance of this polymer. In a more general sense, studies using the broader collection of antibodies against hemicellulosic epitopes have repeatedly demonstrated that a clear and complete picture of the localisation pattern(s) of a given polymer requires the use of mAbs against multiple epitopes in that polymer; use of a single mAb against one epitope is unlikely to yield an accurate or complete assessment.

RECOGNITION OF PECTIN BACKBONES

HOMOGALACTURONAN (HG)-DIRECTED ANTIBODIES

Pectins are the most structurally complex polysaccharides in the primary cell wall and undergo constant physical and chemical remodelling to control wall mechanical properties (Anderson 2016; Qiu et al. 2021). HG is an abundant domain within pectins and consists of an unbranched, linear chain of α-1,4-linked galacturonic acid residues appended with *O*-methyl and *O*-acetyl substituents at the C-6 and *O*-2/*O*-3 positions, respectively (Mohnen 2008, Chapter 5). Pectin methylesterases (PMEs) and pectin acetylesterases (PAEs) precisely regulate HG substitution to control the strength and elasticity of the primary wall, and numerous mAbs have been generated that bind to HG with different degrees of methylesterification (Figure 10.7). These tools have been critical for unravelling the structural complexity of the pectin network, although we have yet to fully understand the mechanisms responsible for its multifaceted influence on plant fitness.

Epitope: Demethylesterified HG

The mAbs CCRC-M38, CCRC-M132, JIM5, LM18, and LM19 preferentially bind HG epitopes with low degrees of methylesterification, although each mAb presents slight variations in tolerance for methyl substitution (Verhertbruggen et al. 2009a; Pattathil et al. 2010). JIM5 was among the first of this group to be extensively characterised (Knox et al. 1990). Competition assays against partially methylesterified hapten inhibitors demonstrated that JIM5 recognises HG epitopes consisting of four unsubstituted α-1,4-linked galacturonic acid residues and tolerates adjacent or flanking methyl-ester substituents (Clausen et al. 2003). Subsequent hierarchical clustering analysis from an ELISA-based polysaccharide screen designated CCRC-M38 and CCRC-M132 to have a similar binding profile to JIM5. However, CCRC-M38 binds to fully de-esterified oligogalacturonides with a degree of polymerisation of at least four (DP≥4) (Pattathil et al. 2010), while JIM5 requires at least a DP of seven to bind fully de-esterified oligogalacturonides. A follow-up study generated an additional set of mAbs, including LM18 and LM19, to potentially identify unesterified HG with a higher degree of specificity (Verhertbruggen et al. 2009a). Like JIM5, both mAbs were able to recognise partially methylesterified HG in several species. However, LM19 was determined to have a higher binding affinity to unesterified HG. Recently published data using glycan microarrays suggest that, like JIM5, LM19 tolerates dispersed methyl-esterification of the galacturonosyl residues (Ruprecht and Pfrengle 2022).

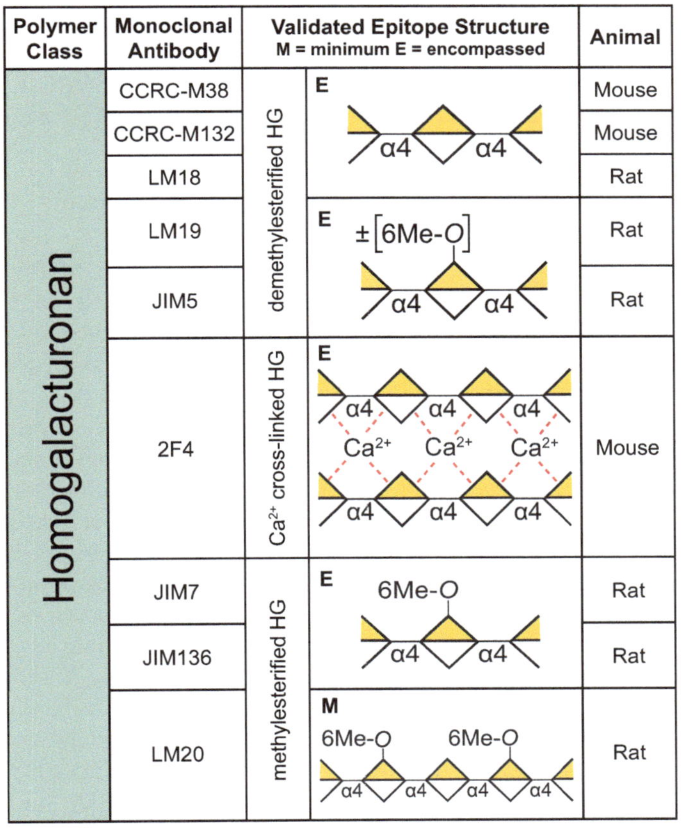

FIGURE 10.7 Homogalacturonan-directed monoclonal antibodies. Experimentally validated epitope structures for homogalacturonan-directed mAbs are shown. The representative schematics of each epitope are labelled with M or E. M indicates that the schematic shown depicts the minimum structure required for recognition. E indicates that the minimum epitope structure is encompassed in the depicted schematic, but additional information is needed for further refinement.

Demethylesterified HG can form cross-links by calcium bridges to support the structural integrity of the pectin network (Marry et al. 2006; Zamil and Geitmann 2017). The mAb 2F4 (Liners et al. 1989) targets a calcium-induced conformation of HG and thus is distinct from the previously described group of mAbs. Initial characterisation of 2F4 was accomplished through ELISA-based screens utilising HG polysaccharides in the presence or absence of divalent cations, which determined the requirement of calcium for pectin recognition (Liners et al. 1989). The influence of the DP and degree of methylesterification on recognition by 2F4 was investigated in a subsequent ELISA-based screen (Liners et al. 1992b), which revealed that two HG oligomers comprised of at least nine galacturonic acid residues with degree of methylesterification <40% are required for dimer formation and recognition by 2F4. While additional specificity screens are needed to confirm the DP and degree of methylesterification requirements of the 2F4 epitope, the binding and immunocytochemical studies performed to date support the idea that calcium ions are necessary to induce the structural conformation recognised by 2F4 but are not directly associated with the epitope structure (Liners and Van Cutsem 1992a).

Epitope: Methylesterified HG

JIM7, JIM136, LM7, and LM20 comprise a set of mAbs that target methylesterified HG and do not bind to unmethylated α-1,4-linked galacturonic acid residues. It is challenging to identify the precise degree of methylesterification that is recognised by each mAb, as the available synthetic HG oligosaccharides used for screening are insufficient representations of the diverse methyl-ester substitution patterns present in the plant cell wall (Willats et al. 2001). The collective use of hierarchical clustering analyses, competition assays, and glycan-based arrays facilitated the partial characterisation of this set of mAbs, and additionally uncovered several differences in epitope specificity (Knox et al. 1990; Clausen et al. 2003; Verhertbruggen et al. 2009a; Pattathil et al. 2010). LM7 was initially characterised to recognise an epitope consisting of both unesterified and methylesterified galacturonic acid residues (Willats et al. 2001). Refinement of this binding profile was accomplished through competitive inhibition assays, in which LM7, contrary to JIM7, preferentially bound to unesterified galacturonic acid residues with adjacent or flanking methylesterified residues (Clausen et al. 2003). Furthermore, JIM7 is characterised to recognise methylesterified residues regardless of the substitution status of adjacent or flanking residues (Clausen et al. 2003; Verhertbruggen et al. 2009a), contributing to its popularity as a probe for broad observations of methylesterification in plant cell walls. However, a comparative analysis of JIM7 and LM20 demonstrated that de-esterification of HG induced a greater loss of binding by LM20, suggesting that JIM7 binds less effectively to methylesterified HG epitopes (Verhertbruggen et al. 2009a). The binding profile of LM20 has been further refined using glycan microarrays against HG oligosaccharides with differential methylation patterns (Ruprecht and Pfrengle 2022). While LM19 recognised both regularly and irregularly spaced methylesterified galacturonic acid residues, LM20 bound only to oligosaccharides wherein every other galacturonosyl residue was methylesterified (Ruprecht and Pfrengle 2022).

Rhamnogalacturonan I (RG-I)-Directed Antibodies

The physical and chemical composition of the pectin network is incredibly fluid and readily reconfigures in response to cellular and environmental signals to support specific architectures across a vast array of plant tissues (Chebli et al. 2021). A critical component of this network involves the pectic polysaccharide RG-I, which is proposed to covalently associate with HG and rhamnogalacturonan II (RG-II) to maintain the structural integrity of the primary cell wall (Cosgrove 2005). The RG-I backbone is based on a repeating disaccharide unit of α-1,4-galacturonic acid and α-1,2-rhamnose residues and displays context-dependent variation in the type and degree of glycosyl substituents, which range from single sugars to branched arabinan, galactan, and arabinogalactan side-chains attached to the rhamosyl residues (Mohnen 2008; Burton et al. 2010). The precise regulation of RG-I structure is highly suggestive of a diverse array of functionalities, but the known complexity of this polymer introduces many obstacles for biochemical and immunohistochemical analyses. To improve our understanding of RG-I and the role it plays in the plant cell wall, several sets of mAbs have been generated to target distinct structural features of RG-I, including the unsubstituted backbone (Figure 10.8) and oligosaccharide side-chains (Figure 10.9).

Epitope: RG-I Backbone

Hierarchical clustering analysis identified the mAbs CCRC-M14, CCRC-M35, CCRC-M36, CCRC-M69, CCRC-M72, and CCRC-M129 as members of a clade that target the pectic backbone (Pattathil et al. 2010). Several HG-directed antibodies displayed similar binding capabilities against the panel of plant polysaccharides screened in this study, but a glycan microarray with synthetic pectic oligosaccharides determined CCRC-M14, CCRC-M35, CCRC-M36, and CCRC-M69 bind specifically to the RG-I backbone (Ruprecht et al. 2017). The exact epitope structure could not be determined because this characterisation was based on the recognition of a single unsubstituted RG-I hexasaccharide composed of three rhamnose-galacturonic acid disaccharide units [→2)-α-L-rhamnose-(1→4)-α-D-galacturonic acid-(1→]$_3$ (DP=6, R_3U_3). Immunolocalisation studies confirmed that the screened mAbs do not strongly bind to arabinogalactans, xylans, XGs, or methylesterified HGs (Young et al. 2008; Voiniciuc et al. 2018). The mAbs INRA-RU1 (RU1) and INRA-RU2 (RU2) also recognised the RG-I backbone oligosaccharide provided in the microarray (DP=6, R_3U_3). These two mAbs had been characterised previously through indirect and competitive ELISA-based screens, which showed that RU1 required a

Polymer Class	Monoclonal Antibody		Validated Epitope Structure M = minimum E = encompassed	Animal
Rhamnogalacturonan I	CCRC-M14	DP-6 (R_3U_3)	E 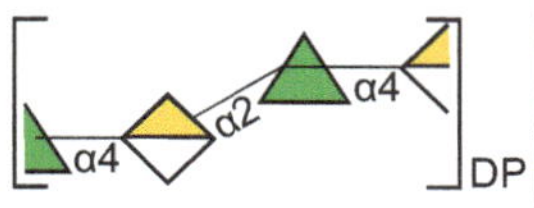	Mouse
	CCRC-M35			Mouse
	CCRC-M36			Mouse
	CCRC-M69			Mouse
	CCRC-M72			Mouse
	CCRC-M129			Mouse
	RU1	DP-12 (R_6U_6)	M 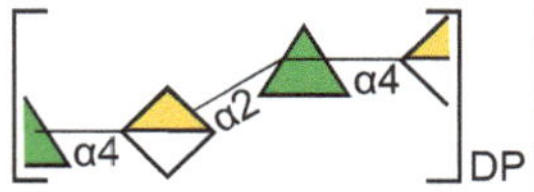	Mouse
	RU2	DP-4 (R_2U_2)		Mouse

FIGURE 10.8 Rhamnogalacturonan I (RG-I)-directed monoclonal antibodies. Experimentally validated epitope structures for RG-I-directed mAbs are shown. DP, or degrees of polymerisation, designates the minimum number of backbone residues (R = rhamnose; U = galacturonic acid) that are required for antibody recognition. The representative schematics of each epitope are labelled with M or E. M indicates that the schematic shown depicts the minimum structure required for recognition. E indicates that the minimum epitope structure is encompassed in the depicted schematic, but additional information is needed for further refinement.

minimum of six rhamnose-galacturonic acid disaccharide units (DP = 12, R_6U_6) for effective binding while only two were needed for RU2 (DP = 4, R_2U_2) (Ralet et al., 2010). Interestingly, the binding efficiency of RU1 decreased dramatically at DP > 14 and RU1 is able to accommodate galactose substitutions (Ralet et al. 2010). Neither of these two mAbs displayed binding to HG, galactan, or arabinan oligosaccharides.

Pectin-Directed Antibodies: Recent Advancements and Prospects

The stability of the pectin network is largely reliant on the regulation of HG methylesterification, as polysaccharides with low degrees of methylesterification are capable of cross-linking with calcium to enhance the mechanical strength of the primary cell wall (Cosgrove 2005; Yang and Anderson 2020). Unesterified HG backbone polymers are additionally subject to degradation by pectin-cleaving enzymes to promote cell wall extensibility during organ development (Peaucelle et al. 2015; Hocq et al. 2017; Voiniciuc et al. 2018). Unfortunately, the enzymes responsible for regulating HG methylesterification are largely unknown (Sénéchal et al. 2014). The current library of HG-directed probes can provide a general insight into the spatial regulation of HG methylesterification patterns during developmental and stress-induced reconfigurations of the cell wall (Sinha and Lynch 1998; Shevell et al. 2000; His et al. 2001; Orfila et al. 2001; Singh et al. 2009; Douchiche et al. 2010; Qi et al. 2017; Wang et al. 2020; Jonsson et al. 2021; Yang et al. 2022). To uncover the specific functional roles of HG heterogeneity, it is essential to develop probes that can distinguish both unique HG glycoforms and pectic polysaccharides that contain HG domains.

The alternating sequence of galacturonic acid and rhamnose residues in the backbone of RG-I plays a significant role in the strength and flexibility of higher order pectin conformations. In comparison to pectic polysaccharides that are based on an HG backbone, RG-I glycostructures are more compact and flexible and can maintain certain structural characteristics regardless of pH (Burton et al. 2010; Alba et al. 2015). While the majority of RG-I is highly substituted in the plant cell wall, the backbone is an important point of focus for understanding RG-I biosynthesis and its involvement in the formation of complex pectic macromolecules (Ulvskov et al. 2005). Immunolocalisation of the RG-I backbone has uncovered a correlation between this polymer and the mechanical strength of various cell walls (Young et al. 2008; Lee et al. 2013), and notably, numerous studies have also suggested the involvement of RG-I in spatial interactions with other major wall components (Tan et al. 2013; Cornuault et al. 2014, 2015). While the nature of these intermolecular interactions has yet to be revealed, such studies demonstrate critical implications of RG-I in the organisation and assembly of a functional cell wall.

Rhamnogalacturonan II (RG-II) is a structurally complex domain of pectin and consists of four oligosaccharide side-chains (A, B, C, and D) and two distinct α-L-Araf substituents (E and F) attached to the (1→4)-linked α-L-GalpA backbone (Funakawa and Miwa 2015). RG-II exists in the plant cell wall as a borate ester cross-linked dimer, in which a single borate diester is formed between the apiosyl residues of side-chain A (O'Neill et al. 1996). Reductions in the stability or content of dimerised RG-II typically results in swollen cell walls with reduced tensile strength, and ultimately reduces a plant's ability to capture, convert, and store energy (O'Neill et al. 2001). Unfortunately, the immunocytochemical tools generated

Polymer Class	Monoclonal Antibody	Validated Epitope Structure	M = minimum E = encompassed	Animal
Pectin- and Protein-linked Arabinogalactans	LM6	5 arabinan DP-2	M 	Rat
	LM13	5 arabinan DP-3		Rat
	LM17	5 arabinan DP-4		Rat
	LM5	β-4 galactan DP-3	M 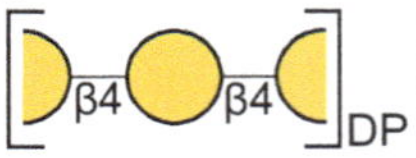	Rat
	CCRC-M107			Mouse
	CCRC-M133	β-4 galactan DP-6	E 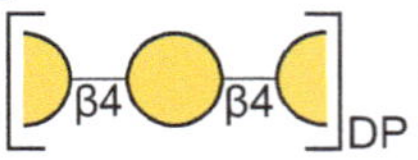	Mouse
	LM26	β-4 galactan w/ β-6 Gal	E 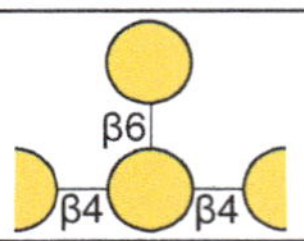	Rat
	LM9	Fer-β-4 galactan	E 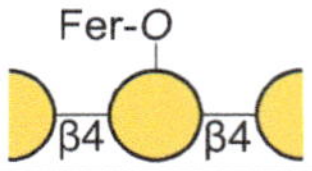	Rat
	CCRC-M8	β-6 galactan-1	M 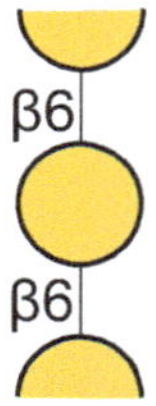	Mouse
	CCRC-M9			Mouse
	CCRC-M15			Mouse
	CCRC-M24			Mouse
	CCRC-M25			Mouse
	CCRC-M32			Mouse
	CCRC-M33			Mouse
	CCRC-M41			Mouse
	CCRC-M44			Mouse
	CCRC-M60			Mouse
	CCRC-M128			Mouse
	CCRC-M17	β-6 galactan-2		Mouse
	CCRC-M77			Mouse
	JIM131			Rat
	CCRC-M7	β-6 galactan-3		Mouse
	CCRC-M12			Mouse
	CCRC-M13			Mouse
	CCRC-M21			Mouse
	CCRC-M22			Mouse
	CCRC-M42			Mouse
	CCRC-M79			Mouse
	CCRC-M80			Mouse
	CCRC-M112			Mouse
	CCRC-M121			Mouse
	CCRC-M122			Mouse
	CCRC-M123			Mouse
	CCRC-M125			Mouse
	CCRC-M126			Mouse
	CCRC-M134			Mouse

FIGURE 10.9 Pectin- and protein-linked arabinogalactan-directed monoclonal antibodies. Experimentally validated epitope structures for pectin- and protein-linked arabinogalactan-directed mAbs are shown. DP, or degrees of polymerisation, designates the minimum number of backbone residues that are required for antibody recognition. The representative schematics of each epitope are labelled with M or E. M indicates that the schematic shown depicts the minimum structure required for recognition. E indicates that the minimum epitope structure is encompassed in the depicted schematic, but additional information is needed for further refinement.

to date are insufficient for target-specific imaging of RG-II. A limited number of polyclonal and monoclonal antibodies have been generated and proposed to target RG-II (Williams et al. 1996; Zhou et al. 2017, 2018). However, the recognised epitope and binding selectivity/specificity of these antibodies are poorly characterised; thus each are insufficient for the visualisation of RG-II in order to gain insight into how, where, and when this polysaccharide is assembled, disassembled, or relocated by plants to influence wall integrity and mechanics. The difficulty in generating bona fide RG-II-directed antibodies is an enigma, given the unique composition and structure of this polysaccharide. The 3D conformation of dimerised RG-II is not yet structurally characterised, and one possible explanation for the apparent lack of immunogenicity is that steric hindrance from this unique structure blocks access to potential binding sites that could be recognised by the immune system of the injected animal. Given the severe, and often lethal, consequences of altering RG-II structure for the plant, there is a critical need for selective and specific probes for RG-II to further our understanding of the roles and functions of this polysaccharide in plant cell wall biology.

RECOGNITION OF PECTIN- AND PROTEIN-LINKED ARABINOGALACTANS

Branched pectic polysaccharides, such as RG-I, contain structurally complex monosaccharide and oligosaccharide side-chains that influence the physical and chemical composition of the pectin network. The most common neutral side-chains of RG-I consist of arabinans, galactans, and types I and II arabinogalactans (AG-I and AG-II), each of which is attached to the rhamnosyl residues in the RG-I backbone (Mohnen 2008).

Arabinan side-chains are based on an α-1,5-linked arabinofuranosyl backbone and can be decorated at *O*-2 or *O*-3 by single arabinosyl residues or additional branched arabinan oligosaccharides. Linear galactan and AG-I substituents consist of a β-1,4-linked galactosyl backbone, in which AG-Is are branched with additional arabinan side-chains. Finally, AG-II side-chains contain a β-1,3-linked galactan backbone with β-1,6-linked galactan or arabinogalactan branches (Mohnen 2008; Yapo 2011).

Pectic side-chains are distinct in chemical composition, structural conformation, and physical properties, and therefore are capable of making extensive, noncovalent binding interactions with surrounding cell wall polymers. By tightly regulating the spatial interactions that form the macromolecular cell wall, plants are able to maintain diverse cell wall arrangements and facilitate processes like cell adhesion and elongation, plant fertility, and plant defence signalling networks (Yang and Anderson 2020; Shin et al. 2021; Forand et al. 2022). Several groups of mAbs have been generated in attempts to target-specific oligosaccharide structures that are commonly associated with RG-I side-chains. However, many of these epitopes are also present on other cell wall and membrane components, specifically glycolipids and arabinogalactan proteins (AGPs) (Chapters 16 and 17), and thus the corresponding mAbs are classified as arabinogalactan-directed rather than RG-I-directed (Figure 10.9).

EPITOPE: α-1,5-ARABINANS

LM6, LM13, and LM17 comprise a set of mAbs that target distinct epitopes within α-1,5-linked arabinans. LM6 was the first mAb of this group to be partially characterised and shows exclusive recognition of α-1,5-linked polyarabinose chains in immunodot binding assays against diverse plant extracts (Willats et al. 1998). High-throughput antibody screens found that the mAbs LM13 and LM17 were similar in binding profile to LM6, indicating that each can recognise epitopes within α-1,5-linked arabinans (Moller et al. 2008; Verhertbruggen et al. 2009b). Comparative epitope analysis of this set of mAbs was accomplished by utilising arabinan-degrading enzymes, including an *endo*-arabinase and multiple α-L-arabinofuranosidases that preferentially cleave branched α-1,2- and/or α-1,3-linked arabinosyl residues or linear α-1,5-arabinans, in combination with immunolabelling experiments on linear and branched arabinans. Enzymatic removal of branched arabinosyl residues, which elevated the abundance of linear arabinan epitopes, led to increased binding of LM6, LM13, and LM17. In contrast, enzymatic cleavage of the α-1,5 backbone appeared to only increase the binding of LM13. Taken together with the results of competitive inhibition assays, all three mAbs were partially characterised to recognise epitopes present in the α-1,5-linked arabinan backbone, with LM13 targeting longer arabinan oligosaccharides. These characterisations were confirmed and refined in subsequent studies utilising glycan-based microarrays (Pedersen et al. 2012; Ruprecht et al. 2017). The mAbs LM6 and LM17 have been established to bind similar epitopes consisting of short α-1,5-linked arabinans, in which the LM6 epitope requires a minimum of two arabinosyl residues (Pedersen et al. 2012). LM17 effectively binds to linear α-1,5-linked arabinans of DP=3 (Ruprecht et al. 2017) and LM13 preferentially binds longer chains of linear α-1,5-linked arabinans of DP=4 or greater (Pedersen et al. 2012).

EPITOPE: β-1,4-LINKED GALACTANS

The mAbs LM5, CCRC-M107, and CCRC-M133 recognise linear chains of β-1,4-linked galactan. LM5 was originally characterised through immunodot and indirect ELISA binding assays to target β-4 galactans in several plant polysaccharides (Jones et al. 1997). Partial characterisation of CCRC-M107 and CCRC-M133 was first established through hierarchical clustering analysis, in which LM5, CCRC-M107, and CCRC-M133 were categorised in a clade of mAbs that bind to linear and branched arabinans, but do not recognise arabinogalactans (Pattathil et al. 2010). Several clear distinctions in the binding profiles of these mAbs have been recently uncovered due to advancements in the chemical synthesis of linear and branched β-1,4-linked galactans. For example, competitive inhibition assays against synthetic β-1,4-galactotetraose analogues has established that LM5 targets the non-reducing end of β-1,4-linked galactans and requires three galactosyl residues for effective binding (Andersen et al. 2016). Multiple studies have used glycan-based microarrays to confirm the specificity of LM5 and refine the characterisation of CCRC-M107 and CCRC-M133, which are both suggested to target internal β-1,4 galactan residues within the side-chain (Pedersen et al. 2012; Ruprecht et al. 2017). CCRC-M107 requires a β-1,4 galactan epitope of at least three contiguous unsubstituted galactosyl residues (Ruprecht et al. 2017). In the same study, CCRC-M133 was capable of binding β-1,4 galactans with DP=6 or greater and did not recognise any substituted compounds, although only a limited number of substituted hexagalactosides were available for screening.

The mAb LM26 (Torode et al. 2018) recognises a β-1,6-galactosyl substitution of β-1,4-galactan oligosaccharides comprised of at least three backbone residues. Immunolocalisation studies demonstrate that the LM26 epitope is associated with a wide range of pectic polysaccharides throughout several plant species and tissues (Torode et al. 2018). LM26 binds with high affinity to cell wall phloem elements and displays a complementary binding pattern to that of LM5, which targets unsubstituted β-1,4-galactan epitopes (Petrova et al. 2022).

The mAb LM9 targets feruloyated β-1,4-linked galactans, a modification that is specific to plants belonging to the Amaranthaceae family (Clausen et al. 2004). The mAb LM12 similarly recognises an epitope containing ferulic acid, but is distinct in its ability to also bind heteroxylan polymers of commelinid monocotyledons (Pedersen et al. 2012), which are often feruloyated. The functional roles of this substitution are not yet well understood, but it has been suggested that feruloyated RG-I side-chains can cross-link together by ferulate diester bonds to contribute to cell-cell adhesion (Yapo 2011; Anderson and Kieber 2020). Ferulate side-chains of xylan may also contribute to cross-linking RG-I and arabinoxylan (Zamil and Geitmann 2017). The continued development and characterisation of selective probes for cell wall-bound phenolics like ferulic acid and coumaric acid will be necessary to uncover the details of such specialised functions in the plant cell wall.

Epitope: β-1,6-Linked Galactans

Hierarchical clustering analyses from high-throughput antibody screens have identified several mAb clades that are specific to arabinogalactans (Moller et al. 2008; Pattathil et al. 2010). One such mAb clade, identified as RG-I/AG, was broadly characterised to bind RG-I polysaccharides isolated from diverse plant materials (Pattathil et al. 2010). Glycan microarray analyses confirmed that the members of this clade do not bind to galactans that consist exclusively of β-1,3-linked or β-1,4-linked backbones, suggesting that the recognised epitope consists of β-1,6-linked galactan present in RG-I, AG-II, and other pectic polysaccharides. The use of diverse, chemically synthesised substituted AG-II oligosaccharides in these screens enabled the refinement of the RG-I/AG clade of mAbs into three distinct subclades, referred to as β-6 galactan-1, -2, and -3 (Ruprecht et al. 2017). Most of the binding profiles of individual mAbs remain partially characterised due to the frequency of β-6 galactan epitopes in the cell wall and the limited number of well-characterised AG oligosaccharides for screening. However, the specificity screens performed to date have uncovered distinctions between the β-6 galactan-1, -2, and -3 binding clades by the degree to which each group tolerates substituents on the β-1,6 galactan backbone. All mAbs in the β-6 galactan-1 and β-6 galactan-3 sets bind to a β-1,6-linked tri- and tetrasaccharides having no other substituents. The use of glycan arrays containing β-1,6-linked galactan chains of varying lengths carrying arabinosyl substituents or attached to either β-1,3-linked or β-1,4-linked galactan backbones demonstrated that the minimal epitope recognised by these antibodies is likely a disaccharide. The β-6 galactan-1 mAbs display the greatest tolerance for side-chain substituents except for 4-linked substituents on the middle galactose of the trisaccharide. The mAbs in β-6 galactan-3 do not tolerate substituents in either the 3- or 4-position of this central galactosyl residue and, like the β-6 galactan-1 antibodies, will bind to a β-6 galactose disaccharides attached to either β-1,3-linked or β-1,4-linked galactan backbones. In contrast, the mAbs in β-6 galactan-2 are the least tolerant of additional substitutions and appear to bind a β-1,6-linked disaccharide that is associated with only a β-1,4-linked galactan backbone. A much larger and more diverse set of galactan oligosaccharides and/or the use of selective hydrolytic enzymes will be required to further refine the epitopes recognised by these three sets of β-6 galactan-directed mAbs.

Epitope: Incomplete Characterisation

Several additional sets of mAbs have been broadly characterised to target protein- and pectin-linked arabinogalactans through hierarchical clustering of their binding profiles to structurally known plant cell wall preparations (Pattathil et al. 2010, 2015). The epitopes recognised by these mAbs are currently unknown due to a lack of structurally characterised arabinogalactan oligosaccharide fragments for specificity screens. Nonetheless, immunolabelling studies have identified distinctions in the epitope abundance and patterning of each mAb category (Bowling and Vaughn 2008; Hall et al. 2013; Méndez et al. 2022) (Figure 10.10).

Three clades of mAbs, referred to as RG-Ia, RG-Ib, and RG-Ic, target unknown epitopes within RG-I from specific plants. The mAbs CCRC-M2 and CCRC-M5 comprise the RG-Ia clade and are selective to gum karaya (Puhlmann et al. 1994; Pattathil et al. 2010, 2015). Members of the RG-Ib clade, CCRC-M30, CCRC-M61, JIM101, and JIM137, share the ability to recognise Camelina seed mucilage and differ slightly in binding to other polysaccharides (Pattathil et al.

Polymer Class	Monoclonal Antibody	Predicted Epitope Structure	Animal
Pectin- and Protein-linked Arabinogalactans	LM16	RG-Ia	Rat
	JIM1		Rat
	JIM16	RG-Ib	Rat
	JIM132		Rat
	JIM101		Rat
	JIM137		Rat
	CCRC-M16	RG-Ic	Mouse
	CCRC-M18		Mouse
	CCRC-M19		Mouse
	CCRC-M23		Mouse
	CCRC-M56		Mouse
	CCRC-M40	Linseed Mucilage RG-I	Mouse
	CCRC-M161		Mouse
	CCRC-M164		Mouse
	JIM3		Rat
	CCRC-M94	Physcomitrella Pectin	Mouse
	CCRC-M98		Mouse
	JIM11	AG-1	Rat
	JIM20		Rat
	JIM93		Rat
	JIM94		Rat
	MAC204		Rat
	JIM12	AG-2	Rat
	JIM14		Rat
	JIM19		Rat
	CCRC-M26	AG-3	Mouse
	CCRC-M31		Mouse
	CCRC-M81		Mouse
	CCRC-M85		Mouse
	PN16.4B4		Mouse
	JIM4		Rat
	JIM8		Rat
	JIM13		Rat
	JIM15		Rat
	JIM17		Rat
	JIM133		Rat
	MAC207		Rat
	MAC266		Rat
	CCRC-M78	AG-4	Mouse
	CCRC-M91		Mouse
	CCRC-M92		Mouse

FIGURE 10.10 Pectin- and protein-linked arabinogalactan-directed monoclonal antibodies. This table lists the remaining pectin- and protein-linked arabinogalactan-directed mAbs. The epitope structures recognised by each set are unknown, thus clades have been broadly characterised through hierarchical clustering of their binding profiles to diverse plant cell wall preparations (Pattathil et al. 2010).

2010). CCRC-M30 is most selective to seed mucilage from Arabidopsis and Linseed and weakly binds to lettuce RG-I (Pattathil et al. 2010). In contrast, CCRC-M61 only binds to Camelina seed mucilage and weakly recognises glucuronoxylans (Pattathil et al. 2010). JIM101 and JIM137 bind with high affinity to okra RG-I and can cross-react with a broad range of plant polysaccharides (Pattathil et al. 2010). The former binds with lower affinity to seed mucilages from *Sinapus alba* and *Camelina sativa*, while the latter binds with lower affinity to citrus and highly methylesterified pectins (Pattathil et al. 2010). Finally, members of the RG-Ic clade, including CCRC-M16, CCRC-M17, CCRC-M18, CCRC-M19, CCRC-M23, and CCRC-M56, have been characterised to recognise an epitope present on RG-I preparations from soybean and Arabidopsis plant cell walls (Pattathil et al. 2010).

Two additional clades have been broadly characterised to recognise unknown epitopes within pectin-linked arabinogalactans. The mAbs JIM3, CCRC-M40, CCRC-M161, and CCRC-M164 share the ability to bind RG-I Linseed seed mucilage, but each mAb greatly differs in the recognition of the remaining plant polysaccharide preparations (Pattathil et al. 2010). For example, in an ELISA-based screen against 54 plant preparations, JIM3 bound to certain monocot xylans and gum arabic polysaccharides while CCRC-M40 bound weakly to nearly every screened preparation (Pattathil et al. 2010) (Figure 10.1). The next clade of mAbs, comprised of CCRC-M94 and CCRC-M98, has been partially characterised to bind with high affinity to a pectic epitope from *Physcomitrella patens*. In comparison to the members of the RG-I Linseed mucilage clade, CCRC-M94 and CCRC-M98 display no cross-reactivity with any other plant polysaccharide among the 54 preparations tested (Figure 10.1).

The members of the AG-1, AG-2, AG-3, and AG-4 clades have been partially characterised to target pectin- and protein-linked arabinogalactans. Unfortunately, competition assays against natural and synthetic hapten inhibitors have provided only limited information regarding the structure of the epitopes recognised by these mAbs (Knox et al. 1989; Pennell et al. 1989; Yates et al. 1996). The current groupings of these mAbs is based on clustering analyses with diverse sets of plant preparations (Pattathil et al. 2010, 2015) and oligosaccharide microarrays (Moller et al. 2008; Pedersen et al. 2012; Ruprecht et al. 2017). The AG-1 clade, consisting of JIM11 (Smallwood et al. 1994), JIM20 (Smallwood et al. 1994; Knox et al. 1995), JIM93, JIM94, and MAC204 (Bradley et al. 1988; Vandenbosch et al. 1989), binds to gum tragacanth and to lettuce and green tomato RG-I preparations. AG-2 comprises JIM12 (Smallwood et al. 1994), JIM14 (Knox et al. 1991; Yates et al. 1996), and JIM19 (Smallwood et al. 1994; Knox et al. 1995; Wang et al. 1995), which bind to linear and branched arabinans and RG-I preparations from diverse plants but do not bind to larch arabinogalactan. AG-3 contains CCRC-M26, CCRC-M31, CCRC-M81, CCRC-M85 (Pattathil et al. 2010, 2015), PN16.4B4 (Norman et al. 1986, 1990), JIM4, JIM8, JIM13, JIM15, JIM17, JIM133 (Knox et al. 1989, 1991; Stacey et al. 1990; Smallwood et al. 1996; Yates et al. 1996), MAC207, and MAC266 (Knox et al. 1989; Pennell et al. 1989; Perotto et al. 1991; Yates et al. 1996). These mAbs bind strongly to gum ghatti, gum arabic, and pectic polysaccharide preparations from tomato and sycamore maple. Finally, AG-4 contains CCRC-M78, CCRC-M91, and CCRC-M92 (Pattathil et al. 2010, 2015) and has been characterised to bind to gum arabic and RG-I preparations from a broad range of plants.

Arabinogalactan-Directed Antibodies: Recent Advancements and Prospects

There is compelling evidence that pectin composition and abundance is modulated throughout developmental and stress-induced reconfigurations of the plant cell wall, although the enzymes responsible for pectin biosynthesis, transport, and modification are only beginning to be identified (Yang and Anderson 2020; Shin et al. 2021). Structural analysis of branched pectic polysaccharides is challenging due to variations in side-chain composition and degree of polymerisation, and as a result it is poorly understood how individual side-chain structures are regulated to contribute to the molecular architecture and biomechanical properties of the pectin network. The available mAbs for pectic arabinan and galactan side-chains are widely used in glycome profiling studies to monitor variations in epitope patterning throughout plant development (Hall et al. 2013; Leroux et al. 2015; Muszyński et al. 2015; Corral-Martínez et al. 2019; Petrova et al. 2022). Immunolocalisation studies using seeds carrying mutations in the enzymes responsible for pectic arabinan degradation have also revealed the functional significance of pectin composition in seed mucilage, as increased levels of branched RG-I in mucilage resulted in inefficient seed coat development and delayed germination (Dean et al. 2007; Arsovski et al. 2009; Kong et al. 2013).

Protein-linked arabinogalactans (AGPs) are ubiquitous throughout the plant cell wall and plasma membrane and have been implicated in a multitude of structural and regulatory pathways (Ellis et al. 2010; Villa-Rivera et al. 2021, Chapters 16 and 17). AGPs are formed through the anchoring of an arabinogalactan structure to a protein core, with the structure and composition of each component displaying a high degree of heterogeneity throughout different plant tissues. It is technically challenging to distinguish between pectin- and protein-linked arabinogalactan epitopes for several reasons (Tan et al. 2012). First, pectin- and protein-linked arabinogalactans are likely to share many structural epitopes, such as arabinan structures recognised by LM6, and thus careful and rigorous analysis is necessary to correctly identify binding patterns. The structures of various classes and subclasses of AGPs are mostly uncharacterised, which additionally limits high-throughput screening approaches to the use of heterogeneous, plant-derived materials, rather than well-characterised, synthetic oligosaccharides. Unfortunately, detailed information regarding the exact structures recognised by arabinogalactan-directed antibodies has remained unattainable. However, the visualisation of dynamic epitope patterns in immunolocalisation studies remains a useful tool for observing the developmental modulation of pectin- and protein-linked arabinogalactans (Bowling and Vaughn 2008; Hall et al. 2013; Moro et al. 2017; Corral-Martínez et al. 2019; Henry and Renzaglia 2021). Changes in the abundance or distribution of several uncharacterised epitopes has revealed a potential functional role of AGPs in low temperature stress response (Yan et al. 2015) and disease resistance (Molina et al. 2021). Several mAbs also display complementary binding patterns at certain developmental stages, including JIM4 and JIM11 (Knox et al. 1989; Smallwood et al. 1994), LM5 and LM26 (Torode et al. 2018), and JIM13 and CCRC-M2 (Freshour et al. 1996). The labelling patterns of both known and unknown glycan epitopes can provide an important insight into the developmental regulation and functional significance of a certain plant cell wall components.

CONCLUSIONS

The ability to both distinguish carbohydrate epitopes on glycoproteins and polysaccharides and to monitor changes in their patterns of deposition is crucial to uncovering the physiological implications of cell wall regulation. The plant cell wall glycan–directed monoclonal antibodies described in this review have provided fundamental insights into the developmental regulation of structurally heterogeneous plant cell wall polysaccharides across various cells, tissues, and organs, as well as identifying important similarities and differences across plant families. The biochemical composition and spatial interactions of cell wall components are precisely regulated during plant growth and development. Thus, mAbs provide us with the capabilities to observe and monitor these changes along developmental gradients. Furthermore, dynamic wall reconfigurations in response to environmental stresses induced by biotic and abiotic factors can be detected and examined using these mAbs. Finally, mAbs have proven useful in targeted genomics efforts to identify cell wall biosynthetic enzymes and investigate the structural and physiological consequences of mutations on cell wall components.

It remains a challenge to further expand and diversify the existing collection of cell wall–directed mAbs, as the complex structure and interconnected nature of cell wall polymers introduces a multitude of obstacles for their generation and characterisation. First, mAbs are generated against epitopes on known plant polysaccharides; the limited availability of many plant cell wall carbohydrates has resulted in the unequal representation of certain classes of cell wall components within current libraries of plant cell wall glycan–directed mAbs (Figure 10.11). The binding profiles of mAbs that target well-characterised polymers, such as XG, are significantly better characterised than those that target poorly understood polymers, such as pectins and AGPs. Also, some polymers (e.g. mannans) have not been targeted for antibody production as much as others (e.g. XGs, xylans). Furthermore, the current library of plant glycan–directed mAbs is limited by the characteristics of the immunogens used. For example, very few antibodies exist that are directed against structural features that include base-labile elements, such as *O*-acetyl and methylester substituents; the polysaccharide preparations used as immunogens are frequently prepared using base extractions and resultant de-esterification of plant cell wall materials. Therefore, alternative strategies for polysaccharide extraction from plant cell walls that preserve, for example, base-labile structural features will need to be employed in order to generate the immunogens needed to further diversify the antibody toolkit and capture more of the epitope space of native plant cell walls.

While mAbs are monospecific in theory, plant cell wall immunogens are heterogeneous and can contain incredibly diverse motifs, thus resulting in probes with broad polymer reactivity or unknown epitope targets. For example, JIM13 and CCRC-M2 were generated against carrot AG-2 and sycamore maple RG-I, respectively. However, the binding profiles of both mAbs remain uncharacterised due to the structural similarities between RG-I and AGP polymers combined with our incomplete understanding of glycan structures present in these two polymers (Figure 10.11). The epitopes recognised by mAbs can also be found in multiple macromolecular contexts. For example, CCRC-M108, CCRC-M109, CCRC-M110, and CCRC-M111 target arabinosylated epitopes in both XGs and xylans (Figure 10.1). Thus, interpretations of experimental results using glycan-directed antibodies must be done with caution. In general, it is safest to view these antibodies as epitope-directed rather than polymer-directed unless orthogonal data are available to indicate that the antibodies used are, in fact, polymer-specific in the plant being studied.

Immunolocalisation of wall epitopes in fixed and sectioned tissues remains the primary use of the library of plant cell wall–directed mAbs to date. There are a few considerations that must be kept in mind when using mAbs for these studies. First, the mAbs will only bind to epitopes exposed on the surface of the section; they cannot

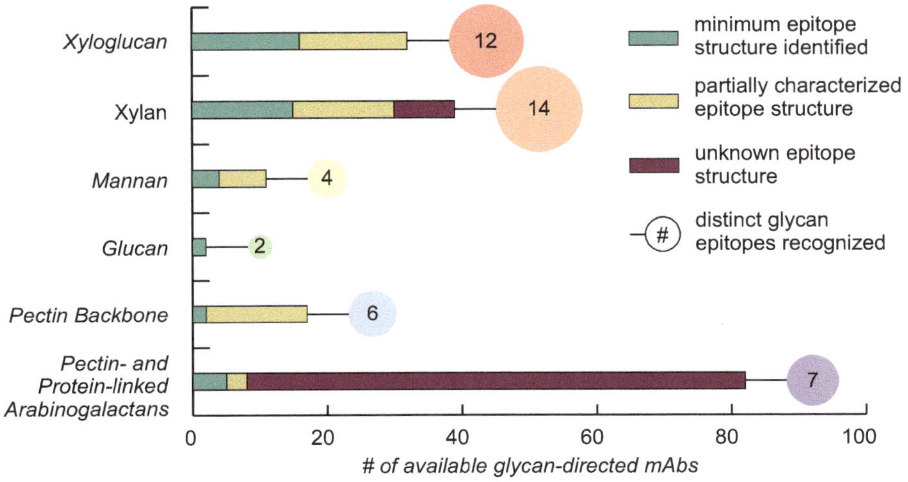

FIGURE 10.11 The quantity and characterisation of glycan-directed mAbs varies between different plant cell wall components.

penetrate either the walls or the resins used to embed the plant tissues. This is also true of enzymes that might be used to pre-treat the sections prior to immunohistochemistry. Thus, with plastic and resin embedding media, the antibodies can only access the cut surfaces/cross-sections of walls. With wax-embedded sections, on the other hand, the wax is removed prior to any immunohistochemistry or other treatments of the sections, thereby exposing additional cell wall surfaces for antibody binding (and enzyme action). Such considerations must be followed when interpreting the results of immunohistochemical experiments. Secondly, the binding of mAbs to their epitopes *in situ* can be compromised by the presence of substituents on the glycans that interfere with antibody binding. For example, methyl and acetyl esters are frequently present on sugars in cell wall glycans *in vivo*. Given that most of the mAbs in the current collection were generated against immunogens that were prepared using base extractions that would remove such groups, the mAbs could be prevented from binding to their epitopes in the native wall. These concerns are particularly relevant to xylans. Indeed, some xylan-directed mAbs do not bind to native cell walls (Pattathil et al. 2010). Thus, it is advisable to treat tissue sections with base before immunolocalising xylans in addition to performing the localisations on untreated sections to obtain a clear picture of xylan distribution patterns. This may be true for other classes of wall polysaccharides as well, though to date there are too few data to assess how critical this is. Thirdly, each class of polysaccharides present in plant cell walls exists as a family of structurally related glycans (glycoforms), and so it is advisable to use mAbs against diverse epitopes in a targeted polysaccharide when seeking information about the distribution of that polysaccharide class in cell walls so as to encompass as many of the glycoforms as possible in the analyses. There is ample evidence in the literature already that different glycoforms of a given polysaccharide class show different localisation patterns in plant tissues, cells, and cell walls, as has been pointed out in several sections of this chapter. Fortunately, the antibody collection is now sufficiently diverse to include mAbs against various epitopes in each polysaccharide class and researchers are well-advised to take full advantage of this diversity.

The size and diversity of the plant glycan–directed antibody collection now permits comprehensive comparative glycomics of plant cell wall preparations in diverse experimental contexts. Two experimentally similar approaches for comparative glycomics of plant cell walls have been described in the literature; both rely on sequential extraction of cell wall preparations with increasingly harsh reagents followed by screening of the extracts using a diverse collection of carbohydrate binding modules (CBMs) and/or antibodies (Pattathil et al. 2012; Kračun et al. 2017). One has been termed Comprehensive Microarray Polymer Profiling (CoMPP) (Moller et al. 2008) and uses printed microarrays of the extracts as the solid support on which to carry out the antibody/CBM screening. The other, termed glycome profiling (Zhu et al. 2010), utilises ELISA for screening of the extracts with the antibody collection. We prefer the term "glycome profiling" because, as has been pointed out throughout this chapter, many of the antibodies in the current library recognise epitopes that are present in multiple macromolecular contexts and thus are not necessarily polymer selective, depending on the plant and tissue being studied. Both experimental approaches for comparative glycomics of plant cell walls provide a moderate throughput, semi-quantitative method that can accommodate large numbers of samples. It is important to remember that the cell wall extraction protocols used in glycome profiling include treatments with strong base that will remove any esterified substituents on cell wall polymers (e.g. methyl-, acetyl-, and phenolic esters); any epitopes including these substituents are therefore lost from the wall extracts prepared in those and subsequent extractions. Glycome profiling provides two important pieces of information about cell wall samples. First, it provides an estimation of the glycan epitope content of a particular cell wall sample; this estimate is only semi-quantitative given the limitations of the extraction techniques used and the difficulties in obtaining absolute quantitation of the amounts of sugar in each extract. Second, glycome profiling provides information about how tightly the various epitopes detected in the screens are bound into the cell wall matrix based on where amongst the extracts a particular epitope is found, thereby giving useful information about overall wall structure. Comparative glycomics of plant cell walls has found broad application to analyse the effects of diverse biomass pretreatment protocols on the extractability of cell wall glycans for biomass utilisation (Alonso-Simón et al. 2010; DeMartini et al. 2011; de Souza et al. 2013; Li et al. 2014; Liu et al. 2014; Socha et al. 2014; Pattathil et al. 2015), for the analysis of plants carrying mutations in cell wall–related genes (Zabotina et al. 2012; Zhao et al. 2013; Sopeña-Torres et al. 2018; Zhang et al. 2023), the comparison of cell wall structure between diverse plants (Lin et al. 2016; da Costa et al. 2017; Wood et al. 2017), and the interactions of plants with other organisms or the environment (Moller et al. 2011; Leite et al. 2017; Molina et al. 2021; Silva et al. 2022), to name just a few.

The summary of information about the epitope specificity of the mAbs in the world-wide collection presented here makes it clear that there is still a lot of work to be done on epitope characterisation for at least a third of the current mAbs, particularly those against arabinogalactan epitopes. Part of the problem is our lack of complete understanding of the detailed structures of these glycans in cell walls, which limits the ability of chemists to design synthetic schemes to enlarge the diversity of oligosaccharides for use in epitope characterisation studies. As an alternative to chemical synthesis, the use of glycoside hydrolases and diverse synthetic enzymes should be explored to generate additional oligosaccharide structures for epitope studies. However, this will require the availability of well-characterised and specific enzymes. Thus, research focused on the enzymes responsible for the degradation of structurally complex polysaccharides will facilitate the expansion of our plant cell wall epitope library and have critical implications in both *in vitro* and *in planta* experiments. Another interesting approach to overcome the technical challenges of chemical synthesis is exemplified by the valorisation of the enzymes responsible for XG biosynthesis to generate novel structural epitopes. It has been shown that *Arabidopsis thaliana* XG xylosyltransferase 2 (AtXXT2) can be used to enzymatically attach additional xylosyl residues to chemically synthesised XG oligosaccharides *in vitro*, providing a route to further diversify these libraries (Dallabernardina et al. 2017). While careful optimisation of this technique is required to generate well-defined and pure plant oligosaccharides, chemoenzymatic strategies show great potential for expanding the current library of epitope structures and identifying additional antibodies that recognise plant cell wall polymers.

Currently, our ability to localise plant cell wall glycans is largely limited to immunohistochemistry in fixed, embedded, and sectioned plant tissue. In other words, in dead tissue! This makes studies of the dynamics of cell wall glycans during plant growth and development and in response to environmental changes/stresses exceedingly difficult. Thus, the development of methods for tagging cell wall glycans *in vivo* is paramount. One approach that has had some success is the incorporation of modified sugars into wall polymers during their synthesis *in vivo*, followed by chemically linking tags to those modified sugars in the wall (so-called click chemistry) (Anderson et al. 2012; Wallace and Anderson 2012; Bukowski et al. 2014). This method has had some success in plants, but very few sugars are unique to a given polysaccharide, making it difficult to assess which polysaccharide has been tagged after clicking. Furthermore, not all clickable sugar derivatives are taken up by plant cells or can be incorporated by the cell's biosynthetic enzymes. In addition, the intrinsic sugar salvage and/or interconversion pathways in plants may cause the alteration of the supplied clickable sugar, further complicating the interpretation of experiments using these reagents. An alternative approach would be to take advantage of currently available cell wall glycan–directed probes, such as the mAbs discussed here and/or CBMs (Gilbert 2010), and express such probes *in planta* in a tagged form, targeting them to the endomembrane system where most cell wall polysaccharides are synthesised. The expectation would be that these expressed and targeted probes would bind to their epitopes during polysaccharide synthesis and remain bound during export to the apoplast and be incorporated into the wall. There are questions about whether the plant cell's vesicular transport system would tolerate such additional "coating" of wall polysaccharides, and whether such coating would affect the biological function(s) of the glycans after incorporation into the wall. Nonetheless, some initial success has been achieved with this approach (Zhang 2014), suggesting that further exploration of this method would be worthwhile.

ACKNOWLEDGEMENTS

B.R.U. and K.J.T are supported by the U.S. Department of Energy, Office of Science, Basic Energy Sciences [DE-SC0008472] and K.J.T. is also supported by the Glycoscience Training Program for Graduate Students at the University of Georgia funded by the National Institutes of Health (T32GM145467), National Institute of General Medical Sciences. Work on plant cell wall–directed monoclonal antibodies by M.G.H. was supported by the National Science Foundation Plant Genome Program (DBI-0421683 and IOS-0923992). The authors thank Sivakumar Pattathil for assistance with the preparation of Figure 10.1.

REFERENCES

Alba, K, Kasapis, S, and Kontogiorgos, V (2015) Influence of pH on mechanical relaxations in high solids LM-pectin preparations. *Carbohydrate Polymers* 127, 182–188.

Alonso-Simón, A, Kristensen, JB, Øbro, J, Felby, C, Willats, WGT, and Jørgensen, H (2010) High-throughput microarray profiling of cell wall polymers during hydrothermal pre-treatment of wheat straw. *Biotechnology and Bioengineering* 105, 509–514.

Amsbury, S, Kirk, P, and Benitez-Alfonso, Y (2018) Emerging models on the regulation of intercellular transport by plasmodesmata-associated callose. *Journal of Experimental Botany* 69, 105–115.

Andersen, MCF, Boos, I, Marcus, SE, Kračun, SK, Rydahl, MG, Willats, WGT, Knox, JP, and Clausen, MH (2016) Characterization of the LM5 pectic galactan epitope with synthetic analogues of β-1,4-D-galactotetraose. *Carbohydrate Research* 436, 36–40.

Anderson, CT (2016) We be jammin': an update on pectin biosynthesis, trafficking and dynamics. *Journal of Experimental Botany* 67, 495–502.

Anderson, CT, and Kieber, JJ (2020) Dynamic construction, perception, and remodeling of plant cell walls. *Annual Review of Plant Biology* 71, 39–69.

Anderson, CT, Wallace, IS, and Somerville, CR (2012) Metabolic click-labeling with a fucose analog reveals pectin delivery, architecture, and dynamics in Arabidopsis cell walls. *Proceedings of the National Academy of Sciences of the USA* 109, 1329–1334.

Anderson, MA, Sandrin, MS, and Clarke, AE (1984) A high proportion of hybridomas raised to a plant extract secrete antibody to arabinose or galactose. *Plant Physiology* 75, 1013–1016.

Arsovski, AA, Popma, TM, Haughn, GW, Carpita, NC, McCann, MC, and Western, TL (2009) AtBXL1 encodes a bifunctional β-D-xylosidase/α-L-arabinofuranosidase required for pectic arabinan modification in Arabidopsis mucilage secretory cells. *Plant Physiology* 150, 1219–1234.

Avci, U, Pattathil, S, Singh, B, Brown, VL, Hahn, MG, and Haigler, CH (2013) Cotton fiber cell walls of *Gossypium hirsutum* and *Gossypium barbadense* have differences related to loosely-bound xyloglucan. *Plos One* 8, e56315.

Barghahn, S, Arnal, G, Jain, N, Petutschnig, E, Brumer, H, and Lipka, V (2021) Mixed linkage β-(1→3),(1→4)-glucan oligosaccharides induce defense responses in *Hordeum vulgare* and *Arabidopsis thaliana*. *Frontiers in Plant Science* 12, 682439.

Benhamou, N (1995) Immunocytochemistry of plant defense mechanisms induced upon microbial attack. *Microscopy Research and* Technique 31, 63–78.

Bowling, AJ, and Vaughn, KC (2008) Immunocytochemical characterization of tension wood: gelatinous fibers contain more than just cellulose. *American Journal of Botany* 95, 655–663.

Bradley, DJ, Wood, EA, Larkins, AP, Galfre, G, Butcher, GW, and Brewin, NJ (1988) Isolation of monoclonal antibodies reacting with peribacteriod membranes and other components of pea root nodules containing *Rhizobium leguminosarum*. *Planta* 173, 149–160.

Brennan, M, Fakharuzi, D, and Harris, PJ (2019) Occurrence of fucosylated and non-fucosylated xyloglucans in the cell walls of monocotyledons: an immunofluorescence study. *Plant Physiology and Biochemistry* 139, 428–434.

Brennan, M, and Harris, PJ (2011) Distribution of fucosylated xyloglucans among the walls of different cell types in monocotyledons determined by immunofluorescence microscopy. *Molecular Plant* 4, 144–156.

Bromley, JR, Busse-Wicher, M, Tryfona, T, Mortimer, JC, Zhang, Z, Brown, DM, and Dupree, P (2013) GUX1 and GUX2 glucuronyltransferases decorate distinct domains of glucuronoxylan with different substitution patterns. *The Plant Journal* 74, 423–434.

Brown, RC, Lemmon, BE, Stone, BA, and Olsen, O-A (1997) Cell wall (1→3)- and (1→3,1→4)-β-glucans during early grain development in rice (*Oryza sativa* L.). *Planta* 202, 414–426.

Broxterman, SE, and Schols, HA (2018) Characterisation of pectin-xylan complexes in tomato primary plant cell walls. *Carbohydrate Polymers* 197, 269–276.

Buckeridge, MS, Pessoa dos Santos, H, and Tiné, MAS (2000) Mobilisation of storage cell wall polysaccharides in seeds. *Plant Physiology and Biochemistry* 38, 141–156.

Buckeridge, MS, Rayon, C, Urbanowicz, B, Tiné, MAS, and Carpita, NC (2004) Mixed linkage (1→3),(1→4)-β-D-glucans of grasses. *Cereal Chemistry* 81, 115–127.

Bukowski, N, Pandey, JL, Doyle, L, Richard, TL, Anderson, CT, and Zhu, Y (2014) Development of a clickable designer monolignol for interrogation of lignification in plant cell walls. *Bioconjugate Chemistry* 25, 2189–2196.

Burton, RA, Gidley, MJ, and Fincher, GB (2010) Heterogeneity in the chemistry, structure and function of plant cell walls. *Nature Chemical Biology* 6, 724–732.

Burton, RA, Wilson, SM, Hrmova, M, Harvey, AJ, Shirley, NJ, Medhurst, A, Stone, BA, Newbigin, EJ, Bacic, A, and Fincher, GB (2006) Cellulose synthase-like CslF genes mediate the synthesis of cell wall (1→3),(1→4)-β-D-glucans. *Science* 311, 1940–1942.

Capek, P, Alföldi, J, and Lišková, D (2002) An acetylated galactoglucomannan from *Picea abies* L. Karst. *Carbohydrate Research* 337, 1033–1037.

Carpita, NC, Defernez, M, Findlay, K, Wells, B, Shoue, DA, Catchpole, G, Wilson, RH, and McCann, MC (2001) Cell wall architecture of the elongating maize coleoptile. *Plant Physiology* 127, 551–565.

Carpita, NC, and McCann, MC (2010) The maize mixed-linkage (1→3),(1→4)-β-D-glucan polysaccharide is synthesized at the golgi membrane. *Plant Physiology* 153, 1362–1371.

Cavalier, DM, Lerouxel, O, Neumetzler, L, Yamauchi, K, Reinecke, A, Freshour, G, Zabotina, OA, Hahn, MG, Burgert, I, Pauly, M, Raikhel, NV, and Keegstra, K (2008) Disrupting two *Arabidopsis thaliana* xylosyltransferase genes results in plants deficient in xyloglucan, a major primary cell wall component. *Plant Cell* 20, 1519–1537.

Chavan, RR, Singh, AP, Azizan, A, and Harris, PJ (2021) Heteromannans are the predominant hemicelluloses in the gametophytic stem of the umbrella moss *Hypnodendron menziesii* and occur in the walls of all cell types. *Planta* 254, 1–14.

Chebli, Y, Bidhendi, AJ, Kapoor, K, and Geitmann, A (2021) Cytoskeletal regulation of primary plant cell wall assembly. *Current Biology* 31, R681–R695.

Chevalier, L, Bernard, S, Ramdani, Y, Lamour, R, Bardor, M, Lerouge, P, Follet-Gueye, M-L, and Driouich, A (2010) Subcompartment localization of the side chain xyloglucan-synthesizing enzymes within Golgi stacks of tobacco suspension-cultured cells. *The Plant Journal* 64, 977–989.

Clausen, MH, Ralet, M-C, Willats, WGT, McCartney, L, Marcus, SE, Thibault, J-F, and Knox, JP (2004) A monoclonal antibody to feruloylated-(1→4)-β-D-galactan. *Planta* 219, 1036–1041.

Clausen, MH, Willats, WGT, and Knox, JP (2003) Synthetic methyl hexagalacturonate hapten inhibitors of anti-homogalacturonan monoclonal antibodies LM7, JIM5 and JIM7. *Carbohydrate Research* 338, 1797–1800.

Cornuault, V, Buffetto, F, Rydahl, MG, Marcus, SE, Torode, TA, Xue, J, Crépeau, M-J, Faria-Blanc, N, Willats, WGT, Dupree, P, Ralet, M-C, and Knox, JP (2015) Monoclonal antibodies indicate low-abundance links between heteroxylan and other glycans of plant cell walls. *Planta* 242, 1321–1334.

Cornuault, V, Manfield, IW, Ralet, M-C, and Knox, JP (2014) Epitope detection chromatography: a method to dissect the structural heterogeneity and interconnections of plant cell-wall matrix glycans. *The Plant Journal* 78, 715–722.

Corral-Martínez, P, Driouich, A, and Seguí-Simarro, JM (2019) Dynamic changes in arabinogalactan-protein, pectin, xyloglucan and xylan composition of the cell wall during microspore embryogenesis in *Brassica napus*. *Frontiers in Plant Science* 10, 332.

Cosgrove, DJ (2005) Growth of the plant cell wall. *Nature Reviews Molecular Cell Biology* 6, 850–861.

Cui, W, and Lee, J-Y (2016) Arabidopsis callose synthases CalS1/8 regulate plasmodesmal permeability during stress. *Nature Plants* 2, 16034.

Cui, X, Shin, H, Song, C, Laosinchai, W, Amano, Y, and Brown, MR (2001) A putative plant homolog of the yeast β-1,3-glucan synthase subunit FKS1 from cotton (*Gossypium hirsutum* L.) fibers. *Planta* 213, 223–230.

da Costa, RMF, Pattathil, S, Avci, U, Lee, SJ, Hazen, SP, Winters, A, Hahn, MG, and Bosch, M (2017) A cell wall reference profile for *Miscanthus* bioenergy crops highlights compositional and structural variations associated with development and organ origin. *New Phytologist* 213, 1710–1725.

Dallabernardina, P, Ruprecht, C, Smith, PJ, Hahn, MG, Urbanowicz, BR, and Pfrengle, F (2017) Automated glycan assembly of galactosylated xyloglucan oligosaccharides and their recognition by plant cell wall glycan-directed antibodies. *Organic & Biomolecular Chemistry* 15, 9995–10000.

de Souza, AP, Leite, DCC, Pattathil, S, Hahn, MG, and Buckeridge, MS (2013) Composition and structure of sugarcane cell wall polysaccharides: implications for second-generation bioethanol production. *BioEnergy Research* 6, 564–579.

Dean, GH, Zheng, H, Tewari, J, Huang, J, Young, DS, Hwang, YT, Western, TL, Carpita, NC, McCann, MC, Mansfield, SD, and Haughn, GW (2007) The *Arabidopsis MUM2* gene encodes a β-galactosidase required for the production of seed coat mucilage with correct hydration properties. *The Plant Cell* 19, 4007–4021.

DeMartini, JD, Pattathil, S, Avci, U, Szekalski, K, Mazumder, K, Hahn, MG, and Wyman, CE (2011) Application of monoclonal antibodies to investigate plant cell wall deconstruction for biofuels production. *Energy & Environmental Science* 4, 4332–4339.

Di Laurenzio, L, Wysocka-Diller, J, Malamy, JE, Pysh, L, Helariutta, Y, Freshour, G, Hahn, MG, Feldmann, KA, and Benfey, PN (1996) The *SCARECROW* gene regulates an asymmetric cell division that is essential for generating the radial organization of the Arabidopsis root. *Cell* 86, 423–433.

Doblin, MS, De Melis, L, Newbigin, E, Bacic, A, and Read, SM (2001) Pollen tubes of *Nicotiana alata* express two genes from different β-glucan synthase families. *Plant Physiology* 125, 2040–2052.

Doering, A, Lathe, R, and Persson, S (2012) An update on xylan synthesis. *Molecular Plant* 5, 769–771.

Donaldson, LA, and Knox, JP (2012) Localization of cell wall polysaccharides in normal and compression wood of Radiata pine: relationships with lignification and microfibril orientation. *Plant Physiology* 158, 642–653.

Douchiche, O, Driouich, A, and Morvan, C (2010) Spatial regulation of cell-wall structure in response to heavy metal stress: cadmium-induced alteration of the methyl-esterification pattern of homogalacturonans. *Annals of Botany* 105, 481–491.

Ebringerová, A (2005) Structural diversity and application potential of hemicelluloses. *Macromolecular Symposia* 232, 1–12.

Ebringerová, A, and Heinze, T (2000) Xylan and xylan derivatives - biopolymers with valuable properties, 1. Naturally occurring xylans structures, isolation procedures and properties. *Macromolecular Rapid Communications* 21, 542–556.

Ebringerová, A, Hromádková, Z, and Heinze, T (2005) Hemicellulose. In *Polysaccharides I: Structure, Characterization and Use*, Heinze T, ed (pp. 1–67). Berlin, Heidelberg: Springer Berlin Heidelberg.

Ellis, M, Egelund, J, Schultz, CJ, and Bacic, A (2010) Arabinogalactan-proteins: key regulators at the cell surface? *Plant Physiology* 153, 403–419.

Escudero, V, Jordá, L, Sopeña-Torres, S, Mélida, H, Miedes, E, Muñoz-Barrios, A, Swami, S, Alexander, D, McKee, LS, Sánchez-Vallet, A, Bulone, V, Jones, AM, and Molina, A (2017) Alteration of cell wall xylan acetylation triggers defense responses that counterbalance the immune deficiencies of plants impaired in the β-subunit of the heterotrimeric G-protein. *The Plant Journal* 92, 386–399.

Fincher, GB (2009) Exploring the evolution of (1→3),(1→4)-β-D-glucans in plant cell walls: comparative genomics can help! *Current Opinion in Plant Biology* 12, 140–147.

Forand, AD, Finfrock, YZ, Lavier, M, Stobbs, J, Qin, L, Wang, S, Karunakaran, C, Wei, Y, Ghosh, S, and Tanino, KK (2022) With a little help from my cell wall: Structural modifications in pectin may play a role to overcome both dehydration stress and fungal pathogens. *Plants* 11, 385.

Freshour, G, Bonin, CP, Reiter, W-D, Albersheim, P, Darvill, AG, and Hahn, MG (2003) Distribution of fucose-containing xyloglucans in cell walls of the *mur1* mutant of Arabidopsis. *Plant Physiology* 131, 1602–1612.

Freshour, G, Clay, RP, Fuller, MS, Albersheim, P, Darvill, AG, and Hahn, MG (1996) Developmental and tissue-specific structural alterations of the cell-wall polysaccharides of *Arabidopsis thaliana* roots. *Plant Physiology* 110, 1413–1429.

Fry, SC, York, WS, Albersheim, P, Darvill, A, Hayashi, T, Joseleau, J-P, Kato, Y, Lorences, EP, Maclachlan, GA, McNeil, M, Mort, AJ, Grant Reid, JS, Seitz, HU, Selvendran, RR, Voragen, AGJ, and White, AR (1993) An unambiguous nomenclature for xyloglucan-derived oligosaccharides. *Physiologia Plantarum* 89, 1–3.

Funakawa, H, and Miwa, K (2015) Synthesis of borate cross-linked rhamnogalacturonan II. *Frontiers in Plant Science* 6, 223.

Gilbert, HJ (2010) The biochemistry and structural biology of plant cell wall deconstruction. *Plant Physiology* 153, 444–455.

Goding, JW (1996) 8- production of monoclonal antibodies. In *Monoclonal Antibodies* (Third Edition), Goding JW, ed (pp. 141–191). London: Academic Press.

Grantham, NJ, Wurman-Rodrich, J, Terrett, OM, Lyczakowski, JJ, Stott, K, Iuga, D, Simmons, TJ, Durand-Tardif, M, Brown, SP, Dupree, R, Busse-Wicher, M, and Dupree, P (2017) An even pattern of xylan substitution is critical for interaction with cellulose in plant cell walls. *Nature Plants* 3, 859–865.

Guedes, FTP, Laurans, F, Quemener, B, Assor, C, Lainé-Prade, V, Boizot, N, Vigouroux, J, Lesage-Descauses, M-C, Leplé, J-C, Déjardin, A, and Pilate, G (2017) Non-cellulosic polysaccharide distribution during G-layer formation in poplar tension wood fibers: abundance of rhamnogalacturonan I and arabinogalactan proteins but no evidence of xyloglucan. *Planta* 246, 857–878.

Guillon, F, Tranquet, O, Quillien, L, Utille, J-P, Ordaz Ortiz, JJ, and Saulnier, L (2004) Generation of polyclonal and monoclonal antibodies against arabinoxylans and their use for immunocytochemical location of arabinoxylans in cell walls of endosperm of wheat. *Journal of Cereal Science* 40, 167–182.

Hahn, MG, Lerner, DR, Fitter, MS, Norman, PM, and Lamb, CJ (1987) Characterization of monoclonal antibodies to protoplast membranes of *Nicotiana tabacum* identified by an enzyme-linked immunosorbent assay. *Planta* 171, 453–465.

Hall, HC, Cheung, J, and Ellis, BE (2013) Immunoprofiling reveals unique cell-specific patterns of wall epitopes in the expanding Arabidopsis stem. *The Plant Journal* 74, 134–147.

Hao, ZY, Avci, U, Tan, L, Zhu, X, Glushka, J, Pattathil, S, Eberhard, SF, Sholes, T, Rothstein, GE, Lukowitz, WL, Orlando, R, Hahn, MG, and Mohnen, D (2014) Loss of Arabidopsis *GAUT12/IRX8* causes anther indehiscence and leads to reduced G lignin associated with altered matrix polysaccharide deposition. *Frontiers in Plant Science* 5, 357.

Henry, JS, and Renzaglia, KS (2021) The placenta of *Physcomitrium patens*: transfer cell wall polymers compared across the three Bryophyte groups. *Diversity* 13, 378.

Hernandez-Gomez, MC, Runavot, JL, Guo, X, Bourot, S, Benians, TA, Willats, WG, Meulewaeter, F, and Knox, JP (2015) Heteromannan and heteroxylan cell wall polysaccharides display different dynamics during the elongation and secondary cell wall deposition phases of cotton fiber cell development. *Plant Cell Physiology* 56, 1786–1797.

His, I, Driouich, A, Nicol, F, Jauneau, A, and Höfte, H (2001) Altered pectin composition in primary cell walls of *korrigan*, a dwarf mutant of *Arabidopsis* deficient in a membrane-bound endo-1,4-β-glucanase. *Planta* 212, 348–358.

Hocq, L, Pelloux, J, and Lefebvre, V (2017) Connecting homogalacturonan-type pectin remodeling to acid growth. *Trends in Plant Science* 22, 20–29.

Jones, L, Seymour, GB, and Knox, JP (1997) Localization of pectic galactan in tomato cell walls using a monoclonal antibody specific to (1→4)-β-D-galactan. *Plant Physiology* 113, 1405–1412.

Jonsson, K, Lathe, RS, Kierzkowski, D, Routier-Kierzkowska, A-L, Hamant, O, and Bhalerao, RP (2021) Mechanochemical feedback mediates tissue bending required for seedling emergence. *Current Biology* 31, 1154–1164.e1153.

Julian, JD, and Zabotina, OA (2022) Xyloglucan biosynthesis: from genes to proteins and their functions. *Frontiers in Plant Science* 13, 920494.

Knox, J (2008) Revealing the structural and functional diversity of plant cell walls. *Current Opinion in Plant Biology* 11, 308–313.

Knox, JP, Day, S, and Roberts, K (1989) A set of cell surface glycoproteins forms an early marker of cell position, but not cell type, in the root apical meristem of *Daucus carota* L. *Development* 106, 47–56.

Knox, JP, Linstead, PJ, Cooper, JPC, and Roberts, K (1991) Developmentally regulated epitopes of cell surface arabinogalactan proteins and their relation to root tissue pattern formation. *The Plant Journal* 1, 317–326.

Knox, JP, Linstead, PJ, King, J, Cooper, C, and Roberts, K (1990) Pectin esterification is spatially regulated both within cell walls and between developing tissues of root apices. *Planta* 181, 512–521.

Knox, JP, Peart, J, and Neill, SJ (1995) Identification of novel cell surface epitopes using a leaf epidermal-strip assay system. *Planta* 196, 266–270.

Kong, Y, Peña, MJ, Renna, L, Avci, U, Pattathil, S, Tuomivaara, ST, Li, X, Reiter, W-D, Brandizzi, F, Hahn, MG, Darvill, AG, York, WS, and O'Neill, MA (2015) Galactose-depleted xyloglucan is dysfunctional and leads to dwarfism in Arabidopsis. *Plant Physiology* 167, 1296–1306.

Kong, Y, Zhou, G, Abdeen, AA, Schafhauser, J, Richardson, B, Atmodjo, MA, Jung, J, Wicker, L, Mohnen, D, Western, T, and Hahn, MG (2013) Galacturonosyltransferase-like5 is involved in the production of arabidopsis seed coat mucilage. *Plant Physiology* 163, 1203–1217.

Kozlova, LV, Ageeva, MV, Ibragimova, NN, and Gorshkova, TA (2014) Arrangement of mixed-linkage glucan and glucuronoarabinoxylan in the cell walls of growing maize roots. *Annals of Botany* 114, 1135–1145.

Kračun, SK, Fangel, JU, Rydahl, MG, Pedersen, HL, Vidal-Melgosa, S, and Willats, WGT (2017) Carbohydrate microarray technology applied to high-throughput mapping of plant cell wall glycans using Comprehensive Microarray Polymer Profiling (CoMPP). In *High-Throughput Glycomics and Glycoproteomics: Methods and Protocols*, Lauc G, Wuhrer M, eds (pp. 147–165). New York, NY: Springer New York.

Lampugnani, ER, Moller, IE, Cassin, A, Jones, DF, Koh, PL, Ratnayake, S, Beahan, CT, Wilson, SM, Bacic, A, and Newbigin, E (2013) *In vitro* grown pollen tubes of *Nicotiana alata* actively synthesise a fucosylated xyloglucan. *Plos One* 8, e77140.

Lazaridou, A, and Biliaderis, CG (2007) Molecular aspects of cereal β-glucan functionality: physical properties, technological applications and physiological effects. *Journal of Cereal Science* 46, 101–118.

Lee, KJD, Cornuault, V, Manfield, IW, Ralet, M-C, and Paul Knox, J (2013) Multi-scale spatial heterogeneity of pectic rhamnogalacturonan I (RG-I) structural features in tobacco seed endosperm cell walls. *The Plant Journal* 75, 1018–1027.

Lee, RC, Burton, RA, Hrmova, M, and Fincher, GB (2001) Barley arabinoxylan arabinofuranohydrolases: purification, characterization and determination of primary structures from cDNA clones. *Biochemical Journal* 356, 181.

Leite, DCC, Grandis, A, Tavares, EQP, Piovezani, AR, Pattathil, S, Avci, U, Rossini, A, Cambler, A, de Souza, AP, Hahn, MG, and Buckeridge, MS (2017) Cell wall changes during the formation of aerenchyma in sugarcane roots. *Annals of Botany* 120, 693–708.

Leroux, O, Sørensen, I, Marcus, SE, Viane, R, Willats, W, and Knox, J (2015) Antibody-based screening of cell wall matrix glycans in ferns reveals taxon, tissue and cell-type specific distribution patterns. *BMC Plant Biology* 15, 56.

Li, H, Bacic, A, and Read, SM (1999) Role of a callose synthase zymogen in regulating wall deposition in pollen tubes of *Nicotiana alata* Link et Otto. *Planta* 208, 528–538.

Li, M, Pattathil, S, Hahn, MG, and Hodge, DB (2014) Identification of features associated with plant cell wall recalcitrance to pretreatment by alkaline hydrogen peroxide in diverse bioenergy feedstocks using glycome profiling. *RSC Advances* 4, 17282–17292.

Li, X, Jackson, P, Rubtsov, DV, Faria-Blanc, N, Mortimer, JC, Turner, SR, Krogh, KB, Johansen, KS, and Dupree, P (2013) Development and application of a high throughput carbohydrate profiling technique for analyzing plant cell wall polysaccharides and carbohydrate active enzymes. *Biotechnology for Biofuels* 6, 94.

Liepman, AH, Nairn, CJ, Willats, WG, Sorensen, I, Roberts, AW, and Keegstra, K (2007) Functional genomic analysis supports conservation of function among cellulose synthase-like a gene family members and suggests diverse roles of mannans in plants. *Plant Physiology* 143, 1881–1893.

Lin, F, Manisseri, C, Fagerström, A, Peck, ML, Vega-Sánchez, ME, Williams, B, Chiniquy, DM, Saha, P, Pattathil, S, Conlin, B, Zhu, L, Hahn, MG, Willats, WGT, Scheller, HV, Ronald, PC, and Bartley, LE (2016) Cell wall composition and candidate biosynthesis gene expression during rice development. *Plant and Cell Physiology* 57, 2058–2075.

Liners, F, Letesson, JJ, Didembourg, C, and Van Cutsem, P (1989) Monoclonal antibodies against pectin: recognition of a conformation induced by calcium. *Plant Physiology* 91, 1419–1424.

Liners, F, Thibault, JF, and Van Cutsem, P (1992b) Influence of the degree of polymerization of oligogalacturonates and of esterification pattern of pectin on their recognition by monoclonal antibodies. *Plant Physiology* 99, 1099–1104.

Liners, F, and Van Cutsem, P (1992a) Distribution of pectic polysaccharides throughout walls of suspension-cultured carrot cells. *Protoplasma* 170, 10–21.

Liu, T, Williams, DL, Pattathil, S, Li, M, Hahn, MG, and Hodge, DB (2014) Coupling alkaline pre-extraction with alkaline-oxidative post-treatment of corn stover to enhance enzymatic hydrolysis and fermentability. *Biotechnology for Biofuels* 7, 48.

Livingston, SJ, Bae, EJ, Unda, F, Hahn, MG, Mansfield, SD, Page, JE, and Samuels, AL (2021) Cannabis glandular trichome cell walls undergo remodeling to store specialized metabolites. *Plant and Cell Physiology* 62, 1944–1962.

Marcotuli, I, Hsieh, YSY, Lahnstein, J, Yap, K, Burton, RA, Blanco, A, Fincher, GB, and Gadaleta, A (2016) Structural variation and content of arabinoxylans in endosperm and bran of durum wheat (*Triticum turgidum* L.). *Journal of Agricultural and Food Chemistry* 64, 2883–2892.

Marcus, SE, Blake, AW, Benians, TA, Lee, KJ, Poyser, C, Donaldson, L, Leroux, O, Rogowski, A, Petersen, HL, Boraston, A, Gilbert, HJ, Willats, WG, and Knox, JP (2010) Restricted access of proteins to mannan polysaccharides in intact plant cell walls. *The Plant Journal* 64, 191–203.

Marcus, SE, Verhertbruggen, Y, Hervé, C, Ordaz-Ortiz, JJ, Farkas, V, Pedersen, HL, Willats, WG, and Knox, JP (2008) Pectic homogalacturonan masks abundant sets of xyloglucan epitopes in plant cell walls. *BMC Plant Biology* 8, 60

Marry, M, Roberts, K, Jopson, SJ, Huxham, IM, Jarvis, MC, Corsar, J, Robertson, E, and McCann, MC (2006) Cell-cell adhesion in fresh sugar-beet root parenchyma requires both pectin esters and calcium cross-links. *Physiologia Plantarum* 126, 243–256.

McCann, MC, Wells, B, and Roberts, K (1992) Complexity in the spatial localization and length distribution of plant cell-wall matrix polysaccharides. *Journal of Microscopy* 166, 123–136.

McCartney, L, Marcus, SE, and Knox, JP (2005) Monoclonal antibodies to plant cell wall xylans and arabinoxylans. *Journal of Histochemistry & Cytochemistry* 53, 543–546.

McCleary, BV, McKie, VA, Draga, A, Rooney, E, Mangan, D, and Larkin, J (2015) Hydrolysis of wheat flour arabinoxylan, acid-debranched wheat flour arabinoxylan and arabino-xylo-oligosaccharides by β-xylanase, α-L-arabinofuranosidase and β-xylosidase. *Carbohydrate Research* 407, 79–96.

Meents, MJ, Motani, S, Mansfield, SD, and Samuels, AL (2019) Organization of xylan production in the Golgi during secondary cell wall biosynthesis. *Plant Physiology* 181, 527–546.

Meikle, PJ, Bonig, I, Hoogenraad, NJ, Clarke, AE, and Stone, BA (1991) The location of (1→3)-β-glucans in the walls of pollen tubes of *Nicotiana alata* using a (1→3)-β-glucan-specific monoclonal antibody. *Planta* 185, 1–8.

Meikle, PJ, Hoogenraad, NJ, Bonig, I, Clarke, AE, and Stone, BA (1994) A (1→3,1→4)-β-glucan-specific monoclonal antibody and its use in the quantitation and immunocytochemical location of (1→3,1→4)-β-glucans. *The Plant Journal* 5, 1–9.

Melida, H, Bacete, L, Ruprecht, C, Rebaque, D, Del Hierro, I, Lopez, G, Brunner, F, Pfrengle, F, and Molina, A (2020) Arabinoxylan-oligosaccharides act as damage associated molecular patterns in plants regulating disease resistance. *Frontiers in Plant Science* 11, 1210.

Méndez, T, Stappung, Y, Moya-León, MA, and Herrera, R (2022) Differential expression of arabinogalactan in response to inclination in stem of *Pinus radiata* seedlings. *Plants* 11, 1190.

Mohnen, D (2008) Pectin structure and biosynthesis. *Current Opinion in Plant Biology* 11, 266–277.

Molina, A, Miedes, E, Bacete, L, Rodríguez, T, Mélida, H, Denancé, N, Sánchez-Vallet, A, Rivière, M-P, López, G, Freydier, A, Barlet, X, Pattathil, S, Hahn, M, and Goffner, D (2021) Arabidopsis cell wall composition determines disease resistance specificity and fitness. *Proceedings of the National Academy of Sciences* 118, e2010243118.

Moller, IE, Licht, HHDF, Harholt, J, Willats, WGT, and Boomsma, JJ (2011) The dynamics of plant cell wall polysaccharide decomposition in leaf-cutting ant fungus gardens. *Plos One* 6, e17506.

Moller, IE, Marcus, SE, Haeger, A, Verhertbruggen, Y, Verhoef, R, Schols, H, Ulvskov, P, Mikkelsen, JD, Knox, JP, and Willats, W (2008) High-throughput screening of monoclonal antibodies against plant cell wall glycans by hierarchical clustering of their carbohydrate microarray binding profiles. *Glycoconjugate Journal* 25, 37–48.

Moller, IE, Sorensen, I, Bernal, AJ, Blaukopf, C, Lee, K, Obro, J, Pettolino, F, Roberts, A, Mikkelsen, JD, Knox, JP, Bacic, A, and Willats, WGT (2007) High-throughput mapping of cell-wall polymers within and between plants using novel microarrays. *Plant Journal* 50, 1118–1128.

Moore, PJ, Darvill, AG, Albersheim, P, and Staehelin, LA (1986) Immunogold localization of xyloglucan and rhamnogalacturonan I in the cell walls of suspension-cultured sycamore cells. *Plant Physiology* 82, 787–794.

Moro, CF, Gaspar, M, da Silva, FR, Pattathil, S, Hahn, MG, Salgado, I, and Braga, MR (2017) *S*-nitrosoglutathione promotes cell wall remodelling, alters the transcriptional profile and induces root hair formation in the hairless *root hair defective 6* (*rhd6*) mutant of *Arabidopsis thaliana*. *New Phytologist* 213, 1771–1786.

Muszyński, A, O'Neill, MA, Ramasamy, E, Pattathil, S, Avci, U, Peña, MJ, Libault, M, Hossain, MS, Brechenmacher, L, York, WS, Barbosa, RM, Hahn, MG, Stacey, G, and Carlson, RW (2015) Xyloglucan, galactomannan, glucuronoxylan, and rhamnogalacturonan I do not have identical structures in soybean root and root hair cell walls. *Planta* 242, 1123–1138.

Nickle, TC, and Meinke, DW (1998) A cytokinesis-defective mutant of Arabidopsis (*cyt1*) characterized by embryonic lethality, incomplete cell walls, and excessive callose accumulation. *The Plant Journal* 15, 321–332.

Nishikawa, S-I, Zinkl, GM, Swanson, RJ, Maruyama, D, and Preuss, D (2005) Callose (β-1,3 glucan) is essential for Arabidopsis pollen wall patterning, but not tube growth. *BMC Plant Biology* 5, 22.

Norman, PM, Kjellbom, P, Bradley, DJ, Hahn, MG, and Lamb, CJ (1990) Immunoaffinity purification and biochemical characterization of plasma membrane arabino-galactan-rich glycoproteins of *Nicotiana glutinosa*. *Planta* 181, 365–373.

Norman, PM, Wingate, VPM, Fitter, MS, and Lamb, CJ (1986) Monoclonal antibodies to plant plasma-membrane antigens. *Planta* 167, 452–459.

O'Neill, MA, Eberhard, S, Albersheim, P, and Darvill, AG (2001) Requirement of borate cross-linking of cell wall rhamnogalacturonan II for Arabidopsis growth. *Science* 294, 846–849.

O'Neill, MA, Warrenfeltz, D, Kates, K, Pellerin, P, Doco, T, Darvill, AG, and Albersheim, P (1996) Rhamnogalacturonan-II, a pectic polysaccharide in the walls of growing plant cell, forms a dimer that is covalently cross-linked by a borate ester. *Journal of Biological Chemistry* 271, 22923–22930.

Ordaz-Ortiz, JJ, Guillon, F, Tranquet, O, Dervilly-Pinel, G, Tran, V, and Saulnier, L (2004) Specificity of monoclonal antibodies generated against arabinoxylans of cereal grains. *Carbohydrate Polymers* 57, 425–433.

Orfila, C, Seymour, GB, Willats, WGT, Huxham, IM, Jarvis, MC, Dover, CJ, Thompson, AJ, and Knox, JP (2001) Altered middle lamella homogalacturonan and disrupted deposition of (1→5)-α-L-arabinan in the pericarp of *Cnr*, a ripening mutant of tomato. *Plant Physiology* 126, 210–221.

Parsons, HT, Stevens, TJ, McFarlane, HE, Vidal-Melgosa, S, Griss, J, Lawrence, N, Butler, R, Sousa, MML, Salemi, M, Willats, WGT, Petzold, CJ, Heazlewood, JL, and Lilley, KS (2019) Separating Golgi proteins from cis to trans reveals underlying properties of cisternal localization. *The Plant Cell* 31, 2010–2034.

Pattathil, S, Avci, U, Baldwin, D, Swennes, AG, McGill, JA, Popper, Z, Bootten, T, Albert, A, Davis, RH, Chennareddy, C, Dong, R, O'Shea, B, Rossi, R, Leoff, C, Freshour, G, Narra, R, O'Neil, M, York, WS, and Hahn, MG (2010) A comprehensive toolkit of plant cell wall glycan-directed monoclonal antibodies. *Plant Physiology* 153, 514–525.

Pattathil, S, Avci, U, Miller, JS, and Hahn, MG (2012) Immunological approaches to plant cell wall and biomass characterization: glycome profiling. In *Biomass Conversion: Methods and Protocols*, Himmel M, ed (pp. 61–72). New York, NY: Springer Science+Business Media, LLC.

Pattathil, S, Avci, U, Zhang, T, Cardenas, CL, and Hahn, MG (2015) Immunological approaches to biomass characterization and utilization. *Frontiers in Bioengineering and Biotechnology* 3, 173.

Pauly, M, and Keegstra, K (2016) Biosynthesis of the plant cell wall matrix polysaccharide xyloglucan. *Annual Review of Plant Biology* 67, 235–259.

Peaucelle, A, Wightman, R, and Höfte, H (2015) The control of growth symmetry breaking in the Arabidopsis hypocotyl. *Current Biology* 25, 1746–1752.

Pedersen, HL, Fangel, JU, McCleary, B, Ruzanski, C, Rydahl, MG, Ralet, M-C, Farkas, V, Von Schantz, L, Marcus, SE, Andersen, MCF, Field, R, Ohlin, M, Knox, JP, Clausen, MH, and Willats, WGT (2012) Versatile high resolution oligosaccharide microarrays for plant glycobiology and cell wall research. *Journal of Biological Chemistry* 287, 39429–39438.

Peña, MJ, Kong, Y, York, WS, and O'Neill, MA (2012) A galacturonic acid-containing xyloglucan is involved in *Arabidopsis* root hair tip growth. *Plant Cell* 24, 4511–4524.

Pennell, RI, Janniche, L, Kjellbom, P, Scofield, GN, Peart, JM, and Roberts, K (1991) Developmental regulation of a plasma membrane arabinogalactan protein epitope in oilseed rape flowers. *Plant Cell* 3, 1317–1326.

Pennell, RI, Knox, JP, Scofield, GN, Selvendran, RR, and Roberts, K (1989) A family of abundant plasma membrane-associated glycoproteins related to the arabinogalactan proteins is unique to flowering plants. *Journal of Cell Biology* 108, 1967–1977.

Peralta, AG, Venkatachalam, S, Stone, SC, and Pattathil, S (2017) Xylan epitope profiling: an enhanced approach to study organ development-dependent changes in xylan structure, biosynthesis, and deposition in plant cell walls. *Biotechnology for Biofuels* 10, 1–13.

Perotto, S, Vandenbosch, KA, Butcher, GW, and Brewin, NJ (1991) Molecular composition and development of the plant with the peribacteroid membrane of pea root nodules. *Development* 112, 763–773.

Perrin, RM, Jia, Z, Wagner, TA, O'Neill, MA, Sarria, R, York, WS, Raikhel, NV, and Keegstra, K (2003) Analysis of xyloglucan fucosylation in Arabidopsis. *Plant Physiology* 132, 768–778.

Petrova, A, Sibgatullina, G, Gorshkova, T, and Kozlova, L (2022) Dynamics of cell wall polysaccharides during the elongation growth of rye primary roots. *Planta* 255, 108.

Pettolino, FA, Hoogenraad, NJ, Ferguson, C, Bacic, A, Johnson, E, and Stone, BA (2001) A (1→4)-β-mannan-specific monoclonal antibody and its use in the immunocytochemical location of galactomannans. *Planta* 214, 235–242.

Prajapati, VD, Jani, GK, Moradiya, NG, Randeria, NP, Nagar, BJ, Naikwadi, NN, and Variya, BC (2013) Galactomannan: a versatile biodegradable seed polysaccharide. *International Journal of Biological Macromolecules* 60, 83–92.

Puhlmann, J, Bucheli, E, Swain, MJ, Dunning, N, Albersheim, P, Darvill, AG, and Hahn, MG (1994) Generation of monoclonal antibodies against plant cell wall polysaccharides. I. Characterization of a monoclonal antibody to a terminal α-(1→2)-linked fucosyl-containing epitope. *Plant Physiology* 104, 699–710.

Qi, J, Wu, B, Feng, S, Lü, S, Guan, C, Zhang, X, Qiu, D, Hu, Y, Zhou, Y, Li, C, Long, M, and Jiao, Y (2017) Mechanical regulation of organ asymmetry in leaves. *Nature Plants* 3, 724–733.

Qiu, D, Xu, S, Wang, Y, Zhou, M, and Hong, L (2021) Primary cell wall modifying proteins regulate wall mechanics to steer plant morphogenesis. *Frontiers in Plant Science* 12, 751372.

Raimundo, SC, Pattathil, S, Eberhard, S, Hahn, MG, and Popper, ZA (2017) β-1,3-glucans are components of brown seaweed (*Phaeophyceae*) cell walls. *Protoplasma* 254, 997–1016.

Ralet, MC, Tranquet, O, Poulain, D, Moïse, A, and Guillon, F (2010) Monoclonal antibodies to rhamnogalacturonan I backbone. *Planta* 231, 1373–1383.

Rebaque, D, Hierro, I, López, G, Bacete, L, Vilaplana, F, Dallabernardina, P, Pfrengle, F, Jordá, L, Sánchez-Vallet, A, Pérez, R, Brunner, F, Molina, A, and Mélida, H (2021) Cell wall-derived mixed-linked β-(1→3),(1→4)-glucans trigger immune responses and disease resistance in plants. *The Plant Journal* 106, 601–615.

Rennie, EA, and Scheller, HV (2014) Xylan biosynthesis. *Current Opinion in Biotechnology* 26, 100–107.

Roberts, K, Phillips, J, Shaw, P, Grief, C, and Smith, E (1985) An immunological approach to the plant cell wall. In *Biochemistry of Plant Cell Walls*, Brett CT, Hillman JR, eds (pp. 125–154). Cambridge, UK: Cambridge University Press.

Ruan, Y-L, Xu, S-M, White, R, and Furbank, RT (2004) Genotypic and developmental evidence for the role of plasmodesmatal regulation in cotton fiber elongation mediated by callose turnover. *Plant Physiology* 136, 4104–4113.

Ruprecht, C, Bartetzko, MP, Senf, D, Dallabernadina, P, Boos, I, Andersen, MCF, Kotake, T, Knox, JP, Hahn, MG, Clausen, MH, and Pfrengle, F (2017) A synthetic glycan microarray enables epitope mapping of plant cell wall glycan-directed antibodies. *Plant Physiology* 175, 1094–1104.

Ruprecht, C, and Pfrengle, F (2022) Synthetic plant glycan microarrays as tools for plant biology. In *Glycan Microarrays: Methods and Protocols*, Kilcoyne M, Gerlach JQ, eds (pp. 115–125). New York, NY: Springer US.

Salmeán, AA, Duffieux, D, Harholt, J, Qin, F, Michel, G, Czjzek, M, Willats, WGT, and Hervé, C (2017) Insoluble (1→3),(1→4)-β-D-glucan is a component of cell walls in brown algae (*Phaeophyceae*) and is masked by alginates in tissues. *Scientific Reports* 7, 2880.

Salnikov, VV, Grimson, MJ, Seagull, RW, and Haigler, CH (2003) Localization of sucrose synthase and callose in freeze-substituted secondary-wall-stage cotton fibers. *Protoplasma* 221, 175–184.

Sandquist, D, Filonova, L, von Schantz, L, Ohlin, M, and Daniel, G (2010) Microdistribution of xyloglucan in differentiating poplar cells. *Bioresources* 5, 796–807.

Scheller, HV, and Ulvskov, P (2010) Hemicelluloses. *Annual Review of Plant Biology* 61, 263–289.

Schlüpmann, H, Bacic, A, and Read, SM (1994) Uridine diphosphate glucose metabolism and callose synthesis in cultured pollen tubes of *Nicotiana alata* Link et Otto. *Plant Physiology* 105, 659–670.

Schmidt, D, Schuhmacher, F, Geissner, A, Seeberger, PH, and Pfrengle, F (2015) Automated synthesis of arabinoxylan-oligosaccharides enables characterization of antibodies that recognize plant cell wall glycans. *Chemistry* 21, 5709–5713.

Selvendran, RR, and Ryden, P (1990) Isolation and analysis of plant cell walls. In *Methods in Plant Biochemistry*, Dey PM, ed (pp. 549–579). Academic Press, Ltd., London, UK.

Sénéchal, F, Wattier, C, Rustérucci, C, and Pelloux, J (2014) Homogalacturonan-modifying enzymes: structure, expression, and roles in plants. *Journal of Experimental Botany* 65, 5125–5160.

Shevell, DE, Kunkel, T, and Chua, N-H (2000) Cell wall alterations in the Arabidopsis emb30 mutant. *The Plant Cell* 12, 2047–2059.

Shin, Y, Chane, A, Jung, M, and Lee, Y (2021) Recent advances in understanding the roles of pectin as an active participant in plant signaling networks. *Plants* 10, 1712.

Silva, C, Adaskaveg, JA, Mesquida-Pesci, SD, Ortega-Salazar, IB, Pattathil, S, Zhang, L, Hahn, MG, van Kan, JAL, Cantu, D, Powell, ALT, and Blanco-Ulate, B (2022) Botrytis cinerea infection accelerates ripening and cell wall disassembly to promote disease in tomato fruit. *Plant Physiology* 191(1), 575–590.

Silva, GB, Ionashiro, M, Carrara, TB, Crivellari, AC, Tine, MA, Prado, J, Carpita, NC, and Buckeridge, MS (2011) Cell wall polysaccharides from fern leaves: evidence for a mannan-rich Type III cell wall in *Adiantum raddianum*. *Phytochemistry* 72, 2352–2360.

Singh, B, Avci, U, Eichler Inwood, SE, Grimson, MJ, Landgraf, J, Mohnen, D, Sørensen, I, Wilkerson, CG, Willats, WGT, and Haigler, CH (2009) A specialized outer layer of the primary cell wall joins elongating cotton fibers into tissue-like bundles. *Plant Physiology* 150, 684–699.

Singh, S, Singh, G, and Arya, SK (2018) Mannans: an overview of properties and application in food products. *International Journal of Biological Macromolecules* 119, 79–95.

Sinha, N, and Lynch, M (1998) Fused organs in the *adherent1* mutation in maize show altered epidermal walls with no perturbations in tissue identities. *Planta* 206, 184–195.

Smallwood, M, Beven, A, Donovan, N, Neill, SJ, Peart, J, Roberts, K, and Knox, JP (1994) Localization of cell wall proteins in relation to the developmental anatomy of the carrot root apex. *The Plant Journal* 5, 237–246.

Smallwood, M, Yates, EA, Willats, WGT, Martin, H, and Knox, JP (1996) Immunochemical comparison of membrane-associated and secreted arabinogalactan-proteins in rice and carrot. *Planta* 198, 452–459.

Smith, E, Roberts, K, Hutchings, A, and Galfre, G (1984) Monoclonal antibodies to the major structural glycoprotein of the Chlamydomonas cell wall. *Planta* 161, 330–338.

Socha, AM, Parthasarathi, R, Shi, J, Pattathil, S, Whyte, D, Bergeron, M, George, A, Tran, K, Stavila, V, Venkatachalam, S, Hahn, MG, Simmons, BA, and Singh, S (2014) Efficient biomass pretreatment using ionic liquids derived from lignin and hemicellulose. *Proceedings of the National Academy of Sciences of the USA* 111, E3587–E3595.

Sopeña-Torres, S, Jordá, L, Sánchez-Rodriguez, C, Miedes, E, Escudero, V, Swami, S, López, G, Pislewska-Bednarek, M, Lassowskat, I, Lee, J, Gu, YN, Haigis, S, Alexander, D, Pattathil, S, Munoz-Barrios, A, Bednarek, P, Somerville, S, Schulze-Lefert, P, Hahn, MG, Scheel, D, and Molina, A (2018) YODA MAP3K kinase regulates plant immune responses conferring broad-spectrum disease resistance. *New Phytologist* 218, 661–680.

Stacey, NJ, Roberts, K, and Knox, JP (1990) Patterns of expression of the JIM4 arabinogalactan-protein epitope in cell cultures and during somatic embryogenesis in *Daucus carota* L. *Planta* 180, 285–292.

Tan, L, Eberhard, S, Pattathil, S, Warder, C, Glushka, J, Yuan, C, Hao, Z, Zhu, X, Avci, U, Miller, JS, Baldwin, D, Pham, C, Orlando, R, Darvill, A, Hahn, MG, Kieliszewski, MJ, and Mohnen, D (2013) An *Arabidopsis* cell wall proteoglycan consists of pectin and arabinoxylan covalently linked to an arabinogalactan protein. *The Plant Cell* 25, 270–287.

Tan, L, Showalter, A, Egelund, J, Hernandez-Sanchez, A, Doblin, M, and Bacic, A (2012) Arabinogalactan-proteins and the research challenges for these enigmatic plant cell surface proteoglycans. *Frontiers in Plant Science* 3, 140.

Torode, TA, O'Neill, R, Marcus, SE, Cornuault, V, Pose, S, Lauder, RP, Kračun, SK, Rydahl, MG, Andersen, MCF, Willats, WGT, Braybrook, SA, Townsend, BJ, Clausen, MH, and Knox, JP (2018) Branched pectic galactan in phloem-sieve-element cell walls: implications for cell mechanics. *Plant Physiology* 176, 1547–1558.

Trethewey, JAK, and Harris, PJ (2002) Location of (1→3)- and (1→3),(1→4)-β-D-glucans in vegetative cell walls of barley (*Hordeum vulgare*) using immunogold labelling. *New Phytologist* 154, 347–358.

Tuomivaara, ST, Yaoi, K, O'Neill, MA, and York, WS (2015) Generation and structural validation of a library of diverse xyloglucan-derived oligosaccharides, including an update on xyloglucan nomenclature. *Carbohydryate Research* 402, 56–66.

Tuomivaara, ST, Popper, ZA, Bootten, TJ, Swennes, AG, Davis, RH, Freshour, G, York, WS, O'Neill, MA, and Hahn, MG (2023) Epitope characterization of monoclonal antibodies with diverse xyloglucan-binding specificities, in preparation.

Ulvskov, P, Wium, H, Bruce, D, Jørgensen, B, Qvist, KB, Skjøt, M, Hepworth, D, Borkhardt, B, and Sørensen, SO (2005) Biophysical consequences of remodeling the neutral side chains of rhamnogalacturonan I in tubers of transgenic potatoes. *Planta* 220, 609–620.

Urbanowicz, BR, Peña, MJ, Moniz, HA, Moremen, KW, and York, WS (2014) Two Arabidopsis proteins synthesize acetylated xylan in vitro. *The Plant Journal* 80, 197–206.

Urbanowicz, BR, Peña, MJ, Ratnaparkhe, S, Avci, U, Backe, J, Steet, HF, Foston, M, Li, H, O'Neill, MA, Ragauskas, AJ, Darvill, AG, Wyman, C, Gilbert, HJ, and York, WS (2012) 4-*O*-methylation of glucuronic acid in Arabidopsis glucuronoxylan is catalyzed by a domain of unknown function family 579 protein. *Proceedings of the National Academy of Sciences of the USA* 109, 14253–14258.

Vandenbosch, KA, Bradley, DJ, Knox, JP, Perotto, S, Butcher, GW, and Brewin, NJ (1989) Common components of the infection thread matrix and the intercellular space identified by immunocytochemical analysis of pea nodules and uninfected roots. *EMBO Journal* 8, 335–341.

Velasquez, SM, Guo, X, Gallemi, M, Aryal, B, Venhuizen, P, Barbez, E, Dünser, KA, Darino, M, Pěnčík, A, Novák, O, Kalyna, M, Mouille, G, Benková, E, Bhalerao, RP, Mravec, J, and Kleine-Vehn, J (2021) Xyloglucan remodeling defines auxin-dependent differential tissue expansion in plants. *International Journal of Molecular Sciences* 22, 9222.

Verhertbruggen, Y, Marcus, SE, Haeger, A, Ordaz-Ortiz, JJ, and Knox, JP (2009a) An extended set of monoclonal antibodies to pectic homogalacturonan. *Carbohydrate Research* 344, 1858–1862.

Verhertbruggen, Y, Marcus, SE, Haeger, A, Verhoef, R, Schols, HA, McCleary, BV, McKee, L, Gilbert, HJ, and Paul Knox, J (2009b) Developmental complexity of arabinan polysaccharides and their processing in plant cell walls. *The Plant Journal* 59, 413–425.

Villa-Rivera, MG, Cano-Camacho, H, López-Romero, E, and Zavala-Páramo, MG (2021) The role of arabinogalactan Type II degradation in plant-microbe interactions. *Frontiers in Microbiology* 12, 730543.

Voiniciuc, C (2022) Modern mannan: a hemicellulose's journey. *New Phytologist* 234, 1175–1184.

Voiniciuc, C, Pauly, M, and Usadel, B (2018) Monitoring polysaccharide dynamics in the plant cell wall. *Plant Physiology* 176, 2590–2600.

Voiniciuc, C, Schmidt, MH, Berger, A, Yang, B, Ebert, B, Scheller, HV, North, HM, Usadel, B, and Gunl, M (2015) MUCILAGE-RELATED10 produces galactoglucomannan that maintains pectin and cellulose architecture in Arabidopsis seed mucilage. *Plant Physiology* 169, 403–420.

Vreeland, V (1970) Localization of a cell wall polysaccharide in a brown alga with labeled antibody. *Journal of Histochemistry & Cytochemistry* 18, 371–373.

Vreeland, V, Slomich, M, and Laetsch, WM (1984) Monoclonal antibodies as molecular probes for cell wall antigens of the brown alga, *Fucus*. *Planta* 162, 506–517.

Wallace, I, and Anderson, CT (2012) Small molecule probes for plant cell wall polysaccharide imaging. *Frontiers in Plant Science* 3, 89.

Wang, M, Heimovaara-Dijkstra, S, Van der Meulen, RM, Knox, JP, and Neill, SJ (1995) The monoclonal antibody JIM19 modulates abscisic acid action in barley aleurone protoplasts. *Planta* 196, 271–276.

Wang, X, Wilson, L, and Cosgrove, DJ (2020) Pectin methylesterase selectively softens the onion epidermal wall yet reduces acid-induced creep. *Journal of Experimental Botany* 71, 2629–2640.

Willats, WGT, Marcus, SE, and Knox, JP (1998) Generation of a monoclonal antibody specific to (1→5)-α-L-arabinan. *Carbohydrate Research* 308, 149–152.

Willats, WGT, Orfila, C, Limberg, G, Buchholt, HC, Van Alebeek, G-JWM, Voragen, AG, Marcus, SE, Christensen, TMIE, Mikkelsen, JD, Murray, BS, and Knox, JP (2001) Modulation of the degree and pattern of methyl-esterification of pectic homogalacturonan in plant cell walls. *Journal of Biological Chemistry* 276, 19404–19413.

Williams, MNV, Freshour, G, Darvill, AG, Albersheim, P, and Hahn, MG (1996) An antibody Fab selected from a recombinant phage display library detects deesterified pectic polysaccharide rhamnogalacturonan II in plant cells. *Plant Cell* 8, 673–685.

Wilson, SM, Burton, RA, Doblin, MS, Stone, BA, Newbigin, EJ, Fincher, GB, and Bacic, A (2006) Temporal and spatial appearance of wall polysaccharides during cellularization of barley (*Hordeum vulgare*) endosperm. *Planta* 224, 655–667.

Wilson, SM, Ho, YY, Lampugnani, ER, Van De Meene, AML, Bain, MP, Bacic, A, and Doblin, MS (2015) Determining the subcellular location of synthesis and assembly of the cell wall polysaccharide (1→3,1→4)-β-D-glucan in grasses. *The Plant Cell* 27, 754–771.

Wood, IP, Pearson, BM, Garcia-Gutierrez, E, Havlickova, L, He, Z, Harper, AL, Bancroft, I, and Waldron, KW (2017) Carbohydrate microarrays and their use for the identification of molecular markers for plant cell wall composition. *Proceedings of the National Academy of Sciences of the USA* 114, 6860–6865.

Xiong, G, Cheng, K, and Pauly, M (2013) Xylan O-acetylation impacts xylem development and enzymatic recalcitrance as indicated by the Arabidopsis mutant *tbl29*. *Molecular Plant* 6, 1373–1375.

Yan, Y, Takáč, T, Li, X, Chen, H, Wang, Y, Xu, E, Xie, L, Su, Z, Šamaj, J, and Xu, C (2015) Variable content and distribution of arabinogalactan proteins in banana (Musa spp.) under low temperature stress. *Frontiers in Plant Science* 6, 353.

Yang, H, Nukunya, K, Ding, Q, and Thompson, BE (2022) Tissue-specific transcriptomics reveal functional differences in floral development. *Plant Physiology* 188, 1158–1173.

Yang, W, Schuster, C, Cherie, Charoensawan, V, Peaucelle, A, Bacic, A, Monika, Wightman, R, and Elliot (2016) Regulation of meristem morphogenesis by cell wall synthases in Arabidopsis. *Current Biology* 26, 1404–1415.

Yang, Y, and Anderson, CT (2020) Biosynthesis, localisation, and function of pectins in plants. In *Pectin: Technological and Physiological Properties*, Kontogiorgos V, ed (pp. 1–15). Springer International Publishing, Cham, Switzerland.

Yapo, BM (2011) Rhamnogalacturonan-I: a structurally puzzling and functionally versatile polysaccharide from plant cell walls and mucilages. *Polymer Reviews* 51, 391–413.

Yates, EA, Valdor, J-F, Haslam, SM, Morris, HR, Dell, A, Mackie, W, and Knox, JP (1996) Characterization of carbohydrate structural features recognized by anti-arabinogalactan-protein monoclonal antibodies. *Glycobiology* 6, 131–139.

Yin, L, Verhertbruggen, Y, Oikawa, A, Manisseri, C, Knierim, B, Prak, L, Jensen, JK, Knox, JP, Auer, M, Willats, WG, and Scheller, HV (2011) The cooperative activities of CSLD2, CSLD3, and CSLD5 are required for normal Arabidopsis development. *Molecular Plant* 4, 1024–1037.

Young, RE, McFarlane, HE, Hahn, MG, Western, TL, Haughn, GW, and Samuels, AL (2008) Analysis of the Golgi apparatus in *Arabidopsis* seed coat cells during polarized secretion of pectin-rich mucilage. *The Plant Cell* 20, 1623–1638.

Yu, L, Lyczakowski, JJ, Pereira, CS, Kotake, T, Yu, X, Li, A, Mogelsvang, S, Skaf, MS, and Dupree, P (2018) The patterned structure of galactoglucomannan suggests it may bind to cellulose in seed mucilage. *Plant Physiology* 178, 1011–1026.

Yu, L, Shi, D, Li, J, Kong, Y, Yu, Y, Chai, G, Hu, R, Wang, J, Hahn, MG, and Zhou, G (2014) CELLULOSE SYNTHASE-LIKE A2, a glucomannan synthase, is involved in maintaining adherent mucilage structure in Arabidopsis seed. *Plant Physiology* 164, 1842–1856.

Zabotina, OA, Avci, U, Cavalier, D, Pattathil, S, Chou, Y-H, Eberhard, S, Danhof, L, Keegstra, K, and Hahn, MG (2012) Mutations in multiple *XXT* genes of Arabidopsis reveal the complexity of xyloglucan biosynthesis. *Plant Physiology* 159, 1367–1384.

Zabotina, OA, van de Ven, WTG, Freshour, G, Drakakaki, G, Cavalier, D, Mouille, G, Hahn, MG, Keegstra, K, and Raikhel, NV (2008) Arabidopsis *XXT5* gene encodes a putative α-1,6-xylosyltransferase that is involved in xyloglucan biosynthesis. *Plant Journal* 56, 101–115.

Zamil, MS, and Geitmann, A (2017) The middle lamella-more than a glue. *Physical Biology* 14, 015004.

Zhang, N., Julian, JD, Yap, CE, Swaminathan, S, and Zabotina, O (2023) The Arabidopsis xylosyltransferases, XXT3, XXT4, and XXT5, are essential to complete the fully xylosylated glucan backbone XXXG-type structure of xyloglucans. *New Phytologist* 238, 1986-1999.

Zhang, T (2014) *Using Carbohydrate-Binding Modules as Tools for in vivo Visualization/Modulation of Plant Cell Wall Polysaccharides* (pp. 1–132). University of Georgia, Athens, GA, USA.

Zhang, X, Rogowski, A, Zhao, L, Hahn, MG, Avci, U, Knox, JP, and Gilbert, HJ (2014) Understanding how the complex molecular architecture of mannan-degrading hydrolases contributes to plant cell wall degradation. *Journal of Biological Chemistry* 289, 2002–2012.

Zhao, Q, Tobimatsu, Y, Zhou, R, Pattathil, S, Gallego-Giraldo, L, Fu, C, Jackson, LA, Hahn, MG, Kim, H, Chen, F, Ralph, J, and Dixon, RA (2013) Loss of function of cinnamyl alcohol dehydrogenase 1 leads to unconventional lignin and a temperature-sensitive growth defect in Medicago truncatula. *Proceedings of the National Academy of Sciences of the USA* 110, 13660–13665.

Zhou, Y, Awano, T, Kobayashi, M, Matoh, T, and Takabe, K (2017) Immunocytochemical detection of rhamnogalacturonan II on forming cell plates in cultured tobacco cells. *Bioscience, Biotechnology, and Biochemistry* 81, 899–905.

Zhou, Y, Kobayashi, M, Awano, T, Matoh, T, and Takabe, K (2018) A new monoclonal antibody against rhamnogalacturonan II and its application to immunocytochemical detection of rhamnogalacturonan II in Arabidopsis roots. *Bioscience, Biotechnology, and Biochemistry* 82, 1780–1789.

Zhu, X, Pattathil, S, Mazumder, K, Brehm, A, Hahn, MG, Dinesh-Kumar, SP, and Joshi, CP (2010) Virus-induced gene silencing offers a functional genomics platform for studying plant cell wall formation. *Molecular Plant* 3, 818–833.

Zola, H (1987) *Monoclonal Antibodies: A Manual of Techniques*. Boca Raton, FL: CRC Press, Inc.

11 Conceptual Models of the Plant Cell Wall

Peter Ulvskov
Copenhagen University

Peter M. Ray
Stanford University

INTRODUCTION

Models guide our thinking and spawn research questions to be addressed experimentally. We therefore consider models of the cell wall important. "On whose shoulders we stand" could have been an appropriate subtitle because we will look into the development of ideas, some outdated, some that have resurfaced and some that deserve to do so.

When cells were discovered by Hooke (1665), it was the walls of cork cells that he observed. While this by definition is the starting point, we will not trace the history of cell walls that far back. We will not even consider the 1830s where research into plant cell wall composition and architecture in the modern sense took its beginning according to Hess (1928). The beginnings of interest in both primary cell wall structure and plant cell growth dates back to the 19th century botanist Carl Nägeli (often spelled Naegeli). In his book *Die Stärkekörner* (The Starch Grains, Nägeli (1858)), he included conclusions about the ultrastructure of cell walls and about their expansion during growth which were further elaborated in several papers that he published in the following years (e.g., Nägeli 1864a,b). Cell wall model building is concerned with establishing an inventory of biopolymers that are its building blocks and understanding cell wall metabolism during growth, differentiation and response to stimuli. A good cell wall model suggests an architecture that can account for, or at least be compatible with, observed cell wall dynamics. Expansion of the primary wall during cell growth is by far the most thoroughly studied and important application of the models that we will discuss. We will therefore deal with cell expansion theories, not in depth but in as much detail as is required to appreciate the thinking that went into the model building. Prior to the discovery of auxin, photo- and gravitropic bending were the primary tools for modulating cell expansion experimentally and these studies led eventually to the identification of the hormone auxin. We are therefore concerned with the level of understanding preceding and following Charles Darwin's remarkable inference, published in his *The Power of Movement in Plants*, of the existence of auxin (Darwin 1880) while studying phototropic bending of canary grass hypocotyls 30 years before the concept of plant growth hormones was coined by Fitting (1910). Auxin provided researchers with a tool to stimulate cell expansion and this led to much increased interest in cell wall architecture: Which cell wall models were compatible with or even could account for the biophysical and biochemical processes observed to accompany auxin-stimulated cell expansion? This question is as pertinent today as it was a century ago.

PRIMARY WALLS AND GROWTH BEFORE AUXIN

The fourth edition of Julius Sachs' classical botany textbook was published 6 years before Darwin's *The Power of Movement in Plants* (Sachs 1874). In it cell expansion is noted to be accompanied by formation of vacuoles that eventually coalesce to form one vacuole that takes up much of cellular volume. The cell wall is a lamellate structure where the water holding capacity of lamellae varies and gives rise to birefringence patterns that can be observed in the light microscope. (The lamellate structure of walls received much attention in the following decades, e.g., by Söding (1934) who noted that the wall shrinks to one third when dried and while the wall is smooth in intact cells, the innermost layer becomes wrinkled in plasmolysed cells). The cell wall is made up of cellulose and minerals; and it is speculated whether material, probably made of carbohydrate, is extruded into the wall from the protoplast as the cell grows. Pectins were soon to be recognised (as were phenolic compounds in secondary walls) (Haberlandt 1896). For the history of pectin discovery and characterisation, see Chapter 5 in this book.

DOI: 10.1201/9781003178309-11

ELASTIC

Deforms *reversibly*, proportionally to stress (load) (when unloaded, recovers original dimensions and shape)

Nägeli (1858) argued that wall expansion during cell growth is caused by the insertion of new wall material (probably intermicellar substance, he thought) into the structure of existing walls, a process that he called intussusception (a term that had been used by Lamarck around 1830 for insertative biosyntheses). Intussusception was accepted as the mode of wall growth by most botanists of that day, including Julius von Sachs, called the "Father of Plant Physiology" and mentor of most of the major 19th-century workers in that field. Stems were considered to have an elongation zone with particularly hydrated walls, a prerequisite for growth by intussusception. Cell expansion was thought to be initiated by water influx, turgor build-up and elastic wall extension. Sachs (1882) and de Vries (1877) are credited with being the first to call attention to an importance of cellular turgor pressure to cell growth. Sachs considered that turgor is necessary to stretch a wall elastically, creating internal spaces into which new wall material can be introduced by intussusception. The role of turgor was appreciated by Darwin and was in line with Strasburger (1882) who pointed out that due to the nature of cellulose, it must be laid down by apposition. He was building on Nägeli who proposed that cell walls are composed of elongated, microscopically invisible, directionally oriented structural units that he called micelles (the term then used for colloidal molecular aggregations). Much later, Frey (1927) reviewed this area compactly and helpfully. Nägeli considered the micelles to run parallel to one another in the plane of the wall and to be separated from one another by a hydrophilic intermicellar substance. The micelles were eventually identified as the cellulose component of the wall. It was Seifritz (1934) who concluded that cellulose has cellobiose as repeating unit and he proposed that cellulose is synthesised in a way not unlike that in *Acetobacter xylinum*, which proved correct and an important clue leading to the discovery of cellulose synthases in higher plants (Pear et al. 1996).

ACETOBACTER XYLINUM

A gram-negative bacterium that is unusual in that it produces cellulose. Its cellulose synthases are not organised as rosettes but in linear terminal complexes. It has been renamed *Komagataeibacter xylinus*.

There was far from consensus at the end of the pre-auxin era. Heyn (1931) listed four theories plus variants but one very notable contribution went unnoticed: Ursprung and Blum (1924) stated the following about what has to happen for a plant cell to begin growth (translated from German, with our parenthetical annotations):

Stages in the Initiation of Plant Cell Growth

1. Active growth of the cell wall by intussusception (this was their belief as to the primary wall-loosening process)
2. Decrease in wall pressure (= stress relaxation)
3. Increase in Saugkraft (their term for the negative of what we now call cell water potential)
4. Water uptake into cell
5. Turgor-driven extension [of the cell wall]

The realisation of the significance of *in vivo* stress relaxation was forgotten and not considered explicitly when Boysen Jensen summarised the main theories in his 1936 book which we shall return to below. Nearly 50 years should pass before it resurfaced (Ray et al. 1972).

THE AUXIN ERA

Auxin was experimentally detected directly first by Peter Boysen Jensen (Boysen Jensen 1911; Boysen Jensen and Nielsen 1925) and Fritz Went (Went 1928). The polymers xylan (of the pentose sugar xylose) and lignin (of phenolic subunits) were added to the cell wall inventory during this period (Lüdkte 1931). The path to identifying indole-3-acetic acid (IAA) as the structure of auxin *in planta*, or heteroauxin, as it was at first called, was extraordinarily tortuous, see pp. 31ff in Audus (1953) for that story. Shortly before 1930, ANJ Heyn joined the Utrecht lab of Fritz Went, wanting to work on and settle, with the help of the newly available auxin, the many issues surrounding primary wall architecture and the physical nature of cell growth. His work eventually influenced this field more than anything that preceded or perhaps followed it until the appearance of the first detailed primary wall structural model by Peter Albersheim's group in 1973. Heyn (1931) introduced a new method for measuring elastic and plastic extensibility of an oat seedling's coleoptile (a sheath-like first leaf, which was Went's principal research material for discovery of auxin). Coleoptiles were excised, cutting off their water supply (from the seedling's roots) so that they could not grow, and were either deprived of auxin by cutting off ("decapitating") their tips (where auxin is produced), or else supplied with auxin either by leaving the tip attached or by replacing it with an auxin-containing agar blocklet. Coleoptiles were held horizontally by their bases and their free, apical ends were loaded with a 250 mg wire rider (weight). After 1 hour or more, the loads were removed and the set was shadowgraphed to record the results. The recovery bending that occurred upon removing the load was a measure of the elastic extensibility, and the curvature that remained after unloading measured the plastic extensibility.

PLASTIC

Deforms elastically up to a yield stress (Y), at which it deforms irreversibly without limit and will not support a higher load

Heyn (1931) found that the occurrence of auxin-induced elongation was always accompanied by increase in wall plasticity, as measured by his method, while changes in elastic extension did not correlate with growth. Boysen Jensen et al. (1936) sums up (p.104 ff) conclusions derived from in particular Heyn, Söding, as well as Pfeffer (1904) and Thimann and Bonner (1933) and proposes three possible scenarios for how cell expansion may be brought about:

1. Elastic extension accompanied by maintenance of wall thickness by apposition or intussusception. Deposition of new wall layers renders previous elastic extension irreversible.
2. Plastic extension. This mechanism is identical with (1) except that extension is irreversible by nature, and limited by turgor and plastic compliance. How the latter might be modulated was not considered.
3. Active growth. This mechanism relies on biosynthetically growing the wall. Turgor does not drive expansion but is merely required to keep the protoplast appressed to the wall and may additionally be required to stretch the wall elastically to allow intussusception, or to drive new wall polymers into the wall so that they will bond with existing ones. The mechanism is irreversible.

Boysen Jensen ruled out Mechanism 1. Went (1928) and Heyn were in favour of 2, i.e., that IAA modulates plastic compliance. Söding (1934), on the other hand, was of the opinion that changes in plastic extensibility were necessary but not sufficient. Söding proposed that extensibility might be modulated by softening of the intermicellar substance (now called *matrix*) combined with intussusception of new material. In other words, he favoured a hybrid 2+3 model in which the intermicellar substances, probably described as being what we would today call viscoelastic, are inserted into the wall in parallel with deposition of new cellulose. Turgor pressure was relegated to a secondary role in driving the intussusception of new intermicellar material. Recycling of matrix polysaccharides would serve two purposes: Reduce the viscosity of the intermicellar material, and provide glucose for cellulose synthesis (Söding 1933). Where and how this all took place were left for speculation and Anderson (1935) made this remark: "It is possible that the plant cell wall may possess some of the characteristics usually attributed to living material during certain phases of its growth".

That low pH is associated with cell expansion was realised early (Bonner 1934a). Rayle & Cleland (1970) experimented on this with oat coleoptiles, and similar experiments were going on in the laboratory of Achim Hager in Heidelberg, Germany. By 1971, both groups had reached the same conclusion, that H^+ ions might be a second-messenger for auxin action on the cell wall. In the last paragraph (p. 217) of his Annual Review article on wall extension, Cleland (1971) wrote, "One explanation [of auxin action] would be that auxin is required for the transport across the [plasma] membrane of [a] wall-loosening factor, perhaps hydrogen ions. This transport would....require respiration to provide an energy source, and if the transport proteins are unstable, continued protein synthesis would be needed to maintain their supply. Such a hypothesis appears to account for all the known facts concerning cell wall extension". Similar German language by Hager et al. (1971) launched the same hypothesis.

It was quickly established that auxin does stimulate H^+ release into auxin-sensitive tissues' cell wall space (Cleland 1973; Rayle 1973). Although these authors showed numerous parallels between auxin-induced H^+ release and elongation, dispute with others (cf. Kutschera and Schopfer 1985) soon arose because of discrepancies, most notably in the timing of the H^+ extrusion response, the above references showing little or no internal acidification until after 30 or 40 minutes of auxin exposure, whereas the full auxin elongation rate response of coleoptiles occurs after only about 15 minutes (Ray and Ruesink 1962), and of other auxin-responsive organs like pea internodes only a little longer (Jacobs and Ray 1976). H^+ excretion is difficult to measure reliably, because of cuticular resistance to H^+ diffusion, which makes a true measurement of internal free-space pH difficult to obtain. A micro pH electrode was, however, successfully inserted into the free space of pea internodes and maize coleoptiles (Jacobs and Ray 1976) and showed definite acidification (to pH <5) within the lag period for auxin stimulation of elongation.

The effect of auxin on proton extrusion is now understood in detail. Proton extrusion is driven by a plasma membrane H^+-ATPase that is regulated by the phosphorylation state of its C-terminus. Auxin enhances the phosphorylation level of that protein's Thr947 in Arabidopsis hypocotyls, but not via affecting a kinase (and without altering the amount of the plasma membrane H^+-ATPase). Instead, auxin induces the expression of SAUR genes (McClure and Guilfoyle 1987), certain of which inhibit a phosphatase that dephosphorylates the ATPase's Thr947 (Spartz et al. 2017). This allows Thr947's phosphorylation level, and thus the ATPase's activity, to increase. This activation occurs fast enough to explain the mechanism of IAA-stimulated proton extrusion; see Fuglsang and Palmgren (2021) for a recent review.

Several objections to the acid-growth theory were based on quantitative comparison of elongations and H^+ extrusions induced in maize coleoptile segments treated either with a near-optimum IAA concentration, or with the fungal drug fusicoccin (FC), an inducer of elongation and of strong trans-plasma membrane H^+ extrusion. Lüthen et al. (1990) demonstrated that wall acidification is required for IAA-induced cell expansion but cannot fully account for it, see Figure 11.1, by demonstrating that using a buffer to lower the apoplastic pH to match that resulting from IAA stimulation only yields half of the effect. Note that this does not disprove the acid-growth theory. It only shows that IAA may have effects other than stimulation of H^+ extrusion that also contribute to stimulating the elongation rate, e.g., stimulation of incorporation of new wall material into the wall interior (intussusception) (Ray, 1967).

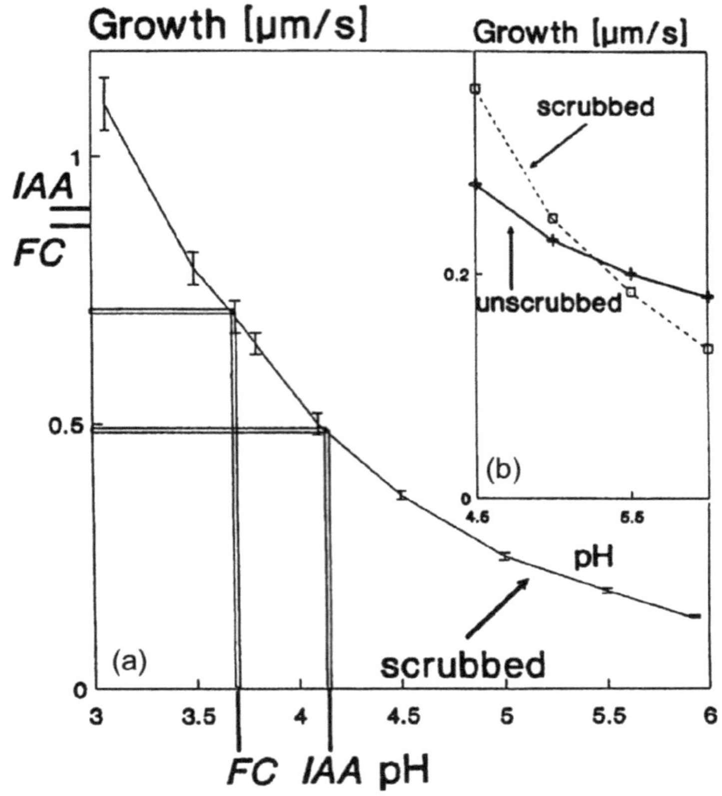

FIGURE 11.1 From Lüthen et al. (1990) comparing IAA and fusicoccin-stimulated growth rates with the growth rates at different pH values determined by buffers (a). Apoplastic pH values due to IAA and fusicoccin-stimulated proton extrusion are indicated on the x-axis. (b) Shows the effect, on elongation, of scrubbing the cuticle with an abrasive to improve H^+ exchange between the tissue's free space and the bathing medium, to improve the buffering and measuring of free-space pH. FC: fusicoccin. (Creative Commons CC BY.)

More than a decade after the acid-growth theory was proposed, Daniel Cosgrove's group at Penn State University developed a simple apparatus by which cell wall skeletons prepared by crushing (or, later, heating) auxin-sensitive tissues such as cucumber seedling hypocotyls, held under a tensile force, could be tested for the occurrence of steady elongation when exposed to different media. They soon found that this would occur simply in an acidic medium (pH 3–4), but did not occur if a skeleton was boiled briefly before testing it. Searching for something that would restore capability for elongation to boiled skeletons, they found that a 2M NaCl extract of elongation-zone cucumber cell walls would do so. So they fractionated this extract by conventional biochemical methods and arrived at an active principle composed of two polypeptides of nearly equal molecular weight, either of which were active (McQueen-Mason et al., 1992). Their activity depended on a low pH, like that required by the acid-growth theory. They were soon named *expansins* (Cosgrove and Li, 1993), and homologous proteins with similar activity were found in a wide range (we now assume all) of land plants.

However, the matter is far from settled and acid-growth theory is still debated, e.g., by Polak and Karcz (2021) in favour and Kutschera and Khanna (2020) arguing against. In any event, that the acid-growth theory is highly relevant, despite the counter-arguments, is shown by the discovery of *expansins* which definitely cause wall-loosening, and which depend upon a low pH to do so, and thus will be mentioned repeatedly in what follows since how they do it should be related to the structural features of the walls that they can act upon.

The acid-growth theory unfolded almost entirely without considering the cell wall architecture that the extruded protons should act upon; so even though there is a lot more to discuss, we will now turn our attention to model building. What is being omitted, for lack of space, are the rest of the mileposts, beyond those discussed just above, in cell wall growth research during the post-auxin period. A glimpse about these is provided in Box 11.1.

BUILDING MODELS

Although cell wall model building has been driven by questions related to cell expansion, it should be appreciated that models of secondary wall predate all four primary wall models that we will look into below.

BOX 11.1 POST-AUXIN MAJOR MILEPOSTS IN CELL WALL ARCHITECTURE AND CELL WALL GROWTH RESEARCH

Dates are mostly approximate, for work extending over several years
Researchers' names separated by commas=collaboration; names separated by semicolon=separate, independent workers

1925–28	Discovery of Auxin (Boysen Jensen; Went)
1931	Wall plasticity as basis for wall expansion & auxin action on it (Heyn)
1934	Auxin I.D. as 3-indolylacetic acid (IAA) (Kögl, Haagen-Smit; Thimann)
1935	Distinctions between primary & secondary walls (Table 11.1) (Bailey)
1950	Transmission electron microscopy (TEM) in biology: cellulose microfibrils; micellar theory ends (Preston; Mühlethaler)
1950	Metabolic (respiratory) basis of plant cell growth (Thimann)
1955	Pectin methylation/Ca^{2+} X-link wall-loosening theory (Bennet-Clark; Carlier, Buffel)
1960	Instron measurements of wall extensibility (Cleland)
1962	High-resolution growth measurements, rapidity of auxin response (Ray, Ruesink)
1965	Lockhart equation: wall expansion/water uptake tradeoff (Box 11.2) (Lockhart)
1967	^{3}H-sugar TEM autoradiography demonstrates that intussusception occurs (Ray)
1970	Acid-growth theory (Cleland, Rayle; Hager)
1971	Tissue tensions, epidermal outer wall control of axial growth (Kutschera)
1972	*In vivo* stress relaxation, growth primary process (Ray, Green, Cleland; Cosgrove)
1973	Covalent network primary wall model (Albersheim, Keegstra)
1975	Xyloglucan as a key matrix player in wall expansion (Albersheim)
1992	Xyloglucan endotransglycosylase discovered, possible wall-loosening enzyme (Smith, Fry; Nishitani, Tominaga)
1992	Expansins detected & identified as wall-loosening proteins (McQueen-Mason, Durachko, Cosgrove)
2001	Reactive oxygen species wall-loosening theory (Fry; Schopfer)
2001–12	Discovery of Cel12A, a creep inducing glucanase leading to Model 4 (Yuan, Wu, Park, Cosgrove)
2002	Yieldins identified (Box 11.2) (Okamoto-Nakazato)
2015	Re-awakened role for pectins, Ca^{2+}X-links and intussusception in *Chara* (Boyer, Proseus)
2015	Addressing cell wall architecture questions with non-invasive NMR techniques (Dupree, Mortimer)

TABLE 11.1
Distinctions between Primary and Secondary Cell Walls

Character	Primary Cell Wall	Secondary Cell Wall
Time of appearance after mitosis	First (after middle lamella)	Second (if at all)
Thickness	Usually thin (≤ 0.5 μm)	Usually thick (1–50 μm)
Composition (roughly), % of dry matter	Cellulose ~30%, matrix polysaccharides > 60%, lignin 0%	Cellulose >50%–60%, matrix polysaccharides ~10%, lignin ≥ 30%
Lamellate structure	No, but many with fine layering of crossed or helicoidal orientations	2–3 major layers, whose microfibrillar orientations differ from one another
Microfibril orientation	Usually loose, diffuse, but often with a definite mean direction	Usually closely parallel in any one major layer
Capable of growth in area	Yes (if appropriately stimulated)	No (with exceptions noted below)

Secondary Wall

Secondary walls were more amenable to study and hence the first illustrations of wall architecture regarded secondary walls. Nägeli employed swelling agents such as strong alkali to delaminate secondary walls, combined with birefringence or X-rays to study the direction of "striations" (Nägeli 1864a,b). These studies and in particular the shifts of microfibril direction in successive lamellae laid the foundations for the models of secondary walls. Making valid inferences from these experiments was riddled with technical problems and remained an issue of discussion for many years (Roelofsen 1951). These difficulties notwithstanding, the model presented in panel (a) of Figure 11.2, from Lüdkte (1931), resembles that shown in modern textbooks, panel (b), and it even acknowledges the templating pectin deposited in between successive lamellae (Yang et al. 2020a). Only the *Querelemente* (transverse elements, labelled e in panel A) analysed in detail by Hess (1928) did not stand the test of time; they turned out to be artefacts as thoroughly documented in the review by Mangenot and Raison (1951).

The classical distinctions between primary and secondary walls, due substantially to the Harvard Biology Department's long-time, widely respected plant anatomist Irving W. Bailey, are roughly as shown in Table 11.1 (cf. (Bailey 1939, 1952)):

Lignin is incorporated into a cell wall after synthesis and deposition of a secondary wall's carbohydrate polymers has been completed. For the relevant enzymology, see Mnich et al. (2020). In wood, the densest lignin is usually not in the secondary walls themselves but in the middle lamellas between the cells, the next dense in the primary walls and least dense in the secondary walls (especially in their middle, thickest "S2" layer). The more strongly lignified wall components are those that contained a larger, water-filled fractional capillary channel volume prior to the insertion of the lignin polymer.

PLASTIC WITH STRAIN HARDENING

If loaded to above Y_o, deforms irreversibly, with Y increasing ("hardening") to equal the then-imposed load

VISCOUS

Deforms irreversibly under any stress, at a rate proportional to stress (load)

VISCOELASTIC

Elastic and viscous elements joined in parallel (time-dependent, reversible strain), or in series (both irreversible viscous, and reversible elastic, strains) or all of these combined

Extension, in the Instron extensometer, of wood (Keckes et al. 2003) and of sclerenchyma (Spatz et al. 1999; Köhler and Spatz 2002) has given evidence of irreversible extension and strain hardening in strain/load diagrams surprisingly similar to those obtained with the Instron from primary-walled structures but with noticeably greater extension/reversal hysteresis loops, implying prominent viscoelasticity. Irreversible Instron extension is of course not biological growth, nor is the irreversible strain frequently observed along with substantial retarded elasticity in bending experiments on woody branches (Ray and Bret-Harte 2019). But this latter irreversible straining seems at least to apply biologically to the branch elevation/angle aspect of multi-year tree crown development. These bending tests actually detected two types of irreversible bending, an instantaneous one with a substantial yield threshold, Y, and a more prominent but slow and time-dependent (viscous) one with little or no Y (Jin et al. 2015; Ray and Bret-Harte 2019), which the above-cited short-term Instron measurements did not detect. Irreversible extensions of the kinds observed seem contrary to the physical implications of the "cannot grow in length" rule that plant biologists still commonly apply to secondary walls.

The Jin et al. (2015) model in Figure 11.2's panel (b) and the textbook model in panel (c) are what we would call multicoat ones (Model 2 in the primary wall model section below). This is corroborated by data from the biomaterials field (Yang et al. 2020b). The figure is based on studies of Gingko, a gymnosperm, and the innermost sleeve of mannan may not apply to all plant families. It has been observed also in sugar beet (Perzon et al. 2020) but xylan is observed to associate with the hydrophilic face of cellulose in both gymnosperms and angiosperms (Busse-Wicher et al. 2016). In grasses, glucuronoarabinoxylan is a major player (Gao et al. 2020), but also here, at least in oat coleoptiles, mannan is more strongly associated with cellulose than any of the xylans are (Ray, 1963).

The unusual behaviour of reaction wood (tension wood in dicots, compression wood in conifers) also lies outside of the "cannot grow" rule, since these secondary-walled xylem cells can make a very stiff inclined branch or tree trunk bend upward in a biological response to lateral gravity. The mechanism of this still seems puzzling, despite considerable research on it (Felten and Sundberg 2013). Recent investigations of tension wood (accompanied by architectural drawings, one of which is shown in panel (d) of Figure 11.2) attempt to understand how a gelatinous tertiary wall (for a discussion of the definition of 'tertiary wall' see Clair et al. (2018)) can build up forces strong enough to bend woody tissue (Gorshkova et al. 2015, 2018). The theory posits that the pectic polysaccharide rhamnogalacturonan-I (RG-I), which is deposited as a spacer molecule between cellulose microfibril bundles in the tertiary wall, will be able to generate tension following galactosidase trimming of its galactan side-chains. This would naively require the molecular weight *decrease* of the RG-I molecules as a result of galactan trimming to be accompanied by an *increase* in hydrodynamic volume. The authors do not state explicitly that this is what the theory posits; it is our interpretation inspired by an analogous phenomenon observed in Arabidopsis seed mucilage in a study of the mucilage-modified2 mutant (*mum2*), a galactosidase knock-out in which mucilage fails to swell (Dean et al. 2007). The observation that mucilage swelling ability is restored in *mum2* if galactose oxidase 'Particles in Mucilage' (RUBY) is also knocked out (Šola et al. 2019) alerts us to how poorly understood the functional properties of RG-I are – see also the discussion in Wolf et al. (2012). It is therefore not surprising that there are several competing theories of how tension is generated in tension wood; see the overview by Alméras and Clair (2016).

We will now turn our attention to primary wall models.

Model 1 "Covalent Network" Model

The first comprehensive model of a primary cell wall was proposed by Keegstra et al. (1973). Xyloglucan (XyG) had been identified as a secreted polysaccharide in cultured sycamore cells (Aspinall et al. 1969). Fine structure elucidation of enzymatically fragmented walls of sycamore suspension cells was an ongoing activity in the Albersheim laboratory and this model was an attempt to put together the pieces

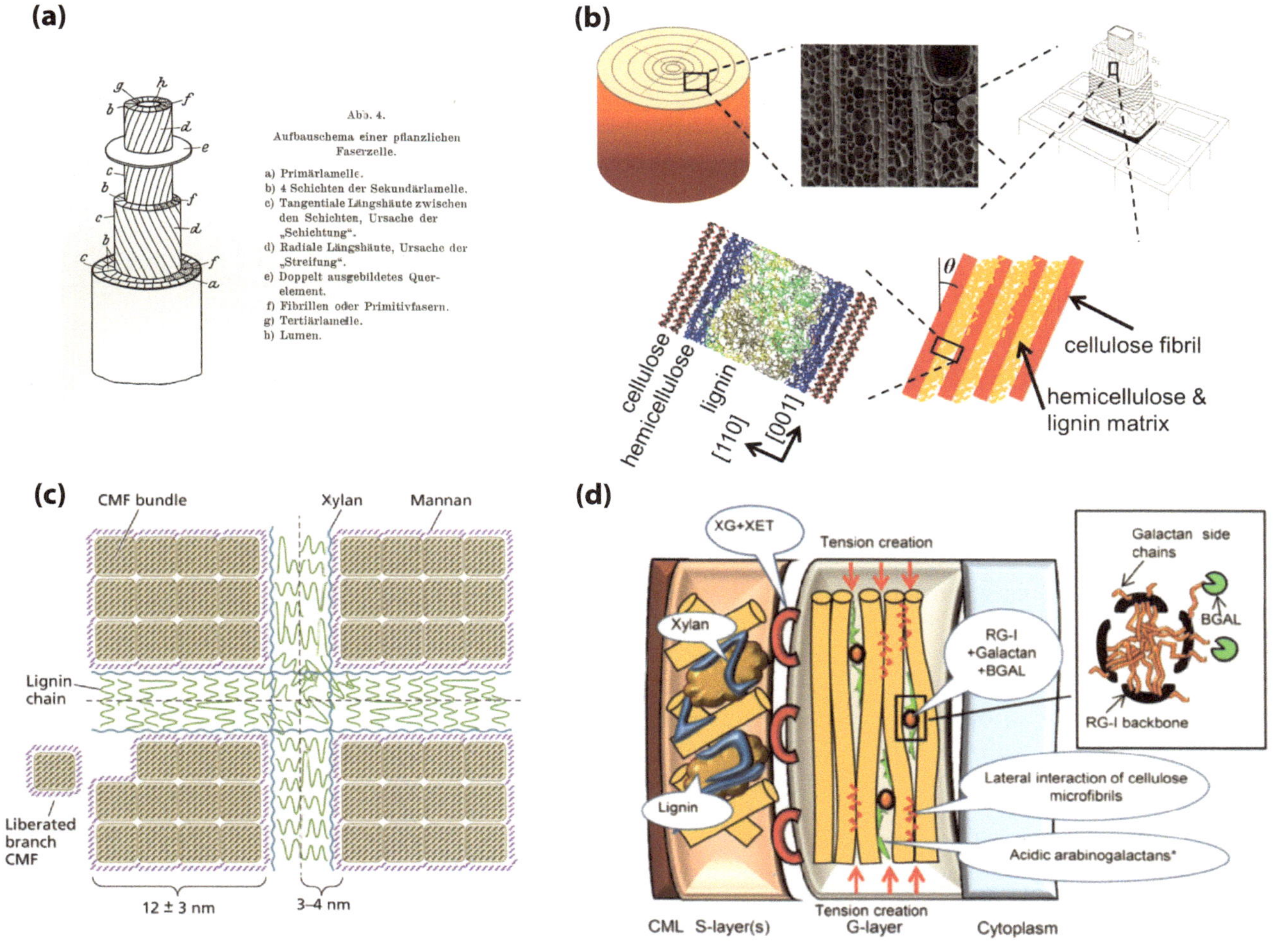

FIGURE 11.2 Secondary cell wall models. (a) Lüdkte (1931), one of the earliest model drawings. (b) The Jin et al. (2015) model. (c) The Terashima et al. (2009) model as redrawn for a widely used textbook (Taiz et al. 2015). (d) A tension wood model drawing featuring a tertiary wall, from Gorshkova et al. (2015). CMF: Cellulose microfibril. XG: Xyloglucan. XET: XG endo-transglucosidase. BGAL: -galactosidase. (Permission to use: (a) off copyright. (b) with kind permission from Elsevier Ltd. (c) with kind permission from Oxford Publishing ltd. (d) creative Commons CC BY.)

in a way that was probably inspired by bacterial walls, i.e., as a cross-linked modular super molecule (Figure 11.3). The experimental strategy relied on enzyme digestion followed by chromatographic separation and then was looking for co-chromatography of constituents of different polymer classes. Fractions were characterised that comprised both XyG and pectin but no reducing ends of XyG were detected in these fractions, only reducing ends featuring galacturonic acid, GalA, a diagnostic sugar for pectins. This led to the proposition that XyG was synthesised onto, or transferred by transglycosylation onto, a galactan side-chain of the pectin, RG-I. The other main cross-link considered was between the glycan part of arabinogalactan protein (AGP) and RG-I. Release of this cross-link relied on sequential digestion with endo-polygalacturonase and protease. An explicit place and role of homogalacturonan (HG) were not considered in the model. Neither was Rhamnogalacturonan-II (RG-II), which was only discovered 5 years later (Darvill et al. 1978). Hydrogen bonding of XyG to cellulose was the most significant non-covalent interaction in the model (Bauer et al. 1973). A revision of the model in 1980 made more room for non-covalent interactions; covalent XyG:pectin cross-links were then regarded as a minor fraction, and hydroxyproline-rich glycoproteins (HRGP) were no longer considered part of the covalent network; a role of XyG in preventing cellulose microfibrils from coalescing into macrofibrils was also presented (Darvill et al. 1980). The latter was an idea that did not deserve to be forgotten, but at the time was indeed forgotten.

Two mechanisms of cell wall expansion were considered: hydrolysis or transglycosylation within the covalently interbonded network, quite similar to ideas proposed much later for the sticky network model (model 3, below), and secondly, sliding of xyloglucan along cellulose microfibrils, in a sense anticipating the mode of action proposed much later for expansins (Cosgrove 2000).

The existence of covalent linkages between polymer classes has been validated in several cases (Fry 2011). Evidence for linkages between cutin (waxy layer that

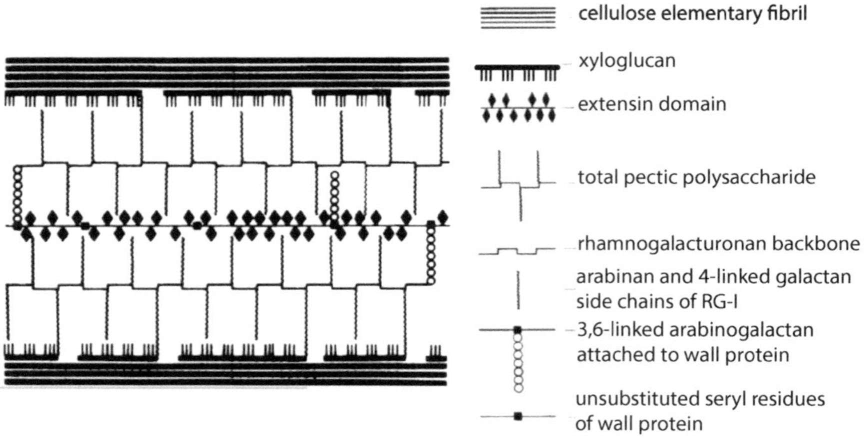

FIGURE 11.3 Covalent cross-linked model. From Keegstra et al. (1973), the cell wall depicted as a covalently inter-linked supramolecular structure (with our slight updates to the nomenclature). (Permission to use: Creative Commons CC BY.)

covers the epidermal outer wall) and XyG has recently been demonstrated (Xin and Fry 2021), and linkages between pectic polymers and XyG are well documented (Thompson and Fry 2000; Popper and Fry 2005); see (Brett et al. 2005) for an overview. It is interesting that there furthermore is evidence that some linkages are formed in the protoplast, and not apoplastically as a wall maturation process (Popper and Fry 2008). The enzymes involved and the linkages formed have not been determined except in the recent demonstration of a barley hetero-transglycosylase that joins XyG with HG, in this case apoplastically (Stratilová et al. 2020).

Cross-links between HRGPs (see previous paragraph) and polysaccharides are known (Qi et al. 1995; Nuñez et al. 2009). Usually, it is not known whether the linkage is glycosidic except in the case of APAP, a supramolecular structure found in Arabidopsis comprising arabinoxylan, pectin and AGP (Tan et al. 2013). The linkage, GlcA-Rha-GalA-Rha…, between a terminal Rha residue of a type II arabinogalactan and the RG-I backbone in APAP was in fact inferred by Keegstra et al. (1973). Glucuronoarabinoxylan as an RG-I side-chain has recently been documented in *Abies sibirica* (Makarova and Shakhmatov 2022).

In addition to these covalent bonds between matrix polymers, there is also evidence for grafting of polysaccharides onto cellulose by transglycosylation. Grafted polysaccharides comprise XyG (Shinohara et al. 2017) as well as both XyG and mixed-linkage glucan (Hrmova et al. 2007; Herburger et al. 2020). Finally, evidence has been presented for pectin linked covalently to cellulose in carrot (Broxterman and Schols 2018).

So the existence of covalent linkages between polymer classes is not questioned. Existence and prevalence are not the same, and the model of the wall as a giant covalent network agrees poorly with the separability of wall components using sequential extraction with solvents that do not break covalent bonds. This discrepancy led to two models that emphasise non-covalent interactions in the wall: The multicoat model and the sticky network model.

Model 2 "Multicoat" Model

The multicoat model regards the wall as a composite in which the cellulose is coated with successive sleeves of polysaccharides and proteoglycans. The model by Talbott and Ray (1992b) was based on the separability features of solvent-extracted wall polymers. Organisation of the various sheaths around microfibrils reflects extractability, with hemicellulosic polymers forming the innermost (closest to the cellulose microfibrils) and hardest to extract polymer class.

Wall self-assembly according to this model can be understood in terms of differential affinities of polysaccharides for the microfibril surface, combined with matrix polymer phase partitioning. Many polysaccharides undergo phase separation even at the quite low concentration, and this has led to the proposition that phase partitioning of polysaccharides plays a role in self-assembly of the wall (Preston 1979). It has been demonstrated *in vitro* for some combinations of wall polysaccharides (MacDougall et al. 1997). Combine this with targeted deposition of polysaccharides and directed secretion of wall-modifying enzymes, and plausible theories for microdomain (local areas of different polymer composition) formation can be formulated (Haas and Peaucelle 2021).

The precise organisation seen in Figure 11.4 with hemicelluloses covering the microfibril surface is in accordance with both the covalent network model and the sticky network model to be considered next. However, populations of very tightly bound pectic populations have been observed (Chapter 5) and Perzon et al. (2020) and NMR studies reveal that cellulose interacts more with pectin than with XyG (Wang and Hong 2016). HRGP may occur in one of the sheaths while later ideas also consider HRGPs inserted orthogonally to the plasma membrane into a pre-existing wall (Kieliszewski et al. 2011).

Stress transmission between matrix and cellulose microfibrils was not discussed explicitly, but it was stated that plant cells have more than one programme for cell expansion and that the multicoat model is compatible with both wall loosening by matrix metabolism and expansion driven by

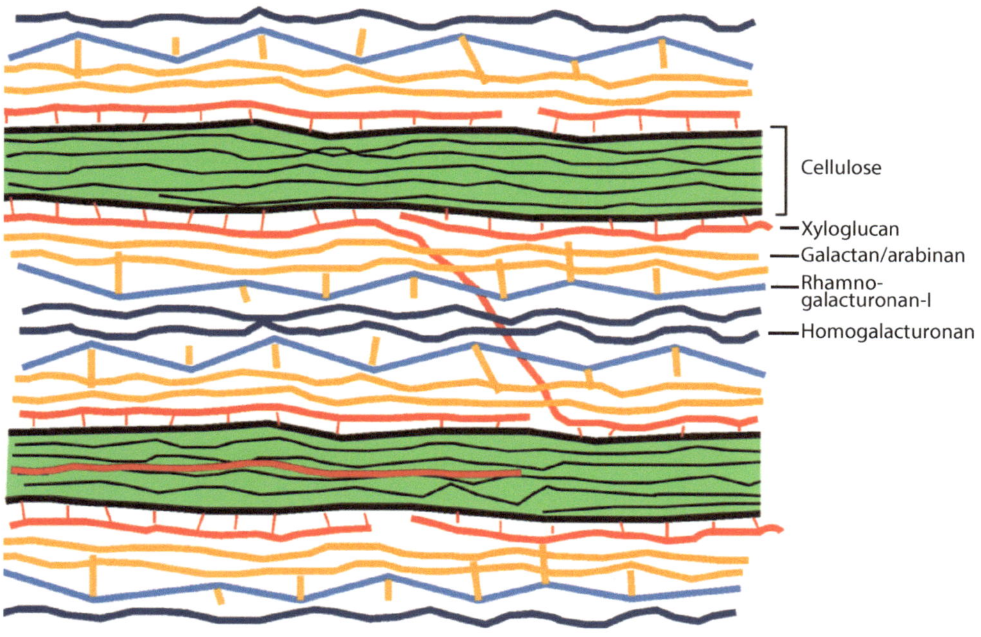

FIGURE 11.4 The multicoat model, developed on the basis of studies with pea (Talbott and Ray 1992b).

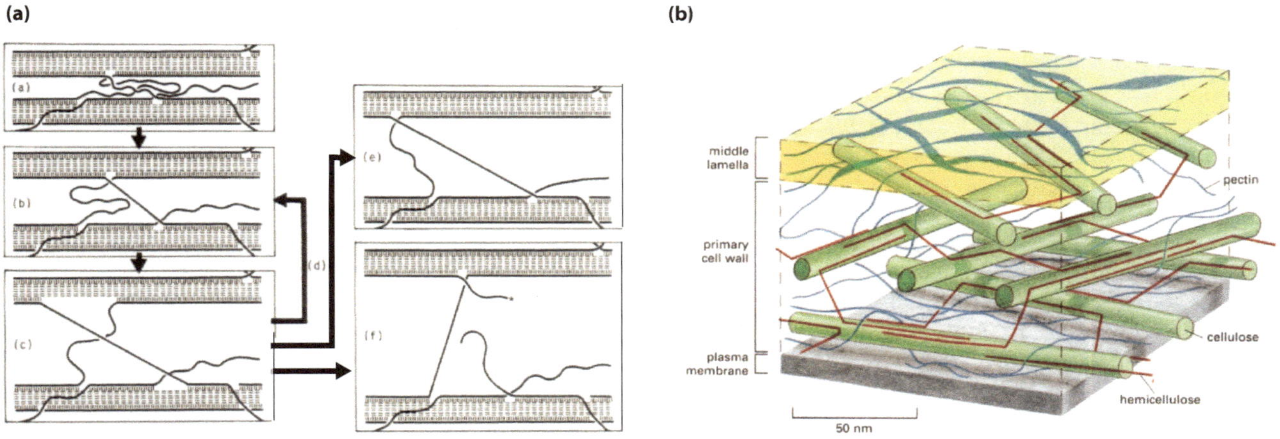

FIGURE 11.5 The sticky network model. (a) Cell expansion by metabolism of XyG tethers, from Fry (1989). (b) The sticky network model drawn to scale by McCann and Roberts (1991). (Permission to use: (a) with kind permission granted by Wiley. (b) courtesy Dr. Maureen McCann).

biosynthesis. Matrix metabolism was suggested both to rely on endo-glucanases and on XET (Fry 1989) (see below). The central role assigned to transglycosylation was prompted by the observation of molecular weight increases of previously deposited XyG following turgor upregulation and reduced XyG sizes following auxin stimulation (Talbott and Ray 1992a). Auxin-mediated changes in XyG size distribution was already well established (Nishitani and Masuda 1983).

Model 3 "Sticky Network" Model

Stress transmission and load-bearing structures in the wall were addressed very succinctly in the sticky network model, which was developed in parallel with the multicoat model. It similarly assumes that non-covalent bonding is an important or the principal element in primary wall and emphasises non-covalent bonding between different polymer classes (thus the "sticky" network, as against a permanent, covalent network in Model 1), as well as transglycosidic covalent bonding between them.

In a lucid paper by Stephen Fry, the idea is presented that cellulose microfibrils are tethered by non-covalent interaction with hemicellulosic chains, where these hemicellulose chains are XyG in type I (non-commelinid monocotyledonous and dicotyledonous) walls (Fry 1989), see panel A of Figure 11.5. Transglycosylation was discussed in this connection a few years before the identification of the first-discovered transglycosylase, XET (Fry et al. 1992; Nishitani and Tominaga 1992). The model furthermore proposes that mixed-linkage glucan plays the role in grasses (type II, commelinid monocotyledonous walls) that XyG plays in type I walls. If xylan were to assume, in non-grass type II walls, the microfibril-tethering role that XyG is thought to play in Type I walls, then a xylan endo-transglycosylase should exist, and that appears to be the case (Johnston et al. 2013). The sticky network model was made precise and explicit by building into it the wall-loosening mechanism in Lockhart's mathematical framework, see Box 11.2, so that wall (kinetic) extensibility (i.e., viscoplastic viscosity) and wall yield threshold become variables dependent on the number of tethers (Passioura and Fry 1992). A lamellated version of the sticky network model can be envisioned in which tethers are confined to individual lamellae as they are deposited (Cosgrove 2000).

VISCOPLASTIC

Elastic below its Y; at loads (L) above Y it deforms irreversibly at a rate proportional to (L – Y), and also supports additional elastic strain in proportion to L

BOX 11.2 LOCKHART

The Lockhart equation (Lockhart 1965)

$$\dot{E} = k\phi L_p A \cdot \frac{\pi - Y}{k\phi + L_p A} \tag{11.1}$$

relates the rate of cell expansion ($\dot{E}$) to the viscoplastic extensibility (1/viscosity) ϕ of the wall, the hydraulic water conductivity L_p limiting water influx, the osmotic potential π of the cell contents, plasma membrane area A and wall yield threshold Y, with k a geometrical constant. That the wall is characterised by a yield threshold, i.e., a threshold of turgor-generated stress below which extension will not take place, was already realised (Cleland 1959). The point is that growth involves coupling together two mechanistically completely separate processes, viz., (i) water uptake into the cell, to enlarge its volume, and (ii) irreversible expansion of the wall to allow this water uptake to occur. (i) Depends, as far as we know, on the physical process of osmosis, which depends upon, and is driven by, a difference in water potential between the contents of the cell and the outside medium.

If auxin is given to a cell, ϕ rises and Y falls (Nakahori et al. 1991), so the wall starts to extend, and thus the turgor P falls, reducing the cell's water potential (Ψ) below the Ψ of the cell's surroundings. This $\Delta\Psi$ starts water uptake. At a certain (usually quickly reached) point, P falls to a value that causes a rate of water volume uptake just equal to the rate of wall chamber volume increase by wall expansion under the force of that P, i.e., a steady state is reached (Cosgrove 1981). At a steady state, P no longer changes. In 1965, Lockhart simply wrote down expressions for (i) and (ii), set them equal to one another, solved that equation for the steady state P, substituted that result for P in the previous expression and rearranged it to get his equation (11.1) above.

Much work since then has shown that normally $k\phi \ll L_pA$ so water enters a cell much more easily than its wall extends. In this case, equation (1) reduces to

$$\dot{E} \approx k\phi(\pi - Y) \approx k\phi(P - Y) \tag{11.2}$$

Auxin causes the yield threshold (Y) to drop because it acidifies the cell wall, and the value of Y is pH-dependent (Nakahori et al. 1991). With cowpea segments, treating wall skeletons with near-boiling temperature inactivates this pH dependence. A certain protein, now named yieldin, has been found that can restore Y's pH-sensitivity (Okamoto-Nakazato 2002). (Measuring Y's pH-sensitivity is not trivial; for an informative discussion, see Suslov et al. (2015).) This restoration is illustrated by the data in Figure 11.6.

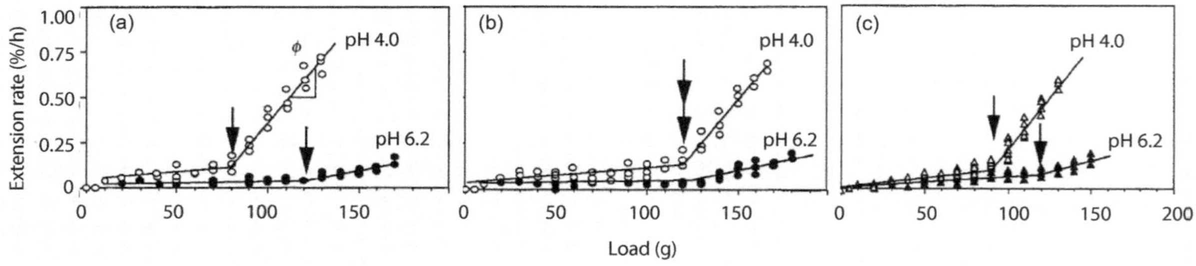

FIGURE 11.6 Demonstration of a yield threshold, its pH-sensitivity and the action of yieldins. (a) Creep assays of living cowpea hypocotyls at pH 4 (open symbols) and 6.2 (closed symbols). Arrows point to yield thresholds. (b) Heat-inactivation abolishes the pH-sensitivity of the yield threshold. (c) pH-sensitivity is restored in heat-inactivated cowpea hypocotyls when a protein extract comprising yieldins is provided. From (Okamoto-Nakazato 2002). (Permission to use: With kind permission from Springer Nature).

This says, *inter alia,* that in the steady state, with water as the external medium, P is almost equal to the cell's osmotic potential, and the growth rate is proportional to the difference between π (or P) and Y.

The sticky network model has several properties that appeal to intuition. The xyloglucan tethers are arranged so that they comprise a portion that is bound to, or even within, a microfibril and are thus inaccessible to XET digestion, contrary to the tethering parts (Pauly et al. 1999), and the biological role of XET may be inferred from the model. Also, expansin action can be rationalised as unzipping of the hydrogen bonds that bind XyG to the microfibril surface. The model is backed by microscopy and a model drawn to scale was prepared by McCann and Roberts (1991); see panel B of Figure 11.5. The authors view the pectic phase as a co-extensive network in agreement with Bailey (1939). The model is compatible with the observation *in vitro* of RG-I side-chains binding to cellulose surfaces (Zykwinska et al. 2007), and it provides a basis for RG-II to play a role in wall strength (O'Neill et al. 2001; Ryden et al. 2003). The model is highly influential and has been redrawn and elaborated upon several occasions, see Somerville et al. (2004) for an example. Carpita and Gibeaut (1993) have developed versions of the model specific for commelinid monocots and for the remainder of the angiosperms, the Type II and Type I, respectively, mentioned above. These models are presented in a widely used textbook by Buchanan et al. (2002).

COMMELINIDS

A clade of monocots featuring palms as the earliest diverging and grasses as the latest. Feruloylated (esterified with ferulic acid) glucuronoarabinoxylan is a signature polysaccharide for this group.

However, there are substantial problems with these inferences from the model. Firstly, the model implies that XET would induce creep in wall specimens held under tension, which experiments confirmed not to be the case (McQueen-Mason et al. 1993). In fact, to the best of our knowledge, creep in wall sections under tension has never been demonstrated for any polysaccharide hydrolase except for Cel12A (to be discussed in relation to the following model). Secondly, the amount of XyG that is surface-bound has been quantified in mung bean: only 3.9% was microfibril-bound, and NMR studies identified pectin, not XyG, as the main component rendered immobile by a tight association with cellulose (Wang and Hong 2016). Using a modelling approach, Thompson (2005) concluded that the hydrogen bonds between XyG and cellulose would not suffice as load-bearing structures. Yi and Puri (2012) also addressed the problem by mathematical modelling and concluded that the strength of the XyG to cellulose binding would not allow the strength of cellulose to be exploited in the wall, leaving cellulose without an evolutionary driver. This conflicts with the findings of Hepworth and Bruce (2000) who were able to infer the strength of cellulose in cylinders of potato tuber from the outside, by small compressions of the tissue cylinders. While it is quite safe to conclude that XyG tethers are not the growth-limiting load-bearing structure in the wall, the existence of tethers *per se* is not questioned. Similar analyses have not been attempted for the two previous models. The most likely outcome would probably also be that stress transmission through the matrix would not be able to exploit the strength of cellulose.

Model 4 "Hot-Spot" Model

Reginald Preston concluded in his 1979 review (Preston 1979) that a model should operate with a series of "interactions from the very weak [...] between hemicelluloses and cellulose (strongest), and directly between the microfibrils themselves (unless these links are formed *de novo* as other compounds are extracted)". Direct intermicrofibril interactions were not considered in any of the models presented. They entered the discussion again in 2010 (Scheller and Ulvskov 2010) and in 2012 the issue was addressed experimentally by Cosgrove and colleagues using the endo-glucanase Cel12A identified earlier (Yuan et al. 2001). This endo-glucanase is active in creep assays albeit with very different kinetics compared with expansins. The latter act immediately while Cel12A action only commences after a lag that cannot be shortened to less than 6 minutes by increasing the enzyme dose. This led the author to propose that it targets a relatively inaccessible substrate that must be degraded for the wall to extend (Yuan et al. 2001). Park and Cosgrove (2012) then established that Cel12A is able to degrade both cellulose and XyG. While a combination of a cellulase and a xyloglucanase will not induce creep, two other enzymes with similar dual specificity will. Finally, it is established that Cel12A will induce creep in wildtype Arabidopsis hypocotyls but not in hypocotyls of the XyG-deficient *xxt1/xxt2* mutant (Cavalier et al. 2008), which is probably surprising given that Cel12A induces creep in composites of bacterial cellulose devoid of XyG. Expansins are active on bacterial cellulose, more so on preparations with high MW XyG than on preparation with shorter XyGs (Darley et al. 2001), but although difficult to measure on the quite stiff pellicles of cellulose with no XyG, expansins are active on this too (Thompson and Islam 2021). We are led to the conclusion that specifically in plant cell walls Cel12A is dependent on XyG for creep induction and that it targets a hard-to-access subpopulation of XyG tightly associated with cellulose.

Park and Cosgrove (2012) propose that these hotspots (panel A of Figure 11.7) are targets for both Cel12A and α-expansins. A solid-state NMR study corroborates the notion that expansin targets a cellulose surface enriched with XyG and with a fine structure that differs from the bulk cellulose (Wang et al. 2013). Confirmation by non-invasive methods like NMR is important. In the Preston quote, above, he raised a concern for artificially induced inter-bonding of cellulose microfibrils, and a similar concern has been expressed regarding the XyG:cellulose interaction (Darley et al. 2001).

(a) (b)

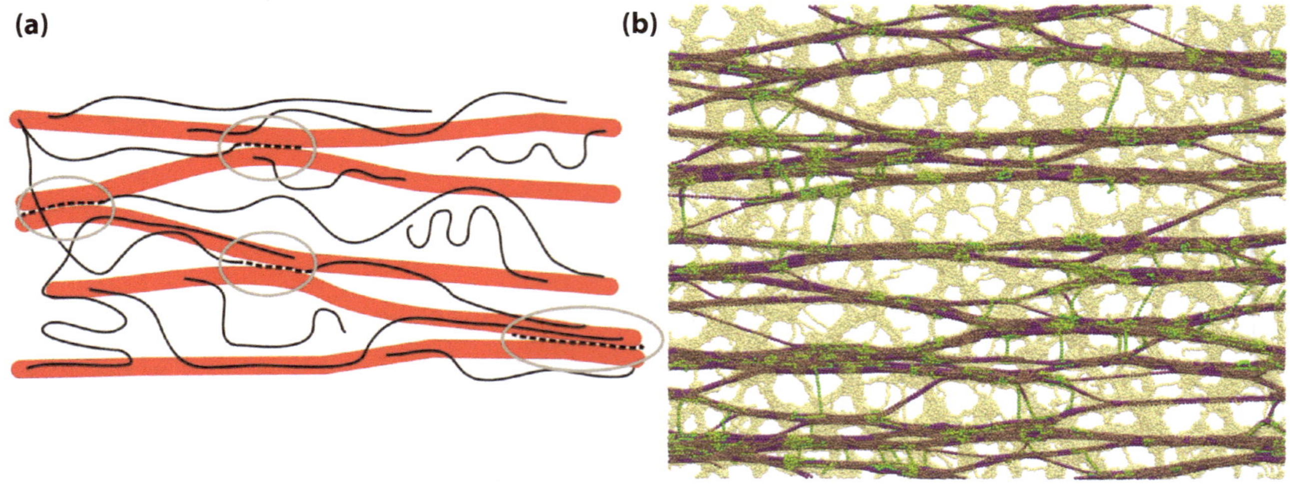

FIGURE 11.7 The hot-spot model. (a) XyG trapped in hotspots where cellulose microfibrils interact. Cellulose microfibrils are red. XyG is shown as a black line, which becomes dashed inside encircled hotspots (from Park and Cosgrove (2012)). (b) Simulation of cell wall self-assembly and wall expansion. Cellulose is purple, XyG is green and pectin is shown in yellow. (Permission to use: (a) Creative Commons CC BY.) (b) This figure is copyrighted by Daniel Cosgrove based on work by Zhang et al. (2021) and is reproduced with permission).

The Cosgrove lab recently presented a molecular simulation of model 4 specifically applied to onion epidermal walls (Zhang et al. 2021), a well-characterised model system (Zhang et al. 2016). The model accounted successfully for both wall assembly and for tensile behaviour. The strengths of matrix polysaccharide:cellulose interactions are mathematically tuneable in the model and simulations indicate that these interactions play very minor roles in wall mechanics, whereas microfibril sliding and interactions between microfibrils were significant.

UNACCOUNTED FOR

The four models each capture some aspects of how cell walls are constructed, but there are many phenomena that are not anticipated by the models and even some that are surprising. We will touch upon some of these in the following.

Cell Wall Expansion Driven by Biosynthesis

Although intussusception was probably favoured by more researchers than any other potential mechanism during the pre-auxin era, it has taken a back seat since its dismissal by Heyn (1931). In most growing organs of intact plants that have been carefully studied, wall synthesis maintains wall thickness approximately constant during wall expansion. Given appropriate feedback control, this may be achieved even with temporal uncoupling of synthesis and cell expansion demonstrated by Refrégier et al. (2004) in etiolated Arabidopsis hypocotyls (with seeds in place to provide carbon) and in light-grown seedlings relying on photosynthesis for carbon supply (Ivakov et al. 2017). In the etiolated state, i.e., without a carbon source, excised segments not only expand without concomitant cell wall biosynthesis but may even recycle wall matrix material, e.g., for respiration or cellulose synthesis (Bret-Harte et al. 1991; Barnes and Anderson 2018). Plants potentially have more than one programme available for wall expansion and this makes it particularly difficult to rule out a proposed mechanism for lack of perfect correlation with growth rate.

Bonner (1934b) reported that oat coleoptiles elongated at +2°C without any wall synthesis at all, and therefore cannot be the basis of wall growth. However, these experiments lacked initial wall mass controls (they included only incubated-minus-IAA controls). When they were repeated including initial controls (Ray 1962), it turned out that wall synthesis did occur at +2°C, but only a little of it, similar to the small elongation at that temperature. Much later, Edelmann et al. (1995) reported an experiment that they thought rules out wall synthesis as contributing to auxin-induced growth. They used cycloheximide (protein synthesis inhibitor) which at a concentration that inhibited both growth and labelled leucine incorporation into the cell wall by 88% did not inhibit incorporation of labelled sugars into wall polymers. They reasoned that since cycloheximide totally inhibits auxin-induced elongation, anything that is on the path from auxin to cell wall loosening should not occur under cycloheximide inhibition. But this logic fails to take into account Ray's (1967) autoradiographic data on coleoptile and pea wall synthesis (see below), which showed that at least half of the total wall synthesis is by apposition, which cannot contribute to wall expansion and could, by this logic, continue when the fraction of wall synthesis (intussusception) that could contribute to wall expansion has been eliminated.

Examination of XyG in charophytic algae revealed that XyG is co-localised with sites of wall expansion (Mikkelsen et al. 2021). The method employed was labelling of living cells with a XyG-specific antibody; hence, it was the external face of the wall that was probed, demonstrating XyG insertion throughout the wall, suggestive of intussusception (a primordial programme for cell expansion?). That intussusception occurs in higher plants was proved by autoradiographs of ^{3}H (tritium)-labelled outer epidermal walls of oat coleoptile and pea stem segments that had synthesised and deposited, into these walls, labelled polysaccharides from precursors that they had produced from the ^{3}H-labelled glucose that was fed to them for 6 hours (Ray 1967). From some of the samples matrix polymers were extracted prior to sectioning and autoradiography. Cellulose was found to be deposited entirely at the wall surface (apposition), but hemicelluloses and pectins were deposited about half by apposition along with the cellulose, and the rest by intussusception throughout the wall's interior. The only question, then, is whether this latter can cause wall-loosening, which cannot be seen by microscopy.

Experimental support for intussusception as a wall-loosening mechanism comes from the parallel inhibition of both elongation and wall synthesis by more than 10 different inhibitors (Ray 1962; Baker and Ray 1965), including monensin (Kutschera and Briggs 1987) and brefeldin A (Schindler et al. 1994), disrupters of the Golgi secretory function which deposits new matrix polymer molecules into the wall (Kim and Brandizzi 2016). Beyond this, we are in the range of logical speculation.

Incorporation of a newly arrived, recently synthesised polysaccharide into the wall structure probably involves either non-covalent bond exchange or trans-glycolytic covalent coupling. We consider the former here. A bond exchange amounts to a new polymer, with unsatisfied bonding capabilities, replacing a non-covalent cross-link between two similar molecules that are already bonded together as part of the wall structure, by bonding similarly to one of them while the other erstwhile partner is freed from its erstwhile connection. This exchange should have a free energy change (ΔG) close to 0 if neither of the previously interbonded polymers is under elastic strain, as would be true at the cell wall's surface facing the plasma membrane, where appositional deposition is occurring.

However, in the wall's interior, many or all of the already-bound polymers that support elastically the load that the cell's turgor is imposing on the wall will be under tensile stress (yielding elastic extension) both within the polymer chains and at cross-links between them. The elastic tensile energy at, and in the vicinity of, the former cross-link would have been an additive part of the formerly cross-linked polymers' free energy (G), but becomes dissipated during the bond exchange because the new partner is under zero elastic strain and the former one must relax when the bond through which it was being strained is severed. This dissipation gives the exchange reaction a negative (downhill) ΔG, i.e., *makes the exchange reaction exergonic* (Ray 1987), creating a driving force tending to make the exchange reaction occur. And, this process constitutes an *in vivo* stress relaxation, resulting from intussusception, which will lead to water uptake and cell wall expansion, both elastic and irreversible. Therefore, it seems impossible that intussusception would *not* cause irreversible cell wall extension.

A Role for Pectic Non-Covalent Bond Exchanges

The consequences may more readily be deduced for simpler or specialised walls. Ca^{2+} ionic "egg-box" cross-links between pectic HG chains are dominant load-bearing structures in the cell wall of the alga *Chara* and also in the sub-apical, expanding region of pollen tubes. Rojas et al. (2011) achieved very good agreement with experimental results when they computationally modelled the elongation zone of pollen tubes and made this observation: "That wall-thickening precedes periods of rapid expansion would be puzzling if thickening acted to curb expansion. However, this observation is consistent with the role of deposition to transiently loosen the cell wall via dissociation of cross-links. Moreover, expansion subsides not when the wall thickens, but when most of the available de-esterified pectin has been incorporated into the load-bearing wall, i.e., when the wall thins". See also the discussion of HG recycling in this process (Haas and Peaucelle 2021).

In at least ten papers published over the decade prior to 2016, John Boyer's group (mainly with Timothy [Ted] Proseus) developed a cyclical model for how the making and breaking of Ca^{2+}/egg-box cross-links by addition of new pectin molecules to the *Chara* wall could lead to *in vivo* stress relaxation and growth (Boyer 2016) and references cited therein). After the first exchange reaction, an additional one (without a new-polymer incorporation) can occur involving the now-unsatisfied bonding site on the freed-up member of the original pair. Repetition of this would lead to a bond-exchange chain-reaction involving no further polymer insertions, or otherwise to a chain-terminating one involving new Ca^{2+} addition and a second new-polymer insertion. The chain-reaction can explain how stress relaxation could reach the wall interior and cause growth even if new-polymer insertions happen only at the wall surface, as Paul Green's (1958) ingenious ^{3}H labelling experiments on the closely related, very similar alga *Nitella* indicated (to a first approximation) was its mode of wall deposition (apposition).

Boyer (2016) argued that their Ca^{2+} bond-exchange model should, or at least might, apply also to land plants. Note, however, that these algae cannot be considered general models for higher plants. They represent a stage prior to the evolutionary advent of expansins (Popper et al. 2011). Important roles for Ca^{2+} in land plants are not to be questioned. A very interesting hypothesis has been advanced regarding a role of certain AGPs in apoplastic Ca^{2+} homeostasis and auxin action (Lamport et al. 2021). Mutant studies in Arabidopsis lend support to several of the aspects of the hypothesis including a role in formation of the Ca^{2+} waves that are important for root growth (Lopez-Hernandez et al. 2020).

This resurgence of interest in a causal connection between wall synthesis and wall expansion was carried further by Ali and Traas' (2016) theoretical paper on incorporation of newly synthesised pectin molecules into a load-bearing pectic wall matrix. Calling it "force-driven polymerisation", in Figure 11.1, they sketch how they think this insertion should occur, stimulated by the stretching of ionic bonds by elastic strain, which should reduce their strength and lower the activation energy for the exchange to occur (as well as contributing to its negative ΔG as noted above).

A Role for Polysaccharide Hydrolases

Auxin stimulates both biosynthesis and turnover of XyG (Labavitch and Ray 1974; Gilkes and Hall 1977). There is evidence for auxin induction of endo-β-glucanase in both dicots (Catalá et al. 2000; Yu et al. 2018a), including in secondary walls (Glass et al. 2015) and in grasses (Kotake et al. 2000). That the enzyme activity is not only correlated with, but contributes to, cell expansion has been investigated by selective inhibition with antibodies against the glucanase both in grasses (Thomas et al. 2000) and in dicots (Hoson et al. 1991; Truelsen and Ulvskov 1995). It is tempting to speculate that there might be a plant equivalent to Cel12A among these enzymes. Arabidopsis Cel1 (At1g70710) is a soluble GH9 endo-glucanase for which antisense knock-down plants have been characterised (Tsabary et al. 2003). The phenotype comprises thinner walls, less cellulose and lignin, and compromised strength. The most striking aspect of the phenotype is the undulating walls observed in roots, both periclinal rhizodermis walls and walls of internal tissues. It has been convincingly shown that undulation can arise in a feed-forward fashion where initial buckling directs localised wall deposition, which further propagates the formation of undulations or lobe formation in epidermal cells in this case (Bidhendi et al. 2019). For the anti-sense Cel1 Arabidopsis, it should also be considered if the undulation may arise due to the failure of elongation of neighbouring cells.

CHEMORHEOLOGICAL

Deforms irreversibly under a load, at a rate that depends on some (bio)chemical reaction occurring within it

The modulation of XyG molecular weight by auxin discussed above (model 2) seems well aligned with a role of hydrolases and XET in cell expansion, a role that should be picked up in creep assays as co-operativity with expansins, but demonstration of this is lacking (McQueen-Mason et al. 1993; Darley et al. 2001; Miedes et al. 2013). Is there something important that creep assays do not pick up, or are we missing an essential feature in all models? The extensiometer creep assay is sensitive to only those chemorheological processes that can be reconstituted *in vitro*. If XET or a β-glucanase depends on concurrent biosynthesis to exert its effect on wall mechanics, then it will go unnoticed in the creep assay. This was addressed in a bioassay: Auxin-stimulated cell expansion in split pea stem segments was impaired by exogenous high MW XyG and promoted by addition of the XyG heptasaccharide XXXG; and the effect was partly inhibited by an antibody against XET. This suggested that XET-mediated XyG incorporation played a role (Takeda et al. 2002).

Colloidal Stabilisation of Cellulose Microfibrils

Hess (1928) pushed the idea that microfibrils are coated with a film of non-cellulosic material and that this has profound impact on the ordering of microfibrils. Newly synthesised cellulose microfibrils in the secondary wall of fibres in *Fagus crenata* are coated with a thin layer of xylan while another population of xylan polymers are secreted into layers of the secondary wall further from the plasma membrane (Awano et al. 2002). Grantham et al. (2017) demonstrated that the pattern of xylan substitution determines its configuration and that an even pattern of side-chains (spaced with an even number of unsubstituted backbone xylosyl residues between side-chains) is required for its binding to the hydrophilic face of cellulose microfibrils. The substitutions in question are glucuronic acid, 4-*O*-methyl-glucuronic acid and acetyl substitutions on the backbone. In rice, it has been shown that acetylation of side-chain arabinosyl residues also affects cellulose interaction (Zhang et al. 2019).

An interesting viewpoint on this was taken by Grant et al. (1969) who used the term "colloidally dispersed" cellulose referring to microfibrils in mucilage. His idea is that what we are observing in mucilage is a repurposed structure and mechanism of controlling cellulose coalescence that is also at play in the wall. Interestingly, a galactoglucomannan with an even side-chain patterning has been shown to stabilise microfibrils in Arabidopsis seed mucilage (Yu et al. 2018b) and knock-out of the galactose oxidase that is assumed to act upon galactan chains of RG-I also affects cellulose organisation in the mucilage (Šola et al. 2019).

Xylan appears to be important for cellulose macrofibrils in secondary walls (Lyczakowski et al. 2019). Cellulose micro- and macrofibril morphology has recently been characterised in irregular xylem (*irx*) mutants that are compromised in the xylan biosynthetic machinery (Crowe et al. 2021). The authors found that changes in microfibril organisation could be observed even in mutants with relatively mild phenotypes (normal levels of xylan but reduced molecular weight). The similarity of *irx* phenotypes caused by mutations in secondary wall CesAs (Turner and Somerville 1997) and in xylan biosynthetic glycosyltransferases (Brown et al. 2005, 2007; Peña et al. 2007) are nicely accounted for if cellulose organisation is a prime role of xylans. The cellulose organising effects are not limited to xylans and glucomannans. Changes in microfibril organisation were observed in the Arabidopsis *xxt1 xxt2* mutant and an interplay with the cytoskeleton was demonstrated, indicating that we are dealing with a rather sophisticated chaperone effect (Xiao et al. 2016).

Swelling and Water as a Plasticiser

The transition from a higher degree of hydration in the stem elongation zone noted by Sachs (1874) to a less hydrated wall in more mature tissue immediately basal to it appears to be accompanied by pectin metabolism, more precisely a reduction in branched pectic species, arabinans in particular (Goldberg et al. 1989). Analysis of relaxation spectra of transformed potato tubers supports a role for arabinans in wall hydration (Hansen et al. 2011). Cell wall models are of particular importance for the understanding of swelling, as it is not the occurrence of various populations of water bound to different polysaccharides; rather it is the 3-dimensional capillarity of the system that is responsible for the behaviour of water molecules in the wall, and this means that surface tension is more important than the water affinity of different polysaccharides (Belton 1997). Its relevance to wall biomechanics is well documented (Evered et al. 2007; Thompson and Islam 2021), but how expansins induce swelling (Yennawar et al. 2006) for example is not understood. And how de-swelling is brought about as a cell moves out of the elongation zone also calls for an explanation. The ionic interaction between extensin and HG (Cannon et al. 2008) causes de-swelling (Valentin et al. 2010) and may be one of the mechanisms at play.

Controlled swelling of pectic nanofilaments discussed by Morris et al. (2009) and now observed by high-resolution confocal microscopy in anticlinal walls of epidermal cells has recently been proposed to be instrumental to lobe formation (Haas et al. 2020). Demonstrating microdomains like these in the wall is at the limit of what light microscopy can do with probes as big as antibodies, but the images are persuasive. The biophysical interpretation is radical, for example, by relegating turgor pressure to play a subordinate role, see the discussion by Cosgrove and Anderson (2020). Better aligned with widely accepted concepts of cell wall-turgor interaction is the study by Bidhendi et al. (2019) which concluded that microtubule-dependent patterning of cellulose fibres in the wall is an essential activity to pattern interdigitated growth and that pectin has a role in the early stages of cell lobe initiation; see also (Belteton et al. 2021). The possible existence of pectic nanofilaments cannot be accounted for by any cell wall model while wall expansion driven by polymer hydration is not unheard of (Ruge 1937). The pectin gel→nanofilament phase changes discovered by Haas et al. (2021) thus also falls in this category of phenomena that future models should accommodate. The proposed mechanism involves enzymatic demethylation of the pectic strands leading to swelling by hydration and electrostatic repulsion, not unlike the mechanisms proposed for tertiary walls in tension wood (see above), and for mucilage secretion.

CONCLUDING REMARKS

The focus of this chapter is on the models that we might refer to as classical. As a result, there are several biomolecules and processes that we have only mentioned *en passant* if at all even though we do not question their relevance to cell wall architecture. These include, e.g., AGPs (Willats and Knox 1996; van Hengel and Roberts 2002) and the class of glycosylphosphatidylinositol-anchored proteins that many of them belong to (Gillmor et al. 2005; Yeats et al. 2018). Lamport, Tan and Kieliszewski's AGP-Calcium capacitor hypothesis mentioned above is obviously relevant and future model building should incorporate AGPs in ways that are compatible with the roles they play in calcium-dependent processes during cell expansion.

We are encouraged to view the cytoskeleton, plasma membrane and cell wall as a continuum as we approach these problems (Chebli et al. 2021), and among processes, cell wall assembly is conspicuous here by its near absence. Four decades of often painstaking research into glycosyltransferases involved in matrix polysaccharide biosynthesis (see Chapter 2) has yielded many mutants in which wall assembly is compromised or compensated for, but little in terms of processes that could yield the ordering and the microdomains that we observe in the walls.

Walls are not the same across the plant kingdom. The discovery of a new hemicellulosic polymer, glucoxylan, which interacts strongly with cellulose and which is confined to Poaceae and 3–4 related families (Little et al. 2019) (and thus not a new feature of the Type II wall in general) rules out, in our opinion, that there will be one unifying cell wall model for flowering plants. The transient use of mixed-linkage glucan during cell expansion in grasses (McCann et al. 2007; Christensen et al. 2010) is just one indication that the mechanisms of wall expansion also differ in profound ways. But defining new cell wall types each with their own model may not be as fruitful as discovering the fundamental design principles. A study of a new linkage in flax mucilage RG-I led Carpita and colleagues to propose that "modification of one polymer may be accompanied by complementary modifications of others to impart functions to heterocomposites not present in either polymer alone" (Naran et al. 2008).

The model builders behind the classical models focused on being compatible with mechanisms of cell expansion. Future model builders will, in addition, appreciate that cell wall biopolymers have self-assembly properties in and by themselves that we need to understand and build models that are compatible with assembly as they try to understand the design principles for what makes a cell wall.

REFERENCES

Ali O, Traas J (2016) Force-driven polymerization and turgor-induced wall expansion. *Trends Plant Science* 21:398–409.

Alméras T, Clair B (2016) Critical review on the mechanisms of maturation stress generation in trees. *Journal of the Royal Society Interface* 13:20160550.

Anderson DB (1935) The structure of the walls of the higher plants. *Botanical Review* 1:52–76.

Aspinall GO, Molloy JA, Craig JW (1969) Extracellular polysaccharides from suspension-cultured sycamore cells. *Canadian Journal of Biochemistry* 47:1063–1070.

Audus JL (1953) *Plant Growth Substances*. Interscience Publishers, New York.

Awano T, Takabe K, Fujita M (2002) Xylan deposition on secondary wall of *Fagus crenata* fiber. *Protoplasma* 219:106–115.

Bailey IW (1939) The microfibrillar and microcapillary structure of the cell wall. *Bulletin of the Torrey Botanical Club* 66:201–213.

Bailey IW (1952) Biological processes in the formation of wood. *Science* 115:255–259.

Baker DB, Ray PM (1965) Relation between effects of auxin on cell wall synthesis and cell elongation. *Plant Physiology* 40:360–368.

Barnes WJ, Anderson CT (2018) Release, recycle, rebuild: cell-wall remodeling, autodegradation, and sugar salvage for new wall biosynthesis during plant development. *Molecular Plant* 11:31–46.

Bauer WD, Talmadge KW, Keegstra K, Albersheim P (1973) The structure of plant cell walls: II. The hemicellulose of the walls of suspension-cultured sycamore cells. *Plant Physiology* 51:174–187.

Belteton SA, Li W, Yanagisawa M, Hatam FA, Quinn MI, Szymanski MK, Marley MW, Turner JA, Szymanski DB (2021) Real-time conversion of tissue-scale mechanical forces into an interdigitated growth pattern. *Nature Plants* 7:826–841.

Belton PS (1997) NMR and the mobility of water in polysaccharide gels. *International Journal of Biological Macromolecules* 21:81–88.

Bidhendi AJ, Altartouri B, Gosselin FP, Geitmann A (2019) Mechanical stress initiates and sustains the morphogenesis of wavy leaf epidermal cells. *Cell Reports* 28:1237-1250.e6.

Bonner J (1934a) The relation of hydrogen ions to the growth rate of the *Avena* coleoptile. *Protoplasma* 21:406–423.

Bonner J (1934b) Studies on the growth hormone of plants: V. The relation of cell elongation to cell wall formation. *Proceedings of the National Academy of Sciences of the United States of America* 20:393–397.

Boyer JS (2016) Enzyme-less growth in *Chara* and terrestrial plants. *Frontiers in Plant Science* 7:866.

Boysen Jensen P (1911) La transmission del'irritation phototropique dans l'avena. *Oversigt over det Danske Kgl Videnskabernes Selskabs Forhandlinger* 1:1–22.

Boysen Jensen P, Avery GS, Burkholder PR, Greichton HR, Scheer BA (1936) *Growth Hormones in Plants*. McGraw-Hill Book Company, London.

Boysen Jensen P, Nielsen N (1925) Studien über die hormonalen Beziehungen zwischen Spitze und Basis der Avenacoleoptile. *Planta* 1:321–331.

Bret-Harte MS, Baskin TI, Green PB (1991) Auxin stimulates both deposition and breakdown of material in the pea outer epidermal cell wall, as measured interferometrically. *Planta* 185:462–471.

Brett CT, Baydoun EA-H, Abdel-Massih RM (2005) Pectin-xyloglucan linkages in type I primary cell walls of plants. *Plant Biosystems* 139:54–59.

Brown DM, Goubet F, Wong VW, Goodacre R, Stephens E, Dupree P, Turner SR (2007) Comparison of five xylan synthesis mutants reveals new insight into the mechanisms of xylan synthesis. *The Plant Journal* 52:1154–1168.

Brown DM, Zeef LA, Ellis J, Goodacre R, Turner SR (2005) Identification of novel genes in *Arabidopsis* involved in secondary cell wall formation using expression profiling and reverse genetics. *Plant Cell* 17:2281–2295.

Broxterman SE, Schols HA (2018) Interactions between pectin and cellulose in primary plant cell walls. *Carbohydrate Polymers* 192:263–272.

Buchanan BB, Gruissem W, Jones RL (2002) *Biochemistry & Molecular Biology of Plants*. Wiley, Somerset, NJ.

Busse-Wicher M, Li A, Silveira RL, Pereira CS, Tryfona T, Gomes TC, Skaf MS, Dupree P (2016) Evolution of xylan substitution patterns in gymnosperms and angiosperms: implications for xylan interaction with cellulose. *Plant Physiology* 171:2418–2431.

Cannon MC, Terneus K, Hall Q, Tan L, Wang Y, Wegenhart BL, Chen L, Lamport DT, Chen Y, Kieliszewski MJ (2008) Self-assembly of the plant cell wall requires an extensin scaffold. *Proceedings of the National Academy of Sciences of the United States of America* 105:2226–2231.

Carpita NC, Gibeaut DM (1993) Structural models of primary cell walls in flowering plants: consistency of molecular structure with the physical properties of the walls during growth. *The Plant Journal* 3:1–30.

Catalá C, Rose JK, Bennett AB (2000) Auxin-regulated genes encoding cell wall-modifying proteins are expressed during early tomato fruit growth. *Plant Physiology* 122:527–534.

Cavalier DM, Lerouxel O, Neumetzler L, Yamauchi K, Reinecke A, Freshour G, Zabotina OA, Hahn MG, Burgert I, Pauly M, Raikhel NV, Keegstra K (2008) Disrupting two *Arabidopsis thaliana* xylosyltransferase genes results in plants deficient in xyloglucan, a major primary cell wall component. *Plant Cell* 20:1519–1537.

Chebli Y, Bidhendi AJ, Kapoor K, Geitmann A (2021) Cytoskeletal regulation of primary plant cell wall assembly. *Current Biology* 31:R681–R695.

Christensen U, Alonso-Simon A, Scheller HV, Willats WG, Harholt J (2010) Characterization of the primary cell walls of seedlings of *Brachypodium distachyon*--a potential model plant for temperate grasses. *Phytochemistry* 71:62–69.

Clair B, Déjardin A, Pilate G, Alméras T (2018) Is the G-layer a tertiary cell wall. *Frontiers in Plant Science* 9:623.

Cleland R (1959) Effect of osmotic concentration on auxin-action and on irreversible and reversible expansion of the *Avena* coleoptile. *Physiologia Plantarum* 12:809–825.

Cleland R (1971) Cell wall extension. *Annual Review of Plant Physiology* 22:197–222.

Cleland R (1973) Auxin-induced hydrogen ion excretion from *Avena* coleoptiles. *Proceedings of the National Academy of Sciences of the United States of America* 70:3092–3093.

Cosgrove DJ (1981) Analysis of the dynamic and steady-state responses of growth rate and turgor pressure to changes in cell parameters. *Plant Physiology* 68:1439–1446.

Cosgrove DJ (2000) Expansive growth of plant cell walls. *Plant Physiology and Biochemistry* 38:109–124.

Cosgrove DJ, Anderson CT (2020) Plant cell growth: do pectins drive lobe formation in arabidopsis pavement cells. *Current Biology* 30:R660–R662.

Cosgrove DJ, Li ZC (1993) Role of expansin in cell enlargement of oat coleoptiles (Analysis of developmental gradients and photocontrol). *Plant Physiology* 103(4):1321–1328.

Crowe JD, Hao P, Pattathil S, Pan H, Ding SY, Hodge DB, Jensen JK (2021) Xylan is critical for proper bundling and alignment of cellulose microfibrils in plant secondary cell walls. *Frontiers in Plant Science* 12:737690.

Darley CP, Forrester AM, McQueen-Mason SJ (2001) The molecular basis of plant cell wall extension. *Plant Molecular Biology* 47:179–195.

Darvill AG, McNeil M, Albersheim P (1978) Structure of plant cell walls: VIII. A new pectic polysaccharide. *Plant Physiology* 62:418–422.

Darvill AG, McNeil M, Albersheim P, Delmer DP (1980) The primary cell walls of flowering plants. In NE Tolbert, ed, *The Biochemistry of Plants, a Comprehensive Tratise*, vol 1, Academic Press, New York, London, pp. 91–162.

Darwin CR (1880) *The Power of Movement in Plants*. John Murrey, London.

de Vries H (1877) *Untersuchungen über die mechanische Ursache der Zellstreckung*. Wilhelm Engelmann, Leipzig.

Dean GH, Zheng H, Tewari J, Huang J, Young DS, Hwang YT, Western TL, Carpita NC, McCann MC, Mansfield SD, Haughn GW (2007) The *Arabidopsis MUM2* gene encodes a β-galactosidase required for the production of seed coat mucilage with correct hydration properties. *Plant Cell* 19:4007–4021.

Edelmann HG, Bergfeld R, Schopfer P (1995) Effect of inhibition of protein glycosylation on auxin-induced growth and the occurrence of osmiophilic particles in maize (*Zea mays* L.) coleoptiles. *Journal of Experimental Botany* 46:1745–1752.

Evered C, Majevadia B, Thompson DS (2007) Cell wall water content has a direct effect on extensibility in growing hypocotyls of sunflower (*Helianthus annuus* L.). *Journal of Experimental Botany* 58:3361–3371.

Felten J, Sundberg B (2013) Biology, chemistry and structure of tension wood. In J Fromm, ed, *Cellular Aspects of Wood Formation*, Springer, Heidelberg, pp. 201–224.

Fitting H (1910) Weitere entwicklungsphysiologische Untersuchungen an Orchidéenblüten. *Zeitschrift Botanik* 2:225–267.

Frey A (1927) Der submikroskopische Feinbau der Zellmembranen. *Naturwissenschaften* 15:760–765.

Fry SC (1989) Cellulases, hemicelluloses and auxin-stimulated growth - a possible relationship. *Physiologia Plantarum* 75:532–536.

Fry SC (2011) Cell wall polysaccharide composition and covalent crosslinking. In P Ulvskov, ed, *Plant Polysaccharides, Biosynthesis and Bioengineering*, Wiley-Blackwell, Oxford, UK, pp. 1–42.

Fry SC, Smith RC, Renwick KF, Martin DJ, Hodge SK, Matthews KJ (1992) Xyloglucan endotransglycosylase, a new wall-loosening enzyme activity from plants. *Biochemical Journal* 282:821–828.

Fuglsang AT, Palmgren M (2021) Proton and calcium pumping P-type ATPases and their regulation of plant responses to the environment. *Plant Physiology* 187:1856–1875.

Gao Y, Lipton AS, Wittmer Y, Murray DT, Mortimer JC (2020) A grass-specific cellulose-xylan interaction dominates in *Sorghum* secondary cell walls. *Nature Communications* 11:6081.

Gilkes NR, Hall MA (1977) The hormonal control of cell wall turnover in *Pisum sativum* L. *New Phytologist* 78:1–15.

Gillmor CS, Lukowitz W, Brininstool G, Sedbrook JC, Hamann T, Poindexter P, Somerville C (2005) Glycosylphosphatidylinositol-anchored proteins are required for cell wall synthesis and morphogenesis in *Arabidopsis*. *Plant Cell* 17:1128–1140.

Glass M, Barkwill S, Unda F, Mansfield SD (2015) Endo-β-1,4-glucanases impact plant cell wall development by influencing cellulose crystallization. *Journal of Integrative Plant Biology* 57:396–410.

Goldberg R, Morvan, C, Hervé du Penhoat C, Michon V (1989) Structure and properties of acidic polysaccharides from mung bean hypocotyls. *Plant Cell Physiology* 30:163–173.

Gorshkova T, Chernova T, Mokshina N, Ageeva M, Mikshina P (2018) Plant 'muscles': fibers with a tertiary cell wall. *New Phytologist* 218:66–72.

Gorshkova T, Mokshina N, Chernova T, Ibragimova N, Salnikov V, Mikshina P, Tryfona T, Banasiak A, Immerzeel P, Dupree P, Mellerowicz EJ (2015) Aspen tension wood fibers contain β-(1→4)-galactans and acidic arabinogalactans retained by cellulose microfibrils in gelatinous walls. *Plant Physiology* 169:2048–2063.

Grant GT, Mcnab C, Rees DA, Skerrett RJ (1969) Seed mucilages as examples of polysaccharide denaturation. *Journal of the Chemical Society D-Chemical Communications* 14:805–806.

Grantham NJ, Wurman-Rodrich J, Terrett OM, Lyczakowski JJ, Stott K, Iuga D, Simmons TJ, Durand-Tardif M, Brown SP, Dupree R, Busse-Wicher M, Dupree P (2017) An even pattern of xylan substitution is critical for interaction with cellulose in plant cell walls. *Nature Plants* 3:859–865.

Green PB (1958) Concerning the site of the addition of new wall substances to the elongating Nitella cell wall. *American Journal of Botany* 45:111–116.

Haas KT, Peaucelle A (2021) From monocots to dicots: the multifold aspect of cell wall expansion. *Journal of Experimental Botany* 72:1511–1513.

Haas KT, Wightman R, Meyerowitz EM, Peaucelle A (2020) Pectin homogalacturonan nanofilament expansion drives morphogenesis in plant epidermal cells. *Science* 367:1003–1007.

Haas KT, Wightman R, Peaucelle A, Höfte H (2021) The role of pectin phase separation in plant cell wall assembly and growth. *Cell Surface* 7:100054.

Haberlandt G (1896) *Physiologische Pflanzenanatomie*. Verlag von Wilhelm Engelmann, Leipzig.

Hager A, Menzel H, Krauss A (1971) Experiments and hypothesis concerning the primary action of auxin in elongation growth. *Planta* 100:47–75.

Hansen SL, Ray PM, Karlsson AO, Jørgensen B, Borkhardt B, Petersen BL, Ulvskov P (2011) Mechanical properties of plant cell walls probed by relaxation spectra. *Plant Physiology* 155:246–258.

Hepworth DG, Bruce DM (2000) A method of calculating the mechanical properties of nanoscopic plant cell wall components from tissue properties. *Journal of Materials Science* 35:5861–5865.

Herburger K, Franková L, Pičmanová M, Loh JW, Valenzuela-Ortega M, Meulewaeter F, Hudson AD, French CE, Fry SC (2020) Hetero-trans-β-glucanase produces cellulose-xyloglucan covalent bonds in the cell walls of structural plant tissues and is stimulated by expansin. *Molecular Plant* 13:1047–1062.

Hess K (1928) Zur Frage des Aufbaues pflanzlicher Membran. *Biochemische Zeitsschrift* 203:409–420.

Heyn ANJ (1931) Der Mechanismus der Zellstreckung. *Recueil des Travaux Botaniques Néerlandais* 28:113–244.

Hooke R (1665) *Micrographia or Some Physiological Descriptions of Minute Bodies Made by Magnifying Glasses*. Courier Corporation, London.

Hoson T, Masuda Y, Sone Y, Misaki A (1991) Xyloglucan antibodies inhibit auxin-induced elongation and cell wall loosening of azuki bean epicotyls but not of oat coleoptiles. *Plant Physiology* 96:551–557.

Hrmova M, Farkas V, Lahnstein J, Fincher GB (2007) A barley xyloglucan xyloglucosyl transferase covalently links xyloglucan, cellulosic substrates, and (1,3;1,4)-β-D-glucans. *Journal of Biological Chemistry* 282:12951–12962.

Ivakov A, Flis A, Apelt F, Fünfgeld M, Scherer U, Stitt M, Kragler F, Vissenberg K, Persson S, Suslov D (2017) Cellulose synthesis and cell expansion are regulated by different mechanisms in growing arabidopsis hypocotyls. *Plant Cell* 29:1305–1315.

Jacobs M, Ray PM (1976) Rapid auxin-induced decrease in free space pH and its relationship to auxin-induced growth in maize and pea. *Plant Physiology* 58:203–209.

Jin K, Qin Z, Buehler MJ (2015) Molecular deformation mechanisms of the wood cell wall material. *Journal of the Mechanical Behavior of Biomedical Materials* 42:198–206.

Johnston SL, Prakash R, Chen NJ, Kumagai MH, Turano HM, Cooney JM, Atkinson RG, Paull RE, Cheetamun R, Bacic A, Brummell DA, Schröder R (2013) An enzyme activity capable of endotransglycosylation of heteroxylan polysaccharides is present in plant primary cell walls. *Planta* 237:173–187.

Keckes J, Burgert I, Frühmann K, Müller M, Kölln K, Hamilton M, Burghammer M, Roth SV, Stanzl-Tschegg S, Fratzl P (2003) Cell-wall recovery after irreversible deformation of wood. *Nature Materials* 2:810–814.

Keegstra K, Talmadge KW, Bauer WD, Albersheim P (1973) The structure of plant cell walls: III. A model of the walls of suspension-cultured sycamore cells based on the interconnections of the macromolecular components. *Plant Physiology* 51:188–197.

Kieliszewski MJ, Lamport DTA, Tan L, Cannon MC (2011) Hydroxyproline-rich glycoproteins: Form and function. In P Ulvskov, ed, *Plant Polysaccharides: Biosynthesis And Bioengineering*, vol 41, Wiley-Blackwell, Oxford, UK, pp. 321–342.

Kim SJ, Brandizzi F (2016) The plant secretory pathway for the trafficking of cell wall polysaccharides and glycoproteins. *Glycobiology* 26:940–949.

Köhler L, Spatz HC (2002) Micromechanics of plant tissues beyond the linear-elastic range. *Planta* 215:33–40.

Kotake T, Nakagawa N, Takeda K, Sakurai N (2000) Auxin-induced elongation growth and expressions of cell wall-bound exo- and endo-β-glucanases in barley coleoptiles. *Plant Cell Physiology* 41:1272–1278.

Kutschera U, Briggs WR (1987) Rapid auxin-induced stimulation of cell wall synthesis in pea internodes. *Proceedings of the National Academy of Sciences of the United States of America* 84:2747–2751.

Kutschera U, Khanna R (2020) Auxin action in developing maize coleoptiles: challenges and open questions. *Plant Signaling & Behavior* 15:1762327.

Kutschera U, Schopfer P (1985) Evidence against the acid-growth theory of auxin action. *Planta* 163:483–493.

Labavitch JM, Ray PM (1974) Relationship between promotion of xyloglucan metabolism and induction of elongation by indoleacetic acid. *Plant Physiology* 54:499–502.

Lamport DTA, Tan L, Kieliszewski MJ (2021) A molecular pinball machine of the plasma membrane regulates plant growth-a new paradigm. *Cells* 10:1935.

Little A, Lahnstein J, Jeffery DW, Khor SF, Schwerdt JG, Shirley NJ, Hooi M, Xing X, Burton RA, Bulone V (2019) A novel (1,4)-β-linked glucoxylan is synthesized by members of the cellulose synthase-like F gene family in land plants. *ACS Central Science* 5:73–84.

Lockhart JA (1965) An analysis of irreversible plant cell elongation. *Journal of Theoretical Biology* 8:264–275.

Lopez-Hernandez F, Tryfona T, Rizza A, Yu XL, Harris MOB, Webb AAR, Kotake T, Dupree P (2020) Calcium binding by arabinogalactan polysaccharides is important for normal plant development. *Plant Cell* 32:3346–3369.

Lüdkte M (1931) Untersuchungen über Aufbau und Bildung der pflanzlichen Zellmembran und ihrer stofflichen Komponenten. *Biochemische Zeitschrift* 233:1–57.

Lüthen H, Bigdon M, Böttger M (1990) Reexamination of the acid growth theory of auxin action. *Plant Physiology* 93:931–939.

Lyczakowski JJ, Bourdon M, Terrett OM, Helariutta Y, Wightman R, Dupree P (2019) Structural imaging of native cryo-preserved secondary cell walls reveals the presence of macrofibrils and their formation requires normal cellulose, lignin and xylan biosynthesis. *Frontiers in Plant Science* 10:1398.

MacDougall AJ, Rigby NM, Ring SG (1997) Phase separation of plant cell wall polysaccharides and its implications for cell wall assembly. *Plant Physiology* 114:353–362.

Makarova EN, Shakhmatov EG (2022) Covalently linked pectin-arabinoglucuronoxylan complex from siberian fir *Abies sibirica* Ledeb. *Carbohydrate Polymers* 277:118832.

Mangenot G, Raison M (1951) Microscopical studies on the swelling of native, bleached and nitrated cellulose fibers. *The Botanical Review* 17:555–628.

McCann MC, Defernez M, Urbanowicz BR, Tewari JC, Langewisch T, Olek A, Wells B, Wilson RH, Carpita NC (2007) Neural network analyses of infrared spectra for classifying cell wall architectures. *Plant Physiology* 143:1314–1326.

McCann MC, Roberts K (1991) Architecture of the primary cell wall. In CW Lloyd, ed, *The Cytoskeletal Basis of Plant Growth and Form*, Academic Press, London, pp. 109–129.

McClure BA, Guilfoyle T (1987) Characterization of a class of small auxin-inducible soybean polyadenylated RNAs. *Plant Molecular Biology* 9:611–623.

McQueen-Mason S, Durachko DM, Cosgrove DJ (1992). Two endogenous proteins that induce cell wall extension in plants. *Plant Cell* 4:1425–1433.

McQueen-Mason SJ, Fry SC, Durachko DM, Cosgrove DJ (1993) The relationship between xyloglucan endotransglycosylase and *in-vitro* cell wall extension in cucumber hypocotyls. *Planta* 190:327–331.

Miedes E, Suslov D, Vandenbussche F, Kenobi K, Ivakov A, Van Der Straeten D, Lorences EP, Mellerowicz EJ, Verbelen JP, Vissenberg K (2013) Xyloglucan endotransglucosylase/hydrolase (XTH) overexpression affects growth and cell wall mechanics in etiolated *Arabidopsis* hypocotyls. *Journal of Experimental Botany* 64:2481–2497.

Mikkelsen MD, Harholt J, Westereng B, Domozych D, Fry SC, Johansen IE, Fangel JU, Łężyk M, Feng T, Nancke L, Mikkelsen JD, Willats WGT, Ulvskov P (2021) Ancient origin of fucosylated xyloglucan in charophycean green algae. *Communications in Biology* 4:754.

Mnich E, Bjarnholt N, Eudes A, Harholt J, Holland C, Jørgensen B, Larsen FH, Liu M, Manat R, Meyer AS, Mikkelsen JD, Motawia MS, Muschiol J, Møller BL, Møller SR, Perzon A, Petersen BL, Ravn JL, Ulvskov P (2020) Phenolic cross-links: building and de-constructing the plant cell wall. *Natural Products Reports* 37:919–961.

Morris VJ, Gromer A, Kirby AR (2009) Architecture of intracellular networks in plant matrices. *Structural Chemistry* 20:255–261.

Nägeli CW (1858) *Die Stärkekörner*. F. Schulthess, Zürich.

Nägeli CW (1864a) Über den inneren Bau der vegetabilischen Zellmembranen. *Sitzungsberichte der königl. bayerischen Akademie der Wissenschaften zu München* 1:282–323.

Nägeli CW (1864b) Über den Inneren Bau der vegetabilischen Zellmembranen. *Sitzungsberichte der königl. bayerischen Akademie der Wissenschaften zu München* 2:114–171.

Nakahori K, Katou K, Okamoto H (1991) Auxin changes both the extensibility and the yield threshold of the cell wall of Vigna hypocotyls. *Plant Cell Physiology* 32:121–129.

Naran R, Chen G, Carpita NC (2008) Novel rhamnogalacturonan I and arabinoxylan polysaccharides of flax seed mucilage. *Plant Physiology* 148:132–141.

Nishitani K, Masuda Y (1983) Auxin-induced changes in the cell wall xyloglucans: effects of auxin on the two different subfractions of xyloglucans in the epicotyl cell wall of *Vigna angularis*. *Plant Cell Physiology* 24:345–355.

Nishitani K, Tominaga R (1992) Endo-xyloglucan transferase, a novel class of glycosyltransferase that catalyzes transfer of a segment of xyloglucan molecule to another xyloglucan molecule. *Journal of Biological Chemistry* 267:21058–21064.

Nuñez A, Fishman M, Fortis L, Cooke P, Hotchkiss A (2009) Identification of extensin protein associated with sugar beet pectin. *Journal of Agricultural and Food Chemistry* 57:10951–10958.

Okamoto-Nakazato A (2002) A brief note on the study of yieldin, a wall-bound protein that regulates the yield threshold of the cell wall. *Journal of Plant Research* 115:309–313.

O'Neill MA, Eberhard S, Albersheim P, Darvill AG (2001) Requirement of borate cross-linking of cell wall rhamnogalacturonan II for Arabidopsis growth. *Science* 294:846–849.

Park YB, Cosgrove DJ (2012) A revised architecture of primary cell walls based on biomechanical changes induced by substrate-specific endoglucanases. *Plant Physiology* 158:1933–1943.

Passioura JB, Fry SC (1992) Turgor and cell expansion: beyond the Lockhart equation. *Functional Plant Biology* 19:565.

Pauly M, Albersheim P, Darvill A, York WS (1999) Molecular domains of the cellulose/xyloglucan network in the cell walls of higher plants. *The Plant Journal* 20:629–639.

Pear JR, Kawagoe Y, Schreckengost WE, Delmer DP, Stalker DM (1996) Higher plants contain homologs of the bacterial CelA genes encoding the catalytic subunit of cellulose synthase. *Proceedings of the National Academy of Sciences of the United States of America* 93:12637–12642.

Peña MJ, Zhong R, Zhou GK, Richardson EA, O'Neill MA, Darvill AG, York WS, Ye ZH (2007) Arabidopsis *irregular xylem8* and *irregular xylem9*: implications for the complexity of glucuronoxylan biosynthesis. *Plant Cell* 19:549–563.

Perzon A, Kračun SK, Jørgensen B, Ulvskov P (2020) Array-based microfibril surface assessment (AMSA) – a method for probing surface-exposed polysaccharides on cellulose nanofibres. *Cellulose* 27:8635–8651.

Pfeffer W (1904) *Pflanzen-Physiologie, Vol. 2: Kraftwechsel.* Wilhelm Engelmann, Leipzig.

Polak M, Karcz W (2021) Some new methodological and conceptual aspects of the "acid growth theory" for the auxin action in maize (*Zea mays* L.) coleoptile segments: do acid- and auxin-induced rapid growth differ in their mechanisms. *International Journal of Molecular Sciences* 22:2317.

Popper ZA, Fry SC (2005) Widespread occurrence of a covalent linkage between xyloglucan and acidic polysaccharides in suspension-cultured angiosperm cells. *Annals of Botany* 96:91–99.

Popper ZA, Fry SC (2008) Xyloglucan-pectin linkages are formed intra-protoplasmically, contribute to wall-assembly, and remain stable in the cell wall. *Planta* 227:781–794.

Popper ZA, Michel G, Hervé C, Domozych DS, Willats WG, Tuohy MG, Kloareg B, Stengel DB (2011) Evolution and diversity of plant cell walls: from algae to flowering plants. *Annual Review of Plant Biology* 62:567–590.

Preston RD (1979) Polysaccharide conformation and cell wall function. *Annual Review of Plant Physiology* 30:55–78.

Qi X, Behrens BX, West PR, Mort AJ (1995) Solubilization and partial characterization of extensin fragments from cell walls of cotton suspension cultures. Evidence for a covalent cross-link between extensin and pectin. *Plant Physiology* 108:1691–1701.

Ray PM (1962) Cell wall synthesis and cell elongation in oat coleoptile tissue. *American Journal of Botany* 49:928–939.

Ray PM (1963) Sugar composition of oat-coleoptile cell walls. *Biochemical Journal* 89(1):144–150.

Ray PM (1967) Radioautographic study of cell wall deposition in growing plant cells. *Journal of Cell Biology* 35:659–674.

Ray PM (1987) Principles of plant cell expansion. In DJ Cosgrove, DP Knievel, eds, *Physiology of Cell Expansion During Plant Growth, American Society of Plant Physiologists*, American Society of Plant Physiologists, Rockville, MD, pp. 1–17.

Ray PM, Bret-Harte MS (2019) Elastic and irreversible bending of tree and shrub branches under cantilever loads. *Frontiers in Plant Science* 10:59.

Ray PM, Green PB, Cleland R (1972) Role of turgor in plant-cell growth. *Nature* 239:163.

Ray PM, Ruesink AW (1962) Kinetic experiments on the nature of the growth mechanism in oat coleoptile cells. *Developmental Biology* 4:377–397.

Rayle DL (1973) Auxin-induced hydrogen-ion secretion in *Avena* coleoptiles and its implications. *Planta* 114:63–73.

Rayle DL, Cleland R (1970) Enhancement of wall loosening and elongation by Acid solutions. *Plant Physiology* 46(2):250–253.

Refrégier G, Pelletier S, Jaillard D, Höfte H (2004) Interaction between wall deposition and cell elongation in dark-grown hypocotyl cells in *Arabidopsis*. *Plant Physiology* 135:959–968.

Roelofsen PA (1951) Contradictory data on spiral structures in the secondary cell wall of fibers of flax, hemp, and ramie. *Textile Research Journal* 21:412–418.

Rojas E, Hotton S, Dumais J (2011) Chemically mediated mechanical expansion of the pollen tube cell wall. *Biophysical Journal* 101:1844–1853.

Ruge U (1937) Untersuchungen über den Einfluß des Heteroauxins auf das Streckungswachstum des Hypokotyls von *Helianthus annuus*. *Zeitschrift für Botanik* 31:1–59.

Ryden P, Sugimoto-Shirasu K, Smith AC, Findlay K, Reiter WD, McCann MC (2003) Tensile properties of *Arabidopsis* cell walls depend on both a xyloglucan cross-linked microfibrillar network and rhamnogalacturonan II-borate complexes. *Plant Physiology* 132:1033–1040.

Sachs J (1874) *Lehrbuch der Botanik nach dem gegenwärtigen Stand der Wissenschaft.* Verlag von Wilhelm Engelmann, Leipzig.

Sachs J (1882) *Vorlesungen über Pflanzenphysiologie.* Wilhelm Engelmann, Leipzig.

Scheller HV, Ulvskov P (2010) Hemicelluloses. *Annual Review of Plant Biology* 61:263–289.

Schindler T, Bergfeld R, Hohl M, Schopfer P (1994) Inhibition of golgi-apparatus function by brefeldin a in maize coleoptiles and its consequences on auxin-mediated growth, cell-wall extensibility and secretion of cell-wall proteins. *Planta* 192:404–413.

Seifritz W (1934) The origin, composition, and structure of cellulose in the living plant. *Protoplasma* 21:129–159.

Shinohara N, Sunagawa N, Tamura S, Yokoyama R, Ueda M, Igarashi K, Nishitani K (2017) The plant cell-wall enzyme AtXTH3 catalyses covalent cross-linking between cellulose and cello-oligosaccharide. *Scientific Reports* 7:46099.

Söding H (1933) Über das Wachstum der Inflorenzenzschäfte. *Jahrbücher für wissenschaftliche Botanik* 77:627–656.

Söding H (1934) Über die Wachstumsmechanik der Haferkoleoptile. *Jahrbücher für wissenschaftliche Botanik* 79:231–255.

Šola K, Gilchrist EJ, Ropartz D, Wang L, Feussner I, Mansfield SD, Ralet MC, Haughn GW (2019) RUBY, a putative galactose oxidase, influences pectin properties and promotes cell-to-cell adhesion in the seed coat epidermis of Arabidopsis. *Plant Cell* 31:809–831.

Somerville C, Bauer S, Brininstool G, Facette M, Hamann T, Milne J, Osborne E, Paredez A, Persson S, Raab T, Vorwerk S, Youngs H (2004) Toward a systems approach to understanding plant cell walls. *Science* 306:2206–2211.

Spartz AK, Lor VS, Ren H, Olszewski NE, Miller ND, Wu G, Spalding EP, Gray WM (2017) Constitutive expression of Arabidopsis SMALL AUXIN UP RNA19 (SAUR19) in tomato confers auxin-independent hypocotyl elongation. *Plant Physiology* 173:1453–1462.

Spatz HC, Köhler L, Niklas KJ (1999) Mechanical behaviour of plant tissues: Composite materials or structures? *Journal of Experimental Botany* 202:3269–3272.

Strasburger E (1882) Über den Bau und das Wachstum der Zellhäute. Verlag von Gustav Fischer, Jena.

Stratilová B, Šesták S, Mravec J, Garajová S, Pakanová Z, Vadinová K, Kučerová D, Kozmon S, Schwerdt JG, Shirley N, Stratilová E, Hrmova M (2020) Another building block in the plant cell wall: barley xyloglucan xyloglucosyl transferases link covalently xyloglucan and anionic oligosaccharides derived from pectin. *The Plant Journal* 104:752–767.

Suslov D, Ivakov A, Boron AK, Vissenberg K (2015) *In vitro* cell wall extensibility controls age-related changes in the growth rate of etiolated Arabidopsis hypocotyls. *Functional Plant Biology* 42:1068–1079.

Taiz L, Zeiger E, Møller IM, Murphy A (2015) *Plant Physiology and Development*. Sinauer Associates Inc., Sunderland, MA.

Takeda T, Furuta Y, Awano T, Mizuno K, Mitsuishi Y, Hayashi T (2002) Suppression and acceleration of cell elongation by integration of xyloglucans in pea stem segments. *Proceedings of the National Academy of Sciences of the United States of America* 99:9055–9060.

Talbott LD, Ray PM (1992a) Changes in molecular size of previously deposited and newly synthesized pea cell wall matrix polysaccharides: effects of auxin and turgor. *Plant Physiology* 98:369–379.

Talbott LD, Ray PM (1992b) Molecular size and separability features of pea cell wall polysaccharides: implications for models of primary wall structure. *Plant Physiol* 98:357–368.

Tan L, Eberhard S, Pattathil S, Warder C, Glushka J, Yuan C, Hao Z, Zhu X, Avci U, Miller JS, Baldwin D, Pham C, Orlando R, Darvill A, Hahn MG, Kieliszewski MJ, Mohnen D (2013) An *Arabidopsis* cell wall proteoglycan consists of pectin and arabinoxylan covalently linked to an arabinogalactan protein. *Plant Cell* 25:270–287.

Terashima N, Kitano K, Kojima M, Yoshida M, Yamamoto H, Westermark U (2009) Nanostructural assembly of cellulose, hemicellulose, and lignin in the middle layer of secondary wall of ginkgo tracheid. *Journal of Wood Science* 55:409–416.

Thimann KV, Bonner J (1933) The mechanism of action of the growth substance in plants. *Proceedings of the Royal Society B: Biological Sciences* 113:126–149.

Thomas BR, Inouhe M, Simmons CR, Nevins DJ (2000) Endo-1,3;1,4-β-glucanase from coleoptiles of rice and maize: role in the regulation of plant growth. *International Journal of Biological Macromolecules* 27:145–149.

Thompson DS (2005) How do cell walls regulate plant growth. *Journal of Experimental Botany* 56:2275–2285.

Thompson DS, Islam A (2021) Plant cell wall hydration and plant physiology: an exploration of the consequences of direct effects of water deficit on the plant cell wall. *Plants* 10:1263.

Thompson JE, Fry SC (2000) Evidence for covalent linkage between xyloglucan and acidic pectins in suspension-cultured rose cells. *Planta* 211:275–286.

Truelsen TA, Ulvskov P (1995) The role of cellulase in hormonal regulation of shoot morphogenesis in tobacco callus. *Planta* 196:727–731.

Tsabary G, Shani Z, Roiz L, Levy I, Riov J, Shoseyov O (2003) Abnormal 'wrinkled' cell walls and retarded development of transgenic *Arabidopsis thaliana* plants expressing endo-1,4-β-glucanase (Cell) antisense. *Plant Molecular Biology* 51:213–224.

Turner SR, Somerville CR (1997) Collapsed xylem phenotype of *Arabidopsis* identifies mutants deficient in cellulose deposition in the secondary cell wall. *Plant Cell* 9:689–701.

Ursprung A, Blum G (1924) Eine Methode zur Messung des Turgor- und Wand-druckes der Zelle, nebst Anwendungen. *Jahrbücher für wissenschaftliche Botanik* 63:1–110.

Valentin R, Cerclier C, Geneix N, Aguié-Béghin V, Gaillard C, Ralet MC, Cathala B (2010) Elaboration of extensin-pectin thin film model of primary plant cell wall. *Langmuir* 26:9891–9898.

van Hengel AJ, Roberts K (2002) Fucosylated arabinogalactan-proteins are required for full root cell elongation in arabidopsis. *The Plant Journal* 32:105–113.

Wang T, Hong M (2016) Solid-state NMR investigations of cellulose structure and interactions with matrix polysaccharides in plant primary cell walls. *Journal of Experimental Botany* 67:503–514.

Wang T, Park YB, Caporini MA, Rosay M, Zhong L, Cosgrove DJ, Hong M (2013) Sensitivity-enhanced solid-state NMR detection of expansin's target in plant cell walls. *Proceedings of the National Academy of Sciences of the United States of America* 110:16444–16449.

Went FW (1928) Wuchsstoff und Wachstum. *Recueil des Travaux Botaniques Néerlandais* 25:1–116.

Willats WG, Knox JP (1996) A role for arabinogalactan-proteins in plant cell expansion: evidence from studies on the interaction of beta-glucosyl Yariv reagent with seedlings of *Arabidopsis thaliana*. *The Plant Journal* 9:919–925.

Wolf S, Hématy K, Höfte H (2012) Growth control and cell wall signaling in plants. *Annual Review of Plant Biology* 63:381–407.

Xiao C, Zhang T, Zheng Y, Cosgrove DJ, Anderson CT (2016) Xyloglucan deficiency disrupts microtubule stability and cellulose biosynthesis in Arabidopsis, altering cell growth and morphogenesis. *Plant Physiology* 170:234–249.

Xin A, Fry SC (2021) Cutin:xyloglucan transacylase (CTX) activity covalently links cutin to a plant cell-wall polysaccharide. *Journal of Plant Physiology* 262:153446.

Yang H, Benatti MR, Karve RA, Fox A, Meilan R, Carpita NC, McCann MC (2020a) Rhamnogalacturonan-I is a determinant of cell-cell adhesion in poplar wood. *Plant Biotechnology Journal* 18:1027–1040.

Yang X, Reid MS, Olsén P, Berglund LA (2020b) Eco-friendly cellulose nanofibrils designed by nature: effects from preserving native state. *ACS Nano* 14:724–735.

Yeats TH, Bacic A, Johnson KL (2018) Plant glycosylphosphatidylinositol anchored proteins at the plasma membrane-cell wall nexus. *Journal of Integrated Plant Biology* 60:649–669.

Yennawar NH, Li LC, Dudzinski DM, Tabuchi A, Cosgrove DJ (2006) Crystal structure and activities of EXPB1 (Zea m 1), a beta-expansin and group-1 pollen allergen from maize. *Proceedings of the National Academy of Sciences of the United States of America* 103:14664–14671.

Yi H, Puri VM (2012) Architecture-based multiscale computational modeling of plant cell wall mechanics to examine the hydrogen-bonding hypothesis of the cell wall network structure model. *Plant Physiology* 160(3):1281–1292.

Yu L, Li Q, Zhu Y, Afzal MS, Li L (2018a) An auxin-induced β-type endo-1,4-β-glucanase in poplar is involved in cell expansion and lateral root formation. *Planta* 247:1149–1161.

Yu L, Lyczakowski JJ, Pereira CS, Kotake T, Yu X, Li A, Mogelsvang S, Skaf MS, Dupree P (2018b) The patterned structure of galactoglucomannan suggests it may bind to cellulose in seed mucilage. *Plant Physiology* 178:1011–1026.

Yuan S, Wu Y, Cosgrove DJ (2001) A fungal endoglucanase with plant cell wall extension activity. *Plant Physiology* 127:324–333.

Zhang L, Gao C, Mentink-Vigier F, Tang L, Zhang D, Wang S, Cao S, Xu Z, Liu X, Wang T, Zhou Y, Zhang B (2019) Arabinosyl deacetylase modulates the arabinoxylan acetylation profile and secondary wall formation. *Plant Cell* 31:1113–1126.

Zhang T, Zheng Y, Cosgrove DJ (2016) Spatial organization of cellulose microfibrils and matrix polysaccharides in primary plant cell walls as imaged by multichannel atomic force microscopy. *The Plant Journal* 85:179–192.

Zhang Y, Yu J, Wang X, Durachko DM, Zhang S, Cosgrove DJ (2021) Molecular insights into the complex mechanics of plant epidermal cell walls. *Science* 372:706–711.

Zykwinska A, Thibault JF, Ralet MC (2007) Organization of pectic arabinan and galactan side chains in association with cellulose microfibrils in primary cell walls and related models envisaged. *Journal of Experimental Botany* 58:1795–1802.

12 The Glycomic Code of the Plant Cell Wall

How Structure Leads to Function

Marcos S. Buckeridge
University of São Paulo

INTRODUCTION

Seen as a Glycomic Code, the plant cell wall holds polymers that produce some heterogeneity, although synthesised with relative precision. Polysaccharides, the principal components of the Glycomic Code, are encoded molecules whose specific functions are associated with forming emergent functional architectures. The architecture here is seen as the physical arrangement of polymers that form the cell wall of a given cell. In the plant kingdom, different architectures are present in type I, II, and III walls, where the main hemicelluloses are xyloglucan, arabinoxylan, and mannan, respectively. Because cell wall architecture results from a combination of interactive polymers, it is helpful to attribute the cell walls to specific allometric proportions. Here, cell wall allometry is defined as the proportions among different polymer domains of the plant cell wall. The term domain in this chapter refers to classes or groups of polymers that confer specific properties to part of the wall. Here, cell wall domains are cellulose, hemicelluloses, pectins, lignin, and proteins. The cellulose–hemicellulose under interaction can also be called domain. The allometry of a given cell wall is associated with the inherent properties of the cell, tissue, or organ. One example is the relatively high proportion of cellulose in wood. The combination of the microfibrils and xylans (or glucomannans) generates the biomechanical properties that allow trees to grow taller than other plants.

The polymer biosynthesis process can be redundant in a group of cells, i.e., being precisely repeated (including its heterogeneity) in the same way in each group of cells. The *redundancy* of this mapping process leads to the emergence of another functional level, giving rise to tissues. Alternatively, cells containing different cell wall allometries (what represents a *degeneracy* of the allometry) may coexist, generating patterns of diversity typical of organs.

To discuss the Glycomic Code in plant cell walls, it will be first explained how the ciphers and codes (Box 12.1) are associated with the semiotics[1] of the genetic code (Zolyan, 2022). The Glycomic Code is the combination of encoded structures that produce cell wall allometry, responsible for the functional diversity of the walls in cells, tissues, and organisms. Short discussions regarding the codes and meanings of some of the main polysaccharides (xyloglucans, mannans, xylans, mixed-linkage beta-glucan, and pectins) attempt to point to evidence of codification in these polymers concerning biological functions and roles. The main elements that can be used to access codification in polysaccharides are their fine structures, i.e., the way decorations are distributed through main chains. Using the Glycomic Code to interpret the plant cell wall helps to understand how biological structure leads to function.

> **BOX 12.1 FROM CIPHERS TO LANGUAGES (WRIXON, 2005)**
>
> - **CIPHER:** method of concealment in which the primary unit is the letter. From Arabic *sifr*, means "nothing"
> - **CODE:** cryptographic system by which a word, number, letter, symbol of phrase is substituted for plaintext words, letter, and phrases. From Latin *codex*
> - **SIGNAL:** signals that rely more upon efficiency of long-distance communication. From Latin *signum*
> - **SCRITPS AND LANGUAGES:** tools to enhance communication between individuals and allow the conveyance of more complex types of information
> - **STEGANOGRAPHY:** messages delivered indirectly by normal language. From Latin *steganos* (cover)

Also discussed will be the evolution of the Glycomic Code, starting from living beings, explaining the existence of extracellular matrices (ECMs) containing polymers with nitrogen and sulfur, and proposing that the selective pressures that led to the chemical and allometric diversities seen in the plant and green algae cell walls have acted on the Glycomic Code.

All biological systems are subjected to what Ernst Mayr (1961) called *dual causality*. In his 2004 book, Mayr defines the term as living beings not being controlled by the natural

DOI: 10.1201/9781003178309-12

laws of physics alone but also by *genetic programs*. Mayr further elaborates the discussion by proposing a difference between *genetic* and *somatic programs*. Although all that happens in a living being is somehow linked to the genetic code, the main emergent property of this code is the result of a *somatic program* on which the selective pressures of evolution work. Following Mayr's definitions, the plant cell walls result from *genetic programs*. However, once the genetically encoded proteins are made, the metabolism in cells, tissues, and organs leads to a cascade of emergent *somatic programs*.

An analogous idea has been recently developed by Marcello Barbieri (2015), who compiled what he called the *codes of life*. Barbieri's Code Biology theory covers mainly the *somatic programs* (in Mayr's sense), proposing a biosemiotics generalisation in which, for any *somatic program*, a group of *signs* is *mapped* into *meanings*, the latter being equivalent to functions. Figure 12.1 contextualises the cell wall in the framework of code Biology.

From a biosemiotics[2] viewpoint, monosaccharides are *signs* that, through metabolism, lead to allometric patterns (generating architectural diversity) that make the shape of cells possible and interfere with their functions.

The emergent cell wall allometries found in nature to date seem to be a result of *dual causality* in the sense that they exist within the limits of the natural physical laws, but at the same time, their functions and biological roles originate from *genetic programs* that culminate in *somatic programs*.

THE GLYCOMIC CODE

The complex polysaccharides' fine structures in plant cell walls are related to crucial plant features: providing shapes to cells (Chapters 11 and 15), participating in cell signaling (Chapter 19), and defending cells against the attack of other organisms (microorganisms, insects, and other herbivores) (Chapter 20).

In the context of Code Biology, a code is a set of rules establishing relations between two worlds (Barbieri, 2015). The Code Biology semiotics defines *worlds* as different entities (Figure 12.1). One is the *world* of *signs*, which are the basic elements of an encrypted message. The other *world* contains the elements represented by the *meanings* (i.e., the functions and roles) that emerge from processing the *signs* into a final function. A third entity—the *adaptor*—acts between both (*signs* and *meanings*), producing functions with their respective biological roles[3] (Tavares and Buckeridge, 2015, Buckeridge, 2018).

In the case of the biological codes, a continuum exists in which the functions and their inherent biological roles become *signs* for subsequent cascades of ciphers and codes that link metabolism to the ecological levels.

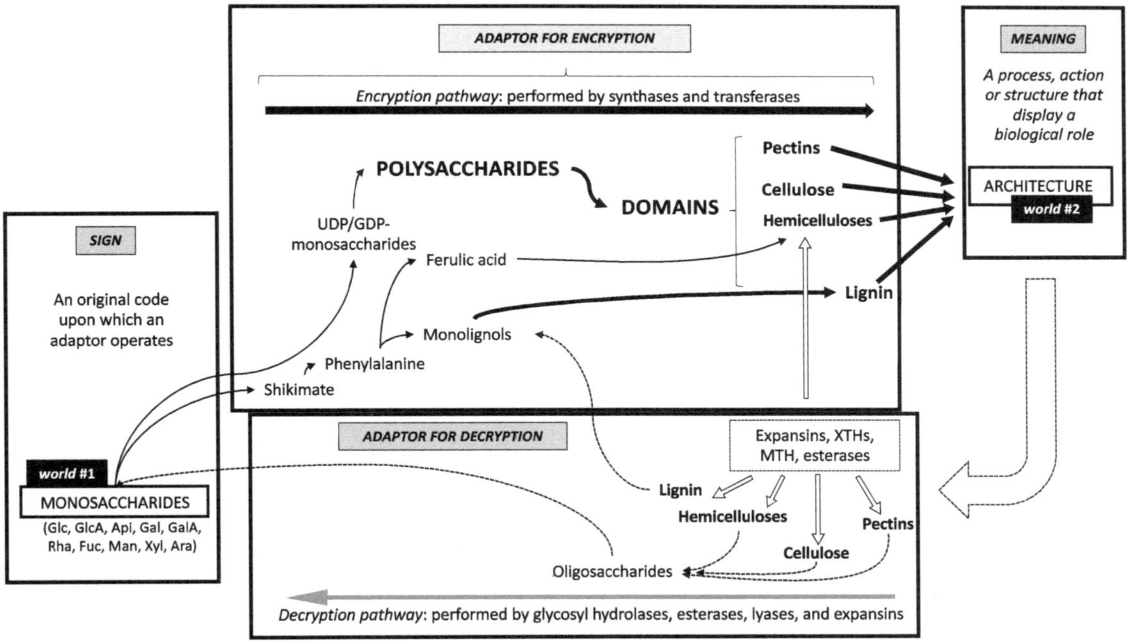

FIGURE 12.1 Biological codes are formed by a group of signs mapped through adaptors (mechanisms) that lead to meaning (represented by biological functions). One example is the DNA code, in which the signs are represented by the triads, the adaptors by the transcription, splicing, and translation mechanisms, and the meaning by the topology of action of the enzymes that act on metabolism. In plant cell walls, the signs are the monosaccharides, amino acids, and phenolic compounds used by the adaptors (polysaccharide and protein biosynthesis and the secondary metabolism that produces the phenolics). The result of the Glycomic Code is cell wall architecture. (Reprinted with permission from Tavares EQP and Buckeridge MS (2015), Do cell walls have a code? Plant Science 241:286–294.) Glc = glucose, GlcA = glucuronic acid, Api = apiose, Gal = galactose, GalA = galacturonic acid, Rha = rhamnose, Fuc = fucose, Man = mannose, Xyl = xylose, Ara = arabinose, XTH = xyloglucan transglycosylase hydrolase, MTH = mannan transglycosylase hydrolase.

Figure 12.1 depicts how the Glycomic Code connects word #1, containing the signs which are the basic elements of cell wall polymers (sugars, amino acids, and phenylpropanoids) through an *adaptor* that encrypts the code into polymers that end up forming an architecture that is reflected as the allometry of the cell wall (see Chapters 3 and 5 for description of part of the encryption system of the Glycomic Code). The architecture and its characteristic allometry can be disassembled by the plants themselves or modified by organisms capable of producing enzymes that can read the code, degrading the polymers, and returning them to their basic sugars.

BIOSEMIOTICS OF THE CELL WALLS

Because the plant cell wall represents most of the biomass on the surface of Earth, the mechanisms behind its assembly process are critical to understanding many aspects of life on Earth. From the biodiversity viewpoint, the food chains in ecosystems that maintain communities' functioning exist due to the capacity of plants to construct their cells, tissues, organs, and a diversity of individuals that are the first step to the biogeochemical processes extant in our planet (Buckeridge, 2023). From the humanity viewpoint, plants are the source of most food, fiber, and energy. They also contribute to a lot of the materials used in industrial systems.

In this context, the plant cell wall is fundamental to the stability of the biosphere because it is quantitatively the most abundant assembly of molecules without which no plant exists. And if plants did not exist, no food chains or ecosystems would be viable.

Viewing Cell Walls as Information

Although DNA is the primary code of all life, the latter is, in fact, part of a succession of ciphers and codes that form proteins, polysaccharides, and other metabolites. They constitute the actual chain of information that produces the connection between photosynthesis and biological roles. For a biological function or role to exist, the parallel work of multiple *adaptors*, increasingly forming complex structures that afford the existence of life, is necessary.

In this chapter, the cell wall is viewed as an encoded complex molecular structure or an assembly of macromolecules that make viable plant cells, tissues, organs, and individuals. The implication of the encoded system that produces single polysaccharides assembled in the ECMs of cells provides the possibility of multicellularity in plants. The capacity to replicate the same code forming identical cells is called *redundancy*, whereas the ability to produce cells with distinct cell wall allometries is called *degeneracy* (Buckeridge, 2006). Thus, the Glycomic Code is considered a result of either *redundancy* or *degeneracy* of an encoding mechanism that ends up with the formation of cell walls, giving rise to different tissues and organs. In this sense, if regarded as the results of a genetic program present in the DNA code, the formation of organs can be seen as a "cascade" of ciphers and codes (Figure 12.2) that diverge toward multiple interconnected adaptors that lead to a plethora of *meanings* (i.e., functions and roles), including the Glycomic Code that is key in the process of plant development and functioning.

Figure 12.2 depicts how information processing leads to the formation of the Glycomic Code, which is responsible for the formation of the ECM of plants. To help understand this succession, Box 12.1 displays definitions of ciphers, codes, signals, languages, and steganography (Wrixon, 2005). As pictured in Figure 12.1, the fundamental elements (or *signs*) of DNA are sugars, nucleotides, and phosphate. These elements are constant and are the basis of both DNA and RNA. However, RNA can be regarded as a cipher of DNA because a single nucleotide is changed along with the RNA molecule's function. Due to complex cellular mechanisms related to translation and protein synthesis, proteins are formed; this is a *code* rather than a *cipher* because the *adaptor* related to the synthesis of the proteins will require a different *sign*, i.e., the amino acids. Among the many aspects of the encoded protein biosynthetic systems is their folding, which determines the interactions among themselves and other molecules. Among the multiple possibilities of functions of proteins that range from signalling metabolism in general, to name but a few, the catalytic function of enzymes will accelerate a series of reactions that will, among other things, result in the Glycomic Code (Chapter 4). It is striking that without this "timing adjustment" given by catalysis made by enzymes, which occurs within cellular compartments and the ECM, life would not be possible on Earth.

Biosemiotics: From Units to Emergent Biological Properties

Figure 12.2 also brings a biosemiotics interpretation (the biochemical processes viewed as information processing) of the Glycomic Code. From encoded information in the DNA, an *object* emerges, the gene. From this *object*, another *sign* appears (the messenger RNA (mRNA)), and from this point onward, proteins can be considered *interpretants*—multiple pathways of information processing that result into biological roles—that will participate in the encoding of metabolism and, finally, the Glycomic Code.

If we now compare Figure 12.2 with the definitions in Box 12.1, the Glycomic Code may be viewed as an encoded structure, but at the same time, it also functions in plant biology as a *signal* or a kind of *language*. As the Glycomic Code hardly functions as a *signal* in displaying long-distance communication, it must probably function as a *language*, since the Glycomic Code displays syntax due to the organization of its terms (see the polysaccharide codes below). Examples are cases where the Glycomic Code *language terms* are sensed by other organisms (especially microorganisms) that can recognise (through their glycomic codes and enzymes) the presence of specific codes that afford or not the penetration of the cells. In other words, microorganisms contain enzymes that display partial ability to *read* the

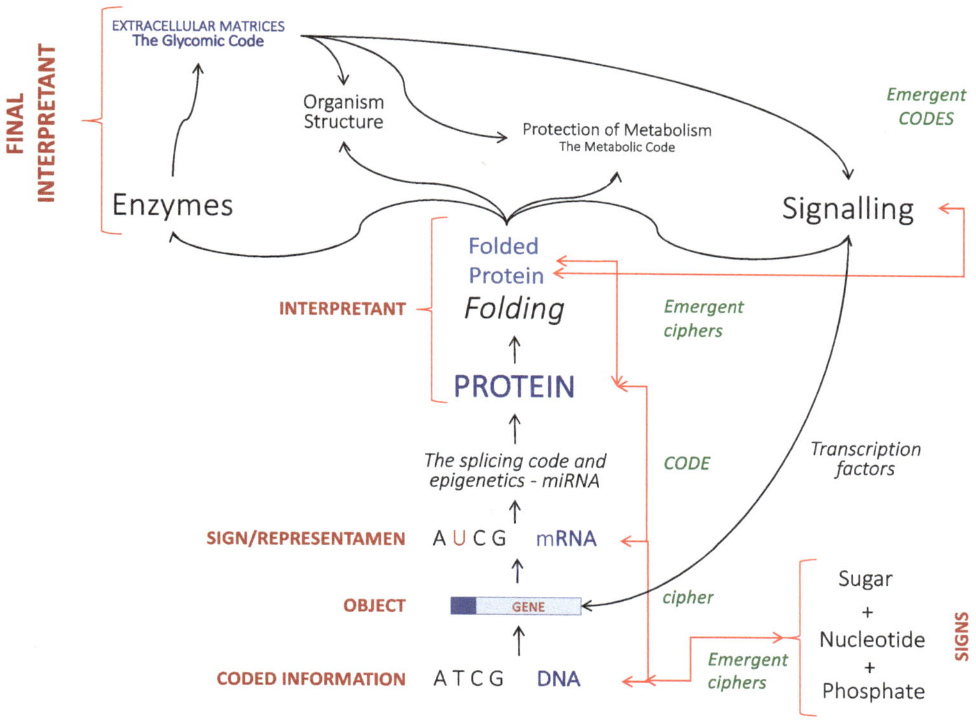

FIGURE 12.2 Biosemiotics and Code Biology of biological molecules taking DNA as a starting code. Ciphers and codes (green) alternate, starting with the DNA genetic code up to the folding of the proteins responsible for enzyme action and structure of the extracellular matrices of living beings. In red, the semiotic terms are pointed at different stages of mapping within the Glycomic Code.

Glycomic Code. In cases of communication among different organisms, one could even classify the phenomenon—based on Box 12.1—as steganography.

EVIDENCE OF THE GLYCOMIC CODE IN PLANT CELL WALLS

Plants deal with environmental and evolutionary pressures through various strategies, producing cell walls with a mixture of polysaccharides displaying distinct properties. The classes of polymers are extremely limited in comparison to the combinatorial possibilities. Three primary cell wall types have been defined to date: I, II, and III depending on the proportions among polymers and cell wall domains. Although they differ regarding the predominance of the type of hemicellulose (xyloglucan for type I, arabinoxylan for type II, and mannan for type III), they share similar biological functions. Subtypes with more subtle modifications (branching types and exchanges of certain sugar moieties in specific positions in the macromolecule) of those types determine biological roles, which are crucial for plant form and function. Examples are the deposition of cellulose in secondary walls of the xylem and fiber cells, the storage walls, the aerenchyma walls, and different types of cells in leaves. Cell wall integrity maintenance mechanisms, examined in Chapter 18, are another piece of evidence that not only does the code exist in plant cell walls but also its integrity must be kept for biological functions and roles to take place.

Using Reductionism to See the Whole

One of the strategies to connect wall form with functions and roles has been the isolation of the polymers and their analyses, followed by strategies to attribute the biological roles to the wall structure. Several procedures for cell wall fractionation with alkali-separated polymers of the plant cell walls with common properties are used (e.g., Beckmann et al., 1923; Carpita, 1984; Sun et al., 1995; De Souza et al., 2013). A weak acid (ammonium oxalate) can extract most of the pectins. Sequential extractions with increasing alkali (NaOH or KOH) concentrations, lead to the separation of hemicelluloses bound to each other or cellulose with different strengths. Alternatively, some polymers are much easier to extract, such as the water-soluble cell wall storage polysaccharides from seeds (galactomannans, xyloglucan, and galactans), for which biological roles have been relatively easier to attribute to structural modifications (i.e., their codes). The cell wall storage polymers are promptly extracted in hot water, followed by precipitation with ethanol. The residue is resolubilised in water in purified form. Furthermore, their structure is relatively more uncomplicated than the primary cell walls. Furthermore, working with seeds and seedlings is more manageable than performing experiments with adult plants.

Multiple techniques have advanced the knowledge about the cell wall polymers, including chemical analyses (monosaccharides, phenylpropanoid, and amino acids of structural proteins), biochemical analysis using specific enzymes that produce oligomers or debranch polymers, and physical techniques such as FTIR), mass spectrometry (that complement oligosaccharide analysis), and nuclear magnetic resonance (Chapter 8). Molecular dynamics, coupled with structural analysis techniques, have also been used to address polymers' interaction and cell wall function (e.g., Martínez-Abad et al., 2017). These techniques have significantly deepened our knowledge about cell wall polymers and led to the proposition of a series of models of polymers assembly in the cell walls that reflect their complexity and diversity.

Progress in cell wall sciences is gradually helping crack the Glycomic Code and make functional attributions to their structural variety. As we gradually acquire more knowledge about plant genomes and the pathways that lead to cell wall biogenesis at the level of chromosomes, strategies could be designed to modify biochemical pathways to change cell wall allometry (e.g., Nawaz et al., 2017, Pagliuso et al., 2022). More systemic approaches have been developed, such as using specific antibodies to localise polymers or groups of polymers (e.g., Buckeridge et al., 2005; Knox, 2008; Avci et al., 2012; De Souza et al., 2013) (Chapter 10) and using molecular biology (genetic transformation) techniques to introduce modifications of genes that produce in vivo structural changes in cell wall polysaccharides (see Wang et al., 2016 for a review). With these and other strategies, researchers have significantly advanced cell wall chemistry and polymer topology so that the cell wall models become progressively more precise (e.g., Martinez-Abad et al., 2017) (Chapter 11).

A parallel can be made between techniques used to address DNA, genetic code, cell walls, and Glycomic Code. Polymer hydrolysis with restriction endonucleases was the first means to access DNA physical maps. By using gel electrophoresis, large DNA fragments were observed. Subsequently, restriction analysis became a valuable tool for accessing DNA characteristics such as single nucleotide polymorphisms, insertions, and deletions (Roberts, 2005).

Cracking the Code: Looking at Longer Polysaccharide Sequences

From the Glycomic Code viewpoint, the use of hydrolases to detect fine structural features of cell wall polysaccharides can be considered analogous to the findings regarding DNA "fine structure" in the 1970s. Peter Kooiman used cellulase digestion of storage xyloglucan in 1961 to find that the polymer is made of four specific blocks. The use of beta-mannanases on the storage galactomannan later in the 1980s showed that the decoration with galactosyl residues is not randomly distributed in the main chain. These features have been shown to affect several polysaccharide properties (McCleary et al., 1985, Dea et al., 1986). For storage xyloglucans, Tiné et al. (2003) showed the same. However, in this case, the authors found that the polymer could be hydrolysed even more precisely under lower cellulase concentrations. It became apparent that the action of cellulase on xyloglucan has analogous properties to the restriction enzymes used to study DNA. In low concentration, cellulase produced larger xyloglucan fragments, which could be analogous to the DNA fragments produced by Danna and Nathans in 1971. The findings related to the fine structure of xyloglucan reinforced the idea that this polymer is encoded. Like DNA, hydrolases could discover some information hidden in the xyloglucan structure. Association between decorations of xylan's fine structure has been linked to function. Jacobs and Dahlman (2001) compared the regularity/irregularity of glucuronic acid decorations distribution in softwood and hardwood. More recently, the regularity in xylans from wood led Martínez-Abad et al. (2017) to find evidence that regular motifs whose substitutions with mGlcA and Arabinose point to functions related to the interaction with cellulose microfibrils.

Codes of Polysaccharides

This chapter will use five groups of polysaccharides to examine the evidence of the Glycomic Code in plant cell walls: xyloglucan, galactomannan, xylans, mixed-linkage beta-glucan (MLG), and pectins. These groups of polymers were chosen to illustrate the idea of the Glycomic Code. The evidence presented does not represent an exhaustive review of codified polymers in plant cell walls, but only examples where they occur.

The polymers chosen as examples are treated as encoded molecules based on evidence correlating structure and function/biological roles. The choice is due to their medium complexity, except for the pectins that constitute the most complex polysaccharides in the plant cell wall. The hemicelluloses were chosen because they have been more intensely studied using hydrolytic enzymes, providing data that afford fine structural examination of the polymers as encoded structures.

Nevertheless, it is essential to highlight that cellulose, lignin, and structural proteins are generally considered part of the Glycomic Code (Tavares and Buckeridge, 2015). Including lignin in the Glycomic Code may sound strange since it is not composed of sugars. However, lignin is covalently bound to hemicellulosic decorations, so the phenolic polymer becomes part of the wall's architecture. Structural proteins, such as hydroxyproline-rich proteins, can also be regarded as part of this code. Relatively less is known about the contribution of structural proteins and arabinogalactan proteins (AGPs) and the architecture of the cell wall, but it is known that such proteins are intrinsic to the cell wall structure. In this way, in Figure 12.1, phenylpropanoids and amino acids are considered signs, and their corresponding parts of the phenylpropanoids and amino acid metabolisms are regarded as adaptors of the Glycomic Code.

The Xyloglucan Code

Xyloglucan hemicellulosic polysaccharide occurs in practically all living plants. It is a branched beta-1,4-linked glucan (the main chain of cellulose) that binds tightly to cellulose (Hayashi, 1989, Levy, 1991). The positions of the galactosyl units are essential to xyloglucan solubility in water (Tiné et al., 2006), degradability (Tiné et al., 2003), the mechanical strength of the cell wall (Peña et al., 2004), and xyloglucan–cellulose interaction (Hayashi, 1989; Lima and Buckeridge, 2001, Lima et al. 2004). In primary walls, the fucose-containing branching of xyloglucan increases binding strength to cellulose (Hayashi, 2006). Several other types of branching exist (Franková and Fry, 2012; Schultink et al., 2014), and their biological roles are slowly being understood.

In the light of Code Biology of the plant cell wall (Figure 12.1), it could be supposed that the biosynthetic machinery follows rules that lead to the positioning for branching with galactosyl and fucosyl residues (some of these rules are described in Chapter 3). Consequently, slightly different *meanings* would be produced, depending on how the *adaptor* processes monosaccharides into polysaccharides. Xyloglucans are made mainly of combinations of repetitive blocks of Glc_4Xyl_3 or $Glc_4Xyl_{2+}Ac$, depending on the botanical group to which a species belongs (Schultink et al., 2014). The preservation of these blocks in the plant kingdom during evolution highlights their essentiality. There is, however, some degree of heterogeneity in the substitution patterns that can be thought as part of the Glycomic Code.

After the finding of the basic constitutive blocks of xyloglucan by Kooiman (1961) until the publication of the nomenclature for xyloglucan fine structure by Fry et al. (1993), the xyloglucan code in storage xyloglucan was explored by Fanutti et al. (1993). In the latter paper, the authors used an endo-beta-transglycosylase to study the mode of action of the enzyme and the combination of limit-digest oligosaccharides in the storage xyloglucan from nasturtium seeds. They showed that the oligosaccharide structure is crucial for the action of the enzyme on the polysaccharide.

Oligosaccharides with five rather than four glucosyl residues in the main chain were discovered in seed storage xyloglucan from the tropical tree species *Hymenaea courbaril* (Buckeridge et al., 1997). A deeper analysis using carefully controlled hydrolysis with cellulase, and mass spectrometry showed that Glc_4Xyl_3 and Glc_5Xyl_4-based oligosaccharides are present in the same polymer molecule. It has been further proposed that the combinations of both oligomers are not arbitrary. Using larger fragments of the *H. courbaril* xyloglucan, it became clear that there are "superblocks" that form another layer of the code. Naming Glc_4Xyl_3 **T** (for tetramer in the main chain) and the Glc_5Xyl_4 **P** (for pentamer in the main chain), it was possible to calculate the probabilities of combinations and find that the most probable motif in the larger molecules would be repetitive sequences of **TPPT**, with the number of **T**s varying in the flanks (Tiné et al., 2006). Such a structure would reflect the distribution of the galactosyl branching in the xyloglucan molecules, changing properties like self-interaction and susceptibility to hydrolysis. A hypothesis to explain the functional meaning of these motifs has been proposed by Buckeridge (2018). The *H. courbaril* xyloglucan was shown to self-interact more strongly than other storage xyloglucans (tamarind, copaifera, and nasturtium). Lima et al. (2004) attributed this feature to the distribution of the galactosyl residue in its molecules. Hypothetically, such distribution would be more accessible to the attack by xyloglucan endo-transglycosylase-hydrolase. These two consequences had been experimentally demonstrated by Lima and Buckeridge (2001), Lima et al. (2004) and Tiné et al. (2006), respectively.

Hymenaea courbaril is a tropical legume tree that has been highly successful during evolution. One of the reasons is the efficiency of seedling establishment, which is one of the significant barriers that control population size (the seedling establishment stage displays a higher mortality rate in the life cycle). Because *H. courbaril* xyloglucan is a storage polysaccharide that is degraded in the cotyledons to provide sugars for the growing seedling, the control of degradation by the enzymes is crucial for seedling performance in the understorey of the rain forests (see details in Buckeridge, 2010 and Buckeridge, 2018). It has been hypothesised that the xyloglucan code in this species provides crucial and highly complex adaptations to storage mobilisation performance, connecting the Glycomic Code with ecological adaptation.

In grasses, the xyloglucan code changed from Glc_4Xyl_3 to $Glc_4Xyl_{2+}Ac$ (Ac for acetyl groups). Quantitatively, xyloglucan is not a significant component in grasses since a distinct wall allometry has evolved in this group of plants, i.e., glucuronoarabinoxylans are the main hemicellulose instead of xyloglucan. From the structural viewpoint, one possible explanation is that, at least for a couple of grass species (maize and sugarcane), there are macro- instead of microfibrils of cellulose in the cell walls (Ding and Himmel, 2006, Buckeridge et al., 2015). The cell wall fractionation of sugarcane cell walls and analyses of the fine structure of the polymers (including xyloglucan) indicated that it is the primary polymer that is more closely associated with cellulose. Thus, one possibility is that the $Glc_4Xyl_{2+}Ac$ could have something to do with a more adaptive assembly of macrofibrils. The vast difference between grasses and eudicots regarding hemicelluloses (glucuronoarabinoxylans, xyloglucan, and beta-glucan) could be associated with developing new biological roles during evolution, such as faster growth rates and more efficient defense mechanisms.

Finding details of the xyloglucan code in storage polymers was possible because these are among the simpler xyloglucans extant. There are several different branching possibilities for xyloglucans and many different contexts are possible (Schultink et al., 2014). Thus, a substantial amount of data regarding the biological functions of these polysaccharides, where each branching type occurs, will be needed to decipher the entire xyloglucan code. The critical point is understanding whether and how different oligosaccharides are combined in single xyloglucan molecules. The second would be attributing specific biological roles associated with the various codes.

The Xylan Code

Non-random arrangements of branching also occur in polysaccharides of the xylan group. For instance, the acetyl esterification positions on the xylan backbones are alternated in the secondary walls of Arabidopsis. According to Busse-Wicher et al. (2014), there seem to exist implications of what is here called the xylan code to the interaction between arabinoxylan and cellulose. Xylan-specific antibodies have been used to show that interfascicular fibers are composed of distinctly substituted arabinoxylan (McCartney et al., 2005). The vascular bundle cells of *Cryptomeria japonica* display xylan polymers with similar substitution degrees in their composition (Kim et al., 2010). A possible explanation for such a fine structure in the interfascicular fibers of *C. japonica* is that the prevalence of substituted arabinoxylan may decrease xylan digestibility. Although this is a different polysaccharide code, a similar function as the one attributed to storage xyloglucan (control of digestibility) can be attributed to acetylated xylan in *C. japonica*.

As proposed in 1993 for xyloglucans, an unambiguous nomenclature has been created for heteroxylans of plant cell walls (Fauré et al., 2009). The 1,4-beta-xylan chain receives the letter X, whereas the arabinosyl decorations should be named A, B, or C, depending on the length of the arabinan chain. To designate in which carbon the arabinosyl residue is linked, the numbers 2 or 3 (the carbons available for branching) should be written as a superscript. The letter U designates a decoration with glucuronic acid. Depending on where it is linked (carbon 2 or 3), a superscript number will be added to the letter U. In case glucuronic acid is methylated, the letter "m" should be added as a superscript. This nomenclature was applied by Martinez-Abad et al. (2017), who used a combination of structural analysis (High Performance Anion Exchange Chromatography with Pulsed Amperometric Detection and mass spectrometry) with molecular dynamics compared xylan oligosaccharides of the types XXA^3XU^mX and $XXXU^mU^mX$. They concluded that depending on the fine structure and the type of cellulose surface (hydrophobic or hydrophilic), the xylan oligosaccharides will bind with different strengths to the microfibril surface. These findings can be used as an example of the codification of glucuronoarabinoxylans related to their structural role in wood cell walls.

Further evidence that the fine structure of xylans is related to susceptibility to hydrolysis is the more uniform distribution of arabino-4-O-methylglucuronoxylans on S1 than S2 tracheid layers compared with O-acetyl-galactoglucomannans (Kim et al., 2011). The outer cell wall layer hypothetically undertakes lower turnover rates. Because it is richer in arabinoxylan, it is probably more recalcitrant to hydrolysis, possibly conferring protection against the attack of microorganisms.

Mutations in various members of the irregular xylem gene family turned the plants deficient in xylan and led to the collapse of the xylem vessels (Brown et al., 2007; Lee et al., 2007). Although xyloglucan has been found to bind efficiently to cellulose, there is evidence that xylan may also cover microfibrils (Awano et al., 2001, Martinez-Abad et al., 2017). Reis and Vian (2004) associated this phenomenon with preventing microfibril aggregation. Later in 2007, Persson et al. (2017) obtained an Arabidopsis mutant deficient in glucuronoxylan and found a phenotype consistent with loss of cell wall integrity. This is consistent with the roles of xylan in maintaining the mechanical properties and low permeability to water and gases that afford water and nutrient conduction by the vascular system.

In the examples explored above, associated with digestibility and the mechanical stability of the fibers and stability of the vascular system, the evidence of regularity and essentiality of polysaccharides of the xylan group might be added as evidence that a xylan code possibly exists (Chapter 18).

It is interesting to note that the xylan and the xyloglucan codes display similar functions in different species. These functions give rise to biological roles (e.g., storage, water conduction, self-interaction control, and hydrolysis resistance). In all cases, the polysaccharide code must likely be precise to provide the adaptation through different biological roles.

The Mannan Code

In plants, the mannan (beta-1,4 linked) polysaccharide family is formed by pure mannans, galactomannans, and glucomannans (Buckeridge, 2010). Seed storage galactomannans are among the most studied polysaccharides because of their industrial applications in the food, medical, and textile sectors, to name but a few. *Cyamopsis tetragonolobus* (guar) and *Ceratonia siliqua* (carob) are examples of species containing galactomannan in their seeds. Galactomannans occur in seeds of many species and families, and significant variation has been found in their degree of galactosylation of the mannan chain (Buckeridge et al., 2000a). Some polymers display the main chain fully branched with alpha-1,6-galactosyl units (mannose: galactose ratio = 1). However, this ratio varies up to 4 depending on the species. Whereas pure mannans produce tissues that are very hard (date palm, for instance), the galactosylation degree significantly changes galactomannan properties, turning the molecule more (man:gal = 1) or less (man:gal = 3) soluble in water. Viscosity also varies, being higher in the former than in the latter.

Some evidence has been produced that corroborates the hypothesis that storage galactomannans display an analogous code to xyloglucan. When products of hydrolysis of galactomannan with endo-beta-mannanase were studied by computer analysis, the modelling indicated that the galactosyl branching distribution was not random. Galactosyl branching in guar occurs at every two or three mannosyl residues (McCleary et al., 1985). It has been demonstrated that the distribution of the decorations with galactose affects galactomannan self-association (Dea et al., 1986).

The use of endo-beta-mannanase as a "restriction" enzyme (similarly to the experiments with very low concentrations of cellulase on xyloglucans) led to the production of higher oligosaccharide blocks that seem to be analogous to the ones found for xyloglucans (Buckeridge and De Souza, 2014). The evidence described above, along with other findings related to the functions of polysaccharides of the mannan group, suggests that they may also display a code like the one found for storage xyloglucans and structural arabinoxylans. However, deeper investigations of more members of the mannan group's fine structure must be made to unveil the higher levels of the mannan code.

Regarding the connection between the polysaccharide code with biological roles, in seeds, pure mannans, whose main chains interact quite strongly (like cellulose), play a role in protecting the embryo development of palm seeds. In legumes, where a range of degrees of substitution with galactose occur, galactomannans could either function as a water retainer, providing not only carbon and energy to the developing embryo but also controlling water transit between the atmosphere and the embryo (Reid and Bewley, 1979, Buckeridge and Dietrich, 1996).

In the case of the glucomannan code, on the one hand, they participate in primary walls playing a structural function. In softwoods, glucomannan appears to replace the biological role of xylan in strengthening microfibrils. In the case of ferns, which possess an entire taxonomic subgroup, mannans and glucomannans are the main hemicelluloses (the Type III walls) (Silva et al., 2011). On the other hand, other species, notably in the groups of orchids and bromeliads, acetylated glucomannans seem to perform functions related to water retention connected to epiphytic habit (Stancato et al., 2001; Goycoolea and Cárdenas, 2003).

The Mixed-Linkage Glucan Code

The MLG is characteristic of the grasses, transiently produced during development (Carpita, 1984). MLG also functions as a storage polysaccharide in cereal seeds, degraded during germination (Buckeridge et al., 2004) (Chapter 9).

Most of the polymer (ca. 90%) is composed of interconnected cellotriosyl (Glc_3) and cellotetraosyl (Glc_4) oligosaccharides, both occurring in the same molecules. In general, the Glc_4:Glc_3 (4:3) ratio ranges from 2 to 3. The larger the 4:3 ratio, the stronger the MLG will bind to cellulose, whereas higher proportions of beta-1,3 linkages will turn the polymer more soluble in water. Larger oligosaccharides from 5 to 11 beta-1,4 linkages have been identified, which led to the discussion about the implications for the three-dimensional conformation of the polymer (Woodward et al., 1983).

In maize coleoptiles, the MLG biosynthetic system is precise in producing a combination of beta-1,3 and -1,4 linkages with a fixed ratio. However, little is known about longer combinations of the linkages, which prevents understanding the diversity of the MLG code at higher levels.

In sugarcane, all plant tissues display MLG to some level in all organs of adult plants. The primary function seems to be structural, but it has been proposed that MLG can also play a storage function (Roulin and Feller, 2001). These authors demonstrated that wheat leaves could start using MLG as a source of carbon for growth when all starch is consumed.

It has been observed that disturbance of the organization of the Golgi apparatus *in vitro* provokes changes in its fine structure. However, no changes can be observed when the biosynthetic machinery is assembled *in vivo* (Buckeridge et al., 1999).

MLG has been introduced in a species that is not genetically equipped to produce it (Arabidopsis) without reporting significant changes in the phenotype. In case this possibly means that when synthesised in a "foreign" system (one that did not develop MLG as a natural component of its cell wall allometry), MLG does not perform its biological function (Doblin et al., 2009), it could be concluded that when changes are made to the Glycomic Code out of context regarding biological roles, they may not produce significant effects on plants. However, a deeper investigation is required to evaluate this hypothesis.

The Pectin Code

Pectins display polysaccharide backbones with alpha-linked acid monosaccharides (Chapter 5). They may be branched with relatively longer polymers of arabinans and galactans (Atmodjo et al., 2013). The degree of methyl esterification (controlling water solubility) and the alpha linkages (that impair flexibility to the main chain) confer the ability to this polymer to fold onto itself under interaction with calcium ions, giving rise to the "egg boxes" (Sundar et al., 2012, Zudnek et al., 2021). Besides forming a store of calcium around the plant cells, this structurally complex polymer forms gels that determine the essential hydrodynamic properties of the middle lamella (Chapters 23 and 24). Pectins also occur in the primary cell wall. In this case, the main galacturonic acid chain alternates with rhamnose residues that may be branched with short polymers of arabinans and galactans (Atmodjo et al., 2013). Elongation of the neutral galactan chain has occurred during evolution. Like galactomannans and xyloglucans, this polymer can function as a carbon reserve for early seeding growth (Buckeridge and Reid, 1994). Arabinan chains have been associated with the rigidity of the wall (Verhertbruggen et al., 2009).

In aquatic plants (duckweeds), pectins of different types (apiogalacturonan and xylogalacturonan) occur, and their configuration in this species is probably related to the habit of living in spring water (Avci et al., 2018, Pagliuso et al., 2020). Due to the apiose-containing pectins, their association with boron retention in the cell wall is thought to adapt to the aquatic condition, where the boron concentration is much lower than in the soil.

With their different shapes determined mainly by the cell walls (the cytoskeleton and the vacuole have critical roles, too), all cells are further interconnected through the middle lamellae (traversed by plasmodesmata), which are therefore part of the Glycomic Code and act as the cement and communication highways among cells (Zamil and Geitmann, 2017). In this sense, as previously proposed (Tavares and Buckeridge, 2015), the middle lamellae could be considered "the wall of the walls" and thence can be interpreted as the actual cell outside. In the endodermis of roots, Casparian strips are regions in the anticlinal walls that are encrusted with lignin and possibly suberin. They block the apoplastic entry of small molecules (e.g., water and ions) into the stele of the root. The only possible path for molecules, including water, is therefore through the endodermic protoplast involving transit across the plasma membrane. Thus, the endodermis helps select nutrients that may reach the plant´s vascular system. Once such substances cross the endodermis cells, a tissue that functions as a type of inner skin (van Fleet, 1961), they may return to the apoplast, entering the vascular system and other tissues potentially staying in the apoplast for the remainder of the path. Together, the middle lamellae of a plant, similarly to the vascular system, permeate the whole body as an inside and, at the same time, outside (apoplastic) network.

WHY A GLYCOMIC CODE? DERIVING FUNCTION FROM STRUCTURE

The examples above (xyloglucan, mannan, xylan, MLG, and pectin) support the hypothesis that cell wall polymers may be produced by encoding mechanisms that are highly sophisticated and accurate programs of polymer assembly. In these cases, biological roles have been attributed to different cell wall allometries and seem related to variations in polysaccharides' fine structure.

To illustrate this suggestion, it has been observed that xylan and mannan synthases are assembled into large protein complexes (Levigne et al., 2004) from several glycosyl transferase families, and further edition of the polysaccharides can be performed *in muro* by transglycosylases and hydrolases (Burr and Fry, 2009). Thus, the result of the role played by the array of synthases, transferases, and hydrolases (i.e., architecture) contains specific combinations of those polysaccharides, giving rise to architectural domains that determine specific biological roles played by a variety of allometries of cell walls.

The polysaccharide codes are straightforward, and their encoding system (an interaction between the genetic and the somatic programs) is being described (photosynthesis>NDP/sugars pathways>Golgi/membrane polymer building>architectural assembly) (see Pagliuso et al., 2021 for a recent review). Thus, the Glycomic Code is manifested through processing the CO_2 that enters the stomata through the metabolism, determining the functional meanings for every wall. Through wall *redundancy* and *degeneration*, the manifestation of the Glycomic Code leads to different architectural designs (allometries) that render fundamental biological roles that originate the main functions of the whole plant. Among the key functions are:

- The vascular system: notable in stems and culms but pervasive to all the plant body so that nutrients, photosynthates, and signal molecules are delivered to all living cells.
- The leaf activities: photosynthesis, light perception, defense, and interaction with the environment.
- The root functions: absorption, transport of nutrients, and protection.
- The reproductive and dispersion tissues (flowers and fruits, spores) are responsible for spreading the individuals of a population throughout the environment.

The persistent and precise execution of the mapping process—with its intrinsic diversity within the plant—determines biological functions and roles selected during evolution. The selection, though, is not random. Only the physically possible structures have appeared in nature since the biological functions depend on natural laws of physics and chemistry that, when followed, provide better adaptation and lead to organisms surviving the competition (Figure 12.3).

THE EVOLUTION OF THE GLYCOMIC CODE

A Brief History of the Genetic Code

In some ways, we can compare the history of the Glycomic Code with one of the Genetic Code. About three decades (from around the 1930s to the 1960s) were necessary for scientists to decipher the Genetic Code. The sequence of discoveries went from the initial finding that nucleic acids are the molecules that transmit information from one generation to another to the findings that triads of nucleotides encode amino acids into proteins (Cobb, 2015). Later in the 1970s, enzymatic fragmentation of DNA by restriction endonucleases became essential tools to access the superior level of the Genetic Code: the genes. The use of restriction enzymes opened the possibility of obtaining DNA genetic maps. Using a combination of specific restriction endonucleases and gel electrophoresis in polyacrylamide gels, Danna and Nathans (1971) observed that endonuclease products could be detected. This paper became one of the classical ones in molecular biology, providing the first description of how to precisely use gel electrophoresis combined with restriction enzymes to detect DNA fragments. Later, genetic mapping became a reality with the advent of continuously improved methods to detect and manipulate DNA fragments.

Here, genetic mapping can be defined as the map of genes in each genome—in this chapter, a plant's genome—responsible for the *mapping* process that transforms *signs* (monosaccharides, amino acids, phenylpropanoids) into a *meaning* (the cell wall polymers). Different architectures and allometries emerge when polymers (polysaccharides, structural proteins, and lignin) are assembled. Semiotically speaking, a script or a language appears (Box 12.1), giving

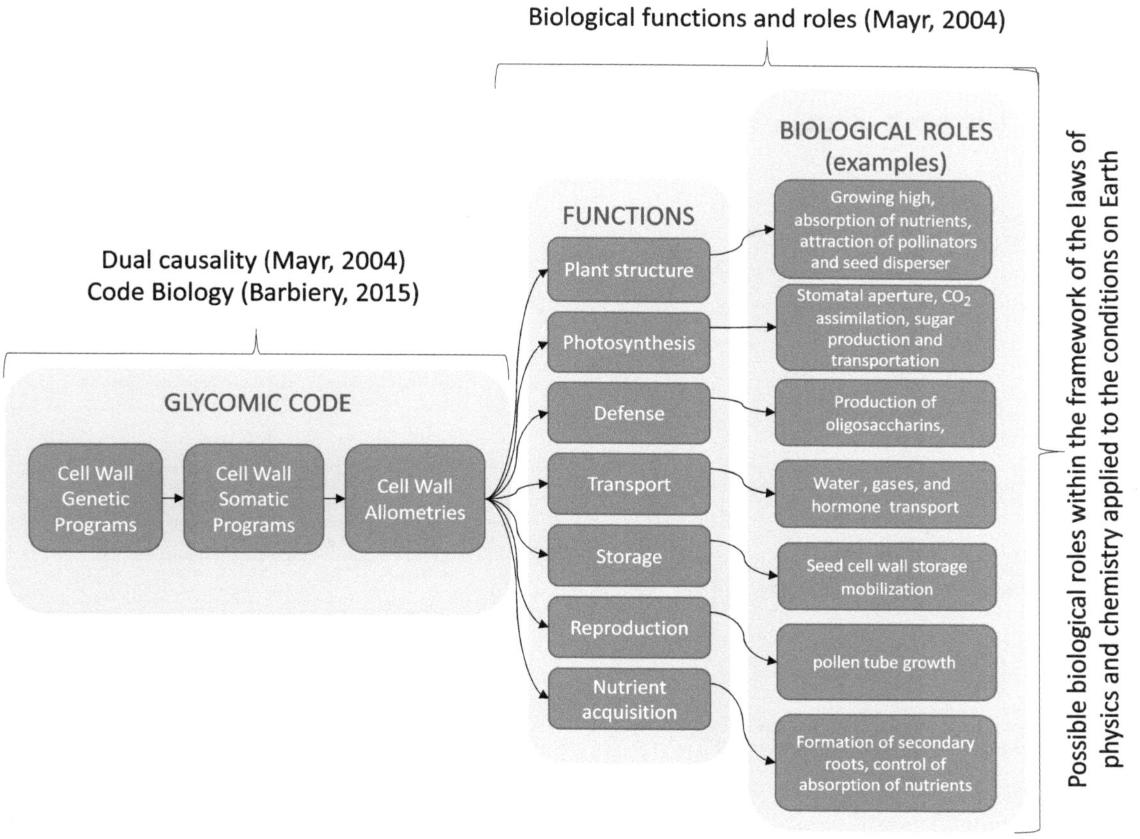

FIGURE 12.3 The Glycomic Code surges in Nature due to dual causality (execution of the genetic code followed by the somatic code). This creates cell wall allometries (combinations of the domains of cellulose, hemicellulose, pectin, and lignin). Different biological functions appear because of cell wall allometry diversity. Each function displays a variety of biological roles. These biological roles constitute the tiny number of possibilities in an enormous universe of possible combinations of signs that could generate allometries (meanings according to the Cold Biology theory). This means that life, as we see it on Earth, is the only possible combination that functions according to the physical and chemical laws of Nature. This figure only gives a few examples of the biological roles of cell walls as generated by the Glycomic Code in plants.

rise to the identity of cells that, when repeated precisely in the same fashion by a group of cells, form a tissue. With the formation of a tissue (e.g., the xylem), new biological roles (transportation of water and nutrients or transportation of hormones) become possible for the plant. The function of long-distance transportation of molecules by plants emerges from different *mapping* processes, also leading to the phloem (transportation of photoassimilates) and the aerenchyma (transportation of gases). It is noteworthy that the *degeneracy* (here defined by deviations in the *mapping* process that lead to different allometries) of the Glycomic Code of a cell having the same genome, but different gene-expression patterns, lead to what we call development, generating a plethora of systems that are embedded in plant biodiversity.

Evolution of the Glycomic Code in Extracellular Matrices

Regarding carbohydrate polymers, the discovery of sugars and the development of methods to reveal the chemical bonds among monosaccharides in polysaccharides (linkage analysis – Hakomori, 1964) led to the modern view of complex carbohydrate polymers.

Using combined physical techniques (e.g., infra-red and Raman spectroscopy, mass spectroscopy, nuclear magnetic resonance) for structural analysis of polysaccharides led to finding a plethora of polymers in nature. Carbohydrate polymers exist in all living organisms' ECMs, except for Archea (Figure 12.4). Polysaccharides can be separated into different classes (Buckeridge, 2018). There are nitrogenated there are nitrogenated (bacteria, animals, and fungi) and non-nitrogenated (red algae, green algae, and plants) polymers. On the other hand, some subgroups contain sulfur in their structures (animals and red algae), the rest being sulfur-free polymers. A Glycomic Code containing sulfur has been associated with cells that live in salty water. Thus, red algae and animals (considered to live in salty water from the inside out) are the sulfate users for their ECMs.[4]

Hypothetically, nitrogenated ECMs seem to be related to organisms that belong to the end of ecological food chains. This may be so because acquiring nitrogen for plants is costly due to the energy spent assimilating it. Being closer to the end of food chains "lowers" the cost of obtaining

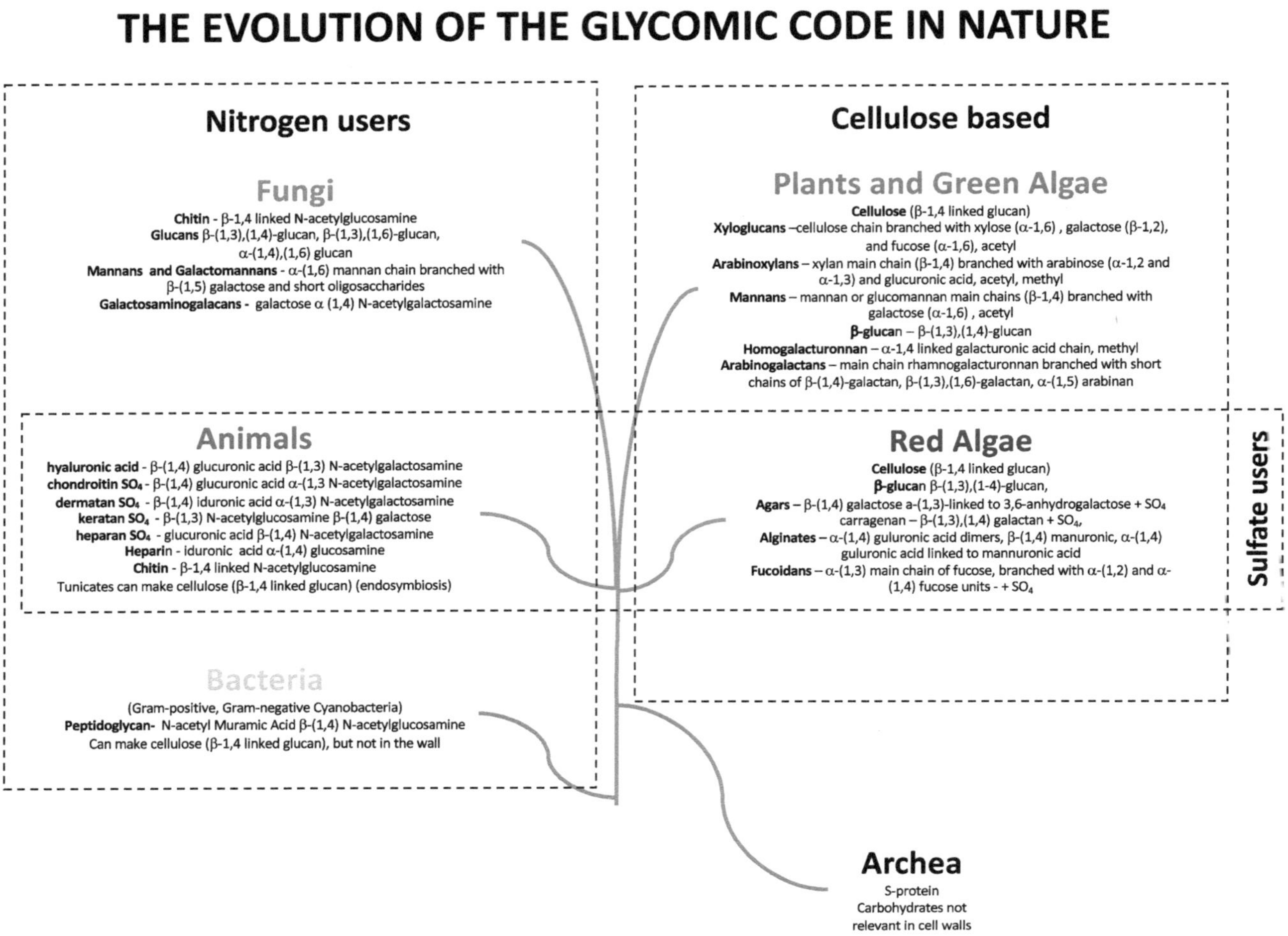

FIGURE 12.4 The evolution of the Glycomic Code in the Extracellular (ECM) matrices of living beings. Except for Archea, all ECMs are composed of carbohydrates (and proteins). Fungi, animals, and Bacteria are considered "N users," whereas plants, green and red algae do not use N for the carbohydrate polymers of their ECM. These are based on cellulose. Another notable structural feature during evolution is a division between groups that use S in their polysaccharides. These are organisms associated with living on the sea or that display cells functioning in high salt concentration (terrestrial animals). The Glycomic Code can be considered the cell walls of photosynthesizing terrestrial organisms. (Reprinted with permission from Buckeridge MS (2018) The evolution of the glycomic codes of extracellular matrices. *Biosystems* 164:112–120.)

nitrogen and increases the possibility of producing an ECM displaying a high proportion of N, which becomes "economically" viable.

Plants and green algae share subclasses of polymers: cellulose and beta-glucan, xyloglucans, xylans, mannans, and the pectic classes formed by homo- and rhamnogalacturonans and arabinogalactans. These ECMs are typically called cell walls.

For the plant cell wall, the knowledge about the structure of the polymers led to the proposition of different models (Chapter 11). From the discovery of mannose and later glucose by Emil Fisher in the 1880s (Hudson, 1941) to the development of methylation analysis by Sen-Itiroh Hakomori (1964) and others, the plant cell wall modeling started to be thought of as an assembly of interactive polymers. The group of Peter Albersheim proposed the first molecularly consistent model of the plant cell wall after publication of a series of three papers (Bauer et al. 1973, Keegstra et al. 1973, Talmadge et al., 1973). The model was conceived not solely based on structure but also on the modifications made on walls from cells growing in a cell suspension, by glycosyl hydrolases. Albersheim's model proposed that some wall polymers were covalently linked, forming an interconnected macrostructure. However, after several years of research, little evidence favored the hypothesis of covalent linkages among polymers within the wall. This led McCann and Roberts to propose, in 1991, a discontinuous model of the cell wall in which hemicelluloses interacted with cellulose microfibrils through hydrogen bonding (a feature already present in Albersheim's model). In the McCann and Roberts model, the pectins form an independent domain in which the cellulose–hemicellulose domain is immersed. Carpita and Gibeaut made further advancements in cell wall modeling in 1993. They proposed that at least two cell wall types existed (type I and II). The former comprises walls containing xyloglucan as the main hemicellulose (occurring mainly in eudicots and non-commelinoid monocots), and the second is present in grasses and contains arabinoxylans as the main hemicelluloses. In 2011, Silva et al. extended this view, proposing the existence of a type III wall in ferns, in which the hemicellulosic domain is composed of mannan.

If we assume that walls are a result of a Glycomic Code, which display different allometries (given by the *somatic programs* proposed by Mayr, 2004) (Figure 12.3), with encoded polysaccharides that, to produce architecture, combine them with lignin and structural proteins, the natural selection process may have acted on domains of the Glycomic Code (domains that are responsible for biological roles) and not specifically on the polymer structure. This is probably because the combination of polymers (architecture), i.e., the result of the *somatic programs*, is "closer" to the biological functions and roles than polymers alone.

Based on the features described above, it can be suggested that selective pressures during the evolution of the cell walls acted on the distinct allometries rather than on isolated genes.

WHY IS THE GLYCOMIC CODE NOT A TRUISM?

A possible criticism of the interpretation of the cell wall as a code could be that explaining the cell wall as the *mapping* (metabolism) process that transforms basic *signs* (monosaccharides, phenylpropanoids, amino acids) leading into *meanings* (functions and roles) would be only another way to explain the same process replacing terms of biochemistry with ones of semiotic. Viewed this way, one could argue that the Glycomic Code idea would be a truism. However, bringing biosemiotics to the discussion of how cell walls are formed and degraded and how they interfere with plant cells' biological functions and roles, tissues, organs, and individuals, denotes the "fabric"[5] of events that lead to the biological functions of the cell wall in plants. In semiotic terms, the "fabric" produces the *interpretants* of the cell wall (Figure 12.3), i.e., the collection of functions and roles resulting from the combination of cell wall allometries. Thinking about the cell wall as an encoded macromolecular structure that results from such a "fabric" of events helps explain the development of tissues, organs, and even different species. If it were not for the Glycomic Code—the set of rules for constructing cell walls with specific allometries—the replication of sister cells with the same characteristics—giving rise to tissues through the development process—would not be possible. Other features are involved in these similarities. Still, the shape-giving, mechanical, and biochemical functions and roles of the cell wall alone can be evoked as essential factors in the formation of tissues. The emergent properties of the groups of cells with similar functions (tissues), under interaction with other cell groups, produce organs. Thus, allometric rules exist for constructing multicellular organisms and plants (Niklas, 1994). In particular, the existence of the Glycomic Code is crucial to achieving the diversity of allometries possible in plants (Figure 12.3).

For instance, in a leaf of a Eudicotyledon species, the epidermis (with pavement cells and stomata—the latter formed by two types of cells) displays functions such as protection against pathogens (defense due to a physical barrier) and connecting (through stomata) the inner tissues of the plant with the atmosphere (exchange of gases and water vapor). However, the leaves of eudicots need a few other cell types to function as an organ. They need the palisade cells, the mesophil, and the cells of the vascular system. Except for the vascular system that permeates the entire plant body, the other cell types derive from developmental modifications in their cell walls that can be viewed as variations of the Glycomic Code. Notably, the vascular system forms a tissue whose Glycomic Code is replicated—with minor variations—in the whole plant body.

Looking at the Vascular System to Understand What a Code is For

The vascular system is an interesting example of the variations in the Glycomic Code, with the development of cells whose biological functions depend on the wall. This is the case of the xylem whose Glycomic Code includes deposition of large proportions of cellulose, glucuronoarabinoxylans, and lignin, and the phloem with its perforated plates with extensive callose deposition.

The vascular system must be efficiently integrated throughout the plant body in all cases, delivering photosynthates and nutrients to all living cells. For the plant body to function adequately and be able to perform and compete in evolution, the control of *mapping* that leads to the variation of the Glycomic Code of the vascular system of plants needs a developmental integration that produces a system obligatorily compatible with the laws of physics and chemistry (West, 2017).

Code Rules

It is now evident that the Glycomic Code has rules, as discovered by the cell wall community of scientists after decades of research. These codes produce the molecules assembled in architectures that become stable under certain physical–chemical conditions. The research community is slowly assembling models of the pathways that produce the cell wall polymers from the primary *signs* of glucose and fructose via UDP-sugar pathways (Bar-Peled et al., 2011, Verbančič et al., 2017, Pagliuso et al., 2021). The NDP-sugars in these pathways are subsequently used in the Golgi and the membrane (for a review, see Zhang et al., 2020). They are further dismantled, separating the sugars from the NDPs as the sugar moieties are assembled into polysaccharides in the Golgi and the plasma membrane.

Another aspect that highlights the necessity of looking at the cell wall as a Glycomic Code is related to the fact that only a few of the possible combinations of elements (*signs*) occur in nature. This means that only some combinations of polymers produce architectures and allometries that function in the context of the plant tissues, organs, and bodies and their relationship with the environment.

A surprising conclusion from a panoramic and summarised view of the cell wall polysaccharides is that the number of combinations existing in nature is smaller than the enormous number of possibilities (1.7 billion for a simple pentamer), given the high number of possible linkages for hydroxyls that carbohydrates could make.

A comparison among the number of possible combinations of five basic units (*signs*) of DNA (bases), protein (amino acids), and polysaccharides (monosaccharides) is instructive to understand the constriction of possibilities that form the Glycomic Code.

To clarify this, one may compare polysaccharides with other macromolecules in living beings. A DNA molecule with four bases could lead to $4^5=1{,}024$ combinations, whereas the possible peptides containing five units constructed from the 20 most common amino acids would produce $20^5=3{,}200{,}000$ combinations. In contrast, the possible polysaccharides made with combinations of the ten different monosaccharides ($10^5=100{,}000$) could be made of hexoses or/and pentoses, which contain three or four free hydroxyls ($3.5^5=525$) and can also be alpha- or beta-linked ($2^5=32$). Multiplying 100,000*525*32, the total number of possibilities of the sugars would be 1,680,700,000. Obviously, the constriction imposed to *mapping* of the Glycomic Code is enormous.

Structure Constricts Function

Despite all possible combinations, most of them can be found in either of the naturally occurring macromolecules (perhaps less so for the DNA). This is possibly related to protein and polysaccharide folding and interaction, and polysaccharides are part of a higher-level encoding system. One possible explanation is that the latter two are much more restricted due to their higher specificity regarding biological function and roles. In the case of the plant cell wall, it is striking that only a dozen combinations occur from the near 1,7 billion possible combinations. In his book *Scale*, Geoffrey West (2017) discusses biological restrictions considering the scaling properties in biology. In the first chapter of his book, he discusses how the discovery of the Navier–Stokes equations was essential to designing modern ships, airplanes, and cars. The Navier–Stokes equations are differential equations that describe the flow of objects in fluids and air. West describes how William Froude provided, in the 1800s, the calculations necessary to improve naval technology dramatically. Since then, the application of the so-called "Froude number" and other developments in engineering have provided the means to design ships, airplanes, and cars that are much more efficient in moving through water and air. West highlights that different cars, ships, and airplane models became increasingly similar since the application spread through the automobile industry. One of the reasons is that most industries learned how to use the Navier–Stokes equations to improve their designs efficiently. The point raised by West later in the book was that, in biology, the allometry is adjusted to performance, and similarly to what happened to the cars, evolution selected the most adapted designs based on what is physically possible. Because genetic programs and somatic programs are part of this process, living beings are the result of dual causality as proposed by Ernst Mayr (2004).

The argument here is that these ideas could be applied to the evolution of the plant cell wall, explaining why a relatively tiny number of polymers, as a proportion of the possible combinations among sugars, led to the known plant cell walls. Hypothetically, there is probably an analogous "Froude number" to building a cell wall. In other words, there should be a mathematical engineering explanation for selecting the combinations of Types I, II, and III walls during evolution because of more adaptable performances. The Glycomic Code is the result of this process. On the one hand, the *signs* existing in abundance (CO_2 being transformed into simple sugars) are available for plants to build their walls. They are the monosaccharides, amino acids, and phenylpropanoids. The possible reason sugars dominate the wall composition is likely related to producing glucose and fructose directly through photosynthesis rather than using amino acids and proteins that require nitrogen, which is more difficult to obtain by plants. Such constriction is probably related to the better-adapted way of *mapping* the Glycomic Code and photosynthesizing capacity of plants. It likely explains why plants and algae display cell walls containing no nitrogen, whereas fungi, animals, and bacteria use it in their ECMs (Buckeridge, 2018).

The Structure Connects to Function

The genetic program behind the execution of the Glycomic Code is "written" in the DNA. Understanding the mapping of monosaccharides into polysaccharides through an encoded system that repeats itself with modifications during the development of the whole plant body expresses a systemic view of the wall that is different from the reductionistic perspective that tends to explain walls through the genes and biochemistry alone. However, the emergent properties of the walls are also a result of the *mapping* system, which is a result of the *somatic program* proposed by Mayr (2004). The *somatic program* consists of the pathways that transform CO_2 into sugars, then metabolise them into nucleotide sugars that are subsequently used by membrane-bound enzymes in the Golgi and the plasma membrane, leading to encoded polysaccharide formation. The final and less studied mechanism is the *in muro* assembly of the complex macromolecular structure that is the manifestation (the interpretant) of the Glycomic Code.

In this way, rather than being a truism or purely philosophical interpretation of molecular biology and biochemistry of the cell wall, the Glycomic Code is the embodiment of the systemic process (the "fabric" mentioned above) that highlights the enormous importance of the cell walls for the biological functions and roles in plants.

One of the consequences of the Glycomic Code is that polysaccharides, cell wall proteins, and lignin should also be encoded because their functions and roles need to be equalised to lead to the architectures and allometries that connect the structure to function.

CONCLUDING REMARKS

The Glycomic Code is central to the issue of the relationship between form and function in plants. This is so because cell walls constitute most of the biomass of plants on the planet. Thus, the allometric diversity of the cell walls can be seen as the main factor behind plant biodiversity. The allometries presently known result from selective pressures that involve adjusting the possible combinations of polymers that make plants functional organisms. The cell wall community has inadvertently been working toward "cracking" the Glycomic Code for several decades. We now have a good understanding of the polymers, their mutual associations, and their relationship with biological functions and roles in plants. The knowledge available about the plant cell walls where biological functions and roles have been pointed out is consistent with the *dual causality* idea. In other words, most of the possible cell wall allometries are limited by natural physical laws. The allometries that can be detected in nature result from *genetic programs* that generate *somatic programs*, which are the elements that undertake selective pressure during evolution.

Accepting that cell walls are encoded structures raises another perspective on manipulating cell wall allometry in the future. In some cases, allometry could be changed to produce plant modifications to make them better adapted for food and fiber production. Another possibility is the creation of "neocodes" (polymers and polymers assemblies not existing yet), which could be used to create novel applications, including adapting plants to different environments on Earth or outside it. Changing cell wall allometry is key to plant engineering. It is becoming possible due to the dramatic increase in knowledge about how the *signs* are *mapped* into *meanings* forming the glycomic code diversity.

ACKNOWLEDGMENTS

I thank my colleagues Adriana Grandis, Bruno Navarro, José Eli da Veiga, Marcelo Zuffo, and Raul Lima for the fascinating discussions about cell walls, evolution and the physical laws applied to engineering. I also thank Anja Geitmann and four reviewers who helped explain the ideas put forward in this chapter much better. I acknowledge the financing support of the National Institute of Science and Technology of Bioethanol (INCT-Bioetanol; FAPESP 2008/57908-6 and 2014/50884-5) and (CNPq 574002/2008-1 and 465319/2014–9).

NOTES

1 Semiotics is a general philosophical theory of signs and symbols that deals primarily with their function in artificially constructed and natural languages and comprises syntactics, semantics, and pragmatics (from Merriam-Webster's dictionary).

2 Biosemiotics is a form of interpretation of biological processes as signs and codes mapped into meanings (i.e., the biological functions and roles). The term biosemiotics was first used by the German scientist Friedrich Rothschild in 1962 and later adopted and extended by the semioticians Thomas Sebeok (Hungarian American) and Thure von Uexkull (German), the latter two considered the fathers of biosemiotics.

3 In this chapter the definition of Ernst Mayr (2004) is used for biological functions and roles. Mayr's example in the book is: *the function of the legs of a rabbit is locomotion, but one of the biological roles of the rabbit's leg is to run away from predators.* For the plant cell wall, one function can be determining the shape of a cell, whereas some possible biological roles could be, for instance, transport of water and nutrients, forming a physical barrier to avoid invasion of microorganisms, or being a reserve of carbon for growth.

4 Many animals, although not aquatic, display sulfur-containing polymers in their extracellular matrices. These matrices are exposed to saline environments inside their bodies. In the case of red algae, the majority occur in the sea and are exposed to high salt levels and a higher concentration of sulfur. They have Glycomic Codes like the plants and green algae, but sulfur-containing polymers probably evolved to adapt to a high-salt (high sulfur) environment.

5 The structure of something; the parts of something that hold it together and make it what it is (from the Cambridge dictionary).

BIBLIOGRAPHY

Atmodjo MA, Hao Z, Mohnen D (2013) Evolving views of pectin biosynthesis. *Annual Reviews of Biology* 64:747–779.

Avci, U, Pattathil S, Hahn MG (2012) Immunological approaches to plant cell wall and biomass characterization: glycome profiling. In: M Himmel (Ed.) *Biomass conversion: methods and protocols.* Methods in molecular biology, vol 908. Humana Press, New York, pp. 73–82.

Avci U, Peña MJ, O'Neill MA (2018) Changes in the abundance of cell wall apiogalacturonan and xylogalacturonan and conservation of rhamnogalacturonan II structure during the diversification of the Lemnoideae. *Planta* 247:953–997.

Awano T, Takabe K, Fujita M (2001) Xylan and lignin deposition on the secondary wall of *Fagus crenata* fibers. *Progress in Biotechnology* 18:137–142.

Bar-Peled M, O'Neill MA (2011) Plant nucleotide sugar formation, interconversion, and salvage by sugar recycling. *Annual Review of Plant Biology* 62:127–155.

Barbieri M (2015) *Code Biology, a New Science in Life* vol. 1. Springer, New York. 224p.

Bauer WD, Talmadge KW, Keegstra K, Albersheim P (1973) The structure of plant cell walls. II the hemicellulose and the walls of suspension-cultured sycamore cells. *Plant Physiology* 51:174–187.

Beckmann E, Liesche O, Lehmann F (1923) Qualitative und quantitative unterschiede der lignin einiger holz- und stroharten. *Biochemisch Zeitshift*, 139: 491–508.

Brown DM, Goubet F, Wong VW, Goodacre R, Stephens E, Dupree P, Turner SR (2007) Comparison of five xylan synthesis mutants reveals new insight into the mechanisms of xylan synthesis, *Plant Journal* 52:1154–1168.

Buckeridge MS, Reid JSG (1994) Purification and properties of a novel ß-galactosidase or exo-ß-(1,4)-galactanase from the cotyledons of germinated Lupinus angustifolius L. seeds. *Planta* 192:502–511.

Buckeridge MS, Dietrich SMC (1996) Mobilisation of the raffinose family oligosaccharides and galactomannan in germinating seeds of Sesbania marginata Benth. (Leguminosae-Faboideae). *Plant Science* 117:33–43.

Buckeridge MS, Crombie HJ, Mendes CJM, Reid JSG, Gidley MJ, Vieira CJ (1997) A new family of xyloglucan oligosaccharides from cotyledons of *Hymenaea courbaril*, *Carbohydrate Research* 303:233–237.

Buckeridge MS, Vergara CE, Carpita NC (1999) The mechanism of synthesis of a cereal mixed-linkage (1→3),(1→4)β-D-glucan: evidence for multiple sites of glucosyl transfer in the synthase complex. *Plant Physiology* 120:1105–1116.

Buckeridge MS, Dietrich SMC, Lima DU (2000) Galactomannans as the reserve carbohydrate in legume seeds In: A.K. Gupta & N. Kaur (Eds.) *Carbohydrate reserves in plants: synthesis and regulation*. Elsevier, Amsterdam p. 283–316.

Buckeridge MS, Rayon, Urbanowicz B, Tiné MAS, Carpita NC (2004). Mixed linkage (1-3),(1-4)-beta-D-glucans of grasses. *Cereal Chemistry* 81:115–127.

Buckeridge MS, Hutcheon I, Reid JSG (2005) The role of exo-(1→4)-ß-galactanase in the mobilisation of polysaccharides from the cotyledonary cell walls of Lupinus angustifolius L. following germination. *Annals of Botany* 96:435–444.

Buckeridge MS (2006) Implications of emergence, degeneracy and redundancy for the modeling of the plant cell wall. In: Takahisa. Hayashi. (Org.). *The Science and the Lore of the Plant Cell Wall: Biosynthesis, Structure and Function*. Boca Raton: BrownWalker Press, Boca Raton: p. 41–47.

Buckeridge MS (2010) Seed cell wall storage polysaccharides: models to understand cell wall biosynthesis and degradation. *Plant Physiology* 154:1017–1023.

Buckeridge MS, De Souza AP (2014) Breaking the "glycomic code" of cell wall polysaccharides may improve second-generation bioenergy production from biomass, *Bioenergy Research* 7:1065–1073.

Buckeridge MS, Dos Santos WD, Tiné MAS, De Souza AP (2015). Chapter 3. The Cell Wall Architecture of Sugarcane and its Implications to Cell Wall Recalcitrance. In: E. Lam, H. Carrer, J.A. da Silva and C. Kole (Eds.). *Compendium of Bioenergy Plants - Sugarcane*. CRC Press. p. 31–50, Print ISBN: 978-1-4987-4338-9, eBook ISBN: 978-1-4987-4339-6.

Buckeridge MS (2018) The evolution of the glycomic codes of extracellular matrices. *Biosystems* 164:112–120.

Buckeridge MS (2023) The diversity of plant carbohydrate hydrolysis in nature and technology. In: Rosana Goldback and Patricia Poletto. (Org). *Polysaccharide-degrading biocatalysts*. Elsevier Inc. p. 55–74.

Burr SJ, Fry SC (2009) Extracellular cross-linking of maize arabinoxylans by oxidation of feruloyl esters to form oligoferuloyl esters and ether-like bonds. *Plant Journal* 58:554–567.

Busse-Wicher M, Gomes TC, Tryfona T, Nikolovski N, Stott K, Granthan NJ, Bolam DN, Skaf M, Dupree P (2014) The pattern of xylan acetylation suggests xylan may interact with cellulose microfibrils as a twofold helical screw in the secondary plant cell wall of *Arabidopsis thaliana*. *Plant Journal* 79:492–506.

Carpita NC (1984) Cell wall development in maize coleoptiles. *Plant Phisiology* 76:205–212.

Carpita NC, Gibeaut DM (1993) Structural models of primary cell walls in flowering plants: consistency of molecular structure with the physical properties of the cell wall during growth. *Plant Journal* 3:1–30.

Cobb M (2015) Life's greatest secret: the race to crack the genetic code. Basic Books, New York.

Danna K, Nathans D (1971) Specific cleavage of Simian virus 40 DNA by restriction endonuclease of hemophilus influenzae. *Proceedings of the National Academy of Science of the United States of America* 68:2913–2917.

De Souza AP, Leite DCC, Pattathil S, Hahn MG, Buckeridge MS (2013) Composition and structure of sugarcane cell wall polysaccharides: implications for second-generation bioethanol production. *Bioenergy Research* 6:564–579.

Dea ICM, Clark AH, McCleary BV (1986) Effect of galactose-substitution-patterns on the interaction properties of galactomannas. *Carbohydrate Research* 147 275-294.

Doblin MS, Pettolino FA, Wilson SM, Campbell R, Burton A., Fincher, GB, Newbigin E, Bacic A (2009) A barley cellulose synthase-like CSLH gene mediates (1,3;1,4)-beta-D-glucan synthesis in transgenic Arabidopsis. Proceedings of the National Academy of Science of the United States of America, 106:5996–6001.

Ding SY, Himmel ME 2006. The maize primary cell wall microfibril: a new model derived from direct visualization. Journal of Agriculture and Food Chemistry 54: 597–606.

Fanutti C, Gidley MJ, Reid JSG (1993) Action of a pure xyloglucan endotransglycosylase formerly called xyloglucan-specific endo-1-4-β-D-glucanase from the cotyledons of germinated nasturtium seeds *Plant Journal* 3:691–700.

Fauré R, Courtin CM, Delcour JA, Dumon C, Faulds CB, Fincher GB, Fort S, Fry SC, Halila S, Kabel MA, Pouvreau L, Quemener B, Rivet A, Saulnier L, Schols HA, Driguez H, O'Donohue J (2009) Brief and informationally rich naming system oligosaccharide motifs of heteroxylans found in plant cell wals. *Australian Journal of Chemistry* 62:533–537.

Franková L, Fry S (2012) Trans-α-xylosidase, a widespread enzyme activity in plants, introduces (1→4)-α-D-xylobiose sid e-chains into xyloglucan structures. *Phytochemistry* 78:29–43.

Fry SC, York WS, Albersheim P, Darvill A, Hayashi T, Joseleau JP, Kato Y, Lorences EP, Maclachlan GA, McNeil M, Mort AJ (1993) An anambiguous nomenclature for xyloglucan-derived oligosaccharides. *Physiologia Plantarum* 89:1–3.

Goycoolea FM, Cárdenas A (2003) Pectins from Opuntia spp.: a short review. *Journal of the Professional Association for Cactus Development* 5:17–19.

Hakomori SI (1964) A rapid permethylation of glycolipid and polysaccharide catalyzed by methylsulfinyl carbanion and dimethyl sufoxide. *Journal of Biochemistry* 55:295–208.

Hayashi T (1989) Xyloglucans in the primary cell wall, *Annual Review of Plant Physiology and Plant Molecular Biology* 40:139–16.

Hayashi T (2006) *The Science and the Lore of the Plant Cell Wall: Biosynthesis, Structure and Function*. BrownWalker Press, Boca Raton: p.41–47.

Hudson CS (1941) Emil Fischer's discovery of the configuration of glucose. *Journal of Chemical Education*, P 252–357.

Jacobs A, Dahlman O (2001) Characterization of the molar masses of hemicelluloses from wood and pulps employing size exclusion chromatography and matrix-assisted laser desorption ionization time-of-flight mass spectrometry. *Biomacromolecules* 2:894–905.

Keegstra K, Talmadge W, Bauer WD, Albersheim P (1973) The structure of the Plant Cell wall. III a model of the walls of suspension-cultured sycamore cells based on the interconnections of the macromolecular components. *Plant Physiology* 51:188–196.

Kooiman P (1961) The constitution of *Tamarindus amyloid*, *Recueil des Travaux Chimiques des Pays-Bas* 80:849–865.

Kim JS, Awano T, Yoshinaga A, Takabe K (2010) Immunolocalization and structural variations of xylan in differentiating earlywood tracheid cell walls of *Cryptomeria japonica*, *Planta* 232:817–824.

Kim JS, Awano T, Yoshinaga A, Takabe K (2011) Occurrence of xylan and mannan polysaccharides and their spatial relationship with other cell wall components in differentiating compression wood tracheids of *Cryptomeria japonica*, *Planta* 233:721–735.

Knox JP (2008) Revealing the structural and functional diversity of plant cell walls, *Current Opinion in Plant Biology* 11:308–313.

Lee CH, Zhon RQ, Richardson EA, Himmelsbach DS, McPhail BT, Ye ZH (2007) The PARVUS gene is expressed in cells undergoing secondary wall thickening and is essential for glucuronoxylan biosynthesis, *Plant Cell Physiology* 48:1659–1672.

Levigne S, Ralet MC, Quéméner B, Thibault JF (2004) Isolation of diferulic bridges ester-linked to arabinan in sugar beet cell walls. *Carbohydrate Research* 339:2315–2319.

Levy S, York WS, Stuike-Prill R, Meyer B, Staehelin LA (1991) Simulations of the static and dynamic molecular conformations of xyloglucan. The role of the fucosylated sidechain in surface-specific sidechain folding. *The Plant Journal* 1:195–215.

Lima DU, Buckeridge MS (2001) Interaction between cellulose and storage xyloglucans: the influence of the degree of galactosylation. *Carbohydrate Polymers* 46:157–163.

Lima DU, Loh W, Buckeridge MS (2004) Xyloglucan-cellulose interaction depends on the side chains and molecular weight of xyloglucan. *Plant Physiology and Biochemistry* 42:389–394.

Martínez-Abad A, Berglund J, Toriz G, Gatenholm P, Henriksson G, Lindstrom M, Wohlert J, Vilaplana F (2017) Regular motifs in xylan modulate molecular flexibility and interactions with cellulose surfaces. *Plant Physiology* 175:1579–1592.

Mayr E (1961) Cause and effect in biology: kinds of causes, predictability, and teleology as viewed by a practicing biologist. *Science* 134:1501–1506.

Mayr E (2004) *What makes biology unique?: considerations on the autonomy of a scientific discipline.* Portuguese translation as Biologia, Ciência Única, in 2005. Editora Schwarcz LTDA. 266p.

McCann MC, Roberts K 1991. Architecture of the primary cell wall. In: CW Lloyd (Ed.). *The cytoskeletal basis of plant growth and form.* Academic Press, London. pp. 109–129.

McCartney L, Marcus SE, Knox JP (2005) Monoclonal antibodies to plant cell wall xylans and arabinoxylans. *Journal of Histochemistry and Cytochemistry* 53:543–546.

McCleary BV, Clark AH, Dea ICM, Rees DA (1985) The fine structures of carob and guar galactomannans. *Carbohydrate Research* 139:237–260.

Nawaz MA, Rehman HM, Imtiaz M, Baloch FS, Lee JD, Yang SH, Lee SI, Chung G (2017) Systems identification and characterization of cell wall reassembly and degradation related genes in *Glycine max* (L.) Merill, a bioenergy legume. *Scientific Reports* 7:10862. doi:10.1038/s41598-017-11495-4.

Niklas K (1994) *Plant Allometry: The scaling of form and process.* Univ of Chicago Press. Chicago and London. 395p.

Pagliuso D, Grandis A, Lam E, Buckeridge MS (2020) High saccharification, low lignin, and high sustainability potential make duckweeds adequate as bioenergy feedstocks. *Bioenergy Research* 14:1082–1092.

Pagliuso D, Navarro BV, Grandi A, Zerillo MM, Lam E, Buckeridge, MS (2022) NDP-sugar pathways overview of Spirodela polyrhiza and their relevance for bioenergy and biorefinery. *Bioenergy Research* 15:1531–1544. doi:10.1007/s12155-021-10355-4.

Peña MJ, Ryden P, Madson M, Smith AC, Carpita NC (2004) The galactose residues of xyloglucans are essential to maintain mechanical strength of the primary cell walls of Arabidopsis during growth. *Plant Physiology* 134:443–451.

Persson S, Cffall KH, Freshour G, Hilley MT, Bauer S, Pondexter P, Hahn MG, Mohnen D, Somerville C (2007) The Arabidopsis irregular xylem8 mutant is deficient in glucuronoxylan and homogalacturonan, which are essential for secondary cell wall integrity. *Plant Cell* 19:237–255.

Reid JSG, Bewley JD (1979) A dual role for the endosperm and its galactomannan reserve in the germinative physiology of fenugreek (Trigonella foenum-graecum L.): an endospermic leguminous seed. *Planta* 147:145–150.

Reis R, Vian B (2004) Helicoidal pattern in secondary cell walls and possible role of xylans in their construction, *Comptes Rendus: Biologies* 327:785–790.

Roberts RJ (2005) How restriction enzymes became the workhorses of molecular biology. *Proceedings of the National Academy of Science of the United States of America* 102:5905–5908.

Roulin S, Feller U (2001). Reversible accumulation of (1→3,1→4)-β-glucan endohydrolase in wheat leaves under sugar depletion. *Journal of Experimental Botany* 52:2323–2332.

Silva GB, Ionashiro M, Carrara TB, Crivellari AC, Tiné MAS, Prado J, Carpita NC, Buckeridg MS (2011) Cell wall polysaccharides from fern leaves: evidence for a mannan-rich type III cell wall in Adiantum raddianum. *Phytochemistry* 72: 2352–2360.

Schultink A, Liu L, Zhu L, Pauly M (2014) Structural diversity and function of xyloglucan sidechain substituents. *Plants* 3:526–542.

Stancato GC, Mazzafera P, Buckeridge MS (2001) Effect of a drought period on the mobilisation of non structural carbohydrates, photosynthetic activity and water status in the epiphytic orchid Cattleya forbesii Lindl. X Laelia tenebrosa Rolfe. *Plant Physiology and Biochemistry* 39:1009–1016.

Sun R, Lawther JM, Banks WB (1995) Influence of alkaline pre-treatments on the cell wall components wheat straw. *Industrial Crops and Products* 4:127–145.

Sundar Raj AA, Rubila S, Jayabalan R, Ranganathan TV (2012) A review on pectin: chemistry due to General Properties of Pectin and its Pharmaceutical Uses. *Scientific Reports* 1:550 doi:10.4172/scientificreports.550.

Talmadge K, Keegstra K, Bauer WD, Albersheim P (1973) The structure of plant cell walls. I. The macromolecular components of the walls of suspension-cultured sycamore cells with a detailed analysis of the pectic polysaccharides. *Plant Physiology* 51:158–173.

Tavares EQP, Buckeridge MS (2015) Do cell walls have a code? *Plant Science* 241:286–294.

Tiné MAS, Lima DU, Buckeridge MS (2003) Galactose branching modulates the action of cellulase on seed storage xyloglucans. *Carbohydrate Polymers* 52:135–141.

Tiné MAS, Silva COS, Lima U, Carpita NC, Buckeridge MS (2006) Fine structure of a mixed-oligomer storage xyloglucan from seeds of *Hymenaea courbaril*. *Carbohydrate Polymers* 66:444–454.

Van Fleet, DS (1961). Histochemistry and function of the endodermis. *The Botanical Review*, 27(2), 165–220.

Verbančič J, Lunn JE, Stitt M, Persson S (2017) Carbon supply and the regulation of cell wall synthesis. *Molecular Plant* 11:75–94.

Verhertbruggen Y, Marcus SE, Haeger A, Verhoef R, Schols HA, McCleary BV, McKee L, Gilbert HJ, Knox JP (2009) Developmental complexity of arabinan polysaccharides and their processing in plant cell walls. *The Plant Journal* 59:413–425.

Wang Y, Fan C, Hu H, Li Y, Sun D, Wang Y, Peng L (2016) Genetic modification of plant cell walls to enhance biomass yield and biofuel production in bioenergy crops. *Biotechnology Advances* 34:997–1017.

West G (2017) *Scale. The universal laws of growth, innovation, sustainability and the pace of life in organisms, cities, economies and companies.* Penguin Press, New York. 527p.

Woodward JR, Fincher GB, Stone BA (1983) Water-soluble (1→3), (1→4)-β-D-glucans from barley (*Hordeum vulgare*) endosperm II. Fine structure. *Carbohydrate Polymers* 3:207–255.

Wrixon FB (2005) *Codes, ciphers, secrets and cryptic communication.* Black Dog and Leventhal Pub. New York. 704p.

Zamil MS, Geitmann A (2017) The middle lamella - more than a glue. *Physical Biology* 14: 015004.

Zhang B, Gao Y, Zhang L, Zhou Y (2020) The plant cell wall: biosynthesis, construction, and functions. *Journal of International Plant Biology* 63:251–272.

Zolyan S (2022) From matter to form: the evolution of the genetic code as semio-poiesis. *Semiotica*, 245:17–61.

Zudnek A, Pieczywek PM, Cybulska J (2021) The primary, secondary and structures of higher levels of pectin polysaccharides. *Comprehensive Reviews in Food Science and Food Safety* 20:1101–1117.

13 We All Stand Together
An Exploration of the Drivers and Constraints That Have Shaped Plant and Algal Cell Wall Diversity

Zoë A. Popper
University of Galway

INTRODUCTION

Possession of polysaccharide-rich cell walls is the one feature that links the phototrophic eukaryotic groups (i.e. brown algae, glaucophytes, red and green algae, and land plants), lineages that last shared a common ancestor more than 1,600 Million Years Ago (MYA) (Yoon et al., 2004) and have diversified significantly since. However, a unifying constant is the importance of cell walls to every aspect of plant and algal physiology (Albersheim et al., 2010).

Many of the prevailing simplified cell wall models (discussed in detailed by Ulvskov and Ray in Chapter 11 of this volume), although challenged over the last 20 years (Zhang et al., 2019; Park and Cosgrove, 2012a), depict a wall composed of cellulose microfibrils that are hydrogen-bonded to polysaccharides, that can act as tethers, and that are embedded in pectins in plants (Silva et al., 2011; Carpita and Gibeaut, 1993; McCann and Roberts, 1992 Keegstra et al., 1973), or alginates, in some brown algae (Deniaud-Bouët et al., 2014). However, covalent links between tethering polysaccharides, such as xyloglucans, and pectins are widespread among flowering plants (Peña et al., 2008; Popper and Fry, 2005; Thompson and Fry, 2000), and enzymes have been characterised that can covalently link xyloglucans to pectins, and other hemicelluloses, including mannans and (1→3, 1→4)-β-D-glucan, and even to cellulose (Franková and Fry, 2021; Herburger et al., 2021; Stratilová et al., 2020a; Herburger et al., 2018). This means that cell wall constituents are physically highly interconnected through a variety of bonds and interactions. The different types of bonds and their relationship to wall mechanical properties and physiological functions such as growth were recently reviewed by Cosgrove (2022). Many wall properties depend on more than one cell wall component. For example, tensile strength, important during expansive cell growth, results from a xyloglucan–cellulose network and from rhamnogalacturonan II-borate complexes (Ryden et al., 2003). With this degree of interconnectivity and complexity, it could be argued that the cell walls of photosynthetic organisms are at least as complex as the vertebrate eye; an organ once considered so complex that the evolutionary processes that generated it were difficult to envisage (Young, 2008). In contrast, the dynamic, diverse, and adaptable nature of cell walls is evident. Changes occur in cell wall composition during development (Chen et al., 2019; Somssich et al., 2016; Seale 2020), in response to different environmental challenges, including water- and temperature- stresses, and wounding by herbivores and pathogens (Chapter 22, Chen et al., 2020; Ahl et al., 2019; Herger et al., 2019; Cosgrove, 2018; Moore and Farrant, 2017; Houston et al., 2016; Le Gall et al., 2015; Voigt, 2014; Franková and Fry, 2013; Moore et al., 2006; Aist, 1976). Specific organs, tissues, cells, and even parts of cells are known to have distinct cell wall compositions (Shtein et al., 2017; Pendle and Benitez-Alfonso, 2015; Eeckhout et al., 2014).

Cell wall diversity has long been used as an important character in algal classifications and, although there are significant limitations in our current understanding, some variation in algal cell wall composition appears to have taxonomic relevance (Stace, 1989). For example, the red algal (Rhodophyta) families that synthesise typical carrageenans in their cell walls form a monophyletic group (Freshwater et al., 1994; Kloareg and Quatrano, 1988). Stebbins (1992) stated that 'even the most superficial survey of ecological plant anatomy points to the adaptive importance of cell wall differentiation' and the evolutionary significance of cell wall diversity has been increasingly explored in land plants and their closest extant ancestors, and has been discussed in many publications in the last ~25 years. For general reviews of cell wall evolution, the reader may like to refer to Fangel et al. (2012), Popper et al. (2011), Popper and Tuohy (2010), Sørensen et al. (2010), Sarkar et al. (2009), and Popper (2008). Recent evidence regarding the occurrence and evolution of xyloglucan, and xyloglucan-modifying enzymes, is detailed in Mikkelsen et al. (2021) and Franková and Fry (2021), and insight into cell wall evolution gained from the genome of *Penium margaritaceum*, a close extant relative of land plants, is explored in Jiao et al. (2020). Therefore, rather than focusing on documenting the many taxonomic differences, some of which are summarised in Figure 13.1, this chapter seeks to explore some of the key drivers and constraints resulting in wall diversity.

DOI: 10.1201/9781003178309-13

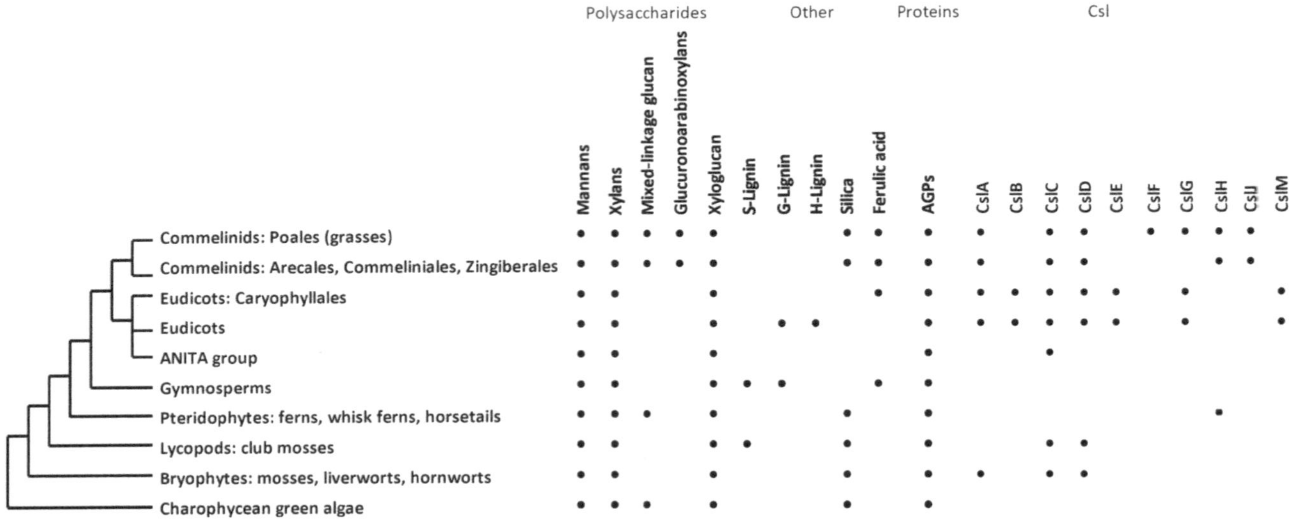

FIGURE 13.1 A simplified phylogeny of the green land plant lineage highlighting the occurrence of selected cell wall components.

The potential for cell walls to be highly diverse is largely derived from being carbohydrate-based; polysaccharides make up over 90% of the dry weight of primary cell walls and over 60% of the dry weight of secondary cell walls (Bacic et al., 1988). There are more than ten major sugar residues that commonly occur in plant cell wall components (Albersheim et al., 2010). However, each sugar residue can be present in multiple forms (e.g. isomers and enantiomers), modified by acetylation and/or methylation, and the glycosidic bonds that link sugar residues within a polymer can be at several positions, and to different sugar residues (Albersheim et al., 2010). The extent to which being carbohydrate-based can facilitate diversity is highlighted by Prestegard et al. (2017) who were able to construct 1,792 distinct tetrasaccharides from a single sugar residue in ring form. Biosynthesis of polysaccharides (and their integration into the wall) is predominantly enzyme-controlled with typically each different glycosidic linkage type synthesised by a specific enzyme. The biosynthesis of pectin alone, although a particularly complex cell wall component, is thought to require the activities of a minimum of 67 distinct enzymes (Atmodjo et al., 2013). Therefore, although the rate of spontaneous mutation in plant genomes is controversial (Monroe et al., 2020), it is likely to play an important role in generating cell wall diversity.

There is an emerging recognition of the primary role of the cell wall in cell mechanics, with the plasma membrane as a mediator between the cell wall, to which it is directly connected, and the cell interior (Codjoe et al., 2022). Cell walls have been described as being "viscoelastoplastic" (Baybrook and Jönsson, 2016), and spatio-temporal differences in the mechanical properties of cell walls are implicated in plant development. Codjoe et al. (2022) recently discussed how the viscoelastoplastic performance of the cell wall, resulting from its component polymers, their interconnections, and metabolism including synthesis and enzymic modifications (see Chapter 4 for a comprehensive discussion of enzymic and non-enzymic wall remodelling), lead to "tuneable" dynamic changes in wall mechanical properties in response to developmental and environmental stress (Anderson and Kieber, 2020). It is tempting to extrapolate from this that plant cell wall composition may have been altered at key points in plant evolution. For example, enabling mechanical homeostasis in a changed environment, such as the altered osmotic environment, experienced in the transition to land. Indeed, cell wall components themselves may act as mechanosensors. Insight into this role was recently shown by Miller et al. (2022) who used a combined approach including mathematical modelling and *in vitro* laboratory experiments and showed that the pollen-specific mechanosensitive ion channel MscS-like (MSL)8, which plays essential roles during pollen hydration, acts by altering pollen cell wall strength in response to ion flux. Osmoregulation is important for many aspects of plant physiology in addition to pollen hydration, including germination. Adapting to a changed osmotic environment was a critical factor in terrestrialisation. It is therefore interesting to speculate whether wall components similar to MSL(8) may have been actors in this transition.

Given that individual cell wall components do not function independently, the degree to which cell wall components are interconnected, and that to maintain cell wall functions it is essential that cell wall integrity is maintained (Gigli-Bisceglia et al., 2020; Rui and Dinneny, 2019; Vaahtera et al., 2019), it is interesting to consider how much freedom individual cell wall components have to evolve and to consider what constraints and drivers might have had the most impact on cell wall composition. One way to do this is to explore the potential evolutionary mechanisms that have resulted in the cell wall compositions of the different extant plant and algal groups, and individual wall components.

What Evolutionary Mechanisms Might Have Shaped Cell Wall Composition?

There are several evolutionary scenarios, each of which leave different hallmarks allowing them (at least in some cases), to be distinguished from each other. The hopeful monster theory (or discontinuous variation) proposes that macromutations leading to significant differences in the organisms that carry them play a major role in evolution (Goldschmidt, 1940). Discontinuous variation can be recognised by the sudden emergence of a novel character in the tree of life. At the time the theory was proposed, it was in startling contrast to the prevailing theory which was that evolutionary change is a slow, steady, gradual, and continuous process brought about by natural selection (the environment) acting on random variation (Darwin, 1859). This type of process can be identified by the continuous presence of a character within a lineage. Diversification may also occur whereby a gene is duplicated and results in a new character. However, the new gene and character tend to share some characteristics with the duplicated gene. Other, more recently recognised mechanisms include Horizontal Gene Transfer (HGT) or Lateral Gene Transfer whereby genetic information is moved between organisms other than the direct transmission of DNA from parent to offspring during reproduction. HGT is well documented as the mechanism by which antibiotic resistance genes are spread between microbes, and it is widely accepted that HGT plays an important role in the evolution of Archea and bacteria (Boto, 2009). However, although it is considered that the transfer of a single or a few genes may give the recipient organisms a new function, and potentially the ability to exploit new niches (Fournier and Gogarten, 2008), the importance of HGT in the evolution of eukaryotes is controversial (Boto, 2009). Ma et al. (2022) recently screened the genomes of a selection of charophytes and land plants with the aim of identifying genes acquired by HGT. Their results suggest that HGT has played an important role in land plant evolution, where it appears to be more abundant in the charophyte and bryophyte genomes than those of the seed plants. HGT has also shaped the cell wall. Specifically, the land plant lineage appears to have acquired a suite of genes involved in cell wall biosynthesis and metabolism including galacturonic acid phosphate kinase, xyloglucan endotransglycosylase, α-L-fucosidase, rhamnogalacturonan lyase, and pectin esterase (Ma et al., 2022). Additionally, convergent evolution, known as homoplasy in cladistics, is the independent evolution of similar or analogous structures that may have similar form and/or function, but were not present in the last common ancestor of those groups.

This chapter discusses the evolutionary mechanisms and selection pressures that may have played important roles in shaping selected cell wall components and the overall composition of the cell walls of extant photosynthetic organisms. It is important to note that more than one evolutionary mechanism may have shaped a specific cell wall component. For example, recent evidence suggests that the pectin esterase gene family, responsible for pectin restructuring, e.g., during development and fruit ripening, was probably acquired by HGT from bacteria and subsequently amplified in land plants (Ma et al., 2022).

BIG DIFFERENCES IN CELL WALL COMPOSITION: THE GHOSTS OF HOPEFUL MONSTERS

Mixed-Linkage Glucan

Although several reports tantalisingly suggested the presence of the polysaccharide mixed-linkage (1→3, 1→4)-β-D-glucan in tobacco protoplasts (Hensel and Franz, 1988), and mung beans (Buchala and Franz, 1974; Franz, 1972), the evidence was unsupported (Takeuchi and Komamine, 1981; Kato and Masuda, 1976) and until recently (Fry et al., 2008a; Sørensen et al., 2008) it was widely thought that among vascular plants the polysaccharide occurred only in the Poales (Trethewey et al., 2005; Smith and Harris, 1999). However, in 2008, two landmark papers (Fry et al., 2008a; Sørensen et al., 2008) concurrently reported that mixed-linkage (1→3, 1→4)-β-D-glucan is present in *Equisetum* (horsetail) cell walls.

Mixed-linkage (1→3, 1→4)-β-D-glucan was shown to occur in extracts prepared from *Equisetum arvense* cell walls by glycan profiling (Sørensen et al., 2008), a high throughput technique (Moller et al., 2007), using a monoclonal antibody, BG1 (BS 400–2, Biosupplies Australia, Parkville, Victoria, Australia), which is highly specific for mixed-linkage (1→3, 1→4)-β-D-glucan (Meikle et al., 1994). Immunolabelling with the same monoclonal antibody showed that the polysaccharide localised to the cell walls of the epidermis and pith parenchyma and in the middle of the stem regions, in the cortical parenchyma, and the sclerenchyma cells that are found at top of the ridges (Sørensen et al., 2008). The vascular bundles did not appear to be labelled (Sørensen et al., 2008). The enzyme lichenase (E.C.3.2.1.73) is specific for mixed-linkage glucan and hydrolyses a (1→4)-linkage where the D-glucose residue is substituted at position 3 (Meikle et al., 1994). Lichenase used as part of an enzyme-based quantification method (McCleary and Glennie-Holmes, 1985) showed that mixed-linkage (1→3, 1→4)-β-D-glucan was not only present in *Equisetum* cell walls but that it was highly abundant, comprising up to 4.6% w/w of the total cell wall material (Fry et al., 2008a; Sørensen et al., 2008). Interestingly, although *Equisetum* mixed-linkage (1→3, 1→4)-β-D-glucan was recognised *in muro* and *in vitro* by both the highly specific anti-mixed-linkage glucan monoclonal antibody (Meikle et al., 1994) and the enzyme lichenase, the polysaccharide is structurally quite different from mixed-linkage (1→3, 1→4)-β-D-glucan from the Poales (Fry et al., 2008a; Sørensen et al., 2008). Mixed-linkage (1→3, 1→4)-β-D-glucan from *Equisetum* and Poales were found to differ in the proportion and distribution of the 1→3 and 1→4 linkages that they possess (Fry et al., 2008a) and may therefore differ in physical characteristics such as their stiffness.

The strong adherence to cellulose microfibrils suggests that *Equisetum* mixed-linkage (1→3, 1→4)-β-D-glucan is structurally important, potentially acting as a tether between cellulose microfibrils (Fry et al., 2008b). Fry et al. (2008b) discovered a novel enzyme, mixed-linkage β-glucan:xyloglucan endotransglucosylase (MXE), present in all species of *Equisetum* that they investigated and that is capable of grafting mixed-linkage (1→3, 1→4)-β-D-glucan to xyloglucan oligosaccharides. In contrast to xyloglucan endotransglucosylase activity, which grafts xyloglucan to xyloglucan oligosaccharides and is typically highest in young, actively growing tissues where it is involved in incorporating newly synthesised xyloglucan into the wall and restructuring the wall as it expands (Fry, 2004), the novel MXE activity was seen to be highest in older tissues where it may play a role in strengthening and repairing mature walls (Fry et al., 2008b).

The discovery that *Equisetum* possesses (1→3, 1→4)-β-D-glucan, and enzymes involved in its metabolism, was surprising because the genus has a very complex evolutionary history that places it within the ferns (Christenhusz et al., 2021). The genus, considered to be the oldest extant vascular plant genus, has a nearly continuous fossil record dating back to the early Carboniferous (~343 MYA). The other main (1→3, 1→4)-β-D-glucan-containing plant group, the Poales (the subject of Chapter 9), is thought to have evolved 70–55 MYA in the Paleocene (Kellogg, 2001; Jacobs et al., 1999), although more recent studies suggest an older origin. For example a recent study based on molecular clock analyses estimated the crown age of Poaceae, one of the families within the class Poales, to be 101 MYA (Huang et al., 2022). However, either scenario places significant time and evolutionary distance between *Equisetum* and the Poales, and suggests that (1→3, 1→4)-β-D-glucan evolved independently in each group, i.e., its presence in both groups is a result of convergent evolution. This is hinted at by the structural and temporal differences in the polysaccharide that have been reported (Fry et al., 2008a,b; Sørensen et al., 2008). Furthermore, whilst it is well documented that the cellulose-synthase-like (Csl) gene families, CslF and CslH, occur in the Poales (Doblin et al., 2009; Burton et al., 2006,2008; Richmond and Somerville, 2000), and are involved in (1→3, 1→4)-β-D-glucan-synthesis, their occurrence in *Equisetum* is open to question since no members of the genus appear to have yet been fully sequenced.

In an effort to determine whether the machinery responsible for (1→3, 1→4)-β-D-glucan biosynthesis, namely CslF and CslH, might occur in *Equisetum*, we currently have to resort to inferences derived from the information available for the most closely related plants, the spikemosses and ferns. The earliest extant vascular plant that has a fully sequenced genome is the spikemoss *Selaginella moellendorffii* and orthologues of CslF and CslH were not identified (Harholt et al., 2012; Banks et al., 2011). Genome-level information for ferns is limited, probably because of their enormous size. *Ceratopteris richardii* is widely regarded as a model fern (Renzaglia and Warne, 1995; Hickok et al., 1995) including for studying fern cell walls (Eeckhout et al., 2014; Leroux et al., 2013a). The *Ceratopteris* transcriptome is not yet complete (Marchant et al., 2019) but has recently been annotated and although several orthologues of Arabidopsis genes relating to cell wall biosynthesis, assembly, and modification are highlighted, none are unambiguously annotated as being either CsLFs or CslHs (Geng et al., 2021). However, Yin et al. (2014) surveyed the genomes and transcriptomes of plants and algae for members of the cellulose synthase superfamily. Their study included transcriptome and Expressed Sequence Tag (EST) data for three ferns, *Ceratopteris richardii*, *Pteridium aquilinum*, and *Adiantum capillus-veneris*. Yin et al. (2014) found that these three ferns have representative genes from CesA, CslA, CslC, CslD, CslE, CslB, and CslH families, but CslF was not identified. This led to the suggestion that CslH diverged after the spikemosses split from the other vascular plants, but before the ferns appeared approximately 400 MYA in the early-mid Devonian (Yin et al., 2014; Pryer et al., 2004).

The likely independent origins of (1→3, 1→4)-β-D-glucan in two distinct groups lead to the question of whether it evolved in *Equisetum* and in the Poales in response to similar selective pressures. The divergences of extant members of *Equisetum* coincide with the breakup of the supercontinent Pangea, approximately 200 MYA, at the end of the Triassic (Christenhusz et al., 2021). It is thought that this dramatically altered the global climate leading to increases in both temperature and humidity (Trappe 2000). Similarly, an abrupt, transient, warm event of ~200,000 years, the Palaeocene-Eocene Thermal Maximum approximately 56 MYA (Kennett and Stott 1991), corresponds with one of the possible time frames for the divergence of the Poales. An earlier divergence of ~100 MYA, in the Cretaceous, as more recently suggested, similarly correlates with a warm climate; the Earth was thought to be ~10°C warmer then than it is today. Many extant grasses thrive in dry open habitats and are drought tolerant (Kellogg 2001). Additionally, C4 photosynthesis, which is more efficient at higher temperatures, has evolved independently several times in the Poales with the grass family, i.e., the Poaceae, accounting for 60% of all C4 plants (Sage, 2004; Kellogg, 2001; Jacobs et al., 1999). The evolution of (1→3, 1→4)-β-D-glucan, at least in *Equisetum* and Poales, could therefore be linked to warmer conditions where its presence may have conferred a survival advantage. It has also been noted that *Equisetum* and the Poales, alongside the presence of (1→3, 1→4)-β-D-glucan, share many morphological characteristics likely arrived at by convergent evolution (Fry et al., 2008b).

It appears that (1→3, 1→4)-β-D-glucan is more widely distributed among monocots than was previously thought. Although it is consistently present in all members of the Poales, in which it is synthesised by members of the CslF, CslH and CslJ families (Little et al., 2018; Farrokhi et al., 2006), immunolabelling with BG1 reveals the presence of the polysaccharide in several commelinid and non-commelinid monocots and is correlated with the broader distribution of CslH and CslJ (Bulone et al., 2019; Little et al., 2018). Monocots are a monophyletic group, meaning that they have a single common ancestor (APG IV 2017). Since (1→3, 1→4)-β-D-glucan and CslH and CslJ have been detected in Acoraceae (Little et al., 2018), this places its origins in this group to ~135–100 MYA, in the late Jurassic to mid-Cretaceous when monocots are thought to have diverged from basal Angiosperms and most of the monocot lineages diverged (Hertweck et al., 2015). Monocots are defined by four key features: (i) sieve tube plastids with wedge-shaped proteinaceous inclusions; (ii) an atactostele stem vasculature, i.e., numerous discrete vascular bundles arranged in rings; (iii) parallel leaf venation (although there are some notable exceptions); and (iv) a single cotyledon (Simpson, 2010). These features are thought to be linked to optimising growth in the marshy or aquatic habitats in which the first monocots are considered to have evolved (Givnish et al., 2018). Since the extant Equisetaceae also live predominantly in marshy habitats, it is tempting to speculate that (1→3, 1→4)-β-D-glucan may offer an advantage in these environments.

However, this is likely not the whole story with regard to (1→3, 1→4)-β-D-glucan evolution. Despite the fact that CslF and CslH were not identified from the *Selaginella* genome, glycan profiling revealed that root, leaf, and stem hemicellulose extracts from *Selaginella* contain epitopes (i.e., the sequence of sugar residues that a monoclonal antibody recognises and binds to) that are recognised by an anti-(1→3, 1→4)-β-D-glucan monoclonal antibody (Harholt et al., 2012). Furthermore, but perhaps less surprisingly given the presence of CslHs in ferns, the same monoclonal antibody is capable of labelling fern cell walls (Leroux et al., 2013b, 2018). Rather than a common evolutionary origin, there is a strong likelihood that (1→3, 1→4)-β-D-glucan has evolved independently several times. This is supported by its occurrence in many different lineages of plants, algae and fungi, including the pathogenic fungus *Rhynchosporium secalis* (Pettolino et al., 2009), lichen-forming Ascomycetes (Gorin et al., 1988; Stone and Clarke, 1992), the lichen *Cetraria islandica* (Honegger and Haisch, 2001), a soil-dwelling freshwater green alga, *Monodus subterraneus* (Stone and Clarke, 1992), charophyte green algae (Sørensen et al., 2011; Eder et al., 2008), and brown algae (Salmeán et al., 2017). An alternative mechanism of biosynthesis has even been suggested by the identification of a glycosyl transferase, Tft1, a glycosyl transferase necessary for the synthesis of cell wall (1→3, 1→4)-β-D-glucan in the fungus *Aspergillus fumigatus* (Samar et al., 2015). Mixed-linkage glucan-like polymers exist that share structural features leading to their digestibility with lichenase, such as those from the liverwort *Lophocolea bidentata* that contains arabinoses as well as glucoses (Popper and Fry, 2003), and label with mixed-linkage glucan specific monoclonal antibodies, such as those from the green seaweed *Ulva* (personal observation). Roberts et al. (2018) also identified a lichenase-sensitive arabinoglucan from the model moss *Physcomitrium patens*, which is synthesised by an enzyme with sequence homology to enzymes present in Ascomycetes but not grasses. This highlights that there are several evolutionary routes leading to synthesis of mixed-linkage glucan and mixed-linkage glucan-like polymers.

Silica

It is interesting to note that correlated with possessing mixed-linkage glucan or mixed-linkage glucan-like polymers as discussed above, liverworts, horsetails, and some ferns, as well as the Poales, are known to accumulate silica in their cell walls (Leroux et al., 2013b, Hodson et al., 2005; Fry et al., 2008a). Analysis of proteins involved in silicon transport and the fossil record suggest that biomineralisation of silica occurred early in land plant evolution, in the Ordovician era, associated with significant changes in plate tectonics, climate and is a period of rapid diversification (Morris et al., 2018; Conley and Carey, 2015; Trembath-Reichert et al., 2015). Antimicrobial properties have also been attributed to silica and its deposition is modulated by reactive oxygen species that are typically produced in response as part of the plant immune response to microbial attack (Zexer and Elbaum, 2022).

Lignin

Lignin, a polymer of phenylpropanoid units, is widely considered to be an innovation of vascular plants. However, the evolutionary history of lignins is far more complex than was initially perceived and is comprehensively reviewed by Novo-Uzal et al. (2012).

The origin of vascular plants approximately 430 MYA, in the Silurian, led to a ten-fold increase in plant species (Harrison and Morris, 2017), with lignins providing the mechanical support allowing plants to grow taller enhancing primary productivity as well as increasing the efficiency of water transport. This era is also associated with diversifications in fungi and insects that may, at least in part, have driven lignin diversification since lignins are also deposited as part of the defence response (Ros Barceló et al., 1997).

Lignins are composed of monolignol units, primarily *p*-coumaryl, coniferyl, and sinapyl alcohols that are precursors of the p-hydroxyphenyl (H), guaiacyl (G), and syringyl (S) lignin units. The classical view is that the lignin found in gymnosperms is composed almost entirely of G-units, with some H units, whereas flowering plant lignins contain G and S units (Novo-Uzal et al., 2012). This view was radically altered in 2013 with the report of newly discovered monomers associated with the seed coats of cacti (Chen et al., 2013), with further diversity in lignin monomers and significant advances in understanding the mechanisms of lignin biosynthesis published in 2019 (Ralph et al., 2019).

The lignins found in the earliest extant vascular plant group, the lycopods, are S-lignins, similarly to those that predominate in flowering plants. However, the presence of S-lignin in these two groups, that diverged ~400 MYA, is an example of convergent evolution as it has been discovered that biosynthesis of S-lignin in *Selaginella* is via a completely different pathway than the S-lignin found in flowering plants (Weng et al., 2008,2010).

The more surprising discovery of HGS lignins in specialised cells of the red alga, *Calliarthron cheilosporioides*, which last shared a common ancestor with vascular plants over a billion years ago is also considered to be an example of convergent evolution (Martone et al., 2009). However, the biosynthetic pathway necessary to confirm this has yet to be identified.

Pectins

Pectins and pectin-like polymers are found in red and green algae as well as land plants (Eder and Lütz-Meindl, 2008; Domozych et al., 2007a,b, Painter, 1983). The major pectins in the land plant lineage include rhamnogalacturonan I (RG-I), rhamnogalcturonan II (RG-II), and homogalcturonan (HG). Monoclonal antibody labelling suggests that all three classes of pectin are present in the cell walls of *Physcomitrium*, the model bryophyte, and members of the vascular plant lineage (Ye and Zhong, 2022). Bioinformatic analysis provides further support for this with gene homologues for galactosyl transferases, β-1,4-galactan β-1,4-galactosyltransferases, RG-II xylosyl transferases, HG methyltransferases, and pectin acetyltransferases identified in the *Physcomitrium* genome, although it should be noted that RG-I arabinosyltransferases and xyloglalacturonan xylosyltransferases appear to be absent (Ye and Zhong, 2022; McCarthy et al., 2014).

The structure of pectins is eloquently discussed in Chapter 5 of this volume. The RG-I backbone consists of a repeating sequence of the disaccharide [2-α-L-Rha(1–4)-α-D-GalA] adorned with side-chains. The side-chains are developmentally and spatially regulated, for example, during fruit ripening (Wang et al., 2019; Panigua et al., 2016; Orfila et al., 2002) and also thought to play an important role in the biomechanical properties of plant tissues (Haas et al., 2020; Cosgrove 2018). However, although RG-I is widespread in the land plant lineage (McCarthy et al., 2014; Kulkarni et al., 2012; Roberts et al., 2012), very little lineage-specific structural variation has been reported, the primary example being the presence of 3-*O*-methyl-D-galactose in pectins from charophycean green algae (O'Rourke et al., 2015). Rhamnogalacturonan I rhamnosyl transferases are involved in biosynthesis of the RG-I backbone and exhibit a significant amount of redundancy (Wachananwat et al., 2020) similarly suggesting conservation of RG-I at a taxonomic level. One RG-I side-chain that might be predicted to have a limited taxonomic occurrence are the arabinans as they have been implicated in stomatal opening (Jones et al., 2003). However, labelling with the monoclonal antibody LM6, which recognises short stretches of arabinosyl residues, suggests they are also part of *Chara* cell walls (Domozych et al., 2009) and that recruitment of arabinans in guard cell function may have been through co-option in function of a pre-existing cell wall component.

Rhamnogalacturonan II (RG-II) also appears to have originated in land plants. RG-II dimerised via its interactions with the element boron, is considered to be important for maintaining cell wall integrity and in plant growth and development; interestingly Boron requirement appears to be correlated with RG-II, with bryophytes, that contain less than 1% of the RG-II quantity found in vascular plants being far less Boron-dependant (Matsunaga et al., 2004). RG-II exhibits a high degree of conservation with the main lineage-specific differences being an increase in content in vascular plants compared to bryophytes, and the presence of 3-*O*-methyl rhamnose in RG-II from some fern species (Matsunaga et al., 2004). Although Pabst et al. (2013) report some species-level differences in monosaccharide side-chains and degree of methylation in RG-II, the overall degree of conservation hints at the fundamental role of pectins. Indeed, enzymic perturbation of pectins including HG in the green alga, *Penium*, has been shown to severely impact physiology (Domozych et al., 2014).

Although pectins are discussed here in relation to 'the ghosts of hopeful monsters', it is important to remember that more than one evolutionary mechanism may shape a single polymer. Pectins are a good example since, as mentioned in the introduction, recent evidence suggests that pectin-modifying enzymes, specifically the pectin esterases, were acquired from bacteria via HGT and subsequently expanded, potentially reflecting diversified roles and new stimuli *in planta* (Ma et al., 2022).

Mannans

The cosmopolitan distribution in red and green algae, and land plants, of wall polysaccharide xylans (Popper and Tuohy, 2010; Domozych et al., 2009a, Carafa et al., 2005; Lahaye et al., 2003) and mannans (Estevez et al., 2009; Dunn et al., 2007; Percival et al., 2001), indicates they also originated in this lineage. Heteromannans are considered to be one of the most ancient cell wall polymers in the plant kingdom and in several green algae, including *Acetabularia*, are the major cell wall polysaccharide (Voiniciuc et al., 2019; Dunn et al., 2007). Mannans, similarly to many, but not all, plant cell wall components, are synthesised by the products of genes resulting from diversification of cellulose synthases (CesAs) to Csl gene families; an expansion that appears to be tightly associated with the land plant lineage (Popper et al., 2011; Yin et al., 2009; Lerouxel et al., 2006). CslAs are the mannan synthases and members of the CslD family who also exhibit mannan synthase activity (Verhertbruggen et al., 2011; Liepman and Cavalier, 2012).

Xylans

Painter (1983) suggested that the red algae indicate an evolutionary branch point for xylans since early diverging lineages are composed of 1→3 linked xylans, whilst later diverging red algal xylans may include 1→3 and 1→4 linkages in the same molecule (Turvey and Williams, 1970), whereas xylans from green algae and land plants are composed of 1→4 linkages. In land plants, xylans and glucuronoxylans are predominantly found in the secondary cell walls of woody and herbaceous eudicots. They are synthesised by members of the glycosyltransferase family, GT43 and GT47, which is an ancient eukaryotic protein family involved in glycosylaminoglycan synthesis in animals and glucuronoxylan elongation in land plants (Lee et al., 2010; Sugahara and Kitagawa, 2002). This family, and the ability to synthesise xylans, appears to have been lost during the evolution of brown algae (Michel et al., 2010). However, a multifaceted approach, including bioinformatics, biochemistry, and immunolocalisation, suggests that plant 1,4-β-xylan originated in the charophytes, with *Klebsormidium* possessing a reduced set of enzymes required for xylan biosynthesis compared to vascular plants, and playing an important role in the transition to life on land (Jensen et al., 2018). Although structural diversity exists in monocot xylans, the presence of arabinofuranose which has long been recognised as a characteristic of commelinids (Albersheim et al., 2010) is the only feature that appears to correlate with a major transition in monocot evolution (Peña et al., 2016). The commelinids are a monophyletic group of monocots that originated ~120 MYA (Hertweck et al., 2015) and also contain ferulic acid in their cell walls which is ester-linked to the glucuronoarabinoxylans and appears to control leaf growth, development, and morphology (de O Buanafina et al., 2017; Harris and Trethewey, 2010). Among land plants, ferulic acid has also been detected in the Caryophyllales, where it is ester-linked to the side-chains of RG-I (Harris and Trethewey, 2010) and in the primary cell walls of gymnosperms (Carnachan and Harris, 2000). It is interesting to note that the Caryophyllales are thought to have originated at a similar time to the commelinids (Yao et al., 2019), suggesting that the presence of ferulic acid in both lineages may be in response to a shared environmental stress.

SOME APPARENT NOVELTIES IN WALL COMPOSITION MAY RESULT FROM HORIZONTAL GENE TRANSFER

As highlighted in 'What evolutionary mechanisms might have shaped the cell wall?', there are several known occurrences of HGT that relate to plant cell walls. Specific cell wall components appear to have been shaped by more than one evolutionary mechanism. Therefore, the example of pectate lyases is discussed within 'Big differences in cell wall composition: the ghosts of hopeful monsters' and the example of xyloglucan endotransglycosylases is discussed in 'Fine-tuning the wall: examples of gradual evolution' as pectins and xyloglucans, respectively, are discussed within these sections.

The Role of Horizontal Gene Transfer in the Transition to Life on Land

The transition to life on land is undoubtedly one of the key events in plant evolution (Rensing, 2018). Water availability is considered a primary driver of plant evolution (Bowles et al., 2022). However, phylogenetic and biochemical analyses of members of the charophycean green algae led Harholt et al. (2016) to argue that land plants 'were terrestrial from the beginning', i.e., that members of the charophycean green algae were already effectively terrestrial, and that it was a change in lifecycle, rather than adaptations to stresses such as desiccation experienced on land, that led to the success of the land plant lineage. Insight into this was gained from the genomes of subaerial Zygnematophyceae, specifically *Spirogloea muscicola*, which is the most closely related charophyte to land plants (Cheng et al., 2019). Cheng et al. (2019) found that gene innovations common to the Zygnematophyceae and embryophytes include genes involved in the biosynthesis of cell wall constituents and their remodelling. Interestingly, genes thought to be involved in desiccation resistance were shown to be gained by HGT from soil bacteria ~580 MYA, diversified to regulate processes including growth and development, leading to the evolution and diversification of land plants (Cheng et al., 2019).

Alginates

Alginates were the first clear example of cell wall polymers known to be synthesised by enzymes acquired through HGT (Michel et al., 2010). As has been previously mentioned, the cell wall compositions of algae are highly diversely related to the fact that 'algae' is an artificial grouping of many very distantly related photosynthetic organisms (Popper and Tuohy, 2010). Some shared features, such as sulphation of carrageenans and alginates (Michel et al., 2010), are linked to the marine habitat, and marine Angiosperms, such as the seagrass *Zostera*, also contain sulphated cell wall polysaccharides (Olsen et al., 2016).

It is interesting to note that the presence of alginates in brown seaweeds appears to be as a result of horizontal gene transfer since the biosynthetic pathway is similar to that identified in another alginate-producing organism, the opportunistic bacterial pathogen *Pseudomonas aeruginosa* (Michel et al., 2010).

Parallels between alginates, in brown algae, and pectins, in land plants, including formation of egg-box structure and gelation (Cao et al., 2020) have led to similar cell wall models for both groups (Deniaud-Bouët et al., 2014; McCann and Roberts, 1992). Furthermore, although a range of monoclonal antibodies have been generated that recognise specific epitopes present in alginates (Torode et al., 2015), some monoclonal antibodies generated against pectins also bind to alginates in brown seaweeds (Raimundo et al., 2016). This points to a potential overall *Bauplan* for cell walls and may suggest constraints for the evolution of cell walls.

FINE-TUNING THE WALL: EXAMPLES OF GRADUAL EVOLUTION

Cellulose

Cellulose, a linear chain of β(1→4) linked D-glucose residues, is the cell wall component that is familiar to most people since it is the primary component of everyday items such as paper and cotton, and has a wide range of biotechnological uses. It is the most abundantly occurring organic polymer on Earth (Hess et al., 1928), and since it was first described by Anselme Payen in 1938 (Phillips, 1961), it has been found to be synthesised by tunicates (Kimura and Itoh, 1995), microorganisms, including bacteria (Roberts et al., 2002) and cyanobacteria (Nobels et al., 2001), as well as algae (Naylor and Russell-Wells, 1934) and plants (Brown, 1985).

Among these lineages, the contribution that cellulose makes to the overall cell wall varies. It typically makes up 20%–50% w/w of the wall in land plants, although in specialised cells such as cotton fibres it may comprise 98% w/w, whereas in red algae it may only account for 1%–8% w/w (Kloareg and Quatrano, 1988). However, greater diversity is observed in cellulose microfibril diameter (Nobels et al., 2001) which Roberts et al. (2002) suggested may derive from differences in cellulose biosynthesis. Cellulose is synthesised by supermolecular complexes of proteins, known as terminal complexes (TCs), in the cell membrane. The proteins are synthesised by cellulose synthases (CesAs) that are widespread in eukaryotes and prokaryotes (Roberts and Roberts, 2002; Roberts et al., 2002; Tsekos, 1999) and are thought to have been attained by red algae during the primary endosymbiosis that led to the origin of this lineage (Roberts and Roberts, 2009), with brown algae acquiring the capacity to synthesise cellulose during the secondary endosymbiosis that led to this lineage (Michel et al., 2010). The CesA genes present in the chlorophyta, and red and brown algae yield products that form linear TCs (Tsekos, 1999). In comparison, the CesAs that form rosette TCs have only been sequenced from the land plant lineage and probably evolved after the divergence of the chlorophycean and charophycean green algae (Yin et al., 2009). Intriguingly, the moss *Physcomitrium* contains a predicted glycosyl transferase with similarity to genes present in *Selaginella* and to cyanobacterial, fungal and red algal genes referred to as 'bacterial-type CesAs' (Roberts et al., 2018). However, although it has not been determined whether the *Physcomitrium* gene synthesises cellulose or linear TCs, it appears that it primarily synthesises a linear polysaccharide composed of unbranched, unsubstituted (1,4)-Glc*p* and (1,3)-Ara*f* residues and bears structural resemblance to mixed-linkage glucan (Roberts et al., 2018).

Since cellulose in land plants is predominantly made by CesAs that form rosette TCs, it might be expected to be relatively homogenous. However, lineage-specific distinctions exist in the level of crystallinity and their cellular distribution. For example, LC-PolScope has revealed that the stomata of different phylogenetic groups exhibit distinct structural patterns of cellulose crystallinity (Shtein et al., 2017).

The CesA protein complexes responsible for cellulose biosynthesis are part of a super family. The cellulose synthase complex is composed of different CesAs in the primary cell wall compared to the secondary cell wall. The primary cell wall of Arabidopsis is synthesised by a complex consisting of CesA1, 3, and either CesA6, 2, 5, or 9, whereas cellulose is synthesised in the secondary cell wall by CesA4, CesA7, and CesA8 (Lampugnani et al., 2019; Kumar and Turner, 2015). CesAs are co-expressed with different genes depending on whether they are involved in primary or secondary cell wall synthesis (Lampugnani et al., 2019). CesAs also exhibit tissue specificity. Lampugnani et al. (2019) reviewed cellulose synthesis in the context of evolution and determined that orthogroups related to cellulose synthesis are first detected in the Zygnematophyceae coincident with the evolution of the preprophase band; they hypothesised that this indicates the requirement for maintaining coordination between cellulose synthesis and microtubules during cell division. Coexpression relationships of the cellulose-related genes suggest that their expression is controlled by a common set of transcription factors that differ between the primary and secondary cell wall (Lampugnani et al., 2019).

Genomic analyses revealed that the CesAs were just one group of a larger group now known as the cellulose gene superfamily consisting of the CesA lineage and several Csl clades including Csl A, B, C, D, E, F, G, H, J, M (Little et al., 2018). The Csls do not synthesise cellulose, but instead form the core biosynthetic machinery for many hemicelluloses. For example, Csl As synthesise mannans, Csl Cs synthesise the backbone of xyloglucan (which is then adorned by the activities of other glycosyltransferases), and Csl F, H, and J synthesise mixed-linkage glucan (Little et al., 2018). The adaptive radiation of CesAs can therefore be considered as the major actor in the evolution of cell walls (Figure 13.1).

Xyloglucan

Xyloglucan has a widespread occurrence in land plants, where it is one of the most quantitatively important polysaccharides in the primary cell wall comprising 20%–25% of the dry mass of dicots and 2%–5% in grasses (Fry, 1989). It was first identified as a storage polysaccharide in plant seeds (e.g. *Tamarindus* and *Tropaeolum*, Le Dizet 1972; Kooiman, 1957) and later recognised to be a key player in expansive cell growth. Xyloglucan hydrogen bonds to cellulose microfibrils and expansive growth is effected by turgor pressure combined with wall loosening brought about by wall-localised proteins belonging to the superfamily known as expansins (Cosgrove, 2000), and the activities of enzymes that cut xyloglucan chains and either attach them to other xyloglucan chains (homo-transglucanases) or other cell wall polysaccharides (heterotransglycanases) (Franková and Fry, 2021; Franková and Fry, 2020). In addition, Herburger et al. (2020) showed that the enzyme hetero-trans-β-glucanase is stimulated by expansins, indicating that there are synergist activities between them in bringing about wall loosening.

Although recent genomic analyses suggest that land plants acquired xyloglucan endotransglycosylases via HGT (Ma et al., 2022), it is clear that following this acquisition there have been modifications in gene number, activity, and stimulus. The action of xyloglucan endotransglucosylases (homo-transglucanases) has been identified in the cell walls of eudicots including Arabidopsis (Vissenberg et al., 2003) as well as in bryophytes and some green algae (Franková and Fry, 2021; Herburger et al., 2018; Van Sandt et al., 2007). Xyloglucan endotransglucosylases include TCH4 from Arabidopsis, which is expressed in response to mechanical stimuli, changes in temperature, light and hormones (Purugganan et al., 1997), thereby conferring adaptive advantages under certain environmental conditions. In the second category, heterotransglycanases, the range of enzyme activities identified is rapidly growing (Franková and Fry, 2021, 2020; Johnston et al., 2013) and includes enzymes that have species-specific distributions, such as hetero-trans-β-glucanase, that target cellulose and is restricted to *Equisetum* (Simmons et al., 2015). Close examination of xyloglucan-acting enzymes within the Glycosidase Hydrolase Family 16 (GH16) of Carbohydrate-Active EnZymes (CAZymes) has revealed that although members of the Endo Glucanase 16 (EG16) and XTH families both originated in green algae, EG16 has not significantly diversified which is in direct contrast with the many duplications identified for XTHs (Behar et al., 2021; Stratilová et al., 2020b, Behar et al., 2018).

There was an apparent mismatch between the ability to detect xyloglucan in charophytes using immune-based techniques including immunohistochemical analyses (Domozych et al., 2010) and glycan profiling (Domozych et al., 2009a) but a seeming lack of biochemical evidence (O'Rourke et al., 2015; Popper and Fry, 2003). This was resolved recently with biochemical, molecular, and immunocytochemical evidence providing support for the presence of xyloglucan in some members of the charophycean green algae and placing the evolution of xyloglucan at approximately 500 MYA in the Cambrian, an era marked by the independent origins of multicellular organisms in many lineages (Mikkelsen et al., 2021). However, taxonomically related structural variation indicates that xyloglucan has been fine-tuned since its origin.

Xyloglucan has a (1→4)-linked-β-D-glucopyranose backbone thought to be synthesised by members of the CslC gene family (Liepman and Cavalier, 2012), and whose structural similarly to cellulose results in it hydrogen-bonding to cellulose. However, the xyloglucan backbone is substituted with side-chains that prevent it hydrogen-bonding to itself and influence the binding to cellulose (Vincken et al., 1995; Hayashi et al., 1994). The extent of structural and compositional variation in xyloglucan side-chains necessitated the development of one-letter-codes to describe them (Franková and Fry, 2012; Fry et al., 1993). Some of the side-chains are developmentally important, e.g., the storage xyloglucans found in dicot seeds contain no fucose (Kooiman, 1961). However, the majority appear to have a taxonomic basis with xyloglucans from many taxa being defined by the presence of unique side-chains. For example, moss and liverwort xyloglucans contain galacturonic acid (Peña et al., 2008) and those from the Asteridae contain a wide-variety of different side-chains that are associated with distinct subgroups (Hoffman et al., 2005). The enormous diversity of structural variation in xyloglucan side-chains has been reviewed from an evolutionary point-of-view by Schultink et al. (2014). However, the driver(s) of this high degree of structural diversity is not known. A hint might be provided by the fact that the fungal lineage that primarily interacts with plants, either beneficially or pathogenically, the Mucoromycota originated and diversified alongside xyloglucans and land plants (Bonfante and Venice, 2020). Many members of the Mucoromycota lineage feed by enzymically digesting plants, primarily their cell walls; some of the different xyloglucan structures may have evolved as defence responses to cell wall degrading pathogenic Mucormycota.

In a similar vein, xyloglucan-derived oligosaccharides have been shown to stimulate plant immunity towards pathogens (Claverie et al., 2018). For a detailed discussion of the role of the plant cell wall in plant immunity, see Chapter 20 of this volume. Interestingly, several wall polymers that have originated and/or diversified in land plants including mixed-linkage glucan and pectins have also been shown to yield immunostimulatory oligosaccharides in response to damage by insect or animal grazers, in which case the oligosaccharides are termed Damage-Associated Molecular Patterns or DAMPs, or in response to pathogens and microbes, in which case they are called Pathogen-Associated Molecular Patterns (PAMPs) or Microbe-Associated Molecular Patterns (MAMPs) (Barghahn et al., 2021; Rebaque et al., 2021; Howlander et al., 2020; Chapter 20). This suggests that diversification of grazers and pathogens are likely to have been an important driver of land plant cell wall diversity.

Arabinogalactan Proteins and Extensins

Arabinogalactan proteins (AGPs) are a large group of highly diverse, wall-localised, proteoglycans. They are composed of two distinct moieties: 1) a carbohydrate moiety that makes up ~90%–98% of an AGP by weight, and is typically rich in arabinose and galactose residues, and 2) a protein moiety, that makes up ~10% of an AGP, and is hydroxyproline-rich. AGPs may be classed as classical or non-classical. Classical AGPs have a hydrophobic transmembrane domain, which appears to be lacking in non-classical AGPs, which also tend to be less heavily glycosylated (Majewska-Sawka and Nothnagel, 2000; Showalter, 2001). AGP biosynthesis is described in illuminating detail by Lopes et al. (Chapter 16). In brief AGP synthesis starts with the translation of an N-terminal signal peptide in the cytosol. This is then guided to the endoplasmic reticulum where it undergoes a range of modifications including hydroxylation of selected proline residues, and attachment of a GPI moiety. AGPs are then transferred to the Golgi apparatus where they undergo *O*-glycosylation. Although many *O*-glycosyltransferases have been identified, including galactosyltransferases, there are several essential enzyme activities that have yet be identified (Chapter 16; Silva et al., 2020).

AGPs have a widespread occurrence in plants, and green and brown algae (Přerovská et al., 2021; Pfeifer et al., 2022; Hervé et al., 2016; Popper and Tuohy, 2010; Domozych et al., 2009; Estevez et al., 2009; Eder et al., 2008; Domozych et al., 2007a, Lee et al., 2005), whereas there appears to be no published evidence for the presence of AGPs in red algae (Johnson et al., 2017). This distribution in combination with similarities in molecular sequence (Hervé et al., 2016), and parallels in the known physiological functions within the different groups, suggest that AGPs may have evolved prior to divergence of the eukaryotic lineages Archaeplastida (which includes the glaucophyte, red algae, green algae, and land plants) and the Telonemids, Stramenopiles, Alveolates, and Rhizaria collectively known as the TSAR supergroup (a group thought to comprise up to half of eukaryotic species diversity, and includes the brown algae) (Burki et al., 2020). Although published dates for the divergence of the main eukaryotic lineages vary greatly, based on molecular and fossil evidence, it is thought that divergence between phototrophic eukaryotic groups occurred approximately 1,600 MYA ago in the Mesoproterozoic era (Yoon et al., 2004).

The functions of AGPs have yet to be fully elucidated but many AGPs are either localised to the cell wall or can diffuse across it acting as soluble signals (Seifert and Roberts, 2007). Evidence for their role as signalling molecules is manifold. For example, recently, two secondary cell wall–specific fasciclin-like AGPs (FLA11 and FLA12) were shown to be important in initiating and regulating lignin and cellulose synthesis and deposition in response to mechanical stress in Arabidopsis (Ma et al., 2022), and in signalling between male and female tissues in flowering plant reproduction (see below). Although multicellularity is thought to have evolved independently in brown algae, green algae, and land plants (De Clerck et al., 2018; Michel et al., 2010) cell–cell signalling enabling coordination of multicellular organisms is essential.

AGPs are known to play major roles in reproduction. In the Fucales, an economically important order of brown algae, AGPs are essential for successful embryogenesis (Hervé et al., 2016). Similarly, in flowering plants, AGPs are involved in embryo and seed development (Costa et al., 2019). In flowering plants, AGPs are also important for pollen tube orientation, gamete delivery, and fertilisation (Pereira et al., 2021; Pereira et al., 2016), which mirrors similar roles involving the male gametes, in this case antheridium development and spermatogenesis, in the green freshwater alga, *Chara corallina* (Domozych et al., 2009). Lopes et al. (Chapter 16) give a detailed discussion of the role of AGPs in abiotic stress, more specifically drought or osmotic stress, and could therefore have been important in events including terrestrialisation, and recolonisation of marine environments by land plants.

Although AGPs have a cosmopolitan occurrence in photosynthetic organisms, it should be highlighted that they exhibit a high degree of structural variation that is linked to specific plant groups and habitats. For example, among the land plants, the AGPs found in the seagrass *Zostera marina* are highly branched, have an unusually high content of terminating 4-*O*-methyl-glucuronic acid residues, and are considered to have evolved in response to the marine environment (Pfeifer et al., 2020). AGPs extracted from ferns (Classen et al., 2019), bryophytes, including three species of moss (Bartels et al., 2017) and the hornwort *Marchantia polymorpha* (Happ and Classen, 2019), gymnosperms (Baumann et al., 2021), as well as charophycean green algae (Pfeifer et al., 2022), and the green seaweed *Ulva lactuca* (Přerovská et al., 2021) each possess unique structural features. For example, bryophyte AGPs contain the sugar residue 3-*O*-methyl-L-rhamnose, which appears to be absent from the AGPs found in flowering plants (Fu et al., 2007) and AGPs from the gymnosperms, *Gingko*, *Cycas* and *Ephedra*, contain the unusual sugar residue acofriose (Baumann et al., 2021). Evidence that AGPs (and extensins) were involved in major transitions in plant evolution including terrestrialisation and the early radiation of flowering plants have been uncovered using bioinformatics-based approaches (Ma et al., 2018; Johnson et al., 2017). Johnson et al. (2017) developed a bioinformatics pipeline based on the amino acid bias of the predicted protein backbone and the characteristic glycomotifs allowing them to identify hydroxy-rich glycoprotein sequences from the 1,000 plants transcriptome project. Interestingly, they found that although AGPs are widespread, GPI-anchored AGPs originate in the green algae, a feature which they suggest is indicative of increased organisation of AGPs within the cell wall–plasma membrane interface (Johnson et al., 2017). The trend towards greater organisation continues with a 2- to 3-fold increase in GPI-AGPs in bryophytes (Johnson et al., 2017), potentially reflecting selection during the transition to land. A similar approach showed that fasciclin-like arabinogalactan proteins can be clustered into distinct groups based on their domain architectures and these map to taxonomic groups (Shafee et al., 2020).

Extensins are another family of wall-localised glycoproteins that have roles including growth, development, and defence. The glycosylation machinery essential for biosynthesis of the diagnostic β-linked Hyp-Ara-Ara-Ara motif appears to be homologous in the green alga *Chlamydomonas*, and in land plants (Møller et al., 2017). However, whereas the *Chlamydomonas* cell wall is primarily composed of extensins, extensins are less quantitatively and functionally important in land plant cell walls. The structures of extensins has diversified in different plant groups. Both the relative abundance of extensins and their structure varies with species. Bioinformatic analysis suggests that their diversification is linked to many crucial points in evolution including terrestrialisation (Liu et al., 2016).

ARE WE MISSING SOMETHING?

Although a wide range of techniques have been developed, enabling characterisation of plant and algal cell walls, each has its strengths and limitations, as discussed below for molecular evidence, biochemical evidence, and microscopy- and immunocytologically derived evidence, and it is likely that there are cell wall components, and physiological mechanisms, that have yet to be identified, as well as those where new evidence may lead to modifications and refinement of current concepts.

Molecular Evidence

Molecular evidence, largely gleaned from whole genome sequences and transcriptomes, has been important for uncovering the presence and distribution of the gene families responsible for the synthesis, integration, and metabolism of various wall components (Little et al., 2018; Yin et al., 2014; Harholt et al., 2012; Liu et al., 2016). However, this is reliant on the availability of sequence information and although land plants comprise over 82% of the biomass on Earth (Bar-On et al., 2018) and are incredibly diverse, only ~0.16% of the ~350,000 extant land plant species have been sequenced (Marks et al., 2021). Although advances in sequencing and annotation are leading to an increased availability of genome-level information, this is predominantly for flowering plants, particularly those that are economically important, and there are large gaps in coverage with the genomes of some groups, such as the gymnosperms, only recently becoming available, whilst genomes for other groups such as the ferns are noticeably lacking (Marks et al., 2021). The first genomes for extant non-flowering seed plants, gymnosperms, despite having considerable commercial importance, only became available in 2013 for *Picea abies* (Norway spruce, Nystedt et al., 2013), and in 2014 for *Pinus tadea* (Loblolly pine, Zimin et al., 2014). The spikemoss, *Selaginella*, genome was published in 2011 (Banks et al., 2011), a partial genome for ferns in 2019 (Marchant et al., 2019), and a quillwort even more recently (*Winkell* et al., 2021), and genomes of non-vascular plants, the mosses and liverworts in 2009 and 2017, respectively (Bowman et al., 2017; Rensing et al., 2008). Notably, this implies that there is insufficient coverage for several land plant groups and in some cases, such as *Equisetum* and hornworts, genome-level information is not yet available. Even fewer algal genomes have been sequenced. The genome of a unicellular green alga, *Penium margaritaceum*, was published in 2020 (Jiao et al., 2020), the filamentous green alga Klebsormidium in 2014 (Hori et al., 2014), the multicellular green alga Chara in 2018 (Nishiyama et al., 2018) and multicellular brown (*Ectocarpus*), red (*Porphyra*), and green seaweeds in 2010, 2017 and 2018 (*Ulva*) (De Clerck et al., 2018; Brawley et al., 2017; Cock et al., 2010).

There are many reasons for the disproportionate coverage including the complexities of sequencing and annotating diverse groups, large genome sizes, polyploidy, and factors such as the socioeconomic impacts on global research that are a legacy of colonialism (Marks et al., 2021).

In addition, sufficient homology between gene sequences is required for proper annotation, and the enormous timespan separating the different photosynthetic groups that possess cell walls indicates that there has been ample opportunity for genes to accumulate a large number of differences making it difficult to place them within particular gene families. This is likely to be exacerbated for the extant photosynthetic marine organisms since they do not represent a monophyletic group and are extremely diverse at the higher taxonomic levels (Raven, 1995). In fact, there is greater variation between individual marine photosynthetic taxa, in terms of their morphology, biochemistry, and genetic information, than there is within the Streptophytina (charophycean green algae and land plants) (Raven, 1995). Furthermore, as has been seen for (1→3, 1→4)-β-D glucan synthesis in *Aspergillus* (Samar et al., 2015), cell wall components may be synthesised by genes that bear no resemblance to the genes that are responsible for making an analogous wall component in other taxa. Although this is an example of convergent evolution, it serves to highlight that genome analysis alone cannot yet provide a full account of all the wall components present in an organism. Perhaps, more importantly, even if all plants and algae were fully sequenced, we could not simply interrogate that information to identify all of the cell wall–related genes. The full complement of enzymes important for the synthesis, integration, and metabolism of wall components has not yet been identified; new enzymes and enzyme activities are still being discovered (Franková and Fry, 2021), and since molecular (gene) data can be equivocal, it is essential to have robust enzyme assays to define specificity.

Biochemical Evidence

The problems associated with gaining molecular evidence for the distribution of cell wall components highlight the importance of the many biochemical methods that are applied in order to characterise cell wall composition. Enormous insight into cell wall structure, composition, and evolution has been gleaned by a multifaceted approach and the development and application of many different methods (Popper, 2020; Fangel et al., 2012; Pettolino et al., 2012; Fry, 2001; Selvendran and O'Neill, 1987) including polysaccharide gel electrophoresis (Goubet et al., 2020), oligosaccharide mass profiling (Lerouxel et al., 2002), and paper chromatography and thin layer chromatography (Fry, 2001). However, the majority of these methods rely on extracting the wall components for detailed characterisation, which depending on the methods used may result in some chemical changes such as deacetylation. Information regarding localisation of the component at tissue, cellular, and subcellular level is also lost.

Microscopy- and Immunocytologically Derived Evidence

Methods that enable *in situ* visualisation of wall components at high resolution yield particularly useful information regarding the specific biological role(s) of wall components. Methods include those that recognise components based on their physical properties such as Fourier Transform Infra-Red Spectroscopy (Mouille et al., 2003), Ramen Imaging (Mateu et al., 2020), and Fluorescence Lifetime Imaging Microscopy (Escamez et al., 2021), as well as those that require recognition of the wall components by specific stains (Sexauer et al., 2021) and antibodies.

The use of monoclonal antibodies (Torode et al., 2016; Pattathil et al., 2010; Moller et al., 2008) and to a lesser extent carbohydrate-binding modules (CBMs, protein domains found in carbohydrate-active enzymes whose ability to bind carbohydrates enables them to be employed as selective labels) (Hervé et al., 2011), either *ex situ*, for example, in glycan profiling, or *in situ*, in immunolabelling experiments (Duffieux et al., 2020; Moore et al., 2020; Raimundo et al., 2016; Leroux et al., 2015; Moller et al., 2007) has been particularly important in identifying and localising cell wall components to specific tissues and taxa. The ability to localise cell wall components at tissue, cell, and even subcellular levels enables many avenues of research that tackle biological questions relating to cell wall function. However, these techniques are dependent on antibody specificity. Monoclonal antibodies may only recognise and bind to polysaccharides that contain a specific sequence of sugar residues, an epitope. This is more complicated than might first be apparent since monoclonal antibodies do not each recognise specific polysaccharides since the epitopes recognised by a single monoclonal antibody may be present in several different polymers. A further complexity is that occasionally an antibody may exhibit an unexpected affinity. For example, at least 12 monoclonal antibodies that were generated against, and have high specificity for, land plant pectins are capable of binding to epitopes present in the cell wall components of *Fucus vesiculosus*, a brown seaweed, even exhibiting subcellular-, cellular-, and tissue-specific localisation patterns (Raimundo et al., 2016). In fact, the cross-reactivity of monoclonal antibodies that bind land plant–derived pectins and brown seaweed–derived alginates has been reported several times (Torode et al., 2015; Raimundo et al., 2016; Larsen et al., 1985). A limitation to the use of antibody-based technologies is that even though there are now several hundred monoclonal antibodies that recognise plant cell wall components, not all are fully characterised with regard to their epitope specificity (Pattathil et al., 2010). Despite the wide range of plant cell wall–directed monoclonal antibodies that are available (see Chapter 10 for more details on cell wall–directed monoclonal antibodies), there are some notable exceptions in the availability of probes towards some wall polymers including cellulose (Głazowska and Mravec, 2021). This stems from the fact that monoclonal antibodies are generated by the response of a host animal to a specific wall component. Additionally, in the case of seaweed polysaccharides, only a select few monoclonal antibodies have been developed and these were predominantly developed towards commercially valuable polysaccharides (Rydahl et al., 2017; Torode et al., 2015,2016). Perhaps, more disappointingly, many of the monoclonal antibodies capable of recognising seaweed cell wall components are either no longer available or have unknown availability (Green et al., 1993; Vreeland et al., 1992; Jones et al., 1988; Vreeland et al., 1984; Vreeland 1970,1972).

New Tools and Avenues of Exploration

Although histological stains (Ruzin, 1999), CBMs, and monoclonal antibodies are extremely useful tools for cell wall studies, they have notable limitations. Most are able to bind to a relatively wide range of polymers. For example, Calcofluor White M2R, also known as Fluorescent Brightener 28, and probably the most widely used stain for cellulose, has the ability to bind to chitin, present in fungi and arthropod cell wall (Anderson et al., 2010). Additionally, as mentioned in 'Alginates', several antibodies raised against pectin are also able to bind to alginates present in brown seaweeds (Raimundo et al., 2016) and as discussed above, the epitope(s) recognised by many monoclonal antibodies may be found within more than one cell wall component.

An important development likely to provide new insights into cell wall microstructure is the transfer of aptamer technologies, which were until recently limited to, although widely used in, biomedical research and diagnostics, and environmental monitoring (Głazowska and Mravec, 2021). Aptamers are short fragments of single-stranded nuclei acid (RNA or DNA) that bind to other molecules with a high degree of selectivity. They can be chemically modified and directly tagged enabling their use as probes to visualise cell wall components in a similar way to CBMs and monoclonal antibodies as well as in direct quantification assays (Głazowska and Mravec, 2021). Another significant advantage of these probes is that the *in vitro* method by which they are developed (Systematic evolution of ligands by exponential enrichment, SELEX) is not reliant on a host animal's immune response meaning that aptamers can be generated for almost any wall component (Głazowska and Mravec, 2021).

The many methods that can be used to visualise cell wall components are elegantly discussed in a recent review by DeVree et al. (2021). One method that stands out as offering the possibility of facilitating new information about cell wall components and their biological and structural functions is click chemistry.

The term click chemistry was coined by Sharpless (Kolb et al., 2001) to describe labelling methods based on the 1,3-dipolar cycloaddition reactions developed by Huisgen working in Munich in the 1950s (Huisgen, 1961). Click chemistry involves feeding biocompatible small markers, many of which are commercially available, to live cells or tissues, where they are metabolically integrated into specific molecules. They are later recognised through reactions that can proceed under biological conditions and do not produce toxic by-products, conjugating the marker to, e.g., a suitable fluorophore. Although click chemistry has been widely used in proteomics and molecular biology, its application in carbohydrate chemistry has been more recent, probably because of the greater complexity of carbohydrates compared to DNA and proteins as well as the reduced availability of suitable markers, and has primarily focused on diagnostics, biomedical applications, and drug discovery (Witczak and Bielski, 2013). However, the transfer of click chemistry to the investigation of plant cell walls in the last 10 years has substantially extended our knowledge regarding pectin delivery and metabolism in Arabidopsis (Anderson et al., 2012) and lignin synthesis and deposition in a range of species (Lion et al., 2017; Simon et al., 2018; Morel et al., 2022). Excitingly, click chemistry offers the tantalising possibility of identifying new wall components, and new roles for previously known wall components, by coupling click chemistry and imaging with the subsequent extraction, and biochemical and chemical characterisation, of the labelled component(s).

CONCLUDING REMARKS

In answer to the question posed at the outset, although only a few select examples have been discussed in detail in this chapter, it is apparent that individual cell wall components have a high degree of freedom to evolve despite being an integral part of a highly interconnected and interdependent entity. This is supported by many studies, including a comparison of the transfer cell walls of bryophyte gametophyte and sporophyte generations carried out by Henry and Renzaglia (2021) whose observations regarding wall diversity led them to conclude that 'similarity in function and morphology of cell walls does not require a common cell wall composition'. Furthermore, it is clear that plant and algal cell walls are mosaic-, or patchwork-like, in being a composite of polymers (see Figure 13.1) that evolved at different times, and in response to different stresses (see Table 13.1). Nevertheless, they are able to act in a highly coordinated way to effect the many essential plant functions that involve cell walls, including growth and development.

Some wall components such as mixed-linkage glucan and lignin appear to have evolved independently more than once, suggesting convergent evolution. Structural and compositional changes present in AGPs, xyloglucan and lignin provide evidence that they have diversified and been fine-tuned in certain species. Therefore, many of the different potential evolutionary scenarios appear to have been actively involved in generating the cell walls of extant plants and algae.

TABLE 13.1
A Summary of the Taxonomic Distribution of Selected Cell Wall Components, Their Potential Origins, and Selection Pressures

Wall Component	Taxa	Putative Date of Origin (Million Years Ago, MYA)	Era	Key Features of the Era
Arabinogalactan proteins (AGPs)	Widespread among photosynthetic eukaryotes	1,600	Mesoproterozoic	
Mannans	Red algae, green algae, land plants	1,500	Archeozoic	Atmosphere lacked free oxygen
Xylans	Red algae, green algae, land plants	1,500	Archeozoic	
Glucuronoarabinoxylans	Commelinids	120	Cretaceous	Relatively warm, significant volcanic activity (causing increased CO_2), high level of diversification of flowering plants
Silica	Select charophycean green algae, bryophytes, ferns, and fern allies including Equisetum, grasses	470–515	Ordovician	Period of rapid diversification High CO_2
Xyloglucan	Charophycean green algae Land plants	~500	Cambrian	Diversification of multicellular organisms
Lignin	Calliarthron cheilosporioides (red alga): HGS lignin	~1,500	Archeozoic	Atmosphere lacked free oxygen
	Vascular plants	430	Silurian	Warm, high atmospheric CO_2, warm shallow seas, diversification of fungi and insects
	• Selaginella: S-lignin	383	Devonian	Warm, arid
	• Gymnosperm (with significant diversification and dominance): G and H units	~300 (~240)	Late Carboniferous (expansion in the Mesozoic)	Warm, high atmospheric CO_2, insect pollination
	• Angiosperm: H and S units lignin	140–270[a]	Early Cretaceous–Permian	

(Continued)

TABLE 13.1 (*Continued*)
A Summary of the Taxonomic Distribution of Selected Cell Wall Components, Their Potential Origins, and Selection Pressures

Wall Component	Taxa	Putative Date of Origin (Million Years Ago, MYA)	Era	Key Features of the Era
Mixed-linkage glucan	Green, red, and brown algae	?	?	
	Ferns; CslH?	400	Early-mid Devonian	Warm, arid
	Equisetum	200	Triassic	Breakup of the supercontinent Pangea, increased temperature and humidity, origin of mammals
	Monocots; CslF, CslJ[b]	135–100	Late Jurassic-Mid Cretaceous	Dinosaurs, rainforest at the South pole
	Grasses (CslF, CslH, CslJ)	56 (expansion ~34[c])	Paleocene–Eocene Thermal Maximum (becoming ecologically dominant in the Cenozoic)	Reduction in atmospheric CO_2 (Cenozoic)
Ferulic acid	Gymnosperms	~300	Late Carboniferous	Potentially lack of seasons
	Commelinids	120[d]	Lower–Mid-Cretaceous	
	Caryophyllales	114[e]	Mid-Cretaceous	

[a] Until recently flowering plants were believed to have evolved in the Cretaceous. However, a newly discovered fossil that has been named Florigerminis jurassica appears to contain a leafy branch, a fruit, and a flower bud. The 164 million-year-old fossil is highly controversial but could push back the evolution of flowering plants to within the Jurassic (Cui et al., 2022). In light of controversies regarding the origin of flowering plants, the best description is currently thought to span 140–270 MYA, i.e., from the Early Cretaceous to the Permian (Sauquet et al., 2022).

[b] Little et al. (2018).

[c] The rise of grasslands was recently linked to a decline in atmospheric CO2 (Palazzesi et al., 2022).

[d] Hertweck et al. (2015).

[e] Yao et al. (2019)

Further support for the independence of individual wall components comes from the finding that despite the supposition that the two most quantitatively and qualitatively important cell wall components in dicotyledonous plants, cellulose and xyloglucan, are considered to have a highly interdependent relationship, as suggested by the tethered network model, xyloglucan is not required for formation and integrity of the cellulose network in Arabidopsis (Kuki et al., 2020). Although there are discernible changes in wall biomechanical properties (Park and Cosgrove, 2012b), the xyloglucan-deficient mutants *xxt1/xxt2* have functional cell walls and cellulose microfibrils are integrated into the wall with a relatively unaltered pattern (Kuki et al., 2020).

Plant and algal cell walls have previously been described as an 'apoplastic intelligence system' (Nishitani and Demura, 2015). Cell walls are highly dynamic and the integration and coordination of the individual polysaccharides, proteoglycans, and phenolic compounds that make up cell walls, and their capacity to alter without, in many cases, compromising biological and biomechanical functions, and overall cell wall integrity are remarkable (Coen and Cosgrove, 2023; Chapter 18). This points to the fundamental importance and potential plasticity of the underlying mechanisms that control and support cross-talk between wall components, plant and algal cell and organism function, and interactions with the environment.

ACKNOWLEDGEMENTS

'We all stand together' is perhaps an odd choice of title for a single-author chapter. However, I hope that it acknowledges and reflects my appreciation of, and the inspiration derived from, the entire cell wall community and imparts the sense of the cell wall in itself as a cohesive community of different components. I thank the editor, Anja Geitmann, and two anonymous reviewers for their kind, thorough, and constructive critique that helped me to improve this chapter. Funding from the Department of Agriculture, Food and Marine (Biocrop) and EU (BlueBio) is also gratefully acknowledged.

REFERENCES

Ahl LI, Mravec J, Jørgensen B, Rudall PJ, Rønsted N, Grace OM (2019) Dynamics of intracellular mannan and cell wall folding in the drought responses of succulent *Aloe* species. *Plant Cell and Environment* 42:2458–2471.

Aist JR (1976) Papillae and related wound plugs of plant cells. *Annual Review of Phytopathology* 14:145–163.

Albersheim P, Darvill AG, Roberts K, Sederoff R, Staehelin A (2010) *Plant cell walls*. Garland Science.

Anderson CT, Carroll A, Akhmetova L, Somerville C (2010) Real-time imaging of cellulose reorientation during cell wall expansion in Arabidopsis roots. *Plant Physiology* 152:787–796.

Anderson CT, Wallace IS, Somerville CR (2012) Metabolic click-labelling with a fucose analog reveals pectin delivery, architecture, and dynamics in Arabidopsis cell walls. *The Proceedings of the National Academy of Sciences* 109:1329–1334.

Anderson CT, Kieber JJ (2020) Dynamic construction, perception and remodelling of plant cell walls. *Annual Review of Plant Biology* 71:39–69.

APG IV (2017) Angiosperm Phylogeny Group classification for the orders and families of flowering plants. *Checklist dataset* doi:10.15468/fzuaam. accessed via GBIF.org on 2021-11-29.

Atmodjo MA, Hao Z, Mohnen D (2013) Evolving views of pectin biosynthesis. *Annual Review of Plant Biology* 64:747–779.

Bacic A, Harris PJ, Stone BA (1988) Structure and function of plant cell walls. In: J. Preiss (ed.), *The biochemistry of plants*. Academic Press.

Banks JA, Nishiyama T, Hasebe M, Bowman JL, Gribskov M, dePamphilis C *et al.* (2011) The *Selaginella* genome identifies genetic changes associated with the evolution of vascular plants. *Science* 20:960–963.

Barghahn S, Arnal G, Jain N, Petutschnig E, Brumer H, Lipka V (2021) Mixed linkage β-1,3/1,4-glucan oligosaccharides induce defense responses in *Hordeum vulgare* and *Arabidopsis thaliana*. *Frontiers in Plant Science* 17:682439. doi: 10.3389/fpls.2021.682439.

Bar-On YM, Phillips R, Milo R (2018) The biomass distribution on Earth. *Proceedings of the National Academy of Sciences* 115:6506–6511.

Bartels D, Baumann A, Maeder M, Geske T, Heise EM, von Schwartzenberg K, Classen B (2017) Evolution of plant cell wall: Arabinogalactan-proteins from three moss genera show structural differences compared to seed plants. *Carbohydrate Polymers* 163:227–235.

Baumann A, Pfeifer L, Classen B (2021) Arabinogalactan-proteins from non-coniferous gymnosperms have unusual structural features. *Carbohydrate Polymers* 261:117831.

Baybrook SA, Jönsson H (2016) Shifting foundations: the mechanical cell wall and development. *Current Opinion in Plant Biology* 29:115–120.

Behar H, Graham SW, Brumer H (2018) Comprehensive cross-genome survey and phylogeny of glycoside hydrolase family 16 members reveals the evolutionary origin of EG16 and XTH proteins in plant lineages. *Plant Journal* 95: 1114–1128.

Behar H, Tamura K, Wagner ER, Cosgrove DJ, Brumer H (2021) Conservation of endo-glucanase 16 (EG16) activity across highly divergent plant lineages. *Biochemical Journal* 478:3063–3078.

Bonfante P, Venice F (2020) Mucoromycota: going to the roots of plant-interacting fungi. *Fungal Biology Reviews* 100–113.

Boto L (2009) Horizontal gene transfer in evolution: facts and challenges. *Proceedings of the Royal Society Series B. Biological Sciences* 277:819–827.

Bowles AMC, Paps J, Bechtold U (2022) Water-related innovations in land plants evolved by different patterns of gene cooption and novelty. *New Phytologist*. 235:732–742.

Bowman JL, Kohchi T, Yamato TT, Jenkins J, Shu S, Ishizaki K, Yamaoka S, Nishihama R, Nakamura Y, Berger F, Adam C, Aki SS, Althoff F, Araki T, Arteaga-Vazquez MA, Balasubrmanian S, Barry K, Bauer D, Boehm CR, Briginshaw L, Caballero-Perez J, Bruno C, Chen F, Chiyoda S, Chovati M, Davies KM, Delmans M, Demura T, Dierschke T, Dolan L, Dorantes-Acosta AE, Eklund DM, Florent SN, Flores-Sandoval E, Fujiyama A, Fukuzawa H, Galik B, Grimanelli D, Grimwood J, Grossniklaus U (2017) Insights into land plant evolution garnered from the *Marchantia polymorpha* genome. *Cell* 171:287–304.

Brawley SH, Blouin NA, Ficko-Blean E, Wheeler GL, Lohr M, Goodson HV, Jenkins JW, Blaby-Haasi CE, Helliwell KE, Chan CX, Marriage TN, Bhattacharyan D, Klein AS, Badis Y, Brodie J, Cao Y, Collén J, Dittamic SM, Gachon CMM, Green BR, Karpowicz SJ, Kim JW, Kudahl UJ, Linu S, Michel G, Mittag M, Olson BJSC, Pangilinan JL, Peng Y, Qiu H, Shuh S, Singer JT, Smith AG, Sprecher BN, Wagner V, Wang W, Wang Z-Y, Yanh Y, Yarish C, Zäuner-Riekaa S, Zhuang Y, Zou Y, Lindquist EA, Grimwood J, Barry KW, Rokhsar DS, Schmutz J, Stiller JW, Grossman AR, Prochnik SE (2017) Insights into the red algae and eukaryotic evolution from the genome of *Porphyra umbilicalis* (Bangiophyceae, Rhodophyta). *Proceedings of the Royal Academy of Sciences* 114:E6361–E370.

Brown RM Jr. (1985) Cellulose microfibril assembly and orientation: recent developments. *Journal of Cell Science* Supplement 2:13–32.

Buchala AJ, Franz G (1974) A hemicellulosic β-glucan from the hypocotyls of *Phaseolus aureus*. *Phytochemistry* 13:1887–1889.

Bulone V, Schwerdt JG, Fincher GB (2019) Co-evolution of enzymes involved in plant cell wall metabolism in grasses. *Frontiers in Plant Science* 10:1009.

Burki F, Roger AJ, Brown MW, Simpson AGB (2020) The new tree of Eukaryotes. *Trends in Ecology and Evolution* 35: 43–55.

Burton RA, Wilson SM, Hrmova M, Harvey AJ, Shirley NJ, Medhurst A *et al.* (2006) Cellulose synthase-like CslF genes mediate the synthesis of cell wall (1,3;1,4)-β-D-glucans. *Science* 311:1940–1942.

Burton RA, Jobling SA, Harvey AJ, Shirley NJ, Mather DE, Bacic A *et al.* (2008) The genetics and transcriptional profiles of the cellulose synthase-like HvCslF gene family in barley (*Hordeum vulgare* L.). *Plant Physiology* 146:1821–1833.

Cao L, Lu W, Mata A, Nishinari K, Fang Y (2020) Egg-box model-based gelation of alginate and pectin: a review. *Carbohydrate Polymers* 242:116389.

Carafa A, Duckett JG, Knox JP, Ligrone R (2005) Distribution of cell-wall xylans in bryophytes and tracheophytes: new insights into basal interrelationships of land plants. *New Phytologist* 168:231–240.

Carnachan SM, Harris PJ (2000) Ferulic acid is bound in the primary cell walls of all gymnosperm families. *Biochemical Systematics and Ecology* 28:865–879.

Carpita NC, Gibeaut DM (1993) Structural models of primary cell walls in flowering plants: consistency of molecular structure with the physical properties of the walls during growth. *Plant Journal*. 3:1–30.

Chen D, Melton LD, Zujovic Z, Harris PJ (2019) Developmental changes in collenchyma cell-wall polysaccharides in celery (*Apium graveolens* L.) petioles. *BMC Plant Biology* 19:81.

Chen F, Tobimatsu Y, Jackson L, Nakashima J, Ralph R, Dixon RA (2013) Novel seed coat lignins in the Cactaceae: structure, distribution and implications for the evolution of lignin diversity. *Plant Journal* 73:201–211.

Chen P, Jung NU, Giarola V, Bartels D (2020) The dynamic responses of cell walls in resurrection plants during dehydration and rehydration. *Frontiers in Plant Science* 10:1698.

Cheng S, Xian W, Fu Y, Marin B, Keller J, Wu T, Sun W, Li X, Xu Y, Zhang Y, Wittek S, Reder T, Günther G, Gontcharov A, Wang S, Li L, Liu X, Wang J, Yang H, Xu X, Delaux PM, Melkonian B, Wong GK, Melkonian M (2019) Genomes of subaerial Zygnematophyceae provide insights into land plant evolution. *Cell* 179:1057–1067.

Christenhusz MJM, Chase MW, Fay MF, Hidalgo O, Leitch IJ, Pellicer J, Viruel J (2021) Biogeography and genome size evolution in the oldest extant vascular plant genus, *Equisetum* (Equisetaceae). *Annals of Botany* 127:681–695.

Classen B, Baumann A, Utermoehlen J (2019) Arabinogalactan-proteins in spore-producing land plants. *Carbohydrate Polymers* 210:215–224.

Claverie J, Balacey S, Lemaître-Guillier C, Brulé D, Chiltz A, Granet L, Noirot E, Daire X, Darblade B, Héloir M-C, Poinssot B (2018) The cell wall-derived xyloglucan is a new DAMP triggering plant immunity in *Vitis vinifera* and *Arabidopsis thaliana. Frontiers in Plant Science* 9:1725. doi.10.3389/fpls.2018.01725.

Cock MJ, Sterck L, Rouzé P, Scornet D, Allen AE, Amoutzias G, Anthouard V, Artiguenave F, Aury JM, Badger JH, Beszteri B, Billiau K, Bonnet E, Bothwell JH, Bowler C, Boyen C, Brownlee C, Carrano CJ, Charrier B, Cho GY, Coelho SM, Collén JM, Corre E, Da Silva C, Delage L, Delaroque N, Dittami SM, Doulbeau S, Elias M, Farnham G, Gachon CMM, Gschloessl B, Heesch S, Jabbari K, Jubin C, Kawai H, Kimura K, Kloareg B, Küpper FC, Lang D, Le Bail A, Leblanc C, Lerouge P, Lohr M, Lopez PJ, Martens C, Maumus F, Michel G, Miranda-Saavedra D, Morales J, Moreau H, Motomura T, Nagasato C, Napoli CA, Nelson DR, Nyvall-Collén P, Peters AF, Pommier C, Potin P, Poulain J, Quesneville H, Read B, Rensing SA, Ritter A, Rousvoal S, Samanta S, Samson G, Schroeder DC, Ségurens B, Strittmatter M, Tonon T, Tregear JW, Valentin K, von Dassow P, Yamagishi T, Van de Peer Y, Wincker P (2010) The *Ectocarpus* genome and the independent evolution of multicellularity in brown algae. *Nature* 465:617–621.

Codjoe JM, Miller K, Haswell ES (2022) Plant cell wall mechanobiology: Greater than the sum of its parts. *Plant Cell* 34:129–145.

Coen E, Cosgrove DJ (2023) The mechanics of plant morphogenesis. Science 379:eade8055.

Conley D, Carey J (2015). Silica cycling over geologic time. *Nature Geoscience* 8:431–432.

Cosgrove DJ (2000) Loosening of plant cell walls by expansins. *Nature* 407:321–326.

Cosgrove DJ (2018) Diffuse growth of plant cell walls. *Plant Physiology* 176:16–27.

Cosgrove DJ (2022) Building an extensible cell wall. *Plant Physiology* 189:1246–1277.

Costa ML, Pereira AM, Pinto S, Silva J, Pereira LG, Coimbra S (2019) In silico and expression analyses of fasciclin-like arabinogalactan proteins reveal functional conservation during embryo and seed development. Plant Reproduction 32:3.

Cui DF, Hou Y, Yin P, Wang X (2022) A Jurassic flower bud from China. In: S.-C. Chang and D. Zheng (Eds.) *Mesozoic Biological Events and Ecosystems in East Asia*. London. Geological Society, Special Publications 521.

Darwin C (1859) *On the origin of species by means of natural selection or the preservation of favoured races in the struggle for life*. London, John Murray.

De Clerck O, Kao S-M, Bogaert K, Blomme J, Foflonker F, Kwantes M, Vancaester E, Vanderstraeten L, Aydogdu E, Boesger J, Califano G, Charrier B, Clewes R, Coates J, Del Cortona A, Fernandez-Pozo N, Gachon C, Hanikenne M, Lattermann L, Leliaert F, Liu X, Maggs C, Popper ZA, Raven J, Sterck L, Van Bel M, Wilhelmsson PKI, Bhattacharya D, Rensing SA, Van Der Straeten D, Vardi A, Vandepoele K, Van de Peer Y, Wichard T, Bothwell J (2018) Insights into the evolution of multicellularity from the sea lettuce genome. *Current Biology* 28:2921–2933.

de O Buanafina MM, Iyer PR, Buanafina MF, Shearer EA (2017) Reducing cell wall feruloylation by expression of a fungal ferulic acid esterase in *Festuca arundinacea* modifies plant growth, leaf morphology and the turnover of cell wall arabinoxylans. *PLoS One* 12:e0185312.

Deniaud-Bouët E, Kervarec N, Michel G, Tonon T, Kloareg B, Hervé C (2014) Chemical and enzymatic fractionation of cell walls from Fucales: insights into the structure of the extracellular matrix of brown algae. *Annals of Botany* 114:1203–1216.

DeVree BT, Steiner LM, Głazowska S, Ruhnow F, Herburger K, Persson S, Mravec J (2021) Current and future advances in fluoresence-based visualization of plant cell wall components and cell wall biosynthetic machineries. *Biotechnology for Biofuels* 14: 78.

Doblin MS, Pettolino FA, Wilson SM, Campbell R, Burton RA, Fincher GB *et al.* (2009) A barley cellulose synthase-like CSLH gene mediates (1,3;1,4)-β-D-glucan synthesis in transgenic Arabidopsis. *Proceeding of the National Academy of Sciences* 14:5996–6001.

Domozych DS, Serfis A, Kiemle SN, Gretz MR (2007a) The structure and biochemistry of charophycean cell walls. *I. Pectins of Penium margaritaceum. Protoplasma* 230:99–115.

Domozych DS, Elliot L, Kiemle SN, Gretz MR (2007b) *Pleurotanium trabecula*, a desmid of wetland biofilms: the extracellular matrix and adhesion mechanisms. *Journal of Phycology* 43:1022–1031.

Domozych DS, Sørensen I, Willats WGT (2009) The distribution of cell wall polymers during antheridium development and spermatogenesis in the Charophycean green alga *Chara corallina. Annals of Botany* 104:1045–1056.

Domozych DS, Sørensen I, Pettolino F, Bacic A, Willats WGT (2010) The cell wall polymers of the charophycean green alga *Chara corallina*: immunobinding and biochemical screening. *International Journal of Plant Sciences* 171:345–361.

Domozych DS, Sørensen I, Popper ZA, Ochs J, Andreas A, Fangel JU, Pielach A, Sacks C, Brechka H, Willats WGT, Rose JKC (2014) Pectin metabolism and assembly in the cell wall of the charophyte green alga *Penium margaritaceum. Plant Physiology* 165:105–118.

Duffieux D, Marcus SE, Knox JP, Hervé C (2020) Monoclonal antibodies, Carbohydrate-Binding Modules, and detection of polysaccharides in cell walls from plants and marine algae. Chapter 20. In: Z. A. Popper (Ed.) *The plant cell wall: methods and protocols*, 2nd edition. Methods in Molecular Biology. New York, Springer.

Dunn EK, Shoue DA, Huang X, Kline RE, Mackay AL, Carpita NC, Taylor IEP, Mandoli DF (2007) Spectroscopic and biochemical analysis of regions of the cell wall of the unicellular 'mannan weed', *Acetabularia acetabulum. Plant and Cell Physiology* 48:122–133.

Eder M, Lütz-Meindl U (2008) Pectin-like carbohydrates in the green alga *Micrasterias* characterized by cytochemical analysis and energy filtering TEM. *Journal of Microscopy* 231:201–214.

Eder M, Tenhaken R, Driouich A, Lütz-Meindl U (2008) Occurrence and characterization of arabinogalactan-like proteins and hemicelluloses in *Micrasterias* (Streptophyta). *Journal of Phycology* 44:1221–1234.

Eeckhout S, Leroux O, Willats WGT, Popper ZA, Viane RLL (2014) Comparative glycan profiling of *Ceratopteris richardii* 'C-Fern' gametophytes and sporophytes links cell-wall composition to functional specialization. *Annals of Botany* 114:1295–1307.

Escamez S, Terryn C, Gandla ML, Yassin Z, Scheepers G, Näsholm T, Sundman O, Jönsson LJ, Lundberg-Felten J, Tuominen H, Niittylä T, Paës G (2021) Fluorescence Lifetime Imaging as an *in situ* and label-free readout for the chemical composition of lignin. *ACS Sustainable Chemistry and Engineering* 9:17381–17392.

Estevez JM, Fernández PV, Kasulin L, Dupree P, Ciancia M (2009) Chemical and in situ characterization of macromolecular components of the cell walls from the green seaweed *Codium fragile. Glycobiology* 19: 212–228.

Fangel JU, Ulvskov P, Knox JP, Mikkelsen MD, Harholt J, Popper ZA, Willats WG (2012) Cell wall evolution and diversity. *Frontiers in Plant Science* 3:152.

Farrokhi N, Burton RA, Brownfield L, Hrmova M, Wilson SM, Bacic A *et al.* (2006) Plant cell wall biosynthesis: genetic, biochemical and functional genomics approaches to the identification of key genes. *Plant Biotechnology Journal* 4:145–167.

Fournier GP, Gogarten JP (2008) Evolution of acetoclastic methanogenesis in *Methanosarcina* via horizontal gene transfer from cellulolytic *Clostridia. Journal of Bacteriology* 190:1124–1127.

Franková L, Fry SC (2012) Trans-α-xylosidase, a widespread enzyme activity in plants, introduces (1→4)-α-D-xylobiose side-chains into xyloglucan structures. *Phytochemistry* 78:29–43.

Franková L, Fry SC (2013) Biochemistry and physiological roles of enzymes that 'cut and paste' plant cell-wall polysaccharides. *Journal of Experimental Botany* 64:3519–3550.

Franková L, Fry SC (2020) Activity and action of cell wall transglycanases. Chapter 10. In: ZA Popper (Ed.) *The plant cell wall: methods and protocols*, 2nd edition. Methods in Molecular Biology. New York, Springer.

Franková L, Fry SC (2021) Hemicellulose-remodelling transglycanase activities from charophytes: towards the evolution of the land-plant cell wall. *Plant Journal* 108:7–28.

Franz G (1972) Polysaccharidmetabolismus in den Zellwänden wachsender Keimlinge von *Phaseolus aureus. Planta* 102:334–347.

Freshwater DW, Frederique S, Butler BS, Hommersand MH, Chase MW (1994) A gene phylogeny of the red algae (Rhodophyta) based on plastid *rbcL. Proceedings of the National Academy Of Sciences* 91:7281–7285.

Fry SC (1989) The structure and functions of xyloglucan. *Journal of Experimental Botany* 40:1–11.

Fry SC (2001) *The Growing Plant Cell Wall: Chemical and Metabolic Analysis.* Reprint edition. New Jersey, The Blackburn Press.

Fry SC (2004) Tansley review: primary cell wall metabolism: tracking the careers of wall polymers in living plant cells. *New Phytologist* 161:641–675.

Fry SC, Nesselrode BHWA, Miller JG, Mewburn BR (2008a) Mixed-linkage (1→3, 1→4)-β-D-glucan is a major hemicellulose of *Equisetum* (horsetail) cell walls. *New Phytologist* 179:104–115.

Fry SC, Mohler KE, Nesselrode BHWA, Franková L (2008b) Mixed-linkage β-glucan : xyloglucan endotransglucosylase, a novel wall-remodelling enzyme from *Equisetum* (horsetails) and charophytic algae. *Plant Journal* 55:240–252.

Fry SC, York WS, Albersheim P, Darvill A, Hayashi T, Joseleau JP, Kato Y, Lorences EP, Maclachlan GA, McNeil M, Mort AJ, Reid JSG, Seitz HU, Selvendran RR, Voragen AGJ, White AR (1993) An unambiguous nomenclature for xyloglucan-derived oligosaccharides. *Physiologium Plantarum* 89:1–3.

Fu H, Yadav MP, Nothnagel EA (2007) *Physcomitrella patens* arabinogalactan proteins contain abundant terminal 3-*O*-methyl-L-rhamnosyl residues not found in Angiosperms. *Planta* 226:1511–1524.

Geng Y, Cai C, McAdam SAM, Banks JA, Wisecaver JH, Zhou Y (2021) A *de novo* transcriptome assembly of *Ceratopteris richardii* provides insights into the evolutionary dynamics of complex gene families in land plants. *Genome Biology and Evolution* 13: 3.

Gigli-Bisceglia N, Engelsdorf T, Hamann T (2020) Plant cell wall integrity maintenance in model plants and crop species-relevant cell wall components and underlying guiding principles. *Cellular and Molecular Life Sciences* 77:2049–2077.

Givnish TJ, Zuluaga A, Spalink D, Soto Gomez M, Lam VKY, Saarela JM, Sass C *et al.* (2018) Monocot plastid phylogenomics, timeline, net rates of species diversification, the power of multi-gene analyses, and a functional model for the origin of monocots. *American Journal of Botany* 105:1888–1910.

Głazowska S, Mravec J (2022) An aptamer highly specific to cellulose enables the analysis of the association of cellulose with matrix cell wall polymers *in vitro* and *in vivo. Plant Journal* 108:579–599.

Goldschmidt R (1940) *The material basis of evolution.* Yale University Press.

Gorin P, Baron M, Iacomini M (1998) Storage products of lichens. p9–23. In *Handbook of lichenology*, ed. M. Galun. CRC Press: Boca Raton, FL, USA.

Goubet F, Dupree P, Salomon Johansen K (2020) Carbohydrate gel electrophoresis. Chapter 2. In *The plant cell wall: methods and protocols,* 2nd edition. Methods in Molecular Biology. New York, Springer.

Green JR, Stafford CJ, Jones JL, Wright PJ, Callow JA (1993) Binding of monoclonal antibodies to vegetative tissue and fucose-containing polysaccharides of *Fucus serratus* L. *New Phytologist* 124:397–408.

Haas KT, Wightman R, Meyerowitz EM, Peaucelle A (2020) Pectin homogalacturonan nanofilament expansion drives morphogenesis in plant epidermal cells. *Science* 367:1003–1007.

Happ K, Classen B (2019) Arabinogalactan-proteins from the liverwort *Marchantia polymorpha* L., a member of a basal land plant lineage, are structurally different to those of Angiosperms. *Plants* 29:460.

Harholt J, Sørensen I, Fangel J, Roberts A, Willats WGT, Scheller HV *et al.* (2012) The glycosyltransferase repertoire of the spikemoss *Selaginella moellendorffii* and a comparative study of its cell wall. *PLoS One* 7:e35846.

Harholt J, Moestrup O, Ulvskov P (2016) Why plants were terrestrial from the beginning. *Trends in Plant Science* 21:96–101.

Harris PJ, Trethewey JAK (2010) The distribution of ester-linked ferulic acid in the cell walls of angiosperms. *Phytochemistry Reviews* 9:19–33.

Harrison CJ, Morris JL (2017) The origin and early evolution of vascular plant shoots and leaves. *Philosophical Transactions of the Royal Society Series B* 373:20160496.

Hayashi T, Takeda T, Ogawa K, Mitsuishi Y (1994) Effects of the degree of polymerization on the binding of xyloglucans to cellulose. *Plant and Cell Physiology* 35:893–899.

Henry JS, Renzaglia KS (2021) The placenta of *Physcomitrium patens*: transfer cell wall polymers compared across the three bryophyte groups. *Diversity* 13:378.

Hensel A, Franz G (1988) A (1→3, 4)-linked-β-D-glucan from the cell walls of regenerating tobacco protoplasts. *Carbohydrate Research* 184:285–287.

Herburger K, Ryan LM, Popper ZA, Holzinger A (2018) Localisation and substrate specificities of transglycanases in charophyte algae relate to development and morphology. *Journal of Cell Science* 131:jcs.203208.

Herburger K, Franková L, Pičmanová M, Loh JW, Valenzuela-Ortega M, Meulewaeter F, Hudson A, French CE, Fry SC (2020) Hetero-trans-β-glucanase produces cellulose-xyloglucan covalent bonds in the cell walls of structural plant tissues and is stimulated by expansin. *Molecular Plant* 13:1–16.

Herburger K, Franková L, Pičmanová M, Xin A, Meulewaeter F, Hudson A, Fry SC (2021) Defining natural factors that stimulate and inhibit cellulose:xyloglucan hetero-transglucosylation. *Plant Journal* 105:1549–1565.

Herger A, Dünser K, Kleine-Vehn K, Ringli C (2019) Leucine-rich repeat extensin proteins and their role in cell wall sensing. *Current Biology* 29:R851–R858.

Hertweck KL, Kinney MS, Stuart SA, Maurin O, Matthews S, Chase MW, Gandolfo MA, Pires JC (2015) Phylogenetics, divergence times and diversification from three genomic partitions in monocots. *Botanical Journal of the Linnean Society* 178:375–393.

Hervé C, Marcus SE, Knox JP (2011) Monoclonal antibodies, carbohydrate-binding modules, and the detection of polysaccharides in plant cell walls. Chapter 7. In: ZA Popper (Ed.) *The plant cell wall: methods and protocols,* 1st edition. Methods in Molecular Biology. New York: Springer.

Hervé C, Siméon A, Jam M, Cassin A, Johnson KL, Salmeán AA, Willats WG, Doblin MS, Bacic A, Kloareg B (2016) Arabinogalactan proteins have deep roots in eukaryotes: identification of genes and epitopes in brown algae and their role in *Fucus serratus* embryo development. *New Phytologist* 209:1428–1441.

Hess K, Haller R, Katz JR (1928) *Die Chemie der Zellulose und ihrer Begleiter*. Leipzig, Akademische Velagsgesellschaft.

Hickok LG, Warne T, Fribourg RS (1995) The biology of the fern *Ceratopteris* and its use as a model system. *International Journal of Plant Sciences* 156:332–345.

Hodson MJ, White PJ, Mead A, Broadley MR (2005) Phylogenetic variation in the silicon composition of plants. *Annals of Botany* 96:1027–1046.

Hoffman M, Jia Z, Peña MJ, Cash M, Harper A, Blackburn II AR, Darvill A, York WS (2005) Structural analysis of xyloglucans in the primary cell walls of plants in the subclass Asteridae. *Carbohydrate Research* 115:1826–1840.

Honegger R, Haisch A (2001) Immunocytochemical location of the (1→3) (1→4)-β-glucan lichenin in the lichen-forming ascomycete *Cetraria islandica* (Icelandic moss). *New Phytologist* 150:739–746.

Hori K, Maruyama F, Fujisawa T, Togashi T, Yamamoto N, Seo M, Sato S, Yamada T, Mori H, Tajma N, Moriyama T, Ikeuchi M, Watanabe M, Wada H, Kobayashi K, Saito M, Masuda T, Sasaki-Sekimoto Y, Mashiguchi K, Awai K, Shimojima M, Masuda S, Iwai M, Nobusawa T, Narise T, Kondo S, Saito H, Sato R, Murakawa M, Ihara Y, Oshima-Yamada Y, Ohtaka K, Satoh M, Sonobe K, Ishii M, Ohtani R, Kanamori-Sato M, Honoki R, Miyazaki D, Mochizuki H, Umetsu J, Higashi K, Shibata D, Kamiya Y, Sato N, Nakamura Y, Tabata S, Ida S, Kurokawa K, Ohta H (2014) Klebsormidium flaccidium genome reveals primary factors for plant terrestrial adaptation. Nature Communications 5: 3978.

Houston K, Tucker MR, Chowdhury J, Shirley N, Little A (2016) The plant cell wall: a complex and dynamic structure as revealed by the responses of genes under stress conditions. *Frontiers in Plant Science* 7:984. doi. 10.3389/fpls.2016.00984.

Howlander P, Bose SK, Zhang C, Jia X, Wang W, Yin H (2020) Pectin oligosaccharide (POS) induce resistance in *Arabidopsis thaliana* through salicylic acid-mediated signalling pathway against *Pseudomonas syringae* pv. Tomato DC3000. *Physiological and Molecular Plant Pathology* 110:101483.

Huang W, Zhang L, Columbus JT, Hu Y, Zhao Y, Tang L, Guo Z, Chen W, McKain M, Bartlett M, Huang C-H, Li D-Z, Ge S, Ma H (2022) A well-supported nuclear phylogeny of Poaceae and implications for the evolution of C4 photosynthesis. *Molecular Plant* 15:755–777.

Huisgen R (1961) Centenary lecture. 1,3-Dipolar cyloadditions. *Proceedings of the Chemistry Society of London* 357.

Jacobs BF, Kingston JD, Jacobs LL (1999) The origin of grass-dominated ecosystems. *Annals of the Missouri Botanical Gardens* 86:590–643.

Jensen JK, Busse-Wicher M, Poulsen CP, Fangel JU, Smith PJ, Yang J-Y, Peña MJ, Dinesen MH, Martens HJ, Melkonian M, Wong GK- S, Moremen KW, Wilkerson CG, Vibe Scheller H, Dupree P, Ulvskov P, Urbanowicz BR, Harholt J (2018) Identification of an algal xylan synthase indicates that there is functional orthology between algal and plant cell wall biosynthesis. *New Phytologist* 218:1049–1060.

Jiao C, Sørensen I, Sun X, Sun H, Behar H, Alseekh S, Philippe G, Palacio Lopez K, Sun L, Reed R, Jeon S, Kiyonami R, Zhang S, Fernie AR, Brumer H, Domozych DS, Fei Z, Rose JKC (2020) The *Penium margaritaceum* genome: hallmarks of the origins of land plants. *Cell* 181:1097–1111.

Johnson KL, Cassin AM, Lonsdale A, Wong GK- S, Soltis DE, Miles NW, Melkonian M, Melkonian B, Deyholos MK, Leebens-Mack J, Rothfels CJ, Stevenson DW, Graham SW, Wang X, Wu S, Pires JC, Edger PP, Carpenter EJ, Bacic A, Doblin MS, Schultz CJ (2017) Insights into the evolution of hydroxyproline-rich glycoproteins from 1000 Plant Transcriptomes. *Plant Physiology* 174:904–921.

Johnston SL, Prakash R, Chen NJ, Kumagai MH, Turano HM, Cooney JM *et al.* (2013) An enzyme activity capable of endotransglycosylation of heteroxylan polysaccharides is present in plant primary cell walls. *Planta* 237:173–187.

Jones JL, Callow JA, Green JR (1988) Monoclonal antibodies to sperm surface antigens of the brown alga *Fucus serratus* exhibit region-, gamete-, species- and genus-preferential binding. *Planta* 176:298–306.

Jones L, Milne JL, Ashford D, McQueen-Mason SJ (2003) Cell wall arabinan is essential for guard cell function. *Proceedings of the National Academy of Sciences* 100:11783–11788.

Kato Y, Masuda K (1976) Presence of a xyloglucan in the cell wall of *Phaseolus aureus* hypocotyls. *Plant and Cell Physiology* 17:1185–1198.

Keegstra K, Talmadge KW, Bauer WD, Albersheim P (1973) The structure of plant cell walls. III. A model of the walls of suspension-cultured sycamore cells based on the interconnections of the macromolecular components. *Plant Physiology* 51:188–196.

Kellogg EA (2001) Evolutionary history of grasses. *Plant Physiology* 125: 1198–1205.

Kennett JP, Stott LD (1991) Abrupt deep-sea warming, palaeoceanographic changes and benthic extinctions at the end of the Palaeocene. *Nature* 353:225–229.

Kimura S, Itoh T (1995) Evidence for the role of the glomerulocyte in cellulose synthesis in the tunicate *Metandrocarpa uedai*. *Protoplasma* 186:24–33.

Kloareg B, Quatrano RS (1988) Structure of the cell walls of marine algae and ecophysiological functions of the matrix polysaccharides. *Oceanography and Marine Biology: An Annual Review* 26:259–315.

Kolb HC, Finn MC, Sharpless KB (2001) Click-chemistry: diverse chemical function from a few good reactions. *Angewandte Chemie International edition* 40: 2004–2021.

Kooiman P (1957) Amyloids of plant seeds. *Nature* 179:107–109.

Kooiman P (1961) The constitution of *Tamarindus* amyloid. *Recueil des Travaux Chimiques des Pays-Bas Belgique* 80:849–856.

Kuki H, Yokoyama R, Kuroha R, Nishitani K (2020) Xyloglucan is not essential for the formation and integrity of the cellulose network in the primary cell wall regenerated from Arabidopsis protoplasts. *Plants* 9: 629.

Kulkarni AR, Peña MJ, Avci U, Mazumder K, Urbanowicz BR, Pattathil S *et al.* (2012) The ability of land plants to synthesize glucuronoxylans predates the evolution of tracheophytes. *Glycobiology* 22:439–451.

Kumar M, Turner S (2015) Plant cellulose synthesis: CESA proteins crossing kingdoms. *Phytochemistry* 112:91–99.

Lahaye M, Rondeau-Mouro C, Deniaud E, Buléon A (2003) Solid-state 13C NMR spectroscopy studies of xylans in the cell wall of *Palmaria palmata (L. Kuntze, Rhodophyta). Carbohydrate Research* 338:1559–1569.

Lampugnani ER, Flores-Sandoval E, Tan QW, Mutwil M, Bowman JL, Persson S (2019) Cellulose synthesis - central components and their evolutionary relationships. *Trends in Plant Science* 24:5.

Larsen B, Vreeland V, Laetsch WM (1985) Assay-dependent specificity of a monoclonal antibody with alginate. *Carbohydrate Research* 143:221–227.

Le Dizet P (1972) Quelques précisions sur la structure de l'amyloïde de capucine. *Carbohydrate Research* 24 :505–509.

Le Gall H, Phillipe F, Domon J-M, Gillet F, Pelloux J, Rayon C (2015) Cell wall metabolism in response to abiotic stress. *Plants* 4:112.

Lee KJD, Sakata Y, Mau SL, Pettolino F, Bacic A *et al.* (2005) Arabinogalactan proteins are required for apical cell extension in the moss *Physcomitrella patens*. *Plant Cell* 17: 3051–3065.

Lee CH, Teng Q, Huang W, Zhong R, Ye ZH (2010) The Arabidopsis family GT43 glycosyltransferases form two functionally non-redundant groups essential for the elongation of glucuronoxylan backbone. *Plant Physiology* 153:526–541.

Leroux O, Eeckhout S, Viane RLL, Popper ZA (2013a) *Ceratopteris richardii* (C-Fern): a model for investigating adaptive modification of vascular plant cell walls. *Frontiers in Plant Science* 4:367.

Leroux O, Leroux F, Mastroberti AA, Santos-Silva F, Van Loo D, Bagniewska-Zadworna A, Van Hoorebeke L, Bals S, Popper ZA, Marianth JEA (2013b) Heterogeneity of silica and glycan epitope distribution in epidermal idioblast cell walls in *Adiantum raddianum* laminae. *Planta* 237:1453–1464.

Leroux O, Sørensen I, Marcus SE, Viane RLL, Willats WGT, Knox JP (2015) Antibody-based screening of cell wall matrix glycans in ferns reveals taxon, tissue and cell-type specific distribution patterns. *BMC Plant Biology* 15:56.

Leroux O, Eder M, Saxe F, Dunlop JWC, Popper ZA, Viane RLL, Knox JP (2018) Comparative *in situ* analysis reveals the dynamic nature of sclerenchyma cell walls of the fern *Asplenium rutifolium*. *Annals of Botany* 121:345–358.

Lerouxel O, Choo TS, Seveno M, Usadel B, Faye L, Lerouge P, Pauly M (2002) Rapid structural phenotyping of plant cell wall mutants by enzymatic oligosaccharide fingerprinting. *Plant Physiology* 130:1754–1763.

Lerouxel O, Cavalier DM, Liepman AH, Keegstra K (2006) Biosynthesis of plant cell wall polysaccharides: a complex process. *Current Opinion in Plant Biology* 9:621–630.

Liepman AH, Cavalier DM (2012) The cellulose synthase-like A and cellulose synthase-like C families: recent advances and future perspectives. *Frontiers in Plant Science* 24;3:109..

Lion C, Simon C, Huss B, Blervacq A-S, Tirot L, Toybou D, Spriet C, Slomianny C, Guerardel Y, Hawkins S, Biot C (2017) BLISS: A bioorthogonal dual-labeling strategy to unravel lignification dynamics in plants. *Cell Chemical Biology* 24:326–338.

Little A, Schwerdt JG, Shirley NJ, Khor SF, Neumann K, O'Donovan LA *et al.* (2018) Revised phylogeny of the Cellulose Synthase gene superfamily: insights into cell wall evolution. *Plant Physiology.* 177:1124–1141.

Liu X, Wolfe R, Welch LR, Domozych DS, Popper ZA, Showalter AM (2016) Bioinformatic identification and analysis of extensins in the plant kingdom. *PloS One* 11(2), e0150177 doi: 10.1371/journal.pone.0150177.

Ma Y, Zeng W, Bacic A, Johnson K (2018) AGPs through space and time. *Annual Plant Reviews* 1 https://doi.org/10.1002/9781119312994.apr0608

Ma J, Wang S, Zhu X, Sun G, Chang G, Li L, Hu X, Zhang S, Zhou Y, Song C-P, Huang J (2022) Major episodes of horizontal gene transfer drove the evolution of land plants. *Molecular Plant* 15:857–871.

Ma Y, Macmillan CP, de Vries L, Mansfield SD, Hao P, Ratcliffe J, Bacic A, Johnson KL (2022) FLA11 and FLA12 glycoproteins fine-tune secondary cell wall properties in response to mechanical stresses. *New Phytologist* 23:1750–1767.

Majewska-Sawka A, Nothnagel EA (2000) The multiple roles of arabinogalactan proteins in plant development. *Plant Physiology* 122:3–9.

Marchant DB, Sessa EB, Wolf PG, Heo K, Barbazuk WB, Soltis PS, Soltis DE (2019) The C-Fern (*Ceratopteris richardii*) genome: insights into plant genome evolution with the first partial homosporous fern genome assembly. *Science Reports* 9:18181.

Marks RA, Hotaling S, Frandsen PB, VanBuren R (2021) Representation and participation across 20 years of plant genome sequencing. *Nature Plants* 7:1571–1578.

Martone PT, Estevez JM, Lu F, Ruel K, Denny MW, Sommerville C, Ralph J (2009) Discovery of lignin in seaweed reveals convergent evolution of cell-wall architecture. *Current Biology* 19:169–175.

Mateu BP, Bock P, Gierlinger N (2020) Ramen Imaging of Plant Cell walls. Chapter 15. In: ZA Popper (ed.) *The Plant Cell Wall: Methods and Protocols, 2nd edition.* Methods in Molecular Biology. New York, Springer.

Matsunaga T, Ishii T, Matsumoto S, Higuchi M, Darvill A, Albersheim P, O'Neill MA (2004) Occurrence of the primary cell wall polysaccharide rhamnogalacturonan II in pteridophytes, lycophytes, and bryophytes: implications for the evolution of vascular plants. *Plant Physiology* 134:339–351.

McCann MC, Roberts K (1992) Architecture of the Primary Cell Wall. In: CW Lloyd. (Ed.) *The cytoskeletal basis of plant growth and form.* London, Academic Press, pp 109–129.

McCarthy TW, Der JP, Honaas LA, de Pamphilis CW, Anderson CT (2014) Phylogenetic analysis of pectin-related gene families in Physcomitrella patens and nine other plant species yields evolutionary insights into cell walls. *BMC Plant Biology* 14:79.

McCleary BV, Glennie-Holmes M (1985) Enzymatic quantification of (1→3)(1→4)-β-D-glucan in barley and malt. *Journal of the Institute of Brewing* 91:285–295.

Meikle PJ, Hoogenraad NJ, Bonig I, Clarke AE, Stone BA (1994) A (1→3,1→4)-beta-glucan-specific monoclonal antibody and its use in the quantitation and immunocytochemical location of (1→3,1→4)-beta-glucans. *Plant Journal* 5:1–9.

Michel G, Tonon T, Scornet D, Cock JM, Kloareg B (2010) Central and storage carbon metabolism of the brown alga *Ectocarpus siliculosus*: insights into the origin and evolution of storage carbohydrates in Eukaryotes. *New Phytologist* 188: 67–81.

Mikkelsen MD, Harholt J, Westereng B, Domozych DS, Fry SC, Johansen IE, Fangel JU, Łężyk M, Feng T, Nencke L, Mikkelsen JD, Willats WGT, Ulvskov P (2021) Ancient origin of fucosylated xyloglucan in charophycean green algae. *Communications Biology* 4:754.

Miller K, Strychalski W, Nickaeen M, Carlsson A, Haswell ES (2022) *In vitro* experiments and kinetic models of Arabidopsis pollen hydration mechanics show that MSL8 is not a simple tension-gated osmoregulatory. *Current Biology* 32:1–14.

Moller I, Sørensen I, Bernal AJ, Blaukopf C, Lee K, Øbro J, Pettolino FA, Roberts A, Mikkelsen JD, Knox JP, Bacic A, Willats WGT (2007) High-throughput mapping of cell-wall polymers within and between plants using novel microarrays. *Plant Journal* 50:1118–1128.

Moller I, Marcus SE, Haeger A, Verhertbruggen Y, Verhoef R, Schols H, Ulvskov P, Mikkelsen JD, Knox JP, Willats W (2008) High-throughput screening of monoclonal antibodies against plant cell wall glycans by hierarchical clustering of their carbohydrate microarray binding profiles. *Glycoconjugate Journal* 25:37–48.

Møller SR, Yi X, Velásquez SM, Gille S, Hansen PLM, Poulsen CP, Olsen CE, Rejzek M, Parsons H, Yang Z, Wandall HH, Clausen H, Field RA, Pauly M, Estevez JM, Harholt J, Ulvskov P, Petersen BL (2017) Identification and evolution of a plant cell wall specific glycoprotein glycosyl transferase, ExAD. *Scientific Reports* 7:45341.

Monroe JG, Srikant T, Carbonell-Bejerano P, Exposito-Alonso M, Weng M-L, Rutter MT, Fenster CB, Weigel D (2020) Mutation bias shapes gene evolution in *Arabidopsis thaliana. bioRxiv* doi: 10.1101/2020.06.17.156752.

Moore JP, Nguema-Ona E, Chevalier L, Lindsey GG, Brandt WF, Lerouge P, Farrant JM, Driouich A (2006) Response of the leaf cell wall to desiccation in the resurrection plant *Myrothamnus flabellifolius*, *Plant Physiology* 141:651–662.

Moore JP, Farrant JM (2017) Arabinose-rich polymers as an evolutionary strategy to plasticize resurrection plant cell walls against desiccation: A case study of the woody angiosperm *Myorthamnus flabellifolia. South African Journal of Botany* 109:354.

Moore JP, Gao Y, Zietsman AYY, Fangel JU, Trygg J, Willats WGT, Vivier M (2020) Analysis of plant cell walls using high-throughput profiling techniques with multivariate methods. Chapter 18. In: ZA Popper (ed.) *The plant cell wall: methods and protocols, 2nd edition*, Methods In Molecular Biology. New York, Springer.

Morel O, Lion C, Neutelings C, Stefanov J, Baldacci-Cresp F, Simon C, Gierlinger N, Biot C, Hawkins S, Spriet C (2022) Mapping lignification dynamics with a combination of chemistry, data segmentation and ratiometric analysis. *Plant Physiology* 188:816–830.

Morris JL, Puttick MN, Clark JW, Edwards D, Kenrick P, Pressel S, Wellman CH, Yang Z, Schneider H, Donoghue PCJ (2018) The timescale of early land plant evolution. *Proceedings of the National Academy of Science* 115:2274–2283.

Mouille G, Robin S, Lecomte M, Pagant S, Höfte H (2003) Classification and identification of Arabidopsis cell wall mutants using Fourier-Transform InfraRed (FTIR) microspectroscopy. *Plant Journal* 35:393–404.

Naylor G, Russell-Wells B (1934) On the presence of cellulose and its distribution in the cell walls of brown and red algae. *Annals of Botany* 48:635–641.

Nishitani K, Demura T (2015) Editorial: An emerging view of plant cell walls as an apoplastic intelligent system. *Plant and Cell Physiology* 56177–56179.

Nishiyama T, Sakayama H, de Vries J, Buschmann H, Saint-Morccux D, Ullrich KK, Haas FB, Vanderstraeren L, Becker B, Lang D, Vosolsobe S, Rombauts S, Wilhelmsson PKI, Janitza P, Kern R, Heyl A, Rümpler F, Calderón Villalobos LIA, Clay JM, Skokan R, Toyoda A, Suzuki Y, Kagoshima H, Schiljlen E, Tajeshwar N, Catarino B, Hetherington AJ, Saltykova A, Bonnot C, Breuninger H, Symeonidi A, Radhakrishnan KG, Hedrich R, Ulvskov P, Glockner G, Delwiche CF, Petrášek J, Van de Peer Y, Friml J, Beilby M, Dolan L, Kohara Y, Sugano S, Fujiyama A, Delaux P-M, Quint M, Theissen G, Hagermann M, Harholt J, Dunand C, Zachgo S, Langdale J, Maumus F, Van der Straeten D, Gould SB, Rensing SA (2018) The Chara genome: secondary complexity and implications for plant terrestrialization. Cell 174:448-464.

Nobels NR, Romanovicz DK, Brown Jr RM (2001) Cellulose in cyanobacteria. Origin of vascular plant cellulose synthase? *Plant Physiology* 12:529–542.

Novo-Uzal E, Pomar F, Gómez Ros LV, Espiñeira JM, Ros Barceló A (2012) Evolutionary history of lignins. *Advances in Botanical Research* 61:311–350.

Nystedt B, Street NR, Wetterbom A, Zuccolo A, Lin YC, Scofield DG *et al.* (2013) The Norway spruce genome sequence and conifer genome evolution. *Nature* 497:579–584.

Olsen J, Rouzé P, Verhelst B, *et al.* (2016) The genome of the seagrass *Zostera marina* reveals angiosperm adaptation to the sea. *Nature* 530:331–335.

Orfila C, Huisman MM, Willats WG, van Albeek GJ, Schols HA, Seymour GB, *et al.* (2002) Altered cell wall disassembly during ripening of *Cur* tomato fruit: implications for cell adhesion and fruit softening. *Planta* 215: 440–447.

O'Rourke C, Gregson T, Murray L, Sadler IH, Fry SC (2015) Sugar composition of the pectic polysaccharides of charophytes, the closest algal relatives of land-plants: presence of 3-*O*-methyl-D-galactose residues. *Annals of Botany* 116:225–236.

Pabst M, Fischl RM, Brecker L, Morelle W, Fauland A, Köfeler H, Altmann F, Léonard R (2013) Rhamnogalacturonan II structure shows variation in the side chains monosaccharide composition and methylation status within and across different plant species. *Plant Journal* 76:61–72.

Painter TJ (1983) Algal polysaccharides. In: GO Aspinall (Ed.). *The Polysaccharides.* 2. New York, Academic Press:p195-285.

Palazzesi L, Hidalgo O, Barrada VD, Forest F, Höhna S (2022) The rise of the grasslands is linked to atmospheric CO_2 decline in the late Palaeogene. *Nature Communications* 13:293.

Panigua C, Blanco-Portales R, Barceló-Muñoz M, García-Gago JA, Waldron MA, Barceló-Muñoz J. *et al.* (2016) Antisense down-regulation of the strawberry β-galactosidase gene FaβGal4 increases cell wall galactose levels and reduces fruit softening. *Journal of Experimental Botany* 67:619–631.

Park YB, Cosgrove DJ (2012a) A revised architecture of primary cell walls based on biomechanical changes induced by substrate-specific endoglucanases. *Plant Physiology* 158:1933–1943.

Park YB, Cosgrove DJ (2012b) Changes in cell wall biomechanical properties in the xyloglucan-deficient *xxt1/xxt2* mutant of Arabidopsis. *Plant Physiology.* 158:465–475.

Pattathil S, Avci U, Baldwin D, Swennes AG, McGill JA, Popper Z, Bootten T, Albert A, Davis RH, Chenareddy C, Dong R, O'Shea B, Rossi R, Leoff C, Freshour G, Narra R, O'Neill MA, York WS, Hahn MG (2010) A comprehensive toolkit of plant cell wall glycan-directed monoclonal antibodies. *Plant Physiology* 153:514–525.

Peña MJ, Darvill AG, Eberhard S, York WS, O'Neill MA (2008) Moss and liverwort xyloglucans contain galacturonic acid and are structurally distinct from the xyloglucans synthesised by hornworts and vascular plants. *Glycobiology* 18:891–904.

Peña MJ, Kulkarni AR, Backe J, Boyd M, O'Neill MA, York WS (2016) Structural diversity of xylans in the cell walls of monocots. *Planta* 244:589–606.

Pendle A, Benitez-Alfonso Y (2015) Immunofluorescence detection of callose deposition around plasmodesmata sites. *Methods in Molecular Biology* 1217:95–104.

Percival E, Rupérez P, Fulgencio SC (2001) Dietary fibre and physicochemical properties of Spanish seaweeds. *European Food Research and Technology* 212:349–354.

Pereira AM, Lopes AL, Coimbra S (2016) Arabinogalactan proteins as interactors along the crosstalk between the pollen tube and the female tissues. *Frontiers in Plant Science* 7:1895.

Pereira AM, Moreira D, Coimbra S, Masiero S (2021) Paving the way for fertilization: the role of the transmitting tract. *International Journal of Molecular Sciences* 22:2603.

Pettolino F, Sasaki I, Turbic A, Wilson SM, Bacic A, Hrmova M, Fincher GB (2009) Hyphal cell walls from the plant pathogen *Rhynchosporium secalis* contain (1,3; 1,6)-β-D-glucans, galacto- and rhamnomannans, (1,3; 1,4)-β-D-glucans and chitin. *FEBS Journal* 276:4122–4133.

Pettolino F, Walsh C, Fincher GB, Bacic A (2012) Determining the polysaccharide composition of plant cell walls. *Nature Protocols* 7:1590–1607.

Pfeifer L, Shafee T, Johnson KL, Bacic A, Classen B (2020) Arabinogalactan-proteins of *Zostera marina* L. contain unique glycan structures and provide insight into adaptation processes to saline environments. *Scientific Reports* 10:8232.

Pfeifer L, Utermöhlen J, Happ K, Permann C, Holzinger A, von Schwartzenberg K, Classen B (2022) Search for evolutionary roots of land plant arabinogalactan-proteins in charophytes: presence of a rhamnogalactan-protein in *Spirogyra pratensis* (Zygnematophyceae). *Plant Journal* 109:568–584.

Phillips M (1961) Anselm Payen. In: E. Forbes (Ed.) *Great Chemists*, New York, Interscience Publishers, Inc. pp. 497–504.

Popper ZA (2008) Evolution and diversity of green plant cell walls. *Current Opinion in Plant Biology* 11:286–292.

Popper ZA, Fry SC (2003) Primary cell wall composition of bryophytes and Charophytes. *Annals of Botany* 91:1–12.

Popper ZA, Fry SC (2005) Widespread occurrence of a covalent linkage between xyloglucan and acidic polysaccharides in suspension-cultured angiosperm cells. *Annals of Botany* 96:91–99.

Popper ZA, Tuohy MG (2010) Beyond the green: understanding the evolutionary puzzle of plant and algal cell walls. *Plant Physiology* 153:373–383.

Popper ZA, Michel G, Hervé C, Domozych DS, Willats WGT, Tuohy MG, Kloareg B, Stengel DB (2011) Evolution and diversity of plant cell walls: from algae to flowering plants. *Annual Review of Plant Biology* 62:567–590.

Popper ZA (2020) *The Plant Cell Wall: Methods and Protocols. Ed.* New York, Springer.

Přerovská T, Henke S, Bleha R, Spiwok V, Gillarová S, Yvin JC, Ferrières V, Nguema-Ona E, Lipovová P (2021) Arabinogalactan-like glycoproteins from *Ulva lactuca* (Chlorophyta) show unique features compared to land plants AGPs. *Journal of Phycology* 57:619–635.

Prestegard JH, Liu J, Widmalm G (2017) Oligosaccharides and Polysaccharides. Chapter 3. In: A Varki, RD Cummings, JD Esko, *et al.* (Eds.) *Essentials of Glycobiology*, 3rd edition, Cold Spring Harbor Laboratory Press.

Pryer KM, Schuettpelz E, Wolf PG, Schneider H, Smith AR, Cranfill R (2004) Phylogeny and evolution of ferns (monilophytes) with a focus on the early leptosporangiate divergences. *American Journal of Botany* 91:1582–1598.

Purugganan MM, Braam J, Fry SC (1997) The Arabidopsis TCH4 xyloglucan endotransglycosylase. Substrate specificity, pH optimum, and cold tolerance. *Plant Physiology* 115:181–190.

Raimundo SC, Avci U, Hopper C, Pattathil C, Hahn MG, Popper ZA (2016) Immunolocalization of cell wall carbohydrate epitopes in seaweeds: presence of land plant epitopes in *Fucus vesiculosus* L. (Phaeophyceae). *Planta* 243:337–354.

Ralph J, Lapierre C, Boerjan W (2019) Lignin structure and its engineering. *Current Opinion in Biotechnology* 56:240–249.

Raven JA (1995) The early evolution of land plants: aquatic ancestors and atmospheric interactions. *Botanical Journal of Scotland* 47:151–175.

Rebaque D, del Hierro I, López G, Bacete L, Vilaplana F, Dallabernardina P, Pfrengel F, Jordá L, Sánchez-Vallet A, Pérez R, Brunner F, Molina F, Mélida H (2021) Cell wall-derived mixed-linked β-1,3/1,4-glucans trigger immune responses and disease resistance in plants. *Plant Journal* 106:601–615.

Rensing SA (2018) Great moments in evolution: the conquest of land by plants. *Current Opinion in Plant Biology* 42:49–54.

Rensing SA, Lang D, Zimmer AD, Terry A, Salamov A, Shapiro H, Nishiyama T, Perroud PF, Lindquist EA, Kamisugi Y, Tanahashi T, Sakakibara K, Fujita T, Oishi K, Shin -I T, Kuroki Y, Toyoda A, Suzuki Y, Hashimoto S, Yamaguchi K, Sugano S, Kohara Y, Fujiyama A, Anterola A, Aoki S, Ashton N, Barbazuk WB, Barker E, Bennetzen JL, Blankenship R, Cho SH, Dutcher SK, Estelle M, Fawcett JA, Gundlach H, Hanada K, Heyl A, Hicks KA, Hughes J, Lohr M, Mayer K, Melkozernov A, Murata T, Nelson DR, Pils B, Prigge M, Reiss B, Renner T, Rombauts S, Rushton PJ, Sanderfoot A, Schween G, Shiu SH, Stueber K, Theodoulou FL, Tu H, Van de Peer Y, Verrier PJ, Waters E, Wood A, Yang L, Cove D, Cuming AC, Hasebe M, Lucas S, Mishler BD, Reski R, Grigoriev IV, Quatrano RS, Boore JL (2008) The *Physcomitrella* genome reveals evolutionary insights into the conquest of land by plants. *Science* 319:64–69.

Renzaglia KS, Warne TR (1995) *Ceratopteris*: An ideal model system for teaching plant biology. *International Journal of Plant Sciences* 156:385–392.

Richmond T, Somerville C (2000) The cellulose synthase superfamily. *Plant Physiology* 124:495–498.

Roberts AW, Roberts E, Delmer DP (2002) Cellulose synthase (CesA) genes in the green alga *Mesotaenium calderiorum*. *Eukaryotic Cell* 1: 847–855.

Roberts AW, Roberts E (2004) Cellulose synthase (CesA) genes in algae and seedless plants. *Cellulose* 11: 419–435.

Roberts AW, Roberts EM, Haigler CH (2012) Moss cell walls: structure and biosynthesis. *Front. Plant Sci.* 3:166.

Roberts AW, Lahnstein J, Hsieh YSY, Xing X, Yap K, Chaves AM, Scavuzzo-Duggan TR, Dimitroff G, Lonsdale A, Roberts E, Bulone V, Fincher GB, Doblin MS, Bacic A, Burton RA (2018) Functional characterization of a glycosyltransferase from the moss *Physcomitrella patens* involved in the biosynthesis of a novel cell wall arabinoglucan. *Plant Cell* 30:1293–1308.

Ros Barceló A (1997) Lignification in plant cell walls. *International Review of Cytology* 176:87–132.

Rui Y, Dinneny J (2019) A wall with integrity: surveillance and maintenance of the plant cell wall under stress. *New Phytologist* 225:1428–1439.

Ruzin SE (1999) *Plant Microtechnique and Microscopy*. Oxford University Press, Oxford, UK.

Rydahl MG, Krac Un SK, Fangel JU, Michel G, Guillouzo A, Génicot S, Mravec J, Harholt J, Wilkens C, Motawia MS, Svensson B, Tranquet O, Ralet MC, Jørgensen B, Domozych DS, Willats WGT (2017) Development of novel monoclonal antibodies against starch and ulvan - implications for antibody production against polysaccharides with limited immunogenicity. *Science Reports* 24:9326.

Ryden P, Sugimoto-Shirasu K, Smith AC, Findlay K, Rieter W-D, McCann MC (2003) Tensile properties of Arabidopsis cell walls depend on both a xyloglucan cross-linked microfibrillar network and rhamnogalacturonan II-borate complexes. *Plant Physiology* 132:1033–1040.

Sage RF (2004) The evolution of C4 photosynthesis. *New Phytologist* 161:341–370.

Salmeán AA, Duffieux D, Harholt J, Qin F, Michel G, Czjzek M, Willats WGT, Hervé C (2017) Insoluble (1→3), (1→4)-β-D-glucan is a component of cell walls in brown algae (Phaeophyceae) and is masked by alginates in tissues. *Science Reports* 7:2880.

Samar D, Kieler JB, Klutts JS (2015) Identification and deletion of Tft1, a predicted glycosyltransferase necessary for cell wall β-1,3;1,4-glucan synthesis in *Aspergillus fumigatus*. *PLoS One* 10:e0117336.

Sarkar P, Bosneaga E, Auer M (2009) Plant cell walls throughout evolution: towards a molecular understanding of their design principles. *Journal of Experimental Botany* 60:3615–3635.

Sauquet H, Ramírez-Barahona S, Magallón S (2022) What is the age of flowering plants? *Journal of Experimental Botany* 73:3840–3853.

Schultink A, Liu L, Zhu L, Pauly M (2014) Structural diversity and function of xyloglucan sidechain substituents. *Plants* 526–542.

Seale M (2020) Cell wall remodeling during wood development. *Plant Physiology* 182:1800–1801.

Seifert GJ, Roberts K (2007) The biology of arabinogalactan proteins. *Annual Review of Plant Biology* 58:137–161.

Selvendran RR, O'Neill MA (1987) Isolation and analysis of cell walls from plant material. In: D Glick (ed.) *Methods of biochemical analysis*. Wiley.

Sexauer M, Shen D, Schön M, Grube Andersen T, Markmann K (2021) Visualizing polymeric components that define distinct root barriers across lineages. *Development* 148:199820.

Shafee T, Bacic A, Johnson K (2020) Evolution of sequence-diverse disordered regions in a protein family: order within the chaos. *Molecular Biology and Evolution*. 37:2155–2172.

Showalter AM (2001) Arabinogalactan proteins: structure, expression and function. *Cellular and Molecular Life Sciences* 58:1399–1417.

Shtein I, Shelef Y, Marom Z, Zelinger E, Schwartz A, Popper ZA, Bar-On B, Harpaz-Saad S (2017) Stomatal cell wall composition: distinctive structural patterns associated with different phylogenetic groups. *Annals of Botany* 119:2021–2033.

Silva GB, Ionashiro M, Carrara TB, Crivellari AC, Tiné MAS, Prado J, Carpita NC, Buckeridge MS (2011) Cell wall polysaccharides from fern leaves: Evidence for a mannan-rich Type III cell wall in *Adiantum raddianum*. *Phytochemistry* 72:2352–2360.

Silva J, Ferraz R, Dupree P, Showalter AM, Coimbra S (2020) Three decades of advances in Arabinogalactan-Protein biosynthesis. *Frontiers in Plant Science* 11: 2014.

Simmons TJ, Mohler KE, Holland C, Goubet F, Franková L, Houston DR *et al.* (2015) Hetero-trans-β-glucanase, an enzyme unique to *Equisetum* plants, functionalizes cellulose. *Plant Journal* 83:753–769.

Simon C, Lion C, Spriet C, Baldacci-Cresp F, Hawkins S, Biot C (2018) One, two, three: a bioorthogonal triple labelling strategy for studying the dynamics of plant cell wall formation *in vivo*. *Angewandte Chemie International Edition* 57:1–8.

Simpson MG (2010) *Plant Systematics*. Second Edition. Amsterdam: Elsevier.

Smith BG, Harris PJ (1999) The polysaccharide composition of Poales cell walls: Poaceae cell walls are not unique. *Biochemical Systematics and Ecology* 27:33–53.

Somssich M, Khan GA, Persson S (2016) Cell wall heterogeneity in root development of Arabidopsis. *Frontiers in Plant Science* 17:1242.

Sørensen I, Pettolino FA, Wilson SM, Doblin MS, Johansen B, Bacic A, Willats WGT (2008) Mixed-linkage (1,3; 1,4)-β-D-glucan is not unique to the Poales and is an abundant component of *Equisetum arvense* cell walls. *Plant Journal* 54:510–521.

Sørensen I, Domozych DS, Willats WGT (2010) How have plant cell walls evolved? *Plant Physiology* 153:366–372.

Sørensen I, Pettolino FA, Bacic A, Ralph J, Lu F, O'Neill MA, Fei Z, Rose JKC, Domozych DS, Willats WG (2011) The charophycean green algae provide insights into the early origins of plant cell walls. *Plant Journal* 68:201–211.

Stace CA (1989) *Plant Taxonomy and Biosystematics*. Second edition. Cambridge University Press..

Stebbins GL (1992) Comparative aspects of plant morphogenesis: a cellular, molecular and evolutionary approach. *American Journal of Botany* 79:589–598.

Stone BA, Clarke AE (1992) *Chemistry and biology of (1,3)*-β-D-*glucan*. Victoria, Australia, La Trobe University Press.

Stratilová B, Šesták S, Mravec J, Garajová S, Pakanová Z, Vadinová K, Kučerová D, Kozmon S, Schwerdt JG, Shirley N, Stratilová E, Hrmova M (2020a) Another building block in the plant cell wall: barley xyloglucan xylosyl transferases link covalently xyloglucan and anionic oligosaccharides derived from pectin. *Plant Journal* 104:752–767.

Stratilová B, Kozmon S, Stratilová E, Hrmova M (2020b) Plant xyloglucan xyloglucosyl transferases and the cell wall structure: subtle but significant. *Molecules* 25:5619.

Sugahara K, Kitagawa H (2002) Heparin and heparan sulfate biosynthesis. *IUBMB Life* 54:163–175.

Takeuchi Y, Komamine A (1981) Glucans in the cell walls regenerated from *Vinca rosea* protoplasts. *Plant and Cell Physiology* 22:1585–1597.

Thompson JE, Fry SC (2000) Evidence for covalent linkage between xyloglucan and acidic pectins in suspension-cultured rose cells. *Planta* 211:275–286.

Torode TA, Marcus SE, Jam M, Tonon T, Blackburn RS, Hervé C, Knox. JP (2015) Monoclonal antibodies directed to fucoidan preparations from brown algae. *PLoS One* 10(2):e0118366.

Torode TA, Siméon A, Marcus SE, Jam M, Le Moigne A-M, Duffieux D, Knox JP, Hervé C (2016) Dynamics of cell wall assembly during early embryogenesis in the brown alga *Fucus*. *Journal of Experimental Botany* 67:6089–6100.

Trappe J (2000) Pangea: extravagant sedimentary resource formation during supercontinent configuration, an overview. *Palaeogeography, Palaeoclimatology, Palaeoecology* 161:35–48.

Trembath-Reichert E, Wilson JP, McGlynn SE, Fischer WW (2015) Evolution of land plant silica biomineralization. Proceedings of the National. Academy of Sciences 112:5449–5454.

Trethewey JAK, Campbell LM, Harris PJ (2005) (1→3),(1→4)-β-D-Glucans in the cell walls of the Poales (*Sensu lato*): An immunogold labelling study using a monoclonal antibody. *American Journal of Botany*. 92:1660–1674.

Tsekos I (1999) The sites of cellulose synthesis in algae: diversity and evolution of cellulose-synthesising enzyme complexes. *Journal of Phycology* 35:635–655.

Turvey JR, Williams EC (1970) The structure of some xylans from red algae *Phytochemistry* 9:2383–2388.

Vaahtera L, Schulz J, Hamann T (2019) Cell wall integrity maintenance during plant development and interaction with the environment. *Nature Plants* 5:924–932.

Van Sandt VST, Stieperaere H, Guisez Y, Verbelen J-P, Vissenberg K (2007) XET activity is found near sites of growth and cell elongation in bryophytes and some green algae: New insights into the evolution of primary cell wall elongation. *Annals of Botany* 99:39–51.

Verhertbruggen Y, Yin L, Oikawa A, Vibe Scheller H (2011) Mannan Synthase activity in the CSLD family. *Plant Signalling and Behavior* 6:1620–1623.

Vincken JP, Dekeizer A, Bledman G, Voragen AGJ (1995) Fractionation of xyloglucan fragments and their interaction with cellulose. *Plant Physiology*. 108:1579–1585.

Vissenberg K, Van Sandt V, Fry SC, Verbelen J-P (2003) Xyloglucan endotransglucosylase action is high in the root elongation zone and in the trichoblasts of all vascular plants from *Selaginella* to *Zea mays*. *Journal of Experimental Botany* 54:335–344.

Voigt CA (2014) Callose-mediated resistance to pathogenic intruders in plant defense-related papillae. *Frontiers in Plant Science* 5:168.

Voiniciuc C, Dama M, Gawenda N, Stritt F, Pauly M (2019) Mechanistic insights from plant heteromannan synthesis in yeast. *Proceedings of the National Academy of Sciences* 116:522–527.

Vreeland V (1970) Localization of a cell wall polysaccharide in a brown alga with labeled antibody. *Journal of Histochemistry and Cytochemistry* 18:371–373.

Vreeland V (1972) Immunocytochemical localization of the extracellular polysaccharide alginic acid in the brown seaweed, *Fucus distichus*. *Journal of Histochemistry and Cytochemistry* 20(5):358–367.

Vreeland V, Slomich M, Laetsch WM (1984) Monoclonal antibodies as molecular probes for cell wall antigens of the brown alga, *Fucus*. *Planta* 162:506–517.

Vreeland V, Zablackis E, Laetsch WM (1992) Monoclonal antibodies as molecular markers for the intracellular and cell wall distribution of carrageenan epitopes in *Kappaphycus* (Rhodophyta) during tissue development. *Journal of Phycology* 28:328–342.

Wachananawat B, Kuroha T, Takenaka Y, Kajiura H, Naramoto S, Yokoyama R, Ishizaki K, Nishitani K, Ishimizi T (2020) Diversity of pectin rhamnogalacturonan I rhamnosyltransferases in Glycosyltransferase Family 106. *Frontiers in Plant Science* 11:997.

Wang D, Samsulrizal NH, Yan C, Allcock NS, Craigon J, Blanco-Ulate B, *et al.* (2019) Characterization of CRISPR mutants targeting genes modulating pectin degradation in ripening tomato. *Plant Physiology*. 170:544–557.

Weng J-K, Akiyama T, Bonawitz ND, Li X, Ralph J, Chapple C (2010) Convergent evolution of syringyl lignin biosynthesis via distinct pathways in the lycophyte *Selaginella* and flowering plants. *Plant Cell* 22:1033–1045.

Weng J-K, Li X, Stout J, Chapple C (2008) Independent origins of syringyl lignin in vascular plants. *Proceedings of the National Academy of Sciences* 105:7887–7892.

Winkell D, Kuo L-Y, Yang H-P, Ashok AM, Irisarri I, Dadras A, de Vries S, de Vries J, Huang Y-M, Li Z, Barker MS, Hartwick NT, Michael TP, Li F-W (2021) Underwater CAM photosynthesis elucidated by Isoetes genome. *Nature Communications* 12:6348.

Witczak ZJ, Bielski R (2013) *Click Chemistry in Glycoscience: New Developments and Strategies*. N.J, John Wiley and Sons Inc.

Yao G, Jin JJ, Li H-T, Yang J-B, Mandala VS, Croley M, Mostow R, Douglas NA, Chase MW, Christenhusz MJM, Soltis DE, Soltis PS, Smith SA, Brockington SF, Moore MJ, Yi TS, Li D-Z (2019) Plastid phylogenomic insights into the evolution of the Caryophyllales. *Molecular Phylogenetics and Evolution* 134:74–86.

Ye Z-H, Zhong (2022) Cell wall biology of the moss *Physcomitrium patens*. *Journal of Experimental Biology* 73:4440–4453.

Yin Y, Huang J, Xu Y (2009) The cellulose synthase superfamily in fully sequenced plants and algae. *BMC Plant Biology* 9:99.

Yin Y, Johns MA, Cao H, Rupani M (2014) A survey of plant and algal genomes and transcriptomes reveals new insights into the evolution and function of the cellulose synthase superfamily. *BMC Genomics* 15: 260.

Yoon HS, Hackett DD, Ciniglia C, Pinto G, Bhattacharya D (2004) A molecular timeline for the origin of photosynthetic eukaryotes. *Molecular Biology and Evolution* 21:809–818.

Young GC (2008) Early Evolution of the Vertebrate Eye-Fossil Evidence. *Evolution: Education and Outreach* 1:427–438.

Zexer N, Elbaum R (2022) Hydrogen peroxide modulates silica deposits in sorghum roots, *Journal of Experimental Botany* 73:1450–1463.

Zhang T, Tang H, Vavylonis D, Cosgrove DJ (2019) Disentangling loosening from softening: insights into primary cell wall structure. *Plant Journal* 100:1101–1117.

Zimin A, Stevens KA, Crepeau MW, Holtz-Morris A, Koriabine M, Marçais G *et al.* (2014) Sequencing and assembly of the 22-gb loblolly pine genome. *Genetics* 196:875–890.

14 Hydrophobisation of Cell Walls by Cutin, Suberin and Associated Waxes

Kiran Suresh, Viktoria V. Zeisler-Diehl, Lukas Schreiber
University of Bonn

INTRODUCTION

Secondary modifications of carbohydrate cell walls of higher plants represent important evolutionary adaptions to life on the land when leaving water. Two main factors can primarily be made responsible for these adaptions. First, the density of the atmosphere is nearly 1,000-times lower than water, which requires additional mechanical stabilisation of land-living plants. This problem of stability was evolutionarily solved by the deposition of the aromatic biopolymer lignin into primary carbohydrate cell walls (Chapter 21). Second, the water potential in the atmosphere surrounding the aboveground plant body is more negative compared to living plant cells in nearly all situations (Nobel 2009). This results in a constant driving force for the transpiration of water from the plant body into the atmosphere (Figure 14.1). To minimise this uncontrolled water loss, carbohydrate cell walls forming plant–atmosphere interfaces are modified by the deposition of the hydrophobic biopolymer cutin with its associated waxes (Riederer and Müller 2006). Thus, two main challenges when conquering the mainland were solved by two different types of apoplastic cell wall biopolymers, an aromatic and an aliphatic polymer during plant evolution.

Besides the protection from an uncontrolled water loss (Riederer and Schreiber 2001), plant cuticles also restrict the leaching of ions from the leaf interior and uptake of substances deposited on the outer side of the cuticle (Heredia 2003). Finally, cuticles represent primary barriers for microorganisms trying to infect the living plant tissue (Ziv et al. 2018). Besides the aboveground part of the plant, roots representing the belowground part of land-living plants also developed hydrophobised interfaces between the root and the surrounding soil environment. Under normal conditions, in well-watered soil, there is no need to protect roots from water loss, since the water potential gradient is directed inwards as long as the soil is not drying out (Figure 14.1). Roots must tightly control the uptake of substances dissolved in soil water, must hinder nutrients from passively diffusing back into the soil from the xylem vessels located in the central cylinder, and they need to be protected from the infection of microorganisms. In roots, it is the hydrophobic biopolymer suberin deposited at the soil root interfaces (Schreiber et al. 1999).

In the following review, we will (i) briefly recapitulate the main steps in the history of cutin and suberin research, (ii) shortly summarise the current state of knowledge of the chemical composition, the biosynthesis, and the function of hydrophobised plant cell walls, (iii) discuss selected relevant methodological and applied aspects related to plant interfaces made of cutin and suberin, and finally (iv) point out important future questions and approaches further improving our knowledge about hydrophobic plant interfaces.

HISTORY OF CUTIN AND SUBERIN RESEARCH

As often in science, major progress is made with new technologies becoming available. The first microscopes allowed insight into the cellular structure of living organisms. Suberised cork oak tissue composed of dead suberised plant cell walls was one of the first biological samples, which led to the concept of the cell as the microscopic unit of living organisms (Hooke 1665). The plant cuticle, representing a thin film that is covering the outer surface of aerial plant organs, was described significantly later (Brongniart 1834) probably because the cuticle in most cases is much thinner than the often multicellular suberised plant tissue. With electron microscopy, the next step of subcellular resolution became possible and this led to the observation that suberin and cutin polymers often are highly organised on the ultrastructural level. Alternating dark and light lamellae, having a thickness in the nanometre range (about 5–10 nm), can be visualised in suberised (Sitte and Rennier 1963) and in some species also in cutinised plant cell walls (Jeffree 2006). Until today, the exact molecular organisation of these dark and light lamellae is not fully clarified, although several models have been suggested (Bernards 2002; Graca 2015). Remarkably, an optically similar organisation (dark and light lamellae) on the ultrastructural level has also been described for epidermal cells (corneocytes) of mammalians forming the interface towards the atmosphere (Madison 2003).

The cutin polymer covering aerial organs always contains cuticular wax, which can easily be extracted with organic solvents (Samuels et al. 2008). The suberin polymer may contain wax, often in comparable amounts as cutin, when suberin is deposited in plant tissues facing the atmosphere (Schreiber et al. 2005), but it can also have only

DOI: 10.1201/9781003178309-14

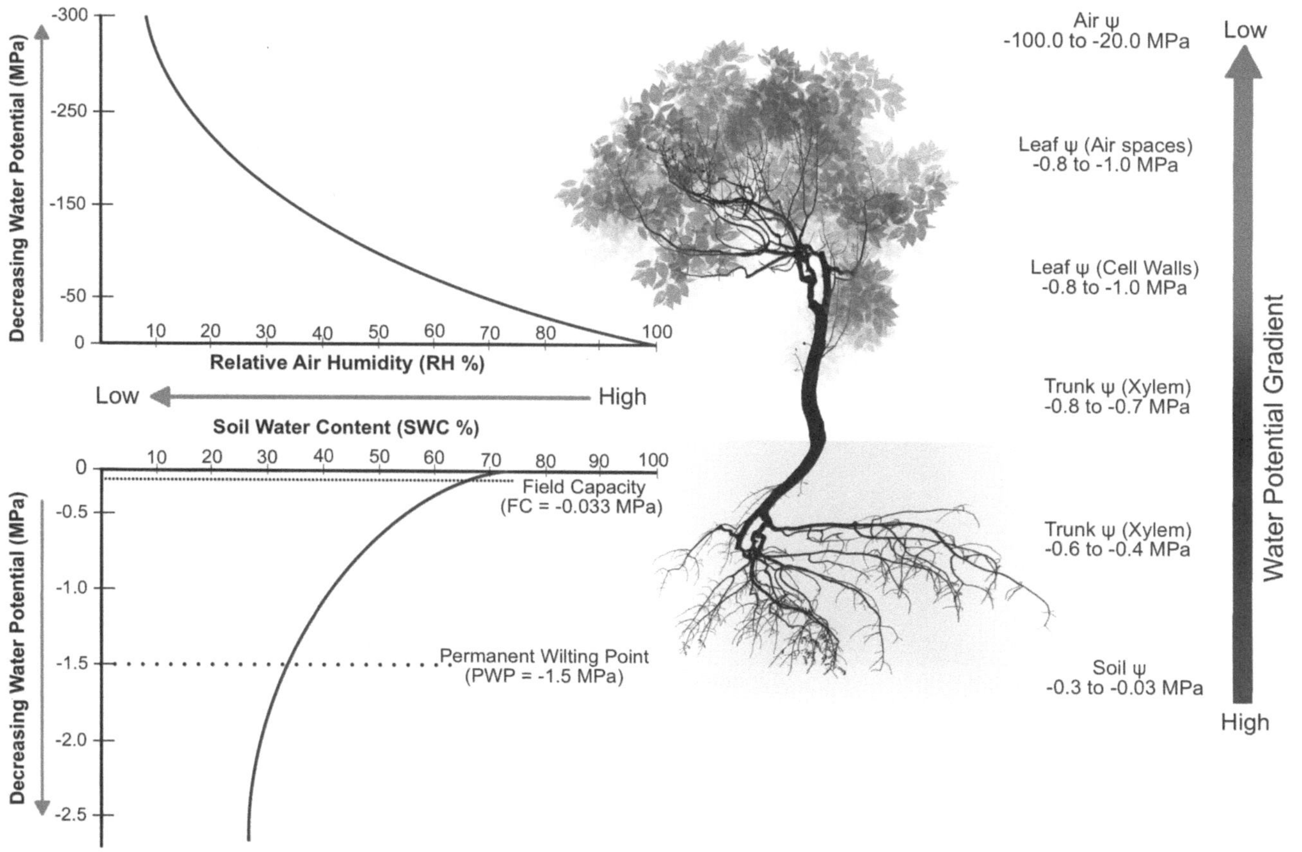

FIGURE 14.1 Schematic diagram representing water potential gradients between soil, plant, and air. The graphs on the left side show the decrease of water potentials at 25°C with decreasing relative air humidity (upper figure) and decreasing relative soil water content (lower figure). The dotted lines in the lower left figure show the water potentials at field capacity (water-saturated soil) and at the permanent wilting point given for most of the economic plants. On the right side, the decreasing gradient in water potentials between soil, root, shoot, leaf, and atmosphere are shown.

traces of wax or no wax at all when occurring belowground or inside the plant tissue (e.g. endodermis) as an internal interface (Schreiber et al. 1999). Cuticular wax is deposited asymmetrically into the outer parts of the cutin polymer (intracuticular wax) and to the outer surface (epicuticular wax). With the availability of scanning electron microscopy, it was shown that epicuticular wax can have a highly ordered crystalline ultrastructure (Koch and Barthlott 2006) and different crystalloids structure in different plants (Figure 14.2). It is well established that epicuticular wax renders leaf surfaces hydrophobic and thus are highly water repellent, which hinders water from penetrating stomatal pores and infiltrating the intercellular gas space of leaves, which would severely interfere with the vital gas exchange. Non-wettability of leaf surfaces has also been described as a self-cleaning effect of the cuticles known as the Lotus effect (Barthlott and Neinhuis 1997). Apart from epicuticular wax, intracuticular wax has been shown to be essential for establishing and optimising the transport-limiting properties of cuticles (Zeisler and Schreiber 2016) since upon wax extraction water permeability across cuticles is increasing by 2 to 3 orders of magnitude (Schönherr 1976). Another vital function of the plant cuticle is that it protects developing leaves and petals from organ fusion which was discovered in the developing era of molecular biological research with the increasing availability of mutants. Cuticle mutants in which the cuticle is missing or defective are often characterised by fusions of leaves and petals during organ development (Sieber et al. 2000).

In the mid of the last century, the cell wall underlying the cuticle was thought to be traversed by ectodesmata (Franke 1961). It was claimed that, similar to plasmodesmatic connections between neighbouring cells, plasmodesmatic files called ectodesmata stretch out from the outer epidermal cell membrane through the cell wall into the cuticle and to the cuticle surface. It was described that they represent preferential pathways for plant nutrients applied to the leaf surfaces and speculated that they might act as "sensors" between the leaf and the surrounding environment. However, some years later, it was shown in convincing experiments that ectodesmata do not exist (Schönherr and Bukovac 1970). Rather, the observed structures represent particular areas in the cuticle, which preferentially allow polar substances (e.g. nutrients) to diffuse through the otherwise hydrophobic cuticle (Schönherr 2006). Thus, the experiments and the observations, which had led to the

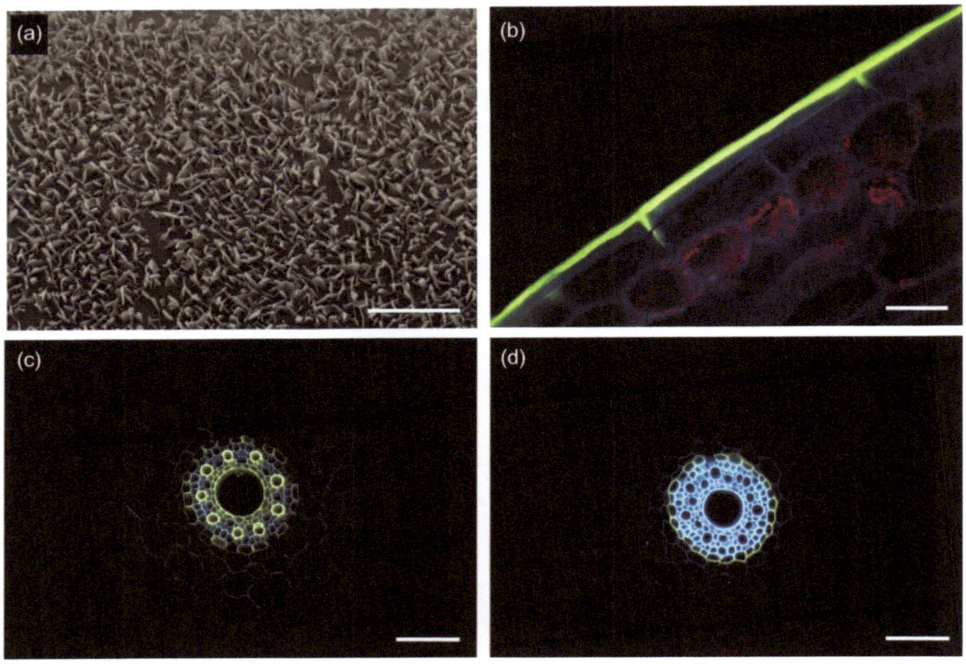

FIGURE 14.2 Images of hydrophobised cell walls in leaves and roots of plants. (**A**) Scanning electron micrograph of the abaxial leaf side of barley (*H. vulgare* cv. Scarlett) showing the morphology of the epicuticular wax crystalloids. (**B**) Fluorescence micrograph of a cross section of the adaxial *C. miniata* leaf after staining with Fluorol Yellow 088. The cuticle is distinguishable by its greenish/yellow fluorescence when viewed under UV excitation (365 nm). (**C**) Fluorescence micrograph of a barley (*H. spontaneum* ICB181243) root cross section after staining with Berberine-Aniline. Casparian strips in the radial endodermal cell walls can be distinguished by their yellow colour when viewed under UV excitation (365 nm). (**D**) Fluorescence micrograph of barley (*H. vulgare* cv. Scarlett) root cross section stained with Fluorol Yellow 088. Suberised endodermal cell wall can be identified by their yellow fluorescence when viewed under UV excitation (365 nm). Size bars: $A = 4$ µm; B, C and $D = 100$ µm.

postulation of ectodesmata, were correct. However, the explanation of the structural basis for these observations of preferential sites of polar paths of transport into the leaf was wrong. At present, it is accepted as a concept that there are particular sites in plant cuticles, predominantly located around guard-cells and trichomes, and over anticlinal cell walls and leaf veins, which allow the cuticular diffusion of polar compounds (Fernandez and Eichert 2009). The exact molecular or structural nature of these polar paths of transport still needs to be clarified (see the Future questions below).

Generally, it is assumed that the interface between roots and the soil that limits the unhindered apoplastic transport of water and dissolved compounds is solely established by the endodermis containing Casparian strips and suberin lamellae (Figure 14.2). Since the endodermis is located deep inside the root, separating the central cylinder from the root cortex, it is surprising that there should be no interface of similar structure and function closer to the outer root surface facing the soil. This is indeed the case with many roots forming an exodermis (Enstone et al. 2003; Perumalla et al. 1990). The exodermis is defined as a root hypodermis equipped with Casparian strips and suberin lamellae similar to the endodermis (Peterson 1989). In fact, the occurrence of an exodermis in roots was described already in 1903 by Kroemer. The reason why the exodermis was more or less ignored and rarely mentioned in textbooks in the following decades until its "rediscovery" is not fully clear. But one reasonable explanation could be that root research, based on physiological and molecular biological methods, mostly investigates hydroponically grown plants due to the easy experimental accessibility of the roots. These growth conditions represent a significantly different environment compared to soil affecting root structure. We know now that many species do not form an exodermis when cultivated in hydroponics. But, already in aeroponic cultivation, with the roots sitting in a gas phase with nearly 100% humidity and the roots being frequently sprayed with nutrient solution, an exodermis is formed (Zimmermann et al. 2000). Clearly, particular physiological and environmental signals are needed to induce an exodermis. Thus, when using artificial cultivation systems like hydroponics, which undoubtedly offer major experimental advantages, it must always be kept in mind that the structure and/or function of plants and their organs are not necessarily identical to those in their natural habitat. A further reason why the concept of an exodermis is often not considered might simply be due to the fact that the majority of physiological and molecular biological plant research conducted today is done on the model species *Arabidopsis thaliana*. Brassicaceae in fact is one of the rare plant families that do not develop an exodermis. As a third reason, it must be mentioned that roots of Eudicotyledons rapidly initiate secondary growth, which leads to the shedding and loss of all root tissues (including the endo- and exodermis) outside the pericycle. The final developmental state of the interface between the soil and the root of dicotyledonous plants is formed by a suberised periderm similar to the shoots with secondary thickening growth (Serra et al. 2022; Wunderling et al. 2018). This multilayered periderm essentially has a protective function forming a highly resistant barrier between roots and shoots and the surrounding environment. Due to its economic importance, cork oak periderm is one of the most intensely investigated periderm structures (Graca 2015).

Cutinised and suberised cell walls both represent hydrophobised plant cell walls (Xin and Herburger 2021). They can also be described as a composite material or composite membrane since they are composed of cutin, waxes, and carbohydrates (Schreiber and Schönherr 1990). The term composite membrane is used to describe technical materials or membranes composed of chemically different types of polymers and substances (Hwang 2011), which are often used in special applications like reverse osmosis. This applies perfectly to plant cuticles that are characterised by a pronounced asymmetry (Schönherr and Riederer 1988) with a gradient increasing in polarity from outside (cuticle/atmosphere interface) to inside (epidermal cell wall). On the inner side of the cuticle, the polar carbohydrate cell wall is in close covalent connection with the hydrophobic cutin polymer (Fernandez et al. 2016). The highly hydrophobic and partially crystalline cuticular waxes are deposited into the outer parts of the hydrophobic cutin polymer and on the cuticle surface facing the atmosphere (Buschhaus and Jetter 2011; Zeisler-Diehl et al. 2018). Suberised tissue-like cork is composed of the aliphatic suberin polymer, substantial amounts of carbohydrates, lignin, and soluble lipophilic waxes (Graca and Santos 2007; Pereira 2013). Suberin is a very special material with unique properties (Silva et al. 2005; Silvestre et al. 2008), which so far have not been able to be obtained by synthetic materials or membranes.

CURRENT STATE OF THE KNOWLEDGE

Both cutin and suberin monomers and associated wax molecules are essentially composed of fatty acid derivatives. In addition, suberin contains aromatic monomers, mainly ester-linked coumaric and ferulic acid and a non-ester-linked aromatic polymer fraction has also been described (Bernards 2002). Glycerol acts in both, suberin and cutin polymers, as a small cross-linker (Graca and Pereira 2000). Several detailed and up-to-date reviews on the chemical composition of these hydrophobic biopolymers and associated waxes, and biosynthesis of the monomers and related genes (Nawrath 2002; Fich et al. 2016; Pollard et al. 2008; Samuels et al. 2008; Shukla and Barberon 2021; Vishwanath et al. 2015; Xin and Herburger, 2021) are available. Therefore, we will present only a concise introduction and refer to the literature for more detailed information. Fatty acids *de novo* synthesis takes place in plastids and further modifications of C16- and C18-fatty acids are performed at the endoplasmic reticulum. Two major steps occurring in the endoplasmatic reticulum, elongation and functionalisation represent the most important processes leading to the characteristic cutin, suberin, and wax compounds. Elongation of C16- and C18-fatty acids leads to so-called very long-chain fatty acids with chain lengths up to C30 or higher. These very long-chain fatty acids are then further oxygenated.

Whereas cutin monomers mostly have chain lengths of C16 and C18, with only lower amounts of higher chain lengths, it is typical for suberin that monomer chain lengths can vary between C16 and C18 up to C30 or even higher depending on the plant species (Kolattukudy 1980). In contrast to cutin and suberin compounds, wax molecules are characterised by chain lengths varying between C20 and C30 or higher with only a minor amount of C16 and C18 molecules (Walton 1990). In wax of certain plant groups, even chain lengths between C40 and C60 are formed by esterification of long-chain alcohols and fatty acids. Besides the wax fraction derived from the elongation of fatty acids, pentacyclic triterpenoids form a second often relevant wax fraction in certain species (Buschhaus and Jetter 2011). The biosynthesis of pentacyclic triterpenoids is facilitated via the biosynthetic terpenoid pathways (McGarvey and Croteau 1995).

Genes of two gene families, ketoacyl-CoA synthases, also called fatty acid elongases, and cytochrome P450 monooxygenases, also called hydroxylases, are coding for the most important key enzymes active in cutin, suberin, and wax biosynthesis (Pollard et al. 2008). Most of the biosynthetic genes including elongases and hydroxylases have successfully been characterised for the first time in *Arabidopsis* within the last 20 years by analysing their tissue and organ-specific expression, their substrates and products, and knockout or overexpression mutants. Based on this valuable information, closely related orthologue genes have been described in other species including a series of important economic plants such as potato (Serra et al. 2009a,b) and rice (Yu et al. 2008). In cuticular wax, elongated primary fatty acids are modified to either aldehydes and then to alcohols or, alternatively, they are decarbonylated to form alkanes. Esters are formed by condensing long-chain alcohols with long-chain fatty acids. These substance classes (fatty acids, aldehydes, alcohols, alkanes, and esters), besides pentacyclic triterpenoids, represent the major constituents found in most plant waxes. However, a wealth of other hydrophobic compounds (ketones, diketones, secondary alcohols, diols, triterpenoid esters, etc.) have also been described to occur in plant waxes, often only in evolutionary closely related groups.

In cutin and suberin polymers, the most relevant substance classes are hydroxy fatty acids and dihydroxy fatty acids (diacids) of varying chain lengths. These substances offer the possibility of efficiently esterifying them among each other but also to glycerol, mainly via glycerol-phosphate-acyltransferases (Yang et al. 2012) to the mid-chain alcohol, and to coumaric and ferulic acids by aliphatic suberin feruloyl transferases (Molina et al. 2009). Thus, a three-dimensional hydrophobic and mechanically stable polymer network can be synthesised, where the hydrophobic wax molecules, if present, can be deposited into the polymer and on the surface in the case of the cuticular membrane. Genes coding for hydroxylases, being involved in further functionalisation of fatty acids leading to hydroxy fatty acids and diacids found in cutin and suberin, have been characterised in the model species *Arabidopsis* (Bak et al. 2011; Pinot and Beisson 2010) and economic plants like potato (Serra et al. 2009a,b). Besides hydroxy fatty acids and diacids, monofunctional primary fatty acids and alcohols are also regularly described as further substance classes occurring in cutin and suberin polymers. Since they are monofunctional, they cannot efficiently contribute to establishing a three-dimensional polymer network. But monofunctional primary fatty acids and alcohols could be esterified to free non-esterified alcoholic and carboxylic groups of the hydroxy fatty acids and diacids and thus polymer hydrophobicity could be improved.

Besides monomer synthesis, further important steps in forming the final functional polymers represent the transport of cutin and suberin monomers from the cytosol to the cell wall, where they must be deposited and finally polymerised. Similarly, hydrophobic wax molecules need to be shuttled across the hydrophilic aqueous outer epidermal cell wall to be deposited within the hydrophobic polymers. There is increasing evidence that ABC-transporters (ATP-binding-cassette transporters) are involved in the trans-membrane transport of cutin and suberin monomers (Do et al. 2018; Elejalde-Palmett et al. 2021; Shiono et al. 2014). In addition, it was shown that lipid transport proteins are involved in the transport of wax compounds across the aqueous cell wall space (DeBono et al. 2009). Nevertheless, there are also some indications that the transport of cutin and suberin monomers and wax molecules travelling from the cytosol to the cell sites of deposition could be mediated by a vesicular transport (Dominguez et al. 2015). Finally, it was convincingly shown that polymerisation of cutin occurs enzymatically via an extracellular acyltransferase (Yeats et al. 2012). It has also been suggested that a non-enzymatic cross-linking of cutin (and thus potentially also suberin) in the apoplast could additionally be involved in polymerisation (Segado et al. 2020). For suberin polymerisation and degradation, lipases have been described as being involved (Ursache et al. 2021). Thus, the transport of cutin, suberin, and wax molecules, as well as cutin and suberin polymerisation, still offers some questions to be solved in detail in the future, whereas biosynthesis of cutin, suberin monomers, and wax molecules is already fairly well understood.

Functionally the hydrophobic polymers cutin and suberin should represent apoplastic barriers. Various functions and scenarios of cutin and suberin acting as barriers need to be distinguished and discussed depending on the polymer and whether the wax is present or not. This includes the following points: barrier for water transport, barrier for the transport of substances dissolved in water, transport barrier for gas, and barrier for microorganisms. Unquestionably, both cutin and suberin polymers forming the interface between the plant and the surrounding above- and belowground environment need to be barriers hindering pathogenic microorganisms from easily colonising the living tissues (Khalil et al. 2019). In roots establishing a mycorrhiza, fungal hyphae need to grow within the apoplast of the cortical root tissue. However, they normally cannot cross the endodermis and enter the central cylinder. Thus, the endodermal cell wall, modified by lignin and suberin, obviously represents an efficient apoplastic barrier hindering microorganisms from entering the central cylinder.

It is also unquestionable that cuticles, composed of the cutin polymer encrusted with waxes, covering the outer leaf surfaces, are very efficient transport barriers protecting plants from uncontrolled water loss when stomata are closed. The water potential of the atmosphere in nearly all situations will be more negative than the water potential in the leaf tissue (Figure 14.1). 99% relative humidity already corresponds to a water potential of −1.35 MPa (Nobel 2009), whereas living well-watered plant tissue has a water potential up to −1.0 MPa. Thus, the driving force for transpiration (water concentration gradient between the leaf and the atmosphere) is normally directed from the leaf interior to the atmosphere. Only when humidity approaches 100%, e.g., during the night, pure water will condense on the cuticle surface and the gradient will be inverse (Berry et al. 2019). The high efficiency of plant cuticles as transpiration barriers has been thoroughly characterised in the past (Schuster et al. 2017). It was shown that plant cuticles are as efficient or even more efficient in retaining water when compared to synthetic polymers such as polyethylene used in our daily lives (Riederer and Schreiber 2001).

Cuticles are also very efficient transport barriers for substances dissolved in water (Baur et al. 1997). As a consequence, the loss of nutrients (leaching) from the leaf interior is limited, and the diffusion of foliar-applied agrochemicals from the leaf surface into the leaf is restricted (Schönherr and Riederer 1989). In addition, cuticles represent efficient barriers to gases in the atmosphere (e.g. O_2, CO_2, and N_2), which are all larger than water molecules (Lendzian and Kerstiens 1991). Structurally, this high impermeability of the plant cuticle is obtained by impregnating the cutin polymer with cuticular waxes on the outer side of the leaf facing the atmosphere. Upon wax extraction, rates of water loss increased by factors between 10 and 1,000-fold, depending on the plant species and organ investigated (Schreiber 2010). Thus, cutin is only a highly efficient transport barrier when impregnated with cuticular waxes. It was shown that it is mostly the intracuticular wax fraction that establishes the transport barrier (Zeisler-Diehl et al. 2018), whereas the epicuticular wax fractions, especially with epicuticular wax crystallites, render leaves highly hydrophobic and unwettable (Barthlott and Neinhuis 1997), and they can help to reflect DNA damaging radiation (Long et al. 2003).

Roots growing in soil face a completely different situation compared to leaves located in the atmosphere. Roots represent organs that must efficiently absorb water and dissolved nutrients from the soil. At the same time, they must prevent other potentially toxic ions (e.g. NaCl) from passively entering and pathogens from successfully colonising the living root tissue. The solution to this problem is best described by the composite transport model (Kim et al. 2018) postulating a symplastic path (diffusion via living cells connected by plasmodesmata), an apoplastic path (diffusion via the aqueous cell walls), and a transcellular path being a combination of both symplastic and apoplastic transport, which requires influx and efflux transporters for compounds to traverse the plasma membrane.

Discussing radial water uptake of roots at first, it is a justified question whether roots actually need or have to develop an apoplastic transport barrier for water. With plants growing in well-watered soil, which will represent the typical condition, the water potential in the soil will be close to zero and only slightly negative (Figure 14.1). With the water potential inside the plant being normally more negative, the water potential gradient driving the diffusion of water will be directed inwards and water will passively follow this gradient. In adaptation to decreasing water potential in the soil, plants need to decrease their own water potential in roots to values lower than the soil water potential. Thus, the water uptake is secured. However, the permanent wilting point of crops is given as an averaged water potential of around −1.5 MPa in soil (Rai et al. 2017). This is fairly mild compared to the atmosphere where water potential easily reaches a value of −100 MPa (about 50% relative humidity). It is highly questionable that an apoplastic barrier in roots, as efficient as in leaves, would help. Establishing such an efficient (cuticular-like) transport barrier in roots in theory might help to limit water loss from the roots to the soil, but at the same time water and nutrient uptake would be severely blocked leading to water and nutrient shortages. Alternatively, it has been suggested that endodermal suberisation is relevant for the optimisation of nutrient uptake in roots since it should mainly affect the apoplastic and transcellular path of transport but most probably not the symplastic path of transport (Barberon et al. 2016).

Water uptake should be secured as long as roots can adjust their water potential to values more negative than in soil. Obviously, water potentials beyond −1.5 MPa cannot easily be achieved by plant cells. Furthermore, compared to leaf cuticles, which are rendered highly impermeable because they are sealed with wax molecules, similar wax amounts and compositions have not been described in enzymatically isolated root suberin (Schreiber et al. 1999). This conclusion is supported when quantitatively comparing water transport of leaf cuticles with roots based on permeances, giving the speed at which water is diffusing radially across hydrophobised cell walls. Root water transport velocities (m s^{-1}) are always 3–4 orders of magnitude higher compared to leaf cuticles (Schreiber 2010). Only when wax molecules of similar amounts and composition are deposited in the suberin polymer, permeances comparable to cuticles can be established with suberised tissues. This was convincingly shown with the periderm of potato tubers (Schreiber et al. 2005). Wax deposition was strongly induced when storing harvested tubers in the ambient atmosphere and water permeability significantly decreased reaching values similar to cuticles.

In some experimental approaches, the apoplastic diffusion of dye molecules is microscopically traced (Varney et al. 1993). When dye molecules are hindered in their free apoplastic diffusion, it is often concluded that there must also be a barrier for free water diffusion. Since all these dye molecules have relatively high molecular weights (300–800 g mol^{-1}) compared to water (18 g mol^{-1}), this conclusion is questionable (Figure 14.3). The same argument of molecular size might also hold true for the diffusion of nutrients, which are larger than water and additionally either negatively or positively charged. A wax-free suberin polymer in roots, which will hardly form a barrier for a polar but uncharged water molecule, could still establish a significantly more pronounced diffusional barrier for ions dissolved in water. Thus, from a functional point of view, apoplastic suberised barriers in roots most probably have the function to hinder nutrients, which have actively been pumped into the central cylinder by the consumption of energy via the symplastic pathway, from passively diffusing back via the apoplast. Only this mechanism allows for establishing the root pressure and the gradient of water potentials from the soil into the central cylinder. Water will passively follow this gradient into the root. Similarly, these suberised barriers will keep potentially toxic ions (e.g. NaCl), which should not enter the xylem vessels in the central cylinder, outside by hindering them from a free apoplastic diffusion into the root along a concentration gradient directed inwards.

A special situation has to be solved by wetland plants like rice. Since the oxygen supply of the roots, especially to the growing root tip, is largely limited underwater in paddy fields, roots develop an aerenchyma (Abiko et al. 2012; Shiono et al. 2014). This allows oxygen to rapidly diffuse from the atmosphere via stomata and the intercellular air space of leaves, shoots, and roots down to the root tip, establishing a high oxygen concentration. As an adaptation to oxygen deficiency, the root exodermis is strongly suberised (Kotula et al. 2009). This efficiently hinders oxygen from being radially lost into the surrounding hypoxic aqueous environment when longitudinally diffusing towards the root tip. At the same time, radial uptake of water in roots is not severely blocked. In response to salt stress, rice roots also show an enhanced suberisation and a reduced uptake of NaCl but no effect on water uptake (Ranathunge et al. 2011). This allows concluding that root function is optimised in the way that radial water uptake is still possible when root apoplastic barriers are significantly reinforced in response to abiotic stress hindering the diffusional loss of oxygen or uptake of NaCl. Since water can use both the apoplastic and the symplastic pathways, this optimisation of radial water uptake might take place by a modulation of the aquaporin activity and/or expression affecting the symplastic pathway.

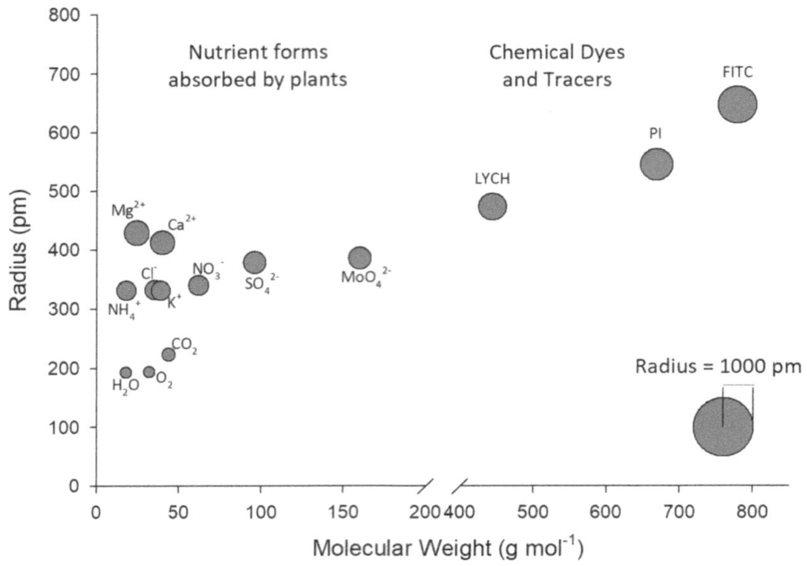

FIGURE 14.3 Relationship between molecular weights (g mol^{-1}) of water, inorganic nutrients, and organic dye molecules used as apoplastic tracers and the respective molecular radii given in picometer (pm). Molecular radii of water, oxygen (O_2), carbon dioxide (CO_2), Lucifer Yellow Carbohydrazide (LYCH), Propidium Iodide (PI), and Fluorescein Isothiocyanate (FITC) were calculated from molar volumes (V_m) estimated according to Abraham and McGown (1978). For nutrients, effective hydrated ionic radii were obtained from Nightingale (1959).

METHODOLOGICAL AND APPLIED ASPECTS

As pointed out above, hydrophobised cell walls can be described as composite membranes or materials composed of a complex mixture of different substance classes (aliphatic and aromatic compounds, traces of proteins, etc.) and even different polymers (cutin or suberin mixed to some extent with the carbohydrate cell wall and sometimes also lignin). Until today, most studies investigating the chemical composition of cutin and suberin are based on analysing the monomers released after depolymerisation (Graca et al. 2015). To some extent, NMR (nuclear magnetic resonance) technologies have been applied to study the intact polymer and this gives insights into the cross-linking between the different monomers and compounds (Dastmalchi et al. 2021; Huang et al. 2017). Nevertheless, methods leading to a detailed and precise picture of how the three-dimensional ultrastructure of cutinised and suberised cell walls exactly looks like on the level of a molecular resolution are still very limited.

This also applies to waxes sealing cutin and suberin, rendering them highly impermeable. Using adequate organic solvents, waxes can easily be extracted and subsequently recrystallised. This approach allowed the development of a fairly precise picture of the ultrastructural organisation, their degree of crystallinity, and the exact crystal lattice (Reynhardt and Riederer 1991, 1994) including the striking three-dimensional epicuticular wax structures such as platelets or tubules (Dora et al. 2018; Koch and Ensikat 2008; Koch et al. 2006), applying methods like NMR, EPR (electron paramagnetic resonance), DSC (differential scanning calorimetry), AFM (atomic force microscope), and X-Ray diffraction. Nevertheless, our knowledge of how and where cuticular waxes, establishing the transport barrier, are exactly deposited within the hydrophobic cutin and/or suberin polymer is still limited. High-resolution electron microscopy can be applied to study the ultrastructure of cutin and suberin. Due to the fairly harsh sample preparation, including dehydration of the samples with organic solvents and embedding in epoxy resins allowing ultrathin sectioning, it is highly improbable that waxes survive this treatment unaffected. It is more realistic that waxes are partially or fully extracted during sample preparation for TEM (transmission electron microscope) investigations due to the treatment with an organic solvent.

Modern confocal microscopy offers the chance to investigate living tissue in three dimensions (x-, y-, and z-axes). In theory, this could allow observing the diffusion of dye molecules of different physicochemical properties (e.g. polar vs. lipophilic) in the z-axis from the cuticle surface through the cuticle into the outer epidermal cell wall via live imaging (Liu and Gaskin 2004). This might enable identifying which pathways molecules are using when diffusing through the cuticle. The resolution of confocal microscopy in the direction of the z-axis is about 500 nm, and this is at the limit for investigating cuticles normally having a thickness of about 1 μm or less. However, since it is the wax layer establishing the transport barrier, which can be estimated to be about ten-times thinner, a direct high-resolution imaging of the transport-limiting barrier in intact living leaves with a non-invasive technique is unfortunately still not possible.

In agro-chemistry, plant protecting agents or nutrients are sprayed on leaf surfaces. This offers the advantage of precisely hitting the leaf surface of the desired species, but since cuticles are water repellent and highly impermeable, care must be taken that spray droplets are not bouncing off from leaf surfaces after impact and that substances are diffusing in substantial amounts through cuticles into the leaves. This is obtained by adding adjuvants and surfactants to the spray solution reducing the surface tension of the aqueous droplets, which enhances leaf surface wetting significantly (Kirkwood 1993). Most interestingly, surfactants can act as plasticisers on the cuticular transport barrier of waxes, since they significantly increase the diffusion of lipophilic organic molecules through the cuticles (Schönherr and Baur 1994). In the future, it could be interesting to optimise the cuticular penetration of polar nutrients when applied during foliar nutrition searching for adequate plasticisers.

Another very interesting idea suggests the possibility of using cutin and suberin monomers as a sustainable source for building synthetic, but bio-based and thus biodegradable, polymers with desired properties replacing the alternatives of polymers synthesised from fossil resources (Heredia-Guerrero et al. 2017; Tedeschi et al. 2018). Besides cellulose and lignin, representing the dominant carbohydrate and aromatic biopolymers in plant cell walls, suberin and cutin occur in plant cell walls of specialised tissues forming apoplastic barriers. Both cutinised and suberised cell walls remain in significant amounts as residues in food (e.g. potatoes and tomatoes) and beverage (e.g. fruits and grapes) production, which are normally used to feed animals or as manure. However, since they represent a rich source of bio-based renewable lipid materials (waxes and oxygenated fatty acids), they could replace classical polymers like polyethylene, which so far are mostly obtained from petrochemistry. The first experimental approach synthesising a bio-based polymer, composed of alginate, oxygenated fatty acids from tomato cutin, and beeswax, looked very promising in terms of hydrophobicity, biomechanical properties, and water vapour transmission (Heredia-Guerrero et al. 2017; Tedeschi et al. 2018). In the future, optimised versions of these types of polymers could be used for packaging food and thus replace petroleum-based synthetic polymers, although a technical realisation will still need some time.

FUTURE QUESTIONS AND APPROACHES

There are still important research aspects related to the structure and function of cutinised and suberised plant cell walls which await to be solved in the future. This is because our knowledge remains limited, often due to methodological limitations, but also due to inconsistency of the available data and/or controversial views and interpretations of the existing data within the scientific community. In the following, some of them will be mentioned and discussed.

Intuitively it is always assumed that thick cuticles should result in lower rates of water permeability. However, most available experimental data do not confirm this. In fact, cuticles with the highest wax coverages had the highest transpiration rates (Becker et al. 1986). Since it is the waxes establishing the transport barrier, it was also postulated that cuticular transpiration should decrease with increasing wax amounts. However, comparing cuticles isolated from different plant species, significantly varying in wax amounts, did not show any reasonable correlation between cuticular transpiration and wax coverage (Figure 14.4). From the classical theory of diffusion of molecules across synthetic polymers, it is predicted that barrier properties of a polymer decrease in direct proportion to an increase in thickness (Crank and Park 1968). This, however, applies to synthetic polymers, which are chemically homogenous. When comparing plant cuticles of different species, which are fairly heterogeneous since they are characterised by pronounced differences in cutin and wax composition and amounts, it should not be too surprising that there is no negative correlation between wax coverage and barrier properties (Figure 14.4). Thus, the important questions of (i) how the barrier properties of cuticles are established on the molecular and ultrastructural level, (ii) how barrier properties are related to the species-specific chemical compositions, and (iii) why plant cuticles of different species are often chemically so different are still unanswered.

Whereas it is widely accepted that polar paths of transport across lipophilic cuticles exist (Fernandez and Eichert 2009; Fernandez et al. 2021), their exact nature has not yet been verified. These polar paths of transport specifically allow polar substances, especially charged ions, to diffuse across hydrophobic cuticles. Microscopy indicates that they are preferentially localised around stomata, at the base of trichomes, and along leaf veins (Schreiber 2005). It has been suggested that these polar paths of transport could be formed by free polar functional groups, like acids and alcohols, which have not been esterified during cutin polymerisation and/or by polar cell wall carbohydrates extending into the cutin polymer and thus providing a polar aqueous environment preferentially used for the diffusion of polar compounds (Schönherr 2006). These polar paths of transport have been named “aqueous polar pores” and average diameters of 0.5 nm have been estimated (Schreiber 2005). However, it must be mentioned that the exact ultrastructural and molecular nature of these polar paths of transport or polar pores have not been identified until now. In principle, their existence has been postulated based on ion permeability across cuticles, which hardly could be explained without assuming the existence of these polar paths of transport. Thus, the elucidation of their exact nature still awaits to be done in the future, e.g., with microscopic techniques offering higher resolution than they have been available up to now.

In recent years, there is an increasing interest in investigating and modelling the potential significance of foliar water uptake also called “inverse transpiration” (Vesala et al. 2017). Since the water potential gradient will mostly be directed towards the atmosphere, having a water potential that is more negative than inside the leaf, water will be lost from the leaf due to stomatal and cuticular transpiration (Figure 14.1). However, in certain environmental situations,

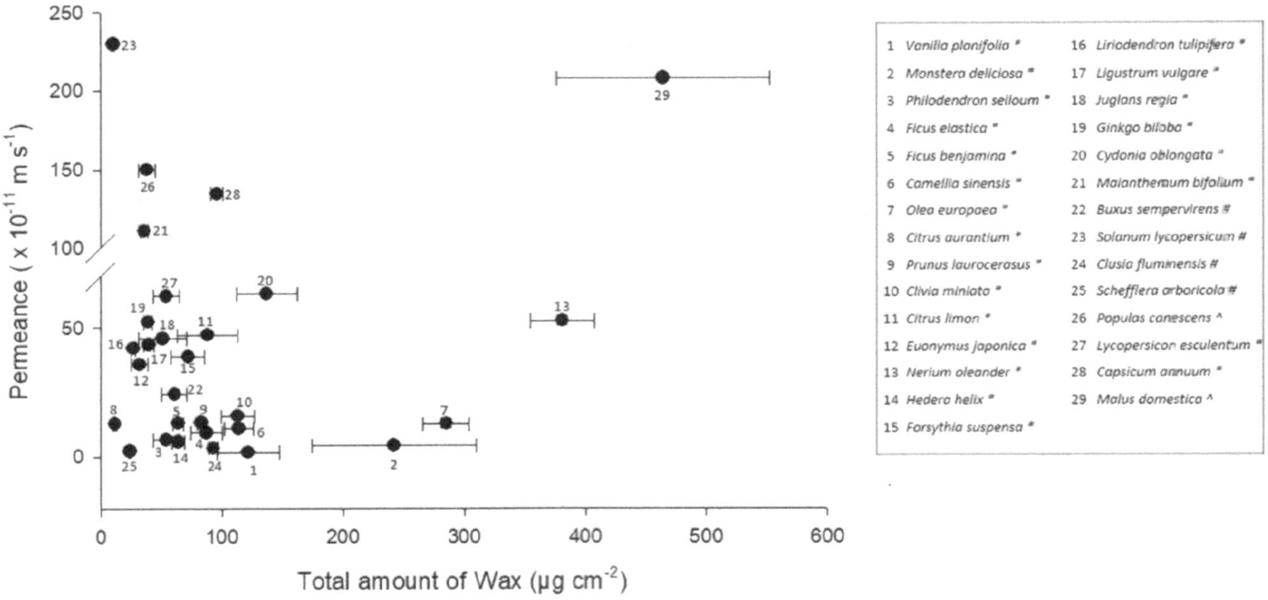

FIGURE 14.4 Correlation between cuticular transpiration of leaves and fruits of several species given as permeances (m s^{-1}) and wax coverages of isolated cuticles (µg cm^{-2}). Numbers 1–26 are values of leaf cuticles and numbers 27–29 are values of fruit cuticles. Values for permeances and wax amounts were obtained from * Schreiber and Riederer (1995a,b), # Zeisler (2013) and ^ Zeisler-Diehl et al. (2018).

e.g., an air humidity of 100% leading to a water potential close to zero and a more negative water potential inside the leaves than outside, reverse transpiration could occur. In parallel, dew deposition on leaf surfaces could lead to water droplets or water films, which could also result in the diffusion of water across the cuticle into the leaf (Berry et al. 2019; Schreel and Steppe 2020). It can be predicted that the preferential trans-cuticular pathway for this water uptake should be the postulated polar paths of transport. Such an asymmetry of water transport across isolated cuticles and synthetic aliphatic polymers, with cellulose nanocrystals incorporated, has been shown to occur when the gradient in water potential is reversed (Kamtsikakis et al. 2021). Although reverse transpiration and cuticular uptake of water can be predicted to occur under special environmental situations, future research needs to determine and quantify whether and to what extent this process would significantly contribute to the water status of plants.

In roots, it is well accepted that the apoplastic barrier is formed by the endodermis and, if present, the exodermis. Nevertheless, its exact function and its precise chemical composition are still to some extent a matter of debate. In its primary developmental state, the endodermis contains only Casparian strips, present as dots in the radial cell walls between two adjacent endodermal cells (Geldner 2013). By chemical analysis of enzymatically isolated Casparian strips, it has been shown that they are carbohydrate cell walls that contain the secondary cell wall modifications lignin (up to 6%) and at lower amounts suberin (up to 2%) (Schreiber et al. 1999). Based on the temporal sequence of the endodermis-specific expression of lignin and suberin genes in roots of *Arabidopsis thaliana*, it was concluded that Casparian strips are exclusively composed of lignin (Naseer et al. 2012). The expression of suberin genes was only observed when the secondary developmental state of the endodermis, with the suberin lamella deposited on all inner surfaces of the primary endodermal cell walls, could be detected microscopically. However, a chemical analysis of the primary and secondary endodermal cell walls of *Arabidopsis* is still missing, but this is technically extremely challenging due to the very fine roots. On the other hand, *in vivo* Raman microscopy indicated that both, lignin and suberin, are simultaneously deposited in Casparian strips in endodermal cell walls in roots of corn (*Zea mays*; Man et al. 2018) and Chinese fir (*Cunninghamia lanceolate*; Song et al. 2019). Thus, future research will have to answer the question of whether different species have different chemical compositions of their Casparian strips or not?

Independent from this question related to the exact chemical composition of Casparian strips, it is also an interesting question why radial cell walls of endodermal cells in the primary developmental state are modified by lignin and suberin leading to the formation of Casparian strips, and to what extent this modification/hydrophobisation with aromatic and aliphatic cell wall polymers will establish an efficient apoplastic transport barrier for water? As already pointed out above (Section 3), large dye molecules and charged inorganic nutrients are blocked by Casparian strips and this makes sense physiologically for the root to function properly. But is there a need to apoplastically block water at the endodermis, as it is often discussed (Geldner 2013)? So far it was never directly shown that small water molecules are retained at Casparian strips, it was so far only deduced from the diffusion of large tracers such as dye molecules

(Varney et al. 1993). If lignin alone would be sufficient to establish an efficient apoplastic barrier for water, then the question arises, why suberin and cutin polymers, including waxes, are needed. From leaf cuticles and suberised potato periderm, we know that they are only efficient barriers for water if they contain waxes in significant amounts. So far, waxes have not been found in similar amounts and not in comparable chemical composition in isolated endodermal cell walls.

Thus, it is questionable whether lignin alone or together with suberin, but without waxes, can form a perfect apoplastic barrier for the diffusion of water. *In situ* studies of the radial water transport in lupine (*Lupinus albus*) roots growing in soil using deuterated water and imaging by rapid neutron tomography (Zarebanadkouki et al. 2019) showed that the apoplastic water flux was nearly 20 times faster compared to the cell-to-cell pathway. Since the apoplast amounted to only 5% of the total radial surface area across which water can diffuse, it was calculated that about 50% of the water is flowing apoplastically. This in fact does not indicate that Casparian strips form perfect apoplastic water barriers. However, lignin is normally deposited in primary plant cell walls to substantially improve their mechanical stability. Thus, it can be speculated that lignin depositions in the radial endodermal cell walls, forming the Casparian strips, support the formation of a biomechanically stable network surrounding the central cylinder of the root. Similar to the plant cell, where the mechanically stable cell wall has the function to withstand the negative osmotic pressure of the cytoplasm and the vacuole, to prevent the cells from unhindered expansion and damage, the negative root pressure is established within the central cylinder. Thus, there should be a pronounced osmotic pressure in the central cylinder acting on the root cortex with the endodermis forming the innermost cell layer of the cortex, which then must be able to withstand this root pressure of the central cylinder. Enzymatically isolated, and thus relaxed, Casparian strips of *Clivia miniata* were characterised by a pronounced undulation (Schreiber et al. 1994), which indicates that they are most probably stretched and under a strong tension *in vivo*. The highly localised reinforcement of the radial endodermal cell walls by lignification could thus be the main reason for lignin depositions in Casparian strips. In addition, they still can form an apoplastic barrier for nutrients, preventing them from diffusing apoplastically back along the concentration gradient directed from the central cylinder towards the surrounding soil environment. This potential biomechanical function of the lignification of the Casparian strip could be an interesting research question to be addressed in the future.

CONCLUSIONS

Hydrophobised plant cell walls, modified by cutin, suberin, and wax, are essential for land-living plants to separate between the inside of the plant and the outside environment. Whereas the chemical composition and the genetic background contributing to the biosynthesis are fairly well studied, important questions related to the functions of hydrophobised plant cell walls are to some extent still not fully answered, and they offer interesting research tasks to be solved in the future.

ACKNOWLEDGEMENTS

This work was supported by a grant from Deutsche Forschungsgemeinschaft (DFG).

REFERENCES

Abiko T, Kotula L, Shiono K, Malik AI, Colmer TD, Nakazono M (2012) Enhanced formation of aerenchyma and induction of a barrier to radial oxygen loss in adventitious roots of *Zea nicaraguensis* contribute to its waterlogging tolerance as compared with maize (*Zea mays* ssp. Mays). *Plant Cell and Environment* 35:1618–1630.

Abraham MH, McGowan JC (1987) The use of characteristic volumes to measure cavity terms in reversed phase liquid chromatography. *Chromatographia* 23:243–246.

Bak S, Beisson F, Bishop G, Hamberger B, Höfer R, Paquette S, Werck-Reichhart D (2011) Cytochromes. *The Arabidopsis Book*, no. 9: e0144, p450. https://www.ncbi.nlm.nih.gov/pmc/articles/PMC3268508/pdf/tab.0144.pdf.

Barberon M, Engelbertus J, Vermeer M, De Bellis D, Wang P, Naseer S, Andersen TG, Humbel BM, Nawrath C, Takano J, Salt DE, Geldner N (2016) Adaptation of root function by nutrient-induced plasticity of endodermal differentiation. *Cell* 164: 447–459.

Barthlott W, Neinhuis C (1997) Purity of the sacred lotus, or escape from contamination in biological surfaces. *Planta* 202:1–8.

Baur P, Buchholz A, Schönherr J (1997) Diffusion in plant cuticles as affected by temperature and size of organic solutes: Similarity and diversity among species. *Plant, Cell and Environment* 20:982–994.

Becker M, Kerstiens G, Schönherr J (1986) Water permeability of plant cuticles: Permeance, diffusion and partition coefficients. *Trees* 1:54–60.

Bernards MA (2002) Demystifying suberin. *Canadian Journal of Botany* 80:227–240.

Berry ZC, Emery NC, Gotsch SG, Goldsmith GR (2019) Foliar water uptake: Processes, pathways, and integration into plant water budgets. *Plant, Cell and Environment* 42:410–423.

Brongniart A (1834) Sur l'épiderme des plantes. Nouvelles récherches sur la structure de l'épiderm des végétaux. *Annales des Sciences Naturelles (Botanique)* 1:65–71.

Buschhaus C, Jetter R (2011) Composition differences between epicuticular and intracuticular wax substructures: How do plants seal their epidermal surfaces? *Journal of Experimental Botany* 62:841–853.

Crank J, Park GS (1968) *Diffusion in polymers*. London, New York: Academic Press.

Dastmalchi K, Chira O, Rodriguez MP, Yoo B, Serra O, Figueras M, Stark RE (2021) A chemical window into the impact of RNAi silencing of the StNAC103 gene in potato tuber periderms: Soluble metabolites, suberized cell walls, and antibacterial defense. *Phytochemistry* 190:112885.

DeBono A, Yeats TH, Rose JK, Bird D, Jetter R, Kunst L, Samuels L (2009) Arabidopsis LTPG is a glycosylphosphatidylinositol-anchored lipid transfer protein required for export of lipids to the plant surface. *The Plant Cell* 21:1230–1238.

Do T, Martinoia E, Lee Y (2018) Functions of ABC transporters in plant growth and development. *Current Opinion in Plant Biology* 41:32–38.

Dominguez E, Heredia-Guerrero JA, Heredia A (2015) Plant cutin genesis: Unanswered questions. *Trends in Plant Science* 20:551–558.

Dora SK, Koch K, Barthlott W, Wandelt K (2018) Kinetics of solvent supported tubule formation of lotus (*Nelumbo nucifera*) wax on highly oriented pyrolytic graphite (HOPG) investigated by atomic force microscopy. *Beilstein Journal of Nanotechnology* 9:468–481.

Elejalde-Palmett C, Martinez San Segundo I, Garroum I, Charrier L, De Bellis D, Mucciolo A, Guerault A, Liu J, Zeisler-Dieh V, Aharoni A, Schreiber L, Bakan B, Clausen MH, Geisler M, Nawrath C (2021) ABCG transporters export cutin precursors for the formation of the plant cuticle. *Current Biology* 31: 2111–2123.

Enstone DE, Peterson CA, Ma F (2003) Root endodermis and exodermis: Structure, function, and responses to the environment. *Journal of Plant Growth Regulation* 21:335–351.

Fernandez V, Eichert T (2009) Uptake of hydrophilic solutes through plant leaves: Current state of knowledge and perspectives of foliar fertilization. *Critical Reviews in Plant Sciences* 28:36–68.

Fernandez V, Guzman-Delgado P, Graca J, Santos S, Gil L (2016) Cuticle structure in relation to chemical composition: Re-assessing the prevailing model. *Frontiers in Plant Science* 7:427.

Fernandez V, Gil-Pelegrin E, Eichert T (2021) Foliar water and solute absorption: An update. *Plant Journal* 105:870–883.

Fich EA, Segerson NA, Rose JKC (2016) The plant polyester cutin: Biosynthesis, structure, and biological roles. *Annual Review of Plant Biology* 67:207–233.

Franke W (1961) Ectodesmata and foliar absorption. *American Journal of Botany* 48:683–691.

Geldner N (2013) The endodermis. *Annual Review of Plant Biology* 64:531–558.

Graca J, Pereira H (2000) Methanolysis of bark suberins: Analysis of glycerol and acid monomers. *Phytochemical Analysis* 11:45–51.

Graca J, Santos S (2007) Suberin: A biopolyester of plants' skin. *Macromolecular Bioscience* 7(2):128–135.

Graca J (2015) Suberin: The biopolyester at the frontier of plants. *Frontiers in Chemistry* 3:62.

Graca J, Cabral V, Santos S, Lamosa P, Serra O, Molinas M, Schreiber L, Kauder F, Franke R (2015) Partial depolymerization of genetically modified potato tuber periderm reveals intermolecular linkages in suberin polyester. *Phytochemistry* 117:209–219.

Heredia A (2003) Biophysical and biochemical characteristics of cutin, a plant barrier biopolymer. *Biochimica et Biophysica Acta* 1620:1–7.

Heredia-Guerrero JA, Heredia A, Dominguez E, Congolani R, Bayer IS, Athanassiou A, Benítez JJ (2017) Cutin from agro-waste as a raw material for the production of bioplastics. *Journal of Experimental Botany* 68:5401–5410.

Hooke R (1665) *Micrographia: or some physiological descriptions of minute bodies made by magnifying glasses with observations and inquiries thereupon*. London: Printed by Martyn J & Allestry J.

Huang WL, Serra O, Dastmalchi K, Jin LQ, Yang LJ, Stark RE (2017) Comprehensive MS and solid-state NMR metabolomic profiling reveals molecular variations in native periderms from four *Solanum tuberosum* potato cultivars. *Journal of Agricultural and Food Chemistry* 65:2258–2274.

Hwang ST (2011) Fundamentals of membrane transport. *Korean Journal of Chemical Engineering* 28:1–15.

Jeffree CE (2006) The fine structure of the plant cuticle. In *Biology of the plant cuticle*, eds. Riederer M, Müller C, 11–110. Oxford: Blackwell Publishing Ltd.

Khalil M, Lerat S, Beaudoin N, Beaulieu C (2019) The plant pathogenic bacterium *Streptomyces scabies* degrades the aromatic components of potato periderm via the β-ketoadipate pathway. *Frontiers in Microbiology* 10:2795.

Kamtsikakis A, Baales J, Zeisler-Diehl VV, Vanhecke D, Zoppe JO, Schreiber L, Weder C (2021) Asymmetric water transport in dense leaf cuticles and cuticle-inspired compositionally graded membranes. *Nature Communications* 12:1267.

Kim YX, Ranathunge K, Lee S, Lee Y, Lee D, Sung J (2018) Composite transport model and water and solute transport across plant roots: An update. *Frontiers in Plant Science* 9:193.

Kirkwood RC (1993) Use and mode of action of adjuvants for herbicides - A review of some current work. *Pesticide Science* 38:93–102.

Koch K, Barthlott W (2006) Plant epicuticular waxes: Chemistry, form, self-assembly and function. *Natural Product Communications* 1:11.

Koch K, Barthlott W, Koch S, Hommes A, Wandelt K, Mamdouh W, De-Feyter S, Broekmann P (2006) Structural analysis of wheat wax (*Triticum aestivum*, c.v. 'Naturastar' L.): From the molecular level to three dimensional crystals. *Planta* 223:258–270.

Koch K, Ensikat HJ (2008) The hydrophobic coatings of plant surfaces: Epicuticular wax crystals and their morphologies, crystallinity and molecular self-assembly. *Micron* 39:759–772.

Kolattukudy PE (1980) Cutin, suberin, and waxes. In *Lipids: Structure and function: A comprehensive treatise*, eds. Stumpf PK, 571–645. New York: Academic Press.

Kotula L, Ranathunge K, Steudle E (2009) Apoplastic barriers effectively block oxygen permeability across outer cell layers of rice roots under deoxygenated conditions: Roles of apoplastic pores and of respiration. *New Phytologist* 184:909–917.

Kroemer H (1903) Wurzelhaube, Hypodermis und Endodermis der Angiospermenwurzel. *Bibliotheca Botanica* 59:1–148.

Lendzian KJ, Kerstiens G (1991) Sorption and transport of gases and vapors in plant cuticles. In *Reviews of environmental contamination and toxicology*, eds. Ware GW, 65–128. New York: Springer.

Liu ZQ, Gaskin RE (2004) Visualisation of the uptake of two model xenobiotics into bean leaves by confocal laser scanning microscopy: Diffusion pathways and implication in phloem translocation. *Pest Management Science* 60:434–439.

Long L, Patel HP, Cory WC, Stapleton AE (2003) The maize epicuticular wax layer provides UV protection. *Functional Plant Biology* 30:75–81.

Madison KC (2003) Barrier function of the skin: "la raison d'etre" of the epidermis. *Journal of Investigative Dermatology* 121:231–241.

Man Y, Zhao Y, Ye R, Lin J, Jing Y (2018) In vivo cytological and chemical analysis of Casparian strips using stimulated raman scattering microscopy. *Journal of Plant Physiology* 220:136–144.

McGarvey DJ, Croteau R (1995) Terpenoid metabolism. *The Plant Cell* 7:1015–1026.

Molina I, Li-Beisson Y, Beisson F, Ohlrogge JB, Pollard M (2009) Identification of an Arabidopsis feruloyl-coenzyme A transferase required for suberin synthesis. *Plant Physiology* 151:1317–1328.

Naseer S, Lee Y, Lapierre C, Franke RB, Nawrath C, Geldner N (2012) Casparian strip diffusion barrier in Arabidopsis is made of a lignin polymer without suberin. *Proceedings of the National Academy of Sciences 109:*10101-10106.

Nawrath C (2002) The biopolymers cutin and suberin. *The Arabidopsis Book*, no.1: e0021. https://www.ncbi.nlm.nih.gov/pmc/articles/PMC3243382/pdf/tab.0021.pdf.

Nightingale ER (1959) Phenomenological theory of ion solvation. Effective radii of hydrated ions. *The Journal of Physical Chemistry A* 63:1381–1387.

Nobel PS (2009) *Physicochemical and environmental plant physiology.* Fourth Edition. Amsterdam: Academic Press.

Pereira H (2013) Variability of the chemical composition of cork. *Bioresources* 8:2246–2256.

Perumalla CJ, Peterson CA, Enstone DE (1990) A survey of angiosperm species to detect hypodermal Casparian bands. I. Roots with a uniserate hypodermis and epidermis. *Botanical Journal of the Linnean Society* 103:93–112.

Peterson CA (1989) Significance of the exodermis in root function. In *Structural and functional aspects of transport in roots*, eds. Loughamn BC, Gašparíková O, Kolek J, 35–40. Dordrecht: Springer.

Pinot F, Beisson F (2010) Cytochrome P450 metabolizing fatty acids in plants: Characterization and physiological roles. *FEBS Journal* 278:195–205.

Pollard M, Beisson F, Li Y, Ohlrogge JB (2008) Building lipid barriers: Biosynthesis of cutin and suberin. *Trends in Plant Science* 13:236–246.

Rai RK, Singh VP, Upadhyay A (2017) Soil Analysis. In *Planning and evaluation of irrigation projects*, eds. Rai RK, Singh VP, Upadhyay A, 505–523. London: Academic Press.

Ranathunge K, Lin J, Steudle E, Schreiber L (2011) Stagnant deoxygenated growth enhances root suberization and lignifications, but differentially affects water and NaCl permeabilities in rice (*Oryza sativa* L.) roots. *Plant Cell and Environment* 34:1223–1240.

Reynhardt EC, Riederer M (1991) Structure and molecular dynamics of the cuticular wax from leaves of *Citrus-aurantium* L. *Journal of Physics D-Applied Physics* 24:478–486.

Reynhardt EC, Riederer M (1994) Structures and molecular dynamics of plant waxes. II Cuticular waxes from leaves of *Fagus sylvatica L. and Hordeum vulgare L. European Biophysics Journal* 23:59–70.

Riederer M, Schreiber L (2001) Protecting against water loss: Analysis of the barrier properties of plant cuticles. *Journal of Experimental Botany* 52:2023–2032.

Riederer M, Müller C (2006) *Biology of the plant cuticle.* Oxford: Annual plant reviews.

Samuels L, Kunst L, Jetter R (2008) Sealing plant surfaces: Cuticular wax formation by epidermal cells. *Annual Review of Plant Biology* 59:683–707.

Schönherr J (1976) Water permeability of isolated cuticular membranes: The effect of cuticular waxes on diffusion of water. *Planta* 131:159–164.

Schönherr J (2006) Characterization of aqueous pores in plant cuticles and permeation of ionic solutes. *Journal of Experimental Botany* 57:2471–2491.

Schönherr J, Baur P (1994) Modelling penetration of plant cuticles by crop protecting agents (CPA) and effects of adjuvants on rates of penetration. *Pesticide Science* 42:185–208.

Schönherr J, Bukovac MJ (1970) Preferential polar pathways in the cuticle and their relationship to ectodesmata. *Planta* 92:189–201.

Schönherr J, Riederer M (1988) Desorption of chemicals from plant cuticles: Evidence for asymmetry. *Archives of Environmental Contamination and Toxicology* 17:13–19.

Schönherr J, Riederer M (1989) Foliar penetration and accumulation of organic chemicals in plant cuticles. In *Reviews of Environmental contamination and toxicology*, eds. Ware GW, 1–70. New York: Springer.

Schreel JDM, Steppe K (2020) Foliar water uptake in trees: Negligible or necessary? *Trends in Plant Science* 25:590–603.

Schreiber L, Schönherr J (1990) Phase transitions and thermal expansion coefficients of plant cuticles. *Planta* 182:186–193.

Schreiber L, Breiner HW, Riederer M, Düggelin M, Guggenheim R (1994) The Casparian strip of *Clivia miniata* reg. roots: Isolation, fine structure and chemical nature. *Botanica Acta* 107:353–361.

Schreiber L, Riederer M (1996a) Determination of diffusion coefficients of octadecanoic acid in isolated cuticular waxes and their relationship to cuticular water permeabilities. *Plant, Cell and Environment* 19:1075–1082.

Schreiber L, Riederer M (1996b) Ecophysiology of cuticular transpiration: Comparative investigation of cuticular water permeability of plant species from different habitats. *Oecologia* 107:426–432.

Schreiber L, Hartmann K, Skrabs M, Zeier J (1999) Apoplastic barriers in roots: Chemical composition of endodermal and hypodermal cell walls. *Journal of Experimental Botany* 50:1267–1280.

Schreiber L (2005) Polar paths of diffusion across plant cuticles: New evidence for an old hypothesis. *Annals of Botany* 95:1069–1073.

Schreiber L, Franke R, Hartmann K (2005) Wax and suberin development of native and wound periderm of potato (*Solanum tuberosum* L.) and its relation to peridermal transpiration. *Planta* 220:520–530.

Schreiber L (2010) Transport barriers made of cutin, suberin and associated waxes. *Trends in Plant Science* 15:546–553.

Schuster AC, Burghardt M, Riederer M. (2017) The ecophysiology of leaf cuticular transpiration: Are cuticular water permeabilities adapted to ecological conditions? *Journal of Experimental Botany* 68:5271–5279.

Segado P, Heredia-Guerrero JA, Heredia A, Dominguez E (2020) Cutinsomes and cutin synthase1 function sequentially in tomato fruit cutin deposition. *Plant Physiology* 183:1622–1637.

Serra O, Pekka Mähönen A, Hetherington AJ, Ragni L (2022) The making of plant armor: The periderm. *Annual Review of Plant Biology* 73:405–432.

Serra O, Soler M, Hohn C, Franke R, Schreiber L, Prat S, Molinas M, Figueras M (2009a) Silencing of StKCS6 in potato periderm leads to reduced chain lengths of suberin and wax compounds and increased peridermal transpiration. *Journal of Experimental Botany* 60:697–707.

Serra O, Soler M, Hohn C, Sauveplane V, Pinot F, Franke R, Schreiber L, Prat S, Molinas M, Figueras M (2009b) CYP86A33-Targeted gene silencing in potato tuber alters suberin composition, distorts suberin lamellae, and impairs the periderm's water barrier function. *Plant Physiology* 149:1050–1060.

Shiono K, Ando M, Nishiuchi S, Takahashi H, Watanabe K, Nakamura M, Matsuo Y, Yasuno N, Yamanouchi U, Fujimoto M, Takanashi H, Ranathunge K, Franke R, Shitan N, Nishizawa N, Takamure I, Yano M, Tsutsumi N,

Schreiber L, Yazaki K, Nakazono M, Kato K (2014) RCN1/OsABCG5, an ATP-binding cassette (ABC) transporter, is required for hypodermal suberization of roots in rice (*Oryza sativa*). *The Plant Journal* 214:40–51.

Shukla V, Barberon M (2021) Building and breaking of a barrier: Suberin plasticity and function in the endodermis. *Current Opinion in Plant Biology* 64:102153.

Sieber P, Schorderet M, Ryser U, Buchala A, Kolattukudy P, Métraux JP, Nawrath C (2000) Transgenic Arabidopsis plants expressing a fungal cutinase show alterations in the structure and properties of the cuticle and postgenital organ fusions. *The Plant Cell* 12:721–737.

Silva SP, Sabino MA, Fernandes EM, Correlo VM, Boesel LF, Reis RL (2005) Cork: Properties, capabilities and applications. *International Materials Reviews* 50:345–365.

Silvestre AJD, Neto CP, Gandini A (2008) Cork and suberins: Major sources, properties and applications. In *Monomers, Polymers and Composites from Renewable Resources*, eds. Belgacem MN, Gandini A, 305–320. Elsevier.

Sitte P, Rennier R (1963) Untersuchungen an cuticularen Zellwandschichten. *Planta* 60:19–40.

Song C, Shen W, Du L, Wen J, Lin J, Li R (2019) Development and chemical characterization of Casparian strips in the roots of Chinese fir (*Cunninghamia lanceolata*). *Trees* 33:827–836.

Tedeschi G, Benitez JJ, Ceseracciu L, Dastmalchi K, Itin B, Stark RE, Heredia A, Athanassiou A, Heredia-Guerrero JA (2018) Sustainable fabrication of plant cuticle-like packaging films from tomato pomace agro-waste, beeswax, and alginate. *ACS Sustainable Chemistry and Engineering* 6:14955–14966.

Ursache R, De Jesus Vieira Teixeira C, Dénervaud Tendon V, Gully K, De Bellis D, Schmid-Siegert E, Andersen TG, Shekhar V, Calderon S, Pradervand S, Nawrath C, Geldner N, Vermeer JEM (2021) GDSL-domain proteins have key roles in suberin polymerization and degradation. *Nature Plants* 7:353–364.

Varney GT, Mccully ME, Canny MJ (1993) Sites of entry of water into the symplast of maize roots. *New Phytologist* 125:733–741.

Vesala T, Sevanto S, Grönholm T, Salmon Y, Nikinmaa E, Hari P, Hölttä T (2017) Effect of leaf water potential on internal humidity and CO_2 dissolution: Reverse transpiration and improved water use efficiency under negative pressure. *Frontiers in Plant Science* 8:54.

Vishwanath SJ, Delude C, Domergue F, Rowland O (2015) Suberin: Biosynthesis, regulation, and polymer assembly of a protective extracellular barrier. *Plant Cell Reports* 34:573–586.

Walton TJ (1990) Waxes, cutin and suberin. *Methods in Plant Biochemistry* 4:105–158.

Wunderling A, Ripper D, Barra-Jimenez A, Mahn S, Sajak K, Mehdi BT, Ragni L (2018) A molecular framework to study periderm formation in Arabidopsis. *New Phytologist* 219:216–229.

Xin A, Herburger K (2021) Mini Review: Transport of hydrophobic polymers into the plant apoplast. *Frontiers in Plant Science* 11:590990.

Yang W, Simpson JP, Li-Beisson Y, Beisson F, Pollard M, Ohlrogge JB (2012) A land-plant-specific glycerol-3-phosphate acyltransferase family in Arabidopsis: Substrate specificity, sn-2 preference, and evolution. *Plant Physiology* 160:638–652.

Yeats TH, Martin LB, Viart HM, Isaacson T, He Y, Zhao L, Matas AJ, Buda GJ, Domozych DS, Clausen MH, Rose JK (2012) The identification of cutin synthase: Formation of the plant polyester cutin. *Nature Chemical Biology* 8:609–611.

Yu D, Ranathunge K, Huang H, Pei Z, Franke R, Schreiber L, He C (2008) Wax Crystal-Sparse Leaf1 encodes a β-ketoacyl CoA synthase involved in biosynthesis of cuticular waxes on rice leaf. *Planta* 228:675–685.

Zarebanadkouki M, Trtik P, Hayat F, Carminati A, Kaestner A (2019) Root water uptake and its pathways across the root: Quantification at the cellular scale. *Scientific Reports* 9:12979.

Zeisler V (2013) *Die pflanzliche Kutikula. PhD*, Rheinische Friedrich-Wilhelms-Universität Bonn, Germany. https://nbn-resolving.org/urn:nbn:de:hbz:5n-34411

Zeisler V, Schreiber L (2016) Epicuticular wax on cherry laurel (*Prunus laurocerasus*) leaves does not constitute the cuticular transpiration barrier. *Planta* 243:65–81.

Zeisler-Diehl V, Müller Y, Schreiber L (2018) Epicuticular wax on leaf cuticles does not establish the transpiration barrier, which is essentially formed by intracuticular wax. *Journal of Plant Physiology* 227:66–74.

Zimmermann HM, Hartmann K, Schreiber L, Steudle E (2000) Chemical composition of apoplastic transport barriers in relation to radial hydraulic conductivity of corn roots (*Zea mays* L.). *Planta* 210:302–311.

Ziv C, Zhao Z, Gao YG, Xia Y (2018) Multifunctional roles of plant cuticle during plant-pathogen interactions. *Frontiers in Plant Science* 9:1088.

15 Shapeshifters

Functional Interplays between Plant Cell Walls and Dynamic, Reversible Shape Changes in Plant Cells and Tissues

Charles T. Anderson
The Pennsylvania State University

INTRODUCTION

Plant cell walls are incredibly strong, flexible, and durable: for example, they support the ability of some trees to tower over 100 metres against the downward pull imposed by gravity while enabling the branch tips of those same trees to whip freely through hurricane-force winds, allowing them to weather storms and endure across many centuries. People have long leveraged the unique physical attributes of plant-derived materials to construct long-lasting clothing, furniture, and buildings: for instance, the Tarkhan Dress was woven from flax fibres at the dawn of Egypt's First Dynasty ~4,500 years ago and survives even today (Stevenson and Michel 2016). Likewise, the Horyu-ji temple in Ikaruga, Nara Prefecture, Japan was constructed from wood over 1300 years ago and still stands, despite experiencing strong earthquakes throughout its history (Nakahara et al. 2000). At a biological level, the patterned deposition and biomechanical properties of plant cell walls underpin cell, tissue, and organ shape, as well as the movements of plant cells and organs that are often slow enough to remain imperceptible to casual observation and only become observable by time-lapse imaging (Stark 2008), but in some cases the movement can be faster than the blink of an eye and require high-speed cameras to capture (Vincent et al. 2011). Very few plant cells remain completely rigid and stationary over the course of the lifetime of a plant, meaning that despite their bricklike patterning in the plant body, almost all plant cells move and flex to some degree; otherwise, plants would be too brittle to grow, heal, and survive the vagaries of changing environments.

Plant cell walls are composed of interwoven networks of biopolymers that include polysaccharides, glycoproteins, and polyphenolic compounds, along with enzymes, metabolites, ions, and water (Anderson and Kieber 2020). With a tensile strength similar to that of steel (Wu et al. 2014), cellulose is the strongest component of the plant cell wall and is composed of β-1,4-linked glucan chains that coalesce soon after polymerisation into cable-like, partially crystalline microfibrils (Yang and Kubicki 2020). Cellulose is laid down in the cell wall by multimeric protein complexes that move in the plasma membrane, driven by the force of cellulose polymerisation, along linear patterns that follow cortical microtubules or pre-existing wall patterning (Chan and Coen 2020). Other polysaccharides, including acidic pectins and branched hemicelluloses, are synthesised by multiple enzymes in mobile Golgi bodies (see Chapters 2 and 17) before being secreted into the wall. Upon secretion, pectins and hemicelluloses interact with the surfaces of cellulose microfibrils and each other to form a hydrated matrix that surrounds, separates, and interlinks the cellulose. Pectins are negatively charged and can thus be highly hydrated but can also crosslink via calcium and boron complexation to form gels (Funakawa and Miwa 2015; Du et al. 2022), whereas hemicelluloses interact with cellulose to help form "biomechanical hotspots" that limit cell wall extensibility (Cosgrove 2022). These configurations of polymers facilitate the expansion of the primary cell walls that surround actively expanding plant cells. The sequential deposition of cell wall layers and the reorganisation of existing wall components during cell expansion are thought to generate a strong but flexible biomaterial, or "flexoskeleton" (Yi et al. 2019), that can expand many-fold while retaining its ability to withstand the high turgor pressure that accompanies cell growth in plants.

The patterning of both the microtubule cytoskeleton and exocytosis events determine the initial nanoscale structure of the cell wall (Gu and Rasmussen 2022); however, this structure changes as plant cells modify their cell walls through the action of lytic and ligating enzymes, regulated ion-based crosslinking, and mechanical and signalling feedback loops that arise during cell growth (Anderson and Kieber 2020). The endogenous enzymes that remodel cell wall polymers include cellulases that hydrolyse cellulose chains and pectin methyl-esterases that remove the methyl groups from pectin backbones, making them susceptible to pectin-cleaving enzymes that include polygalacturonases and pectate lyases while also exposing negatively charged carboxyl groups that can be crosslinked

DOI: 10.1201/9781003178309-15

by calcium. The latter two enzyme classes soften the cell wall and can, along with other pectin-degrading enzymes, facilitate cell separation at the middle lamella layer that adheres most plant cells to one another (Bou Daher and Braybrook 2015; Zamil and Geitmann 2017). In the case of hemicelluloses, endo-hydrolases cleave these polymers, whereas endotransglycosylases can both cleave and relink them, generating new hemicellulose networks. Puzzlingly, pectin and hemicellulose degradation *per se* seem to be insufficient to enable cell wall expansion at the mechanical level; rather, these lytic enzymes appear to act in concert with expansins, non-enzymatic proteins that bind to hemicellulose–cellulose interfaces and facilitate wall "creep" (Wang et al. 2013).

After cell expansion is complete, certain types of plant cells, including xylem and fibre cells, deposit thickened secondary walls that are composed largely of cellulose, hemicellulose, and polyphenolic lignin (Meents et al. 2018). During secondary wall formation, cellulose is laid down in aligned layers that can alternate in orientation to produce an immensely strong wall. Lignin, which polymerises in the wall via radical coupling reactions that are driven by monolignol radicalisation by peroxidases and laccases (Tobimatsu and Schuetz 2019), further strengthens and stiffens the cell wall, allowing fibre cells to support upright posture in plants, and can also alter the wall's hydration status and hydrophilicity, for instance, facilitating water transport up the plant stem by making the inner surface of hollow xylem cells hydrophilic to maintain continuity of water columns while waterproofing the bulk secondary wall to control lateral water diffusion into and out of the water column. The components of the plant cell wall interact dynamically to produce robust biomaterials that are resistant to degradation while, in many cell types, remaining sufficiently flexible to enable plants to move to capture light or carbon dioxide for photosynthesis, find support, minimise predation and abiotic stress, or in some carnivorous plant species even capture animal prey.

The patterned deposition of wall polymers by plant cells, in concert with spatiotemporally precise wall reorganisation and remodelling, helps determine how cell volume and shape can change dynamically during plant growth and development (Saffer 2018). The ability of the cell wall to swell, shrink, bend, and twist in some cell types can also actuate cell and organ movements in response to environmental and intrinsic stimuli (Morris and Blyth 2019). In this chapter, we will peer through the lenses of scientific history and technological advancement to explore how the tunable physical properties of the cell wall, in combination with the mechanical perceptions, responses, and interventions carried out by plant cells, collectively termed their mechanobiology, have been discovered and connect plant cell walls to dynamic, reversible changes in cell shape and facilitate the cellular deformations that help drive many plant movements. For a complementary exploration of how cell walls regulate cell expansion during irreversible growth processes, see Chapter 6.

OBSERVATION AND MEASUREMENT OF 3D CELL SHAPE, WALL STRUCTURE AND MECHANICS, AND CELL AND TISSUE MOVEMENTS IN PLANTS

Observing Plant Cells across Four Dimensions

Plants generate cells in a dazzling variety of shapes (Ivakov and Persson 2013) that range from geometrically simple spheres, cylinders, and rectangular prisms to complex lobed and stellate cells (Figure 15.1). Some of these cell shapes are reminiscent of the complex macroscopic morphologies of plant tissues and organs. Organ shape mirrors cell shape in some cases: for example, stalks and stems often contain long, thin cylindrical cells, whereas flat organs such as leaves and petals contain wide, flat epidermal cells. However, there is not always a direct correspondence between cell and tissue/organ shape; leaves, for example, can also contain columnar cells in palisade parenchyma and spherical cells in spongy mesophyll. In all cases, the cell wall that surrounds each protoplast, or plasma membrane-delimited cell, constrains its shape, moulding the cell into its final form, which scientists and naturalists began to appreciate and analyse with the advent of practical light microscopy during the 17th century, in which cells, the basic units of life that had previously remained invisible, were first revealed to human eyes and minds.

Robert Hooke named not only plant cells, but also their walls, in his 1665 masterpiece *Micrographia* (Hooke 1665), where he used a compound light microscope to observe and draw the round transverse cross sections and rectangular longitudinal cross sections of cells in hand-sectioned cork tissue (*Micrographia*, Observation XVIII, Figure 1). Applying impressive biological insight, Hooke also related the cell morphologies and arrangements of wood and cork to their differing tissue organisation and functions. In the ensuing centuries, many naturalists and scientists made microscopic observations and recordings of cell shapes and sizes, especially at the external surfaces of organs and in cross sections, across a diversity of plant taxa. Histological staining methods, for instance, the use of the polychromatic dye and toluidine blue to label cell walls of differing composition (O'Brien et al. 1964) advanced the ability of observers to discern different cells, and some cell parts such as the nucleus, from one another. However, because it relies on the ability to transmit light through a sample and thus requires sectioning in many cases, and because it cannot achieve optical sectioning and thus leaves out-of-focus objects in the image, brightfield light microscopy is limited in its ability to accurately reveal the 3D shapes of larger plant cells, especially when they have high aspect ratios (i.e., are very long and thin) or are intertwined and interconnected in convoluted patterns. This challenge also applies to methods such as conventional modes of transmission electron microscopy (Hawes 2012), for which very thin sections must be prepared and from which 3D reconstructions

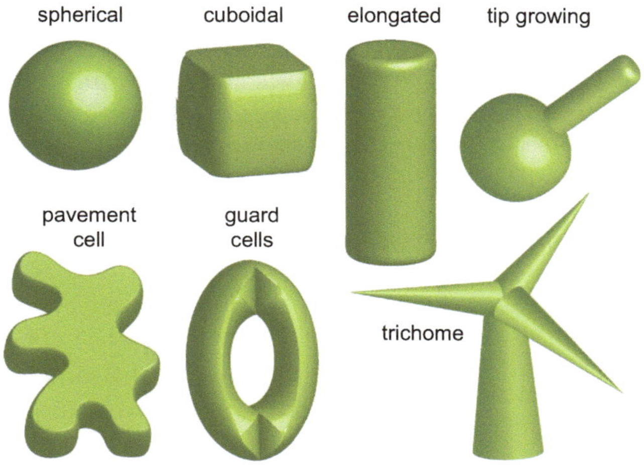

FIGURE 15.1 Examples of plant cell shapes. Bottom row includes examples of epidermal cells.

of cell shape, for example from serial sections (Gay and Anderson 1954), are extremely labour-intensive.

The ability to record and quantify plant cell shapes in 3D has been greatly facilitated by confocal fluorescence microscopy (Hepler and Gunning 1998) and digital imaging and image processing (Moreno et al. 2006), both of which arose late in the last century and have been refined in this one. Confocal microscopy, which encompasses both a laser scanning modality in which a spot of excitation light is rastered over the sample and emitted photons pass through a pinhole to a detector to generate an image of a single focal plane, or a spinning disk modality in which light passing through aligned arrays of microlenses and pinholes is spun across the optical field at high speed and captured by a camera, removes blurring from out-of-focus light, allowing for the collection of sharp, 3D image stacks that can be reconstructed and processed computationally. A related technique, light-sheet fluorescence microscopy, selectively illuminates flat imaging planes and allows for rapid capture of 3D volumes in larger, growing plant tissues (Ovečka et al. 2018). Remaining challenges in the accurate capture of cell shape using light microscopy include spherical aberrations in microscope objectives that can distort the dimensions of the image stack; scattering within the plant tissue that obscures cell boundaries; and autofluorescence that can interfere with the detection of a desired label (Moreno et al. 2006; Shaw and Ehrhardt 2013). In the case of plant cell walls, their thickness is often near or below the diffraction-imposed resolution limit of optical microscopy, defined as the wavelength of the excitation light divided by twice the numerical aperture of the objective (λ/2NA) in the Abbe equation (Komis et al. 2015). This limitation means that precisely capturing the boundaries and thicknesses of cell walls and thus the shapes of the encased protoplasts, which are bound by even thinner plasma membranes ~3 nm in thickness, in 3D in intact tissues is very challenging. Labelling cell boundaries with bright, photostable dyes or genetically encoded markers is a prerequisite for capturing 3D cell shapes precisely. Super-resolution microscopy, which uses optical, detection, and processing methods to surpass the diffraction limit, can help to more precisely map cell walls, their component distributions, and cell boundaries (Fitzgibbon et al. 2010; Komis et al. 2015; Schubert 2017; Shaw et al. 2019; Haas et al. 2020; Peaucelle et al. 2020; Duncombe et al. 2022), but the application of super-resolution microscopy to analysing 3D cell shape is still hampered by challenges relating to sample preparation, autofluorescence, and the scattering of light as it passes through the cell wall, which has a refractive index above 1.4 and can be highly scattering and birefringent due to the presence of ordered cellulose, and the protoplast, which is aqueous but inhomogeneous and has a refractive index closer to 1.3 (Shaw and Ehrhardt 2013).

Optical microscopy has the virtue of being compatible with live cell imaging, allowing the capture of key processes such as cellulose synthesis (Paredez et al. 2006; Duncombe et al. 2020, 2022), but other approaches can also be used to capture plant cell and cell wall structure. Atomic force microscopy (AFM) bypasses the resolution limit of light microscopy and can reveal both the surface topology and mechanical properties of plant cell walls (Marga et al. 2005; Ding and Himmel 2006; Peaucelle et al. 2011; Fernandes et al. 2012; Zhang et al. 2016), but is technically demanding and does not typically reveal chemical information about the sample unless it is combined with a technique such as infrared spectroscopy (Kesari et al. 2021; Bhagia et al. 2022; Bilkey et al. 2022). Other imaging methods can transcend the diffraction limit of light but usually require sample fixation, freezing, embedding, and/or sectioning and are therefore incompatible with living cells. These include transmission electron microscopy, which in combination with immunogold labelling of cell wall components can assist in mapping the nanoscale distribution of those components (Wilson and Bacic 2012), and scanning electron microscopy, which can reveal cell and wall surface shape, topography, and patterning (Zheng et al. 2017). Serial blockface sectioning in combination with scanning electron microscopy (Kittelmann et al. 2016; Czymmek et al. 2020) can reveal the three-dimensional cellular architecture of tissues, but achieving high contrast for cell wall polymers in this and other modalities of electron microscopy is challenging due to the relative electron transparency of polysaccharides. One of the most recent approaches to collecting three-dimensional data for plant cell walls is cryo-electron tomography (cryo-ET), which when combined with ion beam milling can be used to determine the three-dimensional molecular architecture of small regions of cell walls (Nicolas et al. 2022). Further advances in imaging techniques that reveal details of cell wall structure, including polarised light microscopy (Verbelen and Stickens 1995), vibrational spectroscopy-enabled microscopy (Gierlinger 2018), super-resolution light microscopy, electron microscopy, and cryo-microscopy, in combination with immunohistochemical labelling of cell wall components (Hervé et al. 2011; Haas et al. 2020) and ever-expanding data storage and processing capabilities, might allow these imaging approaches to converge to provide 3D molecular maps of the cell walls of whole plant cells or tissues. At a larger scale, X-ray computed tomography (Piovesan et al. 2021), high-resolution tomography and laminography achieved using synchrotron radiation (Verboven et al. 2015), and laser ablation tomography (Strock et al. 2022) can be used to measure 3D cell shape in the context of larger tissues, but coarser resolution and long acquisition times currently limit the ability of these techniques to connect nano- to macro-scale wall, cell, and tissue organisation. For whole tissues, organs, and plants, methods to integrate terabytes of 3D imaging data with mechanical and genetic information into interactive, 4D atlases of tissue growth and development are just beginning to be explored (Refahi et al. 2021).

Image capture is also a realm of accelerating advancement in studies of the morphology and dynamics of plant cell shape. From Hooke's meticulous drawings of cork (Hooke 1665) through Darwins' capture of plant movements using glass plates (Darwin and Darwin 1880) and the early days of photomicrography (Breidbach 2002), perception of plant cell shape and movements were temporally limited by the speed with which the human hand could reproduce what the eye saw or by the relatively high cost, poor sensitivity, and cumbersome capture mechanisms of plate- and film-based imaging. Digital image capture has increased many-fold the number of images that can be collected, and high-speed imaging (Vincent et al. 2011) has enabled researchers to begin to quantitatively analyse the rapid changes in geometry that occur during prey capture by Venus fly traps (*Dionaea muscipula*) and bladderworts (*Utricularia* spp.) or during rapid seed and pollen release (Mitchell et al. 2022). Plants do not always move in the same time signatures as animals, so we must often dramatically slow down or speed up to match their tempos.

As floods of imaging data have overwhelmed the capacity of hand measurements, automated image segmentation and artificial intelligence (AI) are stepping into the gap to assist. Segmentation software like MorphographX (de Reuille et al. 2015) can be used to measure and track changes in 3D cell shape in the context of tissues and organs (Bassel and Smith 2016). However, segmentation of blurry data arising from light scattering remains a challenge for automated analyses (Rui et al. 2016; Yi et al. 2018), but multiphoton and light-sheet microscopy (Heiner et al. 2018; DeVree et al. 2021; Ovečka et al. 2022), automated deconvolution, and computer vision/machine learning approaches (Davaasuren et al. 2022) have the potential to help address this problem.

Experimental Approaches to Studying Plant Cell and Wall Biomechanics and How These Influence Plant Cell Shape and Shape Changes

It has long been appreciated that plants can exert mechanical force via two linked processes, hydration that leads to the establishment of turgor pressure and cell wall expansion (Baskin 2017; Cosgrove 2022). One example of this type of force in action is the ability of roots to widen faults in solid rock via hydraulically driven outward pressure as they grow with the gravity vector. Measurements of whole-plant mechanical properties and forces (Shah et al. 2017) have deepened our understanding of growth phenomena and their regulation, but how they connect to the detailed configuration and nano-mechanics of plant cell walls has remained more obscure. Recently, several experimental methods to measure wall mechanics at the single-cell and subcellular levels have emerged. One imaging-based approach uses sequential deflation and inflation of cells via osmolytes, in combination with confocal data collection, image segmentation, and calculation of volumetric elastic modulus, to estimate wall mechanical properties across individual cells in intact tissues (Kierzkowski et al. 2012). Fluorescence Brillouin microscopy, which exploits the interaction of laser light with thermal vibrations in the sample to derive mechanical properties solely from imaging, is an emerging method for analysing the biomechanics of plant cell walls (Elsayad et al. 2016). For finer-scale estimates of subcellular strain that can be used to calculate mechanical properties, fiducial markers such as fluorescent beads (Yanagisawa et al. 2015) or plasmodesmatal markers (Belteton et al. 2021) can be used.

One approach for directly measuring wall mechanics at the subcellular scale is AFM, in which a probe tip that can vary in size, shape, and hardness and is attached to a flexible cantilever is pressed or tapped against the cell or wall (Peaucelle et al. 2012). The resulting data, when processed using appropriate mechanical models, can reveal both the nanoscale topography and the modulus of the sample being probed. Small, pointed AFM tips can image and probe the mechanics of individual wall polymers (Zhang et al. 2016), whereas larger, round AFM tips have been used to measure mechanical properties across multiple cell layers in meristem tissue (Peaucelle et al. 2011). One caveat of many AFM studies is that they are carried out after cell plasmolysis (Carter et al. 2017), which reduces the confounding factor of turgor pressure in the data, but might lead to artefacts in measuring cell/wall mechanics since there is evidence that cell walls stiffen under turgor pressure, forming a "hydroskeleton" (Bidhendi and Geitmann 2016). Cellular force microscopy can also be used to measure the mechanics of plant cells and tissues at larger forces and spatial scales than AFM (Majda et al. 2019). Nanoindentation, in which a hard probe is used to iteratively indent a sample of interest, has recently been used in combination with finite element modelling to simultaneously estimate both wall modulus and turgor pressure in expanding and contracting guard cells (Chen et al. 2021). All these methods rely on properly calculating cell/wall modulus using parameterised mathematical models, none of which are perfect.

Currently, quantitative modelling of 3D shape changes in plant cells and how these might be accompanied by alterations in wall and/or cell mechanics is facilitated by approaches that include finite element modelling (Bidhendi and Geitmann 2018) and AI-based image segmentation. One challenge is connecting these computational methods to high-quality imaging data, which if an active change in shape for individual cells is being quantified must be collected in living cells (Rui et al. 2017; Yi et al. 2018), to nanoscale and mesoscale mechanics. Although methods such as serial sectioning scanning electron microscopy and cryo-ET can reconstruct wall, cell, and tissue morphology in 3D at high resolution, the cells must first be killed and embedded, precluding the accurate capture of dynamic changes in the shapes or mechanics of single cells.

At a fundamental level, all plant growth patterns and movements are driven by changes in cell shape or volume since cell division does not *per se* drive plant growth but simply involves the subdivision of an existing cell. Elongating hypocotyl cells allow seedlings growing in the dark to rapidly stretch upwards, an adaptation that pushes the top of the seedling towards the light to begin photosynthesising. The lengthening and shortening of stomatal guard cells enable plants to reversibly open and close their stomatal pores to control water vapour release and carbon dioxide absorption at the leaf surface. Cells on opposite sides of the pulvini of sensitive plants grow and shrink to fold their leaflets closed in response to touch or intense heat. Petiole cells can also expand and contract to enable circadian leaf nodding. Contractile cells in the roots of bulb-bearing plants cause the bulbs to be pulled deeper into the soil, avoiding frost damage, and the twining movements of vines are also driven in part by contractile cells. As our ability to detect, locate, and quantify cell wall polymers extends into the nanoscale, the molecular and biomechanical mechanisms underlying these macro- to micro-scale phenomena are starting to be discovered, as detailed below.

DEVELOPMENTAL ORIGINS OF PLANT CELLS WITH REVERSIBLE SHAPE AND SIZE

Most plant somatic cells begin life as cuboidal (e.g., in a tightly packed tissue like the root meristem), spheroidal (e.g., in a loosely packed tissue like spongy mesophyll), or curvilinear prism (e.g., in a growing epidermis) (Figure 15.1), arising from either fertilisation in the case of the zygote or cytokinesis in the case of all other cells. For each cell, the tissue in which it resides and its cell identity influence its morphogenesis: for example, the spongy mesophyll cells in the leaves of many plants arise from divisions of rounded precursors and can either remain round throughout their morphogenesis or adopt stellate morphologies, facilitating contact with the air spaces in the leaf while retaining minimal contacts with their neighbouring cells (Zhang et al. 2021), whereas pavement cells in the leaf epidermis arise from divisions of protodermal cells and can adopt tightly interlocking, flattened morphologies at maturity that resemble puzzle pieces (Zuch et al. 2022).

The developmental signals that promote cell morphogenesis include transcriptional, cell signalling, and biomechanical cues (Zuch et al. 2022). Transcriptional networks ultimately control the expression of the genes that produce the proteins that comprise the cytoskeletal elements, biosynthetic machinery, and endomembrane trafficking systems that initiate both cell expansion and the symmetry-breaking events that lead to anisotropic growth. For example, trichome differentiation, abundance, and shape are determined in part by a network of transcription factors and related signalling pathways (Yang and Ye 2013). The polarised localisation of some proteins (e.g., BREAKING OF ASYMMETRY IN THE STOMATAL LINEAGE, BASL) (Dong et al. 2009) can set up feedback loops that drive asymmetric cell division and morphogenesis, as in the case of the spatially restricted development of stomatal guard cells (Facette et al. 2019). Cell biological cues also include the perception of external stimuli like light (via photoreceptors) and gravity (via statoliths and potentially other sensory mechanisms), which can in turn feed into signalling pathways that result in uneven patterning and remodelling of the cell wall; for example, tip-growing root hairs depend on gravity for their correct morphogenesis (Kwon et al. 2015). Stress perception at the cellular and tissue scales that results in cytoskeletal remodelling can also result in uneven production, reinforcement, and remodelling of the plant cell wall that results in differential growth at the cell or tissue scales, as is hypothesised to be the case for trichomes (Yanagisawa et al. 2015) and for the epidermis at large (Uyttewaal et al. 2012).

However, the extent to which *reversible* cell shapes are determined by the same developmental signalling and transcriptional pathways that underlie irreversible growth has not been fully elucidated. One of the best-studied cell types, in terms of developmental patterning, is stomatal guard cells, which reversibly swell and shrink to open and close the stomatal pores in the leaf epidermis of vascular plants. The molecular players, including extracellular peptides, receptors, MITOGEN ACTIVATED PROTEIN (MAP) kinases, and transcription factors, that drive stomatal differentiation have been extensively characterised, especially in the model plant *Arabidopsis thaliana* (Lee and Bergmann 2019) and, more recently, in the model grass

Brachypodium distachyon (McKown and Bergmann 2020; Spiegelhalder and Raissig 2021). However, how the patterning and modification of the cell wall, in combination with cellular mechanical processes driven by changes in turgor pressure enable sister guard cells to undergo spatially limited separation to form the stomatal pore (Rui et al. 2019) and result in a deformable cell wall that can behave elastically at the micro-scale, in contrast to the walls of irreversibly growing cell walls that display viscoelastic behaviour (Cosgrove 2022), is still incompletely understood.

How cells that undergo elongation into rectangular/cylindrical prisms might differentially attain irreversible or reversible cell shapes is also poorly understood, since in roots and hypocotyls of *Arabidopsis*, where the relationship between cell wall structure and cell elongation has been thoroughly examined, cell elongation is essentially irreversible. Detailed developmental, genetic, cell biological, and biomechanical analyses of contractile cells, such as those that exist in contractile roots or twining vines (Chery et al. 2022), are needed to more fully appreciate the differences between irreversibly and reversibly elongating cells. Even less well understood are the developmental and wall patterning mechanisms that underlie the formation of cells that can dehydrate and rehydrate while maintaining their structural integrity, such as those that exist in resurrection plants (Chen et al. 2020) (see Chapter 22). Further work is needed to disentangle the potentially divergent developmental and morphogenetic pathways that differentiate plant cells with fixed shapes from those that can undergo reversible cell deformation, especially since all cell types that are capable of reversible shape changes also undergo irreversible growth early in their lives.

INFLUENCES AND RESPONSES OF CELL WALLS IN RELATION TO REVERSIBLE SHAPE CHANGES IN PLANT CELLS AND TISSUES

Many more plant cell types can dynamically change shape than is typically appreciated due to the anisotropic nature of their walls that often make them more deformable in one direction than another, in combination with their volumes often being defined in part by turgor pressure. The biomechanics underlying cell shape are determined in the cell wall by the abundance, configurations, and degree of cross-linking of a combination of cellulose/xyloglucan, pectin networks, and glycoproteins (Fukuda 2014). It is assumed that in most cells, turgor pressure, which is intrinsically isotropic, is evenly distributed across the cell, although this might not hold true for long, thin cells or cell protrusions such as pollen tubes and root hairs, where osmotic and thus water potential gradients might exist along the cell axis.

Reversible deformations in plant cells can involve changes in shape with little change in total cell volume (i.e., passive bending or contraction), or changes in both volume and shape that often result from water moving into or out of the cell (Figure 15.2). Passive cell bending, as must occur when the elongated bases of organs, such as leaf petioles, bend in strong winds, necessarily involves the compression of cell wall components on the concave side and the extension or separation of wall components on the convex side. Coordinated cell expansion and compression on opposite sides of organs during some gravitropic, thigmotropic, and phototropic responses involves differential changes in cell shape, with cell compression or bending, which does not always occur, being reversible. Plant cells and the tissues they compose can also shrink in response to abiotic stresses such as salt and drought stress, with corresponding responses that impinge on the cell wall occurring at the cellular and subcellular levels (Colin et al. 2023). The nutation many plants undergo during growth (Darwin and Darwin 1880) is thought to arise from differential, irreversible cell expansion, but in the case of periodic nastic movements in organs that are not rapidly growing, such as leaf nodding, reversible cell expansion and shrinking almost certainly occurs. Reversible, asymmetric shape changes in cells across the pulvini of sensitive plants (e.g., *Mimosa pudica*) (Sleboda et al. 2023) and the hinges of Venus flytraps (*Dionaea muscipula*) ultimately change the shapes of the organs to which they are attached; in *Mimosa*, this change in shape is thought to be driven largely by ion/water flux across the pulvinus (Hagihara and Toyota 2020), whereas in *Dionaea* trap closure is driven partly by buckling (Morris and Blyth 2019). Because they are driven in part by active ion transport, both mechanisms require large amounts of adenosine triphosphate (ATP) and, when frequently induced, can harm plant health; however, whether repeated changes in the shapes of these cells over many cycles of leaflet/trap closure and opening result in structural and/or mechanical changes in their cell walls has not been studied in detail.

One of the most-studied examples of a cell type that changes shape reversibly due to active processes is stomatal guard cells, which elongate and widen slightly (Meckel et al. 2007) while remaining anchored at their ends (Carter et al. 2017) to open the stomatal pore in response to light and other stimuli, and shrink to close it in response to darkness, drought, or pathogen/pest attack (Spiegelhalder and Raissig 2021). Cellulose is synthesised in a radial organisation in the guard cell wall (Fujita and Wasteneys 2014; Rui and Anderson 2016), which also contains localised differences in pectin distribution (Carter et al. 2017; Shtein et al. 2017; Woolfenden et al. 2018). In guard cells induced to open or close, confocal imaging indicates that cellulose unbundles and bundles, respectively (Rui and Anderson 2016), implying micro-scale reversibility in wall structure. However, cell physiological and biomechanical analyses of guard cells in wild-type controls and pectin-modifying mutants of *Arabidopsis*, wherein mutations in a pectin-modifying enzyme result in stepwise stomatal closure but overexpression of the same enzyme accelerates stomatal opening (Rui et al. 2017), and the asymmetric changes observed in wall modulus and turgor pressure by nanoindentation during light-induced stomatal opening and closure (Chen et al. 2021), suggest the presence of structural

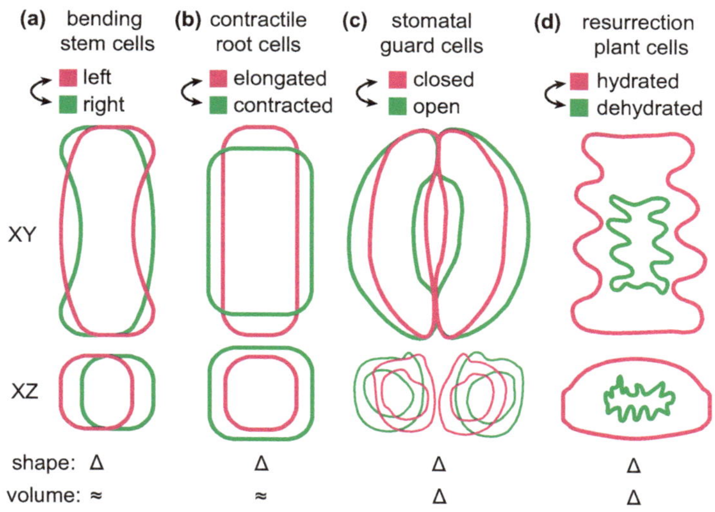

FIGURE 15.2 Examples of dynamic, reversible shape changes in plant cells. Longitudinal (XY) and transverse (XZ) cross sections of cells are shown in two interconvertible states represented by magenta and green. Bending cells in a stem or petiole (**A**) can switch from a leftward to a rightward bend without changing substantially in volume or cross-sectional area. Contractile cells (**B**), for example, in bulb roots, can switch from elongated to contracted, changing aspect ratio and cross-sectional area. Stomatal guard cells (**C**) elongate, distend to widen the stomatal pore, and change slightly in cross-sectional shape between closed and open states. Epidermal cells in a resurrection plant (**D**) shrink and wrinkle in all dimensions upon dehydration and can return to an expanded state upon rehydration. Cells are not drawn to scale.

and/or mechanical hysteresis in the expansion and contraction of guard cells, meaning that opening and closing are not truly symmetrical processes at the nanoscale. Adding weight to the latter concept, expansins can facilitate stomatal opening (Zhang et al. 2011); given their well-known influence on irreversible wall extension (Cosgrove 2022), it is likely that non-covalent bond breakage during stomatal opening and bond re-formation during stomatal closure result in a cell wall that does not perfectly return to its initial state (Figure 15.3). A plethora of mechanical models have been built to help explain the apparently elastic behaviour of guard cells (Bidhendi and Geitmann 2018; Yi et al. 2019), but researchers are only beginning to incorporate molecular-scale details into these models.

The contractile cells of contractile roots and twining vines depend on a specialised wall layer, the gelatinous layer (or G-layer), that is composed mostly of oriented cellulose and can shrink in response to dehydration (Chery et al. 2022). The configuration of longitudinally oriented cellulose and pectins has been proposed to act like a "muscle" during dehydration, wherein tension generated by rhamnogalacturonan-I interspersing with cellulose results in wall shortening (Gorshkova et al. 2018). Such tension, when released, can lead to a widening and shortening of cells (Figure 15.2), although in the epidermis of contractile roots, where cell–cell adhesion is critical for maintaining organ integrity, the tissue itself buckles (Cyr et al. 1988). After initial growth, G-fibre cells of contractile roots and twining vines can anisotropically shrink to accomplish organ shortening, bending, and twisting. Tissue twisting during growth, such as that which occurs in the roots of certain *Arabidopsis* mutants, also requires changes in cell shape from rectilinear to off-axis (Sedbrook et al. 2004). It remains to be seen whether this cell-based twisting arises from defects in the nanoscale patterning of the cell wall, although in the cells of *Arabidopsis* petals that twist in a pectin-deficient mutant this might be the case (Saffer et al. 2017).

Extreme examples of reversible cell shape change are found in resurrection plants (Chen et al. 2020), which shrink as they desiccate in the dry season but can re-expand as the plants are rehydrated by seasonal rains (see Chapter 22). Dehydration involves changes in cell wall–related gene expression and wall composition (Vicré et al. 2004), implying that some wall alterations might be irreversible during dehydration/rehydration cycles, but arabinose-rich polymers, including pectins and protein-linked arabinogalactans, have been implicated in maintaining sufficient flexibility in the cell wall to enable cell re-expansion (Moore et al. 2013). At the tissue level, cell–cell adhesion likely also constrains reversible cell deformations. Although this is seldom discussed in the literature, some plant cells appear not to be tightly adhered to one another; for example, in leatherwood (*Dirca palustris*), a lack of lignification in the middle lamella/cell corners might underlie the remarkable flexibility of its stems and branches (Mottiar et al. 2020).

A conceptual model for how irreversible changes in wall nanostructure might underlie reversible shape changes in plant cells is illustrated in Figure 15.3. Sliding, bending, and detachment of cellulose microfibrils is thought to underly irreversible wall extension (Zhang et al. 2021), a process that is facilitated by mechanical force and the action of wall loosening proteins such as expansins. In reversibly deforming walls, cellulose–xyloglucan binding interactions at biomechanical hotspots and pectin crosslinking via calcium

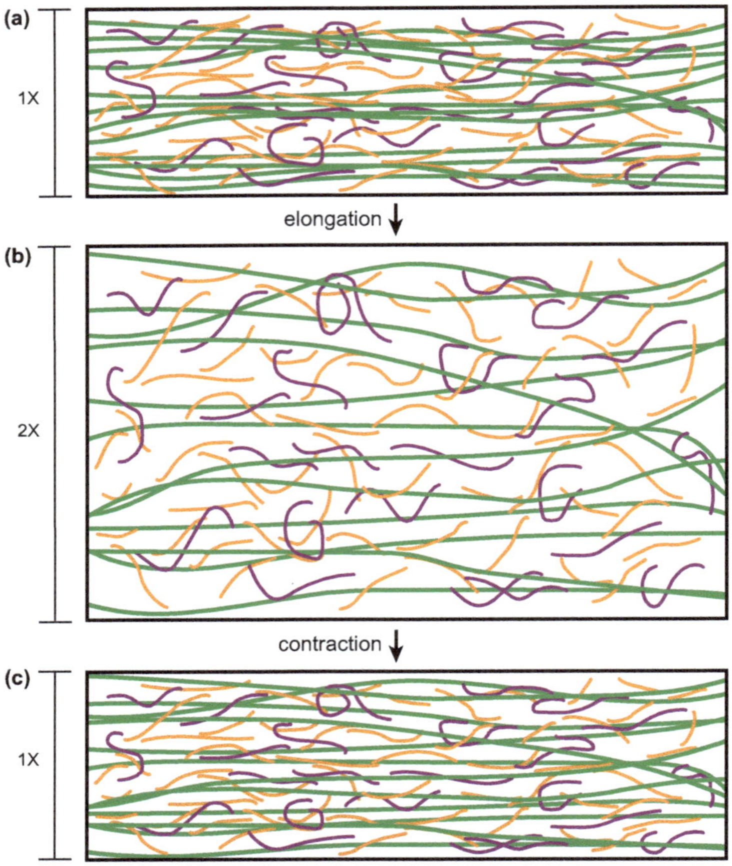

FIGURE 15.3 Rearrangements and deformations of cell wall polysaccharides in a hypothetical wall that reversibly elongates and exhibits hysteresis. In the initial state (**A**), cellulose (green) is present in transverse bundles analogous to those that might exist in closed guard cells; pectins (orange) contact cellulose surfaces or exist in interfibrillar spaces and are crosslinked at some points; and hemicelluloses (purple) contact cellulose, especially at points of cellulose–cellulose contact, or exist in interfibrillar spaces. After an exaggerated twofold elongation (**B**), cellulose is more evenly dispersed, pectin–pectin crosslinks have been disrupted in many cases, and hemicellulose molecules are deformed and/or displaced. Finally, after a twofold contraction (**C**) the wall segment returns to its initial dimensions, but cellulose, pectins, and hemicelluloses are not always in their initial configurations, resulting in altered wall architecture and mechanics; for example, here the cellulose does not return to a bundled configuration. In this simplified example, only one wall layer is shown, polysaccharides are not drawn to scale, cell wall proteins and other components are omitted, and effects of Poisson ratio are not depicted.

and borate diesters might be reversible at the micro-scale while differing at the molecular level, implying that cells can maintain elastic behaviours while undergoing iterative, irreversible rearrangements of their wall polymers.

CURRENT AND FUTURE EXPLORATIONS OF REVERSIBLE CHANGES IN PLANT CELL SHAPE AND THEIR RELATIONS TO CELL WALL ARCHITECTURE AND MECHANICS

Although scientific pioneers like Hooke and the Darwins made astounding leaps of insight about plant tissue architecture and movements based on relatively simple observations, we are now starting to be able to see details that they could not. Our growing ability to observe nanoscale wall organisation in intact cells and tissues, for instance with super-resolution microscopy, allows us to hypothesise new mechanisms for wall expansion, as has recently been done for epidermal pavement cells (Haas et al. 2020), although these can be controversial (Bidhendi and Geitmann 2019; Majda et al. 2019; Cosgrove and Anderson 2020). Rapid, sensitive, and non-damaging methods for capturing 3D cell shape and mechanics simultaneously, which are still in the early stages of development (Chen et al. 2021), would allow for the data collected with such methods to be seamlessly incorporated into computational modelling frameworks, although these must also be constantly refined to accommodate new information. Iterative modelling and experimentation, especially among an expanded range of plant species with unique shape-shifting cells, will help expand our understanding of reversible cell deformation.

How do reversible changes in cell shape facilitate mechanosensation and defence in plants? Damage to leaves such as that caused by insects, which can cause calcium transients that rapidly spread across the whole plant (Toyota et al. 2018), might result in altered supracellular stress patterns that lead to differential growth and wound repair (Uyttewaal et al. 2012; Sampathkumar et al. 2014). Trichomes can temporarily buckle to allow for carnivore trap activation (Di Palma et al. 1961), toxin injection into herbivores by stinging nettles (Jensen et al. 2020), or potential detection of predators or environmental mechanostimuli (Tooker et al. 2010; Liu et al. 2017; Zhou et al. 2017). The links between cell shape changes and tropic responses in plants will be fascinating to study as our ability to detect and connect them flourishes.

Dissecting the temporal order of molecular events during cell wall deposition, reorganisation, and remodelling that influences reversible changes in cell shape should allow us to better understand the assembly principles that allow for reversibility. However, it is not yet clear how truly reversible plant cell shapes might be, in terms of durability and longevity. For example, there is evidence that stomatal function declines with leaf age (Reich and Borchert 1988), but whether this is due to "molecular ageing" of the guard cell wall or other metabolic or physiological changes associated with leaf ageing is unclear. Although it is tempting to speculate that plant cells with reversible shapes could be used for bio-inspired design of flexible, durable, and potentially self-healing materials, structures, and products, fundamental questions such as those above must still be addressed before engineering efforts get fully underway.

The answers to long-standing questions regarding reversible shape changes in plant cells and their influences on plant growth, development, and behaviour feel tantalisingly close, and in answering these questions, current and future scientists will open new lines of inquiry that connect the dynamic metabolism occurring both outside and inside plant cells to the plastic construction of their protective cell walls and their shape-shifting abilities. Who knows what new wonders we will soon see emerging over the horizon, and after doing so seek to understand?

ACKNOWLEDGEMENTS

Writing of this chapter was supported by grant MCB 2015943 from the US National Science Foundation. The author is grateful for fascinating discussions with the late Barbara Pickard regarding plant mechanosensation and trichome cell walls.

REFERENCES

Anderson CT, Kieber JJ (2020) Dynamic construction, perception, and remodeling of plant cell walls. *Annual Review of Plant Biology* 71:39–69.

Baskin TI (2017) Plant cell growth: Cellulose caught slipping. *Nature Plants* 3:1–3.

Bassel GW, Smith RS (2016) Quantifying morphogenesis in plants in 4d. *Current Opinion in Plant Biology* 29:87–94.

Belteton SA, Li W, Yanagisawa M, Hatam FA, Quinn MI, Szymanski MK, Marley MW, Turner JA, Szymanski DB (2021) Real-time conversion of tissue-scale mechanical forces into an interdigitated growth pattern. *Nature Plants* 7:826–841.

Bhagia S, Ďurkovič J, Lagaňa R, Kardošová M, Kačík F, Cernescu A, Schäfer P, Yoo CG, Ragauskas AJ (2022) Nanoscale ftir and mechanical mapping of plant cell walls for understanding biomass deconstruction. *ACS Sustainable Chemistry & Engineering* 10:3016–3026.

Bidhendi AJ, Geitmann A (2016) Relating the mechanics of the primary plant cell wall to morphogenesis. *Journal of Experimental Botany* 67:449–461.

Bidhendi AJ, Geitmann A (2018) Finite element modeling of shape changes in plant cells. *Plant Physiology* 176:41–56.

Bidhendi AJ, Geitmann A (2019) Geometrical details matter for mechanical modeling of cell morphogenesis. *Developmental Cell* 50:117-125 e112.

Bilkey N, Li H, Borodinov N, Ievlev AV, Ovchinnikova OS, Dixit R, Foston M (2022) Correlated mechanochemical maps of arabidopsis thaliana primary cell walls using atomic force microscope infrared spectroscopy. *Quantitative Plant Biology* 3:e31.

Bou Daher F, Braybrook SA (2015) How to let go: Pectin and plant cell adhesion. *Frontiers in Plant Science* 6:523.

Breidbach O (2002) Representation of the microcosm-the claim for objectivity in 19th century scientific microphotography. *Journal of the History of Biology* 35:221–250.

Carter R, Woolfenden H, Baillie A, Amsbury S, Carroll S, Healicon E, Sovatzoglou S, Braybrook S, Gray JE, Hobbs J, Morris RJ, Fleming AJ (2017) Stomatal opening involves polar, not radial, stiffening of guard cells. *Current Biology* 27:2974-2983, e2972.

Chan J, Coen E (2020) Interaction between autonomous and microtubule guidance systems controls cellulose synthase trajectories. *Current Biology* 30:941-947 e942.

Chen P, Jung NU, Giarola V, Bartels D (2020) The dynamic responses of cell walls in resurrection plants during dehydration and rehydration. *Frontiers in Plant Science* 10:1698.

Chen Y, Li W, Turner JA, Anderson CT (2021) Pectate lyase like12 patterns the guard cell wall to coordinate turgor pressure and wall mechanics for proper stomatal function in arabidopsis. *Plant Cell* 33:3134–3150.

Chery JG, Glos RA, Anderson CT (2022) Do woody vines use gelatinous fibers to climb? *New Phytologist* 233:126–131.

Colin L, Ruhnow F, Zhu J-K, Zhao C, Zhao Y, Persson S (2023) The cell biology of primary cell walls during salt stress. *Plant Cell* 35:201–217.

Cosgrove DJ (2022) Building an extensible cell wall. *Plant Physiology* 189(3):1246–1277.

Cosgrove DJ, Anderson CT (2020) Plant cell growth: Do pectins drive lobe formation in arabidopsis pavement cells? *Current Biology* 30:R660–R662.

Cyr RJ, Lin B-L, Jernstedt JA (1988) Root contraction in hyacinth. Ii. Changes in tubulin levels, microtubule number and orientation associated with differential cell expansion. *Planta* 174:446–452.

Czymmek K, Sawant A, Goodman K, Pennington J, Pedersen P, Hoon M, Otegui MS (2020) Imaging plant cells by high-pressure freezing and serial block-face scanning electron microscopy. *Plant endosomes: Methods and protocols.*, pp 69–81.

Darwin C, Darwin FE (1880) The power of movement in plants. John Murray, London.

Davaasuren D, Chen Y, Jaafar L, Marshall R, Dunham AL, Anderson CT, Wang JZ (2022) Automated 3d segmentation of guard cells enables volumetric analysis of stomatal biomechanics. *Patterns* 3:100627.

de Reuille PB, Routier-Kierzkowska A-L, Kierzkowski D, Bassel GW, Schüpbach T, Tauriello G, Bajpai N, Strauss S, Weber A, Kiss A (2015) Morphographx: A platform for quantifying morphogenesis in 4d. *eLife* 4:e05864.

DeVree BT, Steiner LM, Głazowska S, Ruhnow F, Herburger K, Persson S, Mravec J (2021) Current and future advances in fluorescence-based visualization of plant cell wall components and cell wall biosynthetic machineries. *Biotechnology for Biofuels* 14:1–26.

Di Palma JR, Mohl R, Best Jr W (1961) Action potential and contraction of dionaea muscipula (venus flytrap). *Science* 133:878–879.

Ding SY, Himmel ME (2006) The maize primary cell wall microfibril: A new model derived from direct visualization. *Journal of Agricultural and Food Chemistry* 54:597–606.

Dong J, MacAlister CA, Bergmann DC (2009) Basl controls asymmetric cell division in arabidopsis. *Cell* 137:1320–1330.

Du J, Anderson CT, Xiao C (2022) Dynamics of pectic homogalacturonan in cellular morphogenesis and adhesion, wall integrity sensing and plant development. *Nature Plants*:1–9.

Duncombe SG, Barnes WJ, Anderson CT (2020) Imaging the delivery and behavior of cellulose synthases in arabidopsis thaliana using confocal microscopy. *Methods in Cell Biology* 160:201–213.

Duncombe SG, Chethan SG, Anderson CT (2022) Super-resolution imaging illuminates new dynamic behaviors of cellulose synthase. *Plant Cell* 34:273–286.

Elsayad K, Werner S, Gallemí M, Kong J, Sánchez Guajardo ER, Zhang L, Jaillais Y, Greb T, Belkhadir Y (2016) Mapping the subcellular mechanical properties of live cells in tissues with fluorescence emission-brillouin imaging. *Science Signaling* 9:rs5–rs5.

Facette MR, Rasmussen CG, Van Norman JM (2019) A plane choice: Coordinating timing and orientation of cell division during plant development. *Current Opinion in Plant Biology* 47:47–55.

Fernandes AN, Chen X, Scotchford CA, Walker J, Wells DM, Roberts CJ, Everitt NM (2012) Mechanical properties of epidermal cells of whole living roots of arabidopsis thaliana: An atomic force microscopy study. *Physical review. E, Statistical, nonlinear, and soft matter physics* 85:021916.

Fitzgibbon J, Bell K, King E, Oparka K (2010) Super-resolution imaging of plasmodesmata using three-dimensional structured illumination microscopy. *Plant Physiology* 153:1453–1463.

Fujita M, Wasteneys GO (2014) A survey of cellulose microfibril patterns in dividing, expanding, and differentiating cells of arabidopsis thaliana. *Protoplasma* 251:687–698.

Fukuda H (2014) *Plant cell wall patterning and cell shape.* John Wiley & Sons, Hoboken, New Jersey.

Funakawa H, Miwa K (2015) Synthesis of borate cross-linked rhamnogalacturonan ii. *Frontiers in Plant Science* 6:223.

Gay H, Anderson TF (1954) Serial sections for electron microscopy. *Science* 120:1071–1073.

Gierlinger N (2018) New insights into plant cell walls by vibrational microspectroscopy. *Applied Spectroscopy Reviews* 53:517–551.

Gorshkova T, Chernova T, Mokshina N, Ageeva M, Mikshina P (2018) Plant 'muscles': Fibers with a tertiary cell wall. *New Phytologist* 218:66–72.

Gu Y, Rasmussen CG (2022) Cell biology of primary cell wall synthesis in plants. *Plant Cell* 34:103–128.

Haas KT, Rivière M, Wightman R, Peaucelle A (2020) Multitarget immunohistochemistry for confocal and super-resolution imaging of plant cell wall polysaccharides. *Bio-protocol* 10:e3783–e3783.

Haas KT, Wightman R, Meyerowitz EM, Peaucelle A (2020) Pectin homogalacturonan nanofilament expansion drives morphogenesis in plant epidermal cells. *Science* 367:1003–1007.

Hagihara T, Toyota M (2020) Mechanical signaling in the sensitive plant mimosa pudica l. *Plants* 9:587.

Hawes C (2012) *Electron microscopy of plant cells.* Academic Press, Cambridge, MA.

Heiner Z, Zeise I, Elbaum R, Kneipp J (2018) Insight into plant cell wall chemistry and structure by combination of multiphoton microscopy with raman imaging. *Journal of Biophotonics* 11:e201700164.

Hepler PK, Gunning BE (1998) Confocal fluorescence microscopy of plant cells. *Protoplasma* 201:121–157.

Hervé C, Marcus SE, Knox JP (2011) Monoclonal antibodies, carbohydrate-binding modules, and the detection of polysaccharides in plant cell walls. In: *The plant cell wall.* Springer, New York City, pp 103–113.

Hooke R (1665) *Micrographia: Or some physiological descriptions of minute bodies made by magnifying glasses, with observations and inquiries thereupon.* J. Martyn and J. Allestry, London.

Ivakov A, Persson S (2013) Plant cell shape: Modulators and measurements. *Frontiers in Plant Science* 4:439.

Jensen KH, Knoblauch J, Christensen AH, Haaning KS, Park K (2020) Universal elastic mechanism for stinger design. *Nature Physics* 16:1074–1078.

Kesari KK, O'Reilly P, Seitsonen J, Ruokolainen J, Vuorinen T (2021) Infrared photo-induced force microscopy unveils nanoscale features of norway spruce fibre wall. *Cellulose* 28:7295–7309.

Kierzkowski D, Nakayama N, Routier-Kierzkowska AL, Weber A, Bayer E, Schorderet M, Reinhardt D, Kuhlemeier C, Smith RS (2012) Elastic domains regulate growth and organogenesis in the plant shoot apical meristem. *Science* 335:1096–1099.

Kittelmann M, Hawes C, Hughes L (2016) Serial block face scanning electron microscopy and the reconstruction of plant cell membrane systems. *Journal of Microscopy* 263:200–211.

Komis G, Samajova O, Ovecka M, Samaj J (2015) Super-resolution microscopy in plant cell imaging. *Trends in Plant Science* 20:834–843.

Kwon T, Sparks JA, Nakashima J, Allen SN, Tang Y, Blancaflor EB (2015) Transcriptional response of arabidopsis seedlings during spaceflight reveals peroxidase and cell wall remodeling genes associated with root hair development. *American Journal of Botany* 102:21–35.

Lee LR, Bergmann DC (2019) The plant stomatal lineage at a glance. *Journal of Cell Science* 132:jcs228551.

Liu S, Jiao J, Lu TJ, Xu F, Pickard BG, Genin GM (2017) Arabidopsis leaf trichomes as acoustic antennae. *Biophysical Journal* 113:2068–2076.

Majda M, Krupinski P, Jonsson H, Hamant O, Robert S (2019) Mechanical asymmetry of the cell wall predicts changes in pavement cell geometry. *Developmental Cell* 50:9–10.

Majda M, Sapala A, Routier-Kierzkowska A-L, Smith RS (2019) Cellular force microscopy to measure mechanical forces in plant cells. Methods in Molecular Biology 1992: 215–230.

Marga F, Grandbois M, Cosgrove DJ, Baskin TI (2005) Cell wall extension results in the coordinate separation of parallel microfibrils: Evidence from scanning electron microscopy and atomic force microscopy. *Plant Journal* 43:181–190.

McKown KH, Bergmann DC (2020) Stomatal development in the grasses: Lessons from models and crops (and crop models). *New Phytologist* 227:1636–1648.

Meckel T, Gall L, Semrau S, Homann U, Thiely G (2007) Guard cells elongate: Relationship of volume and surface area during stomatal movement. *Biophysical Journal* 92:1072–1080.

Meents MJ, Watanabe Y, Samuels AL (2018) The cell biology of secondary cell wall biosynthesis. *Annals of Botany* 121(6):1107–1125.

Mitchell N, Piatczyc NP, Wang DD, Edwards J (2022) High-speed video and plant ultrastructure define mechanisms of gametophyte dispersal. *Applications in Plant Sciences*:e11463.

Moore JP, Nguema-Ona EE, Vicré-Gibouin M, Sørensen I, Willats WG, Driouich A, Farrant JM (2013) Arabinose-rich polymers as an evolutionary strategy to plasticize resurrection plant cell walls against desiccation. *Planta* 237:739–754.

Moreno N, Bougourd S, Haseloff J, Feijó JA (2006) Imaging plant cells. In *Handbook of biological confocal microscopy.* Springer New York, pp 769–787.

Morris RJ, Blyth M (2019) How water flow, geometry, and material properties drive plant movements. *Journal of Experimental Botany* 70:3549–3560.

Mottiar Y, Gierlinger N, Jeremic D, Master ER, Mansfield SD (2020) Atypical lignification in eastern leatherwood (dirca palustris). *New Phytologist* 226:704–713.

Nakahara K, Hisatoku T, Nagase T, Takahashi Y (2000) Earthquake response of ancient five-story pagoda structure of horyu-ji temple in japan. *Proceedings of the* 12th World Congress on Earthquake Engineering, New Zealand Society for Earthquake Engineering, Auckland, New Zealand..

Nicolas WJ, Fäßler F, Dutka P, Schur FK, Jensen G, Meyerowitz E (2022) Cryo-electron tomography of the onion cell wall shows bimodally oriented cellulose fibers and reticulated homogalacturonan networks. *Current Biology* 32:2375–2389.

O'Brien T, Feder N, McCully ME (1964) Polychromatic staining of plant cell walls by toluidine blue o. *Protoplasma* 59:368–373.

Ovečka M, von Wangenheim D, Tomančák P, Šamajová O, Komis G, Šamaj J (2018) Multiscale imaging of plant development by light-sheet fluorescence microscopy. *Nature Plants* 4:639–650.

Ovečka M, Sojka J, Tichá M, Komis G, Basheer J, Marchetti C, Šamajová O, Kuběnová L, Šamaj J (2022) Imaging plant cells and organs with light-sheet and super-resolution microscopy. *Plant Physiology* 188:683–702.

Paredez AR, Somerville CR, Ehrhardt DW (2006) Visualization of cellulose synthase demonstrates functional association with microtubules. *Science* 312:1491–1495.

Peaucelle A, Braybrook SA, Le Guillou L, Bron E, Kuhlemeier C, Hofte H (2011) Pectin-induced changes in cell wall mechanics underlie organ initiation in arabidopsis. *Current Biology* 21:1720–1726.

Peaucelle A, Braybrook S, Hofte H (2012) Cell wall mechanics and growth control in plants: The role of pectins revisited. *Frontiers in Plant Science* 3:121.

Peaucelle A, Wightman R, Haas KT (2020) Multicolor 3d-dstorm reveals native-state ultrastructure of polysaccharides' network during plant cell wall assembly. *IScience* 23:101862.

Piovesan A, Vancauwenberghe V, Van De Looverbosch T, Verboven P, Nicolaï B (2021) X-ray computed tomography for 3d plant imaging. *Trends in Plant Science* 26:1171–1185.

Refahi Y, Zardilis A, Michelin G, Wightman R, Leggio B, Legrand J, Faure E, Vachez L, Armezzani A, Risson A-E (2021) A multiscale analysis of early flower development in arabidopsis provides an integrated view of molecular regulation and growth control. *Developmental Cell* 56:540–556. e548.

Reich PB, Borchert R (1988) Changes with leaf age in stomatal function and water status of several tropical tree species. *Biotropica* 20:60–69.

Rui Y, Anderson CT (2016) Functional analysis of cellulose and xyloglucan in the walls of stomatal guard cells of arabidopsis. *Plant Physiology* 170:1398–1419.

Rui Y, Yi H, Kandemir B, Wang JZ, Puri VM, Anderson CT (2016) Integrating cell biology, image analysis, and computational mechanical modeling to analyze the contributions of cellulose and xyloglucan to stomatal function. *Plant Signaling & Behavior* 11:e1183086.

Rui Y, Xiao C, Yi H, Kandemir B, Wang JZ, Puri VM, Anderson CT (2017) Polygalacturonase involved in expansion3 functions in seedling development, rosette growth, and stomatal dynamics in *arabidopsis thaliana. Plant Cell* 29:2413–2432.

Rui Y, Chen Y, Yi H, Purzycki T, Puri VM, Anderson CT (2019) Synergistic pectin degradation and guard cell pressurization underlie stomatal pore formation. *Plant Physiology* 180:66–77.

Saffer AM (2018) Expanding roles for pectins in plant development. *Journal of Integrative Plant Biology* 60:910–923.

Saffer AM, Carpita NC, Irish VF (2017) Rhamnose-containing cell wall polymers suppress helical plant growth independently of microtubule orientation. *Current Biology* 27:2248–2259. e2244.

Sampathkumar A, Krupinski P, Wightman R, Milani P, Berquand A, Boudaoud A, Hamant O, Jönsson H, Meyerowitz EM (2014) Subcellular and supracellular mechanical stress prescribes cytoskeleton behavior in arabidopsis cotyledon pavement cells. *eLife* 3:e01967.

Schubert V (2017) Super-resolution microscopy - applications in plant cell research. *Frontiers in Plant Science* 8:531.

Sedbrook JC, Ehrhardt DW, Fisher SE, Scheible WR, Somerville CR (2004) The arabidopsis sku6/spiral1 gene encodes a plus end-localized microtubule-interacting protein involved in directional cell expansion. *Plant Cell* 16:1506–1520.

Shah DU, Reynolds TP, Ramage MH (2017) The strength of plants: Theory and experimental methods to measure the mechanical properties of stems. *Journal of Experimental Botany* 68:4497–4516.

Shaw SL, Ehrhardt DW (2013) Smaller, faster, brighter: Advances in optical imaging of living plant cells. *Annual Review of Plant Biology* 64:351–375.

Shaw SL, Thoms D, Powers J (2019) Structured illumination approaches for super-resolution in plant cells. *Microscopy* 68:37–44.

Shtein I, Shelef Y, Marom Z, Zelinger E, Schwartz A, Popper ZA, Bar-On B, Harpaz-Saad S (2017) Stomatal cell wall composition: Distinctive structural patterns associated with different phylogenetic groups. *Annals of Botany* 119:1021–1033.

Sleboda DA, Geitmann A, Sharif-Naeini R (2023) Multiscale structural anisotropy steers plant organ actuation. *Current Biology* 33:639–646.

Spiegelhalder RP, Raissig MT (2021) Morphology made for movement: Formation of diverse stomatal guard cells. *Current Opinion in Plant Biology* 63:102090.

Stark LA (2008) Plant movements revealed. *CBE-Life Sciences Education* 7:284–287.

Stevenson A, Michel WD (2016) Confirmation of the world's oldest woven garment: The tarkhan dress. *Antiquity Project Gallery* 349.

Strock CF, Schneider HM, Lynch JP (2022) Anatomics: High-throughput phenotyping of plant anatomy. *Trends in Plant Science* 27:520–523.

Tobimatsu Y, Schuetz M (2019) Lignin polymerization: How do plants manage the chemistry so well? *Current Opinion in Biotechnology* 56:75–81.

Tooker J, Peiffer M, Luthe DS, Felton GW (2010) Trichomes as sensors: Detecting activity on the leaf surface. *Plant Signaling & Behavior* 5:73–75.

Toyota M, Spencer D, Sawai-Toyota S, Jiaqi W, Zhang T, Koo AJ, Howe GA, Gilroy S (2018) Glutamate triggers long-distance, calcium-based plant defense signaling. *Science* 361:1112–1115.

Uyttewaal M, Burian A, Alim K, Landrein B, Borowska-Wykręt D, Dedieu A, Peaucelle A, Ludynia M, Traas J, Boudaoud A (2012) Mechanical stress acts via katanin to amplify differences in growth rate between adjacent cells in arabidopsis. *Cell* 149:439–451.

Verbelen JP, Stickens D (1995) In vivo determination of fibril orientation in plant cell walls with polarization cslm. *Journal of Microscopy* 177:1–6.

Verboven P, Herremans E, Helfen L, Ho QT, Abera M, Baumbach T, Wevers M, Nicolaï BM (2015) Synchrotron x-ray computed laminography of the three-dimensional anatomy of tomato leaves. *Plant Journal* 81:169–182.

Vicré M, Lerouxel O, Farrant J, Lerouge P, Driouich A (2004) Composition and desiccation-induced alterations of the cell wall in the resurrection plant craterostigma wilmsii. *Physiologia Plantarum* 120:229–239.

Vincent O, Weißkopf C, Poppinga S, Masselter T, Speck T, Joyeux M, Quilliet C, Marmottant P (2011) Ultra-fast underwater suction traps. *Proceedings of the Royal Society B: Biological Sciences* 278:2909–2914.

Wang T, Park YB, Caporini MA, Rosay M, Zhong L, Cosgrove DJ, Hong M (2013) Sensitivity-enhanced solid-state nmr detection of expansin's target in plant cell walls. *Proceedings of the National Academy of Sciences* 110:16444–16449.

Wilson SM, Bacic A (2012) Preparation of plant cells for transmission electron microscopy to optimize immunogold labeling of carbohydrate and protein epitopes. *Nature Protocols* 7:1716–1727.

Woolfenden HC, Baillie AL, Gray JE, Hobbs JK, Morris RJ, Fleming AJ (2018) Models and mechanisms of stomatal mechanics. *Trends in Plant Science* 23:822–832.

Wu X, Moon RJ, Martini A (2014) Tensile strength of iβ crystalline cellulose predicted by molecular dynamics simulation. *Cellulose* 21:2233–2245.

Yanagisawa M, Desyatova AS, Belteton SA, Mallery EL, Turner JA, Szymanski DB (2015) Patterning mechanisms of cytoskeletal and cell wall systems during leaf trichome morphogenesis. *Nature Plants* 1:15014.

Yang C, Ye Z (2013) Trichomes as models for studying plant cell differentiation. *Cellular and Molecular Life Sciences* 70:1937–1948.

Yang H, Kubicki JD (2020) A density functional theory study on the shape of the primary cellulose microfibril in plants: Effects of c6 exocyclic group conformation and h-bonding. *Cellulose* 27:2389–2402.

Yi H, Chen Y, Wang JZ, Puri VM, Anderson CT (2019) The stomatal flexoskeleton: How the biomechanics of guard cell walls animate an elastic pressure vessel. *Journal of Experimental Botany*.

Yi H, Rui Y, Kandemir B, Wang JZ, Anderson CT, Puri VM (2018) Mechanical effects of cellulose, xyloglucan, and pectins on stomatal guard cells of arabidopsis thaliana. *Frontiers in Plant Science* 9:1566.

Zamil MS, Geitmann A (2017) The middle lamella-more than a glue. *Physical Biology* 14:015004.

Zhang L, McEvoy D, Le Y, Ambrose C (2021) Live imaging of microtubule organization, cell expansion, and intercellular space formation in arabidopsis leaf spongy mesophyll cells. *Plant Cell* 33:623–641.

Zhang T, Zheng Y, Cosgrove DJ (2016) Spatial organization of cellulose microfibrils and matrix polysaccharides in primary plant cell walls as imaged by multichannel atomic force microscopy. *Plant Journal* 85:179–192.

Zhang XQ, Wei PC, Xiong YM, Yang Y, Chen J, Wang XC (2011) Overexpression of the arabidopsis alpha-expansin gene atexpa1 accelerates stomatal opening by decreasing the volumetric elastic modulus. *Plant Cell Reports* 30:27–36.

Zhang Y, Yu J, Wang X, Durachko DM, Zhang S, Cosgrove DJ (2021) Molecular insights into the complex mechanics of plant epidermal cell walls. *Science* 372:706–711.

Zheng Y, Cosgrove DJ, Ning G (2017) High-resolution field emission scanning electron microscopy (fesem) imaging of cellulose microfibril organization in plant primary cell walls. *Microscopy and Microanalysis* 23:1048–1054.

Zhou LH, Liu SB, Wang PF, Lu TJ, Xu F, Genin GM, Pickard BG (2017) The arabidopsis trichome is an active mechanosensory switch. *Plant, Cell & Environment* 40:611–621.

Zuch DT, Doyle SM, Majda M, Smith RS, Robert S, Torii KU (2022) Cell biology of the leaf epidermis: Fate specification, morphogenesis, and coordination. *Plant Cell* 34:209–227.

16 The Biology of Arabinogalactan Proteins in Plant Abiotic Stress and Reproduction

Raquel Figueiredo, Ana Lopes, Ana Marta Pereira, Diana Moreira, Jessy Silva, Maria João Ferreira, and Sílvia Coimbra
LAQV/REQUIMTE - Universidade do Porto

Luís Gustavo Pereira
GreenUPorto - Universidade do Porto

INTRODUCTION – HISTORICAL PERSPECTIVE

The plant cell wall that surrounds and sustains all plant cells is predominantly composed of polysaccharides but about 5–10% by mass accounting for cell wall proteins (Jamet et al., 2008; Anderson and Kieber, 2020). Evidence for the presence of specific proteins on the cell wall was initially revealed 60 years ago, when hydroxyproline (Hyp) was detected in cell wall preparations isolated from cell homogenates. By using cell suspensions from sycamore (*Acer pseudoplatanus* L.) and bean (*Phaseolus vulgaris*), Lamport and Northcote (1960) showed the presence of Hyp-rich proteins in the plant cell walls of these two species. Interestingly, also using a sycamore cell suspension, Aspinall et al. (1969) performed a polysaccharide analysis and found three types of polysaccharides, one of which was an arabinogalactan (AG). One year later, Lamport (1970) used a sycamore cell suspension extract and found that its composition and solubility in 10% trichloroacetic acid (which precipitates most proteins) was novel, producing an "acidic protein-polysaccharide complex" that was then identified as arabinogalactan proteins (AGPs). Furthermore, AGPs were shown to be present between the cell wall and the plasma membrane. These periplasmic location of soluble AGPs versus membrane bound was demonstrated by using the Yariv reagent (Jermyn et al., 1975), which, in turn, enabled the assay and quantitative distribution of AGPs in three cell surface pools: plasma membrane, periplasm and cell wall of cell suspension cultures (Lamport et al., 2006).

The hydrofluoric acid deglycosylation of proteins cleaves the glycosidic links (Mort and Lamport, 1977), leaving the peptide bonds intact for further enzymatic proteolysis and peptide sequencing assays. On the other hand, alkaline hydrolysis cleaves peptide bonds but leaves glycosides largely intact, including *O*-Hyp glycosides (Lamport, 1967). This allowed the isolation not only of Hyp-arabinosides of extensins but also of AGP Hyp-polysaccharides that were essential to their further structural elucidation much more recently (Tan et al., 2010).

AGPs are a highly glycosylated family of proteins, which are included in the Hyp-rich glycoprotein (HRGPs) superfamily (Showalter, 2001; Johnson et al., 2017). AGPs possess a protein core composed mainly of Pro/Hyp, Ala, Ser and Thr residues, ornamented with AG polysaccharides (Showalter et al., 2010). Information regarding the polypeptide backbone of AGPs was mostly obtained by its purification and amino acid analysis (Jermyn and Guthrie, 1985; Gleeson et al., 1989; Komalavilas et al., 1991). However, Chen et al. (1994) and Du et al. (1994) were the first to clone the protein gene of an AGP from cultured *Pyrus communis* cells and from *Nicotiana alata* styles, respectively. On their analysis, both concluded that the AGP backbone was composed of three domains: an N-terminal secretion signal sequence, a central sequence rich in Hyp/Pro residues and a C-terminal hydrophobic region. These criteria were used to classify AGPs as "classical". "Non-classical" AGPs would contain an N-terminal hydrophobic secretion signal sequence followed by one or more Pro/Hyp, Ala, Ser and Thr-rich regions, which can be Hyp-*O*-glycosylated, along with other non-Pro/Hyp, Ala, Ser and Thr-rich regions, such as hydrophilic C-terminal asparagine-rich domains (Mau et al., 1995; Du et al., 1996a). This was the starting point to understand the extent of diversity among AGPs and to decipher the protein moiety of these proteins. In fact, years later, another milestone in the AGP characterisation was achieved by identifying and characterising the glycosylphosphatidylinositol (GPI) anchor in a *P. communis* AGP

DOI: 10.1201/9781003178309-16

(Youl et al., 1998; Oxley and Bacic, 1999). This finding led to hypothesise a possible role for AGPs in signalling events.

Two of the most iconic techniques to study AGP localisation, distribution and expression were those based on the β-Yariv reagent and on monoclonal antibodies. β-Yariv is a chemical that specifically binds to the β-(1→3)-linked D-galactose backbone of AGPs (Yariv et al., 1967) and was first used to precipitate AGPs from soya bean, maize and jack-bean, and later for AGPs' distribution pattern studies and glycan structural analysis (Jermyn et al., 1975; Anderson et al., 1977). As for monoclonal antibodies, they can specifically detect carbohydrate epitopes of AGPs (Knox, 1995). These two approaches have shown that AGPs are ubiquitous in the plant kingdom (Serpe and Nothnagel, 1999; Lee et al., 2005) and are also present in several algae (Hervé et al., 2016).

The genomic era for plant research started in the beginning of the 20th century with the sequencing of the *Arabidopsis thaliana* genome by the Arabidopsis Genome Initiative (2000). This accomplishment impelled the production of an enormous amount of data and, consequently, the development of several bioinformatic tools with great impact on AGP research. Borner et al. (2002) predicted the existence of 210 GPI-anchored proteins on the Arabidopsis genome, from which more than 40% would probably have AG glycosylation modules. At the same time, Schultz et al. (2002) carried out an identification of AGPs and AG peptides based on the biased amino acid composition of AGPs. In this study, 47 proteins were classified as AGPs. A few years later, in an attempt to identify HRGPs in the Arabidopsis genome, considering the amino acid composition and specific motifs of known HRGPs, the BIO OHIO software program was developed (Showalter et al., 2010). From the 166 HRGPs predicted by the programme, 85 were classified as AGPs and 4 as AGP/extensin hybrids. More recently, Ma et al. (2017) developed the Finding-AGP pipeline, which identifies AGP-like sequences based on the definition of successful AG glycosylation. In this study, besides the identification of 151 AGPs in Arabidopsis, the bioinformatics analysis was applied to other 46 plant species, whose genomes had been sequenced. The results showed thousands of AGPs predicted across the plant kingdom, from bryophyte to eudicots.

To date, AGPs are classified based on the protein backbone and the presence or absence of specific domains, namely, classical AGPs, AG peptides, lysine-rich AGPs, fasciclin-like AGPs (FLAs), early nodulin-like AGPs with plastocyanin-like domains, xylogen-like AGPs with nonspecific lipid transfer protein domains, other chimeric AGPs (not included in the mentioned groups) and AGP/Extensin hybrids (Borner et al., 2002; Schultz et al., 2002; Johnson et al., 2003; Borner et al., 2003; Mashiguchi et al., 2009; Ma and Zhao, 2010; Showalter et al., 2010; Kobayashi et al., 2011; Ma et al., 2017). Evidently, AGPs are a very heterogeneous family, making its definition difficult. For that reason, Silva et al. (2020) proposed the following definition for AGPs: a large family of glycoproteins, which shares common features, like the usual protein domain rich in Pro/Hyp, Ala, Ser, and Thr; the occurrence of Ala-Hyp, Ser-Hyp, and/or Thr-Hyp dipeptide repeats; the existence of an N-terminal hydrophobic secretion signal sequence; the presence of type II AGs linked to Hyp residues; the capability to interact with β-Yariv reagent (Yariv et al. 1962, 1967); the possibility to be localised in the plasma membrane by a C-terminal GPI anchor; and the presence of other functional domains.

In this chapter, we will focus on the present knowledge about the biosynthesis and subcellular localisation of AGPs and discuss how AGPs are involved in the abiotic stress response (high temperature, drought and salt stress) and in plant reproduction. The ongoing change in global climate will negatively impact plant reproduction and consequently seed production. A better understanding of the biology of AGPs during abiotic stress is therefore essential to improve seed yield and overall plant health.

AGP BIOSYNTHESIS

AGP biosynthesis involves several steps of addition, removal and modification of amino acids, lipids and carbohydrates. Early steps in the biosynthesis of the polypeptide backbone of AGPs are presumed to occur by the typical eukaryotic pathway for proteins destined for secretion (Walter and Johnson, 1994; Schatz and Dobberstein, 1996), wherein the process initiates in the cytosol with the translation of a N-terminal signal peptide. The signal recognition particle then recognises the signal peptide and removes it, while guiding the nascent polypeptide backbone to the endoplasmic reticulum (ER). Subsequently, prolyl-4-hydroxylases (P4Hs) in the ER lumen convert selected Pro residues of AGP polypeptide backbones into Hyp residues by providing reactive hydroxyl groups (Yuasa et al., 2004; Faye et al., 2005; Hijazi et al., 2014; Nguema-Ona et al., 2014), whose oxygen atom has been shown to be exclusively derived from direct fixation of atmospheric oxygen (Lamport, 1963). According to the extended Pro hydroxylation code, Pro residues that occur after Ala, Gln, Hyp, Pro, Ser, Thr, and Val residues are selected to be hydroxylated (Canut et al., 2016; Duruflé et al., 2017).

AGPs containing the additional sequence for a GPI anchor can be further modified in the ER through glypiation (Schultz et al., 1998; Yeats et al., 2018) (Figure 16.1). Glypiation involves the addition of a GPI anchor to proteins, which in turn allows their attachment to the outer leaflet of the plasma membrane (Borner et al., 2003). The GPI anchor addition sequence (GAS) is an extended C-terminal region of the protein, which directs the cleavage of the polypeptide chain and subsequent linking of a GPI anchor to the new terminal amino acid residue (the ω site). In this C-terminal region, several subregions may be identified, specifically a flexible segment of about 11 polar amino acid residues, followed by the ω region of approximately 4 small residues (one

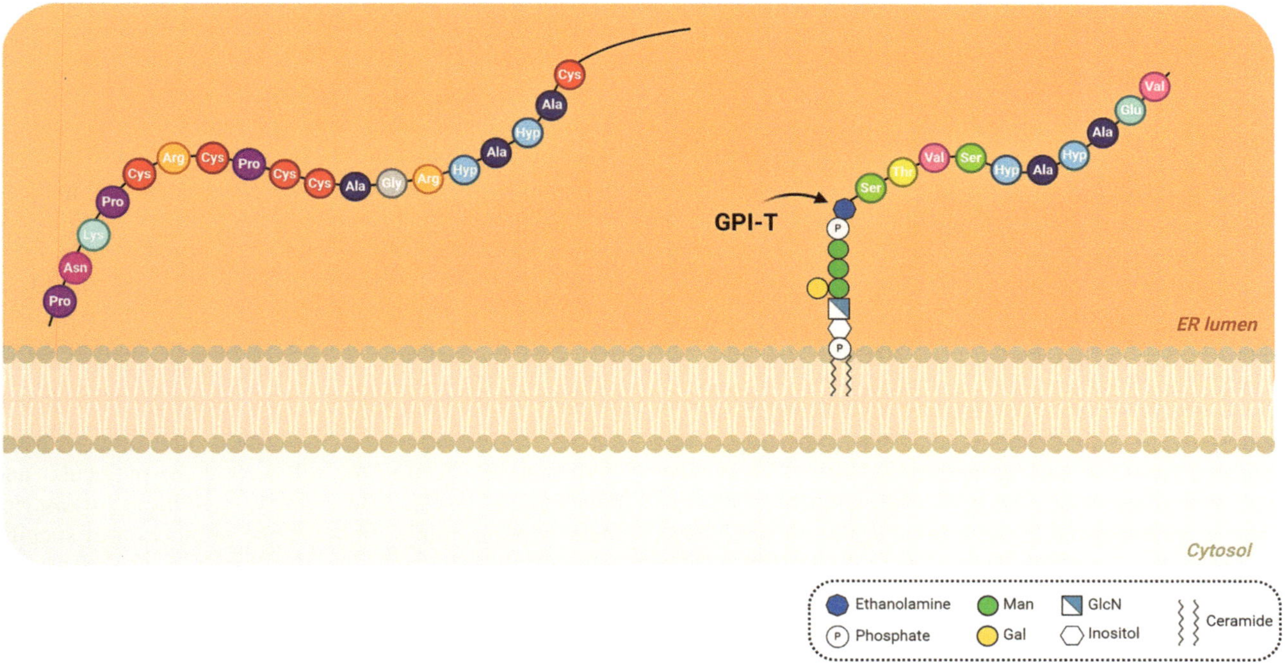

FIGURE 16.1 Detailed steps of the glypiation of AGPs in the ER. The preassembled GPI anchor is attached to the ω site of the mature protein via a transamidation reaction catalysed by the transamidase complex (GPI-T). AtAGP13 (At4g26320; NM_118765) and AtAGP42 (At1g51915; NM_104072), two AG peptides were used as models in the schematic (left and right AGP, respectively). AtAGP13 (Q9STQ3-1) and AtAGP42 (Q8L9S8-1) have the smallest amino acid sequence predicted to have or not have a GPI anchor, respectively (Showalter et al., 2010). The extended Pro hydroxylation code was applied to determine which Pro residues are hydroxylated (Canut et al., 2016; Duruflé et al., 2017). In these cases, only Pro residues after alanine (Ala) residues were converted to Hyp. Signal peptides and C-terminal anchor addition sequence positions were determined using UniProt (The UniProt Consortium, 2019). The GPI model structure was based on the GPI of PcAGP1 (Oxley and Bacic, 1999), which consists of phosphoethanolamine attached to the protein, three mannosyl, one galactosyl residue and glucosamine-inositol linked to phosphoceramide. Created by BioRender.com. Reprinted from Silva et al. (2020) Frontiers in Plant Science 11.

of which is the ω residue), a spacer region of about 6 moderately polar amino acid residues and finally a C-terminal region with 9–24 amino acid residues with high hydrophobic character (Schultz et al., 1998; Eisenhaber et al., 2003; Ellis et al., 2010; Desnoyer and Palanivelu, 2020).

The GPI moiety itself is highly conserved among all organisms. It is initially synthesised on the cytosolic surface of the ER via the sequential addition of glucosamine, three α-linked mannosyl (Man) residues and phosphoethanolamine to a phosphatidylinositol substrate (Figure 16.1). The GPI transamidase complex (GPI-T) recognises and cleaves the GAS between the ω and ω+1 sites. The GPI moiety is added to the ω site by a transamidation reaction (Schultz et al., 1998; Ellis et al., 2010; Yeats et al., 2018; Beihammer et al., 2020; Desnoyer and Palanivelu, 2020).

AGP polypeptide backbones, either free in the ER lumen or tethered by a GPI anchor to the ER membrane, are transferred to the Golgi apparatus where *O*-glycosylation takes place, and several monosaccharides are added successively to them by glycosyltransferases (GTs) (Silva et al., 2020). In plants, *O*-glycosylation occurs mainly on Hyp residues and less often on Ser and Thr residues (Nothnagel, 1997; Faye et al., 2005; Duruflé et al., 2017). The sugars involved in these *O*-glycosylations in AGPs are type II AGs linked to the hydroxyl group of Hyp residues (Faye et al., 2005; Ellis et al., 2010) (Figure 16.2). Type II AGs have a backbone of β-(1→3)-linked galactan substituted at O6 positions with side chains of β-(1→6)-linked galactan. Other sugars can be attached to the side chains, such as arabinose (Ara), and less frequently glucuronic acid (GlcA), rhamnose, fucose (Fuc) or xylose (Clarke et al., 1979; Gaspar et al., 2001; Showalter, 2001; Tan et al., 2004, 2010; Tryfona et al., 2010, 2012). It has been proposed that the carbohydrate moiety of AGPs is composed of more or less defined subunits of about 15 sugar residues, each containing two residues of GlcA, and that these lie at the heart of a Ca^{2+} binding chemical device for calcium storage and release (Lamport and Várnai, 2013). These works have profound implications for the knowledge of calcium signalling networks in higher plants. According to the Hyp contiguity hypothesis (Kieliszewski and Lamport, 1994), AGPs which have clustered non-contiguous Hyp residues are arabinogalactosylated (Shpak et al., 1999; Kieliszewski, 2001; Shpak et al., 2001; Faye et al., 2005).

During AGP *O*-glycosylation, GTs work co-ordinately to control the density, length and sequences of AG chains (Qu et al., 2008). Diverse GTs are involved in the biosynthesis of AGP glycans, such as galactosyltransferases (GALTs), arabinosyltransferases, fucosyltransferases (FUTs), rhamnosyltransferases, xylosyltransferases, glucuronosyltransferases (GLCATs) and GlcA methyltransferases. To the best of our knowledge, 24 relevant enzymes have been identified in *A. thaliana*. GALT2, GALT3, GALT4, GALT5, GALT6 (Basu et al., 2013, 2015a, b; Showalter and Basu, 2016), HPTG1, HPTG2 and HPTG3 (Ogawa-Ohnishi and Matsubayashi, 2015) are eight hydroxyproline-*O*-galactosyltransferases responsible for the first step of AGP glycosylation by adding the first galactose (Gal) to Hyp residues. Subsequently, other GAlTs, such as GALT8 (Narciso et al., 2021), KSN4 (Suzuki et al., 2017) and the At1g77810 gene product (Qu et al., 2008), synthesise the β-(1→3)-galactan backbone, whereas β-(1→6)-galactan side chains are added for instance by GALT31A (Geshi et al., 2013) and GALT29A (Dilokpimol et al., 2014). Other sugars can be added to AGPs, such as Fuc by FUT4 and FUT6 (Wu et al., 2010; Liang et al., 2013; Tryfona et al., 2014). The two GlcA residues present in the β-(1→6)-galactan side chains may be added by GLCAT14A, GLCAT14B, GLCAT14C, GLCAT14D, GLCAT14E or the At4g27480 gene product (Knoch et al., 2013; Dilokpimol and Geshi, 2014; Lopez-Hernandez et al., 2020; Zhang et al., 2020; Xuan et al., 2021) and may be further 4-*O*-methylated by GlcA methyltransferases, such as AGM1 and AGM2 (Temple et al., 2019). Over the last decade, an ambitious global effort led to the discovery of all the above enzymes; however, several GTs remain to be identified, such as α-arabinofuranosyltransferases, α-arabinopyranosyltransferases, α-rhamnosyltransferases, α-xylosyltransferases, α-GALTs and other β-GALTs, β-GLCATs, α-FUTs and GlcA methyltransferases (Silva et al., 2020).

Several chimeric AGPs, like FLAs, contain also the Asn-X-Ser/Thr-specific *N*-glycosylation sequence, where X can be any amino acid except Pro (Du et al., 1996a; Johnson et al., 2003; Faye et al., 2005; Ellis et al., 2010; Duruflé et al., 2017). *N*-glycosylation occurs on Asn residues by addition of the oligosaccharide precursor $Glc_3Man_9GlcNAc_2$ that undergoes a process of maturation in the ER and Golgi (Faye et al., 2005; Nguema-Ona et al., 2014).

Finally, *O*-glycosylated GPI-anchored AGPs are transported to the plasma membrane where they remain temporarily attached to the outer leaflet of the plasma membrane (Showalter, 2001; Schultz et al., 1998), whereas AGPs without anchor are directly released to the cell wall (Silva et al., 2020). GPI-anchored AGPs may be released from the plasma membrane by the action of phosphatidylinositol-specific phospholipases that cleave inositol from the diacylglycerol ceramide moiety. Indeed, the quantity of attached AGPs and their controlled release to the extracellular matrix by phospholipases may establish a signalling network for neighbouring cells (Schultz et al., 1998; Oxley and Bacic, 1999; Gillmor et al., 2005; Desnoyer and Palanivelu, 2020), but this has yet to be established definitively. AGPs not GPI-anchored may also liberate sugars by the action of glycoside hydrolases (GHs), which may also function as signalling molecules in specific signalling pathways (Showalter, 2001). AGPs may be destined to degradation by proteolytic cleavage (Fincher et al., 1983; Faye et al., 2005) or by transport to the vacuole through endocytic multivesicular bodies (Herman and Lamb, 1992; Šamaj et al., 2000).

AGPs IN ABIOTIC STRESS

Plants are exposed to variable and extreme soil and climate factors, namely, high temperatures, salinity or drought, among others, that disturb plant development and ultimately crop productivity. The plant cell wall, a dynamic and heterogeneous polysaccharide matrix, is a key element to induce plant tolerance to environmental stress. The main cell wall components, such as cellulose, pectins, hemicelluloses and a varied group of different glycoproteins, interact with each other and with the environment affecting the cell wall at the molecular and biochemical levels. Under stress conditions, transcriptional reprogramming induces the production of specific cell wall proteins, affecting the synthesis, deposition and reorganisation of cell wall polysaccharides and consequently cell wall architecture. HRGPs are deeply involved in cell wall-mediated signalling cascades, stress tolerance and differentiation processes. Besides being involved in sexual plant reproduction, they are involved in other developmental processes such as seedling growth (Lu et al., 2001), cell expansion and division (Yang et al. 2007), mechanical wounding response (Guan and Nothnagel, 2004), programmed cell death (PCD; Gao and Showalter, 1999) and abiotic stress-response mechanisms such as salt tolerance (Lamport et al., 2006), heat stress (Mareri et al., 2016), drought (Tseng et al., 2013), hypoxia, anoxia (Fragkostefanakis et al., 2012) and dehydration/desiccation (Cui et al., 2012; Moore et al., 2013).

Salt acclimation can promote extensive remodelling of the cell wall, thus priming and enforcing plants to be ready for subsequent salt exposure. The response of plant cells to salt stress involves the release of different AGPs, maybe due to the role of AGPs in controlling cell expansion and osmoregulation (Lamport et al., 2006; Mareri et al., 2016). Immunolabelling assays using different anti-AGP antibodies showed that AGPs accumulate in the culture medium of salt-adapted tobacco cells (Olmos et al., 2017). Moreover, NaCl-tolerant cells displayed a reduced rate of cell elongation and concomitantly a decrease in cell wall extensibility, showing a low level of AGPs on the plasma membrane (Lamport et al., 2006). The decrease of AGPs in the cell wall during salt stress may result in cell wall stiffness, in the activation of signalling mechanisms by releasing extracellular AGPs, or in the accumulation of soluble sugars in the cytoplasm to decrease water loss (Ezquer et al., 2020). Furthermore, non-adapted tobacco cells also accumulate high levels of AGPs when exposed to salt stress but are almost restricted to the plasma membrane, not releasing them to the apoplast (Lamport et al., 2006). Moreover, the AGP FLA4/SOS5 plays a key role during cell expansion under salt stress conditions (Johnson et al., 2011). In *Populus*, 18 FLAs were expressed in root tissues under

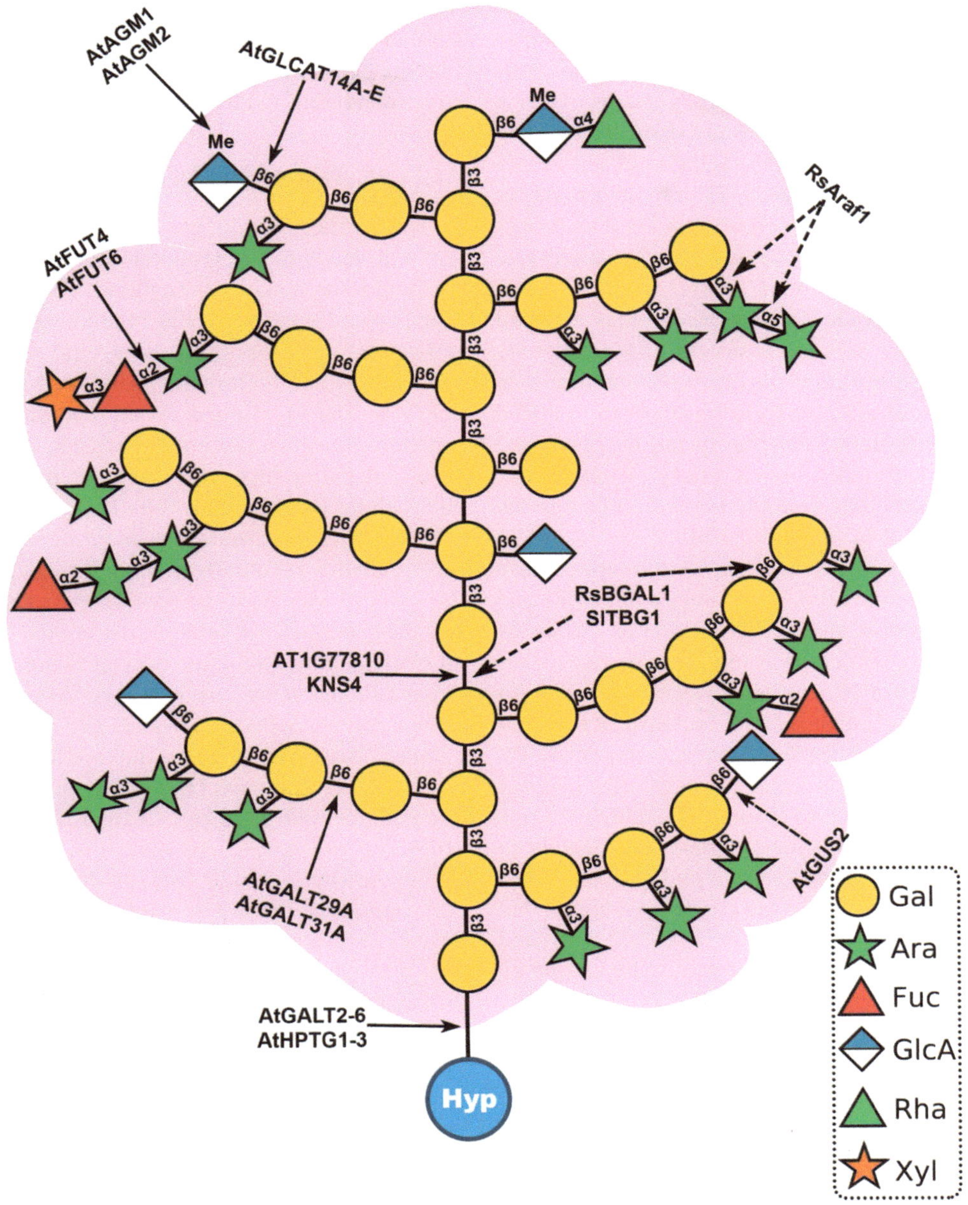

FIGURE 16.2 Model structure of type II arabinogalactan polysaccharides (AGs) and sites of action of known GTs and GHs acting on AGPs. Type II AGs are *O*-glycosidically linked to hydroxyproline (Hyp) and consist of a β-((1→6))-linked backbone of Gal with β-(1→6)-galactan side chains. Further modifications involve the addition of arabinosyl (Ara), fucosyl (Fuc), rhamnosyl (Rha), glucuronosyl (GlcA), 4-*O*-methylglucuronosyl (4-*O*-MeGlcA) and xylosyl (Xyl) residues. Solid line arrows (left) represent sites of action of GTs and dotted line arrows (right) indicate sites of action of GHs. This structure is based on AGs analysed from Arabidopsis leaves (Tryfona et al., 2012). Reprinted from Silva et al. (2020) Frontiers in Plant Science 11.

salt stress (Zang et al., 2015). The increase in FLA gene expression may indicate that FLAs are involved in cell–cell communication (MacMillan et al., 2010) or that the release of oligosaccharides from them may increase the intracellular osmotic potential and decrease the speed of dehydration during the osmotic stress associated with salt exposure (Ma and Zhao, 2010). The AGP glycan structures of *Zostera marina* L. contain unique features, including a high degree of branching and an unusually high content of terminating 4-*O*-methyl-glucuronic acid (4-*O*MeGlcA) residues that may provide insight into the evolution of *Z. marina* from land plants into marine environment (Pfeifer et al., 2020). The increased content of terminating uronic acid residues of *Z. marina* AGPs results in an improved Ca^{2+}-binding ability that may be involved in the reduction of Na^{+}-binding, as an adaptation to the saline marine environment (Lamport and Várnai, 2013; Pfeifer et al., 2020).

In Arabidopsis, among the genes differentially regulated in response to drought, three pectinesterases, a pectin methylesterase, and an AGP were identified (Seki et al.,

2002). In rice, the expression of several AGPs such as OsAGP1, OsAGP15 and OsELA3 were induced in response to drought (Tseng et al., 2013). The expression analysis in roots of drought-resistant versus drought-sensitive rice genotypes, revealed genes differentially expressed among which were identified some AGPs (Rabello et al., 2008). These observations suggest that AGPs may help preserve the mechanical integrity of the cell wall by keeping proper water levels in the cell wall. This reorganisation could also be implicated in the creation of a "buffer zone" that prevents the direct interaction between membranes and cell wall matrix, protecting membrane structure (Mareri et al., 2019). Moreover, also during drought, AGPs could rely on the activation of specific deglycosylation enzymes that lead to the release of polysaccharide chains and the increased content of oligosaccharides which might result in the increase of the intracellular osmotic pressure, reducing the speed of dehydration (Ma and Zhao, 2010).

Under high-temperature conditions, several reports have indicated that AGP expression levels were altered (Li and Showalter, 1996; Lima et al., 2013; Mareri et al., 2016). In tomato (*Solanum lycopersicum* L.), high temperatures transiently down-regulate AGP expression levels in stems (Li and Showalter, 1996). In contrast, an increase in the content of arabinose and Gal, the main components of AGPs, was reported in the leaves of coffee plants (*Coffea arabica* L.) under heat stress conditions (Lima et al., 2013). Plasma membrane AGPs have been shown to have the capacity to be oxidatively cross-linked by H_2O_2 or endogenous agents (Kjellbom et al., 1997), which has led to the hypothesis that such cross-linking might lead to stiffening of cell walls and ultimately enable cells to reduce the loss of water caused by high temperatures (Seifert and Roberts, 2007). In reproductive tissues, AGP levels and distribution are strongly altered after a short high-temperature treatment (3 h at 42 °C) in pistils of *S. lycopersicum* cv. Micro-Tom, with stigmas and ovules being more affected by heat stress than other pistil structures (Mareri et al., 2016). Moreover, the severity of stress seemed to be greater when the heat treatment was applied closer to anthesis, probably because the tissues already developed are not fully able to restore their normal condition after heat stress, whereas those tissues receiving the treatment at the beginning of their development have more time to restore their normal physiological state (Mareri et al., 2016). These results suggest that heat stress profoundly affects AGP expression and distribution in a tissue-specific manner and that different tissues show different AGP pattern under stress conditions: stems and reproductive tissues down-regulate AGP expression while leaf tissues, more exposed to water loss, seem to increase AGP content.

Under abiotic stress conditions, the molecular mechanisms of AGPs are multi-layered: AGPs can be up- or down-regulated depending on the plant tissue, developmental stage or the nature of the stress agent. AGPs have been hypothesised to be involved in the thickening and hardening of cell walls by oxidative cross-linking, and possibly in the release of free sugars that might be transported into the cytosol to osmotically help protect plant cells from stress conditions (Mareri et al., 2019).

AGPs IN PLANT REPRODUCTION

In plants, the reproductive process, essential for species survival, leads to the production of seeds, which are also crucial for the survival of mankind. The study of the molecular players involved in the fundamental processes finely regulating plant reproduction is vital for increasing crop productivity and high-quality seed production in a sustainable way. The angiosperm sexual reproduction begins when a pollen grain reaches a pistil stigma, producing a pollen tube. The pollen tube carries the two male gametes (sperm cells) along the stigmatic cells and the transmitting tract, which is rich in exudates containing glycoproteins. The male gametes eventually reach the ovule, which harbours the embryo sac, with the female gametes, the antipodals and the synergids, the latter attracting the pollen tube. Near an ovule, a pollen tube turns its growing direction into the ovule to enter it by one synergid. After synergid degeneration, sperm cells are released and fuse with the female gametes, developing into the embryo and endosperm, initiating seed development (Higashiyama and Takeuchi, 2015; Dresselhaus and Coimbra, 2016; Zhou and Dresselhaus, 2019).

The entire reproductive process encompasses several signalling pathways involving many molecules that have been described over the years, and AGPs in particular were shown to be involved in many important steps in this process (Table 16.1). The involvement of AGPs in sexual plant reproduction has been demonstrated along the last 30 years in different species, such as *Gladiolus* spp., *Lilium longiflorum*, *Brassica napus*, *Bellis perennis*, *Beta vulgaris*, *Actinidia deliciosa*, *Amaranthus hypochondriacus*, *Catharanthus roseus*, *Trithuria submersa*, *A. thaliana*, *Malus x domestica*, *Quercus suber*, *Annona cherimola* and *S. lycopersicum* (Gleeson and Clarke, 1980; Pennell et al., 1991; Coimbra and Salema, 1997; Coimbra and Duarte, 2003; Majewska-Sawka and Rodriguez-Garcia, 2006; Coimbra et al., 2007; Losada and Herrero, 2012; Costa et al., 2013; Chudzik et al., 2014; Lopes et al., 2016; Lora et al., 2018; Lara-Mondragón and MacAlister, 2021).

By now, evidence supports the involvement of AGPs in somatic embryogenesis (Hengel et al., 2001; Ellis et al., 2010), in female gametogenesis initiation (Acosta-García and Vielle-Calzada, 2004; Demesa-Arévalo and Vielle-Calzada, 2013), pollen grain development and pollen tube germination (Coimbra et al., 2009; Li et al., 2010; Pereira et al., 2016a), in pollen tube targeting the ovules (Costa et al., 2013; Pereira et al., 2014; Hou et al., 2016), in polytubey block (Pereira et al., 2016b), along the entire pathway travelled by the pollen tube until the last interactions

TABLE 16.1
Distribution and roles of AGPs in male and female reproductive tissues and during the double fertilisation process, in the most frequently analysed plant species

Species Name	AGPs ID	Reproductive tissues	Assumed function	References
Actinidia deliciosa		Intine | Exine | Pollen tube wall	Maintenance of cell shape during pollen growth and development	(Coimbra and Duarte, 2003)
		Transmitting tissue | Stigmatic surface | papillae	Intervening in the adhesion of pollen grain to the papillae	
		Obturator	Pollen–pistil interaction	
		Embryo sac	Signalling the ovule entrance	
Amaranthus hypochondriacus		Transmitting tissue | Stigma Papillae cells | Micropylar pole of nucellus	Directing pollen tube growth until the embryo sac	(Coimbra and Salema, 1997; Coimbra and Duarte, 2003)
		Embryo sac | Egg apparatus	Providing directional guides for tube growth inside the ovule	
		Pollen tube pathway		
Annona cherimola		Papillae of the stigma	Supporting role in pollen tube growth	(Lora et al., 2018)
		Egg apparatus		
		Chalazal pole of nucellus | Inner integument	Prevention of polytubey	
Arabidopsis thaliana	AGP6 AGP11	Pollen grain | Pollen tube | Transmitting tissue	Signalling pollent tube guidance into the embryo sac	(Coimbra et al., 2007; Coimbra et al., 2009; Costa et al., 2013)
			Preventing early pollen germination	
		Integuments | Embryo sac wall | Egg apparatus	Markers for gametophyte cell differentiation Involvement in the programmed cell death	(Coimbra et al., 2007)
	AGP23 AGP40 FLA3 FLA14	Pollen grain | Pollen tube	Candidates for male–female exchanges during reproduction Microspore development, affecting cellulose deposition Preventing premature pollen germination in the anthers under high relative humidity	(Qin et al., 2007; Li et al., 2010; Coimbra and Pereira, 2012; Costa et al., 2013; Li et al., 2017; Miao et al., 2021)
	AGP1 AGP12 AGP15 AGP19	Stigma | Style | Transmitting tissue Chalaza and funiculus cells of the ovule Siliques	Distinct steps of pollen tube growth through the pistil Double fertilisation Fruit production	(Pereira et al., 2014; Pereira et al., 2016a; Lopes et al., 2019)
	AGP18 AGP22 AGP24 AGP4/ JAGGER	Stigma surface | Ovary wall Embryo sac wall | Filiform apparatus | Micropylar integuments Integumentary cells in the abaxial ovule region Mature female gametophyte (central cell, synergids, egg cell) Megaspore mother cell | Functional megaspore Synergid	Arabidopsis developmental processes Signalling molecules at the beginning of female gametophyte development Active regulation of selection and survival of viable megaspores Signalling pathway leading polytubey blockage - avoiding polyspermy	(Acosta-García and Vielle-Calzada, 2004; Tucker et al., 2012; Demesa-Arévalo and Vielle-Calzada, 2013; Pereira et al., 2014; Tucker and Koltunow, 2014; Pereira et al., 2016a)

(Continued)

TABLE 16.1 (*Continued*)
Distribution and roles of AGPs in male and female reproductive tissues and during the double fertilisation process, in the most frequently analysed plant species

Species Name	AGPs ID	Reproductive tissues	Assumed function	References
	ENODL11-15	Funiculus | Ovules	Facilitating male–female communication during double fertilisation	(Hou et al., 2016)
		Pollen grain, Pollen tube wall, Transmitting tissue	Pollen tube guidance into the embryo sac Important signalling molecules at the pollen tube apex	(Pereira et al., 2014; Pereira et al., 2016b)
		Tapetum cells	Pollgen grain wall formation Tapetum cell wall degradation	(Li et al., 2017)
		Transmitting tissue	Programmed cell death	(Crawford et al., 2007; Pereira et al., 2021)
Bellis perennis		Anther wall	Dividing microsporocytes and microspores in the protoplasts of tapetal cells	(Chudzik et al., 2014)
		Pollen grain	Anther development	
Beta vulgaris		Primexine and incipient exine Young microspore wall	Post-telophase tetrads stage	(Majewska-Sawka and Rodriguez-Garcia, 2006)
Brassica napus		Tapetal anther layers Sperm cells	Definition of overall shape of sperm cells and identification of sperm cells	(Pennell et al., 1991; Corral-Martínez et al., 2016)
		Microspore cytoplasm | Exine, intine of mature pollen grain Generativel cell wall	Microspore differentiation and tapetal dismantling	
		Mature pollen grain (cytoplasm, wall of generative and sperm cells) | pollen tube Aperture regions	Pollen maturation	
Brassica campestris	BcMF18 BcMF8	Microspore	Intine development and aperture formation	(Lin et al., 2014; Lin et al., 2018)
Capsicum annuum	CaAGP18	Vegetative and reproductive organs	Seedling, flowering, pollen–pistil interactions and fruiting stages	(Verdugo-Perales et al., 2018)
Larix decidua Mill		Male gametophytes Ovule tissues	Pollination Progamic phase After fertilisation	(Rafińska et al., 2021)
Lilium longiflorum		Pollen tube tip region | Transmitting tract in the style	Adhesion during pollination and pollen tube growth	(Jauh and Lord, 1996)
Malus x domestica		Transmitting tract Stigmatic cells	Promotion of heterotrophic pollen tube growth Involvement of female tissues in prezygotic male–female cross-talk	(Losada and Herrero, 2012)

(*Continued*)

TABLE 16.1 (*Continued*)
Distribution and roles of AGPs in male and female reproductive tissues and during the double fertilisation process, in the most frequently analysed plant species

Species Name	AGPs ID	Reproductive tissues	Assumed function	References
Magnolia virginiana				(Losada et al., 2014)
Nicotiana alata		Stigma	Expression of self-incompatibility	(Du et al., 1996)
		Style cells	Promoting the adjusted environment for pollen tube adhesion and germination	
		Stigma exudate	Nutrition of growing pollen tube	
Nicotiana tabacum	TTS	Anther at the pollen mother cell stage | Microspore | Mature pollen grain	Stabiliser of the pollen grain cell wall architecture Intervene in the mechanism of generative cell movement	(Cheung et al., 1995; Wu et al., 2000; Qin et al., 2007; Noyszewski et al., 2017)
		Vesicles, cytoplasmic compartments and wall of generative cell and vegetative cell	Cell–cell interaction during pollen tube growth and development Male gametogenesis	
		Aperture region of pollen grain | pollen tube wall and cytoplasm	Surface molecules in generative and sperm cells	
Olea europaea		Wall of generative cell | Outer exine layer Stigma exudate | Cytoplasm of papillae	Recognition and adhesion of the pollen tube and the stylar transmitting cells, as well as the egg and sperm cells Component of stigmatic exudate	(Suárez et al., 2013)
		Transmitting tract in the style	Implications in pollen–pistil interaction, pistil development	
Oryza sativa	MTR1	Male reproductive cells	Tapetum formation and microspore development	(Tan et al., 2012)
Quercus suber		Anther wall layers Tapetum, pollen mother cell wall Intine wall near the pollen apertures Intine of the vegetative cell and the Generative cell wall	Involvement in micro gametogenesis	(Costa et al., 2015; Lopes et al., 2016)
		Inner wall of the integument Newly formed endosperm cell walls	Signalling female gametogenesis	
Solanum lycopersicum	SlyFLA9	Ovary Pollen–pistil	Pollen–pistil interaction	(Lara-Mondragón and MacAlister, 2021)
Torenia fournieri	AMOR	Surface cell layer of the placenta, ovules	Pollen tube competence Intercellular communication	(Mizukami et al., 2016)
Trithuria submersa		Intine wall of pollen grain	Signalling in pollen grain apertures, enhancing future pollen tube emergence	(Costa et al., 2013)
		Ovary wall | Outer integuments Stigmatic hairs	Essential structural components and probably important signalling molecules	

before fertilisation (Pereira et al., 2014; Lopes et al., 2016; Mizukami et al., 2016; Pereira et al., 2016a,b).

AGPs play important roles very early in reproduction in the female and male germline development. Through immunolocalisation studies with anti-AGP antibodies, Coimbra and co-workers (2007) detected these glycoproteins in the functional megaspore and in the female gametophyte; in a similar study using *Q. suber* flowers, the epitopes of these proteins were recognised in ovules developing megasporocytes (Lopes et al., 2016). Importantly, two studies revealed that AGP18 is essential for the initiation of female gametogenesis in Arabidopsis (Acosta-García and Vielle-Calzada, 2004; Demesa-Arévalo and Vielle-Calzada, 2013). It was shown that this AGP is transcribed especially in the megaspore mother cell, but its glycosylation pattern is defined only in the functional megaspore, confirming the importance of AGPs in cell identity determination. Other studies revealed two other AGPs, AGP22 and AGP24, to be expressed equally in the megaspore mother cell and in the functional megaspore (Tucker et al., 2012; Tucker and Koltunow, 2014). On the "male side", epitopes of AGPs were recognised in different stages of microsporogenesis in several species: *A. thaliana* (Pereira et al., 2006; Qin et al., 2007; Coimbra and Pereira, 2012; Li et al., 2017), *B. perennis* (Chudzik et al., 2014), *Q. suber* (Costa et al., 2015), *B. napus* (Corral-Martínez et al., 2016) and *B. vulgaris* (Majewska-Sawka and Rodriguez-Garcia, 2006). In *A. thaliana*, AGP6, AGP11, AGP23 and AGP40 are pollen-specific and are involved in the pollen tube endosome machinery and in premature pollen germination prevention inside the anthers under high relative humidity (Costa et al., 2013; Pereira et al., 2016b; Lopes et al., 2019). BcMF18 and BcMF8 isolated from *B. campestris* are the orthologues of AtAGP6 and AtAGP11, respectively, and both are fundamental for pollen biogenesis, acting through independent pathways during microspore development (Lin et al., 2014, 2018). In addition to the classical AGPs, FLAs are also recruited during gametophyte development. FLA3 is expressed in pollen grains and pollen tubes and is involved in microspore development, affecting cellulose deposition (Li et al., 2010). In rice, MICROSPORE AND TAPETUM REGULATOR1 (MTR1) encodes a FLA specifically expressed in male reproductive cells, and its mutants showed a male sterile phenotype, indicating that MTR1 is indispensable for tapetum formation and microspore development (Tan et al., 2012). Recently, a role for FLA14 has been determined in pollen development and premature pollen germination prevention inside the anthers under high relative humidity as suggested earlier for Arabidopsis AGP6 and AGP11 (Miao et al., 2021).

One of the most important phases of the reproductive process is double fertilisation, a unique process in angiosperms and a key process for successful seed development. The first interactions between the pollen grain and the female tissues begin at the stigma of the flower. Several studies in phylogenetically diverse plant species revealed AGPs to be involved in this step of the reproductive process. In a study using the monoclonal antibodies JIM8 and JIM13, it was concluded that the acquisition of stigma receptivity in the apple flower is dependent on the secretion of AGPs, and that the disappearance of these AGPs in the stigma after pollination is closely related to pollen tube growth (Losada and Herrero, 2012). The secretion of these AGPs has been correlated with pistil maturity in other species such as *N. alata* (Du et al., 1996b), *T. submersa* (Costa et al., 2013), *Olea europaea* (Suárez et al., 2013), *Magnolia virginiana* (Losada et al., 2014) and *Q. suber* (Lopes et al., 2016). So, AGPs present in stigmatic cells play roles in pollen–stigma interactions and in pollen grain competence to initiate pollen tube growth.

In the next step of this pathway, the pollen tube grows through female tissues of the style until it reaches the ovules. The presence of AGPs along these tissues has been demonstrated in several studies (Coimbra et al., 2007; Pereira et al., 2014, 2016b). One study investigating the distribution of AGPs recognised by the JIM8 antibody in heat-stressed pistils of *S. lycopersicum* cv. Micro-Tom demonstrated that heat stress affects AGPs content and distribution along the pistil tissues (Mareri et al., 2019, 2016). More recently, several AGPs expressed in tomato pistils with dynamic localisations in response to fertilisation were identified. A chimeric FLA (SlyFLA9) demonstrated higher levels of expression in tomato ovary, suggesting a possible role in pollen–pistil interactions (Lara-Mondragón and MacAlister, 2021). A putative sequence encoding AGP18 isolated from *Capsicum annuum* (CaAGP18) was predicted to be important in vegetative and reproductive organs of bell pepper at seedling, flowering, pollen–pistil interactions and fruiting stages (Verdugo-Perales et al., 2018).

One of the female tissues fundamental for pollen tube growth is the transmitting tract, in which regulated programmed cell death facilitates the pollen tubes' journey to the ovules (Crawford et al., 2007; Pereira et al., 2021). In *N. tabacum*, the Transmitting Tissue-Specific (TTS) proteins—specific AGPs—have been identified as related to pollen tube targeting the ovules. This study showed that these proteins stimulate pollen tube growth *in vitro*, attract pollen tubes in *semi-in vivo* culture systems, and are necessary for optimal pollen tube growth *in vivo* by functioning as a source of nutrients and adhesion molecules for growing pollen tubes (Cheung et al., 1995; Wu et al., 2000; Noyszewski et al., 2017).

Studies using genetic approaches and reporter genes have shown that other AGPs, such as AGP1, AGP4 (JAGGER) and AGP19, are abundant in the transmitting tract of *A. thaliana* (Yang et al., 2007; Pereira et al., 2014, 2016a,b). AGP19 is strongly expressed in female tissues, namely, in the style, ovary walls, transmitting tract and siliques. The loss-of-function mutant of AGP19 revealed reduced flower production and smaller siliques compared to wild-type plants, indicating the importance of this AGP in reproduction (Yang et al., 2007). Pereira and colleagues (2014) revealed that AGP1, AGP12 and AGP15 are localised throughout various pistil tissues, suggesting that these may nutritionally contribute to the growth of the pollen tube from the top of the stigma to the base of the pistil, facilitating their movement, guiding them to their targets, or even making them competent for pollen tube reception by the embryo sac. The presence of AGPs in the transmitting tract was also observed in *Lilium longiflorum* (Jauh and Lord, 1996) and *O. europaea* (Suárez et al., 2013) by transmission electron microscopy.

Besides AGPs, many other players act during the growth of the pollen tube along the pistil tissues, one of the most important being the calcium ion. Lamport and collaborators have endorsed that AGPs act as calcium chelators during pollen tube growth based on a novel Hechtian oscillator that integrates a periplasmic glycoprotein-calcium condenser (AGP-Ca^{2+}) with AGPs located at the pollen tube tip as the source of cytosolic Ca^{2+} oscillations (Lamport and Várnai, 2013; Lamport et al., 2018). More recently, a review by Lamport and collaborators clearly highlighted AGPs as essential players involved in novel molecular pinball machines of the plasma membrane controlling cytosolic Ca^{2+} levels that regulate plant metabolism (Lamport et al., 2021). In this case, it will be interesting to find whether specific AGPs expressed in reproductive tissues act in calcium-mediated signalling pathways, given that many of the reproductive events are dependent on the oscillations of this ion (Iwano et al., 2004, 2012). Lopez-Hernandez and colleagues (2020) revealed that the capacity of AGPs to bind Ca^{2+} ions appear to be conferred by the GlcA residues in AGP sugar chains. In this study, mutants of GLCATs, responsible for GlcA addition, demonstrated defects in plant development coming from the reduction in Ca^{2+} binding to AGP sugars (Lopez-Hernandez et al., 2020; Zhang et al., 2020). The study of this GLCATs can be important to understand the possible essential roles of AGP-Ca^{2+} connection during the reproductive processes.

During its growth along the transmitting tract, the pollen tube abruptly turns its growing direction and emerges from the septum in the direction of an ovule. By this time, two phases of pollen tube orientation are defined: a funicular orientation phase—in which the pollen tube adheres to and grows through the funiculus—and a micropylar orientation phase—in which the pollen tube is guided from the funiculus towards the ovule micropyle (Shimizu and Okada, 2000; Mizuta and Higashiyama, 2018). These two orientation phases depend not only on signals from the sporophytic tissues but also on signals from the female gametophyte (Pereira et al., 2016a). In the micropylar orientation phase, the synergids are the preponderant female cells since they are the ones responsible for producing diffusible chemical signals that attract the pollen tube to the entrance of the embryo sac. AGPs are present in the embryo sac, namely, in the synergid cells of different species as possible signalling molecules or as a sink of calcium ions (Coimbra and Salema, 1997; Coimbra and Duarte, 2003; Coimbra et al., 2007; Lopes et al., 2016).

A specific group of chimeric AGPs—ENODL11/12/13/14/15 proteins—are expressed in the funiculus and ovules of Arabidopsis, controlling pollen tube reception. These proteins play a key role in male–female communication facilitating the success of double fertilisation (Hou et al., 2016). JAGGER (AGP4) was the first specific AGP to be proven essential during the micropylar phase of pollen tube growth, revealing a significant role in preventing the entry of more than one pollen tube into the embryo sac in *A. thaliana*. In JAGGER mutants, the persistent synergid does not undergo programmed cell death after successful fertilisation so more than one pollen tube is attracted to the embryo sac, with polytubey phenotype occurrence (multiple pollen tubes growing into one embryo sac). Thus, JAGGER is important for death of the persistent synergid and cessation of pollen tube attraction into ovules, avoiding polyspermy (Pereira et al., 2016b). Lora and collaborators (2018) observed the presence of AGPs in the *A. cherimola* nucellar cells around the micropyle and in its single integument, and correlated this AGPs' abundance with pollen tube attraction by analysing *Annona* and *Arabidopsis* mutants lacking integuments. In this way, they demonstrated the role of the inner integument in pollen tube attraction and a possible role of AGPs in this process and in this species (Lora et al., 2018). A very specific molecule, AMOR, was identified as a bioactive AG sugar chain from derivatives of *Torenia fournieri* ovules. These molecules were shown to be responsible for the establishment of pollen tube competence to identify the ovule attraction signals (Mizukami et al., 2016). Given that the capacity of AGPs to bind Ca^{2+} ions seem to be dependent on its GlcA residues, it is plausible to speculate that AMOR acts together with Ca^{2+} in pollen tube attraction into the ovule. Furthermore, as referred above, several enzymes have been identified that add these GlcA residues to the sugar chains from AGPs. The identification of AMOR as AG sugar chains along with AGP differential localisation in the reproductive tissues suggests that the functions of AGPs and their sugar chains may be locally controlled by AG biosynthesis enzymes (Su and Higashiyama, 2018).

The full activation of AGP functions in different tissues is dependent on the presence of specific glycosylation enzymes at certain sites and times in development, and points to their importance as potential factors in indicating that they may participate in important regulatory signalling pathways.

Recently, it was found that the role of AGPs in reproduction is relevant not only in angiosperms but also in gymnosperms, which points to an ubiquity of these molecules at the reproductive level in different types of plants. The presence of AGPs in *Larix decidua* Mill., a gymnosperm, was demonstrated by immunolocalisation studies, in which the epitopes of AGPs were localised in male gametophytes and ovule tissues during pollination, during the progamic phase, and after fertilisation and in *in vitro* growing pollen tubes (Rafińska et al., 2021).

It is clear that AGPs are involved in all stages of the reproductive process, not only in the model plant Arabidopsis but also in species of great agronomic interest where successful seed production is crucial. The involvement of these glycoproteins in sexual plant reproduction can be considered a sweet love affair. The study of these glycoproteins is fundamental to understanding the signalling pathways and their specific functions in the reproductive process of angiosperms. In the near future, new methodologies emerging from molecular biology, biochemistry, and bioimaging may be crucial tools for deciphering the specific role of each AGP in pollen–pistil communications.

CONCLUSIONS

AGPs are a highly diverse class of heavily HRGP that have been under intense research in the last 60 years, with thousands of AGPs predicted across the plant kingdom, where they are reported to play different functions in both vegetative and reproductive tissues. In this review, we present an overview of the accumulated knowledge on AGP involvement in abiotic stress tolerance and in plant reproduction, as these two physiological processes are essential for plant growth and seed yield, and consequently for sustainable food security. The analyses of the available reports indicate that AGPs are strongly affected by abiotic stress in a tissue-specific manner and that different tissues show different AGP distribution patterns under stress conditions. In those stress responses characterised by an increase in specific classes of AGPs (e.g. drought), the increased AGPs may contribute to the cell wall hardening and thickening, helping plant cells to avoid internal damages, and to cell–cell communication networks that could be important for an efficient transmission of stress-related signalling. On the other hand, the increase in AGP carbohydrate moieties may regulate the intracellular osmotic potential. Curiously, other stress agents may negatively affect the expression of AGPs (e.g. high temperature). How AGPs intervene to allow plant cell walls to cope with different stress factors remains to be further elucidated. Likewise, AGPs are abundant along the pathway followed by the pollen tube from the surface of the stigma towards the ovule. The typical double fertilisation of flowering plants encompasses a complex genetic network of regulatory signals and refined cross-talk in which AGPs intervene either as signalling or nutrient molecules. The breakthrough in understanding AGP function is probably connected with the Ca^{2+} homeostasis. Lamport's hypothesis (Lamport et al., 2018) that plasma membrane AGPs bind Ca^{2+} in the periplasmic space, and when H^+ are pumped out across the plasma membrane lowering the periplasmic pH and titrating the glucuronosyl carboxyl groups on the AGPs, then the Ca^{2+} is released from the AGPs, passing through channels in the plasma membrane to increase the cytosolic Ca^{2+} concentration, and initiate Ca^{2+} signalling, has gained support with recent articles (Lopez-Hernandez et al., 2020; Zhang et al., 2020). This area of research is clearly important and it should be more intensely explored. As we have been predicting for more than 20 years, signals are exchanged at every step along the pollen tube pathway, beginning with the recognition of pollen by cells on the pistil surface and continuing until reaching the embryo sac. AGP abundance characterises the pathway from stigma to synergids demonstrated by immunocytochemical localisation reported in several studies.

ACKNOWLEDGEMENTS

This work received financial support from PT national funds (FCT/MCTES, Fundação para a Ciência e Tecnologia and Ministério da Ciência, Tecnologia e Ensino Superior) through the project UIDB/50006/2020, the Strategic Funding of GreenUPorto (UIDB/05748/2020 and UIDP/05748/2020), and by SeedWheels FCT Project – POCI-01–0145-FEDER-027839. AL (SFRH/BD/115960/2016), DM (SFRH/BD/143557/2019) and MJF (SFRH/BD/143579/2019) received a FCT PhD grant. JS (LCF/BQ/DR20/11790010) received a La Caixa Foundation PhD grant.

REFERENCES

Acosta-García G, Vielle-Calzada J-P (2004) A classical arabinogalactan protein is essential for the initiation of female gametogenesis in arabidopsis. *Plant Cell* 16: 2614–2628.

Anderson R, Clarke A, Jermyn M, Knox R, Stone B (1977) A carbohydrate-binding arabinogalactan-protein from liquid suspension cultures of endosperm from *Lolium multiflorum*. *Functional Plant Biology* 4: 143–158.

Anderson CT, Kieber JJ (2020) Dynamic construction, perception, and remodeling of plant cell walls. *Annual Review of Plant Biology* 71: 39–69.

Aspinall GO, Molloy JA, Craig JW (1969) Extracellular polysaccharides from suspension-cultured sycamore cells. *Canadian Journal Biochemestry* 47: 1063–1070.

Basu D, Liang Y, Liu X, Himmeldirk K, Faik A, Kieliszewski M, Held M, Showalter AM (2013) Functional identification of a hydroxyproline-*O*-galactosyltransferase specific for arabinogalactan protein biosynthesis in Arabidopsis. *Journal of Biological Chemistry* 288: 10132–10143.

Basu D, Tian L, Wang W, Bobbs S, Herock H, Travers A, Showalter AM (2015a) A small multigene hydroxyproline-*O*-galactosyltransferase family functions in arabinogalactan-protein glycosylation, growth and development in Arabidopsis. *BMC Plant Biology* 15: 1–23.

Basu D, Wang W, Ma S, DeBrosse T, Poirier E, Emch K, Soukup E, Tian L, Showalter AM (2015b) Two hydroxyproline galactosyltransferases, GALT5 and GALT2, function in arabinogalactan-protein glycosylation, growth and development in Arabidopsis. *PLoS One* 10: e0125624.

Beihammer G, Maresch D, Altmann F, Strasser R (2020) Glycosylphosphatidylinositol-anchor synthesis in plants: a glycobiology perspective. *Frontiers in Plant Science* 11: 611188.

Borner, GHH, Sherrier, DJ, Stevens, TJ, Arkin, IT, Dupree, P (2002) Prediction of glycosylphosphatidylinositol-anchored proteins in Arabidopsis. A genomic analysis. *Plant Physiology* 129: 486–499.

Borner GHH, Lilley KS, Stevens TJ, Dupree P (2003) Identification of glycosylphosphatidylinositol-anchored proteins in Arabidopsis. A proteomic and genomic analysis. *Plant Physiology* 132: 568–577.

Canut H, Albenne C, Jamet E (2016) Post-translational modifications of plant cell wall proteins and peptides: a survey from a proteomics point of view. *Biochimica et Biophysica Acta - Proteins Proteomics* 1864: 983–990.

Chen CG, Pu ZY, Moritz RL, Simpson RJ, Bacic A, Clarke AE, Mau SL (1994) Molecular cloning of a gene encoding an arabinogalactan-protein from pear (*Pyrus communis*) cell suspension culture. *Proceedings of the National Academy of Sciences* 91: 10305–10309.

Cheung AY, Wang H, Wu HM (1995) A floral transmitting tissue-specific glycoprotein attracts pollen tubes and stimulates their growth. *Cell* 82: 383–393.

Chudzik B, Szczuka E, Zarzyka B, Leszczuk A (2014) Distribution of arabinogalactan proteins during microsporogenesis in the anther of *Bellis perennis* L (Asteraceae). *Acta Biologica Cracoviensia Series Botanica* 56: 49–60.

Clarke AE, Anderson RL, Stone BA (1979) Form and function of arabinogalactans and arabinogalactan-proteins. *Phytochemistry* 18: 521–540.

Coimbra S, Salema R (1997) Immunolocalization of arabinogalactan proteins in *Amaranthus hypochondriacus* L ovules. *Protoplasma* 199: 75–82.

Coimbra S, Duarte C (2003) Arabinogalactan proteins may facilitate the movement of pollen tubes from the stigma to the ovules in *Actinidia deliciosa* and *Amaranthus hypochondriacus*. *Euphytica* 133: 171–178.

Coimbra S, Almeida J, Junqueira V, Costa ML, Pereira LG (2007) Arabinogalactan proteins as molecular markers in *Arabidopsis thaliana* sexual reproduction. *Journal of Experimental Botany* 58: 4027–4035.

Coimbra S, Costa M, Jones B, Mendes MA, Pereira LG (2009) Pollen grain development is compromised in Arabidopsis *agp6 agp11* null mutants. *Journal of Experimental Botany* 60: 3133–3142.

Coimbra S, Pereira LG (2012) Arabinogalactan proteins in *Arabidopsis thaliana* pollen development. In *Transgenic Plants - Advances and Limitations*, ed. Çiftçi YO, 329–352. InTech Publishers.

Corral-Martínez P, García-Fortea E, Bernard S, Driouich A, Seguí-Simarro JM (2016) Ultrastructural immunolocalization of arabinogalactan protein, pectin and hemicellulose epitopes through anther development in *Brassica napus*. *Plant Cell Physiology* 57: 2161–2174.

Costa M, Pereira AM, Rudall PJ, Coimbra S (2013) Immunolocalization of arabinogalactan proteins (AGPs) in reproductive structures of an early-divergent angiosperm, *Trithuria* (Hydatellaceae). *Annals of Botany* 111: 183–190.

Costa ML, Sobral R, Costa MMR, Amorim MI, Coimbra S (2015) Evaluation of the presence of arabinogalactan proteins and pectins during *Quercus suber* male gametogenesis. *Annals of Botany* 115: 81–92.

Crawford BCW, Ditta G, Yanofsky MF (2007) The *NTT* gene is required for transmitting-tract development in carpels of *Arabidopsis thaliana*. *Current Biology* 17: 1101–1108.

Cui S, Hu J, Guo S, Wang J, Cheng Y, Dang X, Wu L, He Y (2012) Proteome analysis of *Physcomitrella patens* exposed to progressive dehydration and rehydration. *Journal of Experimental Botany* 63: 711–726.

Demesa-Arévalo E, Vielle-Calzada JP (2013) The classical arabinogalactan protein AGP18 mediates megaspore selection in Arabidopsis. *Plant Cell* 25: 1274–1287.

Desnoyer N, Palanivelu R (2020) Bridging the GAPs in plant reproduction: a comparison of plant and animal GPI-anchored proteins. *Plant Reproduction* 33: 129–142.

Dilokpimol A, Geshi N (2014) *Arabidopsis thaliana* glucuronosyltransferase in family GT14. *Plant Signaling & Behavior* 9: e28891.

Dilokpimol A, Poulsen CP, Vereb G, Kaneko S, Schulz A, Geshi N (2014) Galactosyltransferases from *Arabidopsis thaliana* in the biosynthesis of type II arabinogalactan: molecular interaction enhances enzyme activity. *BMC Plant Biology* 14: 1–14.

Dresselhaus T, Coimbra S (2016) Plant reproduction: AMOR enables males to respond to female signals. *Current Biology* 26: 321–323.

Du H, Simpson RJ, Moritz RL, Clarke AE, Bacic A (1994) Isolation of the protein backbone of an arabinogalactan-protein from the styles of *Nicotiana alata* and characterization of a corresponding cDNA. *Plant Cell* 6: 1643–1653.

Du H, Clarke AE, Bacic A (1996a) Arabinogalactan proteins: a class of extracellular matrix proteoglycans involved in plant growth and development. *Trends Cell Biology* 6: 411–414.

Du H, Simpson RJ, Clarke AE, Bacic A (1996b) Molecular characterization of a stigma-specific gene encoding an arabinogalactan-protein (AGP) from *Nicotiana alata*. *Plant Journal* 9: 313–323.

Duruflé H, Hervé V, Balliau T, Zivy M, Dunand C, Jamet E (2017) Proline hydroxylation in cell wall proteins: is it yet possible to define rules? *Frontiers in Plant Science* 8: 1–10.

Eisenhaber B, Wildpaner M, Schultz CJ, Borner GHH, Dupree P, Eisenhaber F (2003) Glycosylphosphatidylinositol lipid anchoring of plant proteins. Sensitive prediction from sequence- and genome-wide studies for Arabidopsis and rice. *Plant Physiology* 133: 1691–1701.

Ellis M, Egelund J, Schultz CJ, Bacic A (2010) Arabinogalactan-proteins: Key regulators at the cell surface? *Plant Physiology* 153: 403–419.

Ezquer I, Salameh I, Colombo L, Kalaitzis P (2020) Plant cell walls tackling climate change: biotechnological strategies to improve crop adaptations and photosynthesis in response to global warming. *Plants* 9: 212.

Faye L, Boulaflous A, Benchabane M, Gomord V, Michaud D (2005) Protein modifications in the plant secretory pathway: Current status and practical implications in molecular pharming. *Vaccine* 23: 1770–1778.

Fincher GB, Stone BA, Clarke AE (1983) Arabinogalactan-proteins: structure, biosynthesis, and function. *Annual Review of Plant Biology* 34: 47–70.

Fragkostefanakis S, Dandachi F, Kalaitzis P (2012) Expression of arabinogalactan proteins during tomato fruit ripening and in response to mechanical wounding, hypoxia and anoxia. *Plant Physiology and Biochemistry* 52: 112–118.

Gao M, Showalter AM (1999) Yariv reagent treatment induces programmed cell death in Arabidopsis cell cultures and implicates arabinogalactan protein involvement. *Plant Journal* 19: 321–331.

Gaspar Y, Johnson KL, Mckenna JA, Bacic A, Schultz CJ (2001) The complex structures of arabinogalactan-proteins and the journey towards understanding function. *Plant Molecular Biology* 47: 161–176.

Geshi N, Johansen JN, Dilokpimol A, Rolland A, Belcram K, Verger S, Kotake T, Tsumuraya Y, Kaneko S, Tryfona T, Dupree P, Schelletr HV, Höfte H, Mouille G (2013) A galactosyltransferase acting on arabinogalactan protein glycans is essential for embryo development in Arabidopsis. *Plant Journal* 76: 128–137.

Gillmor SC, Lukowitz W, Brininstool G, Sedbrook JC, Hamann T, Poindexter P, Somerville C (2005) Glycosylphosphatidylinositol-anchored proteins are required for cell wall synthesis and morphogenesis in Arabidopsis. *Plant Cell* 17: 1128–1140.

Gleeson PA, Clarke AE (1980) Arabinogalactans of sexual and somatic tissues of Gladiolus and Lilium. *Phytochemistry* 19: 1777–1782.

Gleeson PA, McNamara M, Wettenhall REH, Stone BA, Fincher GB (1989) Characterization of the hydroxyproline-rich protein core of an arabinogalactan-protein secreted from suspension-cultured *Lolium multiflorum* (Italian ryegrass) endosperm cells. *Biochemical Journal* 264: 857–862.

Guan Y, Nothnagel EA (2004) Binding of arabinogalactan proteins by Yariv phenylglycoside triggers wound-like responses in Arabidopsis cell cultures. *Plant Physiology* 135: 1346–1366.

Hengel AJ, Tadesse Z, Immerzeel P, Schols H, Kammen A, de Vries SC (2001) *N*-acetylglucosamine and glucosamine-containing arabinogalactan proteins control somatic embryogenesis. *Plant Physiology* 125: 1880–1890.

Herman EM, Lamb CJ (1992) Arabinogalactan-rich glycoproteins are localized on the cell surface and in intravacuolar multivesicular bodies. *Plant Physiology* 98: 264–272.

Hervé C, Siméon A, Jam M, Cassin A, Johnson KL, Salmeán AA, Willats WGT, Doblin MS, Bacic A, Kloareg B (2016) Arabinogalactan proteins have deep roots in eukaryotes: identification of genes and epitopes in brown algae and their role in *Fucus serratus* embryo development. *New Phytologist* 209: 1428–1441.

Higashiyama T, Takeuchi H (2015) The mechanism and key molecules involved in pollen tube guidance. *Annual Review of Plant Biology* 66: 393–413.

Hijazi M, Velasquez SM, Jamet E, Estevez JM, Albenne C (2014) An update on post-translational modifications of hydroxyproline-rich glycoproteins: toward a model highlighting their contribution to plant cell wall architecture. *Frontiers in Plant Science* 5: 395.

Hou Y, Guo X, Cyprys P, Zhang Y, Bleckmann A, Cai L, Huang Q, Luo Y, Gu H, Dresselhaus T, Dong J, Qu LJ (2016) Maternal ENODLs are required for pollen tube reception in Arabidopsis. *Current Biology* 26: 2343–2350.

Iwano M, Shiba H, Miwa T, Che FS, Takayama S, Nagai T, Miyawaki A, Isogai A (2004) Ca^{2+} dynamics in a pollen grain and papilla cell during pollination of Arabidopsis. *Plant Physiology* 136: 3562–3571.

Iwano M, Ngo QA, Entani T, Shiba H, Nagai T, Miyawaki A, Isogai A, Grossniklaus U, Takayama S (2012) Cytoplasmic Ca^{2+} changes dynamically during the interaction of the pollen tube with synergid cells. *Development* 139: 4202–4209.

Jamet E, Albenne C, Boudart G, Irshad M, Canut H, Pont-Lezica R (2008) Recent advances in plant cell wall proteomics. *Proteomics* 8: 893–908.

Jauh GY, Lord EM (1996) Localization of pectins and arabinogalactan-proteins in lily (*Lilium longiflorum* L.) pollen tube and style, and their possible roles in pollination. *Planta* 199: 251–261.

Jermyn M, Guthrie R (1985) A final assault on the structure of carrot AGP's. In: *AGP News (Proceedings of the arabinogalactan protein club, vol. 5)* pp. 4–25, Clarke, A.E., ed. Melbourne, Australia.

Jermyn MA, Yeow YM, Woods EF (1975) A class of lectins present in the tissues of seed plants. *Functional Plant Biology* 2: 501–531.

Johnson KL, Jones BJ, Bacic A, Schultz CJ (2003) The fasciclin-like arabinogalactan proteins of Arabidopsis. A multigene family of putative cell adhesion molecules. *Plant Physiology* 133: 1911–1925.

Johnson KL, Kibble NAJ, Bacic A, Schultz CJ (2011) A fasciclin-like arabinogalactan-protein (FLA) mutant of *Arabidopsis thaliana*, *fla1*, shows defects in shoot regeneration. *PLoS One* 6: e25154.

Johnson KL, Cassin AM, Lonsdale A, Bacic A, Doblin MS, Schultz CJ (2017) Pipeline to identify hydroxyproline-rich glycoproteins. *Plant Physiology* 174: 886–903.

Kieliszewski MJ (2001) The latest hype on Hyp-*O*-glycosylation codes. *Phytochemistry* 57: 319–323.

Kieliszewski MJ, Lamport DTA (1994) Extensin: repetitive motifs, functional sites, post-translational codes, and phylogeny. *Plant Journal* 5: 157–172.

Kjellbom P, Snogerup L, Stöhr C, Reuzeau C, McCabe PF, Pennell RI (1997) Oxidative cross-linking of plasma membrane arabinogalactan proteins. *Plant Journal* 12: 1189–1196.

Knoch E, Dilokpimol A, Tryfona T, Poulsen CP, Xiong G, Harholt J, Petersen BL, Ulvskov P, Hadi MZ, Kotake T, Tsumuraya Y, Pauly M, Dupree P, Geshi N (2013) A β-glucuronosyltransferase from *Arabidopsis thaliana* involved in biosynthesis of type II arabinogalactan has a role in cell elongation during seedling growth. *Plant Journal* 76: 1016–1029.

Knox JP (1995) Developmentally regulated proteoglycans and glycoproteins of the plant cell surface. *FASEB J* 9: 1004–1012.

Kobayashi Y, Motose H, Iwamoto K, Fukuda H (2011) Expression and genome-wide analysis of the xylogen-type gene family. *Plant Cell Physiology* 52: 1095–1106.

Komalavilas P, Zhu JK, Nothnagel EA (1991) Arabinogalactan-proteins from the suspension culture medium and plasma membrane of rose cells. *Journal of Biological Chemistry* 266: 15956–15965.

Lamport DTA, Northcote DH (1960) Hydroxyproline in primary cell walls of higher plants. *Nature* 188: 665–666.

Lamport DTA (1963) Oxygen fixation into hydroxyproline of plant cell wall protein. *Journal of Biological Chemistry* 238: 1438–1440.

Lamport DTA (1967) Hydroxyproline-*O*-glycosidic linkage of the plant cell wall glycoprotein extensin. *Nature* 216: 1322–1324.

Lamport DTA (1970) Cell wall metabolism. *Annual Review of Plant Biology* 21: 235–270.

Lamport DTA, Kieliszewski MJ, Showalter AM (2006) Salt stress upregulates periplasmic arabinogalactan proteins: using salt stress to analyse AGP function. *New Phytologist* 169: 479–492.

Lamport DTA, Várnai P (2013) Periplasmic arabinogalactan glycoproteins act as a calcium capacitor that regulates plant growth and development. *New Phytologist* 197: 58–64.

Lamport DTA, Tan L, Held MA, Kieliszewski MJ (2018) Pollen tube growth and guidance: Occam's razor sharpened on a molecular arabinogalactan glycoprotein Rosetta Stone. *New Phytologist* 217: 491–500.

Lamport DTA, Tan L, Kieliszewski MJ (2021) A molecular pinball machine of the plasma membrane regulates plant growth-a new paradigm. *Cells* 10: 1935.

Lara-Mondragón CM, MacAlister CA (2021) Arabinogalactan glycoprotein dynamics during the progamic phase in the tomato pistil. *Plant Reproduction* 34: 131–148.

Lee KJD, Sakata Y, Mau S-L, Pettolino F, Bacic A, Quatrano RS, Knight CD, Knox JP (2005) Arabinogalactan proteins are required for apical cell extension in the moss *Physcomitrella patens*. *Plant Cell* 17: 3051–3065.

Li SX, Showalter AM (1996) Cloning and developmental/stress-regulated expression of a gene encoding a tomato arabinogalactan protein. *Plant Molecular Biology* 32: 641–652.

Li J, Yu M, Geng L-L, Zhao J (2010) The fasciclin-like arabinogalactan protein gene, *FLA3*, is involved in microspore development of Arabidopsis. *Plant Journal* 64: 482–497.

Li D-D, Xue J-S, Zhu J, Yang Z-N (2017) Gene regulatory network for tapetum development in *Arabidopsis thaliana. Frontiers in Plant Science* 8: 1559.

Liang Y, Basu D, Pattathil S, Xu WL, Venetos A, Martin SL, Faik A, Hahn MG, Showalter AM (2013) Biochemical and physiological characterization of *fut4* and *fut6* mutants defective in arabinogalactan-protein fucosylation in Arabidopsis. *Journal of Experimental Botany* 64: 5537–5551.

Lima RB, dos Santos TB, Vieira LG, Ferrarese ML, Ferrarese-Filho O, Donatti L, Boeger MR, Petkowicz CL (2013) Heat stress causes alterations in the cell-wall polymers and anatomy of coffee leaves (*Coffea arabica* L.). *Carbohydrate Polymers* 93: 135–143.

Lin S, Dong H, Zhang F, Qiu L, Wang F, Cao J, Huang L (2014) *BcMF8*, a putative arabinogalactan protein-encoding gene, contributes to pollen wall development, aperture formation and pollen tube growth in *Brassica campestris. Annals of Botany* 113: 777–788.

Lin S, Yue X, Miao Y, Yu Y, Dong H, Huang L, Cao J (2018) The distinct functions of two classical arabinogalactan proteins BcMF8 and BcMF18 during pollen wall development in *Brassica campestris. Plant Journal* 94: 60–76.

Lopes AL, Costa ML, Sobral R, Costa MM, Amorim MI, Coimbra S (2016) Arabinogalactan proteins and pectin distribution during female gametogenesis in *Quercus suber* L. *Annals of Botany* 117: 949–961.

Lopes AL, Moreira D, Ferreira MJ, Pereira AM, Coimbra S (2019) Insights into secrets along the pollen tube pathway in need to be discovered. *Journal of Experimental Botany* 70: 2979–2992.

Lopez-Hernandez F, Tryfona T, Rizza A, Yu XL, Harris MOB, Webb AA, Kotake T, Dupree P (2020) Calcium binding by arabinogalactan polysaccharides is important for normal plant development. *Plant Cell* 32: 3346–3369.

Lora J, Larranaga N, Hormaza JI (2018) Genetics and breeding of fruit crops in the Annonaceae family: Annona spp and Asimina spp. *Advances in Plant Breeding* 651–672.

Losada JM, Herrero M (2012) Arabinogalactan-protein secretion is associated with the acquisition of stigmatic receptivity in the apple flower. *Annals of Botany* 110: 573–584.

Losada JM, Herrero M, Hormaza JI, Friedman WE (2014) Arabinogalactan proteins mark stigmatic receptivity in the protogynous flowers of *Magnolia virginiana* (Magnoliaceae). *American Journal of Botany* 101: 1963–1975.

Lu H, Chen M, Showalter AM (2001) Developmental expression and perturbation of arabinogalactan-proteins during seed germination and seedling growth in tomato. *Physiologia Plantarum* 112: 442–450.

Ma H, Zhao J (2010) Genome-wide identification, classification, and expression analysis of the arabinogalactan protein gene family in rice (*Oryza sativa* L.). *Journal of Experimental Botany* 61: 2647–2668.

Ma Y, Yan C, Li H, Wu W, Liu Y, Wang Y, Chen Q, Ma H (2017) Bioinformatics prediction and evolution analysis of arabinogalactan proteins in the plant kingdom. *Frontiers in Plant Science* 8: 66.

MacMillan CP, Mansfield SD, Stachurski ZH, Evans R, Southerton SG (2010) Fasciclin-like arabinogalactan proteins: specialization for stem biomechanics and cell wall architecture in Arabidopsis and Eucalyptus. *Plant Journal* 62: 689–703.

Majewska-Sawka A, Rodriguez-Garcia MI (2006) Immunodetection of pectin and arabinogalactan protein epitopes during pollen exine formation of *Beta vulgaris* L. *Protoplasma* 228: 41–47.

Mareri L, Faleri C, Romi M, Mariani C, Cresti M, Cai G (2016) Heat stress affects the distribution of JIM8-labelled arabinogalactan proteins in pistils of *Solanum lycopersicum* cv Micro-Tom. *Acta Physiologiae Plantarum* 38: 184.

Mareri L, Romi M, Cai G (2019) Arabinogalactan proteins: actors or spectators during abiotic and biotic stress in plants? *Plant Biosystems* 153: 173–185.

Mashiguchi K, Asami T, Suzuki Y (2009) Genome-wide identification, structure and expression studies, and mutant collection of 22 EARLY NODULIN-LIKE PROTEIN genes in Arabidopsis. *Bioscience, Biotechnology, and Biochemistry* 73: 2452–2459.

Mau S-L, Chen C-G, Pu Z-Y, Moritz RL, Simpson RJ, Bacic A, Clarke AE (1995) Molecular cloning of cDNAs encoding the protein backbones of arabinogalactan-proteins from the filtrate of suspension-cultured cells of *Pyrus communis* and *Nicotiana alata. Plant Journal* 8: 269–281.

Miao Y, Cao J, Huang L, Yu Y, Lin S (2021) FLA14 is required for pollen development and preventing premature pollen germination under high humidity in Arabidopsis. *BMC Plant Biology* 21: 254.

Mizukami AG, Inatsugi R, Jiao J, Kotake T, Kuwata K, Ootani K, Okuda S, Sankaranarayanan S, Sato Y, Maruyama D et al (2016) The AMOR arabinogalactan sugar chain induces pollen-tube competency to respond to ovular guidance. *Current Biology* 26: 1091–1097.

Mizuta Y, Higashiyama T (2018) Chemical signaling for pollen tube guidance at a glance. *Journal of Cell Science* 131: 208447.

Moore JP, Nguema-Ona EE, Vicré-Gibouin M, Sørensen I, Willats WGT, Driouich A, Farrant JM (2013) Arabinose-rich polymers as an evolutionary strategy to plasticize resurrection plant cell walls against desiccation. *Planta* 237: 739–754.

Mort AJ, Lamport DTA (1977) Anhydrous hydrogen fluoride deglycosylates glycoproteins. *Anal Biochem* 82: 289–309.

Narciso JO, Zeng W, Ford K, Lampugnani ER, Humphries J, Austarheim I, van de Meene A, Bacic A, Doblin MS (2021) Biochemical and functional characterization of GALT8, an Arabidopsis GT31 β-(1,3)-galactosyltransferase that influences seedling development. *Frontiers in Plant Science* 12: 886.

Nguema-Ona E, Vicre-Gibouin M, Gotte M, Plancot B, Lerouge P, Bardor M, Driouich A (2014) Cell wall *O*-glycoproteins and *N*-glycoproteins: aspects of biosynthesis and function. *Frontiers in Plant Science* 5: 499.

Nothnagel EA (1997) Proteoglycans and related components in plant cells. *International Review of Cytology* 195–291.

Noyszewski AK, Liu YC, Tamura K, Smith AG (2017) Polymorphism and structure of style-specific arabinogalactan proteins as determinants of pollen tube growth in *Nicotiana. BMC Evolutionary Biology* 17: 186.

Ogawa-Ohnishi M, Matsubayashi Y (2015) Identification of three potent hydroxyproline *O*-galactosyltransferases in Arabidopsis. *Plant Journal* 81: 735–746.

Olmos E, García De La Garma J, Gomez-Jimenez MC, Fernandez-Garcia N (2017) Arabinogalactan proteins are involved in salt-adaptation and vesicle trafficking in tobacco BY-2 cell cultures. *Frontiers in Plant Science* 8: 1092.

Oxley D, Bacic A (1999) Structure of the glycosylphosphatidylinositol anchor of an arabinogalactan protein from *Pyrus communis* suspension-cultured cells. *Proceedings of the National Academy of Sciences* 96: 14246–14251.

Pennell RI, Janniche L, Kjellbom P, Scofield GN, Peart JM, Roberts K (1991) Developmental regulation of a plasma membrane arabinogalactan protein epitope in oilseed rape flowers. *Plant Cell* 3: 1317–1326.

Pereira LG, Coimbra S, Oliveira H, Monteiro L, Sottomayor M (2006) Expression of arabinogalactan protein genes in pollen tubes of *Arabidopsis thaliana*. *Planta* 223: 374–380.

Pereira AM, Masiero S, Nobre MS, Costa ML, Solís M-T, Testillano PS, Sprunck S, Coimbra S (2014) Differential expression patterns of arabinogalactan proteins in *Arabidopsis thaliana* reproductive tissues. *Journal of Experimental Botany* 65: 5459–5471.

Pereira AM, Lopes AL, Coimbra S (2016a) Arabinogalactan proteins as interactors along the crosstalk between the pollen tube and the female tissues. *Frontiers in Plant Science* 7: 1895.

Pereira AM, Nobre MS, Pinto SC, Lopes AL, Costa ML, Masiero S, Coimbra S (2016b) "Love is strong, and you're so sweet": JAGGER is essential for persistent synergid degeneration and polytubey block in *Arabidopsis thaliana*. *Molecular Plant* 9: 601–614.

Pereira AM, Moreira D, Coimbra S, Masiero S (2021) Paving the way for fertilization: the role of the transmitting tract. *International Journal of Molecular Sciences* 22: 2603.

Pfeifer L, Shafee T, Johnson KL, Bacic A, Classen B (2020) Arabinogalactan-proteins of *Zostera marina* L. contain unique glycan structures and provide insight into adaption processes to saline environments. *Scientific Reports* 10: 8232.

Qin Y, Chen D, Zhao J (2007) Localization of arabinogalactan proteins in anther, pollen, and pollen tube of *Nicotiana tabacum* L. *Protoplasma* 231: 43–53.

Qu Y, Egelund J, Gilson PR, Houghton F, Gleeson PA, Schultz CJ, Bacic A (2008) Identification of a novel group of putative *Arabidopsis thaliana* β-(1,3)-galactosyltransferases. *Plant Molecular Biology* 68: 43–59.

Rabello AR, Guimarães CM, Rangel PH, da Silva FR, Seixas D, de Souza E, Brasileiro AC, Spehar CR, Ferreira ME, Mehta Â (2008) Identification of drought-responsive genes in roots of upland rice (*Oryza sativa* L.). *BMC Genomics* 9: 485.

Rafińska K, Niedojadło K, Świdziński M, Niedojadło J, Bednarska-Kozakiewicz E (2021) Spatial and temporal distribution of arabinogalactan proteins during *Larix decidua* Mill. male gametophyte and ovule interaction. *International Journal of Molecular Sciences* 22: 4298.

Šamaj J, Šamajová O, Peters M, Baluška F, Lichtscheidl I, Knox JP, Volkmann D (2000) Immunolocalization of LM2 arabinogalactan protein epitope associated with endomembranes of plant cells. *Protoplasma* 212: 186–196.

Schatz G, Dobberstein B (1996) Common principles of protein translocation across membranes. *Science* 271: 1519–1526.

Schultz C, Gilson P, Oxley D, Youl J, Bacic A (1998) GPI-anchors on arabinogalactan-proteins: implications for signalling in plants. *Trends Plant Science* 3: 426–431.

Schultz CJ, Rumsewicz MP, Johnson KL, Jones BJ, Gaspar YM, Bacic A (2002) Using genomic resources to guide research directions. The arabinogalactan protein gene family as a test case. *Plant Physiology* 129: 1448–1463.

Seifert GJ, Roberts K (2007) The Biology of Arabinogalactan Proteins. *Annual Review of Plant Biology* 58: 137–161.

Seki M et al (2002) Monitoring the expression pattern of around 7,000 Arabidopsis genes under ABA treatments using a full-length cDNA microarray. *Functional & Integrative Genomics* 2: 282–291.

Serpe MD, Nothnagel EA (1999) Arabinogalactan-proteins in the Multiple Domains of the Plant Cell Surface. *Advances in Botanical Research* 30: 207–289.

Shimizu KK, Okada K (2000) Attractive and repulsive interactions between female and male gametophytes in Arabidopsis pollen tube guidance. *Development* 127: 4511–4518.

Showalter AM (2001) Arabinogalactan-proteins: structure, expression and function. *Cellular and Molecular Life Sciences* 58: 1399–1417.

Showalter AM, Keppler B, Lichtenberg J, Gu D, Welch LR (2010) Focus issue on plant cell walls: a bioinformatics approach to the identification, classification, and analysis of hydroxyproline-rich glycoproteins. *Plant Physiology* 153: 485–513.

Showalter AM, Basu D (2016) Extensin and arabinogalactan-protein biosynthesis: glycosyltransferases, research challenges, and biosensors. *Frontiers in Plant Science* 7: 814.

Shpak E, Leykam JF, Kieliszewski MJ (1999) Synthetic genes for glycoprotein design and the elucidation of hydroxyproline-*O*-glycosylation codes. *Proceedings of the National Academy of Sciences* 96: 14736–14741.

Shpak E, Barbar E, Leykam JF, Kieliszewski MJ (2001) Contiguous hydroxyproline residues direct hydroxyproline arabinosylation in *Nicotiana tabacum*. *Journal of Biological Chemistry* 276: 11272–11278.

Silva J, Ferraz R, Dupree P, Showalter AM, Coimbra S (2020) Three decades of advances in arabinogalactan-protein biosynthesis. *Frontiers in Plant Science* 11: 2014.

Su S, Higashiyama T (2018) Arabinogalactan proteins and their sugar chains: functions in plant reproduction, research methods, and biosynthesis. *Plant Reproduction* 31: 67–75.

Suárez C, Zienkiewicz A, Castro AJ, Zienkiewicz K, Majewska-Sawka A, Rodríguez-García MI (2013) Cellular localization and levels of pectins and arabinogalactan proteins in olive (*Olea europaea* L.) pistil tissues during development: Implications for pollen-pistil interaction. *Planta* 237: 305–319.

Suzuki T, Narciso JO, Zeng W, van de Meene A, Yasutomi M, Takemura S, Lampugnani ER, Doblin MS, Bacic A, Ishiguro S (2017) KNS4/UPEX1: a type II arabinogalactan β-(1,3)-galactosyltransferase required for pollen exine development. *Plant Physiology* 173: 183–205.

Tan L, Qiu F, Lamport DTA, Kieliszewski MJ (2004) Structure of a hydroxyproline (Hyp)-arabinogalactan polysaccharide from repetitive Ala-Hyp expressed in transgenic *Nicotiana tabacum*. *Journal of Biological Chemistry* 279: 13156–13165.

Tan H, Liang W, Hu J, Zhang D (2012) MTR1 encodes a secretory fasciclin glycoprotein required for male reproductive development in rice. *Development Cell* 22: 1127–1137.

Tan L, Varnai P, Lamport DTA, Yuan C, Xu J, Qiu F, Kieliszewski MJ (2010) Plant *O*-hydroxyproline arabinogalactans are composed of repeating trigalactosyl subunits with short bifurcated side chains. *Journal of Biological Chemistry* 285: 24575–24583.

Temple H, Mortimer JC, Tryfona T, Yu X, Lopez-Hernandez F, Sorieul M, Anders N, Dupree P (2019) Two members of the DUF579 family are responsible for arabinogalactan methylation in Arabidopsis. *Plant Direct* 3: e00117.

The Arabidopsis Genome Initiative (2000) Analysis of the genome sequence of the flowering plant *Arabidopsis thaliana*. *Nature* 408: 796–815.

Tryfona T, Liang HC, Kotake T, Kaneko S, Marsh J, Ichinose H, Lovegrove A, Tsumuraya Y, Shewry PR, Stephens E, Dupree P (2010) Carbohydrate structural analysis of wheat flour arabinogalactan protein. *Carbohydrate Research* 345: 2648–2656.

Tryfona T, Liang HC, Kotake T, Tsumuraya Y, Stephens E, Dupree P (2012) Structural characterization of Arabidopsis leaf arabinogalactan polysaccharides. *Plant Physiology* 160: 653–666.

Tryfona T, Theys TE, Wagner T, Stott K, Keegstra K, Dupree P (2014) Characterisation of FUT4 and FUT6 α-(1→2)-fucosyltransferases reveals that absence of root arabinogalactan fucosylation increases Arabidopsis root growth salt sensitivity. *PLoS One* 9: e93291.

Tseng I-C, Hong C-Y, Yu S-M, Ho T-HD (2013) Abscisic acid- and stress-induced highly proline-rich glycoproteins regulate root growth in rice. *Plant Physiology* 163: 118–134.

Tucker MR, Okada T, Hu Y, Scholefield A, Taylor JM, Koltunow AM (2012) Somatic small RNA pathways promote the mitotic events of megagametogenesis during female reproductive development in Arabidopsis. *Development* 139: 1399–1404.

Tucker MR, Koltunow AM (2014) Traffic monitors at the cell periphery: the role of cell walls during early female reproductive cell differentiation in plants. *Current Opinion in Plant Biology* 17: 137–145.

Verdugo-Perales M, la Rocha RV, León-Félix J, Osuna-Enciso T, Heredia JB, Osuna-Castro JA, Islas-Osuna MA, Sañudo-Barajas JA (2018) Gene expression of an arabinogalactan lysine-rich protein *Ca*AGP18 during vegetative and reproductive development of bell pepper (*Capsicum annuum* L.). *3 Biotech* 8: 5.

Walter P, Johnson AE (1994) Signal sequence recognition and protein targeting to the endoplasmic reticulum membrane. *Annual Review of Cell and Developmental Biology* 10: 87–119.

Wu HM, Wong E, Ogdahl J, Cheung AY (2000) A pollen tube growth-promoting arabinogalactan protein from *Nicotiana alata* is similar to the tobacco TTS protein. *Plant Journal* 22: 165–176.

Wu Y, Williams M, Bernard S, Driouich A, Showalter AM, Faik A (2010) Functional identification of two nonredundant Arabidopsis α(1,2)fucosyltransferases specific to arabinogalactan proteins. *Journal of Biological Chemistry* 285: 13638–13645.

Xuan L, Lu W, Gluza P, Ebert B, Kotake T, Lu M, Zhang Y, Clausen MH, Johnson KL, Doblin MS, Heazlewood JL, Bacic A, Song L, Zeng W (2021) A pipeline towards the biochemical characterization of the Arabidopsis GT14 family. *International Journal of Molecular Sciences* 22: 1360.

Yang J, Sardar HS, McGovern KR, Zhang Y, Showalter AM (2007) A lysine-rich arabinogalactan protein in Arabidopsis is essential for plant growth and development, including cell division and expansion. *Plant Journal* 49: 629–640.

Yariv J, Rapport MM, Graf L (1962) The interaction of glycosides and saccharides with antibody to the corresponding phenylazo glycosides. *Biochemical Journal* 85: 383–388.

Yariv J, Lis H, Katchalski E (1967) Precipitation of arabic acid and some seed polysaccharides by glycosylphenylazo dyes. *Biochemical Journal* 105: 1C–2C.

Yeats TH, Bacic A, Johnson KL (2018) Plant glycosylphosphatidylinositol anchored proteins at the plasma membrane-cell wall nexus. *Journal of Integrative Plant Biology* 60: 649–669.

Youl JJ, Bacic A, Oxley D (1998) Arabinogalactan-proteins from *Nicotiana alata* and *Pyrus communis* contain glycosylphosphatidylinositol membrane anchors. *Proceedings of the National Academy of Sciences* 95: 7921–7926.

Yuasa K, Toyooka K, Fukuda H, Matsuoka K (2004) Membrane-anchored prolyl hydroxylase with an export signal from the endoplasmic reticulum. *Plant Journal* 41: 81–94.

Zang L, Zheng T, Chu Y, Ding C, Zhang W, Huang Q, Su X (2015) Genome-wide analysis of the fasciclin-like arabinogalactan protein gene family reveals differential expression patterns, localization, and salt stress response in populus. *Frontiers in Plant Science* 6: 1140.

Zhang Y, Held MA, Showalter AM (2020) Elucidating the roles of three β-glucuronosyltransferases (GLCATs) acting on arabinogalactan-proteins using a CRISPR-Cas9 multiplexing approach in Arabidopsis. *BMC Plant Biology* 20: 221.

Zhou L, Dresselhaus T (2019) Friend or foe: signaling mechanisms during double fertilization in flowering seed plants. *Current Topics In Developmental Biology* 453-496.

17 Root Cell Walls and Secretions

A Role in the Plant's Relationship with Its Surrounding Soil Microbiome

Eric Nguema-Ona
CMI-Roullier (Centre Mondial de l'innovation – Groupe Roullier)

Romain Castilleux
Swedish University of Agricultural Sciences

John P. Moore
Stellenbosch University

Maïté Vicré
Université de ROUEN Normandie

Azeddine Driouich
Université de ROUEN Normandie

INTRODUCTION

The plant cell wall is a polysaccharide-rich compartment, which surrounds each plant cell. Cell walls were visualised for the first time in cork by Robert Hooke (1635–1703) using one of the first microscopic devices. Hooke reported on a slice of cork material featuring empty small spaces he termed "cells" (from Latin *cellula*, or monk's room) (Hooke, 1665). The structures he was visualising would later be called cell walls. Plant cell walls have been extensively studied over the past decades, and they consist mainly of three classes of polysaccharides, namely, cellulose, hemicelluloses and pectins, which are invariably found in cell walls of land plants (Lampugnagni et al. 2018; Carpita and Gibeaut. 1993). Cell walls also contain enzymatic proteins and non-enzymatic structural glycoproteins, such as hydroxyproline-rich glycoproteins (HRGPs), specifically arabinogalactan proteins (AGPs) and extensins (EXTs) (Nguema-Ona et al. 2014, 2012; Velasquez et al. 2012; Seifert and Robert 2007). Apart from cellulose and callose, these different cell wall components are synthesised inside the cell, particularly in the Golgi apparatus and the endoplasmic reticulum organelles. They are subsequently secreted and further assembled in the external, apoplastic compartment or in the cell wall (Staehelin et al. 1990). Importantly, structural and compositional differences within these different classes of polysaccharides and glycoproteins are reported across the plant kingdom as well as between plant organs and tissues. These differences often translate into lineage and evolutionary specialisations (Popper 2008; Popper et al. 2011). In this case, the differences arise from the existing repertoire of cell wall–associated biosynthetic genes. These compositional/structural differences can also translate into functional modifications that accompany the activation of various biological processes (Willats et al. 1999; McCartney et al. 2003; Nguema-Ona et al. 2013a; Knox et al. 1991; Knox 1997). Here, the differences are often related to enzymatically driven, pH-dependent, cell wall remodelling activities. These remodelling activities lead to modifications of the composition/structure of wall components, and promote previously impossible molecular interactions, polymer trimming, and physicochemical modifications, which then promote various biological processes. Using immunological, biochemical and molecular approaches, such modifications have been reported for both cell wall–located polysaccharides such as hemicelluloses (Nguema-Ona et al. 2013a) and pectins (Willats et al. 1999; McCartney et al. 2003), as well as for structural HRGPs such as AGPs (Knox et al. 1991; Chapter 16). Today, plant cell walls are known to play a role in almost all aspects of plant biology for they can alternatively be seen as interfaces where signalling molecules, ions or photosynthates transit and as essential compartments for plant cell growth and development (Cosgrove 2016, 2018), for fruit development and ripening (Dheilly et al. 2016; Brummell and Harpster. 2001), as well as for plants' ability to resist various biotic (de Lorenzo et al, 2019; Chapter 20, this volume) and abiotic stresses (Tenhaken 2011; Legall et al. 2015; Chapter 22).

DOI: 10.1201/9781003178309-17

Plant–microbe interactions encompass a range of situations where plant tissues encounter microorganisms. In the soil, roots cohabit with an innumerable quantity and variety of living microorganisms including bacteria, protists, fungi, oomycetes, with which they will develop different types of interactions. The outcome of these encounters is sometimes beneficial for both plants and the microorganisms. In contrast, the outcome can also be detrimental for the plants, or even destructive when the involved plants are crops. When detrimental to the plants, the encounter leads to parasitism and pathogenicity (Newton et al. 2010; Rey and Schornak 2013). When beneficial, it leads to symbiosis. Some of these root–microbe interactions involve direct contact between roots and one or several types of microorganisms at a time. These interactions often occur in the rhizosphere, a micro-environment delimited by the 2–3 millimetres of soil in contact with roots and with root cap–derived cells in the apical region of the root. Plant cells are able by different means to sense and to recognise, chemically or physically, the presence of living microorganisms and to relay the information to the cell, which will in turn respond adequately to the information received, either to favour beneficial interactions, or to mount an appropriate defence response against the invader. How roots interact with soil microorganisms is currently one of the most active fields of investigation in plant science research (He et al. 2021; Bergelson et al. 2019). The cell wall compartment and the apoplastic space which surround plant cells play a role during these interaction events (Delaunay et al. 2014; Thorsten 2012; Nühse, 2012; Hematy et al. 2009). Also, the responses will involve compositional and structural remodelling of the cell walls (de Lorenzo et al. 2019; Nguema-Ona et al. 2013b).

Roots are the underground organs of land plants. They anchor the plant body to the ground and multiply, ramify, and explore vast belowground volumes of soil in order to ensure that the plant has access to water and mineral nutrients which, along with the carbon captured during the photosynthesis, are used to drive plant growth and development, plant reproduction and subsequent seed/fruit production. Each root can be subdivided into different functional zones: namely, the differentiation zone, the elongation zone, the meristematic zone and the root cap zone (Benfey et al. 1993; Figure 17.1). Each zone has specific functional roles. The differentiation zone is the part of the root that forms root hairs, while the meristematic zone and the elongation zone are characterised by cell division and proliferation, and cell elongation, respectively. The root cap zone defines the foremost apical zone of the root and consists of different types of cells including columella cells and lateral root cap cells. The latter cells are programmed to separate from the root cap and to be released into the external environment. Root cap cells that separate from the roots were first named border cells (BCs; Hawes et al. 1990; see also Vicré et al. 2005 and Driouich et al. 2007 for the finding of border-like cells; BLCs), and were shown to be viable, metabolically active and able to synthesise cell wall polysaccharides (e.g., pectins, hemicelluloses) as well as glycoproteins such as AGPs and EXTs (Driouich et al. 2007, 2013). Moreover, BCs and BLCs were shown to release large amounts of these cell wall components into the external environment, forming a thick mucilage embedding them (Driouich et al. 2010, 2019; Kumar and Iyer-Pascuzzi 2020). Here also, cell walls of both BCs and BLCs, and their secreted polysaccharides and glycoproteins were shown to play several biological functions including the mediation of root–microbe interactions (Driouich et al. 2019; Kumar and Iyer-Pascuzzi 2020). The existence of these highly dynamic cells and secreted compounds allowed Driouich and coworkers to define the root extracellular trap (RET), a network of molecules and root cap–derived cells (BCs/BLCs) involved in root immune defence (Driouich et al. 2013, 2019).

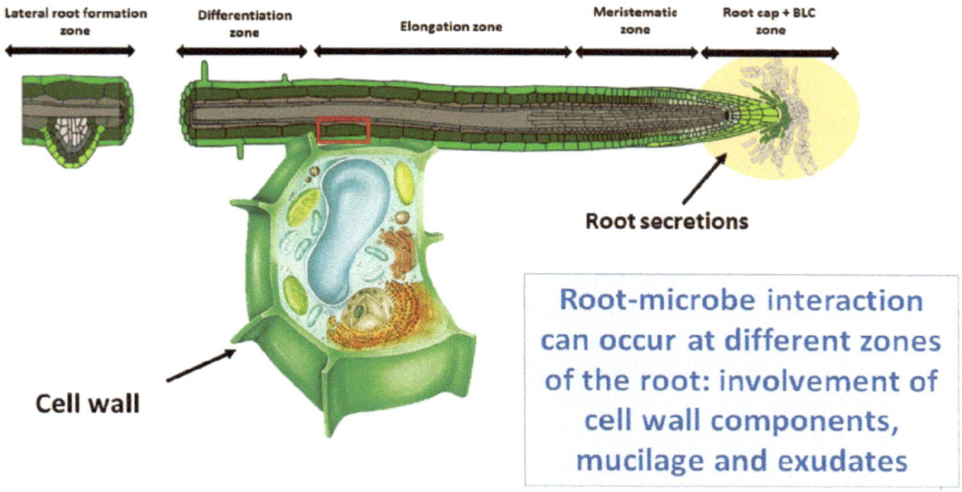

FIGURE 17.1 Root zones, cell wall and microbial interactions. Drawing showing different zones of a root apex and the mucilaginous secretions. The zoomed-in portion represents a root cell with its cell wall, the first barrier against pathogenic microorganisms. Root–microbe interaction can occur at different zones of the root and involves root cell walls, mucilage and exudates. BLC: border-like cells. The root drawing is a modified version from Benfey *et al.*, 2010. *The Plant J.* 61 pp 992–1000. Plant cell model was taken from http://www.vecteezy.com

In this chapter, we aim to recapitulate how progress was made from the very first findings decades ago to the current state-of-the-art and how we came to the current understanding of the role of root cell wall components in mediating root–microbe interactions. Here, we focus mostly on the role of high molecular weight cell wall components with an emphasis on AGPs and EXTs. In addition to cell wall–related secreted components, root secretions also include diverse and abundant low molecular weight molecules. Literature often associates the term "exudate" with these secreted molecules. The importance of these components in the mediation of root–microbe interactions is also very well documented (Bais et al. 2006; Huang et al. 2014; Canarini et al. 2019; Pascale et al. 2020) but will not be developed in this chapter.

ROOT–MICROBE INTERACTIONS MEDIATED BY ROOT SECRETIONS

The central role of the root cap in the root and the whole plant functioning was remarkably highlighted by Darwin as early as 1880.

"We believe that there is no structure in plants more wonderful, as far as its functions are concerned, than the tip of the radicle" Darwin and Darwin (1880).

The accumulating knowledge on the root system collected over the past 140 years has reinforced this assumption on the role of the root apex and has confirmed that this zone is also crucial in root–microbe relationships. Increasing evidence indicates that root–microbe interactions are much more sophisticated than previously thought and that such associations have a significant impact on the health and survival of the whole plant.

Pioneer Plant Cell Biologists and Their Contribution to Our Understanding to the Description of Root Secretions

The presence of adhering soil particles forming sheaths of sand grains on the surface of roots was first observed by Volkens (1887) in desert grasses. It was proposed that these rhizosheaths were part of plant adaptation to the desert dunes. The sand grains were shown to be maintained at the root surface due to mucilage coverings produced by the external layer of cells at the root tip (Price 1911). These observations are among the first reports highlighting the role of root mucilage in maintaining a close functional relationship between roots and soil and a possible role in plant adaptation to their environment.

The definition of rhizosphere was provided by Hiltner in 1904 as the "soil influenced by roots". Later on, Knusdon (1919) reported that the rhizosphere contains dead root-cap cells that have sloughed off from the root outer surface. The term "desquamated-cells" was then commonly used by plant scientists. These cells and the root tip were shown to be covered by a mucigel filling empty spaces between the root surface and the soil. This boundary between root and soil was found to be enriched in microbes but remained poorly investigated. Most studies on the soil-borne microbes from the rhizosphere were restricted to indirect methods based on isolated microorganisms on agar plates but not from direct observation at the root surface. Jenny and Grossenbacher (1963) developed new microscopical techniques to explore this frontier at the root surface and reported the presence of mucilaginous compounds derived from the plant and associated microorganisms. Mucigel was indeed reported to be composed of root cap slime, degraded and degenerating root cap cells, epidermal mucilage as well as microbial-derived mucilage (Morré and Jones 1966, Foster 1982). The polysaccharide-rich slime arising from the root cap cells was described as partly secreted by the Golgi apparatus from the peripheral root cap cells probably using carbohydrate reserves sourced from the disintegrating amyloplasts (Clowes and Woolston 1978). Ultrastructural analysis using electron microscopy revealed the presence of hypertrophied Golgi stacks and vesicles in root peripheral cells and confirmed their role in the synthesis and secretion of polysaccharides (Morré and Jones 1966; Bowles and Northcote 1976; Rougier 1981; Staehelin et al. 1990). It became obvious that a better understanding of the molecular composition of mucigel was necessary in order to gain information regarding interactions between the root system and beneficial as well as pathogenic microorganisms (Jones and Moore 1967). To this end, Chaboud (1983) isolated and characterised the monosaccharide composition of mucigel from maize roots. They found that galactose, fucose and glucose were the main monosaccharides present in root cap slime with significant amounts of xylose and arabinose. All these sugars are commonly present in plant cell wall polysaccharides. At the same time, scientists also started to pay attention to the root-derived cap cells embedded in this mucigel and their function in root–microbe interactions.

The function of the root cap in the perception of and response to gravity has been clearly demonstrated and recognised as of primary importance for a long time (Iverson 1969; Barlow 1974; Wilkins and Waynes 1985; Moore and Evans 1986). Interestingly, Vermer and McCully (1982) reported that root cap–derived cells (BCs/BLCs) released from the maize root cap into the rhizosphere remain alive for a few days following separation. The function of these cells then became a subject of debate among root biologists and a new vision has emerged regarding the significance of these "sloughed off cap cells" within the rhizosphere. Martha Hawes was amongst the first scientists in the 1990s to unravel the hidden potential of these previously neglected cells and the term "BCs" was then proposed to highlight their role at the interface between the root body and the soil (Hawes 1990; Hawes et al. 1998). The production and release of BCs is highly controlled and species-specific (Groot et al. 2004; Hamamoto et al. 2006) and is accompanied by cell-wall remodelling and a loss of cell adhesion involving hydrolysis of homogalacturonan and a specific distribution of EXTs and xyloglucan (XyG) epitopes (Brigham et al. 1995; Wen et al. 1999; Durand et al. 2009; Plancot et al. 2013; Mravec et al. 2017). Different BC populations were identified based on their mode of detachment and release such as root BLCs in flax and *Arabidopsis thaliana* or distinct sheath tissues in Acacia (Vicré et al. 2005; Endo et al. 2011; Plancot et al. 2013). The mechanisms involved in the genetic control of BC production and release are complex and not completely elucidated, but numerous transcription factors, peptides and auxins were found to be key elements in the regulation of root cap cell shedding (del Campillo et al. 2004; Willemsen et al. 2008; Bennett et al. 2010; Kamiya et al. 2016; Karve et al. 2016; Kumar and lyer-Pascuzzi 2020).

Root Secretion-Based Mediation of Root-Microbe Interactions

Root BCs and their associated mucilage (illustrated in Figure 17.2) are described as central elements in root–soil communication and were shown to either attract or repel a wide range of microorganisms, including bacteria, fungi, oomycetes and nematodes to provide root protection (Zhao et al. 2000; Driouich et al. 2010; Hawes et al. 2011; Ropitaux et al. 2020). These interactions at the "frontline" of the root include the involvement of defence proteins, phenolic compounds and polysaccharides mainly produced by BCs (Gunawardena and Hawes 2002; Gunawardena et al. 2005; Wen et al. 2007, 2009; Cannesan et al. 2012; Watson et al. 2015; Weiller et al. 2017; Ropitaux et al. 2020). Interestingly, histones and extracellular DNA were observed in the surrounding mucilage of BCs from pea for the first time by Wen et al. (2007, 2009). In mammals, extracellular DNA is released in blood thus forming the « Neutrophil Extracellular Trap » or « NET » that is important for immunity (Brinkemann et al. 2004). By analogy with the NET, the « Root Extracelluar Trap; RET » model was developed to explain that both BCs and mucilage act in synergy in root protection. The fibrous structure of the mucilage forms a kind of web containing antimicrobial molecules that trap and facilitate the neutralisation of pathogens by BCs (Driouich et al. 2013; Hawes et al. 2016; Tran et al. 2016). Such a structure was shown to provide an efficient physical and/or chemical barrier blocking the entrance of zoospores of *Phytophthora parasitica* and protecting the root cap from infection (Ropitaux et al. 2020). Pathogens such as *Ralstonia solanacearum* that cause bacterial wilt disease can evade this layer of defence and then be able to effectively infect the roots (Tran et al. 2016; Park et al. 2019). As the root system grows, it is tempting to speculate that the different RETs produced by the different root tips of the plants progressively fuse together forming a large « cloud of RETs» surrounding the whole root system of the plant. This hypothesis is termed the "RET cloud model", a large protective network controlling interaction of root with the rhizosphere microbiome as proposed by Driouich et al. (2019).

Root secretions including exudates are essential in mediating interactions between plants and organisms as recently reviewed by Vives-Peris et al. (2020). The authors compiled

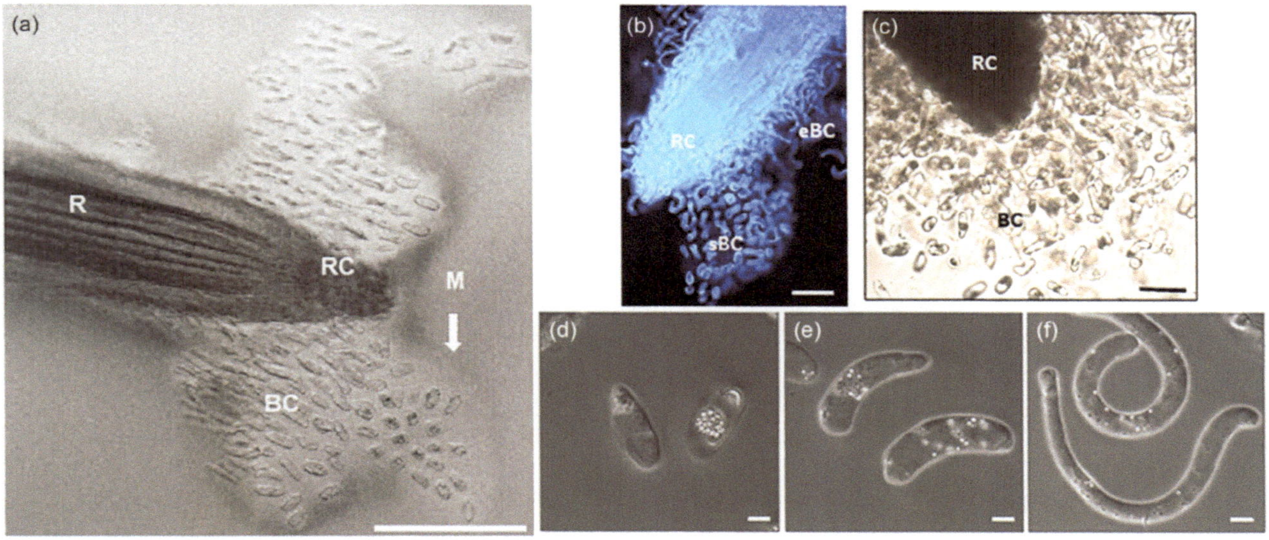

FIGURE 17.2 Diversity of root border cell populations in different crop species. (A) Brightfield micrograph of root tip of tomato (*Solanum lycopersicum*) stained with India Ink showing root border cells and associated mucilage (arrow) forming the « RET » (Root Extracellular Trap). Photo by Dr. Maxime Gotté – University of Rouen Normandie. (B) Fluorescence micrograph showing border cells detaching from the root cap of potato (*Solanum tuberosum*) visualised using calcofluor staining. Two subpopulations of border cells are observed: small and elongated border cells. Reprinted with permission from Koroney et al. 2016, Annals of Botany. 118 pp 797–808. (C) Brightfield micrograph showing a whole population of border cells released by pea root tip (*Pisum sativum*) including small border cells (D), intermediate border cells (E) and elongated border cells (F). Reprinted with permission from Cannesan et al., 2011 Annals of Botany. 108, pp 1658–1670. BC: Border cells; eBC: elongated border cells; M: mucilage; R: root: RC: root cap; sBC: small border cells. Scale bars: 500 μm (A); 100 μm (B); 20 μm (C); 5 μm (D-F).

recent data stating the impact of biological conditions such as the age of the plant, plant genotype and the presence of neighbouring plants (allelopathy) on the amount and composition of root exudates released into the soil. The root system modulates the soil nutrient availability and modifies the soil pH through the composition of its exudates, thereby affecting the recruitment of mycorrhizal fungi and plant growth–promoting rhizobacteria and interfering with pathogenic soil-borne microorganisms. It is thus essential to develop new tools to unravel the fine scale molecular dialogue established between root-secreted molecules and microorganisms and to decipher how root secretions can impact the life cycle of pathogens. To this end, Laloum et al. (2020) developed a novel three-dimensional tool, the "3D *in vitro* choice", to explore chemotaxis responses between root secretions (including BCs) and soil-borne pathogens. Such early events are determinant for root colonisation and subsequent plant infection. Soil-borne fungi such as *Nectria haematococca* were reported to be specifically attracted and neutralised by root BCs of pea (Gunawardena and Hawes (2002); Gunawardena et al. (2005). The perception of chemical gradients by mobile pathogens such as oomycetes and plant parasitic nematodes are key elements in the establishment of root colonisation and further infection and disease. A better understanding of these interactions is thus essential to develop biological agents and/or molecules to control root infection and pathogen populations. Laloum et al. (2021) showed that root exudates from *Faba bean* have a repulsive/repellant effect on zoospores of *Aphanomyces euteiches* and reduce pea root infection opening new perspectives in pea root rot disease management. Flavonoids released within the rhizosphere are well-known candidates to mediate root–microbe communication. Root secretions from pea are particularly enriched in high molecular weight mucilage containing AGPs involved in interactions with several rhizosphere bacteria including *Rhizobium leguminosarum* and *Pseudomonas fluorescens* (Knee et al. 2001). Interestingly, AGPs from pea roots were reported to be part of the molecular actors promoting rhizobium attachment to the root surface (Xie et al. 2011).

More recent data strongly suggest that cell wall polymers such as AGPs are also key actors in root–microbe interaction including soil-borne pathogens (see Figure 17.3). For instance, zoospores of *A. euteiches* are not only attracted by prunetin but also by AGPs produced and secreted by pea root cap and BCs (Cannesan et al. 2012). Also, pectin

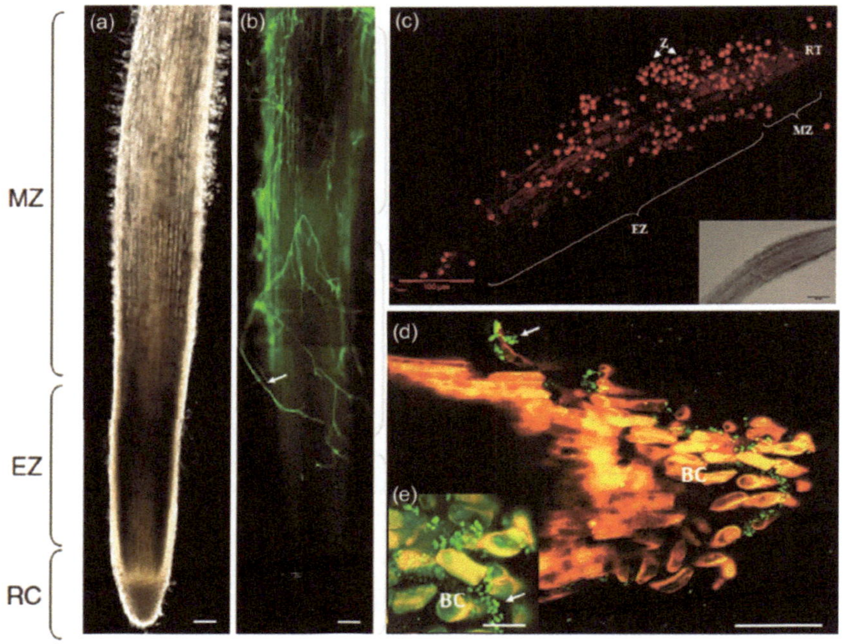

FIGURE 17.3 Interactions of root with soil-borne pathogenic microbes. (A-B) Early infection of pea root (*Pisum sativum*) by the pathogenic oomycete *Aphanomyces euteiches*. (A) Bright field micrograph of a pea root tip at 4 days post-inoculation. (B) FITC-WGA (Wheat Germ Agglutinin) fluorescence staining showing the presence of mycelia (green; arrow) on the root surface of the elongation and maturation zones. Note that the root cap is deprived of infection (no mycelia). Reprinted with permission from Cannesan et al., 2011 Annals of Botany. 108, pp 1658–1670. (C) Fluorescence micrograph showing zoospores of the oomycete *Phytophtora parasitica* accumulating over *Arabidopsis thaliana* root after 3 hours of inoculation. Note that the zoospores (in red; arrows) accumulate in an area covering the end of the meristematic zone and the elongation zone. Zoospores are stained with propidium iodide. Inset: Brightfield micrograph of the same root. Reprinted with permission from Castilleux et al., 2020. Annals of Botany. 125, pp 751–763. (D-E) Fluorescence micrographs showing the association of the bacterium *Pectobacterium atrosepticum* with border cells of potato root tip. The bacteria (arrows) are stained in green (they are GFP-expressing cells). Reprinted with permission from Koroney et al., 2016. Annals of Botany. 118 pp 797–808. BC: Border cells; eBC: elongated border cells; EZ: elongation zone; MZ: meristematic zone; RC: root cap; RT: root tip; Z: zoospores; sBC: small border cells. Scale bars: 100 µm (A, B, C); 86 µm (D); 20 µm (E).

and cellulose seem to be involved in controlling such interactions (Karve et al. 2016). The authors have shown that root BLCs of the *nin-like protein 7-1 (nlp7-1)* Arabidopsis mutant had reduced homogalacturonan and cellulose contents and increased susceptibility to infection by the root-infecting fungal pathogen *Fusarium oxysporum*.

ROOT–MICROBE INTERACTIONS MEDIATED BY ROOT CELL WALL COMPARTMENTS

Pioneering Cell Wall Biologists and Their Contribution to Identification of Key Cell Wall Components

Lamport and Northcote (1960) were the first to detect and measure hydroxyproline in the isolated walls of sycamore and bean suspension cultures. Further evidence accumulated (Lamport 1963) that atmospheric oxygen was incorporated into hydroxyproline-rich cell wall of sycamore cell suspension cultures in the conditions tested. Additional confirmation of a plant cell wall carbohydrate–protein linkage between hydroxyproline and arabinan-derived arabinose was demonstrated by paper electrophoresis (Lamport 1967). The author also discussed the role these linkages may play in cross-linking the cell wall. These milestones established Lamport and co-workers as pioneers in the study of plant cell wall proteins termed HRGPs, which include EXTs and AGPs (reviewed in Lamport 1970). Later, Stephen Fry and colleagues were able to show that a range of different tyrosine cross-links form in EXTs (Fry 1979). Tyrosine residues were shown to participate in dityrosine, isodityrosine and pulcherosine cross-links between these wall proteins *in muro* (Fry 1982; Brady et al. 1997). In parallel, Albersheim and colleagues pioneered the application of chemical and biochemical methods to plant cell wall carbohydrate chemistry (Albersheim et al. 1967; Gelineo-Albersheim et al. 2017). The initial focus was on pectin chemistry but this expanded to other plant cell wall polymers and preparations including hemicelluloses and cellulose (Gelineo-Albersheim et al. 2017). At the time of their discovery, HRGPs were not seen as important players in root–microbe interactions.

The Role of Root Cell Wall Compartments in Mediating Root–Microbe Interactions

Cell wall compartment here refers to the polysaccharide-rich and HRGP complexes and components, which surround living cells in the root rhizosphere region. Keith Roberts and coworkers described clearly the Arabidopsis primary root zone using cell biology techniques (Dolan et al. 1993). They revealed a diversity of tissues and cell shapes, from the apex to the differentiation zone of the Arabidopsis primary root. Using antibodies that recognise specific pectic polysaccharide-associated chains, Knox and coworkers observed developmentally regulated patterns of cell wall organisation in Arabidopsis primary roots (Knox 1997; Willats et al. 1999). Interestingly, many microbes interacting with or invading root tissues have developed specific strategies to infect and enter the root body. These strategies are often targeted at specific root zones (root tips, root hairs initiating lateral roots, elongation zone, etc.). Given that cell wall composition varies among root zones, a link between local cell wall composition and the ability of microorganisms to infect and colonise root tissues can be drawn. For instance, *Ralstonia solanacearum*, a plant pathogenic bacterium that colonises the xylem of a very wide range of host plants, causes bacterial wilt. Initial steps during the infection process involve bacteria accumulation and penetration via the root tips. This penetration was reported in *Arabidopsis thaliana* (Digonnet et al. 2012) and in *Medicago truncatula* (Turner et al. 2009), and is followed by pectin degradation by a type 2 secretion system-dependent (T2SS-dependent), pectin degrading enzyme. Interestingly, root tip cell walls are known to be enriched in pectins (Nagayama et al. 2019; Durand et al. 2009). Pectinolytic enzymes fragment pectin into oligomers, which act as substrates for bacterial growth (Tans-Kersten et al. 1998). Another example is *Verticillium dahlia*, a fungal plant pathogen that causes verticillium wilt in many plant species. The parasitic phase begins when the fungus penetrates the root of a susceptible plant in the region of elongation (root tip or at the sites of lateral roots formation) and colonises the cortex. From the cortex, the hyphae penetrate the xylem vessels where conidia are formed (Palanga et al. 2021). To enter into the vascular (xylem) tissues, the fungus needs to cross the endodermis, which acts as a physical barrier against infection. Therefore, crossing the endodermis may only be achieved when it has not yet developed (at the root tip), or when it is damaged (for instance, by nematode infection) (Bowers et al. 1996). Verticillium species produce enzymes that play a role in plant cell wall degradation (Bidochka et al. 1999). Of all these enzymes, the pectinolytic enzymes are the best studied and thought to play a critical role in pathogenesis (Huang and Mahoney 1999). This is mainly based on the observation that Verticillium has to breach the pectin-containing pit structures between vessel elements and at vessel ends in order to spread (Fradin and Thomma 2006).

In the case of symbiosis including root nodulation symbiosis (RNS), arbuscular mycorrhisation (AM) and ecto-mycorrhisation (Ecto-M), important modifications of the host cell wall compartment, sometimes even triggered by morphogenetic events, have been observed. Driouich and coworkers have discussed the specific role of cell wall HRGPs, AGPs and EXTs, when root tissues of the host body reorganise during the establishment of these symbioses (Nguema-Ona et al. 2012b, 2013a, 2014; Castilleux et al. 2018). It was reported that AGPs play a role in rhizobia interactions, from the initiation of the interaction (Xie at al. 2012; see also Vicré et al. 2005) to the processes that accompany the formation and the establishment of a functional RNS (Rich et al. 2014). Likewise, both AGPs and EXTs were also found to be involved in the different steps leading to the formation of functional AM (van Buuren et al. 1999; Schultz and Harrison 2008). In the case of Ecto-M, a precise role of EXTs and AGPs has not been reported. Only expansin-associated wall remodelling activities have been described (Balestrini and Bonfante 2014).

A recent advance was made in our understanding of the functioning of AGPs: It has been demonstrated that AGPs can play a role as an apoplastic-localised, pH-dependent, versatile sink for calcium and are able to control apoplastic calcium concentrations, and thus fuel several calcium-dependent responses to cell wall mechanical perturbation (Lopez-Hernandez et al, 2020). Such a role might be important during the intrusion of host root tissues during the formation of RNS and AM. It has been reported that the symbiosis signalling pathway, common to both RNS and AM, induces a rhythmic calcium signal, referred to as calcium spiking, which in turn triggers symbiosis-related gene expression and a cascade of morphogenetic events that result in the development of new organ formation (Rich et al. 2014; Balestrini and Bonfante 2014). It is tempting to speculate that AGPs would play such a role of a calcium capacitor that would induce the calcium rhythmic signal. AGPs are located in the apoplast. This compartment gathers and concentrates many molecular players that potentially mediate cellular responses (Delaunois et al. 2014). For instance, apoplast-located proteases termed subtilases were shown to control AM development (Takeda et al. 2009). This mechanism involved the production of apoplast-located peptides. Frenzel et al. (2005) reported on the presence in the apoplast of AM-developing structures of lectins in Medicago. Valot et al. (2005) also identified apoplast-located proteins regulating arbuscular mycorrhizal symbiosis processes. Thus, it is likely that a finely orchestrated sequence of events accompanies AM development in this microenvironment. Would AGPs play a calcium capacitor role there? Further, would AGPs found in secretions including RET and root cap–derived cells (above section) orchestrate a calcium-dependent sequence of events? More research is needed to explore these questions.

Pathogenicity also involves important modifications and a reorganisation of the cell wall polymers within the cell wall compartment. A first report on the role of HRGPs in that context was made in 1974 by Esquerré-Tugaye and Mazau. They noticed an increase in arabinose and hydroxyproline content in the cell wall of melon suffering from anthracnose, laying the foundation for further studies on the modification of cell wall composition and properties of plant material infected by different microbial pathogens. They concluded that following inoculation, the melon cell wall protein–enriched fraction contained quantitatively more tetra-arabinosides substances and the structure of these substances differed from those of untreated plants (Esquerré-Tugaye and Mazau 1974). These results prompted Esquerré-Tugaye and Lamport to investigate globally the role of HRGPs, i.e., EXTs, during muskmelon infection caused by the fungus *Colletotrichum lagenarium* (Esquerré-Tugaye and Lamport 1979a,b). They observed a net increase in hydroxyproline content of infected tissues and for the first time proposed the existence of defence mechanisms involving an accumulation of EXTs. In roots, EXTs were further reported to accumulate in BLCs of Arabidopsis roots following an elicitation with the bacterial peptide flagellin 22 (Plancot et al. 2013). In this study, the authors noticed, following the elicitation, a production of reactive oxygen species that accompanied a modification in the cell wall distribution of EXT epitopes, probably due to hydrogen peroxide–driven EXT crosslinking reaction which has occurred. Castilleux et al. furthered this study by evaluating the response to elicitation of roots and BLCs of a series of Arabidopsis mutants deficient in different glycosyltransferases previously reported to affect EXT arabinosylation (Castilleux et al. 2019, 2018). They confirmed the modification of EXT-associated epitopes following elicitation in wild-type Arabidopsis plants. Interestingly, elicitation of roots and BLCs of Arabidopsis mutants defective in EXT arabinosylation showed a different distribution of EXT-associated epitopes. Moreover, mutants were much more prone to pathogen invasion than the wild-type. This study has given rise to an enthusiastic discussion within plant cell wall community on the importance of this cell wall glycoprotein in defence (see Castilleux et al. 2021; Tan and Mort 2020). Interestingly, a recent study reported on the role of AGPs of apple fruit, following the infection with the fungal pathogen *Penicillium spinulosum* (Leszczuk et al. 2019). Here, the spatial redistribution of AGP-associated epitopes at the cell surface of the fruit was conducted, following inoculation with *Penicillium spinulosum*. The authors showed that epitopes associated with AGPs were indeed redistributed. A further, meaningful experiment, which consisted of pre-incubation of these apple fruits with the Yariv reagent, prior to inoculation with the pathogen, immobilised AGPs at the cell surface (see Nguema-Ona et al. 2007 for Yariv-induced immobilisation of AGPs at the cell surface), but interestingly led to a much more faster colonisation of the fruit. This suggested that in a living system, AGPs are actively redistributed during infection. This also indicates that AGP redistribution is a significant process that contributes to the cellular responses aiming at controlling the disease progression (Leszczuk et al. 2019). Other cell wall polymers were also shown to become modified, or to remodel their structure, thus indirectly modifying the entire plant cell wall architecture. In a study aiming at dissecting molecular changes that take place in the leaves of tobacco lines overexpressing the grapevine (*Vitis vinifera*) polygalacturonase inhibiting protein1 *Vv*PGIP1 (and resistant to *Botrytis cinerea* attacks), it was shown that several genes associated to cell wall metabolism were found to be differentially regulated. A set of XyG endotransglycosylase/hydrolases (XET/XTH) were particularly down-regulated, suggesting that the plant had remodelled its XyG/cellulosic network, in order to slow a potential pathogen attempt to penetrate the host cell wall (Alexandersson et al. 2011). Using an enzymatic XyG oligosaccharide fingerprinting technique, Nguema-Ona et al. (2013) showed that XyG oligomer profiles corresponding to the XyG domain susceptible to the action of endoglucanases was altered. This indicated that an *in muro* remodelling had effectively taken place thus supporting the findings of Alexandersson et al. (2011). Also, in support of this, a recent study by Gavrin and co-workers demonstrated that Medicago was able to developmentally modulate its root cell wall architecture via a controlled exocytosis of XyG polymers and their subsequent incorporation into the wall (Gavrin et al. 2020). Interestingly, this remodelling, which involved the activation of a SCAR/WAVE complex protein API which controls actin cytoskeleton dynamics in roots, led to a root cell wall architecture reorganisation. This ultimately conferred host plant resistance against an oomycete pathogen. Finally, it has also to be remembered that plant cells are equipped with a cell wall integrity system able to monitor changes that would be due to external cues including pathogen ingress (Nühse 2012; Thorsten 2012; Hematy et al. 2009). This system is able to rapidly sense a perturbation and convey a signal in the cytosol, which in turn will orchestrate appropriate defence responses.

To conclude this section, an additional role cell wall components would play during detrimental or pathogenic root microbe interaction is their ability to activate the plant immune system. Here, the arms race–related as well as plant immunity–related subjects are a matter of intense research of practical value in pest management. The literature around this topic is prolific. General aspects of plant immunity have been addressed elsewhere in this book (Chapter 20) and activation of plant immunity in roots specifically was covered by Driouich and co-workers (see Chuberre et al. 2018). The authors highlighted that substantial differences exist compared to aerial plant organs, and that root immune responses are tissue-specific. The authors also suggested a compartmentalisation of defence mechanisms in roots. For foundational papers in this subject, the readers are referred to the following papers: Flor (1971); Jones and Dangl (2006); Huckelhoven (2007); and Böller and Felix (2009).

CONCLUSION AND PERSPECTIVES

The cell wall compartment of root cells and root secretions (enriched in cell wall components) play a major role in mediating root–microbe interactions. During the past decade, the understanding of the role of several cell wall components in mediating these interactions has increased. AGPs were already proposed to influence the relationship of root with the surrounding microorganisms by different means (Nguema-Ona et al. 2013a; see also Hromadová et al. 2021). These would involve different AGP populations, the entire AGP including its glycans or various oligosaccharide fragments that can be released from root AGPs. Recent developments on the biology of AGPs (this chapter; see also comments on AGP biology in MOORE, this volume) further arouse our curiosity to understand their role in root–microbe interactions. Likewise, the role of EXTs during root responses to pathogen attack has also recently stimulated active discussion in the plant biology community (Castilleux et al. 2021; Tan and Mort 2020). And, here again, EXT glycans seem to play a central role in the control of the root–microbe relationship. Also, the AGP-, pectin- and hemicellulose-enriched, RET emerges as an important part of the root, or an extension of the root, localised in the rhizosphere, which actively mediates root interactions with microorganisms. One would wonder now, how would the absence of a given polysaccharide component of the RET would impact its functional role? How would these components or association of components "indifferentially" affect beneficial or detrimental interactions? Insights on the role of homogalacturonan pectin as a cement of the RET (Durand et al. 2009), or on the role of AGPs in interfering with the infection cycle of root pathogenic oomycetes (Cannesan et al. 2012), have started to provide answers to this question.

In addition to their role in controlling individual microorganisms, it would also be interesting to find out if cell wall components found in the RET also contribute to shaping the structure and the relationship between microbial communities in the rhizosphere for instance. It is known that the different soil microorganism populations, while inhabiting the rhizosphere, may develop mutualistic, antagonistic, altruistic and pathogenic behaviours between them (He et al. 2021). Plant genotypes, and particularly the diversity of their secreted metabolites, known to play a role in host physiological processes, including cell expansion and proliferation or defence, probably influence microbial community structure (Bergelson et al. 2019; He et al. 2021). This additional level of interactions adds to the complexity of root–microbe interactions. Understanding these interactions will have implications for the harnessing of microbiota as agents promoting plant stress tolerance, plant productivity, as agents improving the bioavailability of nutrients or preventing invasion by microbial pathogens.

ACKNOWLEDGEMENT

Eric Nguema-Ona is grateful for the invitation to contribute to this book chapter, and also to Agro Innovation International TIMAC AGRO (Saint Malo, France), which provided the environment that made this work possible. He is thankful for the experience and advice he received while a PhD student in Azeddine Driouich laboratory, more particularly from Prof. Tobias I Baskin (University of Massachusetts, USA). He is also thankful for the experience he gained as a postdoc fellow in John P. Moore and Melané Vivier laboratory (University of Stellenbosch, South Africa), Azeddine Driouich and Maite Vicré are grateful to la Fédération de Recherche En Sciences Végétales "Normandie Végétal" FED4277, l'Université de Rouen Normandie et la Région de Normandie for financial support.

REFERENCES

Albersheim P, Darvill A, Roberts K, Sederoff R, Staehelin A (2011) *Plant Cell Walls*. Garland Science ed.

Alexandersson E, Becker JVW, Jacobson D, Nguema-Ona E, Steyn C, Denby CJ, Vivier MA (2011) Constitutive expression of a grapevine polygalacturonase-inhibiting protein affects gene expression and cell wall properties in uninfected tobacco. *BMC Research Notes* 4:493.

Bais HP, Weir TL, Perry LG, Gilroy S, Vivanco JM (2006) The role of root exudates in rhizosphere interactions with plants and other organisms. Annual Reviews in Plant Biology 57:233–266.

Ballestrini R, Bonfante P (2014) Cell wall remodeling in mycorrhizal symbiosis: a way towards biotrophism. *Frontiers in Plant Science* 5:237.

Barlow P (1974) Regeneration of the cap of primary roots of Zea Mays. *New Phytologist* 73:937–954.

Benfey PN, Linstead PJ, Roberts K, Schiefelbein JW, Hauser M-T, Aeschbacher RA (1993) Root development in Arabidopsis: four mutants with dramatically altered root morphogenesis. *Development* 119:57–70.

Bennett T, Scheres B (2010). Root development-two meristems for the price of one? *Current topics in developmental biology* 91:67–102.

Bergelson J, Mittelstrass J, Horton W (2019) characterizing both bacteria and fungi improves understanding of the Arabidopsis root microbiome. *Scientific Reports* 9:24.

Bidochka MJ, Burke S, Ng L (1999) Extracellular hydrolytic enzymes in the fungal genus Verticillium: adaptations for pathogenesis. *Canadian Journal of Microbiology* 45:856–864.

Boller T, Felix G (2009) A renaissance of elicitors: perception of microbe-associated molecular patterns and danger signals by pattern-recognition receptors. *Annual Review Plant Biology* 60:379–406.

Bowers JH, Nameth ST, Riedel RM, Rowe RC (1996) Infection and colonization of potato roots by Verticillium dahliae as affected by *Pratylenchus penetrans* and *P. crenatus*. *Phytopathology* 86:614–621.

Bowles DJ, Northcote DH (1976) The size and distribution of polysaccharides during their synthesis within the membrane system of maize root cells. *Planta* 128:101–106.

Brady JD, Sader IH, Fry SC (1997) Pulcherosine, an oxidatively coupled trimer of tyrosine in plant cell walls: its role in cross-link formation. *Phytochemistry* 47:349–353.

Brigham LA, Woo HH, Nicoll SM, Hawes MC (1995) Differential expression of proteins and mRNAs from border cells and root tips of pea. *Plant Physiology* 109:457–463.

Brinkmann V, Reichard U, Goosmann C, Fauler B, Uhlemann Y, Weiss DS, Weinrauch Y, Zychlinsky A (2004) Neutrophil extracellular traps kill bacteria. *Science* 303:1532–1535.

Brummell DA, Harpster MH (2001) Cell wall metabolism in fruit softening and quality and its manipulation in transgenic plants. *Plant Molecular Biology 2001* 47:311–40.

Campillo ED, Abdel-Aziz A, Crawford D, Patterson SE (2004) Root cap specific expression of an endo-β-1,4--glucanase (cellulase): a new marker to study root development in Arabidopsis. *Plant and Molecular Biology* 56:309–323.

Canarini A, Kaiser C, Merchant A, Richter A, Wanek W (2019) Root Exudation of Primary Metabolites: Mechanisms and Their Roles in Plant Responses to Environmental Stimuli. *Frontiers in Plant Science* 10:157.

Cannesan MA, Durand C, Burel C, Gangneux C, Lerouge P, Ishii T, Laval K, Follet-Gueye ML, Driouich A, Vicre-Gibouin M (2012). Effect of Arabinogalactan Proteins from the Root Caps of Pea and *Brassica napus* on *Aphanomyces euteiches* Zoospore Chemotaxis and Germination. *Plant Physiology* 159:1658–1670.

Carpita NC, Gibeaut DM (1993) Structural models of primary cell in flowering plants: consistency of molecular with the physical properties of the walls during growth. *The Plant Journal* 3:1–30

Castilleux R, Plancot B, Ropitaux M, Carreras A, Leprince J, Boulogne I, Follet-Gueye M-L, Popper Z, Driouich A, Vicré M (2018) Cell wall extensins in root-microbe interactions and root secretions. *Journal of Experimental Botany* 69:4235–4247.

Castilleux R, Plancot B, Gugi B, Attard A, Loutelier-Bourhis C, Lefranc B, Nguema-Ona E, Arkoun M, Yvin J-C, Driouich A, Vicré M (2020) Extensin arabinosylation is involved in root response to elicitors and limits oomycete colonization. *Annals of Botany* 25:751–763.

Castilleux R, Plancot B Vicré M, Nguema-Ona E, Driouich A (2021) Extensin, an underestimated key component of cell wall defence? *Annals of Botany* 127:709–713.

Chaboud A (1983) Isolation, purification and chemical composition of maize root cap slime. *Plant and Soil* 73:395–402.

Chaboud A, Rougier M (1984). Identification and localization of sugar components of rice (Oryza sativa L.) root cap mucilage. *Journal of plant physiology* 116:323–330.

Chuberre C, Plancot B, Driouich A, Moore JP, Bardor M, Gügi B, Vicré M (2018) Plant immunity is compartmentalized and specialized in roots. *Frontiers in Plant Science* 9:1692.

Clowes FAL, Woolston RE (1978) Sloughing of root cap cells. *Annals of Botany* 42:83–89.

Cosgrove DJ (2016) Plant cell wall extensibility: connecting plant cell growth with cell wall structure, mechanics, and the action of wall-modifying enzymes. *Journal of Experimental Botany* 67:463–476.

Cosgrove DJ (2018) Diffuse growth of plant cell walls. *Plant Physiology* 176:16–27.

Delaunois B, Jeandet P, Clément C, Baillieul F, Dorey S, Cordelier S (2014) Uncovering plant-pathogen crosstalk through apoplastic proteomic studies. *Frontiers in Plant Science* 5:249.

Dheilly E, Le Gall S, Guillou M-C, Renou J-P, Bonnin E, Orsel M, Lahaye M (2016) Cell wall dynamics during apple development and storage involves hemicellulose modifications and related expressed genes. *BMC Plant Biology* 16:201.

Digonnet C, Martinez Y, Denance N, Chasseray M, Dabos P, Ranocha P, Marco Y, Jaulneau A, Goffner D (2012) Deciphering the route of *Ralstonia solanacearum* colonization in Arabidopsis thaliana roots during a compatible interaction: focus at the plant cell wall. *Planta* 236:1419–1431.

Driouich A, Durand C, Cannesan MA, Percoco G, Vicre-Gibouin M (2010). Border cells versus border-like cells: are they alike? *Journal of Experimental Botany* 61:3827–3831.

Driouich A, Cannesan M-A, Dardelle F, Durand C, Plancot B, Bernard S, Follet-Gueye M-L, Vicré M (2012) Unity is strength: the power of border cells and border-like cells in relation with plant defense. In *Secretions and Exudates in Biological Systems*, Vivanco J. M., Baluska F. ed. Springer.

Driouich A, Follet-Gueye M-L, Vicré-Gibouin M, Hawes M (2013) Root border cells and secretions as critical elements in plant host defense. *Current Opinion in Plant Biology* 16:489–495.

Driouich A, Smith C, Ropitaux M, Chambard M, Boulogne I, Bernard S, Follet-Gueye M-L, Vicré M, Moore JP (2019) Root extracellular trap versus neutrophil extracellular traps in host defence, a case of functional convergence? *Biological Reviews* 94:1685–1700.

Durand C, Vicré-Gibouin M, Follet-Gueye ML, Duponchel, L, Moreau M, Lerouge P, Driouich A (2009) The organization pattern of root border-like cells of Arabidopsis thaliana is dependent on cell wall homogalacturonan. *Plant Physiology* 150:1411–1421.

Endo I, Tange T, Osawa H (2011) A cell-type-specific defect in border cell formation in the Acacia mangium root cap developing an extraordinary sheath of sloughed-off cells. *Annals of Botany* 108:279–90.

Esquerré-Tugayé MT, Mazau D (1974) Effect of a fungal disease on extensin, the plant cell wall glycoprotein. *Journal of Experimental Botany* 25:509–513.

Esquerré-Tugayé MT, Lamport DTA (1979a) Cell surfaces in plant-microorganism interactions I. A structural investigation of cell wall hydroxyproline-rich glycoproteins which accumulate in fungus-infected plants. *Plant Physiology* 64:314–319.

Esquerré-Tugayé MT, Lamport DTA (1979b) Cell surfaces in plant-microorganism interactions II. Evidence for the accumulation of hydroxyproline-rich glycoproteins in the cell walls od diseased plants as a defense mechanism. *Plant Physiology* 64:320–326.

Flor HH (1971) Current status of the gene-for-gen concept. *Annual Review of Phytopathology* 9:275–296.

Foster RC (1982) The fine structure of epidermal cell mucilages of roots. *New Phytologist* 91:727–40.

Fradin EF, Thomma BPHJ (2006) Physiology and molecular aspects of Verticillium wilt diseases caused by V. dahliae and V. albo-atrum. *Molecular Plant Pathology* 7:71–86.

Frenzel A, Manthey K, Perlick AM, Meyer F, Pühler A, Küster H, Krajinski F (2005) Combined transcriptome profiling reveals a novel family of arbuscular mycorrhizal-specific *Medicago truncatula* lectin genes. *Molecular Plant Microbe Interactions* 18:771–782.

Fry SC (1979) Phenolics components of the primary cell wall and their possible role in the hormonal regulation of growth. *Planta* 146:343–351.

Fry SC (1982) Isodityrosine, a new cross-linking amino acid from plant cell-wall glycoprotein. *Biochemical Journal* 204:343–351.

Gavrin A, Rey T, Torode TA, Toulotte J, Chatterjee A, Kaplan JL, Evangelisti E, Takagi H, Charoensawan V, Rengel D, Journet E-P, Debellé F, de Carvalho-Niebel F, Terauchi R, Braybrook S, Schornack S (2020) Developmental modulation of root cell wall architecture confers resistance to an oomycete pathogen. *Current Biology* 30:4165–4176.

Gelineo-Albersheim I, Darvill A, Howard K (2018) Peter Albersheim 30 March 1934 - 23 July 2017. *Glycobiology* 28:714–718.

Groot EP, Doyle JA, Nichol SA, Rost TL (2004) Phylogenetic distribution and evolution of root apical meristem organization in dicotyledonous angiosperms. *International Journal of Plant Sciences* 165: 97–105.

Gunawardena U, Hawes MC (2002) Role of border cells in localized root infection by pathogenic fungi. *Molecular Plant Microbe Interaction* 15:1128–1136.

Gunawardena U, Rodriguez M, Straney D, Romeo JT, Van Etten HD, Hawes MC (2005) Tissue-specific localization of pea root infection by *Nectria haematococca.* Mechanisms and consequences. *Plant Physiology* 137:1363–1374.

Hamamoto L, Hawes MC, Rost TL (2006) The production and release of living root cap border cells is a function of root apical meristem type in dicotyledonous angiosperm plants. *Annals of Botany* 97:917–923.

Hartmann A, Rothballer M, Schmid M (2008) Lorenz Hiltner, a pioneer in rhizosphere microbial ecology and soil bacteriology research. *Plant Soil* 312:7–14.

Hawes MC (1990). Living plant cells released from the root cap: A regulator of microbial populations in the rhizosphere? *Plant and Soil* 129:19–27.

Hawes MC, Brigham LA, Wen F, Woo HH, Zhu Y (1998) Function of root border cells in plant health: pioneers in the rhizosphere. *Annual Review in Phytopathology* 36:311–27.

Hawes MC, Curlango-Rivera G, Wen F, White GJ, VanEtten HD, Xiong Z (2011). Extracellular DNA: the tip of root defenses? *Plant Science* 180:741–745.

Hawes M, Allen C, Turgeon BG, Curlango-Rivera G, Minh Tran T, Huskey, DA, Xiong Z (2016). Root border cells and their role in plant defense. *Annual Review of Phytopathology* 54:143–161.

He X, Zhang QI, Beibei L, Jin Y, Jiang L, Wu R (2021) Network mapping of root-microbe interactions in *Arabidopsis thaliana. Biofilms and Microbiomes* 7:72.

Hematy K, Cherk C, Somerville S (2009) Host-pathogen warfare at the plant cell wall. *Current Opinion in Plant Biology* 12:406–413.

den Herder G, van Isterdael G, Beeckman T, de Smet I (2010) The roots of a new green revolution. *Trends in Plant Science* 15:600–607.

Hiltner L (1904) Über neuere Erfahrungen und Probleme auf dem Gebiete der Bodenbakteriologie unter besonderer Berücksichtigung der Gründüngung und Brache. *Arb DLG* 98:59–78.

Hooke R (1665) Micrographia, or some physiological descriptions of minute bodies made by magnifying glasses, with observations and inquiries thereupon. *Micrographia. The Royal Society.*

Hromadová D, Soukup A, Tylová E (2021) Arabinogalactan proteins in plant roots - an update on possible functions. *Frontiers in Plant Science* 12:674010.

Huang X, Chaparro JM, Reardon KF, Zhang R, Shen Q,, Vivanco JM (2014) Rhizosphere interactions: root exudates, microbes, and microbial communities. *Botany* 92:267–275.

Huang LK, Mahoney RR (1999) Purification and characterization of an endopolygalacturonase from *Verticillium albo-atrum. Journal of Applied Microbiology* 86:145–156.

Hückelhoven R (2007) Cell wall-associated mechanisms of disease resistance and susceptibility. *Annual Review of Phytopathology* 45:101–127.

Iversen TH (1969) Elimination of geotropic responsiveness in roots of cress (Lepidium sativum) by removal of statolith starch. *Physiologia Plantarum* 22:1251–1262.

Jenny H, Grossenbacher K (1963) Root-soil boundary zones as seen in the electron microscope. *Soil Science Society of America Journal* 27:273–277.

Kamiya M, Higashio SY, Isomoto A, Kim JM, Seki M, Miyashima S, Nakajima K (2016) Control of root cap maturation and cell detachment by BEARSKIN transcription factors in Arabidopsis. *Development* 143:4063–4072.

Karve R, Suárez-Román F, Iyer-Pascuzzi AS (2016) The transcription factor NIN-LIKE PROTEIN7 controls border-like cell release. *Plant Physiology* 171:2101–2111.

Knox JP, Linstead PJ, Peart J, Cooper C, Roberts K (1991) Developmentally-regulated epitopes of cell surface arabinogalactan-proteins and their relation to root tissue pattern formation. *The Plant Journal* 1:317–326.

Knox JP (1997) The use of antibodies to study the architecture and developmental regulation of plant cell walls. *International Review of Cytology* 171:79–120.

Knudson L (1919) Viability of detached root cap cells. *American Journal of Botany* 6:309–310.

Kumar N, Iyer-Pascuzzi AS (2020). Shedding the last layer: Mechanisms of root cap cell release. *Plants* 9:308.

Jones JDL, Dangl JL (2006) The plant immune system. *Nature* 444:323–329.

Laloum Y, Ngala B, Ianszen M, Boulogne I, Plasson C, Fournet S, Gotte M, Nguema-Ona E, Le Roux AC, Gobert V, Driouich A, Vicré M (2020). A novel in vitro tool to study cyst nematode chemotaxis. *Frontiers in Plant Science* 11:1024.

Laloum Y, Gangneux C, Gügi B, Lanoue A, Munsch T, Blum A, Gauthier A, Trinsoutrot-Gattin I, Boulogne I, Vicré M, Driouich A, Laval K, Follet-Gueye M-L (2021). Faba bean root exudates alter pea root colonization by the oomycete *Aphanomyces euteiches* at early stages of infection. *Plant Science* 312:111032.

Lamport DTA, Northcote DH (1960) Hydroxyproline in primary cell walls of higher plants. *Nature* 19:665–666.

Lamport DTA (1962) Oxygen fixation into hydroxyproline of plant cell wall proteins. *Journal of Biological Chemistry* 238:1438–1440.

Lamport DTA (1967) Hydroxyproline-O-glycosidic linkage of the plant cell wall glycoprotein extensin. *Nature* 216:1322–1324.

Lamport DTA (1970) Cell wall metabolism. *Annual Review of Plant Physiology* 21:235–270.

Lampugnani ER, Khan GA, Somssich M, Persson S (2018) Building a plant cell wall at a glance. *Journal of Cell Science* 131:jcs207373.

Le Gall H, Philippe F, Domon JM, Gillet F, Pelloux J, Rayon C (2015) Cell wall metabolism in response to abiotic stress. *Plants* 4:112–166.

Leszczuk A, Pieczywek PM, Gryta A, Frac M, Zdunek A (2019) Immunocytochemical studies on the distribution of arabinogalactan proteins (AGPs) as a response to fungal infection in Malus x domestica fruit. *Scientific Reports* 9:17428.

Lopez-Hernandez F, Tryfona T, Rizza A, Yu X,Harris MO, Webb AA, Kotake T, Dupree P (2020) Calcium binding by arabinogalactan polysaccharides is important for normal plant development. *The Plant Cell* 32:3346–3369.

de Lorenzo G, Ferrari S, Giovannoni M, Mattei B, Cervone F (2019) Cell wall traits that influence plant development, immunity, and bioconversion. *The Plant Journal* 97:134–147.

McCartney L, Steele-King CG, Jordan E, Knox JP (2003) Cell wall pectic beta-1,4-D-galactan marks the acceleration of cell elongation in the Arabidopsis seedling root meristem. *The Plant Journal* 33:447–454.

Morré JD, Jones DD, Mollenhauer HH (1967) Golgi apparatus mediated polysaccharide secretion by outer root cap cells of *Zea mays*: I. Kinetics and secretory pathway. *Planta* 74:286–301.

Moore R, Evans ML (1986) How roots perceive and respond to gravity. *American Journal of Botany* 73:574–87.

Mravec J, Guo X, Hansen AR, Schückel J, Kračun SK, Mikkelsen MD, Mouille G, Johansen IE, Ulvskov P, Domozych DS, Willats WGT (2017) Pea border cell maturation and release involve complex cell wall structural dynamics. *Plant Physiology* 174:1051–1066.

Nagayama T, Nakamura A, Yamaji N, Satoh S, Furukawa J, Iwai H (2019) Changes in the distribution of pectin in root border cells under aluminum stress. *Frontiers in Plant Science* 10:1216

Newton AC, Fitt BDL, Atkins SD, Walters DR, Daniell TJ (2010) Pathogenesis, parasitism and mutualism in the trophic space of microbe-plant interactions. *Trends in Microbiology* 18:365–373.

Nguema-Ona E, Bannigan A, Chevalier L, Baskin TI, Driouich A (2007) Disruption of arabinogalactan proteins disorganizes cortical microtubules in the root of Arabidopsis thaliana. *The Plant Journal* 52:240–251.

Nguema-Ona E, Coimbra S, Vicré-Gibouin M, Mollet J-C, Driouich A (2012a) Arabinogalactan proteins in root and pollen tube cells: distribution and functional aspects. *Annals of Botany* 110:383–404.

Nguema-Ona, E, Moore, JP, Fagerström A, Fangel JU, Willats WGT, Hugo A, Vivier MA (2012b) Profiling the main cell wall polysaccharides of tobacco leaves using high-throughput and fractionation techniques. *Carbohydrate Polymers* 88:939–949.

Nguema-Ona E, Moore JP, Fagerström A, Fangel JU, Willats WGT, Hugo A, Vivier MA (2013a) Overexpression of the grapevine PGIP1 in tobacco results in compositional changes in the leaf arabinoxyloglucan nework in the absence of fungal infection. *BMC Plant biology* 13:46.

Nguema-Ona E, Vicré-Gibouin M, Cannesan M-A, Driouich A (2013b) Arabinogalactan proteins in root-microbe interactions. *Trends in Plant Science* 18:1360–1365.

Nguema-Ona E, Vicré-Gibouin M, Gotté M, Plancot B, Lerouge P, Bardor M, Driouich A (2014) Cell wall O-glycoproteins and N-glycoproteins: aspects of biosynthesis and function. *Frontiers in Plant Science* 5:499.

Nühse TS (2012) Cell wall integrity signaling and innate immunity in plants. *Frontiers in Plant Science* 3:280.

Pascale A, Proietti S, Pantelides IS, Stringlis IA (2020) Modulation of the root microbiome by plant molecules: the basis for targeted disease suppression and plant growth promotion. *Frontiers in Plant Science* 10:1741.

Palanga KK, Liu R, Ge Q, Gong J, Li J, Lu Q, Li P, Yuan Y, Gong W (2021) Current advances in pathogen-plant interaction between *Verticillium dahliae* and cotton provide new insight in the disease management. *J Cotton Res* 4:25.

Park HJ, Wang W, Curlango-Rivera G, Xiong Z, Lin Z, Huskey DA, Hawes MC, VanEtten HD, Turgeon BG (2019) A DNase from a fungal phytopathogen is a virulence factor likely deployed as counter defense against host-secreted extracellular DNA. *mBio* 10:e02805–18.

Plancot B, Santaella C, Jabe, R, Kiefer-Meyer M-C, Follet-Gueye M-L, Leprince J, Gattin I, Souc C, Driouich A, Vicré-Gibouin M (2013) Deciphering the responses of root border-like cells of Arabidopsis and Flax to pathogen-derived elicitors. *Plant physiology* 163:1584–1597.

Popper ZA (2008) Evolution and diversity of green plant cell walls. *Current Opinion Plant Biology* 11:286–292.

Popper ZA, Michel G, Hervé C, Domozych DS, Willats WGT, Tuohy MG, Kloareg B, Stengel DB (2011) Evolution and diversity of plant cell walls: from algae to flowering plants. *Annual Review of Plant Biology* 62:567–590.

Price SR (1911) The roots of some north african desert – grasses. *New Phytologist* 10:328–40.

Rey T, Schornack S (2013) Interactions of beneficial and detrimental root-colonizing filamentous microbes with plant hosts. *Genome Biology* 14:121.

Rich MK, Schorderet M, Reinhardt D (2014) The role of the cell wall in mutualistic symbioses of plants. *Frontiers in Plant Science* 5:238.

Ropitaux M, Bernard S, Schapman D, Follet-Gueye M-L, Vicré M, Boulogne I, Driouich A (2020) Root border cells and mucilage secretions of soybean, *Glycine Max* (Merr) L.: characterization and role in interactions with the oomycete *Phytophthora parasitica*. *Cells* 9:2215.

Rougier M (1981) Secretory activity of the root cap. *Encyclopedia of plant physiology*13:542–574.

Schultz CJ, Harrison MJ (2008) Novel plant and fungal AGP-like proteins in the *Medicago truncatula - Glomus intraradices* arbuscular mycorrhizal symbiosis. *Mycorrhiza* 18:403–412.

Seifert GJ, Roberts K (2007) The Biology of arabinogalactan-proteins. *Annual Review of Plant Biology* 58:137–161.

Staehelin LA, Giddings TH, Kiss JZ, Sack FD (1990). Macromolecular differentiation of Golgi stacks in root tips of Arabidopsis and Nicotiana seedlings as visualized in high pressure frozen and freeze-substituted samples. *Protoplasma* 157:75–91.

Takeda N, Sato S, Asamizu E, Tabata S, Parniske M (2009) Apoplastic plant subtilases support arbuscular arbuscular mycorrhizal development in *Lotus japonicus*. *The Plant Journal* 58:766–777.

Tan L, Mort A (2020) Extensins at the front line of plant defence. A commentary on: 'Extensin arabinosylation is involved in root response to elicitors and limits oomycete colonization. *Annals of Botany* 125:vii–viii.

Tans-Kersten J, Guan Y, Allen C (1998) *Ralstonia solanacearum* pectin methylesterase is required for growth on methylated pectin but not for bacterial wilt virulence. *Applied Environmental Microbiology* 64:4918–4923

Tenhaken R (2015) Cell wall remodeling under abiotic stress. *Frontiers in Plant Science* 5:771.

Thorsten H (2012) Plant cell wall integrity maintenance as an essential component of biotic stress response mechanisms. *Frontiers in Plant Science* 3:77.

Tran TM, MacIntyre A, Hawes M, Allen C (2016) Escaping underground nets: extracellular DNases degrade plant extracellular traps and contribute to virulence of the plant pathogenic bacterium *Ralstonia solanacearum*. *PLoS Pathogens* 12:e1005686.

Turner M, Jauneau A, Genin S, Tavella MJ, Vailleau F, Gentzbittel L, Jardinaud MF (2009) Dissection of bacterial wilt on *Medicago truncatula* revealed two type III secretion system effectors acting on root infection process and disease development. *Plant Physiology* 150:1713–1722.

Valot B, Dieu M, Recorbet G, Raes M, Gianinazzi S, Dumas-Gaudot E (2005) Identification of membrane-associated proteins regulated by the arbuscular mycorrhizal symbiosis. *Plant Molecular Biology* 59:565–580.

van Buuren, ML, Malodonado-Mendoza IE, Trieu AT, Blaylock LA, Harrison MJ (1999) Novel genes induced during an arbuscular myccorrhizal (AM) symbiosis formed between *Medicago truncatula* and *Glomus versiforme*. *Molecular Plant Microbe Interactions* 12:171–181.

Velasquez M, Salter JS, Dorosz JG, Petersen BL, Estevez JM (2012) Recent advances on the posttranslational modifications of EXTs and their roles in plant cell walls. *Frontiers in Plant Science* 3:93.

Vermeer J, McCully ME (1982). The rhizosphere in *Zea*: new insight into its structure and development. *Planta* 156:45–61.

Vicré M, Santaella C, Blanchet S, Gateau A, Driouich A (2005) Root border-like cells of Arabidopsis: microscopical characterization and role in the interaction with rhizobacteria. *Plant Physiology* 138:998–1008.

Vives-Peris V, de Ollas C, Gómez-Cadenas A, Pérez-Clemente RM (2020) Root exudates: from plant to rhizosphere and beyond. *Plant cell reports* 393–17.

Volkens G (1887) *Die Flora der aegyptisch-arabischen Wüste auf Grundlage anatomisch-physiologischer Forschungen*. Gebrüder Borntraeger, Berlin, 156p.

Watson BS, Bedair MF, Urbanczyk-Wochniak E, Huhman DV, Yang DS, Allen SN, Sumner LW (2015). Integrated metabolomics and transcriptomics reveal enhanced specialized metabolism in *Medicago truncatula* root border cells. *Plant physiology* 167:1699–1716.

Weiller F, Moore JP, Young P, Driouich A, Vivier MA (2017). The Brassicaceae species *Heliophila coronopifolia* produces root border-like cells that protect the root tip and secrete defensin peptides. *Annals of botany* 119:803–813.

Wen F, Zhu Y, Hawes MC (1999). Effect of pectin methylesterase gene expression on pea root development. *The Plant Cell* 11:1129–1140.

Wen F, Van Etten HD, Tsaiprailis G, Hawes MC (2007) Extracellular proteins in *Pisum savitum* L. root tip and border cell exudates. *Plant Physiology* 143:773–783.

Wen F, White GJ, VanEtten HD, Xiong Z, Hawes MC (2009) Extracellular DNA is required for root tip resistance to fungal infection. *Plant Physiology* 151:820–829.

Willats WGT, Steele-King CG, Marcus SE, Knox JP (1999) Side chains of pectic polysaccharides are regulated in relation to cell proliferation and cell differentiation. *The Plant Journal* 20:619–628.

Willemsen V, Bauch M, Bennett T, Campilho A, Wolkenfelt H, Xu J, Scheres B (2008) The NAC domain transcription factors FEZ and SOMBRERO control the orientation of cell division plane in Arabidopsis root stem cells. *Developmental cell* 15(6):913–922.

Xia Y, Petti C, Williams MA, DeBolt S (2014) Experimental approaches to study plant cell walls during plant microbe interactions. *Frontiers in Plant Science* 5:540.

Xie F, Williams A, Edwards A, Downie JA (2012) A plant arabinogalactan-like glycoprotein promotes a novel type of polar surface attachment by *Rhizobium leguminosarum*. *Molecular Plant Microbe interactions* 25:250–258.

Zhao X, Schmitt M, Hawes MC (2000) Species-dependent effects of border cell and root tip exudates on nematode behavior. *Phytopathology* 90:1239–1245.

18 Plant Cell Wall Integrity Maintenance

Dhika Amanda

Thorsten Hamann
Norwegian University of Science and Technology

INTRODUCTION

Plants, as sessile organisms, are nearly constantly exposed to stress during growth and development. Therefore, an extensive range of adaptive responses to cope with the stress arising has become established over time. Environmental stressors include abiotic factors, such as temperature (Rizhsky et al. 2004), drought (Ashraf and Foolad 2007), osmotic stress (Zhao et al. 2010; Feng et al. 2014), and salinity (Guo et al. 2012); or biotic ones caused by other organisms (Saijo and Loo 2019). As the primary barrier, the cell wall must adapt and respond to wall damage caused by the stressors. For example, high salinity alters pectin crosslinking in cell walls and microtubules, leading to disruption of cellulose deposition and impairment of wall integrity, illustrating the importance of the CWI maintenance mechanism as shown in the previous study using *Arabidopsis* root (Wang et al. 2007). The study shows that the orientation of cortical microtubules in the *Arabidopsis* root is disrupted after treatment with NaCl (Wang et al. 2007).

In addition to external stimuli, plants also need to respond to rapid changes that occur internally. As plant cells are joined together by their cell walls, cells experience mechanical stress from their neighbours. This occurs during expansive growth when internal pressure and forces from neighbouring cells exert mechanical stress on the cell wall. For example, epidermal cells, highly rigid and compact cells, are under maximal tensile force, while parenchyma cells, forming a low density layer, are under maximal compressive force (Hejnowicz et al. 2000). Consequently, mechanical damage could occur, threatening the functional integrity of the cell wall. Therefore, CWI maintenance sensors must constantly monitor changes in the chemical and mechanical properties of the cell walls, enabling them to initiate processes counteracting the mechanical forces possibly causing cell wall damage (CWD), and repair or compensate for the damage that has already occurred (Schopfer 2006; Hejnowicz et al. 2000).

The concept of the CWI maintenance mechanism was originally established based on research on *Saccharomyces cerevisieae*. The results demonstrated that the yeast cell wall is a highly dynamic structure, which is regularly modifying its composition and structure in response to the physiological needs of the cell and environmental change in a tightly controlled manner (Rodicio and Heinisch 2010). Stimuli indicative of CWI impairment are detected by five dedicated cell wall sensors (WALL STRESS-RESPONSIVE COMPONENT 1 to 3 (WSC1, WSC2, and WSC3)), MATING-INDUCED DEATH2 (MID2), MID TWO-LIKE PROTEIN 1 (MTL1), a stretch-activated, plasma membrane–localised complex consisting of MATING-INDUCED DEATH1 (MID1) and CALCIUM CHANNEL1 (CCH1) or the hyper-osmotic stress sensor HIGH OSMOLARITY GLYCEROL1 (HOG1) (Levin 2011).

In plants, seminal publications providing the first support for the existence of a dedicated plant cell wall integrity (CWI) maintenance mechanism came out at the beginning of the millennium (Cano-Delgado et al. 2000, 2003; Ellis and Turner 2001, 2002; Manfield et al. 2004). The authors were interested in either cell morphogenesis (Cano-Delgado) or plant–pathogen interaction (Ellis), which elegantly highlights the relevance of CWI maintenance both during plant development and defence. They found reducing cellulose production during primary cell wall formation leads to activation defence responses and adaptive changes in wall composition and growth. The next major step was the identification and implication of THESEUS1 (THE1) in detection of reduced cellulose production (Hématy et al. 2007). This work represented a major breakthrough and was followed by more systematic studies of the CWI maintenance mechanism exemplified by Hamann et al. (2009) and Tsang et al. (2011). In parallel, evidence surfaced that another member FERONIA (FER) of the kinase family that THE1 belonged to, also may have a role in CWI maintenance (Kessler and Grossniklaus 2011).

CWI maintenance seems to hold an essential role in growth regulation, hence its conservation throughout the plant kingdom. A recent study using the single-celled alga *Chlamydomonas reinhardtii* indicated the existence of an integrated system of osmo-sensing and CWI maintenance, which regulates the expression of cell wall metabolism genes (Cronmiller et al. 2019). In addition, CWI maintenance during rhizoid growth in an early diverging plant, *Marchantia polymorpha*, is controlled by Malectin-like receptor kinases (MLRs) and a Pto-interacting protein 1 (PTI1)-like signalling module similarly like in *Arabidopsis thaliana*, showing that CWI signalling has already existed in a common ancestor of land plants (Westermann et al. 2019).

In this chapter, we introduce the basic concept of the CWI maintenance mechanism active in plants, followed by presenting evidence regarding CWI maintenance activity in primary and secondary walls. This will include a global

DOI: 10.1201/9781003178309-18

overview of the knowledge available regarding the processes responsible for mechano-sensing and CWD detection in plants as well as molecular components involved. We will proceed to discuss the role of CWI maintenance in regulating the cell cycle as an example of how the mechanism can influence plant development. We will bring the chapter to a close by briefly reviewing the contributions of CWI maintenance to biotic and abiotic stress responses and synthesise the knowledge presented into key take home messages. This short summary also highlights our focus on stimulus perception and early signal transduction. The reason for this being the limited knowledge available regarding downstream targets of early signal transduction processes.

CWI MAINTENANCE IN PRIMARY CELL WALLS

Extensive studies investigating the regulation of CWI maintenance in plants have been performed, with a particular focus on primary cell wall formation. Several stimuli have been proposed to act as indicators for impaired CWI, such as alterations in cell wall components due to disrupted metabolism, crosslinking, or mechanical stress at the cell wall–plasma membrane interface (Rui and Dinneny 2020).

Alterations in Cell Wall Components

The primary wall mainly comprises cellulose, hemicellulose, and pectin, with changes in the content and composition of these constituents capable of triggering the CWI maintenance mechanism. The *Arabidopsis* mutant *procuste1* (*prc1*) displays reduced cell elongation and ectopic lignin deposition due to disruption of cellulose biosynthesis during primary cell wall formation (Delgado-Cerezo et al. 2012; Fagard et al. 2000). The results suggest that lignin deposition could occur as part of a compensatory reaction to replace the missing load-bearing cellulose microfibrils. Similarly, a double-mutant *xyloglucan xylosyltransferase 1* and *2* (*xxt1/xxt2*) with xyloglucan deficiency exhibits reduced anisotropic growth leading to a smaller size plant (Cavalier et al. 2008; Park and Cosgrove 2012; Xiao et al. 2016). However, *xxt1/xxt2* walls have more aligned cellulose microfibrils and more extensibility after treatment with pectin degrading enzymes, suggesting that other wall components are rearranged to re-distribute the mechanical load in the absence of xyloglucan (Park and Cosgrove 2012; Xiao et al. 2016). Pectins have been implicated in CWI maintenance due to their sensitivity towards mechanical deformation (Bacete et al. 2018). Homogalacturonan (HG), the most abundant form of pectin, is secreted into the cell wall in highly methyl-esterified form and undergoes de-methylesterification mediated by PECTIN METHYLESTERASES (PME) to various degrees (Pelloux et al. 2007). PECTIN METHYLESTERASES INHIBITORs (PMEIs) are involved in PME regulation, and the interactions of PME and PMEI influence the physical properties of the polymer network. For example, de-methylesterified HG can form crosslinks with Ca^{2+}, generating "egg-box" structures that promote wall stiffening and porosity (Wolf and Greiner 2012; Ridley et al. 2001; Hocq et al. 2017). Such changes in cell wall structure could be detected by sensors allowing CWD detection (Rui and Dinneny 2020).

The existence of a pectin integrity monitoring system has been proposed before. For example, the brassinosteroid (BR) receptor BRASSINOSTEROID-INSENSITIVE 1 (BRI1) has been implicated in monitoring the HG status in the cell wall (Wolf et al. 2012). A mutant allele of BRI1 was identified as suppressor of the growth defects in plants overexpressing PME1 (Wolf et al. 2012). Impaired HG crosslinking/de-crosslinking, caused by PME activity, triggers BRI1 and the BRI1-ASSOCIATED KINASE1 (BAK1) coreceptor. These, in turn, can activate BR signalling via damage-associated molecular patterns (DAMPs) or pathogen-associated molecular patterns (PAMPs) recognition, resulting in the activation of wall remodelling agents (Wolf et al. 2012; Yasuda et al. 2017). Examples of DAMPs are pectin- or xyloglucan-derived oligosaccharides and plant elicitor peptides (PEPs) (Claverie et al. 2018; Souza et al. 2017; Huffaker et al. 2006: see Chapter 20).

Pectin-derived DAMPs such as oligogalacturonides (OGs) hold a pivotal role in pectin integrity sensing. Upon exposure to pathogen-derived or mechanical damage, OGs are released by the cell wall as a result of HG-digesting enzymes, such as polygalacturonases and pectate lyases (Hou et al. 2019). Similar to xyloglucan-derived oligosaccharides (Claverie et al. 2018), OGs can trigger a wide range of responses, including activation of mitogen-activated protein kinases (MAPKs), callose deposition (Galletti et al. 2008), reactive oxygen species (ROS) production (Bellincampi et al. 2000), and defence responses against pathogens (Ferrari et al. 2007).

Cello-oligomers, including cellotriose, cellobiose, cellotetraose, and other short-chain β-1,4-linked D-glucoses, are generated during the breakdown of cellulose by microbial or plant-derived glucosidases (Hou et al. 2019). Studies show that cello-triose induces elevation of cytoplasmic free calcium ion concentration $[Ca^{2+}]_{cyt}$ in Arabidopsis roots, leading to increased ROS production (Johnson et al. 2018). Furthermore, exogenous application of cello-oligomers induces expression of plant immunity-related genes and activates MAPK signalling cascade, implicating cello-oligomers in CWI monitoring and perception of PAMPs (Johnson et al. 2018; Souza et al. 2017).

PEP1 is a 23-aa long peptide elicitor initially identified in *Arabidopsis* and is derived from the PRECURSOR OF PEPTIDE 1 (PROPEP1) (Huffaker et al. 2006; Yamaguchi et al. 2006). Upon cell damage, elevation of $[Ca^{2+}]_{cyt}$ activates METACASPASE4 (MC4), which cleaves and releases PEP1 from PROPEP1, located in the vacuolar membrane (Hander et al. 2019). PEP1 then binds to the PEP RECEPTOR1 (PEPR1), leading to heterodimerisation of PEPR1 with its coreceptor BAK1 and activates the BAK1-dependent PEPR pathway (Schulze et al. 2010; Tang et al. 2015). Additionally, PEP–PEPR interactions are also involved in the activation of jasmonic acid (JA), salicylic acid in response to CWD as well as ethylene-mediated immune responses (Ma et al. 2012; Flury et al. 2013; Tintor et al. 2013; Ross et al. 2014; Engelsdorf et al. 2018).

A wide range of plant proteins such as extensins, a major class of hydroxyproline-rich glycoproteins (HRGPs), are also involved in modification of the CW structure. Extensins are characterised by a repetitive Ser-$(Hyp)_4$ module in which Ser is being mono-galactosylated (Lamport et al. 1973; Saito et al. 2014) and Hyp residues are being *O*-arabinosylated (Campargue et al. 1998; Held et al. 2004; Tan et al. 2003). Several *in vitro* studies show that complete or partial removal of Hyp-*O*-arabinosides leads to alteration of extensin crosslinking (Chen et al. 2015; Held et al. 2004; Schnabelrauch et al. 1996). Another extensin residue that is also involved in the crosslinking is Tyr. Tyr residues can form covalent bonds with the help of wall-associated peroxidases, leading to cell wall reinforcement and delaying pathogen spread (Fry 1982; Brady and Fry 1997; Bradley et al. 1992; Schnabelrauch et al. 1996).

Mechanical Stress at Cell Wall–Plasma Membrane Interface

Mechanical forces have long been proposed to act as morphogenic factors that shape plants by reprogramming plant development, thus altering growth (Monshausen and Gilroy 2009). Growth heterogeneity is promoted by allowing neighbouring cells to undergo rapid growth changes, facilitating organ emergence (Uyttewaal et al. 2012). Mechanical forces in plants are the result of turgor pressure in a given cell or indirect pressure exerted by adjacent cells leading to mechanical stress (Hejnowicz et al. 2000). This type of stress is usually anisotropic and depends on cell shape and wall thickness, but it can also be superimposed on a wall by surrounding cells depending on the structure of the organ (Hejnowicz et al. 2000).

Plant water status is determining the direction of water fluxes, and the water movement into and out of the cell can be sensed as mechanical signals (Haswell and Verslues 2015). While being exposed to osmotic stress, the extracellular domains of plasma membrane–embedded proteins can undergo conformational changes, which initiate the signalling processes activating downstream responses (Haswell and Verslues 2015). In addition, the plasma membrane is exposed to turgor pressure (0.1–0.6 MPa), which pushes the plasma membrane against the wall, reducing the distance between the two to less than a few tens of nanometres (Wang et al. 2006; Martiniere et al. 2012). Consequently, slight changes in water potential can result in alterations of the cell wall–plasma membrane contact area and wall deformation. Studies showed that decreased water potential, upon dehydration, alters crosslinks between wall polymers and affects microtubule stability, which in turn modifies deposition of new wall materials (Wang et al. 2016). Knowledge regarding the processes responsible for detecting mechanical stress in the wall, displacement of wall versus plasma membrane and the subsequent translation into signals leading to wall remodelling is limited, and certain wall sensors have been implicated and are reviewed in detail in the next sub-chapter (Wolf et al. 2012; Vogel and Sheetz 2006).

PROTEINS INVOLVED IN CELL WALL INTEGRITY MONITORING

The CWI maintenance mechanism is well characterised in *Saccharomyces cerevisiae.* The dedicated pathway involves a set of five membrane-spanning sensors, WSC1 to 3, MID2, and MTL1 (Rodicio and Heinisch 2010; Kock et al. 2015; Neeli-Venkata et al. 2021). In response to changes in the surface tension of the wall, the extracellular domain changes its conformation. This is translated across the plasma membrane, leading to conformational changes in the cytoplasmic region of the membrane-spanning sensors allowing activation of GTPase Rho1 (Vay et al. 2004; Petkova et al. 2012). Rho1-GTP then interacts with PROTEIN KINASE C (PKC1), triggering a MAPKinase-based signalling cascade targeting transcription factors that activate genes required for cell wall synthesis, remodelling, and cytoskeletal reorganisation (Kock et al. 2015). The monitoring mechanism in yeast is generally triggered by three different events (membrane stretch, CWD, and osmo-stress), which each requires a specific set of sensors and signalling mechanism (Figure 18.1). The same classification is also found in plant, and it is discussed below.

Mechano-Perception: The Role of Channels

Stretch-activated ion channels have been proposed to regulate cell wall expansion through a mechano-sensing feedback loop involving cytosolic calcium, which acts as intracellular secondary messenger activating downstream responses (Falke et al. 1988; Cosgrove and Hedrich 1991; Ding and Pickard 1993; Garrill et al. 1994; Kikuyama and Tazawa 2001). During hyper-osmotic conditions, plasmolysis occurs, i.e., the plasma membrane is separated from the cell wall, and thin Hechtian strands connecting the protoplast to the wall become visible (Pont-Lezica et al. 1993). Such a disruption in the contact area between the plasma membrane–cell wall could act as stimulus detected by stretch-activated mechano-sensitive channels, which in turn modulate ion transport to initiate the CWI maintenance process (Novaković et al. 2018; Basu and Haswell 2017; Seifert and Blaukopf 2010; Wolf et al. 2012). Various mechano-sensitive channels from a broad range of organisms have been characterised, yet identifying mechano-sensitive calcium channels in plants has proven difficult.

In yeast, MID1 and Ca^{2+} CHANNEL1 (CCH1) are required for mechano-perception (Figure 18.1). MID1 is a stretch-activated, calcium-permeable, nonselective cation channel located at the plasma membrane and endoplasmic

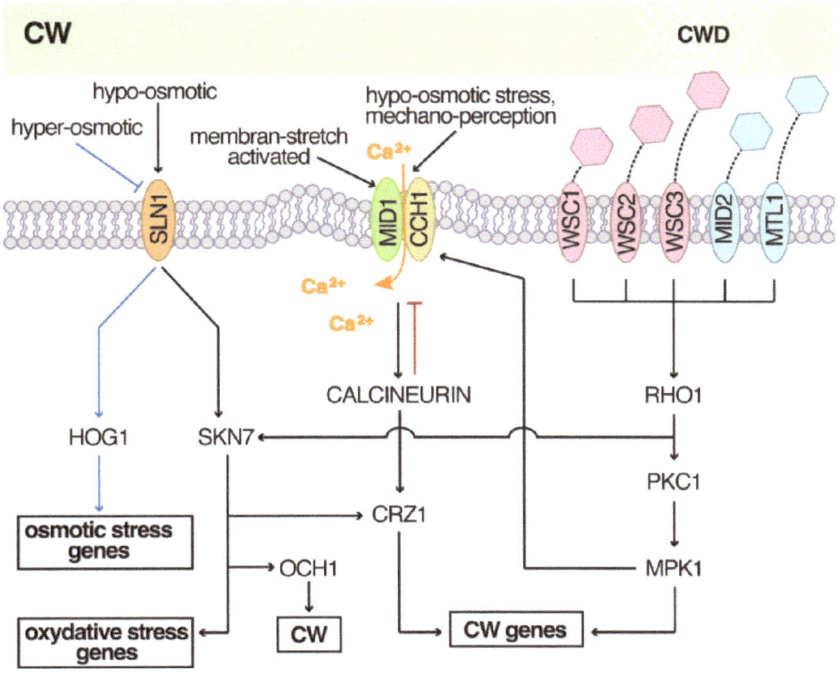

FIGURE 18.1 Schematic overview of yeast CWI monitoring signalling pathways. Inactivation of SLN1 under hyper-osmotic condition activates the HIGH OSMOLARITY GLYCEROL 1 (HOG1), which is responsible for hyper-osmotic stress responses (Jacoby et al. 1997). On the contrary, SLN1 is auto-phosphorylated upon hypo-osmotic stress. It modulates activity of SKN7 transcription factor, which controls expression of oxidative stress–induced genes (Li et al. 1998). SKN7 also activates OUTER CHAIN ELONGATION 1 (OCH1), which is involved in the maturation of cell wall N-glycoproteins (Li et al. 2002). The MID1-CCH1 complex is involved in both osmo-sensing and mechano-perception, controlling expression of CW biosynthesis genes via the calcineurin/calcineurin-responsive zinc finger 1 (CRZ1) pathway. The CWI maintenance pathway is initiated in yeast via five plasma membrane–bound sensors, WSC1-3, MID2, and MTL1, which regulate responses to CWI impairment via the PKC1-MPK1-based signalling cascade.

reticulum (ER) membrane (Iida et al. 1994; Kanzaki et al. 1999; Yoshimura et al. 2004). CCH1 is a voltage-gated calcium channel that monitors membrane tension. CCH1 forms a complex with MID1, facilitating Ca^{2+} influx in response to membrane deformation and promoting cell wall biosynthesis via the calcineurin pathway (Garrett-Engele et al. 1995; Fischer et al. 1997). In the Ascomycete *Botrytis cinerea*, MID1 and CCH1 homologs are required for vegetative growth under low-calcium conditions. However, unlike yeast, MID1 is only found at the ER membrane (Harren and Tudzynski 2013).

To date, homologs of the bacterial MECHANO-RESPONSIVE CHANNEL OF SMALL CONDUCTANCE (MSCs) and MID1 COMPLEMENTING ACTIVITY (MCAs) are the best candidates for plant mechano-sensor channels (Nakagawa et al. 2007). MCA1 and MCA2 are located at the plasma membrane and proposed to act as regulatory components of a mechano-sensitive channel complex. Their function seems to be conserved between yeast and plants, as the expression of *Arabidopsis* MCA1/2 partially rescues MID1-deficient yeast cells (Nakagawa et al. 2007; Yamanaka et al. 2010). MCA1 and MCA2 regulate Ca^{2+} influx through the plasma membrane in cultured cells of *Nicotiana tabacum,* altering cell proliferation and mechanical stress–induced gene expression (Kurusu et al. 2012). More recently it was shown that MCA2 enables Ca^{2+} influx in vitro in response to membrane stretching (Yoshimura et al. 2021). Additionally, overexpression of *MCA1* induces increased expression of the *Arabidopsis TOUCH4* (*TCH4*) gene, also known as (*CALMODULINE-LIKE PROTEIN 12, CML12*), which encodes a calcium-sensitive protein linked to touch responses (Braam 2005, 1992). These results suggest that touch-induced responses can be activated by elevating *MCA1* expression (Yamanaka et al. 2010). Recent studies showed that MCA1 is required to detect gravity vector changes (Nakano et al. 2021). Gravity stimulation causes sedimentation of high-density plastids, which exerts mechanical stress on actin filaments, causing displacement. The displacement activates mechano-sensitive Ca^{2+} channels, such as MCA1, resulting in a novel type of $[Ca^{2+}]_{cyt}$ increase that happens slowly and whose role remains to be elucidated (Nakano et al. 2021).

Six homologs of the MSCs channels have been found in the *Oryza sativa* (rice) genome and ten MSCs-LIKE (*MSL*) genes in *Arabidopsis* (Haswell and Meyerowitz 2006). MSL2 and MSL3 are located in the plastids where they have a role in plastid division, as well as JA production during stress-induced wall remodelling (Engelsdorf et al. 2018). MSL9 and MSL10 are found mainly in the plasma membrane of root cortex cells (Haswell et al. 2008). Both apparently act as mechano-sensors at the plasma membrane, responsible for mechano-sensitive gating of a Cl^- permeable channel (Haswell et al. 2008). However, overexpression of the gene encoding the channel does not disrupt growth or modifies the response to mechanical perturbation (Haswell et al. 2008). It has been proposed that a prolonged release of Cl^- through MSL10, results in plasma membrane depolarisation and activation of the Ca^{2+} channel, leading to $[Ca^{2+}]_{cyt}$ burst (Basu and Haswell 2020; Hedrich 2012). A recent study showed that MSL10 influences ROS production and expression of mechano-sensitive genes when a plant is exposed to hypo-osmotic stress (Basu and Haswell 2020; Hamilton et al. 2015; Zhou et al. 2022). However, it is important to bear in mind that probably other plant-specific mechano-responsive channels exist, which we have not yet identified.

DEFECTIVE KERNEL1 (DEK1), a plant-specific member of the calpain family of Ca^{2+}-activated Cys-proteases, has an important role in mechano-sensing, specification, and maintenance of the epidermal cell layer by indirectly regulating epidermal-specific transcription factors (Galletti et al. 2015). A single, highly conserved *DEK1* gene is found in all plants (Lid et al. 2002) and its role has been investigated in *Zea mays* (Becraft et al. 2002; Lid et al. 2002), *Arabidopsis thaliana* (Lid et al. 2005; Johnson and Ingram 2005; Johnson et al. 2008; Roeder et al. 2012; Galletti et al. 2015), rice (Hibara et al., 2009), *Nicotiana benthamiana* (Ahn et al. 2004), and *Physcomitrella patens* (Perroud et al. 2014). The 240 kDa protein is located in the plasma membrane and has an extracellular loop, a large integral membrane region, and a cytoplasmic calpain domain (Lid et al. 2002). The extracellular loop within the transmembrane domain of DEK1 acts as a sensor to perceive mechanical stimuli, as shown in a study using *Physcomitrella patens* (Demko et al. 2014). Upon perceiving the stimuli, the transmembrane domain of DEK1 enables mechanically mediated Ca^{2+} transport into the cytoplasm (Tran et al. 2017). DEK1 then undergoes autolytic cleavage in two positions: in the region adjacent to the calpain domain and within the juxtamembrane domain, releasing (and activating) the calpain domain into the cytoplasm (Johnson et al. 2008). Currently, the downstream target of calpain remains to be determined. The increased level of cytoplasmic Ca^{2+} also enhances the calpain activity to trigger downstream signalling, which leads to wall remodelling (Johnson et al. 2008). Several studies reported increased wall thickness as well as elevated cellulose and pectin content in plants overexpressing the calpain domain (CALPAIN OE), whereas the opposite effects were observed in knock-down lines (Amanda et al. 2017, 2016). Furthermore, results from transcriptome and gene expression analyses suggest that DEK1 modulates the expression of genes responding to CWD (Johnson et al. 2008; Demko et al. 2014; Amanda et al. 2016, 2017). While these results support the notion that DEK1 can act as cell wall sensor and is possibly involved in CWI maintenance, further studies are required to understand its specific mode of action and characterise the downstream targets regulated by DEK1.

Cell Wall Damage Detection: The Role of Cell Wall Integrity Sensors

Various molecules have been implicated in monitoring of CWI maintenance (Wolf et al. 2012). Changes in the cell wall matrix could activate these molecules and promote direct responses or indirect ones, resulting in wall remodelling (Wolf et al. 2012; Monshausen and Gilroy 2009). Several potential plant CWI sensors have been identified, characterised, and are being discussed below.

Five RECEPTOR-LIKE WAKs and 26 *WAK-like* genes have been found in the *Arabidopsis* genome, with the *WAK-like* ones better known for their prominent role in disease resistance (Verica and He 2002; Zuo et al. 2015). WAK expression levels are elevated in plants upon exposure to biotic and abiotic stresses (Anderson et al. 2001; Wolf et al. 2012). WAKs consist of a highly conserved cytoplasmic kinase domain, a single transmembrane region, and an extracellular domain. WAK1 interacts with glycine-rich proteins in the cell wall, while the external domains of WAK1 and WAK2 bind to pectins with binding tightness mediated by the degree of HG methylesterification (Kohorn et al. 2006; Humphrey et al. 2007; Jung et al. 2019). Binding to de-esterified HG is preferential compared to esterified HG, while binding affinity towards pectin is highly enhanced by calcium. This means that interactions between plasma membrane–anchored WAKs and pectin in the cell walls could allow monitoring of the status of the walls (Decreux and Messiaen 2005). WAKs also play essential roles in pathogen defence, which highlights again the importance of the cell walls during plant–pathogen interactions and the possible involvement of cell wall signalling in defence responses. When pathogens digest plant cell walls and generate OGs, WAK binding to OGs seems to activate MPK3- and MPK6- (mitogen-activated protein kinase) dependent pathways, inducing defence responses (Kohorn 2015; Kohorn et al. 2009, 2014). The binding of OGs to WAK1 seems to be mediated by GLYCINE-RICH PROTEIN 3 (GRP3), a structural wall protein regulating cell expansion negatively (Park et al. 2001; Gramegna et al. 2016).

Based on similar processes in animals (Kapp et al. 2017; Ludwig et al. 2021), an RGD (arginine-glycine-aspartic acid)-like recognition system has been implicated in detecting mechanical stress by physically connecting the plasma membrane to the cell wall. Previous studies in *Arabidopsis* suspension culture cells and onion epidermal cells showed that RGD treatment abolishes the connection of cytoplasm to wall attachment sites (Hechtian strands) (Mellersh and Heath 2001; Canut et al. 1998). Gouget et al. (2006) identified eight potential RECEPTOR-LIKE KINASES (RLKs) as RGD-binding proteins. One of these is an L-TYPE

LECTIN-LIKE RECEPTOR-LIKE KINASE (LeRLK), which binds RGD tripeptides and mediates plasma membrane–cell wall adhesion (Gouget et al. 2006). Lectin domains are known to bind carbohydrates, yet no carbohydrate-binding partner for the LeRLKs has been identified in plants (Wolf et al. 2012). Although the LECTIN-LIKE RLK function is not clearly understood, the cytosolic kinase domain of this protein could signal changes in the status of the cell wall to the cell interior (Gouget et al. 2006). In addition, members of the PROLINE-RICH EXTENSIN-LIKE RECEPTOR KINASE (PERK), PR5-LIKE RECEPTOR KINASE/THAUMATIN, and the LysM families can also interact with carbohydrates and have specific roles in cell wall signalling (Miya et al. 2007; Trudel et al. 1998; Bai et al. 2009; Wolf et al. 2012).

Catharanthus roseus RLK (CrRLK1) proteins belong to another subfamily of RLKs, which has been extensively studied in the context of CWI maintenance. To date, the functions of 15 out of 17 *Arabidopsis* CrRLK1s have been investigated; these include ANXUR1 and 2 (ANX1 and ANX2) in CWI maintenance in pollen tube (Boisson-Dernier et al. 2009; Feng et al. 2019; Lindner et al. 2012; Miyazaki et al. 2009) and tip-growing cells (Boisson-Dernier et al. 2013); BUDDHA'S PAPER SEAL 1 and 2 (BUP1 and 2) in pollen tube growth and integrity maintenance (Ge et al. 2019; Feng et al. 2019); HERCULES RECEPTOR KINASE1 and 2 (HERK1 and HERK2) in cell elongation (Guo et al. 2009); ERULUS (ERU)/ $[Ca^{2+}]_{cyt}$-associated protein kinase (CAP1) in pollen tube growth and root development (Schoenaers et al. 2018); CURVY1 (CVY1) in cell elongation and morphogenesis (Gachomo et al. 2014); ANJEA (ANJ) in pollen tube reception (Galindo-Trigo et al. 2020); and MEDOS1 to 4 (MDS1 to 4) in metal ion stress responses (Richter et al. 2018).

Other prominent members of the CrRLK1-Like family, which have been proposed as CWI sensors, are THE1 and FER. THE1 has been proposed to act as a CWI sensor for several reasons. The THE1 ectodomain comprises two Malectin-like domains, which are known to bind carbohydrates in animals (Schallus et al. 2008). THE1 was initially identified as a suppressor of the inhibition of cell elongation caused by disrupted cellulose biosynthesis (Hématy et al. 2007). THE1 has also been implicated in the modulation of cell expansion involving the motor kinesin AtKINESIN-13A (AtKIN13A) (Fujikura et al. 2014). In parallel, THE1 is also required for CWD-induced ROS production, ectopic lignin deposition, and growth inhibition (Wolf et al. 2012; Hématy et al. 2007). In response to cellulose biosynthesis inhibition (CBI)-induced CWD, THE1 acts upstream of MCA1 and FEI2 to induce production of JA, salicylic acid, and CWI responses (Engelsdorf et al. 2018). Results from a recent study indicate that THE1 regulates root wall stiffness in cellulose-deficient seedling roots and modulates phytohormone production induced by CWD and hyper-osmotic stress (Bacete et al. 2021; Shih et al. 2014; Cornblatt et al. 2021).

In addition, several THE1-interactors have been identified, including ANX1/2 and BUPs1/2 during pollen tube growth and sperm cell release, as well as RAPID ALKALINIZATION FACTOR 34 (RALF34) during cell expansion (Ge et al. 2017). RALF34 binds to the ectodomain of THE1, and its signalling is dependent on THE1 (Gonneau et al. 2018). THE1-RALF34 interactions are pH-sensitive, with no binding at pH 6.5, and the highest affinity being reached at pH 7.8 (Gonneau et al. 2018). RALF34 induces a rapid increase of cytoplasmic Ca^{2+} in wild-type plants, which is absent in *the1-1* plants, indicating how the interactions may generate a Ca^{2+} signal regulating downstream responses (Gonneau et al. 2018; Ge et al. 2017). During PTI, THE1 seems to act upstream of GUANINE EXCHANGE FACTOR4 (GEF4) to mediate defence responses influencing resistance against the necrotrophic fungus *Botrytis cinerea* (Qu et al. 2017). Furthermore, this interaction seems to require RHO-RELATED PROTEIN FROM PLANTS11 (ROP11) (Qu et al. 2017). While these results provide some insight into the signalling processes requiring THE1, most of the signalling cascades involved remain to be elucidated in detail.

FER is another member of the CrRLK1-Like family. It was initially identified because of its role in the regulation of pollen tube elongation and fertilisation (Haruta et al. 2014; Stegmann et al. 2017). Subsequently, it was implicated in nitric oxide accumulation (Duan et al. 2020); root growth and root hair development (Haruta et al. 2014; Zhu et al. 2020; Stegmann et al. 2017); root nutation (Li et al. 2020); leaf epidermis morphogenesis (Li et al. 2015); stomatal movement (Yu and Assmann 2018); flowering time (Wang et al. 2020); and seed size and set (Yu et al. 2014; Li et al. 2016). In the context of CWI maintenance, FER could be involved in detecting changes in pectin structure since it can bind de-methylesterified HG *in vitro* via the extracellular Malectin domain (Feng et al. 2018; Lin et al. 2022). Results from several studies indicated interactions between FER and RALF1, suggesting that RALF1 binds to a plasma membrane complex consisting of FER and LRE-like glycosylphosphatidylinositol-anchored protein GPI-AP1 (LLG1)/ LORELEI (LRE) (Li 2016; Ge et al. 2019). RALF1 binding activates FER, leading to recruitment of the receptor-like cytoplasmic kinase RIPK and phosphorylation of H^+-ATPase. Phosphorylation results in inhibition of proton transport and transient alkalinisation of the apoplast, causing the disassociation of the BRI1-brassinolide (BL)-BAK1 complex (Haruta et al. 2014; Li et al. 2015; Li et al. 2016; Du et al. 2016; Campos et al. 2018; Dressano et al. 2017). Results from experiments with French bean show that apoplastic alkalinisation is affecting the activity of cell wall–associated peroxidases capable of producing ROS, which modulates crosslinking of cell wall components (Bolwell et al. 2002). This would explain the inhibition of root elongation in seedlings treated with RALF1 peptide (Haruta et al. 2014). However, alkalinisation could also influence growth by activating cell wall loosening enzymes (Cosgrove 2015). These observations illustrate that the specific processes responsible for growth regulation remain to be elucidated. In addition, another FER interaction with its ligand, RALF23, is also shown to influence plant immunity by regulating JA signalling via phosphorylation of MYC2 and inhibits accumulation of salicylic acid (Guo et al. 2018).

Interestingly, FER and THE1 loss of function mutants show the opposite phenotypic effect when treated with isoxaben. THE1 loss-function seedlings display JA production and lignin deposition induced by cellulose biosynthesis inhibition, while a FER knock-down allele has enhanced responses (similar to the phenotypic effects of a THE1 gain of function allele) (Engelsdorf et al. 2018). These results implicate both in the response to CWI impairment and indicate that they have distinct functions.

LEUCINE-RICH REPEAT RECEPTOR-LIKE KINASES (LRR-RLKs) have also been implicated in CWI maintenance. One LRR-RLK, MALE DISCOVERER 1-INTERACTING RLK 2 (MIK2), contains 24 LRRs in its extracellular domain, and has a transmembrane and an intracellular kinase domain (Van der Does et al. 2017). The *mik2* mutant displays impaired gene expression, JA accumulation, and lignin deposition in response to CBI-induced CWD. In addition, its activity is dependent on THE1, providing further evidence for its involvement in CWI maintenance (Van der Does et al. 2017). *mik2* also exhibits increased sensitivity to salt stress, implicating MIK2 in salt stress adaptation. However, results from studies using *mik2-1* and *mik2-1 the-1* plants show that MIK2 and THE1 regulate responses to CBI in different ways, supporting the notion that a regulatory network is responsible for CWI maintenance, not a single linear cascade (Van der Does et al. 2017). While MIK2 is also involved in defence responses like THE1, its role in the resistance against the fungal root pathogen *Fusarium oxysporum* is independent of THE1 (Van der Does et al. 2017).

FEI2 is an LRR-RLKs that seems involved in CWI maintenance and acts downstream of THE1 (Engelsdorf et al. 2018). However, further studies are needed to characterise the potential ligand and the specific role of FEI2 in relation to THE1-mediated signalling in detail. Similar to *mik2*, *fei1* and *fei2* mutants show increased sensitivity towards high salt concentrations and CBI, further supporting the connection between salt sensitivity response and CWI. FEI1 and FEI2 seem to be involved in the same signalling process that also modulates SALT OVERLY SENSITIVE5 (SOS5) and GALACTOSYLTRANSFERASE 2 AND 5 (GALT2 and GALT5), which encode arabinogalactan-proteins (AGPs) and AGP-specific galactosyltransferases, respectively, with SOS5 possibly acting as a ligand for FEI1/FEI2 (Basu and Haswell 2017; Xu et al. 2018). AGPs are HRGPs that are covalently linked to cell wall polysaccharides (Tan et al. 2013), modulate cell to cell adhesion (Schultz et al. 1998), reinforce the cell wall with the help of peroxidases (Kjellbom et al. 1997), and maintain wall–plasma membrane integrity (Tan et al. 2018; Chapter 16). The importance of the O-glycosylation of AGPs by GALTs was implied by a previous study using *galt2 galt5* double mutants, which exhibited altered root elongation, root hair growth, pollen tube growth, flowering time, and leaf and inflorescence development (Basu et al. 2015).

FASCICLIN-LIKE ARABINOGALACTAN-PROTEINS (FLAs) form a subgroup of the AGPs, which have as a distinct feature a glycosyl-phosphatidylinositol (GPI) anchor that attaches the proteins to the plasma membrane (Novaković et al. 2018). Fasciclin-1 (FAS1)-like domains are also found in proteins in bacteria, fungi, and animals, where they have specific functions in inter-cell communication or interactions between cells and the extracellular matrix (Kim et al. 2000; Ghatak et al. 2014). In parallel, plant FLAs are also involved in the responses to different stresses exemplified by *FLA7* in maize and *FLA9* in *Arabidopsis*, whose expression patterns are altered in response to drought stress, leading to reduced AGP activity and increased abortion of fertilised ovaries (Cagnola et al. 2018). Additionally, FLAs also participate in the maintenance of the mechanical characteristics of stems (Wang et al. 2017). Increased gibberellin production in the xylem during mechanical or gravity stress triggers degradation of the DELLA protein REPRESSOR OF GIBBERELLIC ACID (RGA1), which results in the activation of *FLAs*, promoting tension wood formation in poplar. Results from a genetic analysis suggested that FLA4/SOS5 is involved in pectin integrity maintenance during seed coat mucilage production (Griffiths et al. 2016; Seifert 2021). In parallel, FLA4/SOS5 seems to activate the response to salt stress via FEI1/2 (Griffiths et al. 2016; Seifert 2021). While the available evidence implicates FLAs in CWI maintenance relevant processes, their specific mode of action and their relationship to damage-induced wall remodelling remain to be determined.

EVIDENCE FOR THE EXISTENCE OF A CWI MAINTENANCE MECHANISM IN SECONDARY WALLS

Lignin, mainly consisting of p-hydroxyphenyl, guaiacyl, and syringyl units, is one of the major components in the secondary walls (Vanholme et al. 2019). Lignin is essential in providing additional mechanical strength and hydrophobicity during growth and inhibiting the spread of pathogen during defence response. The lignification process giving rise to lignin is initiated in the middle lamella, involves pectin, and ensures that lignin is embedded in the secondary cell wall (Lairez et al. 2005; Habrant et al. 2009). Lignin-deficient *Arabidopsis* lines have been generated by reducing the activities of *HYDROXYCINNAMOYL COA: SHIKIMATE/QUINATE HYDROXYCINNAMOYL TRANSFERASE* (*HCT*) or *CINNAMOYL COA REDUCTASE 1* (*CCR1*) (Gallego-Giraldo et al. 2020). In these lignin-deficient lines, the expression of genes encoding pectin degrading enzymes is upregulated, leading to the release of pectin-derived elicitors, which trigger defence responses via the activation of pathogenesis-related genes (Gallego-Giraldo et al. 2020). Voxeur and Höfte (2020) proposed that reduced lignin levels are sensed by CWI sensors and facilitate the release of latent elicitors, resulting in activation of the pectin-hydrolysing enzyme ARABIDOPSIS DEHISCENCE ZONE POLYGALACTURONASE 1 (ADPG1). ADPG1 is needed for the activation of the latent elicitor, thus creating a positive feedback loop. The latent elicitor is then perceived by the OG receptor and initiates cellular responses (Voxeur and Höfte 2016). The data suggest that impaired secondary walls' integrity triggers changes in other cell wall components, which alter cell wall composition and structure. However, the mode of action of the mechanism responsible for lignin monitoring and regulating the responses is still unknown (Gallego-Giraldo et al. 2020).

The CWD-induced CWI maintenance response in secondary walls is mainly driven by external factors, as illustrated by several studies finding increased lignin content and modified composition after plants are exposed to biotic and abiotic perturbation (Tenhaken 2015; Moura et al. 2010). The link between CWD and lignin biosynthesis has been shown in several studies focusing on primary cell walls where normally no lignin is being deposited, illustrating the plasticity of the plant cell walls. Cellulose biosynthesis was inhibited either using mutants such as *CONSTITUTIVE EXPRESSION OF VSP1* (*CEV1*) (Ellis et al. 2002) and *ECTOPIC LIGNIFICATION 1* (*ELI1*) (Caño-Delgado et al. 2003) or inhibitors such as isoxaben (Manfield et al. 2004; Caño-Delgado et al. 2003; Ellis et al. 2002; Hamann et al. 2009). CBI led to altered cell wall composition and formation of ectopic lignin as well as induction of JA production via *ALLENE OXIDE SYNTHASE* (*AOS*) (Denness et al. 2011) and *JASMONIC ACID RESISTANT1* (*JAR1*) (Hamann et al. 2009), ethylene (Ellis et al. 2002; Caño-Delgado et al. 2003), SA, callose (Hamann et al. 2009), and ROS production via RESPIRATORY BURST OXIDASE HOMOLOG D (RBOHD) (Denness et al. 2011). The CWD-induced ectopic lignin deposition is mediated by THE1 (Hématy et al. 2007). Denness et al. (2011) proposed a two-stage response towards CWD, in which ROS and a JA-based feedback loop are needed to modulate CWD-induced lignin production in the root elongation zone.

CWI MAINTENANCE FOR GROWTH CONTROL AND DEVELOPMENT

Knowledge regarding the contributions of CWI maintenance to growth and development in plants is still limited. However, evidence has been accumulating confirming its relevance. Here, two examples serve to illustrate the contributions of CWI maintenance to growth and development.

Reduced cellulose levels in primary walls due to mutations in *CELLULOSE SYNTHASES* (*CESAs*) result in impaired cell division (Sampathkumar et al. 2019). Furthermore, a DEK1 knock-down line of *Nicotiana benthamiana* shows elevated protein levels of the D-type cyclins CYCLIND (CYCD) 2 and 3, and PROLIFERATING CELL NUCLEAR ANTIGEN (PCNA), as well as elevated expression levels of *E2 PROMOTER BINDING FACTOR* (*E2F*), E2F-regulated genes, *RETINOBLASTOMA* (*RB*), and *KNOTTED1* (*KN1*)-type homeobox genes (Ahn et al. 2004). It implies that DEK1 might act as a key regulator of cell proliferation and differentiation by controlling CYCD/RB pathways.

Recently, it was shown that CWD induced by CBI inhibits its cell cycle progression in an osmo-sensitive manner (Gigli-Bisceglia et al. 2018). The inhibition involves NITRATE REDUCTASE 1 (NIA1)- and 2 (NIA2)-mediated processes and apparently enhanced degradation of iP-type cytokinins. Results from gene expression analysis suggest that cytokinin degrading enzymes encoded by *CYTOKININ OXIDASE/DEHYDROGENASE 2* and *3* (*CKX2* and *CKX3*) and regulated by NIA1/2 are responsible (Gigli-Bisceglia et al. 2018). Analysis of cell cycle regulators suggested that the changes in iP-type cytokinins may influence cell cycle progression by modulating the activity of CYCD3;1, an established cell cycle regulator. Interestingly, the alterations of the cell cycle are not mediated by THE1, based on gene expression data for the B-type cyclins CYCLINB1;1 and CYCD3;1 (Gigli-Bisceglia et al. 2018). The results support the notion that THE1-dependent and THE1-independent CWI pathways exist. Intriguingly, a recent study investigating the impact of wounding on cell cycle progression discovered a signalling cascade in *Arabidopsis* and *Rosa rugosa* responsible for mediating the impact, which involves *AtMYB46* and *RrMYB18* (Shi et al. 2021).

The role of CWI in gametophytic development has also been investigated. FER has been shown to ensure correct sperm delivery and block polyspermy in *Arabidopsis* (Li, Wu, and Cheung 2016; Duan et al. 2014). As the first pollen tube reaches the ovule, FER initiates the production of a ROS peak through an RHO GTPase-based signalling process at the filiform apparatus (Duan et al. 2010; Punwani et al. 2007), a thickened cell wall region located at the entrance of the female gametophyte (Li et al. 2016). The local ROS peak causes CWD, weakening the pollen tube walls, inducing pollen tube rupture and sperm release (Duan et al. 2014). Additionally, FER also maintains pectin in a de-esterified state at the filiform apparatus, which contributes later to mediating the accumulation of nitric oxide after the arrival of the first pollen tube (Duan et al. 2020). Nitric oxide leads to modifications of the LURE1 chemoattractant in two positions (cys17 and cys84). These modifications prevent LURE1 from acting as a chemoattractant and interacting with its receptor, thus preventing multiple fertilisation of the female gametophyte by late-arriving sperms (Duan et al. 2020).

CWI MAINTENANCE DURING EXPOSURE TO ABIOTIC STRESS

Research into the processes regulating plant adaptation to abiotic stress has traditionally focused on physiological and general stress signalling mechanisms. While transcriptomics-, proteomics-, and metabolomics-based approaches have generated a more global perspective on processes taking place in plants during exposure to abiotic stress, knowledge regarding adaptive changes in cell wall metabolism and a possible role of CWI maintenance in this context is still limited (Novaković et al. 2018; Wang et al. 2016). Here, the knowledge available regarding the impact of abiotic stress on the cell wall and the relevance of the CWI maintenance mechanism will be discussed.

Water Stress

Climate change exposes plants more frequently to extreme conditions, such as droughts or floods, which lead to physiological changes including reduced leaf water potential and turgor pressure, impaired stomata closure, and reduced growth (Farooq et al. 2009). Although investigating the nature of cell wall structure and composition changes induced by water stress is still limited by imaging and biomechanical tools and techniques available, exciting insights have been generated (Rui and Dinneny 2020). A study using six cultivars of the common bean *Phaseolus vulgaris* L. detected a decreased elasticity modulus of the cell walls after exposure to drought, indicating that plasticity had changed (Martínez et al. 2007). Furthermore, reduced wall thickness has also been observed in Black Spruce (*Picea mariana*) saplings (Balducci et al. 2015). Results from studies using *Arabidopsis*, tobacco, and wheat plants show altered cellulose content after drought stress (Piro et al. 2003; Bray 2004; Iraki et al. 1989). Interestingly, prolonged drought stress can lead to the incorporation of lignin to avoid water loss. Proteomic studies of the elongation zone of rice roots, detected an increased abundance of cell wall loosening enzymes within 16 h of exposure to drought, followed by increased lignin biosynthesis after 48 h (Yang et al. 2006). These observations indicate that cell wall composition and structure are actively adapted to drought stress and imply the possible existence of a regulatory mechanism responsible for these adaptive modifications.

Similar observations have been made in plants exposed to flooding stress. Physiologically, they display reduced hypocotyl and root elongation, as well as impaired lateral root development due to sustained submergence (Le Gall et al. 2015). However, some plants show enhanced stem elongation instead, which contributes to maintaining the apical part above water, suggesting there may be species or context-dependent factors regulating growth adaptation (Cho and Kende 1997; Vreeburg et al. 2005; Voesenek et al. 2006). A recent study of submerged *Arabidopsis* found enhanced ethylene-based signalling induces expression of MICROTUBULE-DESTABILIZING PROTEIN 60 (MDP60), resulting in altered microtubule organisation and increased hypocotyl elongation (Wang et al. 2020). Flooding also decreases pectin content, cell wall sugar content, and lignin deposition (Oourne et al. 2009; Vitorino et al. 2001; Komatsu et al. 2010). In the rosette stage of *Arabidopsis* plants, the NAC transcription factor SPEEDY HYPONASTIC GROWTH (SHYG) and ethylene-based signalling processes regulate the rapid upward (hyponastic growth) leaf movement during waterlogging (Rauf et al. 2013). The available evidence also suggested that SHYG acts upstream of *EXPANSINs* (*EXPs*) and *XYLOGLUCAN ENDOTRANSGLYCOSYLASE/HYDRO-LASE (XTHs)*, contributing to cell wall remodelling (Rauf et al. 2013).

Since both water deficit and waterlogging can cause profound changes in osmolarity, osmo-sensing could play a pivotal role in modifying cell wall composition and structure in response to water stress. In addition to the sensors discussed previously in sub-chapter 3, the Ca^{2+}-permeable channel INDUCED CA^{2+} INCREASE 1 (OSCA1) could be relevant since it seems to act as a hyper-osmotic stress sensor (Yuan et al. 2014). OSCA1 is located at the plasma membrane and enables calcium influx upon sensing changes in plasma membrane tension due to changes in osmolarity (Zhang et al. 2021; Maity et al. 2019; Liu et al. 2018). Increased $[Ca^{2+}]_{cyt}$ concentration then triggers expression of stress-responsive genes via Calcium-dependent protein kinase (CPK)-based signalling processes (Rauf et al. 2013). A study of the loss of function mutant *osca1* found that root growth is hypersensitive to osmotic stress (Yuan et al. 2014). While the results illustrate the importance of OSCA1 in maintaining growth under osmotic stress conditions, it remains to be determined if there are effects on cell wall composition and structure.

Direct evidence implicating the CWI maintenance mechanism in modifications of cell wall metabolism in response to drought or waterlogging stress is limited. Cell wall stiffness in *Arabidopsis* seedling roots is reduced in response to hyper-osmotic stress, and THE1 has an indirect effect on this process (Bacete et al. 2021). Interestingly, abscisic acid (ABA) production induced by drought stress requires intact cell walls in *Arabidopsis* and *Pisum sativum* mesophyll cells, while THE1 is a negative regulator of ABA production in *Arabidopsis* seedlings. While these results implicate THE1-mediated CWI maintenance in the regulation of cell wall stiffness and ABA production upon hyper-osmotic stress, the precise molecular mechanism responsible remains to be determined.

Salt Stress

Soil salinity has a huge impact on plant productivity, which is especially important for agricultural crops. It alters the osmotic pressure of soils and interferes with nutrient uptake, causing plants to have stunted growth, chlorosis, and necrosis of leaves due to ion (Na^{2+}) toxicity (Machado and Serralheiro 2017; Shannon and Grieve 1998; Zhu 2002). Physiologically, plants experiencing salinity stress display defects similar to water-stressed plants, such as reduced wall extensibility (Uddin et al. 2014). Established salt-induced stress responses require the glucuronosyltransferase MONOCATION-INDUCED $[Ca^{2+}]_{cyt}$– INDUCED 1 (MOCA1), which acts as a salt-sensor (Jiang et al. 2019). MOCA1 glycosylates the glycosyl inositol phosphorylceramide (GIPC) sphingolipids, allowing GIPCs to enable Ca^{2+} influx into the cytosol. Elevated $[Ca^{2+}]_{cyt}$ concentration is detected by SOS3 and SOS3-LIKE CALCIUM BINDING PROTEIN8/CALCINEURIN B-LIKE PROTEIN10 (SCaBP8/CBL10) and triggers the activation of the serine/threonine protein kinase SOS2 located in the plasma membrane (Halfter et al. 2000; Quan et al. 2007;

Lin et al. 2009). SOS2 then activates the Na^+/H^+ antiporter SOS1, which facilitates NA^+ efflux to maintain turgor pressure (Lin et al. 2009; Qiu et al. 2002; Shi et al. 2000).

Salinity stress also affects cell wall composition and structure. These effects include reduced cellulose content, impaired pectin crosslinking due to the loss of Ca^{2+}, and induced lignin deposition (Feng et al. 2018; Kesten et al. 2019; Chun et al. 2019; Endler et al. 2015). As CESAs move along cortical microtubules at the cell surface during cellulose biosynthesis with the help of CELLULOSE SYNTHASE INTERACTING 1 (CSI1), cellulose deficiency in plants exposed to salt stress occurs due to the depolymerisation of the cortical microtubules (Endler et al. 2015; Paredez, Somerville, and Ehrhardt 2006; Gu et al. 2010). SPIRAL1 (SPR1) is a microtubule-stabilising protein. Under the high salt condition, SPR1 is degraded by the 26S proteosome, resulting in microtubule depolymerisation, which leads to impaired cellulose biosynthesis (Wang et al. 2011). Additionally, atomic force microscopy (AFM) studies of elongation zone at the root tip region show that the cell wall undergoes softening upon exposure to high salinity (Feng et al. 2018). This occurs due to Na^+ competing with and displacing pectin-bound Ca^{2+}, disrupting the Ca^{2+}-crosslinked pectin network (Feng et al. 2018). The cell walls of wild-type plant roots manage to recover stiffness during the growth recovery phase after experiencing salinity stress and continue to grow. In contrast, FER loss-function seedlings fail to do so. Detailed studies of the *fer* seedlings implicate FER in monitoring pectin crosslinking and initiating rapid increases of $[Ca^{2+}]_{cyt}$ to maintain CWI and prevent cells from bursting (Feng et al. 2018).

Recent studies proposed that callose deposition forms part of the response to salt stress. Under high salinity, Ca^{2+} influx initiates the production of phosphatidic acid, which alters plasma membrane properties, leading to an accumulation of the CYS-RICH RECEPTOR-LIKE KINASE (CRK2) at plasmodesmata. CRK2 then interacts with CALLOSE SYNTHASE 1 (CALS1) to promote callose deposition, which in turn leads to enhanced salt tolerance (Hunter et al. 2019). These results suggest that CWI maintenance is also contributing to cell wall adaptation in response to salt stress but also indicate how little is known about the mode of action of the mechanism.

Temperature Stress

Plants are subject to daily temperature changes, with climate change leading to more dramatic temperature fluctuations. Increased or decreased temperatures beyond a threshold cause heat stress and irreversible damage. Examples of heat stress damage include leaf etiolation, wilting, and reduced crop yields (Chen et al. 2014; Fahad et al. 2017). Physiological effects observed in plants exposed to high temperature are over-production of ROS causing oxidative damage, less-elongated palisade cells with impaired adhesion, and structural damage of chloroplast (Lima et al. 2013). Meanwhile, cold stress causes altered leaf expansion, wilting, chlorosis, and rupture of cells due to the formation of ice crystals in the cell wall and plasma membrane, which leads to necrosis (Pearce 1999; Mahajan and Tuteja 2005; Sanghera et al. 2011; Le Gall et al. 2015).

Cell wall composition and structure are also affected by temperature. Studies of cold acclimation detected increased cell wall thickness and modified composition (Kubacka-Zębalska and Kacperska 1999; Solecka et al. 2008; Weiser et al. 1990; Panter et al. 2020; Tanino et al. 2013; Scholz et al. 2012). These were correlated with changes in biomechanical properties (such as wall stiffness) in several plant species, leading to increased wall stiffness to prevent CWD during freezing. Pectin seems to be one of the wall components that are most responsive to cold stimuli, since increased pectin content was observed in oilseed rape, pea, and bromeliad after cold exposure (Solecka et al. 2008; Weiser et al. 1990; Kubacka-Zębalska and Kacperska 1999; Carvalho et al. 2013). Results from studies of coffee (*Coffea arabica* L.) plants subjected to heat stress indicate decreased pectin content and increased hemicellulose (Lima et al. 2013). When exposed to elevated temperature, PME activity is elevated and triggers the removal of Ca^{2+} bridges from HG, resulting in increased levels of de-methylesterified pectin and apoplastic Ca^{2+} levels (Wu et al. 2010). Elevated Ca^{2+} levels induce the expression of *HEAT-SHOCK PROTEIN* (*HSP*) genes (Gong et al. 1998; Saidi et al. 2009; Zhang et al. 2009). While evidence regarding the impact of temperature stress on cell wall composition and structure is accumulating, knowledge regarding the regulatory mechanisms and a possible specific role for the CWI maintenance mechanism is very limited, and further studies are required.

CONCLUDING REMARKS

Plant cell walls are constantly exposed to a wide range of external and internal stresses, impairing their integrity. Therefore, the CWI maintenance mechanism plays a critical role. Several CWI sensors have been characterised, and their mode of actions can be classified into those which detect alterations in cell wall properties and those which sense changes in plasma membrane tension (Figure 18.2). Although pronounced progress has been made in recent years with respect to understanding the mode of action of the mechanism, it is also evident that there are still significant gaps in our knowledge. CWI monitoring is a very complex system and to date, the majority of the intracellular signalling elements involved are still unknown. In addition, the detailed visualisation of how cell wall polymers change upon exposure to stress and how changes are being perceived with the necessary spatio-temporal resolution by the cell require the development of more advanced *in vivo* imaging tools and techniques. Furthermore, most of the research into the mode of action of the CWI maintenance mechanism has been performed using dicot plant models, such as *Arabidopsis*, and is

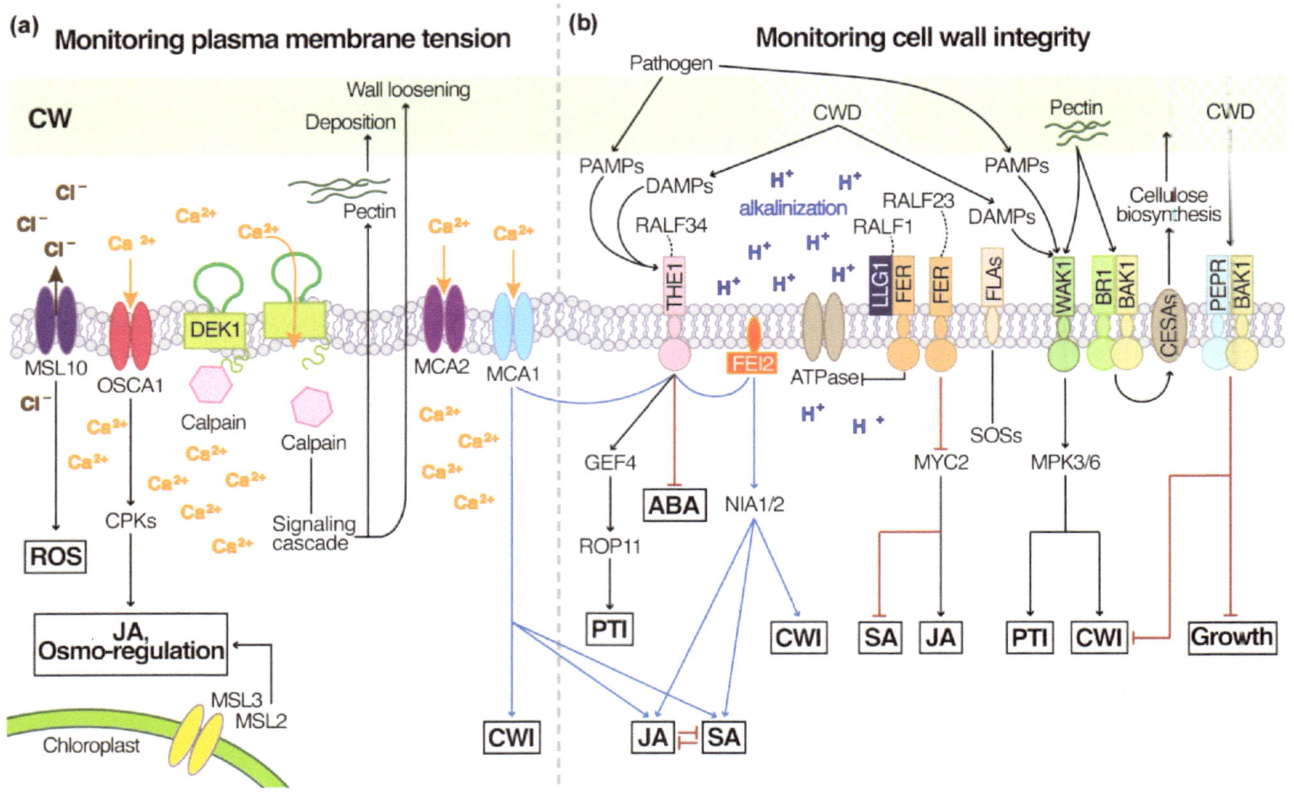

FIGURE 18.2 Overview of the components currently implicated in plant CWI maintenance. (A) CWI ion channels (e.g., MCA1) detect changes in plasma membrane tension. Altered membrane tension can be caused by abiotic stresses such as water deficit, flooding, soil salinity, temperature and osmotic or mechanical stress, leading to alterations in turgor pressure. Activated CWI maintenance channels directly (or indirectly) facilitate Ca^{2+} transport into the cytoplasm, triggering downstream responses including production of ROS and JA as well as changes in turgor, wall composition, and structure, which in turn modify the mechanical characteristics of the walls. (B) CWD can arise during exposure to abiotic and biotic stress as well as during growth and development. CWD impairs crosslinking between wall components or may cause release of DAMPs. The signals generated lead to activation of plasma membrane-located CWI monitoring sensors, which in turn activate downstream responses including phytohormone production and changes in cellular metabolism.

focused on the primary cell wall. Simultaneously the lack of knowledge regarding CWI maintenance in crop species and relevance of the cell wall in monocot species for the response to stress has become more recognised (Gigli-Bisceglia et al. 2020). The latter is exemplified by a recent study on the rice leaf-tip necrosis mutant (*ltn-212*), which found that a damaged CW leads to production of a ROS burst and expression of PR genes (Cao et al. 2021). *POLYGALACTURONASE 1* (*OsPG1*) was shown to be responsible for the necrotic leaf tips, suggesting *OsPG1* is involved in rice defence responses via CW remodelling (Cao et al. 2021). Therefore, there are exciting opportunities to perform cell wall and CWI maintenance research in the monocots model. It is also important to bear in mind that the potential of using targeted cell wall modification to improve crop performance has not been fully realised yet. Using knowledge regarding the mode of action of the CWI maintenance mechanism could create the opportunities to utilise them productively.

REFERENCES

Ahn J W, Kim M, Lim J H, Kim G T, & Pai H S (2004) Phytocalpain controls the proliferation and differentiation fates of cells in plant organ development. *The Plant Journal* 38(6):969–981. https://doi.org/10.1111/j.1365-313X.2004.02102.x

Amanda D, Doblin M S, Galletti R, Bacic A, Ingram G C, & Johnson K L (2016) DEFECTIVE KERNEL1 (DEK1) regulates cell walls in the leaf epidermis. *Plant Physiology* 172(4):2204–2218. https://doi.org/10.1104/pp.16.01401

Amanda D, Doblin M S, Galletti R, Bacic A, Ingram G C, & Johnson K L (2017) Regulation of cell wall genes in response to DEFECTIVE KERNEL1 (DEK1)-induced cell wall changes. *Plant Signaling & Behavior* 12(8):e1345405. https://doi.org/10.1080/15592324.2017.1345405

Amanda D, Doblin M S, MacMillan C P, Galletti R, Golz J F, Bacic A, Ingram G C, & Johnson K L (2017) *Arabidopsis* DEFECTIVE KERNEL1 regulates cell wall composition and axial growth in the inflorescence stem. *Plant Direct* 1(6):e00027. https://doi.org/10.1002/pld3.27

Anderson C M, Wagner T A, Perret M, He Z H, He D, & Kohorn B D (2001) WAKs: Cell wall-associated kinases linking the cytoplasm to the extracellular matrix. *Plant Molecular Biology* 47(1-2):197–206. https://doi.org/10.1023/A:1010691701578

Ashraf M, Foolad M R (2007) Roles of glycine betaine and proline in improving plant abiotic stress resistance *Environmental and Experimental Botany* 59:206–216. https://doi.org/10.1016/j.envexpbot.2005.12.006

Bacete L, Mélida H, Miedes E, & Molina A (2018) Plant cell wall-mediated immunity: cell wall changes trigger disease resistance responses. *The Plant Journal* 93(4):614–636. https://doi.org/10.1111/tpj.13807

Bacete L, Schulz J, Engelsdorf T, Bartosova Z, Vaahtera L, Yan G, Gerhold J, Tichá T, Øvstebø C, Gigli-Bisceglia N, Johannessen-Starheim S, Margueritat J, Kollist H, Dehoux T, McAdam S A M, & Hamann T (2021) THESEUS1 modulates cell wall stiffness and abscisic acid production in *Arabidopsis thaliana. PNAS* 119(1):e2119258119. *https://*doi.org/10.1101/2021.07.22.453418

Bai L, Zhang G, Zhou Y, Zhang Z, Wang W, Du Y, Wu Z, & Song C P (2009) Plasma membrane-associated proline-rich extensin-like receptor kinase 4 a novel regulator of Ca^{2+} signalling is required for abscisic acid responses in *Arabidopsis thaliana. The Plant Journal* 60(2):314–327. https://doi.org/10.1111/j.1365-313X.2009.03956.x

Balducci L, Deslauriers A, Giovannelli A, Beaulieu M, Delzon S, Rossi S, & Rathgeber C B (2015) How do drought and warming influence survival and wood traits of *Picea mariana* saplings?. *Journal of Experimental Botany* 66(1): 377–389.https://doi.org/10.1093/jxb/eru431

Basu D, Wang W, Ma S, DeBrosse T, Poirier E, Emch K, Soukup E, Tian L, & Showalter A M (2015) Two hydroxyproline galactosyltransferases GALT5 and GALT2 function in arabinogalactan-protein glycosylation growth and development in *Arabidopsis. PLoS One* 10:e0125624–e24. https://doi.org/10.1371/journal.pone.0125624

Basu D, & Haswell E S (2017) Plant mechanosensitive ion channels: an ocean of possibilities. *Current Opinion in Plant Biology* 40:43–48. https://doi.org/10.1016/j.pbi.2017.07.002

Basu D, & Haswell E S (2020) The mechanosensitive ion channel MSL10 potentiates responses to cell swelling in *Arabidopsis* seedlings. *Current Biology* 30(14):2716-2728.e2716. https://doi.org/https://doi.org/10.1016/j.cub.2020.05.015

Becraft P W, Li K, Dey N, & Asuncion-Crabb Y (2002) The maize *DEK1* gene functions in embryonic pattern formation and cell fate specification. *Development* 129(22):5217–5225. https://doi.org/10.1242/dev.129.22.5217

Bellincampi D, Dipierro N, Salvi G, Cervone F, & De Lorenzo G (2000) Extracellular H_2O_2 induced by oligogalacturonides is not involved in the inhibition of the auxin-regulated *rolB* gene expression in tobacco leaf explants. *Plant Physiology* 122(4):1379–1386. https://doi.org/10.1104/pp.122.4.1379

Boisson-Dernier A, Roy S, Kritsas K, Grobei M A, Jaciubek M, Schroeder J I, & Grossniklaus U (2009) Disruption of the pollen-expressed FERONIA homologs ANXUR1 and ANXUR2 triggers pollen tube discharge. *Development* 136(19):3279–3288. https://doi.org/10.1242/dev.040071

Boisson-Dernier, A, Kessler S, & Grossniklaus U (2011) The walls have ears: the role of plant CrRLK1Ls in sensing and transducing extracellular signals. *Journal of Experimental Botany* 62(5):1581. https://doi.org/10.1093/jxb/erq445

Boisson-Dernier A, Lituiev D S, Nestorova A, Franck C M, Thirugnanarajah S, & Grossniklaus U (2013) ANXUR receptor-like kinases coordinate cell wall integrity with growth at the pollen tube tip via NADPH oxidases. *PLoS Biology* 11(11):e1001719. https://doi.org/10.1371/journal.pbio.1001719

Bolwell G P, Bindschedler L V, Blee K A, Butt V S, Davies D R, Gardner S L, Gerrish C, & Minibayeva F (2002) The apoplastic oxidative burst in response to biotic stress in plants: a three-component system. *Journal of Experimental Botany* 53(372):1367–1376. https://doi.org/10.1093/jexbot/53.372.1367

Braam J (1992) Regulated expression of the calmodulin-related TCH genes in cultured *Arabidopsis* cells: induction by calcium and heat shock. *PNAS* 89(8):3213–3216. https://doi.org/10.1073/pnas.89.8.3213

Braam J (2005) In touch: Plant responses to mechanical stimuli. *New Phytology* 165(2):373–389. https://doi.org/10.1111/j.1469-8137.2004.01263.x

Bradley D J, Kjellbom P, & Lamb C J (1992) Elicitor- and wound-induced oxidative cross-linking of a proline-rich plant cell wall protein: A novel rapid defense response. *Cell* 70:21–30. https://doi.org/10.1016/0092-8674(92)90530-P

Brady J D, Fry S C (1997) Formation of Di-Isodityrosine and loss of isodityrosine in the cell walls of tomato cell-suspension cultures treated with fungal elicitors or H_2O_2. *Plant Physiology* 115:87–92. https://doi.org/10.1104/pp.115.1.87

Bray E A (2004) Genes commonly regulated by water-deficit stress in *Arabidopsis thaliana. Journal of Experimental Botany* 55(407):2331–2341. https://doi.org/10.1093/jxb/erh270

Cabib E, Roh D, Schmidt M, Crotti L B, & Varma A (2001) The yeast cell wall and septum as paradigms of cell growth and morphogenesis *Journal of Biological Chemistry* 276:19679–19682. https://doi.org/10.1074/jbc.R000031200

Cagnola J I, Dumont de Chassart G J, Ibarra S E, Chimenti C, Ricardi M M, Delzer B, Ghiglione H, Zhu T, Otegui M E, Estevez J M, & Casal J J (2018) Reduced expression of selected *FASCICLIN-LIKE ARABINOGALACTAN PROTEIN* genes associates with the abortion of kernels in field crops of *Zea mays* (maize) and of *Arabidopsis* seeds. *Plant, Cell & Environment* 41(3):661–674. https://doi.org/10.1111/pce.13136

Campargue C, Lafitte C, Esquerré-Tugayé M-T, & Mazau D (1998) Analysis of hydroxyproline and hydroxyproline-arabinosides of plant origin by high-performance anion-exchange chromatography-pulsed amperometric detection. *Analytical Biochemistry* 257:20–25. https://doi.org/10.1006/abio.1997.2526

Campos W F, Dressano K, Ceciliato P H O, Guerrero-Abad J C, Silva A L, Fiori C S, Morato do Canto A, Bergonci T, Claus L A N, Silva-Filho M C, & Moura D S (2018) *Arabidopsis thaliana* rapid alkalinization factor 1-mediated root growth inhibition is dependent on calmodulin-like protein 38. *The Journal of biological chemistry* 293(6):2159–2171. https://doi.org/10.1074/jbc.M117.808881

Cano-Delgado A, Metzlaff K, & Bevan MW (2000) The *eli1* mutation reveals a link between cell expansion and secondary cell wall formation. *Arabidopsis thaliana. Development* 127(15):3395–405. https://doi.org/10.1242/dev.127.15.3395.

Caño-Delgado A, Penfield S, Smith C, Catley M, & Bevan M (2003) Reduced cellulose synthesis invokes lignification and defense responses in *Arabidopsis thaliana. The Plant Journal* 34(3):351–362. https://doi.org/10.1046/j.1365-313X.2003.01729.x

Canut H, Carrasco A, Galaud J P, Cassan C, Bouyssou H, Vita N, Ferrara P, & Pont-Lezica R (1998) High affinity RGD-binding sites at the plasma membrane of *Arabidopsis thaliana* links the cell wall. *The Plant Journal* 16(1):63–71. https://doi.org/10.1046/j.1365-313x.1998.00276.x

Cao Y, Zhang Y, Chen Y, Yu N, Liaqat S, Wu W, Chen D, Cheng S, Wei X, Cao L, Zhang Y, & Liu Q (2021) *OsPG1* encodes a polygalacturonase that determines cell wall architecture and affects resistance to bacterial blight pathogen in rice. *Rice* 14:36. https://doi.org/10.1186/s12284-021-00478-9

Cavalier D M, Lerouxel O, Neumetzler L, Yamauchi K, Reinecke A, Freshour G, Zabotina O A, Hahn M G, Burgert I, Pauly M, Raikhel N V, & Keegstra K (2008) Disrupting two *Arabidopsis thaliana* xylosyltransferase genes results in plants deficient in xyloglucan a major primary cell wall component. *The Plant Cell* 20(6):1519–1537. https://doi.org/10.1105/tpc.108.059873

Carvalho C P, Hayashi A H, Braga M R, & Nievola C C (2013) Biochemical and anatomical responses related to the in vitro survival of the tropical bromeliad *Nidularium minutum* to low temperatures. *Plant Physiology and Biochemistry* 71:144–154. https://doi.org/10.1016/j.plaphy.2013.07.005

Chen W L, Yang W J, Lo H F, & Yeh D M (2014) Physiology anatomy and cell membrane thermostability selection of leafy radish (*Raphanus sativus* var *oleiformis* Pers.) with different tolerance under heat stress. *Scientia horticulturae* 179:367–375. https://doi.org/10.1016/j.scienta.2014.10.003

Chen Y, Dong W, Tan L, Held M A, & Kieliszewski M J (2015) Arabinosylation plays a crucial role in extensin cross-linking. *In vitro Biochemistry insights* 8:1–13. https://doi.org/10.4137/BCI.S31353

Cho H T, Kende H (1997) Expression of expansin genes is correlated with growth in deepwater rice. *The Plant Cell* 9(9):1661–1671. https://doi.org/10.1105/tpc.9.9.1661

Chun H J, Baek D, Cho H M, Lee S H, Jin B J, Yun D J, Hong Y S, & Kim M C (2019) Lignin biosynthesis genes play critical roles in the adaptation of *Arabidopsis* plants to high-salt stress. *Plant Signaling & Behavior* 14(8):1625697. https://doi.org/10.1080/15592324.2019.1625697

Claverie J, Balacey S, Lemaître-Guillier C, Brulé D, Chiltz A, Granet L, Noirot E, Daire X, Darblade B, Héloir M C, & Poinssot B (2018) The cell wall-derived xyloglucan is a new DAMP triggering plant immunity in *Vitis vinifera* and *Arabidopsis thaliana*. *Frontiers in Plant Science* 9:1725. https://doi.org/10.3389/fpls.2018.01725

Cornblatt C.S, Shih H-W, & Monshausen G B (2021) Stimulus-specific processing of the plasma membrane receptor-like kinase FERONIA in *Arabidopsis thaliana*. *BioRxiv*: 2021.10.08.463713. https://doi.org/10.1101/2021.10.08.463713

Cosgrove D J, Hedrich R (1991) Stretch-activated chloride potassium and calcium channels coexisting in plasma membranes of guard cells of *Vicia faba* L. *Planta* 186(1):143–153. https://doi.org/10.1007/BF00201510

Cosgrove D J (2015) Plant expansins: diversity and interactions with plant cell walls. *Current Opinion in Plant Biology* 25:162–172. https://doi.org/10.1016/j.pbi.2015.05.014

Cronmiller E, Toor D, Shao N C, Kariyawasam T, Wang M H, & Lee J H (2019) Cell wall integrity signaling regulates cell wall-related gene expression in *Chlamydomonas reinhardtii*. *Scientific Reports* 9(1):12204. https://doi.org/10.1038/s41598-019-48523-4

Decreux A, Messiaen J (2005) Wall-associated kinase WAK1 interacts with cell wall pectins in a calcium-induced conformation. *Plant and Cell Physiology* 46(2):268–278. https://doi.org/10.1093/pcp/pci026

Delgado-Cerezo M, Sánchez-Rodríguez C, Escudero V, Miedes E, Fernández P V, Jordá L, Hernández-Blanco C, Sánchez-Vallet A, Bednarek P, Schulze-Lefert P, Somerville S, Estevez J M, Persson S, & Molina A (2012) *Arabidopsis* heterotrimeric g-protein regulates cell wall defense and resistance to necrotrophic fungi. *Molecular Plant* 5(1):98–114. https://doi.org/https://doi.org/10.1093/mp/ssr082

Demko V, Perroud P F, Johansen W, Delwiche C F, Cooper E D, Remme P, Ako A E, Kugler K G, Mayer K F X, Quatrano R, & Olsen O A (2014) Genetic analysis of DEFECTIVE KERNEL1 loop function in three-dimensional body patterning in *Physcomitrella patens*. *Plant Physiology* 166(2):903–919. https://doi.org/10.1104/pp.114.243758

Denness L, McKenna J F, Segonzac C, Wormit A, Madhou P, Bennett M, Mansfield J, Zipfel C, & Hamann T (2011) Cell wall damage-induced lignin biosynthesis is regulated by a reactive oxygen species- and jasmonic acid-dependent process in *Arabidopsis*. *Plant Physiology* 156(3):1364–1374. https://doi.org/10.1104/pp.111.175737

Ding J P, & Pickard B G (1993) Mechanosensory calcium-selective cation channels in epidermal cells. *The Plant Journal* 3(1):83–110. https://doi.org/10.1111/j.1365-313x.1993.tb00013.x

Dressano K, Ceciliato P H O, Silva A L, Guerrero-Abad J C, Bergonci T, Ortiz-Morea F A, Bürger M, Silva-Filho M C, & Moura D S (2017) BAK1 is involved in AtRALF1-induced inhibition of root cell expansion. *PLoS Genetics* 13(10):e1007053. https://doi.org/10.1371/journal.pgen.1007053

Du C, Li X, Chen J, Chen W, Li B, Li C, Wang L, Li J, Zhao X, Lin J, Liu X, Luan S, & Yu F (2016) Receptor kinase complex transmits RALF peptide signal to inhibit root growth in *Arabidopsis*. *PNAS* 113(51):E8326–e8334. https://doi.org/10.1073/pnas.1609626113

Duan Q, Kita D, Johnson E A, Aggarwal M, Gates L, Wu H M, & Cheung A Y (2014) Reactive oxygen species mediate pollen tube rupture to release sperm for fertilization in *Arabidopsis*. *Nature Communications* 5:3129. https://doi.org/10.1038/ncomms4129

Duan Q, Kita D, Li C, Cheung A Y, & Wu H M (2010) FERONIA receptor-like kinase regulates RHO GTPase signaling of root hair development. *PNAS* 107(41):17821–17826. https://doi.org/10.1073/pnas.1005366107

Duan Q, Liu M C J, Kita D, Jordan S S, Yeh F L J, Yvon R, Carpenter H, Federico A N, Garcia-Valencia L E, Eyles S J, Wang C S, Wu H M, & Cheung A Y (2020) FERONIA controls pectin- and nitric oxide-mediated male-female interaction. *Nature* 579(7800):561–566. https://doi.org/10.1038/s41586-020-2106-2

Ellis C, & Turner JG (2001) The Arabidopsis mutant *cev1* has constitutively active jasmonate and ethylene signal pathways and enhanced resistance to pathogens. *The Plant Cell* 13(5):1025–1033. https://doi.org/10.1105/tpc.13.5.1025

Ellis C, Karafyllidis I, Wasternack C, & Turner J G (2002) The *Arabidopsis* mutant *cev1* links cell wall signaling to jasmonate and ethylene responses. *The Plant Cell* 14(7):1557–1566. https://doi.org/10.1105/tpc.002022

Elorza M V, Rico H, & Sentandreu R (1983) Calcofluor white alters the assembly of chitin fibrils in *Saccharomyces cerevisiae* and *Candida albicans* cells. *The Journal of General Microbiology* 129:1577–1582. https://doi.org/10.1099/00221287-129-5-1577

Endler A, Kesten C, Schneider R, Zhang Y, Ivakov A, Froehlich A, Funke N, & Persson S (2015) A mechanism for sustained cellulose synthesis during salt stress. *Cell* 162(6):1353–1364. https://doi.org/10.1016/j.cell.2015.08.028

Engelsdorf T, Gigli-Bisceglia N, Veerabagu M, McKenna J F, Vaahtera L, Augstein F, Van der Does D, Zipfel C, & Hamann T (2018) The plant cell wall integrity maintenance and immune signaling systems cooperate to control stress responses in *Arabidopsis thaliana. Science Signaling* 11(536). https://doi.org/10.1126/scisignal.aao3070

Fagard M, Desnos T, Desprez T, Goubet F, Refregier G, Mouille G, McCann M, Rayon C, Vernhettes S, & Höfte H (2000) PROCUSTE1 encodes a cellulose synthase required for normal cell elongation specifically in roots and dark-grown hypocotyls of *Arabidopsis. The Plant Cell* 12(12):2409–2423. https://doi.org/10.1105/tpc.12.12.2409

Fahad S, Bajwa A A, Nazir U, Anjum S A, Farooq A, Zohaib A, Sadia S, Nasim W, Adkins S, Saud S, Ihsan M Z, Alharby H, Wu C, Wang D, & Huang J (2017) Crop production under drought and heat stress: plant responses and management options. *Frontiers in Plant Science* 8(:1147). https://doi.org/10.3389/fpls.2017.01147

Falke L C, Edwards K L, Pickard B G, & Misler S (1988) A stretch-activated anion channel in tobacco protoplasts. *FEBS Letters* 237(1-2):141–144. https://doi.org/10.1016/0014-5793(88)80188-1

Farooq M, Wahid A, Kobayashi N, Fujita D, & Basra S M (2009) Plant drought stress: Effects mechanisms and management. *Agronomy for sustainable development* 29:185–212. https://doi.org/10.1051/agro:2008021

Feng Z T, Deng Y Q, Fan H, Sun Q J, Sui N, & Wang B S (2014) Effects of NaCl stress on the growth and photosynthetic characteristics of *Ulmus pumila* L seedlings in sand culture. *Photosynthetica* 52:313–320. https://doi.org/10.1007/s11099-014-0032-y

Feng W, Kita D, Peaucelle A, Cartwright H N, Doan V, Duan Q, Liu M C, Maman J, Steinhorst L, Schmitz-Thom I, Yvon R, Kudla J, Wu H M, Cheung A Y, & Dinneny J R (2018) The FERONIA receptor kinase maintains cell-wall integrity during salt stress through $Ca^{(2+)}$ signaling. *Current Biology* 28(5):666-675.e665. https://doi.org/10.1016/j.cub.2018.01.023

Feng H, Liu C, Fu R, Zhang M, Li H, Shen L, Wei Q, Sun X, Xu L, Ni B, & Li C (2019) LORELEI-LIKE GPI-ANCHORED PROTEINS 2/3 regulate pollen tube growth as chaperones and coreceptors for ANXUR/BUPS receptor kinases in *Arabidopsis. Molecular Plant* 12(12):1612–1623. https://doi.org/10.1016/j.molp.2019.09.004

Ferrari S, Galletti R, Denoux C, De Lorenzo G, Ausubel F M, & Dewdney J (2007) Resistance to *Botrytis cinerea* induced in *Arabidopsis* by elicitors is independent of salicylic acid ethylene or jasmonate signaling but requires *PHYTOALEXIN DEFICIENT3. Plant Physiology* 144(1):367–379. https://doi.org/10.1104/pp.107.095596

Fischer M, Schnell N, Chattaway J, Davies P, Dixon G, & Sanders D (1997) The *Saccharomyces cerevisiae CCH1* gene is involved in calcium influx and mating. *FEBS Letters* 419(2):259–262. https://doi.org/https://doi.org/10.1016/S0014-5793(97)01466-X

Flury P, Klauser D, Schulze B, Boller T, & Bartels S (2013) The anticipation of danger: microbe-associated molecular pattern perception enhances AtPep-triggered oxidative burst. *Plant Physiology* 161(4):2023–2035. https://doi.org/10.1104/pp.113.216077

Fry S C (1982) Isodityrosine a new cross-linking amino acid from plant cell-wall glycoprotein. *The Biochemical journal* 204:449–455. https://doi.org/10.1042/bj2040449

Fujikura U, Elsaesser L, Breuninger H, Sanchez-Rodriguez C, Ivakov A, Laux T, Findlay K, Persson S, & Lenhard M (2014) Atkinesin-13A modulates cell-wall synthesis and cell expansion in *Arabidopsis thaliana* via the THESEUS1 pathway. *PloS Genetics* 10(9):e1004627. https://doi.org/10.1371/journal.pgen.1004627

Gachomo E W, Jno Baptiste L, Kefela T, Saidel W M, & Kotchoni S O (2014) The *Arabidopsis CURVY1 (CVY1)* gene encoding a novel receptor-like protein kinase regulates cell morphogenesis flowering time and seed production. *BMC Plant Biology* 14(:221). https://doi.org/10.1186/s12870-014-0221-7

Galindo-Trigo S, Blanco-Touriñán N, DeFalco T A, Wells E S, Gray J E, Zipfel C, & Smith L M (2020) CrRLK1L receptor-like kinases HERK1 and ANJEA are female determinants of pollen tube reception. *EMBO Reports* 21(2):e48466. https://doi.org/10.15252/embr.201948466

Gallego-Giraldo L, Liu C, Pose-Albacete S, Pattathil S, Peralta A G, Young J, Westpheling J, Hahn M G, Rao X, Knox J P, De Meester B, Boerjan W, & Dixon R A (2020) ARABIDOPSIS DEHISCENCE ZONE POLYGALACTURONASE 1 (ADPG1) releases latent defense signals in stems with reduced lignin content. *PNAS* 117(6):3281. https://doi.org/10.1073/pnas.1914422117

Galletti R, Denoux C, Gambetta S, Dewdney J, Ausubel F M, De Lorenzo G, & Ferrari S (2008) The AtrbohD-mediated oxidative burst elicited by oligogalacturonides in *Arabidopsis* is dispensable for the activation of defense responses effective against *Botrytis cinerea. Plant Physiology* 148(3):1695–1706. https://doi.org/10.1104/pp.108.127845

Galletti R, Johnson K L, Scofield S, San-Bento R, Watt A M, Murray J A H, & Ingram G C (2015) DEFECTIVE KERNEL 1 promotes and maintains plant epidermal differentiation. *Development* 142(11):1978–1983. https://doi.org/10.1242/dev.122325

Garrett-Engele P, Moilanen B, & Cyert M S (1995) Calcineurin the Ca^{2+}/calmodulin-dependent protein phosphatase is essential in yeast mutants with cell integrity defects and in mutants that lack a functional vacuolar $H^{(+)}$-ATPase. *Molecular and Cellular Biology* 15(8):4103–4114. https://doi.org/10.1128/mcb.15.8.4103

Garrill A, Tyerman S D, & Findlay G P (1994) Ion channels in the plasma membrane of protoplasts from the halophytic angiosperm *Zostera muelleri. The Journal of Membrane Biology* 142(3):381–393. https://doi.org/10.1007/BF00233443

Ge Z, Bergonci T, Zhao Y, Zou Y, Du S, Liu M C, Luo X, Ruan H, García-Valencia L E, Zhong S, Hou S, Huang Q, Lai L, Moura D S, Gu H, Dong J, Wu H M, Dresselhaus T, Xiao J, Cheung A Y, Qu L J (2017) *Arabidopsis* pollen tube integrity and sperm release are regulated by RALF-mediated signaling. *Science* 358(6370):1596–1600. https://doi.org/10.1126/science.aao3642

Ge Z, Zhao Y, Liu M C, Zhou L Z, Wang L, Zhong S, Hou S, Jiang J, Liu T, Huang Q, Xiao J, Gu H, Wu H M, Dong J, Dresselhaus T, Cheung A Y, & Qu L J (2019) LLG2/3 are co-receptors in

BUPS/ANX-RALF signaling to regulate *Arabidopsis* pollen tube integrity. *Current Biology* 29(19):3256-3265.e3255. https://doi.org/10.1016/j.cub.2019.08.032

Ghatak S, Misra S, Norris R A, Moreno-Rodriguez R A, Hoffman S, Levine R A, Hascall V C, & Markwald R R (2014) Periostin induces intracellular cross-talk between kinases and hyaluronan in atrioventricular valvulogenesis. *Journal of Biological Chemistry* 289(12):8545–8561. https://doi.org/10.1074/jbc.M113.539882

Gigli-Bisceglia N, Engelsdorf T, Strnad M, Vaahtera L, Khan G A, Yamoune A, Alipanah L, Novák O, Persson S, Hejatko J, & Hamann T (2018) Cell wall integrity modulates *Arabidopsis thaliana* cell cycle gene expression in a cytokinin- and nitrate reductase-dependent manner. *Development* 145(19). https://doi.org/10.1242/dev.166678

Gong M, van der Luit A H, Knight M R, & Trewavas A J (1998) Heat-shock-induced changes in intracellular Ca^{2+} level in tobacco seedlings in relation to Thermotolerance1. *Plant Physiology* 116(1):429–437. https://doi.org/10.1104/pp.116.1.429

Gonneau M, Desprez T, Martin M, Doblas V G, Bacete L, Miart F, Sormani R, Hématy K, Renou J, Landrein B, Murphy E, Van De Cotte B, Vernhettes S, De Smet I, & Höfte H (2018) Receptor kinase THESEUS1 is a rapid alkalinization factor 34 receptor in *Arabidopsis. Current Biology* 28(15):2452-2458.e2454. https://doi.org/https://doi.org/10.1016/j.cub.2018.05.075

Gouget A, Senchou V, Govers F, Sanson A, Barre A, Rougé P, Pont-Lezica R, & Canut H (2006) Lectin receptor kinases participate in protein-protein interactions to mediate plasma membrane-cell wall adhesions in *Arabidopsis. Plant Physiology* 140(1):81–90. https://doi.org/10.1104/pp.105.066464

Gramegna G, Modesti V, Savatin D V, Sicilia F, Cervone F, & De Lorenzo G (2016) *GRP-3* and *KAPP* encoding interactors of WAK1 negatively affect defense responses induced by oligogalacturonides and local response to wounding. *Journal of Experimental Botany* 67(6):1715–1729. https://doi.org/10.1093/jxb/erv563

Griffiths J S, Crepeau M J, Ralet M C, Seifert G J, & North H M (2016) Dissecting seed mucilage adherence mediated by FEI2 and SOS5. *Frontiers in Plant Science* 7:1073. https://doi.org/10.3389/fpls.2016.01073

Gu Y, Kaplinsky N, Bringmann M, Cobb A, Carroll A, Sampathkumar A, Baskin T I, Persson S, &and Somerville C R (2010) Identification of a cellulose synthase-associated protein required for cellulose biosynthesis. *PNAS* 107: 12866–71. https://doi.org/10.1073/pnas.1007092107

Guo H, Li L, Ye H, Yu X, Algreen A, & Yin Y (2009) Three related receptor-like kinases are required for optimal cell elongation in *Arabidopsis thaliana. PNAS* 106(18):7648–7653. https://doi.org/10.1073/pnas.0812346106

Guo Y, Jia W, Song J, Wang D, Chen M, &and Wang B (2012) *Thellungilla halophila* is more adaptive to salinity than *Arabidopsis thaliana* at stages of seed germination and seedling establishment *Acta Physiologiae. Plantarum* 34:1287–1294. https://doi.org/10.1007/s11738-012-0925-y

Guo H, Nolan T M, Song G, Liu S, Xie Z, Chen J, Schnable P S, Walley J W, & Yin Y (2018) FERONIA receptor kinase contributes to plant immunity by suppressing jasmonic acid signaling in *Arabidopsis thaliana. Current Biology* 28(20):3316-3324.e3316. https://doi.org/10.1016/j.cub.2018.07.078

Habrant A, Gaillard C, Ralet M C, Lairez D, & Cathala B (2009) Relation between chemical structure and supramolecular organization of synthetic lignin–pectin particles. *Biomacromolecules* 10(11):3151–3156. https://doi.org/10.1021/bm900950r

Halfter U, Ishitani M, & Zhu J K (2000) The *Arabidopsis* SOS2 protein kinase physically interacts with and is activated by the calcium-binding protein SOS3. *PNAS* 97(7):3735–3740. https://doi.org/10.1073/pnas.040577697

Hamann T, Bennett M, Mansfield J, & Somerville C (2009) Identification of cell-wall stress as a hexose-dependent and osmosensitive regulator of plant responses. *The Plant Journal* 57(6):1015–1026. https://doi.org/10.1111/j.1365-313X.2008.03744.x

Hamilton E S, Jensen G S, Maksaev G, Katims A, Sherp A M, & Haswell E S (2015) Mechanosensitive channel MSL8 regulates osmotic forces during pollen hydration and germination. *Science* 350:438–441. https://doi.org/10.1126/science.aac6014

Hander T, Fernández-Fernández Álvaro D, Kumpf Robert P, Willems P, Schatowitz H, Rombaut D, Staes A, Nolf J, Pottie R, Yao P, Gonçalves A, Pavie B, Boller T, Gevaert K, Van Breusegem F, Bartels S, & Stael S (2019) Damage on plants activates Ca^{2+}-dependent metacaspases for release of immunomodulatory peptides. *Science* 363(6433):eaar7486. https://doi.org/10.1126/science.aar7486

Harren K, Tudzynski B (2013) CCH1 and MID1 are functionally required for vegetative growth under low-calcium conditions in the phytopathogenic ascomycete *Botrytis cinerea. Eukaryotic Cell* 12(5):712–724. https://doi.org/10.1128/EC.00338-12

Haruta M, Sabat G, Stecker K, Minkoff B B, & Sussman M R (2014) A peptide hormone and its receptor protein kinase regulate plant cell expansion. *Science* 343(6169):408–411. https://doi.org/10.1126/science.1244454

Haswell E S, Meyerowitz E M (2006) MscS-like proteins control plastid size and shape in *Arabidopsis thaliana. Current Biology* 16(1):1–11. https://doi.org/10.1016/j.cub.2005.11.044

Haswell E S, Peyronnet R, Barbier-Brygoo H, Meyerowitz E M, & Frachisse J M (2008) Two MscS homologs provide mechanosensitive channel activities in the *Arabidopsis* root. *Current Biology* 18(10):730–734. https://doi.org/10.1016/j.cub.2008.04.039

Haswell E S, Verslues P E (2015) The ongoing search for the molecular basis of plant osmosensing. *Journal of General Physiology* 145(5):389–394. https://doi.org/10.1085/jgp.201411295

He Z H, Fujiki M, &and Kohorn B D (1996) A cell wall-associated receptor-like protein kinase. *Journal of Biological Chemistry* 271:19789–1993. https://doi.org/10.1074/jbc.271.33.19789

Hedrich R (2012) Ion channels in plants. *Physiological Reviews* 92(4):1777–1811. https://doi.org/10.1152/physrev.00038.2011

Hejnowicz Z, Rusin A, & Rusin T (2000) Tensile tissue stress affects the orientation of cortical microtubules in the epidermis of sunflower hypocotyl. *Journal of Plant Growth Regulation* 19(1):31–44. https://doi.org/10.1007/s003440000005

Held M A, Tan L, Kamyab A, Hare M, Shpak E, &and Kieliszewski M J (2004) Di-isodityrosine is the intermolecular cross-link of isodityrosine-rich extensin analogs cross-linked *in Vitro. Journal of Biological Chemistry* 279:55474–55482. https://doi.org/10.1074/jbc.M408396200

Hématy K, Sado P E, Van Tuinen A, Rochange S, Desnos T, Balzergue S, Pelletier S, Renou J P, & Höfte H (2007) A Receptor-like kinase mediates the response of *Arabidopsis* cells to the inhibition of cellulose synthesis. *Current Biology* 17(11):922–931. https://doi.org/https://doi.org/10.1016/j.cub.2007.05.018

Hocq L, Pelloux J, & Lefebvre V (2017) Connecting homogalacturonan-type pectin remodeling to acid growth. *Trends in Plant Science* 22(1):20–29. https://doi.org/10.1016/j.tplants.2016.10.009

Hou S, Liu Z, Shen H, & Wu D (2019) Damage-associated molecular pattern-triggered immunity in plants. *Frontiers in Plant Science* 10(646). https://doi.org/10.3389/fpls.2019.00646

Huffaker A, Pearce G, & Ryan C A (2006) An endogenous peptide signal in *Arabidopsis* activates components of the innate immune response. *PNAS* 103(26):10098–10103. https://doi.org/10.1073/pnas.0603727103

Humphrey T V, Bonetta D T, & Goring D R (2007) Sentinels at the wall: Cell wall receptors and sensors. *New Phytology* 176(1):7–21. https://doi.org/10.1111/j.1469-8137.2007.02192.x

Hunter K, Kimura S, Rokka A, Tran H C, Toyota M, Kukkonen J P, &and Wrzaczek M (2019) CRK2 enhances salt tolerance by regulating callose deposition in connection with PLDα1. *Plant Physiology* 180:2004–2021. https://doi.org/10.1104/pp.19.00560

Iida H, Nakamura H, Ono T, Okumura M S, & Anraku Y (1994) MID1 a novel *Saccharomyces cerevisiae* gene encoding a plasma membrane protein is required for Ca^{2+} influx and mating. *Molecular and Cellular Biology* 14(12):8259–8271. https://doi.org/10.1128/mcb.14.12.8259-8271.1994

Iraki N M, Bressan R A, Hasegawa P, & Carpita N C (1989) Alteration of the physical and chemical structure of the primary cell wall of growth-limited plant cells adapted to osmotic stress. *Plant Physiology* 91(1):39–47. https://doi.org/10.1104/pp.91.1.39

Iraki N M, Bressan R A, Hasegawa P M, & Carpita N C (1989) Alteration of the physical and chemical structure of the primary cell wall of growth-limited plant cells adapted to osmotic stress. *Plant Physiology* 91:39–47. https://doi.org/10.1104/pp.91.1.39

Jacoby T, Flanagan H, Faykin A, Seto A G, Mattison C, & Ota I (1997) Two protein-tyrosine phosphatases inactivate the osmotic stress response pathway in yeast by targeting the mitogen-activated protein kinase Hog1. *Journal of Biological Chemistry* 272:17749–17755. https://doi.org/10.1074/jbc.272.28.17749

Jiang Z, Zhou X, Tao M, Yuan F, Liu L, Wu F, Wu X, Xiang Y, Niu Y, Liu F, Li C, Ye R, Byeon B, Xue Y, Zhao H, Wang H N, Crawford B M, Johnson D M, Hu C, Pei C, Zhou W, Swift G B, Zhang H, Vo-Dinh T, Hu Z, Siedow J N, Pei Z M (2019) Plant cell-surface GIPC sphingolipids sense salt to trigger Ca^{2+} influx. *Nature* 572(7769):341–346. https://doi.org/10.1038/s41586-019-1449-z

Johnson K L, Ingram G C (2005) Sending the right signals: Regulating receptor kinase activity. *Current Opinion in Plant Biology* 8(6):648–656. https://doi.org/10.1016/j.pbi.2005.09.007

Johnson K L, Faulkner C, Jeffree C E, & Ingram G C (2008) The phytocalpain DEFECTIVE KERNEL 1 is a novel *Arabidopsis* growth regulator whose activity is regulated by proteolytic processing. *The Plant Cell* 20(10):2619–2630. https://doi.org/10.1105/tpc.108.059964

Johnson J M, Thürich J, Petutschnig E K, Altschmied L, Meichsner D, Sherameti I, Dindas J, Mrozinska A, Paetz C, Scholz S S, Furch A C U, Lipka V, Hedrich R, Schneider B, Svato; A, & Oelmüller R (2018) A Poly(A) ribonuclease controls the cellotriose-based interaction between *Piriformospora indica* and its host *Arabidopsis*. *Plant Physiology* 176(3):2496–2514. https://doi.org/10.1104/pp.17.01423

Jung N U, Giarola V, Chen P, Knox J P, & Bartels D (2019) *Craterostigma plantagineum* cell wall composition is remodelled during desiccation and the glycine-rich protein CpGRP1 interacts with pectins through clustered arginines. *The Plant Journal* 100(4):661–676. https://doi.org/https://doi.org/10.1111/tpj.14479

Kanzaki M, Nagasawa M, Kojima I, Sato C, Naruse K, Sokabe M, & Iida H (1999) Molecular identification of a eukaryotic stretch-activated nonselective cation channel. *Science* 285(5429):882–886. https://doi.org/doi:10.1126/science.285.5429.882

Kapp T G, Rechenmacher F, Neubauer S, Maltsev O V, Cavalcanti-Adam E A, Zarka R, Reuning U, Notni J, Wester H-J, Mas-Moruno C, Spatz J, Geiger B, &and Kessler H (2017) A comprehensive evaluation of the activity and selectivity profile of ligands for RGD-binding integrins. *Scientific Reports* 7:39805. https://doi.org/10.1038/srep39805

Kessler SA, & Grossniklaus U (2011) She´s the boss: signaling in pollen tube reception. *Curr Opin Plant Biol.* 14(5):622-7. https://doi.org/10.1016/j.pbi.2011.07.012

Kesten C, Wallmann A, Schneider R, McFarlane H E, Diehl A, Khan G A, van Rossum B J, Lampugnani E R, Szymanski W G, Cremer N, Schmieder P, Ford K L, Seiter F, Heazlewood J L, Sanchez-Rodriguez C, Oschkinat H, & Persson S (2019) The companion of cellulose synthase 1 confers salt tolerance through a Tau-like mechanism in plants. *Nature Communications* 10(1):857. https://doi.org/10.1038/s41467-019-08780-3

Kikuyama M, & Tazawa M (2001) Mechanosensitive Ca^{2+} release from intracellular stores in *Nitella flexilis*. *Plant and Cell Physiology* 42(4):358–365. https://doi.org/10.1093/pcp/pce043

Kim J E, Kim S J, Lee B H, Park R W, Kim K S, & Kim I S (2000) Identification of motifs for cell adhesion within the repeated domains of transforming growth factor-beta-induced gene betaig-h3. *Journal of Biological Chemistry* 275(40):30907–30915. https://doi.org/10.1074/jbc.M002752200

Kjellbom P, Snogerup L, Stöhr C, Reuzeau C, McCabe P F, & Pennell R I (1997) Oxidative cross-linking of plasma membrane arabinogalactan proteins. *The Plant Journal* 12:1189–1196. https://doi.org/10.1046/j.1365-313x.1997.12051189.x

Kock C, Dufrene Y F, & Heinisch J J (2015) Up against the wall: is yeast cell wall integrity ensured by mechanosensing in plasma membrane microdomains?. *Applied and Environmental Microbiology Journal* 81(3):806–811. https://doi.org/10.1128/aem.03273-14

Kohorn B D, Kobayashi M, Johansen S, Friedman H P, Fischer A, & Byers N (2006) Wall-associated kinase 1 (WAK1) is crosslinked in endomembranes and transport to the cell surface requires correct cell-wall synthesis *Journal of Cell Science* 119(11):2282–2290. https://doi.org/10.1242/jcs.02968

Kohorn B D, Kobayashi M, Johansen S, Riese J, Huang L F, Koch K, Fu S, Dotson A, & Byers N (2006) An *Arabidopsis* cell wall-associated kinase required for invertase activity and cell growth. *The Plant Journal* 46:307–316. https://doi.org/10.1111/j.1365-313X.2006.02695.x

Kohorn B D, Johansen S, Shishido A, Todorova T, Martinez R, Defeo E, & Obregon P (2009) Pectin activation of MAP kinase and gene expression is WAK2 dependent. *The Plant Journal* 60(6):974–982. https://doi.org/10.1111/j.1365-313X.2009.04016.x

Kohorn B D, Kohorn S L, Saba N J, & Martinez V M (2014) Requirement for pectin methyl esterase and preference for fragmented over native pectins for wall-associated kinase-activated EDS1/PAD4-dependent stress response in *Arabidopsis*. *Journal of Biological Chemistry* 289(27):18978–18986. https://doi.org/10.1074/jbc.M114.567545

Kohorn B D (2015) The state of cell wall pectin monitored by wall associated kinases: A model. *Plant Signaling & Behavior* 10(7):e1035854. https://doi.org/10.1080/15592324.2015.1035854

Komatsu S, Kobayashi Y, Nishizawa K, Nanjo Y, & Furukawa K (2010) Comparative proteomics analysis of differentially expressed proteins in soybean cell wall during flooding stress. *Amino Acids* 39(5):1435–1449. https://doi.org/10.1007/s00726-010-0608-1

Kubacka-Zębalska M, & Kacperska A (1999) Low temperature-induced modifications of cell wall content and polysaccharide composition in leaves of winter oilseed rape (*Brassica napus* L var oleifera L.) *Plant Science* 148(1):59–67. https://doi.org/10.1016/S0168-9452(99)00122-3

Kurusu T, Yamanaka T, Nakano M, Takiguchi A, Ogasawara Y, Hayashi T, Iida K, Hanamata S, Shinozaki K, Iida H, & Kuchitsu K (2012) Involvement of the putative Ca^{2+}-permeable mechanosensitive channels NtMCA1 and NtMCA2 in Ca^{2+} uptake Ca^{2+}-dependent cell proliferation and mechanical stress-induced gene expression in tobacco (*Nicotiana tabacum*) BY-2 cells. *Journal of Plant Research* 125(4):555–568. https://doi.org/10.1007/s10265-011-0462-6

Lairez D, Cathala B, Monties B, Bedos-Belval F, Duran H, & Gorrichon L (2005) Aggregation during coniferyl alcohol polymerization in pectin solution: a biomimetic approach of the first steps of lignification. *Biomacromolecules* 6(2):763–774. https://doi.org/10.1021/bm049390y

Lamport D T, Katona L, & Roerig S (1973) Galactosylserine in extensin. *The Biochemical journal* 133:125–132.

Le Gall H, Philippe F, Domon J M, Gillet F, Pelloux J, & Rayon C (2015) Cell wall metabolism in response to abiotic stress. *Plants (Basel)* 4(1):112–166. https://doi.org/10.3390/plants4010112

Levin D E (2011) Regulation of cell wall biogenesis in *Saccharomyces cerevisiae*: the cell wall integrity signaling pathway. *Genetics* 189:1145–1175. https://doi.org/10.1534/genetics.111.128264

Li S, Ault A, Malone C L, Raitt D, Dean S, Johnston L H, Deschenes R J, & Fassler J S (1998) The yeast histidine protein kinase Sln1p mediates phosphotransfer to two response regulators Ssk1p and Skn7p. *The EMBO Journal* 17:6952–6962. https://doi.org/10.1093/emboj/17.23.6952

Li S, Dean S, Li Z, Horecka J, Deschenes R J, & Fassler J S (2002) The eukaryotic two-component histidine kinase Sln1p regulates OCH1 via the transcription factor Skn7p. *Molecular Biology of The Cell* 13:412–424. https://doi.org/10.1091/mbc.01-09-0434

Li C, Yeh F L, Cheung A Y, Duan Q, Kita D, Liu M C, Maman J, Luu E J, Wu B W, Gates L, Jalal M, Kwong A, Carpenter H, & Wu H M (2015) Glycosylphosphatidylinositol-anchored proteins as chaperones and co-receptors for FERONIA receptor kinase signaling in *Arabidopsis*. *Elife* 4:06587. https://doi.org/10.7554/eLife.06587

Li C, Wang L, Cui Y, He L, Qi Y, Zhang J, Lin J, Liao H, Lin Q, Yang T, Yu F, & Liu X (2016) Two *FERONIA-like receptor* (*FLR*) genes are required to maintain architecture fertility and seed yield in rice. *Molecular Breeding* 36(11):151. https://doi.org/10.1007/s11032-016-0580-x

Li C, Wu H M, & Cheung A Y (2016) FERONIA and her pals: functions and mechanisms. *Plant Physiology* 171(4):2379–2392. https://doi.org/10.1104/pp.16.00667

Li E, Wang G, Zhang Y L, Kong Z, & Li S (2020) FERONIA mediates root nutating growth. *The Plant Journal* 104(4):1105–1116. https://doi.org/10.1111/tpj.14984

Lid S E, Gruis D, Jung R, Lorentzen J A, Ananiev E, Chamberlin M, Niu X, Meeley R, Nichols S, & Olsen O A (2002) The *DEFECTIVE KERNEL 1* (*DEK1*) gene required for aleurone cell development in the endosperm of maize grains encodes a membrane protein of the calpain gene superfamily. *PNAS* 99(8):5460–5465. https://doi.org/10.1073/pnas.042098799

Lid S E, Olsen L, Nestestog R, Aukerman M, Brown R C, Lemmon B, Mucha M, Opsahl-Sorteberg H G, & Olsen O A (2005) Mutation in the *Arabidopisis thaliana DEK1* calpain gene perturbs endosperm and embryo development while over-expression affects organ development globally. *Planta* 221(3):339–351. https://doi.org/10.1007/s00425-004-1448-6

Lima R B, dos Santos T B, Vieira L G E, Ferrarese M d L L, Ferrarese-Filho O, Donatti L, Boeger M R T, & Petkowicz C L (2013) Heat stress causes alterations in the cell-wall polymers and anatomy of coffee leaves (*Coffea arabica* L.). *Carbohydrate Polymers* 93(1):135–143. https://doi.org/10.1016/j.carbpol.2012.05.015

Lin H, Yang Y, Quan R, Mendoza I, Wu Y, Du W, Zhao S, Schumaker K S, Pardo J M, & Guo Y (2009) Phosphorylation of SOS3-LIKE CALCIUM BINDING PROTEIN8 by SOS2 protein kinase stabilizes their protein complex and regulates salt tolerance in *Arabidopsis*. *The Plant Cell* 21(5):1607–1619. https://doi.org/10.1105/tpc.109.066217

Lin W, Tang W, Pan X, Huang A, Gao X, Anderson C.T, & Yang Z (2022) *Arabidopsis* pavement cell morphogenesis requires FERONIA binding to pectin for activation of ROP GTPase signaling. *Current Biology* 32:497-507.e4. https://doi.org/10.1016/j.cub.2021.11.030

Lindner H, Müller L M, Boisson-Dernier A, & Grossniklaus U (2012) CrRLK1L receptor-like kinases: not just another brick in the wall. *Current Opinion in Plant Biology* 15(6):659–669. https://doi.org/10.1016/j.pbi.2012.07.003

Liu X, Wang J, & Sun L (2018) Structure of the hyperosmolality-gated calcium-permeable channel OSCA1.2. *Nature Communications* 9(1):5060. https://doi.org/10.1038/s41467-018-07564-5

Ludwig B S, Kessler H, Kossatz S, & Reuning U (2021) RGD-binding integrins revisited: how recently discovered functions and novel synthetic ligands (re-) shape an ever-evolving field. *Cancers* 13:1711. https://doi.org/10.3390/cancers13071711

Ma Y, Walker R K, Zhao Y, & Berkowitz G A (2012) Linking ligand perception by PEPR pattern recognition receptors to cytosolic Ca^{2+} elevation and downstream immune signaling in plants. *PNAS* 109(48):19852. https://doi.org/10.1073/pnas.1205448109

Machado R M A, & Serralheiro R P (2017) Soil salinity: effect on vegetable crop growth management practices to prevent and mitigate soil salinization. *Horticulturae* 3(2):30. https://www.mdpi.com/2311-7524/3/2/30

Mahajan S, & Tuteja N (2005) Cold salinity and drought stresses: an overview. *Archives of Biochemistry and Biophysics* 444(2):139–158. https://doi.org/10.1016/j.abb.2005.10.018

Maity K, Heumann J M, McGrath A P, Kopcho N J, Hsu P K, Lee C W, Mapes J H, Garza D, Krishnan S, Morgan G P, Hendargo K J, Klose T, Rees S D, Medrano-Soto A, Saier M H, Jr, Piñeros M, Komives E A, Schroeder J I, Chang G, &

Stowell M H B (2019) Cryo-EM structure of OSCA1.2 from *Oryza sativa* elucidates the mechanical basis of potential membrane hyperosmolality gating. *PNAS* 116(28):14309–14318. https://doi.org/10.1073/pnas.1900774116

Manfield I W, Orfila C, McCartney L, Harholt J, Bernal A J, Scheller H V, Gilmartin P M, Mikkelsen J D, Paul Knox J, & Willats W G (2004) Novel cell wall architecture of isoxaben-habituated *Arabidopsis* suspension-cultured cells: global transcript profiling and cellular analysis. *The Plant Journal* 40(2):260–275. https://doi.org/10.1111/j.1365-313X.2004.02208.x

Martínez J P, Silva H, Ledent J F, & Pinto M (2007) Effect of drought stress on the osmotic adjustment cell wall elasticity and cell volume of six cultivars of common beans (*Phaseolus vulgaris* L.). *European Journal of Agronomy* 26(1):30–38. https://doi.org/https://doi.org/10.1016/j.eja.2006.08.003

Martiniere A, Lavagi I, Nageswaran G, Rolfe D J, Maneta-Peyret L, Luu D T, Botchway S W, Webb S E, Mongrand S, Maurel C, Martin-Fernandez M L, Kleine-Vehn J, Friml J, Moreau P, & Runions J (2012) Cell wall constrains lateral diffusion of plant plasma-membrane proteins. *PNAS* 109(31):12805–12810. https://doi.org/10.1073/pnas.1202040109

Mellersh D G, & Heath M C (2001) Plasma membrane-cell wall adhesion is required for expression of plant defense responses during fungal penetration. *The Plant Cell* 13(2):413–424. https://doi.org/10.1105/tpc.13.2.413

Miya A, Albert P, Shinya T, Desaki Y, Ichimura K, Shirasu K, Narusaka Y, Kawakami N, Kaku H, & Shibuya N (2007) CERK1 a LysM receptor kinase is essential for chitin elicitor signaling in *Arabidopsis*. *PNAS* 104(49):19613–19618. https://doi.org/10.1073/pnas.0705147104

Miyazaki S, Murata T, Sakurai-Ozato N, Kubo M, Demura T, Fukuda H, & Hasebe M (2009) *ANXUR1* and *2* sister genes to *FERONIA/SIRENE* are male factors for coordinated fertilization. *Current Biology* 19(15):1327–1331. https://doi.org/10.1016/j.cub.2009.06.064

Monshausen G B, & Gilroy S (2009) Feeling green: mechanosensing in plants. *Trends in Cell Biology* 19(5):228–235 https://dx. doi.org/10.1016/j.tcb.2009.02.005

Moura J C, Bonine C A, de Oliveira Fernandes Viana J, Dornelas M C, & Mazzafera P (2010) Abiotic and biotic stresses and changes in the lignin content and composition in plants. *Journal of Integrative Plant Biology* 52(4):360–376. https://doi.org/10.1111/j.1744-7909.2010.00892.x

Nakagawa Y, Katagiri T, Shinozaki K, Qi Z, Tatsumi H, Furuichi T, Kishigami A, Sokabe M, Kojima I, Sato S, Kato T, Tabata S, Iida K, Terashima A, Nakano M, Ikeda M, Yamanaka T, & Iida H (2007) *Arabidopsis* plasma membrane protein crucial for Ca^{2+} influx and touch sensing in roots. *PNAS* 104(9):3639–3644. https://doi.org/10.1073/pnas.0607703104

Nakano M, Furuichi T, Sokabe M, Iida H, & Tatsumi H (2021) The gravistimulation-induced very slow $Ca^{(2+)}$ increase in *Arabidopsis* seedlings requires MCA1 a $Ca^{(2+)}$-permeable mechanosensitive channel. *Scientific Reports* 11(1):227. https://doi.org/10.1038/s41598-020-80733-z

Neeli-Venkata R, Diaz C M, Celador R, Sanchez Y, & Minc N (2021) Detection of surface forces by the cell-wall mechanosensor WSC1 in yeast. *Developmental Cell* 56:2856-2870.e7. https://doi.org/10.1016/j.devcel.2021.09.024

Novaković L, Guo T, Bacic A, Sampathkumar A, & Johnson K L (2018) Hitting the wall-sensing and signaling pathways involved in plant cell wall remodeling in response to abiotic stress. *Plants (Basel)* 7(4):89. https://doi.org/10.3390/plants7040089

Ooume K, Inoue Y, Soga K, Wakabayashi K, Fujii S, Yamamoto R, & Hoson T (2009) Cellular basis of growth suppression by submergence in azuki bean epicotyls. *Annals of Botany* 103(2):325–332. https://doi.org/10.1093/aob/mcn198

Panter P, Panter J, & Knight H (2020) Impact of cell-wall structure and composition on plant freezing tolerance. *Annual Plant Reviews online* 3:607–642). https://doi.org/10.1002/9781119312994.apr0746

Paredez A R, Somerville C R, & Ehrhardt D W (2006) Visualization of cellulose synthase demonstrates functional association with microtubules. *Science* 312:1491–1495 https://doi.org/10.1126/science.1126551

Park A R, Cho S K, Yun U J, Jin M Y, Lee S H, Sachetto-Martins G, & Park O K (2001) Interaction of the *Arabidopsis* receptor protein kinase WAK1 with a glycine-rich protein AtGRP-3. *Journal of Biological Chemistry* 276(28):26688–26693. https://doi.org/10.1074/jbc.M101283200

Park Y B, & Cosgrove D J (2012) Changes in cell wall biomechanical properties in the xyloglucan-deficient *xxt1/xxt2* mutant of *Arabidopsis*. *Plant Physiology* 158(1):465–475. https://doi.org/10.1104/pp.111.189779

Pearce R S (1999) Molecular analysis of acclimation to cold. *Plant Growth Regulation* 29(1):47–76. https://doi.org/10.1023/A:1006291330661

Pelloux J, Rustérucci C, & Mellerowicz E (2007) New insight into pectin methylesterase structure and function. *Trends in Plant Science* 12:267–277. https://doi.org/10.1016/j.tplants.2007.04.001

Perroud P F, Demko V, Johansen W, Wilson R C, Olsen O A, & Quatrano R S (2014) DEFECTIVE KERNEL1 (DEK1) is required for three-dimensional growth in *Physcomitrella patens*. *New Phytology* 203(3):794–804. https://doi.org/10.1111/nph.12844

Petkova M I, Pujol-Carrion N, & de la Torre-Ruiz M A (2012) Mtl1 O-mannosylation mediated by both Pmt1 and Pmt2 is important for cell survival under oxidative conditions and TOR blockade. *Fungal Genetics and Biology* 49(11):903–914. https://doi.org/10.1016/j.fgb.2012.08.005

Piro G, Leucci M R, Waldron K, & Dalessandro G (2003) Exposure to water stress causes changes in the biosynthesis of cell wall polysaccharides in roots of wheat cultivars varying in drought tolerance. *Plant Science* 165(3):559–569. https://doi.org/10.1016/S0168-9452(03)00215-2

Pont-Lezica R F, McNally J G, & Pickard B G (1993) Wall-to-membrane linkers in onion epidermis: some hypotheses. *Plant Cell & Environment* 16(2):111–123. https://doi.org/https://doi.org/10.1111/j.1365-3040.1993.tb00853.x

Punwani J A, Rabiger D S, & Drews G N (2007) MYB98 positively regulates a battery of synergid-expressed genes encoding filiform apparatus localized proteins.*The Plant Cell* 19(8):2557–2568. https://doi.org/10.1105/tpc.107.052076

Qiu Q S, Guo Y, Dietrich M A, Schumaker K S, & Zhu J K (2002) Regulation of SOS1 a plasma membrane Na^+/H^+ exchanger in *Arabidopsis thaliana* by SOS2 and SOS3. *PNAS* 99(12):8436–8441. https://doi.org/10.1073/pnas.122224699

Qu S, Zhang X, Song Y, Lin J, & Shan X (2017) THESEUS1 positively modulates plant defense responses against *Botrytis cinerea* through GUANINE EXCHANGE FACTOR4 signaling. *Journal of Integrative Plant Biology* 59(11):797–804. https://doi.org/10.1111/jipb.12565

Quan R, Lin H, Mendoza I, Zhang Y, Cao W, Yang Y, Shang M, Chen S, Pardo J M, & Guo Y (2007) SCABP8/CBL10 a putative calcium sensor interacts with the protein kinase SOS2 to protect *Arabidopsis* shoots from salt stress. *The Plant Cell* 19(4):1415–1431. https://doi.org/10.1105/tpc.106.042291

Rauf M, Arif M, Fisahn J, Xue G P, Balazadeh S, & Mueller-Roeber B (2013) NAC transcription factor speedy hyponastic growth regulates flooding-induced leaf movement in *Arabidopsis*. *The Plant Cell* 25(12):4941–4955. https://doi.org/10.1105/tpc.113.117861

Richter J, Watson J M, Stasnik P, Borowska M, Neuhold J, Berger M, Stolt-Bergner P, Schoft V, & Hauser M T (2018) Multiplex mutagenesis of four clustered *CrRLK1L* with CRISPR/Cas9 exposes their growth regulatory roles in response to metal ions. *Scientific Reports* 8(1):12182. https://doi.org/10.1038/s41598-018-30711-3

Ridley B L, O'Neill M A, & Mohnen D (2001) Pectins: structure biosynthesis and oligogalacturonide-related signaling. *Phytochemistry* 57(6):929–967. https://doi.org/10.1016/s0031-9422(01)00113-3

Rizhsky L, Liang H, Shuman J, Shulaev V, Davletova S, & Mittler R (2004) When defense pathways collide the response of *Arabidopsis* to a combination of drought and heat stress. *Plant Physiology* 134:1683–1696. https://doi.org/10.1104/pp.103.033431

Rodicio R, & Heinisch J J (2010) Together we are strong-cell wall integrity sensors in yeasts. *Yeast* 27(8):531–540. https://doi.org/10.1002/yea.1785

Roeder A H K, Cunha A, Ohno C K, & Meyerowitz E M (2012) Cell cycle regulates cell type in the *Arabidopsis* sepal. *Development* 139(23):4416–4427. https://doi.org/10.1242/dev.082925

Ross A, Yamada K, Hiruma K, Yamashita-Yamada M, Lu X, Takano Y, Tsuda K, & Saijo Y (2014) The *Arabidopsis* PEPR pathway couples local and systemic plant immunity. *The EMBO Journal* 33(1):62–75. https://doi.org/https://doi.org/10.1002/embj.201284303

Rui Y, & Dinneny J R (2020) A wall with integrity: surveillance and maintenance of the plant cell wall under stress. *New Phytologyogist* 225(4):1428–1439. https://doi.org/10.1111/nph.16166

Saidi Y, Finka A, Muriset M, Bromberg Z, Weiss Y G, Maathuis F J M, & Goloubinoff P (2009) The heat shock response in moss plants is regulated by specific calcium-permeable channels in the plasma membrane. *The Plant Cell* 21(9):2829–2843. https://doi.org/10.1105/tpc.108.065318

Saijo Y, Loo E (2019) Plant immunity in signal integration between biotic and abiotic stress responses. *New Phytologyogist* 225:87–104. https://doi.org/10.1111/nph.15989

Saito F, Suyama A, Oka T, Yoko-o T, Matsuoka K, Jigami Y, & Shimma Y (2014) Identification of novel peptidyl serine *galactosyltransferase* gene family in plants. *Journal of Biological Chemistry* 289:20405–20420. https://doi.org/10.1074/jbc.M114.553933

Sampathkumar A, Peaucelle A, Fujita M, Schuster C, Persson S, Wasteneys G O, & Meyerowitz E M (2019) Primary wall cellulose synthase regulates shoot apical meristem mechanics and growth. *Development* 146(10):dev179036. https://doi.org/10.1242/dev.179036

Sanghera G S, Wani S H, Hussain W, & Singh N B (2011) Engineering cold stress tolerance in crop plants. *Current Genomics* 12(1):30–43. https://doi.org/10.2174/138920211794520178

Schallus T, Jaeckh C, Fehér K, Palma A S, Liu Y, Simpson J C, Mackeen M, Stier G, Gibson T J, Feizi T, Pieler T, & Muhle-Goll C (2008) Malectin: a novel carbohydrate-binding protein of the endoplasmic reticulum and a candidate player in the early steps of protein N-glycosylation. *Molecular Biology of The Cell* 19(8):3404–3414. https://doi.org/10.1091/mbc.e08-04-0354

Schnabelrauch L S, Kieliszewski M, Upham B L, Alizedeh H, & Lamport D T (1996) Isolation of pI 4.6 extensin peroxidase from tomato cell suspension cultures and identification of Val-Tyr-Lys as putative intermolecular cross-link site. *The Plant Journal* 9:477–489. https://doi.org/10.1046/j.1365-313X.1996.09040477.x

Schoenaers S, Balcerowicz D, Breen G, Hill K, Zdanic M, Mouille G, Holman T J, Oh J, Wilson M H, Nikonorova N, Vu L D, De Smet I, Swarup R, De Vos W H, Pintelon I, Adriaensen D, Grierson C, Bennett M J, & Vissenberg K (2018) The auxin-regulated CrRLK1L kinase ERULUS controls cell wall composition during root hair tip growth. *Current Biology* 28(5):722-732.e726. https://doi.org/10.1016/j.cub.2018.01.050

Scholz F G, Bucci S J, Arias N, Meinzer F C, & Goldstein G (2012) Osmotic and elastic adjustments in cold desert shrubs differing in rooting depth: coping with drought and subzero temperatures. *Oecologia* 170(4):885–897. https://doi.org/10.1007/s00442-012-2368-y

Schopfer P (2006) Biomechanics of plant growth. *American Journal of Botany* 93(10):1415–1425. https://doi.org/10.3732/ajb.93.10.1415

Schultz C, Gilson P, Oxley D, Youl J, & Bacic A (1998) GPI-anchors on arabinogalactan-proteins: implications for signalling in plants. *Trends in Plant Science* 3:426–431. https://doi.org/10.1016/S1360-1385(98)01328-4

Schulze B, Mentzel T, Jehle A K, Mueller K, Beeler S, Boller T, Felix G, & Chinchilla D (2010) Rapid heteromerization and phosphorylation of ligand-activated plant transmembrane receptors and their associated kinase BAK1. *Journal of Biological Chemistry* 285(13):9444–9451. https://doi.org/10.1074/jbc.M109.096842

Seifert G J, & Blaukopf C (2010) Irritable walls: the plant extracellular matrix and signaling. *Plant Physiology* 153(2):467–478. https://doi.org/10.1104/pp.110.153940

Seifert G J (2021) The FLA4-FEI pathway: A unique and mysterious signaling module related to cell wall structure and stress signaling. *Genes (Basel)* 12(2):145. https://doi.org/10.3390/genes12020145

Shannon M.C, & Grieve C M (1998) Tolerance of vegetable crops to salinity. *Scientia horticulturae* 78:5–38. https://doi.org/10.1016/S0304-4238(98)00189-7

Shi H, Ishitani M, Kim C, & Zhu J K (2000) The *Arabidopsis thaliana* salt tolerance gene *SOS1* encodes a putative Na^+/H^+ antiporter. *PNAS* 97(12):6896–6901. https://doi.org/10.1073/pnas.120170197

Shi J, Zhang Q, Yan X, Zhang D, Zhou Q, Shen Y, Anupol N, Wang X, Bao M, Larkin R M, Luo H, & Ning G (2021) A conservative pathway for coordination of cell wall biosynthesis and cell cycle progression in plants. *The Plant Journal* 106:630–648. https://doi.org/10.1111/tpj.15187

Shih H W, Miller N D, Dai C, Spalding E P, & Monshausen G B (2014) The receptor-like kinase FERONIA is required for mechanical signal transduction in *Arabidopsis* seedlings. *Current Biology* 24:1887–1892. https://doi.org/10.1016/j.cub.2014.06.064

Solecka D, Zebrowski J, & Kacperska A (2008) Are pectins involved in cold acclimation and de-acclimation of winter oil-seed rape plants?. *Annals of Botany* 101(4):521–530. https://doi.org/10.1093/aob/mcm329

Souza C A, Li S, Lin A Z, Boutrot F, Grossmann G, Zipfel C, & Somerville S C (2017) Cellulose-derived oligomers act as damage-associated molecular patterns and trigger defense-like responses. *Plant Physiology* 173(4):2383–2398. https://doi.org/10.1104/pp.16.01680

Stegmann M, Monaghan J, Smakowska-Luzan E, Rovenich H, Lehner A, Holton N, Belkhadir Y, & Zipfel C (2017) The receptor kinase FER is a RALF-regulated scaffold controlling plant immune signaling. *Science* 355(6322):287–289. https://doi.org/10.1126/science.aal2541

Tan L, Leykam J F, & Kieliszewski M J (2003) Glycosylation motifs that direct arabinogalactan addition to arabinogalactan-proteins. *Plant Physiology* 132:1362–69. https://doi.org/10.1104/pp.103.021766

Tan L, Eberhard S, Pattathil S, Warder C, Glushka J, Yuan C, Hao Z, Zhu X, Avci U, Miller J S, Baldwin D, Pham C, Orlando R, Darvill A, Hahn M G, Kieliszewski M J, & Mohnen D (2013) An *Arabidopsis* cell wall proteoglycan consists of pectin and arabinoxylan covalently linked to an arabinogalactan protein. *The Plant Cell* 25:270–287. https://doi.org/10.1105/tpc.112.107334

Tang J, Han Z, Sun Y, Zhang H, Gong X, & Chai J (2015) Structural basis for recognition of an endogenous peptide by the plant receptor kinase PEPR1. *Cell Research* 25(1):110–120. https://doi.org/10.1038/cr.2014.161

Tan L, Tees D, Qian J, Kareem S, & Kieliszewski M J (2018) Intermolecular interactions between glycomodules of plant cell wall arabinogalactan-proteins and extensins. *Cell surface* 1:25–33. https://doi.org/10.1016/j.tcsw.2018.03.001

Tanino K K, Kobayashi S, Hyett C, Hamilton K, Liu J, Li B, Borondics F, Pedersen T, Tse J, Ellis T, Kawamura Y, & Uemura M (2013) *Allium fistulosum* as a novel system to investigate mechanisms of freezing resistance. *Physiologia Plantarum* 147(1):101–111. https://doi.org/10.1111/j.1399-3054.2012.01716.x

Tenhaken R (2015) Cell wall remodeling under abiotic stress. *Frontiers in Plant Science* 5(771). https://doi.org/10.3389/fpls.2014.00771

Tintor N, Ross A, Kanehara K, Yamada K, Fan L, Kemmerling B, Nürnberger T, Tsuda K, & Saijo Y (2013) Layered pattern receptor signaling via ethylene and endogenous elicitor peptides during *Arabidopsis* immunity to bacterial infection. *PNAS* 110(15):6211. https://doi.org/10.1073/pnas.1216780110

Tran D, Galletti R, Neumann E D, Dubois A, Sharif-Naeini R, Geitmann A, Frachisse J M, Hamant O, & Ingram G C (2017) A mechanosensitive Ca^{2+} channel activity is dependent on the developmental regulator DEK1. *Nature Communications* 8(1):1009. https://doi.org/10.1038/s41467-017-00878-w

Trudel J, Grenier J, Potvin C, & Asselin A (1998) Several Thaumatin-Like proteins bind to β-1,3-glucans. *Plant Physiology* 118(4):1431–1438. https://doi.org/10.1104/pp.118.4.1431

Tsang DL, Edmond C, Harrington JL, & Nühse TS (2011) Cell wall integrity controls root elongation via a general 1-aminocyclopropane-1-carboxylic acid-dependent, ethylene-independent pathway. *Plant Physiology* 156(2):596–604. https://doi.org/10.1104/pp.111.175372

Uddin M N, Hanstein S, Faust F, Eitenmüller P T, Pitann B, & Schubert S (2014) Diferulic acids in the cell wall may contribute to the suppression of shoot growth in the first phase of salt stress in maize. *Phytochemistry* 102:126–136. https://doi.org/10.1016/j.phytochem.2014.02.014

Uyttewaal M, Burian A, Alim K, Landrein B, Borowska-Wykret D, Dedieu A, Peaucelle A, Ludynia M, Traas J, Boudaoud A, Kwiatkowska D, & Hamant O (2012) Mechanical stress acts via katanin to amplify differences in growth rate between adjacent cells in *Arabidopsis*. *Cell* 149(2):439–451. https://doi.org/10.1016/j.cell.2012.02.048

Van der Does D, Boutrot F, Engelsdorf T, Rhodes J, McKenna J F, Vernhettes S, Koevoets I, Tintor N, Veerabagu M, Miedes E, Segonzac C, Roux M, Breda A S, Hardtke C S, Molina A, Rep M, Testerink C, Mouille G, Höfte H, Hamann T, Zipfel C (2017) The *Arabidopsis* leucine-rich repeat receptor kinase MIK2/LRR-KISS connects cell wall integrity sensing root growth and response to abiotic and biotic stresses. *PloS Genetics* 13(6):e1006832. https://doi.org/10.1371/journal.pgen.1006832

Vanholme R, De Meester B, Ralph J, & Boerjan W (2019) Lignin biosynthesis and its integration into metabolism. *Current Opinion in Biotechnology* 56:230–239. https://doi.org/10.1016/j.copbio.2019.02.018

Vay H A, Philip B, & Levin D E (2004) Mutational analysis of the cytoplasmic domain of the Wsc1 cell wall stress sensor. *Microbiology* 150(10):3281–3288. https://doi.org/10.1099/mic.0.27264-0

Verica J A,& He Z H (2002) The cell wall-associated kinase (WAK) and WAK-like kinase gene family. *Plant Physiology* 129(2):455–459. https://doi.org/10.1104/pp.011028

Vitorino P G, Alves J D, Magalhães P C, Magalhães M M, Lima L C O, & Oliveira L E M d (2001) Flooding tolerance and cell wall alterations in maize mesocotyl during hypoxia *Pesquisa Agropecuária. Brasileira* 36:1027–1035. https://doi.org/10.1590/S0100-204X2001000800004

Vitorino P G, Alves J D, Magalhães P C, Magalhães M M, Lima, L C O, & Mota de Oliveira L E (2001) Flooding tolerance and cell wall alterations in maize mesocotyl during hypoxia *Pesquisa Agropecuária Brasileira* 36:1027–1035. https://doi.org/10.1590/S0100-204X2001000800004.

Voesenek L A, Colmer T D, Pierik R, Millenaar F F, & Peeters A J (2006) How plants cope with complete submergence. *New Phytology* 170(2):213–226. https://doi.org/10.1111/j.1469-8137.2006.01692.x

Vogel V, & Sheetz M (2006) Local force and geometry sensing regulate cell functions *Nature Reviews Molecular and Cellular Biology* 7(4):265–275. https://doi.org/10.1038/nrm1890

Voxeur A, & Höfte H (2016) Cell wall integrity signaling in plants: "To grow or not to grow that's the question". *Glycobiology* 26(9):950–960. https://doi.org/10.1093/glycob/cww029

Voxeur A, & Höfte H (2020) Pectin-derived immune elicitors in response to lignin modification in plants. *PNAS* 117(9):4442–4444. https://doi.org/10.1073/pnas.2000509117

Vreeburg R A, Benschop J J, Peeters A J, Colmer T D, Ammerlaan A H, Staal M, Elzenga T M, Staals R H, Darley C P, McQueen-Mason S J, & Voesenek L A (2005) Ethylene regulates fast apoplastic acidification and expansin A transcription during submergence-induced petiole elongation in *Rumex palustris*. *The Plant Journal* 43(4):597–610. https://doi.org/10.1111/j.1365-313X.2005.02477.x

Wang L, Hukin D, Pritchard J, & Thomas C (2006) Comparison of plant cell turgor pressure measurement by pressure probe and micromanipulation. *Biotechnology Letters* 28(15):1147–1150. https://doi.org/10.1007/s10529-006-9075-x

Wang C, Li J, & Yuan M (2007) Salt tolerance requires cortical microtubule reorganization in *Arabidopsis. Plant and Cell Physiology* 48:1534–47. https://doi.org/10.1093/pcp/pcm123

Wang S, Kurepa J, Hashimoto T, & Smalle J A (2011) Salt stress-induced disassembly of *Arabidopsis* cortical microtubule arrays involves 26S proteasome-dependent degradation of SPIRAL1. *The Plant Cell* 23(9):3412–3427. https://doi.org/10.1105/tpc.111.089920

Wang T, McFarlane H E, & Persson S (2016) The impact of abiotic factors on cellulose synthesis. *Journal of Experimental Botany* 67(2):543–552. https://doi.org/10.1093/jxb/erv488

Wang H, Jin Y, Wang C, Li B, Jiang C, Sun Z, Zhang Z, Kong F, & Zhang H (2017) Fasciclin-like arabinogalactan proteins PtFLAs play important roles in GA-mediated tension wood formation in *Populus. Scientific Reports* 7(1):6182. https://doi.org/10.1038/s41598-017-06473-9

Wang X, Ma Q, Wang R, Wang P, Liu Y, & Mao T (2020) Submergence stress-induced hypocotyl elongation through ethylene signaling-mediated regulation of cortical microtubules in *Arabidopsis. Journal of Experimental Botany* 71:1067–77. https://doi.org/10.1093/jxb/erz453

Wang L, Yang T, Lin Q, Wang B, Li X, Luan S, & Yu F (2020) Receptor kinase FERONIA regulates flowering time in *Arabidopsis. BMC Plant Biology* 20(1):26. https://doi.org/10.1186/s12870-019-2223-y

Weiser R L, Wallner S J, & Waddell J W (1990) Cell wall and extensin mRNA changes during cold acclimation of pea seedlings. *Plant Physiology* 93(3):1021–1026. https://doi.org/10.1104/pp.93.3.1021

Westermann J, Streubel S, Franck C M, Lentz R, Dolan L, & Boisson-Dernier A (2019) An evolutionarily conserved receptor-like kinases signaling module controls cell wall integrity during tip growth. *Current Biology* 29(22):3899-3908.e3893. https://doi.org/https://doi.org/10.1016/j.cub.2019.09.069

Wolf S, Hématy K, & Höfte H (2012) Growth control and cell wall signaling in plants. *Annual Review of Plant Biology* 63:381–407. https://doi.org/10.1146/annurev-arplant-042811-105449

Wolf S, & Greiner S (2012) Growth control by cell wall pectins. *Protoplasma* 249(2):169–175. https://doi.org/10.1007/s00709-011-0371-5

Wolf S, Mravec J, Greiner S, Mouille G, & Höfte H (2012) Plant cell wall homeostasis is mediated by brassinosteroid feedback signaling. *Current Biology* 22(18):1732–1737. https://doi.org/10.1016/j.cub.2012.07.036

Wu H C, Hsu S F, Luo D L, Chen S J, Huang W D, Lur H S, & Jinn T L (2010) Recovery of heat shock-triggered released apoplastic Ca^{2+} accompanied by pectin methylesterase activity is required for thermotolerance in soybean seedlings. *Journal of Experimental Botany* 61(10):2843–2852. https://doi.org/10.1093/jxb/erq121

Xiao C, Zhang T, Zheng Y, Cosgrove D J, & Anderson C T (2016) Xyloglucan deficiency disrupts microtubule stability and cellulose biosynthesis in *Arabidopsis* altering cell growth and morphogenesis. *Plant Physiology* 170(1):234–249. https://doi.org/10.1104/pp.15.01395

Xu Z, Xin T, Bartels D, Li Y, Gu W, Yao H, Liu S, Yu H, Pu X, Zhou J, Xu J, Xi C, Lei H, Song J, & Chen S (2018) Genome analysis of the ancient tracheophyte *Selaginella tamariscina* reveals evolutionary features relevant to the acquisition of desiccation tolerance. *Molecular Plant* 11(7):983–994. https://doi.org/10.1016/j.molp.2018.05.003

Yamaguchi Y, Pearce G, & Ryan C A (2006) The cell surface leucine-rich repeat receptor for AtPep1 an endogenous peptide elicitor in *Arabidopsis* is functional in transgenic tobacco cells. *PNAS* 103(26):10104. https://doi.org/10.1073/pnas.0603729103

Yamanaka T, Nakagawa Y, Mori K, Nakano M, Imamura T, Kataoka H, Terashima A, Iida K, Kojima I, Katagiri T, Shinozaki K, & Iida H (2010) MCA1 and MCA2 That Mediate Ca^{2+} uptake have distinct and overlapping roles in *Arabidopsis. Plant Physiology* 152(3):1284–1296. https://doi.org/10.1104/pp.109.147371

Yang L, Wang C C, Guo W D, Li X B, Lu M, & Yu C L (2006) Differential expression of cell wall related genes in the elongation zone of rice roots under water deficit. *Russian Journal of Plant Physiology* 53(3):390–395. https://doi.org/10.1134/S1021443706030150

Yasuda S, Okada K, & Saijo Y (2017) A look at plant immunity through the window of the multitasking coreceptor BAK1. *Current Opinion in Plant Biology* 38:10–18. https://doi.org/10.1016/j.pbi.2017.04.007

Yoshimura H, Tada T, & Iida H (2004) Subcellular localization and oligomeric structure of the yeast putative stretch-activated Ca^{2+} channel component MID1. *Experimental Cell Research* 293(2):185–195. https://doi.org/10.1016/j.yexcr.2003.09.020

Yoshimura K, Iida K, & Iida H (2021) MCAs in *Arabidopsis* are Ca^{2+}-permeable mechanosensitive channels inherently sensitive to membrane tension. *Nature Communications* 12:6074. https://doi.org/10.1038/s41467-021-26363-z

Yu F, Li J, Huang Y, Liu L, Li D, Chen L, & Luan S (2014) FERONIA receptor kinase controls seed size in *Arabidopsis thaliana. Molecular Plant* 7(5):920–922. https://doi.org/10.1093/mp/ssu010

Yu Y, & Assmann S M (2018) Inter-relationships between the heterotrimeric Gβ subunit AGB1 the receptor-like kinase FERONIA and RALF1 in salinity response. *Plant, Cell & Environment* 41(10):2475–2489. https://doi.org/10.1111/pce.13370

Yuan F, Yang H, Xue Y, Kong D, Ye R, Li C, Zhang J, Theprungsirikul L, Shrift T, Krichilsky B, Johnson D M, Swift G B, He Y, Siedow J N, & Pei Z M (2014) OSCA1 mediates osmotic-stress-evoked Ca2+ increases vital for osmosensing in *Arabidopsis. Nature* 514(7522):367–371. https://doi.org/10.1038/nature13593

Zhang H, Zhu J, Gong Z, & Zhu J K (2021) Abiotic stress responses in plants. *Nature Reviews Genetics* 23:104–119. https://doi.org/10.1038/s41576-021-00413-0

Zhang W, Zhou R G, Gao Y J, Zheng S Z, Xu P, Zhang S Q, & Sun D Y (2009) Molecular and genetic evidence for the key role of AtCaM3 in heat-shock signal transduction in *Arabidopsis. Plant Physiology* 149(4):1773–1784. https://doi.org/10.1104/pp.108.133744

Zhao K F, Song J, Fan H, Zhou S, &and Zhao M (2010) Growth response to ionic and osmotic stress of NaCl in salt-tolerant and salt-sensitive maize. *Journal of Integrative Plant Biology* 52:468–75. https://doi.org/10.1111/j.1744-7909.2010.00947.x

Zhou X, Zheng Y, Wang L, Li H, Guo Y, Li M, Sun M, &and Zhao P (2022) SYP72 interacts with the mechanosensitive channel MSL8 to protect pollen from hypoosmotic shock during hydration. *Nature Communications* 13:73. https://doi.org/10.1038/s41467-021-27757-9

Zhu J K (2002) Salt and drought stress signal transduction in plants. *Annual Review of Plant Biology* 53:247–273. https://doi.org/10.1146/annurev.arplant.53.091401.143329

Zhu S, Estévez J M, Liao H, Zhu Y, Yang T, Li C, Wang Y, Li L, Liu X, Pacheco J M, Guo H, & Yu F (2020) The RALF1-FERONIA complex phosphorylates eIF4E1 to promote protein synthesis and polar root hair growth. *Molecular Plant* 13(5):698–716. https://doi.org/10.1016/j.molp.2019.12.014

Zuo W, Chao Q, Zhang N, Ye J, Tan G, Li B, Xing Y, Zhang B, Liu H, Fengler K A, Zhao J, Zhao X, Chen Y, Lai J, Yan J, & Xu M (2015) A maize wall-associated kinase confers quantitative resistance to head smut. *Nature Genetics* 47(2):151–157. https://doi.org/10.1038/ng.3170

19 Cell Wall Signalling in Plants

Kay Schneitz
Technical University of Munich

Sebastian Wolf
Eberhard Karls Universität Tübingen

INTRODUCTION

Plant cells are pressurised and encased by a semi-rigid, polysaccharide-rich cell wall. Growth occurs through the regulated increase in cell wall extensibility, which allows the internal turgor pressure to displace cell wall polymers until a new equilibrium is reached and the cell wall is consolidated. Thus, by default, maintaining cell wall integrity of the cell wall is of vital importance for plants. In addition, various abiotic and biotic stresses directly alter cell wall properties to which plant cells are required to respond. Therefore, it has long been assumed that signalling pathways that directly or indirectly monitor cell wall status must exist to ensure cell integrity. However, abundant evidence suggests a much broader function for cell wall–associated signalling.

For example, cell wall cues are required to maintain the differentiation status of mesophyll cells (Davey et al. 2005; Chupeau et al. 2013), instruct cytoskeletal rearrangements (Lin et al. 2022), and are necessary for abscisic acid production under hyperosmotic conditions (Bacete et al. 2022). The cell wall and the protoplast must communicate to enable the dynamic cell wall remodelling that occurs during cell and tissue growth. Cell wall damage caused by abiotic stress or pathogen attack leads to a defence response that frequently results in cell wall remodelling as well (Boller and Felix 2009). At the same time, the cell wall constitutes a natural barrier to the intercellular communication that coordinates such responses between cells in an organ. This obstacle is overcome by signalling mechanisms involving small molecules that move through the cell wall and are recognised by, for example, cell surface receptors. Another general signalling mechanism involves diffusion or controlled movement of factors through plasmodesmata, channels connecting most plant cells. In recent years, functional links between plasma-membrane-localised cell surface receptors and plasmodesmata-controlled intercellular communication have been established (Sager and Lee 2018; Li et al. 2021b). Still, we know very little about the molecular signalling mechanisms that reside at the nexus of monitoring the cell wall status and the control of development and stress responses. It is completely unclear how cell wall remodelling is orchestrated across cells within a growing or mature tissue. On the other hand, overwhelming evidence corroborating the existence of a multitude of cell wall–associated signalling pathways has accumulated over recent years. Here, we aim to summarise this evidence and point out open questions.

In this chapter, we consider as cell wall signalling any signalling event directly responsive to cell wall cues, including pathways monitoring the biochemical composition and/or architecture of the cell wall or those with cell wall–derived polysaccharide ligands. In addition, direct physical contact of the extracellular domain of a cell surface receptor with the cell wall would qualify if changes in this interaction or cell wall alterations affect signalling output. As a rule, the corresponding mechanisms involve feedback between the cell wall and intracellular signalling processes. We also refer the reader to several excellent recent reviews covering various aspects of cell wall signalling (Anderson and Kieber 2020; Rui and Dinneny 2020) as do Chapters 18 (CWI), 15 (CW/cell shape, mechanics), 6 (cell expansion), and 20 (defence) in this book.

THE PLANT CELL WALL

The plant cell wall is a complex structure composed of carbohydrates, phenolic compounds, and a plethora of different cell-wall-bound proteins (Höfte and Voxeur 2017; Lampugnani et al. 2018). Cellulose microfibrils are the main load-bearing elements and are cross-linked to matrix polysaccharides, mainly various hemicelluloses and pectins. Cellulose synthesis is driven by mobile cellulose synthase (CESA) complexes at the plasma membrane which track along cortical microtubules (Paredez et al. 2006). The galacturonic acid (GalA)-containing pectins are synthesised in the Golgi and secreted into the cell wall where they undergo further modifications (Atmodjo et al. 2013). Homogalacturonan (HG), one of the central components of pectin, is a linear polymer of α(1,4)-linked GalA subunits and is initially produced in a neutral methylesterified form. Once in the cell wall, HG becomes de-methylesterified by pectin methylesterases (PMEs) (Jolie et al., 2010), a process that is counteracted by PME inhibitor proteins (PMEIs) (Wormit and Usadel 2018). De-methylesterification of HG leads to complex, context-dependent effects. For example, it can result in the exposure of negative charges, the formation of Ca^{2+}-mediated cross-links, and a more rigid cell wall (Domozych et al. 2014; Jiang et al. 2005). However,

DOI: 10.1201/9781003178309-19

PME activity is often associated with growing tissues and extensible cell walls, affecting tissue morphogenesis (Peaucelle et al. 2008, 2011, 2012, 2015; Atmodjo et al. 2013; Haas et al. 2020). PME activity is a prerequisite for HG degradation by polygalacturonases (PGs) and pectate lyases, resulting in cell wall softening or production of small soluble fragments, so-called oligogalacturonides (OGs). OGs are also generated during pathogen attack because of microbial cell wall degrading activities and represent a class of damage-associated molecular patterns (DAMPs), inducing defence responses (Ferrari et al. 2013). Hemicelluloses are synthesised in the Golgi and secreted into the cell wall. Xyloglucans (XyGs) represent the main component of hemicellulose in dicot primary cell walls and are thought to cross-link microfibrils at distinct hotspot regions (Vissenberg et al. 2005; Park and Cosgrove 2012). Secondary dicot cell walls include xylans as their main hemicellulose and may also contribute to microfibril cross-links. Thus, the control of pectin or hemicellulose properties contributes to organogenesis and plant immunity.

BRIEF HISTORY OF CELL WALL SIGNALLING RESEARCH

Compared to other areas of plant cell wall research, cell wall signalling is a relatively young field. However, it was revealed as early as 1981 that plants respond to fragments of their own cell wall in a similar fashion as to pathogen-derived elicitor molecules. These pioneering studies performed in the Albersheim, West, Ryan, and Wood laboratories established that short carbohydrate fragments, collectively termed oligosaccharides, can exert signalling function and affect plant physiology, immunity and morphogenesis (Bishop et al. 1981; Cervone et al. 1981; Lee and West 1981a, 1981b; Nothnagel et al. 1983; Tran Thanh Van et al. 1985; Eberhard et al. 1989; Marfà et al. 1991; Bellincampi et al. 1993). Later, it became apparent that plants can respond to alteration of their cell wall with surprising plasticity. Shedletzky et al. were the first to show that cultured cells can survive in the presence of high doses of cellulose biosynthesis inhibitors after going through a habituation period. Compensatory modifications of other cell wall components enable growth without the load-bearing capacity of cellulose (Shedletzky et al. 1990; Manfield et al. 2004). This remarkable plasticity of the cell wall entails large-scale transcriptional rearrangements of cell wall biosynthesis and remodelling genes, which are difficult to imagine without the contribution of signalling events. To our knowledge, the first unequivocal conceptualisation of cell wall signalling as a cell wall homeostatic mechanism is provided by Burton et al. (2000), in which the authors postulate the existence of "feedback loops" to explain their observation that knock-down of cellulose synthesis results in increased abundance of pectin as a compensatory mechanism. In agreement with this, plants with altered cell wall composition due to genetic impairment of the biosynthetic machinery show strongly altered development, physiology, and immune responses (Ellis and Turner 2001; Ellis et al. 2002). For example, jasmonate and ethylene responses are altered in the cellulose synthase mutant *cev1*, as revealed by Ellis et al., who, to our knowledge, were the first authors to use the term "cell wall signalling" referring to plants (Ellis et al. 2002).

The receptor kinase WALL-ASSOCIATED KINASE1 (WAK1) represented an early candidate for a cell surface receptor involved in cell wall signalling as a range of biochemical data suggested an association with the cell wall and *in vitro* binding of its extracellular domain to various types of pectin fragments (He et al. 1996; Wagner and Kohorn 2001; Decreux and Messiaen 2005; Decreux et al. 2006). In addition, results from domain-swap experiments were compatible with the notion of WAK1 acting as an OG receptor (Brutus et al. 2010) (see also below). If a final demonstration of signalling processes instructed by cell wall state was needed, it was provided in 2007, when Hématy et al. described the isolation of THESEUS1 (THE1), a plasma-membrane-localised receptor kinase involved in the orchestration of the response to cellulose deficiency (Hématy et al. 2007). The study clearly showed that growth impairments in cellulose synthase mutants are an indirect consequence of cellulose deficiency perception, rather than the result of biomechanical limitations. Accordingly, cellulose synthase mutants show wide-ranging transcriptional rearrangements which are, to a large part, dependent on THE1. To date, cell wall signalling is implicated in a growing number of processes and more and more potential functions are uncovered. For one receptor kinase, FERONIA (FER) cell wall association could be linked to downstream signalling responses, suggesting a bona fide cell wall perception pathway (Lin et al. 2022). However, the full complexity of cell wall signalling pathways and their integration with the regulatory mechanisms of development, physiology, and immune signalling are unknown. In the following, we aim to provide an overview of the current state of knowledge regarding these processes.

CELLULAR REACTIONS TO CELL WALL ALTERATIONS

Identification of cell wall signalling components often requires brute force manipulation of cell wall biosynthesis or modifications such as those found in mutants or after pharmacological treatments. As a result, we know much more about the response to these "artificial" laboratory conditions than about the physiological role of most known cell wall signalling components. The physiological relevance of cell wall signalling pathways can emerge when this knowledge is capitalised upon in different biological contexts. Therefore, we first summarise what is known about the response to cell wall alterations before the potential roles in immune responses, development, and physiology are discussed.

SIGNALLING NETWORK REGULATING THE RESPONSE TO REDUCED CELLULOSE BIOSYNTHESIS

The plant cell responds to changes in cellulose content induced by intrinsic genetic perturbations or by exogenous compounds. The response of Arabidopsis seedlings to cellulose biosynthesis inhibition (CBI) triggered by the herbicide isoxaben represents a particularly well-characterised cell wall stress response (Hamann et al. 2009; Denness et al. 2011; Tsang et al. 2011; Wormit et al. 2012; Engelsdorf et al. 2018; Gigli-Bisceglia et al. 2018; Bacete et al. 2022) (Chapter 18). Exogenous application of isoxaben induces a rapid clearing of CESA complexes from the plasma membrane (Paredez et al. 2006) and thus a block in cellulose biosynthesis. In line with a primary effect of isoxaben on CESA complexes, several isoxaben-resistant mutants carry defects in genes encoding CESA subunits (Scheible et al. 2001; Desprez et al. 2002), and the effects of isoxaben treatment and the phenotypes of *cesA* mutants largely overlap (Cano-Delgado et al. 2000; Fagard et al. 2000; Ellis and Turner 2001; Ellis et al. 2002; Caño-Delgado et al. 2003). Some of these effects, such as lignin accumulation, depend on the presence of sucrose in the medium (Hamann et al. 2009). Sucrose appears to exercise a signalling role rather than a metabolic function as non-metabolisable analogues of sucrose or glucose still promote isoxaben-induced lignification. Interestingly, the response to isoxaben-mediated CBI is also sensitive to turgor pressure as indicated by the observation that much of it is suppressed if isoxaben is supplemented by an osmoticum, such as sorbitol (Hamann et al. 2009; Engelsdorf et al. 2018). In general, a reduction in cellulose biosynthesis may lead to a weakened cell wall, which in turn results in cell bulging due to reduced resistance of the cell wall to turgor pressure. Treatment with sorbitol counteracts this effect.

The response to isoxaben-induced CBI appears to be mediated by a regulatory signalling network, not a single linear pathway. Several cell surface receptor kinases are known to participate in this network, such as THE1, a member of the *Catharanthus roseus* protein-kinase-1-like (CrLRK1L) family of receptor kinases, and several leucine-rich repeat (LRR) receptor kinases, including MALE DISCOVERER1-INTERACTING RECEPTOR-LIKE KINASE 2/ LEUCINE-RICH REPEAT KINASE FAMILY PROTEIN INDUCED BY SALT STRESS (MIK2/LRR-KISS), FEI2, and STRUBBELIG (SUB). THE1 represents the canonical member of this group and was identified in a mutant screen for suppressors of the *prc1-1* mutant, lacking activity of the CESA6 subunit of cellulose synthase (Hématy et al. 2007). In comparison to *prc1-1* mutants, plants lacking both *THE1* and *PRC1* activity maintain reduced cellulose content but show partially restored hypocotyl growth and reduced ectopic lignin accumulation in the hypocotyl and the root of dark-grown seedlings while overexpressing *THE1* enhanced the *prc1-1* phenotype. Moreover, *THE1* is a central regulator of the isoxaben-induced CBI response (Van der Does et al. 2017; Engelsdorf et al. 2018; Bacete et al. 2022). Thus, it was proposed that THE1 may act as a cell wall integrity sensor as it promotes a compensatory response to reduced cellulose synthesis (Hématy et al. 2007).

The LRR receptor kinase gene *FEI2* and *MID-COMPLEMENTING ACTIVITY1* (*MCA1*), encoding a potential mechano-sensitive ion channel, appear to be part of the *THE1* genetic pathway with *THE1* acting upstream of *FEI2* and *MCA1* (Engelsdorf et al. 2018). *FEI2*, together with its close homolog *FEI1*, is also required for the organisation of seed mucilage, cellulose biosynthesis, and cell expansion in the presence of high sugar concentrations and (Xu et al. 2008; Harpaz-Saad et al. 2011). The structure of the FEI extracellular domains, carrying five LRRs, suggests that they possibly act as co-receptors, and thus function similar to BRASSINOSTEROID INSENSITIVE 1-ASSOCIATED KINASE 1 (BAK1), and its close homologs that also feature extracellular domains with five LRRs and act as co-receptors for a wide variety of signalling receptor kinases (Liu et al. 2020). The respective partner receptor kinases of the FEIs are presently unknown. *SALT OVERLY SENSITIVE5* (*SOS5*), encoding an arabinogalactan protein (AGP), is known to affect a range of similar processes than the *FEI*s and may act in the *FEI* signalling pathway (Xu et al. 2008; Harpaz-Saad et al. 2011; Griffiths et al. 2014). However, it still needs to be established if *SOS5* functions in the CBI response.

In the context of the response to CBI, the LRR receptor kinase gene *MIK2* shares overlapping but also distinct functions with *THE1*, indicating that *THE1* and *MIK2* do not act in a simple linear pathway (Van der Does et al. 2017; Coleman et al. 2021). Moreover, *MIK2* turned out to play multiple roles in development and stress responses. Genetic data indicate that *MIK2* contributes to the control of normal root growth direction and salt tolerance as well as the resistance to *Fusarium oxysporum*, a fungal root pathogen (Julkowska et al. 2017; Van der Does et al. 2017; Coleman et al. 2021). In addition, there is evidence that MIK2 constitutes part of a pollen tube receptor complex that is required for pollen tube guidance (Wang et al. 2016). The function of *MIK2* in the regulation of root growth direction and salt tolerance was found to depend on *THE1* while its role in the defence against *F. oxysporum* did not, in line with the notion that *THE1* does not participate in pattern-triggered immunity (Engelsdorf et al. 2018). Current data indicate that MIK2 regulates immune signalling by directly binding to peptides of the SERINE RICH ENDOGENOUS PEPTIDE 12 (SCOOP12) family of secreted peptides (Rhodes et al. 2021). Binding of SCOOP12 peptide to MIK2 induces heterodimerisation of MIK2 with its co-receptor BAK1, eliciting immunity-related downstream responses. It remains to be investigated if a MIK2/BAK1 complex is also involved in the CBI response. However, there is evidence that BAK1 does not play a direct role in promoting this process (Engelsdorf et al. 2018), suggesting that MIK2 may exert this function together with a different co-receptor.

The receptor kinase SUB represents another central factor in the signalling network regulating the CBI response in seedlings (Chaudhary et al. 2020). Genetic evidence suggests that SUB is a direct regulator of isoxaben-induced lignin and callose accumulation in the seedling root and cotyledons, respectively. SUB, also known as SCRAMBLED (SCM), has functions beyond the regulation of the CBI response. It was shown to regulate ovule development, floral morphogenesis, leaf development, and root hair patterning (Chevalier et al. 2005; Kwak et al. 2005; Lin et al. 2012). A detailed genetic analysis indicated that SUB does not provide specificity for the individual developmental processes (Vaddepalli et al. 2011). Thus, it was suggested that SUB functions as a co-receptor that may interact with several separate signalling receptors. In line with this notion, the extracellular domains of SUB carry seven LRRs and thus resemble the short extracellular domains of BAK1 or the FEIs. There is additional genetic evidence supporting this view (see below).

SUB is located at the plasma membrane and at plasmodesmata (Vaddepalli et al. 2014). At plasmodesmata, SUB physically interacts with the 'multiple C2 domains with two transmembrane regions protein' (MCTP) family member QUIRKY (QKY). Current evidence suggests that QKY maintains SUB at the plasma membrane as *qky* loss-of-function mutations lead to reduced levels of SUB at the plasma membrane (Gao et al. 2019; Song et al. 2019; Chaudhary et al. 2021). While QKY is required for all known developmental functions of SUB, such as root hair patterning or ovule development, its role in the isoxaben-induced CBI response is restricted to the promotion of lignin accumulation in the seedling root. This finding suggested that a SUB/QKY complex at plasmodesmata is required for isoxaben-induced lignin accumulation while SUB is part of one or several other cell surface complexes at the plasma membrane that regulate different aspects of the CBI response, such as callose accumulation (Chaudhary et al. 2020). Alternatively, lignin production may require higher levels of SUB at the plasma membrane compared to other CBI-induced downstream responses. The role of other known genetic components of SUB signalling in the isoxaben-induced CBI response remains to be investigated.

THE1, *FEI2*, and *MIK2* appear to function in a different genetic pathway than *SUB*, as *SUB*, *THE1*, and *MIK2* differentially affect expression of marker genes and jasmonate accumulation (Denness et al. 2011; Van der Does et al. 2017; Chaudhary et al. 2020). In addition, *SUB* is required for the inhibition of root growth in *prc1-1* mutants while *THE1* plays such a role in the inhibition of hypocotyl growth (Hématy et al. 2007; Chaudhary et al. 2020). By contrast, *FEI2* appears to negatively affect root growth inhibition in *prc1-1* (Xu et al. 2008). The finding that *THE1* does not affect root growth of *prc1-1* mutants (Chaudhary et al. 2020) suggests that *FEI2* and *THE1* may also have separate functions in the CBI response. Moreover, while *MIK2* controls root skewing in plate-grown seedlings in a *CESA6*-dependent manner, it apparently does not influence root growth of the *prc1-1* mutant (Van der Does et al. 2017). It should be noted that such a global analysis of genetic pathways can be misleading. It is perhaps not unlikely that specific downstream responses to cellulose reduction are regulated by corresponding response-specific pathways and with some potential overlap with respect to participating signalling factors. For example, it is conceivable that *SUB* and *THE1* interact in the control of lignin accumulation in the seedling root upon isoxaben treatment but not in the control of root and hypocotyl growth in *prc1-1* seedlings. It is thus important to precisely define the signalling readout and its potential temporal and cell-type specificity in further biochemical and cell biological studies of this signalling network.

There are likely additional pathways responsive to altered cellulose content or quality, as, for example, mutants carrying a defect in *COBRA*, encoding a GPI-anchored protein required for cellulose biosynthesis (Roudier et al. 2005), show misregulation of ~250 genes (Sorek et al. 2015). A second site mutation in the MEDIATOR complex suppresses the *cobra* phenotype by inducing expression of PMEIs, partially recovering gene expression changes and cellulose content. Pectin in *cobra* is highly de-methylesterified and the PMEI-mediated recovery suggests an intimate connection between cellulose biosynthesis and pectin modification, as well as signalling surveillance of either or both processes (Sorek et al. 2015).

CELL SURFACE RECEPTORS MONITORING PECTIN STATUS

FER stands out among all currently known cell wall signalling components as cell wall binding can be linked to a measurable signalling output, tied to a developmental programme. FER binds to de-methylesterified pectin through its MalectinA region and is required for maintaining cell wall integrity during salt stress (Feng et al. 2018; Lin et al. 2022). Notably, *fer* mutants, which show enhanced cell bursting under ionic stress conditions, can be rescued by pectin cross-linking agents such as calcium ions and borate, suggesting that displacement of calcium ions by monovalent cations could be a primary way in which plants experience salt stress (Feng et al. 2018). Alternatively, PME activity might be affected by those ions directly. A seminal study reported that pectin, most likely in its Ca^{2+}-cross-linked form, can directly activate FER signal transduction. This, in turn, activates ROP activity, organising microtubule bundling and epidermal pavement cell morphogenesis. In agreement with this, *rop*, *fer*, and *pme* mutants as well as PMEI overexpressing plants all share the same pavement cell defect with decreased lobing (Lin et al. 2022). Pectin methylesterification status has also been implicated as a determinant of the mechanical asymmetry between the different cell wall regions in lobed pavement cells (Majda et al. 2017; Bidhendi et al. 2019; Haas et al. 2020). Thus, further studies will have to disentangle the mechanical and signalling contribution of pectin modification, but one attractive hypothesis is that initial pectin asymmetry leads to signalling-mediated wall reinforcement, including microtubule-guided cellulose biosynthesis, to orchestrate and accentuate the complex growth behaviour of interdigitating cells. Notably, FER is also required for the alignment of microtubules to stress patterns in pavement cells (Tang et al. 2022). Therefore, one remaining question is how mechanical force is translated into biochemical activity and whether this requires additional components besides the pectins, for example, RALF peptides.

An independent line of evidence suggests that the state of pectin modification is also monitored by other cell surface receptors. Pharmacological or genetic interference with pectin de-methylesterification leads to an activation of brassinosteroid (BR) signalling, which in turn enacts a compensatory upregulation of pectin biosynthesis and PME activity (Wolf et al. 2012b). This feedback signalling from the cell wall to the BR signalling pathway, which controls a large number of cell wall–related genes (Sun et al. 2010), requires RECEPTOR-LIKE PROTEIN 44 (RLP44), which is sufficient to active BR signalling by interacting directly with the BR receptor BRASSINOSTEROID INSENSITIVE1 (BRI1) and its co-receptor BAK1, thus promoting BR receptor complex formation independently of the BR ligand (Wolf et al. 2014; Holzwart et al. 2018). While it remains to be seen how exactly this cell surface complex senses cell wall signals, it is required for maintaining cell wall integrity under stress conditions and, together with phytosulphokine signalling, is involved in the maintenance of vascular cell fate (Wolf et al. 2014; Holzwart et al. 2018; Holzwart et al. 2020).

Possibly, the earliest cell wall signalling component candidates to be identified (He et al. 1996; Wagner and Kohorn 2001), the wall-associated kinases (WAKs), represent another class of receptor kinases implied in monitoring changes to pectin (Kohorn 2016). The extracellular domains of WAKs contain a linker separating an N-terminal galacturonan binding (GUB) domain that binds pectin *in vitro* from two C-terminal EGF-like repeats that precede the transmembrane domain. The extracellular domains of WAK1 and WAK2 bind long polymers of cross-linked HG and OGs with a DP between 9 and 15 *in vitro* (Decreux and Messiaen 2005; Decreux et al. 2006; Kohorn et al. 2009) and domain-swap experiments suggest WAK1 to function as an OG receptor *in vivo* (Brutus et al. 2010). The five Arabidopsis *WAK* genes are located within a 30 kb genomic interval and genetic studies of *WAK*s have been hampered by lethality and possible redundancy. The current notion of WAK function states that WAKs differentially regulate developmental and stress responses depending on whether WAKs bind longer pectin fragments or OGs, respectively.

CELL WALL SIGNALLING AND THE PLANT'S IMMUNE RESPONSE

Modifications of cell wall composition, such as in cell wall mutants, are often associated with altered disease resistance. Interestingly, the same cell wall mutation can affect susceptibility to different pathogens in opposite ways. However, a recent study found that the abundance of certain cell wall epitopes correlates with and predicts resistance to various pathogens (Molina et al. 2021). While it is possible that altered mechanics or biochemical differences in cell wall–related mutants directly impede or favour pathogen attack, a significant fraction of these cell wall alteration–mediated effects on immunity occurs through signalling. In the simplest scenario, cell wall variability leads to differential production of cell wall breakdown products acting as DAMPs, triggering immune responses (Hou et al. 2019; Pontiggia et al. 2020) (Chapter 20). During recent years, more and more cell wall-derived DAMP activities besides the classical pectic OGs (Hahn et al. 1981; Nothnagel et al. 1983) have been reported. For example, pectic cell wall fractions of a mutant with a defect in the cytokinin response regulator ARR6 trigger calcium spikes, MITOGEN-ACTIVATED PROTEIN KINASE (MAPK) activation, and gene expression changes similar to pathogen-derived elicitors (Bacete et al. 2020). Further biochemical characterisation of these fractions revealed that β-1,4-xylooligosaccharides are sensed as DAMPs and that the particularly potent compound 3^3-α-L-arabinofuranosyl-xylotetraose triggers strong immune responses in Arabidopsis and crop plants when applied externally (Mélida et al. 2020). This nicely shows that studying cell walls of mutant plants can pave the way for fundamental insights and promising applications, despite the signalling-mediated, secondary nature of the associated mutant phenotypes. In line with this, several mutant cell wall fractions were shown to contain so far uncharacterised DAMP activities acting through a variety of signalling pathways (Molina et al. 2021). These DAMPs apparently are of diverse chemical nature. For example, hemicellulose-derived tri- and tetra-saccharides were recently identified to activate immune responses in rice during infection with *Magnaporthe oryzae* (Yang et al. 2021). Other hemicellulose-derived DAMPs include mixed-linkage glucan (β-1,3/1,4-glucan) oligosaccharides (Barghahn et al. 2021; Rebaque et al. 2021; Yang et al. 2021) and xyloglucan oligomers (González-Pérez et al. 2014; Claverie et al. 2018). Breakdown products of cellulose (Aziz et al. 2007; Souza et al. 2017; Johnson et al. 2018; Locci et al. 2019), and callose (Mélida et al. 2018; Wanke et al. 2020), as well as cutin have been shown to trigger immune responses (Fauth et al. 1998; Chassot et al. 2007). By far, the best-characterised DAMPs are pectin HG-derived OGs (Hahn et al. 1981; Nothnagel et al. 1983; Simpson et al. 1998; Decreux and Messiaen 2005; Decreux et al. 2006; Moscatiello et al. 2006; Ferrari et al. 2007; Casasoli et al. 2008; Galletti et al. 2008; Kohorn et al. 2009, 2016; Brutus et al. 2010; Benedetti et al. 2015; Mattei et al. 2016; Davidsson et al. 2017; Voxeur et al. 2019). Active OGs typically have a degree of polymerisation of 3–15 DP and are generated by PGs. A recent study suggests that some pathogens might evade the OG-derived immune response by employing pectin lyase instead of PGs, resulting in OGs associated with reduced virulence (Voxeur et al.

2019). Active OGs can be produced by both pathogen- and plant-derived enzymatic activities but require prior demethylesterification by PME (Wakabayashi et al. 2003; Osorio et al. 2008). On the other hand, PMEIs, inhibitors of PMEs, are required for full resistance to *Botrytis cinerea* (Lionetti et al. 2017), whereas overexpression of PMEIs increases resistance (Lionetti et al. 2007; Volpi et al. 2011). Conversely, PMEs and pectin biosynthesis have been shown to contribute to resistance against *P. syyringae* (Bethke et al. 2014, 2016). Notably, in addition to providing the substrate for PGs, and locally lowering the pH, which might alter cell wall enzyme activities, PME activity also releases methanol, which can have signalling activity in its own right (Hann et al. 2014; Komarova et al. 2014). These findings reflect the complexity of pectin functions and modification and demonstrate the need to understand cell wall signalling responses in order to interpret mutant phenotypes.

Apart from OGs, which have been suggested to be recognised by WAK1 (Brutus et al. 2010), receptors for any of the described cell wall–derived DAMPs have yet to be found. A variety of cell wall mutants show signs of a constitutively active defence response. For example, defence and stress-related genes are overrepresented among THE1-regulated transcripts in cellulose synthase mutants (Hématy et al. 2007). In addition, the *cev1* and *eli1* mutants with a lesion in CESA3, a key enzyme for primary cell wall cellulose biosynthesis, shows increased levels of jasmonate and ethylene, resulting in constitutive stress gene expressions and enhanced resistance against powdery mildew (Ellis and Turner 2001; Ellis et al. 2002; Caño-Delgado et al. 2003). Activation of defence response can also be invoked by isoxaben-induced reduction of cellulose biosynthesis, while mutants in secondary cellulose biosynthetic genes show altered performance under pathogen attack that is jasmonate, salicylic acid, and ethylene independent but requires abscisic acid signalling (Hernández-Blanco et al. 2007). Altered pectin composition can result in increased resistance towards powdery mildew (Vogel et al. 2002; Vogel et al. 2004), and it was hypothesised that the increased availability of oligosaccharide precursors is causative for this effect. Mutants in hemicellulose biosynthesis or modification also show altered defence signalling, as shown for mutants with defects in mixed-linkage glucan content in rice (Vega-Sánchez et al. 2012) and plants with a defect in ESKIMO1, a xylan acetyltransferase (Xiong et al. 2013; Yuan et al. 2013; Urbanowicz et al. 2014; Escudero et al. 2017).

Alteration in lignin, a target of interest for generating optimised bioenergy crops, often leads to growth penalties that are signalling-mediated and can involve defence pathways. For example, dwarfism of Arabidopsis *hct* mutants, affected in the monolignol pathway and exhibiting reduced lignin accumulation, is caused by enhanced salicylic acid levels and can be rescued by interfering with salicylic acid signalling without altering lignin content (Gallego-Giraldo et al. 2011a). Similarly, alfalfa *hct* mutants show enhanced defence gene expression and ROS production through upregulation of salicylic acid, jasmonate, and abscisic acid signalling, resulting in enhanced resistance against fungal infection (Gallego-Giraldo et al. 2011b). This is likely at least in part mediated by release of cell wall breakdown products as mild cell wall extraction released more pectin than in the wildtype and exposing plants to these extracts triggered defence gene activation. Lignin biosynthesis is often induced as a response to pathogen attack (Vance et al. 1980; Lange et al. 1995), presumably to reinforce cell wall mechanics. However, the above-mentioned results show an intimate relationship between lignin content and/or quality and the release of pectic fragments triggering defence responses. This was recently corroborated by observations involving altered expression of *F5H*, which encodes the enzyme that catalyses the first step in S lignin formation. Extracts of *F5H* overexpression plants as well as *f5h* mutants not only differentially triggered defence gene expression in a pectinase-sensitive manner (Gallego-Giraldo et al. 2018) but also contained DAMPs derived from hemicellulose and/or AGPs. In Arabidopsis *hct* and *ccr1* mutants, defective in central biosynthetic enzymes in the monolignol pathway, production of a defence-active OG DAMP could be linked to ectopic expression of a PG, likely induced by an unknown cell wall signalling pathway responsive to the primary effect of lignin alteration (Gallego-Giraldo et al. 2020). Conversely, reduced pectin biosynthesis leads to decreased lignin and decreased recalcitrance (Li et al. 2019). Interestingly, downregulation of pectin biosynthesis in some crop plants is apparently not countered by an extensive signalling response but reduced cell wall integrity is assumed to promote growth and saccharification yield (de Godoy et al. 2013; Biswal et al. 2015). Considering this, the complex interaction between pectin and lignification, which is also reflected by the observation that the latter seems to start in the pectin-rich middle lamella, deserves reviewed attention.

CELL WALL SIGNALLING IN GROWTH AND DEVELOPMENT

During plant development, individual cells deploy diverging cell walls based on their cell fate specification programme. Thus, cell walls are part of the specific differentiation mechanisms instructed by development. However, the cell wall can also provide feedback on developmental programmes. For example, it has long been suggested that oligosaccharides can antagonise auxin signalling (York et al. 1984; Branca et al. 1988; Bellincampi et al. 1993; Bellincampi et al. 1996). More recently, Sinclair et al. have shown that photomorphogenesis can be inhibited by small pectic OGs (Sinclair et al. 2017). In the following, we review the available information on developmental control potentially or evidently exerted by cell wall properties.

CELL FATE CONTROL

Does the cell wall play a role in the regulation of cell fate? In animal systems, the extracellular matrix is known to contribute to cell fate control via biochemical and mechanical signals (Hynes 2009; Watt and Huck 2013). In Fucus brown algae, laser microsurgery of two-celled embryos revealed a function of the cell wall in maintaining cell fate of rhizoid and thallus cells during subsequent early development (Berger et al. 1994). Several observations indicated that the cell wall also plays a role in cell fate control in plants. Protoplasting plant cells results in major dedifferentiation events as indicated by the ability of individual protoplasts to develop into full plants accompanied by changes in gene expression, chromatin status, and reactivation of transposons (Davey et al. 2005; Chupeau et al. 2013). However, the signalling mechanisms underlying the cell wall–mediated control of cell fate remain unknown and the unavoidable deprivation from positional signals represents an important caveat with protoplasting experiments.

The control of root hair fate in the epidermis of the Arabidopsis root meristem has served as a prominent model to study the spatial control of cell fate. Available evidence strongly supports the notion that cell wall signalling contributes to the regulation of this process. At its core, the mechanism regulating root hair patterning results in the differential accumulation of a transcription factor complex (WER-bHLH-TTG) containing WEREWOLF (WER), TRANSPARENT TESTA GLABRA 1 (TTG1), and two basic helix-loop-helix proteins in prospective hair and non-hair cells (Schiefelbein et al. 2014). In developing non-hair cells, the WER-bHLH-TTG complex promotes transcription of the *GLABRA2* (*GL2*) and *CAPRICE* (*CPC*) transcription factor genes. GL2 in turn promotes non-hair development. CPC moves to the neighbouring cells destined to become hair cells where it interferes with WER-bHLH-TTG complex formation. The expression, activity, and movement of some of the involved transcription factors also depends on signalling mediated by the receptor kinases BRI1 and SUB (Kwak et al. 2005; Kuppusamy et al. 2009; Cheng et al. 2014). The available evidence indicates that BRI1 signalling promotes *GL2* expression and non-hair cell formation by removing an inhibitory phosphorylation of TTG1 (Cheng et al. 2014). SUB is thought to repress *WER* expression in hair cells and to promote movement of CPC between epidermal cells (Kwak and Schiefelbein 2007; Kwak and Schiefelbein 2008; Cheng et al. 2014; Song et al. 2019).

Interestingly, recent evidence suggests that the cell wall contributes to the specification of root hair fate by affecting BR- and SUB-mediated signalling. First, treatment of roots with β-glucosyl Yariv, which binds *O*-glycans of the cell wall–localised AGPs leading to AGP aggregation and disruption of native protein distribution (Kitazawa et al. 2013), was reported to result in root hair patterning defects (Borassi et al. 2020). The data indicate the underlying mechanism to involve a not yet fully understood complex crosstalk with BR signalling. The treatment impaired the control of *GL2* expression and non-hair fate and thus phenocopied plants defective in *BRI1* activity. A combination of genetic and pharmacological experiments indicated that *O*-glycosylated AGP21 represents a central component of the mechanism. In line with these findings, a triple mutant lacking activity of three hydroxyproline *O*-galactosyltransferases involved in the biosynthesis of AGPs also showed increased root hair density (Ogawa-Ohnishi and Matsubayashi 2015). Second, exposing Arabidopsis seedlings to sub-lethal doses of isoxaben was observed to result in altered root hair patterning, as indicated by altered spatial expression of a reporter for *GL2* expression and an aberrant distribution of root hairs (Chaudhary et al. 2021). Moreover, isoxaben-induced severe morphological defects in floral and ovule development. Interestingly, similar defects are characteristic for the phenotype of *sub* mutants (Chevalier et al. 2005; Kwak et al. 2005). This led to the hypothesis that the corresponding mechanism involves the downregulation of *SUB* activity. Indeed, it was found that isoxaben-induced changes in the architecture of SUB complexes at the plasma membrane, elevated internalisation of SUB that was independent of QKY, and eventually reduced *SUB* transcript levels. This likely results in altered *WER* expression and CPC movement. Ectopic expression of *SUB* resulted in the attenuation of isoxaben-induced defects, suggesting that the main effect of isoxaben on root hair patterning and tissue morphogenesis is mediated via a control of SUB activity. It remains to be investigated what molecular signal induces this response, but the collective evidence is compatible with a cell wall–derived component (Chaudhary et al. 2021).

PLANT CELL POLARITY

Cell polarity represents an essential aspect of plant growth and physiology. It refers to the asymmetric distribution of individual molecules or higher-order assemblies of molecules at the cellular level. In a tissue context, cell polarity often appears to be coordinated between neighbouring cells. Thus, this type of tissue cell polarity is thought to be interpreted by a coordinate system that provides direction for a vast array of different developmental or physiological processes that are oriented along major body axes, such as the apical-basal (shoot-root) axis, or the proximal-distal axis in leaves. For example, cell polarity may influence the orientation of the cell division plane during asymmetric cell division, leading to the formation of new cell types and tissues, the orientation of growth in anisotropically expanding cells, or the subcellular distribution of importers and exporters required for the directed transport of nutrients (Muroyama and Bergmann 2019; Elliott and Kirchhelle 2020; Ramalho et al. 2021). Prominent examples of asymmetrically distributed proteins include several members of the PIN-FORMED (PIN) family of plasma membrane–localised auxin efflux transporters (Adamowski and Friml 2015) differentially distributed along the apical-basal axis, such as PIN1 or PIN2 (Gälweiler et al. 1998; Müller et al. 1998), the boron exporter REQUIRES HIGH BORON1 (BOR1), a plasma membrane protein that localises to the inner face of root epidermal and endodermal cells (Takano et al. 2010), or members of the SOSEKI (SOK) family of peripheral membrane-associated proteins, each localising to distinct cell corners and edges in a cell-type-specific manner (Yoshida et al. 2019).

The cell wall can exhibit polarity in its organisation. The geometry of the polyhedral cells making up a plant tissue includes faces, edges, and vertices and these geometric entities can be characterised by distinct biochemical properties (Elliott and Kirchhelle 2020). As a rule, anisotropic cell elongation in diffusely growing cells is accompanied by a perpendicular orientation of cellulose microfibrils (Green 1962). Moreover, it has been established that separate faces of a cell can differ in microfibril anisotropy (Chan et al. 2011; Crowell et al. 2011). There is also evidence that different faces can show distinct cell wall matrix composition (Peaucelle et al. 2015) while cell edges frequently accumulate de-methylesterified pectin or phenolic compounds (Elliott and Kirchhelle 2020).

The mechanism regulating tissue cell polarity, also known as axial polarity, in plants is poorly understood (Muroyama and Bergmann 2019; Ramalho et al. 2021). The process of establishing cell polarity can be subdivided into several discrete steps: symmetry breaking, reinforcement, and implementation. Several components thought to play a role in polarity reinforcement have been identified. They constitute complexes of intracellular cortical proteins with asymmetrical localisation. They include a complex containing BREAKING OF ASYMMETRY IN THE STOMATAL LINEAGE (BASL), and members of the BREVIS RADIX (BRX) family that is involved in cell specification during early stomatal development (Dong et al. 2009; Rowe et al. 2019), or a complex containing PROTEIN KINASE ASSOCIATED WITH BRX (PAX), BRX, and phosphatidylinositol-4-phosphate 5-kinases (PIP5K1/2) required for the maintenance of a local PIN minimum in protophloem differentiation (Marhava et al. 2018, 2020). These two distinct protein complexes are considered to belong to polarity mechanisms that are transiently required for the regulation of specific cell fate and cell differentiation processes (Muroyama and Bergmann 2019; Ramalho et al. 2021). By contrast, the SOK and PIN proteins were suggested to be part of an evolutionary ancient universal polarity system that extends beyond individual cell or tissue types and aligns with major body axes (Yoshida et al. 2019; van Dop et al. 2020; Ramalho et al. 2021).

It is conceivable that the structural and biochemical asymmetries of the cell wall contribute to a feedback loop regulating tissue cell polarity. Interestingly, polar subcellular localisation of SOK proteins appeared to be unaffected by impairments in membrane trafficking or cytoskeletal organisation but depended on an undisturbed cell wall as treatment with cell wall degrading enzymes resulting in aberrant SOK localisation (Yoshida et al., 2019). Asymmetric de-methylesterification associated with breaking growth symmetry during the transition from isotropic to anisotropic growth in the hypocotyl epidermis of dark-grown seedlings preceded noticeable changes in cortical microtubule and cellulose microfibril orientation. This process was still occurring upon genetically interfering with cortical microtubule reorientation (Peaucelle et al. 2015). Such observations may point to an instructive role of the cell wall in the control of early steps of cell polarisation, but a detailed understanding of the role of the cell wall in tissue cell polarity is lacking. Most of the available evidence indicates a role of the cell wall in maintaining cell polarity. For example, the BASL protein showed polar localisation in stably transformed BY-2 cell filaments. Interestingly, BASL still underwent dynamic relocalisation patterns and assumed asymmetric distributions in BY-2 cell protoplasts (Chan et al. 2020) and BASL distribution was aligned with the growth axis of regenerating protoplasts. It was proposed that establishing a polarity axis results in asymmetries in cell wall biochemistry and subsequent anisotropic growth, which in turn may reinforce the localisation of a polarised molecular address (Chan et al. 2020). In line with such a notion, the cell wall is required for fixation of the primary developmental axis in Fucus brown algae, the cell walls of which are in many ways similar to those of plants (Kropf et al. 1988).

Present data suggest that the cell wall is required for the asymmetric distribution of at least a subset of cell polarity factors, such as PINs or SOKs. The regulation of subcellular localisation of PINs is very well characterised. Experimental and computational evidence suggested that a combination of super-polar recycling, limited lateral diffusion, and spatially defined internalisation of PINs provides a robust mechanism for maintaining PIN polarity (Kleine-Vehn et al. 2011; Li et al. 2021a). Interestingly, however, the cell wall status can have diverse effects on PIN localisation, ranging from partial or complete removal of PINs from the plasma membrane to distinct defects in PIN polar localisation. For example, apical hook formation in the Arabidopsis seedling is caused by differential cell elongation regulated by auxin (Mazzella et al. 2014). In the current model, a positive feedback loop between auxin distribution and cell wall modifications establishes an asymmetry in cell wall mechanochemical properties and promotes tissue bending in apical hook formation (Aryal et al. 2020; Jonsson et al. 2021). Cells on the concave or inner side of the apical hook were found to exhibit higher levels of HG methylesterification and increased stiffness in comparison to the convex or outer side. The concave side also accumulated PIN1 and a prominent auxin response domain while the convex side showed very little, if any, PIN1. Interfering with asymmetric HG methylesterification by overexpressing PECTINMETHYLESTERASE INHIBITOR 5 (PMEI5) resulted in the absence of PIN1 from the plasma membrane. While the level of other PINs at the plasma membrane, such as PIN3 and PIN4, was reduced, their subcellular distribution appeared unaffected (Jonsson et al. 2021).

The experiments manipulating the cell wall of apical hook cells indicate that increasing cell wall stiffening can lead to PIN1 clearance from the plasma membrane. Cell wall loosening can induce an increase in PIN1 levels at the plasma membrane and affect PIN1 polarity as indicated by the observation that isoxaben-treated shoot apical meristem cells exhibit stronger PIN1 signals and a more focussed subcellular PIN1 localisation in comparison to untreated cells (Heisler et al. 2010). The isoxaben-induced change in the subcellular PIN1 pattern was interpreted to be a consequence of altered mechanical stress (see also below). Alternatively, the cell wall could influence PIN1 localisation in a more direct manner. In yet another set of experiments, it was found that PIN1 showed polar localisation in BY-2 cell filaments but became distributed throughout the entire cell circumference upon protoplasting (Boutté et al. 2006). Apparently, the cell wall can also limit lateral diffusion of PINs in Arabidopsis root epidermal cells as genetic or chemical interference with cellulose biosynthesis, or mechanical disruption of the cell wall, resulted in a lateral spreading of PINs in these cells (Feraru et al. 2011). The authors proposed a model in which the cell wall contributes to the restriction of spatial domains by counteracting the unrestricted spreading of polarity regulators through cellulose-based mechanical cell wall–plasma membrane connections. Such a connection appears to require the integrin-like protein NON-RACE-SPECIFIC DISEASE RESISTANCE1 (NDR1) (Li et al. 2021a) a protein involved in cell wall–plasma membrane adhesion (Knepper et al. 2011). This notion is in line with the observations that proteins with extracellular domains are generally limited in lateral diffusion in a cell-wall-dependent fashion (Martiniere et al. 2011; Martinière et al. 2012). In addition, the cell wall was found to regulate the dynamics and size of nanodomains in Arabidopsis (McKenna et al. 2019). Nanodomains are spatially and functionally distinct domains of the plasma membrane that organise signalling components, such as receptor kinases (Jarsch et al. 2014; Bücherl et al. 2017; Gronnier et al. 2017; Pan et al. 2020). A role of the cell wall in limiting lateral movement appears to be at odds with the noticeably more focussed localisation of PIN1 in isoxaben-treated shoot apical meristem cells in comparison to untreated controls (Heisler et al. 2010). The disparity in subcellular PIN distribution upon altered cell wall biochemistry between the various cell types may be rooted in differences in cell wall treatment or in cell-type-specific variations in the complex regulatory mechanism underlying PIN localisation. Further experiments are required to resolve the apparent discrepancy in the results.

CELL WALL SIGNALLING INDUCED BY MECHANICAL CUES

The cell wall is a major determinant of the cell mechanics and can act as conduit for tissue-level stress patterns. Therefore, sensing of mechanical stress might directly or indirectly involve the cell wall. It has long been known that microtubules re-align around a wounding site (Hush and Overall 1992; Geitmann et al. 1997). Later, it was revealed that microtubules align along stress patterns in many tissues, which presumably will affect cellulose depositioning downstream of mechanical cues

(Hamant et al. 2008; Uyttewaal et al. 2012; Jacques et al. 2013; Sampathkumar et al. 2014; Bidhendi et al. 2019). It is less clear whether the cell wall–cytoskeleton nexus is also involved in orienting microtubules downstream of mechanical stress. Evidence supporting this notion comes from the observation that cellulose synthase mutants and plants treated with isoxaben show altered microtubule alignment (Paredez et al. 2008). In addition, a combination of biochemical and modelling approaches suggests that central changes in the mechanics of the primary cell wall in onion epidermal cells relate to alterations of the cellulose network, not the hemicellulosic or pectic components (Zhang et al. 2021). Simulations involving a coarse-grained model based on polymer physics suggested several microfibril movements during wall stretching including microfibril straightening, curving, or sliding. It is therefore conceivable that mechanisms monitoring such parameters elicit mechanics-related signal transduction orienting cortical microtubules and the synthesis of cellulose microfibrils. Notably, microtubules also align with stress patterns above a certain stress threshold in protoplasts, i.e., in the absence of a fully developed cell wall (Colin et al. 2020), whereas below that threshold, cell geometry seems to be the main determinant of microtubule orientation (Durand-Smet et al. 2020). Another potential scenario in which mechanical signals are transduced through the cell wall involves the receptor kinase FER, which has been suggested to act as a mechano-sensor and was reported, along with its LEUCINE-RICH REPEAT EXTENSIN (LRX) interaction partners to form direct contacts with the cell wall (Shih et al. 2014; Feng et al. 2018; Lin et al. 2022; Tang et al. 2022). FER was recently suggested to mediate rapid microtubule alignment with stress patterns in pavement cells (Tang et al. 2022), whereas another study reported that the FER mechano-signalling pathway is genetically separable from the long-term microtubule stress readout (Malivert et al. 2021), suggesting there is more than one mechanism for microtubule realignment. Other CrRLK1L receptor kinases besides FER have also been implicated in mechano-transduction together with their RALF ligands; however, the contribution of the cell wall to this process is not yet clear (Zhou et al. 2021).

Despite these advances, much remains to be discovered regarding the contribution of the cell wall to the perception of mechanical cues. The search for similarities with the yeast cell wall integrity monitoring pathway yielded the identification of MCA1 and 2, stretch activated channels (Yoshimura et al. 2021) that complement the yeast *MATING PHEROMONE INDUCED DEATH 1* (*MID1*) mutant (Nakagawa et al. 2007; Yamanaka et al. 2010) and regulate Ca^{2+} influx through the plasma membrane

(Kurusu et al. 2012) required under hypo-osmotic stress conditions and upon mechanical stimulus (Kurusu et al. 2013; Hamant and Haswell 2017) as well as for root penetration into hard agar (Nakagawa et al. 2007). While MCA1 is required for the full response to CBI (Engelsdorf et al. 2018), it remains to be seen whether and in which way the cell wall is upstream of MCA1.

Recently, the Fasciclin-like AGPs FLA11 and FLA12 have been implicated in tuning the properties of secondary cell walls in response to mechanical stress (Ma et al. 2022). As the GPI-anchored, heavily glycosylated FLAs can link the cell wall and the plasma membrane and evidence exists for signalling functions of AGPs (Motose et al. 2001, 2004; Xu et al. 2008; Harpaz-Saad et al. 2011, 2012; Griffiths et al. 2014), they have long been regarded as potential cell wall sensor proteins, possibly cooperating with membrane-spanning kinases (Seifert 2018; Turupcu et al. 2018). The report by Ma et al. (2021) is supporting this idea and should renew interest into this enigmatic protein family.

THE CELL WALL AND THE CONTROL OF PLASMODESMATA CONDUCTIVITY

Plasmodesmata mediate intercellular communication by enabling passage of a wide range of factors, including small molecular weight molecules, RNAs, proteins, and viruses (Sager and Lee 2018; Reagan and Burch-Smith 2020; Li et al. 2021b). For example, radial patterning in the Arabidopsis root meristem relies on mobile factors, such as the transcription factor SHORT ROOT (SHR) and the microRNAs 165/6, which move between cell layers through plasmodesmata (Carlsbecker et al. 2010; Vaten et al. 2011). Plasmodesmata traverse the cell wall separating two neighbouring cells and it therefore seems likely that the cell wall affects various aspects of plasmodesmata biology. Very little is known, however, about the possible role of the cell wall in these processes. Arguably, the best understood mechanism relates to the control of plasmodesmata permeability through the regulated turnover of the β-1,3 glucan callose. Callose accumulation occurs between the cell wall and the plasma membrane at the orifice of plasmodesmata and is induced by a variety of stresses and developmental factors. High levels of callose limit the passage of cargo through plasmodesmata while lower levels facilitate transport. Turnover of callose is achieved by the antagonistic activities of two enzymes, callose synthase and β-1,3 glucanase, that synthesise and degrade callose, respectively (Wu et al. 2018). Evidence for a biological function of callose turnover at plasmodesmata has, for example, been obtained in stomatal development. Reduced activity of *CHORUS* (*CHOR*), encoding callose synthase family member CalS10 (also known as GLS8), results in increased plasmodesmata permeability, abnormal movement of the transcription factor SPEECHLESS (SPCH) to neighbouring cells, and excessive proliferation of early stomatal-lineage cells (Guseman et al. 2010).

Interestingly, it has recently been found that addition of callose reduces stiffness and increases elasticity of cellulose hydrogels (Abou-Saleh et al. 2018). The results suggest molecular interactions between the polymers, disruption of the organisation of cellulose, a weakening of intermolecular bonding, and easier sliding of the polymer chains. These findings potentially open exciting new research directions. If the presence of callose in the cell wall at plasmodesmata leads to similar effects on cellulose, it is conceivable that callose may promote cell wall signalling events induced by increased cellulose elasticity that would emanate from plasmodesmata. However, it remains to be demonstrated if callose similarly influences cellulose properties in plants. Moreover, a reduction in cellulose biosynthesis does not appear to affect plasmodesmata permeability during stomatal development (Kong et al. 2012).

The Arabidopsis plasmodesmata proteome is characterised by many signalling proteins, including receptor kinases and GPI-anchored proteins (Fernandez-Calvino et al. 2011; Brault et al. 2019). Several rice receptor kinases isolated from cell wall fractions localised to plasmodesmata, indicating a possible function of plasmodesmata in cell wall signalling (Jo et al. 2011). These findings led to the concept of plasmodesmata acting as hubs for signalling factors (Maule et al. 2011). Some plasmodesmata-localised cell surface proteins are known to affect callose turnover, such as the β-1,3 glucanase PdBG1 involved in lateral root patterning (Benitez-Alfonso et al. 2013). The receptor kinases QSK1 and INFLORESCENCE MERISTEM KINASE2 (IMK2) localise to plasmodesmata upon osmotic stress (Grison et al. 2019). QSK1 is required for the response to mannitol stress, callose synthesis, and the correct timing of lateral root development. Moreover, a complex of the receptor kinase SUB and the MCTP protein QKY at plasmodesmata appears to be required for tissue morphogenesis and cell wall signalling (see above).

Plasmodesmata can form simple or complex branched morphologies (Ehlers and Kollmann 2001). De-novo formation of secondary plasmodesmata through existing cell walls occurs as well and usually results in branched plasmodesmata. The different morphologies are associated with distinct trafficking behaviours (Oparka et al. 1999). These observations raise exciting but largely unexplored questions, such as if and how cell wall remodelling contributes to these fascinating processes. Indeed, apart from differential callose accumulation the cell wall at plasmodesmata can have distinct biochemical characteristics. For example, pectin of the cell wall surrounding pit fields, clusters of plasmodesmata, in tomato fruit pericarps and tobacco trichomes is enriched in α(1,5)-arabinan (Orfila and Knox 2000; Faulkner et al. 2008). There is evidence for an interaction of pectin-modifying factors and the control of viral movement factors (Reagan and Burch-Smith 2020). In tobacco, data obtained from blot overlay binding and yeast two-hybrid assays suggest that a tobacco PME interacts with the MP30 viral movement protein of tobacco mosaic virus (Dorokhov et al. 1999; Chen et al. 2000). In addition, downregulation of this PME in antisense plants indicated a role in the local and systemic movement of tobacco mosaic virus (Dorokhov et al. 1999; Chen et al. 2000; Chen and Citovsky 2003). The molecular and cellular mechanism of this interesting interaction remains to be investigated in more detail, though.

OUTLOOK AND PERSPECTIVES

Over the course of the last decade, much progress has been made in the field of cell wall signalling (Wolf et al. 2012a, 2022) (Figure 19.1), attracting considerable interest and help from developmental biologists, biochemists, physicists, and modellers. As pointed out in this overview, important challenges remain, and more open questions have appeared. Nearly all identified components contributing to cell wall signalling play pleiotropic biological activities, which makes analysis of individual functions very difficult. Conversely, we do not have a firm grasp of how many cell wall signalling pathways exist and, potentially, many more await discovery. The identification of new cell wall signalling components often requires brute force approaches such as cell wall modification of biosynthesis inhibition, which might not represent a natural stress condition. This initial step merely demonstrates the existence of a cell wall surveillance system, the physiological role of which then has to be deciphered in more relevant conditions. Another challenge for the field is the identification of plausible cell wall perception mechanisms, which is greatly hampered by the inherent difficulties of carbohydrate chemistry. A major bottleneck in this regard is the availability of pure, defined carbohydrate oligomers (Pfrengle 2017) necessary for solving, for example, cell wall receptor structures in complex with their carbohydrate ligands. Furthermore, any study of cell wall–mediated signalling events is incomplete without consideration of cell wall mechanics. Unfortunately, our capabilities for measuring mechanical properties at cellular and subcellular resolution are very limited, especially in deeper tissues beneath the outer surface of the plant. Therefore, new techniques such as Brillouin microscopy, which potentially caters to this need, is already widely adopted in other biological systems (Elsayad et al. 2016; Prevedel et al. 2019), and was recently used to assess THE1-dependent cell wall stiffness (Bacete et al. 2022), and it holds great promise for plant biology.

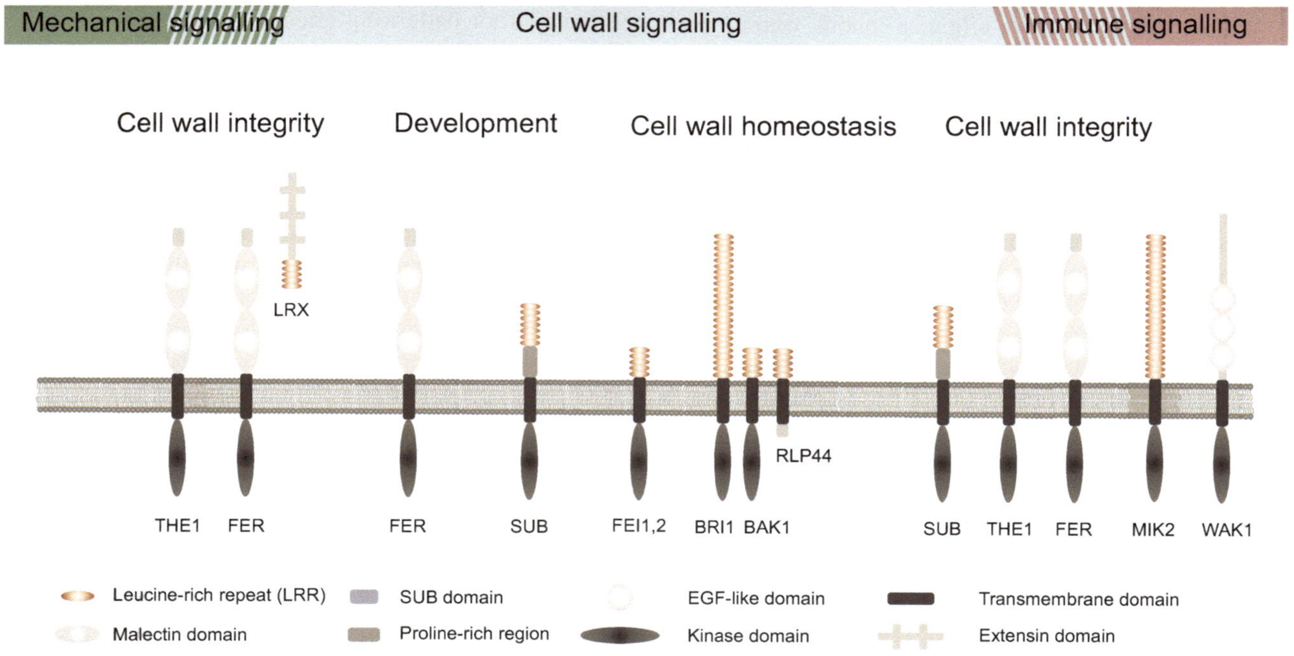

FIGURE 19.1 Schematic representation of cell surface receptors regulating putative and confirmed cell wall–responsive pathways. Note the overlap between cell wall signalling and mechanical or immune signalling. FER= FERONIA, THE1 = THESEUS1, SUB = STRUBBELIG, BRI1 = BRASSINOSTEROID INSENSITIVE1, BAK1 = BRI1-ASSOCIATED KINASE1, RLP44 = RECEPTOR-LIKE PROTEIN44, WAK1 = WALL-ASSOCIATED KINASE1, MIK2 = MALE DISCOVERER1-INTERACTING RECEPTOR-LIKE KINASE 2/LEUCINE-RICH REPEAT KINASE FAMILY PROTEIN INDUCED BY SALT STRESS.

ACKNOWLEDGEMENTS

We thank members of the Schneitz and Wolf labs for fruitful discussions.

FUNDING

Work on cell wall signalling to KS is funded by the German Research Council (DFG) through a SFB924 grant (TP A2) and to SW through the Heisenberg programme (WO 1660/10–1).

REFERENCES

Abou-Saleh RH, Hernandez-Gomez MC, Amsbury S, Paniagua C, Bourdon M, Miyashima S, Helariutta Y, Fuller M, Budtova T, Connell SD, Ries ME, Benitez-Alfonso Y (2018) Interactions between callose and cellulose revealed through the analysis of biopolymer mixtures. *Nat Commun* 9:4538.

Adamowski M, Friml J (2015) PIN-dependent auxin transport: action, regulation, and evolution. *Plant Cell* 27:20–32.

Anderson CT, Kieber JJ (2020) Dynamic construction, perception, and remodeling of plant cell walls. *Annu Rev Plant Biol* 71:39–69.

Aryal B, Jonsson K, Baral A, Sancho-Andres G, Routier-Kierzkowska AL, Kierzkowski D, Bhalerao RP (2020) Interplay between cell wall and auxin mediates the control of differential cell elongation during apical hook development. *Curr Biol* 30:1733-1739.e3.

Atmodjo MA, Hao Z, Mohnen D (2013) Evolving views of pectin biosynthesis. *Annu Rev Plant Biol* 64:747–779.

Aziz A, Gauthier A, Bézier A, Poinssot B, Joubert JM, Pugin A, Heyraud A, Baillieul F (2007) Elicitor and resistance-inducing activities of beta-1,4 cellodextrins in grapevine, comparison with beta-1,3 glucans and alpha-1,4 oligogalacturonides. *J Exp Bot* 58:1463–1472.

Bacete L, Mélida H, López G, Dabos P, Tremousaygue D, Denancé N, Miedes E, Bulone V, Goffner D, Molina A (2020) *Arabidopsis* Response Regulator 6 (ARR6) modulates plant cell-wall composition and disease resistance. *Mol Plant Microbe Interact* 33:767–780.

Bacete L, Schulz J, Engelsdorf T, Bartosova Z, Vaahtera L, Yan G, Gerhold JM, Tichá T, Øvstebø C, Gigli-Bisceglia N, Johannessen-Starheim S, Margueritat J, Kollist H, Dehoux T, McAdam SAM, Hamann T (2022) THESEUS1 modulates cell wall stiffness and abscisic acid production in *Arabidopsis thaliana*. *Proc Natl Acad Sci U S A* 119:e2119258119.

Barghahn S, Arnal G, Jain N, Petutschnig E, Brumer H, Lipka V (2021) Mixed linkage β-1,3/1,4-glucan oligosaccharides induce defense responses in *Hordeum vulgare* and *Arabidopsis thaliana*. *Front Plant Sci* 12:682439.

Bellincampi D, Cardarelli M, Zaghi D, Serino G, Salvi G, Gatz C, Cervone F, Altamura MM, Costantino P, Lorenzo GD (1996) Oligogalacturonides prevent rhizogenesis in *rolB*-transformed tobacco explants by inhibiting auxin-induced expression of the *rolB* gene. *Plant Cell* 8:477–487.

Bellincampi D, Salvi G, De Lorenzo G, Cervone F, Marfà V, Eberhard S, Darvill A, Albersheim P (1993) Oligogalacturonides inhibit the formation of roots on tobacco explants. *The Plant Journal* 4:207–213.

Benedetti M, Pontiggia D, Raggi S, Cheng Z, Scaloni F, Ferrari S, Ausubel FM, Cervone F, De Lorenzo G (2015) Plant immunity triggered by engineered in vivo release of oligogalacturonides, damage-associated molecular patterns. *Proc Natl Acad Sci U S A* 112:5533–5538.

Benitez-Alfonso Y, Faulkner C, Pendle A, Miyashima S, Helariutta Y, Maule A (2013) Symplastic intercellular connectivity regulates lateral root patterning. *Dev Cell* 26:136–147.

Berger F, Taylor A, Brownlee C (1994) Cell fate determination by the cell wall in early *Fucus* development. *Science* 263:1421–1423.

Bethke G, Grundman RE, Sreekanta S, Truman W, Katagiri F, Glazebrook J (2014) Arabidopsis *pectin methylesterases* contribute to immunity against *Pseudomonas syringae*. *Plant Physiol* 164:1093–1107.

Bethke G, Thao A, Xiong G, Li B, Soltis NE, Hatsugai N, Hillmer RA, Katagiri F, Kliebenstein DJ, Pauly M, Glazebrook J (2016) Pectin biosynthesis Is critical for cell wall integrity and immunity in *Arabidopsis thaliana*. *Plant Cell* 28:537–556.

Bidhendi AJ, Altartouri B, Gosselin FP, Geitmann A (2019) Mechanical stress initiates and sustains the morphogenesis of wavy leaf epidermal cells. *Cell Rep* 28:1237-1250.e6.

Bishop PD, Makus DJ, Pearce G, Ryan CA (1981) Proteinase inhibitor-inducing factor activity in tomato leaves resides in oligosaccharides enzymically released from cell walls. *Proc Natl Acad Sci U S A* 78:3536–3540.

Biswal AK, Hao Z, Pattathil S, Yang X, Winkeler K, Collins C, Mohanty SS, Richardson EA, Gelineo-Albersheim I, Hunt K, Ryno D, Sykes RW, Turner GB, Ziebell A, Gjersing E, Lukowitz W, Davis MF, Decker SR, Hahn MG, Mohnen D (2015) Downregulation of *GAUT12* in *Populus deltoides* by RNA silencing results in reduced recalcitrance, increased growth and reduced xylan and pectin in a woody biofuel feedstock. *Biotechnol Biofuels* 8:41.

Boller T, Felix G (2009) A renaissance of elicitors: perception of microbe-associated molecular patterns and danger signals by pattern-recognition receptors. *Annu Rev Plant Biol* 60:379–406.

Borassi C, Gloazzo Dorosz J, Ricardi MM, Carignani Sardoy M, Pol Fachin L, Marzol E, Mangano S, Rodríguez Garcia DR, Martínez Pacheco J, Rondón Guerrero YDC, Velasquez SM, Villavicencio B, Ciancia M, Seifert G, Verli H, Estevez JM (2020) A cell surface arabinogalactan-peptide influences root hair cell fate. *New Phytol* 227:732–743.

Boutté Y, Crosnier MT, Carraro N, Traas J, Satiat-Jeunemaitre B (2006) The plasma membrane recycling pathway and cell polarity in plants: studies on PIN proteins. *J Cell Sci* 119:1255–1265.

Branca C, Lorenzo GD, Cervone F (1988) Competitive inhibition of the auxin-induced elongation by α-D-oligogalacturonides in pea stem segments. *Physiologia Plantarum* 72:499–504.

Brault ML, Petit JD, Immel F, Nicolas WJ, Glavier M, Brocard L, Gaston A, Fouché M, Hawkins TJ, Crowet JM, Grison MS, Germain V, Rocher M, Kraner M, Alva V, Claverol S, Paterlini A, Helariutta Y, Deleu M, Lins L, Tilsner J, Bayer EM (2019) Multiple C2 domains and transmembrane region proteins (MCTPs) tether membranes at plasmodesmata. *EMBO Rep* 20:e47182.

Brutus A, Sicilia F, Macone A, Cervone F, De Lorenzo G (2010) A domain swap approach reveals a role of the plant wall-associated kinase 1 (WAK1) as a receptor of oligogalacturonides. *Proc Natl Acad Sci U S A* 107:9452–9457.

Bücherl CA, Jarsch IK, Schudoma C, Segonzac C, Mbengue M, Robatzek S, MacLean D, Ott T, Zipfel C (2017) Plant immune and growth receptors share common signalling components but localise to distinct plasma membrane nanodomains. *Elife* 6:e25114.

Burton RA, Gibeaut DM, Bacic A, Findlay K, Roberts K, Hamilton A, Baulcombe DC, Fincher GB (2000) Virus-induced silencing of a plant cellulose synthase gene. *Plant Cell* 12:691–706.

Caño-Delgado A, Penfield S, Smith C, Catley M, Bevan M (2003) Reduced cellulose synthesis invokes lignification and defense responses in *Arabidopsis thaliana. Plant J* 34:351–362.

Cano-Delgado AI, Metzlaff K, Bevan MW (2000) The *eli1* mutation reveals a link between cell expansion and secondary cell wall formation in *Arabidopsis thaliana. Development* 127:3395–3405.

Carlsbecker A, Lee JY, Roberts CJ, Dettmer J, Lehesranta S, Zhou J, Lindgren O, Moreno-Risueno MA, Vaten A, Thitamadee S, Campilho A, Sebastian J, Bowman JL, Helariutta Y, Benfey PN (2010) Cell signalling by microRNA165/6 directs gene dose-dependent root cell fate. *Nature* 465:316–321.

Casasoli M, Spadoni S, Lilley KS, Cervone F, De Lorenzo G, Mattei B (2008) Identification by 2-D DIGE of apoplastic proteins regulated by oligogalacturonides in *Arabidopsis thaliana. Proteomics* 8:1042–1054.

Cervone F, Andebrhan T, Coutts RHA, Wood RKS (1981) Effects of french bean tissue and leaf protoplasts on *Colletotrichum lindemuthianum* polygalacturonase. *Journal of Phytopathology* 102:238–246.

Chan J, Eder M, Crowell EF, Hampson J, Calder G, Lloyd C (2011) Microtubules and CESA tracks at the inner epidermal wall align independently of those on the outer wall of light-grown *Arabidopsis* hypocotyls. *J Cell Sci* 124:1088–1094.

Chan J, Mansfield C, Clouet F, Dorussen D, Coen E (2020) Intrinsic cell polarity coupled to growth axis formation in tobacco BY-2 cells. *Curr Biol* 30:4999-5006.e3.

Chassot C, Nawrath C, Métraux JP (2007) Cuticular defects lead to full immunity to a major plant pathogen. *Plant J* 49:972–980.

Chaudhary A, Chen X, Gao J, Leśniewska B, Hammerl R, Dawid C, Schneitz K (2020) The Arabidopsis receptor kinase strubbelig regulates the response to cellulose deficiency. *PLoS Genetics* 16:e1008433.

Chaudhary A, Chen X, Leśniewska B, Boikine R, Gao J, Wolf S, Schneitz K (2021) Cell wall damage attenuates root hair patterning and tissue morphogenesis mediated by the receptor kinase strubbelig. *Development* 148:dev199425.

Chen MH, Citovsky V (2003) Systemic movement of a tobamovirus requires host cell pectin methylesterase. *Plant J* 35:386–392.

Chen MH, Sheng J, Hind G, Handa AK, Citovsky V (2000) Interaction between the tobacco mosaic virus movement protein and host cell pectin methylesterases is required for viral cell-to-cell movement. *EMBO J* 19:913–920.

Cheng Y, Zhu W, Chen Y, Ito S, Asami T, Wang X (2014) Brassinosteroids control root epidermal cell fate via direct regulation of a MYB-bHLH-WD40 complex by GSK3-like kinases. *Elife* 3:e02525.

Chevalier D, Batoux M, Fulton L, Pfister K, Yadav RK, Schellenberg M, Schneitz K (2005) *STRUBBELIG* defines a receptor kinase-mediated signaling pathway regulating organ development in *Arabidopsis. Proc Natl Acad Sci U S A* 102:9074–9079.

Chupeau MC, Granier F, Pichon O, Renou JP, Gaudin V, Chupeau Y (2013) Characterization of the early events leading to totipotency in an *Arabidopsis* protoplast liquid culture by temporal transcript profiling. *Plant Cell* 25:2444–2463.

Claverie J, Balacey S, Lemaître-Guillier C, Brulé D, Chiltz A, Granet L, Noirot E, Daire X, Darblade B, Héloir MC, Poinssot B (2018) The cell wall-derived xyloglucan is a new damp triggering plant immunity in *Vitis vinifera* and *Arabidopsis thaliana. Front Plant Sci* 9:1725.

Coleman AD, Maroschek J, Raasch L, Takken FLW, Ranf S, Hückelhoven R (2021) The Arabidopsis leucine-rich repeat receptor-like kinase MIK2 is a crucial component of early immune responses to a fungal-derived elicitor. *New Phytol* 229:3453–3466.

Colin L, Chevallier A, Tsugawa S, Gacon F, Godin C, Viasnoff V, Saunders TE, Hamant O (2020) Cortical tension overrides geometrical cues to orient microtubules in confined protoplasts. *Proc Natl Acad Sci U S A* 117:32731–32738.

Crowell EF, Timpano H, Desprez T, Franssen-Verheijen T, Emons AM, Höfte H, Vernhettes S (2011) Differential regulation of cellulose orientation at the inner and outer face of epidermal cells in the *Arabidopsis* hypocotyl. *Plant Cell* 23:2592–2605.

Davey MR, Anthony P, Power JB, Lowe KC (2005) Plant protoplasts: status and biotechnological perspectives. *Biotechnol Adv* 23:131–171.

Davidsson P, Broberg M, Kariola T, Sipari N, Pirhonen M, Palva ET (2017) Short oligogalacturonides induce pathogen resistance-associated gene expression in *Arabidopsis thaliana. BMC Plant Biol* 17:19.

de Godoy F, Bermúdez L, Lira BS, de Souza AP, Elbl P, Demarco D, Alseekh S, Insani M, Buckeridge M, Almeida J, Grigioni G, Fernie AR, Carrari F, Rossi M (2013) Galacturonosyltransferase 4 silencing alters pectin composition and carbon partitioning in tomato. *J Exp Bot* 64:2449–2466.

Decreux A, Messiaen J (2005) Wall-associated kinase WAK1 interacts with cell wall pectins in a calcium-induced conformation. *Plant Cell Physiol* 46:268–278.

Decreux A, Thomas A, Spies B, Brasseur R, Van Cutsem P, Messiaen J (2006) In vitro characterization of the homogalacturonan-binding domain of the wall-associated kinase WAK1 using site-directed mutagenesis. *Phytochemistry* 67:1068–1079.

Denness L, McKenna JF, Segonzac C, Wormit A, Madhou P, Bennett M, Mansfield J, Zipfel C, Hamann T (2011) Cell wall damage-induced lignin biosynthesis is regulated by a reactive oxygen species- and jasmonic acid-dependent process in Arabidopsis. *Plant Physiol* 156:1364–1374.

Desprez T, Vernhettes S, Fagard M, Refrégier G, Desnos T, Aletti E, Py N, Pelletier S, Höfte H (2002) Resistance against herbicide isoxaben and cellulose deficiency caused by distinct mutations in same cellulose synthase isoform CESA6. *Plant Physiol* 128:482–490.

Domozych DS, Sørensen I, Popper ZA, Ochs J, Andreas A, Fangel JU, Pielach A, Sacks C, Brechka H, Ruisi-Besares P, Willats WG, Rose JK (2014) Pectin metabolism and assembly in the cell wall of the charophyte green alga *Penium margaritaceum. Plant Physiol* 165:105–118.

Dong J, Macalister CA, Bergmann DC (2009) BASL controls asymmetric cell division in *Arabidopsis. Cell* 137:1320–1330.

Dorokhov YL, Mäkinen K, Frolova OY, Merits A, Saarinen J, Kalkkinen N, Atabekov JG, Saarma M (1999) A novel function for a ubiquitous plant enzyme pectin methylesterase: the host-cell receptor for the tobacco mosaic virus movement protein. *FEBS Lett* 461:223–228.

Durand-Smet P, Spelman TA, Meyerowitz EM, Jönsson H (2020) Cytoskeletal organization in isolated plant cells under geometry control. *Proc Natl Acad Sci U S A* 117:17399–17408.

Eberhard S, Doubrava N, Marfa V, Mohnen D, Southwick A, Darvill A, Albersheim P (1989) Pectic cell wall fragments regulate tobacco thin-cell-layer explant morphogenesis. *Plant Cell* 1:747–755.

Ehlers K, Kollmann R (2001) Primary and secondary plasmodesmata: structure, origin, and functioning. *Protoplasma* 216:1–30.

Elliott L, Kirchhelle C (2020) The importance of being edgy: cell geometric edges as an emerging polar domain in plant cells. *J Microsc* 278:123–131.

Ellis C, Karafyllidis I, Wasternack C, Turner JG (2002) The Arabidopsis mutant *cev1* links cell wall signaling to jasmonate and ethylene responses. *Plant Cell* 14:1557–1566.

Ellis C, Turner JG (2001) The Arabidopsis mutant *cev1* has constitutively active jasmonate and ethylene signal pathways and enhanced resistance to pathogens. *Plant Cell* 13:1025–1033.

Elsayad K, Werner S, Gallemí M, Kong J, Sánchez Guajardo ER, Zhang L, Jaillais Y, Greb T, Belkhadir Y (2016) Mapping the subcellular mechanical properties of live cells in tissues with fluorescence emission-Brillouin imaging. *Sci Signal* 9:rs5.

Engelsdorf T, Gigli-Bisceglia N, Veerabagu M, McKenna JF, Vaahtera L, Augstein F, Van der Does D, Zipfel C, Hamann T (2018) The plant cell wall integrity maintenance and immune signaling systems cooperate to control stress responses in *Arabidopsis thaliana*. *Sci Signal* 11:eaao3070.

Escudero V, Jordá L, Sopeña-Torres S, Mélida H, Miedes E, Muñoz-Barrios A, Swami S, Alexander D, McKee LS, Sánchez-Vallet A, Bulone V, Jones AM, Molina A (2017) Alteration of cell wall xylan acetylation triggers defense responses that counterbalance the immune deficiencies of plants impaired in the β-subunit of the heterotrimeric G-protein. *Plant J* 92:386–399.

Fagard M, Desnos T, Desprez T, Goubet F, Refregier G, Mouille G, McCann M, Rayon C, Vernhettes S, Höfte H (2000) *PROCUSTE1* encodes a cellulose synthase required for normal cell elongation specifically in roots and dark-grown hypocotyls of Arabidopsis. *Plant Cell* 12:2409–2424.

Faulkner C, Akman OE, Bell K, Jeffree C, Oparka K (2008) Peeking into pit fields: a multiple twinning model of secondary plasmodesmata formation in tobacco. *Plant Cell* 20:1504–1518.

Fauth, Schweizer, Buchala, Markstadter, Riederer, Kato, Kauss (1998) Cutin monomers and surface wax constituents elicit H_2O_2 in conditioned cucumber hypocotyl segments and enhance the activity of other H_2O_2 elicitors. *Plant Physiol* 117:1373–1380.

Feng W, Kita D, Peaucelle A, Cartwright HN, Doan V, Duan Q, Liu MC, Maman J, Steinhorst L, Schmitz-Thom I, Yvon R, Kudla J, Wu HM, Cheung AY, Dinneny JR (2018) The FERONIA receptor kinase maintains cell-wall integrity during salt stress through Ca^{2+} signaling. *Curr Biol* 28:666-675. e5.

Feraru E, Feraru MI, Kleine-Vehn J, Martiniere A, Mouille G, Vanneste S, Vernhettes S, Runions J, Friml J (2011) PIN polarity maintenance by the cell wall in *Arabidopsis*. *Curr Biol* 21:338–343.

Fernandez-Calvino L, Faulkner C, Walshaw J, Saalbach G, Bayer E, Benitez-Alfonso Y, Maule A (2011) Arabidopsis plasmodesmal proteome. *PLoS One* 6:e18880.

Ferrari S, Galletti R, Denoux C, De Lorenzo G, Ausubel FM, Dewdney J (2007) Resistance to *Botrytis cinerea* induced in Arabidopsis by elicitors is independent of salicylic acid, ethylene, or jasmonate signaling but requires *PHYTOALEXIN DEFICIENT3*. *Plant Physiol* 144:367–379.

Ferrari S, Savatin DV, Sicilia F, Gramegna G, Cervone F, Lorenzo GD (2013) Oligogalacturonides: plant damage-associated molecular patterns and regulators of growth and development. *Front Plant Sci* 4:49.

Gallego-Giraldo L, Escamilla-Trevino L, Jackson LA, Dixon RA (2011a) Salicylic acid mediates the reduced growth of lignin down-regulated plants. *Proc Natl Acad Sci U S A* 108:20814–20819.

Gallego-Giraldo L, Jikumaru Y, Kamiya Y, Tang Y, Dixon RA (2011b) Selective lignin downregulation leads to constitutive defense response expression in alfalfa (*Medicago sativa* L.). *New Phytol* 190:627–639.

Gallego-Giraldo L, Liu C, Pose-Albacete S, Pattathil S, Peralta AG, Young J, Westpheling J, Hahn MG, Rao X, Knox JP, De Meester B, Boerjan W, Dixon RA (2020) ARABIDOPSIS DEHISCENCE ZONE POLYGALACTURONASE 1 (ADPG1) releases latent defense signals in stems with reduced lignin content. *Proc Natl Acad Sci U S A* 117:3281–3290.

Gallego-Giraldo L, Posé S, Pattathil S, Peralta AG, Hahn MG, Ayre BG, Sunuwar J, Hernandez J, Patel M, Shah J, Rao X, Knox JP, Dixon RA (2018) Elicitors and defense gene induction in plants with altered lignin compositions. *New Phytol* 219:1235–1251.

Galletti R, Denoux C, Gambetta S, Dewdney J, Ausubel FM, De Lorenzo G, Ferrari S (2008) The AtrbohD-mediated oxidative burst elicited by oligogalacturonides in Arabidopsis is dispensable for the activation of defense responses effective against *Botrytis cinerea*. *Plant Physiol* 148:1695–1706.

Gälweiler L, Guan C, Müller A, Wisman E, Mendgen K, Yephremov A, Palme K (1998) Regulation of polar auxin transport by AtPIN1 in *Arabidopsis* vascular tissue. *Science* 282:2226–2230.

Gao J, Chaudhary A, Vaddepalli P, Nagel MK, Isono E, Schneitz K (2019) The Arabidopsis receptor kinase STRUBBELIG undergoes clathrin-dependent endocytosis. *J Exp Bot* 70:3881–3894.

Geitmann A, Hush JM, Overall RL (1997) Inhibition of ethylene biosynthesis does not block microtubule re-orientation in wounded pea roots. *Protoplasma* 198:135–142.

Gigli-Bisceglia N, Engelsdorf T, Strnad M, Vaahtera L, Khan GA, Yamoune A, Alipanah L, Novák O, Persson S, Hejatko J, Hamann T (2018) Cell wall integrity modulates Arabidopsis thaliana cell cycle gene expression in a cytokinin- and nitrate reductase-dependent manner. *Development* 145:dev166678.

González-Pérez L, Perrotta L, Acosta A, Orellana E, Spadafora N, Bruno L, Bitonti BM, Albani D, Cabrera JC, Francis D, Rogers HJ (2014) In tobacco BY-2 cells xyloglucan oligosaccharides alter the expression of genes involved in cell wall metabolism, signalling, stress responses, cell division and transcriptional control. *Mol Biol Rep* 41:6803–6816.

Green PB (1962) Mechanism for plant cellular morphogenesis. *Science* 138:1404–1405.

Griffiths JS, Tsai AY, Xue H, Voiniciuc C, Sola K, Seifert GJ, Mansfield SD, Haughn GW (2014) SALT-OVERLY SENSITIVE5 mediates Arabidopsis seed coat mucilage adherence and organization through pectins. *Plant Physiol* 165:991–1004.

Grison M, Kirk P, Brault M, Wu XN, Schulze WX, Benitez-Alfonso Y, Immel F, Bayer EMF (2019) Plasma membrane-associated receptor like kinases relocalize to plasmodesmata in response to osmotic stress. *Plant Physiol* 181:142–160.

Gronnier J, Crowet JM, Habenstein B, Nasir MN, Bayle V, Hosy E, Platre MP, Gouguet P, Raffaele S, Martinez D, Grelard A, Loquet A, Simon-Plas F, Gerbeau-Pissot P, Der C, Bayer EM, Jaillais Y, Deleu M, Germain V, Lins L, Mongrand S (2017) Structural basis for plant plasma membrane protein dynamics and organization into functional nanodomains. *Elife* 6: e26404.

Guseman JM, Lee JS, Bogenschutz NL, Peterson KM, Virata RE, Xie B, Kanaoka MM, Hong Z, Torii KU (2010) Dysregulation of cell-to-cell connectivity and stomatal patterning by loss-of-function mutation in *Arabidopsis CHORUS* (*GLUCAN SYNTHASE-LIKE 8*). *Development* 137:1731–1741.

Haas KT, Wightman R, Meyerowitz EM, Peaucelle A (2020) Pectin homogalacturonan nanofilament expansion drives morphogenesis in plant epidermal cells. *Science* 367:1003–1007.

Hahn MG, Darvill AG, Albersheim P (1981) Host-pathogen interactions: XIX. The endogenous elicitor, a fragment of a plant cell wall polysaccharide that elicits phytoalexin accumulation in soybeans. *Plant Physiol* 68:1161–1169.

Hamann T, Bennett M, Mansfield J, Somerville C (2009) Identification of cell-wall stress as a hexose-dependent and osmosensitive regulator of plant responses. *Plant J* 57:1015–1026.

Hamant O, Haswell ES (2017) Life behind the wall: sensing mechanical cues in plants. *BMC Biol* 15:59.

Hamant O, Heisler MG, Jonsson H, Krupinski P, Uyttewaal M, Bokov P, Corson F, Sahlin P, Boudaoud A, Meyerowitz EM, Couder Y, Traas J (2008) Developmental patterning by mechanical signals in *Arabidopsis*. *Science* 322:1650–1655.

Hann CT, Bequette CJ, Dombrowski JE, Stratmann JW (2014) Methanol and ethanol modulate responses to danger- and microbe-associated molecular patterns. *Front Plant Sci* 5:550.

Harpaz-Saad S, McFarlane HE, Xu S, Divi UK, Forward B, Western TL, Kieber JJ (2011) Cellulose synthesis via the FEI2 RLK/SOS5 pathway and cellulose synthase 5 is required for the structure of seed coat mucilage in Arabidopsis. *Plant J* 68:941–953.

Harpaz-Saad S, Western TL, Kieber JJ (2012) The FEI2-SOS5 pathway and CELLULOSE SYNTHASE 5 are required for cellulose biosynthesis in the Arabidopsis seed coat and affect pectin mucilage structure. *Plant Signal Behav* 7:285–288.

He ZH, Fujiki M, Kohorn BD (1996) A cell wall-associated, receptor-like protein kinase. *J Biol Chem* 271:19789–19793.

Heisler MG, Hamant O, Krupinski P, Uyttewaal M, Ohno C, Jonsson H, Traas J, Meyerowitz EM (2010) Alignment between PIN1 polarity and microtubule orientation in the shoot apical meristem reveals a tight coupling between morphogenesis and auxin transport. *PLoS Biol* 8:e1000516.

Hématy K, Sado PE, Van Tuinen A, Rochange S, Desnos T, Balzergue S, Pelletier S, Renou JP, Höfte H (2007) A receptor-like kinase mediates the response of *Arabidopsis* cells to the inhibition of cellulose synthesis. *Curr Biol* 17:922–931.

Hernández-Blanco C, Feng DX, Hu J, Sánchez-Vallet A, Deslandes L, Llorente F, Berrocal-Lobo M, Keller H, Barlet X, Sánchez-Rodríguez C, Anderson LK, Somerville S, Marco Y, Molina A (2007) Impairment of cellulose synthases required for *Arabidopsis* secondary cell wall formation enhances disease resistance. *Plant Cell* 19:890–903.

Höfte H, Voxeur A (2017) Plant cell walls. *Curr Biol* 27:R865–R870.

Holzwart E, Huerta AI, Glöckner N, Garnelo Gómez B, Wanke F, Augustin S, Askani JC, Schürholz AK, Harter K, Wolf S (2018) BRI1 controls vascular cell fate in the *Arabidopsis* root through RLP44 and phytosulfokine signaling. *Proc Natl Acad Sci U S A* 115: 11838–11843.

Holzwart E, Wanke F, Glöckner N, Höfte H, Harter K, Wolf S (2020) A mutant allele uncouples the brassinosteroid-dependent and independent functions of BRASSINOSTEROID INSENSITIVE 1. *Plant Physiol* 182:669–678.

Hou S, Liu Z, Shen H, Wu D (2019) Damage-associated molecular pattern-triggered immunity in plants. *Front Plant Sci* 10:646.

Hush JM, Overall RL (1992) Re-orientation of cortical F-actin is not necessary for wound-induced microtubule re-orientation and cell polarity establishment. *Protoplasma* 169:97–106.

Hynes RO (2009) The extracellular matrix: not just pretty fibrils. *Science* 326:1216–1219.

Jacques E, Verbelen JP, Vissenberg K (2013) Mechanical stress in Arabidopsis leaves orients microtubules in a 'continuous' supracellular pattern. *BMC Plant Biol* 13:163.

Jarsch IK, Konrad SS, Stratil TF, Urbanus SL, Szymanski W, Braun P, Braun KH, Ott T (2014) Plasma membranes are subcompartmentalized into a plethora of coexisting and diverse microdomains in *Arabidopsis* and *Nicotiana benthamiana*. *Plant Cell* 26:1698–1711.

Jiang L, Yang SL, Xie LF, Puah CS, Zhang XQ, Yang WC, Sundaresan V, Ye D (2005) VANGUARD1 encodes a pectin methylesterase that enhances pollen tube growth in the Arabidopsis style and transmitting tract. *Plant Cell* 17:584–596.

Jo Y, Cho WK, Rim Y, Moon J, Chen XY, Chu H, Kim CY, Park ZY, Lucas WJ, Kim JY (2011) Plasmodesmal receptor-like kinases identified through analysis of rice cell wall extracted proteins. *Protoplasma* 248:191–203.

Johnson JM, Thürich J, Petutschnig EK, Altschmied L, Meichsner D, Sherameti I, Dindas J, Mrozinska A, Paetz C, Scholz SS, Furch ACU, Lipka V, Hedrich R, Schneider B, Svatoš A, Oelmüller R (2018) A poly(A) ribonuclease controls the cellotriose-based interaction between *Piriformospora indica* and its host Arabidopsis. *Plant Physiol* 176:2496–2514.

Jolie RP, Duvetter T, Van Loey AM, Hendrickx ME (2010) Pectin methylesterase and its proteinaceous inhibitor: a review. *Carbohydrate Res* 345(18): 2583–2595.

Jonsson K, Lathe RS, Kierzkowski D, Routier-Kierzkowska A-L, Hamant O, Bhalerao RP (2021) Mechanochemical feedback mediates tissue bending required for seedling emergences. *Curr Biol* 31:1154-1164.e3.

Julkowska MM, Koevoets IT, Mol S, Hoefsloot H, Feron R, Tester MA, Keurentjes JJB, Korte A, Haring MA, de Boer GJ, Testerink C (2017) Genetic components of root architecture remodeling in response to salt stress. *Plant Cell* 29:3198–3213.

Kitazawa K, Tryfona T, Yoshimi Y, Hayashi Y, Kawauchi S, Antonov L, Tanaka H, Takahashi T, Kaneko S, Dupree P, Tsumuraya Y, Kotake T (2013) β-galactosyl Yariv reagent binds to the β-1,3-galactan of arabinogalactan proteins. *Plant Physiol* 161:1117–1126.

Kleine-Vehn J, Wabnik K, Martiniere A, Langowski L, Willig K, Naramoto S, Leitner J, Tanaka H, Jakobs S, Robert S, Luschnig C, Govaerts W, Hell SW, Runions J, Friml J (2011) Recycling, clustering, and endocytosis jointly maintain PIN auxin carrier polarity at the plasma membrane. *Mol Syst Biol* 7:540.

Knepper C, Savory EA, Day B (2011) Arabidopsis NDR1 is an integrin-like protein with a role in fluid loss and plasma membrane-cell wall adhesion. *Plant Physiol* 156:286–300.

Kohorn BD (2016) Cell wall-associated kinases and pectin perception. *J Exp Bot* 67:489–494.

Kohorn BD, Hoon D, Minkoff BB, Sussman MR, Kohorn SL (2016) Rapid oligo-galacturonide induced changes in protein phosphorylation in Arabidopsis. *Mol Cell Proteomics* 15:1351–1359.

Kohorn BD, Johansen S, Shishido A, Todorova T, Martinez R, Defeo E, Obregon P (2009) Pectin activation of MAP kinase and gene expression is WAK2 dependent. *Plant J* 60:974–982.

Komarova TV, Sheshukova EV, Dorokhov YL (2014) Cell wall methanol as a signal in plant immunity. *Front Plant Sci* 5:101.

Kong D, Karve R, Willet A, Chen M-K, Oden J, Shpak ED (2012) Regulation of plasmodesmatal permeability and stomatal patterning by the glycosyltransferase-like protein KOBITO1. *Plant Physiol* 159:156–168.

Kropf DL, Kloareg B, Quatrano RS (1988) Cell wall is required for fixation of the embryonic axis in *Fucus* zygotes. *Science* 239:187–190.

Kuppusamy KT, Chen AY, Nemhauser JL (2009) Steroids are required for epidermal cell fate establishment in *Arabidopsis* roots. *Proc Natl Acad Sci USA* 106:8073–8076.

Kurusu T, Kuchitsu K, Nakano M, Nakayama Y, Iida H (2013) Plant mechanosensing and Ca2+ transport. *Trends Plant Sci* 18:227–233.

Kurusu T, Nishikawa D, Yamazaki Y, Gotoh M, Nakano M, Hamada H, Yamanaka T, Iida K, Nakagawa Y, Saji H, Shinozaki K, Iida H, Kuchitsu K (2012) Plasma membrane protein OsMCA1 is involved in regulation of hypo-osmotic shock-induced Ca2+ influx and modulates generation of reactive oxygen species in cultured rice cells. *BMC Plant Biol* 12:11.

Kwak SH, Schiefelbein J (2007) The role of the SCRAMBLED receptor-like kinase in patterning the *Arabidopsis* root epidermis. *Dev Biol* 302:118–131.

Kwak SH, Schiefelbein J (2008) A feedback mechanism controlling SCRAMBLED receptor accumulation and cell-type pattern in Arabidopsis. *Curr Biol* 18:1949–1954.

Kwak SH, Shen R, Schiefelbein J (2005) Positional signaling mediated by a receptor-like kinase in *Arabidopsis*. *Science* 307:1111–1113.

Lampugnani ER, Khan GA, Somssich M, Persson S (2018) Building a plant cell wall at a glance. *J Cell Sci* 131:jcs207373.

Lange BM, Lapierre C, Sandermann H (1995) Elicitor-induced spruce stress lignin (structural similarity to early developmental lignins). *Plant Physiol* 108:1277–1287.

Lee SC, West CA (1981a) Polygalacturonase from *Rhizopus stolonifer*, an elicitor of casbene synthetase activity in castor bean (*Ricinus communis* L.) seedlings. *Plant Physiol* 67:633–639.

Lee SC, West CA (1981b) Properties of *Rhizopus stolonifer* polygalacturonase, an elicitor of casbene synthetase activity in castor bean (*Ricinus communis* L.) seedlings. *Plant Physiol* 67:640–645.

Li H, von Wangenheim D, Zhang X, Tan S, Darwish-Miranda N, Naramoto S, Wabnik K, De Rycke R, Kaufmann WA, Gütl D, Tejos R, Grones P, Ke M, Chen X, Dettmer J, Friml J (2021a) Cellular requirements for PIN polar cargo clustering in *Arabidopsis thaliana*. *New Phytol* 229:351–369.

Li M, Yoo CG, Pu Y, Biswal AK, Tolbert AK, Mohnen D, Ragauskas AJ (2019) Downregulation of pectin biosynthesis gene *GAUT4* leads to reduced ferulate and lignin-carbohydrate cross-linking in switchgrass. *Commun Biol* 2:22.

Li ZP, Paterlini A, Glavier M, Bayer EM (2021b) Intercellular trafficking via plasmodesmata: molecular layers of complexity. *Cell Mol Life Sci* 78:799–816.

Lin L, Zhong SH, Cui XF, Li J, He ZH (2012) Characterization of temperature-sensitive mutants reveals a role for receptor-like kinase SCRAMBLED/STRUBBELIG in coordinating cell proliferation and differentiation during Arabidopsis leaf development. *Plant J* 72:707–720.

Lin W, Tang W, Pan X, Huang A, Gao X, Anderson CT, Yang Z (2022) *Arabidopsis* pavement cell morphogenesis requires FERONIA binding to pectin for activation of ROP GTPase signaling. *Curr Biol* 32:497-507.e4.

Lionetti V, Fabri E, De Caroli M, Hansen AR, Willats WG, Piro G, Bellincampi D (2017) Three pectin methylesterase inhibitors protect cell wall integrity for Arabidopsis immunity to *Botrytis*. *Plant Physiol* 173:1844–1863.

Lionetti V, Raiola A, Camardella L, Giovane A, Obel N, Pauly M, Favaron F, Cervone F, Bellincampi D (2007) Overexpression of pectin methylesterase inhibitors in Arabidopsis restricts fungal infection by *Botrytis cinerea*. *Plant Physiol* 143:1871–1880.

Liu J, Li J, Shan L (2020) SERKs. *Curr Biol* 30:R293–R294.

Locci F, Benedetti M, Pontiggia D, Citterico M, Caprari C, Mattei B, Cervone F, De Lorenzo G (2019) An Arabidopsis berberine bridge enzyme-like protein specifically oxidizes cellulose oligomers and plays a role in immunity. *Plant J* 98:540–554.

Ma Y, MacMillan CP, de Vries L, Mansfield SD, Hao P, Ratcliffe J, Bacic A, Johnson KL (2022) FLA11 and FLA12 glycoproteins fine-tune stem secondary wall properties in response to mechanical stresses. *New Phytol* 233:1750–1767.

Majda M, Grones P, Sintorn IM, Vain T, Milani P, Krupinski P, Zagórska-Marek B, Viotti C, Jönsson H, Mellerowicz EJ, Hamant O, Robert S (2017) Mechanochemical polarization of contiguous cell walls shapes plant pavement cells. *Dev Cell* 43:290-304.e4.

Malivert A, Erguvan Ö, Chevallier A, Dehem A, Friaud R, Liu M, Martin M, Peyraud T, Hamant O, Verger S (2021) FERONIA and microtubules independently contribute to mechanical integrity in the *Arabidopsis* shoot. *PLoS Biol* 19:e3001454.

Manfield IW, Orfila C, McCartney L, Harholt J, Bernal AJ, Scheller HV, Gilmartin PM, Mikkelsen JD, Paul Knox J, Willats WG (2004) Novel cell wall architecture of isoxaben-habituated *Arabidopsis* suspension-cultured cells: global transcript profiling and cellular analysis. *Plant J* 40:260–275.

Marfà V, Gollin DJ, Eberhard S, Mohnen D, Danill A, Albersheim P (1991) Oligogalacturonides are able to induce flowers to form on tobacco explants. *Plant J* 1:217–225.

Marhava P, Aliaga Fandino AC, Koh SWH, Jelínková A, Kolb M, Janacek DP, Breda AS, Cattaneo P, Hammes UZ, Petrášek J, Hardtke CS (2020) Plasma membrane domain patterning and self-reinforcing polarity in *Arabidopsis*. *Dev Cell* 52:223-235.e5.

Marhava P, Bassukas AEL, Zourelidou M, Kolb M, Moret B, Fastner A, Schulze WX, Cattaneo P, Hammes UZ, Schwechheimer C, Hardtke CS (2018) A molecular rheostat adjusts auxin flux to promote root protophloem differentiation. *Nature* 558:297–300.

Martiniere A, Gayral P, Hawes C, Runions J (2011) Building bridges: formin1 of Arabidopsis forms a connection between the cell wall and the actin cytoskeleton. *Plant J* 66:354–365.

Martinière A, Lavagi I, Nageswaran G, Rolfe DJ, Maneta-Peyret L, Luu DT, Botchway SW, Webb SE, Mongrand S, Maurel C, Martin-Fernandez ML, Kleine-Vehn J, Friml J, Moreau P, Runions J (2012) Cell wall constrains lateral diffusion of plant plasma-membrane proteins. *Proc Natl Acad Sci U S A* 109:12805–12810.

Mattei B, Spinelli F, Pontiggia D, De Lorenzo G (2016) Comprehensive analysis of the membrane phosphoproteome regulated by oligogalacturonides in *Arabidopsis thaliana*. *Front Plant Sci* 7:1107.

Maule AJ, Benitez-Alfonso Y, Faulkner C (2011) Plasmodesmata - membrane tunnels with attitude. *Curr Opin Plant Biol* 14:683–690.

Mazzella MA, Casal JJ, Muschietti JP, Fox AR (2014) Hormonal networks involved in apical hook development in darkness and their response to light. *Front Plant Sci* 5:52.

McKenna JF, Rolfe DJ, Webb SED, Tolmie AF, Botchway SW, Martin-Fernandez ML, Hawes C, Runions J (2019) The cell wall regulates dynamics and size of plasma-membrane nanodomains in *Arabidopsis*. *Proc Natl Acad Sci U S A* 116:12857–12862.

Mélida H, Bacete L, Ruprecht C, Rebaque D, Del Hierro I, López G, Brunner F, Pfrengle F, Molina A (2020) Arabinoxylan-oligosaccharides act as damage associated molecular patterns in plants regulating disease resistance. *Front Plant Sci* 11:1210.

Mélida H, Sopeña-Torres S, Bacete L, Garrido-Arandia M, Jordá L, López G, Muñoz-Barrios A, Pacios LF, Molina A (2018) Non-branched β-1,3-glucan oligosaccharides trigger immune responses in Arabidopsis. *Plant J* 93:34–49.

Molina A, Miedes E, Bacete L, Rodríguez T, Mélida H, Denancé N, Sánchez-Vallet A, Rivière MP, López G, Freydier A, Barlet X, Pattathil S, Hahn M, Goffner D (2021) *Arabidopsis* cell wall composition determines disease resistance specificity and fitness. *Proc Natl Acad Sci U S A* 118:e2010243118.

Moscatiello R, Mariani P, Sanders D, Maathuis FJ (2006) Transcriptional analysis of calcium-dependent and calcium-independent signalling pathways induced by oligogalacturonides. *J Exp Bot* 57:2847–2865.

Motose H, Sugiyama M, Fukuda H (2001) An arabinogalactan protein(s) is a key component of a fraction that mediates local intercellular communication involved in tracheary element differentiation of zinnia mesophyll cells. *Plant Cell Physiol* 42:129–137.

Motose H, Sugiyama M, Fukuda H (2004) A proteoglycan mediates inductive interaction during plant vascular development. *Nature* 429:873–878.

Müller A, Guan C, Gälweiler L, Tänzler P, Huijser P, Marchant A, Parry G, Bennett M, Wisman E, Palme K (1998) *AtPIN2* defines a locus of *Arabidopsis* for root gravitropism control. *EMBO J* 17:6903–6911.

Muroyama A, Bergmann D (2019) Plant cell polarity: creating diversity from inside the box. *Annu Rev Cell Dev Biol* 35:309–336.

Nakagawa Y, Katagiri T, Shinozaki K, Qi Z, Tatsumi H, Furuichi T, Kishigami A, Sokabe M, Kojima I, Sato S, Kato T, Tabata S, Iida K, Terashima A, Nakano M, Ikeda M, Yamanaka T, Iida H (2007) *Arabidopsis* plasma membrane protein crucial for Ca2+ influx and touch sensing in roots. *Proc Natl Acad Sci U S A* 104:3639–3644.

Nothnagel EA, McNeil M, Albersheim P, Dell A (1983) Host-pathogen interactions: XXII. A galacturonic acid oligosaccharide from plant cell walls elicits phytoalexins. *Plant Physiol* 71:916–926.

Ogawa-Ohnishi M, Matsubayashi Y (2015) Identification of three potent hydroxyproline O-galactosyltransferases in Arabidopsis. *Plant J* 81:736–746.

Oparka KJ, Roberts AG, Boevink P, Santa Cruz S, Roberts I, Pradel KS, Imlau A, Kotlizky G, Sauer N, Epel B (1999) Simple, but not branched, plasmodesmata allow the non-specific trafficking of proteins in developing tobacco leaves. *Cell* 97:743–754.

Orfila C, Knox JP (2000) Spatial regulation of pectic polysaccharides in relation to pit fields in cell walls of tomato fruit pericarp. *Plant Physiol* 122:775–781.

Osorio S, Castillejo C, Quesada MA, Medina-Escobar N, Brownsey GJ, Suau R, Heredia A, Botella MA, Valpuesta V (2008) Partial demethylation of oligogalacturonides by pectin methyl esterase 1 is required for eliciting defence responses in wild strawberry (*Fragaria vesca*). *Plant J* 54:43–55.

Pan X, Fang L, Liu J, Senay-Aras B, Lin W, Zheng S, Zhang T, Guo J, Manor U, Van Norman J, Chen W, Yang Z (2020) Auxin-induced signaling protein nanoclustering contributes to cell polarity formation. *Nat Commun* 11:3914.

Paredez AR, Persson S, Ehrhardt DW, Somerville CR (2008) Genetic evidence that cellulose synthase activity influences microtubule cortical array organization. *Plant Physiol* 147:1723–1734.

Paredez AR, Somerville CR, Ehrhardt DW (2006) Visualization of cellulose synthase demonstrates functional association with microtubules. *Science* 312:1491–1495.

Park YB, Cosgrove DJ (2012) A revised architecture of primary cell walls based on biomechanical changes induced by substrate-specific endoglucanases. *Plant Physiol* 158:1933–1943.

Peaucelle A, Braybrook S, Höfte H (2012) Cell wall mechanics and growth control in plants: the role of pectins revisited. *Front Plant Sci* 3:121.

Peaucelle A, Braybrook SA, Le Guillou L, Bron E, Kuhlemeier C, Hofte H (2011) Pectin-induced changes in cell wall mechanics underlie organ initiation in Arabidopsis. *Curr Biol* 21:1720–1726.

Peaucelle A, Louvet R, Johansen JN, Höfte H, Laufs P, Pelloux J, Mouille G (2008) Arabidopsis phyllotaxis is controlled by the methyl-esterification status of cell-wall pectins. *Curr Biol* 18:1943–1948.

Peaucelle A, Wightman R, Höfte H (2015) The control of growth symmetry breaking in the *Arabidopsis* hypocotyl. *Curr Biol* 25:1746–1752.

Pfrengle F (2017) Synthetic plant glycans. *Curr Opin Chem Biol* 40:145–151.

Pontiggia D, Benedetti M, Costantini S, De Lorenzo G, Cervone F (2020) Dampening the DAMPs: How plants maintain the homeostasis of cell wall molecular patterns and avoid hyper-immunity. *Front Plant Sci* 11:613259.

Prevedel R, Diz-Muñoz A, Ruocco G, Antonacci G (2019) Brillouin microscopy: an emerging tool for mechanobiology. *Nat Methods* 16:969–977.

Ramalho JJ, Jones VAS, Mutte S, Weijers D (2021) Pole position: How plant cells polarize along the axes. *Plant Cell* 34: 174–192.

Reagan BC, Burch-Smith TM (2020) Viruses reveal the secrets of plasmodesmal cell biology. *Mol Plant Microbe Interact* 33:26–39.

Rebaque D, Del Hierro I, López G, Bacete L, Vilaplana F, Dallabernardina P, Pfrengle F, Jordá L, Sánchez-Vallet A, Pérez R, Brunner F, Molina A, Mélida H (2021) Cell wall-derived mixed-linked β-1,3/1,4-glucans trigger immune responses and disease resistance in plants. *Plant J* 106:601–615.

Rhodes J, Yang H, Moussu S, Boutrot F, Santiago J, Zipfel C (2021) Perception of a divergent family of phytocytokines by the Arabidopsis receptor kinase MIK2. *Nat Commun* 12:705.

Roudier F, Fernandez AG, Fujita M, Himmelspach R, Borner GH, Schindelman G, Song S, Baskin TI, Dupree P, Wasteneys GO, Benfey PN (2005) COBRA, an Arabidopsis extracellular glycosyl-phosphatidyl inositol-anchored protein, specifically controls highly anisotropic expansion through its involvement in cellulose microfibril orientation. *Plant Cell* 17:1749–1763.

Rowe MH, Dong J, Weimer AK, Bergmann DC (2019) A plant-specific polarity module establishes cell fate asymmetry in the Arabidopsis stomatal lineage. *BioRxiv* 614636.

Rui Y, Dinneny JR (2020) A wall with integrity: surveillance and maintenance of the plant cell wall under stress. *New Phytol* 225:1428–1439.

Sager RE, Lee JY (2018) Plasmodesmata at a glance. *J Cell Sci* 131:jcs209346.

Sampathkumar A, Krupinski P, Wightman R, Milani P, Berquand A, Boudaoud A, Hamant O, Jonsson H, Meyerowitz EM (2014) Subcellular and supracellular mechanical stress prescribes cytoskeleton behavior in *Arabidopsis* cotyledon pavement cells. *Elife* 3:e01967.

Scheible WR, Eshed R, Richmond T, Delmer D, Somerville C (2001) Modifications of cellulose synthase confer resistance to isoxaben and thiazolidinone herbicides in *Arabidopsis Ixr1* mutants. *Proc Natl Acad Sci U S A* 98:10079–10084.

Schiefelbein J, Huang L, Zheng X (2014) Regulation of epidermal cell fate in Arabidopsis roots: the importance of multiple feedback loops. *Front Plant Sci* 5:47.

Seifert GJ (2018) Fascinating fasciclins: a surprisingly widespread family of proteins that mediate interactions between the cell exterior and the cell surface. *Int J Mol Sci* 19:E1628.

Shedletzky E, Shmuel M, Delmer DP, Lamport DT (1990) Adaptation and growth of tomato cells on the herbicide 2,6-dichlorobenzonitrile leads to production of unique cell walls virtually lacking a cellulose-xyloglucan network. *Plant Physiol* 94:980–987.

Shih HW, Miller ND, Dai C, Spalding EP, Monshausen GB (2014) The Receptor-like kinase FERONIA Is required for mechanical signal transduction in *Arabidopsis* seedlings. *Curr Biol* 24:1887–1892.

Simpson SD, Ashford DA, Harvey DJ, Bowles DJ (1998) Short chain oligogalacturonides induce ethylene production and expression of the gene encoding aminocyclopropane 1-carboxylic acid oxidase in tomato plants. *Glycobiology* 8:579–583.

Sinclair SA, Larue C, Bonk L, Khan A, Castillo-Michel H, Stein RJ, Grolimund D, Begerow D, Neumann U, Haydon MJ, Krämer U (2017) Etiolated seedling development requires repression of photomorphogenesis by a small cell-wall-derived dark signal. *Curr Biol* 27:3403-3418.e7.

Song JH, Kwak SH, Nam KH, Schiefelbein J, Lee MM (2019) QUIRKY regulates root epidermal cell patterning through stabilizing SCRAMBLED to control CAPRICE movement in *Arabidopsis*. *Nat Commun* 10:1744.

Sorek N, Szemenyei H, Sorek H, Landers A, Knight H, Bauer S, Wemmer DE, Somerville CR (2015) Identification of MEDIATOR16 as the *Arabidopsis* COBRA suppressor MONGOOSE1. *Proc Natl Acad Sci U S A* 112:16048–16053.

Souza CA, Li S, Lin AZ, Boutrot F, Grossmann G, Zipfel C, Somerville SC (2017) Cellulose-derived oligomers act as damage-associated molecular patterns and trigger defense-like responses. *Plant Physiol* 173:2383–2398.

Sun Y, Fan XY, Cao DM, Tang W, He K, Zhu JY, He JX, Bai MY, Zhu S, Oh E, Patil S, Kim TW, Ji H, Wong WH, Rhee SY, Wang ZY (2010) Integration of brassinosteroid signal transduction with the transcription network for plant growth regulation in Arabidopsis. *Dev Cell* 19:765–777.

Takano J, Tanaka M, Toyoda A, Miwa K, Kasai K, Fuji K, Onouchi H, Naito S, Fujiwara T (2010) Polar localization and degradation of *Arabidopsis* boron transporters through distinct trafficking pathways. *Proc Natl Acad Sci U S A* 107:5220–5225.

Tang W, Lin W, Zhou X, Guo J, Dang X, Li B, Lin D, Yang Z (2022) Mechano-transduction via the pectin-FERONIA complex activates ROP6 GTPase signaling in *Arabidopsis* pavement cell morphogenesis. *Curr Biol* 32:508-517.e3.

Tran Thanh Van K, Toubart P, Cousson A, Darvill AG, Gollin DJ, Chelf P, Albersheim P (1985) Manipulation of the morphogenetic pathways of tobacco explants by oligosaccharins. *Nature* 314:615–617.

Tsang DL, Edmond C, Harrington JL, Nühse TS (2011) Cell wall integrity controls root elongation via a general 1-aminocyclopropane-1-carboxylic acid-dependent, ethylene-independent pathway. *Plant Physiol* 156:596–604.

Turupcu A, Almohamed W, Oostenbrink C, Seifert GJ (2018) A speculation on the tandem fasciclin 1 repeat of FLA4 proteins in angiosperms. *Plant Signal Behav* 13:e1507403.

Urbanowicz BR, Peña MJ, Moniz HA, Moremen KW, York WS (2014) Two Arabidopsis proteins synthesize acetylated xylan in vitro. *Plant J* 80:197–206.

Uyttewaal M, Burian A, Alim K, Landrein B, Borowska-Wykret D, Dedieu A, Peaucelle A, Ludynia M, Traas J, Boudaoud A, Kwiatkowska D, Hamant O (2012) Mechanical stress acts via katanin to amplify differences in growth rate between adjacent cells in *Arabidopsis*. *Cell* 149:439–451.

Vaddepalli P, Fulton L, Batoux M, Yadav RK, Schneitz K (2011) Structure-function analysis of STRUBBELIG, an Arabidopsis atypical receptor-like kinase involved in tissue morphogenesis. *PLoS One* 6:e19730.

Vaddepalli P, Herrmann A, Fulton L, Oelschner M, Hillmer S, Stratil TF, Fastner A, Hammes UZ, Ott T, Robinson DG, Schneitz K (2014) The C2-domain protein QUIRKY and the receptor-like kinase STRUBBELIG localize to plasmodesmata and mediate tissue morphogenesis in *Arabidopsis thaliana*. *Development* 141:4139–4148.

Van der Does D, Boutrot F, Engelsdorf T, Rhodes J, McKenna JF, Vernhettes S, Koevoets I, Tintor N, Veerabagu M, Miedes E, Segonzac C, Roux M, Breda AS, Hardtke CS, Molina A, Rep M, Testerink C, Mouille G, Höfte H, Hamann T, Zipfel C (2017) The Arabidopsis leucine-rich repeat receptor kinase MIK2/LRR-KISS connects cell wall integrity sensing, root growth and response to abiotic and biotic stresses. *PLoS Genet* 13:e1006832.

van Dop M, Fiedler M, Mutte S, de Keijzer J, Olijslager L, Albrecht C, Liao CY, Janson ME, Bienz M, Weijers D (2020) DIX domain polymerization drives assembly of plant cell polarity complexes. *Cell* 180:427-439.e12.

Vance CP, Kirk TK, Sherwood RT (1980) Lignification as a mechanism of disease resistance. *Annu Rev Phytopathology* 18:259–288.

Vaten A, Dettmer J, Wu S, Stierhof YD, Miyashima S, Yadav SR, Roberts CJ, Campilho A, Bulone V, Lichtenberger R, Lehesranta S, Mahonen AP, Kim JY, Jokitalo E, Sauer N, Scheres B, Nakajima K, Carlsbecker A, Gallagher KL, Helariutta Y (2011) Callose biosynthesis regulates symplastic trafficking during root development. *Dev Cell* 21:1144–1155.

Vega-Sánchez ME, Verhertbruggen Y, Christensen U, Chen X, Sharma V, Varanasi P, Jobling SA, Talbot M, White RG, Joo M, Singh S, Auer M, Scheller HV, Ronald PC (2012) Loss of *Cellulose synthase-like F6* function affects mixed-linkage glucan deposition, cell wall mechanical properties, and defense responses in vegetative tissues of rice. *Plant Physiol* 159:56–69.

Vissenberg K, Fry SC, Pauly M, Höfte H, Verbelen JP (2005) XTH acts at the microfibril-matrix interface during cell elongation. *J Exp Bot* 56:673–683.

Vogel JP, Raab TK, Schiff C, Somerville SC (2002) *PMR6*, a pectate lyase-like gene required for powdery mildew susceptibility in Arabidopsis. *Plant Cell* 14:2095–2106.

Vogel JP, Raab TK, Somerville CR, Somerville SC (2004) Mutations in *PMR5* result in powdery mildew resistance and altered cell wall composition. *Plant J* 40:968–978.

Volpi C, Janni M, Lionetti V, Bellincampi D, Favaron F, D'Ovidio R (2011) The ectopic expression of a pectin methyl esterase inhibitor increases pectin methyl esterification and limits fungal diseases in wheat. *Mol Plant Microbe Interact* 24:1012–1019.

Voxeur A, Habrylo O, Guénin S, Miart F, Soulié MC, Rihouey C, Pau-Roblot C, Domon JM, Gutierrez L, Pelloux J, Mouille G, Fagard M, Höfte H, Vernhettes S (2019) Oligogalacturonide production upon *Arabidopsis thaliana-Botrytis cinerea* interaction. *Proc Natl Acad Sci U S A* 116:19743–19752.

Wagner TA, Kohorn BD (2001) Wall-associated kinases are expressed throughout plant development and are required for cell expansion. *Plant Cell* 13:303–318.

Wakabayashi K, Hoson T, Huber DJ (2003) Methyl de-esterification as a major factor regulating the extent of pectin depolymerization during fruit ripening: a comparison of the action of avocado (*Persea americana*) and tomato (*Lycopersicon esculentum*) polygalacturonases. *J Plant Physiol* 160:667–673.

Wang T, Liang L, Xue Y, Jia PF, Chen W, Zhang MX, Wang YC, Li HJ, Yang WC (2016) A receptor heteromer mediates the male perception of female attractants in plants. *Nature* 531:241–244.

Wanke A, Rovenich H, Schwanke F, Velte S, Becker S, Hehemann JH, Wawra S, Zuccaro A (2020) Plant species-specific recognition of long and short β-1,3-linked glucans is mediated by different receptor systems. *Plant J* 102:1142–1156.

Watt FM, Huck WT (2013) Role of the extracellular matrix in regulating stem cell fate. *Nat Rev Mol Cell Biol* 14:467–473.

Wolf S (2022) Cell wall signaling in plant development and defense. *Annu Rev Plant Biol* 73:16.1-16.31.

Wolf S, Hématy K, Höfte H (2012a) Growth control and cell wall signaling in plants. *Annu Rev Plant Biol* 63:381–407.

Wolf S, Mravec J, Greiner S, Mouille G, Höfte H (2012b) Plant cell wall homeostasis is mediated by brassinosteroid feedback signaling. *Curr Biol* 22:1732–1737.

Wolf S, van der Does D, Ladwig F, Sticht C, Kolbeck A, Schurholz AK, Augustin S, Keinath N, Rausch T, Greiner S, Schumacher K, Harter K, Zipfel C, Hofte H (2014) A receptor-like protein mediates the response to pectin modification by activating brassinosteroid signaling. *Proc Natl Acad Sci U S A* 111:15261–15266.

Wormit A, Butt SM, Chairam I, McKenna JF, Nunes-Nesi A, Kjaer L, O'Donnelly K, Fernie AR, Woscholski R, Barter MC, Hamann T (2012) Osmosensitive changes of carbohydrate metabolism in response to cellulose biosynthesis inhibition. *Plant Physiol* 159:105–117.

Wormit A, Usadel B (2018) The multifaceted role of pectin methylesterase inhibitors (PMEIs). *Int J Mol Sci* 19:E2878.

Wu SW, Kumar R, Iswanto ABB, Kim JY (2018) Callose balancing at plasmodesmata. *J Exp Bot* 69:5325–5339.

Xiong G, Cheng K, Pauly M (2013) Xylan O-acetylation impacts xylem development and enzymatic recalcitrance as indicated by the Arabidopsis mutant *tbl29*. *Mol Plant* 6:1373–1375.

Xu SL, Rahman A, Baskin TI, Kieber JJ (2008) Two leucine-rich repeat receptor kinases mediate signaling, linking cell wall biosynthesis and ACC synthase in *Arabidopsis*. *Plant Cell* 20:3065–3079.

Yamanaka T, Nakagawa Y, Mori K, Nakano M, Imamura T, Kataoka H, Terashima A, Iida K, Kojima I, Katagiri T, Shinozaki K, Iida H (2010) MCA1 and MCA2 that mediate Ca^{2+} uptake have distinct and overlapping roles in Arabidopsis. *Plant Physiol* 152:1284–1296.

Yang C, Liu R, Pang J, Ren B, Zhou H, Wang G, Wang E, Liu J (2021) Poaceae-specific cell wall-derived oligosaccharides activate plant immunity via OsCERK1 during *Magnaporthe oryzae* infection in rice. *Nat Commun* 12:2178.

York WS, Darvill AG, Albersheim P (1984) Inhibition of 2,4-dichlorophenoxyacetic acid-stimulated elongation of pea stem segments by a xyloglucan oligosaccharide. *Plant Physiol* 75:295–297.

Yoshida S, van der Schuren A, van Dop M, van Galen L, Saiga S, Adibi M, Möller B, ten Hove CA, Marhavy P, Smith R, Friml J, Weijers D (2019) A SOSEKI-based coordinate system interprets global polarity cues in *Arabidopsis*. *Nature Plants* 5:160–166.

Yoshimura K, Iida K, Iida H (2021) MCAs in Arabidopsis are Ca^{2+}-permeable mechanosensitive channels inherently sensitive to membrane tension. *Nat Commun* 12:6074.

Yuan Y, Teng Q, Zhong R, Ye ZH (2013) The Arabidopsis DUF231 domain-containing protein ESK1 mediates 2-O- and 3-O-acetylation of xylosyl residues in xylan. *Plant Cell Physiol* 54:1186–1199.

Zhang Y, Yu J, Wang X, Durachko DM, Zhang S, Cosgrove DJ (2021) Molecular insights into the complex mechanics of plant epidermal cell walls. *Science* 372:706–711.

Zhou X, Lu J, Zhang Y, Guo J, Lin W, Van Norman JM, Qin Y, Zhu X, Yang Z (2021) Membrane receptor-mediated mechano-transduction maintains cell integrity during pollen tube growth within the pistil. *Dev Cell* 56:1030-1042.e6.

20 The Ancient Battle between Plants and Pathogens

Resilience of the Plant Cell Wall and Damage-Associated Molecular Patterns (DAMPs) Drive Plant Immunity

Daniela Pontiggia, Sarah Giulietti, Giovanna Gramegna, Vincenzo Lionetti, Riccardo Lorrai, Lucia Marti, Simone Ferrari, Giulia De Lorenzo, and Felice Cervone
Sapienza Università di Roma

INTRODUCTION

The plant cell wall is a strong and resilient barrier against environmental stresses. It is built to resist pathogenic attacks of microbes and insects with its recalcitrance to enzymatic disruption and is the place where important signals and mechanisms of innate immunity are activated. Plant immunity may have initially evolved as a response to potentially dangerous situations caused by mechanical wounds, which can become an easy passageway for microbes into plant tissues (Savatin et al., 2014). It is also likely that innate immune responses of other organisms along the tree of life share this same evolutionary origin, based on the necessity of protecting and repairing wounded tissues, even in the absence of microbes (De Lorenzo et al., 2018). Immunity relies on the capacity of perceiving the presence of "self" and "non-self" danger signals, referred to as Damage-Associated Molecular Patterns (DAMPs) and Microbe-Associated Molecular Patterns (MAMPs), respectively (Tanaka and Heil, 2021; De Lorenzo and Cervone, 2022). MAMPs are highly conserved microbial molecules derived from molecular structures that usually characterise pathogens of different classes, as, for example, bacterial flagellin and chitooligomers from fungal cell walls and exoskeletons of insects (Choi and Klessig, 2016). In the case of plants, many characterised DAMPs are derived from the fragmentation of the cell wall components (Pontiggia et al., 2020).

In the plant kingdom, immunity to potential pathogens is the rule and disease is the exception. Upon pathogen perception, plants counteract the infection through the activation of two layers of defence, pattern-triggered immunity (PTI) and effector-triggered immunity (ETI). In the widely accepted "zig-zag model" proposed several years ago (Jones and Dangl, 2006), PTI is considered the first line of defence activated by MAMPs through specific pattern-recognition receptors (PRR) localised on the plasma membrane, whereas ETI is a second layer of defence, triggered by microbial genotype–specific pathogenicity effectors released into the plant cell and recognised by intracellular plant genotype–specific receptor proteins that function as resistance (R) factors. The effects of PTI include the rapid activation of a broad spectrum of responses such as an oxidative burst, Ca^{2+} influx, activation of protein kinases and transcriptional reprogramming leading to hormonal and metabolic alterations (Bigeard et al., 2015; Peng et al., 2018). ETI is characterised by a more robust and longer response than PTI, which in many cases culminates in localised programmed cell death known as the "hypersensitive response" (Cui et al., 2015). PTI and ETI are inter-dependent and mutually enhance each other to produce a highly effective immune response to a wide variety of pathogens (Ngou et al., 2021; Pruitt et al., 2021; Tian et al., 2021; Yuan et al., 2021). Moreover, local activation of immunity can confer a systemic protection against pathogens in distal tissues mediated by the hormone salicylic acid and referred to as systemic acquired resistance (SAR) (Fu and Dong, 2013). Local mechanical damage and activation of immunity by necrotrophs and insects as well as beneficial microbes may also lead to a systemic protective response, where jasmonic acid plays an important role (Vega-Munoz et al., 2020).

Many MAMPs and a plethora of effectors and their receptors have been discovered in the last 20 years in a large number of plant–pathogen combinations, reflecting the result of the long co-evolutionary battle between plants and pathogens. However, before the evolutionary acquisition of the sophisticated mechanisms of recognition and

DOI: 10.1201/9781003178309-20

response to MAMPs and effectors as depicted in the zig-zag model, a more ancient battle for survival must have occurred at the level of the plant cell wall, with plants evolving cell wall components that resist pathogen attack and microbes continually evolving new cell wall degrading enzymes (CWDEs). In this context, conceivably, plants evolved the capacity to recognise and respond to cell wall–derived molecules, i.e., DAMPS, as indicators of a breach formation in the wall and molecular signals of danger (Figure 20.1).

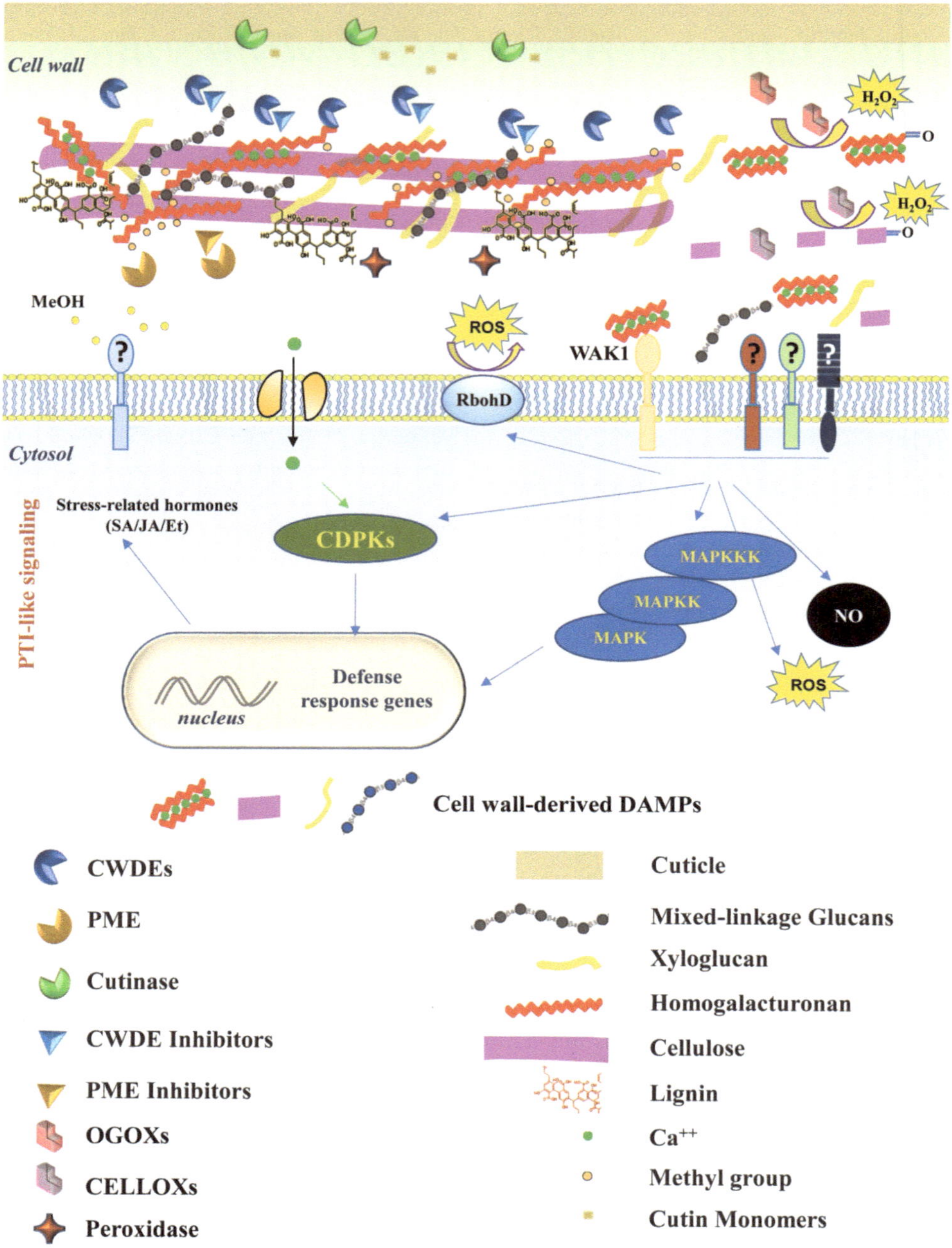

FIGURE 20.1 CWDEs, cell wall–derived DAMPs and pattern-triggered immunity (PTI). Schematic representation of the events occurring during the cell wall degradation by plant pathogens. Cell wall modifying and degrading enzymes, the activity of which is regulated by specific inhibitors, act on cell wall polysaccharides, releasing fragments that may act as DAMPs. These are likely recognised by plasma membrane receptors (PRRs), as in the case of OGs and its receptor WAK1. Responses downstream of receptor activation include ROS production, both extracellular (mediated by the NADPH oxidase RbohD) and intracellular, production of nitric oxide (NO), activation of CDPKs and of the MAPK cascade, synthesis of hormones and transcriptional reprogramming. Methanol released by PME may also act as a signal. The signalling activity of OGs and CDs is dampened by specific oxidases belonging to the family of Berberine-Bridge Enzyme-like family (BBE-l), which also produce H_2O_2.

CWDES AND THEIR INHIBITION

The penetration and spreading of many microbes, nematodes and insects in plant tissue is facilitated by an arsenal of CWDEs, which also generate necessary nutrients for the invading organisms (Toth and Birch, 2005; Calderón-Cortés et al., 2012; Kubicek et al., 2014). Fungi produce a remarkable number and variety of CWDEs, mostly represented by glycosyl hydrolases (Kubicek et al., 2014). Soft rot bacteria also synthesise significant amounts of CWDEs that are key virulence factors (Charkowski et al., 2012) and parasitic nematodes and phytophagous insects have acquired the ability to produce CWDEs apparently *via* horizontal transfer of CWDE-encoding genes from bacteria (Shelomi et al., 2016; Haeger et al., 2020; Haeger et al., 2021). The capacity of producing a diverse array of CWDEs coupled to the correct spatio-temporal regulation of their synthesis is critical for obtaining efficient cell wall degradation and is the key to the success of many pathogens (Bellincampi et al., 2014; Lorrai and Ferrari, 2021). The specific CWDEs secreted by a particular pathogen during infection most likely reflect the adaptation to specific cell wall compositions of different hosts. The concurrent production of multiple CWDEs is also necessary for full virulence (Kubicek et al., 2014; Lorrai and Ferrari, 2021). For example, *Fusarium graminearum* needs the combined action of xylanase, polygalacturonase (PG) and cellulase production to display full virulence (Paccanaro et al., 2017). Moreover, inactivation of a transcription factor required for the coordinated expression of cutinase, cellulase and pectate lyase significantly reduces virulence of *Alternaria alternata* on tangerine (*Citrus reticulata*) (Ma et al., 2019).

CWDEs not only cause cell wall degradation but also induce necrosis of the host tissue, as observed with xylanases (Furman-Matarasso et al., 1999; Noda et al., 2010), cellulases (Ma et al., 2015), cutinases (Gui et al., 2018) and pectinases (Zuppini et al., 2005; Wang et al., 2009; Zhang et al., 2014; Zhang et al., 2021). Defence and cell death–inducing activity of microbial CWDEs may not depend on enzymatic activity (De Lorenzo et al., 1997; Poinssot et al., 2003). For example, some defence responses of Arabidopsis require the PRR RLP42 that senses a conserved 9-amino acid fragment of fungal PGs (Zhang et al., 2021). In some cases, therefore, CWDEs may be regarded to be typical MAMPs (Gui et al., 2018). In addition, CWDEs can indirectly trigger plant defence responses through the release of DAMPs, as discussed below.

Besides canonical CWDEs, lytic polysaccharide monooxygenases (LPMOs), which utilise H_2O_2 or molecular oxygen for the oxidative degradation of polysaccharides, have drawn the attention of researchers as possible pathogenicity factors that act in cell wall degradation. Their possible role as pathogenicity factors is supported by the observation that many *LPMO* genes are present in the genome of several phytopathogenic fungi and oomycetes (Vaaje-Kolstad et al., 2010; Kim et al., 2014; Bissaro et al., 2017; Labourel et al., 2020).

In contrast to pathogens, much less is known about the production of CWDEs by beneficial microbes. The lifestyle of symbionts like mycorrhizal fungi is not compatible with the production of destructive amounts of CWDEs, but rather with their controlled and limited activity (Cervone et al., 1988; Sillo et al., 2016; Bonfante, 2018; Perotto et al., 2018). Consistent with this notion is the combined analysis of 135 fungal genomes, which shows that transitions from saprotrophy to symbiosis involve widespread losses of CWDEs (Miyauchi et al., 2020; Mesny et al., 2021).

CWDEs Produced by Microbes

A wide range of pectinases, secreted by fungi and bacteria, can be classified according to their site of cleavage (endo- and exo-), preferred substrate (pectin or pectate) and mode of cleavage (hydrolases or lyases) (Schols et al., 2009). PGs hydrolyse the α-(1–4) linkages of homogalacturonan (HG), either internally (endo-PGs) or at the non-reducing end (exo-PGs). Pectate lyases (endo-PLs and exo-PLs) and pectin lyases (PNLs) catalyse the β-elimination of non-esterified and highly methyl-esterified pectin, respectively. PLs, but not PNLs, require calcium for optimal activity. Crystallographic studies on bacterial PLs allowed the first identification of a novel protein domain characterised by a right-handed parallel β-helix (Yoder et al., 1993; Pickersgill et al., 1994). Notably, PGs and PNLs also fold into a parallel β-helix and share many of the structural features of PLs, despite limited amino acid sequence identity (Mayans et al., 1997; Federici et al., 1999; Federici et al., 2001; Bonivento et al., 2008).

PGs, prevalently produced by fungi, and PLs, prevalently produced by bacteria, are among the first microbial CWDEs secreted during infection. This suggests that the activity of pectinases is essential for the subsequent action of other CWDEs to achieve a massive tissue maceration (De Lorenzo et al., 1997). In these terms, pectinases can be useful in the pretreatment of plant biomass to ameliorate saccharification and bioconversion efficiency (Lionetti et al., 2010; Francocci et al., 2013). The impact of pectin degradation is most evident during attack by soft rot microbes, which cause the dissolution of the middle lamella and consequent maceration of the host tissues (Hugouvieux-Cotte-Pattat et al., 2014).

The degradation of pectin and other cell wall polysaccharides was suggested as an important factor of pathogenicity and specificity of microbes more than 50 years ago (Bateman and Millar, 1966; Albersheim et al., 1969; Cooper and Wood, 1973), but supporting genetic evidence could not be obtained easily for a long time due to gene redundancy and the existence of many enzyme isoforms produced by the same pathogen. Many years later, targeting single microbial genes for disruption (Scott-Craig et al., 1990; Walton, 1994) revealed in some cases the role of specific PGs (ten Have et al., 1998; Isshiki et al., 2001; Oeser et al., 2002) and PLs (Yakoby et al., 2001; Ben-Daniel et al., 2012; Yang et al., 2018) as important virulence factors.

The activity of microbial PGs and PLs is strictly dependent on the status of methylation and acetylation of their substrates. Indeed, many pathogens secrete pectin methylesterases (PMEs) to remove methyl groups from pectin, which consequently becomes less recalcitrant to degradation (Fan et al., 2017; Lionetti et al., 2017). Pathogens can also recruit the plant PMEs to assist the infection process, as suggested by the observation that *Pectobacterium carotovorum* and *Botrytis cinerea* induce rapid expression of AtPME3 in Arabidopsis as a susceptibility factor required for the colonisation of the host tissue (Raiola et al., 2011). Other fungal pectin–degrading enzymes such as xylogalacturonan hydrolases, rhamnogalacturonan lyases and rhamnogalacturonan acetylesterases have been characterised but their importance as virulence factors has not been yet fully demonstrated (Pagès et al., 2003; Williams et al., 2014; Mandelc and Javornik, 2015).

Recently, it has been reported that silencing a HG-degrading LPMO produced by the oomycete *Phytophthora infestans* causes reduced virulence on potato (Sabbadin et al., 2021). The HG fragments generated by this enzyme are oxidised oligogalacturonides (OGs), which, unlike non-oxidised OGs, are unable to trigger host defences (Benedetti et al., 2018). Therefore, LPMOs, unlike PGs and PLs (see later), contribute to deconstruct the host cell wall while simultaneously avoiding the activation of plant immunity. It is noteworthy that an LPMO active on chito-oligosaccharides is highly expressed in the haustoria of *Podosphaera xanthii* and contributes to the suppression of chitin-induced plant immunity (Polonio et al., 2021).

Endoglucanases, exoglucanases and β-glucosidases acting on cellulose are thought to contribute to the virulence of some pathogenic fungi (Van Vu et al., 2012) and bacteria (Hwang et al., 2019). On the other hand, disruption of single cellulase genes in *Cochliobolus carbonum* and *Botrytis cinerea* failed to significantly impair their virulence (Sposato et al., 1995; Espino et al., 2005), likely reflecting enzyme redundancy.

The synergy of glucanases and LPMOs for an efficient pathogenicity is suggested by the expansion of the LPMO family in aggressive fungal pathogens and their elevated levels of expression during infection, especially during the transformation of hemibiotrophs from biotrophic to necrotrophic growth (reviewed in Jagadeeswaran et al., 2021). As in the case of the enzymes acting on HG, the release of oxidised cellulose fragments by LPMOs is not expected to induce immunity since oxidised cellodextrins (CDs) are inactive as elicitors (Locci et al., 2019). Nevertheless, in contrast with the results of Locci et al. (2019), several LPMO-oxidised cellulose oligosaccharides have been reported to evoke immunity in Arabidopsis, conferring resistance to the necrotrophic fungus *Botrytis cinerea* (Zarattini et al., 2021).

Enzymes such as endo-1,4-β-xylanases, endo-1, 3-β-xylanases, β-xylosidases, α-arabinofuranosidases, α-glucuronidases and acetylxylanases act on hemicelluloses (Moreira and Filho, 2016; Bajaj and Mahajan, 2019). Early attempts to determine their role in pathogenesis by gene disruption were not conclusive likely due to their functional redundancy (Apel et al., 1993; Apel-Birkhold and Walton, 1996; Wu et al., 1997). More recent work indicates that simultaneous silencing of ten xylanase genes significantly reduces virulence of *Magnaporthe grisea* (Nguyen et al., 2011). In some cases, a major role of single xylanase isoforms has been related to their ability to induce cell death (Noda et al., 2010; Yang et al., 2018).

Lignin-degrading enzymes are mainly produced by saprophytes or weak parasites and also by some phytopathogenic fungi (Pollegioni et al., 2015). A recent paper reports that a chemical inhibitor of laccases provides field protection against infection by *Gaeumannomyces graminis* var. *tritici* (Wang et al., 2020).

Finally, the role of cutinases as virulence factors of pathogenic fungi is well established and discussed in several papers (Rogers et al., 1994; Serrano et al., 2014; Aragón et al., 2017; Gui et al., 2018; Ziv et al., 2018) (Figure 20.1).

CWDEs Produced by Nematodes and Insects

Pectinases, cellulases and xylanases produced by cyst and root-knot nematodes contribute to their pathogenicity on plant roots. Notably, in contrast to pathogenic microbes, which usually produce pectinases before the other CWDEs, nematodes secrete cocktails of various CWDEs at the same time through the stylet orifice (Chen et al., 2005; Mitreva-Dautova et al., 2006; Vanholme et al., 2007; Abad et al., 2008; Davis et al., 2011). Transcriptomics and genomics studies indicate that at least some groups of CWDEs are produced and play important roles in the insect feeding capacity. Interestingly, pectinases from insects are similar to bacterial pectinases, suggesting their possible acquisition through horizontal gene transfer (Watanabe and Tokuda, 2010; Shelomi et al., 2016; Haeger et al., 2020; Haeger et al., 2021).

Inhibitors of CWDEs

Plants counteract CWDEs by producing protein inhibitors that slow down the cell wall degradation (Juge, 2006; Lagaert et al., 2009). Well characterised inhibitors are the PG-inhibiting proteins (PGIPs) active against both microbial and insect PGs (De Lorenzo et al., 2001; Kalunke et al., 2015; Rathinam et al., 2020). After the first report of such inhibitors (Albersheim and Anderson, 1971), it was shown much later that these proteins favour the formation *in vitro* of elicitor-active OGs (Cervone et.al., 1987, 1989). Today, more than 170 PGIPs have been described (Kalunke et al., 2015). A PGIP from *Phaseolus vulgaris* was the first plant leucine-rich repeat (LRR) protein to have its crystal structure solved. It shows a helical elongated shape characterised by parallel β-sheets and a negatively charged surface on the concave side that is crucial for the interaction and inhibition of PGs (Di Matteo et al., 2003; Benedetti et al., 2011). Notably, the LRR structure found in PGIPs was later found to be widespread among the receptors and co-receptors involved in plant immunity, including PRRs that perceive microbial MAMPs and R proteins mediating the gene-for-gene resistance (Figure 20.2) (Di Matteo et al., 2003; Chakraborty et al., 2019; Song et al., 2021). Gene families encoding PGIPs in Arabidopsis and crop plants have been extensively characterised and their role in defence against specific pathogens is well documented (Kalunke et al., 2015). Several isoforms vary in their ability to inhibit PGs from different fungi or even different PGs from the same fungal strain. For instance, PvPGIP2 from *P. vulgaris* inhibits PGs from *Fusarium phyllophilum* and *Aspergillus niger*, whereas another isoform from the same plant, PvPGIP1, is effective only against the *A. niger* enzyme (Leckie et al., 1999). In Arabidopsis, two closely related genes, *AtPGIP1* and *AtPGIP2*, encode inhibitors with similar activity against a PG of *B. cinerea*, but with differential activity against a PG from *Colletotrichum acutatum*. Notably, expression of *AtPGIP2* is induced during pathogen infection by jasmonic acid but not by OGs, whereas *AtPGIP1* displays a jasmonic acid-independent induction by both fungal infection and OGs (Ferrari et al., 2003).

Overexpression of *PGIP* genes increases resistance to fungi and bacteria in several plants (Powell et al., 2000; Ferrari et al., 2003; Aguero et al., 2005; Joubert et al., 2006; Janni et al., 2008; Ferrari et al., 2012), whereas silencing of endogenous PGIPs increases susceptibility to *B. cinerea* in Arabidopsis (Ferrari et al., 2006) and to *F. graminearum* in Arabidopsis and wheat (Ferrari et al., 2012). Grapevine plants over-expressing a pear PGIP display reduced infection by the bacterium *Xylella fastidiosa*, which expresses a PG encoded by a gene that is essential for virulence (Aguero et al., 2005; Roper et al., 2007). Inhibitory activity of PGIPs towards PGs derived from mirids (D'Ovidio et al., 2004; Frati et al., 2006) and leaf beetles (Haeger et al., 2020) has been reported. The importance of PGIPs in defence is emphasised by the recent finding that PGIP-inactivating effectors are produced by necrotrophic fungi such as *Sclerotinia sclerotiorum* to enhance their virulence (Wei et al., 2022).

Besides PGIPs, xylanase inhibitors such as the wheat proteins XIP-I (xylanase inhibitor protein I) (Payan et al., 2003) and TAXI (*Triticum aestivum* xylanase inhibitor) (Debyser and Delcour, 1997; Debyser et al., 1999) have been characterised. The crystal structure of TAXI in complex with a xylanase from *Aspergillus niger* reveals the direct interaction of the inhibitor with the active site of the enzyme and its competition with the enzyme substrate (Sansen et al., 2004).

Xyloglucan endoglucanase inhibitor proteins (XEGIPs) have been characterised in a few dicots (Qin et al., 2003; York et al., 2004; Naqvi et al., 2005; Habrylo et al., 2013). The crystal structure of a XEGIP in complex with an endoglucanase shows that, similarly to TAXI, XEGIP inhibits the enzyme by competing with the enzyme substrate (Yoshizawa et al., 2012).

In the interaction between the oomycete *Phytophthora sojae* and soybean, a plant inhibitor (GmGIP1) of the microbial xyloglucan endoglucanase (XEG1) partially blocks pathogen virulence. As an illustration of an ongoing arms race, a paralogous PsXEG1-like protein, PsXLP1, is secreted by the oomycete that has no enzyme activity but interacts more tightly with GmGIP1 than XEG1, acting as a decoy and thereby preventing the inhibition of XEG1 (Ma et al., 2017). Interestingly, XEG1 from *P. sojae* is also recognised by a receptor-like protein (RXEG1) from *Nicotiana benthamiana* and the crystal structure of the complex reveals that binding of XEG1 promotes association of RXEG1 with the LRR-type co-receptor BAK1, a known player in plant immunity (Sun et al., 2022).

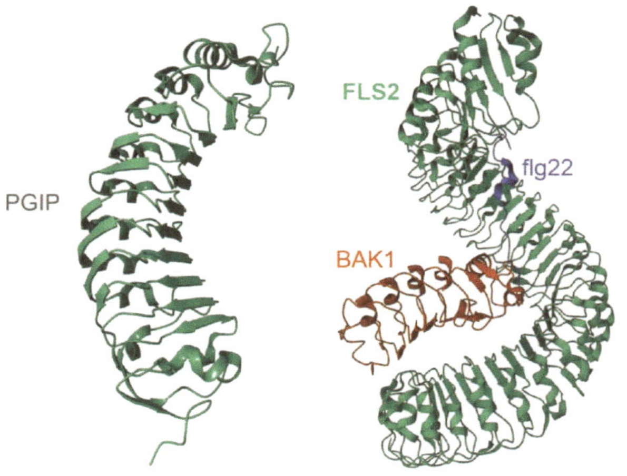

FIGURE 20.2 Structure of LRR domains. Structure of PGIP2 of *P. vulgaris* (left) and the LRR domains of FLS2 and its co-receptor BAK1 interacting with the bacterial MAMP ligand flg22 (right).

CELL WALL MODIFICATIONS AFFECT THE PLANT IMMUNITY

Cell wall modifications can be obtained through natural, induced or targeted mutations, manipulation of the expression of plant cell wall–related genes, or through the expression *in planta* of microbial CWDEs. The concept that plant resistance to pathogens is affected by the cell wall structure is well-accepted. Cell wall modifications may not only result in a more recalcitrant territory for the enzyme-assisted microbial invasion of the plant tissue but may also prime more complicated mechanisms of response not yet elucidated (Bacete et al., 2018).

A lower content of de-methyl-esterified homogalacturonan, obtained by transgenic expression of a fungal PG (Capodicasa et al., 2004; Ferrari et al., 2008) or by mutation (Francocci et al., 2013), results in increased resistance to *B. cinerea* and reduced growth. Some mutants with a reduced galacturonic acid content are more susceptible to the bacterium *Pseudomonas syringae* and the fungus *B. cinerea* (Bethke et al., 2016). Other mutants with defects in pectin composition are impaired in resistance against the fungal pathogen *Colletotrichum higginsianum* (Engelsdorf et al., 2017).

The degree of HG methyl esterification is regulated by PMEs and PME inhibitors (PMEIs) that dynamically orchestrate the overall methylation and distribution of methyl groups. Arabidopsis plants with a reduced AtPME3 activity or which over-express PMEIs are more resistant to *B. cinerea* (Lionetti et al., 2007). Conversely, PMEI mutants exhibit compromised immunity to the fungus (Lionetti et al., 2017). Specific *pme* single, double, triple and quadruple mutants are more susceptible to *P. syringae* with no decrease of total PME activity, suggesting that the specific mode of action of PMEs is more relevant than their amounts (Bethke et al., 2014). Indeed, unlike AtPME3, AtPME17, which performs a blockwise pattern of pectin de-methylesterification that favours the formation of egg-box structures between HG, strongly contributes to resistance against *B. cinerea* (Del Corpo et al., 2020). During plant–pathogen interactions, both the amount and the pattern of pectin methylesterification can be altered (Lionetti et al., 2012; Bellincampi et al., 2014; Bethke et al., 2014). The perturbation of Arabidopsis and *Brachypodium distachyon* cell wall acetylation by transgenic expression of a pectin-specific fungal acetylesterase also leads to increased resistance to the fungal pathogens *B. cinerea* and *Bipolaris sorokiniana* but not to the bacteria *P. syringae* and *Xanthomonas oryzae* (Pogorelko et al., 2013).

Reduced synthesis of cellulose or mutations of the cellulose synthase genes enhance resistance to many pathogens. It is possible that the alteration of the territory to be invaded is not properly recognised by the microbes (Ellis and Turner, 2001; Ellis et al., 2002; Caño-Delgado et al., 2003; Hernandez-Blanco et al., 2007). On the other hand, alterations of cellulose synthesis are induced by isoxaben and lead to activation of defence responses (the so-called cell wall damage syndrome) (Gigli-Bisceglia et al., 2020). Arabidopsis mutants with an altered xyloglucan structure are more resistant to the necrotrophic fungus *Plectosphaerella cucumerina* (Delgado-Cerezo et al., 2012). The overexpression of the glycosyltransferases involved in xylan synthesis in barley increases the resistance to *Blumeria graminis* (Chowdhury et al., 2017). The reduced level of acetylation of hemicellulose in Arabidopsis improves resistance to different pathogens such as *B. cinerea*, *Hyaloperonospora arabidopsidis, Erysiphe cichoracearum* and *Erysiphe orontii* (Vogel et al., 2004; Manabe et al., 2011). The overexpression in Arabidopsis and *Brachipodium* of a hemicellulose acetylesterase from *A. nidulans* enhances resistance to *B. cinerea* and *B. sorokiniana* (Pogorelko et al., 2013). Many different kinds of cell wall modifications have recently been shown to have a significant impact on disease resistance and on the overall defensive ability of plants (Molina et al., 2021).

One of the plant responses against fungi is the deposition of callose by glucan synthase-like (GSL) enzymes localised on the plasma membrane (Schneider et al., 2016; Wang et al., 2021). Callose also typically accumulates in the cell wall region close to the neck zone of plasmodesmata where it works as a tissue barrier against invading viruses (Amsbury et al., 2017; Liu et al., 2021). Arabidopsis plants over-expressing a GSL enzyme accumulate elevated levels of callose that almost completely block the penetration of powdery mildew fungi (Ellinger et al., 2013; Ellinger and Voigt, 2014). Accordingly, a barley mutant silenced in a GSL enzyme exhibits a slower callose deposition and is more susceptible to *Blumeria graminis* (Chowdhury et al., 2016).

Lignification of the cell wall often occurs in response to mechanical damage or infection and contributes to plant resistance to bacteria (Tronchet et al., 2010; Ma et al., 2017; Lee et al., 2019) and fungi (Xu et al., 2011; Eynck et al., 2012; Shi et al., 2012; Yang et al., 2017; Liu et al., 2018; Liu et al., 2021). Lignin accumulation has been reported to contribute to the resistance of rice to insects such as the brown planthopper (Jannoey et al., 2017), leafrollers (Tianpei et al., 2015), aphids (Wang et al., 2017) and root-knot nematodes (Fujimoto et al., 2015). Interestingly, alfalfa and Arabidopsis plants impaired in lignin biosynthesis display constitutive expression of defence responses (Gallego-Giraldo et al., 2011). In this case, it is thought that cell wall remodelling causes the release of an endogenous PG likely generating DAMPs not yet structurally characterised (Gallego-Giraldo et al., 2020).

Recalcitrance of cell wall to enzymatic degradation during fungal infections can be enhanced by cross-links formed between phenolics such as ferulic acid or between polysaccharides and phenolics (Mnich et al., 2020). Feruloyl esterases have been reported to contribute to virulence of some fungal pathogens (Xu et al., 2018; Thaker et al., 2021) and, accordingly, transgenic expression of a fungal feruloyl esterase in Arabidopsis and *Brachypodium* increases susceptibility to fungal necrotrophs (Reem et al., 2016). Moreover, plant peroxidases can form cross-links between glycoproteins such as extensins and contribute to restrict pathogen invasion (Castilleux et al., 2018; Castilleux et al., 2021).

CELL WALL DAMPs

Upon attack by pathogenic microbes, nematodes or insects, the plant cell wall may release danger signals that activate plant immunity. Research in this field started with basic biochemical tools many years ago, when attempts were made in several laboratories to detect how pathogenicity factors such as PGs interact and bind to cell wall and membranes in relation to activation of immunity (Cervone et al., 1978; Bishop et al., 1981; Cervone et al., 1981; Hahn et al., 1981; Lee and West, 1981, 1981). The hypothesis was that plants could have evolved the ability of sensing key pathogenicity factors like PGs to activate resistance responses.

Danger signals released from cell wall were originally named "endogenous elicitors" (Hahn et al., 1981), but nowadays are referred to as DAMPs (Land, 2003; De Lorenzo et al., 2018; Tanaka and Heil, 2021). The high structural complexity of the cell wall allows to predict the potential formation of a variety of fragments, some of which have been demonstrated to act as DAMPs (Hou et al., 2019; Mélida et al., 2020). An emerging notion is that not only pathogen attack but also mechanical lesions, cell wall remodelling, cell wall manipulation operated by transgenesis or cell wall mutations can lead to the activation of endogenous CWDEs that generate DAMPs (Benedetti et al., 2015; Gallego-Giraldo et al., 2020; Pontiggia et al., 2020; Molina et al., 2021). More generally, a complex cell wall integrity system is responsible for the perception of alterations in the cell wall structure and consequent induction of the appropriate cellular responses (Engelsdorf et al., 2018; Bacete and Hamann, 2020; Gigli-Bisceglia et al., 2020). If the fragments released during cell wall alterations display DAMP activity, likely they are subjected to homeostatic control to avoid a hyper-immune response that often compromises plant fitness and survival (Pontiggia et al., 2020).

DAMPs may originate both from the fibrillar and matrix cell wall components as well as from the cuticle. Pectin and particularly HG are well known substrates of both microbial and plant-derived pectinases. The enzymatic hydrolysis of HG by PGs and PLs transiently generates unmethylated OGs capable of activating an immune response (Figure 20.1) (Ferrari et al., 2013; Hou et al., 2019). OGs composed of 10–15 residues of α-(1–4)-linked D-galacturonic acid display a marked DAMP activity in several plants. In Arabidopsis, shorter OGs with a degree of polymerisation between two and six have a similar but milder activity (Davidsson et al., 2017; De Lorenzo et al., 2018). OGs are perceived by the wall-associated kinase (WAK1) (Brutus et al., 2010) and likely by other receptors still to be characterised (Gravino et al., 2017). The longevity of OGs transiently formed during HG hydrolysis by microbial and insect PGs is enhanced by plant-derived PGIPs (Figure 20.3) (Cervone et al., 1987; Cervone et al., 1989; D'Ovidio et al., 2004). Evidence is lacking to support the hypothesis that other

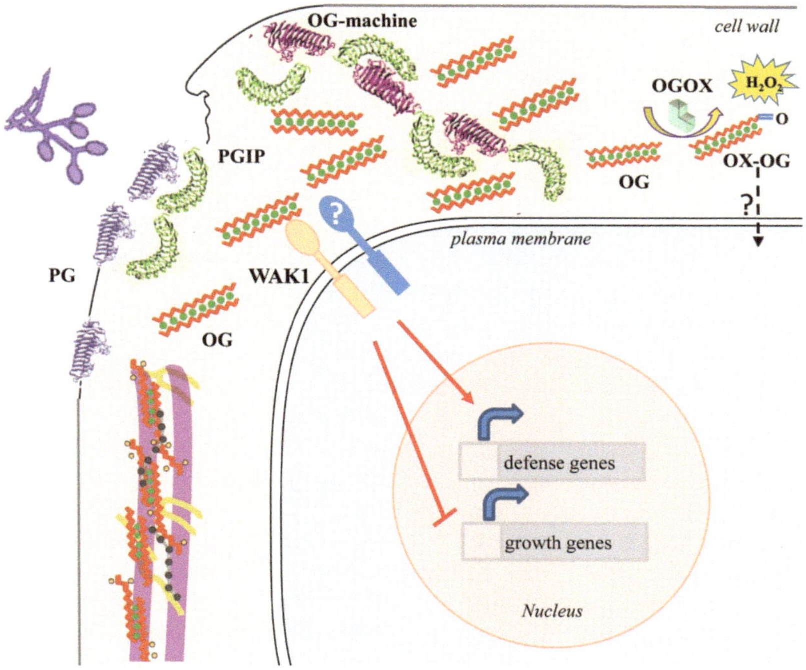

FIGURE 20.3 PG, PGIP and OGs in plant–pathogen interactions. Interaction of microbial and insect PGs with PGIPs favours the formation of OGs that activate immunity. In Arabidopsis, the homeostasis of OGs is maintained by OGOX that oxidises OGs impairing their elicitor activity. The expression under an inducible promoter of a PG–PGIP fusion protein, named "OG machine", leads to the accumulation of OGs on command and consequent immunity.

kinds of pectic fragments such as methylated or oxidised OGs found in plant tissues upon long incubation with macerating pathogens or enzymes act as DAMPs (Voxeur et al., 2019). It is well known that higher concentrations of OGs elicit defences while lower concentrations regulate growth and development processes (Ferrari et al., 2013; Pontiggia et al., 2020; Yang et al., 2021). Treatment with OGs induces Ca^{2+} influx, ROS production and activation of mitogen-activated protein kinases (MAPKs) and calcium-dependent protein kinases (CDPKs) (Bellincampi et al., 2000; Galletti et al., 2008; Gravino et al., 2015). Intracellular transduction of OGs is followed by expression of immune-related genes, increase of stress-related hormones and accumulation of defence molecules such as phytoalexins and callose (Ferrari et al., 2007; Denoux et al., 2008; Gravino et al., 2015).

Cellulose fragments, i.e., CDs, activate plant immunity in both Arabidopsis and grapevine, and among them, cellotriose exhibits a marked induction of defence responses (Aziz et al., 2007; Souza et al., 2017; Johnson et al., 2018; Locci et al., 2019). Hemicellulose fragments also act as DAMPs. Mannan oligosaccharides from locust bean gum have been shown to protect rice and tobacco against different pathogens (Zang et al., 2019). Xyloglucan-derived fragments trigger an immune response comparable to that of OGs in Arabidopsis and grapevine (Claverie et al., 2018; Heloir et al., 2019). Fragments of arabinoxylan activate defences in Arabidopsis, tomato and pepper (Mélida et al., 2020).

Linear β-1,3-glucan fragments induce plant immunity and, if released from a fungal cell wall, are considered to be typical MAMPs (Mélida et al., 2018; Wanke et al., 2020). However, they may be released from plant-derived callose of papillae by pathogen-derived enzymes and, in this case, may be regarded to be typical DAMPs (Voigt, 2014; Mélida et al., 2018). The possible role of β-1,3-glucans as DAMPs is further suggested by plant virology studies. During viral infection, callose accumulates in the plasmodesmata to reduce viral spreading into the plant tissues. However, some viruses such as the tobacco mosaic virus avoid this defence mechanism by inducing the host production of β-1,3-glucanase that potentially forms β-1,3-glucan fragments by degrading callose. β-1,3-glucan fragments may consequently act as a DAMP against other kinds of microbes (Kozieł et al., 2021).

Recently, the pathogen of rice *Magnaporthe oryzae* has been reported to secrete a glycosyl hydrolase capable of degrading the β-1,3-1,4-glucans and releasing a mixed linkage β-1,3-1,4-glucan trisaccharides and tetrasaccharides that activate rice immunity (Yang et al., 2021). Mixed linkage β-1,3-1,4-glucan fragments activate immunity not only in rice but also in the monocot barley (Barghahn et al., 2021) and the dicot Arabidopsis (Barghahn et al., 2021; Rebaque et al., 2021). While linear β-1,3- and β-1,4-glucans are characteristic of callose and cellulose, respectively, and are present in both monocots and dicots, mixed linkage β-1,3-1,4-glucans are found in both plants and microbes (Burton and Fincher, 2009). In plant, they are present in *Poaceae*, such as rice and barley, but are absent in dicotyledons (Gigli-Bisceglia et al., 2020; Yang et al., 2021). Therefore, these oligosaccharides can be considered as typical DAMPS in terms of their ability to activate immunity in *Poaceae* (Barghahn et al., 2021; Yang et al., 2021), while in Arabidopsis the activated immunity is better regarded as a canonical MAMP-mediated activity (Barghahn et al., 2021; Rebaque et al., 2021).

Fragments derived from the degradation of the plant cuticle by fungal enzymes have also been shown to activate plant defences and may be regarded as DAMPs (Arya et al., 2021).

The Interplay between DAMPs and Hormones

In addition to being generated during pathogen attack, cell wall fragments are also generated during growth and development due to cell wall modifications and remodelling (Barnes and Anderson, 2018; Yang et al., 2021). Both defence and growth involve a high energy cost and maintaining an appropriate growth-defence trade-off requires the correct distribution of resources (Li et al., 2020; Figueroa-Macías et al., 2021). Cell wall fragments are candidate regulators of the most advantageous growth-defence trade-offs. OGs, for example, are known to stimulate immunity in a dose-dependent manner (Denoux et al., 2008) and have a negative effect on growth, also in a dose-dependent manner (Branca et al., 1988; Ferrari et al., 2008). Transgenic plants expressing a chimeric protein formed by a plant PGIP and a fungal PG, referred to as the "OG machine", produce OGs on command (Figure 20.3) and show a dose-dependent increase of salicylic acid and immunity, concomitantly with a reduction of plant growth (Benedetti et al., 2015). Some auto-immune dwarf mutants characterised by loss-of-function mutations of negative immunity regulators or gain-of-function mutations of immunity receptors have elevated salicylic acid levels and enhanced immunity (van Wersch et al., 2016). An effect of OGs depleting energy reserves from general plant metabolism has also been described (Krzyzaniak et al., 2018).

The balance between physiological and pathological responses is controlled, on the one hand, by the interplay of stress-related hormones such as salicylic acid, jasmonic acid, ethylene, abscisic acid and, on the other hand, growth- and development-related hormones, such as auxin, gibberellins, cytokinins and brassinosteroids (Figueroa-Macías et al., 2021). Several DAMPs and MAMPs negatively affect the activity of growth hormones and positively affect the activity of stress-related hormones (Branca et al., 1988; Navarro et al., 2006; Wang et al., 2007; Savatin et al., 2011). OGs, for example, up-regulate genes involved in the biosynthesis and signalling of stress hormones such as ethylene, not only at the local level but also systemically in Arabidopsis and tomato (Simpson et al., 1998; Gravino et al., 2015; Davidsson et al., 2017; Gamir et al., 2021). Treatments with OGs increase resistance of Arabidopsis to *B. cinerea*, *Pseudomonas syringae*, and *Pectobacterium carotovorum* and concomitantly increase the biosynthesis of salicylic acid and jasmonic acid (Benedetti et al., 2015; Davidsson et al., 2017; Howlader et al., 2020). Other putative cell wall DAMPs like cellobiose induce the biosynthesis and signalling of salicylic acid, jasmonic acid and ethylene in Arabidopsis (Souza et al., 2017), while xyloglucans induce the accumulation of *PR1*, a marker gene of salicylic acid (Claverie et al., 2018). Treatment of *Nicotiana benthamiana* or *Oryza sativa* with mannan enhances the expression of salicylic acid– and jasmonic acid–related genes (Zang et al., 2019).

Conversely, OGs antagonise some processes induced by auxin such as cell expansion in pea hypocotyls (Branca et al., 1988), the induction of lateral root formation in tobacco (Bellincampi et al., 1993) and the induction of several auxin-responsive genes in Arabidopsis (Mauro et al., 2002; Savatin et al., 2011). At the same time, auxin has a negative effect on OG-dependent responses, compromising induced resistance both in Arabidopsis and in tobacco (Savatin et al., 2011).

The Interplay between DAMPs and MAMPs

Cell wall–derived DAMPs trigger a PTI-like response using similar signal transduction mechanisms (Aziz et al., 2007; Souza et al., 2017; Claverie et al., 2018; Heloir et al., 2019; Zang et al., 2019; Mélida et al., 2020; Barghahn et al., 2021; Yang et al., 2021). However, there are also some marked differences with the PTI response induced by MAMPs. For example, as compared to the synthetic peptides flg22 and elf18, corresponding to bacterial flagellin and elongation factor-TU, respectively, OGs elicit different early responses and a limited overlap of downstream responses, which are weaker and shorter than those elicited by flg22 and elf18 and which are not compensated by increasing their concentrations (Denoux et al., 2008; Savatin et al., 2014; Gravino et al., 2015; Kohorn et al., 2016; Mattei et al., 2016; Souza et al., 2017; Giovannoni et al., 2021; Marti et al., 2021). The first evidence of a synergy between DAMPs and MAMPs was obtained by Davis et al. (1986). Co-treatment with a DAMP and a MAMP enhances PTI-related responses leading to higher levels of immunity as compared with treatments with single elicitors (Davis et al., 1986; van Aubel et al., 2016; Souza et al., 2017; Moreau et al., 2020). It is conceivable that cell wall degradation and release of DAMPs play an auxiliary role to the action of MAMPs and are necessary to activate a stronger and more efficient immune response. A strict DAMP–MAMP connection in the immune responses of Arabidopsis roots has recently been suggested. A cooperation between cellular damage, likely releasing DAMPs, and perception of MAMPs is necessary for a strong activation of immunity in the presence of pathogenic microbes. On the other hand, the simple recognition of MAMPs, which are also present in beneficial soil microorganisms, is not expected to be sufficient for activation of a strong resistance (Zhou et al., 2020) (Figure 20.4).

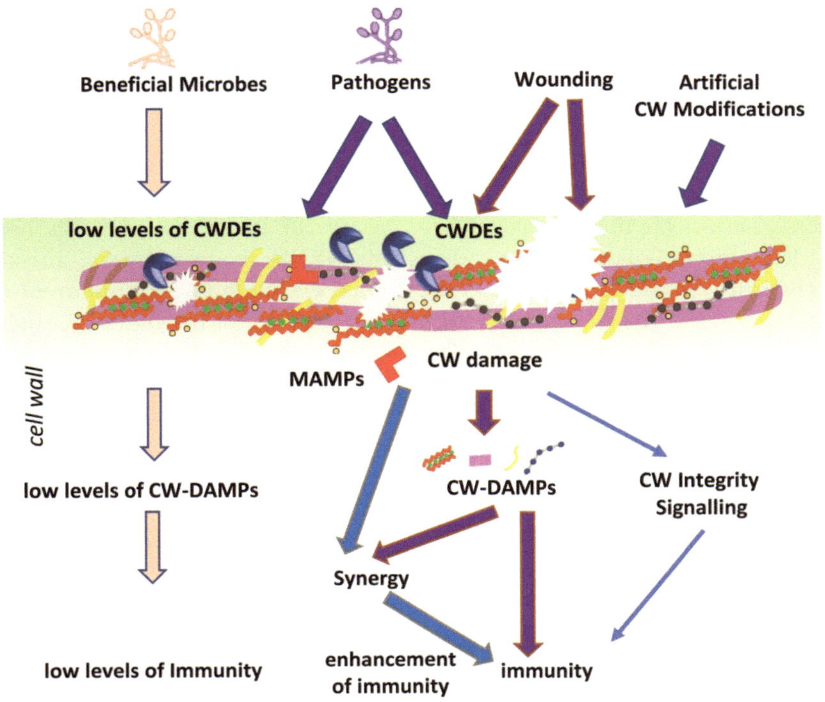

FIGURE 20.4 The interplay between wounding, DAMPs and MAMPs. Cell wall degradation, wounding and release of DAMPs may function additively or synergistically with MAMPs, in order to activate a strong and efficient immune response. The simple recognition of MAMPs, which are present in beneficial microorganisms unable of deconstructing the plant cell wall, is not sufficient to activate a strong resistance.

An interesting question is whether MAMPs evolved to exploit immunity mechanisms and pathways originally traced by DAMPs. The cell wall is the first obstacle that pathogens have to overcome and it seems likely that there must have been a great deal of selective pressure to evolve an appropriate response to wounds. It is well known that several CWDEs are induced upon wounding and that cell wall–derived DAMPs such as OGs are involved in the self-healing of wounds (Savatin et al., 2014; Gramegna et al., 2016). The transient nature of DAMP-related responses and the partial overlap of the signal transduction mechanisms of DAMPs and MAMPs suggest that they function in additive or synergic manner to achieve a strong immunity. The concomitant recognition of pathogenic MAMPs together with the perception of cellular damage can increase the duration, extension and strength of the processes that lead to the activation of strong responses to infections.

The Homeostasis of DAMPs

A hyper-immunity response caused by the accumulation of high levels of DAMPs causes deleterious effects such as reduced growth and cell death, which can be prevented by homeostatic control of DAMP levels. Several berberine-bridge enzyme-like (BBE-like) proteins were discovered to be capable of oxidising DAMPs such as OGs and CDs (Benedetti et al., 2018; Locci et al., 2019; Pontiggia et al., 2020; Eggers et al., 2021). It is likely that BBE-like proteins capable of oxidising other kinds of DAMPs and MAMPs will be characterised in the future since the involvement of this protein family in immunity seems to be one of their main functions (Rodrigues Oblessuc et al., 2019; Pontiggia et al., 2020). The oxidation of OGs and CDs impairs their elicitor activity in Arabidopsis. However, the impairment of DAMP activity appears not to affect the overall resistance response against pathogens such as *B. cinerea*, likely because oxidised oligosaccharides are more recalcitrant to further hydrolysis and less efficiently utilised as a carbon source by pathogens (Benedetti et al., 2018; Locci et al., 2019).

In the oxidation reaction of OGs, CDs and possibly other DAMPs by BBE-like proteins, H_2O_2 is produced, which may further reinforce the cell wall recalcitrance to degradation. On the other hand, if a limited amount of H_2O_2 is generated in a limited number of cells where DAMPs are generated, the produced H_2O_2 may act as a positional signal for short-range cell-cell communication in both immunity and development (Figure 20.1) (Pontiggia et al., 2020; Scortica et al., 2022).

CONCLUSIONS AND PERSPECTIVES

Elucidation of the function of DAMPs in the innate immunity has become a primary focus of interest for understanding the basis of important biological phenomena, not only in the plant kingdom, where DAMPs were first discovered, but in the animal kingdom as well. Insights into the biology of both the plant-defence and immunity-hyper-immunity trade-offs in both plants and animals are expected to be gained from the current studies on DAMPs. Many events related to growth, cell proliferation and positional signalling as well as root formation and morphogenesis may be linked to the activity of DAMPs (Bellincampi et al., 1996; Pontiggia et al., 2020; Yang et al., 2021; Scortica et al., 2022).

Cell wall–derived DAMPs, such as OGs, CDs and others, alone or in combination with MAMPs, are already used to develop new agricultural products to thwart pathogen attack that are more respectful of the environment than conventional agrochemicals. For instance, a mixture of chito-oligosaccharides and OGs is registered in the EU as a low-risk fungicide to protect tomato, cucumber and grapevine against powdery mildew (van Aubel et al., 2016). Cell wall–derived DAMPs, therefore, can potentially be used in organic agriculture with the only limitations related to possible detrimental effects if they are added in excess or to their time of persistence in the environment (Trouvelot et al., 2014). A more sophisticated biotech approach based on chimeric proteins may generate cell wall–derived DAMPs *in planta* at the site of infection. In transgenic plants, when the PG–PGIP chimeric protein ("OG machine") is controlled by the *PR-1* promoter, which is induced by microbial infection, it generates OGs without obvious detrimental effects on plant growth (Figure 20.3) (Benedetti et al., 2015). The use of other promoters and DAMP-generating "machines" can be envisioned in the future. Also, the engineering of both native and chimeric DAMP receptors is expected in the future to improve the arsenal of technological tools against crop diseases (De Lorenzo et al., 2011).

Some approaches based on the current knowledge of how plants counteract the cell wall degradation have already been successfully applied in several crop plants. Overexpression of protein inhibitors of CWDEs improves the resistance against many pathogenic fungi, bacteria and oomycetes (reviewed in Kalunke et al., 2015; Rathinam et al., 2020). In the case of PGIPs, not only native PGIPs but also PGIPs modified by site-directed mutagenesis or artificial evolution to gain improved recognition capabilities may be used in the future. A study using the two hybrid systems showed that a truncated PGIP from French beans, which is only one-third the size of the full length protein, exhibits an efficient inhibition activity against PGs from *A. niger* and *B. cinerea* (Chiu et al., 2021). The screening of a library to isolate peptides that specifically bind and inhibit the PG of *Agrobacterium vitis* yielded peptides with sequence similarity to PGIPs (Warren et al., 2016). Truncated and PGIP-derived peptides could potentially expand our tools against microbial pathogens and be the basis for engineering proteins with enhanced antimicrobial activity. Besides PGIPs, inhibitors of xylanases (Moscetti et al., 2013; Moscetti et al., 2015; Tundo et al., 2020) and of pectin methylesterases (Lionetti et al., 2007, 2017) have been used to improve crop resistance to pathogens. Crop plants such as tobacco, wheat, and cotton over-expressing PMEIs have been shown to be more resistant to both fungal and viral infections (An et al., 2008; Volpi et al., 2011; Lionetti et al., 2014; Liu et al., 2018).

In conclusion several biotechnological opportunities to engineer durable resistance and protect crops against diseases are foreseen to derive from the current studies on cell wall degradation and cell wall–derived DAMPs. The knowledge of how pathogens deconstruct the plant cell wall, how plants perceive cell wall perturbations of integrity and how cell wall DAMPs are released and regulate cellular activity is a fertile field of research with increasing expectations of new discoveries.

ACKNOWLEDGEMENTS

Research in the laboratory of the authors is supported by Sapienza University of Rome (Ateneo Grant RG12117A898EABE0) and Ministry of University and Research (2017ZBBYNC_PRIN 2017).

REFERENCES

Abad P, Gouzy J, Aury JM, Castagnone-Sereno P, Danchin EG, Deleury E, Perfus-Barbeoch L, Anthouard V, Artiguenave F, Blok VC, Caillaud MC, Coutinho PM, Dasilva C, De Luca F, Deau F, Esquibet M, Flutre T, Goldstone JV, Hamamouch N, Hewezi T, Jaillon O, Jubin C, Leonetti P, Magliano M, Maier TR, Markov GV, McVeigh P, Pesole G, Poulain J, Robinson-Rechavi M, Sallet E, Ségurens B, Steinbach D, Tytgat T, Ugarte E, van Ghelder C, Veronico P, Baum TJ, Blaxter M, Bleve-Zacheo T, Davis EL, Ewbank JJ, Favery B, Grenier E, Henrissat B, Jones JT, Laudet V, Maule AG, Quesneville H, Rosso MN, Schiex T, Smant G, Weissenbach J, Wincker P (2008) Genome sequence of the metazoan plant-parasitic nematode *Meloidogyne incognita*. *Nature Biotechnology* 26:909–915.

Aguero CB, Uratsu SL, Greve C, Powell ALT, Labavitch JM, Meredith CP, Dandekar AM (2005) Evaluation of tolerance to Pierce's disease and Botrytis in transgenic plants of *Vitis vinifera L.* expressing the pear PGIP gene. *Molecular Plant Pathology* 6:43–51.

Albersheim P, Jones TM, English PD (1969) Biochemistry of the cell wall in relation to infective processes. *Annu Rev Phytopathol* 7:171–194.

Albersheim P, Anderson AJ (1971) Proteins from plant cell walls inhibit polygalacturonases secreted by plant pathogens. *Proceedings of the National Academy of Sciences* 68:1815–1819.

Amsbury S, Kirk P, Benitez-Alfonso Y (2017) Emerging models on the regulation of intercellular transport by plasmodesmata-associated callose. *Journal of Experimental Botany* 69: 105–115.

An SH, Sohn KH, Choi HW, Hwang IS, Lee SC, Hwang BK (2008) Pepper pectin methylesterase inhibitor protein CaPMEI1 is required for antifungal activity, basal disease resistance and abiotic stress tolerance. *Planta* 228:61–78.

Apel-Birkhold PC, Walton JD (1996) Cloning, disruption, and expression of two endo-b-1, 4-xylanase genes, XYL2 and XYL3, from *Cochliobolus carbonum*. *Appl Environ Microbiol* 62:4129–4135.

Apel PC, Panaccione DG, Holden FR, Walton JD (1993) Cloning and targeted gene disruption of XYL1, a a-1,4-xylanase gene from the maize pathogen *Cochliobolus carbonum*. *Molecular Plant-Microbe Interactions* 6:467–473.

Aragón W, Reina-Pinto JJ, Serrano M (2017) The intimate talk between plants and microorganisms at the leaf surface. *Journal of Experimental Botany* 68:5339–5350.

Arya GC, Sarkar S, Manasherova E, Aharoni A, Cohen H (2021) The plant cuticle: An ancient guardian barrier set against long-standing rivals. *Frontiers in Plant Science* 12:663165.

Aziz A, Gauthier A, Bézier A, Poinssot B, Joubert JM, Pugin A, Heyraud A, Baillieul F (2007) Elicitor and resistance-inducing activities of beta-1,4 cellodextrins in grapevine, comparison with beta-1,3 glucans and alpha-1,4 oligogalacturonides. *Journal of Experimental Botany* 58:1463–1472.

Bacete L, Mélida H, Miedes E, Molina A (2018) Plant cell wall-mediated immunity: cell wall changes trigger disease resistance responses. *Plant Journal* 93:614–636.

Bacete L, Hamann T (2020) The role of mechanoperception in plant cell wall integrity maintenance. *Plants (Basel)* 9:574.

Bajaj P, Mahajan R (2019) Cellulase and xylanase synergism in industrial biotechnology. *Applied Microbiology and Biotechnology* 103:8711–8724.

Barghahn S, Arnal G, Jain N, Petutschnig E, Brumer H, Lipka V (2021) Mixed-linkage b-1,3/1,4-glucan oligosaccharides induce defense responses in *Hordeum vulgare* and *Arabidopsis thaliana*. *Frontiers in Plant Science* 12:682439.

Barnes WJ, Anderson CT (2018) Release, recycle, rebuild: cell-wall remodeling, autodegradation, and sugar salvage for new wall biosynthesis during plant development. *Molecular Plant* 11:31–46.

Bateman DF, Millar RL (1966) Pectic enzymes in tissue degradation. *Annual Review of Phytopathology* 4:119–144.

Bellincampi D, Cardarelli M, Zaghi D, Serino G, Salvi G, Gatz C, Cervone F, Altamura MM, Costantino P, De Lorenzo G (1996) Oligogalacturonides prevent rhizogenesis in rolB-transformed tobacco explants by inhibiting auxin-induced expression of the *rolB* gene. *Plant Cell* 8:477–487.

Bellincampi D, Cervone F, Lionetti V (2014) Plant cell wall dynamics and wall-related susceptibility in plant-pathogen interactions. *Frontiers in Plant Science* 5:228.

Bellincampi D, Salvi G, De Lorenzo G, Cervone F, Marfa V, Eberhard S, Darvill A, Albersheim P (1993) Oligogalacturonides inhibit the formation of roots on tobacco explants. *Plant Journal* 4:207–213.

Bellincampi D, Dipierro N, Salvi G, Cervone F, De Lorenzo G (2000) Extracellular H_2O_2 induced by oligogalacturonides is not involved in the inhibition of the auxin-regulated *rolB* gene expression in tobacco leaf explants. *Plant Physiology* 122:1379–1385.

Ben-Daniel BH, Bar-Zvi D, Tsror Lahkim L (2012) Pectate lyase affects pathogenicity in natural isolates of Colletotrichum coccodes and in pelA gene-disrupted and gene-overexpressing mutant lines. *Molecular Plant Pathology* 13:187–197.

Benedetti M, Leggio C, Federici L, De Lorenzo G, Pavel NV, Cervone F (2011) Structural resolution of the complex between a fungal polygalacturonase and a plant polygalacturonase-inhibiting protein by small-angle X-ray scattering. *Plant Physiology* 157:599–607.

Benedetti M, Pontiggia D, Raggi S, Cheng Z, Scaloni F, Ferrari S, Ausubel FM, Cervone F, De Lorenzo G (2015) Plant immunity triggered by engineered in vivo release of oligogalacturonides, damage-associated molecular patterns. *Proceedings of the National Academy of Sciences* 112:5533–5538.

Benedetti M, Verrascina I, Pontiggia D, Locci F, Mattei B, De Lorenzo G, Cervone F (2018) Four Arabidopsis berberine bridge enzyme-like proteins are specific oxidases that inactivate the elicitor-active oligogalacturonides. *Plant Journal* 94:260–273.

Bethke G, Grundman RE, Sreekanta S, Truman W, Katagiri F, Glazebrook J (2014) Arabidopsis PECTIN METHYLESTERASEs contribute to immunity against Pseudomonas syringae. *Plant Physiology* 164:1093–1107.

Bethke G, Thao A, Xiong G, Li B, Soltis NE, Hatsugai N, Hillmer RA, Katagiri F, Kliebenstein DJ, Pauly M, Glazebrook J (2016) Pectin biosynthesis is critical for cell wall integrity and immunity in *Arabidopsis thaliana*. *Plant Cell* 28:537–556.

Bigeard J, Colcombet J, Hirt H (2015) Signaling mechanisms in pattern-triggered immunity (PTI). *Molecular Plant* 8:521–539.

Bishop PD, Makus DJ, Pearce G, Ryan CA (1981) Proteinase inhibitor-inducing factor activity in tomato leaves resides in oligosaccharides enzymically released from cell walls. *Proceedings of the National Academy of Sciences* 78:3536–3540.

Bissaro B, Rohr AK, Muller G, Chylenski P, Skaugen M, Forsberg Z, Horn SJ, Vaaje-Kolstad G, Eijsink VGH (2017) Oxidative cleavage of polysaccharides by monocopper enzymes depends on H_2O_2. *Nature Chemical Biology* 13:1123–1128.

Bonfante P (2018) The future has roots in the past: the ideas and scientists that shaped mycorrhizal research. *New Phytologist* 220:982–995.

Bonivento D, Pontiggia D, Di Matteo A, Fernandez-Recio J, Salvi G, Tsernoglou D, Cervone F, De Lorenzo G, Federici L (2008) Crystal structure of the endopolygalacturonase from the phytopathogenic fungus Colletotrichum lupini and its interaction with polygalacturonase-inhibiting proteins. *Proteins-Structure Function and Bioinformatics* 70:294–299.

Branca C, De Lorenzo G, Cervone F (1988) Competitive inhibition of the auxin-induced elongation by alpha-D-oligogalacturonides in pea stem segments. *Physiologia Plantarum* 72:499–504.

Brutus A, Sicilia F, Macone A, Cervone F, De Lorenzo G (2010) A domain swap approach reveals a role of the plant wall-associated kinase 1 (WAK1) as a receptor of oligogalacturonides. *Proceedings of the National Academy of Sciences* 107:9452–9457.

Burton RA, Fincher GB (2009) (1,3;1,4)-b-D-glucans in cell walls of the poaceae, lower plants, and fungi: a tale of two linkages. *Molecular Plant* 2:873–882.

Calderón-Cortés N, Quesada M, Watanabe H, Cano-Camacho H, Oyama K (2012) Endogenous plant cell wall digestion: a key mechanism in insect evolution. *Annual Review of Ecology, Evolution, and Systematics* 43:45–71.

Caño-Delgado A, Penfield S, Smith C, Catley M, Bevan M (2003) Reduced cellulose synthesis invokes lignification and defense responses in Arabidopsis thaliana. *Plant Journal* 34:351–362.

Capodicasa C, Vairo D, Zabotina O, McCartney L, Caprari C, Mattei B, Manfredini C, Aracri B, Benen J, Knox JP, De Lorenzo G, Cervone F (2004) Targeted modification of homogalacturonan by transgenic expression of a fungal polygalacturonase alters plant growth. *Plant Physiology* 135:1294–1304.

Castilleux R, Plancot B, Ropitaux M, Carreras A, Leprince J, Boulogne I, Follet-Gueye ML, Popper ZA, Driouich A, Vicré M (2018) Cell wall extensins in root-microbe interactions and root secretions. *Journal of Experimental Botany* 69:4235–4247.

Castilleux R, Plancot B, Vicré M, Nguema-Ona E, Driouich A (2021) Extensin, an underestimated key component of cell wall defence? *Annals of Botany* 127:709–713.

Cervone F, Scala A, Scala F (1978) Polygalacturonase from Rhizoctonia fragariae : further characterization of two isozymes and their action towards strawberry tissue. *Physiological Plant Pathology* 12:19–26.

Cervone F, Aldebrhan T, Coutts RHA, Wood RKS (1981) Effects of French Bean tissue and leaf protoplasts on *Colletotrichum lindemuthianum* polygalacturonase. *Phytopath Z* 102:246.

Cervone F, De Lorenzo G, Degra L, Salvi G, Bergami M (1987) Purification and characterization of a polygalacturonase-inhibiting protein from *Phaseolus-vulgaris* L. *Plant Physiology* 85:631–637.

Cervone F, Castoria R, Spanu P, Bonfante-Fasolo P (1988) Pectinolytic activity in some ericoid mycorrhizal fungi. *Transactions of the British Mycological Society* 91:537–539.

Cervone F, Hahn MG, De Lorenzo G, Darvill A, Albersheim P (1989) Host-pathogen interactions. XXXIII. A plant protein converts a fungal pathogenesis factor into an elicitor of plant defense responses. *Plant Physiology* 90:542–548.

Chakraborty S, Nguyen B, Wasti SD, Xu G (2019) Plant leucine-rich repeat receptor kinase (LRR-RK): structure, ligand perception, and activation mechanism. *Molecules* 24.

Charkowski A, Blanco C, Condemine G, Expert D, Franza T, Hayes C, Hugouvieux-Cotte-Pattat N, Lopez Solanilla E, Low D, Moleleki L, Pirhonen M, Pitman A, Perna N, Reverchon S, Rodriguez Palenzuela P, San Francisco M, Toth I, Tsuyumu S, van der Waals J, van der Wolf J, Van Gijsegem F, Yang CH, Yedidia I (2012) The role of secretion systems and small molecules in soft-rot Enterobacteriaceae pathogenicity. *Annual Review of Phytopathology* 50:425–449.

Chen Q, Rehman S, Smant G, Jones JT (2005) Functional analysis of pathogenicity proteins of the potato cyst nematode *Globodera rostochiensis* using RNAi. *Molecular Plant Microbe Interactions* 18:621–625.

Chiu T, Behari A, Chartron JW, Putman A, Li Y (2021) Exploring the potential of engineering polygalacturonase-inhibiting protein as an ecological, friendly, and nontoxic pest control agent. *Biotechnology and Bioengineering* 118:3200–3214.

Choi HW, Klessig DF (2016) DAMPs, MAMPs, and NAMPs in plant innate immunity. *Bmc Plant Biology* 16:232.

Chowdhury J, Schober MS, Shirley NJ, Singh RR, Jacobs AK, Douchkov D, Schweizer P, Fincher GB, Burton RA, Little A (2016) Down-regulation of the glucan synthase-like 6 gene (HvGsl6) in barley leads to decreased callose accumulation and increased cell wall penetration by Blumeria graminis f. sp. hordei. *New Phytologist* 212:434–443.

Chowdhury J, Lück S, Rajaraman J, Douchkov D, Shirley NJ, Schwerdt JG, Schweizer P, Fincher GB, Burton RA, Little A (2017) Altered expression of genes implicated in xylan biosynthesis affects penetration resistance against powdery mildew. *Frontiers in Plant Science* 8:445.

Claverie J, Balacey S, Lemaitre-Guillier C, Brule D, Chiltz A, Granet L, Noirot E, Daire X, Darblade B, Heloir MC, Poinssot B (2018) The cell wall-derived xyloglucan is a new DAMP triggering plant immunity in *Vitis vinifera* and *Arabidopsis thaliana*. *Frontiers in Plant Science* 9:1725.

Cooper RM, Wood RKS (1973) Induction of synthesis of extracellular cell-wall degrading enzymes in vascular wilt fungi. *Nature* 246:309–311.

Cui H, Tsuda K, Parker JE (2015) Effector-triggered immunity: from pathogen perception to robust defense. *Annual Review of Plant Biology* 66:487–511.

D'Ovidio R, Raiola A, Capodicasa C, Devoto A, Pontiggia D, Roberti S, Galletti R, Conti E, O'Sullivan D, De Lorenzo G (2004) Characterization of the complex locus of bean encoding polygalacturonase-inhibiting proteins reveals subfunctionalization for defense against fungi and insects. *Plant Physiology* 135:2424–2435.

Davidsson P, Broberg M, Kariola T, Sipari N, Pirhonen M, Palva ET (2017) Short oligogalacturonides induce pathogen resistance-associated gene expression in *Arabidopsis thaliana. Bmc Plant Biology* 17:19.

Davis KR, Darvill AG, Albersheim P (1986) Host-pathogen interactions. XXXI. Several biotic and abiotic elicitors act synergistically in the induction of phytoalexin accumulation in soybean. *Plant Molecular Biology* 6:23–32.

Davis EL, Haegeman A, Kikuchi T (2011) Degradation of the plant cell wall by nematodes. *In* J Jones, G Gheysen, C Fenoll, eds, Genomics and Molecular Genetics of Plant-Nematode Interactions, 10.1007/978-94-007-0434-3_12. Springer, Dordrecht, pp 255–272.

De Lorenzo G, Castoria R, Bellincampi D, Cervone F (1997) Fungal invasion enzymes and their inhibition. *In* GC Carroll, P Tudzynski, eds, The Mycota. V. Plant Relationships, Part B. Springer-Verlag, Berlin, pp 61–83.

De Lorenzo G, D'Ovidio R, Cervone F (2001) The role of polygalacturonase-inhibiting proteins (PGIPS) in defense against pathogenic fungi. *Annual Review of Phytopathology* 39:313–335.

De Lorenzo G, Brutus A, Savatin DV, Sicilia F, Cervone F (2011) Engineering plant resistance by constructing chimeric receptors that recognize damage-associated molecular patterns (DAMPs). *(FEBS) Letters* 585:1521–1528.

De Lorenzo G, Ferrari S, Cervone F, Okun E (2018) Extracellular DAMPs in plants and mammals: immunity, tissue damage and repair. *Trends in Immunology* 39:937–950.

De Lorenzo G, Cervone F (2022) Plant immunity by damage-associated molecular patterns (DAMPs). *Essays Biochem* 66:459–469.

Debyser W, Delcour JA (1997) Inhibitors of xylanolytic and beta-Glucanolytic enzymes. *In*,.

Debyser W, Peumans WJ, Van Damme EJM, Delcour JA (1999) Triticum aestivum xylanase inhibitor (Taxi), a new class of enzyme inhibitor affecting breadmaking performance. *Journal of Cereal Science* 30:39–43.

Del Corpo D, Fullone MR, Miele R, Lafond M, Pontiggia D, Grisel S, Kieffer-Jaquinod S, Giardina T, Bellincampi D, Lionetti V (2020) AtPME17 is a functional *Arabidopsis thaliana* pectin methylesterase regulated by its PRO region that triggers PME activity in the resistance to *Botrytis cinerea. Molecular Plant Pathology* 21:1620–1633.

Delgado-Cerezo M, Sánchez-Rodríguez C, Escudero V, Miedes E, Fernández PV, Jordá L, Hernández-Blanco C, Sánchez-Vallet A, Bednarek P, Schulze-Lefert P, Somerville S, Estevez JM, Persson S, Molina A (2012) Arabidopsis heterotrimeric G-protein regulates cell wall defense and resistance to necrotrophic fungi. *Molecular Plant* 5:98–114.

Denoux C, Galletti R, Mammarella N, Gopalan S, Werck D, De Lorenzo G, Ferrari S, Ausubel FM, Dewdney J (2008) Activation of defense response pathways by OGs and Flg22 elicitors in Arabidopsis seedlings. *Molecular Plant* 1:423–445.

Di Matteo A, Federici L, Mattei B, Salvi G, Johnson KA, Savino C, De Lorenzo G, Tsernoglou D, Cervone F (2003) The crystal structure of polygalacturonase-inhibiting protein (PGIP), a leucine-rich repeat protein involved in plant defense. *Proceedings of the National Academy of Sciences, USA* 100:10124–10128.

Eggers R, Jammer A, Jha S, Kerschbaumer B, Lahham M, Strandback E, Toplak M, Wallner S, Winkler A, Macheroux P (2021) The scope of flavin-dependent reactions and processes in the model plant *Arabidopsis thaliana. Phytochemistry* 189:112822.

Ellinger D, Naumann M, Falter C, Zwikowics C, Jamrow T, Manisseri C, Somerville SC, Voigt CA (2013) Elevated early callose deposition results in complete penetration resistance to powdery mildew in Arabidopsis. *Plant Physiology* 161:1433–1444.

Ellinger D, Voigt CA (2014) Callose biosynthesis in Arabidopsis with a focus on pathogen response: what we have learned within the last decade. *Annals of Botany* 114:1349–1358.

Ellis C, Turner JG (2001) The Arabidopsis mutant *cev1* has constitutively active jasmonate and ethylene signal pathways and enhanced resistance to pathogens. *Plant Cell* 13:1025–1033.

Ellis C, Karafyllidis I, Wasternack C, Turner JG (2002) The Arabidopsis mutant cev1 links cell wall signaling to jasmonate and ethylene responses. *Plant Cell* 14:1557–1566.

Engelsdorf T, Gigli-Bisceglia N, Veerabagu M, McKenna JF, Vaahtera L, Augstein F, Van der Does D, Zipfel C, Hamann T (2018) The plant cell wall integrity maintenance and immune signaling systems cooperate to control stress responses in *Arabidopsis thaliana. Sci. Signal.* 11:eaao3070.

Engelsdorf T, Will C, Hofmann J, Schmitt C, Merritt BB, Rieger L, Frenger MS, Marschall A, Franke RB, Pattathil S, Voll LM (2017) Cell wall composition and penetration resistance against the fungal pathogen *Colletotrichum higginsianum* are affected by impaired starch turnover in Arabidopsis mutants. *Journal of Experimental Botany* 68:701–713.

Espino JJ, Brito N, Noda J, Gonzalez C (2005) *Botrytis cinerea* endo-ß-1, 4-glucanase Cel5A is expressed during infection but is not required for pathogenesis. *Physiological and Molecular Plant Pathology* 66.

Eynck C, Seguin-Swartz G, Clarke WE, Parkin IAP (2012) Monolignol biosynthesis is associated with resistance to Sclerotinia sclerotiorum in *Camelina sativa. Molecular Plant Pathology* 13:887–899.

Fan H, Dong H, Xu C, Liu J, Hu B, Ye J, Mai G, Li H (2017) Pectin methylesterases contribute the pathogenic differences between races 1 and 4 of Fusarium oxysporum f. sp. cubense. *Sci Rep* 7:1–16.

Federici L, Mattei B, Caprari C, Savino C, Cervone F, Tsernoglou D (1999) Crystallization and preliminary X-ray diffraction study of the endo- polygalacturonase from *Fusarium moniliforme. Acta Crystallogr.D.Biol.Crystallogr.* 55 (Pt 7):1359–1361.

Federici L, Caprari C, Mattei B, Savino C, Di Matteo A, De Lorenzo G, Cervone F, Tsernoglou D (2001) Structural requirements of endopolygalacturonase for the interaction with PGIP (polygalacturonase-inhibiting protein). *Proceedings of the National Academy of Sciences, USA* 98:13425–13430.

Ferrari S, Savatin DV, Sicilia F, Gramegna G, Cervone F, Lorenzo GD (2013) Oligogalacturonides: plant damage-associated molecular patterns and regulators of growth and development. *Frontiers in Plant Science* 4:49.

Ferrari S, Galletti R, Vairo D, Cervone F, De Lorenzo G (2006) Antisense expression of the Arabidopsis thaliana AtPGIP1 gene reduces polygalacturonase-inhibiting protein accumulation and enhances susceptibility to Botrytis cinerea. *Molecular Plant-Microbe Interactions* 19:931–936.

Ferrari S, Galletti R, Denoux C, De Lorenzo G, Ausubel FM, Dewdney J (2007) Resistance to *Botrytis cinerea* induced in Arabidopsis by elicitors is independent of salicylic acid, ethylene, or jasmonate signaling but requires PHYTOALEXIN DEFICIENT3. *Plant Physiology* 144:367–379.

Ferrari S, Galletti R, Pontiggia D, Manfredini C, Lionetti V, Bellincampi D, Cervone F, De Lorenzo G (2008) Transgenic expression of a fungal endo-polygalacturonase increases plant resistance to pathogens and reduces auxin sensitivity. *Plant Physiology* 146:669–681.

Ferrari S, Sella L, Janni M, De Lorenzo G, Favaron F, D'Ovidio R (2012) Transgenic expression of polygalacturonase-inhibiting proteins in Arabidopsis and wheat increases resistance to the flower pathogen Fusarium graminearum. *Plant Biology* 14:31–38. doi: 10.1111/j.1438-8677.2011.00449.x.

Ferrari S, Vairo D, Ausubel FM, Cervone F, De Lorenzo G (2003) Tandemly duplicated arabidopsis genes that encode polygalacturonase-inhibiting proteins are regulated coordinately by different signal transduction pathways in response to fungal infection. *Plant Cell* 15:93–106.

Figueroa-Macías JP, García YC, Núñez M, Díaz K, Olea AF, Espinoza L (2021) Plant growth-defense trade-offs: molecular processes leading to physiological changes. *International Journal of Molecular Sciences* 22:693.

Francocci F, Bastianelli E, Lionetti V, Ferrari S, De Lorenzo G, Bellincampi D, Cervone F (2013) Analysis of pectin mutants and natural accessions of Arabidopsis highlights the impact of de-methyl-esterified homogalacturonan on tissue saccharification. *Biotechnology for Biofuels* 6:163.

Frati F, Galletti R, De Lorenzo G, Salerno G, Conti E (2006) Activity of endo-polygalacturonases in mirid bugs (Heteroptera : Miridae) and their inhibition by plant cell wall proteins (PGIPs). *European Journal of Entomology* 103:515–522.

Fu ZQ, Dong X (2013) Systemic acquired resistance: turning local infection into global defense. *Annual Review of Plant Biology* 64:839–863.

Fujimoto T, Mizukubo T, Abe H, Seo S (2015) Sclareol induces plant resistance to root-knot nematode partially through ethylene-dependent enhancement of lignin accumulation. *Molecular Plant Microbe Interactions* 28:398–407.

Furman-Matarasso N, Cohen E, Du Q, Chejanovsky N, Hanania U, Avni A (1999) A point mutation in the ethylene-inducing xylanase elicitor inhibits the beta-1-4-endoxylanase activity but not the elicitation activity. *Plant Physiology* 121:345–351.

Gallego-Giraldo L, Jikumaru Y, Kamiya Y, Tang YH, Dixon RA (2011) Selective lignin downregulation leads to constitutive defense response expression in alfalfa (*Medicago sativa* L.). *New Phytologist* 190:627–639.

Gallego-Giraldo L, Liu C, Pose-Albacete S, Pattathil S, Peralta AG, Young J, Westpheling J, Hahn MG, Rao X, Knox JP, De Meester B, Boerjan W, Dixon RA (2020) ARABIDOPSIS DEHISCENCE ZONE POLYGALACTURONASE 1 (ADPG1) releases latent defense signals in stems with reduced lignin content. *Proc Natl Acad Sci USA* 117:3281–3290.

Galletti R, Denoux C, Gambetta S, Dewdney J, Ausubel FM, De Lorenzo G, Ferrari S (2008) The AtrbohD-mediated oxidative burst elicited by oligogalacturonides in Arabidopsis is dispensable for the activation of defense responses effective against *Botrytis cinerea*. *Plant Physiology* 148:1695–1706.

Gamir J, Minchev Z, Berrio E, Garcia JM, De Lorenzo G, Pozo MJ (2021) Roots drive oligogalacturonide-induced systemic immunity in tomato. *Plant Cell Environ* 44:275–289.

Gigli-Bisceglia N, Engelsdorf T, Hamann T (2020) Plant cell wall integrity maintenance in model plants and crop species-relevant cell wall components and underlying guiding principles. *Cellular and Molecular Life Sciences* 77:2049–2077.

Giovannoni M, Marti L, Ferrari S, Tanaka-Takada N, Maeshima M, Ott T, De Lorenzo G, Mattei B (2021) The plasma membrane-associated Ca^{2+}- binding protein PCaP1 is required for oligogalacturonide and flagellin-induced priming and immunity. *Plant, Cell & Environment* 44:3078–3093.

Gramegna G, Modesti V, Savatin DV, Sicilia F, Cervone F, De Lorenzo G (2016) GRP-3 and KAPP, encoding interactors of WAK1, negatively affect defense responses induced by oligogalacturonides and local response to wounding. *Journal of Experimental Botany* 67:1715–1729.

Gravino M, Savatin DV, Macone A, De Lorenzo G (2015) Ethylene production in *Botrytis cinerea*- and oligogalacturonide-induced immunity requires calcium-dependent protein kinases. *Plant Journal* 84:1073–1086.

Gravino M, Locci F, Tundo S, Cervone F, Savatin DV, De Lorenzo G (2017) Immune responses induced by oligogalacturonides are differentially affected by AvrPto and loss of BAK1/BKK1 and PEPR1/PEPR2. *Molecular Plant Pathology* 18:582–595.

Gui YJ, Zhang WQ, Zhang DD, Zhou L, Short DPG, Wang J, Ma XF, Li TG, Kong ZQ, Wang BL, Wang D, Li NY, Subbarao KV, Chen JY, Dai XF (2018) A *Verticillium dahliae* extracellular cutinase modulates plant immune responses. *Mol Plant Microbe Interact* 31:260–273.

Habrylo O, Forster A, Jeltsch JM, Phalip V (2013) The characterisation of xyloglucanase inhibitors from *Humulus lupulus*. *Phytochemistry* 90:70–77.

Haeger W, Henning J, Heckel DG, Pauchet Y, Kirsch R (2020) Direct evidence for a new mode of plant defense against insects via a novel polygalacturonase-inhibiting protein expression strategy. *Journal of Biological Chemistry* 295:11833–11844.

Haeger W, Wielsch N, Shin NR, Gebauer-Jung S, Pauchet Y, Kirsch R (2021) New players in the interaction between beetle polygalacturonases and plant polygalacturonase-inhibiting proteins: Insights from proteomics and gene expression analyses. *Frontiers in Plant Science* 12:660430.

Hahn MG, Darvill AG, Albersheim P (1981) Host-pathogen interactions. XIX. The endogenous elicitor, a fragment of a plant cell wall polysaccharide that elicits phytoalexin accumulation in soybeans. *Plant Physiology* 68:1161–1169.

Heloir MC, Adrian M, Brule D, Claverie J, Cordelier S, Daire X, Dorey S, Gauthier A, Lemaitre-Guillier C, Negrel J, Trda L, Trouvelot S, Vandelle E, Poinssot B (2019) Recognition of elicitors in grapevine: From MAMP and DAMP perception to induced resistance. *Frontiers in Plant Science* 10:1117.

Hernandez-Blanco C, Feng DX, Hu J, Sanchez-Vallet A, Deslandes L, Llorente F, Berrocal-Lobo M, Keller H, Barlet X, Sanchez-Rodriguez C, Anderson LK, Somerville S, Marco Y, Molina A (2007) Impairment of cellulose synthases required for Arabidopsis secondary cell wall formation enhances disease resistance. *The Plant Cell* 19:890–903.

Hou S, Liu Z, Shen H, Wu D (2019) Damage-Associated Molecular Pattern-triggered immunity in plants. *Frontiers in Plant Science* 10:646.

Howlader P, Bose SK, Jia X, Zhang C, Wang W, Yin H (2020) Oligogalacturonides induce resistance in *Arabidopsis thaliana* by triggering salicylic acid and jasmonic acid pathways against Pst DC3000. *International Journal of Biological Macromolecules* 164:4054–4064.

Hugouvieux-Cotte-Pattat N, Condemine G, Shevchik VE (2014) Bacterial pectate lyases, structural and functional diversity. *Environmental Microbiology Reports* 6:427–440.

Hwang IS, Oh E-J, Lee HB, Oh C-S (2019) Functional characterization of two cellulase genes in the Gram-positive pathogenic bacterium *Clavibacter michiganensis* for wilting in tomato. *Molecular Plant-Microbe Interactions* 32:491–501.

Isshiki A, Akimitsu K, Yamamoto M, Yamamoto H (2001) Endopolygalacturonase is essential for citrus black rot caused by Alternaria citri but not brown spot caused by Alternaria alternata. *Molecular Plant-Microbe Interactions* 14:749–757.

Jagadeeswaran G, Veale L, Mort AJ (2021) Do lytic polysaccharide monooxygenases aid in plant pathogenesis and herbivory? *Trends Plant Science* 26:142–155.

Janni M, Sella L, Favaron F, Blechl AE, De Lorenzo G, D'Ovidio R (2008) The expression of a bean polygalacturonase-inhibiting proteins in transgenic wheat confers increased resistance to the fungal pathogen *Bipolaris sorokiniana*. *Molecular Plant-Microbe Interactions* 21:171–177.

Jannoey P, Channei D, Kotcharerk J, Pongprasert W, Nomura M (2017) Expression analysis of genes related to rice resistance against brown planthopper, *Nilaparvata lugens*. *Rice Science* 24:163–172.

Johnson JM, Thurich J, Petutschnig EK, Altschmied L, Meichsner D, Sherameti I, Dindas J, Mrozinska A, Paetz C, Scholz SS, Furch ACU, Lipka V, Hedrich R, Schneider B, Svatos A, Oelmuller R (2018) A Poly(A) ribonuclease controls the cellotriose-based interaction between *Piriformospora indica* and its host Arabidopsis. *Plant Physiology* 176:2496–2514.

Jones JDG, Dangl JL (2006) The plant immune system. *Nature* 444:323–329.

Joubert DA, Slaughter AR, Kemp G, Becker JV, Krooshof GH, Bergmann C, Benen J, Pretorius IS, Vivier MA (2006) The grapevine polygalacturonase-inhibiting protein (VvPGIP1) reduces Botrytis cinerea susceptibility in transgenic tobacco and differentially inhibits fungal polygalacturonases. *Transgenic Research* 15:687–702.

Juge N (2006) Plant protein inhibitors of cell wall degrading enzymes. *Trends in Plant Science* 11:359–367.

Kalunke RM, Tundo S, Benedetti M, Cervone F, De Lorenzo G, D'Ovidio R (2015) An update on polygalacturonase-inhibiting protein (PGIP), a leucine-rich repeat protein that protects crop plants against pathogens. *Frontiers in Plant Science* 6:146.

Kim S, Stahlberg J, Sandgren M, Paton RS, Beckham GT (2014) Quantum mechanical calculations suggest that lytic polysaccharide monooxygenases use a copper-oxyl, oxygen-rebound mechanism. *Proceedings of the National Academy of Sciences* 111:149–154.

Kohorn BD, Hoon D, Minkoff BB, Sussman MR, Kohorn SL (2016) Rapid Oligo-Galacturonide Induced Changes in Protein Phosphorylation in Arabidopsis. *Molecular & Cellular Proteomics* 15:1351–1359.

Kozieł E, Otulak-Kozieł K, Bujarski JJ (2021) Plant cell wall as a key player during resistant and susceptible plant-virus interactions. *Frontiers in Microbiology* 12:656809–656809.

Krzyzaniak Y, Negrel J, Lemaitre-Guillier C, Clément G, Mouille G, Klinguer A, Trouvelot S, Héloir MC, Adrian M (2018) Combined enzymatic and metabolic analysis of grapevine cell responses to elicitors. *Plant Physiology Biochemistry* 123:141–148.

Kubicek CP, Starr TL, Glass NL (2014) Plant cell wall-degrading enzymes and their secretion in plant-pathogenic fungi. *Annual Review of Phytopathology* 52:427–451.

Labourel A, Frandsen KEH, Zhang F, Brouilly N, Grisel S, Haon M, Ciano L, Ropartz D, Fanuel M, Martin F, Navarro D, Rosso MN, Tandrup T, Bissaro B, Johansen KS, Zerva A, Walton PH, Henrissat B, Leggio LL, Berrin JG (2020) A fungal family of lytic polysaccharide monooxygenase-like copper proteins. *Nature Chemical Biology* 16:345–350.

Lagaert S, Belien T, Volckaert G (2009) Plant cell walls: Protecting the barrier from degradation by microbial enzymes. *Seminars in Cell and Developmental Biology* 20:1064–1073.

Land WG (2003) Allograft injury mediated by reactive oxygen species: from conserved proteins of Drosophila to acute and chronic rejection of human transplants. Part III: interaction of (oxidative) stress-induced heat shock proteins with toll-like receptor-bearing cells of innate immunity and its consequences for the development of acute and chronic allograft rejection. *Transplantation Reviews* 17:67–86.

Leckie F, Mattei B, Capodicasa C, Hemmings A, Nuss L, Aracri B, De Lorenzo G, Cervone F (1999) The specificity of polygalacturonase-inhibiting protein (PGIP): a single amino acid substitution in the solvent-exposed beta-strand/beta-turn region of the leucine-rich repeats (LRRs) confers a new recognition capability. *EMBO Journal* 18:2352–2363.

Lee SC, West CA (1981) Polygalacturonase from *Rhizopus stolonifer*, an elicitor of casbene synthetase activity in castor bean (Ricinus communis L.) seedlings. *Plant Physiology* 67:633–639.

Lee SC, West CA (1981) Properties of Rhizopus stolonifer polygalacturonase, an elicitor of casbene synthetase activity in castor bean seedlings. *Plant Physiology* 67:640–645.

Lee MH, Jeon HS, Kim SH, Chung JH, Roppolo D, Lee HJ, Cho HJ, Tobimatsu Y, Ralph J, Park OK (2019) Lignin-based barrier restricts pathogens to the infection site and confers resistance in plants. *EMBO Journal* 38:e101948.

Li P, Lu YJ, Chen H, Day B (2020) The lifecycle of the plant immune system. *Critical Reviews in Plant Sciences* 39:72–100.

Lionetti V, Cervone F, Bellincampi D (2012) Methyl esterification of pectin plays a role during plant-pathogen interactions and affects plant resistance to diseases. *The Journal of Plant Physiology* 169:1623–1630.

Lionetti V, Francocci F, Ferrari S, Volpi C, Bellincampi D, Galletti R, D'Ovidio R, De Lorenzo G, Cervone F (2010) Engineering the cell wall by reducing de-methyl-esterified homogalacturonan improves saccharification of plant tissues for bioconversion. *Proceedings of the National Academy of Sciences, USA* 107:616–621.

Lionetti V, Raiola A, Camardella L, Giovane A, Obel N, Pauly M, Favaron F, Cervone F, Bellincampi D (2007) Overexpression of pectin methylesterase inhibitors in Arabidopsis restricts fungal infection by *Botrytis cinerea*. *Plant Physiology* 143:1871–1880.

Lionetti V, Raiola A, Cervone F, Bellincampi D (2014) Transgenic expression of pectin methylesterase inhibitors limits tobamovirus spread in tobacco and Arabidopsis. *Molecular Plant Pathology* 15:265–274.

Lionetti V, Fabri E, De Caroli M, Hansen AR, Willats WG, Piro G, Bellincampi D (2017) Three pectin methylesterase inhibitors protect cell wall integrity for Arabidopsis immunity to *Botrytis*. *Plant Physiology* 173:1844–1863.

Liu Q, Luo L, Zheng L (2018) Lignins: Biosynthesis and biological functions in plants. *International Journal of Molecular Sciences* 19:E335.

Liu N, Sun Y, Pei Y, Zhang X, Wang P, Li X, Li F, Hou Y (2018) A pectin methylesterase inhibitor enhances resistance to Verticillium wilt. *Plant Physiology* 176:2202–2220.

Liu D, Wu J, Lin L, Li P, Li S, Wang Y, Li J, Sun Q, Liang J, Wang Y (2021) Overexpression of cinnamoyl-CoA reductase 2 in *Brassica napus* increases resistance to *Sclerotinia sclerotiorum* by affecting lignin biosynthesis. *Frontiers in Plant Science* 12:732733.

Liu J, Zhang L, Yan D (2021) Plasmodesmata-involved battle against pathogens and potential strategies for strengthening hosts. *Frontiers in Plant Science* 12:644870.

Locci F, Benedetti M, Pontiggia D, Citterico M, Caprari C, Mattei B, Cervone F, De Lorenzo G (2019) An Arabidopsis berberine bridge enzyme-like protein specifically oxidizes cellulose oligomers and plays a role in immunity. *Plant Journal* 98:540–554.

Lorrai R, Ferrari S (2021) Host cell wall damage during pathogen infection: mechanisms of perception and role in plant-pathogen interactions. *Plants (Basel)* 10.

Ma Y, Han C, Chen J, Li H, He K, Liu A, Li D (2015) Fungal cellulase is an elicitor but its enzymatic activity is not required for its elicitor activity. *Molecular Plant Pathology* 16:14–26.

Ma QH, Zhu HH, Han JQ (2017) Wheat ROP proteins modulate defense response through lignin metabolism. *Plant Science* 262:32–38.

Ma Z, Zhu L, Song T, Wang Y, Zhang Q, Xia Y, Qiu M, Lin Y, Li H, Kong L, Fang Y, Ye W, Dong S, Zheng X, Tyler BM (2017) A paralogous decoy protects *Phytophthora sojae* apoplastic effector PsXEG1 from a host inhibitor. *Science* 355:710–714.

Ma H, Zhang B, Gai Y, Sun X, Chung KR, Li H (2019) Cell-wall-degrading enzymes required for virulence in the host selective toxin-producing necrotroph. *Frontiers in Microbiology* 10:2514.

Manabe Y, Nafisi M, Verhertbruggen Y, Orfila C, Gille S, Rautengarten C, Cherk C, Marcus SE, Somerville S, Pauly M, Knox JP, Sakuragi Y, Scheller HV (2011) Loss-of-function mutation of Reduced Wall Acetylation2 in Arabidopsis leads to reduced cell wall acetylation and increased resistance to *Botrytis cinerea*. *Plant Physiology* 155:1068–1078.

Mandelc S, Javornik B (2015) The secretome of vascular wilt pathogen Verticillium albo-atrum in simulated xylem fluid. *Proteomics* 15:787–797.

Marti L, Savatin DV, Gigli-Bisceglia N, de Turris V, Cervone F, De Lorenzo G (2021) The intracellular ROS accumulation in elicitor-induced immunity requires the multiple organelle-targeted Arabidopsis NPK1-related protein kinases. *Plant, Cell & Environment* 44:931–947.

Mattei B, Spinelli F, Pontiggia D, De Lorenzo G (2016) Comprehensive analysis of the membrane phosphoproteome regulated by oligogalacturonides in *Arabidopsis thaliana*. *Frontiers in Plant Science* 7:1107.

Mauro ML, De Lorenzo G, Costantino P, Bellincampi D (2002) Oligogalacturonides inhibit the induction of late but not of early auxin-responsive genes in tobacco. *Planta* 215:494–501.

Mayans O, Scott M, Connerton I, Gravesen T, Benen J, Visser J, Pickersgill R, Jenkins J (1997) Two crystal structures of pectin lyase A from Aspergillus reveal a pH driven conformational change and striking divergence in the substrate-binding clefts of pectin and pectate lyases. *Structure* 5:677–689.

Mélida H, Bacete L, Ruprecht C, Rebaque D, Del Hierro I, Lopez G, Brunner F, Pfrengle F, Molina A (2020) Arabinoxylan-oligosaccharides act as damage associated molecular patterns in plants regulating disease resistance. *Frontiers in Plant Science* 11:1210.

Mélida H, Sopena-Torres S, Bacete L, Garrido-Arandia M, Jorda L, Lopez G, Munoz-Barrios A, Pacios LF, Molina A (2018) Non-branched beta-1,3-glucan oligosaccharides trigger immune responses in Arabidopsis. *Plant Journal* 93:34–49.

Mesny F, Miyauchi S, Thiergart T, Pickel B, Atanasova L, Karlsson M, Hüttel B, Barry KW, Haridas S, Chen C, Bauer D, Andreopoulos W, Pangilinan J, LaButti K, Riley R, Lipzen A, Clum A, Drula E, Henrissat B, Kohler A, Grigoriev IV, Martin FM, Hacquard S (2021) Genetic determinants of endophytism in the Arabidopsis root mycobiome. *Nature Communications* 12:7227.

Mitreva-Dautova M, Roze E, Overmars H, de Graaff L, Schots A, Helder J, Goverse A, Bakker J, Smant G (2006) A symbiont-independent endo-1,4-β-xylanase from the plant-parasitic nematode *Meloidogyne incognita*. *Molecular Plant Microbe Interactions* 19:521–529.

Miyauchi S, Kiss E, Kuo A, Drula E, Kohler A, Sanchez-Garcia M, Morin E, Andreopoulos B, Barry KW, Bonito G, Buee M, Carver A, Chen C, Cichocki N, Clum A, Culley D, Crous PW, Fauchery L, Girlanda M, Hayes RD, Keri Z, LaButti K, Lipzen A, Lombard V, Magnuson J, Maillard F, Murat C, Nolan M, Ohm RA, Pangilinan J, Pereira MF, Perotto S, Peter M, Pfister S, Riley R, Sitrit Y, Stielow JB, Szollosi G, Zifcakova L, Stursova M, Spatafora JW, Tedersoo L, Vaario LM, Yamada A, Yan M, Wang P, Xu J, Bruns T, Baldrian P, Vilgalys R, Dunand C, Henrissat B, Grigoriev IV, Hibbett D, Nagy LG, Martin FM (2020) Large-scale genome sequencing of mycorrhizal fungi provides insights into the early evolution of symbiotic traits. *Nature Communications* 11:5125.

Mnich E, Bjarnholt N, Eudes A, Harholt J, Holland C, Jørgensen B, Larsen FH, Liu M, Manat R, Meyer AS, Mikkelsen JD, Motawia MS, Muschiol J, Møller BL, Møller SR, Perzon A, Petersen BL, Ravn JL, Ulvskov P (2020) Phenolic cross-links: building and de-constructing the plant cell wall. *Natural Product Reports* 37:919–961.

Molina A, Miedes E, Bacete L, Rodríguez T, Mélida H, Denancé N, Sánchez-Vallet A, Rivière MP, López G, Freydier A, Barlet X, Pattathil S, Hahn M, Goffner D (2021) Cell wall composition determines disease resistance specificity and fitness. *Proceedings of the National Academy of Sciences, USA* 118:e2010243118.

Moreau S, van Aubel G, Janky R, Van Cutsem P (2020) Chloroplast electron chain, ROS production, and redox homeostasis are modulated by COS-OGA elicitation in tomato (*Solanum lycopersicum*) leaves. *Frontiers in Plant Science* 11:597589.

Moreira LR, Filho EX (2016) Insights into the mechanism of enzymatic hydrolysis of xylan. *Applied Microbiology and Biotechnology* 100:5205–5214.

Moscetti I, Tundo S, Janni M, Sella L, Gazzetti K, Tauzin A, Giardina T, Masci S, Favaron F, D'Ovidio R (2013) Constitutive expression of the xylanase inhibitor TAXI-III delays *Fusarium* head blight symptoms in durum wheat transgenic plants. *Molecular Plant Microbe Interactions* 26:1464–1472.

Moscetti I, Faoro F, Moro S, Sabbadin D, Sella L, Favaron F, D'Ovidio R (2015) The xylanase inhibitor TAXI-III counteracts the necrotic activity of a *Fusarium graminearum* xylanase in vitro and in durum wheat transgenic plants. *Molecular Plant Pathology* 16:583–592.

Naqvi SM, Harper A, Carter C, Ren G, Guirgis A, York WS, Thornburg RW (2005) Nectarin IV, a potent endoglucanase inhibitor secreted into the nectar of ornamental tobacco plants. Isolation, cloning, and characterization. *Plant Physiology* 139:1389–1400.

Navarro L, Dunoyer P, Jay F, Arnold B, Dharmasiri N, Estelle M, Voinnet O, Jones JD (2006) A plant miRNA contributes to antibacterial resistance by repressing auxin signaling. *Science* 312:436–439.

Ngou BPM, Ahn H-K, Ding P, Jones JDG (2021) Mutual potentiation of plant immunity by cell-surface and intracellular receptors. *Nature* 592:110–115.

Nguyen QB, Itoh K, Van Vu B, Tosa Y, Nakayashiki H (2011) Simultaneous silencing of endo-b-1,4 xylanase genes reveals their roles in the virulence of *Magnaporthe oryzae*. *Molecular Microbiology* 81:1008–1019.

Noda J, Brito N, Gonzalez C (2010) The *Botrytis cinerea* xylanase Xyn11A contributes to virulence with its necrotizing activity, not with its catalytic activity. *Bmc Plant Biology* 10:38.

Oeser B, Heidrich PM, Muller U, Tudzynski P, Tenberge KB (2002) Polygalacturonase is a pathogenicity factor in the *Claviceps purpurea* /rye interaction. *Fungal Genetics and Biology* 36:176–186.

Paccanaro MC, Sella L, Castiglioni C, Giacomello F, Martinez-Rocha AL, D'Ovidio R, Schafer W, Favaron F (2017) Synergistic effect of different plant cell wall-degrading enzymes is important for virulence of *Fusarium graminearum*. *Mol Plant Microbe Interact* 30:886–895.

Pagès S, Valette O, Abdou L, Bélaïch A, Bélaïch J-P (2003) A rhamnogalacturonan lyase in the *Clostridium cellulolyticum* cellulosome. *Journal of Bacteriology* 185:4727–4733.

Payan F, Flatman R, Porciero S, Williamson G, Juge N, Roussel A (2003) Structural analysis of xylanase inhibitor protein I (XIP-I), a proteinaceous xylanase inhibitor from wheat (*Triticum aestivum*, var. Soisson). *Biochemical Journal* 372:399–405.

Peng Y, van Wersch R, Zhang Y (2018) Convergent and Divergent Signaling in PAMP-Triggered Immunity and Effector-Triggered Immunity. *Molecular Plant Microbe Interactions* MPMI06170145CR.

Perotto S, Daghino S, Martino E (2018) Ericoid mycorrhizal fungi and their genomes: another side to the mycorrhizal symbiosis? *New Phytologist* 220:1141–1147.

Pickersgill R, Jenkins J, Harris G, Nasser W, Robert-Boudouy J (1994) The structure of Bacillus subtilis pectate lyase in complex with calcium. *Nature Structural & Molecular Biology* 1:717–723.

Pogorelko G, Lionetti V, Fursova O, Sundaram RM, Qi M, Whitham SA, Bogdanove AJ, Bellincampi D, Zabotina OA (2013) Arabidopsis and *Brachypodium distachyon* transgenic plants expressing *Aspergillus nidulans* acetylesterases have decreased degree of polysaccharide acetylation and increased resistance to pathogens. *Plant Physiology* 162:9–23.

Poinssot B, Vandelle E, Bentejac M, Adrian M, Levis C, Brygoo Y, Garin J, Sicilia F, Coutos-Thevenot P, Pugin A (2003) The endopolygalacturonase 1 from *Botrytis cinerea* activates grapevine defense reactions unrelated to its enzymatic activity. *Molecular Plant-Microbe Interactions* 16:553–564.

Pollegioni L, Tonin F, Rosini E (2015) Lignin-degrading enzymes. *Febs Journal* 282:1190–1213.

Polonio Á, Fernández-Ortuño D, de Vicente A, Pérez-García A (2021) A haustorial-expressed lytic polysaccharide monooxygenase from the cucurbit powdery mildew pathogen *Podosphaera xanthii* contributes to the suppression of chitin-triggered immunity. *Molecular Plant Pathology* 22:580–601.

Pontiggia D, Benedetti M, Costantini S, De Lorenzo G, Cervone F (2020) Dampening the DAMPs: how plants maintain the homeostasis of cell wall molecular patterns and avoid hyper-immunity. *Frontiers in Plant Science* 11:613259.

Powell ALT, van Kan J, ten Have A, Visser J, Greve LC, Bennett AB, Labavitch JM (2000) Transgenic expression of pear PGIP in tomato limits fungal colonization. *Molecular Plant-Microbe Interactions* 13:942–950.

Pruitt RN, Locci F, Wanke F, Zhang L, Saile SC, Joe A, Karelina D, Hua C, Frohlich K, Wan WL, Hu M, Rao S, Stolze SC, Harzen A, Gust AA, Harter K, Joosten M, Thomma B, Zhou JM, Dangl JL, Weigel D, Nakagami H, Oecking C, Kasmi FE, Parker JE, Nurnberger T (2021) The EDS1-PAD4-ADR1 node mediates Arabidopsis pattern-triggered immunity. *Nature* 598:495–499.

Qin Q, Bergmann CW, Rose JK, Saladie M, Kolli VS, Albersheim P, Darvill AG, York WS (2003) Characterization of a tomato protein that inhibits a xyloglucan- specific endoglucanase. *Plant Journal* 34:327–338.

Raiola A, Lionetti V, Elmaghraby I, Immerzeel P, Mellerowicz EJ, Salvi G, Cervone F, Bellincampi D (2011) Pectin methylesterase is induced in Arabidopsis upon infection and is necessary for a successful colonization by necrotrophic pathogens. *Molecular Plant-Microbe Interactions* 24:432–440.

Rathinam M, Rao U, Sreevathsa R (2020) Novel biotechnological strategies to combat biotic stresses: polygalacturonase inhibitor (PGIP) proteins as a promising comprehensive option. *Applied Microbiology and Biotechnology* 104:2333–2342.

Rebaque D, Del Hierro I, López G, Bacete L, Vilaplana F, Dallabernardina P, Pfrengle F, Jordá L, Sánchez-Vallet A, Pérez R, Brunner F, Molina A, Mélida H (2021) Cell wall-derived mixed-linked β-1,3/1,4-glucans trigger immune responses and disease resistance in plants. *Plant Journal* 106:601–615.

Reem NT, Pogorelko G, Lionetti V, Chambers L, Held MA, Bellincampi D, Zabotina OA (2016) Decreased polysaccharide feruloylation compromises plant cell wall integrity and increases susceptibility to necrotrophic fungal pathogens. *Frontiers in Plant Science* 7:630.

Rodrigues Oblessuc P, Vaz Bisneta M, Melotto M (2019) Common and unique Arabidopsis proteins involved in stomatal susceptibility to *Salmonella enterica* and *Pseudomonas syringae*. *FEMS Microbiology Letters* 366:fnz197.

Rogers LM, Flaishman MA, Kolattukudy PE (1994) Cutinase gene disruption in *Fusarium solani* f sp pisi decreases its virulence on pea. *Plant Cell* 6:935–945.

Roper MC, Greve LC, Warren JG, Labavitch JM, Kirkpatrick BC (2007) Xylella fastidiosa requires polygalacturonase for colonization and pathogenicity in *Vitis vinifera* grapevines. *Molecular Plant-Microbe Interactions* 20:411–419.

Sabbadin F, Urresti S, Henrissat B, Avrova AO, Welsh LRJ, Lindley PJ, Csukai M, Squires JN, Walton PH, Davies GJ, Bruce NC, Whisson SC, McQueen-Mason SJ (2021) Secreted pectin monooxygenases drive plant infection by pathogenic oomycetes. *Science* 373:774–779.

Sansen S, De Ranter CJ, Gebruers K, Brijs K, Courtin CM, Delcour JA, Rabijns A (2004) Structural basis for inhibition of Aspergillus niger xylanase by *Triticum aestivum* xylanase inhibitor-I. *Journal of Biological Chemistry* 279:36022–36028.

Savatin DV, Ferrari S, Sicilia F, De Lorenzo G (2011) Oligogalacturonide-auxin antagonism does not require post-transcriptional gene silencing or stabilization of auxin response repressors in Arabidopsis. *Plant Physiology* 157:1163–1174.

Savatin DV, Gigli-Bisceglia N, Marti L, Fabbri C, Cervone F, De Lorenzo G (2014) The Arabidopsis NUCLEUS- AND PHRAGMOPLAST-LOCALIZED KINASE1-related protein kinases are required for elicitor-induced oxidative burst and immunity. *Plant Physiology* 165:1188–1202.

Savatin DV, Gramegna G, Modesti V, Cervone F (2014) Wounding in the plant tissue: the defense of a dangerous passage. *Frontiers in Plant Science* 5:470.

Schneider R, Hanak T, Persson S, Voigt CA (2016) Cellulose and callose synthesis and organization in focus, what's new? *Current Opinion in Plant Biology* 34:9–16.

Schols HA, Visser RGF, Voragen AG (2009) Pectins and pectinases. Wageningen Academic Publishers.

Scortica A, Giovannoni M, Scafati V, Angelucci F, Cervone F, De Lorenzo G, Benedetti M, Mattei B (2022) Berberine bridge enzyme-like oligosaccharide oxidases act as enzymatic transducers between microbial glycoside hydrolases and plant peroxidases. *Mol Plant Microbe Interact* 10.1094/MPMI-05-22-0113-TA.

Scott-Craig JS, Panaccione DG, Cervone F, Walton JD (1990) Endopolygalacturonase is not required for pathogenicity of *Cochliobolus carbonum* on maize. *Plant Cell* 2:1191–1200.

Serrano M, Coluccia F, Torres M, L'Haridon F, Metraux JP (2014) The cuticle and plant defense to pathogens. *Frontiers in Plant Science* 5:274.

Shelomi M, Danchin EG, Heckel D, Wipfler B, Bradler S, Zhou X, Pauchet Y (2016) Horizontal gene transfer of pectinases from bacteria preceded the diversification of stick and leaf insects. *Sci Rep* 6:26388.

Shelomi M, Heckel DG, Pauchet Y (2016) Ancestral gene duplication enabled the evolution of multifunctional cellulases in stick insects (Phasmatodea). *Insect Biochemistry and Molecular Biology* 71:1–11.

Shi H, Liu Z, Zhu L, Zhang C, Chen Y, Zhou Y, Li F, Li X (2012) Overexpression of cotton (*Gossypium hirsutum*) *dirigent1* gene enhances lignification that blocks the spread of *Verticillium dahliae*. *Acta Biochimica et Biophysica Sinica (Shanghai)* 44:555–564.

Sillo F, Fangel JU, Henrissat B, Faccio A, Bonfante P, Martin F, Willats WG, Balestrini R (2016) Understanding plant cell-wall remodelling during the symbiotic interaction between *Tuber melanosporum* and *Corylus avellana* using a carbohydrate microarray. *Planta* 244:347–359.

Simpson SD, Ashford DA, Harvey DJ, Bowles DJ (1998) Short chain oligogalacturonides induce ethylene production and expression of the gene encoding aminocyclopropane 1-carboxylic acid oxidase in tomato plants. *Glycobiology* 8:579–583.

Song W, Forderer A, Yu D, Chai J (2021) Structural biology of plant defence. *New Phytologist* 229:692–711.

Souza CA, Li S, Lin AZ, Boutrot F, Grossmann G, Zipfel C, Somerville SC (2017) Cellulose-derived oligomers act as damage-associated molecular patterns and trigger defense-like responses. *Plant Physiology* 173:2383–2398.

Sposato P, Ahn JH, Walton JD (1995) Characterization and disruption of a gene in the maize pathogen Cochliobolus carbonum encoding a cellulase lacking a cellulose binding domain and hinge region. *Molecular Plant-Microbe Interactions* 8:602–609.

Sun Y, Wang Y, Zhang X, Chen Z, Xia Y, Wang L, Sun Y, Zhang M, Xiao Y, Han Z, Wang Y, Chai J (2022) Plant receptor-like protein activation by a microbial glycoside hydrolase. *Nature* 610: 335–342.

Tanaka K, Heil M (2021) Damage-Associated Molecular Patterns (DAMPs) in plant innate immunity: Applying the danger model and evolutionary perspectives. *Annual Review of Phytopathology* 59:53–75.

ten Have A, Mulder W, Visser J, van Kan JA (1998) The endopolygalacturonase gene Bcpg1 is required for full virulence of Botrytis cinerea. *Molecular Plant Microbe Interactions* 11:1009–1016.

Thaker A, Mehta K, Patkar R (2021) Feruloyl esterase Fae1 is required specifically for host colonisation by the rice-blast fungus *Magnaporthe oryzae*. *Current Genetics* Sep 15. doi: 10.1007/s00294-021-01213-z. Epub ahead of print.

Tian H, Wu Z, Chen S, Ao K, Huang W, Yaghmaiean H, Sun T, Xu F, Zhang Y, Wang S, Li X, Zhang Y (2021) Activation of TIR signalling boosts pattern-triggered immunity. *Nature* 598:500–503.

Tianpei X, Li D, Qiu P, Luo J, Zhu Y, Li S (2015) Scorpion peptide LqhIT2 activates phenylpropanoid pathways via jasmonate to increase rice resistance to rice leafrollers. *Plant Science* 230:1–11.

Toth IK, Birch PR (2005) Rotting softly and stealthily. *Current Opinion in Plant Biology* 8:424–429.

Tronchet M, Balagué C, Kroj T, Jouanin L, Roby D (2010) Cinnamyl alcohol dehydrogenases-C and D, key enzymes in lignin biosynthesis, play an essential role in disease resistance in Arabidopsis. *Molecular Plant Pathology* 11:83–92.

Trouvelot S, Héloir MC, Poinssot B, Gauthier A, Paris F, Guillier C, Combier M, Trdá L, Daire X, Adrian M (2014) Carbohydrates in plant immunity and plant protection: roles and potential application as foliar sprays. *Frontiers in Plant Science* 5:592.

Tundo S, Paccanaro MC, Elmaghraby I, Moscetti I, D'Ovidio R, Favaron F, Sella L (2020) The xylanase inhibitor TAXI-I increases plant resistance to *Botrytis cinerea* by inhibiting the BcXyn11a xylanase necrotizing activity. *Plants (Basel)* 9.

Vaaje-Kolstad G, Westereng B, Horn SJ, Liu Z, Zhai H, Sorlie M, Eijsink VG (2010) An oxidative enzyme boosting the enzymatic conversion of recalcitrant polysaccharides. *Science* 330:219–222.

van Aubel G, Cambier P, Dieu M, Van Cutsem P (2016) Plant immunity induced by COS-OGA elicitor is a cumulative process that involves salicylic acid. *Plant Science* 247:60–70.

Van Vu B, Itoh K, Nguyen QB, Tosa Y, Nakayashiki H (2012) Cellulases belonging to glycoside hydrolase families 6 and 7 contribute to the virulence of *Magnaporthe oryzae*. *Molecular Plant-Microbe Interactions* 25:1135–1141.

van Wersch R, Li X, Zhang Y (2016) Mighty dwarfs: arabidopsis autoimmune mutants and their usages in genetic dissection of plant immunity. *Frontiers in Plant Science* 7:1717.

Vanholme B, VAN Thuyne W, Vanhouteghem K, DE Meutter J, Cannoot B, Gheysen G (2007) Molecular characterization and functional importance of pectate lyase secreted by the cyst nematode *Heterodera schachtii*. *Molecular Plant Pathology* 8:267–278.

Vega-Munoz I, Duran-Flores D, Fernandez-Fernandez AD, Heyman J, Ritter A, Stael S (2020) Breaking bad news: dynamic molecular mechanisms of wound response in plants. *Frontiers in Plant Science* 11:610445.

Vogel JP, Raab TK, Somerville CR, Somerville SC (2004) Mutations in PMR5 result in powdery mildew resistance and altered cell wall composition. *Plant Journal* 40:968–978.

Voigt CA (2014) Callose-mediated resistance to pathogenic intruders in plant defense-related papillae. *Front Plant Sci* 5:168.

Volpi C, Janni M, Lionetti V, Bellincampi D, Favaron F, D'Ovidio R (2011) The ectopic expression of a pectin methyl esterase inhibitor increases pectin methyl esterification and limits fungal diseases in wheat. *Molecular Plant Microbe Interactions* 24:1012–1019.

Voxeur A, Habrylo O, Guenin S, Miart F, Soulie MC, Rihouey C, Pau-Roblot C, Domon JM, Gutierrez L, Pelloux J, Mouille G, Fagard M, Hofte H, Vernhettes S (2019) Oligogalacturonide production upon *Arabidopsis thaliana-Botrytis cinerea* interaction. *Proceedings of the National Academy of Sciences* 116:19743–19752.

Walton JD (1994) Deconstructing the cell wall. *Plant Physiology* 104:1113–1118.

Wang D, Pajerowska-Mukhtar K, Culler AH, Dong X (2007) Salicylic acid inhibits pathogen growth in plants through repression of the auxin signaling pathway. *Current Biology* 17:1784–1790.

Wang X, Li Q, Niu X, Chen H, Xu L, Qi C (2009) Characterization of a canola C2 domain gene that interacts with PG, an effector of the necrotrophic fungus *Sclerotinia sclerotiorum*. *Journal of Experimental Botany* 60:2613–2620.

Wang Y, Sheng L, Zhang H, Du X, An C, Xia X, Chen F, Jiang J, Chen S (2017) CmMYB19 over-expression improves aphid tolerance in *Chrysanthemum* by promoting lignin synthesis. *International Journal of Molecular Sciences* 18.

Wang Z, Peng Q, Gao X, Zhong S, Fang Y, Yang X, Ling Y, Liu X (2020) Novel fungicide 4-chlorocinnamaldehyde thiosemicarbazide (PMDD) inhibits laccase and controls the causal agent of take-all disease in wheat, *Gaeumannomyces graminis* var. tritici. *Journal of Agricultural and Food Chemistry* 68:5318–5326.

Wang Y, Li X, Fan B, Zhu C, Chen Z (2021) Regulation and function of defense-related callose deposition in plants. *International Journal of Molecular Sciences* 22.

Wanke A, Rovenich H, Schwanke F, Velte S, Becker S, Hehemann JH, Wawra S, Zuccaro A (2020) Plant species-specific recognition of long and short β-1,3-linked glucans is mediated by different receptor systems. *Plant J* 102:1142–1156.

Warren JG, Kasun GW, Leonard T, Kirkpatrick BC (2016) A phage display-selected peptide inhibitor of *Agrobacterium vitis* polygalacturonase. *Molecular Plant Pathology* 17:480–486.

Watanabe H, Tokuda G (2010) Cellulolytic systems in insects. *Annual Review of Entomology* 55:609–632.

Wei W, Xu L, Peng H, Zhu W, Tanaka K, Cheng J, Sanguinet KA, Vandemark G, Chen W (2022) A fungal extracellular effector inactivates plant polygalacturonase-inhibiting protein. *Nat Commun* 13:2213.

Williams HL, Sturrock RN, Islam MA, Hammett C, Ekramoddoullah AK, Leal I (2014) Gene expression profiling of candidate virulence factors in the laminated root rot pathogen *Phellinus sulphurascens*. *BMC Genomics* 15:603.

Wu SC, Ham KS, Darvill AG, Albersheim P (1997) Deletion of two endo -alfa-1,4-xylanase genes reveals additional isozymes secreted by the rice blast fungus. *Molecular Plant-Microbe Interactions* 10:700–708.

Xu L, Zhu L, Tu L, Liu L, Yuan D, Jin L, Long L, Zhang X (2011) Lignin metabolism has a central role in the resistance of cotton to the wilt fungus *Verticillium dahliae* as revealed by RNA-Seq-dependent transcriptional analysis and histochemistry. *Journal of Experimental Botany* 62:5607–5621.

Xu M, Gao X, Chen J, Yin Z, Feng H, Huang L (2018) The feruloyl esterase genes are required for full pathogenicity of the apple tree canker pathogen Valsa mali. *Molecular Plant Pathology* 19:1353–1363.

Yakoby N, Beno-Moualem D, Keen NT, Dinoor A, Pines O, Prusky D (2001) *Colletotrichum gloeosporioides* pelB is an important virulence factor in avocado fruit-fungus interaction. *Molecular Plant-Microbe Interactions* 14:988–995.

Yang Q, He Y, Kabahuma M, Chaya T, Kelly A, Borrego E, Bian Y, El Kasmi F, Yang L, Teixeira P, Kolkman J, Nelson R, Kolomiets M, L Dangl J, Wisser R, Caplan J, Li X, Lauter N, Balint-Kurti P (2017) A gene encoding maize caffeoyl-CoA O-methyltransferase confers quantitative resistance to multiple pathogens. *Nature Genetics* 49:1364–1372.

Yang C, Liu R, Pang J, Ren B, Zhou H, Wang G, Wang E, Liu J (2021) Poaceae-specific cell wall-derived oligosaccharides activate plant immunity via OsCERK1 during *Magnaporthe oryzae* infection in rice. *Nature Communications* 12:2178.

Yang Y, Yang X, Dong Y, Qiu D (2018) The *Botrytis cinerea* xylanase BcXyl1 modulates plant immunity. *Frontiers in Microbiology* 9:2535.

Yang Y, Zhang Y, Li B, Yang X, Dong Y, Qiu D (2018) A *Verticillium dahliae* pectate lyase induces plant immune responses and contributes to virulence. *Frontiers in Plant Science* 9:1271.

Yang Y, Anderson CT, Cao J (2021) Polygalacturonase45 cleaves pectin and links cell proliferation and morphogenesis to leaf curvature in *Arabidopsis thaliana*. *The Plant Journal* 106:1493–1508.

Yoder MD, Keen NT, Jurnak F (1993) New domain motif: The structure of pectate lyase C, a secreted plant virulence factor. *Science* 260:1503–1507.

York WS, Qin Q, Rose JK (2004) Proteinaceous inhibitors of endo-b-glucanases. *Biochimica et Biophysica Acta (BBA)* 1696:223–233.

Yoshizawa T, Shimizu T, Hirano H, Sato M, Hashimoto H (2012) Structural basis for inhibition of xyloglucan-specific endo-β-1,4-glucanase (XEG) by XEG-protein inhibitor. *Journal of Biological Chemistry* 287:18710–18716.

Yuan M, Jiang Z, Bi G, Nomura K, Liu M, Wang Y, Cai B, Zhou JM, He SY, Xin XF (2021) Pattern-recognition receptors are required for NLR-mediated plant immunity. *Nature* 592:105–109.

Yuan M, Ngou BPM, Ding P, Xin XF (2021) PTI-ETI crosstalk: an integrative view of plant immunity. *Current Opinion in Plant Biology* 62:102030.

Zang H, Xie S, Zhu B, Yang X, Gu C, Hu B, Gao T, Chen Y, Gao X (2019) Mannan oligosaccharides trigger multiple defence responses in rice and tobacco as a novel danger-associated molecular pattern. *Molecular Plant Pathology* 20:1067–1079.

Zarattini M, Corso M, Kadowaki MA, Monclaro A, Magri S, Milanese I, Jolivet S, de Godoy MO, Hermans C, Fagard M (2021) LPMO-oxidized cellulose oligosaccharides evoke immunity in Arabidopsis conferring resistance towards necrotrophic fungus *Botrytis cinerea*. *Communications biology* 4:1–13.

Zhang L, Kars I, Essenstam B, Liebrand TWH, Wagemakers L, Elberse J, Tagkalaki P, Tjoitang D, van den Ackerveken GFJM, van Kan JAL (2014) Fungal endopolygalacturonases are recognized as microbe-associated molecular patterns by the Arabidopsis receptor-like protein RESPONSIVENESS TO BOTRYTIS POLYGALACTURONASES1. *Plant Physiology* 164:352–364.

Zhang L, Hua C, Pruitt RN, Qin S, Wang L, Albert I, Albert M, van Kan JAL, Nürnberger (2021) Distinct immune sensor systems for fungal endopolygalacturonases in closely related Brassicaceae. *Nature Plants* 7.

Zhou F, Emonet A, Denervaud Tendon V, Marhavy P, Wu D, Lahaye T, Geldner N (2020) Co-incidence of damage and microbial patterns controls localized immune responses in roots. *Cell* 180:440-453 e418.

Ziv C, Zhao Z, Gao YG, Xia Y (2018) Multifunctional roles of plant cuticle during plant-pathogen interactions. *Frontiers in Plant Science* 9:1088.

Zuppini A, Navazio L, Sella L, Castiglioni C, Favaron F, Mariani P (2005) An endopolygalacturonase from *Sclerotinia sclerotiorum* induces calcium-mediated signaling and programmed cell death in soybean cells. *Mol Plant Microbe Interact* 18:849–855.

21 Cell Wall Polysaccharide Matrix Dynamics during Wood Development

Pramod Sivan and Francisco Vilaplana
KTH Royal Institute of Technology

Ewa J. Mellerowicz
Swedish University of Agricultural Sciences

INTRODUCTION

The cell walls in wood comprise the most abundant renewable resource on earth for sustainable production of eco-friendly materials, chemicals and energy (Keegan et al. 2013; Bar-On et al. 2018; Martínez-Abad et al. 2018). The functional properties, and both economic and environmental significance of this material, are mainly due to its chemical composition and structural integrity. Chemically, wood is composed of cellulose and lignin interconnected by matrix polysaccharides that include polymers of pectins, hemicelluloses, glycosylated structural proteins and callose (Figure 21.1). Due to their dynamic interactive features, the matrix polysaccharides play major roles in wood formation and its functional properties from initial development to maturity (Whitney et al. 1999; Åkerholm and Salmen 2001; Ryden et al. 2003; Martínez-Abad et al. 2017). Wood has great cellular and chemical complexity, varying both within trees, in response to environmental cues and among taxonomic groups. This review highlights the historical perspectives and our current understanding of matrix polysaccharides' roles in this complexity. The first five sections cover matrix dynamics during the successive stages of xylogenesis, cell division in the vascular cambium, xylem cell expansion, biosynthesis of secondary wall layers, formation of pits and perforations and post-synthetic matrix modification in secondary walls. Further sections address the variability of matrix polysaccharides among wood cell types in normal wood (NW), compare hardwoods and softwoods and discuss interactions among the cell wall polysaccharides. The last two sections cover changes in matrix polysaccharides in reaction wood and heartwood.

CELL WALL MATRIX POLYSACCHARIDES IN THE VASCULAR CAMBIUM

Vascular cambium is a lateral secondary meristem that is responsible for radial growth of higher vascular plants through formation of secondary xylem (wood) and secondary phloem (Larson 1994). The cambial meristem concept began with description of vital sap between bark and wood recorded in the fourteenth century (Larson 1994; Samuels et al. 2006). The fluid-like nature of the cambium observed in these early studies is due to the highly vacuolated character of cambial cells, as documented by observations of live cambial cells by Irving Bailey in the early 20th century (Bailey 1959), and hydrophilic nature of their cell walls, which are rich in pectins (Figures 21.1 and 21.2). Understanding of cambium's meristematic nature, first discussed by Marcello Malpighi in the 17th century, was later developed by application of the cell theory of Theodore Schwan and Matthias Jakob Schleiden published in 1839 by several contemporary botanists. This resulted in a view of the cambium as a cell layer between phloem and wood (Larson 1994; Samuels et al. 2006). Careful microscopic analyses of active cambium in conifers led Karl Gustav Sanio to develop cambial initial theory in the 1850s. This postulates that each initial gives rise to two daughter cells, one of which retains cambial initial identity while the other becomes committed to either xylem or phloem fate as a xylem or phloem mother cell, respectively (Larson 1994; Iqbal and Ghouse 1990; Samuels et al. 2006). These mother cells divide periclinally one or more times before entering terminal differentiation. Approximate locations of initials within the cambial meristem are indicated by high frequencies of anticlinal cell divisions that allow expansion of the meristem's circumference (Larson 1994). Apart from that, there are no clear differences between initials and mother cells in either ultrastructure or cell wall chemistry, and the zone of meristematic activity is jointly called "vascular cambium" or "cambial zone" (Catesson 1964; Larson 1994; Schrader et al. 2004).

Unlike in other meristems, there are two morphologically distinct cell types in the vascular cambium: axially elongated fusiform cambial cells (FCCs) that give rise to axial systems in secondary xylem and phloem, and isodiametric ray cambial cells (RCCs) that give rise to rays (Mellerowicz et al. 2001). Transcriptomic analysis of laser-dissected RCCs and FCCs from tangential freeze-dried sections of aspen vascular cambium have indicated that RCCs are enriched in homogalacturonan (HG)-related transcripts while FCCs have more galactan and xyloglucan (XG)-related transcripts (Goué et al. 2008). This suggests a differential matrix distribution of these cell types, which has been confirmed by immunolocalisation of the galactan LM5 epitope (Goué et al. 2008; Figure 21.2d).

DOI: 10.1201/9781003173309-21

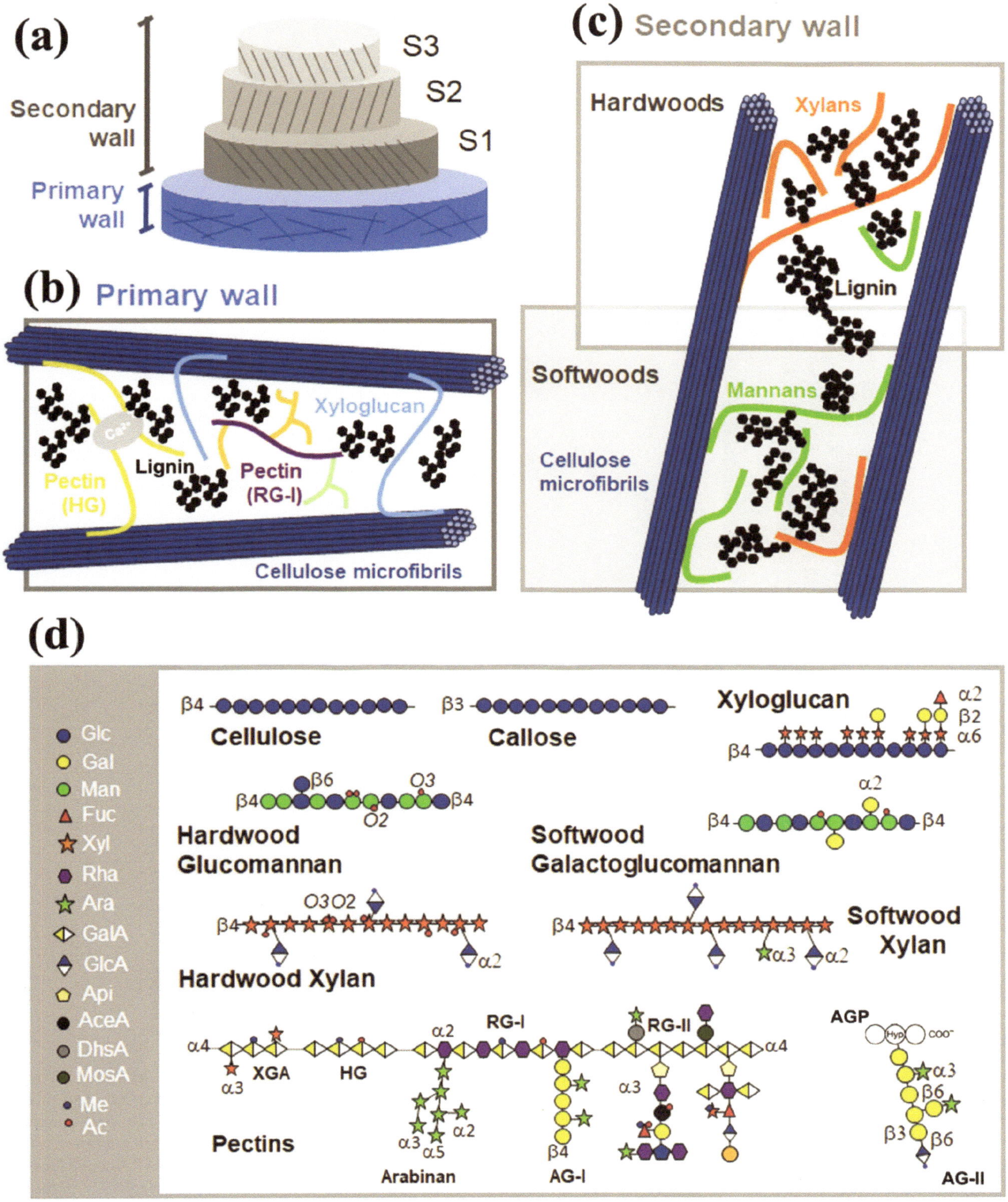

FIGURE 21.1 Wood cell wall architecture and composition. **(a)** Layered structure of wood cell walls. Supramolecular architecture of **(b)** primary cell wall and **(c)** secondary cell layers. **(d)** Molecular structure of plant cell wall polysaccharides (Burton et al. 2010), showing general monosaccharides and glycosidic linkages of each subfamily are described. The substitution pattern of each polysaccharide depends on the plant type, tissue and developmental stage. HG, XGA, RG-I, AG-I and RG-II, respectively, refer to Homogalacturonan, Xylogalacturonan, Rhamnogalacturonan I, Arabinogalactan I and Rhamnogalacturonan II pectin domains, while AG-II (Arabinogalactan II) is a part of glycosylation of arabinogalactan proteins (AGPs).

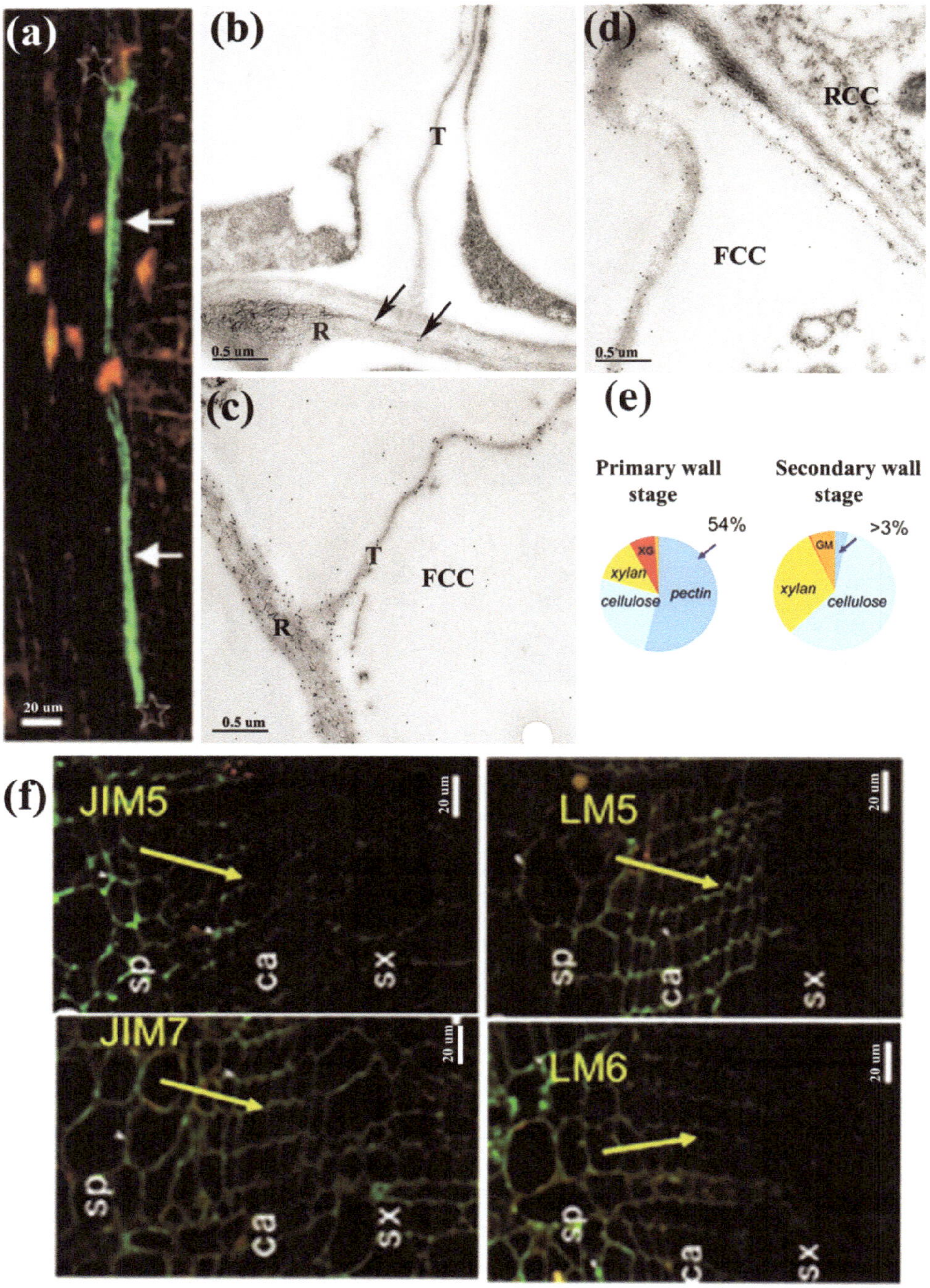

FIGURE 21.2 Variability of the cell wall matrix composition among cells of the vascular cambium in aspen. **(a)** Radial section of periclinally dividing fusiform cambial cell (FCC) showing callose (green signal, arrows) deposited on the growing cell plate between the phragmoplast poles (stars). (Reprinted with permission from Chaffey and Barlow, 2002, Planta 214: 526–536.) **(b)** Acidic homogalacturonan (HG) labelled with JIM5 antibody is found in the middle lamella of cambial radial (R) walls (arrows) but not tangential (T) walls. **(c)** RG-I β-1,4-galactan labelled with LM5 is present in R and cambial walls (Mellerowicz EJ and Immerzeel P, unpublished). **(d)** LM5 immunolabelling showing that RG-I β-1,4-galactan epitopes are more abundant in FCC than ray cambial cell (RCC). (Reprinted with permission from Goué et al. 2008 New Phytologist 180: 45–56.) **(e)** Carbohydrate composition of cell walls in the cambium of *Populus* showing predominance of pectins, contrasting with the carbohydrate composition of the secondary wall layers. Based on data from Simson and Timell (1978a; 1978b; 1978c; 1978d). **(f)** Immunolocalisation of the following pectin epitopes in cambium: JIM5, lightly methylesterified (acidic) HG; JIM7, highly methylesterified HG; LM5, β-1,4-galactan; LM6, α-1,5-arabinan. (Reprinted with permission from Liu et al. 2017, International Journal of Molecular Science 18: 1716).

FCCs, RCCs and their derivative secondary xylem and secondary phloem cells are organised in radial files, each formed by one initial dividing periclinally. The deposition of the periclinal cell plate in FCCs is guided by a special organelle called the phragmosome (Sinnote and Bloch 1940), which consists of cytoplasm with longitudinally oriented microtubules (Goosen-de et al. 1984). Perpendicularly oriented phragmoplast microtubules extending between two daughter nuclei guide vesicles containing cell wall matrix material to the equatorial plane where the vesicles coalesce and release their content, thereby forming the periclinal cell plate (Bailey 1959). The cell plate, which is rich in callose (Figure 21.2a), grows to the tips and sides of an FCC following the phragmoplast (Chaffey and Barlow 2002), and the primary wall (PW) layers are deposited on both sides of the cell plate. In contrast to radial walls, the nascent periclinal walls are resistant to extraction with ethylene diamine tetra acetic acid (EDTA), indicating that they contain little calcium-bound acidic pectin (Catesson 1994). This has been confirmed by immunolabelling experiments showing that labelling by JIM5 antibodies, which recognise acidic HG, predominantly occurs in the middle lamella of radial walls, while LM5 and JIM7 antibodies (which, respectively, recognise β-1,4 galactan and methylesterified HG) abundantly label radial and tangential walls (Figure 21.2b, c, f; Liu et al. 2017). Each time a periclinal division occurs, a new cell wall is formed, not only between the daughter cells but also around the entire cells. Thus, the radial walls of FCCs are several folds thicker than the tangential walls, as well as have a different cell wall composition, the radial walls being enriched in acidic HG (Figure 21.2b, c, f).

Besides pectins, the PWs of vascular cambium contain xylan and mannan, as well as arabinogalactan proteins and extensins. PW xylan is not labelled by LM10 and LM11 antibodies, which label SW xylan, either in cambial and expanding xylem cells or the compound middle lamella (CML) of mature xylem cells (Figure 21.4c–e; Hervé et al. 2009; Kim and Daniel 2012). In Arabidopsis, glucuronoxylan (GX) with largely unmethylated Glc*p*A spaced every six xylosyl units has been detected in PWs (Mortimer et al. 2015), with some substitution of Glc*p*A by a pentose (Mortimer et al. 2015; Chong et al. 2015) and similar form of xylan might be present in the PW of vascular cambium.

Gene expression profiles across the developing xylem of aspen have helped to identify specific *Populus* homologs of glycosyl transferases with demonstrated functions in biosynthesis of HG, RGI, RGII, XG, mannan, callose, AGP and extensins, which are putatively involved in biosynthesis of PW matrix in the cambium and radial expansion zone (Table 21.1; Sundell et al. 2017; Kumar et al. 2019). Few of the candidate genes have been functionally studied in developing woody tissues. One of them, *PtGAUT4* was shown to encode a HG galacturonosyl transferase involved in biosynthesis of HG (Biswal et al. 2018). Its downregulation substantially increased poplar growth. Candidate genes putatively involved in PW xylan biosynthesis are broadly expressed across the wood-forming zone and are most

TABLE 21.1
Genes with predicted functions in polysaccharide matrix biosynthesis upregulated in the cambial zone and radial expansion zone of Populus. Data extracted from Kumar et al. (2019).

CAZy family	*P. trichocarpa* ID		*A. thaliana* orthologue		Biosynthetic function
GT31	• Potri.011G151400		• AT1G32930	• GALT31A	β-1,6-galactan β-1,6-galactosyl transferase
GT31	• Potri.011G004200		• AT1G05170		• Galactosyl transferase
GT31	• Potri.001G450200		• AT1G32930	• GALT31A	• β-1,6-galactan β-1,6-galactosyl transferase
GT31	• Potri.002G089800		• AT1G77810		• Putative galactosyl transferase
GT31	• Potri.010G151600		• AT3G06440	• GALT3	• Galactosyl transferase
GT77	• Potri.010G046100		• AT1G70630	• RAY1	• Arabinofuranosyl transferase
GT77	• Potri.002G134400		• AT1G19360	• RRA3	• Extensin β-1,2-L-arabinosyl transferase
GT95	• Potri.003G195600		• AT5G13500	• HPAT3	• Extensin β-1,2-L-arabinosyl transferase
GT48	• Potri.001G012200		• AT5G13000		• Callose synthase
GT2	• Potri.006G116900	• CSLA3	• AT5G03760	• CSLA9	• Glucomannan synthase
GT2	• Potri.004G189000	• CSLA4	• AT5G22740	• CSLA2	• Glucomannan synthase
GT2	• Potri.002G200300	• CSLD1	• AT1G02730	• CSLD5	• Tip growth, mannan synthase
GT2	• Potri.014G125100	• CSLD2	• AT1G02730	• CSLD5	• Tip growth, mannan synthase
GT2	• Potri.001G136200	• CSLD4	• AT3G03050	• CSLD3	• Tip growth, mannan synthase
GT2	• Potri.013G082200	• CSLD6	• AT3G03050	• CSLD3	• Tip growth, mannan synthase
GT106	• Potri.012G064200		• AT5G15740	• RRT1	• RGI rhamnosyl transferase
GT106	• Potri.004G113500		• AT5G15740	• RRT1	• RGI rhamnosyl transferase
GT77	• Potri.014G092400	• RGTX1-B	• AT4G01220	• MPG4	• RGII α-1,3-xylosyltransferase

(*Continued*)

TABLE 21.1 (*Continued*)
Genes with predicted functions in polysaccharide matrix biosynthesis upregulated in the cambial zone and radial expansion zone of Populus. Data extracted from Kumar et al. (2019).

CAZy family	*P. trichocarpa* ID		*A. thaliana* orthologue		Biosynthetic function
GT8	• Potri.008G116900	• GATL3-A	• AT1G13250	• GATL3	• Putative galacturonosyl transferase
GT8	• Potri.010G129400	• GATL3-B	• AT1G13250	• GATL3	• Putative galacturonosyl transferase
GT8	• Potri.013G146200	• GAUT10-A	• AT2G20810	• GAUT10	• Putative galacturonosyl transferase
GT8	• Potri.019G108900	• GAUT10-B	• AT2G20810	• GAUT10	• Putative galacturonosyl transferase
GT8	• Potri.019G105100	• GAUT10-C	• AT2G20810	• GAUT10	• Putative galacturonosyl transferase
GT8	• Potri.016G001700	• GAUT4-A	• AT5G47780	• GAUT4	• HG galacturonosyl transferase
GT8	• Potri.002G246500	• GAUT8	• AT3G25140	• GAUT8/QUA1	• HG/xylan galacturonosyl transferase
GT8	• Potri.014G142000	• GAUT15	• AT3G58790	• GAUT15	• HG/xylan galacturonosyl transferase
GT92	• Potri.005G258900	• GALS1	• AT2G33570	• GALS1	• β-1,4-galactan synthase
GT92	• Potri.003G156600	• GALS2-A	• AT5G44670	• GALS2	• β-1,4-galactan synthase
GT92	• Potri.001G074600	• GALS2-B	• AT5G44670	• GALS2	• β-1,4-galactan synthase
GT2	• Potri.018G009300	• CSLC4-A	• AT3G28180	• CSLC5	• XG β-1,4-glucan synthase
GT2	• Potri.006G270900	• CSLC4-B	• AT3G28180	• CSLC5	• XG β-1,4-glucan synthase
GT2	• Potri.014G190900	• CSLC6-B	• AT3G07330	• CSLC6	• XG β-1,4-glucan synthase
GT2	• Potri.002G248400	• CSLC-R/	• AT3G07330	• CSLC6	• XG β-1,4-glucan synthase
GT34	• Potri.008G208000	• XXT1-A	• AT3g62720	• XXT1	• XG 6-xylosyl transferase
GT34	• Potri.010G025300	• XXT1-C	• AT3g62720	• XXT1	• XG 6-xylosyl transferase
GT34	• Potri.012G068300	• XXT3-A	• AT5G07720	• XXT3	• XG 6-xylosyl transferase

Current gene model in v3.1 *Populus trichocarpa* genome assembly replacing the former model Potri.002G113100.

TABLE 21.2
CAZyme genes with predicted functions in primary wall xylan biosynthesis upregulated in the PW/SW zone of Populus. Data extracted from Kumar et al. (2019).

CAZy family	*P. trichocarpa* ID		*A. thaliana* orthologue		Biosynthetic function
GT8	• Potri.004G124500	• GAUT13-A	• AT3G01040	• GAUT13	• galacturonosyl transferase
GT8	• Potri.017G090800	• GAUT13-B	• AT3G01040	• GAUT13	• galacturonosyl transferase
GT8	• Potri.017G106800	• GAUT9-B	• AT3G02350	• GAUT9	• galacturonosyl transferase
GT43	• Potri.007G047500	• GT43C	• AT5G67230	• IRX14	• Xylan backbone biosynthesis
GT43	• Potri.002G107300	• GT43E	• AT1G27600	• IRX9L	• Xylan backbone biosynthesis
GT47	• Potri.003G162000	• GT47A-2	• AT1G27440	• IRX10	• Xylan β-1,4 xylosyl transferase
GT47	• Potri.015G116700		• AT5G61840	• IRX10L	• Xylan β-1,4 xylosyl transferase
GT61	• Potri.012G051500		• AT3G18170		• pentose addition
GT8	• Potri.005G187900	• GUX3	• AT1G77130	• GUX3	• Xylan glucuronosyl transferase
CE6	• Potri.001G300800		• AT4G34215		• Xylan acetyl esterase

strongly expressed during primary to secondary wall transition (PW/SW) (Table 21.2; Kumar et al. 2019).

The mitotic activity of the cambium alternates with dormancy periods, which in temperate regions coincide with winter, and in tropical arid regions with the dry season or endogenous apical growth rhythms. After cessation of mitotic activity, the cell walls of FCCs become thick outside the primary pit fields (Chen et al. 2010; Rao et al. 2011). Ultrastructural studies have revealed microfibrillar structures with helicoidal patterns in the dormant cambium of *Acer pseudoplatanus* (Catesson 1994) and *Aesculus hippocastanum* (Chaffey et al. 1998), suggesting that dense arrays of cellulose microfibrils could be components of the thickened cell walls. In dormancy, cambium cell wall has a denser fibrillar network, smaller pores and fewer cross-links between microfibrils than in active states (Chen et al. 2010), suggesting that the activity-dormancy transitions involve alterations of the matrix polysaccharides. In poplar, xylan and XG accumulate in cell walls of the dormant cambial cells (Baïer et al. 1994; Viëtor et al. 1995; Ermel et al. 2000) along with acidic HG (Ermel et al. 2000; Chen et al. 2010). However, the xylan of dormant cambium walls differs from the xylan of SWs as it does not react with LM10 antibodies (Figure 21.4c; Kim and Daniel 2012). During cambial reactivation in spring, cell wall material in the middle lamella, including the calcium-bound pectin, is removed from radial walls prior to or concomitant with resumption of mitotic activity by cambial cells (Funada and Catesson 1991).

MATRIX DYNAMICS IN GROWING XYLEM CELLS

The ultimate size of a xylem cell depends on both its type and a balanced mechanism of intrusive and symplastic growth (Larson 1994; Mellerowicz et al. 2001; Mellerowicz 2006). In hardwood species, vessel elements greatly increase in diameter, mainly through extension of their radial walls via symplastic growth (together with adjacent cells) and intrusive growth (with vessel elements intruding sideways between cells of adjacent radial files). The fibres expand radially by symplastic growth of radial walls while their tangential walls remain attached to neighbouring cells within a radial file and do not extend. The radial expansion of vessel elements is much greater than that of other axial elements and is an early diagnostic trait. In contrast to vessel elements and axial parenchyma cells, the fibres elongate through intrusive apical growth. Their middle part, corresponding to the length of FCCs, remains in contact with adjacent cells and does not elongate, while new cell wall material is deposited at the tips that intrude into the space between the adjacent cells (Sinnott and Bloch 1939; Esau 1953; Larson 1994; Lev-Yadun 2001). Almost 50 years ago, Wenham and Cusick (1975) proposed that the mechanism involved turgor pressure causing the separation of middle lamellae and rounding of the cells, thus creating free space for intrusive growth. This idea has been recently confirmed by modelling (Majda et al. 2021). In conifers, the expansion of tracheids involves only the radial wall expansion in the radial direction without the apical intrusive growth; hence, the tracheids keep the original length of FCCs (Larson 1994). Similarly, axial parenchyma cells of hardwoods and softwoods expand only through radial diffuse growth without cell elongation (Esau 1953; Larson 1994). These cells sometimes divide transversely during differentiation, forming short axial strands of parenchyma, which have similar lengths to their progenitor FCCs. Ray parenchyma cells typically elongate in the radial direction, through mechanisms that likely involve diffuse symplastic growth.

Matrix polysaccharides are involved in both symplastic and intrusive growth. The orientation of cellulose microfibrils in the PW of FCCs is random to longitudinal (Chaffey et al. 1999; Derba-Maceluch et al. 2015), thus allowing radial cell expansion by diffuse growth when the matrix polymers cross-linking cellulose microfibrils are modified. The polysaccharide matrix of PWs of cambial cells of both hardwoods and softwoods consists of pectins, hemicelluloses and various glycosylated structural proteins, which, respectively, account for 30–50, 20–25 and up to 10% of the total mass of the PWs (Figures 21.1b and d, 21.2e; Mellerowicz et al. 2001; Scheller and Ulvskov 2010). The pectins include HG, rhamnogalacturonan I (RGI) and rhamnogalacturonan II (RGII). The hemicelluloses are largely XG, with some xylan and mannan, while the proteins include arabinogalactan proteins and extensins. Their modification, which is required for growth, mainly depends on enzymatic activities of apoplastic enzymes and cell wall–residing expansins. All known apoplastic proteins belong to large CAZy families. Recently, the protein family members specifically upregulated during xylem cell expansion have been identified in aspen wood-forming tissues (Kumar et al. 2019). Functions of few of these proteins have been determined, but some are α-expansins, which have prime responsibility for cell wall stress relaxation during diffuse growth by pH-dependent disruption of hydrogen bonds between cellulose microfibrils and cross-linking glycans (McQueen-Mason and Cosgrove 1995). They include a vascular-specific group A (Gray-Mitsumune et al. 2004), also known as clade IV (Sampedro et al. 2006), conserved in hardwoods and softwoods. *Pt*EXPA1 of this group promotes creep of cellulose-XG composites *in vitro* and enhances radial fibre expansion but not radial vessel expansion or fibre elongation when overexpressed in hybrid aspen (Gray-Mitsumune et al. 2008). This suggests that *Pt*EXPA1 limits xylem cell's symplastic growth but not their intrusive growth. Its expression is a marker of xylem cell expansion (Sundell et al. 2017).

The most abundant polysaccharide-active enzymes expressed during xylem cell expansion are the pectin-active enzymes (Kumar et al. 2019). Primary cell wall plasticity is known to be mediated by the methylesterification of HG (Peaucelle et al. 2011), changes in which have been demonstrated to precede cellulose reorientation connected to anisotropic growth (Peaucelle et al. 2015). HG methylesterification is regulated synthetically by pectin methyl transferases in the Golgi, and post-synthetically by cell wall–localised pectin methylesterases (PMEs), which are regulated by PME inhibitors (PMEIs), subtilisin-type Ser proteases (SBTs) and at least one E3 ubiquitin ligase (Levesque-Tremblay et al. 2015). Differing effects of HG methylesterification on cell growth have been reported and mechanisms involved are far from clear (Derbyshire et al. 2007; Peaucelle et al. 2011; Raggi et al. 2015; Chapters 5 and 6). However, immunolocalisation of HG in developing wood tissues has detected an increase in de-esterification at the onset of growth (Siedlecka et al. 2008). During cell wall development, regions of unesterified Gal*p*A residues can be either cross-linked via Ca^{2+} ions (forming "egg box" structures) that increase wall stiffness or cleaved by pectin-degrading enzymes (e.g., polygalacturonases) that promote wall loosening. According to Siedlecka et al. (2008), overexpression of hybrid aspen (*P. tremula x tremuloides*) *PME1* (*PtxtPME1*), which is similar to *AtPME3*, generates ester-free HG stretches, and hinders intrusive and diffuse growth of fibres and vessel elements, while suppression of *PtxtPME1* has opposite effects. These findings suggest that *Ptxt*PME1 activity may play an important role in cell wall strengthening by suppressing symplastic and intrusive growth during and following the process. Recent analysis of intrusively growing aspen fibres has revealed that tips of these fibres are highly labelled by JIM5 and JIM7 antibodies, indicating that the newly deposited wall at the tips is rich in esterified and de-esterified HG (Majda et al. 2021). The HG-rich tips are subtended by a sub-apical region with transverse, quickly changing to helical, cellulose microfibrils, which supports the tip growth at the apex and diffuse growth in a short sub-apical zone. Further from the tips, the wall becomes thicker, with more cross-linked cellulose microfibrils and more non-fucosylated XG. Thus, in addition to pectin metabolism, XG metabolism appears to be involved in fibres' intrusive growth.

Xyloglucan endotransglycosylase (XET) enzymes were first reported in 1992 (Farkas et al. 1992; Fry et al. 1992; Nishitani and Tominaga 1992; Okazawa et al. 1993; reviewed in Chapter 4), and since then have been implicated in cell growth and XG cross-linking in cell walls. XETs mediate endo-cleavage of a XG donor, followed by ligation to a XG acceptor (Fry et al. 1992; Nishitani and Tominaga 1992; Darley et al. 2001; Sulová et al. 2001). Expression of genes encoding XETs has been detected in growing vascular tissues and they putatively mediate cell expansion (Xu et al. 1995; Antosiewicz et al. 1997). *In situ* tests in fresh sections through developing aspen wood, performed to detect XET activity and the presence of the donor substrate XG, revealed the strongest signals in cell walls of cambial and expanding xylem cells (Bourquin et al. 2002). XET activity and XG were detected in both radial and tangential cell walls, and were more prominent in expanding fibres than expanding vessel elements. This closely followed the pattern of fucosylated XG content detected with CCRC-M1 antibody and hybrid aspen XTH34 (*Ptxt*XTH34) immunolocalisation. Overexpression of *PtxtXTH34* in hybrid aspen has been found to increase XG incorporation into the tightly bound XG fraction of cell walls of primary-walled cambial and developing xylem cells, and increase radial expansion of vessel elements but reduce radial expansion of fibres (Nishikubo et al. 2011). In addition, differential effects of *Ptxt*XTH34 on cell expansion in these cell types have been found, and are attributed to differences in their XG content and native XET activity (Bourquin et al. 2002). Thus, XET activity could limit expansion in vessel elements that have little native XET activity and little XG in their PWs, while fibres have high XET activity and XG content. In the fibres, increase in XET activity could raise levels of XG cross-linking and thus play a key role in strengthening of their walls. Recently, two *XTH* genes expressed in vascular tissues that are phylogenetically closely related to *PtxtXTH34* (*AtXTH4* and *AtXTH9*) were functionally tested in Arabidopsis (Kushwah et al. 2020). The single and double mutants had shorter fibres and thinner vessel elements than the wild type. Moreover, lengths of tails of vessel elements were highly affected in the mutants. The results show that these *XTH* genes have overlapping positive functions in mediating symplastic and intrusive growth of vessel elements and fibre elongation in Arabidopsis.

BIOSYNTHESIS OF SECONDARY WALL LAYERS

The evolution of secondary cell wall was a key adaptation that enabled the diversification of terrestrial plants and augmentation of their above-ground biomass. Secondary wall layers are deposited at the inner side of the PW (Figures 21.1 and 21.3) when xylem cell expansion is completed. The formation of SWs in Arabidopsis is triggered by a distinctive genetic programme controlled by master-inducers and downstream factors that activate transcription of genes encoding glycosyl

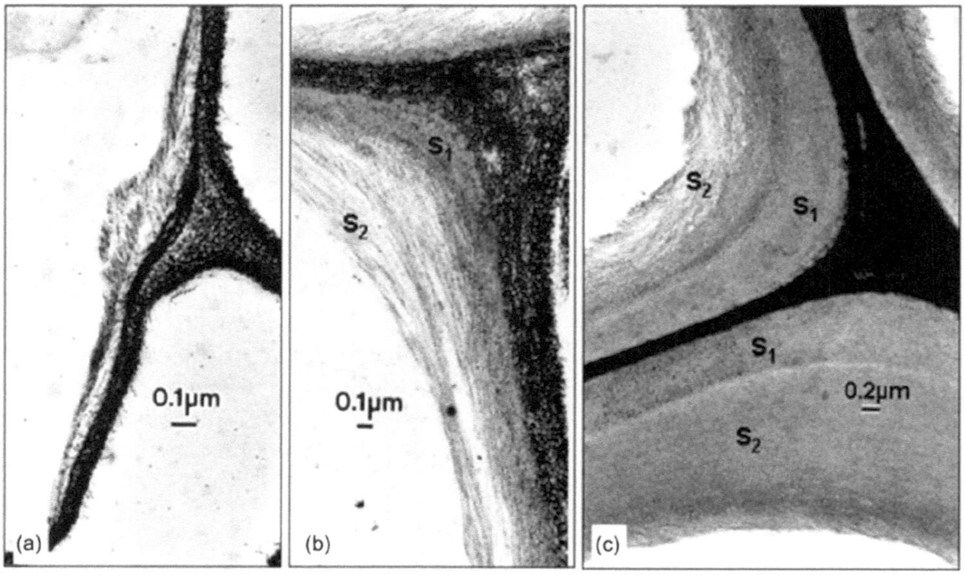

FIGURE 21.3 Assembly of secondary cell wall in developing xylem fibres of poplar. **(a)** Early stage of deposition of cellulose microfibrils in S1. **(b)** Nascent cellulose microfibrils in S2 with a loose arrangement. **(c)** Upper part, fibre with incomplete and loose S2; bottom part, mature fibre with compact S2. (Reprinted with permission from Ruel et al., 2006, Maderas. Ciencia y Tecnología 8: 107–116.)

transferases synthesising secondary cell wall polymers (Kubo et al. 2005; Zhong et al. 2008; Taylor-Teeples et al. 2015). In *Populus*, SW-associated NAC domain (SND) and vascular-related NAC domain (VND) proteins of NAC Class IIB clade have been shown to act as master switchers of SW formation that induce expression of downstream genes involved in the biosynthesis of SW cellulose, xylan and lignin (Ohtani et al. 2011; Zhong et al. 2011; Chen et al. 2019). In gymnosperms, the SW regulatory programme might be different, because Jokipii-Lukari et al. (2017) identified no NAC Class IIB co-expression networks in wood-forming tissues of spruce, but detected a set of separate CesA genes similar to SW CesA genes of angiosperms.

The timing of SW development varies among different cell types. In hardwoods, SW deposition ends and subsequent lignification occurs earlier in vessel elements and contact parenchyma cells (vessel-associated cells) than in fibres and isolation cells (Murakami et al. 1999). In conifers, tracheids develop SWs rapidly and close to the cambium, whereas deposition of SWs in ray parenchyma cells is delayed until late in the growing season, the next season, or even until the heartwood is formed (Yamamoto et al. 1978; Nakaba et al. 2006; Zheng et al. 2014). Moreover, the SW development is not synchronous among cells within a ray. But the molecular bases for these differences have not yet been discovered.

Most woody plants deposit SW with a tripartite structure with successive S1, S2 and S3 layers (Figures 21.1 and 21.3), characterised by alternating orientation of cellulose microfibrils (Abe et al. 1995; Funada 2001). While the PWs have plastic, dynamic, and highly hydrated architecture that enables cell growth, the SWs have a denser and thicker network of cellulose microfibrils with a different combination of matrix polysaccharides (Figure 21.1). Deposition of SW layers limits further radial cell expansion but provides xylem cells with mechanical strength and rigidity. The first formed (S1) layer has a flat, almost transverse microfibril array with an S-type of helix that gradually changes clockwise to a longitudinal arrangement and steep Z-type helix in the thickest (S2) layer. Finally, the S3 layer is formed after an abrupt reorientation of the microfibrils back to the more transverse S-type helix (Figure 21.1; Abe et al. 1995; Funada 2001). The changes in microfibril orientation during formation of successive cell wall layers are paralleled by the reorientation of cortical microtubules (Abe et al. 1995; Chaffey et al. 1999; Kaneda et al. 2010).

Freshly deposited cellulose microfibrils observed in the innermost developing SW layers form loose parallel arrays that appear to consolidate with time (Figure 21.3; Ruel et al. 2006). Although not experimentally proven, this loose structure could be related to a diurnal pattern of cellulose and matrix deposition. According to scanning electron microscopy (SEM) observations of the developing SW layers in tracheids of *Cryptomeria japonica*, a distinct array of newly deposited cellulose microfibrils can be seen during the daytime when the cells are under water stress, whereas this array appears to be covered by amorphous matrix material consisting of mannan and other compounds, possibly lignin precursors, during the night when the cells are turgid (Hosoo et al. 2002). This led to a hypothesis that while cellulose is deposited during the day, the hemicelluloses and lignin are deposited at night. Recent analyses of the diurnal pattern of incorporation of radiolabelled carbon from $^{13}CO_2$ into different cell wall compounds in aspen confirmed diurnal patterns of incorporation of lignin, cellulose, xylan, and proteins, and that the ratio of nighttime to daytime rates of lignin incorporation are particularly high (Mahboubi et al. 2015). Thus, lignin polymerisation could occur later than cellulose deposition, resulting in an incomplete cell wall assembly and loose cellulose network in the inner layer. However, it is still not clear if rates of hemicellulose deposition substantially differ, as suggested by the findings of Hosso et al. (2002). Careful ultrastructural analysis of developing xylem cells depositing SWs using high-pressure freezing fixation has revealed the presence of both nascent cellulose microfibrils aligned with cortical microtubules and secretory vesicles fusing with plasmamembrane in pine and poplar (Samuels et al. 2002; Kaneda et al. 2010). These results indicate that cellulose and hemicellulose deposition in SW coincide temporally, but we cannot exclude the possibility that rates of these processes may differ during diurnal cycles, resulting in deposition of a hemicellulose-poor cell wall layer during the day. Another possibility is that the loose nascent cellulose network could require mechanical compression by the high turgor pressure during the night to acquire mature architecture.

The polysaccharide composition of SW layers profoundly differs from that of PWs (Figures 21.1 and 21.2e; Mellerowicz et al. 2001; Scheller and Ulvskov 2010; Mellerowicz and Gorshkova 2012). Pectins are not present in SW layers, although they are the main polysaccharides in PWs, and the layers have dramatically different hemicellulose composition, with replacement of the XG by xylans and/or mannans. Moreover, different biosynthetic enzymes are involved in matrix polymer biosynthesis of the PW and SW layers. In angiosperms, the SW biosynthetic genes are induced by the dedicated SW programme master switchers of NAC family clade IIB discussed above.

In hardwoods, biosynthesis of secondary cell wall xylan is mediated by different isoforms of transferases from those participating in PW xylan biosynthesis (Tables 21.2 and 21.3; Ratke et al. 2015; Chen et al. 2019; Kumar et al. 2019). For example, in Arabidopsis, xylan backbones are synthesised by a xylan synthase complex that includes a xylan xylosyl transferase of the GT47 family (IXR10/IRX10L) and two scaffolding proteins of the GT43 family (IRX9/IRX9L and IRX14/IRX14L) (Jensen et al. 2014; Urbanowicz et al. 2014; Mortimer et al. 2015; Zeng et al. 2016). The scaffolding protein *At*IRX9L and xylosyl transferase *At*IRX10L are found in the PW xylan synthase complex, but they are replaced by *At*IRX9 and *At*IRX10 in the SW xylan synthase complex (Mortimer et al. 2015). Biochemical functions of these proteins are conserved between primary and secondary synthase complexes, but their expression patterns

TABLE 21.3
CAZyme genes with predicted function in secondary wall matrix biosynthesis upregulated in the SW zone of Populus. Data extracted from Kumar et al., 2019.

CAZy family	*P. trichocarpa* ID		*A. thaliana* orthologue		Biosynthetic function
GT14	• Potri.013G066200	• GT14A	• AT3G03690	• GLCAT14A	β-1,6-/β-1,3-galactan β-1,6-glucuronosyl transferase
GT14	• Potri.015G107100		• AT3G03690	• GLCAT14A	• β-1,6-/β-1,3-galactan β-1,6-glucuronosyl transferase
• GT14	• Potri.006G263000		• AT2G37585	• GlcAT14C	• β-1,3-/β-1,6-galactan β-1,6-glucuronosyl transferase
• GT29	• Potri.004G184200		• AT1G08280	• GALT29A	• β-1,3-/β-1,6-galactan β-1,6-galactosyl transferase
• GT31	• Potri.002G228900	• GT31B-1	• AT1G05170		• Putative galactosyl transferase
• GT31	• Potri.014G153700	• GT31B-2	• AT1G05170		• Putative galactosyl transferase
• GT31	• Potri.006G228900		• AT2G26100		• Putative galactosyl transferase
• GT31	• Potri.006G011300		• AT4G21060	• GALT2	• Hydroxyproline-β-O-galactosyl transferase
• GT31	• Potri.003G136000		• AT5G53340	• HPGT1	• Hydroxyproline O-galactosyl transferase
• GT95	• Potri.006G258800		• AT5G25265	• HPAT1	• Extensin β-1,2-L-arabinosyl transferase
• GT95	• Potri.001G028100		• AT5G13500	• HPAT3	• Extensin β-1,2-L-arabinosyl transferase
• GT48	• Potri.015G089300		• AT2G36850	• GSL8	• Callose synthase
• GT48	• Potri.005G058300		• AT2G13680	• GSL2/CALS5	• Callose synthase
• GT48	• Potri.003G214200		• AT5G13000	• GSL12/CALS3	• Callose synthase
• GT106	• Potri.010G250600		• AT1G51630	• MSR2	• Mannan biosynthesis
• GT2	• Potri.019G049700	• CSLD5	• AT3G03050	• CSLD3	• Tip growth, mannan synthase
• GT2	• Potri.008G026400	• CSLA1 /GT2A	• AT5G03760	• CSLA9	• Glucomannan synthase
• GH5_7	• Potri.006G109900	• MAN4	• AT5G01930	• MAN6	• Mannanase
• GH5_7	• Potri.016G138600	• MAN6	• AT5G01930	• MAN6	• Mannanase
• GH5_7	• Potri.013G130400	• MAN8	• AT2G20680	• MAN2	• Mannanase
• GT106	• Potri.017G101300		• AT5G15740	• RRT1	• RGI rhamnosyl transferase
• GT106	• Potri.015G048100		• AT5G15740	• RRT1	• RGI rhamnosyl transferase
• GT8	• Potri.002G200200	• GATL7-A/ GT8H	• AT3G62660	• GATL7	• Putative galacturonosyl transferase
• GT8	• Potri.014G125000	• GATL7-B	• AT3G62660	• GATL7	• Putative galacturonosyl transferase
• GT8	• Potri.T138200	• GAUT7-C	• AT2G38650	• GATL7	• Putative galacturonosyl transferase
• GT75	• Potri.019G051700		• AT5G16510	• RGP5	• Putative mutase
• GT2	• Potri.002G114200	• CSLC12A	• AT4G07960	• CSLC12	• XG synthase
• GT2	• Potri.005G146900	• CSLC12B	• AT4G07960	• CSLC12	• XG synthase
• GT34	• Potri.015G061800	• XXT3-B/ GT34B	• AT5G07720	• XXT3	• XG xylosyl transferase
• GT37	• Potri.001G034100		• AT2G03220	• FUT1/MUR2	• XG fucosyl transferase
• GT43	• Potri.006G131000	• GT43A	• AT2G37090	• IRX9	• Xylan backbone biosynthesis

(Continued)

TABLE 21.3 (*Continued*)
CAZyme genes with predicted function in secondary wall matrix biosynthesis upregulated in the SW zone of Populus. Data extracted from Kumar et al., 2019.

CAZy family	*P. trichocarpa* ID		*A. thaliana* orthologue		Biosynthetic function
• GT43	• Potri.016G086400	• GT43B	• AT2G37090	• IRX9	• Xylan backbone biosynthesis
• GT43	• Potri.005G141500	• GT43D	• AT5G67230	• IRX14L	• Xylan backbone biosynthesis
• GT43	• Potri.006G240200	• GT43G	• AT1G27600	• IRX9L	• Xylan backbone biosynthesis
• GT47	• Potri.001G068100	• GT47A-1	• AT1G27440	• IRX10	• Xylan xylosyl transferase
• GT47	• Potri.012G109600	• GT47D-1	• AT5G61840	• IRX10L	• Xylan xylosyl transferase
• GT47	• Potri.015G107200		• AT5G61840	• IRX10L	• Xylan xylosyl transferase
• GT47	• Potri.012G109200		• AT5G61840	• IRX10L	• Xylan xylosyl transferase
• GT8	• Potri.001G416800	• GAUT12-A/ GT8D1	• AT5G54690	• GAUT12/IRX8	• Xylan reducing end/pectin galacturonosyl transferase
• GT8	• Potri.011G132600	• GAUT12-B/ GT8D2	• AT5G54690	• GAUT12/IRX8	• Xylan reducing end/pectin galacturonosyl transferase
• GT8	• Potri.004G111000	• GAUT9-A	• AT3G02350	• GAUT9	• Xylan/pectin galacturonosyl transferase
• GT47	• Potri.009G006500	• GT47C	• AT2G28110	• FRA8/IRX7	• Xylan reducing end biosynthesis
• GT8	• Potri.014G040300	• GATL1-A/ GT8F	• AT1G19300	• PARVUS/GATL1	• Xylan reducing end biosynthesis
• GT8	• Potri.002G132900	• GATL1-B/GATL1.2	• AT1G19300	• PARVUS/GATL1	• Xylan reducing end biosynthesis
• GT8	• Potri.007G031700	• GATL2-A/ GT8E	• AT3G50760	• GATL2	• Putative galacturonosyl transferase involved in xylan biosynthesis
• GT8	• Potri.007G107200	• GUX1-A	• AT3G18660	• GUX1	• Xylan glucuronosyl transferase
• GT8	• Potri.005G061600	• GUX1-B/ GT8B	• AT3G18660	• GUX1	• Xylan glucuronosyl transferase
• GT8	• Potri.014G029900	• GUX2/ GT8A	• AT4G33330	• GUX2	• Xylan glucuronosyl transferase
• GH10	• Potri.002G113132*	• Xyn10A	• AT1G58370	• XYN1	• Trans-xylanase
• CE6	• Potri.014G022600	•	• AT3G53010		• Acetyl xylan esterase

differ. In *Populus*, *At*IRX9L and *At*IRX9 correspond to *Pt*GT43E and *Pt*GT43A/B, respectively (Lee et al. 2011; Ratke et al. 2015; Kumar et al. 2019). Several *Populus* SW xylan biosynthetic genes appear to have the same functions and regulation as homologs in Arabidopsis. Promoters of *PtGT43A* and *PtGT43B* are activated by master SW switchers *Ptxt*MYB021/*At*MYB46 and *Pt*NAC085/*At*SND1 (Ratke et al. 2015). *PtGT43A*, *PtGT43B* and *PtGT43E* have been shown to have the same function as *AtIRX9*, whereas *PtGT43C* and *PtGT43D* have the same function as *AtIRX14* (Lee et al. 2011). Knockdown of *PtGT43A* and *B* in *Populus* reportedly reduces xylan content, xylan chain length and SW thickness (Lee et al. 2011; Ratke et al. 2015).

Separate sets of PW- and SW-specific xylan glucuronosyl transferases have also been found in Arabidopsis, designated *At*GUX3 and *At*GUX1/*At*GUX2, respectively (Mortimer et al. 2015), and their homologs in *Populus* show differential expression during PW and SW biosynthesis in differentiating wood cells (Tables 21.2 and 21.3; Kumar et al. 2019). These transferases catalyse addition of Glc*p*A with different spacing on the xylosyl backbone, resulting in differences between PW and SW xylan structures, and different xylan domains in SWs. GUX1 and GUX3 may mediate glucuronosylation every other Xyl*p* but GUX3 typically mediates glucuronosylation every sixth Xyl*p* (in PWs) and GUX1 every eight Xyl*p* (in the major domain of SW xylan), whereas GUX2 mediates glucuronosylation with irregular patterns (in the minor domain of SW xylan) (Bromley et al. 2013; Mortimer et al. 2015). Conifers were recently found to have genes belonging to *AtGUX1*/*AtGUX3* and *AtGUX2* clades, called *GUX1* and *GUX2* clades, respectively (Lyczakowski et al. 2021). Members of the GUX1 clade typically add a glucuronosyl side chain to every sixth Xyl*p*, while members of the GUX2 clade can add Glc*p*A to consecutive Xyl*p* units making motifs, which have been detected in conifer xylans (Martínez-Abad 2017). The more frequent spacing pattern of Glc*p*A in gymnosperms than in angiosperms (Busse-Wicher et al. 2016) could be due to a lack of acetylation in the former since acetylating and glucuronosylating enzymes compete for the same substrate: the second C of Xyl*p* motifs (Chong et al. 2014; Lee et al. 2014; Lyczakowski et al. 2021). Most glucuronate in SW xylan is methylated at the *O*4 position (Teleman 2009) via a reaction catalysed by Golgi-localised glucuronoxylan methyl transferases (GXMs/GXMTs) of the DUF579 family (Urbanowicz et al. 2012; Lee et al. 2012). Four GXMs have been characterised in poplar (Youan et al. 2014a; 2014b; Song et al. 2016).

Biosynthesis of a reducing end tetrasaccharide sequence first discovered in birch xylan (Johansson and Samuelson 1977) is needed for xylan backbone biosynthesis. This sequence is known to be synthesised by GT8 family transferases, *At*PARVUS and *At*GAUT12//*At*IRX8, and a GT47 family transferase, *At*FRA8/*At*IRX7, although details of biochemical reactions catalysed by these enzymes are still unclear (Rennie and Scheller 2014). *Populus* homologs of *AtPARVUS* (*PtGT8E* and *F*; Lee et al. 2009a), *AtFRA8 (PtGT47C*; Lee et al. 2009b) and two homologs of *AtGAUT12* (*PtGAUT12-1/GT8D* and *PtGAUT12-2/GT8E*; Li et al. 2011; Biswal et al. 2015) seem to be involved in xylan biosynthesis according to knockdown phenotypes of *Populus* and/or complementation analyses of Arabidopsis mutants.

Glucomannan (GM) is present in both PW and SW layers in hardwoods and softwoods, but is much more abundant in the latter. There could be a structural difference between the GMs synthesised in the two cell wall layers since GM-specific antibodies that label SW do not react with the PW layer (Figure 21.4ab; Maeda et al. 2000; Kim et al. 2010b). GM and mannan backbone biosynthesis activities have been reported in cellulose synthase–like (CSL) family A (Dhugga et al. 2004; Liepman et al. 2005; Suzuki et al. 2006) and CSLD (Yin et al. 2011; Verhertbruggen et al. 2011), respectively. Expression of *AtCSL9A* is controlled by the *At*MYB46 master SW regulator (Kim et al. 2014). In *Populus*, different *CSLA* and *CSLD* genes are expressed during PW and SW biosynthesis (Tables 21.1 and 21.3; Suzuki et al. 2006; Kumar et al. 2019), suggesting they are differentially regulated by the secondary cell wall programme, and the SW-upregulated *PtCSLA1* has been shown to participate in GM synthesis (Suzuki et al. 2006). Addition of 1,6-galactosyl side chains in galactoglucomannans (GGMs) is known to be mediated by members of the GT34 family (Edwards et al. 1999). Deposition of GM in softwood tracheids is first detected in cell corners (Kim et al. 2010b) and enlarged, darkly stained Golgi bodies are apparently involved in GM deposition (Samuels et al. 2002).

The presence of XG in mature S layers has not been demonstrated, but deposition of XG has been immunohistochemically observed in the cell wall during SW development in *Populus* (Bourquin et al. 2002; Baba et al. 2009; Sandquist et al. 2010; Nishikubo et al. 2011). XG signals have been detected, using the CCRC-M1 antibody and XG-specific carbohydrate-binding module (CBM) FXG-14b, at the surface of nascent SW layers (Baba et al. 2009; Sandquist et al. 2010), with increasing intensity in the PW-middle lamella region, jointly called the CML, during SW-layer biosynthesis (Bourquin et al. 2002; Nishikubo et al. 2011). XET *in situ* activity incorporating XG acceptor into SW layers has also been detected in developing S layers (Baba et al. 2009). These results indicate that XG biosynthesis continues during SW deposition, in agreement with the expression of genes associated with XG biosynthesis in SW-forming xylem cells (Table 21.3; Sundell et al. 2017; Kumar et al. 2019). Moreover, alterations of XG structure in cell walls by overexpressing xyloglucanase in poplar (Park et al. 2004; Yamamoto et al. 2011) or knocking down the major *XTH* genes (*AtXTH4* and *AtXTH9*) involved in vascular tissue differentiation in *Arabidopsis* (Kushwah et al. 2020) have been found to significantly affect the cellulose structure and architecture of SWs. However, such alterations in SW could be mediated by cell wall damage perception mechanisms triggered by xylogluco-oligosaccharide mobilisation from primary cell walls and not necessarily by XG modification in SWs (Kushwah et al. 2020). It has also been suggested that XG may help to keep S layers together (Bourquin et al. 2002), but this requires experimental validation.

Almost all matrix polysaccharides are modified by acetylation, which affects their properties and interactions in cell walls. Conifer xylan is a notable exception as it is not acetylated (Figure 21.1). Acetyl-CoA, which serves as substrate for acetylation, is delivered to the Golgi apparatus through channels encoded by *REDUCED WALL ACETYLATION (RWA)* genes and the acetyl groups are transferred to nascent polysaccharides by polysaccharide-specific acetyl transferases of the TBL family (Rennie and Scheller 2014; Donev et al. 2018). In *Populus*, four *RWA* genes of two clades exhibiting differential expression patterns during wood development contribute to xylan acetylation (Pawar et al. 2017) and several TBL members add acetyl groups at specific Xyl*p* positions. *Ptr*XOAT11 and 12 monoacetylate Xyl*p* at *O*-2 and *O*-3, *Ptr*XOAT1, 2 and 3 both mono- and diacetylate Xyl*p* at these positions, while *Ptr*XOAT8 and 10 acetylate Xyl*p* decorated with glucuronosyl residues (Zhong et al. 2018). Other members of the TBL family are involved in mannan and pectin acetylation (Gille and Pauly 2012).

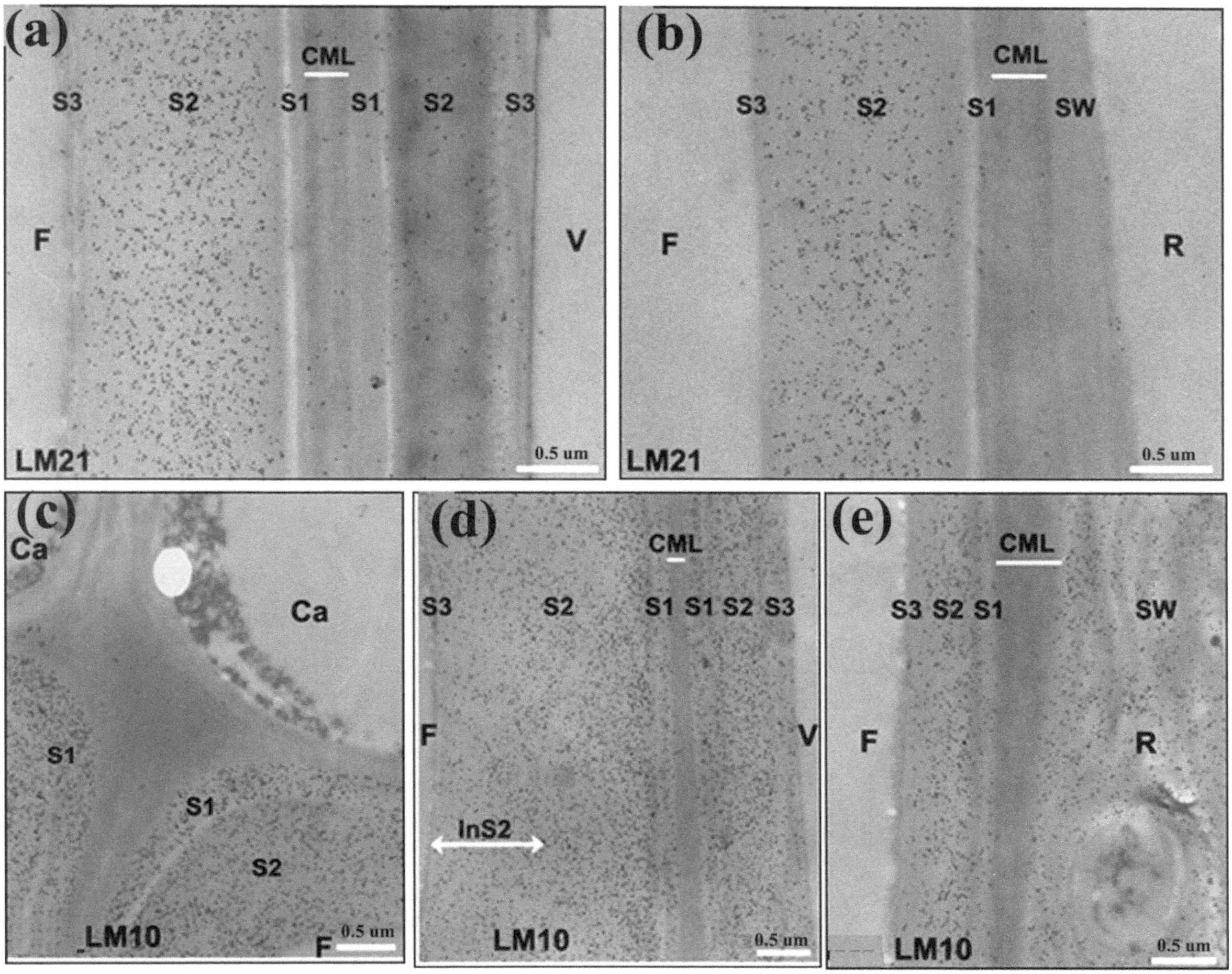

FIGURE 21.4 Immunogold localisation of glucomannan with LM21. **(a, b)** Xylan with LM10, **(c–e)** monoclonal antibodies in secondary walls of aspen fibres (F), vessel elements (V) and ray cells (R). Note the much weaker LM21 labelling in vessel elements and ray cells than in fibres and the stronger signals from S2 than S1 and S3 layers **(a, b)**. In contrast, there are no obvious differences among fibres, vessel elements and ray cells in labelling with LM10 antibody **(c–e)**. Fibres exhibit stronger labelling in the outer than the inner S2 layer. Cambial cell walls in dormant state (Ca) remain unlabelled. CML, compound middle lamella; InS2, inner S2 layer; SW, secondary wall. Bar=500nm. (Reprinted with permission from Kim and Daniel, 2012, Planta 236: 35–50.)

ROLES OF MATRIX POLYSACCHARIDES IN FORMATION OF PITS AND PERFORATIONS

SWs are not deposited over pits and perforations. These regions can be first identified as microtubule-depleted zones positioned by components of the ROP11 signalling pathway (Funada et al. 1997; Chaffey et al. 1999; Oda et al. 2010; Oda and Fukuda 2012). A separate ROP signalling pathway then organises biosynthesis of secondary cell wall over the pit regions (Sugiyama et al. 2019). Callose is synthesised in a ring at the sites of pits and perforations (Chaffey and Barlow 2002), which likely affects cellulose deposition. Moreover, xylan is deposited at the edge of SWs hanging over the pits, regulating pit size and cellulose deposition in the SWs over pits (Wang et al. 2022). The cortical microtubules and actin microfilaments form circular patterns over pits (Chaffey et al. 1997a,b; 1999; 2000), matching the cellulose microfibril orientation around the pits and forming overhanging SW structures that form borders of the pits.

Pit membrane corresponding to the PW is modified during development, which varies among cell types (Plavcová and Hacke 2011; Kim and Daniel 2013). In vessel-parenchyma pit membranes, HG epitopes are abundant and XG epitopes have been detected. These pit membranes are asymmetrical, with an amorphous layer only present at the parenchyma cell side (discussed in Section "Variability of Matrix Polysaccharides Among Wood Cell Types"). In the intervessel pits, esterified and de-esterified HG epitopes are initially abundant along with XG epitopes. During wall development, XG epitopes become sparse and HG epitopes become restricted to a narrow annulus zone of the pit membrane (outer pit membrane ring) in mature xylem (Plavcová and Hacke 2011; Kim and Daniel 2013; Herbette et al. 2015). Weak signals from RG-I arabinan have also been detected, but no xylan or mannan signals, in accordance with pit membrane's primary cell wall nature. In mature conducting vessel elements, pit membranes contain crystalline cellulose and lignin (Herbette et al. 2015). Extensins with their diverse glycan epitopes have also been associated with intervessel pit membranes, and many extensin-encoding genes are expressed during wood development (Abedi et al. 2021). Based on immunolocalisation experiments with four hardwood species, Plavcová and Hacke (2011) proposed that pit membrane annulus and main membrane parts represent two functionally distinct domains that might regulate cavitation thresholds by determining the extent of tension-induced membrane deflection. Perforations are formed in end walls of vessel elements devoid of SW layers by modification and complete removal of CML. The CML at the site of a future perforation becomes thicker initially and subsequently thinner than in adjacent areas, and its final removal coincides with cell death (Jacobsen et al. 2015). Active hydrolytic processes are putatively involved in cell wall removal according to ultrastructural evidence (Murmanis 1978), but the enzymes involved have not yet been identified.

POST-SYNTHETIC MATRIX MODIFICATION IN SECONDARY WALLS

Following their discharge by Golgi vesicles at the plasma membrane, matrix polysaccharides diffuse in the apoplast and become associated with the cell wall either spontaneously or enzymatically. Enzymes involved in cell wall assembly and extracellular polymer modification are important determinants of cell wall properties. *In situ* enzymatic activity assays with live SWs have revealed cellulase, exo-glucanase, galactosidase and xyloglucanase (XEH) activities that have been monitored in real time by confocal microscopy (Figure 21.5a, b, d; Ibatullin et al. 2009; Banasiak et al. 2014). These activities have been detected in developing and mature SWs and in CML, suggesting that some cell wall remodelling occurs during biosynthesis and some *post-mortem,* probably until the cell walls are fully lignified. Interestingly, XET activity has been detected in the gelatinous cell wall layer that does not lignify even 4 years after its deposition (Figure 21.5c; Nishikubo et al. 2007).

Studies of the role of cell wall modification by reverse genetics have shown that PtMAN6 protein, which has endo-mannanase activity, is associated with the plasma membrane of differentiating vessel elements in *Populus* (Zhao et al. 2013). Its overexpression inhibited formation and lignification of SWs not only of developing vessels but also other wood cells, whereas its suppression increased SW thickness and lignification not only in wood but also in pith. This activity seems to have a regulatory role, coordinating SW formation via oligosaccharide signalling. In addition, suppression of *PtxtXYN10A*, which encodes a xylan transglycosylase that mediates the cutting and re-ligation of xylan in SWs (Derba-Maceluch et al. 2015) has been found to stimulate plants' primary and secondary growth, and alter the cellulose microfibril angle (MFA) in wood. The activity may hypothetically release mechanical tension in the growing SWs, while its suppression may stimulate mechanosensing pathways that activate growth and alter cortical microtubules, thereby affecting cellulose orientation. Similar phenotypes have been observed following mild suppression of xylan biosynthesis in transgenic aspen (Ratke et al. 2018).

Although SWs do not contain pectins, pectin modification in the CML appears to affect mature xylem cell walls. Biswal et al. (2014; 2018) found that overexpression of the pectate lyase *PtxtPL1–27* or suppression of the pectin biosynthetic gene *PtGAUT4* enhanced wood saccharification potential. They also found that *Ptxt*PL1–27 mobilised xylan and XG from wood cell walls, indicating that pectins might be interconnected to these hemicelluloses (Biswal et al. 2014). Even stronger effects on saccharification yields were detected when HG methylesterification was suppressed in Arabidopsis by expression of a fungal PME inhibitor (Lionetti et al. 2010). In agreement, the modification of CML is nowadays considered as a vital strategy for decreasing recalcitrance of lignocellulose (Chapter 24).

VARIABILITY OF MATRIX POLYSACCHARIDES AMONG WOOD CELL TYPES

Immunolabelling experiments have clearly revealed high heterogeneity in distributions of SW matrix polymers in secondary xylem. In *Populus*, xylan labelling patterns with LM10 antibody are similar in fibres, vessel elements and ray cells (Figures 21.4c-d; Kim and Daniel 2012), while GM labelling with LM21 antibody is much stronger in fibres than in vessel elements and ray cells (Figure 21.4a, b; Kaneda et al. 2010; Kim and Daniel 2012). In accordance with these observations, Shi et al. (2017) found that *PtCSLA1* transcripts encoding a GM synthase were more abundant in transcriptomes of laser-dissected developing fibres than in those of vessel elements.

Ray and axial parenchyma cells develop a specialised tertiary cell wall layer ("the amorphous layer") over the lignified SW layers, composed of a 'protective layer' of contact cells (also known as vessel-associated cells) and chemically similar 'isotropic layer' of isolation cells (cells with no pits to vessel elements (Chafe 1974)). The matrix composition of the amorphous/isotropic layer resembles that of the PW (Morris et al. 2018), as it is abundant in pectins and contains XG and arabinogalactan proteins (Plavcová and Hacke 2011; Kim and Daniel 2013; Herbette et al. 2015). This layer of contact cells may eventually lignify.

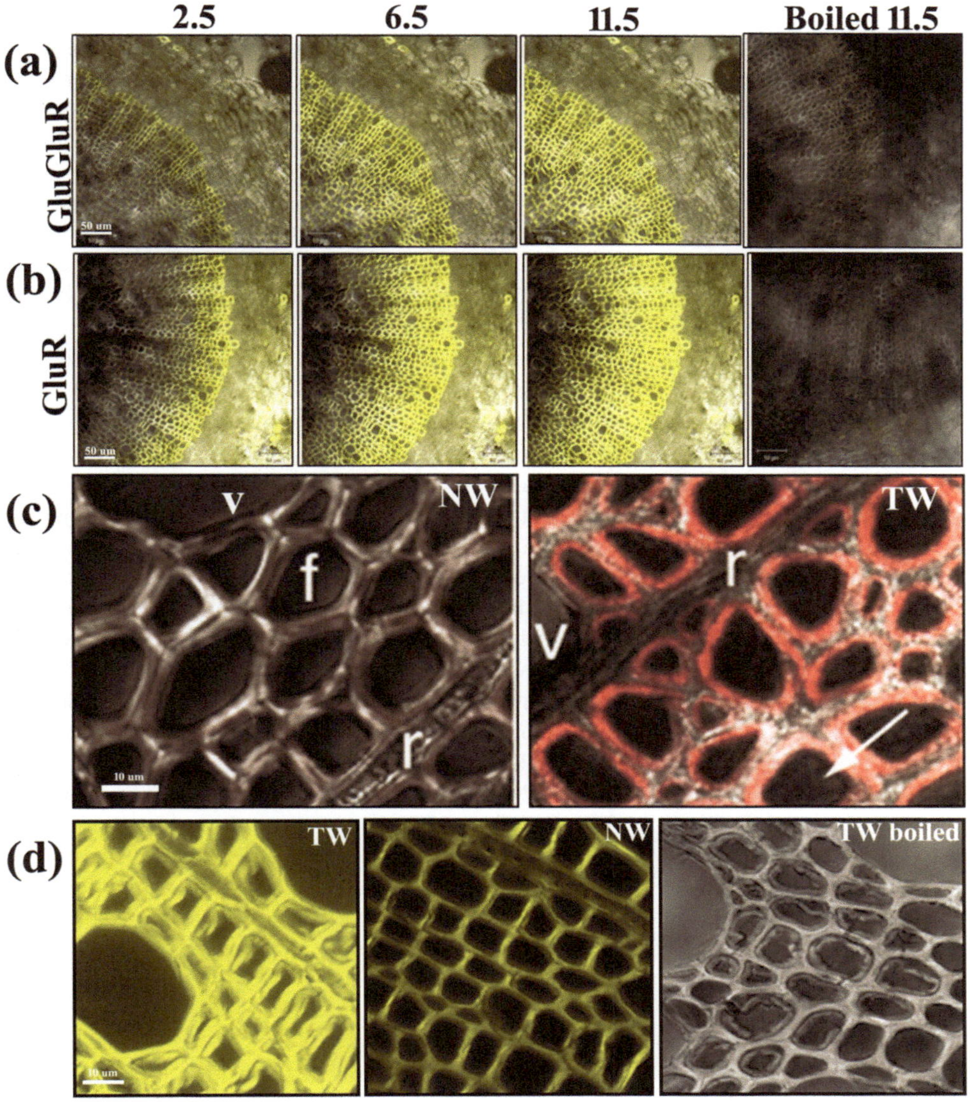

FIGURE 21.5 *In situ* enzymatic activities detected in secondary and tertiary walls. **(a, b)** Enzyme activities in developing and mature secondary walls in the secondary xylem of 8-week-old *Arabidopsis thaliana* (Columbia) hypocotyl. **(a)** Cellulase or glucosidase activity detected with the substrate Glu-Glu-β-resorufin (GluGluR) and **(b)** glucosidase activity with the substrate Glu-β-resorufin (GluR), which yields yellow fluorescence signals when hydrolysed by an endogenous activity. The pictures were taken after incubation with the substrate for 2.5, 6.5 and 11.5 min. No signal was detected in boiled hypocotyl sections (negative controls) incubated with the substrates for 11.5 min. Scale bars = 50 μm. (Banasiak A and Mellerowicz EJ, unpublished). **(c)** *In situ* XET activity in a 4-year-old normal wood (NW) and tension wood (TW) of a field-grown aspen demonstrated by incorporation of XXXG-sulphorhodamine (red signals). XET activity is detected specifically in TW gelatinous fibres in the secondary wall layers adjacent to G-layers (arrow). Note that activity is present 4 years after G-fibre biosynthesis. V, vessel element; f, fibre; r, ray cell. (Reprinted with permission from Nishikubo et al., 2007, Plant and Cell Physiology 48: 843–855.) **(d)** β-1,4-galactosidase activity in TW and NW of aspen detected with Gal-resorufin substrate. Note the prominent activity in TW. No signal was detected in boiled TW samples (negative controls). The activity was proposed to participate in generation of tensile stress in TW G-layers. (Reprinted with permission from Gorshkova et al., 2015, Plant Physiology 169: 2048–2063.)

STRUCTURE OF MATRIX POLYSACCHARIDES IN FULLY DIFFERENTIATED WOOD

The material properties of wood have fascinated and inspired scientists for centuries. Research spanning more than 200 years generated the current view of wood as a biocomposite of cellulose microfibrils embedded in a tightly connected matrix of hemicelluloses and polyphenolic lignins, with well-defined ordering from nanoscale to macroscale (Cosgrove and Jarvis 2012; Burgert and Keplinger 2013). Fundamental wood chemistry research may have started in the early 19th century when a French chemist, Anselme Payen, discovered a "resistant fraction" in tissues of herbs and hardwoods composed of carbon, hydrogen and oxygen, then subsequently determined the atomic composition of purified cellulose (Sjöström 1992; Young 2008). The polymeric structure of cellulose was discovered by the German organic chemist Hermann Staudinger in the 1920s. Extensive research on carbohydrates by Walter Haworth elucidated the structure of glycopyranose units in cellulose. The German chemist Karl Freudenberg provided evidence of the presence of β-glycosidic linkages between these sugar moieties (Young 2008). The presence of sugars other than glucose in wood was detected before the end of the 19th century and the term "hemicelluloses" was introduced by von Ernst Schulze in 1891 (Timell 1964). In pioneering work on hemicelluloses in the early 20th century, Margaret O'Dwyer detected approximately one O-methylhexuronic acid residue in every 10 xylose residues of xylan isolated from oak wood (O'Dwyer 1934; 1939). Sugar motifs were also isolated from aspen wood and identified (Jones and Wise 1952), but a complete structure of wood GX was first determined by Aspinall et al. (1954) in European beech. In a series of studies on the occurrence of *O*-acetyl groups in wood hemicelluloses, Lindberg and coworkers recorded 7.1 ester linkages per 10 xylose residues in xylan of silver beech (Hägglund et al. 1956; Lindberg et al. 1973: Timell 1964). Locations of *O*-acetyl groups on the C2 and C3 position of xylose residue in hardwood xylan were revealed by Bouveng et al. (1961). Jones et al. (1957) isolated GMs from wood hemicelluloses, and subsequently Timell and coworkers extensively studied the diversity of hemicelluloses in softwoods and hardwoods (Timell 1964; Bouveng et al. 1960; Lindberg et al. 1973; Rosell and Svensson 1975). Based on their studies, it was recognised that the major hemicelluloses of SWs of vascular plants are xylans and mannans (Timell 1967; Willför et al. 2005a,b; McKee et al. 2016). In conifers (softwoods), GGM is the most abundant hemicellulose (20–25% dry weight (d.w.)) and lower amounts of arabinoglucuronoxylan (AGX) are present (10–15% d.w.). In contrast, hemicelluloses of eudicot SWs are dominated by acetylated GX and have relatively little GM (Scheller and Ulvskov 2010; Huang et al. 2016), suggesting that xylan may have acquired GM functions in hardwoods (Dammström et al. 2009).

Xylans are ubiquitous hemicelluloses in the SW of vascular plants (Carafa et al. 2005). Their general molecular structure consists of a backbone of β-(1→4) D-xylopyranose (Xyl*p*) units, which can be decorated by various glycosyl substitutions and chemically modified by acetylation depending on the plant type, tissue and developmental stage (Busse-Wicher et al. 2016; Martínez-Abad et al. 2017). These substitutions are necessary to maintain xylan solubility (Mikkelsen et al. 2015), as unsubstituted xylan can form crystalline chain structures with a three-fold screw helix (Nieduszynski and Marchessault 1971). In conifers, xylans are of the AGX type, with α-(1→2) D-Glc*p*A in 4-*O*-methylated form (MeGlc*p*A) and α-(1→3) L-arabinofuranose (Ara*f*) as the main decorations of the xylosyl backbone and no reported acetylations (Escalante et al. 2012; Busse-Wicher et al. 2016; McKee et al. 2016). Busse-Wicher et al. (2016) investigated xylan substitution patterns in four gymnosperm lineages (Coniferophyta, Cycadophyta, Ginkgophyta and Gnetophyta) and found similar spacing of MeGlc*p*A substitution in them on every sixth xylosyl residue. They also found higher levels of arbinosylation in Ginkgophyta and cycads than in conifers, and an absence of arabinose and presence of acetyl substitution in Gnetophyta, similar to features of eudicot xylan. Xylan decorations may have co-evolved with lignin since Gnetophyta have the G-S-type of lignin, like eudicots (Nawari et al. 2016). Ara*f* substitution in conifer AGX is evenly spaced two Xyl*p* residues from the ones substituted with MeGlc*p*A (Busse-Wicher et al. 2016). In subsequent analyses, Martínez-Abad et al. (2017) found a major domain with MeGlc*p*A substitutions every six Xyl*p* units and Araf decorations evenly spaced from MeGlc*p*A units (not only two units away), and a minor domain with consecutive glucuronation. The role of the consecutive glucuronation in AGX is still unclear, but it may putatively influence interaction with lignin (Martínez-Abad et al. 2017). Molecular dynamic simulations of the identified xylan motifs have suggested that the flexibility of the three-fold screw conformation of softwood xylan in solution is enhanced by the presence of arabinofuranosyl decorations. Moreover, the occurrence of glycosyl substitutions in AGX influences steric tolerance during adsorption on both the hydrophilic and hydrophobic surfaces of cellulose microfibrils, favouring stabilisation of the two-fold screw conformation. This suggests that not only hydrogen bonding but also nonpolar driving forces play important roles in xylan's adsorption onto cellulose surfaces (Martínez-Abad et al. 2017; Pereira et al. 2017).

Eudicots contain acetylated GX with distinct structural motifs in terms of relative frequencies of acetylation and glucuronidation. The decorations by acetate and glucuronic acid groups make the hydrophilic xylan surface more hydrophobic and acidic, respectively (Simmons et al. 2016). The GX populations have different domains with distinct substitution patterns, a major easily extractable domain with even MeGlc*p*A spacing and acetylation with respect to MeGlc*p*A side chains (Bromley 2013; Busse-Wicher et al. 2014; Chong et al. 2014; Martínez-Abad et al. 2018), and a more recalcitrant minor domain characterised by tighter glucuronidation spacing and sparser acetylation (Bromley 2013; Martínez-Abad et al. 2018). On a xylan backbone in the ribbon-like two-fold helical screw conformation, all the decorations that are evenly spaced as in the major domain will face the same side, thus creating a compatible region for hydrogen bonding to the hydrophilic surface of cellulose (Busse-Wicher et al. 2014; 2016; Simmons et al. 2016). NMR analyses of GX populations from birch wood (Martínez-Abad et al. 2018) have revealed that easily extractable GX interacts with lignin through γ-ester linkages between MeGlc*p*A and the γ-carbon of lignin, while the recalcitrant GX populations correlate with higher amounts of benzyl ether lignin-carbohydrate complexes (LCCs). This indicates that the xylan substitutions may play a distinct role in the interactions with lignin during the lignification process, and hence their recalcitrance to processing.

Acetylated GGMs are the most abundant hemicelluloses in the mature wood of conifers. GGM consists of a backbone of β-(1→4)-linked D-mannopyranosyl (Man*p*) and D-glucopyranosyl (Glc*p*) units, with both α-(1→6) D-galactopyranosyl (Gal*p*) decorations and acetylations at *O*-2 and/or *O*-3 positions of the Man*p* units (Hannuksela and du Penhoat 2004; Lundqvist et al. 2002). The occurrence of an even pattern of galactosylation has been previously reported in a seed mucilage GGM from *Arabidopsis thaliana*, which again suggests that it may contribute to stable binding to the hydrophilic and hydrophobic faces of cellulose microfibrils (Yu et al. 2018).

In spruce, sequential subcritical water extractions and analyses have revealed the presence of two separate GGM populations, with distinct substitution patterns and hence potentially distinct effects on the supramolecular architecture of lignocellulose (Martínez-Abad et al. 2020). One was easily extractable and had a higher acetylation content, with lower frequencies of glucosyl and galactosyl motifs, than the other more recalcitrant mannan population, which also had a more complex glucosylation pattern of the backbone (with consecutive Glc*p*-Man*p* motifs) and richer galactosylation profile. Molecular dynamic simulations revealed that both the contents and distributions of Man*p* and Glc*p* in the GM backbone influence the binding strength of GGM to cellulose (Yu et al. 2018; Martínez-Abad et al. 2020). The axial conformation of the C2 hydroxyl groups in mannose hinders their adsorption in a cellulose-like conformation onto the microfibrils. However, even spacing of Man*p* residues alternating with Glc*p* units enables strong adsorption of the GM backbone to the cellulose, similarly to a pure β-glucan chain (Martínez-Abad et al. 2020). These results suggest that the recalcitrant GGM population in spruce could be directly interacting with the cellulose surfaces, mediated by the higher content of Glc*p* in the backbone, and the presence of even motifs of alternating Man*p* units with higher content of Gal*p* substitutions (Martínez-Abad et al. 2020).

ROLE OF MATRIX POLYSACCHARIDES IN DETERMINING LIGNOCELLULOSE PROPERTIES

The most important biological role of the diverse molecular structures of hemicelluloses is believed to be modulation of their interaction with cellulose microfibrils (Bromley et al. 2013; Busse-Wicher et al. 2014, 2016; Martínez-Abad et al. 2017) and covalent linkages with lignins in LCCs (Martínez-Abad et al. 2018). The spatial organisation of the cellulose, hemicellulose and lignin components in lignocellulosic biomass is still largely unknown. However, recent studies involving advanced imaging and solid scattering techniques have started to reveal the exquisite supramolecular architecture of microfibrillar and macrofibrillar domains (Hafrén et al. 1999; Terashima et al. 2009; Fernandes et al. 2011; Kang et al. 2019; Terrett et al. 2019; Lyczakowski et al. 2019). SWs of softwoods and hardwoods have large aggregates known as macrofibrils that have average diameters of 10–20 nm (Fernandes et al. 2011; Jarvis 2018), but can reach up to 60 nm as revealed by cryo-SEM (Lyczakowski et al, 2019). In these macrofibrillar domains, rigid mannan and xylan domains putatively react with individual cellulose microfibrils with 3–4 nm diameters, and unbound matrix mannan and xylan populations are also reportedly present (Terrett et al. 2019). In SWs of *Arabidopsis thaliana*, xylan and lignin contribute to macrofibrils (Lyczakowski et al. 2019), and both rigid xylan domains tightly adsorbed to the cellulose microfibrils and more flexible xylan domains are present (Terrett et al. 2019). The presence of both rigid and flexible hemicellulose domains in secondary cell walls suggests that hemicelluloses may act as an interphase that extends in a controlled manner the rigid and semicrystalline organisation of the cellulose microfibrils towards a more mobile and disordered phase. This combination of rigid and flexible organisation of wood hemicelluloses seems to be required for modulation of cellulose fibrillar networks' biomechanical properties (Berglund et al. 2020).

The covalent linkages between hemicelluloses and lignin that form LCCs are probably among the most controversial aspects of wood chemistry (Gimmarella et al. 2019a; Terrett and Dupree 2019). There are three major types of LCCs, mediated by sugars such as Xyl*p*, Glc*p*A, Glc*p*, Man*p*, Gal*p* and Ara*f*: phenyl glycosides that link the phenolic C-4 to the anomeric C1-*O* in sugars; γ-esters that link γ-carbons in lignin to the carboxyl group in mGlc*p*A (C6-*O*); and benzyl ethers that link α-carbons in lignin to either primary hydroxyl groups (*O*-6 in hexoses and *O*-5 in Ara*f*) or secondary hydroxyls (*O*-2 or *O*-3 in different sugars) (Tarasov et al. 2018; Giummarella et al. 2019a; Terret and Dupree 2019). For a long time, it was not known if LCCs are present in natural wood or they are introduced during wood processing and/or isolation of LCCs. More recently, their natural occurrence has been supported by analyses of lignification in spruce cell lines forming extracellular lignin (Giummarella et al. 2019b). Further corroboration has been provided by discovery of glucuronoyl esterases that specifically hydrolyse γ-ester linkages between GX and lignin in wood-decaying microorganisms (Špániková and Biely 2006) and solid state NMR-based detection of close contacts between lignin and acidic xylan in intact secondary walls of maize (Kang et al. 2019).

CELL WALL MATRIX POLYSACCHARIDES IN REACTION WOOD

Spatial positions of trees' stem axes are controlled by active growth responses of secondarily thickened stems to changes in the direction of gravity or light, or longitudinal mechanical stresses, which regulate stem position by generating growth stresses. This involves changes in cambial growth patterns and xylem cell differentiation, resulting in a type of wood with distinct structure and cell wall chemistry, called reaction wood (Clair et al. 2013). Reaction wood typically forms in leaning stems and branches. In gymnosperms, it forms on the lower stem side and is known as compression wood, while in angiosperms it forms on the upper side and is called tension wood. Compression wood generates longitudinal pressure, while tension wood (TW) generates longitudinal tension in the wood. Despite these positional and operational differences, the formation of both types of reaction wood results in higher lignification of the lower side of a leaning stem than the upper side. Due to this similarity, several pioneering researchers attempted to apply a common model of generation of growth stresses within tree trunks to explain stem bending (Boyd 1985). The two main competing theories extensively used by researchers to model the observed strains in the wood were "lignin swelling" and "cellulose crystallisation". It soon became apparent that whereas compression generation by compression wood correlated with high lignin content, tension generation correlated with cellulose content and low MFA (Yamamoto 1998). Current research focuses on fine structures of the two types of reaction wood to elucidate how the pressure and the tension are generated (Felten and Sundberg 2013; Fagerstedt et al. 2014). Matrix polysaccharides appear to play essential roles in these processes.

Tension Wood

Most temperate and approximately 40% of tropical hardwoods form TW with gelatinous (G) fibres characterised by the presence of a translucent G-layer inner to the SW layer(s), thus forming a tertiary wall layer (Figure 21.6; Wardrop and Dadswell 1948; Fisher and Stevenson 1981). In some species, a multilayered cell wall develops with lignified SW layers interspersed with non-lignified G-layers (Clair et al. 2006; Ruelle et al. 2006; Ghislain et al. 2016; Higaki et al. 2021). The composition of the G-layer dramatically differs from that of SW layers, as reviewed by Mellerowicz and Gorshkova (2012). Norberg and Meier (1966) concluded from analyses of the chemical and physical properties of the G-layer in aspen TW that G-fibres contain almost pure crystalline cellulose with low (almost axial) MFA. Subsequent analyses of isolated G-layers indicated the presence of other polysaccharides, with crystalline cellulose and matrix polysaccharides accounting for about 70–80% and 15–20% d.w., respectively (Furuya et al. 1970; Nishikubo et al. 2007; Kaku et al. 2009; Guedes et al. 2017). No lignin was detected in isolated G-layers by chemical analysis reported by Kaku et al. (2009), but sometimes lignin-like phenolics have been identified with lignin antibodies (reviewed by Pilate et al. 2004) and Raman spectroscopy imaging (Gierlinger and Schwanninger 2006). The G-layer also contains proteins, accounting for about 3% d.w. (Kaku et al. 2009). G-layers have very peculiar architecture, as their cellulose forms four times larger crystallites than those in SW, with much larger mesoporosity and almost axial orientation (Mellerowicz and Gorshkova 2012). The high porosity hypothetically drives the lateral swelling of the G-layer, its longitudinal contraction (Chang et al. 2015) and lateral pressure on SW (Goswami et al. 2008). The SW in G-fibres has relatively high lignin and xylan contents, and high MFA (Timel 1969; Clair et al. 2011; Rüggeberg et al. 2013); thus, the outward pressure on SW layers is predicted to induce longitudinal strain in SWs (Goswami et al. 2008). The second most abundant sugar of the G-layer after Glc*p* is reportedly either Gal*p*, accounting for about 20% of neutral sugars according to Kaku et al. (2009) and Guedes et al. (2017), or Xyl*p*, accounting for 1.5–18.9% of neutral sugars according to Norberg and Meier (1966), Furuya et al. (1970) and Nishikubo et al. (2007).

Gal has been associated with two types of polymers: β-1,4-galactan of RGI and type II acidic arabinogalactan (AG-II) associated with arabinogalactan proteins. The presence of β-1,4-galactan in G-layers is supported by convergent evidence from linkage analyses, polymer fingerprinting and immunolocalisation studies (Figure 21.6e, f; Gustafsson et al. 1952; Meir 1962; Kuo and Timell 1969; Bowling and Vaughn 2008; Arend 2008; Gorshkova et al. 2015; Guedes et al. 2017). During G-layer maturation RG-I, β-1,4-galactan chains are trimmed (Guedes et al. 2017), probably by β-1,4-galactosidase(s) that are associated with the G-layer, and reportedly highly active in developing G-fibres (Figure 21.5d; Gorshkova et al. 2015). The trimmed β-1,4-galactan enclosed within cellulose macrofibrils by entrapment mechanisms is predicted to generate longitudinal tensile stress within the G-layers (Mellerowicz et al. 2008; Mellerowicz and Gorshkova 2012; Gorshkova et al. 2015), but this hypothesis still requires functional testing *in planta*.

Acidic arabinogalactans comprise another abundant component of G-layers. They have been identified by Yariv reagent (Lafarguette et al. 2004), immunolocalisation of different AG-II components (Bowling and Vaughn 2008; Gerttula et al. 2015; Guedes et al. 2017), polysaccharide profiling (Gorshkova et al. 2015) and proteomic analyses of isolated G-layers (Kaku et al. 2009; Bygdell et al. 2017). Moreover, transcriptional induction of many AGPs of the *AtFLA12* clade is a hallmark of TW transcriptomes (Lafarguette et al. 2004; Andersson-Gunnerås et al. 2006). FLAs have been shown to influence cellulose orientation and cell walls' mechanical properties (MacMillan et al. 2010; 2015), but the mechanism of their action is not understood. *PtFLA6*, which belongs to the same clade as *AtFLA12*, mediates TW formation downstream of gibberellins (Wang et al. 2017).

The identity of Xyl*p*-containing polymers is still being debated. Xyl*p* has been interpreted as evidence of either xylan (Guedes et al. 2017) or XG (Kaku et al. 2009; Nishikubo et al. 2007). The G-layers do not contain xylan that is labelled with antibodies recognising SW xylan structures, such as AX1, LM10, LM11 and UX1 (Figure 21.6d; Kim and Daniel 2012; Guedes et al. 2017; Pramod et al. 2019). However, trace amounts of GX with a distinct structure are supported by recent oligomer profiling showing that xylanase digestion of isolated G-layers yields fragments including two types of glucuronosylated and methylated

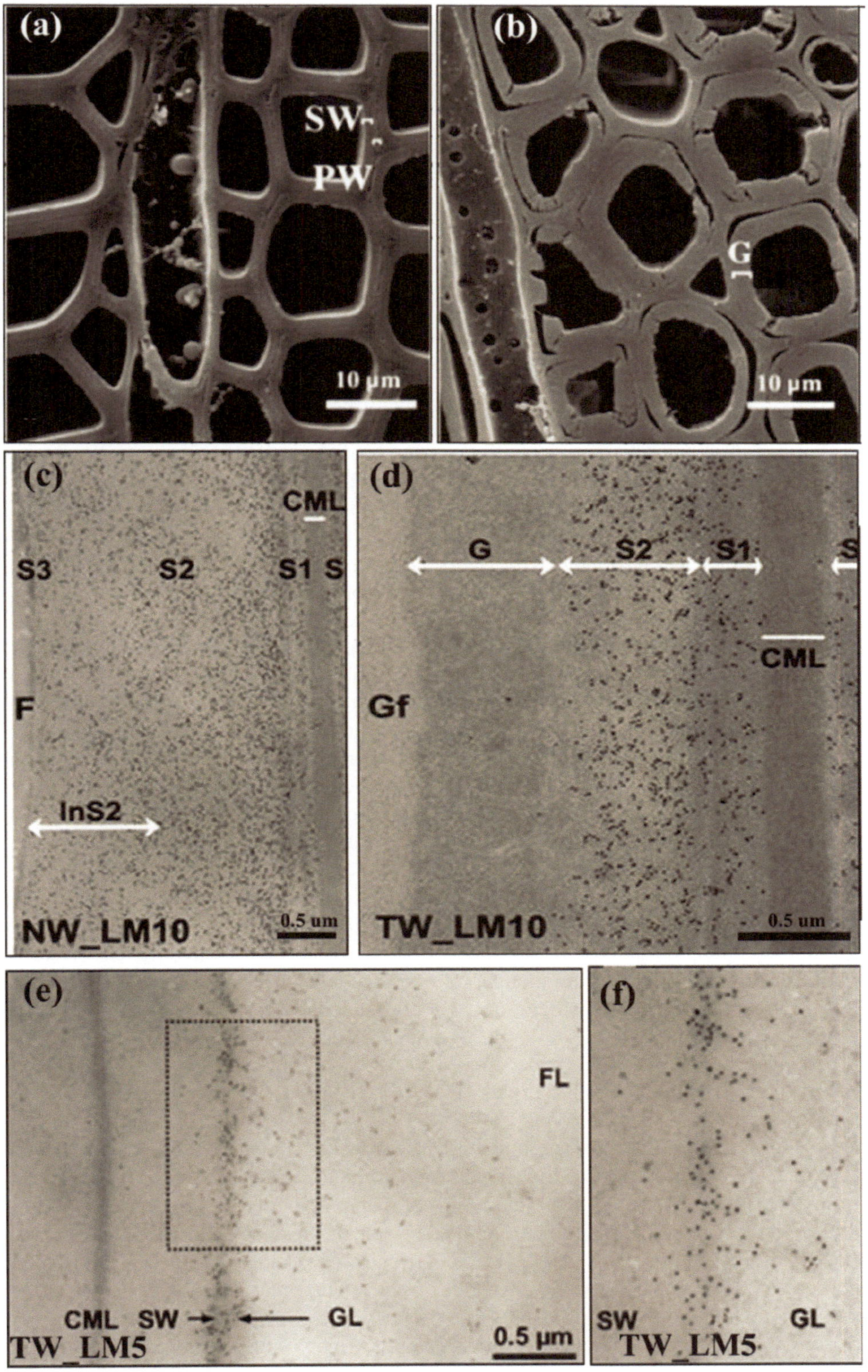

FIGURE 21.6 Tension wood (TW) of poplar. **(a, b)** Scanning electron micrographs of opposite wood (a) and TW (b). Note an additional thick layer (G-layer) inwards of the secondary cell wall layer. PW, primary cell wall; SW, secondary cell wall; and G, gelatinous layer. (Reprinted with permission from Pilate et al., 2004, New Phytologist 163:63–72.) **(c–f)** Immunogold localisation of xylans by LM10 antibodies **(c, d)** and (1,4)-β-galactan **(e, f)** by LM5 antibodies in normal wood (NW) and TW. NW fibres (F) show stronger LM10 labelling in the outer than the inner S2 layer (c), while TW fibres have uniform labelling in S1 and S2 layers but no labelling in the G-layer. (Reprinted with permission from Kim and Daniel, 2012, Planta 236:35–50.) **(e, f)** LM5 labelling of a TW fibre showing intense signals in the G-layer, especially the contact zone between the G-layer and adjacent S2 layer. (Reprinted with permission from Arend, 2008, Tree Physiology 28:1263–1267.)

residues (Guedes et al. 2017). No other residues typically present in xylan fingerprints were reported, suggesting that the xylan's structure strongly differs from that of SW xylan. Moreover, the anti-xylan antibodies CCRC-M119 and CCRC-M120 that detected signals in developing G-layers in G-fibers did not recognise their SW layers. However, CCRC-M120 labelled their CML, suggesting that a polymer similar to PW xylan could be synthesised during G-layer development (Guedes et al. 2017). Other studies have found associations between Xyl*p* and the presence of XG in G-layers. These associations are corroborated by the presence of relatively high amounts of 4,6-linked Glc*p*, terminal Xyl*p*, 2-linked Gal*p* and terminal Fuc*p* (Nishikubo et al. 2007, Kaku et al. 2009); immunolocalisation of fucosylated XG using CCRC-M1 and other antibodies (Nishikubo et al. 2007; Meloche et al. 2007; Bowling and Vaughn 2008; Baba et al. 2009; Sanquist et al. 2010); and *in situ* XET activity (Nishikubo et al. 2007; Baba et al. 2009; Gerttula et al. 2015). Some authors have found that XG is abundant in developing G-layers, but detected no XG signals in mature G-layers, which was attributed to the tight architecture of mature G-layers and XG entrapment within macrofibrils (Nishikubo et al. 2007; Baba et al. 2009). XG has also been immunolocalised between S- and G-layers (Baba et al. 2009; Sandquist et al. 2010), and interestingly XET activity seems to follow XG signals, moving from the G-layer to adjacent SW layers during G-fibre development (Figure 21.5c; Nishikubo et al., 2007). Several XTH genes are reportedly expressed more strongly in TW than in NW (Nishikubo et al. 2007) and *Pt*XET16E/*Pt*XTH6 has been identified in proteomic analyses of isolated G-layers (Kaku et al. 2009). In addition, XG is apparently essential for the righting movement in inclined stems since Baba et al. (2009) found that expression of fungal xyloglucanase, which reduced XG contents, totally inhibited this movement in poplar. XG and XET activities in G-fibres are thought to assist the binding of different cell wall layers (Mellerowicz et al. 2008).

Roughly half of species examined to date make functional TW without G-fibres (Onaka 1949; Bailliers et al. 1995; Clair et al. 2006, Ruelle et al. 2006; Pramod et al. 2021). The matrix composition of such wood is unknown, but it has been shown to have lower MFAs, higher cellulose to lignin ratios and higher S to G ratios than NW (Sugiyama et al. 1993; Okuyama et al. 1994; Bailleres et al. 1995; Yoshizawa et al. 2000; Yoshida et al. 2002; Ruelle et al. 2007; Pramod et al. 2021). More research is needed to determine if cell wall porosity is increased in such TW walls, or polymers are trapped within cellulose macrofibrils, or other features of cell wall architecture contribute to longitudinal stresses in such TW (Clair et al. 2006).

Compression Wood

Compression wood is characterised by the presence of rounder than normal tracheids, with a thicker than normal SW containing a S1 layer and a highly modified S2 layer with distinct characteristics. It can be divided into an outer lignified S2L and inner S2i region containing helical cavities (Donaldson and Singh 2013; Zhang et al. 2016). The S3 layer is absent in severe compression wood (Wardrop and Dadswell 1950; Zhang et al. 2016).

Analysis of the cell wall chemistry of compression wood has revealed that it has much higher lignin contents and lower cellulose contents than NW (Donaldson and Singh 2013). Cellulose microfibrils of its S2 layer have higher MFAs (Wardrop and Dadswell 1950; Gorišek and Torelli 1999; Donaldson et al. 2004; Yeh et al. 2005, 2006) and lower crystallinity (Lee 1961; Scurfield 1973; Timell 1986) than those in NW. Advanced techniques such as fluorescence spectroscopy, microautoradiography, and time-of-flight secondary ion mass spectroscopy have shown that the middle lamellae and S2L layer of compression wood tracheids have higher amounts of *p*-hydroxyphenyl lignin units than other layers (Fukushima and Terashima 1991; Donaldson et al. 2010; Tokareva et al. 2007; Fagerstedt et al. 2014).

Immunolocalisation experiments have shown that the S2L and S2i layers have lower mannan and xylan contents than the S2 layer in NW (Kim et al. 2011), but the S2L layer contains β-1,4-galactan, which is not present in SWs of NW (Schmitt et al. 2006; Altaner et al. 2007; Mast et al. 2009; Kim et al. 2010a; Zhang et al. 2016). Because of a high correlation of galactan with lignin contents in compression wood tracheids, galactan putatively plays a role in S2L lignification (Donaldson and Knox 2012). Callose or laricinan (β-1,3-glucan) is another unusual matrix polysaccharide found in compression wood and localised in helical cavities of the S2i layer of several conifer species (Brodzki 1972; Waterkeyn et al. 1982; Altaner et al. 2007; Zhang et al. 2016). Although compression wood has clearly characterised and prominent differences in cell wall composition and cell wall architecture from NW, it is still not known how they generate longitudinal pressure in tracheids (Donaldson and Singh 2013).

FORMATION OF HEARTWOOD

Heartwood is formed through a sequence of physiological, anatomical, cellular and chemical changes in the inner layer of wood, resulting in death of all the living cells and conversion of substances stored in them into heartwood extractives (Magel and Höll 1993; Hillis 1999). The extractives usually make the heartwood darker than the surrounding sapwood. The amounts and composition of extractives vary among plant species and under different environmental conditions, but they usually have antifungal and antibacterial properties, repel water and increase wood's mechanical strength, thereby changing its appearance, increasing its biotic resistance and enhancing its durability (Hillis 1999).

Ray parenchyma cells, which are not lignified in earlier phases, become lignified in processes preceding the cell death associated with heartwood formation. This involves activation of enzymes involved in the pentose phosphate pathway, phenyl propanoid pathway and tricarboxylic acid (TCA) cycle, together with accumulation of secondary metabolites such as cinnamic acids, lignin, terpenoids, and flavonoids (Higuchi 1976; Hillis 1999). After the death of parenchyma cells, their organelles autolyse and all accumulated compounds are converted by non-specific oxidation and polymerisation (Hillis 1999). Their death is preceded by the occlusion of adjacent embolised tracheary elements, so that their embolism can no longer be repaired. Contact parenchyma cells secrete gels and gums, composed of AG-II and pectins, which diffuse into adjacent tracheids (softwoods) or vessel elements (hardwoods) (De Micco et al. 2016). In some species, tyloses occluding tracheary elements are also formed. In this process, the protective pecto-cellulosic layer in the contact parenchyma cells overlying the pit membranes (Czaninski 1977; Gregory 1978; Evert 2006; De Micco et al. 2016) is stretched into the vessel lumen like a balloon with active deposition of cell wall materials including cellulose, hemicellulose, pectins, suberin and lignin (Foster 1967; Pearce and Holloway 1984; Schaffer and Wisniewski 1989). When tylosis expansion is complete, additional layers are formed: an outer amorphous layer that may be smooth or granular and an inner layer with microfibrillar structure resembling a thickened PW (Foster 1964; Schmitt et al. 1997). In *Eucalyptus*, *Kalopanax*, *Quercus* and *Robinia*, the outer layer reportedly has randomly oriented cellulose microfibrils while the inner layer has parallel microfibril orientation (Foster 1967; Sachs et al. 1970; Shibata et al. 1982). The vessels can be occluded further with gels or gums secreted by previously formed tyloses (Chattaway 1949), walls of which can become impregnated with suberin or lignin (Sachs et al. 1970; Obst et al. 1988; Evert 2006).

Occlusion of vessel elements by gels and tyloses can also be induced soon after their differentiation without heartwood formation. Wounding or pathogen attack induces such occlusions, via responses mediated by ethylene and jasmonate signalling pathways (Shibata et al. 1982; Leśniewska et al. 2017). In *Vitis*, the occlusion follows a short period of vessel hydraulic activity (Jacobsen et al. 2015).

Comparative studies have indicated that heartwood has lower cellulose and higher pectin content than sapwood, but similar lignin contents, in several hardwood species (Willför et al. 2005a; Çetinkol et al. 2012). However, relative amounts of matrix polysaccharides in heartwood and sapwood vary substantially. For example, more matrix polysaccharides have been detected in heartwood than sapwood in *Fagus sylvatica*, but less in *Eucalyptus globulus* and *Populus deltoides* (Willför et al. 2005a). Similar amounts of matrix have been detected in both wood types in other species. In most species, much more water-soluble polysaccharides have been extracted from sapwood than from heartwood.

Lower cellulose contents have also been detected in heartwood of softwoods than in their sapwood (Willför et al. 2005b). Norway spruce heartwood also contains more lignin, AGX and pectin than its sapwood (Bertaud and Holmbom 2004), and larch heartwood has very high AG-II contents (Sjöström 1992; Coté and Timell 1967). In other softwood species, similar contents of matrix polysaccharides have been found in sapwood and heartwood (Willför et al. 2005b). These results show that the cell wall matrix composition can be altered in heartwood compared to sapwood, with some species exhibiting more pronounced changes than others. However, neither the mechanisms responsible for these changes nor their significance have been elucidated so far.

POST SCRIPTUM

In this review, we attempt to summarise current knowledge of wood matrix polysaccharides, with a historical perspective. It focusses on their structures and biosynthesis, the areas of the field where there have been the most substantial advances in the past 20 years. We also attempt to provide a broad overview of matrix polysaccharides' variability in different types of wood, in different species of higher vascular plants and at different developmental stages. The lengths of the section reflect the state of knowledge of the respective topics. While there is clear evidence that matrix polysaccharides play key roles in wood, there are still large gaps in our knowledge of the interactions among these polymers and other wood cell wall components, and understanding of how they mediate diverse cell wall functions in the wood.

REFERENCES

Abe H, Funada R, Imaizumi H, Ohtani J, Fukazawa K (1995) Dynamic changes in the arrangement of cortical microtubules in conifer tracheids during differentiation. *Planta* 197:418–421.

Abedi T, Castilleux R, Nibbering P, Niittylä T (2021) The spatiotemporal distribution of cell wall-associated glycoproteins during wood formation in *Populus*. *Frontiers in Plant Sciences* 11:611607.

Åkerholm M, Salmén L (2001) Interactions between wood polymers studied by dynamic FT-IR spectroscopy. *Polymer* 42:963–969.

Altaner C, Hapca AI, Knox JP, Jarvis MC (2007) Detection of β-1-4-galactan in compression wood of Sitka spruce [*Picea sitchensis* (Bong.) Carrière] by immunofluorescence. *Holzforschung* 61:311–316.

Andersson-Gunnerås S, Mellerowicz, EJ, Love J, Segerman, B, Ohmiya Y, Coutinho PM, Nilsson P, Henrissat B, Moritz T, Sundberg B (2006) Biosynthesis of cellulose-enriched tension wood in *Populus*: global analysis of transcripts and metabolites identifies biochemical and developmental regulators in secondary wall biosynthesis. *Plant Journal* 45:144–165.

Antosiewicz DM, Purugganan MM, Polisensky DH, Braam J (1997) Cellular localization of Arabidopsis xyloglucan endotransglycosylase-related proteins during development and after wind stimulation. *Plant Physiology* 115:1319–1328.

Arend M (2008) Immunolocalization of (1,4)-β-galactan in tension wood fibers of poplar. *Tree Physiology* 28: 1263–1267.

Aspinall GO, Hirst EL, Mahomed RS (1954) Hemicellulose A of beechwood (*Fagus sylvatica*). *Journal of the Chemical Society (Resumed)* 1954: 1734–1738.

Baba K, Park YW, Kaku T, Kaida R, Takeuchi M, Yoshida M, Hosoo Y, Ojio Y, Okuyama T, Taniguchi T, Ohmiya Y, Kondo T, Shani Z, Shoseyov O, Awano T, Serada S, Norioka N, Norioka S, Hayashi T (2009) Xyloglucan for generating tensile stress to bend tree stem. *Molecular Plant* 2:893–903.

Baïer, M, Goldberg R, Catesson AM, Liberman M, Bouchemal N, Michon V, Du-Penhoat CH (1994) Pectin changes in samples containing poplar cambium and inner bark in relation to the seasonal cycle. *Planta* 193: 446–454.

Bailey IW (1959) *Contributions to Plant Anatomy*. Waltham, Massachusetts, USA: Chronica Botanica Co.

Baillères H, Chanson B, Fournier M, Tollier MT, Monties B (1995) Wood structure, chemical composition and growth strains in *Eucalyptus* clones. *Annales des Sciences Forestières* 52: 157–172.

Banasiak A, Ibatullin FM, Brumer H, Mellerowicz EJ (2014) Glycoside hydrolase activities in cell walls of sclerenchyma cells in the inflorescence stems of *Arabidopsis thaliana* visualized *in situ*. *Plants* 3: 513–525.

Bar-On YM, Phillips R, Milo R (2018) The biomass distribution on earth. *Proceedings of the National Academy of Sciences of the United States of America* 115: 6506–6511.

Berglund J, Mikkelsen D, Flanagan BM, Dhital S, Gaunitz S, Henriksson G, Lindström ME, Yakubov GE, Gidley MJ, Vilaplana F (2020) Wood hemicelluloses exert distinct biomechanical contributions to cellulose fibrillar networks. *Nature Communications* 11:4692.

Bertaud F, Holmbom B (2004) Chemical composition of earlywood and latewood in Norway spruce heartwood, sapwood and transition zone wood. *Wood Science and Technology* 38: 245–256.

Biswal AK, Atmodjo MA, Li M, Baxter HL, Yoo CG, Pu Y, Lee YC, Mazarei M, Black IM, Zhang JY, Ramanna H, Bray AL, King ZR, LaFayette PR, Pattathil S, Donohoe BS, Mohanty SS, Ryno D, Yee K, Thompson OA, Rodriguez M Jr, Dumitrache A, Natzke J, Winkeler K, Collins C, Yang X, Tan L, Sykes RW, Gjersing EL, Ziebell A, Turner GB, Decker SR, Hahn MG, Davison BH, Udvardi MK, Mielenz JR, Davis MF, Nelson RS, Parrott WA, Ragauskas AJ, Neal Stewart C Jr, Mohnen D (2018) Sugar release and growth of biofuel crops are improved by downregulation of pectin biosynthesis. *Nature Biotechnology* 36:249–257.

Biswal AK, Soeno K, Latha Gandla M, Immerzeel P, Pattathil S, Lucenius J, Serimaa R, Hahn MG, Moritz T, Jönsson LJ, Israelsson-Nordström M, and Mellerowicz EJ (2014) Aspen pectate lyase *Ptxt*PL1-27 mobilizes matrix polysaccharides from woody tissues and improves saccharification yield. *Biotechnology for Biofuels* 7:11.

Biswal AK, Hao Z, Pattathil S, Yang X, Winkeler K, Collins C, Mohanty SS, Richardson EA, Gelineo-Albersheim I, Hunt K, Ryno D, Sykes RW, Turner GB, Ziebell A, Gjersing E, Lukowitz W, Davis MF, Decker SR, Hahn MG, Mohnen D (2015) Downregulation of *GAUT12* in *Populus deltoides* by RNA silencing results in reduced recalcitrance, increased growth and reduced xylan and pectin in a woody biofuel feedstock. *Biotechnology for Biofuels* 8:41.

Bourquin V, Nishikubo N, Abe H, Brumer H, Denman S, Eklund, M, Christiernin M, Teeri, TT, Sundberg B, Mellerowicz EJ (2002) Xyloglucan endotransglycosylases have a function during the formation of secondary cell walls of vascular tissues. *Plant Cell* 14:3073–3088.

Bouveng HO, Gregg PJ, Lindberg B (1960) Position of *O*-acetyl groups in birch xylan. *Acta Chemica Scandinavica* 14:742–748.

Bouveng HO (1961) Phenylisocyanate derivatives of carbohydrates. II. Location of the *O*-acetyl groups in birch xylan. *Acta Chemica Scandinavica* 15: 96–100.

Bowling AJ, Vaughn, KC (2008) Immunocytochemical characterization of tension wood: gelatinous fibers contain more than just cellulose. *American Journal of Botany* 95:655–663.

Boyd JD (1985) The key factor in growth stress generation in trees lignification or crystallisation? *IAWA Journal* 6:139–150.

Brodzki P (1972) Callose in compression wood tracheids. *Acta Societatis Botanicorum Poloniae* 41:321–327.

Bromley JR, Busse-Wicher M, Tryfona T, Mortimer, JC, Zhang, Z, Brown DM, Dupree P (2013) GUX1 and GUX2 glucuronyltransferases decorate distinct domains of glucuronoxylan with different substitution patterns. *Plant Journal* 74:423–434.

Burgert I, Keplinger T (2013) Plant micro- and nanomechanics: experimental techniques for plant cell-wall analysis. *Journal of Experimental Botany* 64:4635- 4649.

Burton RA, Gidley MJ, Fincher GB (2010) Heterogeneity in the chemistry, structure and function of plant cell walls. *Nature Chemical Biology* 6:724–732.

Busse-Wicher M, Gomes TCF, Tryfona T, Nikolovski N, Stott K, Grantham NJ, Bolam DN, Skaf MS, Dupree P (2014) The pattern of xylan acetylation suggests xylan may interact with cellulose microfibrils as a twofold helical screw in the secondary plant cell wall of *Arabidopsis thaliana*. *Plant Journal* 79:492–506.

Busse-Wicher M, Li A, Silveira RL, Pereira CS, Tryfona T, Gomes TCF, Skaf MS, Dupree P (2016) Evolution of xylan substitution patterns in gymnosperms and angiosperms: implications for xylan interaction with cellulose. *Plant Physiology* 171:2418–2431.

Bygdell J, Srivastava V, Obudulu O, Srivastava MK, Nilsson R, Sundberg B, Trygg J, Mellerowicz EJ, Wingsle G (2017) Protein expression in developing tension wood in *Populus*, monitored at high tissue resolution. *Journal of Experimental Botany* 68:3405–3417.

Carafa A, Duckett JG, Knox JP, Ligrone R (2005) Distribution of cell-wall xylans in bryophytes and tracheophytes: new insights into basal interrelationships of land plants. *New Phytologist* 168:231–240.

Catesson A-M (1964) Origine, fonctionnement et variations cytologiques saisonnières du cambium de *l'Acer pseudoplatanus* L. (Aceraceae). *Annales des Sciences Naturelles Botanique* Paris 12e serie, 5: 229–498.

Catesson AM (1994) Cambial ultrastructure and biochemistry: changes in relation to vascular tissue differentiation and the seasonal cycle. *International Journal of Plant Sciences* 155:251–261.

Çetinkol ÖP, Smith-Moritz AM, Cheng G, Lao J, George A, Hong K, Henry R, Simmons BA, Heazlewood JL, Holmes BM (2012) Structural and chemical characterization of hardwood from tree species with applications as bioenergy feedstocks. *PLoS One* 7:12.

Chafe SC (1974) Cell wall formation and "protective layer" development in the xylem parenchyma of trembling aspen. *Protoplasma* 80:335–354.

Chaffey N, Barnett J, Barlow P (1997) Cortical microtubule involvement in bordered pit formation in secondary xylem vessel elements of *Aesculus hippocastanum* L. (Hippocastanaceae): a correlative study using electron microscopy and indirect immunofluorescence microscopy. *Protoplasma* 197:64–75.

Chaffey N, Barlow PW, J.R. Barnett JR (1998) A seasonal cycle of cell wall structure is accompanied by a cyclical rearrangement of cortical microtubules in fusiform cambial cells within taproots of *Aesculus hippocastanum* L. (Hippocastanaceae). *New Phytologist* 139:623–635.

Chaffey N, Barnett JR, Barlow P (1999) A cytoskeletal basis for wood formation in angiosperm trees: the involvement of cortical microtubules. *Planta* 208:19–30.

Chaffey N, Barnett J, Barlow P (2000) A cytoskeletal basis for wood formation in angiosperm trees: the involvement of microfilaments. *Planta* 210:890–896.

Chaffey N, Barlow, P (2002) Myosin, microtubules, and microfilaments: co-operation between cytoskeletal components during cambial cell division and secondary vascular differentiation in trees. *Planta* 214:526–536.

Chattaway MM (1949) The development of tyloses and secretion of gum in heartwood formation. *Australian Journal of Biological Sciences* 2:227–240.

Chen H, Wang JP, Liu H, Li H, Lin YJ, Shi R, Yang C, Gao J, Zhou C, Li Q, Sederoff RR, Li W, Chiang VL (2019) Hierarchical transcription factor and chromatin binding network for wood formation in black cottonwood (*Populus trichocarpa*). *Plant Cell* 31:602–626.

Chen HM, Han JJ, Cui, KM He XQ (2010) Modification of cambial cell wall architecture during cambium periodicity in *Populus tomentosa* Carr. *Trees - Structure and Function* 24:533–540.

Chong SL, Virkki L, Maaheimo H, Juvonen M, Derba-Maceluch M, Koutaniemi S, Roach M, Sundberg B, Tuomainen P, Mellerowicz EJ, Tenkanen M (2014) *O*-acetylation of glucuronoxylan in *Arabidopsis thaliana* wild type and its change in xylan biosynthesis mutants. *Glycobiology* 24:494–506.

Chong SL, Koutaniemi S, Juvonen M, Derba-Maceluch M, Mellerowicz EJ, Tenkanen M (2015) Glucuronic acid in *Arabidopsis thaliana* xylans carries a novel pentose substituent. *International Journal of Biological Macromolecules* 79:807–812.

Clair B, Ruelle J, Beauchêne J, Prévost MF, Fournier M (2006) Tension wood and opposite wood in 21 tropical rain forest species. 1. Occurrence and efficiency of the G-layer. *IAWA Journal* 27:329- 338.

Clair B, Alméras T, Pilate G, Jullien D, Sugiyama J, Riekel C (2011) Maturation stress generation in poplar tension wood studied by synchrotron radiation microdiffraction. *Plant Physiology* 155:562–570.

Clair B, Alteyrac J, Gronvold A, Espejo J, Chanson B, Alméras T (2013) Patterns of longitudinal and tangential maturation stresses in *Eucalyptus nitens* plantation trees. *Annals of Forest Science* 70:801–811.

Cosgrove DJ, Jarvis MC (2012) Comparative structure and biomechanics of plant primary and secondary cell walls. *Frontiers in Plant Science* 3:204.

Coté Jr WA, Timell TE (1967) Studies on larch arabinogalactan. III. Distribution of arabinogalactan in tamarack. *Tappi* 50:285–289.

Czaninski Y (1977) Vessel-associated cells. *IAWA Bull.* 3:51–55.

Dammström S, Salmén L, Gatenholm P (2009) On the interactions between cellulose and xylan, a biomimetic simulation of the hardwood cell wall. *BioResources* 4:3–14.

Darley CP, Forrester AM, McQueen-Mason SJ (2001) The molecular basis of plant cell wall extension. *Plant Molecular Biology*. 47:171–187.

De Micco V, Balzano A, Wheeler EA, Baas P (2016) Tyloses and gums: a review of structure, function and occurrence of vessel occlusions. *IAWA Journal* 37:186–205.

Derba-Maceluch M, Awano T, Takahashi J, Lucenius J, Ratke C, Kontro I, Busse-Wicher M, Kosik O, Tanaka R, Winzéll A, Kallas Å, Leśniewska J, Berthold F, Immerzeel P, Teeri TT, Ezcurra I, Dupree P, Serimaa R, Mellerowicz EJ (2015) Suppression of xylan transglycosylase *Ptxt*Xyn10A affects cellulose microfibril angle in secondary wall in aspen wood. *New Phytologist* 205:666–681.

Derbyshire P, McCann MC, Roberts K (2007) Restricted cell elongation in Arabidopsis hypocotyls is associated with a reduced average pectin esterification level. *BMC Plant Biology* 7:31.

Dhugga KS, Barreiro R, Whitten B, Stecca K, Hazebroek J, Randhawa GS, Dolan M, Kinney AJ, Tomes D, Nichols S, Anderson P (2004) Guar seed beta-mannan synthase is a member of the cellulose synthase super gene family. *Science* 303:363–366.

Donaldson LA, Grace J, Downes GM (2004) Within-tree variation in anatomical properties of compression wood in radiata pine. *IAWA Journal* 25:253–271.

Donaldson L, Radotić K, Kalauzi A, Djikanović D, Jeremić M (2010) Quantification of compression wood severity in tracheids of *Pinus radiata* D. Don using confocal fluorescence imaging and spectral deconvolution. *Journal of Structural Biology* 169:106–115.

Donaldson LA, Knox JP (2012) Localisation of cell wall polysaccharides in normal and compression wood of radiata pine - relationships with lignification and microfibril orientation. *Plant Physiology* 158:642–653.

Donaldson LA, Singh AP (2013) Formation and structure of compression wood. In *Cellular Aspects of Wood Formation*. ed. Fromm J. Plant Cell Monographs 225-256. Springer-Verlag.

Donev E, Gandla ML, Jönsson, LJ, Mellerowicz EJ (2018). Engineering non-cellulosic polysaccharides of wood for the biorefinery. *Frontiers in Plant Sciences* 9:1537.

Dünisch O, Richter HG, Koch G (2010) Wood properties of juvenile and mature heartwood in *Robinia pseudoacacia* L. *Wood Science and Technology* 44:301–313.

Edwards ME, Dickson CA, Chengappa S, Sidebottom C, Gidley MJ, Reid JS (1999) Molecular characterisation of a membrane-bound galactosyltransferase of plant cell wall matrix polysaccharide biosynthesis. *Plant Journal* 19:691–697.

Ermel FF, Follet-Gueye ML, Cibert C, Vian B, Morvan C, Catesson AM, Goldberg R (2000) Differential localization of arabinan and galactan side chains of rhamnogalacturonan 1 in cambial derivatives. *Planta* 210:732–740.

Esau K (1953) *Plant anatomy*. New York, NY, USA: John Wiley & Sons.

Escalante A, Gonçalves A, Bodin A, Stepan A, Sandström C, Toriz G, Gatenholm P (2012) Flexible oxygen barrier films from spruce xylan. *Carbohydrate Polymers* 87: 2381–2387.

Evert RF (2006) *Esau's Plant Anatomy: Meristems, Cells, and Tissues of the Plant Body: Their Structure, Function, and Development*. Hoboken, New Jersey: John Wiley & Sons.

Fagerstedt KV, Mellerowicz E, Gorshkova T, Ruel K and Joseleau J-P (2014) Cell wall polymers in reaction wood. In *The Biology of Reaction Wood*, ed. Gardiner B, Barnett J, Saranpää P, Gril J. 37–106. Springer Series in Wood Science, Berlin: Springer-Verlag.

Farkas V, Sulova Z, Stratilova E, Hanna R, Maclachlan G (1992) Cleavage of xyloglucan by *Nasturtium* seed xyloglucanase and transglycosylation to xyloglucan subunit oligosaccharides. *Archives of Biochemistry and Biophysics* 298: 365–370.

Felten, J, Sundberg B (2013) Biology, chemistry and structure of tension wood. In *Plant Cell Monographs*, ed Fromm J, 203–224. Berlin, Heidelberg: Springer-Verlag.

Fernandes AN, Thomas LH, Altaner CM, Callow P, Forsyth VT, Apperley DC, Kennedy CJ, Jarvis MC (2011) Nanostructure of cellulose microfibrils in spruce wood. *Proceedings of the National Academy of Sciences of the United States of America* 108:E1195–E1203.

Fisher JB, Svenson JW (1981) Occurrence of reaction wood in branches of dicotyledons and its role in tree architecture. *Botanical Gazette* 142:82–95.

Foster RC (1964) Fine structure of tyloses. *Nature* 204: 494–495.

Foster RC (1967) Fine structure of tyloses in three species of the Myrtaceae. *Australian Journal of Botany* 1:25–34.

Fry SC, Smith RC, Renwick KF, Martin DJ, Hodge SK, Matthews KJ (1992) Xyloglucan endotransglycosylase, a new wall-loosening enzyme activity from plants. *Biochemical Journal* 282: 821–828.

Fukushima K, Terashima N (1991) Heterogeneity in formation of lignin - Part XV: Formation and structure of lignin in compression wood of *Pinus thunbergii* studied by microautoradiography. *Wood Science and Technology* 25:371–381.

Funada R, Catesson AM (1991) Partial cell wall lysis and the resumption of meristematic activity in *Fraxinus excelsior* cambium. *IAWA Bulletin* n.s. 12:439–444.

Funada R, Abe H, Furusawa O, Imaizumi H, Fukazawa K, Ohtani J (1997) The orientation and localization of cortical microtubules in differentiating conifer tracheids during cell expansion. *Plant and Cell Physiology* 38:210–212.

Funada R (2001) The role of cortical microtubules in wood formation. In *Progress in Biotechnology* 18:127–135, Elsevier.

Furuya N, Takahashi S, Miyazaki M (1970) The chemical composition of gelatinous layer from the tension wood of *Populus euroamericana*. *Mokuzai Gakkaishi* 16:26–30.

Gerttula S, Zinkgraf M, Muday GK, Lewis DR, Ibatullin FM, Brumer H, Hart F, Mansfield SD, Filkov V, Groover A (2015) Transcriptional and hormonal regulation of gravitropism of woody stems in *Populus*. *Plant Cell* 27:2800–2813.

Ghislai B., Nicolini E-A, Romain R, Ruelle J, Yoshinaga A, Alford MH, Clair B (2016). Multilayered structure of tension wood cell walls in Salicaceae *sensu lato* and its taxonomic significance. *Botanical Journal of the Linnean Society* 182:744–756.

Gierlinger N, Schwanninger M (2006) Chemical Imaging of poplar wood cell walls by confocal Raman microscopy. *Plant Physiology* 140:1246–1254.

Gille S, Pauly M (2012) *O*-acetylation of plant cell wall polysaccharides. *Frontiers in Plant Sciences* 3:12.

Giummarella N, Pu Y, Ragauskas AJ, Lawoko M (2019a) A critical review on the analysis of lignin carbohydrate bonds. *Green Chemistry 21*: 1573–1595.

Giummarella N, Balakshin M, Koutaniemi S, Kärkönen A, Lawoko M (2019b) Nativity of lignin carbohydrate bonds substantiated by biomimetic synthesis. *Journal of Experimental Botany 70*:5591–5601.

Goosen-de Roo L, Bakhuizen R, van Spronsen PC, Libbenga KR (1984) The presence of extended phragmosomes containing cytoskeletal elements in fusiform cambial cells of *Fraxinus excelsior* L. *Protoplasma* 122: 145–152.

Gorišek Ž, Torelli N (1999) Microfibril angle in juvenile, adult and compression wood of spruce and silver fir. *Phyton* 39:129–132.

Gorshkova T, Mokshina N, Chernova T, Ibragimova N, Salnikov V, Mikshina P, Tryfona T, Banasiak A, Immerzeel P, Dupree P, Mellerowicz EJ (2015) Aspen tension wood fibers contain β-(1→4)-galactans and acidic arabinogalactans retained by cellulose microfibrils in gelatinous walls. *Plant Physiology* 169:2048–2063.

Goswami L, Dunlop JWC, Jungnikl K, Eder M, Gierlinger N, Coutand C, Jeronimidis G, Fratzl P, Burgert I (2008) Stress generation in the tension wood of poplar is based on the lateral swelling power of the G-layer. *Plant Journal* 56:531–538.

Goué N, Lesage-Descauses MC, Mellerowicz EJ, Magel E, Label P, Sundberg B (2008) Microgenomic analysis reveals cell type-specific gene expression patterns between ray and fusiform initials within the cambial meristem of *Populus*. *New Phytologist* 180:45–56.

Gray-Mitsumune M, Mellerowicz EJ, Abe H, McQueen-Mason S, Winzéll A, Sterky F, Blomqvist K, Schrader J, Teeri TT, Sundberg B (2004) Expansins abundant in secondary xylem belong to Subgroup A of the a-expansin gene family. *Plant Physiology* 135:1552–1564.

Gray-Mitsumune M, Blomquist K, McQueen-Mason S, Teeri TT, Sundberg B, Mellerowicz EJ (2008) Ectopic expression of a wood-abundant expansin *PttEXPA1* promotes cell expansion in primary and secondary tissues in aspen. *Plant Biotechnology Journal* 6:62–72.

Gregory RA (1978) Living elements of the conducting secondary xylem of sugar maple (*Acer saccharum* Marsh.) *IAWA Bulletin* 4:65–69.

Guedes FTP, Laurans F, Quemener B, Assor C, Lainé-Prade V, Boizot N, Vigouroux J, Lesage-Descauses MC, Leplé JC, Déjardin A, Pilate G (2017) Non-cellulosic polysaccharide distribution during G-layer formation in poplar tension wood fibers: abundance of rhamnogalacturonan I and arabinogalactan proteins but no evidence of xyloglucan. *Planta* 246:857–878.

Gustafsson C, Ollinmaa PJ, Saarnio J, Finsnes E, Sörensen JS, Sörensen NA (1952) The carbohydrates in birchwood. *Acta Chemica Scandinavica* 6:1299–1300.

Hafrén J, Fujino T, Itoh T (1999) Changes in cell wall architecture of differentiating tracheids of *Pinus thunbergii* during lignification. *Plant and Cell Physiology* 40:532–541.

Hägglund E, Lindberg B, McPherson J (1956). Dimethyl sulfoxide, a solvent for hemicellulose. *Acta Chemica Scandinavica* 10: 1160–1164.

Hannuksela T, Hervé du Penhoat C (2004) NMR structural determination of dissolved *O*-acetylated galactoglucomannan isolated from spruce thermomechanical pulp. *Carbohydrate Research* 339:301–312.

Herbette S, Bouchet B, Brunel N, Bonnin E, Cochard H, Guillon F (2015) Immunolabelling of intervessel pits for polysaccharides and lignin helps in understanding their hydraulic properties in *Populus tremula* × *alba*. *Annals of Botany* 115:187–199.

Hervé C, Rogowski A, Gilbert HJ, Knox JP (2009) Enzymatic treatments reveal differential capacities for xylan recognition and degradation in primary and secondary plant cell walls. *Plant Journal* 58:413–422.

Higaki A, Kadowaki Y, Yoshinaga A, Takabe K (2021) Xylan deposition and lignification in differentiating tension wood fibers in *Mallotus japonicus* (Euphorbiaceae) with multilayered structure. *Holzforschung* 75:1–12.

Higuchi I (1976) Biochemical aspects of lignification and heartwood formation. *Wood Research: Bulletin of the Wood Research Institute Kyoto University* 59/60:180–199.

Hillis WE (1999) The formation of heartwood and its extractives. In *Phytochemicals in Human Health Protection, Nutrition, and Plant Defense,* ed. Romeo JT 215–253. New York: Kluwer Academic/Plenum Publishers.

Hosoo Y, Yoshida M, Imai T, Okuyama T (2002) Diurnal difference in the amount of immunogold-labeled glucomannans detected with field emission scanning electron microscopy at the innermost surface of developing secondary walls of differentiating conifer tracheids. *Planta* 215:1006–1012.

Huang Y, Wang L, Chao Y, Nawawi DS, Akiyama T, Yokoyama T, Matsumoto Y (2016) Relationships between hemicellulose composition and lignin structure in woods. *Journal of Wood Chemistry and Technology* 36:9–15.

Ibatullin FI, Banasiak A, Baumann MJ, Greffe L, Takahashi J, Mellerowicz EJ, Brumer H (2009) A real-time fluorogenic assay for the visualization of glycoside hydrolase activity *in planta*. *Plant Physiology* 151:1741–1750.

Iqbal M, Ghouse AKM (1990) Cambial concept and organisation. In *The Vascular Cambium*, ed. M. Iqbal, 1–36. Taunton, England: Research Studies Press.

Jacobsen AL, Rodrigues-Zaccaro FD, Lee TF, Valdovinos J, Toschi HS, Martinez JA, Pratt RB (2015) Grapevine xylem development, architecture and function. In *Functional and Ecological Xylem Anatomy*, ed. Hacke U, 133–162. Switzerland: Springer International Publishing.

Jarvis MC (2018) Structure of native cellulose microfibrils, the starting point for nanocellulose manufacture. *Philosophical Transactions of the Royal Society A Mathematical, Physical and Engineering Sciences* 376:20170045.

Jensen JK, Johnson NR, Wilkerson CG (2014) *Arabidopsis thaliana* IRX10 and two related proteins from psyllium and *Physcomitrella patens* are xylan xylosyltransferases. *Plant Journal* 80:207–215.

Johansson MH, Samuelson O (1977) Reducing end groups in birch xylan and their alkaline degradation. *Wood Science Technology.* 11: 251–263.

Jokipii-Lukkari S, Sundell D, Nilsson O, Hvidsten TR, Street NR, Tuominen H (2017) NorWood: a gene expression resource for evo-devo studies of conifer wood development. *New Phytologist* 216:482–494.

Jones JKN, Wise LE (1952). The hemicelluloses present in aspen wood (*Populus tremuloides*). Part II. *Journal of the Chemical Society* 750:3389–3993.

Jones JKN, Merler E, Wise LE (1957) The hemicelluloses present in aspen wood (*Populus tremuloides*). Part I. The constitution of pentosan and hexosan fractions. *Canadian Journal of Chemistry* 311:634–645.

Kaku T, Serada S, Baba K, Tanaka F, Hayashi T (2009) Proteomic analysis of the G-layer in poplar tension wood. *Journal of Wood Science* 55:250–257.

Kaneda M, Rensing K, Samuels L (2010) Secondary cell wall deposition in developing secondary xylem of poplar. *Journal of Integrative Plant Biology* 52:234–243.

Kang X, Kirui A, Dickwella Widanage MC, Mentink-Vigier F, Cosgrove DJ, Wang T (2019) Lignin-polysaccharide interactions in plant secondary cell walls revealed by solid-state NMR. *Nature Communications* 10:347.

Keegan D, Kretschmer B, Elbersen B, Panoutsou C (2013) Cascading use: a systematic approach to biomass beyond the energy sector. *Biofuels, Bioproducts and Biorefining.*

Kim JS, Awano T, Yoshinaga A, Takabe K (2010a) Immunolocalization of β-1-4-galactan and its relationship with lignin distribution in developing compression wood of *Cryptomeria japonica. Planta* 232:109–119.

Kim JS, Awano T, Yoshinaga A, Takabe K (2010b) Temporal and spatial immunolocalization of glucomannans in differentiating earlywood tracheid cell walls of *Cryptomeria japonica. Planta* 232:545–554.

Kim JS, Awano T, Yoshinaga A, Takabe K (2011) Occurrence of xylan and mannan polysaccharides and their spatial relationship with other cell wall components in differentiating compression wood tracheids of *Cryptomeria japonica. Planta* 233:721–735.

Kim JS, Daniel G (2012) Distribution of glucomannans and xylans in poplar xylem and their changes under tension stress. *Planta* 236:35–50.

Kim JS, Daniel G (2013) Developmental localization of homogalacturonan and xyloglucan epitopes in pit membranes varies between pit types in two poplar species. *IAWA J.* 34:245–262.

Kim WC, Reca IB, Kim Y, Park S, Thomashow MF, Keegstra K, Han KH (2014) Transcription factors that directly regulate the expression of *CSLA9* encoding mannan synthase in *Arabidopsis thaliana. Plant Molecular Biology* 84:577–587.

Kubo M, Udagawa M, Nishikubo N, Horiguchi G, Yamaguchi M, Ito J, Mimura T, Fukuda H, Demura T (2005) Transcription switches for protoxylem and metaxylem vessel formation. *Genes and Development* 19:1855–1860.

Kumar V, Hainaut M, Delhomme N, Mannapperuma C, Immerzeel P, Street NR, Henrissat B, Mellerowicz EJ (2019) Poplar carbohydrate-active enzymes: whole-genome annotation and functional analyses based on RNA expression data. *Plant Journal* 99:589–609.

Kuo CM, Timell TE (1969) Isolation and characterization of a galactan from tension wood of American beech (*Fagus grandifolia* Ehrl). *Svensk Papperstidning-Nordisk Cellulosa* 72:703–716.

Kushwah S, Banasiak A, Nishikubo N, Derba-Maceluch M, Majda M, Endo S, Kumar V, Gomez L, Gorzsás A, McQueen-Mason S, Braam J, Sundberg B, Mellerowicz EJ (2020) Arabidopsis *XTH4* and *XTH9* contribute to wood cell expansion and secondary wall formation. *Plant Physiology* 182:1946–1965.

Lafarguette F, Leplé JC, Déjardin A, Laurans F, Costa G, Lesage-Descauses MC, Pilate G (2004) Poplar genes encoding fasciclin-like arabinogalactan proteins are highly expressed in tension wood. *New Phytologist* 164:107–121.

Larson PR (1994) *The Vascular Cambium: Development and Structure*. Springer Series in Wood Science, Berlin, Heidelberg: Springer-Verlag.

Lawoko M, Henriksson G, Gellerstedt G (2005) Structural differences between the lignin-carbohydrate complexes present in wood and in chemical pulps. *Biomacromolecules* 6:3467–3473.

Lee CL (1961) Crystallinity of wood cellulose fibres studied by X-ray methods. *Forest Products Journal* 11:108–112.

Lee C, Teng Q, Huang W, Zhong R, Ye Z-H (2009a). The poplar GT8E and GT8F glycosyltransferases are functional orthologs of Arabidopsis PARVUS involved in glucuronoxylan biosynthesis. *Plant and Cell Physiology* 50:1982–1987.

Lee C, Teng Q, Huang W, Zhong R, Ye, Z-H (2009b) Downregulation of *PoGT47C* expression in poplar results in a reduced glucuronoxylan content and an increased wood digestibility by cellulase. *Plant and Cell Physiology* 50:1075–1089.

Lee C, Teng Q, Huang W, Zhong R, Ye ZH (2011) Molecular dissection of xylan biosynthesis during wood formation in poplar. *Molecular Plant* 4:730–747.

Lee C, Teng Q, Zhong R, Ye ZH (2014) Alterations of the degree of xylan acetylation in Arabidopsis xylan mutants. *Plant Signaling and Behavior* 9:e27797.

Lee C, Teng Q, Zhong R, Yuan Y, Haghighat M, Ye ZH (2012) Three Arabidopsis DUF579 domain-containing GXM proteins are methyltransferases catalyzing 4-*O*-methylation of glucuronic acid on xylan. *Plant and Cell Physiology* 53:1934–1949.

Leśniewska J, Öhman D, Krzesłowska M, Kushwah S, Barciszewska-Pacak M, Kleczkowski LA, Sundberg B, Moritz T, Mellerowicz EJ (2017) Defense responses in aspens with altered PME activity reveal the hormonal inducers of tyloses. *Plant Physiology* 173:1409–1419.

Lev-Yadun S (2001) Intrusive growth - the plant analog of dendrite and axon growth in animals. *New Phytologist* 150:508–512.

Levesque-Tremblay G, Pelloux J, Braybrook SA, Müller K (2015) Tuning of pectin methylesterification: consequences for cell wall biomechanics and development. *Planta* 242:791–811.

Li Q, Min D, Wang JP, Peszlen I, Horvath L, Horvath B, Nishimura Y, Jameel H, Chang H-M, Chiang VL, Li C (2011) Downregulation of glycosyltransferase 8D genes in *Populus trichocarpa* caused reduced mechanical strength and xylan content in wood. *Tree Physiology* 31:226–236.

Liepman AH, Wilkerson CG, Keegstra K (2005) Expression of cellulose synthase-like (Csl) genes in insect cells reveals that CslA family members encode mannan synthases. *Proceedings of the National Academy of Sciences of the United States of America* 102:2221–2226.

Lindberg B, Rosell KG, Svensson S (1973) Position of the *O*-acetyl groups in birch xylan. *Svensk Papperstidning-Nordisk Cellulosa* 76: 30–32.

Lionetti V, Francocci F, Ferrari S, Volpi C, Bellincampi D, Galletti R, D'Ovidio R, De Lorenzo G, Cervone F (2010) Engineering the cell wall by reducing de-methyl-esterified homogalacturonan improves saccharification of plant tissues for bioconversion. *Proceedings of the National Academy of Sciences of the United States of America* 107:616–621.

Liu J, Hou J, Chen H, Pei K, Li Y, He XQ (2017) Dynamic changes of pectin epitopes in cell walls during the development of the procambium-cambium continuum in poplar. *International Journal of Molecular Sciences* 18:1716.

Liu J, Leppänen AS, Kisonen V, Willför S, Xu C, Vilaplana F (2018) Insights on the distribution of substitutions in spruce galactoglucomannan and its derivatives using integrated chemo-enzymatic deconstruction, chromatography and mass spectrometry. *International Journal of Biological Macromolecules* 112:616–625.

Lundqvist J, Teleman A, Junel L, Zacchi G, Dahlman O, Tjerneld F, Stålbrand H (2002) Isolation and characterization of galactoglucomannan from spruce (*Picea abies*). *Carbohydrate Polymers* 48:29–39.

Lyczakowski JJ, Bourdon M, Terrett OM, Helariutta Y, Wightman R, Dupree P (2019) Structural imaging of native cryo-preserved secondary cell walls reveals the presence of macrofibrils and their formation requires normal cellulose, lignin and xylan biosynthesis. *Frontiers in Plant Sciences* 10:1398.

Lyczakowski JJ, Yu L, Terrett OM, Fleischmann C, Temple H, Thorlby G, Sorieul M, Dupree P (2021) Two conifer GUX clades are responsible for distinct glucuronic acid patterns on xylan. *New Phytologist* 231:1720–1733.

MacMillan CP, Mansfield SD, Stachurski ZH, Evans R, Southerton SG (2010) Fasciclin-like arabinogalactan proteins: specialization for stem biomechanics and cell wall architecture in Arabidopsis and Eucalyptus. *Plant Journal* 62:689–703.

MacMillan CP, Taylor L, Bi Y, Southerton SG, Evans R, Spokevicius A (2015) The fasciclin-like arabinogalactan protein family of *Eucalyptus grandis* contains members that impact wood biology and biomechanics. *New Phytologist* 206:1314–1327.

Maeda Y, Awano T, Takabe K, Fujita M (2000) Immunolocalization of glucomannans in the cell of differentiating tracheids in *Chmaecyparis obtusa*. *Protoplasma* 213:148–156.

Magel EA, Höll W (1993) Storage carbohydrates and adenine nucleotides in trunks of *Fagus Sylvatica* L. in relation to discolored wood. *Holzforschung* 47:19–24.

Mahboubi A, Linden P, Hedenström M, Moritz T, Niittylä T (2015) ^{13}C tracking after $^{13}CO_2$ supply revealed diurnal patterns of wood formation in aspen. *Plant Physiology* 168:478–89.

Majda M, Kozlova L, Banasiak A, Derba-Maceluch M, Iashchishyn IA, Morozova-Roche LA, Smith RS, Gorshkova T, Mellerowicz EJ (2021) Elongation of wood fibers combines features of diffuse and tip growth. *New Phytologist* 232: 673–691.

Martínez-Abad A, Berglund J, Toriz G, Gatenholm P, Henriksson G, Lindström M, Wohlert J, Vilaplana F (2017) Regular motifs in xylan modulate molecular flexibility and interactions with cellulose surfaces. *Plant Physiology* 175:1579–1592.

Martínez-Abad A, Giummarella N, Lawoko, M Vilaplana F (2018) Differences in extractability under subcritical water reveal interconnected hemicellulose and lignin recalcitrance in birch hardwoods. *Green Chemistry* 20:2534–2546.

Martínez-Abad A, Jiménez-Quero A, Wohlert J (2020) Influence of the molecular motifs of mannan and xylan populations on their recalcitrance and organization in spruce softwoods. *Green Chemistry* 22:3956–3970.

Mast SW, Donaldson LA, Torr K, Phillips L, Flint H, West M, Strabala TJ, Wagner A (2009) Exploring the ultrastructural localization and biosynthesis of beta (1,4)-galactan in *Pinus radiata* compression wood. *Plant Physiology* 150:573–583.

McKee LS, Sunner H, Anasontzis GE, Toriz G, Gatenholm P, Bulone V, Vilaplana, F Olsson L (2016) A GH115 α-glucuronidase from *Schizophyllum commune* contributes to the synergistic enzymatic deconstruction of softwood glucuronoarabinoxylan. *Biotechnology for Biofuels* 9:2.

McQueen-Mason SJ, Cosgrove DJ (1995) Expansin mode of action on cell walls: analysis of wall hydrolysis, stress relaxation, and binding. *Plant Physiology* 107:87–100.

Meier H, Dahlén M, Frisell C (1962) Studies on a galactan from tension wood of beech (*Fagus silvatica* L.). *Acta Chemica Scandinavica* 16:2275–2283.

Mellerowicz EJ (2006) Xylem cell expansion - lessons from poplar. In *The Science and Lore of the Plant Cell Wall,* ed. Hayashi T. 267–275, Boca Raton: Universal Publishers Brown Walker Press.

Mellerowicz EJ, Gorshkova TA (2012) Tensional stress generation in gelatinous fibres: a review and possible mechanism based on cell-wall structure and composition. *Journal of Experimental Botany* 63:551–565.

Mellerowicz EJ, Baucher M, Sundberg B, Boerjan W (2001) Unravelling cell wall formation in the woody dicot stem. *Plant Molecular Biology* 47:239–274.

Mellerowicz EJ, Immerzeel P, Hayashi T (2008) Xyloglucan - the molecular muscle of trees. *Annals of Botany* 101:659–665.

Meloche CG, Knox JP, Vaughn KC (2007) A cortical band of gelatinous fibers causes the coiling of redvine tendrils: a model based upon cytochemical and immunocytochemical studies. *Planta* 225:485–498.

Mikkelsen D, Flanagan BM, Wilson SM, Bacic A, Gidley MJ (2015) Interactions of arabinoxylan and (1,3)(1,4)-β-glucan with cellulose networks. *Biomacromolecules* 16:1232–1239.

Min DY, Li Q, Chiang V, Jameel H, Chang HM, Lucia L (2014a) The influence of lignin-carbohydrate complexes on the cellulase-mediated saccharification I: Transgenic black cottonwood (western balsam poplar, California poplar) *P. trichocarpa* including the xylan down-regulated and the lignin down-regulated lines. *Fuel 119*:207–213.

Min DY, Yang C, Chiang V, Jameel H, Chang HM (2014b) The influence of lignin-carbohydrate complexes on the cellulase-mediated saccharification II: Transgenic hybrid poplars (*Populus nigra* L. and *Populus maximowiczii* A.). *Fuel 116*: 56–62.

Morris, H, Plavcová L, Gorai M, Klepsch MM, Kotowska M, Schenk HJ, Jansen S (2018) Vessel-associated cells in angiosperm xylem: Highly specialized living cells at the symplast-apoplast boundary. *American Journal of Botany* 105:1–10.

Mortimer JC, Faria-Blanc N, Yu X, Tryfona T, Sorieul M, Ng YZ, Zhang Z, Stott K, Anders N, Dupree P (2015) An unusual xylan in Arabidopsis primary cell walls is synthesised by GUX3, IRX9L, IRX10L and IRX14. *Plant Journal* 83:413–426.

Murakami Y, Funada R, Sano Y, Ohtani J (1999) The differentiation of contact cells and isolation cells in the xylem ray parenchyma of *Populus maximowiczii*. *Annals of Botany* 84:429–435.

Murmanis LL (1978) Breakdown of end walls in differentiating vessels of secondary xylem in *Quercus rubra* L. *Annals of Botany* 42:679–682.

Nakaba S, Sano Y, Kubo T, Funada R (2006) The positional distribution of cell death of ray parenchyma in a conifer, *Abies sacchalinensis*. *Plant Cell Reports* 25:1143–1148.

Nawari DS., Syafii W, Akiyama T, Matsumoto Y (2016) Characteristics of guaiacyl-syringyl lignin in reaction wood in the gymnosperm *Gnetum gnemon* L. *Holzforschung.* 70: 593–602.

Nieduszynski I, Marchessault RH (1971) Structure of β-D-(1→4′) xylan hydrate. *Nature* 232:46–47.

Nishikubo N, Awano T, Banasiak A, Bourquin V, Ibatullin F, Funada R, Brumer H, Teeri TT, Hayashi T, Sundberg B & Mellerowicz EJ (2007) Xyloglucan endo-transglycosylase (XET) functions in gelatinous layers of tension wood fibers in poplar - a glimpse into the mechanism of the balancing act of trees. *Plant and Cell Physiology* 48:843–855.

Nishikubo N, Takahashi J, Roos AA, Derba-Maceluch M, Piens K, Brumer H, Teeri TT, Stålbrand H, Mellerowicz EJ (2011) Xyloglucan endo-transglycosylase-mediated xyloglucan rearrangements in developing wood of hybrid aspen. *Plant Physiology* 155:399–413.

Nishitani, K, Tominaga R (1992) Endo-xyloglucan transferase, a novel class of glycosyltransferase that catalyzes transfer of a segment of xyloglucan molecule to another xyloglucan molecule. *Journal of Biological Chemistry* 267:21058–21064.

Norberg PH, Meier H (1966) Physical and chemical properties of the gelatinous layer in tension wood fibres of aspen (*Populus tremula* L.). *Holzforschung* 20:174–178.

Obst JR, Sachs, IB Kuster TA (1988) The quantity and type of lignin in tyloses of bur oak (*Quercus macrocarpa*). *Holzforschung* 42:229–231.

Oda Y, Iida Y, Kondo Y, Fukuda H (2010) Wood cell-wall structure requires local 2D-microtubule disassembly by a novel plasma membrane-anchored protein. *Current Biology* 20:1197–202.

Oda Y, Fukuda H (2012) Initiation of cell wall pattern by a Rho- and microtubule-driven symmetry breaking. *Science* 337:1333–1336.

O'Dwyer MH (1934) The hemicelluloses of the wood of English oak: The composition and properties of hemicellulose A, isolated from samples of wood dried under various conditions. *Biochemical Journal* 28: 2116–2124.

O'Dwyer MH (1939) The hemicelluloses of the wood of English oak: The structure of hemicellulose A. *Biochemical Journal* 33:713–717.

Ohtani M, Nishikubo N, Xu B, Yamaguchi, M, Mitsuda N, Goué N, Shi F, Ohme-Takagi M, Demura T (2011) A NAC domain protein family contributing to the regulation of wood formation in poplar. *Plant Journal* 67:499–512.

Okazawa K, Sato Y, Nakagawa T, Asada K, Kato I, Tomita E, Nishitani K (1993) Molecular cloning and cDNA sequencing of endoxyloglucan transferase, a novel class of glycosyltransferase that mediates molecular grafting between matrix polysaccharides in plant cell walls. *Journal of Biological Chemistry* 44:83–89.

Okuyama T, Yamamoto H, Yoshida M, Hattori Y, Archer RR (1994) Growth stresses in tension wood: role of microfibrils and lignification. *Annales des Sciences Forestières* 51:291–300.

Onaka F (1949) Studies on compression and tension wood. *Wood Research, Bulletin of the Wood Research Institute, Kyoto University* 24:1–88.

Park YW, Baba K, Furuta Y, Iida I, Sameshima K, Arai M, Hayashi T (2004) Enhancement of growth and cellulose accumulation by overexpression of xyloglucanase in poplar. *FEBS Letters* 564:183–187.

Pawar P M-A, Ratke C, Balasubramanian VB, Chong SL, Gandla ML, Adriasola M, Sparrman T, Hedenström M, Szwaj K, Derba-Maceluch M, Ezcurra I, Tenkanen M, Jönsson LJ, Mellerowicz EJ (2017) Downregulation of *RWA* genes in hybrid aspen affects xylan acetylation and wood processing properties. *New Phytologist* 214:1491–1505.

Pearce RB, Holloway PJ (1984) Suberin in the sapwood of oak (*Quercus robur* L.): its composition from a compartmentalization barrier and its occurrence in tyloses in undecayed wood. *Physiological Plant Pathology* 24: 71–81.

Peaucelle A, Braybrook SA, Le Guillou L, Bron E, Kuhlemeier C, Höfte H (2011) Pectin induced changes in cell wall mechanics underlie organ initiation in Arabidopsis. *Current Biology* 21:1720–1726.

Peaucelle, A, Wightman R, Höfte H (2015) The control of growth symmetry breaking in the arabidopsis hypocotyl. *Current Biology* 25:1746–1752.

Pereira CS, Silveira RL, Dupree P, Skaf MS (2017) Effects of xylan side-chain substitutions on xylan-cellulose interactions and implications for thermal pretreatment of cellulosic biomass. *Biomacromolecules* 18:1311–1321.

Pilate G, Chabbert B, Cathala B, Yoshinaga A, Leplé JC, Laurans F, Lapierre C, Ruel K (2004) Lignification and tension wood. *Comptes Rendus - Biologies* 327:889–901.

Plavcová L, Hacke UG (2011) Heterogeneous distribution of pectin epitopes and calcium in different pit types of four angiosperm species. *New Phytologist* 192:885–897.

Pramod S, Rajput KS, Rao KS (2019) Immunolocalization of β-(1-4)-D-galactan, xyloglucans and xylans in the reaction xylem fibres of *Leucaena leucocephala* (Lam.) de Wit. *Plant Physiology and Biochemistry* 142:217–223.

Pramod S, Rao KS, Rajput KS (2021) Anatomy and cell wall chemistry of tension wood in *Hibiscus cannabinus*. *Bosque* 42:99–110.

Raggi S, Ferrarini A, Delledonne M, Dunand C, Ranocha P, De Lorenzo G, Cervone F, Ferrari S (2015) The Arabidopsis class III peroxidase *At*PRX71 negatively regulates growth under physiological conditions and in response to cell wall damage. *Plant Physiology* 169:2513–2525.

Rao KS, Kim, JS, Kim YS (2011) Early changes in the radial walls of storied fusiform cambial cells during fibre differentiations. *IAWA J.* 32:333–340.

Ratke C, Pawar PMA, Balasubramanian VK, Naumann M, Duncranz ML, Derba-Maceluch M, Gorzsás A, Endo S, Ezcurra I, Mellerowicz EJ (2015) Populus *GT43* family members group into distinct sets required for primary and secondary wall xylan biosynthesis and include useful promoters for wood modification. *Plant Biotechnology Journal* 13:26–37.

Ratke C, Terebieniec BK, Winestrand S, Derba-Maceluch M, Grahn T, Schiffthaler B, Ulvcrona T, Özparpucu M, Rüggeberg M, Lundqvist SO, Street NR, Jönsson LJ, Mellerowicz EJ (2018) Downregulating aspen xylan biosynthetic GT43 genes in developing wood stimulates growth via reprograming of the transcriptome. *New Phytologist* 219:230–245.

Rennie EA, Scheller HV (2014) Xylan biosynthesis. *Current Opinion in Biotechnology* 26:100–107.

Rosell KG, Svensson S (1975). Studies of the distribution of the *O*-methyl- α-D-glucuronic acid residues in birch xylan. *Carbohydrate Research* 42:297–304.

Ruel K, Chevalier-billosta V, Guillemin F, Sierra JB, Joseleau J-P (2006) The wood cell wall at the ultrastructural scale - formation and topochemical organization. *Maderas.Ciencia y Tecnología* 8:107–116.

Ruelle JB, Clair B, Beauchene, J MF, Fournier M (2006) Tension wood and opposite wood in 21 tropical rain forest species 2. Comparison of some anatomical and ultrastructural criteria. *IAWA Journal* 27:341–376.

Ruelle J, Yoshida M, Clair B, Thibaut B (2007) Peculiar tension wood structure in *Laetia procera* (Poepp.) Eichl. (Flacourtiaceae). *Trees - Structure and Function* 21:345–355.

Rüggeberg M, Saxe F, Metzger TH, Sundberg B, Fratzl P, Burgert I (2013) Enhanced cellulose orientation analysis in complex model plant tissues. *Journal of Structural Biology* 183:419–428.

Ryden P, Sugimoto-Shirasu K, Smith AC, Findlay K, Reiter WD, McCann MC (2003) Tensile properties of Arabidopsis cell walls depend on both a xyloglucan cross-linked microfibrillar network and rhamnogalacturonan II-borate complexes. *Plant Physiology* 132:1033–1040.

Sachs I, Kuntz J, Ward J, Nair G, Schultz N (1970) Tyloses structure. *Wood and Fiber* 2:259–268.

Sampedro J, Carey RE, Cosgrove DJ (2006) Genome histories clarify evolution of the expansin superfamily: new insights from the poplar genome and pine ESTs. *Journal of Plant Research* 119:11–21.

Samuels AL, Rensing KH, Douglas CJ, Mansfield SD, Dharmawardhana DP, Ellis BE (2002) Cellular machinery of wood production: differentiation of secondary xylem in *Pinus contorta* var. latifolia. *Planta* 216:72–82.

Samuels AL, Kaneda M, Rensing KH (2006) The cell biology of wood formation: from cambial divisions to mature secondary xylem. *Canadian Journal of Botany* 84:631–639.

Sandquist D, Filonova L, von Schantz L, Ohlin M, Daniel G (2010) Microdistribution of xyloglucan in differentiating poplar cells. *BioResources* 5:796–807.

Schaffer K, Wisniewski M (1989) Development of the amorphous layer (protective layer) in xylem parenchyma of cv. golden delicious apple, cv. loring peach, and willow. *American Journal of Botany* 76:1569–1582.

Scheller HV, Ulvskov P (2010) Hemicelluloses. *Annual Review of Plant Biology* 61:263–289.

Schmitt U, Richter HG, Muche C (1997) TEM study of wound-induced vessel occlusions in European ash (*Fraxinus excelsior* L.). *IAWA Journal* 18:401–404.

Schmitt U, Singh A, Frankenstein C, Möller R (2006) Cell wall modifications in woody stems induced by mechanical stress. *New Zealand Journal of Forestry Science* 36:72–86.

Schrader J, Nilsson J, Mellerowicz E, Berglund A, Nilsson P, Hertzberg M, Sandberg G (2004) A high-resolution transcript profile across the wood-forming meristem of poplar identifies potential regulators of cambial stem cell identity. *Plant Cell* 16: 2278–2292.

Scurfield G (1973) Reaction wood: its structure and function. *Science* 179:647–655.

Shi R, Wang JP, Lin YC, Li Q, Sun YH, Chen H, Sederoff RR, Chiang VL (2017) Tissue and cell-type co-expression networks of transcription factors and wood component genes in *Populus trichocarpa*. *Planta* 245:927–938.

Shibata N, Harada, H Saiki H (1982) Development and structure of traumatic tyloses in *Quercus serrata* Thunb. II. Developing process of the traumatic tylosis wall. *Mokuzai Gakkaishi* 28:1–9.

Siedlecka A, Wiklund S, Péronne MA, Micheli F, Leśniewska J, Sethson I, Edlund U, Richard L, Sundberg B, Mellerowicz EJ (2008) Pectin methyl esterase inhibits intrusive and symplastic cell growth in developing wood cells of *Populus*. *Plant Physiology* 146:554–65.

Simmons TJ, Mortimer JC, Bernardinelli OD, Pöppler AC, Brown SP, DeAzevedo ER, Dupree R, Dupree P (2016) Folding of xylan onto cellulose fibrils in plant cell walls revealed by solid-state NMR. *Nature Communications* 7:13902.

Simson BW, Timell TE (1978a) Polysaccharides in cambial tissues of *Populus tremuloides* and *Tilia americana*. 1. Isolation, fractionation, and chemical composition of the cambial tissues. *Cellulose Chemistry and Technology* 12:39–50.

Simson BW, Timell TE (1978b) Polysaccharides in cambial tissues of *Populus tremuloides* and *Tilia americana*. II. Isolation and structure of a xyloglucan. *Cellulose Chemistry and Technology* 12:51–62.

Simson BW, Timell TE (1978c) Polysaccharides in cambial tissues of *Populus tremuloides* and *Tilia americana*. IV. 4-*O*-methylglucuronoxylan and pectin. *Cellulose Chemistry and Technology* 12:79–84.

Simson BW, Timell TE (1978d) Polysaccharides in cambial tissues of *Populus tremuloides* and *Tilia americana*. V. *Cellulose Chemistry and Technology* 12:137–141.

Sinnott EW, Bloch R (1939) Changes in intercellular relationships during the growth and differentiation of living plant tissues. *American Journal of Botany* 26:625–634.

Sinnott EW, Bloch R (1940) Cytoplasmic behavior during division of vacuolate plant cells. *Proceedings of the National Academy of Sciences* 26:223–227.

Sjöström E (1992) *Wood Chemistry: Fundamentals and Applications*. San Diego, California: Academic Press, Inc.

Song D, Gui J, Liu C, Sun J, Li L (2016). Suppression of *PtrDUF579-3* expression causes structural changes of the glucuronoxylan in *Populus*. *Frontiers in Plant Sciences* 7:493.

Špániková S, Biely P (2006) Glucuronoyl esterase-novel carbohydrate esterase produced by *Schizophyllum commune*. *FEBS letters 580*:4597–4601.

Sugiyama K, Okuyama T, Yamamoto H, Yoshida M (1993) Generation process of growth stresses in cell walls: relation between longitudinal released strain and chemical composition. *Wood Science and Technology* 27:257–262.

Sugiyama Y, Nagashima Y, Wakazaki M, Sato M, Toyooka K, Fukuda H, Oda Y (2019) A Rho-actin signaling pathway shapes cell wall boundaries in Arabidopsis xylem vessels. *Nature Communications* 10:468.

Sulová Z, Baran R, Farkaš V (2001) Release of complexed xyloglucan endotransglycosylase (XET) from plant cell walls by a transglycosylation reaction with xyloglucan-derived oligosaccharides. *Plant Physiology and Biochemistry* 39:927–932.

Sundell D, Street NR, Kumar M, Mellerowicz EJ, Kucukoglu M, Johnsson C, Kumar V, Mannapperuma C, Delhomme N, Nilsson O, Tuominen H, Pesquet E, Fischer U, Niittylä T, Sundberg B, Hvidsten TR (2017) Aspwood: high-spatial-resolution transcriptome profiles reveal uncharacterized modularity of wood formation in *Populus tremula*. *Plant Cell* 29:1585–1604.

Suzuki S, Li L, Sun YH, Chiang VL (2006) The cellulose synthase gene superfamily and biochemical functions of xylem-specific cellulose synthase-like genes in *Populus trichocarpa*. *Plant Physiology* 142:1233–1245.

Tarasov D, Leitch M, Fatehi P (2018) Lignin-carbohydrate complexes: properties, applications, analyses, and methods of extraction: a review. *Biotechnology for Biofuels* 11:1–28.

Taylor-Teeples M, Lin L, de Lucas M, Turco G, Toal TW, Gaudinier A, Young NF, Trabucco GM, Veling MT, Lamothe R, Handakumbura PP, Xiong G, Wang C, Corwin J, Tsoukalas A, Zhang L, Ware D, Pauly M, Kliebenstein DJ, Dehesh K, Tagkopoulos I, Breton G, Pruneda-Paz JL, Ahnert SE, Kay SA, Hazen SP, Brady SM (2015) An Arabidopsis gene regulatory network for secondary cell wall synthesis. *Nature* 517:571–575.

Teleman A (2009) Hemicelluloses and pectins. In *Wood Chemistry and Wood Biotechnology*, eds Gellerstedt G, Ek M, Henriksson G, 101–120, Berlin: De Gruyter.

Terashima N, Kitano, K, Kojima M, Yoshida M, Yamamoto H, Westermark U (2009) Nanostructural assembly of cellulose, hemicellulose, and lignin in the middle layer of secondary wall of ginkgo tracheid. *Journal of Wood Science* 55: 409–416.

Terrett OM, Dupree P (2019) Covalent interactions between lignin and hemicelluloses in plant secondary cell walls. *Current Opinion in Biotechnology 56*:97–104.

Terrett OM, Lyczakowski JJ, Yu L, Iuga D, Franks WT, Brown SP, Dupree R, Dupree P (2019) Molecular architecture of softwood revealed by solid-state NMR. *Nature Communications* 10:4978.

Timell TE (1964) Wood hemicelluloses: Part 1. *Advances in Carbohydrate Chemistry* 19:247–302.

Timell TE (1967) Recent progress in the chemistry of wood hemicelluloses. *Wood Science and Technology* 1:45–70.

Timell TE (1969) The chemical composition of tension wood. *Svensk Papperstidning-Nordisk Cellulosa* 72:173–181.

Timell TE (1986) *Compression Wood in Gymnosperms*. Springer, Berlin.

Tokareva EN, Pranovich AV, Fardim P, Daniel G, Holmbom B (2007) Analysis of wood tissues by time-of-flight secondary ion mass spectrometry. *Holzforschung* 61:647–655.

Urbanowicz BR, Peña MJ, Moniz HA, Moremen, KW, York WS (2014). Two Arabidopsis proteins synthesize acetylated xylan in vitro. *Plant Journal* 80:197–206.

Urbanowicz BR, Pena MJ, Ratnaparkhe S, Avci U, Backe J, Steet HF, Foston M, Li H, O'Neill MA, Ragauskas AJ (2012) 4-*O* methylation of glucuronic acid in Arabidopsis glucuronoxylan is catalyzed by a domain of unknown function family 579 protein. *Proceedings of the National Academy of Sciences of the United States of America* 109:14253–14258.

Verhertbruggen Y, Yin L, Oikawa A, Scheller HV (2011) Mannan synthase activity in the CSLD family. *Plant Signaling and Behavior* 6:1620–1623.

Viëtor RJ, Renard CMGC, Goldberg R, Catesson AM (1995) Cell-wall polysaccharides in growing poplar bark tissue. *International Journal of Biological Macromolecules* 17: 341–344.

Wang H, Jin Y, Wang C, Li B, Jiang C, Sun Z, Zhang Z, Kong F, Zhang H (2017) Fasciclin-like arabinogalactan proteins, PtFLAs, play important roles in GA-mediated tension wood formation in *Populus*. *Scientific Reports* 7:6182.

Wang H, Yang H, Wen Z, Gao C, Gao Y, Tian Y, Xu Z, Liu X, Persson S, Zhang B, Zhou Y (2022) Xylan-based nanocompartments orchestrate plant vessel wall patterning. *Nature Plants* 8:295–306.

Wardrop AB, Dadswell HE (1948) The nature of reaction wood: I. The structure and properties of tension wood fibres. *Australian Journal of Scientific Research* 1:1–16.

Wardrop AB, Dadswell HE (1950) The nature of reaction wood: II. The cell wall organization of compression wood tracheids. *Australian Journal of Scientific Research* 3:1–13.

Waterkeyn L, Caeymaex S, Decamps E (1982) La callose des trachéides du bois de compression chez *Pinus silvestris* et *Larix decidua*. *Bulletin de la Société Royale de Botanique de Belgique* 115:149–155.

Wenham MW, Cusick F (1975) The growth of secondary wood fibres. *New Phytologist* 74: 247–261.

Whitney SEC, Gothard MGE, Mitchell JT, Gidley MJ (1999) Roles of cellulose and xyloglucan in determining the mechanical properties of primary plant cell walls. *Plant Physiology* 121: 657–664.

Willför S, Sundberg A, Hemming J, Holmbom B (2005a) Polysaccharides in some industrially important hardwood species. *Wood Science and Technology* 39:601–617.

Willför S, Sundberg A, Hemming J, Holmbom B (2005b) Polysaccharides in some industrially important softwood species. *Wood Science and Technology* 39:245–258.

Xu W, Purugganan MM, Polisensky DH, Antosiewicz DM, Fry SC, Braam J (1995) Arabidopsis *TCH4*, regulated by hormones and the environment, encodes a xyloglucan endotransglycosylase. *Plant Cell* 7:1555–1567.

Yamamoto H (1998) Generation mechanism of growth stresses in wood cell walls: roles of lignin deposition and cellulose microfibril during cell wall maturation. *Wood Science and Technology* 32:171–182.

Yamamoto K, Fukazawa K, Ishida S (1978) Study on the cell wall development of ray parenchyma in genus *Pinus*: Report 2. The season of cell wall thickening and lignification in *Pinus strobus*. *Research Bulletins of the College Experiment Forests, Hokkaido University* 35:99–108.

Yamamoto M, Saito T, Isogai A, Kurita M, Kondo T, Taniguchi T, Kaida R, Baba K, Hayashi T (2011) Enlargement of individual cellulose microfibrils in transgenic poplars overexpressing xyloglucanase. *Journal of Wood Science* 57:71–75.

Yeh T-F, Goldfarb B, Chang H-M, Peszlen I, Braun JL, Kadla JF (2005) Comparison of morphological and chemical properties between juvenile wood and compression wood of loblolly pine. *Holzforschung* 59:669–674.

Yeh TF, Braun JL, Goldfarb B, Chang HM, Kadla JF (2006) Morphological and chemical variations between juvenile wood, mature wood, and compression wood of loblolly pine (*Pinus taeda* L.). *Holzforschung* 60:1–8.

Yin L, Verhertbruggen Y, Oikawa A, Manisseri C, Knierim B, Prak L, Jensen JK, Knox JP, Auer M, Willats WG, Scheller HV (2011) The cooperative activities of CSLD2, CSLD3, and CSLD5 are required for normal Arabidopsis development. *Molecular Plant* 4:1024–1037.

Yoshida M, Ohta H, Yamamoto H, Okuyama T (2002) Tensile growth stress and lignin distribution in the cell walls of yellow poplar, *Liriodendron tulipifera* Linn. *Trees - Structure and Function* 16:457–464.

Yoshizawa N, Inami A, Miyake S, Ishiguri F, Yokota S (2000) Anatomy and lignin distribution of reaction wood in two *Magnolia* species. *Wood Science and Technology* 34:183–196.

Young RA (2008). Historical developments in wood chemistry. *Turkish Journal of Forestry* 9:1–15.

Yu L, Lyczakowski JJ, Pereira CS, Kotake T, Yu X, Li A, Mogelsvang S, Skaf MS, Dupree P (2018) The patterned structure of galactoglucomannan suggests it may bind to cellulose in seed mucilage. *Plant Physiology* 178:1011–1026.

Yuan Y, Teng Q, Zhong R, Ye Z-H (2014a) Identification and biochemical characterization of four wood-associated glucuronoxylan methyltransferases in *Populus*. *PLoS One* 9:e87370.

Yuan Y, Teng Q, Lee C, Zhong R, Ye Z-H (2014b) Modification of the degree of 4-*O*-methylation of secondary wall glucuronoxylan. *Plant Science* 219-220:42–50.

Zeng W, Lampugnani ER, Picard KL, Song L, Wu, AM, Farion IM, Zhao J, Ford K, Doblin MS, Bacic, A (2016) Asparagus IRX9, IRX10, and IRX14A are components of an active xylan backbone synthase complex that forms in the Golgi apparatus. *Plant Physiology* 171:93–109.

Zhang M, Chavan RR, Smith BG, McArdle BH, Harris PJ (2016) Tracheid cell-wall structures and locations of (1→4)-β-D-galactans and (1→3)-β-D-glucans in compression woods of radiata pine (*Pinus radiata* D. Don). *BMC Plant Biology* 16:194.

Zhao Y, Song D, Sun J, Li L (2013) *Populus* endo-beta-mannanase *Ptr*MAN6 plays a role in coordinating cell wall remodeling with suppression of secondary wall thickening through generation of oligosaccharide signals. *Plant Journal* 74:473–485.

Zheng P, Aoki D, Yoshida M, Matsushita Y, Imai T, Fukushima K (2014) Lignification of ray parenchyma cells in the xylem of *Pinus densiflora*. Part I: Microscopic investigation by POM, UV microscopy, and TOF-SIMS. *Holzforschung* 68:897–905.

Zhong R, Lee C, Zhou J, McCarthy RL, Ye ZH (2008) A battery of transcription factors involved in the regulation of secondary cell wall biosynthesis in Arabidopsis. *Plant Cell* 20:2763–2782.

Zhong R, McCarthy RL, Lee C, Ye ZH (2011) Dissection of the transcriptional program regulating secondary wall biosynthesis during wood formation in poplar. *Plant Physiology* 157:1452–1468.

Zhong R, Cui D, Phillips DR, Ye ZH (2018) A novel rice xylosyltransferase catalyzes the addition of 2-*O*-xylosyl side chains onto the xylan backbone. *Plant and Cell Physiology* 59:554–565.

22 Drying Out Walls
How Do the Cell Walls of Resurrection Plants Survive Desiccation?

John P. Moore
Stellenbosch University

Maïté Vicré
Université de ROUEN Normandie

Eric Nguema-Ona
Agro Innovation International TIMAC AGRO

Azeddine Driouich
Université de ROUEN Normandie

Jill M. Farrant
University of Cape Town

INTRODUCTION

Vegetative desiccation tolerance is the remarkable ability of some plants to survive drying to or below an absolute water potential of −100 MPa, or loss of >90% cellular water, without dying (Bewley, 1979; Oliver et al., 2020). Although rare in angiosperms (ca. 240 out of ca. 3,00,000 known species), this trait is present in every major lineage of land plants, except gymnosperms. It is an ancestral adaptation that facilitated the colonisation of terrestrial habitats by land plants. It is postulated that this trait was lost (or suppressed) in the ancestors of vascular plants as a trade-off for more complex water management and transport systems, but reappeared in the lycophytes, ferns, and at least 13 angiosperm lineages (Gaff and Oliver 2013). Despite this widespread taxonomic distribution, vegetative desiccation tolerance is a relatively rare trait confined to select lineages that tend to occur in very xeric habitats, unsuitable for most other organisms, where they are subjected to long and/or frequent periods in a desiccated state (Alpert 2000; Porembski and Barthlott 2000). This extreme phenotype has gained much attention in recent years, not least because "resurrection plants", as they are commonly called, serve as models for the production of crops with extreme tolerance to extended droughts (Farrant et al. 2015; Hilhorst and Farrant 2018; Farrant and Hilhorst 2021). Recent reviews of the field not only show considerable increase in understanding of the complex multigenic traits underlying desiccation tolerance in different species but also point to the large gaps in our understanding of the phenomenon (e.g. Challabathula et al. 2016; Blomstedt et al. 2018; Zhang and Bartels 2018; Oliver et al. 2020). It is clear, for example, that there are core similarities among desiccation-tolerant plants that have in common ancestral features (Oliver et al. 2020), and in angiosperms many of the similarities are seed associated (Costa et al. 2017; VanBuren et al. 2017; Lyall and Gechev 2020) leading to the postulation that in vegetative tissues of trachyophytes their occurred a rewiring of ancestral desiccation tolerance pathways maintained in seeds and spores (Illing et al. 2005; Farrant and Moore 2011; Costa et al. 2017; VanBuren 2017).

It is predicted that by 2050, southern Africa could face a major reduction in arable land due to long-term drought (Dai 2013). Such dire water shortages are expected to periodically increase over the next decade and many strategies, political, economic and scientific, will be needed to meet the imbalance of increasing global human population and decreasing available natural resources as expounded by the Intergovernmental Panel on Climate Change. Southern Africa is an arid region; much of the arable land occurs along the southern and eastern coasts, which shifts from a temperate-Mediterranean climate to sub-tropical climate. Southern Africa also supports some of the most biologically diverse regions on the planet, with plant species in abundance and richness only matched by the Amazon basin. An example of this biodiversity is found in the unusually named 'resurrection plants', which may hold solutions to crop survival of extreme drought stress (Farrant et al. 2015; Hilhorst and Farrant 2018; Farrant and Hilhorst 2021). They are so named because of their ability to survive desiccation of their vegetative tissues to an air-dry state, whereby they survive in a quiescent/dormant state until water is re-supplied

DOI: 10.1201/9781003178309-22

(in nature by rain), and they 'miraculously' rehydrate and resume full metabolic activity within hours (see Moore et al. 2009 and Farrant and Hilhorst 2021 for online videos). Southern Africa particularly is home to the most genetically diverse group of angiosperm resurrection plant species, e.g., *Myrothamnus* spp., *Xerophyta* spp., *Craterostigma* spp., *Sporobolus* spp., *Eragrostis* spp., etc., found anywhere on the planet (Farrant and Moore 2011) and much research has focused on these species. It is also clear that there are considerable differences among resurrection plants in the processes associated with preventing damage associated with extreme water deficit and in the precise nature of ultimate quiescence obtained (Oliver et al. 2020; Farrant and Hilhorst 2021). Given the significant degree of tissue compaction that occurs during dehydration of resurrection plants and that cell walls make up a considerable proportion of plant biomass (e.g. as cellulose, hemicellulose and pectin polymers commonly), it is logical to ask how these plants achieve this reversible dehydration feat. In this review, we seek to present a comprehensive account of studies that have investigated the cell wall changes that occur in resurrection plants during dehydration and upon rehydration and the consequences thereof, which are linked not only to the signalling and detection of water stress *via* mechanoperception but also the structural adaptations that the plant cell walls of these species must undergo in order to 'resurrect' when water is imbibed.

HOW DO THE CELL WALLS OF ANGIOSPERM RESURRECTION PLANTS UNDERGO REVERSIBLE DESICCATION? – A HISTORICAL PERSPECTIVE

Webb and Arnott (1982) were the first, as far as we are aware, to consider the cell wall folding occurring in seeds to be linked to wall ultrastructural and compositional changes. Cell wall folding and unfolding during the dehydration–rehydration cycle of angiosperm resurrection plants was first studied in depth in the dicot species *Craterostigma wilmsii* by Vicré et al. (1999). Such folding is believed to prevent disruption of the plasma membrane by minimising the tension between it and the cell wall at plasmodesmata junctions and thereby preventing plasmolysis-induced mechanical stress (Vicré et al. 1999, 2004). Vicré et al. were also able to show that cell wall folding occurs simultaneously with a modification in the organisation of xyloglucan-associated epitopes within the wall. Such modification in the cell wall architecture appears at the onset of dehydration and subsequently increases upon severe water loss to remain constant to air dryness. Xyloglucans are essential in cell wall strengthening and Vicré et al. hypothesised that a partial hydrolysis could occur in order to make the cell wall more flexible (Vicré et al. 1999, 2004). In *C. wilmsii*, additional modifications of the pectin matrix are believed to occur, notably strengthening *via* increased cross-linkages with calcium ions (e.g. *via* egg-boxes) at least at the end of dehydration (Vicré et al. 1999, 2004). Hence, the major contribution by Vicré et al. was the idea that xyloglucan reorganisation and calcium deposition were important factors in the desiccation tolerance of resurrection plant cell walls. Jones and McQueen-Mason (2004) were also able to show the expression and activity of expansin (EXP) encoding genes (EXPs are proteins that lack detectable enzyme activities but still loosen hemicellulose networks in cell walls) in the dehydration and rehydration process in the related species *Craterostigma plantagineum*. The second major study on resurrection cell walls was performed by Moore et al. (2006) on the woody angiosperm resurrection plant *Myrothamnus flabellifolia*, which does not fold walls as dramatically as observed in the Craterostigma species. *M. flabellifolia* shows an unusual plant cell wall composition being particularly rich in pectin-associated arabinans and arabinogalactan polymers (Moore et al. 2006, 2008). When plant cell walls lose water, the structural polysaccharides of the wall move into close proximity and irreversibly adhere; this is a major challenge that resurrection plants have to overcome (discussed in Moore et al. 2006, 2008, 2013). Moore et al. proposed that this species has evolved an arabinan-rich cell wall, which affords it the capacity to be constitutively prepared for extreme water loss by preventing polymer adhesion and irreversible associations to occur (Moore et al. 2006, 2008). This pectin plasticising role for arabinans and AGPs has been expounded and discussed in detail by Moore et al. The most recent study by Moore et al. (2013) surveyed a range of South African resurrection plants (i.e. *Mohria cafforum*, *Craterostigma plantagineum*, *Myrothamnus flabellifolia*, *Xerophyta viscosa*, *Xerophyta schlecterii*, *Xerophyta viscosa* and *Eragrostis nindensis*) and confirmed a pectin plasticising strategy, characterised by a high concentration of arabinans and arabinogalactan polymers associated to pectin. The study, modelled in Figure 22.1, concluded arabinosylation of key polymers, i.e., pectin arabinans, arabinogalactan proteins and arabinoxylans, as a means to achieve plasticity/flexibility in diverse families and genera of angiosperm resurrection plants (modelled in Figure 22.1, Moore et al. 2013). Further in-depth studies by Bartels et al. focusing on wall signalling mechanisms in *Craterostigma plantagineum* will be discussed on in more depth later in this chapter. Despite the employment of differing general desiccation tolerance strategies, what appears to be a *leitmotiv* in both monocot and dicot resurrection plants (accounting for anatomical differences and morphology across plant lineages) is cell wall flexibility, the degree of which can vary among species but which collectively confers the ability to reversibly fold upon the extreme water loss associated with desiccation-prone environments.

SOME INSIGHTS INTO CELL WALLS OF DESICCATION-TOLERANT AND SENSITIVE SPIKE MOSSES

Like angiosperm resurrection plants, certain Selaginella species such as the well-known *Selaginella lepidophylla* (also named dinosaur plant, as it survived massive plant extinction) are also able to tolerate and survive desiccation for extended periods. Being evolutionary ancestral to

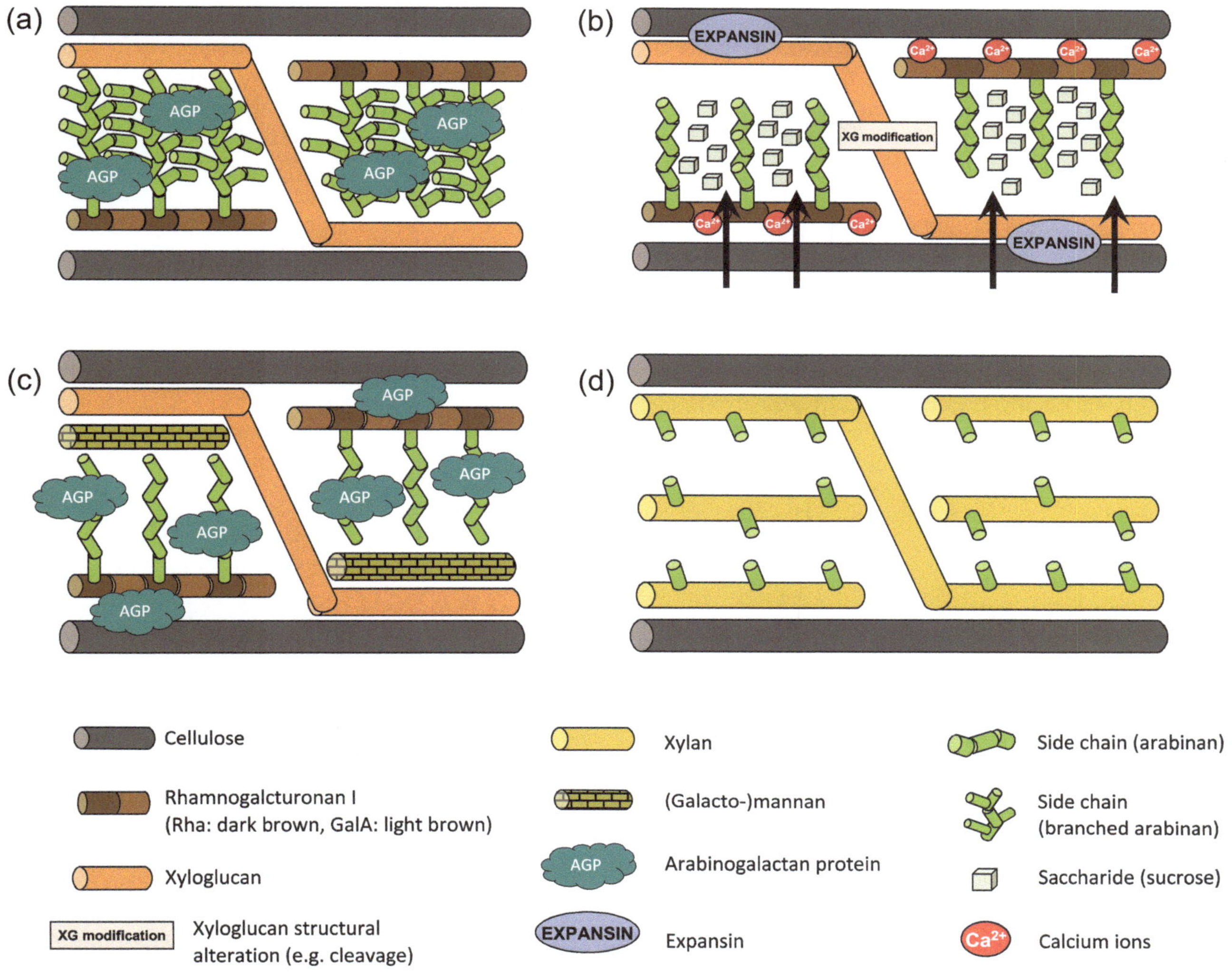

FIGURE 22.1 Cell wall molecular models proposed in Moore et al. (2013). These referred to the resurrection plants *Myrothamnus flabellifolia* (a), *Craterostigma wilmsii* and *Craterostigma plantagineum* (b), *Mohria caffrorum* (c) and Xerophyta spp. and *Eragrostis nindensis* (d). Reproduced from Moore et al. (2013) with permission.

angiosperms, insights into wall composition and changes during desiccation and recovery therefrom can inform strategies selected for by higher plants. Plancot et al. (2019) have provided a detailed structural characterisation of leaf cell walls of the desiccation-tolerant *Sellaginella involvens* and desiccation-sensitive relatives, *Sellaginella moellendorffii, and Sellaginella kraussiana*. Unsurprisingly, the study demonstrates that desiccation causes significant alterations of the cell wall composition and structure of these species; however, the exact nature thereof differed in the tolerant vs sensitive species. Using scanning electron microscopy, the authors first showed that pronounced folding of the cell walls occurred during desiccation of the tolerant *S. involvens*. Such reversible desiccation-induced cell wall folding has been reported for many resurrection angiosperm species (Moore et al. 2009; Farrant et al. 2017) and is a process that depends on the molecular composition and architecture of the cell wall (Moore et al. 2009). The authors went on to show that the hemicelluloses, xyloglucan, arabinoxylan and mannan, are the most abundant polysaccharides of leaf cell walls of the studied Selaginella species. Structural analyses of arabinoxylan showed modification of the structure of this polysaccharide (in terms of branching with arabinose residues) in *S. involvens* as compared to the intolerant *S. moellendorfii* and *S. kraussiana*. Xylan is also abundant in the cell walls of resurrection grasses and xylan branching was also shown to be modified during dehydration (Plancot et al. 2014; Moore et al. 2013). The difference in xylan substitution is thus believed to affect the mechanical properties of the cell wall (e.g. modification of cell wall extensibility) possibly by interacting with other wall components such as cellulose. Analysis of the xyloglucan structures in the Selaginella species revealed that this polysaccharide is of a XXXG-type that can be galactosylated, fucosylated and also arabinosylated. Analysis of the level of different oligosaccharides under desiccation conditions showed that

XXXG fragments increased, whereas the XXFG, XXLG and XLFG fragments decreased in the desiccation-tolerant species *S. involvens*, suggesting significant structural alterations of the polysaccharide. The structure of xyloglucan has been shown to be modified during desiccation in *C. wilmsii* and *M. flabellifolia* and is implicated as a major cell wall process associated with desiccation tolerance (see Vicré et al. 2004; Moore et al. 2006, 2009). Interestingly, the structural analysis also revealed the presence of arabinose-containing fragments (XDDG, XDEG and XLEG/XELG) in the cell wall of *S. kraussina*, rarely found in the cell wall of angiosperm plants. However, no significant change in the level of these oligosaccharides occurred under desiccation conditions. Finally, mannan structure was investigated in that study but the relative abundance of the oligomannan fragments did not change between hydrated and desiccated treatments for the three studied species. However, the O-acetylation status of the fragments was higher in the desiccation-tolerant species *S. involvens*. The level of O-acetylation could alter the interaction of mannan with other polymers like cellulose or pectins thus influencing the extensibility of the cell wall during water stress. Interestingly, the O-acetylation status of the cell wall polysaccharide xylan has been associated with increased drought tolerance of the Arabidopsis mutant *esk1* (Lefebvre et al. 2011). Reid and Bewley (1979) have also suggested that mannan polysaccharides are able to retain water and they can therefore be involved in maintaining cell wall hydration during desiccation.

THE MODEL RESURRECTION PLANT *CRATEROSTIGMA PLANTAGINEUM*

The model *C. plantagineum* Hoch. has now been studied for well over two decades (Bartels 2005). This research included insights into cell wall and membrane signalling was not yet reported on in other angiosperm resurrection plants. For example, it was as early as 2005 that phospholipase-mediated signalling has been reported for this species (Bartels 2005) and other research integral to the model of wall responses to desiccation and rehydration we propose below stems from molecular studies done on *C. plantigineum*. We thus draw heavily on data from this species. However, having said this, due to inherent differences observed among dicot and monocot resurrection plants in terms of mechanisms of desiccation tolerance instituted and wall structure, we acknowledge that similar wall-associated work should be conducted on the monocot resurrection plants, such as *Xerophyta schlecterii* and *Eragrostis nindensis*, for which there has been wall characterisation (discussed above; Moore et al. 2013; Plancot et al. 2014) and for whom there is considerable physiological and omic data available. We thus make extrapolations from such data where relevant.

Whichever model is used, it is clear that plant cell walls and the plasma membranes are physically affected upon water loss by losing turgor pressure, and subsequent breakage of membranes, resulting in leakage and ultimately tissue/organ death. In this context, resurrection plants must possess the innate genetic programming required to tolerate such extreme desiccation stress and therefore need to be deciphered with a view to elucidating these survival mechanisms. The cell wall glycome and glycoproteome and its higher order polysaccharide-rich architecture must enable resurrection plants (such as *C. plantagineum*) to withstand mechanical tension (due to polymer adhesion, hydraulic failure and turgor collapse) caused by an extreme water loss and during the folding/unfolding mechanisms observed during repeated cycles of dehydration and rehydration cycles experienced in the field. In parallel, membrane and metabolic (i.e. osmolytes and re-organised sugar, phenolic and amino acid metabolism) adjustments must occur to ensure that the integrity of the cell membranes is maintained in spite of oxidative and mechanical stressors.

THE PROCESS OF DESICCATION-INDUCED CELL WALL FOLDING AND PLASTICITY IN *CRATEROSTIGMA* SPECIES

The evolution of desiccation tolerance within the Linderniaceae has been expounded upon in Phillips et al. (2008). A centre of endemism and diversification for these plants appears to occur (and have occurred) in ancient tropical montane rainforests of east Africa in Kenya and Tanzania. *Lindernia subracemosa*, a desiccation-sensitive relative of *C. plantagineum*, occurs in these rainforests and experiences adequate rainfall and thus does not require desiccation tolerance. However, sharing this habitat, another species *Lindernia brevidens* does retain vegetative desiccation tolerance. The Craterostigma species on the other hand (*C. plantagineum* and *C. wilmsii*) are found in numerous arid inselberg regions including South Africa. Evaluation of cell wall changes during desiccation and rehydration of *C. wilmsii* (by Vicré et al. 1999, 2004) and of *C. plantagineum* (by Jones and McQueen-Mason 2004; Moore et al. 2013) has not led to a clear reason (or hypothesis) for how this is achieved by these particular plants. This is in contrast to *M. flabellifolia*, *Xerophyta viscosa* and *E. nindensis* where arabinosylation of cell wall polymers seems to be an effective evolutionary mechanism (as reported by Moore et al. 2013). In *C. wilmsii* and *C. plantagineum*, the process appears far more subtle; the cell wall composition is not particularly unusual, and this seems to suggest a remodelling approach of employing genes/proteins such as xyloglucan transhydrolases (XTHs) and EXPs with calcium ion redistribution to plasticise leaf cell walls (fully discussed in Moore et al. 2013). In both species, there is (i) cell wall loosening involving XyG reorganisation and (ii) cell wall stiffening for stabilising and locking of the cell wall structure during desiccation: with egg-box formation (Ca^{2+} increase in *C. wilmsii*) proposed based on experimental evidence to occur in *C. plantagineum* (Giarola et al. 2016).

Together, these findings suggest that desiccation-tolerant *Craterostigma* species must employ a much more subtle approach, and therefore requires more detailed studies. In the most recent work on *C. plantagineum*, a glycine-rich protein

(GRP) was isolated and investigated from a functional perspective (Gariola et al. 2016). These authors demonstrated that this GRP, termed CpGRP1, was able to interact with CpWAK1 (a wall-associated kinase) to mediate a signalling response and presumably acting as a cell wall sensor in *C. plantagineum* for detection of water deficit. They proposed a model in which the CpGRP1-CpWAK1 system interacts with calcium and pectin at the cell wall and plasma membrane interface to mediate signalling and desiccation tolerance gene regulation (Gariola et al. 2016). Interestingly, BhGRP1, a GRP, has been implicated in the Chinese angiosperm (and importantly dicot) resurrection plant *Boea hygrometrica* (Wang et al. 2009; Mitra et al. 2013). However, a number of factors remain unknown; most importantly we do not have sufficient information on the plant cell wall compositional and architectural changes that occur in either species to prevent irreversible polymer changes from happening in the first place.

Despite this, Chen et al. (2020) have recently proposed a molecular model of cell wall integrity and consequent signalling in *C. plantagineum* (see Figure 22.2). They integrate and expand on the observations of the CpGRP1-CpWAK1 system outlined above and draw on work from Jung et al. (2019) who showed that upon desiccation, type II *Cp*GRP1, which exhibits arginine-rich domains, are able to interact with *Cp*WAK1and pectic homogalacturonans (HGs). Of interest is that such a GRP–WAK interaction has also been described in *A. thaliana* (Park et al. 2001) and is central to depict, and further model a system where external, desiccation-induced (for instance) perturbations are being sensed in the apoplastic compartment and transduced in the cytosolic compartment in order to trigger cellular and physiological responses. This is taken into consideration in the model, which further points to the importance of methylation status of HGs, apoplastic/cytoplasmic calcium influxes and the presence of phosphatidic acid as essential to propose a unifying view of mechanisms that occur at the cell wall interface with the plasma membrane, upon dehydration and rehydration (Chen et al. 2020, 2021). This model is congruent with work on this and other angiosperm resurrection plants, which report on different cell wall polymers playing a similar role. These include hemicellulosic xyloglucans (Vicré et al. 1999, 2004; Jones and McQueen-Mason 2004), xylan polysaccharides (Plancot et al. 2019), pectic HGs (Vicré et al. 1999) and rhamnogalacturonan-I (RG-I) galactan and arabinan side chain polysaccharides (Moore et al. 2006, 2013).

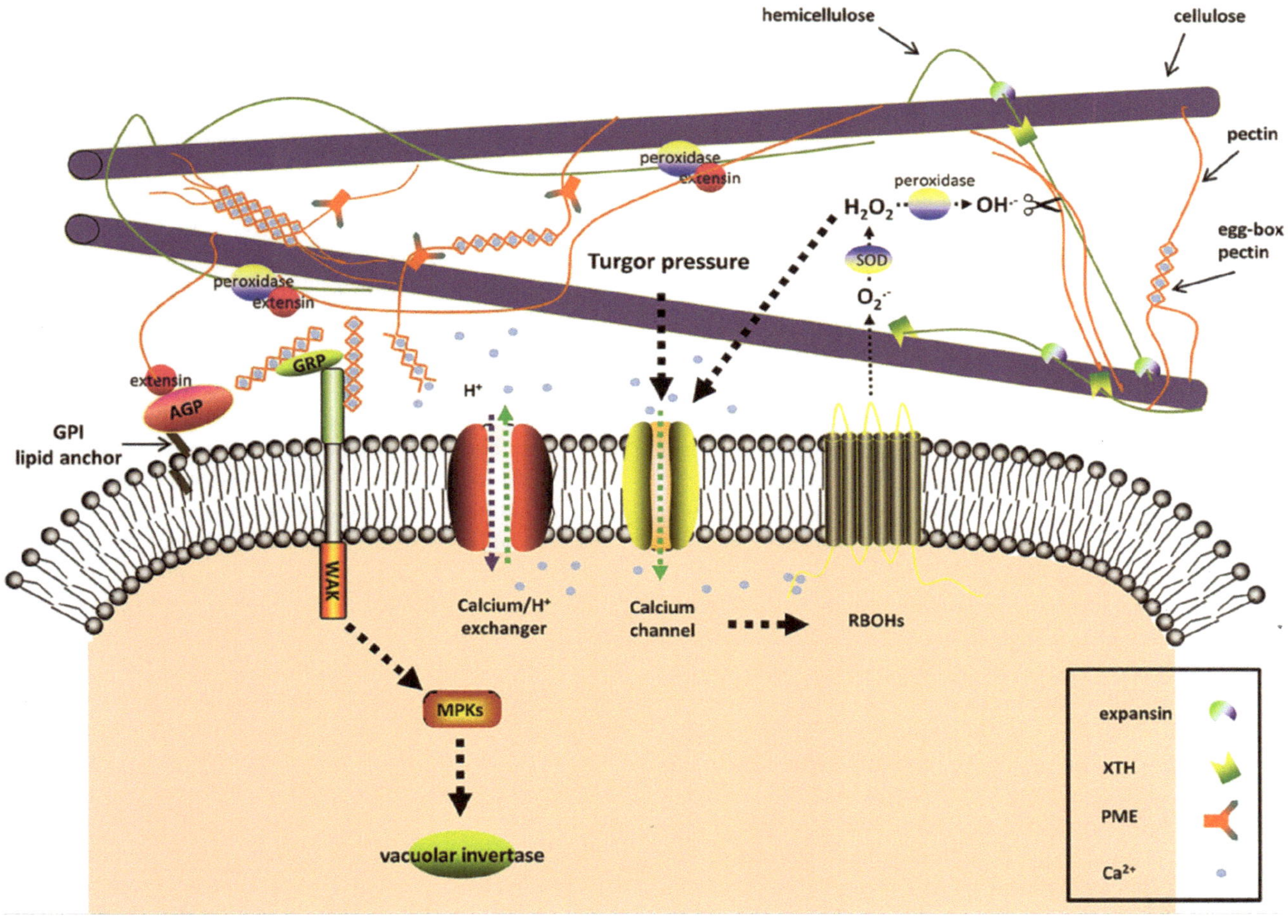

FIGURE 22.2 This model proposed and predicted interactions between apoplastic proteins and signalling molecules that would involve a change in pectin methylation patterning, a calcium signal transduction and a change in gene expression upon dehydration of *Craterostigma plantagineum* cell walls as highlighted in Chen et al. 2020. Reproduced from Chen et al. (2020) with permission.

ARABINOGALACTAN PROTEINS AS PROMISING CANDIDATES FOR MEDIATING DESICCATION-INDUCED CELL WALL SIGNALLING

Cell wall polymers including HG (see Chapter 5) and arabinogalactan proteins (AGPs; see Chapter 16) are implicated in modifying the mechanical properties of the cell wall but are also suggested to play a major role in cell signalling. Calcium influx is amongst the first signals to be triggered in response to perception of modified turgor pressure, and calcium ions (Ca^{2+}) act as a second messenger in water stress responses (Chen et al. 2020). We hypothesise that AGPs could be promising candidates due to their capacities to reversibly bind Ca^{2+} through the glucuronic acid residues of their side chains (Lamport et al. 2014). As a consequence, AGPs represent a major site of Ca^{2+} capacitor activity and a substantial source of cytosolic Ca^{2+} due to their close vicinity to the plasma membrane (Lamport et al. 2014). One might speculate that upon dehydration, AGPs reversibly release their bound Ca^{2+}, thus contributing to the signalling pathways of desiccation responses in resurrection plants. Recent advances in the understanding of the role cell wall– and plasma membrane–located, glycosyl-phosphatidyl-inositol (GPI)-anchored proteins, such as HRGP-related arabinogalactan proteins (Lopez-Hernandez et al. 2020), have indicated their ability to play a role as an apoplastic, reservoir of pH-dependent available calcium (Zhang, 2020; Lamport et al. 2021, 2013). In addition, a better understanding of the role of HRGP-related extensins in the plant cell wall (Castilleux et al. 2021; Nguema-Ona et al. 2014) has allowed us to revisit the model proposed in Chen et al. (2020). Indeed, this mechanical perception is also orchestrated at the cell wall interface with the plasma membrane, by plant cells in general when exposed to biotic stress– or abiotic stress–related mechanical perturbations that occur both in the wall and also the apoplastic compartments. Below we attempt to extrapolate from such available information and present a revised and updated model of resurrection plant cell wall characteristics and potential mode(s) of operation during desiccation and rehydration that in turn facilitate vegetative desiccation tolerance (see Figure 22.3 for a proposed model). We do, however, recognise that such mechanisms may vary with the specific plant species concerned as well as tissue and organ (root or shoot) type, so this is a provisional model based on what we can glean from the literature thus far.

Our understanding of the biology of plasma membrane and apoplastic AGPs has recently taken a tremendous step forward. Using Arabidopsis mutants deficient in the ability to correctly glucuronidate the side chains of AGP-associated type II arabinogalactans (which further impact the methylation of glucuronic acid residues), Lopez-Hernandez et al. (2020) demonstrated that AGPs exhibit an ability to sequester calcium ions in the apoplast (Zhang, 2020). Such sequestration properties (also termed the calcium capacitor hypothesis) have been promoted by Lamport and co-workers (Lamport et al. 2021, 2014, 2013), who highlighted an important role for AGPs in controlling Ca^{2+} ions influxes occurring in the apoplast and accompanying signalling events taking place across the plasma membrane. These pH-dependent properties rely on the methylation status of the terminal glucuronic acid residues (Lopez-Hernandez et al. 2020). This finding, together with the extensive literature on the genetic, compositional, structural and functional aspects of AGPs (Perez-Perez et al. 2018; Mareri et al. 2019; Ma et al. 2017, 2018; Nguema-Ona et al. 2012; Seifert and Roberts 2007; Showalter 2001; Schultz et al. 2000), allows a reconsideration of the function of these glycoproteins as functional actors located at the interface of the plasma membrane, the apoplast, and the cell compartment. In our proposed model, AGPs are postulated to operate as a novel plasma membrane and apoplastic, located sink, which is able to capture Ca^{2+} ions, *via* a GlcA-Ca^{2+}-GlcA bridging. Where two GlcA residues from type II arabinogalactan chains linked to AGPs couple to Ca^{2+} ions during desiccation. This is rendered possible by the pH-dependent removal of methyl groups on AGP-containing glucuronic acid residues. This local and available source of calcium is thus easily mobilisable by the tissues upon rehydration and *via* a rapid transduction of a signal through Ca^{2+} channels and transporters. Interestingly, the binding of acidic HG pectic polysaccharides to type II GRPs (glycine-rich proteins) in the apoplast of dehydrated leaves, renders this micro-environment acidic. This in turn might make this space prone to an additional association between an acidic GlcA residue derived from AGPs, a Ca^{2+} ion, and an acidic GalA residue derived from the pectic HG. This would amplify the Ca^{2+} capacitor effect at the vicinity of the membrane and Ca^{2+} channel transporters (Lamport et al. 2021, 2013). Our explanation better positions the role of Ca^{2+}, which was until recently seen as unclear (Jung et al. 2019; see also Vicré et al. 1999).

Notable progress in the understanding of molecular basis of wall-sensing mechanisms was made within the last 20 years. Indeed, several literature surveys already highlighted a number of plasma membrane–located, molecular players including GRPs, AGPs and GPI-anchored proteins; wall-associated kinases (WAKs); receptor-like kinases (RLK such as THEUSEUS); lectin-receptor kinases; extensin-like receptor kinases (PERKs); and all those involved in cell wall integrity maintenance (Ringli 2010; Hamann 2012; Vaahtera et al. 2019; Bacete and Hamann 2020, Bacete et al. 2022). Here also, an important step forward was made with the understanding of the role of the FERONIA receptor-like kinase in maintaining cell wall integrity during salt stress through Ca^{2+} signalling (Shih et al., 2014; Feng et al. 2018). FERONIA is a transmembrane protein that harbours an apoplastic malectin-like domain, known to physically bind to demethylesterified HGs, and a cytoplasmic kinase domain. The protein is able to perceive a loss of wall integrity, due to salinity-induced dismantling of the apoplastic, pectic network. It was demonstrated that

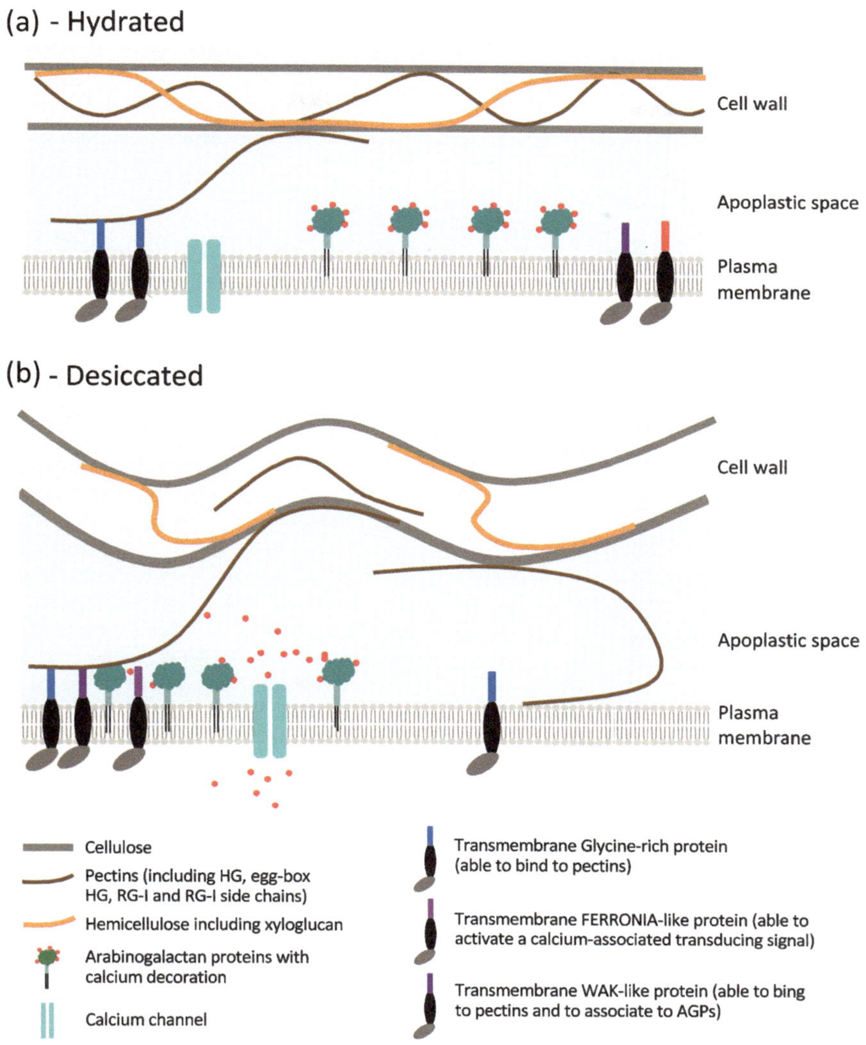

FIGURE 22.3 A proposed hypothetical model incorporating a role for arabinogalactan proteins (AGPs) as important contributors to cell wall integrity signalling for vegetative desiccation tolerance in resurrection plants; the hydrated state is shown in (A) and the desiccated state in (B), and a legend indicating the various cell wall macromolecules is provided. This extends the model proposed by Chen et al. (2020) by highlighting a role for AGPs in calcium ion–mediated signalling. Upon dehydration, membrane-anchored AGPs, along with transmembrane wall-associated kinases (WAKs) and additional transmembrane kinases such as FERONIA with glycine-rich proteins (GRPs), migrate to the plasma membrane surface and form a GRP–WAK–AGP–FERONIA–pectin complex. The pH-dependent release of calcium ions from AGPs would upon rehydration lead to the activation of signalling pathways, which facilitate the rehydration process in resurrection plants.

upon salt-induced mechanical stress, FERONIA plays a role in mediating calcium signalling to prevent plant cell bursting, and like its close homolog THEUSEUS (Hematy et al. 2007; Gonneau et al. 2018) is central in helping the plant to overcome mechanical stress due to water loss. A parallel can be drawn with dehydration–rehydration events that occur during vegetative desiccation tolerance cycles. Interestingly, these studies also allow extending the initially GRP–WAK–pectin association to additional players involving calcium and FERONIA proposing a novel GRP–WAK–FERONIA–pectin–calcium complexing taking place. It is interesting to note that AGPs have been previously shown to be associated with WAKs (Gens et al. 2000), where this association was shown to display, a calcium-dependent, control of mechanical perturbance in the plasmalemmal reticulum (Pickard and Fujiki 2005).

This, along with the fact that AGPs can be GPI-anchored to the plasma membrane (Borner et al. 2003), allows us to, tentatively, depict the possible occurrence of membrane-located microdomains, where, and under the influence of the pH as well as events of dehydration/rehydration, AGPs could define a calcium-enriched microdomain, at the vicinity of the GRP–WAK–FERONIA–pectin, apoplastic- and membrane-located domains. Upon dehydration events, the GRP–WAK complexes would bind to acidic, pectic HG polysaccharides. In parallel, a local formation of GRP–WAK–FERONIA–pectin–AGP complexes enriched with calcium ions chelated between two acidic glucuronic acids belonging to two consecutive AGP-AGP molecules or plausibly between an acidic glucuronic acid belonging to an AGP and an acidic galacturonic acid–containing pectic HG polysaccharides could take place. Pectins represent easily 50% of the cell wall materials in plant leaves (Zablackis et al. 1995). It is possible that the predominant form of pectic HG polysaccharides would be egg-boxes. Upon dehydration, which is accompanied by a pH drop, the acidic, calcium-free, pectic HG polysaccharides formed would then predominate and establish physical connections with the different receptor kinases (THEUSEUS, FERONIA, WAK), while calcium would migrate towards acidic, glucuronic acid- residues associated to AGP and galacturonic acid associated to pectic HG polysaccharides, yielding the formation of calcium stores. Finally, it has been reported that the receptor-like kinase THEUSEUS regulates the activity of several genes having a role in protection of plant tissues against reactive oxygen species, as well as cell wall, related peroxidases and extensins (Hematy et al. 2007). Extensins are known to crosslink in the walls, thus playing a role in cell wall architecture (Nguema-Ona et al. 2014; Tan and Mort 2020; Castilleux et al. 2021).

CONCLUSIONS AND FUTURE PERSPECTIVES

In this chapter, we have highlighted the knowledge gained to date on cell wall changes upon desiccation and rehydration in resurrection plants. A number of cell wall players have been implicated in earlier studies of resurrection plants such as *Craterostigma wilmsii* and *Myrothamnus flabellifolia* including pectic arabinans, AGPs, arabinosylation of xylans, xyloglucan endotransglycoylases, EXPs, and calcium ion redistribution across a wide range of plant species (Moore et al. 2013). A focus on *Boea hygrometrica* and *Craterostigma plantagineum* has centred on the role for GRPs in desiccation tolerance and has led to the plasma membrane–cell wall interface where evidence has been presented for interactions with WAKs and pectin domains of various states of methylation/demethylation, which in turn provided strong evidence for wall localised GRPs in cell wall integrity signalling pathways and subsequent gene expression. However, it is still not clear how the wall itself is able to survive the mechanical and macromolecular changes associated with the absence of water during desiccation. To this end, we focused on the accumulating evidence for AGPs to be active calcium stores that form complexes with GRP–WAK–FERONIA–pectin enriched in calcium ions which we propose to occur in desiccated cell walls. These complexes would then discharge their calcium ions upon rehydration and protect cell wall polymers from irreversible polymer adhesion during desiccation. Resurrection plants are able to extensively reorganise their cell wall constitutive polymers upon dehydration, in a highly controlled manner. These plants are then able to activate the fine-tuning of a calcium-driven resurgence of tissue growth and development, when water is again available. To what extent this reorganisation is rendered possible by remodelling and/or *de novo* synthesis of wall polysaccharides remains to be clarified. We hope that our model will spur further research into this important area of plant biology.

REFERENCES

Alpert P (2000) The discovery, scope, and puzzle of desiccation tolerance in plants. *Plant Ecology*, 151(1):5–17.

Anderson CM, Wagner TA, Perret M, He Z-H, He D, Kohorn BD (2001) WAKs: cell wall-associated kinases linking the cytoplasm to the extracellular matrix. *Plant Molecular Biology* 47:197–206.

Asgari M, Brulé V, Western TL, Pasini D (2020) Nano-indentation reveals a potential role for gradients of cell wall stiffness in directional movement of the resurrection plant *Selaginella lepidophylla*. *Scientific Reports* 10(1):1–10.

Bacete L, Hamann T (2020) The role of mechanoperception in plant cell wall integrity maintenance. *Plants*, *9*(5):574.

Bacete L, Schulz J, Engelsdorf T, Bartosova Z, Vaahtera L, Yan G, Gerhold JM, Tichá T, Øvstebø C, Gigli-Bisceglia N, Johannessen-Starheim S, Margueritat J, Kollist H, Dehoux T, McAdam S, Hamann T (2022) THESEUS1 modulates cell wall stiffness and abscisic acid production in *Arabidopsis thaliana*. *Proceedings of the National Academy of Sciences of the United States of America*, 119(1):e2119258119.

Bartels D (2005) Desiccation tolerance studied in the resurrection plant *Craterostigma plantagineum*. *Integrative and Comparative Biology* 45:696–701.

Bewley JD (1979) Physiological aspects of desiccation tolerance. *Annual Review of Plant Physiology*, *30*(1):195–238.

Blomstedt CK, Griffiths CA, Gaff DF, Hamill JD, Neale AD (2018) Plant desiccation tolerance and its regulation in the foliage of resurrection 'flowering-plant' species. *Agronomy* 8:146.

Borner GHH, Sherrier DJ, Stevens TJ, Dupree P (2003) Identification of glycosylphosphatidylinositol-anchored proteins in Arabidopsis. *Plant Physiology* 137:104–116.

Castilleux R, Plancot B, Gugi B, Attard A, Loutelier-Bourhis C, Lefranc B, Nguema-Ona E, Arkoun M, Yvin J-C, Driouich A, Vicré M (2020) Extensin arabinosylation is involved in root response to elicitors and limits oomycete colonization. *Annals of Botany* 25:751–763.

Castilleux R, Plancot B, Vicré M, Nguema-Ona E, Driouich A (2021) Extensin, an underestimated key component of cell wall defence? *Annals of Botany* 127:709–713.

Challabathula D, Puthur JT, Bartels D (2016) Surviving metabolic arrest: photosynthesis during desiccation and rehydration in resurrection plants. *Annals of the New York Academy of Sciences 1365*(1):89–99.

Chen P, Jung NU, Giarola V, Bartels D (2020) The dynamic responses of cell walls in resurrection plants during dehydration and rehydration. *Frontiers in Plant Science*, *10*:1698.

Chen P, Giarola V, Bartels D (2021) The *Craterostigma plantagineum* protein kinase CpWAK1 interacts with pectin and integrates different environmental signals in the cell wall. *Planta* 253: 92.

Costa MCD, Artur MA, Maia J, Jonkheer E., Derks MF, Nijveen H., Williams B, Mundree SG, Jiménez-Gómez JM, Hesselink T, Schijlen EG (2017) A footprint of desiccation tolerance in the genome of *Xerophyta viscosa*. *Nature Plants*, *3*(4):1–10.

Dai A (2013) Increasing drought under global warming in observations and models. *Nature Climate Change* 3:52–58.

Farrant JM, Hilhorst HW (2021) What is dry? Exploring metabolism and molecular mobility at extremely low water contents. *Journal of Experimental Botany 72*(5):1507–1510.

Farrant JM, Cooper K, Dace HJWS, Bentley J, Hilgart A (2017) Desiccation tolerance. Chapter 9. pp 217–252. In *Plant Stress Physiology*, 2nd edition. Ed. S Shabala. CAB International, Wallingford, United Kingdom.

Farrant JM, Cooper K, Hilgart A, Abdalla KO, Bentley J, Thomson JA, Dace HJ, Peton N, Mundree SG, Rafudeen MS (2015) A molecular physiological review of vegetative desiccation tolerance in the resurrection plant *Xerophyta viscosa* (Baker). *Planta 242*(2):407–426.

Farrant JM, Moore JP (2011) Programming desiccation-tolerance: from plants to seeds to resurrection plants. *Current Opinion in Plant Biology*. 14:1–6.

Feng W, Kita D, Peaucelle A, Cartwright HN, Doan V, Duan Q, Liu M-C, Maman J, Steinhorst L, Schmitz-Thom I, Yvon R, Kudla J, Wu H-M, Cheung AY, Dinneny JR (2018) The FERONIA receptor kinase maintains cell-wall integrity during salt stress through calcium signaling. *Current Biology* 28:666–675.

Gaff DF (1971) Desiccation-tolerant flowering plants in Southern Africa. *Science* 174:1033–1034.

Gaff DF, Oliver M (2013) The evolution of desiccation tolerance in angiosperm plants: a rare yet common phenomenon. *Functional Plant Biology 40*(4):315–328.

Gens JS, Fujiki M, Pickard BG (2000) Arabinogalactan protein and wall-associated kinase in a plasmalemmal reticulum with specialized vertices. *Protoplasma* 212:115–134.

Giarola V, Krey S, von den Driesch B, Bartels D (2016) The *Craterostigma plantagineum* glycine-rich protein CpGRP 1 interacts with a cell wall-associated protein kinase 1 (CpWAK1) and accumulates in leaf cell walls during dehydration. *New Phytologist* 210(2):535–550.

Giarola V, Chen P, Dulitz SJ, König M, Manduzio S, Bartels D (2020) The dehydration- and ABA-inducible germin-like protein CpGLP1 from *Craterostigma plantagineum* has SOD activity and may contribute to cell wall integrity during desiccation. *Planta* 252:84.

Gonneau M, Desprez T, Martin M, Doblas VG, Bacete L, Miart F, Sormani R, Hematy K, Renou J, Landrein B, Murphy E, van de Cotte B, Vernhettes S, de Smet I, Höfte H (2018) Receptor kinase THESEUS1 is a rapid alkalinization factor 34 receptor in Arabidopsis. *Current Biology* 28:2452–2458.

Hamann T (2012) Plant cell wall integrity maintenance as an essential component of biotic stress response mechanisms. *Frontiers in Plant Science* 3:77.

He Z-H, Fujiki M, Kohorn BD (1996) A cell wall-associated, receptor-like protein kinase. *Journal of Biological Chemistry* 271:19789–19793.

Hematy K, Cherk C, Somerville S (2009) Host-pathogen warfare at the plant cell wall. *Current Opinion in Plant Biology*, 12:406–413.

Hematy K, Sado P-E, van Tuinen A, Rochange S, Desnos T, Balzergue S, Pelletier S, Renou, J-P, Höfte H (2007) A receptor-like kinase mediates the response of Arabidopsis cells to the inhibition of cellulose synthesis. *Current Biology* 17:922–931.

Hilhorst HW, Farrant JM (2018) Plant desiccation tolerance: a survival strategy with exceptional prospects for climate-smart agriculture. *Annual Plant Reviews Online* 327–354.

Illing N, Denby KJ, Collett H, Shen A, Farrant JM (2005) The signature of seeds in resurrection plants: a molecular and physiological comparison of desiccation tolerance in seeds and vegetative tissues. *Integrative and Comparative Biology* 45(5):771–787.

Jones L, McQueen-Mason S (2004) A role for expansins in dehydration and rehydration of the resurrection plant *Craterostigma plantagineum*. *FEBS Letters* 559:61–65.

Jung NU, Giarola V, Chen P, Knox JP, Bartels D (2019) *Craterostigma plantagineum* cell wall composition is remodelled during desiccation and the glycine-rich protein CpGRP1 interacts with pectins through clustered arginines. *The Plant Journal* 100(4):661–676.

Kohorn SL, Kohorn BD (2012) The cell wall-associated kinases, WAKs, as pectin receptors. *Frontiers in Plant Science* 8:88.

Kohorn BD (2016) Cell wall-associated kinases and pectin perception. *Journal of Experimental Botany* 67:489–494.

Lamport DTA, Varnai P (2013) Periplasmic arabinogalactan glycoproteins act as a calcium capacitor that regulates plant growth and development. *New Phytologist* 197:58–64.

Lamport DTA, Varnai P, Seal CE (2014) Back to the future with the AGP - calcium flux capacitor. *Annals of Botany* 114:1069–1085..

Lamport DTA, Tan L, Kieliszewski M (2021) A molecular pinball machine of the plasma membrane regulates plant growth - a new paradigm. *Cells* 10:1935.

Lefebvre V, Fortabat M-N, Ducamp A, North HM, Maia-Grondard A, Trouverie J, Durand-Tardif M (2011) ESKIMO1 Disruption in *Arabidopsis* alters vascular tissue and impairs water transport. *PLOS ONE*, 6(2): e16645.

Lopez-Hernandez F, Tryfona T, Rizza A, Yu X, Harris MO, Webb AA, Kotake T, Dupree P (2020) Calcium binding by arabinogalactan polysaccharides is important for normal plant development. *Plant Cell* 32:3346–3369.

Lyall R, Nikoloski Z, Gechev T (2020) Comparative analysis of ROS network genes in extremophile eukaryotes. *International Journal of Molecular Sciences*, 21 (23):9131.

Ma Y, Yan C, Li H, Wu W, Liu Y, Wang Y, Chen Q, Ma H (2017) Bioinformatics prediction and evolution analysis of arabinogalactan proteins in the plant kingdom. *Frontiers in Plant Science* 8:66.

Ma Y, Zeng W, Bacic A, Johnson K (2018) AGPs through time and space. In Roberts, J. A. Ed. *Annual Plant Review Online*. John Wiley and Sons, Ltd., Chichester, UK, 767–804.

Mangeon A, Junqueira RM, Sachetto-Martins G (2010) Functional diversity of the plant glycine-rich proteins superfamily. *Plant Signaling and Behavior* 5:99–104.

Mareri L, Romi M, Cai G (2019) Arabinogalactan proteins: actors or spectactors during abiotic and biotic stress in plants. *Plants Biosystems* 153:173–185.

Marks RA, Farrant JM, Nicholas McLetchie D, VanBuren R (2021). Unexplored dimensions of variability in vegetative desiccation tolerance. *American Journal of Botany* 108(2):346–358.

Mitra J, Xu G, Wang B, Li M, Deng X (2013) Understanding desiccation tolerance using the resurrection plant *Boea hygrometrica* as a model system. *Frontiers in Plant Science* 4:446.

Moore JP, Nguema-Ona E, Chevalier L, Lindsey GG, Brandt WF, Lerouge P, Farrant JM, Driouich A (2006) Response of the leaf cell wall to desiccation in the resurrection plant *Myrothamnus flabellifolius*. *Plant Physiology* 141:651–662.

Moore JP, Farrant JM, Driouich A (2008) A role for pectin-associated arabinans in maintaining the flexibility of the plant cell wall during water deficit stress. *Plant Signaling and Behavior* 3:102–104.

Moore JP, Le NT, Brandt WF, Driouich AD, Farrant JM (2009) Towards a systems-based understanding of plant desiccation tolerance. *Trends in Plant Science* 14:110–117.

Moore JP, Nguema-Ona EE, Vicré-Gibouin M, Sørensen I, Willats WGT, Driouich A, Farrant JM (2013) Arabinose-rich polymers as an evolutionary strategy to plasticize resurrection plant cell walls against desiccation. *Planta* 237:739–754.

Nguema-Ona E, Coimbra S, Vicré-Gibouin M, Mollet J-C, Driouich A (2012) Arabinogalactan proteins in root and pollen tube cells: distribution and functional aspects. *Annals of Botany* 110:383–404.

Nguema-Ona E, Vicré-Gibouin M, Gotté M, Plancot B, Lerouge P, Bardor M, Driouich A (2014) Cell wall O-glycoproteins and N-glycoproteins: aspects of biosynthesis and function. *Frontiers in Plant Science* 5:499.

Oliver MJ, Farrant JM, Hilhorst HW, Mundree S, Williams B, Bewley JD (2020) Desiccation tolerance: avoiding cellular damage during drying and rehydration. *Annual Review of Plant Biology*, 71:435–460.

Park AR, Cho SK, Yun UJ, Jin MY, Lee SH, Sachetto-Martins G, Park OK (2001) Interaction of the Arabidopsis receptor protein kinase Wak1 with Glycine-rich protein AtGRP-3. *Journal of Biological Chemistry* 276: 26688–26693.

Perez-Perez Y, Carneros E, Berenguer E, Solis M T, Barany I, Pintos, B, Gómez-Garay A, Risueño MC, Testillano PS (2018) Pectin de-methylesterification and AGP increase promote cell wall remodeling and are required during somatic embryogenesis of *Quercus suber*. *Frontiers in Plant Science* 9: 1915.

Phillips JR, Fischer E, Baron M, van den Dries N, Facchinelli F, Kutzer M, Rahmanzadeh R, Remus D, Bartels D (2008) *Lindernia brevidens*: a novel desiccation-tolerant vascular plant, endemic to ancient tropical rainforests. *Plant Journal* 54:938–948.

Pickard BG, Fujiki M (2005) Calcium pulsation in BY-2 cells and evidence for control of mechanosensory calcium-selective channels by the plasmalemmal reticulum. *Functional Plant Biology* 32:863–879.

Plancot B, Vanier G, Maire F, Bardor M, Lerouge P, Farrant JM, Moore J, Driouich A, Vicré-Gibouin M, Afonso C, Loutelier-Bourhis C (2014) Structural characterization of arabinoxylans from two African plant species *Eragrostis nindensis* and *Eragrostis tef* using various mass spectrometric methods. *Rapid Communications in Mass Spectrometry* 28(8):908–916.

Plancot B, Gügi B, Mollet JC, Loutelier-Bourhis C, Lerouge P, Follet-Gueye ML, Vicré M, Alfonso C, Nguema-Ona E, Bardor M, Driouich A (2019) Desiccation tolerance in plants: Structural characterization of the cell wall hemicellulosic polysaccharides in three Selaginella species. *Carbohydrate Polymers*. 208:180–190.

Porembski S, Barthlott W (2000) Granitic and gneissic outcrops (inselbergs) as centers of diversity for desiccation-tolerant vascular plants. *Plant Ecology* 151(1):19–28.

Reid JSG, Bewley JD (1979) A dual rôle for the endosperm and its galactomannan reserves in the germinative physiology of fenugreek (*Trigonella foenum-graecum* L.), an endospermic leguminous seed. *Planta*, 147(2):145–150.

Ringli C (2010) Monitoring the outside: cell wall-sensing mechanisms. *Plant Physiology* 153:1445–1452.

Rodriguez MCS, Edsgärd D, Hussain SS, Alquezar D, Rasmussen M, Gilbert T, Nielsen BH, Bartels D, Mundy J (2010) Transcriptomes of the desiccation-tolerant resurrection plant *Craterostigma plantagineum*. *Plant Journal* 63:212–228.

Shih H-W, Miller ND, Dai C, Spalding EP, Monshausen GB (2014) The receptor-like kinase FERONIA is required for mechanical signal transduction in Arabidopsis seedlings. *Current Biology* 24:1887–1892.

Schultz CJ, Johnson KL, Currie G, Bacic A (2000) The classical arabinogalactan protein gene family of Arabidopsis. *Plant Cell* 12:1751–1767.

Seifert GJ, Roberts K (2007) The biology of arabinogalactan proteins. *Annual Reviews in Plant Biology* 58:137–161.

Showalter AM (2001) Arabinogalactan proteins: structure, expression and function. *Cellular and Molecular Life Science* 58:1399–1417.

Suguiyama VF, Sanches RFE, Meirelles ST, Centeno DC, Da Silva EA, Braga MR (2016) Physiological responses to water deficit and changes in leaf cell wall composition as modulated by seasonality in the Brazilian resurrection plant *Barbacenia purpurea*. *South African Journal of Botany* 105:270–278.

Tan L, Mort A (2020) Extensins at the front line of plant defence A commentary on: 'Extensin arabinosylation is involved in root response to elicitors and limits oomycete colonization. *Annals of Botany* 125:vii–viii.

Tebele SM, Marks RA, Farrant JM (2021) Two decades of desiccation biology: A systematic review of the best studied angiosperm resurrection plants. *Plants* 10:2784.

Vaahtera L, Schulz J, Hamann T (2019) Cell wall integrity maintenance during plant development and interaction with the environment. *Nature Plants* 5:924–932.

VanBuren R (2017). Desiccation tolerance: seedy origins of resurrection. *Nature Plants 3*(4):1–2.

VanBuren R, Wai CM, Zhang Q, Song X, Edger PP, Bryant D *et al.* (2017) Seed desiccation mechanisms co-opted for vegetative desiccation in the resurrection grass *Oropetium thomaeum*. *Plant Cell & Environment*. 40:2292–2306.

Vander Willigen C, Pammenter NW, Jaffer MA, Mundree SG, Farrant JM (2003). An ultrastructural study using anhydrous fixation of *Eragrostis nindensis*, a resurrection grass with both desiccation-tolerant and-sensitive tissues. *Functional Plant Biology* 30(3):281–290.

Vicré M, Farrant JM, Driouich A (2004). Insights into the cellular mechanisms of desiccation tolerance among angiosperm resurrection plant species. *Plant, Cell & Environment 27*(11):1329–134.

Vicré M, Sherwin HW, Driouich A, Jaffer MA, Farrant JM (1999) Cell wall characteristics and structure of hydrated and dry leaves of the resurrection plant *Craterostigma wilmsii*, a microscopical study. *Journal of Plant Physiology* 155:719–726.

Vicré M, Lerouxel O, Farrant J, Lerouge P, Driouich A (2004) Composition and desiccation-induced alterations of the cell wall in the resurrection plant *Craterostigma wilmsii*. *Physiologia Plantarum* 120:229–239.

Wang L, Shang H, Liu Y, Zheng M, Wu R, Phillips J, Bartels D, Deng X (2009) A role for a cell wall localized glycine-rich protein in dehydration and rehydration of the resurrection plant *Boea hygrometrica*. *Plant Biology* 11:837- 848.

Webb MA, Arnott HJ (1982) Cell wall conformation in dry seeds in relation to the preservation of structural integrity during desiccation. *American Journal of Botany* 69:1657–1668.

Xiao L, Yang G, Zhang L, Yang X, Zhao S, Ji Z, Zhou Q, Hu M, Wang Y, Chen M, Xu Y, Jin H, Xiao X, Hu G, Bao F, Hu F, Wan P, Li L, Deng X, Kuang T, Xiang C, Zhu J-K, Oliver MJ, He Y (2015) The resurrection genome of Boea hygrometrica: A blueprint for survival of dehydration." *Proceedings of the National Academy of Sciences United States of America* 112:5833–5837.

Zablackis E, Huang J, Muller B, Darvill AG, Albersheim P (1995) Characterization of the cell-wall polysaccharides of *Arabidopsis thaliana* leaves. *Plant Physiology* 107:1129–1138.

Zhang T (2020) Peripheral? Not really! The extracellular arabinogalactan proteins function in calcium signaling. *Plant Cell* 32:3057–3058.

Zhang Q, Bartels D (2018) Molecular responses to dehydration and desiccation in desiccation-tolerant angiosperm plants. *Journal of Experimental Botany* 69(13):3211–3222.

23 The Middle Lamella between Anthropic and Evolutionary Contexts

Rawen Ben Malek and Grégory Mouille
Université Paris-Saclay

INTRODUCTION

We present this book chapter as a critical review of the concept of the middle lamella based on the experimental and theoretical framework throughout history that starts from its identification in the 19th century all the way to the current research in the field. Our motivation is to assess the consistency of the methods and assumptions that led to today's definition of the middle lamella.

The middle lamella is often defined in association with the primary cell wall. The plant primary cell wall is defined as a fibre-reinforced material where cellulose fibres are embedded in a gel-like matrix of pectin. The cellulose–hemicellulose network is thought to provide tensile strength to the primary cell wall while the cross-linked pectin network provides compression and shear resistance. The middle lamella is then presented as an adhesive layer that forms an interface between the primary walls. Biologically, it should act as a cement between neighbouring cells (Albersheim et al. 2010). The middle lamella is considered as the extension of the pectin mesh, i.e., a layer of the primary cell wall deprived of fibre reinforcement. The adhesive function is often related to a calcium-mediated cross-bridging mechanism of the de-methylesterified Homogalacturonans (HG) (one of the components of the pectin). During cytokinesis, the middle lamella emerges from the plate formed during cell division and is therefore the first intercellular layer that forms during cell division.

The middle lamella is therefore defined with regard to the following elements: (i) structural properties visible in microscopy, (ii) biochemical nature and (iii) biological function. However, what if one or more of these features is missing? In meristems, for instance, the middle lamella lacks the optical properties as it is not observed at all, but its presence is generally assumed (Zamil and Geitmann 2017). In some organisms related to plants such as the Volvocaceae, some authors refer to the presence of a middle lamella despite a major biochemical difference. In plants, the middle lamella is a carbohydrate-based gel-like structure while in Volvocaceae proteins are the major constituents of the gel-like matrix (Domozych and Domozych 2014). Finally, the middle lamella has been identified in zones where cell separation has occurred, mainly in tri-cellular junctions (Sobry et al. 2005). It seems then, that the paradigm of the middle lamella somehow drives an "accommodation" to the exceptions that one might encounter, without questioning the concept itself or why such abnormalities occur. In fact, within the framework of "normal science" or day-to-day science, the research is conducted within the prevailed paradigm that remains unquestioned in an overall research consensus and as the abnormalities and errors accumulate, they create a sense of crisis (Kuhn 1962). Lakatos in 1978 described these degenerating programmes as theories that are "fabricated only in order to accommodate known facts".

In this review, we wanted to look deeper into the epistemic aspects that historically led to the definition of the middle lamella considering the abovementioned abnormalities to reassess these definitions and eventually suggest some mechanisms that might explain the occurrence of such exceptions.

THE HISTORY OF CREATING AND REINFORCING THE MODEL OF THE MIDDLE LAMELLA

The Epistemic Practices Defining the Nature of the Middle Lamella

The identification of the first cork cells by Robert Hooke in 1665 (Hooke 1665) followed by the improvement of the microscope by Antonie van Leeuwenhoek have paved the way for the establishment of cell theory and the beginning of cytology (Dröscher 2002; Sekeres and Zarsky 2018). The extensive work on plant cytology, which marked the 19th century, led to the conceptualisation of the middle lamella. The existence of a middle lamella was first reported by the German botanist Von Mohl in 1835 who identified a refractive intercellular substance in the xylem and other tissues (Kerr and Bailey 1934). It was described as a bounding layer that differed from the primary cell wall by its solubility and optical properties. This layer was, then, named middle lamella by Hofmeister (1867) and Sachs (1874). As it has been thoroughly described in Charles Allen's review on the origin and nature of the middle lamella (Allen, 1901), later studies on plant cell divisions by Strasburger (Strasburger 1876) came to confirm the existence of this layer. Strasburger was able to observe the cell plate at magnifications ranging from 600 to 900 diameters which correspond approximatively to the 10X and 16X objectives today (considering that the highest field view achieved in the 19th century was about 18 mm).

DOI: 10.1201/9781003178309-23

In that period, there was no clear distinction between the middle lamella as a component of the primary cell wall and the compound middle lamella (Chapter 24) that resulted from the compression of the primary cell wall/middle lamella continuum between two layers of the secondary wall. In fact, between 1867 and 1869, Dippel defined the middle lamella as a continuum of two anisotropic cell walls bounding an isotropic intercellular substance. However, in 1884, Strasburger identified this layer as distinct from the cell wall. In his experiment, Strasburger demonstrated that after the treatment of endosperm cells with sulphuric acid, the last component to remain was the middle lamella which, in that time, was qualified as *Theilplatte*, i.e., separation plate that he distinguishes from the primary cell wall (Strasburger 1884). The middle lamella layer was then described as the first-layed layer appearing after the splitting process or the deposition of the new cell wall (Strasburger 1898).

A clear distinction between the middle lamella and the compound middle lamella was suggested by Ritter 1925 who attributed the middle lamella term to the isotropic part only and Dippel's middle lamella was described by Van Itterson (1927) as the compound middle lamella. This distinction was supported by the existence of a primary and secondary cell wall (Kerr and Bailey 1934).

In the late 19th century, Mangin undertook the first biochemical studies, exclusively focusing on the identification of the biochemical nature of the middle lamella. He first identifies "pectose" as the major constituent of the middle lamella and later states that it is formed of a "cement" of insoluble pectates and more specifically of calcium pectates (Conrad 1926). At that time, the identification, staining and extraction techniques related to the pectic substance as well as the potential association of pectin with calcium also described in lime pectates have already been identified and performed by Louis Nicolas Vauquelin in 1790, Henri Braconnot in 1825 and Edmond Frémy 1840 who suggested that pectin had the role of cementing cells together (Mangin 1888)[1]. The name pectin is derived from the Greek word πηκτός / pêktós meaning thick, as a description of the obtained substance/gel that results from its extraction (Ropartz and Ralet 2020). To identify pectin in plant tissues, Mangin was the first to suggest the use of ruthenium red in 1893, after the identification of different ruthenium-sesquichloride-based dyes by Joly in 1892. Mangin was the first to use this staining technique to explore the biochemical nature of the middle lamella[2] (Mangin 1893).

Mangin also demonstrated, using various extraction methods and differential staining techniques, that the pectic compounds associated with the cellulose were neutral by opposition to their acidic nature in the middle lamella. Furthermore, he studied multiple types of tissues, demonstrating the existence of a biochemical "diversity" of middle lamella. For instance, he found that in meristematic tissues, the middle lamella (or intercellular layer as he named it) did not stain with phenosaphranin or methylene blue in opposition to the cellulose-bound pectin in the cell wall. In 1917, Ehrlich defined the acid pectin as Galacturonic Acid (GalA), an uronic acid that results from the oxidation of Galactose. In 1914, Fallenberg demonstrated that the removal of methyl (CH_3) from carboxylic groups (COOH) results in the acidic form of pectin (Conrad 1926).

Noteworthy, even at the time of its discovery, the term middle lamella itself was questioned with regard to its origin, the terms median layer and intercellular space were often used to refer to this component (Allen 1901) but it appears that a general settlement has been set on the term middle lamella. Overall, in less than half a century, the fundamental bases of the middle lamella were defined with regard to its staining capacities, its biological role as a cement between the cells and the nature of calcium pectates that compose it.

The Consequence of the Paradigm

With the development of electron microscopy techniques, the observation of the middle lamella by transmission electron microscopy (TEM) was made possible using different fixation and staining techniques, mainly the periodic acid–thio carbohydrazide silver proteinate (PATAg) staining. On TEM pictures, the middle lamella appears as a ~50 nm-thick layer that stains darker than the primary cell wall tissues (Albersheim et al. 1960). According to the authors, the staining of the middle lamella is mainly related to the presence of low methylesterified HGs. The positively charged ions such as iron or silver precipitate and result in the observed electron-dense layer (Figure 23.1). This observation confirmed the nature of the polymers present in the middle lamella as it was already shown by Mangin, at the end the 19th century using ruthenium red staining.

In order to improve the methods of visualising and characterising the biochemical nature of the middle lamella, antibodies oriented towards polysaccharides were developed in the 1970s. The purification and identification of the cell wall components led the development of polysaccharide-binding antibodies (Knox et al. 1970; Vreeland 1970). The main goal was to help localise these components *in muro* and to evaluate the role of the constituents of the cell wall in growth and tissue morphogenesis (Willats et al. 2001). Various antibodies that target de-methylesterified pectin, highly methylesterified pectin, low methylesterified pectin and calcium bridges are traditionally used to study the middle lamella (Bou Daher and Braybrook 2015; Chapter 10). The use of these antibodies has confirmed the diversity in the nature of pectic substances in different biological tissues. This diversity of chemical substances has already been observed by Mangin, Frémy, Ehrlich and Fallenberg by the use of various dyes and extraction methods as stated above. More than one century after the discovery of the middle lamella and the demonstration of its enrichment in low methylesterified pectin, the major technical breakthrough in cytology was TEM and immunocytology that ultimately came to confirm the work of the 19th century.

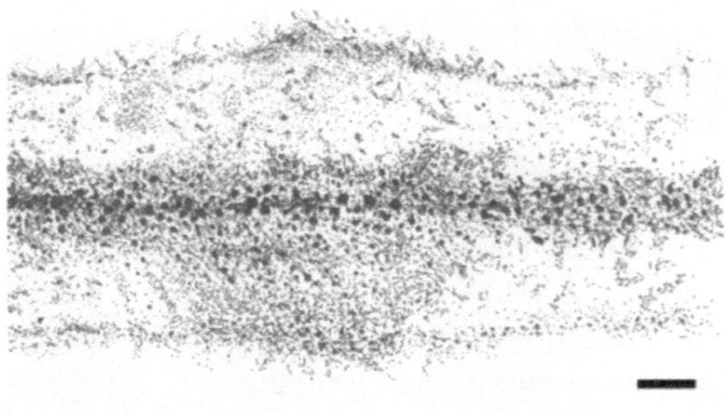

FIGURE 23.1 Drawing of "an enlarged axillary wall of onion roots demonstrating the increased size of the iron particles in the region of the middle lamella." This pattern may be due to the ionic bonds formed between iron cations and COO^- groups of GalA which accumulate more in the region of the middle lamella. Hand-drawn by Rawen Ben Malek after Albersheim et al. (1960). Scale bar = 100 nm.

These experimental methods might in some sense bias our current interpretation of the middle lamella for the following reasons:

The calcium-mediated cross-bridging of the anionic HGs has, so far, only been demonstrated *in vitro*. The capacity of monoclonal antibodies such as 2F4 to identify the egg-box structure is only putative as Liners et al. (1989) mention in the abstract: "The epitope recognised is probably part of the dimers of pectin chains associated according to the 'egg-box' model," and there is no definitive evidence of a strong specificity to such structures *in muro*. In fact, the experimental protocol, which generated the monoclonal 2F4 antibody, was conducted using *in vitro* gels made only of the pectin water-soluble fraction of oranges and a relatively high concentration of calcium (if we compare it to the physiological concentrations of calcium present in the cell wall). This might seem of minor importance, but until today, the native structure of pectin is still unknown as its structural characterisation depends highly on extraction methods that inevitably alter its *in muro* conformation (Atmodjo et al. 2013). Furthermore, some microdomains of the cell wall might present some significant charge or steric hindrance effect, which might alter the binding of antibodies (Sutherland et al. 2009). As a consequence, immunocytology might be a good tool to identify putative structures, but a more specific *in muro* study targeted towards microdomains in the cell wall needs to be conducted using different techniques.

These recent advancements on the characterisation of the middle lamella are addressed within the framework of the general settlement established upon its discovery. This seems to be restricting the new breakthroughs as the methodological routine. The subsequent results do not fall far from those that have already been established in the 19th century. A different viewpoint of the problem might inspire new methods to study the middle lamella. In the next part, we suggest an evolutionary view of the middle lamella and more generally of the ECM to understand the mechanisms related to the appearance of the middle lamella and how it has been maintained through the evolutionary history of plants.

FROM ECM TO MIDDLE LAMELLA

Throughout evolution, the capability of cells to adhere was made possible by the outermost part appressed to the cell wall and membrane known as the ECM (Helm and Potts 2012). The cell ECM provides a dynamical framework in the tissue organisation. The adhesive property of the ECM is due to the cross-links between structural polymers or fibres leading to a reticulated and hydrated gel. The mechanical properties of ECM compounds are assumed to be non-linear due to the constant remodelling of permanent or transient cross-links under the influence of mechanical loadings that trigger a dynamic change in the stiffness of the ECM gel (Mak 2020). The chemical nature of the cross-link bonds itself is not a fixed parameter, and the bounds can be hydrogen, ionic or covalent. What determines the persistence of the adhesion is the stability, the strength and the dynamics of the cross-links. Accordingly, an ECM capable of satisfying these features would facilitate the transition to multicellularity. Therefore, the main biological phenomena that progressively determined the emergence of multicellular organism are first the occurrence of cell adherence, which paved the path for the emergence of aggregative multicellular organism and secondly the symplasmic continuity and intercellular communication that leads to the emergence of complex multicellular organism. As it has been described by Bich et al. (2019), multicellularity is a common phenomenon that is "neither unique nor rare". Accordingly, all the tools needed to acquire multicellularity: cell adhesion, communication, differentiation and coordination have been identified in unicellular organisms, which facilitate the transition multicellularity either in aggregative or more complex form (Grosberg and Strathmann 2007).

The middle lamella is the cell wall–appressed part of the plant ECM. The biochemical nature of the middle lamella shows its potential ability to form a reticulated gel. The general term attributed to the adhesive part of their ECMs is "mucilage". The mucilage, by opposition to the middle lamella, shows a thick, light layer with dispersed polymers. However, due to its chemical nature and biological functions, a similarity between the middle lamella and mucilage may probably exist. In fact, the middle lamella may be an evolved form of the mucilage. In the next part, we suggest to explore this question to see whether the mucilage can shift from its thick and light aspect to a thin and electron-dense layer. If so, what would drive such a shift?

Volvocaceae: The Evolutionary Edge between Mucilage and Middle Lamella

As Chlamydomonas and plants share common ancestors and structural similarities in their cell wall organisation, this unicellular algae is highly suitable as a model for plant metabolism/physiology (Nishii and Miller 2010). The multicellular form of Chlamydomonas is thought to be the Volvocaceae, which contain about 40 multicellular species close to unicellular *Chlamydomonas reinhardtii*. In fact, cytology, physiology and phylogenetic studies support the hypothesis that Volvox evolved from Chlamydomonas. This multicellularity can be explained by two hypotheses (Kaiser 2001):

i. Chlamydomonas sister cells tend to cohere to each other after division surrounding themselves with mucilage which spontaneously generates a multicellular array of non-motile cells which is also known as the palmelloid stage.
ii. In Volvox, we observe failure of daughter cells to separate which result in cytoplasmic bridges.

Both cases point out the fact that the ECM, also called mucilage, along with cytoplasmic bridges play an important role in the biological cycles of Volvocaceae. These two structures are thought to be fundamental during the inversion mechanism, a developmental step present in all Volvocaceae (Green et al. 1981; Sumper et al. 2000).

The micrographs from Volvox show the mucilaginous structure either as a light thick surface as in *V. aureus* (Figure 23.2) or as a thin electron-dense layer crossing the mucilage also as in *V. globator* (Figure 23.3). This thin layer has been qualified as the middle lamella (Ikushima and Maruyama 1968). The presence of such a structure remains elusive, as it is clear that the term middle lamella has only been attributed due to the staining properties of this layer.

Interestingly, this so-called middle lamella appeared in *Volvox* species that preserved their protoplasmic connections after the inversion process (Hoops et al. 2006). Before the inversion, the *Volvox* embryo is formed of a colony of germ cells each possessing two flagella. In the embryonic stage, the flagella are oriented inside the sphere. The inversion occurs in the next generation so that the flagella are exposed to the outside to allow the organism to swim (Haas et al. 2018). The post-inversion-protoplasmic-connections seem to be reinforced by an electron-dense material. From these observations, two hypotheses can be made. Either the emergence of a middle lamella has somehow contributed to the maintenance of the protoplasmic bridges or the reinforcement of the protoplasmic bridges has created a mechanical compression effect that modifies the rheology of the gel and results in the dark-stained-aspect of the middle layer. To further explore these questions, we will examine two cases. In the first case, we will look at an example where the compression effect is removed but the middle lamella is observable. The second case will tackle the putative compression effect where no middle lamella is observed.

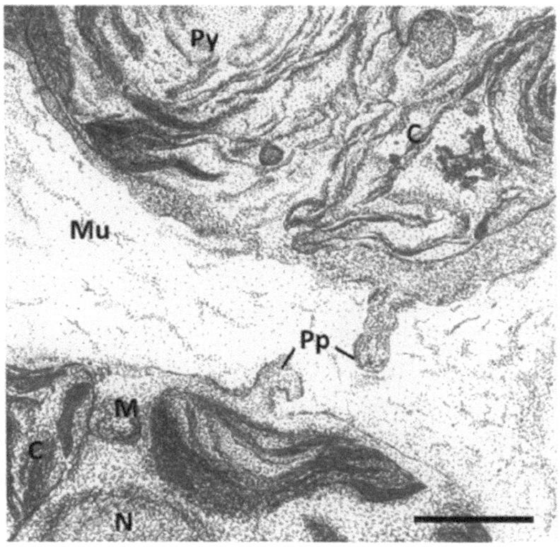

FIGURE 23.2 Drawing of an "electron micrograph of a section through two neighbouring protoplasts in coenobium of the *aureus* type showing protoplasmic projections. C-chloroplast, M-Mitochondria, Mu-Mucilaginous portion, N-Nucleus, Py-pyrenoid." Note how the protoplasmic bridges (called Pp) are ruptured after inversion and the intercellular space is filled with mucilage. The authors emphasise the fact that no middle lamella is observed in this case. Hand-drawn by Rawen Ben Malek after Ikushima and Maruyama (1968). Scale bar = 1 μm.

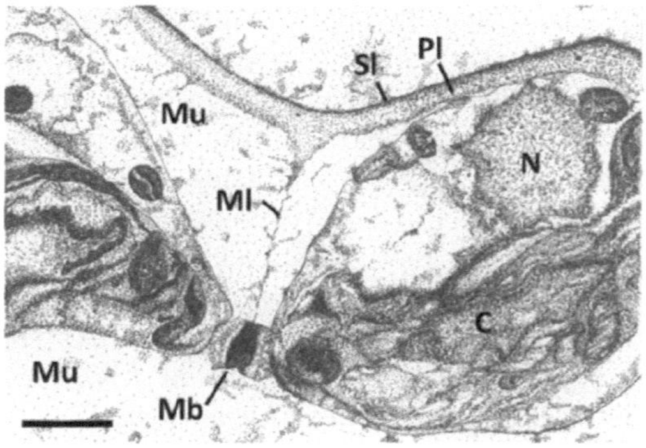

FIGURE 23.3 Drawing of "electron micrograph of a section through the peripheral part of a free-swimming coenobium of the *globator* type of Volvox. Sl-Superficial lamella, Pl-Peripheral lamella, Mb-Medial body, Ml-Middle lamella, C-chloroplast, M-Mitochondria, Mu-Mucilaginous portion, N-Nucleus". The ECM between the two cells appears as a lightly stained thick surface crossed by thin electron-dense layer referred to as the middle lamella between the two cells. The protoplasmic bridge seems to be reinforced by a thick electron-dense material. Hand-drawn by Rawen Ben Malek after Ikushima and Maruyama (1968). Scale bar = 1 μm.

Post-Cytokinesis *In Vitro* Middle Lamella

The middle lamella, as defined in the introduction, emerges after cytokinesis as the first cell wall layer. In that process, vesicles transport methylesterified HG to the level of the phragmoplast (Gu and Rasmussen 2022). The proper optical features of the middle lamella appear after the HGs are de-methylesterified. In this part, we present an example of a post-cytokinesis-induced middle lamella synthesised in a cell suspension. Leppard and Ross Colvin (1971a) tried to recreate an aggregation mechanism similar to algae from plant protoplasts of *Daucus carota* and *Phaseolus vulgaris*. The motivation of this study was to characterise the structure of the intercellular spaces by recreating multicellular organisms. Leppard and Colvin used heavy metals to emphasise the structure of the intercellular substance. The morphologies observed corresponded to electron-opaque fibres, vesicles, electron-transparent rods, myelin-like figures and lacunes or gaps. The emerging frequency of these forms was quantified to show their independence from variability between cell clamps. The vesicular material along with the electron-transparent rods and opaque fibres were considered as an inherent part of the middle lamella as the vesicular aspect has already been reported by Halperin and Jensen in 1967. It was hypothesised that the electron-opaque fibres (Figure 23.4) are probably lignin-like (Leppard and Colvin 1971a). In a later study (Leppard and Colvin 1971b), further extraction methods, infrared spectroscopy and X-Ray, showed a very low content of lignin and the fibrillar substance was identified as a pectin-like polymer. In fact, hydrochloric and carbazole methods showed that this substance was composed of 75% linear polygalacturonic acid (PGA). Similar structures of fibrillar pectins are encountered in nature and more specifically in root–soil interfaces, also called root mucilage (Ramamoorthy and Leppard 1977). These structures have been reported as a cation-exchange system with the soil nutrients and present a high affinity for heavy metal. This capacity of pectin to act as a cation-exchange resin is also present in the middle lamella as heavy metals accumulate in that interface. A question that might come to one's mind is why PGA behaves differently in these cases and forms fibrillar structures rather than the egg-box model in the presence of calcium. Gessa et al. in 1990 have been able to answer this question by synthesising similarly organised gel structures. It seems that the fibrillar pectin is formed when the PGA has a high dissociation degree in an aqueous solution (0.8). When an inorganic calcium substrate fixed on an inert plastic support is put in contact with the PGA solution, a membrane forms with the shape of regular rings. Further acidic treatments and calcium saturation reveal a fibrillar structure. It was hypothesised that this type of arrangement resulted from the dissociation degree of the PGAs which made them arrange as distantly as possible from each other reducing the energy of the system to a minimum. This drives a positioning of the monomers at an angle of 120° in the chain and the carboxylic groups to set out in the same position with a periodicity of 13.1 Å. This type of organisation gives rise to the fibrillar orientation and creates 'free spaces', i.e., void microdomains that have been hypothesised in the cell wall and which drive the ionic mobilisation, diffusion and selective adsorption. The concept of 'free spaces' in the cell wall/middle lamella will be treated below.

Finding the Middle Lamella in the Meristem: Expecting the Unexpected

The meristem is a compact tissue of undifferentiated cells. Despite this compact structure, two cell wall interactions are observed: cell wall corners also known as junction zones marked by cell separation and cohesive zones where supposedly high adhesion forces are observed. However, this statement is based on the sole observation of cells and on the assumptions of polymer interactions. But to date, no satisfactory method to measure cell adhesion forces in

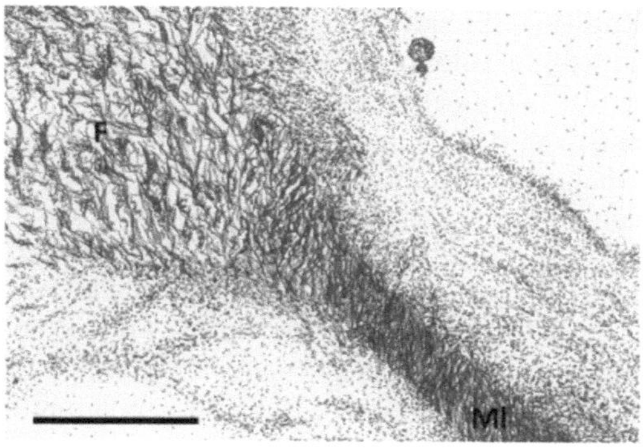

FIGURE 23.4 Drawing of "a cross-section of two cell walls of *Daucus,* which illustrates how the electron-opaque fibres (F) of the cell coat merge gradually into the middle lamella (Ml) between cells. This material received the pre-embedment heavy-atom stain treatment". Hand-drawn by Rawen Ben Malek after Leppard and Colvin (1971a). Scale bar = 500 nm.

plant tissues has been developed (Jarvis 1998; Zamil and Geitmann 2017). Some studies took interest in defining the nature of the polymers in different zones of the meristem using immuno-localisation methods. Although the general idea was to observe the localisation of these antibodies, the micrographs seem to be exploitable to also observe the differential staining of the middle lamella. High-methylesterified pectin was identified mainly in the cohesive zones of meristematic cells from *S. alba* (Sobry et al. 2005). This goes on with the fact that HGs are released in their methylesterified form upon their deposition in the cell wall and are then de-methylesterified by the action of plant pectin methylesterase (Knox 1992; Knox et al. 1990). It would then be logical to observe a very light, if non-existent, staining of the middle lamella in the cohesive zone as the negative charges are masked with methyl or other groups that prevent the binding of metals. Here, there is no optical evidence of the presence of the middle lamella, but its presence is still assumed. The corner junctions, on the other hand, show a high electron density in the micrographs and the use of antibodies such as 2F4 or LM7 (Willats et al. 2001) seems to confirm the acidic nature of the pectin in these zones. It seems that a structure with optical and biochemical properties similar to those of the middle lamella is observed at the interface of cell separation.

When comparing the acidic pectin in junction zones to the pectin in cohesive zones, Sobry et al. (2005) suggested that calcium pectate is more implicated at the level of cell separation by contributing to a minimal tissue cohesion and to the creation of appropriate ionic and porosity environments. According to the authors, calcium was more enriched in less cohesive zones (i.e. junctions) but absent in the region of the middle lamella.

In order for such structures to persist, a reinforcement mechanism is necessary. Plants have been able to adapt to these conditions due to the differential distribution of pectin and other components that are able to form cross-links within the cell wall. Why pectic acids are preferentially present in these specific junction zones remains elusive. As the contribution of the mechanical growth-generated-forces is forging to explain the consequent aspect of the middle lamella, some questions remain unsolved and mechanics alone cannot bring a satisfying answer. For instance, how can compression drive the middle lamella to stain differently? If the primary cell wall also contains HG, why do staining metals preferentially accumulate in the region of the middle lamella? To assess these questions, we suggest exploring the contribution of the Donnan effect on the cell wall at sub-parietal levels.

THE DONNAN–GIBBS EQUILIBRIUM IN THE ECM

It seems that the emergence of the middle lamella could be due to two factors: (i) A volume reduction of the gel due to tension or compression forces and (ii) the anionic nature of the gel. In (i), the compression of the cells against one another (as previously seen in *Volvox*) or the tension produced by the turgor pressure of the cell directly causes a volume reduction of the gel-like matrix. Since the matrix is a porous material, the pores of the matrix also shrink leading to a loss of water. A similar mechanism in the cell wall has been described by Bidhendi and Geitmann (2016). In addition, due to the anionic nature of the gel (ii), it can thermodynamically be considered as an osmotic membrane that generates a Donnan–Gibbs potential (Laktionov et al. 2022). The Donnan–Gibbs equilibrium is the result of an uneven distribution of charged particles in two solutions separated by a semi-permeable membrane. The semi-permeable membrane constrains the system by preventing the transfer of one or more ionic species (Safronov et al. 2006).

The following equation models Donnan's law:

$$\frac{C_{Cl-}^{gel}}{Cs} = \frac{Cs}{C_{Na+}^{gel}} = \sqrt{1+\left(\frac{\alpha}{2Cs\nu_{gel}}\right)^2} \pm \frac{\alpha}{2Cs\nu_{gel}}$$

where C^{gel}_{Cl-} and C^{gel}_{Na+} are the concentrations of monovalent anions and cations, respectively, in the internal solution, Cs is the salinity of the external solution, v_{gel} is the gel molar volume (inverse density of gel segments) and α is the ionisation degree. The "±" sign in the formula accounts for the sign change in polyanion vs. polycation gels (Laktionov et al. 2022).

For example, in a polyanionic gel such as the middle lamella formed of anionic HG, a volume reduction first leads to a loss of water and a reduction in the pore size. The presence of negatively charged carboxylic groups (COO^-) in the volume-reduced pores triggers a non-even distribution of cations. The above equation shows that a volume reduction (v_{gel}) induces the accumulation of cations inside the gels and a reduction of salinity in the rejected solution. In fact, the mechanical squeezing of polyelectrolyte hydrogels has been suggested as a water desalination technique where the charged salts are accumulated in the gel and the desalted water is released after a squeezing step (Li et al. 2011; Arens et al. 2017).

The Donnan effect has previously been reported in the cell walls and in the ECMs in general (Kloareg et al. 2021; Thibault and Ralet 2003) but has not been described in the subdomains of the cell wall such as the middle lamella. Considering that the cell wall is composed of cellulose, hemicellulose and pectin, we should expect to see a binding of metal contrasting agentin parts other than the middle lamella. The fact that these cations preferentially accumulate in the middle lamella zone suggests the presence of microdomains that have a higher affinity for cations. Hope and Stevens introduced the concept of Apparent Free Space, in 1952, which distinguishes two types of free spaces in the cell wall: the water free space (WFS) and the Donnan Free Space (DFS) (Hope and Stevens 1952). The nature of the free space depends on the presence of negatively charged particles and on the size of the pores. Accordingly, macropores are WFS where solvants and molecules can diffuse freely. Micropores are DFS that selectively accumulate cations in the preferential order of their charges (White 2011). Consequently, trivalent cations have priority over divalent cations, which bind more easily than monovalent cations. The compression effect as well as the reticulation of the acidic pectin can therefore trigger the formation of these micropores and create DFSs in the zone of the middle lamella. This would therefore explain the preferential staining of the middle lamella with trivalent/divalent cations in comparison with the cell wall (Albersheim et al. 1960). It would also explain why when negative charges are masked as it is the case for meristems, the middle lamella does not stain as usual.

Therefore, we can assume that the reinforcement of protoplasmic connections in Volvox would maintain this compression effect and create DFSs at the interface of the middle lamella. This microdomain organisation is even more obvious in the case of ripening fruits.

The fruit ripening process is thought to be accompanied by a dynamic remodelling of the middle lamella or the hydrolytic activity of enzymes that lead to the swelling and splitting of the cell wall. This results in the loss of cell adhesion which translates into an overall decrease in tissue firmness (Brummell 2020). Interestingly, these events do not seem to occur in the regions surrounding plasmodesmata and the structure of the middle lamella between two plasmodesmata is maintained. Many investigations that analysed plasmodesmata have found that the region surrounding them stains differently and has denser fibrillar structures (Ben-Arie et al. 1979; Roy et al. 1997). Therefore, regions surrounding the plasmodesmata may differ in composition. Cationic colloidal gold was used by Roy et al. (1997) to recognise the COO^- groups of GalA in de-methylesterified pectin. Interestingly, the results show that cationic gold bounds poorly to the plasmodesmata-wall complex even after de-metylesterification and the use of chelation agents. The middle lamella however stains properly in the micrographs meaning that the metallic dyes were able to bind to negatively charged groups. The poor fixing of the cationic gold might be due to the size of the compressed microdomains. The potential presence of cations (metallic stain or putative Ca^{2+} ions) may create a cationic environment that prevents cationic gold from fixing and/or trigger a steric congestion that prevents cationic gold from entering the microdomain.

CONCLUSION

In this review, we examined our current definition of the middle lamella by looking at the different experimental protocols/data and theoretical hypotheses that built this concept. As we look through examples of the middle lamella in plants and other related organisms, we identify some aspects that question this paradigm. Some cases were treated with regard to the optical properties, the biochemical nature and the biological function. It turns out that building a middle lamella that fits both the current paradigm and the exceptions, described in this chapter, requires specific developmental, mechanical and chemical conditions. First, the middle lamella—either in plants or related green organisms—necessarily requires the presence of anionic polymers that have the capacity to cross-link. In the case of plants, these polymers correspond to de-methylesterified HG where the anionic function comes from the COO^- groups. In other related organisms, such as Volvox, similar functions are ensured by the HRGPs. Secondly, a volume reduction either by tension or compression is needed to build DFSs within the polyanionic gel. In the absence of mechanical constrictions in a calcium-saturated environment, the anionic polymers with a high degree of dissociation organise in maximal distance with respect to each other and adopt a fibrillar conformation. This type of conformation is counterintuitive as we expect HGs to form a gel according to the egg-box model in the presence of calcium. Consequently, we suggest that volume reduction of the polyanionic gels is necessary to create the middle lamella. This can be explained by the Donnan–Gibbs equilibrium. In the case of compression of cells one against the other, DFSs are created. In the DFS, a charge effect is built. As a consequence, heavy metals and cationic salts are selectively accumulated. Therefore, exploring the concept of microdomains or 'free spaces' in the cell wall/middle lamella as well as the rheology of the gels under mechanical constraints similar to those occurring in tissues from plants and related green organisms might constitute a good track for further research on the middle lamella.

ACKNOWLEDGEMENTS

The authors thank Stéphanie Robert and Stéphane Verger for critical reading.

NOTES

1 Mangin, L. (1888) Sur la constitution de la membrane des végétaux. Comptes rendus hebdomadaires des séances de l'Académie des sciences. Tome 107, page 144.

"On savait cependant que, dans les cellules agrégées en tissus, la lamelle moyenne est formée par un corps qui n'a pas de réactions de la cellulose. Les tissus des fruits, des racines, renferment en effet d'après M. Frémy, un principe immédiat qu'il a nommé pectose; ce principe qui n'a pas pu être séparé de la cellulose, produit par une série de transformations complexes, les composés pectiques qu'on rencontre dans les fruits mûrs. D'autre part, M. Maudet a montré que la moelle de certains arbres renferme de la pectose et du pectate de chaux; ces substances formeraient, suivant M. Frémy, un ciment destiné à relier les cellules entre elles, comme cela a lieu probablement dans les fibres libériennes du lin et du chanvre. "

"We knew that in tissues of aggregated plant cells, the middle lamella is formed by a body that does not react with cellulose. Tissues from fruits and roots hold, according to Mr Frémy, an immediate principal he called *pectose*; this principal did not separate from cellulose. It produces by a series of complex transformations, the pectic components that we find in ripened fruits. Furthermore, Mr Maudet demonstrated that the pith of certain trees contains pectose and lime pectates; these substances could constitute, according to M. Frémy, a cement destined to stick cells together, as it probably occurs in fibers from the secondary phloem of flax or hemp.

2 Mangin, L. (1893) sur l'utilisation du rouge de ruthénium, Comptes rendus hebdomadaires des séances de l'Académie des sciences. Tome 116 pages 653–655.

"Le rouge de ruthénium jouit encore d'une propriété non moins importante : il se fixe énergétiquement sur les gommes et les mucilages dérivés des composés pectiques, sans colorer les mucilages formés par la cellulose ou par les produits de liquéfaction de callose."

"Ruthenium red has another and not least important property: it fixes energetically on gums and mucilages derived from pectic components, and doesn't stain mucilages which are composed of cellulose and other products derived from the liquefication of callose."

REFERENCES

Arens L, Albrecht JB, Höpfner J, Schlag, K, Habicht A, Seiffert S, Wilhelm, M (2017) Energy Consumption for the desalination of salt water using polyelectrolyte hydrogels as the separation agent. *Macromolecular Chemistry and Physics* 218:1700237.

Albersheim P, Darvill A, Roberts K, Sederoff R, Staehelin A (2010) *Plant Cell Walls*. Garland Science.

Albersheim P, Mühlethaler K, Frey-Wyssling A (1960) Stained pectin as seen in the electron microscope. *The Journal of Biophysical and Biochemical Cytology* 8:501–506. https://doi.org/10.1083/jcb.8.2.501

Allen CE (1901) On the origin and nature of the middle lamella. *In Gazette* 32 (1). https://www.jstor.org/stable/2464904

Atmodjo MA, Hao Z, Mohnen D (2013) Evolving views of pectin biosynthesis. *Annual Review of Plant Biology* 64:747–779. https://doi.org/10.1146/annurev-arplant-042811-105534

Ben-Arie R, Kislev N, Frenkel C (1979) Ultrastructural changes in the cell walls of ripening apple and pear fruit. *Plant Physiology* 64:197–202. https://doi.org/10.1104/pp.64.2.197

Bich L, Pradeu T, Moreau JF (2019) Understanding multicellularity: The functional organization of the intercellular space. *Frontiers in Physiology* 10:1170. https://doi.org/10.3389/fphys.2019.01170

Bidhendi AJ, Geitmann A (2016) Relating the mechanics of the primary plant cell wall tomorphogenesis. *Journal of Experimental Botany* 67:449–461. doi:10.1093/jxb/erv535

Bou Daher F, Braybrook SA (2015) How to let go: Pectin and plant cell adhesion. *Frontiers in Plant Science* 6:523. https://doi.org/10.3389/fpls.2015.00523

Brummell DA (2020) Sensing when the wall comes tumbling down. *Journal of Experimental Botany*. 71:6865–6868. https://doi.org/10.1093/jxb/eraa436

Conrad CM (1926) A biochemical study of the insoluble pectic substances in vegetables. *American Journal of Botany* 13:531. https://doi.org/10.2307/2435283

Domozych DS, Domozych CE (2014) Multicellularity in green algae: Upsizing in a walled complex. *Frontiers in Plant Science*. 18:649. doi:10.3389/fpls.2014.00649

Dröscher A (2002) Edmund B. Wilson's The Cell and Cell Theory between 1896 and 1925. *History and Philosophy of the Life Science* 24:357–389.

Ehrlich F (1917) Die Pektinstoffe, ihre Konstitution und Bedeutung. *Chemistry Zeitung*. 41:197–200.

Fellenber, TV (1914) Zur Kenntnis des Pektins. *Mitteilungen aus Lebensmitteluntersuchung und Hygiene* 5:225.

Green KJ, Viamontes GI, Kirk DL (1981) Mechanism of formation, ultrastructure, and function of the cytoplasmic bridgesystem duringmorphogenesis in Volvox. *The Journal Cell of Biology* 91. https://rupress.org/jcb/article-pdf/91/3/756/1075648/756.pdf

Grosberg RK, Strathmann RR (2007) The evolution of multicellularity: A minor major transition? *Annual Review of Ecology, Evolution, and Systematics* 38:621–654. https://doi.org/10.1146/annurev.ecolsys.36.102403.114735

Gu Y, Rasmussen CG (2022) Cell biology of primary cell wall synthesis in plants. *Plant Cell* 34:103–128. doi:10.1093/plcell/koab249

Haas PA, Goldstein RE (2018) Embryonic inversion in *Volvox carteri*: The flipping and peeling of elastic lips. *Physical Review E* 98 doi:10.1103/PhysRevE.98.052415

Halperin W, Jensen WA (1967) Ultrastructural changes during growth and embryogenesis in carrot cell cultures. *Journal of Ultrastructure Research* 18:428–443. https://doi.org/10.1016/S0022-5320(67)80128-X

Helm RF, Potts M (2012) Extracellular matrix (ECM). *Ecology of Cyanobacteria II: Their Diversity in Space and Time*. 461–480. Springer Netherlands. https://doi.org/10.1007/978-94-007-3855-3_18

Hofmeister W (1867) *Die Lehre von der Pflanzenzelle*. Leipzig.

Hooke R (1665) *Micrographia, or, Some Physiological Descriptions of Minute Bodies Made by Magnifying Glasses: With observations and inquiries thereupon*. Printed by J. Martyn and J. Allestry, London.

Hoops HJ, Nishii I, Kirk D L (2006) Cytoplasmic bridges in Volvox and its elatives. In *Cell-Cell Channels,* 65–84. Springer New York. https://doi.org/10.1007/978-0-387-46957-7_4

Hope AB, Stevens P G (1952) Electric potential differences in bean roots and their relation to salt uptake. *Australian Journal of Scientific Research*. Ser. B:Biological Sciences *5*:335–343. https://doi.org/10.1071/bi9520335

Ikushima N, Maruyama S (1968) The protoplasmic connection in Volvox. *The Journal of Protozoology*. 15:136–140. https://doi.org/10.1111/j.1550-7408.1968.tb02098.x

Jarvis MC (1998) Intercellular separation forces generated by intracellular pressure. The *Plant, Cell and Environment*. 21:1307–1310. https://doi.org/10.1046/j.1365-3040.1998.00363.x

Kaiser D (2001) Building a multicellular organism. *Annual Review of Genetics* 35:103–123. https://doi.org/10.1146/annurev.genet.35.102401.090145

Kerr T, Bailey IW (1934) The cambium and its derivative tissues: No.X. Structure, optical properties and chemical composition of the so-called middle lamella. *Journal of the Arnold Arboretum 15*:327–349. https://www.jstor.org/stable/43782261

Kloareg B, Badis Y, Cock JM, Michel G (2021) Role and evolution of the extracellular matrix in the acquisition of complex multicellularity in Eukaryotes: A macroalgal perspective. *Genes.* 12(7) https://doi.org/10.3390/genes12071059

Knox JP (1992) Cell adhesion, cell separation and plant morphogenesis. *Plant Journal* 2:137–141. https://doi.org/10.1111/j.1365-313X.1992.00137.x

Knox RB, Heslop-Harrison J, Read C (1970) Localization of antigens associated with the pollen grain wall by immunofluorescence. *Nature*. 225:1067–1068.

Knox JP, Linstead PJ, King J, Cooper C, Roberts K (1990) Pectin esterification is spatially regulated both within cell walls and between developing tissues of root apices. *Planta.* 181:512–521. https://doi.org/10.1007/BF00193004

Kuhn TS (1962) *The Structure of Scientific Revolutions*. University of Chicago Press, Chicago.

Laktionov M, Nová L, Rud OV (2022) Water desalination using polyelectrolyte hydrogel: Gibbs ensemble modeling. *Gels 8*:656. https://doi.org/10.3390/gels8100656

Lakatos I (1978) *The methodology of scientific research programmes*. In *The Methodology of Scientific Research Programmes*. Cambridge University Press. https://doi.org/10.1017/cbo9780511621123

Leppard GG, Colvin JR (1971a) Fine structure of the middle lamella of aggregates of plant cells in suspension culture. *Journal of Cell Biology*. 50:237–246. https://doi.org/10.1083/jcb.50.1.237

Leppard GG, Colvin JR (1971b) Fibrillar lignin or fibrillar pectin? *Journal of Polymer Science*, Part C Polymer symposia 36:321–326. https://doi.org/10.1002/polc.5070360123

Leppard G G, Colvin JR, Rose D, Martin SM (1971) Lignofibrils on the external cell wall surface of cultured plant cells. *Journal of Cell Biology.* 50:63–80.

Li D, Zhang X, Yao J, Simon GP, Wang H (2011) Stimuli-responsive polymer hydrogels as a new class of draw agent for forward osmosis desalination. *Chemical Communication* 47:1710–1712.

Liners F, Letesson JJ, Didembourg C, Van Cutsem P (1989) Monoclonal antibodies against pectin: Recognition of a conformation induced by calcium. *Plant Physiology* 91:1419–1424. doi:10.1104/pp.91.4.1419. PMID:16667195

Mak M (2020) Impact of crosslink heterogeneity on extracellular matrix mechanics and remodeling. *Computational and Structural Biotechnology Journal* 18:3969–3976. https://doi.org/10.1016/j.csbj.2020.11.038

Mangin L (1888) Sur la constitution de la membrane des végétaux. *Comptes rendus hebdomadaires des séances de l'Académie des sciences*. Tome 107:144.

Mangin L (1893) Sur l'utilisation du rouge de ruthénium. *Comptes rendus hebdomadaires des séances de l'Académie des sciences*. Tome 116:653–655.

Mohl H (1835) *Vox. Ueber die Verbindung der Zellen untereinander*. PhD Dissertation, Tübingen.

Nishii I, Miller SM (2010) Volvox: Simple steps to developmental complexity? *Current Opinion in Plant Biology* 13:646–653. https://doi.org/10.1016/j.pbi.2010.10.005

Ritter GJ (1925) Distribution of lignin in wood. *Industrial and Engineering Chemistry* 17:1194–1203.

Ropartz D, Ralet M-C (2020) *Pectin structure. Pectin: Technological and Physiological Properties* 17–36. Springer International Publishing. https://doi.org/10.1007/978-3-030-53421-9_2

Roy S, Watada AE, Wergin, WP (1997) Characterization of the cell wall microdomain surrounding plasmodesmata in apple fruit. *Plant Physiology* 114:539–547. https://doi.org/10.1104/pp.114.2.539

Sachs J (1874) *Lehrbuch der Botanik*. 4th ed. Leipzig.

Safronov AP, Shklyar TF, Borodin VS, Smirnova YA, Sokolov SY, Pollack GH, Blyakhman FA (2006) Donnan potential in hydrogels of poly(methacrylic acid) and its potassium salt. *Water and the Cell.* 273-284. Springer Netherlands. https://doi.org/10.1007/1-4020-4927-7-13

Sekeres J, Zarsky V (2018) 180 Years of the cell: From Matthias Jakob Schleiden to the cell biology of the twenty-first century. *Plant Cell Monograph* 23:7–37. https://doi.org/10.1007/978-3-319-69944-8_2

Sobry S, Havelange A, Van Cutsem P (2005) Immunocytochemistry of pectins in shoot apical meristems: Consequences for intercellular adhesion. *Protoplasma.* 225:15–22. https://doi.org/10.1007/s00709-005-0089-3

Strasburger E (1876) Über Zellbildung und Zelltheilung. Hermann Dabis, Jena.

Strasburger E (1884) *Das kleine botanische Practicum für Anfänger.* G. Fischer, Jena.

Strasburger E (1898) *Die pflanzlichen Zellhäute*. Gebrüder Borntraeger, Berlin.

Sumper M, Nink J, Wenzl S (2000) Self-assembly and cross-linking of Volvox extracellular matrix glycoproteins are specifically inhibited by Ellman's reagent. *European Journal of Biochemistry*. 267:2334–2339. https://doi.org/10.1046/j.1432-1327.2000.01241.x

Sutherland P, Hallett I, Jones M (2009) Probing cell wall structure and development by the use of antibodies: A personal perspective. *New Zealand Journal of Forestry Science* 39. https://nzjfs.scionresearch.com

Thibault J-F, Ralet M-C (2003) Physico-chemical properties of pectins in the cell walls and after extraction. *Advances in Pectin and Pectinase Research* 1:91–105. Springer, Netherlands. https://doi.org/10.1007/978-94-017-0331-4_7

Van Iterson G (1927) De wording van den plantaardigen celwand. *Chemical Weekblad.* 24:166–187.

Vreeland V (1970) Localization of a cell wall polysaccharide in brown alga with labelled antibody. *Journal of Histochemistry and Cytochemistry* 18:371–373.

White PJ (2011) Ion uptake mechanisms of individual cells and roots: Short-distance transport. *Marschner's Mineral Nutrition of Higher Plants*, Third Edition. 7–47. Elsevier Inc. https://doi.org/10.1016/B978-0-12-384905-2.00002-9

Willats WGT, Orfila C, Limberg G, Buchholt HC, Van Alebeek GJWM, Voragen AGJ, Marcus SE, Christensen TMIE, Mikkelsen JD, Murray BS, Knox JP (2001) Modulation of the degree and pattern of methyl-esterification of pectic homogalacturonan in plant cell walls: Implications for pectin methyl esterase action, matrix properties, and cell adhesion. *Journal of Biological Chemistry* 276:19404–19413. https://doi.org/10.1074/jbc.M011242200

Zamil MS, Yi H, Puri VM (2014) Mechanical characterization of outer epidermal middle lamella of onion under tensile loading. *American Journal of Botany* 101:778–787. https://doi.org/10.3732/ajb.1300416

Zamil MS, Geitmann A (2017) The middle lamella - more than a glue. *Physical Biology* 14: 015004. https://doi.org/10.1088/1478-3975/aa5ba5

Zdunek A, Gancarz M, Cybulska J, Ranachowski Z, Zgórska K (2008) Turgor and temperature effect on fracture properties of potato tuber (*Solanum tuberosum* cv. Irga). *International Agrophysics* 22:89–97.

24 The Compound Middle Lamella as a Target for Improved Deconstruction of Woody Biomass

Nicholas C. Carpita and Maureen C. McCann
National Renewable Energy Laboratory

INTRODUCTION

In fruits and some vegetables, the middle lamella is a pectin-rich matrix that is primarily responsible for cell–cell adhesion through calcium and ester cross-links (Chapter 23). All that is needed to disrupt cell–cell adhesion and release single cells from the parenchymatous tissues is a Ca^{2+}-chelator and dilute alkali (Marry et al. 2006; Ordaz-Ortiz et al. 2009; Zamil and Geitmann 2017). Pectins comprise two major classes of polysaccharides: homogalacturonan (HG), polymers with backbones of only (1→4)-α-linked GalA residues, and rhamnogalacturonan-I (RG-I), polymers with backbones of (1→2)-α-Rha-(1→4)-α-GalA disaccharide residues. Some HGs, particularly those in flowers and fruits, bear branches of single Xyl residues linked (1→3)-α- to the HG backbone to form xylogalacturonans (Zandleven et al. 2007). To the RG-I backbone can be attached side chains of (1→4)-β-galactans, branched and unbranched 5-linked α-arabinans, and type I arabinogalactans. Rhamnogalacturonan-II (RG-II) is misnamed as it comprises a distinct domain of HG, substituted with complex Rha-containing chains, and can dimerise through the branch-chain residue, apiose, by borate di,di-ester cross-links (O'Neill et al. 2004). Calcium cross-linked HGs are enriched in middle lamellae, but as HG might be covalently linked to RG-I and RG-II, these molecules might form an interaction that is the basis of cell–cell adhesion in primary walls (Vincken et al. 2002; Tan et al. 2013).

Although the molecular basis of cell–cell adhesion between primary walls of parenchymatous tissues of fruits and vegetables has been characterised, the molecular determinants of adhesion in woody tissues and organs, such as stems, are less well understood. Chelators and dilute alkali are insufficient to disrupt cell–cell adhesion in these organs. Within primary walls of certain cell types are elaborated thick secondary wall composites of cellulose, glucuronoxylans, and other hemicelluloses, co-extensive with networks of acidic phenylpropanoids and lignin. Secondary-walled cells constitute the majority of harvested biomass (Chanoca et al. 2019; Ralph et al. 2019). In fibre cells of wood, the middle lamella and adjacent primary wall form a compound middle lamella (CML) containing lignin and other phenolic substances in addition to pectins, cellulose, and hemicelluloses (Westermark 1985; Donaldson 2001; Fromm et al. 2003) (Figure 24.1) Lignin polymerisation is initiated in the middle lamella, where it might interact with pectins and hemicelluloses to form strong zones of cell–cell adhesion (Donaldson 2001; Zamil and Geitmann 2017). Inter- and intra-species variation in lignin composition in the secondary wall might be equally reflected in the CML, although systematic comparisons have not yet been made.

The degree to which molecular interactions between lignin and cellulose and the crystallinity of cellulose reduce the yields of sugar by cellulolytic enzymes is termed biomass recalcitrance (Himmel et al. 2007). However, at the cellular and organ level, differing proportions of cell types, their sizes, shapes, and densities, and the degrees and nature of fibre cell–cell adhesion all impact sugar yields by enzymatic hydrolysis. Thus, we have broadly defined biomass recalcitrance as those features that disproportionately increase energy requirements, increase cost and complexity of biorefinery operations, and/or reduce the conversion of biomass carbon into desired products (McCann and Carpita 2015). In this chapter, we examine the structural components of the CMLs and provide insights into their relative contributions to recalcitrance.

Determining the molecular basis of cell–cell adhesion in tree species and bioenergy grasses could identify potential targets for genetic modifications that enhance cell–cell separation during shredding of crop biomass to small particles for catalytic conversion of lignocellulosic biomass to fuels or chemicals (Yang et al. 2021). Particle size is a determining factor in efficient heat transfer and catalyst accessibility for thermochemical and/or biochemical conversions (Viamajala et al. 2009). The mechanical reduction of intact biomass to particles, termed comminution, is an energy-intensive process to reduce particle size for more efficient conversion processes. However, the smaller the particle desired, the greater the required energy input (Miao et al. 2011). Engineering biomass properties to reduce the energy inputs for comminution will increase efficiencies for thermo-catalytic or biochemical pathways to biofuels

DOI: 10.1201/9781003178309-24

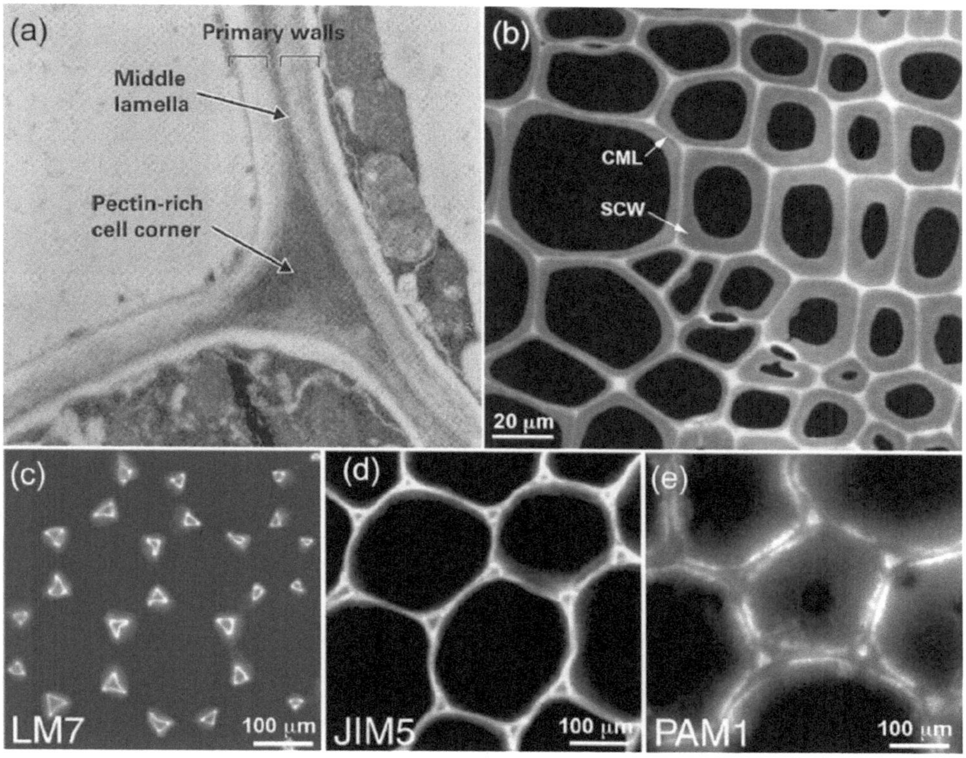

FIGURE 24.1 The middle lamellae of herbaceous and woody plants. (**a**) Herbaceous species with only primary cell walls produce a pectin-rich middle lamella and cell corners (Micrograph by M. McCann) (**b**) Woody species form a CML between the primary walls of SCW-rich xylem cells. Lignin synthesis is initiated in the CML, shown by the fluorescence of the aromatic residues (adapted from Donaldson 2001). (**c–e**) Primary cell walls, immunolabelled with three monoclonal antibodies, to show three types of methyl-esterification of HG in young pea stems or roots (adapted from Willats et al. 2001). (**c**) Leeds monoclonal7 (LM7) detects HGs with non-blockwise de-esterification to about 30%. Intense labelling is found only lining the cell corners. (**d**) John Innes monoclonal5 (JIM5) detects a broad range of both block-wise and non-block-wise HGs de-esterified to between 15 and 70%. (**e**) Phage-display monoclonal1 (PAM1) detects HGs with 0–15% esterification.

and bioproducts. We also review genetic engineering approaches that have been employed to facilitate breakage of the chemical bonds of the CML and reduce energy inputs of biomass processing.

HISTORICAL STUDIES OF THE CML

Over a century ago, the middle lamella was identified by its optical and staining properties as a distinct layer between the cell walls of two daughter cells formed during cell division (Allen 1901). The 'cell plate' was defined as the incipient wall, and an 'intercellular substance' was observed as a new layer between recently divided cells of pine (Strasburger 1875). Although Strasburger (1884) referred to the middle lamella as 'primary wall', he characterised it as a distinctly thin, cutinised, acid-resistant layer exterior to the primary wall. Treub (1878) first suggested that the middle lamella was formed *de novo* by secreted substances between two daughter cells and was not derived from the cell plate.

In dividing cells of onion (*Allium cepa*) root tips, the initial division plane could be stained for carbohydrate, but the staining disappeared during maturation of the cell plate and primary wall, which was then localised to the middle lamella (Timberlake 1900). The middle lamellae of most tissues could be stained with ruthenium red (Mangin 1889, 1893), still routinely used to stain de-esterified pectins. In histochemical studies of secondary xylem formation at the cambium, an intercellular substance, described as the *Zwischensubstanz*, was characterised as comprising calcium pectates and callose (Dippel 1898).

Meshitsuka and Nakano (1985) were first to describe the middle lamellae of woody tissues as a 'compound middle lamella', as differentiated from the pectin-rich middle lamellae of herbaceous tissues by the presence of lignin (Ritter 1925). The term 'lignin (*lignine*)', from the Latin for 'wood', was coined by one of the first plant chemical taxonomists to describe the acid-resistant aromatic network that remained after acid digestion of wood (de Candolle 1813). Because it was the only material remaining after complete dissolution of the cell wall in sulphuric acid, Ritter (1925) determined that 75% of the total lignin of the cell was in the middle lamella, and none was associated with cellulose in woody tissues of alder, basswood (*Tilia* spp.), and pine. A gentle acid treatment showed that lignin was indeed present in secondary walls, but concentrated in the middle lamellae of both hardwood and softwood species (Harlow 1932). A quantitative analysis on small amounts of lamellae isolated from woody sections of Douglas fir gave a lignin value of 72%, leaving 28% uncharacterised but assumed to be carbohydrate (Bailey 1936a). Bailey discounted a major contribution by pectin in woody tissues, but he suggested pentose-containing hemicelluloses comprising about one-half of the uncharacterised material (Bailey 1936b).

The chemical structure of lignin from various softwoods (gymnosperms), such as fir (*Abies* spp.) and hemlock (*Tsuga* spp.), was determined to comprise primarily guaiacyl (G-) units, whereas those of hardwoods (angiosperms), such as birch, poplar, oak (*Quercus* spp.), and maple (*Acer* spp.), were made of both G- and syringyl (S-) units (Hawkins et al. 1937). Additional evidence for this distinction came from use of the Mäule reagent, which stains G-lignin of gymnosperms yellow, whereas the S-lignins of angiosperms stain red (Srivastava 1966). The same monolignols found in the cellulose-rich secondary wall are present in the CML. Consistent with bulk lignin profiles of woody angiosperms and the high methoxy content of the lignin, the middle lamella of birch xylem was found by UV microscopy to be enriched in both G- and S-lignin (Fergus and Goring 1970). Although only 18% of the total lignin was traced to the CML, mechanically isolated CMLs were estimated to be 72–85% lignin by weight. Using nitrobenzene and permanganate oxidation staining, distributions of S- and G-units varied during the development of birch wood, with G-units enriched in the CML, and S-units enriched in the secondary walls (Hardell et al. 1980). Using nitrobenzene oxidation and acidolysis, Westermark (1985) quantified lignin units by GC-MS of whole-wood, mechanically isolated CML, and compression wood in spruce, finding small amounts of unmethoxylated *p*-hydroxyphenyl (H-) units in CML compared to the abundant G-units. Westermark (1985) also found enrichment of the H-units during the early stages of xylem development and in compression wood.

With improvements in interference and confocal laser scanning microscopy (CLSM), lignin localisations could be imaged at xylem cell corners, CML, and the secondary cell walls of pine (Singh and Donaldson 1999) and poplar and grasses (Donaldson et al. 2001). Lignin was highly enriched in the cell corners and CML compared to the secondary walls of these woody species. Similar results were obtained for cell corners and CML of sugarcane (Sant'Anna et al. 2013). Consistent with studies of Westermark (1985), antibodies that recognise H-lignin units detected lignins in the earliest stages of xylem development (Ruel et al. 1994), and monoclonal antibodies against H-, G-, and mixed G-/S-lignins revealed further heterogeneity of lignin type in woody tissues by transmission electron microscopy (TEM) (Ruel et al. 1999). However, compared to direct measurements of enrichment of G- and S-lignin in the CML and cell corners by many researchers, Ruel et al. (1999) found much greater enrichment of the mixed G-/S-lignin in the secondary walls of both poplar and grass walls, with only low abundance of these epitopes in the CML.

Lignin heterogeneity was explored further in several species, including poplar, using radiolabelled monolignols supplied in feeding experiments (Fukushima and Terashima 1990). These exogenous monolignols could be detected by autoradiography of lignified wall domains in sectioned materials. Deposition of H-lignins in the CML occurred almost exclusively at very early stages of xylem development. If developing woody tissues were radiolabelled with coniferin for G-units, and syringin for S-units, then more intense labelling of the CML was detected during secondary wall development. Similar results were obtained in pine (Fukushima and Terashima 1991), but abundant signal from H-units during early xylem formation was followed by intense signal from G-units, and then S-units during induction of compression wood. Results of both studies are consistent with the appearance of H-lignin early in CML development. A more recent sophisticated approach employed click chemistry to associate H-, G-, and S-analogues with their wall domains, finding that the H-units were slightly enriched in cell corners and CML, whereas the G- and S-analogues were associated with the secondary cell wall (SCW) (Morel et al. 2022). How exogenous substrates partition relative to endogenous monolignol substrates is not known. Nevertheless, the preponderance of evidence in the literature indicates that H-units initiate lignification of the pectin-rich cell corners and CML, which then become heavily impregnated with G- units in gymnosperms, and with G- and S-units in angiosperms.

FORMATION OF THE CML DURING WOOD DEVELOPMENT

To reduce recalcitrance of woody biomass, a better understanding is needed of the molecular events of fibre formation leading up to formation of the CML. The vascular cambium produces cells by both periclinal and anticlinal divisions that give rise to cylinders of cells that produce xylem initials on the inner surface and phloem initials to the exterior (Bossinger & Spokevicius 2018). Growth of the xylem initials proceeds by two distinct patterns. A combination of axial and longitudinal expansion of the main body of the xylem initially establishes the ultimate diameter of the fibre or xylary element by symplastic growth (Déjardin et al. 2010). The primary cell walls of poplar xylem are typical of dicots, with xyloglucan, glucuroarabinoxylans (GAXs), and (gluco)mannans as the major hemicelluloses, embedded in a pectin matrix of HG and RG-I (Mellerowicz and Gorshkova 2012). The second pattern of elongation of developing fibres is at their tips, which narrow during elongation and insert, by what is termed 'intrusive growth', into the middle lamellae of the cells above and below the main body of the fibre cells, forming new adhesive connections (Gorshkova et al. 2012). During this process, the middle lamellae transition from pliant structures that allow cells to slip past each other to ones that become stiffened and lignified structures that resist intrusion.

During intrusive growth, immunocytochemical studies show a concentration of HG epitopes at the tapered tips, with xyloglucan epitopes more predominant in the walls of the main body of the cell (Majda et al. 2021). The HG is highly methylated, perhaps contributing to the pliable state of the middle lamella, as highly methylated pectins do not cross-link to form rigid gels. HG is synthesised and secreted in its highly methyl esterified form (Me-HG). Pectin methylesterases (PMEs) can de-esterify the Me-HG in two ways. Some PMEs exhibit a non-blockwise cleavage of methyl esters (Willats et al. 2001). Most PMEs are thought to be 'processive', i.e., they migrate along the backbone in a block-wise fashion to cleave methyl groups from GalA residues from long stretches of HG to form Ca^{2+}-cross-linked zones of tight adhesion (Tanhatan-Nasseri et al. 2011). High-resolution transcript profiling by microarray analysis identified two distinct patterns of PME expression, one upregulated early in initiation and another at the beginning of radial expansion (Schrader et al. 2004). Reduced expression of PMEs stimulated intrusive growth in transgenic aspen (Siedlecka et al. 2008). By contrast, overexpression of recombinant PME interferes with both expansion of the main body and tapered ends of the intrusively growing fibre cells (Siedlecka et al. 2008). Thus, PMEs expressed during the terminal events of growth could de-esterify HG and enable transition of a state from one that is permissive for cell intrusions to one that forms a rigid gel-like matrix by cross-linking with Ca^{2+} (Jarvis 1984). De-esterification of HG by PMEs renders the backbones susceptible to digestion by pectate lyases and polygalacturonases (PGases), so a balance of difference types of PMEs and PGases/pectate lyases is required for growth stasis and termination (Micheli et al. 2000).

Other cross-links have also been implicated in cell–cell adhesion. For example, cell–cell adhesion in sugar beet (*Beta vulgaris*) requires both ester cross-links and Ca^{2+}-cross-linking of HG (Marry et al. 2006). The presence of Ca^{2+} binding sites concentrated in sugar beet middle lamellae and cell corners was revealed by image-electron energy loss spectroscopy (EELS) of sections prepared from TEM (Marry et al. 2006). However, only after sequential treatment with chelator and dilute alkali was separation along the middle lamellae evident.

Another aspect of the formation of the CML is the intricate structure of the cell corners, i.e., the junction of three or more cells. The three-way or four-way junction zones of cells are tightly controlled with respect to pectin composition and complement of cell wall modifying enzymes. Immunocytochemical approaches have shown that structural proteins, such as extensin-like hydroxyproline-rich (HRGP), proline-rich (PRP), and glycine-rich (GRP) proteins, accumulate in the cell corners and in the cell walls of protoxylem (Ye et al. 1991). GRPs form a network around xylem thickenings and in the CML of xylem and parenchyma cells (Ryser 2003). In some instances, the corner is filled but in others large air gaps define a triangular zone, the size of which is tightly controlled by special zones of adhesion (Figures 24.1C-E). Using immunocytochemistry, xyloglucans were also enriched in the newly formed cross walls of clover (*Trifolium pratense*) root tip cells, whereas HG/RG-I epitopes were concentrated in the cell corners (Lynch and Staehelin 1992). The PME inhibitor (PMEI) class of enzymes can partner with PMEs to generate different patterns of partial de-esterification (Sénéchal et al. 2015). Monoclonal antibodies that recognise these different forms of esterified pectins reveal the presence of distinct epitopes in different domains in the primary wall and middle lamella (Figures 24.1C-E). Leeds Monoclonal7 (LM7), which detects HGs with non-blockwise de-esterification to about 30%, exhibits abundant labelling only in cell corners and cell junctions, whereas John Innes Monoclonal5 (JIM5) detects a broad range of both block-wise and non-block-wise HGs de-esterified to between 15 and 70% and also labels middle lamellae. A Phage-display Monoclonal1 (PAM1) detects HGs with 0–15% esterification throughout the middle lamella. RG-I is also located at the cell corners. Both methyl-esterified and unesterified HGs are distributed more or less equally in the cell walls of most cells, and RG-I and its arabinan and galactan side-chains are uniformly distributed along cells walls and cell corners of potato (*Solanum tuberosum*) tubers (Bush and McCann 1999). Potato tubers expressing a fungal RG-lyase, an enzyme that cleaves the RG-I backbone, had deformed cell corner junctions (Oomen et al. 2002). Redistribution of arabinan and galactan side-chains from the primary walls to the middle lamella is observed in these transgenic tubers. Similarly, the Arabidopsis (*Arabidopsis thaliana*) rol1–2 mutant, defective in *Rhamnose synthase1* (RHM1), results in structural defects in RG-I and RG-II that impact normal development of cell corners, leading to severe suppression of root and shoot development (Schumacher et al. 2021). As the LM7 epitope is so precisely localised, polysaccharides containing these epitopes must be specifically targeted to the cell corners, and points of contact for cell–cell adhesion must occur, or enzymes are targeted to polysaccharide substrates to generate these epitope *in situ*. One potential regulator of targeted secretion could be the Rat Sarcoma (Ras)-like GTPase binding protein (Rab), which is a signal transduction element integral to regulation of Golgi-mediated transport in plants (Nielsen et al. 2008). A Rab-A5c protein is associated with vesicle trafficking specifically at the plasma membrane at cell junctions, and a mutation in the *Rab-A5c* gene resulted in severe developmental defects in leaves, lateral roots, and root hairs (Kirchhelle et al. 2016).

The sequential dynamics of polysaccharides can be probed indirectly by differential transcriptional analysis of the machinery of biosynthesis and metabolism. Large-scale RNAseq analysis of thin cryo-sections of young aspen stems gave representative profiles of transcripts expressed differentially by cells of the cambium, early xylem development, and mature xylem (Sundell et al. 2017). Transcripts encoding primary wall polysaccharides dominate profiles in cambial cells and early xylem development, including glycosyl transferases involved in biosynthesis of HG (Caffall and Mohnen 2009), RG-I (Uehara et al. 2017), RG-II (Egelund et al. 2006), the hemicellulose xyloglucan (Zabotina 2012), and primary cell wall cellulose synthase (CesA) genes (Song et al. 2010). By contrast, CesAs associated with secondary wall synthesis become strongly expressed in developing xylem cells and persist through maturation stages (Sundell et al. 2017). Xylans are synthesised by subgroups of synthases homologous to the Arabidopsis Irregular Xylem9/9L (IRX9/9L) and IRX14/14L family Glycosyltransferase43 (GT43) (Wu et al. 2010), and IRX10 family GT47 glycosyl transferases (Brown et al. 2009), which, like CesAs, have isoforms highly expressed during either primary wall or secondary wall synthesis in poplar (Sundell et al. 2017). Although the secondary wall isoforms of cellulose and xylan synthesis define the xylem maturation phase of xylogenesis, several of the primary-wall-related genes, including those of pectin and hemicellulose synthesis, continue expression during this maturation phase (Sundell et al. 2017). Expression of these genes may correspond to the biosynthesis of primary walls of parenchymatous cells and ray cells, but they might also be associated with maturation of the CML. Genes of cell-wall metabolism and rearrangement are also expressed, such as PMEs and pectate lyases, which could potentially function in the transition from pliable to rigid CML during intrusive growth.

CELL–CELL ADHESION IN BIOENERGY GRASSES

Grasses are commelinid monocots, which make a Type II primary cell wall (Carpita and Gibeaut 1993). The polysaccharide structural framework comprises a cellulose microfibril network coated and tethered by glucuronoarabinoxylans (GAXs), and this framework is embedded in matrix polymers of highly substituted GAXs and some pectins (Carpita 1996). Depending on the spacing patterns of acetyl and (4-*O*-methyl)-glucuronosyl (MeGlcA) groups on the xylan backbone, a two-fold or a three-fold helical axis can form (Busse-Wicher et al. 2016). With evenly spaced alternating substitutions, a two-fold axis is favoured and predicted to hydrogen bond to the hydrophilic faces of cellulose microfibrils. Among the commelinid monocots, grasses and other members of the Poales make a (1→3),(1→4)-β-D-glucan that appears transiently during rapid stages of cell elongation and is hydrolysed as elongation ceases (McCann et al. 2007). During primary wall development, cross-linking phenylpropanoids enriched in hydroxycinnamic acids, such as ferulate and *p*-coumarate, are deposited as cells mature.

The Type II secondary walls are composites of cellulose and relatively unsubstituted GAX interconnected with a network of acidic phenylpropanoids and lignin (Carpita 1996; Sarkar et al. 2009). Multi-dimensional magic angle spinning solid-state ^{13}C-nuclear magnetic resonance (^{13}C-NMR) spectra show a majority of xylans in sorghum secondary walls to be mainly in a three-fold screw conformation due to high degrees of substitutions with arabinosyl units (Gao et al. 2020). The patterns of secondary wall formation in grasses are distinct from those of the woody dicots. Primary and secondary wall growth occurs in the internodes represented in a developmental gradient from the base to the apex of the stalk (Penning et al. 2019). Stem segments differentiate from the apical meristem, expanding into leaf nodes, and elongate an internode that constitutes most of the height and biomass of the plant (Penning et al. 2019). Cell division occurs at intercalary meristems confined to just above the node at the base of each internode. Cell elongation occurs in a narrow band above the intercalary meristem, and secondary wall growth continues in a broad band extending the entire length of the internode. By contrast to woody dicots, grasses have no lateral secondary meristematic growth from a cambium, and the secondary wall is confined mostly to the stiff and dense rind tissue. However, woody dicots and grasses will undergo similar transitions of primary wall growth, middle lamella extension, and formation of a CML. A key difference in the composition of the CML is the formation of the hydroxycinnamate-rich phenylpropanoid network throughout the primary wall and extending into the middle lamellae.

RNAseq analysis performed on developing rind tissues from the youngest internodes to the most mature at the base revealed several distinct patterns of expression (Penning et al. 2019). One pattern, with highest expression in the elongating young internodes that decreased in lower internodes, was represented by genes typically expressed during primary wall formation. A second pattern, with expression that increased to a peak in the most mature internodes, was represented by genes expressed during secondary wall formation. Other patterns showed gene expression that peaked in the middle internodes and either fell precipitously or maintained expression in the older, maturing internodes. In general, the younger internodes expressed genes associated with cellulose synthesis, xylan, and pectin synthesis, as well as wall remodelling enzymes known to be involved in primary wall synthesis, whereas genes of monolignol synthesis were associated with secondary wall formation. Nevertheless, distinct isoforms of CesAs, xylan synthesis, and even pectin synthesis were also in the secondary wall pattern (Penning et al. 2019). However, as in poplar, certain CesAs, IRX genes of xylan synthesis, and galacturonosyltransferase (GAUT) genes of HG synthesis comprise the intermediate patterns, where expression of genes thought to be associated with primary wall synthesis extends into internodes engaged mostly in secondary wall formation. Although the vast majority of PGases are highly expressed during primary wall synthesis, PGaseA11, PGaseA12, and PGaseG1 are expressed predominantly during secondary wall formation (Penning et al. 2019). Some of these genes commonly associated with primary wall synthesis, whose expression either initiates or continues into mature internodes, might be involved in the formation of the grass CML. Expression of certain members of benzyl alcohol, anthocyanin, *N*-hydroxycinnamoyl, deacetylvindoline (BAHD) class of acyl-CoA transferases have been implicated in the formation of ferulate- and *p*-coumarate-GAX esters in brachypodium (*Brachypodium distachyon*) (Molinari et al. 2013). Similar to their findings, expression of several *BAHD* genes in maize (*Zea mays*) stems follow an intermediate pattern of both primary wall and secondary wall, but two genes, *BAHD9* and *BAHD2a*, show exceptionally high expression during secondary wall formation (Penning et al. 2019).

LIGNIN COMPOSITION AND RECALCITRANCE

A major component of biomass recalcitrance has been attributed to strong interactions between lignin and cellulose that limit access by saccharifying enzymes to the glucan chains of cellulose (Chen and Dixon 2007). Reducing lignin content and altering lignin composition have been targets for engineering biomass plants with reduced recalcitrance to enzymatic digestion. Because the CML is an initiation site of lignin in all angiosperms, and these cross-linked aromatic networks might be components of cell–cell adhesion, resistance to diffusion of cellulases and other enzyme across and along the CML are likely to be key elements in recalcitrance (Yang et al. 2021). The extensive acidic phenylpropanoid networks in primary walls of grasses could contribute additional factors in cell–cell adhesion.

Monolignol synthesis begins with deamination of phenylalanine to *trans*-cinnamic acid by the ammonia lyase (PAL), subsequent hydroxylation to *p*-coumaric acid by cinnamate 4-hydroxylase (C4H), and its coupling to CoA by 4-coumarate:CoA ligase (4CL) (Figure 24.2). The *p*-coumaroyl-CoA conjugate is converted to its respective aldehyde by CoA-dependent cinnamoyl reductase (CCR), and to its respective monolignol, *p*-coumaroyl alcohol, which is a precursor to H-lignin synthesis. *p*-Coumaroyl-CoA is also converted by several hydroxylase and methyl transferase enzymes to feruloyl-CoA, and then reduced to coniferaldehyde by a CCR, and further to coniferyl alcohol by a cinnamyl alcohol dehydrogenase (CAD), to generate the precursor to G-lignin synthesis. Hydroxylation by ferulate 5-hydroxylase (F5H), subsequent methylation by caffeic acid *O*-methyltransferase (COMT), and conversion of the aldehydes to alcohols by CAD yields sinapyl alcohol, the major lignol substrate for synthesis of S-lignin (Figure 24.2).

The lignin heteropolymer is produced via the oxidative coupling of monolignol subunits corresponding to H-, G-, and S-lignins (Figures 24.2). Monolignols are synthesised at the cytoplasmic surface of the endoplasmic reticulum (ER), but how they are exported across the plasma membrane into the wall has not been firmly established. Inhibitor studies implicated certain classes of ATP-binding cassette (ABC) transporters (Miao and Liu 2010), but only *p*-coumaryl alcohol substrates of H-lignin were found to be actively transported by an Arabidopsis ABCG29 transporter, and not the coniferyl or sinapyl alcohol substrates of G- and S-lignin (Alejandro et al. 2012). An alternative 'sink-driven' mechanism was proposed by Perkins et al. (2022), whereby the process of lignification creates a concentration gradient for monolignols to diffuse through the plasma membrane to sites of polymerisation in the apoplast. This diffusion could be facilitated by close association of ER and plasma membrane to channel the substrates more efficiently (Carpita 2012). The 'sink-driven' mechanism is also consistent with the ability to incorporate unnatural phenolic compounds into the lignin structure without needing to express specific transporters (Perkins et al. 2022).

Overall, lignin is structurally heterogeneous, and its composition is determined by the relative proportions of the exported monolignols and other aromatics. Knowledge of the genes encoding the enzymes of lignin biosynthesis has enabled engineering of lignin (Chanoca et al. 2019; Ralph et al. 2019). A multi-omics approach in *Populus trichocarpa* illustrated several strategies of control of lignin content and composition that can result in increased saccharification yields (Wang et al. 2018). In poplar (*Populus* spp.), disruption of genes encoding certain enzymes of monolignol synthesis, such as C4H (Bjurhager et al. 2010), 4CL (Li et al. 2003; Xiang et al. 2017), cinnamate 3'-hydroxylase (C3'H) (Coleman et al. 2008; Mansfield et al. 2012; Ralph et al. 2012), or CCR (Leplé et al. 2007; Smith et al. 2015), resulted in drastic decreases in lignin content. In some instances, down-regulation of early enzymes in monolignol synthesis was correlated with increased saccharification yields of glucose and xylose (Mansfield et al. 2012; Van Acker et al. 2014). Down-regulation of CAD in poplar resulted in enhanced saccharification, with only modest reduction of lignin (Van Acker et al. 2017). Mutations in CAD block the last step in reduction of hydroxycinnamaldehydes to monolignols, resulting in incorporation of the aldehydes (Van Acker et al. 2017). A monolignol 4-*O*-methyltransferase (MOMT) gene overexpressed in poplar resulted in formation of 4-*O*-methyl coniferyl and sinapyl alcohols that terminated lignin polymerisation, reducing lignin content and increasing saccharification yield (Cai et al. 2016). Mutations of a *COMT* gene resulted in blockage of the S-lignin pathway at 5-hydroxyguaiacyl (5-OH-G), but these units are readily polymerised into lignin at the expense of S-units (Bonawitz and Chapple 2010; Yang et al. 2019). The 5-OH-G units are indistinguishable from G units by conventional lignin analysis, such as the Derivatisation followed by Reductive Cleavage (DFRC) method (Lu and Ralph 1997) or NMR, but catalytic delignification and reduction of monolignol derivatives showed by electrospray ionisation mass spectrometry that 5-OH-G is a significant lignin component in native poplar lignin (Yang et al. 2019).

Many of the early attempts to down-regulate steps in the monolignol synthesis pathway resulted in growth penalties, probably related to the water stress resulting from xylem defects. Alternative strategies that took advantage of the plasticity of lignin synthesis were employed to improve saccharification yield without compromising growth and biomass quality (Muro-Villanueva et al. 2019). For example, down-regulation of the *F5H* gene in poplar produced lignin that is enriched in G units at the expense of S-units, but overexpression of *F5H* resulted in plants with more than 90% S-lignin content (Stewart et al. 2009; Yang et al. 2019). Numerous studies in woody biomass species have now shown that high S-lignin results in enhanced pulping efficiency (Huntley et al. 2003) and saccharification yield (Li et al. 2010; Studer et al. 2011; Yang et al. 2019).

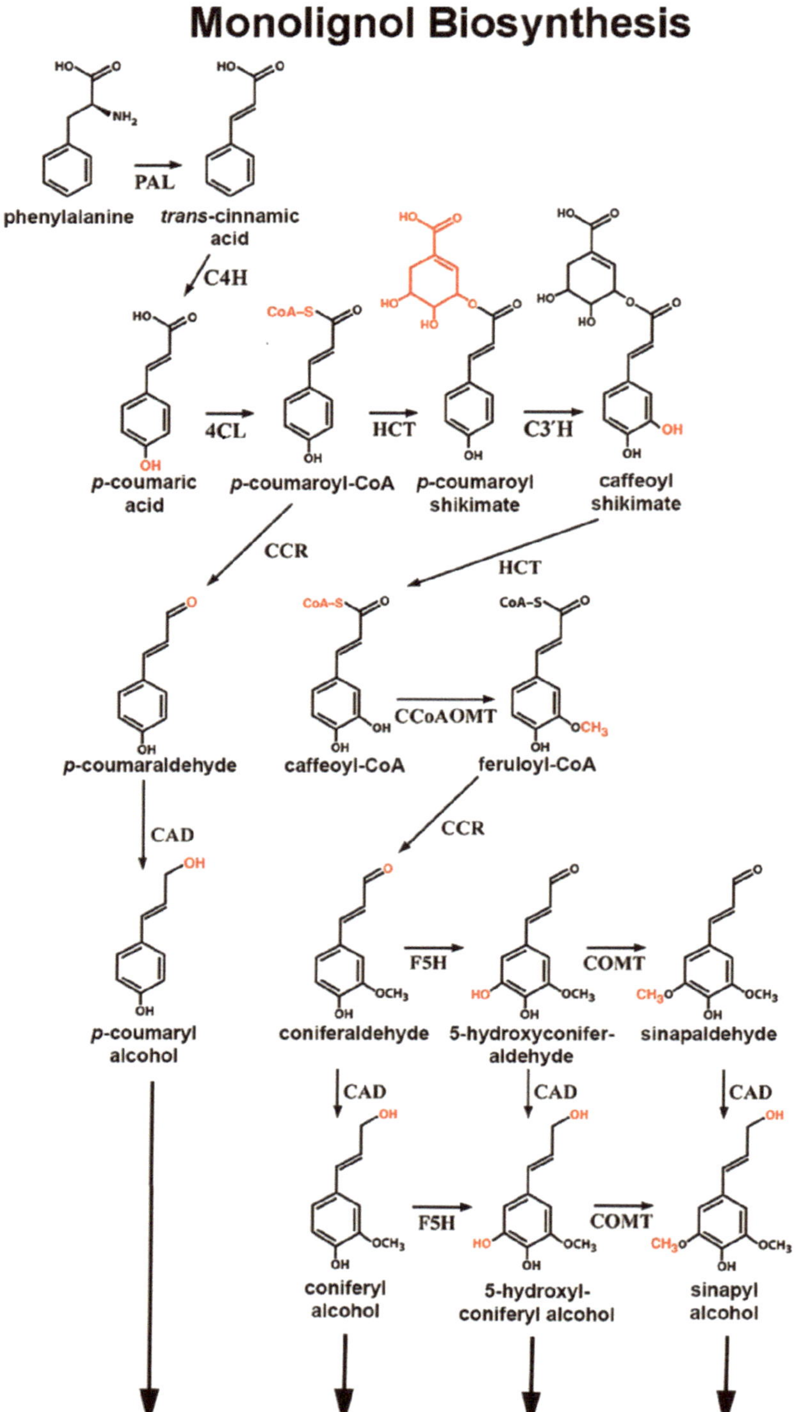

FIGURE 24.2 The monolignol biosynthesis pathways. Synthesis begins with deamination of phenylalanine by the phenylalanine ammonia lyase (PAL). A cinnamoyl 4-hydroxylase (C4H) converts *trans*-cinnamic acid to *p*-coumaric acid. Tyrosine can also be converted to *p*-coumaric acid by tyrosine ammonia lyase and C4H (not shown). The typical route to monolignols is by the formation of *p*-coumaroyl-CoA by 4CL and to feruloyl-CoA by sequential reactions of a *p*-hydroxycinnamoyl-CoA:quinate/shikimate-*p*-hydroxycinnamoyltransferase (HCT), a C3'H, a second HCT, and a caffeoyl-CoA *O*-methyltransferase (CCoAOMT). The *p*-coumaroyl-CoA can be directly reduced to *p*-coumaraldehyde by CCR and to *p*-coumaryl alcohol by CAD to make the monolignol substrate for *H-lignin*. CCR and CAD also reduce feruloyl-CoA to coniferaldehyde and to coniferyl alcohol substrate for *G-lignin*. A combination of F5H and COMT, before or after reduction by CAD, converts coniferaldehyde or coniferyl alcohol to sinapaldehyde or sinapyl alcohol, the substrate of *S-lignin*. Down-regulation of COMT can block formation of sinapaldehyde and sinapyl alcohol, resulting in the accumulation of 5-hydroxylconiferaldehyde and 5-hydroxyl-coniferyl alcohol which are also exported and incorporated into *OH-G-lignin*. For each of these reactions, the change in molecular structure is indicated in red. Adapted from Penning et al. (2009), as modified from Chanoca et al. (2019).

As in poplar, mutations in CAD (Saballos et al. 2009; Sattler et al. 2009), COMT (Sattler et al. 2012), and 4CL (Saballos et al. 2012), representing the classical *brown-midrib* mutants of sorghum, have been found to have enhanced saccharification yield. Suppression of the early genes in monolignol synthesis in grass species also substantially reduces lignin content and increases enzymatic digestibility, often with growth penalty (Halpin 2019). However, the concept that high S-lignin, i.e., high S:G ratios, improves saccharification yield is not so obvious in bioenergy grasses. Down-regulation of COMT, the key enzyme in the formation of sinapyl alcohol of S-lignin in switchgrass (*Panicum virgatum*), resulted in reduced lignin content but showed increased saccharification yields despite a severe reduction in S:G ratios (Fu et al. 2011; Baxter et al. 2014). Up- or down-regulation of F5H and COMT in switchgrass, in parallel or antagonistically, can alter the S:G ratio (Wu et al. 2019). Up-regulating F5H expression in a transgenic switchgrass line with severely down-regulated COMT substantially increased saccharification efficiency of cell walls compared to wild type (WT). However, no correlation was observed between cell wall saccharification efficiency, 5-OH-G-lignin level, and S/G ratio (Wu et al. 2019). Down-regulation of COMT also knocked down S:G ratios in sugarcane (*Saccharum officinarum*), but the low-lignin content was also accompanied by reductions in ferulate and *p*-coumarate esters (Jung et al. 2013).

Interpretation of modifications to lignin structure is more complicated in grasses because of the phenylpropanoid-rich Type II primary cells. Because of the preponderance of hydroxycinnamic acids in grasses, their synthesis and esterification to xylans became the new foci for bioenergy crop improvement. Down-regulation of Hydroxycinnamoyl-Coenzyme A shikimate/quinate hydroxycinnamoyl transferase (*HCT*) genes in brachypodium resulted in only modest changes in lignin content and composition and had little impact on biomass yield, but substantial increases in saccharification yield were observed (Serrani-Yarce et al. 2021). In some grass species, monolignol–*p*-coumarate units can comprise up to 40% of the lignin; *p*-coumarate does not participate in radical-mediated coupling reactions, but is linked by an enzyme, *p*-coumarate monolignol transferase (Withers et al. 2012). *p*-Coumaroyl-CoA:monolignol transferase (PMT) acts specifically in the lignin biosynthetic pathway in Brachypodium (Petrik et al. 2014). In brachypodium, BAHD acyl-CoA transferases are implicated in ferulate and *p*-coumarate esterification to GAX (Molinari et al. 2013). Overexpression of mutated forms of rice BAHD greatly reduced ferulate content but with concomitant increases in *p*-coumarate (Bartley et al. 2013). Little to no impact was observed in vegetative development or lignin content or composition, but biomass prepared from these materials exhibited a 20–40% increase in saccharification yield. Suppression of a BAHD acyltransferase also decreased *p*-coumaroyl substitutions on arabinoxylan and improved biomass digestibility in green foxtail (*Setaria viridis*) (de Souza et al. 2018) and sugarcane (de Souza et al. 2019); the *p*-coumarate in grass lignin is mainly attached to acylated S-units intracellularly by a BAHD PMT (Mota et al. 2021).

LIGNIN AND THE STRUCTURE OF THE CML

Lignin genetic variants provide tools to probe the impact of altered aromatic networks in the CML. The high S-lignin produced by overexpression of F5H is infrequently branched and the chain lengths are shorter (Stewart et al. 2009). This finding is consistent with a hypothesis that, compared to G-lignins, S-lignins are less able to shield cellulose microfibrils from enzymatic attack, resulting in increased saccharification yields. However, modification of lignin in secondary walls would necessarily impact fibre adhesion zones in the CML at the nanoscale as well as at the atomic scale of a lignin–cellulose interaction. In most instances, the enhanced saccharification yields in high S-lignin variants are not observed without pretreatment. The high S-lignin variants might also interfere with normal cell–cell adhesion in the CML, and this is indeed the case. In grasses, combined acid- and steam-expansion treatment resulted in splitting along the CML of Miscanthus (*Miscanthus* spp.) and wheat (*Triticum aestivum*) biomass (Auxenfans et al. 2017). The pretreatments preferentially cleaved lignin β-*O*-4 linkages, resulting in melting and redistribution of lignin in the treated materials. Xylan hydrolysis, reduction in the cross-linking phenolic acids and the redistribution of lignin during pretreatment are each correlated with increased saccharification yield (Auxenfans et al. 2017). Scanning electron microscopy (SEM) and TEM imaging of maize stover after thermochemical pretreatments showed that lignins coalesce into large molten bodies that migrate within and out of the cell wall (Donohoe et al. 2008). Consistent with these observations, cell separation at the CML occurred uniquely in genetic variants with high S-lignin during maleic acid–catalysed hydrolysis of xylans from Arabidopsis floral stems, (Figure 24.3; Ciesielski et al. 2014). Thus, high proportions of β-*O*-4 linkages in S-lignin might be located specifically in the CML, where cleavage of these bonds increased relative surface area and thereby increased accessibility to hydrolytic enzymes, as the basis of enhanced saccharification yield.

To explore further the factors that impact cell–cell adhesion in the CML, a cell–cell separation assay was developed for finely milled wood particles to evaluate impacts of chemical and enzymatic treatments sequentially to remove lignin, xylan, and/or pectin from the CML in genetic variants of hybrid poplars (Yang et al. 2021). Employing transgenic lines with various S:G ratios, neither lignin oxidation by acidic sodium chlorite, nor extraction of pectins and weakly associated hemicelluloses by dilute alkali, was alone sufficient to disrupt cell–cell adhesion, regardless of lignin composition, but a combination of the two treatments resulted in cell separation (Figure 24.4A). Other treatments to remove lignin using Zn^{2+}/Pd (Parsell et al. 2015) or Ni^{+} chemical catalysts (Luo et al. 2016) also did not disrupt cell–cell

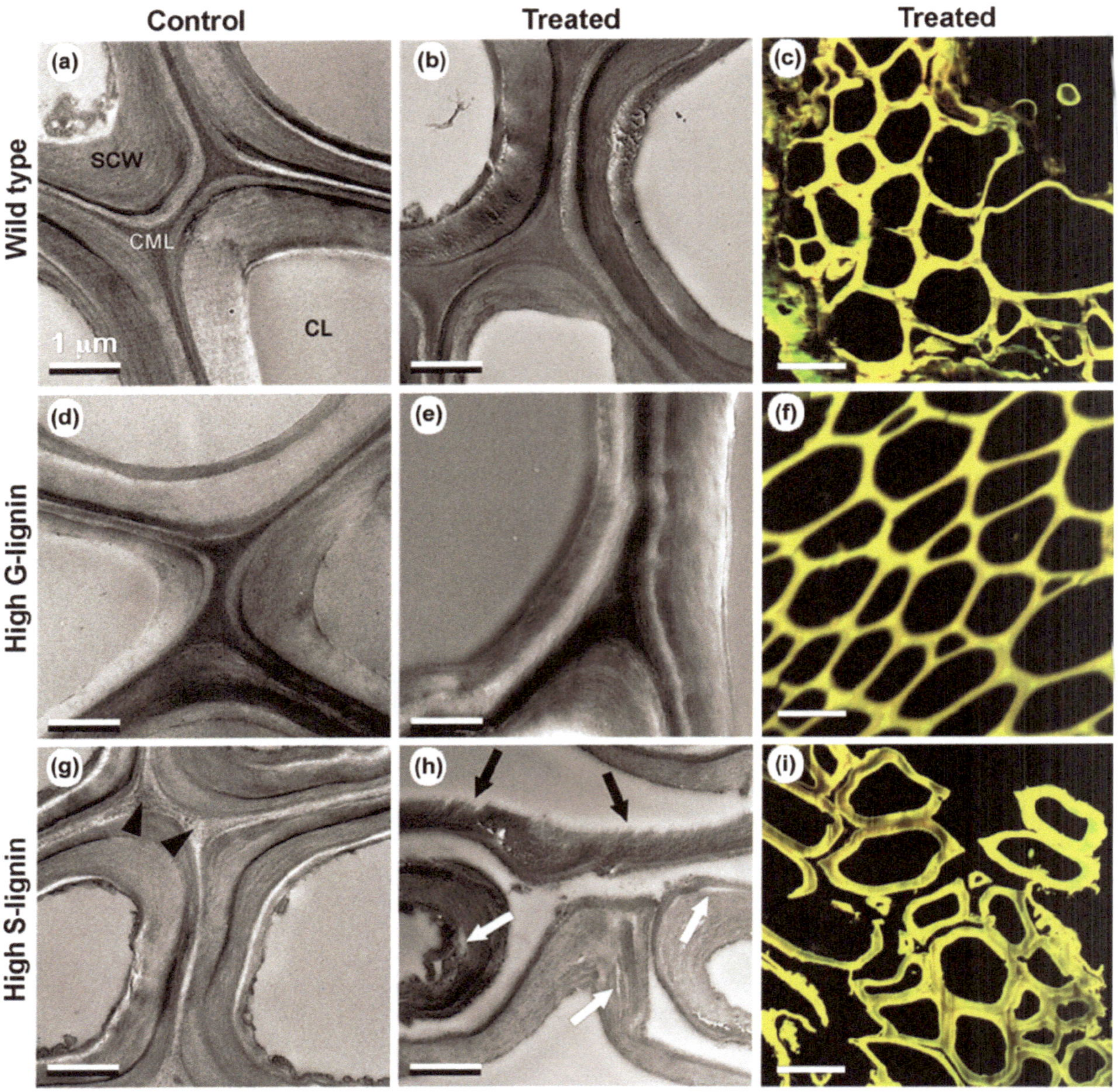

FIGURE 24.3 Electron micrographs of the CML in untreated and treated vascular tissues of Arabidopsis floral stems from lignin variants. Images taken before (Control) and after maleic acid–catalysed hydrolysis of xylans (Treated). Cross-sections of untreated wild-type (WT) by TEM show cell lumen (CL), secondary cell wall (SCW), and compound middle lamella (CML) in untreated WT (**a**), high G-lignin (**d**), and high S-lignin (**g**) lines. Untreated high S-lignin lines displayed much lower electron density in the CML. After treatment with maleic acid, little change in structure is observed in WT (**b**) and high G-lignin (**e**) lines, but substantial disruption of the CML is observed in the high S-lignin (**h**) lines. Confocal laser scanning microscopy shows the absence of cell–cell separation in WT (**C**) and high G-lignin (**f**) lines, but extensive separation in the CML region is evident in the high S-lignin lines (**i**). The SCWs also show extensive delamination (white arrows) and 'nano-fibrillation' of cellulose microfibrils at the interface with the cell lumen (black arrows) (adapted from Ciesielski et al. 2014).

adhesion in woody particles, indicating that lignin alone is not sufficient to maintain adhesion (Yang et al. 2021). Nevertheless, high S-lignin genotypes fragmented to single cells and small cell clusters more easily than did wild-type or high G-lignin genotypes (Figure 24.4B). Acidic chlorite treatments are also known to increase sugar release, upon digestion with hydrolytic enzymes, from Zinnia (*Zinnia elegans*) tracheary elements differentiated *in vitro* (Lacayo et al. 2013), from sugarcane bagasse (Siqueira et al. 2013), and from a variety of other woody grasses (Reeves 1987), most likely because of increased accessibility to glucanases along the cell boundaries.

In poplar (*Populus tremula* x *P. alba*), the acidic chlorite treatment resulted in some loss of material from cell corners, yet the middle lamellae remained intact and were only removed after subsequent extraction with dilute alkali (Figure 24.5; Yang et al. 2021). Genetic modification of lignin has caused severe growth and developmental

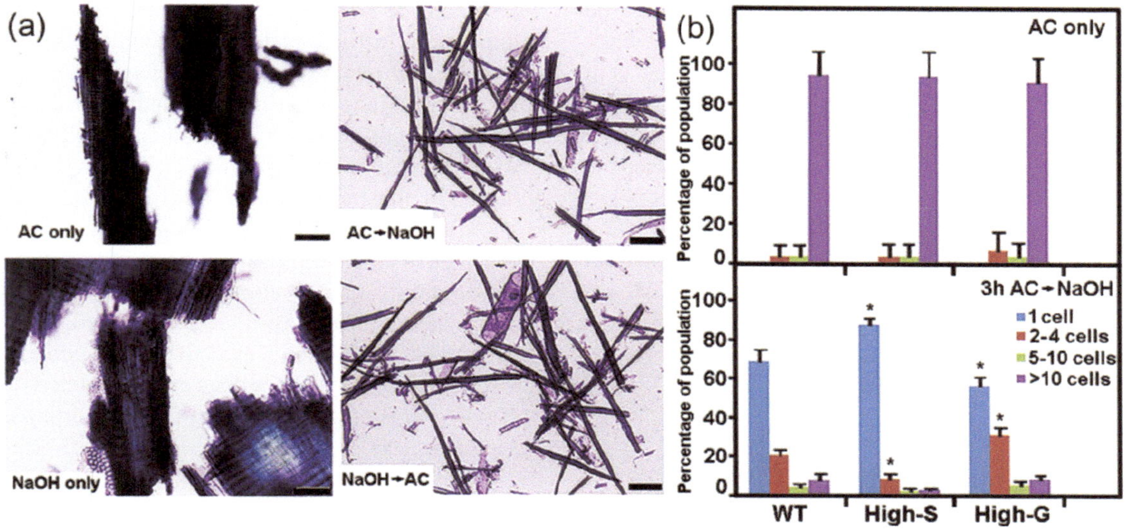

FIGURE 24.4 Cell–cell separation of fibre cells from poplar wood particles requires a combination of acidic sodium chlorite and 0.1 M NaOH. (**a**) Wood particles treated with acidic chlorite (AC) or NaOH alone remain intact. Regardless of order of treatment, a combination of treatments with either AC → NaOH or NaOH → AC results in cell separation. Wood particles and fibre cells were visualised with Toluidine Blue. (**b**) Treatment of wood particles from wild type (WT), a high S-lignin variant (High-S), and a high G-lignin variant (High-G) with acidic sodium chlorite (AC) alone results in little cell–cell separation (*upper panel*). Limited treatment of wood particles for 3 hours in acidic chloride (AC) followed by 24 hours in 0.1 M NaOH results in significantly greater yield of single cells from the high S-lignin particles (*lower panel*). Wood particles from 3-year-old WT poplar (*Populus tremula* x *P. alba*, cv. INRA 717-1B4) were transformed with Arabidopsis F5H or F5H-RNAi constructs to produce high S-lignin (93% S) and high G-lignin (47%), respectively, compared to WT (S-lignin 66%; G-lignin 33%) (adapted from Yang et al. 2021).

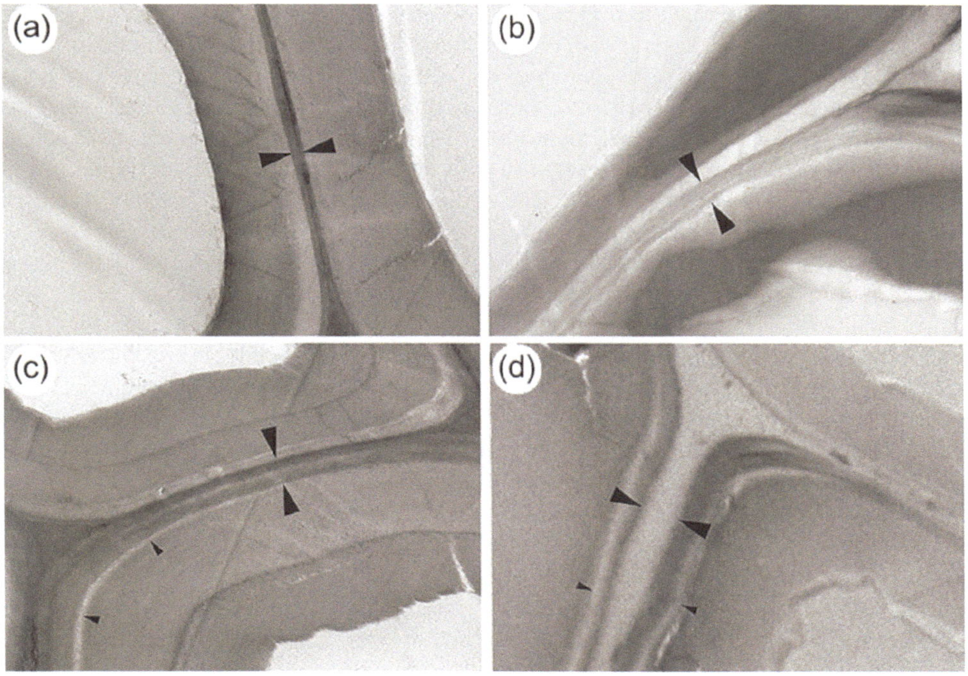

FIGURE 24.5 Electron micrographs of the CML in untreated and treated wild type (WT) poplar particles. (**a**) Electron micrographs of untreated WT particles. Large arrowheads indicate electron-dense CML. (**b**) WT particles after treatment with acidic chlorite. Arrowheads indicate a zone of reduced electron density in the CML and secondary wall: the primary and secondary wall. (**c**) WT particles after treatment with 0.1 M NaOH for 24 hours. Large arrowheads indicate some loss of electron density in the CML; small arrowheads show loss of electron density at interface of CML and secondary wall. (**d**) WT particles after treatment with 3 hours acidic chlorite, followed by 24 hours dilute alkali. Large arrowheads indicate region of cell–cell separation and loss of electron density in the CML; small arrowheads show loss of electron density at interface of CML and secondary wall. Bars = 2 μm (adapted from Yang et al. 2021).

phenotypes (Jones et al. 2001; Schilmiller et al. 2009), but high S-lignin variants exhibit little or no disruption of normal growth in Arabidopsis (Meyer et al. 1998; Marita et al. 1999; Franke et al. 2000), or in the poplar variants (Yang et al. 2019, 2021). High S-lignin poplar, with similar total lignin content as WT, displayed increased cell–cell separation relative to wild-type and high G-lignin variants, with 3 h acidic chlorite and 24 h dilute alkali treatments (Yang et al. 2021). The highly methylated structure of S-lignin might contribute to less cross-linking within the CML, or changes in lignin structure might reduce cross-linking of other cell wall components.

A ROLE FOR GLUCURONOXYLAN IN CELL–CELL ADHESION IN THE CML

After acidic chlorite and dilute alkali induced cell separation, electron-dense staining of wood particles was reduced in patches of secondary wall, between primary and secondary wall, and from cell corners (Yang et al. 2021). As over 90% of the non-cellulosic sugar was Xyl, xylans were a major polysaccharide extracted by these treatments. However, de-lignification treatments followed by xylanase were insufficient to effect complete cell separation in wood particles. In grass species, GAXs are cross-linked into the polyphenyl network via di-ferulic acid cross-bridges, representing one of several types of lignin-carbohydrate linkages (Iiyama et al. 1994). These ester linkages are easily broken and cannot alone constitute a stable junction. However, arabinosyl residues of GAX can initiate several patterns of ester, ether, and phenyl–phenyl linkages to form phenylpropanoid networks (Scalbert et al. 1985; Carpita 1986; Iiyama et al. 1994). In maize coleoptile cell walls, acidic chlorite renders GAX more easily extracted in subsequent treatment with dilute alkali (Carpita 1984). Dilute-acid pretreatments caused a redistribution of maize xylans from primary wall to middle lamella (Brunecky et al. 2009). Partial dissolution of the CML and increased porosity were observed after ammonia-fibre expansion (AFEX) pretreatment of maize stems (Chundawat et al. 2011). Dilute-acid and steam-expansion pretreatments of poplar lignocellulosic biomass also resulted in increased porosity and surface area by disruption of the CML (Meng et al. 2013). Dilute-acid pretreatment of poplar increased celluloytic enzyme absorption into the CML (Kim and Lee 2019).

In poplar wood particles, both RG-I and xylan are associated with lignin, and possibly with each other. A complex of GAX and RG-I was isolated from tomato (*Solanum lycopersicum*) cell walls (Broxterman and Schols 2018), and GAX and RG-I/HG are covalently linked in the Arabinoxylan Pectin Arabinogalactan Protein1 in Arabidopsis cell walls (Tan et al. 2013). Antibody probes specific to the RG-I backbone label the middle lamellae of potato tuber cell walls (Buffetto et al. 2015) and in tobacco (*Nicotiana benthianum*) seed endosperm cells, but only after enzymatic digestion of an abundant heteromannan polymer (Lee et al. 2013). Changes in the distribution of xyloglucan, (1→4)-β-galactan, xylan, and glucomannan epitopes were all associated with changes in lignification in normal and compression wood of pine (Donaldson and Knox 2012). Mannans were excluded from the CML, but xylan was found throughout. Although excluded from the CML, mannans were specifically involved in cell–cell adhesion in pine but not poplar (Liu et al. 2022). Enzymatic or chemical removal of xylan (Cheng et al. 2012) or pectin (Pakarinen et al. 2012) enhances saccharification yield, and genetic modification of enzymes involved in biosynthesis of the xylan backbone results in decreased stem strength (Brown et al. 2005; Brown et al. 2007; Persson et al. 2007). A distinct role for xylans as determinants of cell–cell adhesion in poplar is not clear, but they might be complexed with lignin or pectins.

PECTINS ARE ESSENTIAL COMPONENTS OF THE CML IN WOODY SPECIES

In separate treatments, the wood particles from poplar variants were subjected to Ni^+-mediated catalytic delignification, a treatment that completely removes lignin without destruction of cellulose and other carbohydrates (Luo et al. 2016). When the cell–cell adhesion assay was applied after oxidation of the lignin backbones with acidic chlorite treatment, 0.1 M alkali yielded both pectin and loosely bound hemicelluloses from the wood particles (Figure 24.6). Xylan comprised over 90% of the carbohydrate extracted during the cell–cell separation assays, but sugar and methylation analyses indicated that RG-I was also removed (Yang et al. 2021). Only dilute alkali treatment was needed to achieve complete cell separation after catalytic delignification (Yang et al. 2021). Enzymatic treatments were then used to determine which linkages should be cleaved to achieve cell separation. Among various polysaccharide hydrolases and lyases, a combination of xylanase, which cleaves the (1→4)-β-D-xylan backbone, and RG-lyase, which cleaves the backbone of RG-I (Mutter et al. 1998; Oomen et al. 2002), resulted in cell separation (Figure 24.6).

Neither PGases nor pectate lyases were able to substitute for RG-lyase in the cell-separation assay, indicating a specific role for RG-I in cell–cell adhesion. As proof of concept, transgenic poplar lines that expressed the Arabidopsis *RG-LYASE6* (*AtRGIL6*) cDNA were generated, and these lines showed increased cell separation phenotypes in chemical, enzymatic, and mechanical assays (Yang et al. 2021). A higher proportion of single cells was released from milled wood particles upon limit assays with chlorite and dilute alkali, or with catalytically delignified particles treated with only xylanase and RG-lyase. Woody particles from transgenic materials were fragmented more easily for the same energy inputs, in a proxy assay for comminution (Yang et al. 2021). Genetic control of RG-I content provides a new strategy to increase catalyst accessibility and saccharification yields at the molecular level, and to increase fragmentation of woody biomass at the tissue and organ level. Taken together, poplar requires an amalgam of lignin,

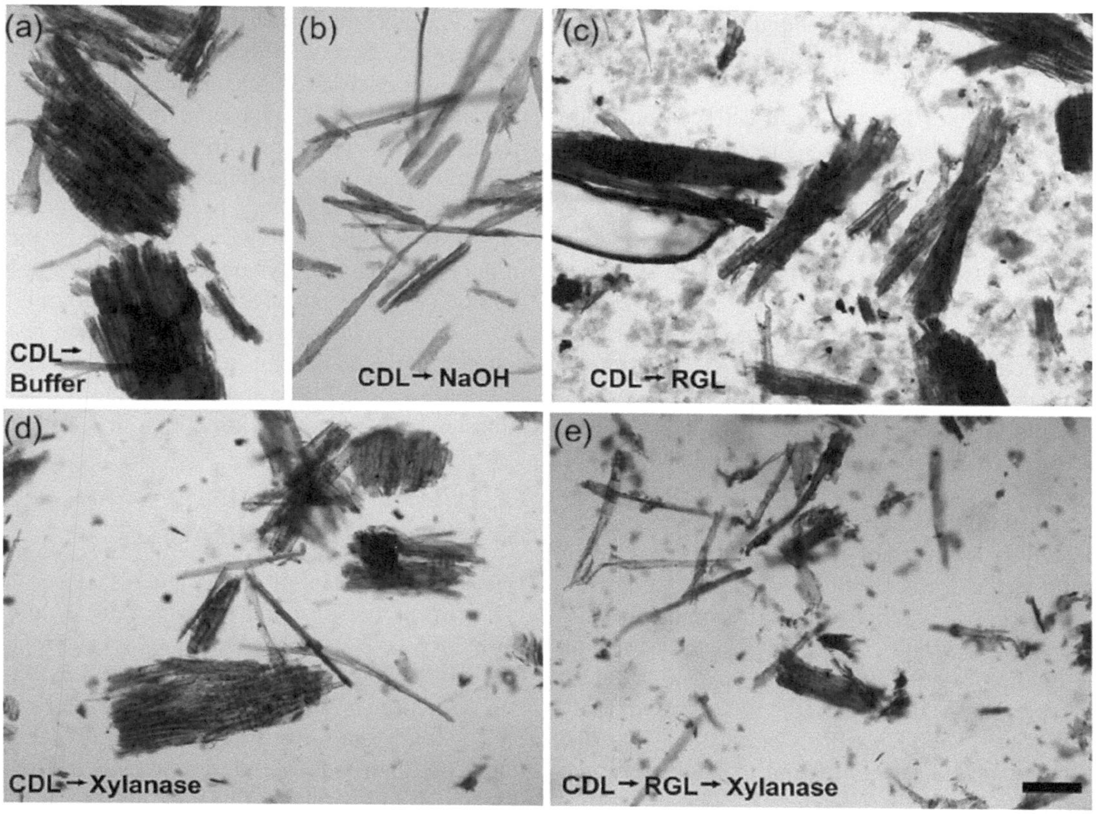

FIGURE 24.6 Cell–cell separation of Catalytically De-Lignified (CDL) wild type (WT) poplar wood particles after dilute alkali or enzyme treatments. (**a**) Bright-field micrographs of Toluidine Blue-stained particles of WT poplar wood following treatment with Ni/C catalyst to remove lignin and then incubation in buffer (CDL→buffer). (**b**) CDL followed by dilute alkali treatment for 24 hours (CDL→NaOH) results in almost complete cell separation. (**c**) and (**d**) CDL treatment followed by endo-(1→4)-α-L- RG-I lyase (CDL→RGL) or with endo-(1→4)-β-D-xylanase (CDL→xylanase) resulted in some disruption of the wood particles. (**e**) CDL treatment followed by a combination of endo-(1→4)-α-L- RG-I lyase and then endo-(1→4)-β-D-xylanase (CDL→RGL→xylanase) resulted in more extensive cell separation. Bars =100 μm (adapted from Yang et al. 2021).

xylan, and RG-I to form the adhesive matrix. A specific role for RG-I in cell–cell adhesion has also been shown for parenchymatous tissues of fruits. (Oomen et al. 2002). In strawberry (*Fragaria×ananassa*), knockdown of RG-lyase expression is correlated with reduced dissolution of the middle lamella during fruit ripening (Molina-Hidalgo et al. 2013). However, overexpression of a tomato RG-lyase homolog resulted in increased tomato fruit firmness, indicating that we still lack understanding of cell–cell adhesion even in well-characterised systems, such as tomato fruit ripening (Ochoa-Jiménez et al. 2018).

Various side chains of RG-I have also been implicated in the mechanical properties in both primary- and secondary-walled cells. Digestion of the RG-I backbone by expression of a fungal RG-lyase resulted in cell swelling in the cortex and periderm of the tuber and relocation of arabinan and galactans to the middle lamellae and cell corners. RG-I isolated from flax (*Linum usitatissimum*) gelatinous fibres forms hydrogels with hyperelastic properties (Mikshina et al. 2015). The removal of galactan and arabinan side chains of RG-I is correlated with the softening and cell separation in apples (*Pyrus malus)* during fruit cold storage (Peña and Carpita 2004). In tobacco (*Nicotiana plumbaginifolia*), loss of RG-I arabinans is associated with increased callus friability, a term for more facile separation of cells in culture (Iwai et al. 2001). Wall porosity can be modified by overexpression of galactanase in tomato and potato (Orfila and Knox 2000; Oxenbøll-Sørensen et al. 2000), and in potato the RG-I side chain status can alter cell elongation and proliferation (Bush et al. 2001). Transgenic potato tubers, in which fungal arabinanase or galactanase genes have been overexpressed to degrade arabinan or galactan side chains, are as strong as control tubers but they lose elasticity and become brittle (Ulvskov et al. 2005). In solution, neutral sugar side chains bind water and increase solubility (Belton 1997) and reduce intrinsic viscosity, but contribute to polymer entanglement, resulting in higher viscosity at higher RG-I concentrations (Hwang et al. 1993). In principle, different micro-environments might be created within cell walls by modification of the side chain structure of RG-I to regulate wall properties, such as porosity, water-binding properties, or resistance to compressive forces. In the poplar cell separation assay, the proportion of RG-I in the acetic chlorite and dilute alkali extracts was only about 2% of the total mass, but an increased propensity of the tissue to fragment was observed when the abundance of RG-I was further reduced in transgenic RG-lyase over-expressing lines (Yang et al. 2021).

By determining roles for pectins in cell–cell adhesion in poplar wood, additional targets for genetic modifications have been identified that could substantially reduce energy inputs for biomass processing at the point of use. Consistent with a role for the contribution of pectin to biomass recalcitrance, down-regulation of HG synthesis enhanced sugar yield in saccharification assays (Biswal et al. 2018a). Despite the paucity of pectins in grasses compared to dicots, down-regulation of GAUT4 results in reduced ferulate and lignin cross-linking in switchgrass, as well as a greater lignin migration after steam pretreatment (Li et al. 2019). By contrast, overexpression of GAUT12 to increase HG content decreased saccharification yield (Biswal et al. 2018b). Deletion of a gene cluster encoding a suite of pectolytic enzymes of *Caldicellulosiruptor bescii* reduced sugar yields in saccharification assays (Chung et al. 2014), whereas controlled expression of recombinant pectinases enhanced digestibility of Arabidopsis (Lionetti et al. 2010). The binding of a pectin backbone around cellulose is thought to play a role in maintaining structural integrity within the primary wall (Dick-Pérez et al. 2012), and the addition of arabinan and galactan chains to RG-I strengthen binding to cellulose (Zykwinska et al. 2007). Galactan content increased by about 50% in secondary walls of transgenic Arabidopsis stem tissue by co-expression of genes of UDP-Gal synthesis, its Golgi transporter, and a galactan synthase1 (GALS1), and floral stems of these plants exhibited enhanced saccharification yield (Aznar et al. 2018). The increases in binding of galactans might result in a loosened wall structure, much like in the gelatinous wall layers in tension wood (Gorshkova et al. 2015).

In Arabidopsis floral stems, cell separation occurred uniquely in genetic variants with high S-lignin after removal of xylan using maleic acid catalysis (Figure 24.5; Ciesielski et al. 2014). These results suggest that both lignin and xylan, and perhaps their interactions, have roles in cell–cell adhesion. Cell–cell adhesion in poplar wood particles was disrupted entirely by treating delignified particles with xylanase and an RG-lyase that cleaved the backbone of the RG-I polymer. By contrast, hydrolysis or lysis of HG did not affect the extent of cell–cell separation. If HG and RG-I were covalently linked together, then lysis of either component would be expected to disrupt cell–cell adhesion in mature wood. Xylan and RG-I might mediate cell–cell adhesion before the initiation of lignification in the CML.

CELLULOSE CRYSTALLINITY AS A RECALCITRANCE FACTOR

Despite the substantial improvements of saccharification yields in the poplar wood particles after catalytic delignification, a considerable amount of recalcitrance remains that can only be attributed to cellulose crystallinity (Figure 24.7). Swelling of the cellulose in trifluoroacetic acid (TFA) breaks this crystallinity, resulting in almost complete digestion of the amorphous cellulose generated in a few hours. An illustration of the inherent recalcitrance of cellulose in the absence of lignin can be seen in the slow digestion of cotton fibre cellulose in a commercial enzyme cocktail called Ctec II (Novozymes) (Figure 24.7). When cellulose is suspended in TFA at freezing temperatures, it swells to a thick gel that upon heating to 55°C is completely dissolved into a pourable liquid (Shiga et al. 2017). When subsequently crashed into ethanol, an amorphous gel is recovered that the CtecII cocktail digests to completion in a few hours (Figure 24.7).

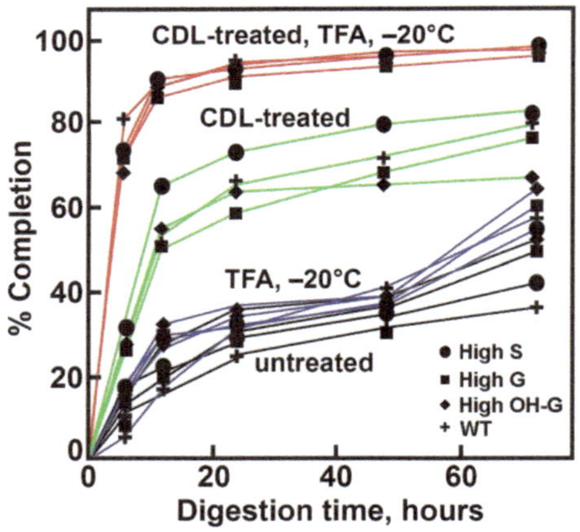

FIGURE 24.7 Rates of enzymatic hydrolysis of milled wood particles from wild type (WT) and transgenic poplar lines before or after CDL. Cellic® Ctec2 (Novozymes) was added to 5 mg of milled particles to initiate hydrolysis, and incubations at 37°C were for up to 72 hours. The lignified wood particles do not swell in TFA at –20°C, and digestion (blue lines) is not improved compared with digestion of untreated particles (black lines), regardless of lignin composition. After CDL-treatment, particles showed enhanced saccharification yields regardless of lignin composition (green lines). However, initial rates of digestion were substantially higher in the high S-lignin transgenic lines, despite the absence of lignin. After CDL treatment, all materials readily swell in TFA at –20°C, and the resulting improvement in digestibility illustrates the recalcitrance of cellulose crystallinity (red lines). WT, high G-lignin, and high S-lignin were generated as described in Figure 24.4. The high 5-OH-G-lignin lines (High OH-G) were generated by *COMT-RNAi* constructs that increased 5-OH-G-lignin from 19% in WT to 33% in the high OH-G-lignin lines (from McCann and Carpita 2020, as modified from Yang et al. 2019).

Newly synthesised cellulose microfibrils of flowering plants are commonly described as *para*-crystalline arrays, which by solid-state ^{13}C-NMR spectroscopy (Jarvis 2013), and the size of the particle rosette CesA complexes (Nixon et al. 2016; Haigler and Roberts 2019), contain about 18 to 24 glucan chains in the crystalline core domain. Regardless of the numbers of chains in microfibrils made by CesA complexes, microfibrils also aggregate into macrofibril bundles (Ding et al. 2014). Deuterium exchange experiments with plant native cellulose showed phases of rapid and much slower exchange, interpreted as exchange with hydrogens in amorphous vs. crystalline domains (Frilette et al. 1948). Deuterium exchange followed by rehydrogenation revealed an intermediate fraction of slow deuteration distinct from deuterium-resistant domains (Jeffries 1963; Lindh and Salmén 2017). Fourier transform infrared spectroscopy of these exchange properties indicated that cotton cellulose is a large aggregate of three bundles, each of about 144 chains (Lindh and Salmén 2017).

A cellulose oxidation protocol was used to determine empirically the relative proportions of amorphous chains, surface chains of crystalline cellulose microfibrils, and the crystalline core chains (Shiga et al. 2021). When catalysed by 2,2,6,6- tetramethylpyperidine-1-oxyl (TEMPO), the *O*-6 primary alcohols of glucosyl residues exposed to water are oxidised by hypochlorite to yield glucuronic acid (Saito and Isogai 2004; Okita et al. 2010). In this protocol, glucuronosyl oxidation products were activated with a water-soluble diimide and then reduced with $NaBD_4$, creating 6-deutero- and 6,6-dideuteroglucosyl residues that could be distinguished from undeuterated glucose by electrospray ionisation-mass spectrometry (ESI–MS). Cellobiose units from the anhydrous domains of microfibrils are undeuterated, whereas only single deuterations per cellobiose unit from surface chains would be expected to be oxidised. Thus, from quantitation of the ratios of surface to anhydrous chains, the relative cellulose bundle size can be estimated. The surprising result was that cellulose microfibrils were generally bundled to the same extent in primary and secondary walls (Shiga et al. 2021).

Lignin is considered the key obstacle for efficient saccharification of woody biomass. Its action at the atomic scale is proposed as binding or shielding cellulose microfibrils from cellulolytic enzymes, but a considerable body of evidence now indicates that the lignified CML plays a role in blocking the diffusion of the enzymes to the secondary wall. By contrast to cotton fibre cellulose, wood particles of the poplar lignin variants were unable to swell in TFA (Yang et al. 2019), and recovered material showed no improvement in saccharification yield, regardless of lignin composition (Figure 24.7). However, materials from catalytic delignification show enhanced saccharification rates, revealing a doubling of the initial rates of digestion in high S-lignin lines, even though no lignin remains in the particles (Figure 24.7). These results indicate that factors impacting recalcitrance are not restricted to lignin–cellulose interactions at the atomic scale but can involve a broad range of factors that vary among biomass types, such as the microscale cell–cell interactions of the CML (Carpita and McCann 2020). In lignin genetic variants, between 67 and 72% of the microfibrils in wood particles were in an anhydrous domain (Shiga et al. 2021). Although catalytic delignification of poplar wood particles more than doubles saccharification yield (Yang et al. 2019), this treatment increases the proportion of anhydrous domains from 67 to 72% to 74 to 79%, indicating that chemical delignification increases, rather than decreases, the proportion of inaccessible cellulose chains. Thus, the increase in saccharification yield observed in delignified poplar wood particles did not result from generation of more amorphous cellulose by the delignification treatment. Given the apparent increased crystallinity of cellulose, these results lend additional evidence that the enhancement of saccharification centres on dissolution of the CML facilitating diffusion of cellulolytic enzymes.

FUTURE PROSPECTS

Woody biomass is a complex, self-assembling, hierarchical material that is sustainable and abundant. Although more recent work focused on improving sugar yield from biomass for biofuel production, a greater emphasis is now on redesigning wood properties using physical and chemical treatments to alter mechanical, optical, thermal, and electrical properties (Hongli et al. 2016). For example, delignification and densification of wood have been demonstrated to produce a material ten-fold stronger than the original wood with mechanical properties that surpass steel and titanium alloys, termed "super wood" (Chen et al. 2020). An alternative strategy is to redesign wood for novel functions using genetic engineering (Carpita and McCann 2020), reducing the carbon intensity of subsequent thermal or chemical treatments for various end-uses. Our overarching hypothesis is that specific cell wall domains, including the CML and secondary thickenings in xylem tissue, contribute disproportionately to biomass structural integrity and recalcitrance. We further propose that by engineering controls to regulate the molecular compositions and nanoscale architectures of heterogeneous extracellular domains, we will gain critical knowledge of the macroscale properties of biomass and the ability to design its properties for end-uses.

ACKNOWLEDGEMENTS

This work was supported by the National Renewable Energy Laboratory (NREL), Laboratory Directed Research and Development (LDRD) program.

REFERENCES

Alejandro S, Lee Y, Tohge T, Sudre D, Osorio S, Park J, Bovet L, Lee Y, Geldner N, Fernie AR, et al. (2012) AtABCG29 is a monolignol transporter involved in lignin biosynthesis. *Current Biology* 22:1207–1212.

Allen CE (1901) The origin and nature of the middle lamella. *Botanical Gazette* 32:1–34.

Aznar A, Chalvin C, Shih PM, Maimann M, Ebert B, Birdseye DS, Loqué D, Scheller HV (2018) Gene stacking of multiple traits for high yield of fermentable sugars in plant biomass. *Biotechnology for Biofuels* 11:2.

Auxenfans T, Crônier D, Chabbert B, Paës G (2017) Understanding the structural and chemical changes in plant biomass following explosion pretreatment. *Biotechnology for Biofuels* 10:36.

Bailey AJ (1936a) Lignin in Douglas fir. Composition of the middle lamella. *Industrial and Engineering Chemistry* 8:52–55.

Bailey AJ (1936b) Lignin in Douglas fir. The pentosan content of the middle lamella. *Industrial and Engineering Chemistry.* 8:389–391.

Bartley LE, Peck ML, Kim S-R, Ebert B, Manisseri C, Chiniquy DM, Sykes R, et al. (2013) Overexpression of a BAHD Acyltransferase, *OsAt10*, alters rice cell wall hydroxycinnamic acid content and saccharification. *Plant Physiology* 161:1615–1633.

Baxter HL, Mazarei M, Labbe N, Kline LM, Cheng Q, Windham MT, Mann DGJ, et al. (2014) Two-year field analysis of reduced recalcitrance transgenic switchgrass. *Plant Biotechnology Journal* 12:914–924.

Belton PS (1997) NMR and the mobility of water in polysaccharide gels. *International Journal of Biological Macromolecules* 21:81–88.

Biswal AK, Atmodjo MA, Li M, Baxter HL, Yoo CG, Pu Y, Lee YC, et al. (2018a) Sugar release and growth of biofuel crops are improved by downregulation of pectin biosynthesis. *Nature Biotechnology* 36:249–257.

Biswal AK, Atmodjo MA, Pattathil S, Amos RA, Yang X, Winkeler K, Collins C, et al. (2018b) Working towards recalcitrance mechanisms: Increased xylan and homogalacturonan production by overexpression of GAlactUronosylTransferase12 (GAUT12) causes increased recalcitrance and decreased growth in Populus. *Biotechnology for Biofuels* 11:1.

Bjurhager I, Olsson A-M, Zhang B, Gerber L, Kumar M, Berglund LA, Burgert I, Sundberg B, Salmén L (2010) Ultrastructure and mechanical properties of *Populus* wood with reduced lignin content caused by transgenic down-regulation of cinnamate 4-hydroxylase. *Biomacromolecules* 11:2359–2365.

Bonawitz ND, Chapple C (2010) The genetics of lignin biosynthesis: connecting genotype to phenotype. *Annual Review of Genetics* 44:337–363.

Bossinger G, Spokevicius AV (2018) Sector analysis reveals patterns of cambium differentiation in poplar stems. *Journal of Experimental Botany* 69:4339–4348.

Brown DM, Zeef LAH, Ellis J, Goodacre R, Turner SR (2005) Identification of novel genes in Arabidopsis involved in secondary cell wall formation using expression profiling and reverse genetics. *Plant Cell* 17:2281–2295.

Brown DM, Goubet F, Wong VW, Goodacre R, Stephens E, Dupree P, Turner SR (2007) Comparison of five xylan synthesis mutants reveals new insight into the mechanisms of xylan synthesis. *Plant Journal* 52:1154–1168.

Brown DM, Zhang Z, Stephens E, Dupree P, Turner SR (2009) Characterization of IRX10 and IRX10-like reveals an essential role in glucuronoxylan biosynthesis in Arabidopsis. *Plant Journal* 57:732–746.

Broxterman SE, Schols HA (2018) Characterisation of pectin-xylan complexes in tomato primary plant cell walls. *Carbohydrate Polymers* 197:269–276.

Brunecky R, Vinzant TB, Porter SE, Donohoe BS, Johnson DK, Himmel ME (2009) Redistribution of xylan in maize cell walls during dilute acid pretreatment. *Biotechnology and Bioengineering* 102:1537–1543.

Buffetto F, Cornuault V, Rydahl MG, Ropartz D, Alvarado C, Echasserieau V, Le Gall S, et al. (2015) The deconstruction of pectic rhamnogalacturonan I unmasks the occurrence of a novel arabinogalactan oligosaccharide epitope. *Plant Cell Physiology* 56:2181–2196.

Bush MS, McCann MC (1999) Pectic epitopes are differentially distributed in the cell walls of potato (*Solanum tuberosum*) tubers. *Physiologia Plantarum* 107:201–213.

Bush MS, Marry M, Huxham IM, Jarvis MC, McCann MC (2001) Developmental regulation of pectic epitopes during potato tuberisation. *Planta* 213:869–880.

Busse-Wicher M, Li A, Silveira RL, Pereira CS, Tryfona T, Gomes TCF, Skaf MS, Dupree P (2016) Evolution of xylan substitution patterns in gymnosperms and angiosperms: implications for xylan interaction with cellulose. *Plant Physiology* 171:2418–2431.

Caffall KH, Mohnen D (2009) The structure, function, and biosynthesis of plant cell wall pectic polysaccharides. *Carbohydrate Research* 344:1879–1900.

Cai Y, Zhang K, Kim H, Hou G, Zhang X, Yang H, Feng H, et al. (2016) Enhancing digestibility and ethanol yield of *Populus* wood via expression of an engineered monolignol 4-*O*-methyltransferase. *Nat Communications* 7:11989

Carpita NC (1984) Fractionation of hemicelluloses from maize cell walls with increasing concentrations of alkali. *Phytochemistry* 23:1089–1093.

Carpita NC (1986) Incorporation of proline and aromatic amino acids into cell walls of maize coleoptiles. *Plant Physiology* 80:660–666.

Carpita NC (1996) Structure and biogenesis of the cell walls of grasses. *Annual Review of Plant Physiology and Plant Molecular Biology* 47:445–476.

Carpita NC (2012) Progress in the biological synthesis of the plant cell wall: new ideas for improving biomass for bioenergy. *Current Opinion in Biotechnology* 23:330–337.

Carpita NC, Gibeaut DM (1993) Structural models of primary cell walls in flowering plants: Consistency of molecular structure with the physical properties of the walls during growth. *Plant Journal* 3:1–30..

Carpita NC, McCann MC (2020) Redesigning plant cell walls for the biomass-based bioeconomy. *Journal of Biological Chemistry* 295:15144–15157.

Chanoca A, de Vries L, Boerjan W (2019) Lignin engineering in forest trees. *Frontiers in Plant Science* 10:912.

Chen F, Dixon RA (2007) Lignin modification improves fermentable sugar yields for biofuel production. *Nature Biotechnology* 25:759–761.

Chen C, Kuang Y, Zhu S, Burgert I, Keplinger T, Gong A, Li T, et al. (2020) Structure-property-function relationships of natural and engineered wood. *Nature Reviews Materials* 5:642–666.

Cheng F, Sheng J, Dong R, Men Y, Gan L, Shen L (2012) Novel xylanase from a Holstein cattle rumen metagenomic library and its application in xylooligosaccharide and ferulic acid production from wheat straw. *Journal of Agriculture and Food Chemistry* 60:12516–12524.

Chundawat SPS, Donohoe BS, da Costa Souza L, Elder T, Agrawal UP, Lu F, Ralph J, Himmel ME, Balan V, Dale BE (2011) Multi-scale visualization and characterization of lignocellulosic plant cell wall deconstruction during thermochemical pretreatment. *Energy and Environmental Science 4*:973–984.

Chung D, Pattathil S, Biswal AK, Hahn MG, Mohnen D, Westpheling J (2014) Deletion of a gene cluster encoding pectin degrading enzymes in *Caldicellulosiruptor bescii* reveals an important role for pectin in plant biomass recalcitrance. *Biotechnology for Biofuels* 7:1–12.

Ciesielski PN, Resch MG, Hewetson B, Killgore JP, Curtin A, Anderson A, Chiaramonti AN, et al. (2014) Engineering plant cell walls: tuning lignin monomer composition for deconstructable biofuel feedstocks or resilient biomaterials. *Green Chemistry* 16:2627–2635.

Coleman HD, Park J-Y, Nair R, Chapple C, Mansfield SD (2008) RNAi-mediated suppression of *p*-coumaroyl-CoA 3-hydroxylase in hybrid poplar impacts lignin deposition and soluble secondary metabolism. *Proceedings of the National Academy of Sciences USA* 105:4501–4506.

de Candolle AP (1813) *Théorie Élémentaire de la Botanique: Exposition des principes de la classification naturelle et de l'art de décrire et d'étudier les végétaux*. Paris, Déterville.

Déjardin A, Laurans F, Arnaud D, Breton C, Pilate G, Leplé J-C (2010) Wood formation in Angiosperms. *Comptes Rendus Biologies* 333:325–334.

de Souza WR, Martins PK, Freeman J, Pellny TK, Michaelson LV, Sampaio BL, Vinecky F, et al. (2018) Suppression of a single *BAHD* gene in *Setaria viridis* causes large, stable decreases in cell wall feruloylation and increases biomass digestibility. *New Phytologist* 218:81–93.

de Souza WR, Pacheco TF, Duarte KE, Sampaio BL, Molinari PAO, Martins PK, Santiago TR, et al. (2019) Silencing of a BAHD acyltransferase in sugarcane increases biomass digestibility. *Biotechnology for Biofuels* 12:111.

Dick-Pérez M, Wang T, Salazar A, Zabotina OA, Hong M (2012) Multidimensional solid-state NMR studies of the structure and dynamics of pectic polysaccharides in uniformly ^{13}C-labeled Arabidopsis primary cell walls. *Magnetic Resonance in Chemistry* 50:539–550..

Ding S-Y, Zhao S, Zeng YN (2014) Size, shape, and arrangement of native cellulose fibrils in maize cell walls. *Cellulose* 21:863–871.

Dippel L (1898) *Das Mikroskop und seine Anwendung*. Zweite umgearb. Aufl. Braunschweig.

Donaldson LA (2001) Lignification and lignin topochemistry - An ultrastructural view. *Phytochemistry* 57:859–873.

Donaldson LA, Knox JP (2012) Localization of cell wall polysaccharides in normal and compression wood of radiata pine: Relationships with lignification and microfibril orientation. *Plant Physiology* 158:642–653.

Donaldson L, Hague J, Snell R (2001) Lignin distribution in coppice poplar, linseed and wheat straw. *Holzforschung* 55:379–385.

Donohoe BS, Decker SR, Tucker MP, Himmel ME, Vinzant TB (2008) Visualizing lignin coalescence and migration through maize cell walls following thermochemical pretreatment. *Biotechnology and Bioenergy* 101:913–925.

Egelund J, Petersen BL, Motawia MS, Damager I, Faik A, Olsen CE, Ishii T, et al. (2006) *Arabidopsis thaliana* RGXT1 and RGXT2 encode Golgi-localized (1,3)-α-D-xylosyltransferases involved in the synthesis of pectic rhamnogalacturonan-II. *Plant Cell* 18:2593–2607.

Fergus BJ, Goring DAI (1970) The distribution of lignin in birch wood as determined by ultraviolet microscopy. *Holzforschung* 24:118–124.

Franke R, McMichael CM, Meyer K, Shirley AM, Cusumano JC, Chapple C (2000) Modified lignin in tobacco and poplar plants overexpressing the Arabidopsis gene encoding ferulate 5-hydroxylase. *Plant Journal* 22:223–234.

Frilette VJ, Hanle J, Mark H (1948) Rate of exchange of cellulose with heavy water. *Journal of the American Chemical Society* 70:1107–1113.

Fromm J, Rockel B, Lautner S, Windeisen E, Wanner G (2003) Lignin distribution in wood cell walls determined by TEM and backscattered SEM techniques. *Journal of Structural Biology* 143:77–84.

Fu C, Mielenz JR, Xiao X, Ge Y, Hamilton CY, Rodriguez MJr, Chen F, et al. (2011) Genetic manipulation of lignin reduces recalcitrance and improves ethanol production from switchgrass. *Proceedings of the National Academy of Sciences USA* 108:3803–3808.

Fukushima K, Terashima N (1990) Heterogeneity in formation of lignin. XIII. Formation of *p*-hydroxyphenyl lignin in various hardwoods visualized by microautoradiography. *Journal of Wood Chemistry and Technology* 10:413–433.

Fukushima K, Terashima N (1991) Heterogeneity in formation of lignin Part XV: Formation and structure of lignin in compression wood of *Pinus thunbergii* studied by microautoradiography. *Wood Science and Technology* 25:371–381.

Gao Y, Lipton AS, Wittmer Y, Murray DT, Mortimer JC (2020) A grass-specific cellulose-xylan interaction dominates in sorghum secondary cell walls. *Nature Communications* 11:6081.

Gorshkova T, Brutch N, Chabbert B, Deyholos M, Hayashi Y, Lev-Yadun S, Mellerowicz EJ, et al. (2012) Plant fiber formation: state of the art, recent and expected progress, and open questions. *Critical Reviews in Plant Sciences* 31:201–228.

Gorshkova T, Mokshina N, Chernova T, Ibragimova N, Salnikov V, Mikshina P, Tryfona T, et al. (2015) Aspen tension wood fibers contain β-(1→4)-galactans and acidic arabinogalactans retained by cellulose microfibrils in gelatinous walls. *Plant Physiology* 169:2048–2063.

Haigler CH, Roberts AW (2019) Structure/function relationships in the rosette cellulose synthesis complex illuminated by an evolutionary perspective. *Cellulose* 26:227–247.

Halpin C (2019) Lignin engineering to improve saccharification and digestibility in grasses *Current Opinion in Biotechnology* 56:223–229.

Hardell H-L, Leary GJ, Stoll M, Westermark U (1980) Variations in lignin structure in defined morphological parts of birch. *Svensk Papperstidning-Nordisk Cellulosa* 83:71–74.

Harlow WM (1932) Contributions to the chemistry of the plant cell wall vi. further studies on the location of lignin in the cell walls of wood. *American Journal of Botany* 19:729–739.

Hawkins WL, Wright GF, Hibbert H (1937) Studies on lignin and related compounds. XXXII. The guaiacyl and syringyl groups in various woods. *Journal of the American Chemical Society* 59:2447–2448.

Himmel ME, Ding S-Y, Johnson DK, Adney WS, Nimlos MR, Brady JW, Foust TD (2007) Biomass recalcitrance: engineering plants and enzymes for biofuels production. *Science* 315:804–807.

Hongli Z, Luo W, Ciesielski PN, Fang Z, Zhu ZY, Henriksson G, Himmel ME, Hu L (2016) Wood-derived materials for green electronics, biological devices, and energy applications. *Chemical Reviews* 116:9305–9374..

Huntley SK, Ellis D, Gilbert M, Chapple C, Mansfield SD (2003) Significant increases in pulping efficiency in C4H-F5H-transformed poplars: Improved chemical savings and reduced environmental toxins. *Journal of Agriculture and Food Chemistry* 51:6178–6183.

Hwang J, Pyun YR, Kokinin JL (1993) Side chains of pectin: some thoughts on their role in plant cell walls and foods. *Food Hydrocolloids* 7:39–53.

Iiyama K, Lam, TB- T, Stone B (1994) Covalent cross-links in the cell wall. *Plant Physiology* 104:315–320.

Iwai H, Ishii T, Satoh S (2001) Absence of arabinan in the side chains of the pectic polysaccharides strongly associated with cell walls of *Nicotiana plumbaginifolia* non-organogenic callus with loosely associated constituent cells. *Planta* 213:907–915.

Jarvis ML (1984) Structure and properties of pectin gels in plant cell walls. *Plant Cell and Environment* 7:153–164.

Jarvis ML (2013) Cellulose biosynthesis: counting the chains. *Plant Physiology* 163:1485–1486.

Jeffries R (1963) An infra-red study of the deuteration of cellulose and cellulose derivatives. *Polymer* 4:375–389.

Jones L, Ennos AR, Turner SR (2001) Cloning and characterization of *irregular xylem4* (*irx4*): a severely lignin-deficient mutant of Arabidopsis. *Plant Journal* 26:205–216.

Jung JH, Vermerris W, Gallo M, Fedenko JR, Erickson JE, Altpeter F (2013) RNA interference suppression of lignin biosynthesis increases fermentable sugar yields for biofuel production from field-grown sugarcane. *Plant Biotechnology Journal* 11:709–716.

Kim JE, Lee J-W (2019) Microstructural changes in the cell wall and enzyme adsorption properties of lignocellulosic biomass subjected to thermochemical pretreatment. *Cellulose* 26:1111–1124.

Kirchhelle C, Chow MC, Foucart C, Neto H, Stierhof YD, Kalde M, Walton C, et al. (2016) The specification of geometric edges by a plant rab GTPase is an essential cell-patterning principle during organogenesis in Arabidopsis. *Developmental Cell* 36:386–400.

Lacayo CI, Hwang MS, Ding S-Y, Thelen MP (2013) Lignin depletion enhances the digestibility of cellulose in cultured xylem cells. *PLoS One* 8:e68266.

Lee KJD, Cornuault V, Manfield IW, Ralet M-C, Knox JP (2013) Multiscale spatial heterogeneity of pectic rhamnogalacturonan I (RG-I) structural features in tobacco seed endosperm cell walls. *Plant Journal* 75:1018–1027.

Leplé J-C, Dauwe R, Morreel K, Storme V, Lapierre C, Pollet B, Naumann A, et al. (2007) Downregulation of cinnamoyl-coenzyme A reductase in poplar: multiple-level phenotyping reveals effects on cell wall polymer metabolism and structure. *Plant Cell* 19:3669–3691.

Li L, Zhou Y, Cheng X, Sun J, Marita JM, Ralph J, Chiang VL (2003) Combinatorial modification of multiple lignin traits in trees through multigene cotransformation. *Proceedings of the National Academy of Sciences USA* 100:4939–4944.

Li X, Ximenes E, Kim, Y, Slininger M, Meilan R, Ladisch M, Chapple C (2010) Lignin monomer composition affects *Arabidopsis* cell-wall degradability after liquid hot water pretreatment. *Biotechnology for Biofuels* 3:27.

Li M, Yoo CG, Pu Y, Biswal AK, Tolbert AK, Mohnen D, Ragauskas AJ (2019) Downregulation of pectin biosynthesis gene *GAUT4* leads to reduced ferulate and lignin-carbohydrate cross-linking in switchgrass. *Communications Biology* 2:22.

Lindh EL, Salmén L (2017) Surface accessibility of cellulose fibrils studied by hydrogen-deuterium exchange with water. *Cellulose* 24:21–33.

Lionetti V, Francocci F, Ferrari S, Volpi C, Bellincampi D, Galletti R, D'Ovidio R, De Lorenzo G, Cervone F (2010) Engineering the cell wall by reducing de-methyl-esterified homogalacturonan improves saccharification of plant tissues for bioconversion. *Proceedings of the National Academy of Sciences, USA* 107:616–621.

Liu D, Qian J, Xu J, Carpita NC, McCann MC, Yang H (2022) Pectin-associated mannans and xylans play distinct roles in cell-cell adhesion in pine and poplar wood. *Industrial Crops and Products* 184:115054.

Lu F, Ralph J (1997) Derivatization followed by reductive cleavage (DFRC method), a new method for lignin analysis: protocol for analysis of DFRC monomers. *Journal of Agriculture and Food Chemistry* 45:2590–2592.

Luo H, Klein IM, Jiang Y, Zhu H, Liu B, Kenttämaa HI, Abu-Omar MM (2016) Total utilization of Miscanthus biomass, lignin and carbohydrates, using earth abundant nickel catalyst. *ACS Sustainable Chemicstry and Engineering* 4:2316–2322.

Lynch MA, Staehelin LA (1992) Domain-specific and cell type-specific localization of two types of cell wall matrix polysaccharides in the clover root tip. *Journal of Cell Biology* 118:467–479.

Majda M, Kozlova L, Banasiak A, Derba-Maceluch M, Iashchishyn IA, Morozova-Roche LA, Smith RS, et al. (2021) Elongation of wood fibers combines features of diffuse and tip growth. *New Phytology* 232:673–691.

Mangin L (1889) Sur la presence des composes pectiques dans les végétaux. *Comptes Rendus* 109:579.

Mangin L (1893) Sur l'emploi du rouge de ruthenium en anatomie végétale. *Comptes Rendus* 116:653.

Mansfield SD, Kang K-Y, Chapple C (2012) Designed for deconstruction - poplar trees altered in cell wall lignification improve the efficacy of bioethanol production. *New Phytology* 194:91–101.

Marita JM, Ralph J, Hatfield RD, Chapple C (1999) NMR characterization of lignins in Arabidopsis altered in the activity of ferulate 5-hydroxylase. *Proceedings of the National Academy of Sciences USA* 96:12328–12332.

Marry M, Roberts K, Jopson SJ, Huxham IM, Jarvis MC, Corsar J, Robertson E, et al. (2006) Cell-cell adhesion in fresh sugar-beet root parenchyma requires both pectin esters and calcium cross-links. *Physiologia Plantarum* 126:243–256.

McCann MC, Defernez M, Urbanowicz BR, Tewari JC, Langewisch T, Olek A, Wells B, Wilson RH, Carpita NC (2007) Neural network analyses of infrared spectra for classifying cell wall architectures. *Plant Physiology* 143:1314–1326.

McCann MC, Carpita NC (2015) Biomass recalcitrance: a multi-scale, multi-factor, and conversion-specific property. *Journal of Experimental Botany* 66:4109–4118.

Mellerowicz EJ, Gorshkova TA (2012) Tensional stress generation in gelatinous fibres: a review and possible mechanism based on cell-wall structure and composition. *Journal of Experimental Bot* 63:551–565.

Meng X, Foston M, Leisen J, DeMartini J, Wyman CE, Arthur J. Ragauskas AJ (2013) Determination of porosity of lignocellulosic biomass before and after pretreatment by using Simons' stain and NMR techniques. *Bioresource Technology* 144:467–476.

Meshitsuka G, Nakano J (1985) Structural characteristics of compound middle lamella lignin. *Journal of Wood Chemistry and Technology* 5:391–404.

Meyer K, Shirley AM, Cusumano JC, Bell-Lelong DA, Chapple C (1998) Lignin monomer composition is determined by the expression of a cytochrome P450-dependent monooxygenase in Arabidopsis. *Proceedings of the National Academy of Sciences USA* 95:6619–6623.

Miao Y-C, Liu C-J (2010) ATP-binding cassette-like transporters are involved in the transport of lignin precursors across plasma and vacuolar membranes. *Proceedings of the National Academy of Sciences USA* 107:22728–22733.

Miao Z, Grift TE, Hansen AC, Ting KC (2011) Energy requirement for comminution of biomass in relation to particle physical properties. *Industrial Crops and Products* 33:504–513.

Micheli F, Sundberg B, Goldberg R, Richard L (2000) Radial distribution pattern of pectin methylesterases across the cambial region of hybrid aspen at activity and dormancy. *Plant Physiology* 124:191–199.

Mikshina PV, Petrova AA, Faizullin DA, Zuev YF, Gorshkova TA (2015) Tissue-specific rhamnogalacturonan I forms the gel with hyperelastic properties. *Biochemistry (Moscow)* 80:1088–1098.

Molina-Hidalgo FJ, Franco AR, Villatoro C, Medina-Puche L, Mercado JA, Hidalgo MA, Monfort A, et al. (2013) The strawberry (*Fragaria x ananassa*) fruit-specific *rhamnogalacturonate lyase1* (*FaRGLyase1*) gene encodes an enzyme involved in the degradation of cell-wall middle lamellae. *Journal of Experimental Botany* 64:1471–1483.

Molinari HBC, Pellny TK, Freeman J, Shewry PR, Mitchell RAC (2013) Grass cell wall feruloylation: Distribution of bound ferulate and candidate gene expression in *Brachypodium distachyon*. *Frontiers in Plant Sciences* 4:50.

Morel O, Lion C, Neutelings G, Stefanov J, Baldacci-Cresp F, Simon C, Biot C, Hawkins S, Spriet C (2022) REPRISAL: mapping lignification dynamics using chemistry, data segmentation, and ratiometric analysis. *Plant Physiology* 188:816–830.

Mota TR, de Souza WR, Oliveira DM, Martins PK, Sampaio BL, Vinecky F, Ribeiro AP, et al. (2021) Suppression of a BAHD acyltransferase decreases *p*-coumaroyl on arabinoxylan and improves biomass digestibility in the model grass *Setaria viridis*. *Plant Journal* 105:136–150.

Muro-Villanueva F, Mao X, Chapple C (2019) Linking phenylpropanoid metabolism, lignin deposition, and plant growth inhibition. *Current Opinion in Biotechnology* 56:202–208.

Mutter M, Colquhoun IJ, Beldman G, Schols H, Bakx EJ, Voragen AGJ (1998) Characterization of recombinant rhamnogalacturonan α-L-rhamnopyranosyl-(1→4)-α-D-galactopyranosyluronide lyase from *Aspergillus aculeatus*. *Plant Physiology* 117:141–152.

Nielsen E, Cheung AY, Ueda T (2008) The regulatory RAB and ARF GTPases for vesicular trafficking. *Plant Physiology* 147:1516–1526.

Nixon BT, Mansouri K, Singh A, Du J, Davis JK, Lee J-G, Slabaugh E, et al. (2016) Comparative structural and computational analysis supports eighteen cellulose synthases in the plant cellulose synthesis complex. *Scientific Reports* 6:28696.

Ochoa-Jiménez V-A, Berumen-Varela G, Burgara-Estrella A, Orozco-Avitia J-A, Ojeda-Contreras, A-J, Trillo-Hernández E-A, Rivera-Domínguez M, et al. (2018) Functional analysis of tomato rhamnogalacturonan lyase gene *Solyc11g011300* during fruit development and ripening. *Journal of Plant Physiology* 231:31–40.

Okita Y, Saito T, Isogai A (2010) Entire surface oxidation of various cellulose microfibrils by TEMPO-mediated oxidation. *Biomacromolecules* 11:1696–1700.

O'Neill MA, Ishii T, Albersheim P, Darvill AG (2004) Rhamnogalacturonan-II: structure and function of a borate cross-linked pectic polysaccharide. *Annual Review of Plant Biology* 55:109–139.

Oomen RJFJ, Doeswijk-Voragen CHL, Bush MS, Vincken J-P, Borkhardt B, van den Broek LAM, Corsar J, et al. (2002) *In muro* fragmentation of the rhamnogalacturonan I backbone in potato (*Solanum tuberosum* L.) results in a reduction and altered location of the galactan and arabinan side-chains and abnormal periderm development. *Plant Journal* 30:403–413.

Ordaz-Ortiz JJ, Marcus SE, Knox JP (2009) Cell wall microstructure analysis implicates hemicellulose polysaccharides in cell adhesion in tomato fruit pericarp parenchyma. *Molecular Plant* 2:910–921.

Orfila C, Knox JP (2000) Spatial regulation of pectic polysaccharides in relation to pit fields in cell walls of tomato fruit pericarp. *Plant Physiology* 122:775–781.

Oxenbøll-Sørensen S, Pauly M, Bush M, Skjøt M, McCann MC, Borkhardt B, Ulvskov P (2000) Pectin engineering: modification of potato pectin by in vivo expression of an endo-1,4-b-D-galactanase. *Proceedings of the National Academy of Sciences USA* 97:7639–7644.

Pakarinen A, Zhang J, Brock T, Maijala P, Viikari L (2012) Enzymatic accessibility of fiber hemp is enhanced by enzymatic or chemical removal of pectin. *Bioresource Technology* 107:275–281.

Parsell T, Yohe S, Degenstein J, Jarrell T, Klein I, Gençer E, Hewetson B, *et al.* (2015) A synergistic biorefinery based on catalytic conversion of lignin prior to cellulose starting from lignocellulosic biomass. *Green Chemistry* 17:1492–1499.

Peña MJ, Carpita NC (2004) Loss of highly branched arabinans and debranching of rhamnogalacturonan I accompany loss of firm texture and cell separation during prolonged storage of apple. *Plant Physiology* 135:1305–1313.

Penning BW, Hunter CT, Tayengwa R, Eveland AL, Dugard CK, Olek AT, Vermerris W, et al. (2009) Genetic resources for maize cell wall biology. *Plant Physiology* 151:1703–1728.

Penning BW, Shiga TM, Klimek JF, SanMiguel PJ, Shreve J, Thimmapuram J, Sykes RW, et al. (2019) Expression profiles of cell-wall related genes vary broadly between two common maize inbreds during stem development. *BMC Genomics* 20:785.

Perkins ML, Schuetz M, Unda F, Chen KT, Bally MB, Kulkarni JA, Yan Y, Pico J, Simone D Castellarin SD, et al. (2022) Monolignol export by diffusion down a polymerization-induced concentration gradient. *Plant Cell* 34:2080–2095.

Persson S, Caffall KH, Glenn Freshour G, Hilley MT, Bauer S, Poindexter P, Hahn MG, et al. (2007) The Arabidopsis *irregular xylem8* mutant is deficient in glucuronoxylan and homogalacturonan, which are essential for secondary cell wall integrity. *Plant Cell* 19:237–255.

Petrik DL, Karlen SD, Cass CL, Padmakshan D, Lu FC, Liu S, Le Bris P, et al. (2014) *p*-Coumaroyl-CoA:monolignol transferase (PMT) acts specifically in the lignin biosynthetic pathway in *Brachypodium distachyon*. *Plant Journal* 77:713–726.

Ralph J, Akiyama T, Coleman HD, Mansfield SD (2012) Effects on lignin structure of coumarate 3-hydroxylase downregulation in poplar. *BioEnergy Research* 5:1009–1019.

Ralph J, Lapierre C, Boerjan W (2019) Lignin structure and its engineering. *Current Opinion in Biotechnology* 56:240–249.

Reeves JB (1987) Sodium chlorite treatment of plant materials: Fiber and lignin composition, digestibility, and their interrelationships. *Journal of Dairy Science* 70:2534–2549.

Ritter GJ (1925) Distribution of lignin in wood. Journal of *Industrial and Engineering Chemistry* 17:1194–1197.

Ruel K, Faix O, Joseleau JP (1994) New immunogold probes for studying the distribution of the different lignin types during plant cell wall biogenesis. *Journal of Trace and Microprobe Techniques* 12:247–265.

Ruel K, Burlat V, Joseleau J-P (1999) Relationship between ultrastructural topochemistry of lignin and wood properties. *International Association of Wood Anatomists Journal* 20: 203–2011.

Ryser U (2003) Protoxylem: the deposition of a network containing glycine-rich cell wall proteins starts in the cell corners in close association with the pectins of the middle lamella. *Planta* 216:854–864.

Saballos A, Ejeta G, Sanchez E, Kang C, Vermerris W (2009) A genome-wide analysis of the cinnamyl alcohol dehydrogenase family in sorghum [*Sorghum bicolor* (L.) Moench] identifies *SbCAD2* as the *Brown midrib6* gene. *Genetics* 181:783–795.

Saballos A, Sattler SE, Sanchez E, Foster TP, Xin Z, Kang C-H, Pedersen JF, Vermerris W (2012) *Brown midrib2* (*Bmr2*) encodes the major 4-coumarate:coenzyme A ligase involved in lignin biosynthesis in sorghum (*Sorghum bicolor* (L.) Moench). *Plant Journal* 70:818–830.

Saito T, Isogai A (2004) TEMPO-mediated oxidation of native cellulose. The effect of oxidation conditions on chemical and crystal structures of the water-insoluble fractions. *Biomacromolecules* 5:1983–1989.

Sant'Anna C, Costa LT, Abud Y, Biancatto L, Miguens FC, de Souza W (2013) Sugarcane cell wall structure and lignin distribution investigated by confocal and electron microscopy. *Microscopy Research and Technique* 76:829–834.

Sarkar P, Bosneaga E, Auer M (2009) Plant cell walls throughout evolution: towards understanding of their design principles. *Journal of Experimental Botany* 60: 3615–3635.

Sattler SE, Saathoff AJ, Haas EJ, Palmer NA, Funnell-Harris DL, Sarath G, Pedersen JF (2009) A nonsense mutation in a cinnamyl alcohol dehydrogenase gene is responsible for the sorghum *brown midrib6* phenotype. *Plant Physiology* 150:584–595.

Sattler SE, Palmer NA, Saballos A, Greene AM, Xin ZG, Sarath G, Vermerris W, Pedersen JF (2012) Identification and characterization of four missense mutations in *Brown midrib12* (*Bmr12*), the caffeic acid *O*-methyltransferase (COMT) of sorghum. *BioEnergy Research* 5:855–865.

Scalbert A, Monties B, Lallemand J, Guittet E, Rolando C (1985) Ether linkage between phenolic acids and lignin fractions from wheat straw. *Phytochemistry* 24:1359–1362.

Schilmiller AL, Stout J, Weng J-K, Humphreys J, Ruegger MO, Chapple C (2009) Mutations in the cinnamate 4-hydroxylase gene impact metabolism, growth and development in Arabidopsis. *Plant Journal* 60:771–782.

Schrader J, Nilsson J, Mellerowicz E, Berglund A, Nilsson P, Hertzberg M, Sandberg G (2004) A high-resolution transcript profile across the wood-forming meristem of poplar identifies potential regulators of cambial stem cell identity. *Plant Cell* 16:2278–2292.

Schumacher I, Fabrice TN, Abdou M-T, Kuhn BM, Voxeur A, Herger A, Roffler S, et al. (2021) Defects in cell wall differentiation of the Arabidopsis mutant *rol1-2* Is dependent on cyclin-dependent kinase CDK8. *Cells* 10:685.

Sénéchal F, L'Enfant M, Domon J-M, Rosiau E, Crépeau M-J, Surcouf O, Esquivel-Rodriguez J, et al. (2015) Tuning of pectin methylesterification: Pectin methylesterase inhibitor7 modulates the processive activity of co-expressed pectin methylesterase 3 in a pH-dependent manner. *Journal of Biology Chemistry* 290:23320–23335.

Serrani-Yarce JC, Escamilla-Trevino L, Barros J, Gallego-Giraldo L, Pu Y, Ragauskas A, Dixon RA (2021) Targeting hydroxycinnamoyl CoA: shikimate hydroxycinnamoyl transferase for lignin modification in *Brachypodium distachyon*. *Biotechnology for Biofuels* 14:50.

Shiga TM, Xiao W, Xiao H, Zhang X, Olek AT, Donohoe BS, Liu J, et al. (2017) Enhanced rates of enzymatic saccharification and catalytic synthesis of biofuel substrates in gelatinized cellulose generated by trifluoroacetic acid. *Biotechnology for Biofuels* 10:310.

Shiga TM, Yang H, Penning BW, Olek AT, McCann MC, Carpita NC (2021) A TEMPO-catalyzed oxidation-reduction method to probe surface and anhydrous crystalline-core domains of cellulose microfibril bundles. *Cellulose* 28:5305–5319.

Siedlecka A, Wiklund S., Péronne M-A, Micheli F, Lesniewska J, Sethson I, Edlund U, et al. (2008) Pectin methyl esterase inhibits intrusive and sympastic cell growth in developing wood cells of *Populus*. *Plant Physiology* 146:554–565.

Singh AP, Donaldson LA (1999) Ultrastructure of tracheid cell walls in radiata pine (*Pinus radiata*) mild compression wood. *Canadian Journal of Botany* 77:32–40.

Siqueira G, Varnai A, Ferraz A, Milagres AMF (2013) Enhancement of cellulose hydrolysis in sugarcane bagasse by the selective removal of lignin with sodium chlorite. *Applied Energy* 102:399–402.

Smith RA, Gonzales-Vigil E, Karlen SD, Park J-Y, Lu F, Wilkerson CG, Samuels L, et al. (2015) Engineering monolignol *p*-coumarate conjugates into poplar and Arabidopsis lignins. *Plant Physiology* 169:2992–3001.

Song D, Shen J, Li L (2010) Characterization of cellulose synthase complexes in *Populus* xylem differentiation. *New Phytologist* 187:777–790.

Srivastava LM (1966) Histochemical studies on lignin. *Technical Association of the Pulp and Paper Industry* 49:173–183.

Stewart JJ, Akiyama T, Chapple C, Ralph J, Mansfield SD (2009) The effects on lignin structure of overexpression of ferulate 5-hydroxylase in hybrid poplar. *Plant Physiology* 150:621–635.

Strasburger E (1875) Über *Zellbildung und Zelltheilung*. Jena Dabis, Germany.

Strasburger E (1884) *Das Botanische Practicum*. Jena Dabis.

Studer MH, DeMartini JD, Davis MF, Sykes RW, Davison B, Keller M, Tuskan GA, Wyman CE 2011. Lignin content in natural Populus variants affects sugar release. *Proceedings of the National Academy of Sciences USA* 108:6300–6305.

Sundell D, Street NR, Kumar M, Mellerowicz EJ, Kucukoglu M, Johnsson C, Kumar V, et al. (2017) AspWood: High-spatial-resolution transcriptome profiles reveal uncharacterized modularity of wood formation in *Populus tremula*. *Plant Cell* 29:1585–1604.

Tan L, Eberhard S, Pattathil S, Warder C, Glushka J, Yuan C, Hao Z, et al. (2013) An Arabidopsis cell wall proteoglycan consists of pectin and arabinoxylan covalently linked to an arabinogalactan protein. *Plant Cell* 25:270–287.

Tanhatan-Nasseri A, Crepeau MJ, Thibault JF, Ralet MC (2011) Isolation and characterization of model homogalacturonans of tailored methylesterification patterns. *Carbohydrate Polymers* 86:1236–1243.

Timberlake HG (1900) The development and function of the cell plate in higher plants. *Botanical Gazette* 30:73–99.

Treub M (1878) *Quelques Recherches sur le Rôle du Noyau dans la Division des Cellules Végétales*. Amsterdam, The Netherland.

Uehara Y, Tamura S, Maki Y, Yagyu K, Mizoguchi T, Tamiaki H, Imai T, et al. (2017) Biochemical characterization of rhamnosyltransferase involved in biosynthesis of pectic rhamnogalacturonan I in plant cell wall. *Biochemical and Biophysical Research Communications* 486:130–136.

Ulvskov P, Wium H, Bruce D, Jorgensen B, Qvist KB, Skjot M, Hepworth D., et al. (2005) Biophysical consequences of remodelling the neutral side chains of rhamnogalacturonan I in tubers of transgenic potatoes. *Planta* 220:609–620.

Van Acker R, Leplé J-C, Aerts D, Storme V, Goeminne G, Ivens B, Légée F, et al. (2014) Improved saccharification and ethanol yield from field-grown transgenic poplar deficient in cinnamoyl-CoA reductase. *Proceedings of the National Academy of Sciences USA* 111:845–850.

Van Acker R, Déjardin A, Desmet S, Hoengenaert L, Vanholme R, Morreel K, Laurans F, et al. (2017) Different routes for conifer- and sinapaldehyde and higher saccharification upon deficiency in the dehydrogenase CAD1. *Plant Physiology* 175:1018–1039.

Viamajala S, McMillan JD, Schell DJ, Elander RT (2009) Rheology of corn stover slurries at high solids concentrations - Effects of saccharification and particle size. *Bioresource Technology* 100:925–934.

Vincken J-P, Schols HA, Oomen RJF, McCann MC, Ulvskov P, Voragen AGJ, Visser RGF (2002) If homogalacturonan were a side chain of rhamnogalacturonan I. Implications for cell wall architecture. *Plant Physiology* 132:1781–1789.

Wang JP, Matthews ML, Williams CM, Shi R, Yang C, Tunlaya-Anukit S, Chen HC, et al. (2018) Improving wood properties for wood utilization through multi-omics integration in lignin biosynthesis. *Nature Communications* 9:1579–1595.

Westermark U (1985) The occurrence of *p*-hydroxyphenylpropane units in the middle-lamella lignin of spruce (*Picea abies*). *Wood Science and Technology* 19:223–232.

Willats WGT, Orfila C, Limberg G, Buchholt HC, van Alebeek G-JWM, Voragen AGJ, Marcus SE, et al. (2001) Modulation of the degree and pattern of methyl-esterification of pectic homogalacturonan in plant cell walls: implications for pectin methyl esterase action, matrix properties, and cell adhesion. *Journal of Biological Chemistry* 276:19404–19413.

Withers S, Lu FC, Kim H, Zhu YM, Ralph J, Wilkerson CG (2012) Identification of grass-specific enzyme that acylates monolignols with *p*-coumarate. *Journal of Biological Chemistry* 287:8347–8355.

Wu AM, Hörnblad E, Voxeur A, Gerber L, Rihouey C, Lerouge P, Marchant A (2010) Analysis of the Arabidopsis IRX9/IRX9-L and IRX14/IRX14-L pairs of glycosyltransferase genes reveals critical contributions to biosynthesis of the hemicellulose glucuronoxylan. *Plant Physiology* 153:542–554.

Wu Z, Wang N, Hisano H, Cao Y, Wu F, Liu W, Bao Y, et al. (2019), Simultaneous regulation of *F5H* in *COMT*-RNAi transgenic switchgrass alters effects of *COMT* suppression on syringyl lignin biosynthesis. *Plant Biotechnology Journal* 17:836–845.

Xiang ZY, Sen SK, Min DY, Savithri D, Lu FC, Jameel H, Chang V, Chang H-M (2017) Field-grown transgenic hybrid poplar with modified lignin biosynthesis to improve enzymatic saccharification efficiency. *ACS Sustainable Chemistry and Engineering* 5:2407–2414.

Yang H, Zhang X, Luo H, Liu B, Shiga TM, Li X, Kim JI, et al. (2019) Overcoming cellulose recalcitrance in woody biomass for the lignin-first biorefinery. *Biotechnology for Biofuels* 12:171.

Yang H, Benatti MR, Karve RA, Fox A, Meilan R, Carpita NC, McCann MC (2021) Rhamnogalacturonan I is a determinant of cell-cell adhesion in poplar wood. *Plant Biotechnology* 18:1027–1040.

Ye ZH, Song YR, Marcus A, Varner JE (1991) Comparative localization of three classes of cell wall proteins. *Plant Journal* 1:175–183.

Zabotina OA (2012) Xyloglucan and its biosynthesis. *Frontiers in Plant Sciences* 3:134.

Zamil MS, Geitmann A (2017) The middle lamella - more than a glue. *Physical Biology* 14:015004.

Zandleven J, Sørensen SO, Harholt J, Beldman G, Schols HA, Scheller HV, Voragen AGJ (2007) Xylogalacturonan exists in cell walls from various tissues of *Arabidopsis thaliana*. *Phytochemistry* 68:1219–1226.

Zykwinska A, Thibault J-F, Ralet M-C (2007) Organization of pectic arabinan and galactan side chains in association with cellulose microfibrils in primary cell walls and related models envisaged, *Journal of Experimental Botany* 58:1795–1802.

25 Cytokinesis and the Formation of a New Cell Wall

Rosalie Sinclair, Mingqin Chang, and Georgia Drakakaki
University of California Davis

PLANT CYTOKINESIS, A UNIQUE AND FUNDAMENTAL PROCESS

Eduard Strasburger, one of the founders of modern plant cell biology and director of the Botany Institute and the Botanical Garden at the University of Bonn from 1881 to 1912, first elucidated the principles of plant cytokinesis. Early in his career, Strasburger characterised and named the distinct stages of mitosis. Using the green algae *Cladophora glomerata*, due to its large cell size, Strasburger illuminated the differences between plant and animal cytokinesis in his book *Über Zellbildung und Zelltheilung* (1876; "On Cell Formation and Cell Division"). While cytokinesis in vacuolated cells has been described since the 1940s with the description of a phragmosome (Sinnott and Bloch 1940), the formation of the cell plate was first illustrated at the ultrastructural level in the 1960s (Hepler and Newcomb 1967). For several decades, major discoveries were based on microscopy techniques followed by in-depth molecular analyses towards the end of the 20th century.

Cytokinesis is the final step in cell division, in which the cytoplasm of two newly formed daughter cells separates. Regulated cell division along with coordination of cell expansion is essential for plant development and cell shape determination (Rasmussen and Bellinger 2018; Sablowski and Gutierrez 2022; Yi and Goshima 2022). Plant cytokinesis is a fundamental process of plant life that is distinct from animal cytokinesis, which occurs through the formation of a cleavage furrow. Plant cells separate through the centrifugal formation of a cell plate that grows outwards and develops into the new cell wall, partitioning the cytoplasm of the dividing cell starting from the centre of the cell.

Cell plate formation is a tightly regulated process of choreographed vesicle delivery via the phragmoplast, followed by vesicle fusion and membrane transformation, which occurs concurrently with the time-specific deposition of polysaccharides (Drakakaki 2015; Smertenko et al. 2017; Sinclair et al. 2022). In plant cytokinesis, a phragmoplast structure serves as a scaffold for the assembly of the cell plate. In contrast, during animal cytokinesis, an actomyosin contractile ring is formed to separate the two daughter cells. Furthermore, while animal cytokinesis is characterised by the formation of a cleavage furrow inwards from the cell periphery, in plants, the construction of a cell plate begins from the centre of the dividing cell at the division plane, expanding centrifugally outwards until fused with the parental cell wall.

Despite their stark differences, during late cytokinesis the midzone/midbody in animal cells (Glotzer 2005) shares some features with the plant phragmoplast/cell plate (Otegui et al. 2005). Membrane delivery is required for both the growth of the midbody in animal cytokinesis and cell plate formation in plants. While the midbody fills a remaining gap in the ingrowing furrow (Glotzer 2005; Steigemann and Gerlich 2009) and the plant cell plate expands radially, both structures ultimately enable the separation of the two daughter cells (Jürgens 2005; Otegui et al. 2005). How these structures have evolved to fulfil their specific functions (Farmer and Prekeris 2022) in different types of eukaryotic cells remains unknown.

This chapter provides an overview of plant cytokinesis, focusing on cell plate development, by first describing the different stages and the overall progression from the point of view of both the prevailing membrane and cytoskeletal arrays, as informed by different microscopy techniques. Cytoskeletal arrangement, followed by molecular components of the endomembrane system and polysaccharide deposition that together contribute to the formation of the cell plate and the new cell wall are discussed. The contribution of technological advances to our current knowledge in cell plate development is emphasised along with current limitations and emerging technologies that can be used to further decipher the mysteries of this fundamental aspect of plant life.

CELL PLATE FORMATION: A MULTISTAGE DYNAMIC PROCESS

Electron microscopy was a major tool that contributed to advancing our understanding of the details of plant cytokinesis (Pickett-Heaps and Northcote 1966; Hepler and Newcomb 1967; Palevitz and Hepler 1974; Gunning et al. 1978). It allowed for the discovery of several aspects of phragmoplast development and cell plate orientation. During the mid-1990s, the Staehelin group used high pressure freezing (HPF) freeze substitution (Samuels et al. 1995) along with electron tomography to capture details on both cytoskeletal and membrane conformations in three dimensions thus informing models of cell plate

DOI: 10.1201/9781003178309-25

formation (Otegui et al. 2001; Segui-Simarro et al. 2004). Fluorescence microscopy using labelled tubulin provided early insights into microtubule polymerisation and translocation of the expanding phragmoplast (Asada et al. 1991). The development of fluorescence and confocal laser scanning microscopy, along with the generation of distinct markers for membrane and cytoskeleton components, allowed for a dynamic illustration of the process (reviewed in Lee and Liu 2013; McMichael et al. 2013; Boruc and Van Damme 2015; Jurgens et al. 2015; Smertenko et al. 2017, 2018; Buschmann and Müller 2019; Sinclair et al. 2022). The *Nicotiana tabacum* Bright-Yellow-2 (BY-2) cell culture system has been an excellent tool to observe larger cells and synchronisation of the cell cycle (Kakimoto and Shibaoka 1989; Samuels et al. 1995; Kumagai-Sano et al. 2006). In summary, cell plate development has been described in the literature through two perspectives: first, cytokinesis can be staged based on cytoskeletal development, which mainly focuses on the microtubule array; and the second is based on the membrane conformations of the cell plate structure during cell plate development, characterised by four developmental stages that can exist simultaneously in different areas of a given cell plate.

Here, we will discuss centrifugal growth of the cell plate primarily through morphological changes of the membrane with reference to the corresponding cytoskeletal conformations (Figures 25.1 and 25.2). We should note that morphological changes of the membrane have been primarily defined using electron microscopy, while the cytoskeletal and overall phragmoplast conformations illustrated in the figures have been described using fluorescence imaging; hence, there are differences in the spatial resolution of the experimental data underlying these concepts.

Centrifugal Growth of the Cell Plate

During plant cytokinesis, a specialised structure known as the phragmoplast delivers cytokinetic vesicles to the ribosome-excluding cell plate assembly matrix (CPAM). The phragmoplast originates from microtubules of the central spindle that separates the chromosomes (Euteneuer and McIntosh 1980; Lee et al. 2001). The microtubules orient with plus ends

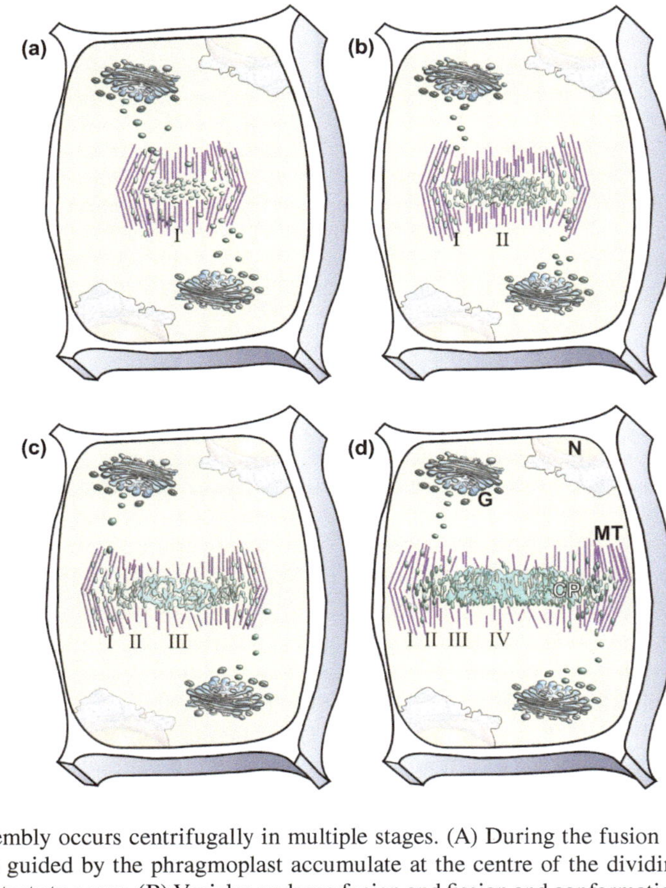

FIGURE 25.1 Cell plate assembly occurs centrifugally in multiple stages. (A) During the fusion of the Golgi-derived vesicle stage (FVS) (I), cytokinetic vesicles guided by the phragmoplast accumulate at the centre of the dividing cell, at the cell plate assembly matrix (CPAM), where fusion starts to occur. (B) Vesicles undergo fusion and fission and conformational changes, resulting in the tubular-vesicular network (TVN) (Stage II). (C) Interconnected membrane structures transition to a tubular network (TN) (Stage III). The membrane network further expands to a continuous planar fenestrated sheet (PFS) (Stage IV). (D) Deposition of additional polysaccharides facilitates transition to a new cell wall, separating the two daughter cells. Note that different stages can occur simultaneously; structures are not to scale. This simplified representation emphasises cell plate membranes. Note that many subcellular compartments are omitted for simplicity and that the location of the nucleus is not exact. For a more accurate representation, see Videos S1, S2 in Sinclair et al. (2022). CP, Cell Plate; MT, Microtubules; G, Golgi; N, Nucleus.

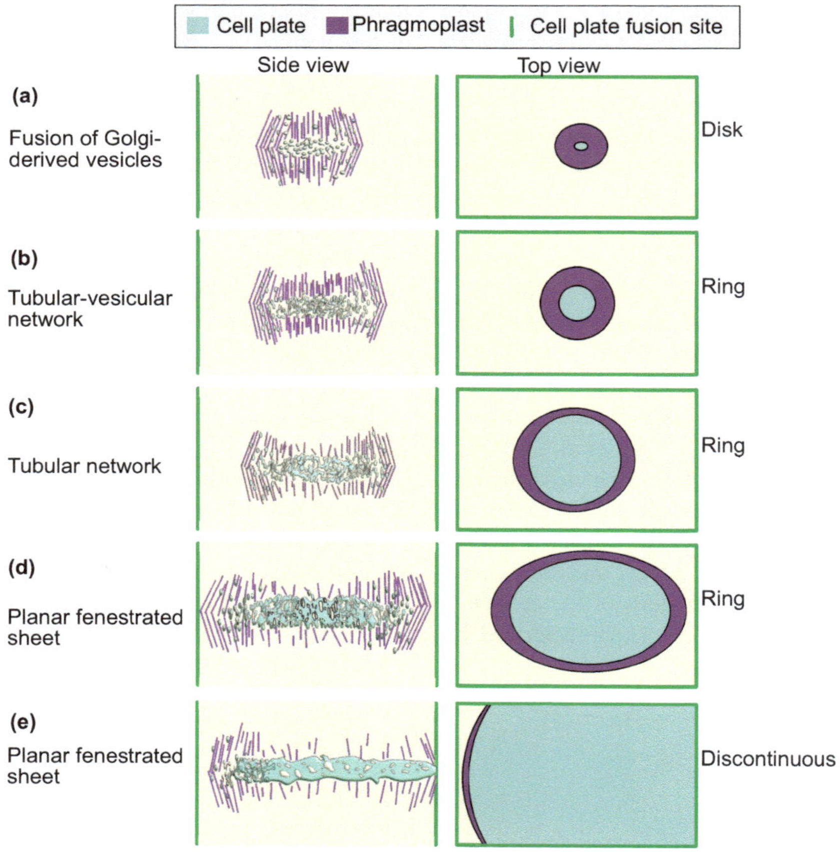

FIGURE 25.2 Schematic representation of the cell plate development through two perspectives. Cell plate assembly occurs centrifugally in simultaneously occurring stages termed by the mature state of the centre-most material. On the left, a view of membrane fusion staging nomenclature; on the right, a two-dimensional phragmoplast at the stages shown as the top view of the corresponding cell plate. (A) During the FVS stage, cytokinetic vesicles guided by the phragmoplast accumulate at the centre of the dividing cell, at the CPAM, where fusion starts to occur. This represents a disk-like tight bundling of the phragmoplast microtubules. (B) Vesicles undergo fusion and fission and conformational changes resulting in tubular-vesicular network (TVN). The phragmoplast continues to expand with the membrane material and shows a ring conformation. (C) Interconnected membrane structures transition to a tubular network (TN). The ring phragmoplast persists. The membrane network further expands to a continuous PFS (D). The microtubules begin to depolymerise in the centre but do not disperse until a connection to the parental cell wall is formed in (E), where they change to a discontinuous conformation. Occasionally, the cell plate connects unevenly to the parental cell wall, causing a discontinuous phragmoplast, while the remaining end continues to mature until it fuses to the parental cell wall completing cytokinesis (E).

towards the division plane maintaining the same orientation as the spindle; the two arrays are on either side of the division plane that was previously established by the preprophase band (PPB) (Lloyd and Buschmann 2007; Schaefer et al. 2017). These microtubules run in a similar orientation to actin microfilaments (Kakimoto and Shibaoka 1987; Palevitz 1987; Traas et al. 1987; Schmit and Lambert 1990).

During the early stages, polymerisation and alignment of microtubules shape the phragmoplast into a disk, facilitating the delivery of vesicles at the cell division plane. When vesicles reach the division plane, they immediately start to fuse together, forming dumbbell and hourglass-shaped conformations with tubular connections (Figures 25.1 and 25.3). The tubulated vesicles continue to morph into a network and transition to a smooth yet wavy membrane disk. Cell plate expansion is radial with a small diameter disk eventually expanding across the entire plane of division (Samuels et al. 1995). The microtubules depolymerise at regions where vesicles have already formed a smooth network at the centre, while they continue to polymerise at the edge of the cell plate (Murata et al. 2013) to form a microtubular ring. This ring continues to expand and grow along the direction of cell plate expansion. This is known as a ring stage phragmoplast (Figures 25.2B, C and D). Thus, the cell plate itself contains a gradient of stages ranging from mature at the cell plate centre to newly accumulated membranes at the leading edge. As portions of the cell plate reach the parental cell wall, they disturb the ring conformation, resulting in a discontinuous ring. This stage is known as a discontinuous ring phragmoplast (Figure 25.2E). In vacuolate shoot cells, instead of symmetrical cell plate expansion, polarised cytokinesis has been observed, in which the cell plate initially anchors on one side of the cell, followed by growth of the cell plate that ultimately joins the opposite site (Cutler and Ehrhardt 2002). This type of cell plate growth is proposed to provide mechanical support during the building of the cell wall in large cells (Cutler and Ehrhardt 2002).

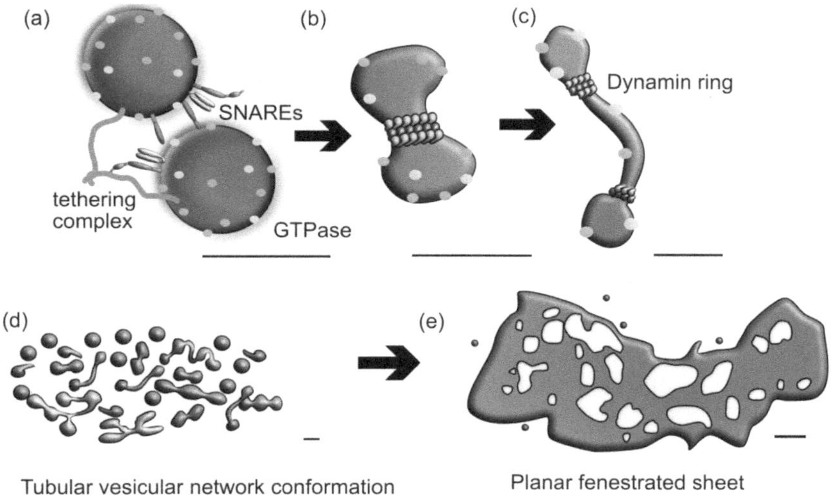

FIGURE 25.3 Cell plate membrane transitions from vesicles to planar fenestrated sheet (PFS). During the early stages of cell plate assembly, delivery of cytokinetic vesicles, regulated by molecular switches, such as RAB GTPases, undergo initial fusion and fission. Vesicle fusion is assisted by tethering complexes and SNAREs (A). The Dynamin DRPC1 wraps around the necks of fused vesicles and causes fission and tubulation (B). Tubular vesicular stretching is accomplished by dynamin rings assembled around the vesicles creating dumbbell-like structures (C) that increase in complexity as more membrane is added and transitions to the tubular-vesicular network (TVN) stage (D). With the presence of a spreading force, TVN transitions to the PFS (E). Schematic description of vesicle fusion is adapted from data collected using electron microscopy (Samuels et al., 1995; Segui-Simarro et al. 2004). Scale bars 50 nm (A) to (D) 0.25 µm (E).

Description of Cell Plate Developmental Stages Based on Membrane Conformations

Despite the dynamic nature of cell plate development, electron microscopy of high pressure cryofixed cells has been the preferred method for the characterisation of vesicle fusion and membrane transition due to the excellent sample preservation and spatial resolution that the method can provide. Electron micrographs allowed for 2D resolution of the cell plate and established foundational models using synchronised BY-2 cells (Samuels et al. 1995). The models were refined with the advancement of this technique into high resolution electron tomography, with 3D renderings of 2D tilt series images of serial sections that covered a large area of the cell plate. Montaged frame images could cover approximately a 4.5×4.5-µm cell plate area, at a nanometre resolution (Otegui et al. 2001; Segui-Simarro et al. 2004). This enabled a deeper understanding of the structure of vesicle fusion and transitions of membrane conformations during cell plate development, while overcoming the limitation posed by artefacts caused by conventional chemical fixation. Based on these excellent studies, we know that distinct membrane transition stages can exist simultaneously.

During the initial *fusion of Golgi-derived vesicles stage* (FVS), vesicular fusion and fission events start with the delivery of cytokinetic vesicles via phragmoplast microtubules. Upon vesicle fusion and fission, dumbbell structures are formed. The morphology of the cell plate membrane changes as more vesicle materials are added (Samuels et al. 1995; Segui-Simarro et al. 2004). Cytokinetic vesicles regulated by GTPases, such as RABA2a (Chow et al. 2008), are fused via the activity of SNARE complexes. A key cytokinetic-specific SNARE is KNOLLE, which was first discovered in 1997 (Lauber et al. 1997). Tethering complexes also assist in this process (Rybak et al. 2014). For details, see below and Figures 25.1A and 25.2.

The morphology of the cell plate membrane changes as more membrane materials are added to the initial cell plate, which transitions to a *tubular-vesicular network* (TVN) stage. As the vesicles fuse, they tubulate largely due to the activity of dynamin-related proteins (Otegui et al. 2001; Segui-Simarro et al. 2004; Konopka and Bednarek 2008). The dynamin springs wrap around the necks of fused vesicles, leading to both fission and tubulation, by stretching the vesicle membranes into long tubes, hourglass, and dumbbell conformations to create a membrane network (Otegui et al. 2001; Segui-Simarro et al. 2004) Figure 25.3. The membrane conformation at this stage is uneven with rough tubulations. The deposition of the polysaccharide callose is reported to begin at this stage (Samuels et al. 1995). Other cell wall polysaccharides, delivered via Golgi-derived vesicles, including hemicellulose and pectin, are suggested to be deposited in small amounts at this stage (Samuels et al. 1995; Drakakaki 2015) (Figure 25.1B).

Further morphological membrane changes occur that transition the TVN into a smoother membrane while still maintaining the tubular connectivity known as the *tubular network* (TN) stage (Samuels et al. 1995). The membrane network now encloses a large lumen where polysaccharides can be deposited. Callose is prominently deposited at this stage, while cellulose deposits at a later stage (Samuels et al. 1995). Clathrin-coated vesicles (CCVs) recycle excess cell plate materials from the maturing cell plate and allow the cell plate to begin to thin into a smoother structure (Segui-Simarro et al. 2004) (Figure 25.1C).

As the cell plate nears full maturation at the *planar fenestrated sheet* (PFS) stage, a plasma membrane (PM) sheet with small pits can now be seen. These pits originated from the large cavities of the TN and will shrink with membrane expansion (Samuels et al. 1995; Segui-Simarro et al. 2004). It is thought that most pits are closed, but some remain open for plasmodesmata development and contain strands of endoplasmic reticulum (ER) (Hepler 1982; Segui-Simarro et al. 2004) (Figure 25.1D).

Though it is not considered a stage of cell plate maturation, cell plate fusion with the parental cell wall and PM is followed by flattening of the cell plate and is an important process to finalise cytokinesis (Mineyuki and Gunning 1990). Membrane "finger-like fusion tubes" extend from the cell plate to meet the parental cell wall (Samuels et al. 1995). Notably, callose has been observed at the junction of the cell plate and the parental cell wall (Samuels et al. 1995). A recently developed animation provides an overview of plant cytokinesis (Videos S1, S2 in Sinclair et al. 2022).

MOLECULAR COMPONENTS INVOLVED IN CYTOKINESIS AND CELL PLATE DEVELOPMENT

As outlined above, cytokinesis and the formation of the cell plate is a dynamic process requiring the coordination of the cytoskeleton, endomembrane trafficking, polysaccharide deposition, and assembly into a new cell wall (Sinclair et al. 2022). At the turn of the century, technology allowed for the sequencing of different genomes, starting with the model plant Arabidopsis (Meinke et al. 1998; Tabata et al. 2000). Further, confocal microscopy became a routine tool for analysis in plants, allowing for live imaging of events using both fluorescent markers and dyes (Livanos et al. 2017; Komis et al. 2018). The development of molecular and genetic tools enabled the study of gene function in mutants using either loss or gain of function phenotypes and contextualised the subcellular behaviour of cytokinetic markers (Cheung et al. 2022), providing a better handle on the functional analysis of the studied molecular components and their role in cytokinesis (McMichael and Bednarek 2013; Boruc and Van Damme 2015; Smertenko et al. 2017; Gu and Rasmussen 2022). The sections below briefly explore selected, well-studied critical components of this process and emphasise plant endomembranes and polysaccharide deposition during cytokinesis (Figure 25.4).

Cytoskeleton and Division-Plane Orientation

Division-plane determination controls the cellular patterns created in meristems that will ultimately shape plant development. During cytokinesis, the cytoskeleton is the backbone of the phragmoplast, a multicomponent machinery that facilitates the construction of the cell plate from membranes and oligosaccharides (Lee and Liu 2013; Smertenko 2018). For details on cytoskeletal and division-plane orientation, we refer readers to these excellent reviews (Rasmussen and Bellinger 2018; Smertenko 2018; Smertenko et al. 2018; Buschmann and Müller 2019; Lee and Liu 2019; Müller 2019; Gu and Rasmussen 2022). The cytoskeleton is the primary system on which the vesicles are delivered to the CPAM. This requires the coordination of microtubules, microfilaments, microtubule-associated proteins (MAPs), and actin binding proteins (ABPs) including kinesin and myosin motor proteins.

The involvement of the cytoskeleton in cytokinesis begins much earlier than the establishment of the phragmoplast. The cell division-plane and phragmoplast orientation are determined by the PPB early in the G2 stage of mitosis (Wick and Duniec 1983; Wick and Duniec 1984; Mineyuki et al. 1988). The PPB consists of a cortical ring of microtubules, actin filaments, cytoskeleton-interacting proteins, and other components that underlie the cortical division zone (CDZ), which is disassembled during prometaphase (Wick and Duniec 1983, 1984). The PPB deposits landmark proteins, aiding in the determination of the CDZ, which supports the alignment of the phragmoplast. The phragmoplast builds upon the remaining mitotic spindle in the centre of the division plane and expands towards the cell cortex during cytokinesis (Lee and Liu 2013; Smertenko 2018). Early ancestors of land plants did not divide through the PPB/phragmoplast system, suggesting that during the course of evolution the PPB and phragmoplast were gradually modified (Buschmann and Zachgo 2016). One hypothesis is that instead of a fully developed PPB, a few cortical microtubules are sufficient to establish a CDZ (Buschmann and Müller 2019). The enigmatic role of PPB is an active area of research awaiting future discoveries. Several critical proteins for phragmoplast positioning accumulate at the division site prior to phragmoplast cell cortex interaction (Rasmussen and Bellinger 2018). One of the earliest discovered proteins associated with the cortical division plane is TANGLED1 (TAN1), a microtubule-cross-linking protein regulating phragmoplast positioning (Smith et al. 1996; Cleary and Smith 1998; Smith et al. 2001; Martinez et al. 2020). Confocal fluorescence microscopy allowed for the understanding that AtTAN1 remains behind at the division site after the PBB is disassembled, providing cortical guidance cues for the phragmoplast (Walker et al. 2007). During telophase, TAN1 along with other division site associated proteins, may organise the cortical microtubule array, which is then incorporated into the phragmoplast and guides it to the division site (Bellinger et al. 2021). In Arabidopsis, mutations in TAN1 display aberrant phenotypes caused by division-plane defects and short roots but only when combined with mutants of AUXIN INDUCED IN ROOT CULTURES9 (AIR9), a microtubule binding protein accumulating at the division site and capable of associating with the PPB memory (Buschmann et al. 2006; Mir et al. 2018).

Additionally, the kinesin12-related proteins Phragmoplast Orienting Kinesins (POKs), POK1 and POK2, core members of division-plane orientation, localise to both the phragmoplast midline and the division site and are involved in phragmoplast positioning. Their function can be dependent on the other core member TAN1 (Lipka et al. 2014; Martinez et al.

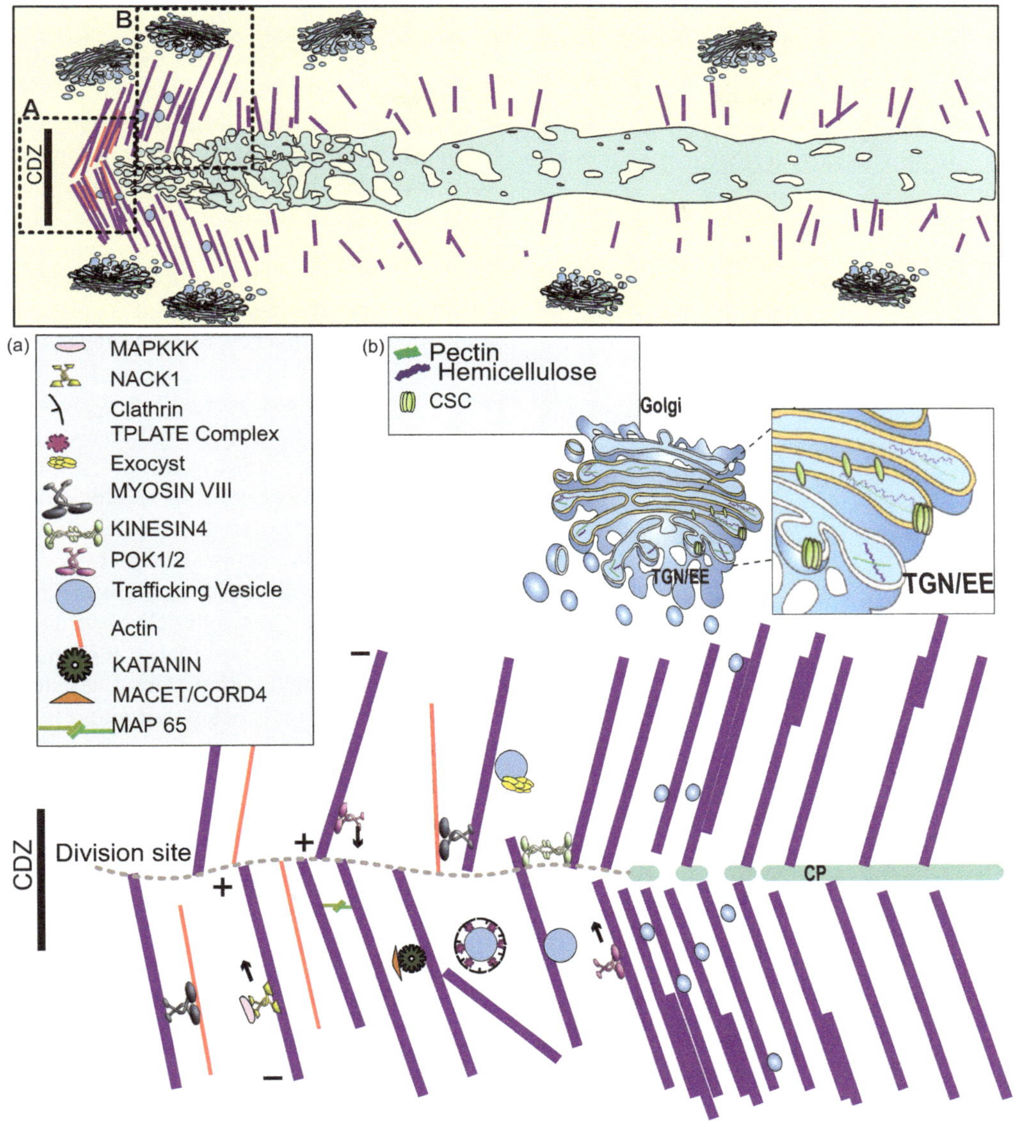

FIGURE 25.4 Conceptual model of plant cytokinesis with a focus on cytoskeletal elements. (A) Key components of the cell cortex underlying the phragmoplast, membrane, and cell wall structures that form the cell plate (teal) from left to right maturation. (B) Pectin and hemicellulose are synthesised in the Golgi and are transported via cytokinetic vesicles along microtubules in a spatiotemporally regulated manner to the plasma membrane, supporting the construction and assembly of the cell plate.

2017; Herrmann et al. 2018; Buschmann and Müller 2019; Gu and Rasmussen 2022; Mills et al. 2022). Absence of POKs leads to phragmoplast guidance defects and abnormal cell plate insertion that deviates from the positions marked by PBB (Müller et al. 2006; Lipka et al. 2014; Martinez et al. 2017). The affinity of POKs with MAP65-3/PLE that cross-links interdigitating microtubules and other members of the Microtubule-associated protein 65 (MAP65) family suggests that such potential interactions regulate phragmoplast guidance and balance phragmoplast organisation (Ho et al. 2011; Herrmann et al. 2018; Buschmann and Müller 2019; Müller 2019). Three IQ67 DOMAIN (IQD) proteins physically interact with POKs and PHGAPs, likely providing a scaffold for division-plane maintenance (Kumari et al. 2021) (Figure 25.4).

During cytokinesis, microtubules in the phragmoplast transition through the disk, ring, and discontinuous ring conformation by the addition of new microtubules nucleating on the leading edge of the phragmoplast. The plant gamma–tubulin ring complex nucleates microtubules on the sides of extant microtubules and is a necessary component for microtubule recruitment and phragmoplast organisation during cell plate formation (Nakamura et al. 2010; Ho et al. 2011; Hotta et al. 2012). When cells enter mitosis, the nuclear envelope serves as a microtubule nucleation site on the surface (Wick and Duniec 1983), with the gamma–tubulin ring complex marking sites where microtubules emanate. Notably, the nuclear envelope also generates new microtubules towards the end of the cell division cycle (Hasezawa et al. 1991).

MAP65 isoforms cross-link anti-parallel microtubules in the phragmoplast midzone, thereby stabilising the phragmoplast structure. Their loss of function displays wider midzones and cytokinetic defects (Smertenko et al. 2000; Hussey et al. 2002; Müller et al. 2004, 2019; Ho et al. 2011; Kosetsu et al. 2013; Smertenko et al. 2018).

Phragmoplast function is regulated via signalling pathways that involve different kinases and their interactions with downstream targets. Among those, the MAPK cascade is the best characterised (Smertenko et al. 2018). The turnover of the phragmoplast microtubules can be controlled by the kinesin-like protein NACK1 (Sasabe et al. 2015). This motor protein functions in a MAP kinase signalling pathway, regulating phragmoplast expansion by controlling the depolymerisation of microtubules (Nishihama et al. 2001, 2002). Further, the microtubule-severing complex KATANIN and its interactors function in phragmoplast organisation and expansion (Sasaki et al. 2019; Panteris et al. 2021) (Figure 25.4A).

Kinesins are motor proteins that primarily bind to microtubules but have also been shown to associate with actin (Klotz and Nick 2012; Tian et al. 2015). They are critical factors in spindle assembly (Liu and Lee 2022) and are responsible for vesicle and organelle transport (Nebenführ and Dixit 2018). During cytokinesis, kinesins direct Golgi-derived vesicles towards the developing cell plate; however, which kinesins are responsible remains elusive. Both plus and minus end-directed kinesins have been implemented in cytokinesis (Nebenfuhr and Dixit 2018). It is proposed that a number of kinesins play a fundamental and non-redundant role in cytokinesis and phragmoplast positioning, demonstrating the complexity of cell plate formation (Gu and Rasmussen 2022). This remains a rich area of research requiring attention and unique adaptation of tools to elucidate cargo delivery, phragmoplast signalling, organisation, and dynamics.

Regulated cell division contributes to the formation of the entire plant body. While the majority of divisions are symmetrical, asymmetrical cell divisions contribute to the development of new cell types (Rasmussen and Bellinger 2018). The orientation of the cell plate has been studied from a cell geometrical perspective since the 1800s (Hofmeister 1863; Sachs 1878; Errera 1886) and has been recently revisited (Besson and Dumais 2014; Martinez et al. 2018; Rasmussen and Bellinger 2018). Geometric properties can be used to generate predictions of cell plate orientation (Martinez et al. 2018) from a set of minimal area configurations (Errera 1886). However, simple geometry is likely not the only contributing factor, as local triggers or environmental stresses, such as wounding, can alter division-plane orientation (Rasmussen and Bellinger 2018). Continuing studies in mutants with cell division orientation defects, such as *tan1* (Martinez et al., 2017), can put the current models to test and uncover the interaction between molecular factors and stochasticity in cell plate orientation.

Endomembrane Components

The plant endomembrane system coordinates the production and delivery of membrane and secretory cargo required for cell plate formation; it also coordinates the recycling of excess materials. The conventional secretory pathway—ER–Golgi–*trans*-Golgi Network (TGN)–PM—is the dominant cytokinetic vesicle pathway; however, endocytic pathways also contribute to cell plate formation (Richter et al. 2014). Delivery of cytokinetic vesicles is highly regulated and involves RAB GTPases and tethering complexes, bringing the cytokinetic vesicles in close proximity to the target membrane at the cell plate, while vesicle fusion is facilitated by other complexes (Figure 25.3A-C). Of RAB GTPases, the RABA clade is well-studied in plant cytokinesis, although a complete roadmap of GTPases is yet to be elucidated. RABA2 and RABA3 are implicated in the delivery and incorporation of new membrane material to the cell plate based on their localisation at the leading edge (Chow et al. 2008), which is persistent through cell plate development (Park et al. 2014). During cytokinesis, RABA2a and RABA1e are likely associated with different vesicle populations, as demonstrated by their behaviour under treatment with the inhibitor of cytokinesis-specific callose Endosidin7 (ES7; Davis et al. 2016). In addition to RABAs, RABEs also show cell plate localisation and interact with the Stomatal Cytokinesis-Defective (SCD) complex that regulates trafficking at the cell plate (Speth et al. 2009; Ahn et al. 2013; Mayers et al. 2017).

The late-secretory ADP-ribosylation factor guanine-nucleotide exchange factors ARF GEFs (BIG1–4) regulate trafficking of newly synthesised proteins and endocytosed cargo from the TGN/EE to the forming cell plate (Richter et al. 2014). Transport Protein Particle (TRAPP) tethering complexes act as GEFs (Jones et al. 2000; Wang et al. 2000; Pinar et al. 2015). TRAPPII acts upstream of RABA1c to facilitate the membrane trafficking from TGN to the cell plate (Qi et al. 2011). Recent extensive protein interaction assays showed that TRAPPII functions upstream of RABA2a, postulating that TRAPPII is a GEF for RABA2a (Kalde et al. 2019). In addition to TRAPPs, the exocyst tethering complex also functions in cytokinesis. The localisation of the two complexes is complementary with the exocyst being present at the onset and towards the end of cytokinesis (Rybak et al., 2014). However, when overlapping, subunits of the exocyst complex interact with TRAPP subunits acting synergistically in regulation of cell plate development (Rybak et al. 2014; Vukasinovic and Zarsky 2016).

The fusion of cytokinetic vesicles is mediated by soluble N-ethylmaleimide-sensitive factor protein attachment protein receptor (SNARE) complexes (El Kasmi et al. 2013). The timeline of their discovery also brings perspective on the development of approaches that have been used to unravel the molecular components in cytokinesis. Early studies on these genes started with the genetic analysis of embryo pattern formation (Mayer et al. 1991), which uncovered two members of the SNARE complex: the syntaxin KNOLLE (Lukowitz et al. 1996) and the accessory protein SEC1/Munc18 protein KEULE (Assaad et al. 2001). The phenotype of unfused vesicles at the cell plate in the corresponding mutants as well as the mitotic expression of KNOLLE supported the importance of these proteins in cytokinesis (Lauber et al. 1997; Waizenegger et al. 2000).

The initial discovery in genetics was followed by protein interaction screens and mutant phenotypes that eventually allowed for the identification of two functional complexes in Arabidopsis (El Kasmi et al. 2013; Cheung et al. 2022). These complexes consist of (i) the Q-SNARE KNOLLE and its partners Qbc-SNARE SNAP33 and either R-SNAREs VAMP721/VAMP722 or (ii) KNOLLE and its partners Qb-SNARE NPSN11, Qc-SNARE SYP71, and also R-SNAREs VAMP721/VAMP722 (Lauber et al. 1997; Heese et al. 2001; Zheng et al. 2002; Zhang et al. 2011; El Kasmi et al. 2013; Jurgens et al. 2015). Sec1/Munc18 (SM) proteins, such as KEULE, assist in SNARE complex formation, which is essential for vesicle fusion during cytokinesis (Assaad et al. 2001). The significance of having two functional complexes involving KNOLLE still remains to be determined. Interestingly, among the SNARE proteins involved in cytokinesis, only KNOLLE is cell cycle–specific and thus is widely accepted as a cytokinesis marker (Figures 25.3A–C, 25.5, and 25.6A). The formation of cis-SNARE complexes takes place in the ER and are disassembled when they reach the cell plate (Karnahl et al. 2017). From an evolutionary point of view, Qa-SNARE SYP132, a non-specialised ancient Qa-SNARE, has counterparts in lower plants and functionally overlaps with the flowering plant–specific KNOLLE, showing a transition from ancient to flowering plant cytokinesis complexes (Park et al. 2018).

Ultrastructural studies support the estimate that 70% of the overall membrane materials accumulated at the CPAM are recycled and removed as the membranes mature and flatten during the transition from the TVN to PFS stages (Samuels et al. 1995; Otegui et al. 2001). Membrane recycling is therefore as important as vesicle delivery. Different components of CCVs, such as the clathrin chain, adaptor protein complexes, and dynamin-related proteins that help the scission of CCVs, are present at the cell plate (Fujimoto et al. 2008; Konopka et al. 2008; Konopka and Bednarek 2008; Fujimoto et al. 2010; Song et al. 2012; McMichael and Bednarek 2013). The role of clathrin-mediated recycling in cell plate maturation is further supported by the presence of TPLATE, a peripherally associated complex on CCVs that helps generate membrane curvature (Van Damme et al. 2011; Johnson et al. 2021) (Figure 25.5 and 25.6A). Cell wall biosynthetic complexes, such as cellulose synthases or SNAREs, can be recycled from the cell plate once their function is complete (Boutté et al. 2010; Gadeyne et al. 2014; Sanchez-Rodriguez et al. 2018; Dahhan and Bednarek 2022).

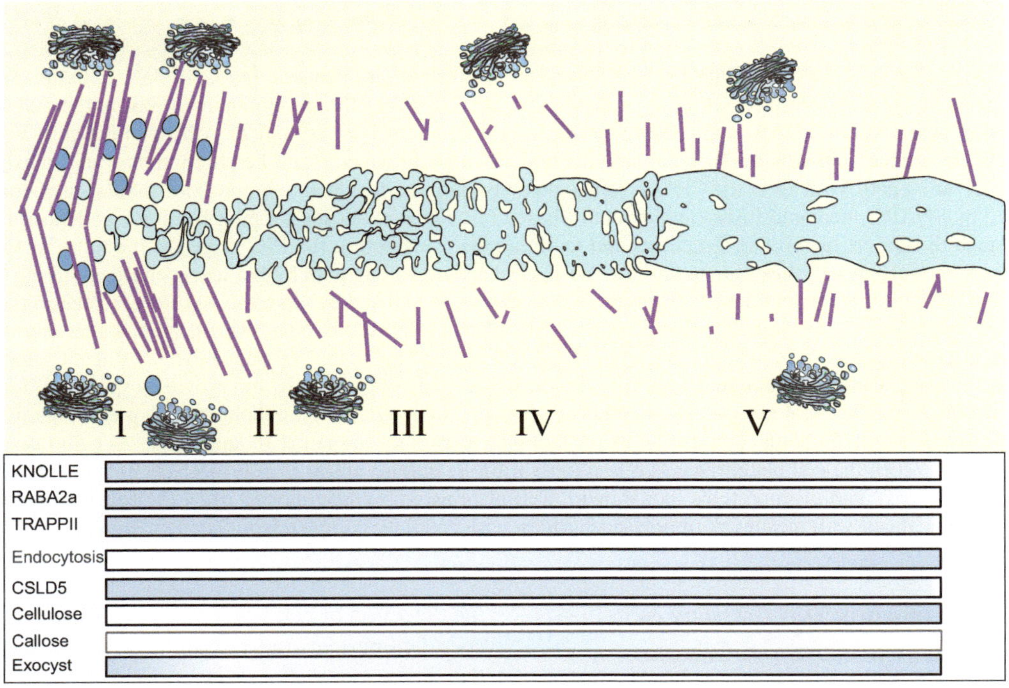

FIGURE 25.5 Visual representation of the temporal activity of selected components involved in cell plate formation. Temporal distribution of tethering complexes Exocyst and TRAPPII, SNARE-KNOLLE, RABA2a, membrane recycling and endocytosis, and polysaccharide deposition (cellulose and callose, blue) throughout the stages of cell plate formation. (I) Fusion of Golgi-derived vesicles stage FVS, (II) tubular-vesicular network (TVN) stage, (III) tubular network (TN) stage, (IV) planar fenestrated sheet (PFS), and (V) cross wall. Structures are not to scale.

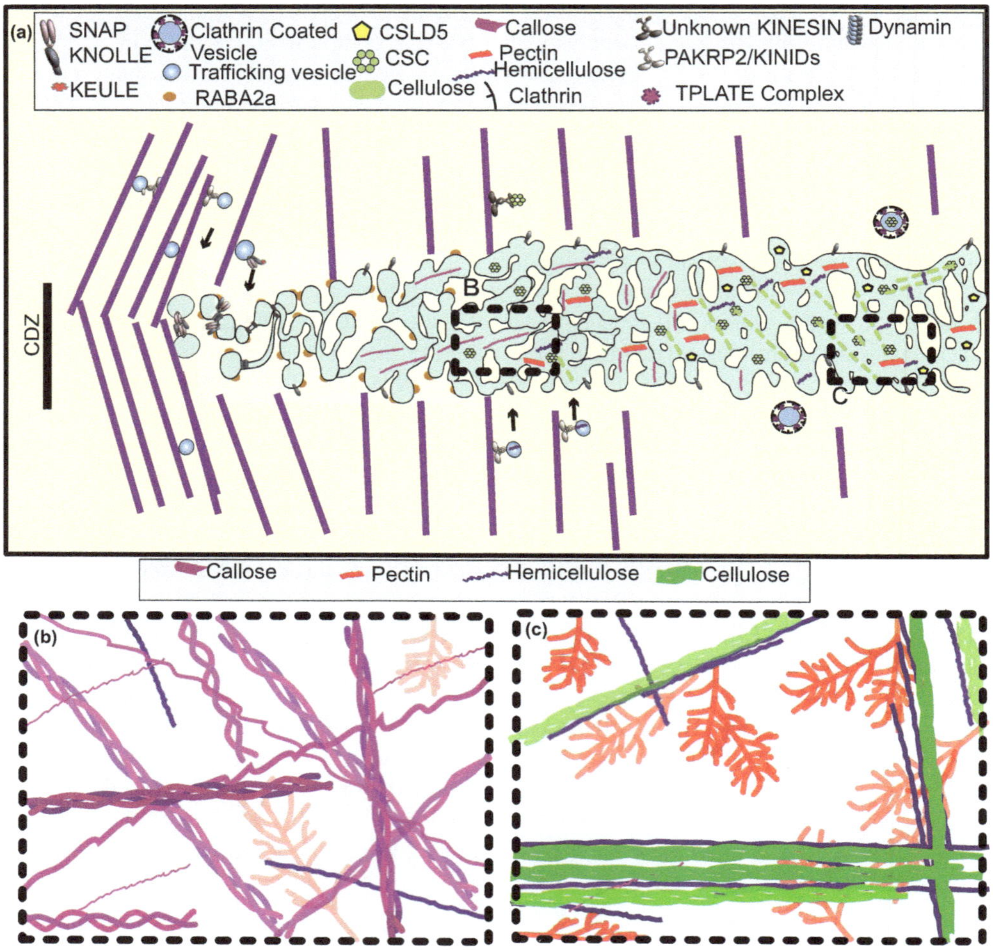

FIGURE 25.6 Proposed model of cell plate maturation and key contributing elements. (A) Cell plate initiation stage comprising tubular-vesicular network (TGN)-derived secretory vesicles that migrate along the phragmoplast towards the cell equator. Vesicle docking and fusion facilitated by SNAREs and the tethering complex such as Exocyst and TRAPPII. In the early stage of cell plate formation, RABA2a (orange) and KNOLLE (blue) labelled vesicles with guidance from the phragmoplast vigorously accumulate at the centre of the dividing cell. Vesicles at the CPAM change their structure by fusion and fission (blue tubular membrane structures in TVN) facilitated by SNARE-mediated fusion and dynamin-based restriction. Callose (hot pink lines) starts depositing at the tubular network (TN) stage to support membrane structures and facilitate the transition to the PFS stage. At the more mature stages, additional polysaccharides, such as cellulose (green dashed lines), are deposited along with higher levels of cross-linking polysaccharides such as hemicellulose (purple) and pectin (red). Membrane remodelling is assisted by the removal of excess material at the centre via clathrin-mediated recycling. Microtubule depolymerisation takes place at the centre. Note that different stages can occur simultaneously. Structures are not to scale. (B) Cartoon representation of cell plate composition starting at the TN stage. Callose (pink) deposition starts at the TN stage. Callose is thought to stabilise the membrane network structure and provide a spreading force for cell plate expansion and transition to a fenestrated sheet. Hemicellulose (purple) and pectin (red) are thought to be present but in low abundance. (C) Cartoon representation of cell plate composition as the cell plate continues to mature in later stages as cellulose (green) begin to replace callose (pink). When the cell plate transitions to a cross wall, it contains cell wall polysaccharides such as cellulose, hemicellulose (purple), pectin (red), and glycoproteins (not shown). Cell plate assembly is complete when the cell wall polysaccharides are organised into the typical cell wall meshwork of polysaccharides, with cellulose completely replacing callose.

Polysaccharide Deposition and Assembly of the New Cell Wall: Cellulose and Callose

Unlike the cytoskeleton and endomembrane trafficking components during cytokinesis, little is known about polysaccharide deposition and assembly during cell plate maturation. Most of the knowledge in the field is derived from the detection of cell wall components using polysaccharide-specific epitopes and immunocytochemistry. Variations in the immunocytochemistry approaches and/or tissue types/preparations contribute to the challenge of distinguishing the various stages of the developing cell plate, thus understating the time-dependent transition of the process. A detailed study has been impeded by the technical limitations to perform live imaging of polysaccharides. Though useful inhibitors are increasingly available to bypass the common lethality phenotypes of cytokinesis mutants, many challenges remain in understanding cell plate–specific polysaccharide deposition events.

Callose, a β-(1,3)-linked glucan, and cellulose, a β-(1,4)-linked glucan, are the two glucan polymers that are synthesised directly at the cell plate membrane. Our early understanding of callose deposition at the cell plate is based on a study by Samuels et al. (1995) using immuno electron microscopy on cryofixed samples. The deposition of callose in the cell plate is transient; it first appears at the TVN stage and is the dominant polysaccharide until the PFS stage (Samuels et al. 1995). It is also found at the junction of the parental cell wall connected with the cell plate, though its function at the cell junction remains unknown. Callose is the prominent component of the cell plate but, unlike cellulose, is rarely present in the fully mature cell wall except in plasmodesmata (Wu et al. 2018) and during transient responses such as response to external stressors (Chen and Kim 2009; Zaveska Drabkova and Honys 2017).

Although callose deposition at the cell plate has been well documented, the molecular understanding of its biosynthesis and removal remains elusive. The callose synthase (CalS) family, also known as the glucan synthase-like (GSL) family, comprises 12 members in Arabidopsis (Chen and Kim 2009). Among the 12 members, CalS10/GSL8 is likely responsible for callose deposition at the cell plate as *gsl8* mutants display typical cytokinetic effects that are seedling-lethal (Chen et al. 2009; Thiele et al. 2009; Guseman et al. 2010). Ectopically expressed CalS1/GSL6 has been localised to the cell plate of BY-2 cells (Hong et al. 2001); however, no cytokinesis defects in mutants have been reported thus far, suggesting that it is unlikely a major isoform contributing to the specificity of callose to cytokinesis. The major hurdles in studying the regulation of callose biosynthesis at the cell plate is the difficulty in generating functional fluorescently tagged CalS lines and conditional mutants that can be monitored in a temporal manner to determine the impact of loss of function in the development of the cell plate.

It has been suggested that CalSs work in a complex at the cell plate (Schneider et al. 2016; Zaveska Drabkova and Honys 2017). *gsl8* mutants show canonical cytokinesis defects and continue to be the only one in the family to show significant cytokinesis failures (Chen et al. 2009; Thiele et al. 2009; Guseman et al. 2010). Genetic and biochemical interactions between GSL8 and GSL10 make the latter a potential interactor of GSL8 at the cell plate. One plausible hypothesis is that callose synthases form complexes similar to cellulose synthases (Saatian et al. 2018). However, if a specific complex of GSL8 exists at the cell plate, its composition remains to be determined. The *in situ* biosynthesis of callose at the plasma membrane is supported by the direct appearance at the cell plate and its absence in secretory vesicles. Callose synthases are likely trafficked through TGN/early endosome (EE) pathways and have been identified in TGN/EE proteomes such as that of SYP61 (Drakakaki et al. 2012), RABF2b/ARA7, and RABD2a/ARA5 (Heard et al. 2015). GSLs are also trafficked via unconventional trafficking pathways, with the involvement of either multivesicular bodies (Bohlenius et al. 2010) or exosomes (Ellinger et al. 2013). Such trafficking has been shown for the GSL isoform GSL5/PMR4 to sites of callose accumulation in response to the plant pathogen *Blumeria graminis* f. sp. *Hordei* (Ellinger et al. 2013), but information about the trafficking of GSL8 is lacking.

The biosynthesis of callose requires high concentrations of calcium (Him et al. 2001), suggesting that a calcium gradient proximal to the cell plate is required for its regulation (Wolniak et al. 1980; Aidemark et al. 2009). This has been a long-standing hypothesis based on observations made in the early 1970s describing the inhibition of cytokinesis resulting from reduced calcium levels after caffeine treatments (Paul and Goff 1973). In support of this hypothesis, a study found that cell plate formation in tobacco BY-2 cells is inhibited under treatment with BAPTA, a calcium chelating buffer (Jurgens M 1994). Similarly, the *Physcomitrella patens sabre* mutant has cytokinesis defects that are caused by abnormal ER–cell plate connections, and delayed/aberrant callose accumulation (Cheng and Bezanilla 2021). A tempting hypothesis is that the irregular callose pattern in the *sabre* mutant is due to abnormal calcium levels at the vicinity of the cell plate. In addition to synthesis and dynamic structure, the transient accumulation of callose requires regulated hydrolysis once the cell plate transforms into a new cell wall. Callose is synthesised at the plasma membrane and is thought to remain near the surface instead of embedding deeply into the cell plate lumen. This facilitates easy access to hydrolytic enzymes to remove the polymers once the cell plate is stabilised by additional polysaccharide deposition deep inside the structure as it matures (Albersheim et al. 2011). The (1,3)-β-glucanase family contains 50 members in Arabidopsis; at present, no specific member has been assigned to callose degradation at the cell plate (Doxey et al. 2007; Levy et al. 2007).

Callose is rapidly deposited in the cell plate lumen (Samuels et al. 1995). Callose polymers are known to self-assemble into different structures (Falch and Stokke 2001). Further, single, double, and triple stranded loops can be formed, giving the polymer a dynamic rigidity and possibly contributing to its modularity in response to different triggers (Falch and Stokke 2001) (Figure 25.6B). With the advancement of imaging modules and the development of recombinantly expressed CalSs, different callose-formed structures, their conformation, and physicochemical properties can be better resolved.

The dominant and transient deposition of callose during the early delicate membrane network stage was postulated to mechanically stabilise the young cell plate until a more permanent cell wall is assembled (Samuels et al. 1995). In addition, the specific cytokinetic callose chemical inhibitor ES7 has been used to characterise the necessity for callose at the cell plate and its dynamics (Park et al. 2014). Interestingly, ES7 does not inhibit wound or stress-induced callose accumulation. The inhibition of cytokinesis-specific callose accumulation with ES7 demonstrates a temporal dependency on callose in expansion and maturation of the cell plate in both land plants and algae, uncovering the evolutionarily conserved role of the polymer in the completion of cytokinesis (Park et al. 2014; Davis et al. 2020).

A recent study by Jawaid et al. (2022) used biophysical modelling to dissect the transition between the TVN stage to PFS and a mature cell plate. Cell plate sub-structures were approximated with testable shapes, and the Helfrich free energy model for membranes was adopted to understand the transition from a vesicular network to a fenestrated sheet and mature cell plate (Figure 25.3 D,E). To reach a mature cell plate(Figure 25.3 D,E). the proposed model requires an additional element: the onset of a two-dimensional spreading/stabilising force coupled with a concurrent loss of spontaneous curvature. The absence of a spreading/stabilising force predicts failure of maturation (Jawaid et al. 2022). The transient accumulation of callose, coinciding with the predicted stages requiring this spreading force, is consistent with the proposed model's concept, lending support for the role of callose in cell plate maturation (Jawaid et al. 2022).

Cellulose is the second polysaccharide directly synthesised at the cell plate (Chen et al. 2018). A distinguishing feature of cellulose deposition in higher-plant cell walls, in contrast to other cellulosic organisms, is its organisation into paracrystalline microfibrils, the deposition of which is associated with the orientation of the cortical microtubules (Ledbetter and Porter 1963). Cellulose microfibrils consist of β-(1,4)-linked glucans synthesised at the plasma membrane by cellulose synthase (CESA) complexes (CSCs) and are produced at the plasma membrane during late to completed cytokinesis stage. For detailed information about cellulose, please refer to Chapter 1 by Doblin et al. Though cellulose is the main load-bearing polymer in parental plant cell walls, cellulose appears to be only a minor component of the early stage cell plate (Samuels et al. 1995). For their biosynthetic function, each cellulose synthase complex requires a minimum of three catalytic cellulose synthase proteins (Desprez et al. 2007; Persson et al. 2007; Wilson et al. 2021). Although not present until the late stages of cytokinesis, specific cellulose-deficient mutants exhibit cytokinesis-deficient phenotypes, suggesting an essential role for cellulose in cell plate maturation (Zuo et al. 2000).

The temporal dynamics of cellulose at the cell plate was first observed by electron microscopy in tobacco BY-2 and Arabidopsis root tip cells (Samuels et al. 1995). During the late TN stage, cellulose deposition begins. The deposition of cellulose coincides with the flattening and stiffening of the cell plate as its content increases steadily through the remaining stages, mainly the PSF, seemingly replacing callose (Samuels et al. 1995). In 2014, *in vivo* imaging of cellulose synthase reported its accumulation at the TVN stage, concurrent with the presence of cellulose (Miart et al. 2014). This study also suggested that CSC delivery occurs through phragmoplast-associated vesicles during the TVN stage or directly by the Golgi apparatus to the CPAM (Miart et al. 2014) (Figure 25.6C). Variations in immunocytochemistry approaches and/or tissues paired with the challenge to distinguish the different maturation stages that simultaneously occur at the developing cell plate could account for the contrasting information on the timing of cellulose biosynthesis at the cell plate.

Recently, it has been shown that one member of the Cellulose Synthase-Like D family, CSLD5, is targeted to newly forming cell plates of dividing cells (Park et al. 2011; Gu et al. 2016). The direct product of CSLD5 has not yet been identified, but similar to the product of CSLD3, it is likely a β-(1,4)-glucan polysaccharide (Yang et al. 2020), providing a framework for crystalline cellulose (Figure 25.6A).

Golgi-Synthesised Polysaccharides and Structural Proteins

The matrix polysaccharides hemicellulose and pectin are synthesised in the Golgi apparatus and transported to the cell plate via Golgi-derived vesicles. *Hemicellulose* in dicotyledonous plants is most abundant as xyloglucan (XyG), a β-1,4 glucan, featuring a regular pattern of xylosylations (Keegstra 2010; Scheller and Ulvskov 2010; Pauly and Keegstra 2016; Julian and Zabotina 2022). The molecular machinery for its biosynthesis in the Golgi is well understood (Julian and Zabotina 2022) and described in detail in Chapter 2. Notably, no studies have been conducted focusing on the localisation and behaviour of biosynthetic proteins specifically during cytokinesis, except the presence of endoxyloglucan transferase (EXGT), a cellulose/xyloglucan network modifying enzyme (Yokoyama and Nishitani 2001). Most of our current knowledge of matrix cell wall polysaccharide transport is derived from a very limited number of immunohistochemical electron microscopy analyses using antibodies raised against polysaccharide epitopes (Moore et al. 1986; Moore and Staehelin 1988; Moore et al. 1991; Lynch and Staehelin 1992; McFarlane et al. 2008; Young et al. 2008; Kang et al. 2011). An early study in sycamore maple (*Acer pseudoplatanus*) cells suggests an "assembly line" model consisting of the initial biosynthesis of the XyG backbone followed by the addition of side chains in Golgi sub-compartments, with the TGN largely containing fully substituted XyG glycans (Zhang and Staehelin 1992). However, recent vesicle glycomic studies using a large-scale antibody ELISA of TGN/EE-isolated vesicles marked by the syntaxin SYP61 suggest that diverse glycans are generated in Golgi/TGN and are transported to the apoplast in their final assembly (Wilkop et al. 2019). Studies using red clover roots have shown that during cell division, XyG is transported through TGN-derived secretory vesicles to the forming cell plate. The polysaccharide is detected at the TVN stage, most likely delivered by secretory vesicles, with its presence most abundant upon completion of the new cell wall (Moore and Staehelin 1988).

The presence of a cellulose/xyloglucan network modifying enzyme at the cross wall of dividing BY-2 cells suggests polysaccharide modification during cell wall assembly at the cell plate (Yokoyama and Nishitani 2001).

While the type of secretory vesicles that carry polysaccharides to the cell plate and their cargo is still largely unknown, glycomic analysis of SYP61 TGN/EE vesicles has shown that this pathway is involved in the traffic of both XyG and pectin oligosaccharides. The ECHIDNA (ECH) and the YIP family of RAB GTPase interacting proteins associated with SYP61 are implicated in TGN polysaccharide trafficking (Gendre et al. 2011; Kang et al. 2011; Gendre et al. 2013; McFarlane et al. 2013). Therefore, it is plausible that TGN/EE subdomains, including the aforementioned components and the SNARE proteins SYP42/SYP43 (Uemura et al. 2012), are involved in polysaccharide secretion. During cytokinesis, the whole secretory machinery is recruited towards the development of the cell plate; thus, it is likely that similar to the SYP61 pathway, cytokinetic vesicles carry xyloglucan oligosaccharides in different forms of substitution for their assembly into the nascent cell wall. A daunting task remains in the glycomic analysis of cytokinetic vesicles. Specifically, when isolated with the help of cytokinetic markers, it will be necessary to dissect the forms of hemicellulose oligosaccharides that are transported to the cell plate.

Pectin, first isolated in 1825 by the French chemist and pharmacist Henri Braconnot, is another Golgi-synthesised polysaccharide. The pectin structure comprises a backbone with galacturonic acid residues, with subgroups determining its diversification and classification into the following main types: homogalacturonan (HG), rhamnogalacturonan I (RG-I), and rhamnogalacturonan II (RG-II), and substituted galacturonans. Pectin accounts for up to 35% of the primary cell wall in dicotyledonous plants, and the other polysacharides are considered to be embedded in it (Mohnen 2008; Caffall and Mohnen 2009; Atmodjo et al. 2013; Du et al. 2022). For a more in-depth discussion on pectin structure and biosynthesis, refer to Chapter 5. The backbones of all pectins are synthesised in the Golgi with side group modifications thought to be added in different Golgi cisternae, which are presumably transported to the cell plate through secretory vesicles. Non-esterified RG-I and HG are present in the *cis* and medial Golgi, while arabinose-containing side chains of RG-I are detected only in the TGN, demonstrating specificity of Golgi cisternae in pectin biosynthesis (Zhang and Staehelin 1992). While pectins are synthesised and secreted in a methyl esterified form at the cell wall, they are de-esterified by pectin methylesterases (PMEs), which leads to their acidification and allows for further cross-linking (Caffall and Mohnen 2009; Albersheim et al. 2011). The presence of methyl esterified pectin at the cell plate has been detected with the aid of the JIM7 monoclonal antibody at the cell plate of Arabidopsis and BY-2 dividing cells (Toyooka et al. 2009; Rybak et al. 2014). The absence of RG-I in the cell plate using RG-I-specific polyclonal antibodies, but presence in the middle lamella of red clover (*Trifolium pretense L.*), suggests that acidification of the polysaccharide occurs at later stages (Moore and Staehelin 1988).

PMEs have also been shown to be transported to the cell plate through a Golgi-derived, but TGN-independent pathway in tobacco BY-2 cells, as demonstrated by Brefeldin A (BFA) insensitivity (Wang et al. 2016). RG-II, a more complex form of pectin, serves as the site of borate cross-linking within pectin to facilitate gel formation. Notably, RG-II has also been detected at the cell plate, which was illustrated using immunoelectron microscopy in tobacco BY-2 cells (Zhou et al. 2017).

The association of specific molecular components of endomembrane trafficking and the corresponding pathways that are specialised in polysaccharide delivery to the cell plate are currently limited to a small number of proteins involved in vesicle trafficking. These include the Secretory Carrier Membrane Protein 2 (SCAMP2), which is associated with the transport of methyl esterified pectin to the cell plate (Toyooka et al. 2009) or tethering complexes, such as the Exocyst and TRAPPII, associated with pectin delivery at the cell plate (Rybak et al. 2014).

The few studies in this field combined with variation within cell types and species dependencies account for our limited understanding of vesicle-mediated polysaccharide transport at the cell plate (Drakakaki 2015; Sinclair et al. 2022). The recent glycomic analysis of SYP61 TGN/EE vesicles showed that both matrix oligosaccharide glycans can be present in the same vesicle population (Wilkop et al. 2019). Interestingly, both methyl esterified pectin and the partially esterified pectin HG backbone were included in the SYP61 glycome, suggesting that both forms of pectin may exist throughout the endomembrane system (Wilkop et al. 2019). It is likely that a combination of pectin glycans are also transported to the cell plate via cytokinetic vesicles, which should be further investigated via glycomic analysis. Given that secretory traffic is heavily recruited towards cell plate formation (Richter et al. 2014) and that the abundance of structural polysaccharides increases in later stages of cell plate formation, a question remains on the regulation of polysaccharide deposition during cytokinesis. One plausible hypothesis is that the regulation is directly conducted on the biosynthesis level at the Golgi apparatus before transport and sorting takes place.

During the fusion of the cross wall with the mother cell wall, the cell plate membrane fuses with the plasma membrane of the mother cell. Initially, the cross wall is generally homogeneous and is continuously developed via the deposition of cellulose and other matrix polysaccharides on each side. This generates a central layer rich in pectin known as the *middle lamella*. Dissolution of the region within the mother cell wall at the junction with the cross wall allows for the rearrangement of the cell wall, its development, and the generation of a three-way junction (Albersheim et al. 2011). The pectin-rich middle lamella differentiates as the adjacent cell wall develops in sequence via a mechanism that it is not well understood. De-esterification of HG along with abundance of RG-I in the middle lamella (Moore and Staehelin 1988) suggest a maturation step and stiffening process of the negatively charged pectin molecules. The middle lamella cements the two daughter cells together and provides a source of adhesion and cell separation as well as helps to transfer load distribution. Further, it accommodates cell to cell communication, changes in cell geometry, and is involved in modification of cell wall chemistry in response to abiotic stress (reviewed in Zamil and Geitmann 2017). Interestingly, in the onion epidermis, cell fracture does not occur directly at the middle lamella but instead at the cell wall (Zamil et al. 2014, 2015), emphasising its mechanical properties and supporting the load-bearing role of pectin in the cell wall (Dick-Pérez et al. 2011; Peaucelle et al. 2012; Cosgrove 2022). Further, the spatial distribution of pectin modifications in the middle lamella reflects their mechanical role against load-bearing forces (Zamil and Geitmann 2017). With the combination of modelling and tools to study the spatiotemporal biochemical transition and the physicochemical properties of this layer, we will gain a better understanding of how the middle lamella differentiates and responds to the diverse stresses that it can face. For further information regarding aspects of the middle lamella and its applications, please see Chapters 23 and 24.

Structural proteins of the cell wall contribute to its physical properties but have no described catalytic activity. Structural proteins constitute roughly 5–10% of the cell wall dry matter (Jamet et al. 2008). These include glycine-rich proteins (GRPs), proline-rich proteins (PRPs), and hydroxyproline-rich glycoproteins (HRGPs), which are so named after the amino acids they enrich. A fourth class, arabinogalactan proteins (AGPs), has been implicated in diverse roles such as cell–cell recognition, cell fate, embryogenesis, and xylem development within a number of species (Albersheim et al. 2011; Showalter and Basu 2016; Silva et al. 2020). An in-depth review of AGPs is provided in Chapter 16.

A member of HRPG, EXTENSIN3 (EXT3), is specifically associated with cell plate development. EXTENSINs self-assemble into dendritic networks (Cannon et al. 2008). This form of self-assembly structure has the potential to react with negatively charged pectin, affecting the organisation of the new cell wall material and the overall cell wall assembly during late cytokinesis (Cannon et al. 2008). AtEXT3 was enriched in the junctions of the mother cell wall and the leading edge of the cell plate, and was shown to be critical in the completion of cell wall formation in Arabidopsis (Hall and Cannon 2002; Cannon et al. 2008). Interestingly, recent studies showed that AtEXT3 is not essential for embryogenesis and plant development in Arabidopsis, suggesting that its function is likely redundant with other related proteins (Doll et al. 2022). In addition to AtEXT3, AGPs labelled by the monoclonal antibody LM14 have been detected at the cell plate, emphasising the overall importance of AGPs at the cell plate (Yu and Zhao 2012).

The expansion and maturation of the cell plate associates with microtubule depolymerisation. The communication of the expanding cell plate with the phragmoplast is likely carried by complex networks that include the aforementioned kinase pathways as well as other proteins (Smertenko et al. 2018). Many discoveries are awaiting with regard to the identification of these signalling components.

CHALLENGES, EMERGING TECHNOLOGIES, AND FUTURE PERSPECTIVES

Cell plate development is a process that requires both spatial and temporal resolution. Compared to the large number of publications on vesicle trafficking during plant cytokinesis that describe several molecular components, such as SNARE complexes, GTPases, and tethering factors (McMichael and Bednarek 2013; Jurgens et al. 2015; Smertenko et al. 2017; Gu and Rasmussen 2022), the literature related to polysaccharide deposition during cell plate formation is limited. Formative studies based on electron microscopy and tomography coupled with antibody labelling of polysaccharides provide excellent resolution and have shaped our current conceptual models (Moore and Staehelin 1988; Samuels et al. 1995; Segui-Simarro et al. 2004). However, electron microscopy is a laborious and destructive process, limiting the number of studies that can be undertaken. To capture the different cytokinetic stages that occur simultaneously in an expanding cell plate (see Video S1 in Sinclair et al. 2022), a four-dimensional (4D) approach is required to depict long-term imaging in a biologically relevant (i.e., multicellular) system. This poses several challenges and opportunities for development.

Historically, ultrastructural and immunofluorescence studies were followed by genetic screenings, which allowed for the identification of several key proteins and interesting mutants in cytokinesis. However, in-depth investigations are hampered by lethal phenotypes (Sollner et al. 2002; Strompen et al. 2002; McMichael and Bednarek 2013). Live imaging using fluorescent proteins and other probes has informed dynamics of endomembrane components that are amenable to tagging. However, several proteins, some of which have enzymatic activity and are multi-spanning membrane proteins, are recalcitrant to fluorescent tagging or recombinant expression. CRISPR-Cas9 editing, inducible systems that can be cell cycle-induced, conditional mutants, such as the temperature-sensitive *tplate* mutant (Wang et al. 2021), and recombineering toolsets for direct chromosomal tagging (Brumos et al. 2020) can allow for functional analyses that overcome many of the challenges faced with canonical cytokinesis mutants. The advancement of single cell omics technologies may prove instrumental in combination with cell cycle markers to establish expression and biochemical profiles. In addition, it may lead to gene discovery in regulatory components. Further, *in silico* protein structure analysis, such as AlphaFold (Callaway 2022), can expedite the functional characterisation of identified proteins.

Because it overcomes lethality challenges, chemical biology has played an important role in dissecting cell plate development. For example, by specifically inhibiting cytokinetic callose deposition, the small molecule ES7 informed the role of this polymer in cytokinesis, which was tested using a biophysical model that dissected the behaviour of endomembrane components during cell plate development (Park et al. 2014; Davis et al. 2016; Jawaid et al. 2022). Similarly, SMIFH2 informed the role of formins in cytokinesis (Zhang et al. 2021). Furthermore, the use of nanotechnology in plants as sensors, probes, and delivery vehicles holds promise in both imaging, chemical, and genetic manipulation of cell plate development (Kwak et al. 2017; Lew et al. 2018; Dong et al. 2019; Sanzari et al. 2019).

Extensive biochemical analysis on the polysaccharide content of a cell plate is challenging due to the diverse stages of cytokinesis in plants. Further, comparisons of reported studies on diverse sets of plant systems is an added complexity. The ever-increasing number of antibodies to visualise specific polysaccharide epitopes are useful in understanding the *de novo* formation of the new cell wall (Pattathil et al. 2010; Hervé et al. 2011; Lee et al. 2011; Avci et al. 2012; Duffieux et al. 2020). These antibodies can be coupled with electron tomography analysis of dividing cells (Otegui 2020). In addition, they can facilitate large-scale analysis of subcellular compartments, as shown for Golgi or specialised TGN populations, to better understand transport routes of polysaccharide cargo of cytokinetic vesicles (Okekeogbu et al. 2019; Parsons et al. 2019; Wilkop et al. 2019). While these antibodies can be informative enough to provide a biochemical profile of isolated cell walls, subcellular compartments, and hopefully cytokinetic vesicles, their incorporation in a live imaging system still awaits development. It is potentially worth exploring the isolation of specific antibody fragments that can be genetically encoded and targeted at the cell plate.

Recent light microscopy advances allow for a better understanding of plant cell dynamics (Komis et al. 2018; DeVree et al. 2021). Super resolution techniques can provide more insight into the specificity of cytokinetic vesicles and the different cell plate stages, while light sheet microscopy provides information about cell plate development dynamics under longer temporal imaging of events. Structured illumination microscopy has been instrumental to better understand CESA dynamics at the PM (Duncombe et al. 2022) and can potentially be applied to dissect protein dynamics at different cell plate stages. Lattice light sheet microscopy (LLSM) affords imaging at much faster rates, with less light exposure of the sample, minimising photobleaching and phototoxic effects (Chen et al. 2014); this can extend the observation time necessary to capture cytokinesis events. LLSM has been used to dissect mitosis and endomembrane dynamics in a variety of systems (Chen et al. 2014; Aguet et al. 2016; Sen et al. 2021). Plant research is slowly embracing the benefits that light sheet microscopy and vertical imaging have to offer (Vyplelová et al. 2017; Glanc et al. 2018; Ovečka et al. 2021). However, the potential of LLSM has not been explored to understand mechanisms that regulate cytokinesis in plants.

Additionally, there has been a recent increase in both qualitatively and quantitatively analysing microscopy results, which in turn can be used to build cell plate development models and interrogate specific components. Quantitative analysis can help dissect characteristic patterns of vesicle motion during developmental stages by using spatiotemporal image correlation spectroscopy (STICS), an imaging domain extension of fluorescence correlation spectroscopy, overcoming the diffraction limits of light microscopy (van Oostende-Triplet et al. 2017). Further, the development of custom image analysis software, such as Fluorescence Morphological Operators Software (FluMOS), allows for the automated measurement of the dimensions of the expanding cell plate and the distinction of different phases during cell plate expansion (van Oostende-Triplet et al. 2017). Artificial intelligence can be used to analyse large datasets by image segmentation, classification, and restoration, overcoming many of the challenges of microscopy (von Chamier et al. 2019). The development of biophysical and mathematical models for cell plate development (Jawaid et al. 2022) can provide insights into the contribution of individual components in cytokinesis and can be used to interrogate the biomechanical contribution of polysaccharides in cell plate maturation. Similarly, 3D cell shape segmentation coupled with modelling can facilitate the prediction of division-plane orientations and aid in the characterisation of contributing proteins (Martinez et al. 2018). Such synergy of information can help inform how cell plate development is regulated.

The generation of probes that allow live imaging of the cell wall, such as Pontamine Fast Scarlet 4B for cellulose, aniline blue fluorochrome for callose, fluorophore-functionalised chitosan oligosaccharides (COS), and metabolic click-mediated labelling for pectin (Anderson et al. 2012; DeVree et al. 2021; Ropitaux et al. 2022), are necessary to understand the spatiotemporal transition of cell plate development and maturation (DeVree et al. 2021). To date, the combination of imaging both cell wall biosynthetic enzymes together with their products remains a challenge but promises to uncover how cell wall assembly is regulated during cytokinesis.

Biomechanical readouts of the cell plate are necessary to interrogate biophysical models. Once adapted to study cytokinesis, biomechanical methods using Brillouin microscopy, which assess stiffness through the thickness of the cell wall (Prevedel et al. 2019, Altartouri et al. 2019), or atomic force microscopy, which can assess external topography and elasticity of plant cells (Braybrook 2015), will be able to better describe the changes in the transition of the cell plate to a new cell wall and how this is regulated in different genetic backgrounds. Focused ion beam scanning electron microscopy (Czymmek et al. 2020; Weiner et al. 2021; Zechmann et al. 2022) has become more accessible through the development of commercial instrumentation that can complement studies using serial sectioning transmission electron microscopy, and thus contributes to an increased pool of 3D cell plate reconstructions that can be used to inform developing models. Biochemical analysis of isolated compartments (Wilkop et al. 2019; Dahhan et al. 2022) has been instrumental in identifying both protein and biochemical cargo, and its adaptation for isolating cytokinetic vesicles can reveal the nature of diverse cargo transported at the cell plate.

Over the last 50 years, the characterisation of cell plate development and identification of several molecular components has been achieved; however, many questions remain unanswered. Some of these related to cell plate expansion are as follows: (i) How many types of cytokinetic vesicles exist and what is their cargo? (ii) Which motor proteins specifically regulate cytokinetic vesicles and cargo delivery to the CPAM? (iii) What are the signalling mechanisms regulating cell plate orientation, cell plate expansion, and phragmoplast organisation? (iv) How is phragmoplast guidance achieved mechanistically? (v) What are the physicochemical properties of the assembled polysaccharide deposits during cell plate formation? (vi) How do specific polysaccharides, such as callose, contribute to the two-dimensional spreading force necessary for cell plate expansion and maturation? (vii) What regulates the fusion of the cell plate with the parental cell wall? (viii) What polysaccharides are deposited in each stage of cell plate development and how is this regulated? (ix) What is the function of the PPB? With all these tools in place, exciting discoveries in plant cytokinesis are awaiting, which together will form a complete picture of the formation of such an essential and dynamic structure for plant development—the new cell wall.

ACKNOWLEDGEMENTS

We apologise to colleagues whose work could not be included in this review due to space limitations. This work was supported by NSF, Grant MCB 1818219 and by a Hatch grant from the United States Department of Agriculture (CA-D-PLS-2132-H). We would like to thank Dr. Buschmann for the thorough and constructive comments on this chapter. This chapter is dedicated in the memory of Prof. Andrew Staehelin for his ground breaking discoveries in plant cytokinesis and cell walls as well as his dedication to training of plant cell biologists.

REFERENCES

Aguet F, Upadhyayula S, Gaudin R, Chou YY, Cocucci E, He K, Chen BC, Mosaliganti K, Pasham M, Skillern W, Legant WR, Liu TL, Findlay G, Marino E, Danuser G, Megason S, Betzig E, Kirchhausen T (2016) Membrane dynamics of dividing cells imaged by lattice light-sheet microscopy. *Mol Biol Cell* 27:3418–3435.

Ahn CS, Han JA, Pai HS (2013) Characterization of in vivo functions of Nicotiana benthamiana RabE1. *Planta* 237:161–172.

Aidemark M, Andersson CJ, Rasmusson AG, Widell S (2009) Regulation of callose synthase activity in situ in alamethicin-permeabilized Arabidopsis and tobacco suspension cells. *BMC Plant Biol* 9:27.

Albersheim P, Darvill A, Roberts K, Sederoff R, Staehelin LA (2011) Plant Cell Walls. Garland Science, New York.

Altartouri B, Bidhendi AJ, Tani T, Conrad C, Chebli Y, Liu N, Karunakaran C, Scarcelli G, Geitmann A (2019) Pectin chemistry and cellulose crystallinity govern pavement cell morphogenesis in a multi-step mechanism. *Plant Physiology* 181:127–141.

Anderson CT, Wallace IS, Somerville CR (2012) Metabolic click-labeling with a fucose analog reveals pectin delivery, architecture, and dynamics in Arabidopsis cell walls. *Proc Natl Acad Sci U S A* 109:1329–1334.

Asada T, Sonobe S, Shibaoka H (1991) Microtubule translocation in the cytokinetic apparatus of cultured tobacco cells. *Nature* 350:238–241.

Assaad FF, Huet Y, Mayer U, Jurgens G (2001) The cytokinesis gene KEULE encodes a Sec1 protein that binds the syntaxin KNOLLE. *J Cell Biol* 152:531–543.

Atmodjo MA, Hao ZY, Mohnen D (2013) Evolving views of pectin biosynthesis. *Annual Review of Plant Biology* 64:747-+.

Avci U, Pattathil S, Hahn MG (2012) Immunological approaches to plant cell wall and biomass characterization: immunolocalization of glycan epitopes. *Methods Mol Biol* 908:73–82.

Bellinger MA, Uyehara AN, Martinez P, McCarthy MC, Rasmussen CG (2021) Cell cortex microtubules contribute to division plane positioning during telophase in maize. *BioRxiv preprint server for biology*.

Besson S, Dumais J (2014) Stochasticity in the symmetric division of plant cells: when the exceptions are the rule. *Front Plant Sci* 5:538.

Bohlenius H, Morch SM, Godfrey D, Nielsen ME, Thordal-Christensen H (2010) The multivesicular body-localized GTPase ARFA1b/1c is important for callose deposition and ROR2 syntaxin-dependent preinvasive basal defense in barley. *Plant Cell* 22:3831–3844.

Boruc J, Van Damme D (2015) Endomembrane trafficking overarching cell plate formation. *Curr Opin Plant Biol* 28:92–98.

Boutté Y, Frescatada-Rosa M, Men S, Chow CM, Ebine K, Gustavsson A, Johansson L, Ueda T, Moore I, Jürgens G, Grebe M (2010) Endocytosis restricts Arabidopsis KNOLLE syntaxin to the cell division plane during late cytokinesis. *Embo j* 29:546–558.

Braybrook SA (2015) Measuring the elasticity of plant cells with atomic force microscopy. *Methods Cell Biol* 125:237–254.

Brumos J, Zhao C, Gong Y, Soriano D, Patel AP, Perez-Amador MA, Stepanova AN, Alonso JM (2020) An improved recombineering toolset for plants. *Plant Cell* 32:100–122.

Buschmann H, Chan J, Sanchez-Pulido L, Andrade-Navarro MA, Doonan JH, Lloyd CW (2006) Microtubule-associated AIR9 recognizes the cortical division site at preprophase and cell-plate insertion. *Curr Biol* 16:1938–1943.

Buschmann H, Zachgo S (2016) The evolution of cell division: from streptophyte algae to land plants. *Trends Plant Sci* 21:872–883.

Buschmann H, Müller S (2019) Update on plant cytokinesis: rule and divide. *Curr Opin Plant Biol* 52:97–105.

Caffall KH, Mohnen D (2009) The structure, function, and biosynthesis of plant cell wall pectic polysaccharides. *Carbohydr Res* 344:1879–1900.

Callaway E (2022) What's next for AlphaFold and the AI protein-folding revolution. *Nature* 604:234–238.

Cannon MC, Terneus K, Hall Q, Tan L, Wang Y, Wegenhart BL, Chen L, Lamport DT, Chen Y, Kieliszewski MJ (2008) Self-assembly of the plant cell wall requires an extensin scaffold. *Proc Natl Acad Sci U S A* 105:2226–2231.

Chen XY, Kim JY (2009) Callose synthesis in higher plants. *Plant Signal Behav* 4:489–492.

Chen XY, Liu L, Lee E, Han X, Rim Y, Chu H, Kim SW, Sack F, Kim JY (2009) The arabidopsis callose synthase gene GSL8 is required for cytokinesis and cell patterning. *Plant Physiology* 150:105–113.

Chen BC, Legant WR, Wang K, Shao L, Milkie DE, Davidson MW, Janetopoulos C, Wu XS, Hammer JA, 3rd, Liu Z, English BP, Mimori-Kiyosue Y, Romero DP, Ritter AT, Lippincott-Schwartz J, Fritz-Laylin L, Mullins RD, Mitchell DM, Bembenek JN, Reymann AC, Böhme R, Grill SW, Wang JT, Seydoux G, Tulu US, Kiehart DP, Betzig E (2014) Lattice light-sheet microscopy: imaging molecules to embryos at high spatiotemporal resolution. *Science* 346:1257998.

Chen HW, Persson S, Grebe M, McFarlane HE (2018) Cellulose synthesis during cell plate assembly. *Physiol Plant* 164:17–26.

Cheng X, Bezanilla M (2021) SABRE populates ER domains essential for cell plate maturation and cell expansion influencing cell and tissue patterning. *Elife* 10:e65166.

Cheung AY, Cosgrove DJ, Hara-Nishimura I, Jürgens G, Lloyd C, Robinson DG, Staehelin LA, Weijers D (2022) A rich and bountiful harvest: Key discoveries in plant cell biology. *Plant Cell* 34:53–71.

Chow CM, Neto H, Foucart C, Moore I (2008) Rab-A2 and Rab-A3 GTPases define a trans-golgi endosomal membrane domain in Arabidopsis that contributes substantially to the cell plate. *Plant Cell* 20:101–123.

Cleary AL, Smith LG (1998) The Tangled1 gene is required for spatial control of cytoskeletal arrays associated with cell division during maize leaf development. *Plant Cell* 10:1875–1888.

Cosgrove DJ (2022) Building an extensible cell wall. *Plant Physiology* 189(3):1246–1277.

Cutler SR, Ehrhardt DW (2002) Polarized cytokinesis in vacuolate cells of Arabidopsis. *Proc Natl Acad Sci U S A* 99:2812–2817.

Czymmek K, Sawant A, Goodman K, Pennington J, Pedersen P, Hoon M, Otegui MS (2020) Imaging Plant Cells by High-Pressure Freezing and Serial block-face scanning electron microscopy. *Methods Mol Biol* 2177:69–81.

Dahhan DA, Bednarek SY (2022) Advances in structural, spatial, and temporal mechanics of plant endocytosis. *FEBS Lett* 596(17):2269–2287.

Dahhan DA, Reynolds GD, Cárdenas JJ, Eeckhout D, Johnson A, Yperman K, Kaufmann WA, Vang N, Yan X, Hwang I, Heese A, De Jaeger G, Friml J, Van Damme D, Pan J, Bednarek SY (2022) Proteomic characterization of isolated Arabidopsis clathrin-coated vesicles reveals evolutionarily conserved and plant-specific components. Plant Cell 34(6):2150–2173.

Davis DJ, McDowell SC, Park E, Hicks G, Wilkop TE, Drakakaki G (2016) The RAB GTPase RABA1e localizes to the cell plate and shows distinct subcellular behavior from RABA2a under Endosidin 7 treatment. *Plant Signal Behav* 11:e984520.

Davis DJ, Wang M, Sørensen I, Rose JKC, Domozych DS, Drakakaki G (2020) Callose deposition is essential for the completion of cytokinesis in the unicellular alga Penium margaritaceum. *J Cell Sci* 133.

Desprez T, Juraniec M, Crowell EF, Jouy H, Pochylova Z, Parcy F, Hofte H, Gonneau M, Vernhettes S (2007) Organization of cellulose synthase complexes involved in primary cell wall synthesis in Arabidopsis thaliana. *Proc Natl Acad Sci U S A* 104:15572–15577.

DeVree BT, Steiner LM, Głazowska S, Ruhnow F, Herburger K, Persson S, Mravec J (2021) Current and future advances in fluorescence-based visualization of plant cell wall components and cell wall biosynthetic machineries. *Biotechnol Biofuels* 14:78.

Dick-Pérez M, Zhang Y, Hayes J, Salazar A, Zabotina OA, Hong M (2011) Structure and interactions of plant cell-wall polysaccharides by two-and three-dimensional magic-angle-spinning solid-state NMR. *Biochemistry-Columbus* 50:989.

Doll NM, Berenguer E, Truskina J, Ingram G (2022) AtEXT3 is not essential for early embryogenesis or plant viability in Arabidopsis. *New Phytol* 236:1629–1633.

Dong R, Li Y, Li W, Zhang H, Liu Y, Ma L, Wang X, Lei B (2019) Recent developments in luminescent nanoparticles for plant imaging and photosynthesis. *Journal of Rare Earths* 37:903–915.

Doxey AC, Yaish MW, Moffatt BA, Griffith M, McConkey BJ (2007) Functional divergence in the Arabidopsis beta-1,3-glucanase gene family inferred by phylogenetic reconstruction of expression states. *Mol Biol Evol* 24:1045–1055.

Drakakaki G (2015) Polysaccharide deposition during cytokinesis: Challenges and future perspectives. *Plant Sci* 236:177–184.

Drakakaki G, van de Ven W, Pan S, Miao Y, Wang J, Keinath NF, Weatherly B, Jiang L, Schumacher K, Hicks G, Raikhel N (2012) Isolation and proteomic analysis of the SYP61 compartment reveal its role in exocytic trafficking in Arabidopsis. *Cell Res* 22:413–424.

Du J, Anderson CT, Xiao C (2022) Dynamics of pectic homogalacturonan in cellular morphogenesis and adhesion, wall integrity sensing and plant development. *Nat Plants* 8:332–340.

Duffieux D, Marcus SE, Knox JP, Hervé C (2020) Monoclonal Antibodies, Carbohydrate-Binding Modules, and Detection of Polysaccharides in Cell Walls from Plants and Marine Algae. *Methods Mol Biol* 2149:351–364.

Duncombe SG, Chethan SG, Anderson CT (2022) Super-resolution imaging illuminates new dynamic behaviors of cellulose synthase. *Plant Cell* 34:273–286.

El Kasmi F, Krause C, Hiller U, Stierhof YD, Mayer U, Conner L, Kong L, Reichardt I, Sanderfoot AA, Jurgens G (2013) SNARE complexes of different composition jointly mediate membrane fusion in Arabidopsis cytokinesis. *Mol Biol Cell* 24:1593–1601.

Ellinger D, Naumann M, Falter C, Zwikowics C, Jamrow T, Manisseri C, Somerville SC, Voigt CA (2013) Elevated early callose deposition results in complete penetration resistance to powdery mildew in Arabidopsis. *Plant Physiol* 161:1433–1444.

Errera L (1886) Sur une condition fondamentale d'équilibre des cellules vivantes. *Annales de la Société belge de microscopie* 13:12–16.

Euteneuer U, McIntosh JR (1980) Polarity of midbody and phragmoplast microtubules. *J Cell Biol* 87:509–515.

Falch BH, Stokke BT (2001) Structural stability of (1→3)-β-D-glucan macrocycles. *Carbohydr Polym* 44:113–121.

Farmer T, Prekeris R (2022) New signaling kid on the block: the role of the postmitotic midbody in polarity, stemness, and proliferation. *Mol Biol Cell* 33:pe2.

Fujimoto M, Arimura S, Nakazono M, Tsutsumi N (2008) Arabidopsis dynamin-related protein DRP2B is co-localized with DRP1A on the leading edge of the forming cell plate. *Plant Cell Rep* 27:1581–1586.

Fujimoto M, Arimura S, Ueda T, Takanashi H, Hayashi Y, Nakano A, Tsutsumi N (2010) Arabidopsis dynamin-related proteins DRP2B and DRP1A participate together in clathrin-coated vesicle formation during endocytosis. *Proc Natl Acad Sci U S A* 107:6094–6099.

Gadeyne A, Sanchez-Rodriguez C, Vanneste S, Di Rubbo S, Zauber H, Vanneste K, Van Leene J, De Winne N, Eeckhout D, Persiau G, Van De Slijke E, Cannoot B, Vercruysse L, Mayers JR, Adamowski M, Kania U, Ehrlich M, Schweighofer A, Ketelaar T, Maere S, Bednarek SY, Friml J, Gevaert K, Witters E, Russinova E, Persson S, De Jaeger G, Van Damme D (2014) The TPLATE adaptor complex drives clathrin-mediated endocytosis in plants. *Cell* 156:691–704.

Gendre D, McFarlane HE, Johnson E, Mouille G, Sjodin A, Oh J, Levesque-Tremblay G, Watanabe Y, Samuels L, Bhalerao RP (2013) Trans-Golgi network localized ECHIDNA/Ypt interacting protein complex is required for the secretion of cell wall polysaccharides in Arabidopsis. *Plant Cell* 25:2633–2646.

Gendre D, Oh J, Boutte Y, Best JG, Samuels L, Nilsson R, Uemura T, Marchant A, Bennett MJ, Grebe M, Bhalerao RP (2011) Conserved Arabidopsis ECHIDNA protein mediates trans-Golgi-network trafficking and cell elongation. *Proc Natl Acad Sci U S A* 108:8048–8053.

Glanc M, Fendrych M, Friml J (2018) Mechanistic framework for cell-intrinsic re-establishment of PIN2 polarity after cell division. *Nature Plants* 4:1082–1088.

Glotzer M (2005) The molecular requirements for cytokinesis. *Science* 307:1735–1739.

Gu F, Bringmann M, Combs JR, Yang J, Bergmann DC, Nielsen E (2016) Arabidopsis CSLD5 functions in cell plate formation in a cell cycle-dependent manner. *Plant Cell* 28:1722–1737.

Gu Y, Rasmussen CG (2022) Cell biology of primary cell wall synthesis in plants. *Plant Cell* 34:103–128.

Gunning BE, Hardham AR, Hughes JE (1978) Evidence for initiation of microtubules in discrete regions of the cell cortex in Azolla root-tip cells, and an hypothesis on the development of cortical arrays of microtubules. *Planta* 143:161–179.

Guseman JM, Lee JS, Bogenschutz NL, Peterson KM, Virata RE, Xie B, Kanaoka MM, Hong Z, Torii KU (2010) Dysregulation of cell-to-cell connectivity and stomatal patterning by loss-of-function mutation in Arabidopsis chorus (glucan synthase-like 8). *Development* 137:1731–1741.

Hall Q, Cannon MC (2002) The cell wall hydroxyproline-rich glycoprotein RSH is essential for normal embryo development in Arabidopsis. *Plant Cell* 14:1161–1172.

Hasezawa S, Marc J, Palevitz BA (1991) Microtubule reorganization during the cell cycle in synchronized BY-2 tobacco suspensions. *Cell motility and the cytoskeleton* 18:94–106.

Heard W, Sklenar J, Tome DF, Robatzek S, Jones AM (2015) Identification of regulatory and cargo proteins of endosomal and secretory pathways in Arabidopsis thaliana by proteomic dissection. *Mol Cell Proteomics* 14:1796–1813.

Heese M, Gansel X, Sticher L, Wick P, Grebe M, Granier F, Jurgens G (2001) Functional characterization of the KNOLLE-interacting t-SNARE AtSNAP33 and its role in plant cytokinesis. *J Cell Biol* 155:239–249.

Hepler PK (1982) Endoplasmic reticulum in the formation of the cell plate and plasmodesmata. *Protoplasma* 111:121–133.

Hepler PK, Newcomb EH (1967) Fine structure of cell plate formation in the apical meristem of Phaseolus roots. *J Ultrastruct Res* 19:498–513.

Herrmann A, Livanos P, Lipka E, Gadeyne A, Hauser MT, Van Damme D, Müller S (2018) Dual localized kinesin-12 POK2 plays multiple roles during cell division and interacts with MAP65-3. *EMBO Rep* 19:e46085.

Hervé C, Marcus SE, Knox JP (2011) Monoclonal antibodies, carbohydrate-binding modules, and the detection of polysaccharides in plant cell walls. *Methods Mol Biol* 715:103–113.

Him JLK, Pelosi L, Chanzy H, Putaux JL, Bulone V (2001) Biosynthesis of (1,3)-beta-D-glucan (callose) by detergent extracts of a microsomal fraction from Arabidopsis thaliana. *European Journal of Biochemistry* 268: 4628–4638.

Ho C-MK, Hotta T, Guo F, Roberson RW, Lee Y-RJ, Liu B (2011) Interaction of antiparallel microtubules in the phragmoplast is mediated by the microtubule-associated protein MAP65-3 in Arabidopsis *The Plant Cell* 23:2909–2923.

Ho CM, Hotta T, Guo F, Roberson RW, Lee YR, Liu B (2011) Interaction of antiparallel microtubules in the phragmoplast is mediated by the microtubule-associated protein MAP65-3 in Arabidopsis. *Plant Cell* 23:2909–2923.

Hofmeister W (1863) Zusatze und Berichtigungen zu den 1851 veroffentlichen Untersuchungen der Entwicklung hoherer Kryptogamen. *Jahrb Wiss Bot* 3:259–293.

Hong Z, Delauney AJ, Verma DP (2001) A cell plate-specific callose synthase and its interaction with phragmoplastin. *Plant Cell* 13:755–768.

Hotta T, Kong Z, Ho CM, Zeng CJ, Horio T, Fong S, Vuong T, Lee YR, Liu B (2012) Characterization of the Arabidopsis augmin complex uncovers its critical function in the assembly of the acentrosomal spindle and phragmoplast microtubule arrays. *Plant Cell* 24:1494–1509.

Hussey PJ, Hawkins TJ, Igarashi H, Kaloriti D, Smertenko A (2002) The plant cytoskeleton: recent advances in the study of the plant microtubule-associated proteins MAP-65, MAP-190 and the Xenopus MAP215-like protein, MOR1. *Plant Mol Biol* 50:915–924.

Jamet E, Albenne C, Boudart G, Irshad M, Canut H, Pont-Lezica R (2008) Recent advances in plant cell wall proteomics. *Proteomics* 8:893–908.

Jawaid MZ, Sinclair R, Bulone V, Cox DL, Drakakaki G (2022) A biophysical model for plant cell plate maturation based on the contribution of a spreading force. *Plant Physiol* 188:795–806.

Johnson A, Dahhan DA, Gnyliukh N, Kaufmann WA, Zheden V, Costanzo T, Mahou P, Hrtyan M, Wang J, Aguilera-Servin J, van Damme D, Beaurepaire E, Loose M, Bednarek SY, Friml J (2021) The TPLATE complex mediates membrane bending during plant clathrin-mediated endocytosis. *Proc Natl Acad Sci U S A* 118:e2113046118.

Jones S, Newman C, Liu F, Segev N (2000) The TRAPP complex is a nucleotide exchanger for Ypt1 and Ypt31/32. *Mol Biol Cell* 11:4403–4411.

Julian JD, Zabotina OA (2022) Xyloglucan biosynthesis: from genes to proteins and their functions. *Front Plant Sci* 13:920494.

Jürgens G (2005) Cytokinesis in higher plants. *Annual Review of Plant Biology* 56:281–299.

Jurgens G, Park M, Richter S, Touihri S, Krause C, El Kasmi F, Mayer U (2015) Plant cytokinesis: a tale of membrane traffic and fusion. *Biochem Soc Trans* 43:73–78.

Jurgens M HL, Rivers BA, Hepler PK (1994) BAPTA-calcium buffers modulate cell plate formation in stamen hairs of Tradescantia-evidence for calcium gradients. *Protoplasma* 183:86–99..

Kakimoto T, Shibaoka H (1987) Actin filaments and microtubules in the preprophase band and phragmoplast of tobacco cells. *Protoplasma* 140:151–156.

Kakimoto T, Shibaoka H (1989) Cytoskeletal ultrastructure of phragmoplast-nuclei complexes isolated from cultured tobacco cells. *In* M Tazawa, ed, Cell Dynamics: Molecular Aspects of Cell Motility Cytoskeleton in Cellular Structure and Activity. Springer Vienna, Vienna, pp 95–103.

Kalde M, Elliott L, Ravikumar R, Rybak K, Altmann M, Klaeger S, Wiese C, Abele M, Al B, Kalbfuß N, Qi X, Steiner A, Meng C, Zheng H, Kuster B, Falter-Braun P, Ludwig C, Moore I, Assaad FF (2019) Interactions between Transport Protein Particle (TRAPP) complexes and Rab GTPases in Arabidopsis. *Plant J* 100:279–297.

Kang BH, Nielsen E, Preuss ML, Mastronarde D, Staehelin LA (2011) Electron tomography of RabA4b- and PI-4Kbeta1-labeled trans Golgi network compartments in Arabidopsis. *Traffic* 12:313–329.

Karnahl M, Park M, Mayer U, Hiller U, Jürgens G (2017) ER assembly of SNARE complexes mediating formation of partitioning membrane in Arabidopsis cytokinesis. *Elife* 6:e25327.

Keegstra K (2010) Plant cell walls. *Plant Physiol* 154:483–486.

Klotz J, Nick P (2012) A novel actin-microtubule cross-linking kinesin, NtKCH, functions in cell expansion and division. *New Phytol* 193:576–589.

Komis G, Novák D, Ovečka M, Šamajová O, Šamaj J (2018) Advances in imaging plant cell dynamics. *Plant Physiol* 176:80–93.

Konopka CA, Backues SK, Bednarek SY (2008) Dynamics of Arabidopsis dynamin-related protein 1C and a clathrin light chain at the plasma membrane. *Plant Cell* 20:1363–1380.

Konopka CA, Bednarek SY (2008) Comparison of the dynamics and functional redundancy of the Arabidopsis dynamin-related isoforms DRP1A and DRP1C during plant development. *Plant Physiol* 147:1590–1602.

Kosetsu K, de Keijzer J, Janson ME, Goshima G (2013) Microtubule-associated Protein65 is essential for maintenance of phragmoplast bipolarity and formation of the cell plate in Physcomitrella patens. *Plant Cell* 25:4479–4492.

Kumagai-Sano F, Hayashi T, Sano T, Hasezawa S (2006) Cell cycle synchronization of tobacco BY-2 cells. *Nature Protocols* 1:2621–2627.

Kumari P, Dahiya P, Livanos P, Zergiebel L, Kölling M, Poeschl Y, Stamm G, Hermann A, Abel S, Müller S, Bürstenbinder K (2021) IQ67 DOMAIN proteins facilitate preprophase band formation and division-plane orientation. *Nat Plants* 7:739–747.

Kwak SY, Wong MH, Lew TTS, Bisker G, Lee MA, Kaplan A, Dong J, Liu AT, Koman VB, Sinclair R, Hamann C, Strano MS (2017) Nanosensor Technology Applied to Living Plant Systems. *Annu Rev Anal Chem (Palo Alto Calif)* 10:113–140.

Lauber MH, Waizenegger I, Steinmann T, Schwarz H, Mayer U, Hwang I, Lukowitz W, Jurgens G (1997) The Arabidopsis KNOLLE protein is a cytokinesis-specific syntaxin. *J Cell Biol* 139:1485–1493.

Ledbetter MC, Porter KR (1963) A "Microtubule" in Plant Cell Fine Structure. *J Cell Biol* 19:239–250.

Lee KJ, Marcus SE, Knox JP (2011) Cell wall biology: perspectives from cell wall imaging. *Mol Plant* 4:212–219.

Lee YJ, Liu B (2019) Microtubule nucleation for the assembly of acentrosomal microtubule arrays in plant cells. *New Phytol* 222:1705–1718.

Lee YR, Giang HM, Liu B (2001) A novel plant kinesin-related protein specifically associates with the phragmoplast organelles. *Plant Cell* 13:2427–2439.

Lee YR, Liu B (2013) The rise and fall of the phragmoplast microtubule array. *Curr Opin Plant Biol* 16:757–763.

Levy A, Guenoune-Gelbart D, Epel BL (2007) beta-1,3-Glucanases: plasmodesmal gate keepers for intercellular communication. *Plant Signal Behav* 2:404–407.

Lew TTS, Wong MH, Kwak SY, Sinclair R, Koman VB, Strano MS (2018) Rational design principles for the transport and subcellular distribution of nanomaterials into plant protoplasts. *Small* 14:e1802086.

Lipka E, Gadeyne A, Stöckle D, Zimmermann S, De Jaeger G, Ehrhardt DW, Kirik V, Van Damme D, Müller S (2014) The phragmoplast-orienting kinesin-12 class proteins translate the positional information of the preprophase band to establish the cortical division zone in Arabidopsis thaliana. *Plant Cell* 26:2617–2632.

Liu B, Lee YJ (2022) Spindle assembly and mitosis in plants. *Annu Rev Plant Biol* 73:227–254.

Livanos P, Chugh M, Müller S (2017) Analysis of phragmoplast kinetics during plant cytokinesis. *Methods Mol Biol* 1662:137–150.

Lloyd C, Buschmann H (2007) Plant division: remembering where to build the wall. *Curr Biol* 17:R1053–1055.

Lukowitz W, Mayer U, Jurgens G (1996) Cytokinesis in the Arabidopsis embryo involves the syntaxin-related KNOLLE gene product. *Cell* 84:61–71.

Lynch MA, Staehelin LA (1992) Domain-specific and cell type-specific localization of two types of cell wall matrix polysaccharides in the clover root tip. *J Cell Biol* 118:467–479.

Martinez P, Allsman LA, Brakke KA, Hoyt C, Hayes J, Liang H, Neher W, Rui Y, Roberts AM, Moradifam A, Goldstein B, Anderson CT, Rasmussen CG (2018) Predicting division planes of three-dimensional cells by soap-film minimization. *Plant Cell* 30:2255–2266.

Martinez P, Dixit R, Balkunde RS, Zhang A, O'Leary SE, Brakke KA, Rasmussen CG (2020) TANGLED1 mediates microtubule interactions that may promote division plane positioning in maize. *Journal of Cell Biology* 219.

Martinez P, Luo A, Sylvester A, Rasmussen CG (2017) Proper division plane orientation and mitotic progression together allow normal growth of maize. *Proc Natl Acad Sci U S A* 114:2759–2764.

Mayer U, Ruiz RAT, Berleth T, Miséra S, Jürgens G (1991) Mutations affecting body organization in the Arabidopsis embryo. *Nature* 353:402–407.

Mayers JR, Hu T, Wang C, Cardenas JJ, Tan Y, Pan J, Bednarek S (2017) SCD1 and SCD2 Form a Complex that Functions with the Exocyst and RabE1 in Exocytosis and Cytokinesis. Plant Cell.

McFarlane HE, Watanabe Y, Gendre D, Carruthers K, Levesque-Tremblay G, Haughn GW, Bhalerao RP, Samuels L (2013) Cell wall polysaccharides are mislocalized to the vacuole in echidna mutants. *Plant Cell Physiol* 54:1867–1880.

McFarlane HE, Young RE, Wasteneys GO, Samuels AL (2008) Cortical microtubules mark the mucilage secretion domain of the plasma membrane in Arabidopsis seed coat cells. *Planta* 227:1363–1375.

McMichael CM, Bednarek SY (2013) Cytoskeletal and membrane dynamics during higher plant cytokinesis. *New Phytol* 197:1039–1057.

McMichael CM, Reynolds GD, Koch LM, Wang C, Jiang N, Nadeau J, Sack FD, Gelderman MB, Pan J, Bednarek SY (2013) Mediation of clathrin-dependent trafficking during cytokinesis and cell expansion by Arabidopsis stomatal cytokinesis defective proteins. *Plant Cell* 25:3910–3925.

Meinke DW, Cherry JM, Dean C, Rounsley SD, Koornneef M (1998) Arabidopsis thaliana: a model plant for genome analysis. *Science* 282:662, 679–682.

Miart F, Desprez T, Biot E, Morin H, Belcram K, Hofte H, Gonneau M, Vernhettes S (2014) Spatio-temporal analysis of cellulose synthesis during cell plate formation in Arabidopsis. *Plant J* 77:71–84.

Mills AM, Morris VH, Rasmussen CG (2022) Localization of PHRAGMOPLAST ORIENTING KINESIN1 at the division site depends on the microtubule binding proteins TANGLED1 and AUXIN-INDUCED IN ROOT CULTURES9 in Arabidopsis. *Plant Cell* 34(11):4583–4599.

Mineyuki Y, S. Gunning BE (1990) A role for preprophase bands of microtubules in maturation of new cell walls, and a general proposal on the function of preprophase band sites in cell division in higher plants. *Journal of Cell Science* 97:527–537.

Mineyuki Y, Wick SM, Gunning BE (1988) Preprophase bands of microtubules and the cell cycle: Kinetics and experimental uncoupling of their formation from the nuclear cycle in onion root-tip cells. *Planta* 174:518–526.

Mir R, Morris VH, Buschmann H, Rasmussen CG (2018) Division plane orientation defects revealed by a synthetic double mutant phenotype. *Plant Physiol* 176:418–431.

Mohnen D (2008) Pectin structure and biosynthesis. *Curr Opin Plant Biol* 11:266–277.

Moore PJ, Darvill AG, Albersheim P, Staehelin LA (1986) Immunogold localization of xyloglucan and rhamnogalacturonan I in the cell walls of suspension-cultured sycamore cells. *Plant Physiol* 82:787–794.

Moore PJ, Staehelin LA (1988) Immunogold localization of the cell-wall-matrix polysaccharides rhamnogalacturonan I and xyloglucan during cell expansion and cytokinesis inTrifolium pratense L.; implication for secretory pathways. *Planta* 174:433–445.

Moore PJ, Swords KM, Lynch MA, Staehelin LA (1991) Spatial organization of the assembly pathways of glycoproteins and complex polysaccharides in the Golgi apparatus of plants. *J Cell Biol* 112:589–602.

Müller S (2019) Plant cell division - defining and finding the sweet spot for cell plate insertion. *Current Opinion in Cell Biology* 60:9–18.

Müller S, Han S, Smith LG (2006) Two kinesins are involved in the spatial control of cytokinesis in Arabidopsis thaliana. *Current Biology* 16:888–894.

Müller S, Smertenko A, Wagner V, Heinrich M, Hussey PJ, Hauser MT (2004) The plant microtubule-associated protein AtMAP65-3/PLE is essential for cytokinetic phragmoplast function. *Curr Biol* 14:412–417.

Murata T, Sano T, Sasabe M, Nonaka S, Higashiyama T, Hasezawa S, Machida Y, Hasebe M (2013) Mechanism of microtubule array expansion in the cytokinetic phragmoplast. *Nat Commun* 4:1967.

Nakamura M, Ehrhardt DW, Hashimoto T (2010) Microtubule and katanin-dependent dynamics of microtubule nucleation complexes in the acentrosomal Arabidopsis cortical array. *Nat Cell Biol* 12:1064–1070.

Nebenführ A, Dixit R (2018) Kinesins and myosins: Molecular motors that coordinate cellular functions in plants. *Annu Rev Plant Biol* 69:329–361.

Nishihama R, Ishikawa M, Araki S, Soyano T, Asada T, Machida Y (2001) The NPK1 mitogen-activated protein kinase kinase kinase is a regulator of cell-plate formation in plant cytokinesis. *Genes Dev* 15:352–363.

Nishihama R, Soyano T, Ishikawa M, Araki S, Tanaka H, Asada T, Irie K, Ito M, Terada M, Banno H, Yamazaki Y, Machida Y (2002) Expansion of the cell plate in plant cytokinesis requires a kinesin-like protein/MAPKKK complex. *Cell* 109:87–99.

Okekeogbu IO, Pattathil S, Gonzalez Fernandez-Nino SM, Aryal UK, Penning BW, Lao J, Heazlewood JL, Hahn MG, McCann MC, Carpita NC (2019) Glycome and proteome components of golgi membranes are common between two Angiosperms with distinct cell-wall structures. *Plant Cell* 31:1094–1112.

Otegui MS (2020) Electron Tomography and Immunogold Labeling as Tools to Analyze De Novo Assembly of Plant Cell Walls. In ZA Popper, ed, The Plant Cell Wall: Methods and Protocols. Springer New York, New York, NY, pp 365–382.

Otegui MS, Mastronarde DN, Kang BH, Bednarek SY, Staehelin LA (2001) Three-dimensional analysis of syncytial-type cell plates during endosperm cellularization visualized by high resolution electron tomography. *Plant Cell* 13:2033–2051.

Otegui MS, Verbrugghe KJ, Skop AR (2005) Midbodies and phragmoplasts: analogous structures involved in cytokinesis. *Trends Cell Biol* 15:404–413.

Ovečka M, Sojka J, Tichá M, Komis G, Basheer J, Marchetti C, Šamajová O, Kuběnová L, Šamaj J (2021) Imaging plant cells and organs with light-sheet and super-resolution microscopy. *Plant Physiology* 188:683–702.

Palevitz B (1987) Accumulation of F-actin during cytokinesis inAllium. Correlation with microtubule distribution and the effects of drugs. *Protoplasma* 141:24–32.

Palevitz BA, Hepler PK (1974) The control of the plane of division during stomatal differentiation in Allium. *Chromosoma* 46:297–326.

Panteris E, Kouskouveli A, Pappas D, Adamakis IS (2021) Cytokinesis in fra2 Arabidopsis thaliana p60-katanin mutant: defects in cell plate/daughter wall formation. *Int J Mol Sci* 22.

Park E, Diaz-Moreno SM, Davis DJ, Wilkop TE, Bulone V, Drakakaki G (2014) Endosidin 7 specifically arrests late cytokinesis and inhibits callose biosynthesis, revealing distinct trafficking events during cell plate maturation. *Plant Physiol* 165:1019–1034.

Park M, Krause C, Karnahl M, Reichardt I, El Kasmi F, Mayer U, Stierhof YD, Hiller U, Strompen G, Bayer M, Kientz M, Sato MH, Nishimura MT, Dangl JL, Sanderfoot AA, Jürgens G (2018) Concerted action of evolutionarily ancient and novel SNARE complexes in flowering-plant Cytokinesis. *Dev Cell* 44:500-511.e504.

Park S, Szumlanski AL, Gu F, Guo F, Nielsen E (2011) A role for CSLD3 during cell-wall synthesis in apical plasma membranes of tip-growing root-hair cells. *Nat Cell Biol* 13:973–980.

Parsons HT, Stevens TJ, McFarlane HE, Vidal-Melgosa S, Griss J, Lawrence N, Butler R, Sousa MML, Salemi M, Willats WGT, Petzold CJ, Heazlewood JL, Lilley KS (2019) Separating golgi proteins from cis to trans reveals underlying properties of cisternal localization. *Plant Cell* 31:2010–2034.

Pattathil S, Avci U, Baldwin D, Swennes AG, McGill JA, Popper Z, Bootten T, Albert A, Davis RH, Chennareddy C, Dong R, O'Shea B, Rossi R, Leoff C, Freshour G, Narra R, O'Neil M, York WS, Hahn MG (2010) A comprehensive toolkit of plant cell wall glycan-directed monoclonal antibodies. *Plant Physiol* 153:514–525.

Paul DC, Goff CW (1973) Comparative effects of caffeine, its analogues and calcium deficiency on cytokinesis. *Exp Cell Res* 78:399–413.

Pauly M, Keegstra K (2016) Biosynthesis of the plant cell wall matrix polysaccharide Xyloglucan. *Annu Rev Plant Biol* 67:235–259.

Peaucelle A, Braybrook S, Höfte H (2012) Cell wall mechanics and growth control in plants: the role of pectins revisited. *Front Plant Sci* 3:121.

Persson S, Paredez A, Carroll A, Palsdottir H, Doblin M, Poindexter P, Khitrov N, Auer M, Somerville CR (2007) Genetic evidence for three unique components in primary cell-wall cellulose synthase complexes in Arabidopsis. *Proc Natl Acad Sci U S A* 104:15566–15571.

Pickett-Heaps JD, Northcote DH (1966) Organization of microtubules and endoplasmic reticulum during mitosis and cytokinesis in wheat meristems. *J Cell Sci* 1:109–120.

Pinar M, Arst HN, Jr., Pantazopoulou A, Tagua VG, de los Rios V, Rodriguez-Salarichs J, Diaz JF, Penalva MA (2015) TRAPPII regulates exocytic Golgi exit by mediating nucleotide exchange on the Ypt31 ortholog RabERAB11. *Proc Natl Acad Sci U S A* 112:4346–4351.

Prevedel R, Diz-Munoz A, Ruocco G, Antonacci G (2019) Brillouin microscopy: an emerging tool for mechanobiology. *Nat Methods* 16:969–977.

Qi X, Kaneda M, Chen J, Geitmann A, Zheng H (2011) A specific role for Arabidopsis TRAPPII in post-Golgi trafficking that is crucial for cytokinesis and cell polarity. *Plant J* 68:234–248.

Rasmussen CG, Bellinger M (2018) An overview of plant division-plane orientation. *New Phytol* 219:505–512.

Richter S, Kientz M, Brumm S, Nielsen ME, Park M, Gavidia R, Krause C, Voss U, Beckmann H, Mayer U, Stierhof YD, Jurgens G (2014) Delivery of endocytosed proteins to the cell-division plane requires change of pathway from recycling to secretion. *Elife* 3:e02131.

Ropitaux M, Hays Q, Baron A, Fourmois L, Boulogne I, Vauzeilles B, Lerouge P, Mollet JC, Lehner A (2022) Dynamic imaging of cell wall polysaccharides by metabolic click-mediated labeling of pectins in living elongating cells. *Plant J* 110(3):916–924.

Rybak K, Steiner A, Synek L, Klaeger S, Kulich I, Facher E, Wanner G, Kuster B, Zarsky V, Persson S, Assaad FF (2014) Plant cytokinesis is orchestrated by the sequential action of the TRAPPII and exocyst tethering complexes. *Dev Cell* 29:607–620.

Saatian B, Austin RS, Tian G, Chen C, Nguyen V, Kohalmi SE, Geelen D, Cui Y (2018) Analysis of a novel mutant allele of GSL8 reveals its key roles in cytokinesis and symplastic trafficking in Arabidopsis. *BMC Plant Biol* 18:295.

Sablowski R, Gutierrez C (2022) Cycling in a crowd: Coordination of plant cell division, growth, and cell fate. *Plant Cell* 34:193–208.

Sachs J (1878) Uber die Anordnung der Zellen in jungsten Pflanzentheilen. *Arb Bot Inst Wurzburg* 2:46–104.

Samuels AL, Giddings TH, Jr., Staehelin LA (1995) Cytokinesis in tobacco BY-2 and root tip cells: a new model of cell plate formation in higher plants. *J Cell Biol* 130:1345–1357.

Sanchez-Rodriguez C, Shi Y, Kesten C, Zhang D, Sancho-Andres G, Ivakov A, Lampugnani ER, Sklodowski K, Fujimoto M, Nakano A, Bacic A, Wallace IS, Ueda T, Van Damme D, Zhou Y, Persson S (2018) The cellulose synthases are cargo of the TPLATE adaptor complex. *Mol Plant* 11:346–349.

Sanzari I, Leone A, Ambrosone A (2019) Nanotechnology in plant science: to make a long story short. *Frontiers in Bioengineering and Biotechnology* 7:120.

Sasabe M, Ishibashi N, Haruta T, Minami A, Kurihara D, Higashiyama T, Nishihama R, Ito M, Machida Y (2015) The carboxyl-terminal tail of the stalk of Arabidopsis NACK1/HINKEL kinesin is required for its localization to the cell plate formation site. *J Plant Res* 128:327–336.

Sasaki T, Tsutsumi M, Otomo K, Murata T, Yagi N, Nakamura M, Nemoto T, Hasebe M, Oda Y (2019) A novel katanin-tethering machinery accelerates cytokinesis. *Curr Biol* 29:4060–4070.e4063.

Schaefer E, Belcram K, Uyttewaal M, Duroc Y, Goussot M, Legland D, Laruelle E, de Tauzia-Moreau ML, Pastuglia M, Bouchez D (2017) The preprophase band of microtubules controls the robustness of division orientation in plants. *Science* 356:186–189.

Scheller HV, Ulvskov P (2010) Hemicelluloses. *Annu Rev Plant Biol* 61:263–289.

Schmit AC, Lambert AM (1990) Microinjected fluorescent phalloidin in vivo reveals the F-actin dynamics and assembly in higher plant mitotic cells. *Plant Cell* 2:129–138.

Schneider R, Hanak T, Persson S, Voigt CA (2016) Cellulose and callose synthesis and organization in focus, what's new? *Curr Opin Plant Biol* 34:9–16.

Segui-Simarro JM, Austin JR, White EA, Staehelin LA (2004) Electron tomographic analysis of somatic cell plate formation in meristematic cells of arabidopsis preserved by high-pressure freezing. *Plant Cell* 16:836–856.

Sen O, Harrison JU, Burroughs NJ, McAinsh AD (2021) Kinetochore life histories reveal an Aurora-B-dependent error correction mechanism in anaphase. *Dev Cell* 56:3082-3099.e3085.

Showalter AM, Basu D (2016) Extensin and arabinogalactan-protein biosynthesis: glycosyltransferases, research challenges, and biosensors. *Front Plant Sci* 7:814.

Silva J, Ferraz R, Dupree P, Showalter AM, Coimbra S (2020) Three decades of advances in arabinogalactan-protein biosynthesis. *Front Plant Sci* 11:610377.

Sinclair R, Hsu G, Davis D, Chang M, Rosquete M, Iwasa JH, Drakakaki G (2022) Plant cytokinesis and the construction of new cell wall. *FEBS Lett* 596(17):2243–2255.

Sinnott EW, Bloch R (1940) Cytoplasmic behavior during division of vacuolate plant cells. *Proc Natl Acad Sci U S A* 26:223–227.

Smertenko A (2018) Phragmoplast expansion: the four-stroke engine that powers plant cytokinesis. *Curr Opin Plant Biol* 46:130–137.

Smertenko A, Assaad F, Baluska F, Bezanilla M, Buschmann H, Drakakaki G, Hauser MT, Janson M, Mineyuki Y, Moore I, Muller S, Murata T, Otegui MS, Panteris E, Rasmussen C, Schmit AC, Samaj J, Samuels L, Staehelin LA, Van Damme D, Wasteneys G, Zarsky V (2017) Plant cytokinesis: terminology for structures and processes. *Trends Cell Biol* 27:885–894.

Smertenko A, Hewitt SL, Jacques CN, Kacprzyk R, Liu Y, Marcec MJ, Moyo L, Ogden A, Oung HM, Schmidt S, Serrano-Romero EA (2018) Phragmoplast microtubule dynamics - a game of zones. *J Cell Sci* 131.

Smertenko A, Saleh N, Igarashi H, Mori H, Hauser-Hahn I, Jiang CJ, Sonobe S, Lloyd CW, Hussey PJ (2000) A new class of microtubule-associated proteins in plants. *Nat Cell Biol* 2:750–753.

Smith LG, Gerttula SM, Han S, Levy J (2001) Tangled1: a microtubule binding protein required for the spatial control of cytokinesis in maize. *J Cell Biol* 152:231–236.

Smith LG, Hake S, Sylvester AW (1996) The tangled-1 mutation alters cell division orientations throughout maize leaf development without altering leaf shape. *Development* 122:481–489.

Sollner R, Glasser G, Wanner G, Somerville CR, Jurgens G, Assaad FF (2002) Cytokinesis-defective mutants of Arabidopsis. *Plant Physiol* 129:678–690.

Song K, Jang M, Kim SY, Lee G, Lee GJ, Kim DH, Lee Y, Cho W, Hwang I (2012) An A/ENTH domain-containing protein functions as an adaptor for clathrin-coated vesicles on the growing cell plate in Arabidopsis root cells. *Plant Physiol* 159:1013–1025.

Speth EB, Imboden L, Hauck P, He SY (2009) Subcellular localization and functional analysis of the Arabidopsis GTPase RabE. *Plant Physiol* 149:1824–1837.

Steigemann P, Gerlich DW (2009) Cytokinetic abscission: cellular dynamics at the midbody. *Trends in Cell Biology* 19:606–616.

Strompen G, El Kasmi F, Richter S, Lukowitz W, Assaad FF, Jurgens G, Mayer U (2002) The Arabidopsis HINKEL gene encodes a kinesin-related protein involved in cytokinesis and is expressed in a cell cycle-dependent manner. *Curr Biol* 12:153–158.

Tabata S, Kaneko T, Nakamura Y, Kotani H, Kato T, Asamizu E, Miyajima N, Sasamoto S, Kimura T, Hosouchi T, Kawashima K, Kohara M, Matsumoto M, Matsuno A, Muraki A, Nakayama S, Nakazaki N, Naruo K, Okumura S, Shinpo S, Takeuchi C, Wada T, Watanabe A, Yamada M, Yasuda M, Sato S, de la Bastide M, Huang E, Spiegel L, Gnoj L, O'Shaughnessy A, Preston R, Habermann K, Murray J, Johnson D, Rohlfing T, Nelson J, Stoneking T, Pepin K, Spieth J, Sekhon M, Armstrong J, Becker M, Belter E, Cordum H, Cordes M, Courtney L, Courtney W, Dante M, Du H, Edwards J, Fryman J, Haakensen B, Lamar E, Latreille P, Leonard S, Meyer R, Mulvaney E, Ozersky P, Riley A, Strowmatt C, Wagner-McPherson C, Wollam A, Yoakum M, Bell M, Dedhia N, Parnell L, Shah R, Rodriguez M, See LH, Vil D, Baker J, Kirchoff K, Toth K, King L, Bahret A, Miller B, Marra M, Martienssen R, McCombie WR, Wilson RK, Murphy G, Bancroft I, Volckaert G, Wambutt R, Düsterhöft A, Stiekema W, Pohl T, Entian KD, Terryn N, Hartley N, Bent E, Johnson S, Langham SA, McCullagh B, Robben J, Grymonprez B, Zimmermann W, Ramsperger U, Wedler H, Balke K, Wedler E, Peters S, van Staveren M, Dirkse W, Mooijman P, Lankhorst RK, Weitzenegger T, Bothe G, Rose M, Hauf J, Berneiser S, Hempel S, Feldpausch M, Lamberth S, Villarroel R, Gielen J, Ardiles W, Bents O, Lemcke K, Kolesov G, Mayer K, Rudd S, Schoof H, Schueller C, Zaccaria P, Mewes HW, Bevan M, Fransz P (2000) Sequence and analysis of chromosome 5 of the plant Arabidopsis thaliana. *Nature* 408:823–826.

Thiele K, Wanner G, Kindzierski V, Jurgens G, Mayer U, Pachl F, Assaad FF (2009) The timely deposition of callose is essential for cytokinesis in Arabidopsis. *Plant Journal* 58:13–26.

Tian J, Han L, Feng Z, Wang G, Liu W, Ma Y, Yu Y, Kong Z (2015) Orchestration of microtubules and the actin cytoskeleton in trichome cell shape determination by a plant-unique kinesin. *Elife* 4:e09351.

Toyooka K, Goto Y, Asatsuma S, Koizumi M, Mitsui T, Matsuoka K (2009) A mobile secretory vesicle cluster involved in mass transport from the Golgi to the plant cell exterior. *Plant Cell* 21:1212–1229.

Traas JA, Doonan JH, Rawlins DJ, Shaw PJ, Watts J, Lloyd CW (1987) An actin network is present in the cytoplasm throughout the cell cycle of carrot cells and associates with the dividing nucleus. *The Journal of cell biology* 105:387–395.

Uemura T, Kim H, Saito C, Ebine K, Ueda T, Schulze-Lefert P, Nakano A (2012) Qa-SNAREs localized to the trans-Golgi network regulate multiple transport pathways and extracellular disease resistance in plants. *Proc Natl Acad Sci U S A* 109:1784–1789.

Van Damme D, Gadeyne A, Vanstraelen M, Inze D, Van Montagu MC, De Jaeger G, Russinova E, Geelen D (2011) Adaptin-like protein TPLATE and clathrin recruitment during plant somatic cytokinesis occurs via two distinct pathways. *Proc Natl Acad Sci U S A* 108:615–620.

van Oostende-Triplet C, Guillet D, Triplet T, Pandzic E, Wiseman PW, Geitmann A (2017) Vesicle dynamics during plant cell cytokinesis reveals distinct developmental phases. *Plant Physiol* 174:1544–1558.

von Chamier L, Laine RF, Henriques R (2019) Artificial intelligence for microscopy: what you should know. *Biochem Soc Trans* 47:1029–1040.

Vukasinovic N, Zarsky V (2016) Tethering complexes in the arabidopsis endomembrane system. *Front Cell Dev Biol* 4:46.

Vyplelová P, Ovečka M, Šamaj J (2017) Alfalfa root growth rate correlates with progression of microtubules during mitosis and cytokinesis as revealed by environmental light-sheet microscopy. *Front Plant Sci* 8:1870.

Waizenegger I, Lukowitz W, Assaad F, Schwarz H, Jürgens G, Mayer U (2000) The Arabidopsis KNOLLE and KEULE genes interact to promote vesicle fusion during cytokinesis. *Curr Biol* 10:1371–1374.

Walker KL, Muller S, Moss D, Ehrhardt DW, Smith LG (2007) Arabidopsis TANGLED identifies the division plane throughout mitosis and cytokinesis. *Curr Biol* 17:1827–1836.

Wang H, Zhuang X, Wang X, Law AH, Zhao T, Du S, Loy MM, Jiang L (2016) A distinct pathway for polar exocytosis in plant cell wall formation. *Plant Physiol* 172:1003–1018.

Wang J, Yperman K, Grones P, Jiang Q, Dragwidge J, Mylle E, Mor E, Nolf J, Eeckhout D, De Jaeger G, De Rybel B, Pleskot R, Van Damme D (2021) Conditional destabilization of the TPLATE complex impairs endocytic internalization. *Proc Natl Acad Sci U S A* 118:e2023456118.

Wang W, Sacher M, Ferro-Novick S (2000) TRAPP stimulates guanine nucleotide exchange on Ypt1p. *J Cell Biol* 151:289–296.

Weiner E, Pinskey JM, Nicastro D, Otegui MS (2021) Electron microscopy for imaging organelles in plants and algae. *Plant Physiology* 188:713–725.

Wick SM, Duniec J (1983) Immunofluorescence microscopy of tubulin and microtubule arrays in plant cells. I. Preprophase band development and concomitant appearance of nuclear envelope-associated tubulin. *J Cell Biol* 97:235–243.

Wick SM, Duniec J (1984) Immunofluorescence microscopy of tubulin and microtubule arrays in plant cells. II. Transition between the pre-prophase band and the mitotic spindle. *Protoplasma* 122:45–55.

Wilkop T, Pattathil S, Ren G, Davis DJ, Bao W, Duan D, Peralta AG, Domozych DS, Hahn MG, Drakakaki G (2019) A hybrid approach enabling large-scale glycomic analysis of post-golgi vesicles reveals a transport route for polysaccharides. *Plant Cell* 31:627–644.

Wilson TH, Kumar M, Turner SR (2021) The molecular basis of plant cellulose synthase complex organisation and assembly. *Biochem Soc Trans* 49:379–391.

Wolniak SM, Hepler PK, Jackson WT (1980) Detection of the membrane-calcium distribution during mitosis in Haemanthus endosperm with chlorotetracycline. *J Cell Biol* 87:23–32.

Wu SW, Kumar R, Iswanto ABB, Kim JY (2018) Callose balancing at plasmodesmata. *J Exp Bot* 69:5325–5339.

Yang J, Bak G, Burgin T, Barnes WJ, Mayes HB, Peña MJ, Urbanowicz BR, Nielsen E (2020) Biochemical and genetic analysis identify CSLD3 as a beta-1,4-glucan synthase that functions during plant cell wall synthesis. *Plant Cell* 32:1749–1767.

Yi P, Goshima G (2022) Division site determination during asymmetric cell division in plants. Plant Cell. 34:2120–2139

Yokoyama R, Nishitani K (2001) Endoxyloglucan transferase is localized both in the cell plate and in the secretory pathway destined for the apoplast in tobacco cells. *Plant and Cell Physiology* 42:292–300.

Young RE, McFarlane HE, Hahn MG, Western TL, Haughn GW, Samuels AL (2008) Analysis of the Golgi apparatus in Arabidopsis seed coat cells during polarized secretion of pectin-rich mucilage. *Plant Cell* 20:1623–1638.

Yu M, Zhao J (2012) The cytological changes of tobacco zygote and proembryo cells induced by beta-glucosyl Yariv reagent suggest the involvement of arabinogalactan proteins in cell division and cell plate formation. *BMC Plant Biol* 12:126.

Zamil MS, Geitmann A (2017) The middle lamella-more than a glue. *Phys Biol* 14:015004.

Zamil MS, Yi H, Puri VM (2014) Mechanical characterization of outer epidermal middle lamella of onion under tensile loading. *Am J Bot* 101:778–787.

Zamil MS, Yi H, Puri VM (2015) The mechanical properties of plant cell walls soft material at the subcellular scale: the implications of water and of the intercellular boundaries. *Journal of Materials Science* 50:6608–6623.

Zaveska Drabkova L, Honys D (2017) Evolutionary history of callose synthases in terrestrial plants with emphasis on proteins involved in male gametophyte development. *PLoS One* 12:e0187331.

Zechmann B, Möstl S, Zellnig G (2022) Volumetric 3D reconstruction of plant leaf cells using SEM, ion milling, TEM, and serial sectioning. *Planta* 255:118.

Zhang GF, Staehelin LA (1992) Functional compartmentation of the golgi apparatus of plant cells: immunocytochemical analysis of high-pressure frozen- and freeze-substituted sycamore maple suspension culture cells. *Plant Physiol* 99:1070–1083.

Zhang L, Smertenko T, Fahy D, Koteyeva N, Moroz N, Kuchařová A, Novák D, Manoilov E, Smertenko P, Galva C, Šamaj J, Kostyukova AS, Sedbrook JC, Smertenko A (2021) Analysis of formin functions during cytokinesis using specific inhibitor SMIFH2. *Plant Physiol* 186:945–963.

Zhang L, Zhang H, Liu P, Hao H, Jin JB, Lin J (2011) Arabidopsis R-SNARE proteins VAMP721 and VAMP722 are required for cell plate formation. *PLoS One* 6:e26129.

Zheng H, Bednarek SY, Sanderfoot AA, Alonso J, Ecker JR, Raikhel NV (2002) NPSN11 is a cell plate-associated SNARE protein that interacts with the syntaxin KNOLLE. *Plant Physiol* 129:530–539.

Zhou Y, Awano T, Kobayashi M, Matoh T, Takabe K (2017) Immunocytochemical detection of rhamnogalacturonan II on forming cell plates in cultured tobacco cells. *Biosci Biotechnol Biochem* 81:899–905.

Zuo J, Niu QW, Nishizawa N, Wu Y, Kost B, Chua NH (2000) KORRIGAN, an Arabidopsis endo-1,4-beta-glucanase, localizes to the cell plate by polarized targeting and is essential for cytokinesis. *Plant Cell* 12:1137–1152.

Index

Note: **Bold** page numbers refer to tables; *italic* page numbers refer to figures and page numbers followed by "n" denote endnotes.